AF522329

Frontiers of Fungal Diversity in India

(Prof. Kamal Festschrift)

Frontiers of Fungal Diversity in India

(Prof. Kamal Festschrift)

G.P. Rao
C. Manoharachari
D.J. Bhat
R.C. Rajak
T.N. Lakhanpal

INTERNATIONAL BOOK DISTRIBUTING CO.
(Publishing Division)

Published by

INTERNATIONAL BOOK DISTRIBUTING CO.

(Publishing Division)
Chaman Studio Building, 2nd Floor,
Charbagh, Lucknow 226 004 U.P. (INDIA)
Tel. : Off. : 450004, 450007, 459058 Fax : 0522-458629
E-Mail : ibdco@sancharnet.in

First Edition. 2003

ISBN 81-85860-92-0

©Publisher

No part of this publication may be reproduced, stored in a retrieval system, or transmitted, in any form or by any means, electronic, mechanical, photocopying, recording or otherwise, without the prior written permission of the publisher.

Composed & Designed at :

Panacea Computers
33, Nehru Road,
Sadar Cantt. Lucknow-226 002
Phone : 481164, 480546
E-mail : prasgupt@rediffmail.com

Printed at:

Army Printing Press
33, Nehru Road,
Sadar Cantt. Lucknow-226 002
Phone : 481164, 480546
E-mail : armypress@sify.com

Contents

Preface

The abundant variations in the types of living organisms in many geophysical areas is called biodiversity. With reference to large organisms plants, animals etc., the distinctive visible anatomical and morphological features enable us to identify the different genera and species and, therefore, we can easily assess the extent of morphotaxonomic diversity. As regards microorganisms, the diversity is to be studied not only from morphotaxonomy standpoint but also with a view to their functional manifestations. Amongst the microbes, the fungi are of great importance on account of their functional manifestations in dairy, bakery, brewery, agriculture, forestry, medicine and deteriorating various artifacts, recycling of elements, natural scavengers and many other natural phenomena responsible for maintaining the environmental balance. However, most fungi are not visible to the naked eye and they need to be studied under the microscope, the culturable fungi can be identified, provided they sporulate but the biotrophs which are not amenable to culture media go unnoticed. It is therefore, imperative that there is a gap in our knowledge of fungal diversity. Fungi are better known than bacteria, viruses and other smaller forms of life such as viroids, prions etc. There are about, 72,000 named species of fungi and new species are regularly being added at the rate of about 1500 each year. The total number of fungal species, both known and unknown, is estimated as, 1.5 M. The 72,000 known species exist in different culture collection centres in the form of several subspecies, pathovars and strains, the total cultures going upto 1,70,000. However, the majority of described strains are not deposited in culture collection centers and are not accessible.

The idea of compiling this volume is in respect of Prof. Kamal, one of the leading mycologists of the country. Prof. Kamal has contributed significantly to morphotaxonomic fungal diversity with reference to foliar forms. On account of peculiar life style, unique biochemistry, immense adaptability and mode of nutrition, the fungal organisms have occupied the position of a separate Kingdom. These organisms have been so important since time immemorial, that hardly a day passes without the effect of these organisms on one or the other walks of human life. The diversity in these organisms is immense and still afoot.

The publication of this volume is purposeful in bringing together the information available till date on various aspects of fungal diversity in Indian sub-continent. The Indian sub continent is characterized with diverse topography, climatic conditions and turn over of green vegetation all the year round in its one or the other parts and correspondingly this sub-continent is adorned with immense fungal diversity. As yet, very little is known about different aspects of fungal diversity in Indian sub-continent. It was thought to be the right time to bring together the information whatever, gathered by resident mycologists so far in different disciplines of fungal diversity, so as to discuss the major prospects and future strategies.

This compilation focuses light on the diversity and bioremediation, biosynthesis of gold nanoparticles employing fungal organisms, biotechnological and molecular approaches in characterization of fungal diversity, biological control, new approaches of fungal diversity including heavy metal toxicities and resistance in fungi, alkaloid biosynthesis and complexone siderophores as iron transport compounds in some group of fungal organisms. An account is also presented on fungal cellulases in novel industrial application, aflatoxin contamination, toxicosis and preventive measures.

A major part of this book is devoted to discuss the recent trends and current status of different fungal forms or groups there of including thermophilic fungi, keratinolytic fungi, rust fungi, Agaricales, acellular slime fungi, fleshy fungi, *Alternaria* species associated with cucurbitaceous vegetables and fungal diversity in forest ecosystems.

A great deal of morphotaxonomic diversity is presented including Coelomycetes, Loculoascomycetes and certain individual fungal genera viz., *Phoma, Ravanelia, Vermiculariopsiella, Phaeoramularia, Gonatophragmium, Stenella, Corynespora, Colletotrichum and Ceratocytis species*. Habitat diversity has also been discussed with reference to indoor and outdoor environments. Some chapters are also devoted to discuss the biodiversity of fungal forms as biocontrol agents and their commercial application in agriculture and human life.

This book is certainly not a comprehensive account of all the aspects of fungal diversity from all over the Indian sub-continent .However, the content of this book will give the idea of all the current researches being carried out in most of the aspects of fungal diversity in Indian sub-continent. Hopefully, it will prove to be modest and useful attempt in accelerating the pace of growth of researchers working on different aspects of fungal diversity in India. All the important aspects of fungal diversity have been discussed at length .

Most sincerely, we extend our thanks to all the authors for readily agreeing to contribute articles and for their timely cooperation in preparation of the manuscripts. We are also thankful to Dr Farhat Abbasi, Dr (Mrs.) Parveen Abbasi, Dr (Mrs.) Maneesha Singh, Mr Rajarshi Gaur, Mr Ashutosh Singh for their help in reading the galley proofs and finalization of the manuscripts. We are also grateful to M/s International Book Distributing Company, Lucknow for their determined efforts to publish the volume on schedule.

We hope this volume will serve as an important reference for students and scientists involved in fungal diversity of agricultural crops, forest ecosystems, indoor and outdoor environments, fungal diseases associated with plants, animals and human beings and it will stimulate research and extension work in all aspects of fungal diversity, by new means to step up research prospects in fungal diversity and its exploitation for the good of mankind in the new millennium.

Editors

About the Editors

Dr. GOVIND P. RAO is working as Head and Senior Scientific Officer (Pl. Pathology) at Sugarcane Research Station, Gorakhpur, India. He did his Ph. D. in Plant Virology in 1987 from Gorakhpur Univ. Dr. Rao is a fellow of Indian Phytopathological Society and Sugarcane Technologists Association of India. He has published 5 edited books to his credit. He is receipient of Young Scientist Award (DST, New Delhi, 1998), National Associateship Award (DBT, New Delhi, 1991) and Best Agricultural Scientist Award, UP (UPCAR, Lucknow, 2001). He is Editor-in Chief of 'Sugar Tech' an international journal of sugar crop and related industries and Associate Editor of 'Journal of Basic & Applied Mycology'. He is Secretary of Society for Sugar Research & Promotion. He is also regular member of Indian Phytopathological Society, American Phytopathological Society, British Society of Plant Pathology and Society for General Microbiology, UK. He has been awarded several national and overseas awards for postdoctoral work and attending international workshops and conferences. He has worked and visited many scientific laboratories of USA, France, Scotland, UK, Yugoslavia, Sri Lanka, China, Nepal, Singapore, Australia and Thailand. He has been involved in the research on sugarcane pathology for the last 15 years. Currently, he is working on characterization of red-rot and smut fungal pathogens of sugarcane through biotechnolgoical and molecular approaches. In addition, he is currently working on the diagnosis, characterization and management of fungal, bacterial, viral and phytoplasma diseases of sugarcane in India.

Prof. C. MANOHARACHARY presently working as Head, Department of Botany and Co-ordinator, UGC - COSIST and UGC- SAP in Botany, Osmania University, Hyderabad has been proud product of Osmania University with M. Sc., Ph. D., Degrees obtained from Osmania University and D. Sc. from University of Mysore. Prof. C. Manoharachary with his excellent academic record has guided 40 Ph. D. students and published 400 research papers in National/ International journals. He is recipient of Young Scientist award (A. P. Science Academy, 1979) and Best Teacher award (by Government of A.P. in 1995). He has served as Chairman, Board of Studies in Botany, Vice-Principal and Principal, University College of Science, In-Charge Dean, Faculty of Science and Convenor P.G. Admissions in Science, Osmania University. He has been the member of many academic and professional bodies of National and International repute. He is also Fellow of Indian Botanical Society, Fellow of A.P. Academy of Sciences and Indian Phytopathological Society. He has organized nine National Symposia /

Workshops and one International Conference. He did his Post-doctoral research work with Prof. J.Webster at U.K. under British Council Fellowship from 1977-78, with Prof. Kreisel at Germany under Cultural Exchange Programme from 1989-90 and with Prof. R.D. Goods at U.S.A. in 1990. He is the Panel member of Plant and Animal Sciences, UGC, and Member Research Degree Committee of various Universities and is also subject expert to UGC, CSIR, DBT and DST. He is associated as an Executive Editor of the Indian Journal of Microbial Ecology, Advisory Member of various Journals, Executive Council member of the Indian Botanical Society and other State and Central bodies. He has successfully completed five major research projects sponsored by DBT, DST, CSIR, UGC and others. Prof. C. Manoharachary is the first Osmanian to get D. Sc. degree in the whole of Science Faculty. Prof. C.Manoharchary is an internationally reputed scientist with excellent academic record and rich experience of administration besides having 32 years of experience as reputed teacher, researcher and dynamic administrator. He has specialized in Mycology, Plant Pathology and Fungal Biotechnology. Recently he has been awarded Dr. Birbal Sahani Medal for his significant contributions in the field of Mycology, Plant Pathology and Microbiology. Prof. Chary has brought huge funding from different funding agencies to develop infrastructure and advanced laboratory, equipment for the Department of Botany, Osmania University. He has guided 18 AVRC Programmes in Botany. Recipient of Andhra Pradesh Scientist Award, 2002, Sectional President of Botany, 89th Indian Science Congress, Lucknow, 2002, President, Indian Phytopathological Society 2002-2003.

Prof. D. J. BHAT, Ph. D. is presently Professor in the Department of Botany, Goa University, Goa, India. He did his M. Sc. (Botany) in 1973 from University of Mysore and M. Phil. (Mycology and Plant Pathology) in 1974 and Ph. D. (Botany: Mycology) in 1979 from University of Madras, India. Dr. Bhat has worked earlier as Associate Professor in the Department of Biology, Asmara University, Ethiopia (1981-87), Research Scientist in the Department of Applied Botany, Mangalore University, Karnataka (1988-90), and Reader in the Department of Microbiology, Goa University (1990-94). A Specialist on fungi, Dr. Bhat is teaching Mycology, Plant Pathology and Agriculture Microbiology to U.G. and P.G. students of Botany since 1981. He has guided several M. Sc. and Ph. D. students on biology of fungi. Presently, Dr. Bhat is the Head, Department of Botany and Dean, Faculty of Life Sciences and Environment, in Goa University. Dr. Bhat has worked on several national and international sponsored research projects aimed at isolation, documentation and conservation of microfungi of the forests of highlands of Ethiopia and the Western Ghats in southern India. He has published more than 60 research papers on floristics, taxonomy and biology of microfungi, both in national and international journals. Dr. Bhat has built up a state-of –the –art facility to work on and repository to deposit original and pure cultures of fungi of India in the Department of Botany, Goa University, present collection of which includes fungi recovered from aquatic, coprophilous, endophytic, entomogenous, litter and soil habitats. Dr. Bhat

has received the Developing Country Fellowship of Royal Society of London and visited the I.M.I., Kew, and University of Exester, U.K. during 1985-86. He has visited the University of Waterloo, Ontario, Canada during 1993 and Novartis, Basel, Switzerland, in 1996. He was also the recipient of Late Shome Memorial Award of the Mycological Society of India. Dr. Bhat has conducted the 22nd and 24th Annual Meetings of the Mycological Society of India and an International Symposium on Ecology of Fungi in Goa University. He held the position of President, Goa Botanical Society during 1996-98 and Treasurer, Mycological Society of India between 1995-2001. Presently, Dr. Bhat is the Secretary of Mycological Society of India.

Dr. R.C.RAJAK is working as Professor and Head, Department of Biological Science, R.D. University, Jabalpur. He did his M.Sc. in botany with first position. He is fellow of Plant Pathological Society of India, Society of Mycology & Plant Pathology and Indian Botanical Society. He has been councilor and president of Annual Meeting of Indian Phytopathological Society and Indian Science congress. He has been also excellent sport achievements in football and tennis in National events. He has been involved in research and teaching of botany especially mycology for the last 34 years. He has been guided 33 students for Ph.D degrees and 2 students for D.Sc. degrees. He has published 165-research article in national and International journals. He has also edited 5 books to his credit. He has successfully completed 11 research projects and 3 ongoing projects from DBT, New Delhi and CSIR, New Delhi.

Prof.T.N.LAKHANPAL is presently Head of Dept. of Bio-Sciences, H.P.University, Shimla. He did his M.Sc (Hons.) with gold medal in 1967 from Punjab University and Ph.D. from Delhi University in 1975. He has been recipient of Fullbright Felowship at Dept. of Forest Science, Oregon State Univ., Coravillis OR, USA in 1992-1993. He has also been offered Professorship in the same department. He is fellow of Phytopathological Society of India and Society of Mycology and Plant Pathology. He is chief editor to 'Indian Journal of Mushrooms'. He has published six books and monographs and over 140 research publications to his credit. He has widely travelled in the universities and institutions of USA, Japan, England and Thailand. He has successfully completed research collaboration with USA, Edinburg and Nigeria. He has guided 21 research students in Ph.D. programmes. His laboratory is serving as a training center for ectomycorrhizal research. He did excellent work on mushroom of Himalayas, taxonomy of Indian mynomycetes, the amanitaceae, boletaceae, Canthrellaceae and Russulaceae of India.

Prof. Kamal

Prof. Kamal : A Life Sketch

Professor Kamal was born on September 2, 1941 to the parents of lower middle class peasantry in Kushinagar District, the then District Deoria of Gorakhpur Commissionery of Uttar Pradesh. He graduated in 1962 and post-graduated in 1964 from the University of Gorakhpur. After post-graduation he was posted at FRI, Dehra Dun to work on 'Fungal Decay and Durability of Indian Timbers' in the Department of Forest Pathology. He obtained Ph.D. Degree in Botany (soil fungi) from Gorakhpur University in 1968 under supervision of Professor K. S. Bhargava, the then Head, Botany department of the University.

As student he had been very attentive, indulgent and single minded. Dr. Kamal joined the Botany Department in Gorakhpur University in 1965 as a research student and in 1967 as lecturer. Since then he has been actively engaged in research in different aspects of biodiversity with particular reference to fungal taxonomy (Hyphomycetes). He received a short time advanced training in taxonomy of fungi in the Department of Special Botany at Swiss Federal Institute of Technology, Zurich, Switzerland in 1979/80 under the supervision of Prof. Emil Muller, a known mycologist of the world. He was INSA visiting Fellow to Royal Nepal Academy of Science and Technology, Kathmandu, Nepal, hosted by Central Department of Botany, Tribhuan University, Kathmandu from September 1993 to February 1994 and U.G.C. visiting fellow in the Department of Botany, Dr. H.S. Gaur University, Sagar (M.P.) in 1990 (for a short duration). He has attended several National/International Symposia/Conferences and delivered invited or keynote lectures. He has chaired sessions in Mycology and Plant Pathology in various seminars and symposia. He was elected as Councillor of Indian Phytopathological Society of India. In 1992 he was elected as President of Indian Phytopathological Society, M.E. Zone. In 2001, he was elected as Vice-President of Mycologial Society of India. He is presently the Editor-in-Chief of 'Society for Basic and Applied Mycology'.

Professor Kamal has already published more than 200 research papers on various aspects of Mycology and Plant Pathology, particularly in taxonomy of foliar fungi. During his research endeavour, he described over 300 new species including recombinations and redispositions and more than a dozen of new genera to his credit. Over sixty students have been awarded Ph.D. Degree under the supervision of Professor Kamal. He has authored/co authored four books on various aspects of Mycology and Microbiology.

Professor Kamal has made his presence marked as dedicated, devoted and disciplined teacher, researcher and guide. His simplicity and sagacity always endeared him to his students and colleagues. He has been a man of few words, frank, straightforward, enlightened, forward looking, mentally alert and agile, always full of new ideas. His planning and execution of research projects have been exemplary. Coming from a very modest farmers family he scaled glorious academic heights by shear hard work, resilience and catholic devotion.

Students Guided for Ph.D. Degree

Student's Name	Year	Title of the Thesis
1.Dr. Mohan Lal Gupta	1975	Studies on Aspergilli from forest soils of Gorakhpur.
2.Dr. Narendra Prasad Singh	1976	An Investigation field at Gorakhpur Air spora over sugarcane field at Gorakhpur.
3. Dr. Krishna Shankar Srivastava	1978	Citric acid Production by Aspergilli.
4.Dr. Sudama Singh	1978	Plant Parasitic Fungi of Gorakhpur.
5.Dr. Rana Pratap Singh	1978	Taxonomic Studies on Plant Parasitic Fungi of Gorakhpur.
6.Dr. Prafulla Kumar	1979	Studies on Plant parasitic Fungi of Gorakhpur.
7.Dr. Ajay Kumar Verma	1980	Studies on Fungal Airspora over Arhar {*Cajanus cajan* (L.) Mill sp. field at Gorakhpur.
8.Dr. Farhat Abbasi	1980	Studies on Plant-parasitic Fungi of Gorakhpur with special reference to Gall forming Fungi.
9.Dr. Rakesh Pratap Singh	1981	Comparative Role of nodal and stalk infections in the epidemiology of Red Rot of Sugarcane.
10. Dr. Bishram Rai	1981	Taxonomic Studies on the Fungi parasitising the Forest vegetation of Terai belt of North-Eastern U. P. with reference to Hyphomycetes.
11. Dr. (Mrs.) Parveen Abbasi	1982	Studies on Plant parasitic Fungi of Gorakhpur.
12. Dr. Akhila Nand Rai	1983	Studies on Fungi parasitising Forest flora of Terai belt of North-Eastern U. P.
13. Dr. Anil Kumar Singh	1983	Taxonomic Studies on Plant-parasitic Fungi of Gorakhpur region.
14. Dr. Prem Narayan	1986	Foliicolous Hyphomycetes of North-Eastern U. P.
15. Dr. Brijendra Kumar Gupta	1986	Taxonomic Studies of parasitic Fungi of Gorakhpur division with reference to leaf-spotting Hyphomycetes.
16. Dr. Ram Keerti Verma	1986	A Taxonomic treatment of the Fungi causing leaf diseases in Plants of North-Eastern U. P.
17. Dr. (Mrs.) Usha Budathoki	1988	Studies on Foliicolous Fungi from Kathmandu Valley.
18. Dr. Ramesh Chand	1989	Taxonomic Studies of Powdery Mildew Fungi of North-Eastern Terai belt of U. P.
19. Dr. Chaturbhuji Gupta	1989	Taxonomic Treatment to the Hyphomycetes of Angiospanic Plants from North-Eastern U. P.
20. Dr. Nasim Ahmed	1989	Taxonomic Studies on Powdery Mildews of Northern India.
21. Dr. Lakshmi Kant	1989	Biodegradation of Celluloses from Agricultural wastes by certain Fungi.
22. Dr. (Mrs.) Poonam	1990	Studies on Seed Mycoflora some Medicinal Plants.
23. Dr. Rakesh Kumar	1990	Studies on Foliicolous Dematiaceious Hyphomycets of U. P.
24. Dr. (Miss) Anima Mishra	1990	Studies on Cellulytic Fungi with reference to Aspergillus.
25. Dr. Sudir Singh Bisht	1991	Studies on Post harvest Fungal Diseases of some fruit and vegetable crops of Kumaon.
26. Dr. Lalji Srivastava	1991	Fungitoxic Evaluation of Volatile constituents of some Higher Plants.

27.	Dr. Ram Vyas Chaturvedi	1991	Evaluation of Essential oil of some Higher Plants against Fungal deterioration of Pulses during Storage.
28.	Dr. Arun Salic Moses	1991	Taxonomy and Biology of *Cercospora* complex from North-Eastern Indian forest.
29.	Dr. Mohd. Khursheed Khan	1991	Taxonomic Studies on Foliicolous fungi from Forest flora of North-Eastern U. P. & Bihar.
30.	Dr. Satish Chandra	1991	Taxonomy and Biology of Foliicolous Hyphomycetes with special reference to some *Pseudocercospora.*
31.	Dr. V. P. Pandey	1992	Taxonomy and Biology of Foliicolous Hyphomycetes with special reference to some *Pseudocercospora.*
32.	Dr. S. N. Joshi	1992	Studies on *Cercospora complex* with special reference to Phaeoisariopsis.
33.	Dr. Rajesh Kumar	1992	An Investigation on Ergot Alkaloid production by *Cleviceps paspali* steve. Hal in Sub-merged culture.
34.	Dr. Rajesh Kumar Srivastava	1992	Taxonomy of Foliicolous Hyphomycetes with reference to *Cercospora* sensu-stricto.
35.	Dr. Aditya Naryan Singh	1992	Taxonomy and Biology of Foliicolous Fungi of South Gorakhpur Forest division with reference to Black mildews.
36.	Dr. Neeraj Srivastava	1992	Taxonomy of Foliicolous Fungi from North-Eastern U. P. with reference to Hyphomycetes.
37.	Dr. J. K. Mathur	1993	Tissue culture Studies on Threatened Medicinal and Aromatic Plants of *Valerianaceae Nardostachys jatamassi* and *Valleriana wallichii.*
38.	Dr. Shivakar Mishra	1993	Isolation and Immuro-chemical *characterigation of Prosopsis jutoflora* Pollen Allergens & Alkaloiles.
39.	Dr. S. P. Chaudhary	1993	Studies on combining ability, gene action and other different genetic parameters of Sesame. (*Sesamum indicum L.*).
40.	Dr. M. D. Kushwaha	1994	Studies on Fungi causing Fish Diseases.
41.	Dr. (Miss.) Poonam Srivastava	1994	Taxonomy of Foliicolous Hyphomycetes with reference to *Cercosporidium complex.*
42.	Dr. H. S. G. Rao	1994	Taxonomy and Biology of Foliicolous Hyphomycetes from Forest of North-India with reference to *Pseudocercospora.*
43.	Dr. Satya Naryan	1996	Taxonomy and Biology of Foliicolous Hypomycetes with special reference to *Cercospora* Sensu-Stricto.
44.	Dr. (Mrs.) Meenu	1996	Studies on Taxonomy of some Foliicolous Hypomycetes with reference to *Corynespora.*
45.	Dr. Sanjay Kumar Singh	1996	Taxonomy of Foliicolous Hypomycetes with special reference to *Phaeoramularia.*
46.	Dr. R. N. Kharwar	1997	Taxonomy of Foliicolous Hypomycetes with special reference to *Mycovellosiella' Senella* and allied forms.
47.	Dr. Archana Singh	1997	Morphotaxonomic Studies of leaf inhabiting Hypomycetes of Terai Forests U. P.
48.	Dr. Archana Singh	1998	Taxonomy of some leaf spot Fungi.
49.	Dr. Ramesh Chandra Verma	1999	Morphotaxonomic Studies on Black Mildews from Nepal Himalayas.

50.	Dr. H. D. Bhartiya	1999	Morhotaxonomic Studies on *Cercospora Fres. Sensu-Stricto*."
51.	Dr. Ravindra Nath Chaube	1999	Studies on Biodiversity of some fleshy Fungi of Ethanobotanical food use.
52.	Dr. Harish Pd. Mishra	1999	Cultivation and Nutritive value of some Mushrooms with reference to uncommon edible forms.
53.	Dr. Madhava Sharan Tripathi	2001	Studies on Taxonomy of Foliicolous Fungi causing Tar-spots and Allied Symptoms in Indian Sub-Continent.
54.	Dr. Km. Nisha Kumari	2001	Morphotaxonomic Bidiversity of Foliicolous cladosporia in Indian Sub-continent.
55.	Dr. Sanjay Kr. Sharma	2002	Studies on combining ability and genetic parameters in Bread Wheat (*Triticum aestivum* L.)
56.	Km. Rita Singh	2002	"Detailed Taxonomic Treatment to Foliicolous form genus *Pseudocercospora* with reference to infrageneric taxa from Indian Sub-continent. (Viva Voce Awaited).
57.	Km. Anuradha	2002	Morphotaxonomic Studies on *Cercosporella* and *Pseudocercosporella* from Indian Sub- continent. (Viva Voce Awaited).
58.	Km. Sapna Chaudhary	2002	Systematic study of Cercosporoid Fungi with special reference to Sirosporium and Stenella. (Viva Voce Awaited).
59.	Km. Nidhi Sharma	2002	Morphotaxonomic Studies of Foliicolous Hyphomycetes with special reference to *Corynespora* and *Drechslera* (Viva Voce Awaited).
60.	Mr. Balwant Kr. Chaudhary	2002	Effect of different spacings and levels of nitrogen on growth, yield and quality of Rajmash (Viva Voce Awaited)
61.	Ms. Swati Singh	2002	Morphotaxonomy of Follicolous Hyphomycetes with reference to *Phaeoisariopsis* and *Phaeoramularia* (Summitted)

Research Papers Published by Prof. Kamal

1. Bhargava, K. S., R. D. Joshi and Kamal (1966). Occurrence of *Cephaleuros* in Gorakhpur. Indian Forester, 92: 589-590.
2. Bhargava, K. S., R. D. Joshi and Kamal (1960). Additional hosts of *Cephaleuros* in Gorakhpur Univ. Res. Jour., 2: 41.
3. K.S.Bhargava and Kamal (1969). Occurrence of *Helminthosporium pedicellatum* in Gorakhpur, India. Plant. Dis. Reptr., 52: 477-478.
4. Kamal and K. S. Bhargava (1969). Two new species of *Aspergillus* from soils, Trans. Mycol. Soc., 52: 336-340.
5. Kamal and C. S. Singh (1969). Rhizosphere mycoflora of certain ornamental Plants, Sci. & Cult., 35: 635-637.
6. Studies on soil fungi from teak froests of Gorakhpur III. Some Aspergilli new to India. Indian Phytopath., 32: 376-381. (1969).
7. Kamal and K. S. Bhargava (1968). Studies on soil fungi from teak forests of Gorakhpur. III. Additions to our knowledge of Indian Aspergilli. Indian Phytopath., 22: 376-381.
8. Kamal and K. S. Bhargava (1970). Studies on soil fungi from Teak forests of Gorakhpur IV. Isolation of fungi by soil plate and soil dilution on plate methods. Sydowia, 24: 233-240.
9. Kamal and K. S. Bhargava (1970). Studies on soil fungi from teak forests of Gorakhpur. V. A. contribution to Indian Torulaceae. India Phytopath., 23: 558-561.
10. Kamal and K. S. Bhargava (1970). Studies on soil fungi from Teak forests of Gorakhpur VI. A study on Penicillia from three teak stands of different ages. Proc. Nat. Acad. Sci. India, 40 (B): 191-195.
11. Kamal and C. S. Singh (1970). Studies on soil fungi from Teak forests of Gorakhpur VIII. A comparative account of fungi of Earthworm casts, Termitarium and Surrounding soil from a teak stand. Annales Inst. Pasteur., 119: 249-259.
12. Kamal and C. S. Singh (1970). On Rhizosphere fungal flora of some ferns from-Darjeeling (India), Annales Inst. Pasteur, 119: 360-368.
13. Kamal and C. S. Singh (1970). A study on Rhizosphere mycopopulation of some pteridophytes. Annales Inst. Pasteur, 119: 468-482.
14. Kamal and C. S. Singh (1970). Succession of fungi on decaying leaves of some pteridophytes. Annales Inst. Pasteur, 119: 468-482.
15. Kamal and C. S. Singh (1970). Rhizosphere mycoflora of some Bryophytes. Annales Inst. Pasteur, 119: 752-755.
16. Kamal and C. S. Singh (1970). Rhizosphere mycoflora of a virus Infected grass (Cynodon dactylon Linn.). Port. Acta. Bio., 11: 359-364.
17. Singh C. S. and Kamal (1970). A study of some Micro-organic contents of a Dwaba soil. Sydowia, 24: 217-222.
18. Mishra, R. R. and Kamal (1970). Rhizosphera mycoflora of virus infected plants. Indian Phytopath., 23: 27-31.
19. Kamal and K. S. Bhargava (1971). Studies on soil fungi from Teak forests of Gorakhpur VII. Gorund vegetation and distribution of fungi in three Teak stands. Proc. Nat. Acad. Sci. India, 41 (B): 395-403.
20. Kamal and R. L. Chaudhary (1971). Rhizosphere fungal flora of two medicinal plants. Sci. & Cult., 37: 489-490.

21. Kamal and K. S. Bhargava (1971). Studies on soil fungi from Teak forests of Gorakhpur IX. Seasonal distribution. Proc. Nat. Acad. Sci. India, 41 (B): 427-436.

22. Mishra, R. R. and Kamal (1971). Aeromycology Air of Gorakhpur. III. Seasonal variation in fungal spora. Mycopath. Mycol. Appl., 45: 301-310.

23. Mishra, R. R. and Kamal (1971). Aeromycology Air of Gorakhpur. II. Seasonal variation in fungal spora. Mycopath. Mycol. Appl., 45: 310-316.

24. Mishra, R. R. and Kamal (1972). Rhizosphere fungal flora of certain Euphorbiaceous plants. Mycopath. Mycol. Apple., 46: 73-79.

25. Kamal and K. S. Bhargava (1972). Studies on soil fungi from Teak forests of Gorakhpur. II. A contribution to Indian Aspergilli. Mycopath. Mycol. Appl., 47: 59-72.

26. Kamal and A. B. Sinha (1972). Anomaly in the germination behavior of Gram (*Cicer arietinum* L.) seeds. Sci. & Cult., 38: 404-405.

27. Kamal and M. L. Gupta (1972). Studies on Aspergilli from different soils of Gorakhpur. I. Effect of Maleic Hydrazide on some Aspergilli. Sci. & Cult., 38: 212-214.

28. Kamal and M. L. Gupta (1973). Studies on Aspergilli from different soils of Gorakhpur II. Effect 2, 4-D on the members of *Aspergillus flavipes* group. Sci. & Cult., 39: 320-323.

29. Kamal and M.L. Gupta (1973). Studies on Aspergilli from soils of Gorakhpur I. Effect of maleic hydrazide on the growth of *Aspergillus flavipes* group. Sci. & Cult., 28: 212-214.

30. Kamal and K. S. Bhargava (1973). Studies on soil fungi from Teak forests of Gorakhpur. X. Edaphic factors and distribution of soil microfungi in teak stands of different ages. Proc. Nat. Acad. Sci. India, 43(B): 16

31. Kamal and M. L. Gupta (1973). Studies on Aspergilli from soils of Gorakhpur IV. The genus *Aspergillus* in relation to other microfungi in Usar soils. Proc. Nat. Acad. Sci. India., 43 (B): 229-236.

32. Kamal and M. L. Gupta (1973). Permeability of *Chenopodium amaranticolor* Coste. Seeds to water as affected by temperature. Planta, 3: 17-18.

33. Kamal and M. L. Gupta (1974). Studies on Aspergilli from soils of Gorakhpur. III. Some Aspergilli new to India. Indian Phytopath., 27: 29-32.

34. Kamal and A. B. Sinha (1974). Studies on Aspergilli from soils of Gorakhpur. IV. Effect of culture filtrates of Certain Aspergilli on germination of *Abelmoschus esculentus* seeds. Tropical Ecol., 15: 11-15.

35. Kamal and A. K. Verma (1974). Studies on Aspergilli from soil of Gorakhpur. V. Effect to culture filtrate of some Aspergillus species of *Helmintosporium rostratum*. An approach to Biological control. Proc. Nat. Acad. Sci. India, 44: 221-224.

36. Kamal and N.P. Singh (1974). On Microfungi from root region of ten sugarcane varieties. Indian Phytopath., 27: 347-354.

37. Kamal, K. S. Srivastava and S. N. Pandey (1974). On Mycoorganic components of Soyabean Rhizosphere part I-Rhizosphere Fungi as influenced by soil application of some NPK Fertilizers. Fert. Technol., 11: 289-291.

38. Kamal and K. S. Srivastava (1974). Rhizosphere microfungi of some weeds. Fert. Technol., 11: 440-442.

39. Kamal, S. N. Pandey and R. A. Sharma (1974). Studies on Soyabean (Glycine max (linn.) Merr.) Part I. Effect of some NPK Fertilizers on growth on flowering. Fert. Technol., 11: 429-431.

40. Kamal, T. P. Mall, R. P. Srivastava (1974). Microfungi at he root soil interface of Arhar (Cajanus cajan Linn. Mill sp.) Infected by Pigeon Pea Sterility Mosaic visus. Fert. Technol., 11: 434-436.

41. Kamal and N. P. Singh (1974). An investigation on mycoorganic content of Air over sugarcane field at Gorakhpur (U. P.) I. Aspergilli. Proc. Nat. Acad. Sci. India., 44 (B): 134-138.

42. Kamal and N. P. Singh (1974). An investigation on mycoorganic content of air over sugar field at Gorakhpur (U. P.). II. *Penicillium, Paecilomyces* and *Scopulariopsis* Proc. Nat. Acad. Sci. India, 44 (B): 156-160.

43. Kamal, T. P. Mall and R. P. Srivastava (1974). Rhizosphere mycoflora of some visus infected cultivars of arhar (*Cajanus cajan* Linn.) Mill sp.-Fert. Technol., 11: 234-238.

44. On myco-organic components of Soyabean rhizosphere Part- III. Effect of Soybean mosaic virus infection on rhizosphere fungal flora. Technol., 12: 67-69. (1974).

45. Kamal and Lalji Srivastava (1975). Succession of micro-fungi on a decaying liverwort (*Cythodium* sp.) Proc. Nat. Acad. Sci. India, 45 (B): 185-191.

46. Kamal, K. S. Srivastva and S. Singh (1975). On Myco-organic components of soyabean Rhizosphere part. II. Rhizosphere effect as influenced by soil application of some NPK fertilizers. Fert. Technol., 12: 135-138.

47. Kamal, U. P. Gupta and K. S. Srivastava (1975). On Myco-organic component of soyabean Rhizosphere part. III. Effect of soyabean mosaic visu infection on Rhizosphere fungal flora. Fert. Technol., 12: 67-69.

48. Kamal, S. N. Pandey and R. A. Sharma and S. Singh (1975). Studies of Soyabean (Glycine Max Linn.) Merr. Part. II-Effect of some NPK Fertilizers on yiels. Fet. Technol., 12: 323-328.

49. Kamal and N. P. Singh (1975). Fungal air spora over cowpea and maize fields at Gorakhpur. Proc. Nat. Acad. Sci. India., 45 (B): 53-60.

50. Kamal and N. P. Singh (1975). Succession of fungi on decomposing *Ophioglossum* fronds. Proc. Nat. Acad. Sci. India, 45 (B): 68-74.

51. Kamal and S. Singh (1975). Wet rot of Castor plants. Indian Phytopath., 28: 410-411.

52. Kamal and S. Singh (1975). *Euphorbia geniculata* Linn. A new host record for *Sclerotium rolfsii* Sacc. Curr. Sci., 44: 747.

53. Kamal and K. S, Srivastava (1975). *Aspergillus lanosus*-A new mould producing citric acid. Curr. Sci., 44: 789.

54. Microfungi from the root region of *Vigna mungo* (Roxb.) Linn. Technol., 13: 311-312. (1976).

55. Kamal and S. Singh (1975). Incidence of groundnut rust in North India. Fert. Technol., 12: 73-74.

56. Kamal and S. Singh (1976). *Ophioglossum vulgtum* Linn. A new host of *Curvularia lunata* Wakk. Boed. Curr. Sci., 45-65.

57. Kamal and S. Singh (1976). *Sida acuta*-an addition to the host list of *Sclerotium rolfsii* Sacc. Curr. Sci., 45: 58-59.

58. Kamal and K. S. Srivastava (1976). A new species of *Aspergillus* producing citric acid. Indian J. Mircobiol., 16: 49-50.

59. Kamal, M. L. Gupta and V. Nath (1976). Studies on Aspergilli from different soils of Gorakhpur. VIII. Effect of light and dark on the growth performance of *Aspergillus niveus* Bloch. Sci. & Cult., 42: 163-164.

60. Kamal and S. Singh (1976). Some new hosts of *Sclerotium rolfsii* from India. Indian Phytopath., 29. 188.

61. Kamal and Ajay Kumar Verma (1976). Micro fungal flora in the root region of arhar (*Cajanus cajan* (L.) (Mill). Tehnol., 13: 155-157.
62. Kamal and S. Singh (1976). Sclerotial wilt of *Ocimum canum* Sims. A new record from India. Curr. Sci., 45: 235.
63. Kamal and S. Singh (1976). Wet rot of -Sesame seedlings in India Sci. & Cult., 42: 269-270.
64. Kamal and P. Kumar (1976). Micro-fungal flora in the root region of Arhar (*Cajanus cajan* (L.) Mill. Sp.). Fert. Technol., 13: 155-157.
65. Kamal and Ajay Kumar Verma (1977). Collar rot of cowpea caused by *Rhizoctonia solanii* Kuhn-New to India. Geobios, 4: 282.
66. Kamal, S. Singh and R. P. Singh (1977). Fungi of Gorakhpur. I. Indian Phytopath., 30: 186-188.
67. Kamal and Ajay Kumar Verma (1977). Fungal air spora over Urd (*Vigna radiata* Roxb.) Linn.) at Gorakhpur as obtained by settle plate method. Proc. Nat. Acad. Sci. India, 47 (B): 241-246.
68. Kamal and R. P. Singh (1978). A new species of Verrucisporota (Hyphomcete) from India. Can. J. Bot., 56: 2785-2786.
69. Kamal S. Singh and R. P. Singh (1978). Fungi of Gorakhpur. II. Fert. Techonol, 15: 71-72.
70. Singh, S. and Kamal (1978). Fungi of Gorakhpur. III. Indian Phytopath., 31: 203-205.
71. Kamal, R. P. Singh and K. S. Bhargava (1978). Sour rot of lemon-New to India. Indian Phytopath., 31: 76-77.
72. Singh, S. and Kamal (1978). Fungi of Gorakhpur. V. Indian Phytopath., 31: 204-206.
73. Singh, S. and Kamal (1978). Fungi of Gorakhpur. VI. Indian Phytopath., 31: 315-317.
74. Kamal and R. P. Singh (1978). Fungi of Gorakhpur. IV. Some new host records from India. Acta Bot. Indica, 6: 188-189.
75. Kamal, S. Singh and R. P. Singh (1979). A. new species of *Periconiella* from Indian. Curr., 48: 211-212.
76. Kamal and Ajay Kumar Verma (1979). Air spora studies over Arhar field-Periodicity of some fungal spores, *Acremonium*. *Acrophialophora* and *Paecilomyces*. Biol. Bull. Indian, 1: 47-52.
77. Kamal and R. P. Singh (1979). Fungi of Gorakhpur. VII. Indian J. Mycol. Pl. Pathol., 9: 170-172.
78. Kamal and S. Singh (1979). Fungi of Gorakhpur. VIII. Indian J. Mycol. Pl. Pathol., 9: 173-176.
79. Kamal, S. Singh and R. P. Singh (1979). Fungi of Gorakhpur. IX. Indian J. Mycol. Pl. Pathol., 9: 245-246.
80. Kamal and S. Singh (1979). Fungi of Gorakhpur. IX. B. Indian Phytopath., 32: 46-50.
81. Kamal and S. Singh (1979). Fungi of Gorakhpur. IX. Indian Phytopath., 32: 216-220.
82. Kamal, R. P. Singh and P. Kumar (1979). Fungi of Gorakhpur. X.Indian Phytopath., 32: 216-220.
83. Kamal, R. P. Singh and P. Kumar (1979). Fungi of Gorakhpur. XI. Indian Phytopath., 32: 229.
84. Kamal, P. Kumar and Rekesh Pratap Singh (1979). New host records of *Collectotrichum* species. Indian Phytopath., 32: 461-462.
85. Kamal and Ajay Kumar Verma (1979). Seed borne Mycoflora of Arhar (T-21)-Effect of culture filtrates of some isolates on seed germination and fungicidal treatment. Indian. J. Mycol. Pl. Pathol., 9: 41-45.
86. Kamal, M. L. Gupta and P. Kumar (1979). Studies on Aspergilli from soils of Gorakhpur VIII-Distribution in four soil types as against plant cover. Indian. J. Mycol. Pl. Pathol., 9 (1): 66-74.
87. Kamal, M. L. Gupta and P. Kumar (1979). Studies on Aspergilli from soils of Gorakhpur X. Seasonal Distribution in four soil types. Indian. J. Mycol. Pl. Pathol., 9: 150-155.

88. Kamal, M. L. Gupta and P. Kumar (1979). Aspergilli from soils of Gorakhpur. IX. Edaphic factors and distribution of Aspergilli- in four soil types. Indian J. Mycol. Pl. Pathol., 9: 56-65.

89. Srivastava, K. S. and Kamal (1979). Production of Citric acid by *Aspergillus niger*. Indian J. Mircrobiol., 19: 145-149.

90. Srivastava, K. S. and Kamal (1979). Effect of some chemical factors on the growth and citric acid production by *Aspergillus niger*. J. Indian. Bot. Soc., 58: 208-214.

91. Kamal and G. Chandra (1979). On sclerotial rot of *Nigella sativa* and *Alternaria* leaf spot of *Annona squamosa*-New to India. Indian J. Mycol. Pl. Pathol., 9: 88.

92. Kumar, P. and Kamal (1979). A new species of *Periconniella* from India. Curr. Sci., 48: 402-403.

93. Kumar, P. and Kamal (1979). A new species of *Pseudocercospora*. Curr. Sci., 48: 654.

94. Kumar, P. and Kamal (1979). *Pseudocercospora*. Contraria (H. & P. Sydow) Deighton-A new record from India. Curr. Sci., 48: 654.

95. Kumar, P. and Kamal (1979). A new species of *Phyllosticta* from India. Curr. Sci., 48: 1082-1083.

96. Kamal and R. P. Singh (1979). Knife cut disease of Sugarcane in India Sugarcane Pathol. News Letter, 23: 34.

97. Singh, R. P. and Kamal (1979). Growth of *Colletotrichum falcatum* on media containing juice of different sugarcane varieties. Sugarcane Pathol., 23: 27-28.

98. Kamal, B. Rai and P. Kumar (1980). A new species of *Stenella* from India. Curr. Sci., 49: 234-235.

99. Kamal and P. Kumar (1980). Fungi of Gorakhpur. XIII. Indian. Phytopath., 33: 265-269.

100. Singh, S., Kamal and K. S. Srivastava (1980). Plant parasitic fungi of Gorakhpur. XXIII. Indian J. Mycol. Pl. Pathol., 10: 104-105.

101. Kamal, P. Kumar and D. N. Shukla (1980). Fungi of Gorakhpur-XIII. Indian Phytopath., 33: 265-269.

102. Kamal, Rai A. N. and Kumar P. (1980). A new species of Sterellar from India. Curr. Sci., 49: 234-235.

103. Kamal, P. Kumar and D. N. Shukla (1980). Fungi of Gorakhpur. XIV. Indian Phytopath., 33: 462-464.

104. 100. Kamal, P. R. Singh and P. Kumar (1980). Fungi of Gorakhpur. XIX. *Stenella*. Sydowia, 33: 162-166.

105. 101. Kamal and P. Kumar (1980). Fungi of Gorakhpur. XVII. *Sarcinella*. Sydowia, 33: 152-156.

106. 102. Kamal and P. Kumar (1980). Fungi of Gorakhpur. XXVA. Environment India., 3: 1-4.

107. 103. Kamal and R. P. Singh (1980). Fungi of Gorakhpur. XXVI. Environment India, 3: 14-17.

108. Kamal, P. Kumar and R. P. Singh (1980). Fungi of Gorakhpur. XVI. Indian Phytopath., 33: 551-556.

109. Srivastava, K. S. and Kamal (1980). Citric acid production by Aspergilli. II. Strain selection. Indian J. Mycol. Pl. Pathol., 10: 23-29.

110. Kamal, P. Kumar, R. P. Singh and D. N. Shukla (1980). Fungi of sugarcane (*Saccharum officinarum*) as influenced by knife cut disease. Bio. Bull. India., 2: 56-59.

111. Singh, R. P. and Kamal (1980). Natural incidene of red-rot in Eastern tract of Utter Pradesh. Cooperative sugar, 11: 1.

112. Kamal and R. P. Singh (1980). Fungi of Gorakhpur. XVIII. *Pseudocercospora* Sydowia, 33: 157-161.

113. Rai, A. N., B. Rai and Kamal (1980). *Walkeromyces* Thaung and *Parapithomyces* Thanung-Two new generic records from India. Curr. Sci., 49: 480-483.

114. Kamal and R. P. Singh (1980). Fungi of Gorakhpur. XXVb. Indian Phytopath., 33: 594-596.

115. Verma A. K. and Kamal (1980) Airspora studies over arhar (Cajanus cajan (L.) Millsp.) field periodicity in *Drechsiera, Epicoccum* and *Penicillium* Nat. Acad. Sci. Letter, 3: 13-17.

116. Kumar, P. and Kamal (1981). Tow new species of *Pseudocercospora* from India. Curr. Sci., 50: 1029-1031.

117. Singh, R. P., Kamal and P. Abbasi (1981). A new species of *Veronaea*. Curr. Sci., 50: 236-237.

118. Kumar, P. and Kamal (1981). Two new species of *Mycovellosiella* Rangel from Gorakhpur. Biol. Bull India, 3: 117-120.

119. Kamal, P. Kumar and B. Rai (1981). A new interesting Hyphomycetes from India. Indian J. Mycol. Pl. Pathol., 11: 144-145.

120. Kamal and R. P. Singh (1981). Fungi of Gorakhpur. XXVII. Indian J. Mycol. Pl. Pathol., 11: 1-4.

121. Kamal, R. P. Singh and P. Kumar (1981). Fungi of Gorakhpur. XXVIII. Indian J. Mycol. Pl. Pathol., 11: 35-39.

122. Kaml, R. P. Singh and K. P. Verma (1981). Effect of Red-rot of Sugarcane on the yield and juice quality. Coop. Sugar, 12: 1-2.

123. Kumar, P. and Kamal (1981). A new species of *Mycovellosiella* from India Curr. Sci., 50: 137-138.

124. Kamal, R. P. Singh and Anil K. Singh (1981). Fungi of Gorakhpur. IVb, Environment India., 4: 43-50.

125. Kamal, A. K. Verma and R. P. Verma (1981). Effect of Culture filterates of certain Aspergilli on germination of Arhar (Cajanus cajan Linn.) Mill Sp. Seeds, Bio. Bull. India, 3: 182-186.

126. Kumar, P. and Kamal (1981). Two new species of fungus genus *Cercospora* Fresh. Curr. Sci., 50: 508-510.

127. Verma, A. K., B. K. Gupta and Kamal (1981). Airspora studies over arhar seasonal distribution of fungal airspora as obtained by gravity plate method. Environment India, 4: 56-61.

128. Shukla, D. N., Anil K. Singh, P. Kumar and Kamal (1982). Fungi of Gorakhpur. XV. Indian Phytopath., 35: 86-91.

129. Singh, S. K., Kamal and A. K. Singh (1982). *Septoria crepidis* Vest. A new record from India. Curr. Sci., 51: 144-145.

130. Rai, A. N., S. K. Singh and Kamal (1982). Addition to the new host records for *Sclerotium rolfsii* (Sacc.) Curz. From India. Environment India, 5: 55-56.

131. Rai, A. N., Kamal and S. K. Singh (1982). A new species of *Mycovellosiella* Rangel. Curr. Sci., 51: 781-782.

132. Kumar, P. and Kamal (1982). A new species of *Mycovellosiella* from India Curr. Sci., 51: 846-847.

133. Rai, A. N. and Kamal (1982). A new species of *Pseudocercospora* Speg. Curr. Sci., 51: 287-288.

134. Verma, A. K., R. P. Verma and Kamal(1982). Airspora over Arhar field. Indian Phytopath., 35: 126-130.

135. Kamal and A. N. Rai (1982). *Pseudocercospora carissae* synonym of *Sirosporium carissae* Indian Phytopath., 35: 316-317.

136. Kamal, B. K. Gupta and P. Kumar (1982). Studies on Aspergilli from soils of Gorakhpur. XII. Effect of Chemical factors on distribution of Aspergilli in four soil types. Environment India, 5: 47-50.

137. Kamal, R. P. Singh and D. N. Shukla (1982). Fungi of Gorakhpur. XXXVIII. Indian J. Mycol. & Pl. Pathol., 12: 160-164.

138. Kamal, R. P. Singh and P. Kumar (1982). Fungi of Gorakhpur. XXX. Indian J. Mycol. Pl. Pathol., 12: 156-159.

139. Kamal, C. Gupta and A. N. Rai (1982). Two new species of *Pseudocercospora* from Indian Phytopath., 35: 316-317.

140. Verma, A. K. and Kamal (1982). Air spora studies over an arhar (Cajanus cajan) field- the diurnal and seasonal periodicity of certain plant pathogens. Indian J. Mycol. Pl. Pathol., 12: 40-45.

141. Kamal, S. Singh and Rakesh P. Singh (1982). Fungi of Gorakhpur. XXX. Indian J. Mycol. Pl Pathol., 12: 156-159.

142. Verma, A. K. and Kamal (1982). Air spora studies over an Arhar (Cajanus cajan) Field-The Diurnal and seasonal periodicity of certain plant pathogens. Indian J. Mycol. Pl Pathol., 12: 40-45.

143. Kamal, S. Singh and rakesh. P. Singh (1982). Fungi of Gorakhpur. XX. Indian J. Mycol. Pl. Pathol., 12: 71-73.

144. Singh, A. K. and Kamal (1982). Fungi of Gorakhpur. Ivc. Awadh Univ. J. Sci., 1: 17-23.

145. Kamal, K. S. Srivastava and B. K. Gupata (1982). Citric acid Production by Aspergilli-Some cultural conditions (Hydrogen ion concentration. Incubation period and Temperature) affecting citric acid production by *Aspergillus niger* starin. Life Sci. Adv., 1B: 89-93.

146. Rai, A. N. and Kamal (1983). Sone new host records for *Acromenium zonatum* from India. Indian Phytopath., 36: 160-161.

147. Kamal, Rakesh Pratap and Anil K. Singh (1983). Effect of extracts from different aerial parts of the susceptible sugarcane variety on growth of *Colletotrichum falcatum* Went. Indian Sugar Crop, J. 9: 19.

148. Kamal, A. N. Rai and G. Morgan-Jones (1983). Notes on Hyphomycetes. XLV. Neopericonea. A new *Phaeodictyosperous* genus from India. Mycotaxon 18: 15-18.

149. Kamal, A. N. Rai and G. Morgan-Jones (1983). Notes on Hyphomycetes. XLVI. *Parafulvia*. A new Foliicolous Phaeophragmosporous genus with catenates coidia. Mycotaxon, 18: 67-71.

150. Singh, R. P., Anil K. Singh and Kamal (1983). Fungi in the Rhizosphere and Rhizoplane of sugarcane seedlings. J. Indian Bot. Soc., 62: 419-420.

151. Singh, Rakesh P., A. K. Singh and Kamal (1984). Effect of nodal and Stalk infections of red-rot on quality of sugarcane juice. Indian Sugar Crop., J. 10: 15-16.

152. Kamal, A. N. Rai and G. Morgan-Jones (1984). *Pseudochuppia*. A new foliicolous, dictyosporic hyphomycete genus from India. Mycologia, 76: 163-165.

153. Singh, Anil K. and Kamal (1984). *Pseudocercosporella pogostemoniae* sp. nov. A new Hyphomycets from India. Curr. Sci., 53: 544-545.

154. Abbasi, P. and Kamal (1984). Fungi of Gorakhpur. XXXVI. Indian J. Mycol. & Pl. Pathol., 14: 126-130.

155. Kamal, A. N. Rai and G. Morgan-Jones (1984). Notes on Hyphomycetes. XLVII. New species of Sarcinella and Sirosporium. Mycotaxon, 20: 589-594.

156. Morgan-Jones, G. and Kamal (1984). Notes on Hyphomycetes. XLVIII. Fuligomyces. A new foliicolous, anamorphic sooty mould genus from Uttar Pradesh. Mycotaxon, 20: 595-598.

157. Singh, A. K., Kamal and S. K. Singh (1985). A new species of *Pseudocercospora* Speg. Curr. Sci., 54: 144-145.

158. Singh, A. K. and Kamal (1985). Fungi of Gorakhpur. XXXVI. J. Indian Bot. Soc., 64: 144-147.

159. Kamal, B. Rai and G. Morgan-Jones (1985). Notes on Hyphomycetes. L. Two new species of *Sirosporium* from India. Mycotaxon, 24: 313-318.

160. Singh, A. K. and Kamal (1985). New fungus record from India. Indian Phytopath., 38: 388.

161. Narayan, P. and Kamal (1985). Some new fungal records from India. Indian Phytopath., 38: 389.

162. Rai, B. and Kamal (1985). A new species of *Parapyricularia* from India. Can J. Bot., 63: 988-989.

163. Kamal, C. Gupta and A. N. Rai (1985). Two new species of *Pseudocercospora* from India. Indian Phytopath., 38: 704-706.

164. Rai, B. and Kamal (1985). New Hyphomycetes from India. Trans. Brit. Mycol. Soc., 85: 566-570.

165. Kamal, R. K. Verma and G. Morgan Jones (1986). Notes on Hyphomycetes. II. *Kameshwaromyces*. A new foliicolous shooty mould like genus from Madhya Pradesh

166. Morgan-Jones, G., Kamal and R. K. Verma (1986). *Goosiella*. A new Pteridocolous, helicosporous Hyphomycete genus from Madhya Pradesh. India. Mycologia, 78: 496-500.

167. Rai, B. and Kamal (1986). New Hyphomycetes from India. Trans. Brit. Mycol. Soc., 85: 566-570.

168. Singh, Anil K. and Kamal (1966). A new species of *Stenella* from India. Indian Phtopath., 39: 274-276.

169. Rai, A. N., B. Rai and Kamal (1986). A new species of *Mycovellosiella* Rangel from India. Curr. Sci., 55: 412-413.

170. Verma, R. K. and Kamal (1986). *Paraaoria himalayana* gen. et. sp.nov. A foliicolous coelomycete from Nepal. Trans. Br. Mycol. Soc., 87: 645-647.

171. Narayan, P. and Kamal (1986). New foliicolous Hyphomycetes from India. Can. J. Bot., 64: 201-207.

172. Kamal and P. Narayan (1986). Fungi from Hilly tracts of Uttar Pradesh l. Indian Phytopath., 39: 198-203.

173. Kamal and P. Narayan (1986). Fungi from Hilly tracts of Uttar Pradesh I. Indian Phytopath., 39: 198-203.

174. Kamal, Narayan, P. and R. P. Verma (1986). Fungi from Hilly tracts of U. P. II. Indian Phytopath., 39: 453-458.

175. Rai, A. N. and Kamal (1986). A new species of *Mycovellosiella* Rangel, Curr. Sci., 55: 798-799.

176. Kumar, R. and Kamal (1986a). New host records of some foliicolous fungi from India II. Indian Phytopath., 39: 627.

177. Kamal and Gupta, B. K. (1986). Fungi of Gorakhpur. XL. Indian J. Mycol. Pl. Pathol., 16: 245-248.

178. Singh A. K. and Kamal (1986). Fungi of Gorakhpur. XXXIX. Kavak, 13-14: 25-24.

179. Rai, A. N. and Kamal (1986). Fungi of Gorakhpur. XIII. Kavak, 14: 31-36.

180. Cannon, P. F., Minter, D. W. and Kamal (1986). *Vladracula* P. Cannon, Minter and Kamal gen. nov. in : The Rhytismataceae of the Indian sub-continent. Mycol., Papers No. 155: 117-119.

181. Verma R. K. and Kamal (1986). *Paraaoria himalayana* gen. et. sp. nov. A foliicolous coelomycete in citrus from Nepal. Trans. Br. Mycol. Soc., 87: 645-647.

182. Rai, A. N. and Kamal (1987). New *Cercospora* species from India. Trans. Br. Mycol. Soc., 89: 124-126.

183. Kamal, B. K. Gupta and R. K. Verma (1987). New species of *Cercospora* and *Cerosporodium* from Uttar Pradesh. Can. J. Bot., 65: 1259-1261.

184. Kamal, P. Narayan and R.K Verma (1987). Fungi from hilly tracts of Uttar Pradesh. II. Indian Phytopathology., 39: 453-458.

185. Verma, R.K. and Kamal (1987a). Studies on Foliicolous Ascomycetes. II. Genus. Actinomycbe Hobnel. Indian Phytopathology, 40: 410-413.

186. Verma, R.K. and Kamal (1987b) Studies on Foliicolous Ascomycetes. III,. Some Interesting bitunicates. Indian Phytoplathology, 40: 459-465.

187. Abbasi, P.R., Kumar and Kamal (1987). Some new foliicolous Cercoseptona from Uttar Pradesh, India. Pers. Mycol. Res., 1. Prof. Agrawal Festchrift: 35-41.

188. Verma. R.K. and Kamal (1987). Agarwalomyces indicus. Gen et sp. nov. Fructicolous synnematous genus from Uttar Pradesh, India. Trans Br. Mycol Soc., 89: 596-599.

189. Gupta, B.K. and Kamal (1987). Some new species of *Pseudocercospora* from India. Pers Mycol. Res. Prof Agrawal festschrift: 19-23.

190. Verma, R.K and Kamal (1987). New species of *Pseudocercospora, Phaeoisariopsois, Sarcenillea* and *Stenella* from Uttar Pradesh and Bihar. Trans Br. Mycol. Soc., 89: 67-72.

191. Gupta, C., abbasi. P.and Kamal (1987). Some new foliicolous Hyphomycetes from North-Eastern U.P. Pers Mycol. Res.-1 Prof. Agrawal Festschrift. 7: 17.

192. Rai, A.N. and Kamal (1987). A new *Stenella* species from India. Mycol. Res, 83: 398-399.

193. Rai and Kamal (1988) A new species *Phaeoisariopsis* Mun. Curr. Sci., 57: 670-672.

194. Verma, R.K., Kamal. M.K. Khan (1988). Some new species of *Septoria* Sacc. From India. Indian Phytopathol., 41: 25-29.

195. Rai, A. N. and Kamal (1989). Two new species of *Psedocercospora* from India. Curr Sci., 58: 1327-1330.

196. Rai, A, N, and Kamal (1989). New *Cercospora* species from India. Trans Br. Mycol. Soc., 89: 124-126.

197. Verma, R.K. and Kamal (1989) A new species and combination of *Mycovellosiella*. Mycol Res., 93: 230-232.

198. Kamal, M.K. Khan and U. budathoki (1989) Peculiar foliicolous Hyphomycetes from Kathmandu Valley, Nepal Ana Net, 15: 5.

199. Budathoki, U., R. K. Verma and Kamal (1989). New species of *Phaeoramularia. Pseudocercospora* and *Stenella* from Nepal. Indian Phytopath., 42: 53-60.

200. Gupta, B. K. and Kamal (1989). New species of *Pseudocercosporella*. Indian Phytopath., 42: 388-392.

201. Verma, R. K., M. K. Khan and Kamal (1989). Some new species and combinations of. *Pseudocercospora*. Mycol. Res., 92: 347-353.

202. Verma, R. K. and Kamal (1989). Studies on Foliicolous Ascomycotina. IV. New species of *Aphanostigma* and *Phyllachora*. Indian Phytopath., 42: 561-565.

203. Budathoki, U. Verma, R. K. and Kamal (1989). New species of *Phaeoramularia. Pseudocercospora* and *Stenella* from Nepal. Indian Phytopath., 42: 53-60.

204. Verma, R. K., R. Chand and Kamal (1990). Two new species of *Uncinula* from Nepal. Mycol. Res., 94: 128-130.

205. Kamal, M. K. Khan and R. K. Verma. (1990). New species and Combinations in. *Pseudocercospora* from India. Mycol. Res., 94: 240-242.

206. Kamal, B. K. Gupta and C. Gupta (1990). Three new species of. *Pseudocercospora* from Uttar Pradesh. Indian Phytopath., 43: 143-149.

207. Verma, R. K. and Kamal (1990). *Pyrenula corticola* sp. nov ; a bark inhibiting ascomycetes from Uttar Pradesh. Indian Phytopath., 43: 568-569.

208. Kamal, Moses, A. S. and Choudhary, R. (1990). Two new species and a new combination in. *Pseudocercospora* from U. P., India. Mycol. Res., 94: 714-717.

209. Rai, A. N. and Kamal (1990). A new species of *Stenella* from India. Indian Phytopath., 43: 214-216.

210. Lakshmi Kant, Kamal and S. N. Mathur (1990). Cellulolytic activities of *Chaetomium globosum* on different cellulosic substrates. World J. Microbiol. Biotechnol., 6: 23-26.

211. Kamal, A. N. Rai and A. S. Moses (1991). Some new species and new combinations in *Pseudocercospora* from India. Mycol. Res., 95: 403-406.

212. Chaudhary, R., C. Gupta and Kamal (1991). New species of *Hotoroconium. Pseudocercospora* and *Stenella* from India. Mycol. Res., 95: 1070-1073.

213. Verma, R. K. and Kamal (1991). Some new species of *Pseudocercospora* Indian Phytopath., 94: 440-447.

214. Bist, S. S. and Kamal (1991). Two new storage rots of Gaderi (*Colocasia antiquorum*) from Kumaon. Indian Phytopath., 44: 444-445.

215. Khan, M. K., U. Budathoki and Kamal (1991). New foliicolous Hyphomycetes from Kathmandu valley, Nepal. Indian Phytopath., 44: 21-29.

216. Khan, M. K., R. K. Verma and Kamal (1992). New species of *Cercospora, Mycovellosiella* and *Phaeiosariospsis*. Indian Phytopath., 45: 26-34.

217. Rai, A. N. and Kamal (1992). Two new species of *Ascochyta* from India. Indian Phytopath., 45: 23-25.

218. Srivastava, K., A. K. Srivastava and Kamal (1993). *Stenellopsis xeromphigena* sp. nov. from North-Eastern, Uttar Pradesh. Indian Phytopath., 46: 243-244.

219. Khan, M. K., Kamal and G. Morgan-Jones (1993). Notes on Hyphomycetes. IXII. Novel species of *Fuligomces* and *Tretospora* from India, with comments on the genera. Mycotaxon, 69: 477-450.

220. Kumar, R., A. N. Rai. and Kamal (1994). Two new Hyphomycetes from Indian sub-continent. Indian Phytopath., 47: 127-132.

221. Kamal (1994). Biological status of Fungi-An Appraisal. Current Trends in Life Sciences, 20: 13-21.

222. Srivastava, K., A. K. Srivastava and Kamal (1994). New species of *Stenella* from India. Mycol. Res., 98: 516-520.

223. Srivastava, N., A. K. Srivastava and Kamal (1994). New synnematous foliicolous Hyphomycetes from India. Mycol. Res., 98: 521-524.

224. Khan, M. K., M. S. Khan and Kamal (1994). New species of *Pseudocercospora Stenella* from India. Mycol. Res., 98: 1277-1283.

225. Srivastava, K., A. K. Srivastava and Kamal (1994). Additions to *Stenella* Sydow. From India. Mycol. Res., 99: 233-236.

226. Rao, H. S. G., S. Chandra and Kamal (1995). Additions to *Pseudocercospora* from India. Mycol. Res., 99: 707-710.

227. Rao, H. S. G., A. Singh and Kamal (1995). New species of *Pseudocercospora* associated with leaf spots of forest flora of India and Nepal. Indian Phytopath., 48: 254-259.

228. Srivastava, N., Kamal and G. Morgan Jones (1995). Notes on Hyphomycetes. LXVI. Two new *Ramularia* species from India. Mycotaxon, 54: 49-55.

229. Khan, M. K., Kamal and G. Morgan-Jones (1995). Notes on Hyphomycetes. LXVI. New species of *Mycovellosiella Phaeiosariopsis, Sirosporium* and *Stenella* from India. Mycotaxon, 54: 27-36.

230. Kamal and A. K. Srivastava (1995). Taxonomic treatment to phtopathogenic fungi and managemtn of diseases with reference to Cercospora complex. Proceeding of the Symposium on Managemnt of Microbes in the Service of Mankind, Allahabad, 1992.

231. Bhall K., Srivastava N and Kamal (1996). New Species of *Mycovellosiella* from North-Eastern Uttar Pradesh, India. Mycol. Res., 100: 1333-1336.

232. Rao, H. S. G., A. S. Moses and S. Narayan (1998). Some new species of *Pseudocercospora* associated with leaf spot in forest flora of North-Eastern Uttar Pradesh. Indian Phytopath., 49: 136-141.

233. Rao, H. S. G., A. S. Moses and P. N. Singh (1996). Some new species of *Pseudocercospora* from India. Mycol. Res., 100: 1071-1076.

234. Meenu & Kamal (1998). New species of *Corynespora* Mycol. Res, 102: 344-346.

235. Singh S. K., A. Singh and Kamal (1997). Additions to *Phaeoramularia* and *Stenella* from Indian sub-continent. Mycol. Res., 101: 530-534.

236. Dubey, R., S. K. Singh and Kamal (1999). Taxonomic treatment to *Alternaria* spp. From North-Eastern U. P. Microbiol. Res., 154: 1-4.

237. Singh, S. K., K. Bhalla and Kamal (1999). New Pathogenic *Phaeoramularia* spp. On forest plants Mycol. Res., 103: 271-274.

238. Singh, A., and S. K. and Kamal (2000). Two new species of *Cercospora* from India. J. Mycol.Res., 36: 44-49.

239. S. Chaudhary, Sharma N. and Kamal (2002). The new species of *Stenella* from North-Eastern Uttar Pradesh India. Indian Phytopath, 55(1): 57-60.

Dr. (Mrs.) K. Shukla
Dr. (Mrs.) P. Abbasi

1

Fungal Diversity and Bioremediation

S.B. Sullia

Department of Microbiology and Biotechnology
Bangalore University, Bangalore-560056

The rich diversity of fungi in the tropical areas is now being explored with greater emphasis than before because of the numerous applications of fungi in medicine,industry, agriculture and in cleansing the environment. The fungi are now being increasingly employed in bioremediation, a pollution control technology which uses biological systems for the degradation of toxic chemicals into less harmful ones. The fungi are unique in producing a variety of extracellular enzymes which can degrade natural polymers such as cellulose and lignin as well as some xenobiotic compounds. The fungi are endowed with aggressive growth potential and being filamentous in nature have a greater reach in the soil environment than bacteria. The white rot fungus *Phanerochaete chrysosporium* has been used for the degradation of aromatic hydrocarbons, chlorinated organics, nitrogen aromatics and several miscellaneous compounds such as sulfonated azodyes. Other fungi such as *Pleurotus ostreatus, Agaricus bisporus, Lentinus oloides, Corius versicolor,* and some yeasts. More work is required in this new area of research to enable us to effectively use fungi in pollution alleviation strategies.

Key words : Fungal diversity, bioremediation, pollution alleviation

The abundant variation in the types of living organisms in any geophysical area is called biodiversity. With reference to large organisms (plants, animals etc.) the distinctive visible anatomical and morphological features enable us to identify the different genera and species and, therefore, we can easily assess the extent of biodiversity. However, most fungi are not visible to the naked eye and they need to be studied under the microscope. The culturable fungi can be identified provided they sporulate but the biotrophs which are not amenable to culture media go unnoticed. It is, therefore, imperative that there is a gap in our knowledge of the diversity of all microbes including fungi. Fungi are better known than bacteria, viruses and other smaller forms of life such as viroids. There are about 72,000 named species of fungi and new species are being added at the rate of about 1500 each year. The total number of fungal species both known and unknown is estimated at 1.5 M. The 72,000 known species exist in different culture collection centres in the form of several subspecies, pathovars and strains, the total cultures going upto 1,70,000 (Bennett and Faison, 1997). However, the majority of described strains are not deposited in culture collection centres and are not accessible.

Functional Diversity

While dealing with microorganisms it is appropriate to give equal emphasis to structural (morphological) and functional diversity. Based on their activities, the fungi can be grouped as cellulase producers, lignin degraders, nitrtate reducers, pectin degraders, phosphate solubilizers, iron chelaters, Penicillin producers, etc.. The common traits between a set fungi implies the presence of certain common gene clusters which have value in a particular ecological niche (Watts *et al.*, 1999).

Habitat Diversity

The mycoflora of unique ecological niches have some common features and it will be very rewarding to explore fungal diversity in thermophilic environments , such as, hot springs, thermal vents, sun-heated soils compost pits, self-heated coal refuse piles, steam line discharge sites etc. The thermophilic fungi and bacteria growing at temperatures between 40-60° C are the sources of thermostable enzymes. Other ecological niches to be explored are the Western Ghats, the Himalayan ranges, marine ecosystems, mangroves, coral reefs, sand dunes, industrial effluent contaminated soils, ant hills, refineries, activated sludge, insects and several other natural sources. Endophytes of plants and animals are very meagerly understood.

Bioprospecting

The process of looking for new forms of life is called bioprospecting. The techniques of fungal exploration have remained by and large conventional. The methods have mainly included the use of different recipes for culture media along with techniques such as dilution plating soil plating leaf shred inoculation, etc. Direct incubation and baiting techniques have also been in use. The emphasis is now being laid on enrichment techniques wherein the medium used mimics the natural habitat in nutritional status. One can isolate nitrate reducers using a medium enriched with nitrate and minimal carbon source. Reduction of carbon source also a means of keeping the contaminants away. Addition of antibiotics in enrichment cultures is another way of keeping away the weedy fungi and bacteria. Enrichment cultures have often been used to isolate fungi capable of degrading certain specific xenobiotics. The xenobiotic in question may be used as the sole carbon source in the medium so that only the fungi with degradative capacities will grown. In liquid culture, competition for a particular substrate will lead to the enrichment of the microbial strain that is able to grow faster than others. One flaw in the enrichment technique is that it leads often to growth of bacteria even when fungi with good potential are present. It is because of the capacity of bacteria to grow faster than fungi. Besides, some fungi may not utilize the degraded product as the sole source of carbon or as supplementary source of carbon, as in the case of the white rot fungi (Jefries *et al.*, 1981). It may be therefore necessary to use bacteriostatic compounds and supplementary carbon sources to isolate zenobiotic degrading fungi from a natural source.

When it is not possible to grow a fungus by conventional means, isolation of the gene pool from the environment may be very rewarding. Environmental DNA is now being harnessed for microorganism. The gene of interest separated from environmental DNA can be tagged onto a vector and cloned into a suitable expression host to yield the product. More work has been done this area with bacteria than fungi (Seow *et al.*, 1997; Sullia, 2001).

Bioprospecting is important because fungi and other microorganisms are the largest future sources for various products and processes. They are of critical importance to the sustainability of life in nature because they are involved in the recycling of elements and in producing and consuming gases such as oxygen and carbon dioxide. Fungi have been employed for industrial production of compounds of intrinsic value such as antibiotics, enzymes, hormones, immunosuppressants, food additives and colorants. They have also been employed as biopesticides and biofertilizers. They have been sources of microbial protein and oil and animal feed. Edible fungi have been known as long as human civilization. The exploration of the extent of fungal diversity on planet earth is indeed a stupendous, but surely a very rewarding task.

Bioremediation

Bioremediation is a pollution control technology that uses biological systems to catalize the degradation or transformation of various toxic chemicals to less harmful forms. The general approaches to bioremediation are to enhance natural biodegradation by native organisms (intrinsic bioremediation), to carry out environmental modification by applying nutrients or aeration (biostimulation) or through addition of microorganisms (bioaugmentation). Unlike conventional technologies, bioremediation can be carried out on-site. Bioremediation is limited in the number of toxic materials it can handle (Hart, 1996), but where applicable, it is cost-effective (Atlas and Unterman, 1999).

Biodegradation, mineralization, bioremediation, biodterioration, biotransformation, bioaccumulation and biosorption are some terms with minor subtle differences but often overlappingly used. Biodegradation is the general term used for all biologically mediated breakdown of chemical compounds and complete biodegradation leads to mineralization. Biotransformation is a step in the biochemical pathway which leads to the conversion of a molecule (precursor) into a product. A series of such steps are required for a biochemical pathway. In environmental science, it is of importance whether the product is less harmful or not (Bennett and Faison, 1997). Biodeterioration refers usually to the breakdown of economically useful compounds but often the term has been used to refer to the degradation of normally resistant substances such as metals, plastics, drugs, cosmetics, painting, sculpture, wood products and equipment (Rose, 1981). Bioremediation refers to the use of biological systems to degrade toxic compounds in the environment. Bioaccumulation or biosorption is the accumulation of the toxic compounds inside the cell or on surface without any degradation of the toxic molecule. This method can be effective in aquatic environments where the organisms can be removed after being loaded with the toxic substance.

The fungi are unique among microorganisms in that they secrete a variety of extracellular enzymes. The decomposition of lignocellulose is rated as the most important degradative event in the carbon cycle of earth (Bennett and Faison, 1997). Enormous literature exists on role of fungi in the carbon and nitrogen cycles of nature (Frankland *et al.*, 1982; Cooke and Rayner, 1984; Carroll and Wicklow, 1992). The role of fungi in the degradation of complex carbon compounds such as starch, cellulose, pectin, lignin, lignocellulose, inulin, xylan, araban etc. is well known. *Trichoderma reesei* is known to possess the complete set of enzymes required to breakdown cellulose to glucose. Degradation of lignocellulose is the characteristic of several basidiomycetous fungi.

Fungi in bioremediation

Fungi are good in the accumulation of heavy metals such as cadmium, copper, mercury, lead and zinc. Systems using *Rhizopus arrhizus* have been developed for treating uranium and thorium.

The ability of fungi to transform a wide variety of hazardous chemicals has aroused interest in using them in bioremediation. (Alexander, 1994). The white rot fungi are unique among eukaryotes for having evolved nonspecific methods for the degradation of lignin; curiously they do not use lignin as a carbon source for their growth (Kirk *et al.*, 1976). Lignin degradation is, therefore, essentially a secondary metabolic process, not required for the main growth process. Lamar *et al.*, 1993) compared the abilities of three lignin-degrading fungi, *Phanerochaete chrysosporium*, *P. sordida* and *Tramates hirsuta* to degrade PCP (Pentachlorphenyl) and creosote in soil. Inoculation of soil with 10% (wt/wt) *Phanerochaete sordia* resulted in the greatest decrease of

PCP and creosote. *P. sordida* was also most useful in the degradation of PAHs (Polycyclic aromatic hydrocarbons) from soil. Davis *et al.* (1993) showed that *P. sordida* was capable of degrading efficiently the three ring PAHs, but less efficiently the four-ring PAHs.

Phanerochaete chrysosporium has been shown to degrade a number of toxic xenobiotics such as aromatic hydrocarbons (Benzo alpha pyrene, Phenanthrene, Pyrene) chlorinated organics (Alkyl halide insecticides, Chloroanilines, DDT, Pentachlorophenols, Trichlorophenol, Polychlorinated biphenyls, Trichlorophenozyacetic acid), nitrogen aromatics (2,4-dinitrotoluene, 2,4,6-trinitrotoluene-TNT) and several miscellaneous compounds such as sulfonated azodyes. Several enzymes which are released such as laccases, polyphenol oxidases, lignin peroxidases etc. play a role in the degradative process. In addition, a variety of intracellular enzymes such as reductases, methyl transferases and cytochrome oxygenases are known to play a role in xenobiotic degradation (Barr and Aust, 1994).

Phanerochaete chrysosporium has been shown to effect the bioleaching of organic dyes (Nigam *et al.*, 1995). Pauli Ollikka *et al.* (1993) have also shown the decolorization of azo-triphynyl methane dyes by lignin peroxidase produced by *P. chrysosporium.* Sami and Radhaune (1995) have demonstrated the role of lignin peroxidase and manganese peroxidase from *P. chrysosporium* in the decolorization of olive mill waste water. The work carried out in our laboratory (Ashoka *et al.*, 2000, 2002) has shown that *Phanerochaete chrysosporium* and microbial consortia were effective in color removal from textile dye effluents. The fungus casued 80% decolorization in broth containing 2.5% of effluent. There was reduction in BOD and COD values. A local isolate of *Fusarium sp.* caused various degress of decolorization ranging from 35 to 85%.

Among the fungal systems, *Phanerochaete chrysosporium* is emerging as the model system for bioremediation. The basidiomycetous fungus *Pleurotus ostreatus* bas been shown to produce an extracellular hydrogen peroxide dependent lignolytic enzyme which removes the color due to remozol brilliant blue. Oxidative enzymes play a very major role in biodegradation. Other fungi which can be used in bioremediation are obviously the members of Zygomycetes e.g., the mucoraceous fungi and the arbuscular mycorrhizal fungi. Aquatic fungi and anaerobic fungi are also candidates for bioremediation.

Among other fungi used in bioremediation, the yeasts, e.g., *Candida tropicalis, Saccharomyces cerevisiae, S. carlbergensis* and *Candida utilis* are important in clearing industrial effluents of unwanted chemicals. *Agaricus bisporus* and *Lentinus oloides* are important in lignocellulose decomposition.

Corius versicolor is important in cleaning up pulp and paper mill wastes. Consortia of fungi and bacteria (usually uncharacterised) are used in composting, the most useful waste disposal practice. Phenolic azo dyes have been shown to be oxidized by the enzyme laccase produced by *Pyricularia oryzae* (Chivukula and Renganathan, 1995).

Future Outlook

Whenever bioremediation figures as the topic of discussion, bacterial agents come into focus and fungi are much less studied. One should realize, however, the greater potential of fungi by virtue of their aggressive growth, greater biomass production and extensive hyphal reach in soil. More research will be focused in future on using the diverse fungal flora for bioremediation. The work on the microbial diversity in the electroplating industrial effluents is going on in our laboratory. There are several promising fungi that can degrade zinc, cyanide and chromium in these effluents.

Future work will be more fucused on the biotechnological aspects. It may be possible to clone the highly efficient degradative enzyme producing genes into bacteria and conversely, baterial genes can be transferred to fungi which are suitable. The high surface-to-cell ratio of filamentous fungi makes them better degraders under certain niches like contaminated soils. Fungi have been shown to even solubilize partially coal, a highly polymeric substance more complex than lignin. There is no doubt, therefore, regarding fungi being harnessed more and more in environmental bioremediation in future.

References

Alexander, M., (1994). Biodegradation and Bioremediation. Acad. Press, San Diego, Calif.

Ashoka, C., Manjunath, K. and Sullia, S.B. (2000). Biological treatment of combined textile dye effluent, In : Ecology of Fungi, D.J. Bhat and S. Raghukumar (eds.), 63-67.

Ashoka, C., Geetha, M.S. and Sullia, S.B. (2002). Bioleaching of composite textile dye effluent using bacterial consortia. Asian Jr. Microbial Biotechn. Env. Sci., **4**:65-68.

Atlas, R.M. and Unterman, R. (1999). Bioremediation. In: Industrial Microbiology & Biotechnology, 2nd Ed., ASM Press, Washington, 666-681.

Barr, D.P. and Aust, S.D. (1994). Mechanisms the white rot fungi use to degrade pollutants. Env. Sci. Technol., **28**: 79A-87A.

Bennett, J.W. and Faison, B.D. (1997). Use of Fungi in Biodegradation. In: Environmental Microbiology, ASM Press, Washington.

Carroll, G.C. and Wicklow, D.T. (1992). The Fungal Community. Its Organization and Role in the Ecosystem. 2nd ed. Marcel Dekker Inc., N.Y.

Chivukula, M. and Renganathan, V. (1995). Phenolic azo dye oxidation by laccase from *Pyricularia oryzae*. Appl. Environ. Microbiol., **61** : 4374-4377.

Cooke, R.C. and Rayner, A.D.M. (1984). Ecology of Saprophytic Fungi. Longman, London.

Davis, M.W., Glaser, J.A., Evans, J.W. and Lamar, R.T. (1993). Field evaluation of lignin degrading fungu *Phanerochaete sordida* to treat creosote contaminated soil. Environ. Sci. Technol., **27**: 2572-76.

Frankland, J.C., Hedger, N.H. and Swift, J.J. (Eds.) (1982). Decomposer Basidiomycetes. Their Biology and Ecology. Cambridge Univ. Press, Cambridge.

Kirk, T.K., Connors, W.J. and Zeikus, J.G. (1976). Requirement of growth substrate during lignin degradation by two wood rotting fungi., Appl. Eviron. Microbiol., **32**: 192-194.

Lamar, R.T. Evans, J.W. and Glaser, J. (1993). Solid phase treatment of pentachlorophenol-contaminated soil using lignin degrading fungi. Environ. Sci. Technol., **27**: 2566-71.

Nigam, P., Banat, I.M., McMullan, G., Dalel Singh, and Marchant, R. (1995). Microbial degradation of textile effluent containing Azo, Diazo and reactive dyes by aerobic and anaerobic bacterial and fungal cultures. 36th Annu. Conf, AMI, Hisar, 37-38.

Pauli, O., Kirsi, A., Veli-Matti, L., Tuomo, G., Timo, R. and Hari, S. (1993). Decolorization of Azo Triphenyl Methane Heterocyclic and Polymeric dyes by lignin peroxidase, isoenzymes from *Phanerochaete chrysosporium*. Appl. Environ. Microbiol., **59** : 4010-4016.

Rose, A.H. (1981). Microbial Biodeterioration. Vol 6, Economic Microbiology. Academic Press, London.

Sami, G. and Radhaune, R. (1995). Role of lignin peroxidase and manganese peroxidase from *Phanerochaete chrysosporium* in the decolorization of olive mill waste water. Appl. Envron. Microbiol., **61**: 1098-1103.

Seow, K.T., Meurer, G., Gerlitz, M., Pienkowski, E.W., Hutchinson, C.R. and Davies, J. (1997). A study of iterative type-II polyketide synthases, using bacterial genes from uncultured microorganisms. J. Bacteriol, **179** : 7360-68.

Sullia, S.B. (2001). How do we preserve our micorbial diversity ? Science and Technology, D.H. Supplement, page 3, July 3, 2001.

Watts, J.E.M., Huddleston-Anderson, A.S. and Wellington, E.M.H. (1999). Bioprospecting. In 'Mammal of Industrial Microbiology & Biotechnology' Eds. Demain, A.L. and Davies, J.E., ASM Press, Washington, D.C. 631-641.

2

Biosynthesis of Gold Nanoparticles Employing Fungi

Absar Ahmad,[1] M. Islam Khan,[1] Rajiv Kumar[2] and Murali Sastry[3]

[1]Biochemical Sciences, [2]Catalysis and [3]Materials Chemistry Divisions National Chemical Laboratory, Pashan Road, Pune – 411 008

There is little doubt that nanomaterials and nanotechnology will play a key role in many technologies of the future. Immediate and potentially exciting applications in medicine, electronics industry and the chemical industry have are close to commercialization. One key aspect of nanotechnology concerns the development of protocols for the synthesis of nanomaterials over a range of chemical compositions, sizes and controlled monodispersity. In the context of the current drive to develop green technologies in the field of advanced materials synthesis, this aspect of nanotechnology assumes considerable importance. In this chapter, we outline some of the strategies developed. Various genera and species of fungi have been systematically screened and particular species of fungi have been identified for both intra-cellular and extra-cellular biosynthesis of gold nanoparticles. Some aspects of immobilization of gold nanoparticles synthesized extra-cellularly in thin film form are also described.

Key words: **Nanoparticles; gold; biosynthesis; green chemistry; immobilization**

There is tremendous current excitement in the study of nanoscale matter (matter having nanometer dimensions, 1 nm = 10^{-7} cm) with respect to their fundamental properties, organization to form superstructures and applications. The unusual physicochemical and optoelectronic properties of nanoparticles arise primarily due to confinement of electrons within particles of dimensions smaller than the bulk electron delocalization length, this process being termed quantum confinement (Brus, 1984; Fendler, 1992; Henglein, 1993, 1995; Weller, 1993; Alivisatos, 1996; Colvin *et al.*, 1994). The exotic properties of nanoparticles have been considered in applications such as optoelectronics (Colvin *et al.*, 1994; Wang and Herron, 1991), catalysis (Schmid, 1992; Hoffman *et al.*, 1992), reprography (Hamilton and Baetzod, 1979), single-electron transistors (SETs) and light emitters (Klein *et al.*, 1997; Weller, 1998), non-linear optical devices (Wang, 1991; Yoffe, 1993) and photoelectrochemical applications (Mansure *et al.*, 1995). Magnetic nanoparticles are being viewed with interest from a fundamental point of view (superparamagnetism in the nanoparticles) (Gider *et al.*, 1996) as well as in applications such as magnetic memory storage devices (Symonds, 1995), magnetic resonance image enhancement (Bulte *et al.*, 1994.) and magnetic refrigeration (Shull *et al.*, 1993). The ability to tune the optical absorption/emission properties of semiconductor nanoparticles (the so-called "quantum dots") by simple variation in nanoparticle size is particularly attractive in the facile band-gap engineering of materials (Fendler and Meldrum, 1995) and the growth of quantum dot lasers (Ledentvos *et al.*, 1996). More recently, nanoscale matter has been looked at with interest for potential application in nanocomputers, synthesis of advanced materials, energy storage devices, electronic and optical displays, chemical and biosensors as well as biomedical devices (Dagani, 2000). As we enter the new millennium, it is becoming increasingly clear that nanoparticles and technologies related to their synthesis, organization and application (which may collectively be termed "nanotechnology") will play a key role in different areas and will directly/indirectly impact the quality of life world-wide. Recognizing the importance of nanomaterials in ushering in the next industrial revolution , many countries have launched major initiatives into the development of a strong fundamental and applied knowledge base in the area of nanotechnology. As an improved understanding of the fundamental properties and thereby, applications of nanoscale matter emerges, it indeed does appear that "there's plenty of room at the bottom" (APSCIT, 1959) in this fascinating area.

Recent development of bio-based nanomaterials synthesis protocols

An important area of research in nanotechnology is the synthesis of nanoparticles of different chemical compositions, sizes and controlled monodispersity. Currently, there is an ever-growing need to develop environmentally benign nanoparticle synthesis processes that do not use toxic chemicals in the synthesis protocol. As a result, researchers in the field of nanoparticle synthesis and assembly have turned to biological systems for inspiration. This is not surprising given that many organisms, both unicellular and multicellular, are known to produce inorganic materials either intra- or extra-cellularly (Smkiss and Wilbur, 1989; Loveley *et al.*, 1987; Schleifer, 1995; Dickson, 1999). Some well-known examples of bio-organisms synthesizing inorganic materials include magnetotactic bacteria (which synthesize magnetite nanoparticles) (Mann, 1992; Oliver *et al.*, 1995; Kroger *et al.*, 1999) diatoms (which synthesize siliceous materials) (Pum and Sleytr, 1999; Sleyter *et al.* 1999) and S-layer bacteria (which produce gypsum and calcium carbonate layers) (Stephen and Maenaughton, 1999). The secrets gleaned from nature have lead to the development of biomimetic approaches for the growth of advanced nanomaterials.

Even though many biotechnological applications such as remediation of toxic metals employ micro-organisms such as bacteria (Southom and Bereridge, 1996; Bereridge and Murray, 190; Fortin and Beveridge, 2000) and yeast (Mehra and Winge, 1991), (the detoxification often occurring via reduction of the metal ions/formation of metal sulfides) it is only relatively recently that materials scientists have been viewing with interest such micro-organisms as possible eco-friendly nanofactories (Klaus *et al.*, 1999; Mukherjee *et al.* 2001). Beveridge and co-workers have demonstrated that gold particles of nanoscale dimensions may be readily precipitated within bacterial cells by incubation of the cells with Au^{3+} ions (Southom and Beveridge, 1996). Klaus-Joerger and co-workers have shown that the bacteria *Pseudomonas stutzeri* AG259 isolated from a silver mine when placed in a concentrated aqueous solution of $AgNO_3$, resulted in the reduction of the Ag^+ ions and formation of silver nanoparticles of well-defined size and distinct morphology within the periplasmic space of the bacteria (Klaus *et al.*, 1999, 2001; Jeorger *et al.*, 2000). As can be seen from the above,, the use of biological organisms in the deliberate synthesis of nanoparticles is a relatively new and exciting area of research with considerable potential for development.

In a break from tradition, which has hitherto relied on the use of prokaryotes such as bacteria in the synthesis of nanoparticles, we have recently embarked on a programme aimed at investigating whether eukaryotic organisms such as fungi may be used to grow nanoparticles of different chemical compositions and sizes. A number of different species of fungi have been investigated in this two-year effort and it has been shown that fungi are extremely good candidates in the synthesis of gold (Mukherjee *et al.*, 2001, 2002), silver (Mukherjee *et al.*, 2001) and indeed quantum dots of the technologically important CdS as well (Ahamd *et al.*, communicated). In the remaining sections of this chapter, we briefly outline the use of the fungi *Verticillium* and *Fusarium oxysporum* in the intra- and extra-cellular synthesis of gold nanoparticles. In the concluding section, the electrostatic entrapment of gold nanoparticles synthesized extra-cellularly in thermally evaporated lipid films is demonstrated. Ultimately, the use of nanoparticles synthesized by a biochemical route in the electronics industry would require their assembly in the form of thin films and consequently, our focus has also been directed to developing nanoparticle assembly protocols (Sasry, 2001).

Biosynthesis of gold nanoparticles employing fungi

Intra-cellular synthesis of gold nanoparticles using Verticillium sp.

The acidophillic fungus, *Verticillium*, was isolated from the *Taxus* plant and maintained on potato-dextrose agar slants at 25 ^{0}C. The fungus was grown in 500 mL Erlenmeyer flasks each containing 100 mL MGYP medium, composed of malt extract (0.3%), glucose (1.0%), yeast extract (0.3%), and peptone (0.5%) at 25-28^{0}C under shaking condition (200 rpm) for 96 h. After 96 h of fermentation, mycelia were separated from the culture broth by centrifugation (5000 rpm) at 10 °C for 20 min and the settled mycelia were washed thrice with sterile distilled water. Ten grams of the harvested mycelial mass was then re-suspended in 100 mL sterile distilled water in 500 mL conical flasks at pH = 5.5 - 6.0 and to this suspension, 100 mL of an aqueous solution of 10^{-4} M chloroauric acid ($HAuCl_4$) was added. The whole mixture was thereafter put into a shaker at 28 °C (200 rpm) and the reaction carried out for a period of 72 h. The bio-transformation was routinely monitored by visual inspection of the biomass as well as measurement of the UV-vis spectra from the fungal cells.

Thin sections of the *Verticillium* cells after reaction with $AuCl_4^-$ ions in the manner described above were prepared for transmission electron microscopy (TEM) analysis as follows. Approximately 1 mm^3 bits of the Au nano-*Verticillium* biomass were taken and fixed in 2.5 % gluteraldehyde in distilled water for 2 h at room temperature. After fixation, the cells were sedimented (1,500 rpm, 10 min) and washed thrice with distilled water. Without postfixation, the pellet was subjected to dehydration with 30, 50, 70 and 90 % ethanol for 15 min at each concentration followed by two changes in absolute ethanol. Since ethanol does not possess good miscibility with epoxy resins, propylene oxide was used as a linking agent. The dehydrated pellet was kept in propylene oxide for 15 min following which the infiltration of the resin was done by placing the pellet in a 1 : 1 mixture of propylene oxide and Epon 812 overnight at room temperature. Embedding was carried out using a mixture of the resin (Epon 812) and hardeners (DDSA – dodecynyl succinic anhydride and MNA – methyl nadic anhydride) in the ratio 1 : 1.5. To this, two drops of tridimethylaminomethyl phenol (DMP30) was added to accelerate the polymerization process. Polymerization was carried out using this mixture at 60 °C for 3 days. Ultrathin sections were cut using an ultramicrotome (Leica Ultracut UCT) and were taken on copper TEM grids (40 μmX 40 μm mesh size). The sections were slightly stained with uranyl acetate and lead citrate prior to TEM analysis. TEM measurements were carried out on a JEOL Model 1200EX instrument operated at an accelerating voltage of 60 kV. A low operating voltage was used to minimize damage to the thin sections by electron beam heating.

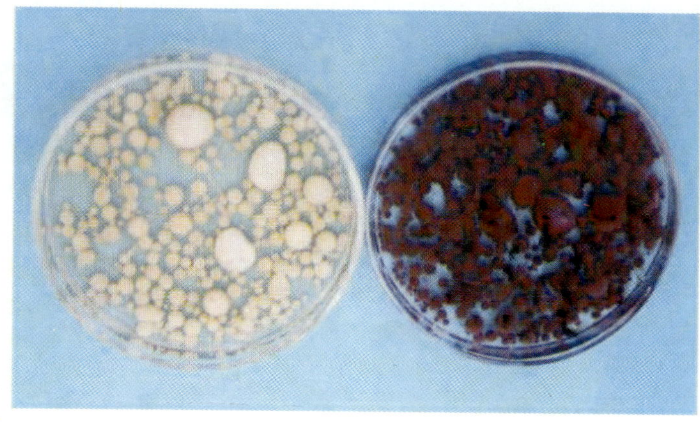

Figure 1 : A) Picture of Petridishes containing the *Verticillium* sp. biomass in aqueous solution of 10^{-3} M $HAuCl_4$ at the beginning of the reaction (Petridish to the right) and after 72 h of reaction (Petridish to the left).

A picture of the *Verticillium* biomass before (Petridish to the right) and after exposure to aqueous chloroauric acid for a period of 72 h (Petridish to the left) is shown in Fig.1. The appearance of the dark purple colouration in the fungal biomass after reaction with $AuCl_4^-$ ions is a clear indicator of the reduction of the Au^{3+} ions and formation of gold nanoparticles in the fungal biomass. It is well known that the lovely colour of gold nanoparticles, ranging from pink to blue, is due to excitation of surface plasmon vibrations in the nanoparticles (Henglein, 1993). Incidentally, this very property has been used for centuries wherever gold nanoparticles were employed as colouring matter in glasses etc. The growth of the nanoparticles occurred only *within the fungal biomass* and not extra-cellularly and is an interesting feature of this particular fungus.

As mentioned in the introduction, the biochemical methods for the synthesis of gold nanoparticles would need to address size control issues as well. Consequently, we analyzed thin sections of the *Verticillium* cells after growth of the gold nanoparticles by TEM. Fig.2 shows TEM pictures of one of the cells in the thin section at different magnifications. The assembly of gold nanoparticles on the surface of the *Verticilium* cells is clearly seen in Fig.2A and is indicated by arrows. Occasionally, the gold nanoparticles formed with the cell (Fig.2A) and is possibly due to diffusion of the metal ions across the cell wall and their reduction within the cell by enzymes present there (Mukherjee *et al.,* 2001). At higher magnification (Fig.2B), the presence of individual nanoparticles can clearly be seen. The particles range in size from 5 to 50 nm and are uniformly dispersed on the cell wall surface. We believe the reduction of the $AuCl_4^-$ ions occurs through enzymes present in the cell wall. It is possible that once formed, the gold nanoparticles are capped by the protein and thus, their aggregation is prevented. It is well known that protein

A

500 nm

B

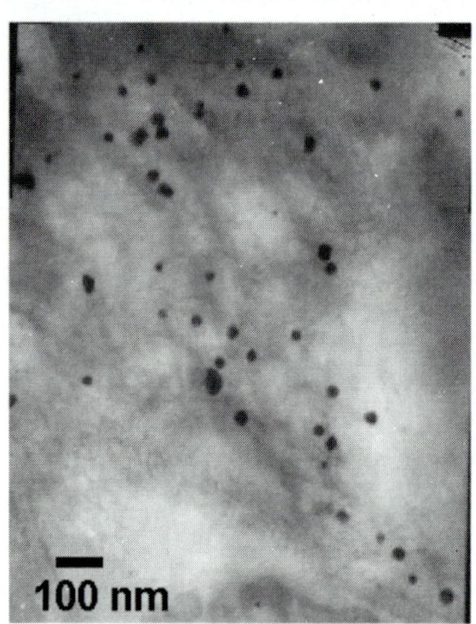

Figure 2 : TEM micrographs recorded from a thin section of the *Verticillium* biomass after reaction with aqueous $AuCl_4^-$ ions for 72 h at different magnifications (A and B). In A, the presence of gold nanoparticles on the cell wall surface as well as occasionally within the cell is indicated by arrows. At a higher magnification, the individual gold nanoparticles on the cell-wall surface are seen (B).

molecules bind strongly to gold nanoparticles and therefore, this is a likely mechanism (Gole *et al.*, 2001).

From the application point of view, it would be important to ascertain whether the fungal cells die under exposure to the $AuCl_4^-$ ions, i.e., if the ions are toxic to the fungus. Fig.3 shows a Petridish containing small portions of the *Verticillium* biomass after formation of gold nanoparticles (dark cores, identified by arrows) and exposed to nutrients. It is clear the fungus grows to cover the Petridish indicating that the metal ions are not toxic to them. This is an interesting aspect of the work and throws open many questions related to

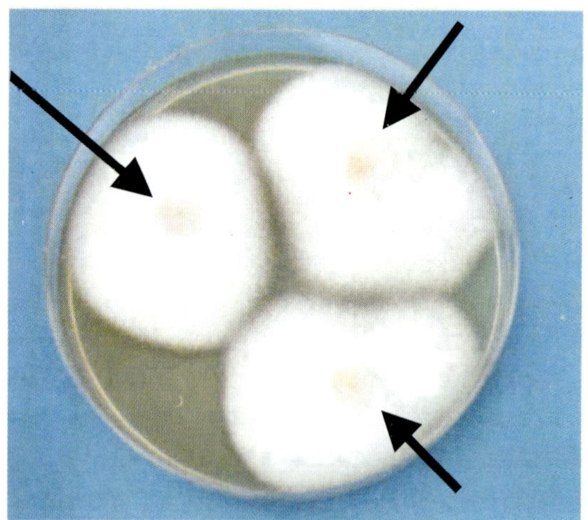

Figure 3. Petridish containing small portions of the gold nanoparticle-*Verticillium* biomass (indicated by arrows) after maintaining in a nutrient medium for 1 week.

the evolution of such organisms. Attempts are currently underway to control the size and polydispersity of the gold nanoparticles formed using *Verticilium* as well as to understand better the enzymatic process leading to the metal nanoparticle formation.

Extra-cellular synthesis of gold nanoparticles using *Fusarium oxysporum*

In a typical experiment, 10 g of the *Fusarium oxysporum* biomass was immersed in 100 mL of water for a period of 72 h following which the solution was filtered and reacted with 10^{-3} M $AuCl_4^-$ ions in the dark. It was observed that the metal ions were slowly reduced in solution as evidenced by a gradual and steady increase in intensity of purple colouration in the solution. Fig.4 shows a picture of a test tube containing the water exposed to the fungal biomass after reaction of $AuCl_4^-$ ions for 12 h. In this case as well, a very dark purple colour is seen in the test tube clearly indicating the presence of gold nanoparticles in the solution (Mukherjee *et al.*, 2002). Unlike in the case of *Verticillium,* the gold nanoparticles using *Fusarium oxysoporum* form extra-cellularly and is a very important development insofar as nanomaterials synthesis is concerned. We would like to mention here that the gold nanoparticles synthesized in solution using *Fusarium oxysporum* were extremely stable in time wit no evidence for aggregation months after their synthesis. This indicates that the nanoparticles are stabilized by a coating of protein molecules, possibly electrostatically. The protein coating may be used to also extract the nanoparticles from solution and assemble them as will be shown below.

The test tube on the right of the picture in Fig.4 shows the gold nanoparticle solution after centrifugation. The pink mass at the bottom of the test tube is the aggregated nanoparticle pellet coated with proteins. This pellet could be washed and the gold nanoparticles readily redissolved in water for further immobilization measurements. The gold nanoparticles observed by TEM are all nearly spherical and in the size range 8 – 40 nm (Mukherjee *et al.*, 2002). As in the case of *Verticillium,* we are attempting to control the size of the gold nanoparticles and

Figure 4. Test to the left showing water exposed to the *Fusarium oxysporum* biomass for 72 h followed by reaction with 10-3 M AuCl4- ions for a further period of 12 h. The test tube to the right is the gold nanoparticle solution shown on the left after centrifugation.

conjugating them with biologically important molecules such as antibodies, DNA etc. The extra-cellular synthesis of gold nanoparticles using *Fusarium oxysporum* is thus tailor-made for such bio-based applications. Attempts are also being made to extend the fungus-based method to the synthesis of nanoparticles of other chemical compositions such as CdS, PbS, ZnS (Ahmad *et al.,* communicated) oxides etc.

Immobilization of gold nanoparticles in thin lipid films

As mentioned previously, the gold nanoparticles synthesized in solution would need to be immobilized in thin film form for further application, especially in the electronics industry and as supported catalysts. In this laboratory, some of us have developed a process for the deposition of thin ionizable lipid films for the immobilization of inorganic ions (Damle *et al.*, 2002), charged nanoparticles (Sastry, 2002), proteins (Gole and Sastry, 2001; Sastry, 2002) and DNA (Sastry, 2002; Ramakrishnan *et al.*, 2001). This method uses electrostatic interactions between the intercalant (the ions/nanoparticles/biomacromolecule) and charged lipid molecules in the film. Since the gold nanoparticles synthesized using *Fusarium oxysporum* are capped with a protective coating of protein, it should be possible to entrap the nanoparticles in a suitable lipid film.

Thin films of octadecylamine (ODA, cationic lipid, 500 Å thick) were thermally vacuum deposited in an Edwards E306 vacuum coating unit

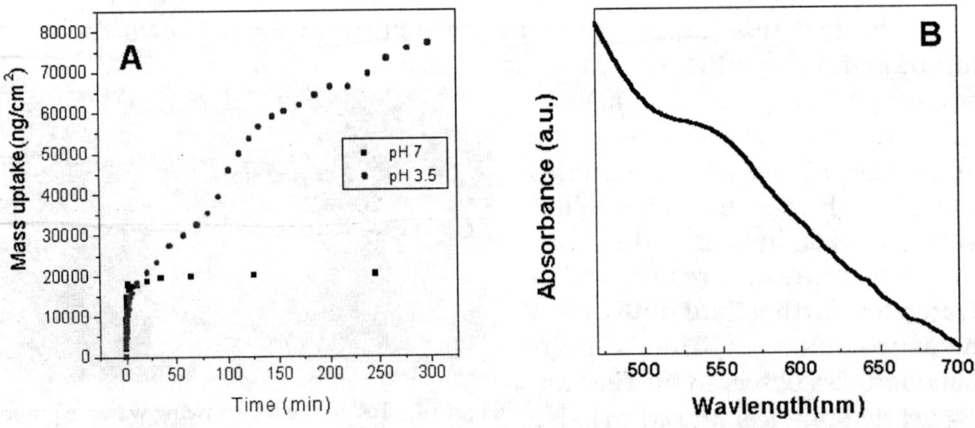

Figure 5. A) QCM mass uptake measured as a function of time of immersion of a 500 Å thick ODA film in gold nanoparticles synthesized extra-cellularly using *Fusarium oxysporum*. The pH values of the gold nanoparticle solution are given in the figure. B) UV-vis spectrum recorded from a 500 Å thick ODA on quartz after immersion in the gold nanoparticle solution synthesized using *Fusarium oxysporum* at pH 7.

operated at a pressure of 1 x 10^{-7} Torr onto carbon-coated TEM grids, gold-coated AT-cut quartz crystals for quartz crystal microgravimetry (QCM) measurements and quartz substrates for UV-vis spectroscopy measurements. The film thickness and deposition rate were monitored *in-situ* using an Edwards QCM. The ODA films on the QCM crystal were immersed in the gold nanoparticle solution (prepared after careful centrifugation and washing of the gold nanoparticle pellet and redispersion in deionized water, Fig.4) and the mass uptake recorded as function if time of immersion. Fig.5A shows the QCM mass uptake data recorded from the ODA film during immersion in the gold nanoparticle solution held at two different pH values (the pH of the solution was adjusted using ammonia and dilute HCl). In the case of the pH 7 solution, the mass uptake is extremely large and reaches a limiting value of ca. 80,000 nm/cm^2. This indicates considerable loading of the film with gold nanoparticles. Indeed, a reddish colour could clearly be observed in the film on the QCM. This experiment was also performed by immersion of the ODA film in the gold nanoparticle solution at pH 3 and it is seen that the final mass loading due to diffusion of the gold nanoparticles into the ODA matrix is considerably smaller than at pH 7 (Fig.5A, 20,000 ng/cm^2 vs 80,000 ng/cm^2). It is well known that electrostatic interactions drive the movement of the gold nanoparticles into the lipid matrix (Sastry, 2001, 2002; Damle *et al.*, 2002, Gole and Sastry, 2001; Rama Krishnan *et al.*, 2001). In the case of gold nanoparticles synthesized using *Fusarium oxysporum*, the protein coating is expected to stabilize the nanoparticles in solution electrostatically. This simple experiment provides indirect proof for the presence of the protein and also enables one to roughly estimate the pI of the protein on the nanoparticle surface. At both pH 7 and 3, the ODA matrix is fully positively charged (pK_B of ODA = 10.8) and therefore the smaller

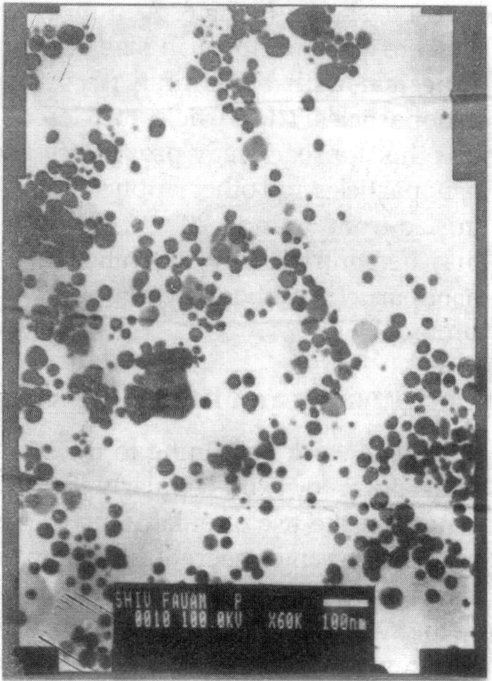

Figure 6. TEM picture recorded from a 500 Å thick ODA film deposited on a carbon-coated TEM grid after immersion in the gold nanoparticle solution synthesized using *Fusarium oxysporum* for 6 h.

protein loading at pH 3 indicates that the charge on the gold nanoparticle-bound protein at this pH is lower than at pH 7 or it changes sign in this pH interval. We believe the pI of the protein is quite small indicating that it would be fully negatively charged at pH 7 and perhaps fully uncharged at pH 3, thereby leading to smaller mass loading. The QCM measurements also provide information on the optimum time of immersion of the lipid film in the nanoparticle solution for other experiments.

The presence of gold nanoparticles in the ODA matrix is conveniently determined by UV-vis spectroscopy due to the lovely colour of the particles. Fig.5B shows the UV-vis spectrum recorded from a 500 Å thick ODA film on quartz after immersion in the gold nanoparticle solution at pH 7 for 6 h. The clear absorption band at ca. 540 nm is a strong indicator of the presence of gold nanoparticles in the film. As mentioned earlier, the strong purple colour in the film on quartz could clearly be seen by the naked eye.

The gold nanoparticles prepared using *Fusarium oxysporum* and entrapped in the ODA matrix was imaged by TEM and one representative image is shown in Fig.6. A number of gold nanoparticles can clearly be seen and are very uniform in size and shape. The particles are predominantly spherical with occasional presence of the well-known triangular gold nanoparticles. The particles are densely packed in the lipid matrix and indicate that this approach may prove very useful in assembling bio-synthesized gold nanoparticles for other applications. The particles in the ODA matrix range in size from 3 nm to 20 nm. The density of nanoparticles in the lipid matrix may be controlled by a number of factors such as time of immersion in the nanoparticle solution, the pH of the solution etc. and add to the power of this protocol.

Conclusions and future directions

The work pertaining to the synthesis of gold nanoparticles using fungi, either intra- or extra-cellularly is, to the best of our knowledge, the first such report on the use of eukaryotic organisms in materials synthesis. The use of specific enzymes secreted by organisms such as fungi in the synthesis of nanoparticles is exciting for the following reasons. The process can be extended to the synthesis of nanoparticles of different chemical compositions and indeed, different shapes and sizes by suitable identification of enzymes secreted by the fungi. This opens up the possibility of creating a rational approach for the synthesis of nanoparticles using fungi with tremendous advantages. This fungal-mediated green chemistry approach towards the synthesis of nanoparticles has many advantages such as ease with which the process can be scaled up, economic viability, possibility of easily covering large surface areas by suitable growth of the mycelia etc. The shift from bacteria to fungi as a means of developing natural "nano-factories" has the added advantage that down-stream processing and handling of the biomass would be much simpler. Further, compared to bacteria, fungi are known to secrete much higher amounts of proteins thereby significantly increasing the productivity of this biosynthetic approach. Application of such bio-nanocomposites in catalysis and other electronic applications are currently being pursued.

References

Ahmad, A., Mukherjee, P., Senapati, S., Mandal, D. Khan, M.I., Kumar, R. and Sastry, M. (2002). Enzyme- mediated extracellular synthesis of Ag and CdS nanoparticles by the fungus *Fusarium oxysporum*. Angew.Chem. Int.Ed. Engl. (communicated).

Alivisatos, A. P. (1996). Semiconductor clusters, nanocrystals, and quantum dots. Science., **271** : 933.

APSCIT (1959). The classic talkby Richard Ferynman entitled "There's plenty of room at the bottom. (available on the website http://www.zyvex.com/nanotech/feynanm.htm).

Beveridge, T. J. and Murray, R. G. E. (1996). Sites of metal deposition in the cell wall of *Bacillus subtilis*. J. Bacteriol., **141 :** 876.

Brus, L. E. (1984). Electron-electron and electron-hole interactions in small semiconductor crystallites: The size dependence of the lowest excited electronic state. J.Chem.Phys., **80** : 4403.

Bulte, J. W. M., Douglas, T., Mann, S., Frankel, R. B., Moskovitz, B. M. Brooks, R. A. Baumgarner, C. D. Vymazal, J. and Frank, J. A. (1994). Magnetoferitin : Biomineralization as a novel molecular approach in the design of iron-oxide - based magnetic resonance contrast agents. Invest. Radiol., **29** : 5214.

Colvin, V. L., Schlamp, M. C. and Alivisatos, A. P. (1994). Light-emitting diodes made from cadmium selenide nanocrystals and a semiconducting polymer. Nature., **370** : 354.

Dagoni, R. (2002). Coverage of the new applications envisaged for nano materials. Chem. Eng. News.

Damle, C., Kumar, A. and Sastry, M. (2002). "Synthesis of Ag/Pd Nanoparticles and Their Low-Temperature Alloying within Thermally Evaporated Fatty Acid Films"J.Phys.Chem.B., **106** : 297.

Fendler, J. H. and Meldrum, F. C. (1995). The colloid chemical approach to nanostructured materials. Adv.Mater., **7** : 607.

Fendler, J. H. (1992). Membrane Mimetic Chemistry Approach to Advanced Materials; Springer-Verlag : Berlin,

Fortin, D. and Beveridge, T.J. (1980). In Biomineralization. From Biology to Biotechnology and Medical Applications (Ed. E. Baeuerien), Wiley-VCH, Weinheim, **7.**

Gole, A. and Sastry, M. (2001) A new method for the generation of pattterened protein films by encapsulation in arrays of thermally evaporated lipids. Biotech.Bioeng., **74** : 172.

Gole, A., Dash, C., Ramakrishnan, V., Sainkar, S.R., Mandale, A.B., Rao, M. and Sastry, M. (2001) "Pepsin-Gold Colloid Conjugates: Preparation, Characterization, and Enzymatic Activity". Langmuir., **17** : 1674.

Hamilton, J. F. and Baetzold, R. C. (1979). Catalysis by small metal clusters. Science. **205**, 1213.

Henglein, A. (1993). Electronics of colloidal nanometer particles. Ber. Bunsen-Ges. **99**, 903 (1995). 3. H. Weller, Colloidal semiconductor Q-Particles: Chemistry in the transition region between solid state and molecules. Angew.Chem. Int.Ed.Engl., **32** : 41.

Henglein, A.(1993). Physicochemical properties of small metal particles in solution : "microelectrode " reactions, chemisorption, compositemetal particles and the atom to metal transition. J.Phys.Chem., **97** : 5457.

Hoffman, A. J. Mills, G., Yee, H. and Hoffman, M. R. (1992). Q-sized cadmium sulfide: synthesis, characterization, and efficiency of photoinitiation of polymerization of several vinylic monomers. J.Phys.Chem., **96** : 5546.

Joerger, R., Klaus, T. and Granqvist, C.G. (2000) "Biologically Produced Silver-Carbon Composite Materials for Optically Functional Thin-Film Coatings". Adv.Mater., **12** : 407.

Klaus-Joerger, T. , Joerger, R., Olsson, E. and Granqvist, C.G. (1999). Bacteria as workers in the living factory: metal-accumulating bacteria and their potential for material science. Trends Biotech. **19,** 15(2000).

Klaus, T., Joerger, R., Olsson, E. and Granqvist, C. G. (2000). Silver-based crystalline nanoparticles microbially fabricated . Proc.Nat.Acad.Sci. **96** : 13611

Klein, D.L., Roth, R., Lim, A. K. L., Alivisatos, A. P. and McEuen, P. L. (1997). "A single-electron transistor made from a cadmium selenide nanocrystal". Nature., **389** : 699

Kroger, N., Deutzmann, R. and Sumper, M. (1995). Polycationic peptides from Diatom Biosilican that direct silica Nanoparticles formation. Science., **286** : 1129.

Ledentsov, N N., Grundmann, M., Kirstaedter, N., Schmidt, O., Heitz, R., Bohrer, J., Bimberg, D., Ustinov, V. M., Shchukin, V. A., Yu Egorov, A., Zhukov, A. E., Zaitsev, S., Kop'ev, P. S., Alferov, Zh. I., Ruvimov, S. S., Kosogov, A. O. Werner, P., Gosele, U. and Heydenrich, (1996). Ordered arrays of quantum dots:formation electronic spectra, relaxation phenomena, lasing. Solid State Electron., **40** : 785

Loveley, D. R. Stolz, J. F., Nord, G. L. and Phillips, E. J. P. (1999). Anaerobic production of magnetite by a dissimilatory iron-reducing microorganism. Nature, **330**: 252 (1987).

Mann, S. (1999). Molecular tectonics in biomineralization and biomimetic materials chemistry. Nature, **365** : 499 (1993).

Mansur, H. S., Grieser, F., Marychurch, M. S. Biggs, S., Urquhart, R. S. and Furlong, D. N. (1995). Photoelectrochemical properties of 'Q-state' CDS particles in Arachidic acid Langmuir-Blodgett films. J.Chem.Soc. Faraday Trans., **91** : 665.

Mehra, R. K. and Winge, D. R. (1991). Metal ion resistance in fungi molecular mechanisms and their regulated expression. J.Cell.Biochem., **45** : 30.

Mukherjee, P., Ahmad, A., Mandal, D., Senapati, S., Sainkar, S.R., Khan, M.I., Ramani, R., Parischa, R., Ajaykumar, P.V., Alam, M., Sastry, M. and Kumar, R. (2001) "Bioreduction of $AuCl_4^-$ Ions by the Fungus, *Verticillium sp.* and Surface Trapping of the Gold Nanoparticles Formed" Angew.Chem.Int.Ed.Engl., **40** : 3585.

Mukherjee, P., Ahmad, A., Mandal, D., Senapati, S. Sainkar, S.R., Khan, M.I., Parischa, R., Ajayakumar, P.V., Alam, M., Kumar, R. and Sastry, M. (2001) "Fungus-Mediated Synthesis of Silver Nanoparticles and Their Immobilization in the Mycelial Matrix: A Novel Biological Approach to Nanoparticle Synthesis". Nano Lett., **1** : 515

Mukherjee, P., Senapati, S., Mandal, D., Ahmad, A., Khan, M.I., Kumar, R. and Sastry, M. (2002) Extracelular synthesis of Gold nanoparticles by the fungus *Fusarium oxysporum* Chem. Bio. Chem., **3** : 461.

Oliver, S. Kupermann, A., Coombs, N., Lough, A. and Ozin, G. A. (1993).Lamellar aluminophopates with surface patterns that mimic diatoms and radiolarian microskeletons. Nature. **378** : 47.

Pum, D. and Sleytr, U. B. (1999). The application of bacterial S-layers in molecular nanotechnology. Trends Biotechnol., **17** : 8.

Ramakrishnan, V., Sable, M., D'Costa, M., Ganesh, K.N. and Sastry, M. (2001). Sequential entrapment of RNA and DNA in lipid bilayers stacks. Chem. Commun. 2622.

Sastry, M. (2000). Nanostructured thin films by self-assembly of surface modified colloidal particles. Current Science, **78** : 1089.

Sastry, M. (2001). *Nanoparticle thin films : an approach based on self-assembly*, Handbook of Surfaces and Interfaces of Materials, (Ed. H. Nalwa, Academic Press, San Diego,), **Vol.3**, Ch.2, 89.

Sastry, M. (2002). Entrapment of proteins and DNA in thermally evapoated lipid films. Trends in Biotech., **20** : 185.

Schmid, G. (1992). Large clusters and colloids, metals in the embryonic state. Chem.Rev., **92** : 1709.

Shi, J., Gider, S., Babcock, K. and Awschalom, D. D. (1996). Magnetic clusters in molecular beams, metals, and semiconductors. Science., **271** : 937.

Shull, R. D., McMichael, R. D. and Ritter, J. J. (1993). Magnetic nanocomposites for magnetic refrigeration. Nanostruct. Mater., **2** : 205.

Simkiss, K. and Wilbur, K. M. (1989). Biomineralization (Academic Press, New York,). S. Mann, (ed.), Biomimetic Materials Chemistry (VCH Publishers, 1996).

Sleytr, U. B., Messner, P., Pum, D. and Sara, M. (1999). Crystalline Bacterial Cell Surface Layers (S Layers): From Supramolecular Cell Structure to Biomimetics and Nanotechnology. Angew. Chem. Int. Ed., **38 :** 1034.

Southam, G. and Beveridge, T. J. (2000). The occurrence of sulphur and phosphorus within bacterially derived crystalline and pseudocrystalline octahedral gold formed invitro. Geochim .Cosmochim.Acta., **60** : 4369

Spring, H. and Schleifer, K. H. (1995). System.Appl.Microbiol., **18 :** 147

Stephen, J. R. and Maenaughton, S. J. (1999). Developments in terrestrial bacterial remediation of metals. Curr.Opin.Biotechnol., **10** : 230.

Symonds, J. L. (1995). Phys.Today., **48** : 26.

Wang, Y. (1993). Nonlinear optical properties of Nanometer- sized semiconductor clusters. Acc.Chem.Res., **24** : 133.

Wang, Y. and Herron, N. (1991). Nanometer-Sized Semiconducter Clusters: Materials Synthesis,Quantum Size effects and photophysical properties. J.Phys.Chem., **95** : 525.

Weller, H. (1998). Transistors and Light Emitters from Single Nanoclusters. Angew. Chem.Int.Ed.Engl., **37** : 1658.

Yoffe, A.D. (1993). Low-dimensional systems: quantum size effects and electronic properties of semiconductor microcrystallites (zero-dimensional systems) and some quasi-two-dimensional systems. Adv. Phys., **42** : 173.

3

Immunodetection of Plant Pathogenic Fungi

B. N. Chakraborty* and U. Chakraborty

Immuno-Phytopathology Laboratory
Department of Botany, University of North Bengal
Siliguri – 734430

Recent trends in detection of plant pathogens include the development of more rapid diagnostics techniques which combine the ability of high specificity for the target organism with detection in very low amounts and at very early stages of infection. Most of these are the immunoassays including ELISA, immunofluorescence, dot blot and western blot. ELISA is now routine practice and various formats have been developed including indirect (PTA), direct (DAS) and dot blots. The pathogen most commonly detected by ELISA have been discussed. Immunofluorescence has also been used to identify various fungal microorganisms. The potentiality of immunodetection techniques for the development of routine assays involving large number of organisms is also immense.

Key words : Immunoassay, Enzyme linked immunosorbent assay (ELISA), Immunflourescence, Dot blot

Disease diagnosis is an art as well as a science. We use the scientific method to perform and interpret tests for the detection of pathogens. The art lies in synthesizing information on symptom development, case history, and results of laboratory tests to determine the most likely cause(s) of disease. Understanding the difference between the terms 'diagnosis' and 'detection', which are often mistakenly used interchangeably, is crucial. Detection of a pathogen does not necessarily prove that it caused the disease at hand. For example, one can usually culture, or induce by incubation, several pathogens from a plant that are not actively causing the observed symptoms. Considering complex etiology of plant disease it can be stated that the 'phenomenon of "one cause-one disease" is rare in nature'. The diagnostician must identify the various causal factors and determine their relative importance. This task can be difficult, as different factors may cause similar

*Corresponding Author e-mail : bnc_nbu@rediffmail.com

symptoms and different symptoms may be caused by the same factor. Just as a positive test result does not prove that the pathogen detected caused the disease, a negative result does not definitely rule out the presence of a particular pathogen. The test itself may fail, or improper selection of plant tissues can lead to erroneous conclusions. Attempts to use culture techniques to recover a vascular pathogen, such as *Fusarium oxysporum* f.sp. *lycopersici*, from wilted tomato foliage rather than vascular tissues would give erroneous negative results. False negative results can occur with any assay, and the diagnostician must be aware of the likelihood of their occurrence.

Obviously, the choice of diagnostic test(s) can greatly influence the diagnosis. Many factors influence the interpretation of test results and the value of the final diagnosis. The quality and case-history of the sample provided, available resources, and expertise of the diagnostician all play a role in the accuracy of the diagnosis. Sample quality is of paramount importance and is probably the most common limiting factor to accurate diagnosis. The best diagnosis would be one for which the affected plant(s) could be observed both in the field and in the laboratory. Unfortunately, expense, time and logistics usually preclude field visits by the diagnostician. Case history information provided with a specimen can be more important to the diagnosis than the specimen itself. This is particularly true when biotic agents play a minor role in the disease. Indeed, many plant specimens submitted to diagnostic labs are afflicted with abiotic problems, such as nutrition imbalances, chemical injury, temperature or moisture extremes, or air pollution. Most abiotic problems must be diagnosed from case-history information and symptom descriptions in the literature. Inadequate or incorrect information often leads to an inaccurate diagnosis.

The competent diagnostician must be able to synthesize information about many different aspects of plant health and know the plant in health as well as disease. The diagnostician has historically been a generalist rather than a specialist, well versed not only in plant pathology, but also in agronomy, horticulture, entomology, soil science and weed science. To understand and interpret modern diagnostic techniques, he or she must also have working knowledge of immunology and molecular biology. In addition to familiarity with pathology, the diagnostician must develop expertise in production practices, response of plants to environmental factors, and characteristics of specific cultivars or hybrids of many different crops. To this extent the individual must become a crop specialist. The difficulties inherent in becoming both a specialist and a generalist in a field that deals with hundreds of plant species and exponentially more pathogens or abiotic agents are obvious. Pathogens most readily detected or identified by various techniques are as follows :

Technique	Pathogens
Serology	
Enzyme linked immunosorbent assay (ELISA)	
Multiwell	Fungi, bacteria, viruses
Flow-through	Fungi, bacteria
Dipstick	Fungi
Dot blot	Fungi, bacteria, viruses
Tissue print	Fungi, viruses
Immunofluorescence	Fungi, bacteria, virus inclusions
Serologically specific electron microscopy (SSEM)	Viruses
Agglutination	Viruses, bacteria
Nucleic acid hybridization	
Dot blot / squash blot	Fungi, bacteria, viruses, viroids
Polymerase chain reaction (PCR)	Fungi, bacteria, viruses, viroids
Tissue print hybridization	Viruses
Fatty acid analysis	Bacteria
dsRNA analysis	Viruses
Polyacrylamide gel electrophoresis (PAGE)	Viroids, viruses
Nutritional test kits	Bacteria
Light microscopy	Fungi, bacteria, nematodes, virus inclusions
Culture	Fungi, bacteria
Baiting	Fungi
Host inoculation	Viruses, viroids, fungi, bacteria, Nematodes
Leaf dips	Viruses
Extraction/identification	Nematodes

Recent trends in detection of plant pathogens include the development of more rapid diagnostic techniques with high specificity for the target organism (Chakraborty, 1988). These techniques can be used to detect fungi, bacteria and viruses present in low amounts in and on plant tissue and, therefore, in many cases the pathogen can be detected at an earlier stage of disease development than was previously possible. Some of these rapid, sensitive techniques are enzyme linked immunosorbent assay (ELISA), immunofluorescence (IF) and the polymerase chain reaction (PCR). Commercial development of such techniques are expensive.

Although most plant pathogenic fungi can be detected by microscopy or other conventional means, but serological techniques have extra advantage where:

(1) the fungus in question is not readily identified by morphological characteristics;

(2) species identification is important and difficult by conventional means;

(3) detection of root pathogens prior to development of foliar symptoms is necessary;

(4) large numbers of samples must be processed for a particular disease for which conventional methods are time-consuming;

(5) rapid, on-site detection is necessary for making disease management decisions for high-value crops;

(6) regulations governing the use of pesticides require demonstration of the presence of a particular pathogen;

(7) the fungus causes disease at low, difficult-to-detect populations in plant tissue; or

(8) plant material is subject to quarantine regulation.

Before the development of serological techniques, laborious and time-consuming assays such as transmission to indicator hosts were used for the routine detection of viruses in many diagnostic labs. Early serological techniques were less sensitive and more time-consuming than those more widely used today. Conventional techniques for virus detection have largely been replaced by the more rapid and sensitive modern techniques, however, some of the conventional techniques still have a place in the diagnostic lab.

The diagnosis of bacterial diseases has been aided by many contributions from both basic and applied research, directed to the rapid detection and identification of plant pathogenic bacteria. Many selective and semiselective media have been developed. Serology continues to become more useful as sensitivity increases and assay time decreases. Several test formats similar to those developed for fungi and viruses have been developed for bacteria, including multiwell ELISA, dot blot, and immunofluorescence. Fatty acid analysis has also proven to be a reliable method of identification, and new techniques, such as nucleic acid probes and the PCR, have extended the limits of specificity.

Many techniques that differ in sensitivity, specificity, reliability and cost are available for the detection of plant pathogens. The most desirable methods for diagnosticians are those that give the least number of false negative or false positive results over time and among workers, are relatively rapid, cost-effective, and detect the broadest range of pathogens. Sensitivity is not as high a priority in the diagnostic lab as it is in regulatory labs or indexing programmes, although a technique must be sensitive enough to detect a pathogen in symptomatic tissue. A diagnostician depends heavily on reliable, accurate, standardized assays. Rapidity is important, although techniques that take an entire day to complete but do not require the diagnostician's full attention are preferably acceptable. Cost is always a consideration, and diagnostic labs will invest in technology that is applicable to the greatest number of pathogens. Analysis of the interpretation of techniques currently available for pathogen detection will serve as a guide to those who must use the art as well as the science for disease diagnosis. The diagnostician must not lose sight of the fact that the main goal is to interpret results. With difficult cases, the diagnostician must integrate many lines of 'hard' and 'soft' empirical evidence in order to make a reliable diagnosis.

Immunological methods are being used increasingly in agricultural research and ELISA is now routine practice for plant virus detection and diagnosis. Application to the study of bacteria and fungi came much later presumably because they are generally much more complex in their make up. Routine testing of large numbers of samples will only be possible when specific, sensitive, easy and reproducible methods of diagnosis are available. Various modifications and amplifications, as used in ELISA for viruses and bacteria can also be used to increase the specificity and sensitivity for fungi. Background reactions can be reduced or eliminated by the addition of different blocking agents which saturate non-specific binding areas. These include for fungi skimmed milk, gelatine, bovine serum albumin, albumin from chicken protein and serum of non-immunised animals. Modifications and amplifications can only improve the results of ELISA when sufficient specific antibodies are available for the particular application. Therefore of all the variables in ELISA procedure, antiserum quality is undoubtedly the most important.

Most of the currently available serological tests for fungi rely on direct or indirect ELISA with either polyclonal or monoclonal antibodies. ELISAs for fungi have been developed in a variety of formats, some of which are more appropriate for diagnostic labs than others. Available formats include: multiwell, flow-through, dipstick, dot blot and tissue print assays. In multiwell ELISA, the wells of commercial kits are precoated with antibodies specific to the target organism. Addition of tissue homogenate to the wells allows binding of antigen to the antibodies, which can be detected by antibody-enzyme conjugate. Multiwell ELISA has been used successfully to detect a wide range

Table 1 : ELISA formats used with plant pathogenic fungi

Fungus	Sample	ELISA type
Acremonium coenophialum	Tall fescue	Protein A Sandwich
Armillaria mellea	Tea	PTA indirect, Dot blot
Bipolaris carbonum	Tea	PTA indirect
Botrytis cinerea	Grapes	PTA indirect
Colletotrichum sp.	Anemones	PTA indirect
Colletrotrichum falcatum	Sugrcane	DAC
Corticium theae	Tea	PTA indirect, Dot blot
Erysiphe graminis f.sp. *hordei*	Barley	PTA indirect
Exobasidium vexans	Tea	PTA indirect, DAS, Dot blot
Fomes lamaoensis	Tea	PTA indirect
Fusarium culmorum	Wheat	PTA indirect, DAS
Fusarium oxysporum	Soybean	PTA indirect
Fusarium graminearum	Soybean	PTA indirect
Gauemannomyces graminis	Wheat	DAS
Glomerella cingulata	Tea	PTA indirect, DAS
Leptosphaeria korrae	Turf grass	PTA indirect
Ophiostoma ulmi	Elm	PTA indirect
Penicillium islandicum	Rice	PTA indirect
Pestalotiopsis theae	Tea	PTA indirect, DAS
Phialophora mutabilis	Pine, Birch	PTA indirect
Phoma exigua	Potato	PTA indirect, DAS
Phomopsis longicolla	Soybean	PTA indirect, DAS
Phytophthora fragariae	Strawberry	PTA indirect
Phytophthora infestans	Potato	PTA indirect
Phytophthora megasperma	Soybean	DAS
Phytophthora nicotianae	Tomato	PTA indirect
Phytophthora parasitica	Soil	DAS
Poria hypobrumea	Tea	PTA indirect, Dot blot
Pseudocercosporella herpotrichoides	Wheat	PTA indirect, DAS
Pythium aphanidermatum	Turf grass	Competetion
Rosellinia arcuata	Tea	PTA indirect, Dot blot
Sclerotium rolfsii	Soybean	PTA indirect
Sclerotinia sclerotiorum	Sunflower	DAS
Sphearostilbe repens	Tea	PTA indirect
Thielaviopsis basicola	Cotton	PTA indirect
Ustilago scitaminea	Sugarcane	DAC, DAS
Ustulina zonata	Tea	PTA indirect, DAS, Dot blot

of plant pathogenesis fungi (Table-1). Sensitivity of multiwell ELISA is high. The main advantage of multiwell ELISA over conventional techniques is rapidity. In flow-through ELISA diagnostics assay is performed on a device composed of absorbent plastic material with antibodies immobilized on the surface. The sample, added dropwise to the top surface, flows past immobilized antibodies that trap the antigen. Antigen is detected by enzyme conjugate. The flow through kits, which are rapid, easy to use, and self-explanatory are particularly suited to field diagnosis. The membrane on the dipstick, which is precoated with specific antibodies, is soaked in buffered crude plant extract to allow antigen-antibody binding. After rinsing the dipstick, bound antigens are detected with enzyme conjugate. The dot-blot format differs from DAS-ELISA in that antigen or plant sap is spotted directly onto a nitrocellulose or polyvinylidene membrane that has not been pretreated with antibodies. For tissue print ELISA a cross-section of plant tissue is pressed directly onto a membrane, thus eliminating the need for an extraction step. If the target pathogen is present, the antigen is transferred from the plant tissue to the membrane. Membrane-bound antigen is detected by antibody-enzyme conjugate. Tissue print ELISA is simple to perform, requires minimal equipment, and is faster than dot blot. Prints of many different sections of plant tissue can be done rapidly for a higher probability of detection. As with dot blot, high background colour from healthy plant sap may occur, but modifiaction of the ELISA procedure can reduce background colour.

Verticillium sp.

A soluble protein extract of *Verticillum dahliae* was used to prepare a specific rabbit antiserum. The reaction of this rabbit antibody to the hyphae of *V.dahliae* was detected with an alkaline phosphatase antirabbit IgG conjugate that hydrolyzed the substrate, napthol-AS-phosphate, to a product that reacted with a diazonium salt, yielding a coloured precipitate outlining the fungal hyphae. The hyphae were readily observed on and in the root cortex of the host usage a dissecting microscope (Gerik *et al.*, 1987). Sundaram *et al.* (1991) also prepared antiserum against purified mycelial proteins from *Verticillium dahliae,* the predominant fungus species in the potato early dying complex. The antisera tested against crude mycelial preparations of *Verticillium* spp. using indirect ELISA reacted positively with 11 out of 12 *V.dahliae* isolates from potato and cotton and soil, but negatively with one isolate from tomato. The isolate did not react with a *Colletotrichum* sp. from potato, or with one isolate of *Rhizoctonia solani* from sugarbeet. Double antibody sandwich ELISA, using polyclonal antisera, detected *V.dahliae* and *V.albo-atrum* in infected roots and stems of potato.

Phytophthora sp.

Phytophthora fragariae was detected by ELISA in roots of strawberry cv.Teniva. Because of high sensitivity of ELISA, presence of fungal antigen was demonstrated before symptoms could be detected in microscopical observations (Werres and Casper, 1987).Antiserum raised against pooled mycelial suspensions from five isolates representing five physiologic races of *P.fragariae*, used in an ELISA detected soluble antigens at protein concentrations as low as 2 ng/ml. Fungal antigens could also be detected in extracts of strawberry plants infected with *P.fragariae*. Root extracts prepared from the alpine strawberry *Fragaria vesca* and *F.ananassa* cv. Cambridge Favourite infected with any of the five isolates studied, produced strong reactions in ELISA (Mohan, 1988). Development of an agri diagnostic *Phytophthora* multiwell ELISA kit for detection of *Phytophthora* in plant tissue, which also readily detected *Phytophthora* in soil where soybeans were damaged by *P.megasperma* f.sp. *glycinea* was also reported by Schmitthenner (1988). A polyclonal antiserum prepared in rabbit immunized with a mycelium extract of *Phytophthora infestans*, reacted in ELISA with mycelium extracts of two *Phytophthora* species but not with those of 10 other microorganisms found on potato. Benson (1991) compared two commercial serological assay kits to a culture plate method for detection of *Phytophthora cinnamomi* in root samples from inoculated azaleas. Both the multiwell E kit and the rapid assay F kit detected *P.cinnamomi* in azalea roots beginning one week after inoculation. Agreement between immunoassay kits and culture plate results for detection of *P.cinnamomi* was most consistent beginning 3-4 weeks after inoculation. Root symptoms, but not foliar symptoms, of *Phytophthora* root rot were evident at this time. Although colour reactions in the rapid assay detectors became increasingly darker after completion of the test, results after 5 min. were as reliable as those after 60 min., since readings for uninoculated controls used to determine test thresholds also increased with time. The multiwell kit detected *P.cinnamomi* in root samples containing as little as 1% infected root tissue. In a commercial nursery survey, 5 and 15% of the azalea root samples at two nurseries had positive ELISA values that were unconfirmed by culture plate. The rapid assay kit dtected *P.cinnamomi*, was easy to use and gave results in a short time. Pscheidt *et al.* (1992) tested the sensitivity of a *Phytophthora* specific immunoassay on 17 species of *Phytophthora* collected throughout the world, including 18 isolate each of *P.cinnamomi* and *P.cactorum*. Kits were also used in the diagnosis of plant specimens with symptoms characteristic of *Phytophthora* infection. *Phyophthora* isolates tested produced a positive reaction with the immunoassay kit. The lowest absorbance relative to other species were obtained from *P.cinnamomi* and *P.megasperma*.Variation in absorbance was high among isolates of *P.cinnamomi* but low among *P.cactorum*. Clinic samples with typical symptoms of *Phytophthora* infection produced a positive reaction with the immunoassay, as did pure cultures of *Phytophthora* sp. isolated from these samples. Cross reaction occurred with several *Pythium* sp. isolated from clinic samples and with several specimens infected with *Peronospora* sp. Cross- reactivity with some of the *Pythium* sp. made interpretation difficult, but when kit results were combined with field histories and symptomatology, the immunoassays proved to be a useful tool in clinical diagnosis. A dipstick immunoassay that was specific for *Phytophthora cinnamomi* was developed for use in soils. Azo dye detection of monoclonal antibody-labelled cysts attached to a nylon membrane provided a rapid, sensitive assay suitable for field use. There was no cross reaction with other *Phytophthora* and *Pythium* species in controlled environment assays or with soil or with other organic matter that adhered to the membrane. The assay was used successfully to detect *P.cinnamomi* in a wide range of soil samples collected from beneath a diverse range of host species. According to the authors, the dipstick assay offers several advantages compared with traditional procedures- ie., familiarity with *Phytophthora* taxonomy is not required, the assay can be performed by unskilled personnel; and soil, rather than infected plant tissues can be assayed. Field testing of the assay showed that in kit form, it could be used as a reliable diagnostic tool to replace or augment current isolation and detection methods. ELISA can also be used for the quantitative determination of fungal biomass in infected tissues. A polyclonal ELISA of *Phytophthora infestans* has been developed by Beckman *et al.*(1994) for use in the determination of fungal biomass during the early stages of infection of tuber discs of *Solanum tuberosum*. By optimizing the dilution of sample extracts and the dilution of primary anti-*P.infestans* antiserum, quantification of the biomass of *P.infestans* in zoospore inoculated tuber disc could be achieved by 8-18 h after inoculation. Differences in growth between avirulent and virulent isolates of *P.infestans* on the resistant potato cv. Kennebec were quantified by 32-48 h after inoculation. Together with a comparison of growth of the same isolates on the susceptible cv. King Edward, these results comprised an ELISA of the Quadrate check. On the resistant host the growth of the avirulent host was invariably arrested by 16h after inoculation, whereas that of the virulent isolate continued throughout the time course. On the susceptible host, however, the avirulent isolate appeared more aggressive than the virulent isolate. Author suggested that ELISA, which is often simpler to perform than other procedures for estimating fungal growth may be used to complement biochemical studies of rapidly induced plant defense response.

Fusarium sp.

Competitive types of two novel ELISAs for *Fusarium* species were developed by Kitagawa *et al.* (1989).Antiserum against a strain (F504) of *F.oxysporum* was elicited in rabbits, and a highly specific, sensistive, and accurate ELISA for the homologous strains was developed by using the antiserum with ß-D-galactosidase-labelled anti-rabbit IgG as the secondary antibody and cell fragments of the strain attached to Amino-Dylark balls as the solid-phase antigen. All other microorganisms tested, including nine other strains of *Fusarium,* showed little cross reactivity. When cell fragments of *F.oxysporum* F501 attached to the balls were used as as a solid phase antigen in a heterologous competitive ELISA, the modified assay was a general assay for 10 strains of four *Fusarium* species with a high sensitivity.The specificity of the heterologous competitive ELISA was shown to be limited to *Fusarium* species. Detection of fusarial toxins using ELISA has also been reported by some authors. Phelps *et al.*(1990) developed an ELISA for detection of isomarticin, a naphthazarin toxin produced by *Fusarium solani.* A carbodimide procedure was used to couple the hapten isomarticin to BSA for the immunogen and to alkaline phosphatase for the enzyme-linked tracer.The resulting assay had a detection limit of 2ng/ml. for isomarticin; other naphthazarin toxins were detectable at less than 10 ng per well in ELISA plates. The assay was specific for naphthazarins. The cross reactivity for a number of phenolic compounds, including the closely related naphthoquinones, was three orders of magnitude less sensitive. Naphthazarin toxins of *Fusarium solani* were also detected by competitive ELISA analysis in xylem fluid of roots rotted by *F.solani* and symptomless scaffold roots and branches of healthy-appearing and diseased citrus trees (Nemee and Charest,1991). In healthy appearing roots of trees with blight symptoms, 11.4 times more toxin was detected than roots of healthy trees; rotted roots contained significantly higher toxins than other roots. Using indirect ELISA, detection of *Fusarium oxysporum* in soybean root tissue was demonstrated by Chakraborty *et al.* (1997b)

Colletotrichum sp.

Infection of sugarcane stalks by *Colletrotrichum falcatum* could be very rapidly detected by an indirect antigen coating enzyme-lined immunosorbent assay (DAC-ELISA) technique using polyclonal antiserum raised against the pathogen, well before symptom expresion. Among the diffetrent host tissues, root eyes, buds and internal white spots wer determined by ELISA to show high pathogen colonization. Differential host pahtogen interaction caused by differetn isolates of the pathogen could also be effectively and more rapidly determined by ELISA procedures. The same technique was also utilized to differentiate the host reaction to pathogen as resistant, moderately resistant, and susceptible based on the pathogen colonization in the cane stalks (Viswanathan *et al.*, 1998, 1999).

Pythium sp.

DAS-ELISA with a monoclonal antibody directed against *Pythium* sp. was developed to detect this pathogen in turfgrass. The monoclonal antibody was the product of hybridoma cell line PA5, produced using *P.aphanidermatum* as immunogen. The antibody bound *P.aphanidermatum, P.graminicola, P.myriotylum* and *P.ultimum* all of which were involved with *Pythium* blight (Miller *et al.*, 1986). Isolation of *Pythium* sp. from different soils by conventional methods revealed *P.violae* to be most common, while *P.sulcatum* was isolated less frequently. Competition ELISA using polyclonal antibodies against *P.violae* and *P.sulcatum* confirmed the results of conventional techniques. With cavities developed on the field grown carrots, ELISA confirmed the predominance of *P.violae*. In one sample, *P.sulcatum* was also isolated from a small number of lesions and was not detected in ELISA. The competition ELISA did not indicate presence of either *Pythium* in a range of non-cavity spot lesions from which attempts at isolation were negative (Lyons and White, 1992).

Penicillium sp.

Antisera obtained from a rabbit immunized with *Penicillium verrucosum* var. *verrucosum* were characterized by immunofluorescence and by direct ELISA for their reactivity with 44 strains of moulds (Fuhrmann and Roqueber, 1989). Antigenically, *P.verrucosum* var. *verrucosum* (subgenus *Penicillium*) appeared to be similar to strains belonging to subgenus *Furcatum*, but antigens confirmed the existence of similar biochemical structures on *Penicillium frequentans, Aspergillus versicolor* and *Aspergillus fumigatus*.

Leptosphaeria sp.

Monoclonal antibody (Mab) LK50 developed against *Leptosphaeria korrae* strain ATCC 56289 was capable of detecting *L. korrae* in cultures and in the naturally infected blue grass samples from three states (Nameth *et al.*, 1990). In cross reactivity tests using indirect ELISA, Mab reacted positively to all 24 isolates of *L.korrae*. Mab reacted negatively with 38 of 42 isolates of related and non related fungi and negatively to apparently healthy grass. The limit of detection was less than 2ng/ml of lyophilized mycelia.

Botrytis sp.

Ricker (1991) detected water soluble antigens produced by *Botrytis cinerea* in naturally infected grape juice by using enzyme immunoassay with an indirect format of antibody HRPO conjugates bound to polyclonal rabbit antibodies directed against *B.cinerea* (anti Bc IgG). Protein-A purified gamma globulin from an early bled antiserum (803-7), which reacted primarily with low molecular weight compounds present only in extracts of *B.cinerea*, was used to specifically detect *B.cinerea* and quantify levels of infection in juice from infected grape berries.

Pseudocercosperella sp.

A double antibody sandwich ELISA test has been developed by Priestley and Dewey (1993) for the detection of *Pseudocercosperella herpotrichoides* using a highly specific monoclonal antibody PH-10 as the capture antibody and genus specific polyclonal rabbit antisera as test antibody. The assay recognized extracts from plants both artificially and naturally infected with *P. herpotrichoides* giving at least three fold higher absorbance values with extracts of *P. herpotrichoides* infected tissue than with extracts from healthy tissues.

Phomopsis sp.

Brill *et al.* (1994) analysed two ELISA formats and antigen preparations against *Phomopsis longicolla*. The polyclonal antibodies (Pabs) were purified to IgG frcation and tested in indirect ELISA and in DAS-ELISA. The Pabs raised to culture filtrate were more specific but less active in binding to members of the *Diaporthe-Phomopsis* complex than were those raised to the mycelial extract immunogen preparation. DAS-ELISA was more specific and 100 fold more sensitive in detecting members of the complex than was indirect ELISA.

Macrophomina sp.

Serological relationship between *Macrophomina phaseolina* and soybean root was demonstrated by Chakraborty and Purkayastha (1983). Major cross reactive antigens were located in soybean root tissue mainly in the susceptible variety.

Thielaviopsis sp.

Soluble protein extracts of chlamydospores and mycelium of *Thielaviopsis basicola* infecting *Gossypium hirsutum* were used to raise polyclonal mouse ascites antibodies. In a fungal capture ELISA, using the purified and biotin labelled IgG fraction both brown and grey cultural types of *T.basicola* were detected, while negligible cross reactivity was observed with other common soil borne fungi. The minimum detection limit of ELISA was between 1 and 20 ng of *T.basicola* proteins depending on the assay. The pathogen could be detected in cotton roots two days after inoculation (Holetz *et.al*, 1994)

Sclerotinia sp.

Jamaux and Spire (1994) developed a serological test that allows the early detection of infection of young petals of *Sclerotinia sclerotiorum*, an important pathogen of rapeseed. Soluble mycelial extrcats of *S.sclerotiorum* were used to produce the first generation antiserum. This was not specific for *S. sclerotiorum* in DAS-ELISA and allowed the screening of cross reacting fungal species such as *Botrytis cinerea*, a pathogen commonly present on rapeseed petals. Use of a polyclonal anti-*B.cinerea* serum enabled the absorption by serial cycles, of *S.sclerotiorum* antigens common to *B.cinerea*. Residual antigens were used as immunogens for the production of the second generation antisera (S1 and S2) which were then tested by DAS-ELISA. Cross reactions with *B.cinerea* decreased with purification cycles of the immunogen whereas cross-reactions with some unrelated fungi slightly increased.

Bipolaris sp.

Major cross reactive antigens shared between *Bipolaris carbonum* and tea were demonstrated by Chakraborty and Saha (1994) using indirect ELISA and its cellular location was also detected in epidermal and mesophyll tissues of tocklai varieties.

Plasmodiophora sp.

Wakeham and White (1996) raised polyclonal antisera to whole and sonicated resting spores of *Plamodiophora brassicae* and two soluble components prepared by filtration and ultracentrifugaion. Cross reactivity of all three antisera with a range of soil fungi, including *Spongospora subteranea* was low. Test formats including western blotting, dipstick, dot blot and indirect ELISA was assessed for their potential to detect resting spores of *P.brassicae* in soil.

Pestalotiopsis sp.

Polyclonal antisera were raised against mycelia and cell wall antigens of *Pestalotiopsis theae* and IgG were further purified by ammonium sulfate fractionation and chromatography on DEAE-Sephadex and the immunoglobin fractions were used for DAC-ELISA, DAS-ELISA and competition ELISA tests (Chakraborty *et al.*, 1995, 1996a). Pathogen was detected in tea leaves as early as 12h after inoculation with *P.theae*. At antiserum dilution upto 1:125, the pathogen could be detected in inoculated leaf extracts up to antigen concentration of 2mg/L. The antiserum reacted with two other isolates of *P.theae* but not with the antigens from mycelial extracts of *Glomerella cingulata* and *Corticium invisum* or with tea leaves inoculated with these pathogens.

Glomerella sp.

Serological cross reactivity between *Glomerella cingulata* and *Camellia sinensis* has been demonstrated by Chakraborty *et al.* (1996b). Pab raised against antigen preparation from mycelia and cell wall of *G.cingulata* were used for detection of pathogen in tea leaf tissue. Antisera raised against cell wall preparations gave better recognition than that against mycelial preparation as observed in DAC-ELISA and DAS-ELISA test (Chakraborty *et al.*, 2002)

Exobasidium sp.

Early detection of *Exobasidium vexans* causal agent of blister blight of tea were achieved using immunoassays by Chakraborty *et al.* (1997) in order to develop field management strategies. DAC-ELISA formats were developed for early detection of the pathogen in tea leaf tissues.

Ustilago sp.

Enzyme linked Immuno-sorbent Asay Techniques were used to detect and quantify the smut pathogen (*Ustilago sutaminea* Syd.) in inoculated sugarcane plants before symptoms expression and pathogen colonization as determined. ELISA was related to resistance or susceptibility of the genotypes. Both Direct ELISA (DAS) using a penicillinase system and Indirect (DAC) using an alkaline phosphatase system were found useful to detect the smut fungus in inoculated plants at a very early stage and the extent of pathogen colonisation was found to be directly related to the susceptibility of the genotype as determined by field evaluations. Teliospores of the fungus as well its dikaryotic mycelium served well as antigen sources. The techniques are suitable for rapidly and precisely screening sugarcane genotypes for smut resistane (Padmanaban and Mohanraj, 1999).

Ustulina sp.

Chakraborty *et al.* (2001a) raised polyclonal antibodies against mycelial and cell wall antigens of *Ustulina zonata* , causing charcoal stump rot disease of tea. Two different ELISA formats such as DAC and DAS were tested to detect the pathogen in soil and artificially inoculated tea root tissues. Such immunological methods developed for early detection of *U.zonata* from soil and tea root tissue will be useful for the proper management of this disease.

Fomes sp.

Polyclonal antibody (PAb) against mycelial antigen of *Fomes lamaoensis*, causal agent of brown root rot disease of tea was raised and analysed following DAC-ELISA formats. Such format was very useful in detecting the pathogen in infested soil which enable disease prevention at an early stage (Chakraborty *et al.*, 2001b).

ELISA has become established as a standard diagnostic procedure in laboratories throughout the world. The versatility of the technique, the sensitivity, speed, and precision with which results can be obtained, and the scale of operation that is possible are attributes of ELISA which, when combined, cannot be matched by any other serodiagnostic method. Undoubtedly, the impact of ELISA has been greatest in the large-scale diagnosis of diseases of perennial and vegetatively propogated crops such as trees, bulbs and ornamentals and possibly of most significance, potatoes and foliar and root diseaes of tea (*Camellia sinensis*). The potential of the ELISA technique for obtaining quantitative data has been exploited for various purposes, including studies on host-pathogen interactions, epidemiological investigations and characterization studies.With increasing interest being shown in the possible use of immunosorbent assays by plant health inspection and certification agencies the need to standardize reagents and operational procedures, as well as methods for the analysis and presentation of results, becomes correspondingly greater. International agreement is difficult to achieve in any field, and the formulation of certification standards, based on immunosorbent assay procedures, is not likely to be easy. However, the relative ease and speed with which plant populations can be assayed for specific pathogens, at the same time enabling quantitative determination of the proportion of infected individuals, at least should facilitate classification of the disease status of the population, if not the formulation of nationally or internationally acceptable standards. There are two principal technical aspects to be considered- (a) the provision of calibrated biological standards (antiserum and antigen) against which test samples can be measured and classified, and (b) the establishment of accepted experimental protocols and analytical procedures.

Immunofluorescence (IF) has also been used to identify various micro-organisms. Plant samples applied to microscope slides as thin tissue sections or soil samples are fixed by a short period of drying or by treatment with chemicals. The optimal fixing method for each sample (antigen) must be determined empirically. To date, indirect immunofluorescence has been preferred for *in situ* analyses of phytopathogenic fungi. It has proved more suitable for detection in plant material or soil than direct immunofluorescence. Phytopathogenic fungi which have been detected in soil and plant by the process (IF) have been listed in Table-2. IF is particularly suitable for detection and localisation of antigens in plant tissue and soil because the cells can be

Table 2 : *In situ* analyses of phytopathogenic fungi using Indirect immunofluorescence (IF)

Alternaria sp.	Barley (grain)	Indirect
Aspergillus sp.	Barley (grain)	Indirect
Bipolaris carbonum	Tea (leaf)	Indirect
Corticium theae	Tea (leaf)	Indirect
Eutypa armeniacea	Grapevines (wood)	Direct, Indirect
Exobasidium vexans	Tea (leaf)	Indirect
Fomes lamaoensis	Tea (root), soil	Indirect
Fusarium graminearum	Soybean (root) , soil	Indirect
Fusarium oxysporum	Soybean (root), soil	Indirect
Fusarium vasinfectum	Cotton (root), Soil	Direct
Ganoderma lucidum	Cashew (root)	Direct
Glomerella cingulata	Tea (leaf)	Indirect
Ophiostoma ulmi	Elm (wood)	Indirect
Pestalotiopsis theae	Tea (leaf)	Indirect
Penicillium sp.	Barley (grain)	Indirect
Penicillium cyclopium	Barley (grain), Soil	Indirect
Phaeolus schweinitzii	Soil	Indirect
Phoma exigua	Potato (tuber)	Indirect
Phytophthora megasperma	Soil	Indirect
Phytophthora cinnamomi	Soil	Direct, Indirect
Plasmodiophora brassicae	Cabbage (root), Soil	Indirect
Sclerotium rolfsii	Soybean (root), soil	Indirect
Sphaerostilbe repens	Tea (root), soil	Indirect
Ustulina zonata	Tea (root), soil	Indirect

viewed directly. Thus, in plants latently infected with phytopathogenic fungi, the pathogen could be detected by IF. Not only identification of the pathogen but quantitative assessments are possible by measuring and counting hyphae as well as resting spores (Wakeham and White, 1996). The availability of sufficient specific antibodies is vital to the success of IF. Interpretation of IF is frequently made more difficult by other sources of fluorescence, such as cross-reactions of other organisms with the antibody, autofluorescence of the plant material and fungicides in soil samples, and non-specific adsorption of marked antibodies to soil particles. The importance of such interference has to be individually determined for each system and requires repeated observations of suitable controls, non-coloured samples or possibly healthy tissue. Different techniques have been tested to eliminate interfering fluorescence in fungi.

Immunological techniques will become increasingly important for the detection, diagnosis and epidemiology of pathogens and non-pathogenic fungi. Of primary importance for immunological studies is selection and preparation of an appropriate antigen and the production of antibody. For most host-pathogen combinations, selection and specific modification of test method will be essential. Choice of a fungal antigen will be determined by the relevant structures or metabolic products to be detected for the pathogen under investigation. This requires appropriate knowledge of the biology and epidemiology of the pathogen. The antigen must be present in sufficient quantity, stable and have the highest possible immunogenicity. Generally fungal antigenic determinants are concentrated in young hyphal tips and those metabolites in the category of surface components may produce specific antibodies of high affinity. For routine use with large sample numbers, time saving and easy to handle methods are essential. ELISA is a good example, since to a large extent the process may be automated. Even more convenient are methods such as dot blot and dip stick assays. Where it is necessary to process large numbers of samples, as in the plant protection services, immunological techniques will become increasingly important. Some methods which are at present have limited use will undoubtedly become more important in work with fungi.

References

Beckman, K. B., Harrison, J. G. and Ingram, D. S. (1994). Optimization of a polyclonal enzyme-linked immunosorbent assay (ELISA) of fungal biomass for the use in studies of plant defense responses. Physiol. Mol. Plant Pathol., **44:**19-32.

Benson, D. M. (1991). Detection of *Phytophthora cinnamoni* in Azalea with commercial serological assay kits. Plant Disease, **75(5):**478-482.

Brill, L. M., McClary, R. D. and Sinclair, J. B. (1994). Analysis of two ELISA formats and antigen preparations using polyclonal antibodies against *Phomopsis longicolla*. Phytopathol, **84:**173-179.

Chakraborty, B. N. (1988). Antigenic Disparity. In: Experimental and Conceptual Plant Pathology (Eds. R.S. Singh, U. S. Singh, W. M. Hess and D. J. Webber), pp. 477-484. Oxford & I. B. H. Publishing, New Delhi, India.

Chakraborty, B. N. and Purkayastha, R. P. (1983). Serological relationship between *Macrophomina phaseolina* and soybean cultivars. Physiol. Plant Pathol., **23:**197-205.

Chakraborty, B. N. and Saha, A. (1994). Detection and cellular location of cross-reactive antigens shared by *Camellia sinensis* and *Bipolaris carbonum*. Physiol. Mol. Plant Pathol., **44:**403-416.

Chakraborty, B. N., Basu, P., Das, R., Saha, A. and Chakraborty, U. (1995). Detection of cross-reactive antigens between *Pestalotiopsis theae* and tea leaves and their cellular location. Annals of Applied Biology, **207**:11-21.

Chakraborty, B. N., Chakraborty, U., Basu, P., Das, R. and Saha, A. (1996b). Serological relationship between *Glomerella cingulata* (Stoneman) Spaulid and Schrenk and *Camellia sinensis* (L.) O. Kuntze. Journal of Plantation Crops, **24**:205-211.

Chakraborty, B. N., Chakraborty, U., Saha, A., Das and R., Basu, P. (1997a). Detection and management of blister blight of tea. In: Management of Threatening Plant Diseases of National Importance (Eds. V. P. Agnihotri, A. K. Sarbhoy and D. V. Singh), pp. 227-235, Malhotra Publishing House, New Delhi.

Chakraborty, B. N., Chakraborty, U., Sengupta, D., Deb, D. and Das, J. (2001a). Development of immuno-diagnostic kits for detection of *Ustulina zonata* in the soil and tea root tissues. Proceedings of International Symposium on Frontiers of Fungal Diversity and Diseases in South-East Asia. Department of Botany, D. D. U. Gorakhpur University, Gorakhpur: pp66.

Chakraborty, B. N., Sarkar, B. and Chakraborty, U. (1997b). Detection of cross-reactive antigens shared by *Fusarium oxysporum* and *Glycine max* by indirect ELISA and their cellular location in root tissues. Folia Microbiol., **42**:607-612.

Chakraborty, U., Basu, P., Das, R., Saha, A. and Chakraborty, B. N. (1996a). Evaluation of antiserum raised against *Pestaloptiosis theae* for detection of grey blight disease of tea by ELISA. Folia Microbiol., **41**:413-418.

Chakraborty, U., Das, R., Basu, P., Guha, S. and Chakraborty, B. N. (2002). Serological cross reactivity between *Glomerella cingulata* and *Camellia sinensis*. Indian Phytopath., **55(1)**:1-7.

Chakraborty, U., Deb, D., Das, J., Sengupta, D. and Chakraborty, B. N. (2001b). Development of polyclonal antibody based immunoassay for detecting *Fomes lamoaensis* causing brown root rot disease of tea. J. Mycol. Pl. Pathol., **31(1)**:118-119.

Fuhrmann, J., Roquebert, M. F., Van Hoegaerden, M. and Strossberg, A. D. (1989). Immunological differentiation of *Penicillium* species. Can. J. Microbiol., **35(4)**:1043-1047.

Gerik, J. S., Lommel, S. A., Huisman, O. C. (1987). A specific serological staining procedure for *Verticillium dahliae* in cotton root tissue. Phytopathol., **77(2)**:261-265.

Holtz, B. A., Karu, A. E. and Weinhold, A. R. (1994). Enzyme-linked immunosorbent assay for detection of *Thielavopsis basicola*. Phytpathol., **84**:977-983.

Jamaux, I. and Spire, D. (1994). Development of a polyclonal antibody-based immunoassay for the detection of *Sclerotinia sclerotiorum* in rapeseed petals. Plant Pathol., **43**:847-862.

Kitagawa, T., Sakamoto, Y., Furumi, K. and Ogura, H. (1989). Novel enzyme immunoassay for specific detection of *Fusarium oxysporum* f. sp. *cucumerinum* and for general detection of various *Fusarium* species. Phytopathol., **79**:162-165.

Lyons, N. F. and White, J. G. (1992). Detection of *Pythium violae* and *Pythium sulcatum* in carrots with cavity spot using competition ELISA. Annals of Applied Biology, **120**:235-244.

Miller, S. A., Grothus, G. D., Peterson, F. P. and Papa, S. L. (1986). Detection of *Pythium* blight in turfgrass using a monoclonal antibody-based diagnostic test. Phytopathol., **76**:1057 (Abstr.).

Mohan, S. B. (1988). Evaluation of antisera raised against *Phytophthora fragariae* for detecting the red core disease of strawberries by enzyme-linked immunosorbent assay (ELISA). Plant Pathol., **37**:206-216.

Nameth, S. T., Shane, W. W. and Stier, J. C. (1990). Development of a monoclonal antibody for detection of *Leptosphaeria korrae*, the causal agent of necrotic ringspot disease of turfgrass. Phytopathol, **80(11)**:1208-1211.

Nemec, S., Jabaji-Hare, S. and Charest, P. M. (1991). ELISA and immunocytochemical detection of *Fusarium solani*-produced naphthazarin toxins in citrus trees in Florida. Phytopathol., **81(12)**:1497-1503.

Padmanaban, P. and Mohanraj, D. (1999). Early and rapid detection of sugarane smut by immanological methods. pp. 46-52 In : Sugarcane Pathology (Ed : G.P. Rao, A. Bergamin Fillo, R.C. Magarcy, L.J.C. Autrey), Science Publ. Inc., New Hampshire, USA.

Phelps, D. C., Nemec, S., Baker, R. and Mansell, R. (1990). Immunoassay for naphthazarin phytotoxin produced by *Fusarium solani*. Phytopathol., **80(3)**:298-302.

Priestly, R. A. and Dewey, F. M. (1993). Development of a monoclonal antibody immunoassay for the eyespot pathogen *Pseudocercosporella herpotrichoides*. Plant Pathol., **42**:403-412.

Pscheidt, J. W., Burket, J. Z., Fischer S. L. and Hamm, P. B. (1992). Sensitive and clinical use of *Phytophthora*-specific immunoassay kits. Plant Disease, **76(9)**:928-932.

Ricker, R. W., Marois, J. J., Dlott, J. W., Bostock, R. M. and Morrison, J. C. (1991). Immunodetection and quantification of *Botrytis cinerea* on harvested wine grapes. Phytopathol., **81(4)**: 404-411.

Schmitthenner, A. F. (1988). ELISA detection of *Phytophthora* from soil. (Abstr.) Phytopathol., **78**:1576.

Sundaram, S., Plasencia, J. and Banttari, E. E. (1991). Enzyme-linked immunosorbent assay for detection of *Verticillium* spp. using antisera produced to *V. dahliae* from potato. Phytopathol., **81(12)**:1485-1489.

Viswanathan, R., Samiyappan, R. and Padmanaban, P. (1998). Specific detection of *Collectrotridum falcatum* in sugarcane by seorlogial technique. Sugarcane, **3** : 18-23.

Viswanathan, R., Padmanaban, P., Mohanraj, D. and Jothi, R. (1999). Rapid detection of red rot in sugarcane and Estmiation of host reaction by ELISA. pp. 35-45. In Sugarcane Pathology (Ed: G.P. Rao, A. Bergamin Fillo, R.C. Magarcy, L.J.C. Autrey), Science Publ. Inc., New Hampshire, USA.

Wakeham, A. J. and White, J. G. (1996). Serological detection in soil of *Plasmodiophora brassicae* resting spores. Physiol. Mol. Plant Pathol., **48**:289-303.

Werres, S. and Casper, R. (1987). Nachweis von *Phytophthora fragariae* Hickman in Wurzeln der Erdbeersorte 'Tenira' mit Hilfe des enzyme-linked immunosorbent assay (ELISA). Journal of Phytopathology, **118(4)**:367-369.

4

Application of Molecular Markers for the Analysis of Genetic Diversity and Systematics of Fungi

K.K. Sabu*, N.S. Pradeep and P. Padmesh
Tropical Botanic Garden and Research Institute (TBGRI), Thiruvananthapuram-695562, Kerala

Analysis of genetic diversity is carried out using a variety of tools and techniques, however, due to its accuracy, the use of molecular markers find much popularity. Such studies often hold diverse applications in the fields of utilisation of genetic resources, molecular taxonomy and conservation of biological materials. There are many reports available which deal with the use of protein/DNA markers in edible mushrooms. Oyster mushrooms (*Pleurotus* sp.) are one of the popular edible mushrooms in the tropics thoroughly studied mainly to scrutinise speculated position of many species and genotypic classes. The present review discusses on the application of isozyme and RAPD markers for the analysis of genetic diversity and systematics of *Pleurotus* sp.

Key words : Genetic diversity, *Pleurotus*, oyster mushroom, isozyme, RAPD

Definitive evidence for genetic diversity within and among populations became available only in late 1960s. This was provided by artificial selection experiments in cultivated crops such as rice, wheat etc. and laboratory model organisms like *Drosophila*. The genetic diversity can be analysed and estimated using a variety of techniques, among which use of molecular markers find greater utility due to its ability to reveal polymorphism at the DNA level. There are many reviews available in the literature dealing with the use of molecular markers for variety of applications on various groups of fungi, the present attempt however, focuses on *Pleurotus*, an important genus of edible mushrooms.

There are hundreds of mushroom varieties grown in the world today. Of these oyster mushrooms (*Pleurotus* (Fr.) P. Kumm.), one of the choice as edible mushrooms in the tropics has gained much importance in the last decade in many countries including India. Due to its growing economic importance and spectra of both intraspecific as well as interspecific genetic variability,

*Corresponding Author E-mail : kksabu@rediffmail.com

several studies were undertaken world over during the past two decades mainly to scrutinise speculated position of many species and genotypic classes. Isozymes and RAPD markers are the two widely used tools in these studies and the technical details of which are discussed in this article along with significant findings on the diversity and systematics of the genus *Pleurotus* using these techniques.

Isozymes

In an elaborate review on isozyme analysis, Micales *et al.* (1986) has summarized utility of the technique in fungal taxonomy and genetics. They ruled out the possible objections raised by classical geneticists on the isozyme banding patterns, which are now being confirmed by crossing experiments. Even today isozyme markers are one of the best tools to answer many research questions in the analysis of genetic polymorphism, taxonomic infrastructure, construction of linkage map, etc. in many fungal taxa (Zervakis *et al.*, 2001; Larraya *et al.*, 2000).

Markert and Moller coined the term 'isozyme' in 1959 to describe the multiple molecular forms of enzymes that exhibit the same enzymatic specificity but different electrophoretic mobility. This is usually arising as a result of mutation, where at the minimum, a base sequence is deleted or replaced by another. If the mutation affects a coding region, the enzyme or protein translated from the DNA sequence will differ slightly from the one formed before mutation. If the product of such sequence is an enzyme having altered amino acid sequence but showing similar substrate specificity, then it will migrate to a different position in an electric field. The different forms of the gene that produce structurally altered proteins are called alleles. The enzyme variants at a single locus are termed alloenzymes or allozymes, while the enzyme products formed from genes at different loci are termed isoenzymes or isozymes (Fig. 1).

Even though molecular weight and size will be similar to the allozymes of an isozyme, they differ each other in net surface charges which resulted in their varied migration under an electric field. The technical utility of electrophoresis coupled with the zymogram technique (Hunter and Markert, 1957), has been considered as the best tool of choice for studies of heritable variations by geneticists, systematists, and population biologists. Electrophoresis is the process by which charged particles move through an electric field depending on their net electric charge. Usually, some inert matrix (e.g. acrylamide, agarose, starch, etc.) and buffer solutions are used to support the particles as well as passage of electric current. The matrix in general mimics a molecular sieve which allows particles having smaller size and weight faster.

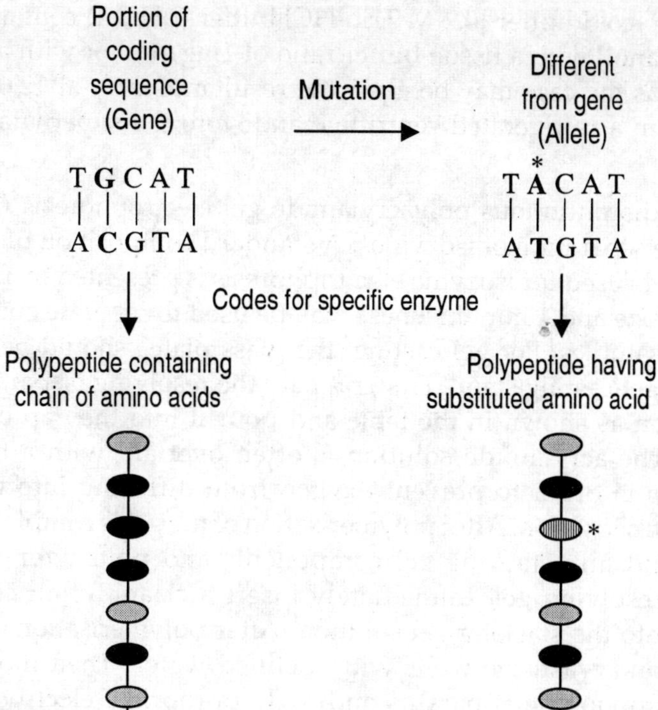

Fig. 1. Pictorial representation of origin of genetic variation leading to the formation of allozymes. The * indicates location of mutation.

Here utility of polyacrylamide gel for enzyme and agarose gel for DNA separation are discussed.

Extraction and separation of isozymes : Enzymes are present in almost all living cells and are even abundant in actively dividing cells. The selection of tissues as the source of isozymes solely depends on the objectives of the study as well as the enzymes to be assayed. In studies involving activity staining or assaying of enzymes, one must extract enzymes at its intact condition from whatever be the tissues analysed. Normally it is carried out using an extraction buffer containing various antioxidants as well as enzyme stabilisers that help prevent destruction of enzyme molecules. The process may be very difficult for mature tissues (particularly higher plant tissues) as it often studded with highly complex mixture of secondary metabolites. Immediately after sampling, the enzyme source must be cleaned properly without damaging the cells, which may otherwise activate phenolic substances and deteriorate the quality of enzymes. It should be stored in a -70^{0}C deep freezer or kept in liquid nitrogen, if enzyme extraction is carried out later.

For extraction of enzymes, the sample is mashed in a pre-chilled mortar in a suitable ice-cold buffer [0.2 M Trio-HCl buffer (pH 8.4) containing 0.05% β-mercaptoethanol] with a tissue-buffer ratio of 1mg/4 μl or with appropriate modifications as the case may be. Spin the resultant slurry at 12,000 rpm for 10 min at 4^0C in a refrigerated centrifuge and store the supernatant at -70^0C before use.

Vertical discontinuous polyacrylamide gel electrophoresis (PAGE) can be used for the separation of isozymes. A standard composition of the gel and electrode buffers used for isozyme electrophoresis is presented in Table 1. Gels of 16 x 18 cm^2 size and 1 mm thickness may be used to separate enzymes from at least 10-15 samples. For gel casting, the glass plates should be assembled and fixed in a gel-casting stand. First, prepare the resolving (separating) gel of the composition as shown in the table and pour it into the gap between the glass plates. The acrylamide solution is often overlaid with a thin film of distilled water in order to prevent oxygen from diffusing into the gel and inhibiting polymerisation. After polymerisation of the gel is completed, remove the overlay and mix stacking gel components and pour over the already polymerised resolving gel. Immediately insert a clean Teflon comb as per requirement into the stacking gel solution. After polymerisation, remove the Teflon comb and wash the wells with distilled water. Then mount the gel onto the electrophoretic apparatus and add appropriate electrode (or tank) buffer solutions to sufficiently immerse the electrodes. Keep the electrophoresis unit inside a fridge, which is able to keep the enzymes native during electrophoresis. Loading of the enzyme extract is done after mixing it with sample buffer (containing 0.5% bromophenol blue and 30% sucrose dissolved in distilled water) in 9:1 ratio. The higher density of the sample buffer help to

Table 1. Details of gel and electrode buffer composition used for isozyme electrophoresis.

Item	Components	Volume (ml) required[1]	
		Stacking gel 4% gel	Separating gel 7% gel
A. Gel buffer	30% Acrylamide mix[2]	0.980	5.800
	Tris buffer[3]	0.940	6.250
	10% APS	0.075	0.250
	H_2O (distilled/deionised)	5.510	12.680
	TEMED	0.005	0.020
B. Electrode buffer	25 mM Tris HCl, 250 mM Glycine (pH 8.3)	2000	

keep the extract inside the well. About 20-30ml of the extract is sufficient to resolve most of the enzymes. Estimation of soluble proteins in the extract needs to be carried out only when quantitative inference is required from the zymogram.

Complete separation of the enzymes may take 4-5 hrs. The power unit of choice is set at 150V (8 v/cm) while the enzymes are moving through the stacking gel and 200V (15 v/cm) at the separating gel zone. After electrophoresis, the gel can be stained for detection of the required isozymes (Table 2) and documented either by appropriate gel documentation system or manually by photography.

Interpretation of zymogram and data analysis : Genetic interpretation of enzyme phenotypes is based on observed differences in zymogram profile after activity staining of PAGE. The standard nomenclature of isozymes and allozymes is based on Wendel and Percy (1990) and Surve Iyer (1995). Locus encoding the most anodally migrating isozyme for each enzyme system is assigned the numerical designation 1, with additional loci numbered sequentially (2, 3...) in order of decreasing electrophoretic mobility. Similarly, the fastest moving allozyme at each locus is given character designation A, with following alleles (B, C...) named in order of decreasing electrophoretic mobility.

POPGENE® Version 1.0 computer program (Yeh and Boyle, 1997) is a user-friendly Microsoft® windows based package for the analysis of genetic variation among and within natural populations. It can be used for the analysis of isozyme variation using both haploid and diploid data sets. It is a freeware distributed over the Internet from the POPGENE® home page at www.ualberta.ca/~fyeh. Standard genetic variability measures including the number and percentage of polymorphic loci, observed number of alleles, allele frequency, gene diversity, observed and expected heterozygocity, genetic distance and identity can be detected using this package. Dendrogram may be drawn based on Nei's measures of distances (Nei, 1973, 1972) using UPGMA method.

RAPD markers

Random amplified polymorphic DNA (RAPD; Williams *et al.*, 1990) markers are a modification of polymerase chain reaction (PCR; Mullis and Faloona, 1987), which provides a means by which billions of copies of a particular target DNA fragment can be made from a complex mixture of genomic DNA. RAPD analysis is widely being used in fungal strain identification, ecophysiological studies, breeding and mapping programmes.

Table 2. Composition of reagents used for activity staining of commonly analysed isozymes

Enzyme	Staining solution[1]	References
Alcohol dehydrogenase (*Adh*) E.C.1.1.1.1	Tris 0.1 M, pH 7.5 (100), NAD^+ (30), MTT (20), PMS (4), Ethanol (6) (add just before incubation). Incubate in dark at 30ºC for 15 – 60 min depending on enzyme activity. Discard solution, stop reaction and fix	Tanksley, 1979
Esterase (*Est*) E.C.3.1.1.2	Sodium phosphate 0.1 M, pH 6.2 (100), α-Naphthyl butyrate 1%(in acetone), Fast blue RR salt (100). Incubate in dark at 30ºC for 4 – 8 h. Discard solution, stop reaction and fix.	Tanksley and Rick, 1980
Glutamate dehydrogenase (*Gdh*) E.C.1.4.1.2	Tris 0.1 M, pH 7.5 (100), $CaCl_2$ 10 mM, Sodium glutamate (800), NAD^+ (30), NBT (20), PMS (4). Incubate in dark at 30ºC for 30 min to 2 h or until bands appear in the gel. Discard solution, stop reaction and fix.	Hartmann *et al.*, 1973
Glutamate oxaloacetate transaminase (*Got*) E.C.2.6.1.1	A. Tris 0.1 M, pH 8.5 (100), α -Ketoglutarate (100), Aspartic acid (200). B. Pyridoxal-5-phosphate (10), Fast blue RR salt (150). Incubate in dark at 30ºC for 2 – 4 h or until blue bands appear. Discard solution, stop reaction and fix.	Decker and Ran, 1963
Malate dehydrogenase (*Mdh*) E.C.1.1.1.37	Tris 0.1 M, pH 7.5 (100), DL-Malate 1 M, pH 7.5 (3), NAD^+(30), MTT (20), PMS (4). Incubate in dark at 30ºC for 15 – 60 min or until blue bands appear. Discard solution, stop reaction and fix.	Brown *et al.*, 1978
Peroxidase (*Per*) E.C.1.11.1.7	Acetate buffer 0.2 M, pH 4.6 (28), Ethanol (68), O-Dianisidine (100), 3% H_2O_2 (2). Incubate in dark at 30ºC for 15 – 60 min. Discard solution, stop reaction and fix.	Brewbaker *et al.*, 1968
Succinic dehydrogenase (*Sdh*) E.C.1.3.99.1	Na phosphate 50 mM, pH 7.0 (100), Sodium EDTA (400), Sodium Succinate (250), ATP Na_2 (50), NAD^+ (70), MTT (40), PMS (2). Incubate in dark at 30ºC for 30 min. Discard solution, stop reaction and fix.	Brewer, 1970
Superoxide dismutase (*Sod*) E.C.1.15.1.1	A. Sodium phosphate 50 mM, pH 7.5 (100), MTT (200) B. Sodium phosphate 50 mM, pH 7.5 (100), TEMED (0.4), Riboflavin (1). First, soak gels in solution A for 20 min in the dark at 30ºC. Then, drain off solution A and pour solution B and incubate under illumination for 15 min or until the achromatic bands are easily distinguishable. Stop reaction and fix.	Vallejos, 1983
Polyphenol oxidase (*Ppo*) EC 1.10.3.2	100mM phosphate buffer, pH 6.0; DOPA (50). Incubate in light at 30ºC for 8 h.	Wang *et al.*, 1989
Acid phosphatase (*Acp*) EC 3.1.3.2	50mM sodium acetate, pH 5.5; 1M $MgCl_2.6H_2O$ (1); Fast black K salt (100); Fast blue RR salt (100); α & β Naphthyl acid phosphate, 1% (3) in 50% acetone. Incubate in dark at 30ºC for 2 h.	Vallejos, 1983
Lactate dehydrogenase (*Ldh*) EC 1.1.1.27	100mM tris-Cl buffer, pH 8.4; 200mM sodium lactate (15); NAD (25); NBT (15); PMS (3). Incubated in dark at 37ºC for 1h.	Wang *et al.*, 1989

1 7 ml of stacking gel and 25 ml of separating gel was required to cast a single gel of 1 mm thickness in the Hoefer Electrophoresis Unit. The volume may vary depending on the make.
2 Acrylamide mix: acrylamide, 29 and bisacrylamide, 1
3 For preparing stacking gel, 0.5 M Tris (pH 6.8) and for separating gel, 1.5 M Tris (pH 8.8) can be used.
4 Quantity expressed in ml for buffers or solvents and mg for fine chemicals.

However, its application in *Pleurotus* research programmes is still not that much popular as evidenced from the review of literature.

Extraction of genomic DNA : Total genomic DNA from the samples may be isolated following modified Murray and Thompson (1980) method using CTAB (Table 3; Fig 2). The extraction buffer may be prepared by adding 1.2% PVP-40T (MW 40,000, Sigma, USA) to remove high phenolic contaminants while double $CHCl_3$ extraction at 10,000 rpm help to remove polysaccharides. After ethanol precipitation, DNA is resuspended in 100 ml of 1x TE buffer (pH 8.0) and store at -20^{0}C till assayed.

PCR assay : Polymerase chain reaction may be carried out in 25 µl reaction mixture containing 2.5 µl 10x amplification buffer (500 mM KCl, 100 µM Tris HCl, 1.0% Triton X-100 and 15 mM $MgCl_2$), 100 µl each of dATP, dGTP, dCTP and dTTP, 1.0 U of *Taq* DNA polymerase, 15 pM of decamer random primer and 50 ng of genomic DNA. Amplification reactions can be performed in a DNA Thermal Cycler. The sequential steps to be programmed in thermal cycler are: 1 cycle of 2 min at 93°C, 2 min at 35°C and 2 min at 72°C followed by 38 cycles of 1 min at 93°C, 1 min at 36°C and 2 min at 72°C. The last cycle may be followed by 10 min extension at 72°C. However, the duration and number of cycles need to be optimized according to the sample/ species analysed.

Agarose gel electrophoresis : The amplified products are resolved in 1.2% agarose gel (1x TBE) followed by ethidium bromide staining and the bands detected are documented using an image analysis and documentation system. For manual documentation, one must place the gel over an UV transilluminator and photograph may be taken using a camera loaded with B/W film and the lens covered with a blue filter. Amplified products that are reproducible and consistent in performance are chosen for data analysis.

RAPD data analysis : POPGENE® can be used for RAPD data analysis too. Most of the variability statistics that could be performed using isozyme data and described earlier can also be analysed using RAPD data. Another interesting computer program is PCO3D® package which is to be used for the analyses of pattern data - geographic variation, variation among individuals, classification of species, etc. The package consists of 3 parts: computation of similarities, principal coordinate analysis and 3-dimensional plotting. The program is available with Dr. Robert P. Adams, Plant Biotechnology Centre, Baylor University, TX 79040, USA.

In order to perform the data analysis, the bands appearing in the documented images should be converted to binary data (0 and 1) in such a way that the presence of a band scores as 1 and its absence as 0. This can be

Table. 3. Stock solutions and reagents used for genomic DNA extraction required for PCR assay.

Reagent	Ingredients[1]	Quantity
Tris buffer	Tris buffer 1 M, pH 8.0 (pH adj. with conc. HCl)	12.11 gm
EDTA	Na_2 EDTA (pH adj. with NaOH beads)	18.61 gm
CTAB Extraction buffer	CTAB 2% w/v	2 gm
	Tris buffer 100 mM, pH 8.0	10 ml (from stock)
	Na_2 EDTA 20 mM, pH 8.0	4 ml (from stock)
	NaCl 1.4 M	8.2 gm
	PVP 1.2%	1.2 gm
	β–Mercaptoethanol 1%	1 ml
CTAB Precipitation buffer	CTAB 1% w/v	1 gm
	Tris buffer 50 mM, pH 8.0	5 ml (from stock)
	Na_2 EDTA 10 mM, pH 8.0	2 ml (from stock)
CTAB-NaCl solution	CTAB 10%	10 gm
	NaCl	4.1 gm
High Salt TE buffer	Tris buffer 10 mM, pH 8.0	1 ml
	Na_2 EDTA 0.1 mM, pH 8.0	20 μl
	NaCl 1 M	5.85 gm
TE buffer	Tris buffer 10 mM, pH 8.0	1 ml
	Na_2 EDTA 1 mM, pH 8.0	0.2 ml
TBE buffer (10x)	Tris base	10.8 gm
	Boric acid	5.5 gm
	Na_2 EDTA 0.5 mM, pH 8.0	20 ml
Phenol saturation buffer	Phenol	100 gm
	Add Tris buffer to get a final pH 8.0	
Sodium acetate	Sodium acetate 3 M	40.8 gm
Gel loading buffer (6x)	Bromophenol blue 0.25%	0.25 gm
	Sucrose 40%	40 gm
Agarose (1.2%)	Agarose	1200 mg
	TBE (10x)	10 ml
	H_2O (Distilled / Deionised)	90 ml
	EtBr	0.8 μl (from 10 mg/ml stock)

[1] All reagents are prepared in 100 ml distilled/deionised water.

Sample tissue, 2 gm

↓

Pulverise tissue to a fine powder with liquid nitrogen, add warm β-ME-CTAB extraction buffer and mix thoroughly, incubate for 2 h at 65°C

↓

Extract with equal volume of $CHCl_3$, spin at 10000 rpm for 5 min

↓

Collect supernatant, add 1 ml CTAB-NaCl buffer (65°C), invert mix, extract with equal volume of $CHCl_3$, mix well, spin at 10000 rpm for 5 min

↓

Collect supernatant, add 1 volume CTAB precipitation buffer, invert mix, incubate at 37°C overnight

↓

Spin at 8000 rpm for 10 min, collect pellet, add 1 ml high salt TE buffer to dissolve the pellet, then add 1 ml isopropanol, keep at -20°C for 30 min

↓

Spin at 10000 rpm for 15 min, collect pellet, wash with 80% EtOH, dry, re-suspend in 0.5 ml TE buffer

↓

Add 4 μl RNAse to the sample, incubate at 55°C for 25 min, then add 0.5 ml $CHCl_3$, invert mix, spin at 10000 rpm for 5 min

↓

Collect supernatant, add 50 μl of 3M Sodium acetate, slightly move, add 1 ml of 100% EtOH, spin at 10000 rpm for 15 min

↓

Collect pellet, add 1 ml of 70% EtOH, spin at 10000 rpm for 5 min, decant and dry

↓

Dissolve the pellet in 100 μl of 1xTE

Fig. 2. Schematic representation of DNA extraction protocol.

used as the input file for data analysis and then to detect the standard genetic diversity statistics.

Pleurotus, a versatile group of basidiomycetes

Pleurotus species are one of the causative agents for wood decay in terrestrial habitats worldwide and are widely collected and cultivated as edible fungi in the East and increasingly in the West. The oyster mushrooms are good sources of non-starchy carbohydrates, having a high content of dietary fiber and contain moderate quantities of good quality protein with most of the essential amino acids, minerals, and vitamins. In terms of amino acid index and protein sources, *Pleurotus* species fall between high-grade vegetable and low-grade meat (Opletal, 1993; Stamets, 1993; Breene, 1990; Bano and Rajarathanam, 1988).

Pleurotus enjoy a worldwide distribution in nature and are met within both temperate and tropical zones; however, they are more prevalent in the sub-tropics. Moreover, they have a wide range of ecological adaptations to the climatic conditions, as well as the ability to colonize a variety of substrates.

Isozyme analysis has been widely used in different species/strains of *Pleurotus*. While in the beginning attention was given to the banding pattern alone (Blaich and Esser, 1975), later the trend has changed by including numerical methods to the allelic data (Larraya *et al.*, 2000). Recently, RAPD assay is being done to differentiate many *Pleurotus* species (Zervakis *et al.*, 2001; Pradeep, 2002).

i) Use of isozymes in the analysis of genetic variability and systematics of *Pleurotus*

In revealing the extent of genetic variability in many *Pleurotus* species, utility of isozymes are well documented (Iracabal and Labarere, 1993; Zervakis and Labarere, 1992; Magae *et al.*, 1990; May and Royse, 1988; May and Royse, 1982). Studies of Iracabal and Labarere (1993) using five enzymes has showed considerable variability among the wild strains of *P. cornucopiae* largely in accordance with their geographic origin. It is also interesting to note that the extent of variability is different in various *Pleurotus* species. While in *P. ostreatus*, isozyme analysis showed higher level (93%) of polymorphism (Kulkarni *et al.*, 1986), the same in *P. citrinopileatus* has showed only 21.05% (Jacob *et al.*, 2000). In a comprehensive study, Zervakis *et al.* (1994) has analysed both the intraspecific diversity and interspecific relatedness of 11 *Pleurotus* species viz., *P. abalones, P. columbinus, P. cornucopiae, P. cystidiosus, P. dryinus, P. eryngii, P. flabellatus, P. ostreatus, P. pulmonarius, P. sajor-caju* and *P. sapidus*. They identified *P. cystidiosus* and *P. eryngii* as species having higher intraspecific genetic variability while *P. cornucopiae, P. dryinus* and *P. sajor-caju* showed little variability.

Differences in isozyme banding patterns have been widely used in many *Pleurotus* species to separate or identify species (Zervakis *et al.*, 1994; Zervakis and Labarere, 1992; Iracabal *et al.*, 1991; May and Royse, 1988; Kulkarni *et al.*, 1986; Boisselier-Dubayle, 1983). For example, distinguishing species by the similarity index of the seven enzymes offered reasonable basis for classifying the *Pleurotus* species in Japan (Magae *et al.*, 1990). The study of Zervakis *et al.* (1994) also offers among other things to the issues related to phylogeny and demarcation of various *Pleurotus* species.

Genetic relationships among the various ecotypes or genotypic classes in many *Pleurotus* species remain ambiguous. The populations of *P. eryngii* species complex, for example, growing in the greater Mediterranean area are in close association with different genera of plants of the family Apiaceae. Their distinct host-specialization served as the principal criterion for the discrimination of several taxa until Zervakis *et al.* (2001) studied this problem using molecular markers. As a result, host-exclusive zymograms for the *Aph* (acid phosphatase) and *Phe-1* (dopa-phenoloxidase) loci obtained from *Pleurotus* strains associated with *Cachrys ferulacea* enables accurate strain identification. It is interesting to note that allele frequencies, genetic variability and mean diversity are high for isolates from *Eryngium* species and *Ferula communis*. In one of the historic studies involving strain identification, May and Royse (1988) has classified several isolates of *Pleurotus* species using isozyme banding patterns originally described as one species based on morphology.

ii) Application of RAPD markers in *Pleurotus* research

Though techniques such as restriction fragment length polymorphism (RFLP) has been widely employed (Iracabal *et al.*, 1995; Matsumoto and Fukumasa-nakai, 1995; Iracabal and Labarere, 1994; Labarere *et al.*, 1993), there are only few reports available on the utility of RAPD markers in revealing systematic relationship and estimating genetic diversity in *Pleurotus*. Zervakis *et al.* (2001) using random primers analysed the populations of *P. eryngii* species complex growing in the greater Mediterranean area. The 42 RAPD patterns independently permitted the separation of the fungal populations into five clusters in accordance with their host-specificity. During the analysis, total of 45 (out of 48) polymorphic bands are scored, and four molecular markers specific for the identification of *Pleurotus* strains growing on *Eryngium asclepium* subspecies *asclepium* and *Cachrys ferulacea* are obtained.

Probably for the first time in India, in a comprehensive and systematic approach, Pradeep (2002) has studied 15 *Pleurotus* species of diverse habit and habitat comprising *P. sajor-caju, P. florida, P. flabellatus* and *P. citrinopileatus* and those species such as *P. cystidiosus, P. eryngii, P. pulmonarius, P. ostreatus, P. sapidus, P. membranaceus, P. platypus, P. fossulatus, P. opuntiae, P. eous, P. djamour* which are cultivated with a regional restriction in various parts of the world. He has used 20 random primers and RAPD data analysis of all the 15 species studied showed higher levels of genetic variation. The mean coefficient of similarity based on Nei and Lei's method is 0.62 which ranged from 0.41 to 0.81 among various species. Cluster analysis based on UPGMA reveals two major groups in these species. Cluster-I comprising of eight species (*P. pulmonarius, P. sajor-caju, P. membranaceus, P. platypus, P. florida, P. djamour, P. ostreatus* and *P. sapidus*) have coefficient of similarity values ranging from 0.50 to 0.81 with a mean value of 0.65. The species *P. pulmonarius, P. sajor-caju* and *P. membranaceus* forms a sub group with an average coefficient value of 0.74, while *P. platypus, P. florida* and *P. djamour* having a mean similarity value of 0.69 forms another group. Seven species (*P. opuntiae, P. eryngii, P. fossulatus, P. citrinopileatus, P. cystidiosus, P. flabellatus* and *P. eous*) in cluster-II shows coefficient of similarity values that ranges from 0.43 to 0.66 with a mean value of 0.53. Here four out of the seven species viz., *P. opuntiae, P. eryngii, P. fossulatus* and *P. citrinopileatus* having higher degree of similarity are grouped together while the other three form a loose group. Among the 15 species studied, *P. ostreatus* and *P. sapidus* shows more similarity (0.67) while *P. eous* is the least similar (0.45) to others.

Conclusion

The commercial production of edible basidiomycetes is increasing every year as a response to the growing public demand probably because of their nutritional properties. Notwithstanding, this production is threatened by biotic and abiotic factors that make it necessary to improve the fungal strains currently used in industry. Breeding of edible basidiomycetes, however, has been mainly empirical and slow since the genetic tools useful in the selection of the new genetic material to be introduced in the commercial strains have not been developed for these fungi as it was for other organisms (Ramirez *et al.*, 2000). Hence, both molecular systematics and genetic improvement of fungi demand appropriate use of molecular markers and a good knowledge of their heredity. It is also important to consider that the meager fossil record for fungi provides little data on the origin and dispersal of species, molecular phylogenetic studies may be the only means for reconstructing evolutionary history within a biogeography context (Vilgalys and Sun, 1994). *Pleurotus* being an important genus comprising both popularly cultivated species/strains across the world, a better understanding of the genetic diversity, molecular systematics and the identification of genes involved in the control of quantitative traits could possibly contribute much towards designing future genetic improvement strategies with respect to widening of genetic base and development of superior genotypes/strains.

References

Bano, Z. and Rajarathanam, S. (1988). *Pleurotus* mushrooms. Part II. Chemical composition, nutritional value, post harvest physiology, preservation and role as human food. Crit. Rev. Food Sci. Nutr., **27**: 87-158.

Blaich, R. and Esser, K. (1975). Function of enzymes in wood destroying fungi. II. Multiple forms of laccase in white rot fungi. Arch Microbiol, **103** : 271-277.

Boisselier-Dubayle, M.C. (1983). Taxonomic significance of enzyme polymorphism among isolates of *Pleurotus* (Basidiomycetes) from umbelliferae. Trans. Br. Mycol. Soc., **81**: 121-127.

Breene, W.M. (1990). Nutritional and medicinal value of speciality mushrooms. J. Food Protection, **53**: 883-894.

Brewbaker, J.L., Upadhya, M.D., Makinen, Y. and MacDonald, T. (1968). Isozyme polymorphism in flowering plants. III. Gel electrophoretic methods and applications. Physiologia Plantarum, **21**: 930-940.

Brewer, G.J. (1970). An introduction to isozyme techniques, Academic Press, New York, London.

Brown, A.H.D. (1978). Isozymes, plant population genetic structure and genetic conservation. Theor. Appl. Genet., **52**: 145-157.

Decker, L.E. and Rau, E.M. (1963). Multiple forms of glutamic-oxaloacetic transaminase in tissues, Proc. Soc. Exp. Biol. Med., **112**: 144-149.

Hartmann, T., Nagel, M. and Ilert, H.T. (1973). Organ specific multiple forms of glutamic dehydrogenase in *Medicago sativa.* Planta, **111**: 119-128.

Hunter, R.L. and Markert, C.L. (1957). Histochemical demonstration of enzymes separated by zone electrophoresis in starch gels. Science, **125**: 1294-1295.

Iracabal, B. and Labarere, J. (1993). Comparison of polymorphism and phenetic variability as determined by the study of hydrolases and oxidoreductases in two cultivated mushrooms *Agaricus bisporus* and *Pleurotus cornucopiae.* Exp. Mycol., **17**: 90-102.

Iracabal, B., Roux, P.H. and Labarere, J. (1991). Study on enzyme polymorphism in *Agaricus* and *Pleurotus* species for characterisation and genetic improvement. In: *Science and cultivation of Edible fungi* (Maher MJ., Ed), 1, 37-41. Academic Press, Rotterdam.

Jacob, J.P., Sabu, K.K. and Abraham, T.K. (2000). Genetic variability of mutant strains of *Pleurotus citrinopileatus* based on isozyme electrophoresis. Mush Res., **9**: 79-84.

Kulkarni, R.K., Kamerath, C.D. and Allard, K.L. (1986). Genetic diversity between isolates of *Pleurotus ostreatus* as revealed by isozyme analysis. In: *Cultivating edible fungi.* (Eds. Wuest PJ, Royse DJ and Beelman RB) Dev. Crop Sci., **10**: 171-181. Elsevier, Amsterdam.

Larraya, L.M., Perez, G., Ritter, E., Pisabarro, A.G. and Ramirez, L. (2000). Genetic linkage map of the edible basidiomycete *Pleurotus ostreatus.* Appl. Environ. Microbiol., **66**: 5290-300.

Magae, Y., Haga, K., Taniguchi, H. and Sasaki, T. (1990). Enzymes of strains of *Pleurotus* species (Basidiomycetes) compared by electrophoresis. J. Gen Appl. Microbiol., **36**: 69-80.

Markert, C.L. and Moller, F. (1959). Multiple forms of enzymes: tissue, ontogenetic and species specific patterns. Proc. Natl. Acad. Sci., **45**: 753-763.

May, B. and Royse, D.J. (1982). Genetic variation and joint segregation of biochemical loci in the common medow mushroom, *Agaricus campestris*. Biochem. Genet., **20**: 1165-1173.

May, B. and Royse, D.J. (1988). Interspecific allozyme variation within the fungal genus *Pleurotus.* Trans. Br. Mycol. Soc., **90**: 29-36.

Micales, J.A., Bonde, M.R. and Peterson, J,L. (1986). The use of isozyme analysis in fungal taxonomy and genetics. Mycotaxon, **27**: 405-449.

Mullis, K.B. and Faloona, F.A. (1987). Specific synthesis of DNA *in vitro* via a polymerase-catalyzed chain reaction. Methods in Enzymology, **155**: 335-350.

Murray, M.G. and Thomson, W.F. (1980). Rapid isolation of high-molecular weight plant DNA. Nucleic Acid Res., **8**: 4321-4325.

Nei, N. (1972). Genetic distance between populations. Amer. Natur., **106**: 283-292.

Nei, M. 1973. Analysis of gene diversity in subdivided populations. Proc. Nat. Acad. Sci. USA., **70**: 3321-3323.

Opletal, L. (1993). Phytotherapeutic aspects of diseases of the circulatory system. 2. The oyster mushrooms and its potential use. Cheskoslovenska Farmacia, **42**: 160-166.

Pradeep, N.S. (2002). Bioprospecting of some species of *Pleurotus.* Ph.D. thesis submitted to Gandhigram Rural Institute - Deemed University. Gandhigram, India.

Ramirez, L., Larraya, L.M. and Pisabarro, A.G. (2000). Molecular tools for breeding basidiomycetes. Int. Microbiol, **3**: 147-52.

Stamets, P. (1993). *Growing Groumets and medicinal mushrooms.* Ten Speed Press, Berkely, CA. Pp 551.

Surve-Iyer, R.S., Adams, G.C., Iezzoni, A.F. and Jones, A.L. (1995). Isozyme detection and variation in *Leucostoma* species from *Prunus* and *Malus.* Mycologia., **87**: 471- 482.

Tanksley, S.D. (1979). Linkage, Chromosomal association and expression of *Adh-1* and *Pgm-2* in tomato. Biochem. Genet., **17**: 1159-1167.

Tanksley, S.D. and Rick, C.M. (1980). Genetics of esterases in species of *Lycopersicon.* Ther. Appl. Genet., **56**: 209-219.

Vallejos, E. (1983). Enzyme activity staining. In *Isozymes in Plant Genetics and breeding,* Part A, (ed. Tanksley SD and Orton TJ), Elsevier Science Publishers BV, Amsterdam, 469-516.

Vilgalys, R. and Sun, B.L. (1994). Ancient and recent patterns of geographic speciation in the oyster mushroom *Pleurotus* revealed by phylogenetic analysis of ribosomal DNA sequences. Proc. Natl. Acad. Sci. USA, **91**: 4599-4603.

Wang, H.C. and Wang, Z.S. (1989). The prediction of strain characteristics of *A. bisporus* by the application of isozyme electrophoresis. *Mushroom Science* **XII** Part I. In : Proc. of 12th Inter. Congress on the Science and Cult. Edible Fungi, 87-100. Braunschweig, Germany. 1987.

Wendel, J.F. and Percy, R.G. (1990). Allozyme diversity and introgression in the Galapagos Islands endemic *Gossypium darwinii* and its relationship to continental *G. barbadense.* Biochem. Syst. Ecol., **18** : 517-528.

Williams, J.G., Kubelik, A.R., Kenneth, J.L., Rafalski, J.A. and Scott, V.T. (1990). DNA polymorphism amplified by arbitrary primers are useful as genetic markers. Nucleic Acid Res., **18**: 6531-6535.

Yeh, I.C. and Boyle, T.J.B. (1997). Population genetic analysis of co-dominant and dominant markers and quantitative traits. Belgian J. Bot., **129**: 157.

Zervakis, G and Labarere, J. (1992). Taxonomic relation ship within the fungal genus *Pleurotus* as determined by isoelectric focussing analysis of enzyme pattern. J. Gen. Microbiol., **138**: 635-645.

Zervakis, G., Sourdis, J. and Ballis, C. (1994). Genetic variability and systematics of eleven *Pleurotus* species based on isozyme analysis. Mycol Res., **98**: 329-341.

Zervakis, G.I., Venturella, G. and Papadopoulou, K. (2001). Genetic polymorphism and taxonomic infrastructure of the *Pleurotus eryngii* species-complex as determined by RAPD analysis, isozyme profiles and ecomorphological characters. Microbiology, **147**: 3183-94.

5

Molecular Mechanisms Involved in Biological Control of Fungal Plant Pathogens of Major Agricultural and Horticultural Crops

R. Samiyappan*, S. Nakkeeran, V. Ramamoorthy, A. Ramanathan, P. Balasubramanian and T. Marimuthu,

Dept. of Plant Pathology, Tamil Nadu Agricultural University, Coimbatore 641 003

Beneficial soil microbes belonging to fluorescent pseudomonads and *Trichoderma* spp are commonly employed for the management of soil-borne diseases and to some extent foliar pathogens of agricultural and horticultural crops. These antagonistic microbes suppress the pathogen's propagules by various modes of actions *viz.*, colonization of infection niche, production of antibiotics, siderophores and lytic enzymes. Microbial genes regulating spatial and temporal synthesis of these secondary antifungal metabolites play an important role in suppression of disease development. Moreover, certain isolates of fluorescent pseudomonads and *Trichoderma* spp increase the plant growth and yield by production of various plant growth-promoting substances. In addition to direct antagonism and plant growth promoting activity, many strains of fluorescent pseudomanads and certain *Trichoderma* spp induce systemic resistance in response to infection by fungal pathogens by activating various defense-related genes encoding proteins which prevent the further colonization of pathogens in plant tissue.

Key Words : Biological control, Fungal plant pathogens, molecular mechanisms

Of late there is increasing public concern regarding the continued use of chemical pesticides which are damaging human health or the environment. Such concerns are driving us to search for ecologically sound methods to control several plant diseases which are responsible for significant reduction in yield both in major Agricultural and Horticultural crops every year in India. Biocontrol of plant diseases, therefore, is currently needed for Indian Agriculture in order to sustain the food production without polluting agriculture ecosystem with chemical pesticides. This has to be done to keep pace with the increasing Indian population in the coming decades. Although few biopesticides are in the Indian market for management of plant diseases they constitute less than 1% of the pesticides market in India.

* Corresponding Author e-mail: rsamiyappan@hotmail.com

Biopesticides are preferred by many than chemical pesticides, because they a) do not leave harmful residues b) are target specific and do not destroy beneficial organisms and c) promote the growth of crop plants. Main agents employed as biocontrol agents include bacteria viz., fluorescent pseudomonads (*Pseudomonas fluorescens* and *P. putida*) and *Bacillus* spp. and fungi viz., *Trichoderma* spp. In India, scope for exploitation of fluorescent pseudomonads (also called as rhizobacteria) and *Trichoderma* spp. as biopesticides is greater when compared to other groups of bioagents in the coming decades. The present review paper mainly focuses on the molecular and biochemical mechanisms associated with the suppression of plant pathogens by major biocontrol agents.

Rhizobacteria

In the rhizosphere region of the plants rhizobacteria are abundantly present and many of these bacteria are not only benefitted from the nutrients secreted by the plant roots but also influence the plants by direct or by indirect way and lead to the stimulation of plant growth. Large scale application of plant growth promoting rhizobacteria (PGPR) to crop plants as inoculants would be attractive as it would substantially reduce the use of chemical pesticides and increases the crop yield, thereby helping to feed the growing Indian population. Worldwide, there are several PGPR products being commercially produced. These strains can be used optimally only if the molecular basis of their beneficial effects and the ways by which these traits are influenced by biotic and abiotic factors are known. There are several mechanisms involved in the biocontrol efficacy of several PGPR strains and mainly they include production of anti-microbial metabolites and inducing systemic resistance in host plants.

Biocontrol by PGPR

Suppressive soils harbour several rhizobacterial strains which are capable of suppressing plant pathogens. Major mechanisms include competition for nutrients, niche exclusion, induced systemic resistance (ISR) and production of antifungal or antimicrobial metabolites. Most of the bacterial strains which belong to PGPR group are fluorescent pseudomonads (FP). Many of the FP strains produce antimicrobial metabolites viz., 2,4-Diacetylphloroglucinol (DAPG), phenazines, pyrrolnitrin and viscosinamide. The genetic basis of the biosynthesis of the more frequently detected antimicrobial metabolites in *Pseudomonas* spp. has been elucidated. Recently Raaijmakers and Weller (2001) reported that genetically diversed population of DAPG producing *Pseudomonas* spp. occurring on the roots of wheat cultivated soil were naturally suppressive to take all disease of wheat. Among 101 isolates of fluorescent pseudomonads studied, 16 different groups were identified by random amplified polymorphic DNA (RAPD) analysis. One RAPD group made up 50% of the total population of DAPG-producing *Pseudomonas* spp. Both short and long term studies indicated that this dominant genotype, exemplified by *P. fluorescens* Q8r1-96, is highly adapted to the wheat rhizosphere.

Regulation of Antimicrobial Metabolite Production

Production of antimicrobial metabolites in *Pseudomonas* spp. is subjected to complex regulation. Key factors in the regulation of the biosynthesis of most antimicrobial metabolites are global regulation and quorum sensing. Global regulation is directed by the *gacS* / *gacA* genes, which encode a two component regulatory system that senses an unknown signal(s). Quorum sensing involves production of N-acyl homoserine lactone (AHL) signal molecules by AHL synthase such as Lux 1 (Bassler, 1999). At a threshold concentration of AHL, which is reached only when a certain density of bacterial cells is present, the AHL will sufficiently bind to and activates a transcriptional regulator, such as LuxR. The activated form of the transcriptional regulator then stimulates gene expression.

A recent report on transgenic plants that produce various AHL's (Fray *et al.*, 1999) presented novel perspectives on biocontrol and on optimizing the application of biocontrol strains. It was suggested that plants can produce and secrete substances that mimic AHL activity and could therefore influence the density – dependent behaviour of rhizobacteria (Tepiitski *et al.*, 2000). A striking new finding in *P. fluorescens* CHAO is the regulation of DPAG production by autoinduction and its repression by salicylate and pyoluteorin produced by the same cells, as in the repression of DPAG production by the fungal metabolite fusaric acid (Schnider *et al.*, 2000).

Induced Systemic Resistance

Several rhizobacterial strains have the ability to induce a state of systemic resistance in plants which provides protection against a broad spectrum of phytopathogenic organisms including fungi, bacteria and viruses (Van Loon *et al.*, 1998; Ramamoorthy *et al.*, 2001). ISR pathway is induced when the plant is challenged by pathogenic organisms (Nandakumar *et al.*, 2001). The list of rhizobacterial *Pseudomonas* spp. known to induce an ISR is growing rapidly. A degree of dependence on plant genotype is observed in the generation of these ISR's. Bacterial determinants that claimed to produce ISRs include siderophores, the O-antigen of lipopolysaccharides and salicylic acid. The latter compound has even been indicated to cause an ISR when present in nanogram amounts (De Meyer *et al.*, 1999). Elucidation of the plant factors involved in the pathways leading to ISR and SAR has shown that induced disease resistance can be enhanced by the simultaneous activation of ISR and SAR pathways (Van Wees *et al.*, 2000). Induction of pathogenesis related proteins and phenolics is noticed in tomato treated with *P. fluorescens* in response to infection by *Fusarium oxysporum* f. sp. *lycopersici* (Ramamoorthy *et al.*, 2002).

Rhizosphere Competence

Rhizobacterial inoculants are often applied in seed coatings. After sowing, the bacterial inoculants must be able to establish themselves in the rhizosphere at population densities sufficient to produce a beneficial effect. Hence, efficient bacterial strains should survive in the rhizosphere, make use of nutrients exuded by the plant root, proliferate, be able to efficiently colonize the root system and be able to compete with indigenous micro organisms. Inadequate biocontrol in field experiments has often been correlated to poor root colonization. Identification of the genes and traits involved in the processes of inoculation and root colonization will give a detailed insight into plant-bacterial interactions and leads to the more efficient application of bacterial strains. *P. fluorescens* genes that are specifically expressed in the rhizosphere (i.e. *rhi* genes) have been identified using *in vivo* expression technology (IVET) (Rainey, 1999). More than twenty *rhi* genes have been identified, of which fourteen showed significant homology to genes that are involved in nutrient acquisition, stress response or secretion. Efficient scavenging for iron using siderophores makes *Pseudomonas* strains more competitive. On the basis of identified colonization genes and traits, colonization mutants of *P. chlororaphis* strain, PCL 1391 have been constructed. These mutants have lost their ability to control tomato foot and root rot, showing for the first time that root colonization is an essential trait for biocontrol (Chin *et al.*, 2000). In the rhizosphere, bacterial inoculants compete for nutrients and ecological niches with indigenous bacteria, fungi and actinomycetes. Recently, genes of *Pseudomonas* biocontrol strains have been identified that can be induced or repressed by the presence of phytopathogenic fungi. IVET technology has been used to show that the presence of *Phytophthora parasitica* can induce various genes in *P. putida*, including genes encoding diacylglycerol kinase (Lee and Cooksey, 2000).

Genetic Modification of Rhizobacteria

The identification of genes involved in the ability of rhizobacterial strains to improve plant growth creates the potential to improve the performance of biocontrol strains or to construct novel biocontrol strains by genetic modification. Complete operons, as well as single genes under the control of their own regulatory genes or regulated by the constitutive expression of the *tac* or *lac* promoters, have been transferred to rhizobacterial strains. There are some studies in which the introduction of genes into rhizobacterial strains has enhanced biocontrol and/or plant growth promotion. These genes include the cry-toxin encoding cry1AC7 gene of *Bacillus thuringiensis* (Downing *et al.*, 2000), the chitinase – encoding chiA gene of *Serratia marcescens* (Downing and Thomson, 2000) and endochitinase of *Manduca sexta* against *Rhizoctonia solani* (Samiyappan and Muthukrishnan, 1992; Samiyappan *et al.*, 1999).

Whole Genome Analysis

Recent technological developments in molecular biological techniques have resulted in the publication of many whole genome sequences. Sequencing of approximately 50 microbial genomes has been completed and those of another one hundred and fifty are in progress. Among these, several are rhizosphere inhabiting bacteria viz., *P. fluorescens, P. putida, P. aeruginosa* and *Bacillus subtilis*. The sequence data will, by comparative and functional genomics, facilitate the identification of genes that are specifically expressed on the seed or in the rhizosphere, that are involved in the regulation and production of secondary metabolites or anti-microbial metabolites. Further, the construction of bacterial artificial chromosome (BAC) libraries for the study of gene expression and to identify genes of interest is of great value, especially in the study of bacteria whose genome has not been sequenced.

TRICHODERMA

Trichoderma interacts with pathogen and host in soils either directly or indirectly. In indirect interactions plant responds to the presence of the antagonist, resulting in induced resistance or plant growth promotion. Based on the interaction, the mechanism of action may be direct or indirect.

Direct Mechanism of Action

Competition

Most of the soil-borne pathogens are controlled by competition for space or court of infection on roots and seeds. Biological control of silver leaf disease caused by *Chondrostereum purpureum* was controlled by *Trichoderma viride* through early colonization of fresh wounds. Competition also seems to be the most potent mechanism employed by *T. harzianum* T-35 in the control of *F.*

oxysporum. f. sp. *vasinfectum* and *F. oxysporum*. f. sp. *melonis* in the cotton rhizosphere and melon (Sivan and Chet, 1989).

Antibiosis

Antibiotics are low molecular weight (<1kDa) secondary metabolites produced during nutrient limiting conditions. *Trichoderma* species produce more than 43 secondary metabolites which exhibit antibiotic activity (Sivasithambaram and Ghisalberti, 1998). Among these compounds alkyl pyrones, isonitriles, polyketides, petaibols, diketopiperazines, sesquiterpenes and steroids have often found to be associated with biocontrol activity (Howell, 1998).

Volatile antibiotic produced by *Trichoderma harzianum* was inhibitory to *Rhizoctonia solani* (Dennis and Webster, 1971) and *Macrophomina phaseolina* (Tamimi and Hutchinson, 1975). The volatile metabolites effective against fungal pathogens were identified as Pentyl analogues namely 6-n-pentyl-2H-pyran- 2-one and 6-n-pentenyl-2H-pyran- 2-one (Claydon *et al.*, 1987). Production of volatile harzianolide by *T. harzianum* [3- (2- hydroxyl-propyl)-4 (hexa – 2″ - dienyl – 2(5H) furanone] was found to be antifungal (Claydon *et al.*, 1991). *T. viride* produces number of antifungal metabolites such as trichodermin (Gotfredson and Vangedal, 1965), dermadin (Pyke and Dietz,1966), trichoviridin (Yamano *et al.*, 1970)and sesquiterpene heptalic acid (Itoh *et al.*, 1980).

Gliotoxin an epidiothidiketopiperazine produced by *G. virens* G20 is a key factor in reducing damping off of Zinnia (Wilhitte *et al.*, 1994). *G. Virens* pertaining to 'Q' group produced gliotoxin and dimethyl gliotoxin, but 'P' group produced gliovirin (diketopiperazine) and heptelidic acid. Gliovirin was inhibitory to *P. ultimum* causing damping-off in cotton than 'Q' strains, which was effective against *R. solani* (Howell *et al.*, 1993).

The UV mutants of *T. harzianum* produced trichorzianines, which are hydrophobic peptides, interact with phospholipid membrane and induce membrane permeability (Merlier *et al.*, 1984; Boda *et al.*, 1985). *T. hamatum* produced trichoviridin 3-(3-isocyano-6-oxabicyclo (3, 10) hex-2-ch-5yl) acrylic acid and 3-(3-isocyano cyclopet-2-enzylidne) propionic acid (Baldwin *et al.*, 1981). In addition, it produced three isonitriles among which isonitrin A was broad spectrum in activity (Edenborough and Hebert, 1988). *T. longibrachiatum* produced trichodermin (Watts *et al.*, 1988) and trichobrachin (Bruckner *et al.*, 1990). The cells exposed to antibiotics of *Trichoderma* spp. create ultra structural changes like retraction of plasmalemma, breakdown of organelles, disintegration of cytoplasm, loss of turgor, weakening and death of host cells.

Cellulases

Trichoderma species were potential bio-control agents and producers of powerful cellulases (Papavizas, 1985). Mandels *et al.* (1971) irradiated *T. viride* and produced new strain that secreted twice as much cellulase as its wild parent. Improved strains of *T. reesei* were isolated through mutation and selection. It produced more cellulase than the wild strains did (Montenecourt and Eveleigh, 1979). Baker (1991) found a perfect correlation between increased cellulase production and rhizosphere competence by mutants of *T. harzianum.* It is suggested that mutants use the mucigel on roots as a substrate. Mutants with rhizosphere competence were more efficient bio-control agent and induced increased growth responses from seed treatment than wild type parent. The quality and quantity of cellulases from *Trichoderma* spp. were improved by mutagenesis (Cuevas *et al.*, 1994).

Lytic Enzymes

Proteases

Trichoderma species are well known producers of proteases. Proteolytic activity of *T.viride* was claimed to be involved in biocontrol of *Sclerotium rolfsii* (Rodriguez- Kabana et al., 1978). Serine proteases produced by *T. harzianum* played an important role in mycoparasitism of plant pathogens (Geremia *et al.*, 1993). Autoclaved mycelium, fungal cell wall or chitin induced this enzyme and its production was repressed by glucose. Proteases are suggested to degrade cell wall, membranes and proteins released by the lysis of the pathogen, thus making the nutrients available for the mycoparasite (Goldman *et al.*, 1994). Protease production by *T. harzianum* was induced by the presence of live and dead conidia of *B. cinerea*. Germinability of *B. cinerea* was affected by the culture filtrate containing protease. The action of *Trichoderma* protease on the infection of *B. cinerea* on the surface of bean leaves was due to the adverse effect on the germination of *Botrytis* conidia and to the inactivation of hydrolytic enzymes of pathogen responsible for plant tissue necrosis. Foliar application of *T. harzianum* controls *B. cinerea, Pseudoperonospora cubensis, Sclerotinia sclerotiorum,* and *Sphacelotheca fuliginea* in cucumber. Enzymes like pectinase, cutinase, glucanase and chitinase secreted by *B. cinerea* were suppressed by protease produced by the antagonist and was responsible for the biocontrol of foliar pathogens (Katan *et al.*, 2000).

β-1, 3 glucanases

Lytic enzymes play a vital role in mycoparasitism and degrade the pathogen cell wall. The enzymes such as chitinases, and β-1, 3 glucanases lyse the host wall and leads to the leakage of protoplasmic contents, which are in turn used as food material for the multiplication of antagonist (Cherif and Benhamou, 1990; Tronsmo *et al.*, 1993). Strains of *T. harzianum* produced a mixture of lytic enzymes comprising of chitinolytic and gluconolytic enzymes; which involved in mycoparasitism (Sivan and Chet, 1989). Excretion of β-1,3 glucanase into growth medium and to the soil by *T. harzianum* was observed (Elad *et al.*, 1982). Carbon sources such as laminarin, pustulan, cell walls of *R. solani* and mycelia of several fungi incorporated in the enzyme production medium induced the secretion of β-1, 3 glucanase by *T. harzianum*. The purified endochitinase, chitobiosidase, N-acetyl-(-glucosaminidase) and glucan, -1,3-β-glucosidase and their combinations inhibited the spore germination and germ tube elongation of *Botrytis cinerea*. Combinations of the purified enzymes resulted in the synergistic increase in antifungal activity (Lorito *et al.*, 1994). High synergistic activity was observed by combining 1,3-β-glucosidases and chitinolytic enzymes (Tronsmo *et al.*, 1993). This corresponds to the effect observed with combination of β-1,3 glucanase and endochitinase isolated from plants. *T. harzianum* also produced N-acetylglucosaminidase, endochitinase, chitobiosidase and endo β-1, 3 glucanases in soybean rhizosphere active against *R. solani* (Dalsoglio *et al.*, 1998).

Chitinases

Chitinase, the basic protein catalyses the hydrolysis of chitin. Chitin, a linear polymer of β-1,4-N-acetyl glucosamine, is often considered to be the second most abundant polysaccharide in nature. Relationship between mycolytic enzymes, chitinases and β-1,3 glucanases produced by mycoparasitic fungi and their significance in fungal cell wall lysis and degradation have been well established (Elad *et al.*, 1980, 1999). Assay of chitinase could be used as a basis for screening potential bio-control agents (Elad *et al.*, 1982). *Trichoderma* spp. attack several plant pathogenic fungi by excreting lytic enzymes including β-1, 3 glucanase, proteinases and chitinases enabling them to degrade host cell walls and thus reduce disease incidence (Geremia *et al.*, 1993; Lorito *et al.*, 1993; Goldman *et al.*, 1994). *T. harzianum* strain Tm had six distinct chitinolytic enzymes. Two were identified as β-1, 4-N-acetyl glucosaminidases. It hydrolyzes chitin to N-acetyl glucosamine (Glc NAc) monomers. Four were endochitinases that randomly cleave at internal sites over the entire length of the chitin microfibril (CHIT 52, CHIT 42, CHIT 33 and CHIT31) (Haran *et al.*, 1995). *T. harzianum* – P1 released chimeric units from chitin and termed as chitobiosidase (40 kDa) (Harman *et al.*, 1993).

The involvement of *Trichoderma* chitinases in mycoparasitism was performed in liquid cultures supplemented with different carbon sources e.g.

chitin, purified fungal cell walls, glucose or GlcNAc. Complete degradation of mycelial walls of pathogenic fungi depends on major extent on chitinolysis and hence the degrading ability to chitin by *Trichoderma* spp. assumes a greater significance in antagonistic process than the other mechanisms / enzymes involved in the process (Cherif and Benhamou, 1990). Chitinases and glucanases were produced by *Trichoderma* isolates, in soil, which had live mycelium of *S. rolfsii* and *R. solani*. High levels of β-1,3 glucanase and chitinase activities were also detected by dual agar cultures of *T. harzianum* with *S. rolfsii* and *R. solani* (Elad *et al.*, 1983). Direct interaction of *T. harzianum* with *R. solani*, strongly enhanced the endochitinase CHIT 42 (Carsolio *et al.*, 1994). The spore germination and hyphal elongation of *B. cinerea, F. solani, F. graminearum, Ustilago avenae* and *Uncinula necator* were inhibited substantially by the combination of chitinolytic and gluconolytic enzymes purified from *T. harzianum* (Harman *et al.*, 1993). The chitinolytic enzymes from *T. harzianum* were more active than that of the enzyme from other sources such as plant and bacteria were effective against a wider range of fungi (Harman *et al.*, 1993).

Diversity of *Trichoderma* Chitinases

Exteracellular secretion of chitinases from different species of *Trichoderma* reveals the presence of endochitinase with different molecular weights (40, 42, 46,37 and 33 kDa), chitobiosidase (40 kDa), N –acetyl-ß-D-glucosaminidase (64, 69,73 kDa), and exochitinase (28kDa) (De La Cruz *et al.*, 1992; Harman *et al.*, 1993; Lorito *et al.*,1994; Ulhoa and Peberdy, 1991., 1992; Deane *et al.*, 1998). Among various forms of chitinase endochitinase of 42 kDa and N–acetyl-ß-D-glucosaminidase of 70 and 73 kDa have been isolated frequently. *T. atroviride, T. harzianum, T. virens, T. hamatum* and *T. cf. harzianum* produced endochitinase of 42 kDa. But the gene ech42 of *T. atroviride* differed in their nucleotide sequence up to 35% and up to 17 % in the amino acid sequence. It correlated well with the phylogenetic distance of the investigated species (Kullnig *et al.*, 2001).

Antifungal Activity of Chitinase

High level induction of extra-cellular chitinase was obtained by growing *Trichoderma* on purified chitin, fungal cell walls or mycelia as a sole carbon sources but the same was not induced when chitosan, cellulose, unpurified chitin or laminarin were used as carbon sources (Lorito *et al.*, 1994). The 42kDa endo chitinase, the 40 kDa chitobiosidase and 73kDa N-acetyl b-D glucosidase were highly inhibitory to the spore germination, hyphal elongation of several fungal pathogens namely *Botrytis cinerea, Fusarium* spp., *Alternaria* spp. *Ustilago avenae, Uncinula necator* and against the fugus containing chitin as their cell wall make up (Lorito *et al* 1996; Schirmbock *et al.*, 1994). The endo-chitinases of *Trichoderma* were effective in its antifungal action than plant, bacterial and other fungal chitinases (Lorito *et al.*, 1996). These enzymes have the potentiality to lyse soft structure of the hyphal tip, hard chitinous wall of matured hyphae, conidia, chlamydospores and sclerotia (Benhamou and Chet, 1993; Rousseau *et al.*, 1996).

Synergism of chitinase with other antifungal compounds

Chitinase action is synergistic with other chitinase component and other antibiotics putatively involved in biocontrol (Haran.*et al.*, 1995; Lorito *et al.*,1996). This type of synergism with other gluconolytic enzymes enhances the lytic and antifungal action even in cases where the enzymes have little or no activity when applied alone (Lorito *et al.*, 1996). The synergism between chitinase and peptaibol antibiotics explains that the antibiotics form voltage gated ion channels in black liquid membranes and modifiy the permeability of liposomes in the absence of applied voltage (El Hadjji *et al.*, 1989). Hence, while the chitinase reduces the rigidity of the cell wall, antibiotics inhibit the membrane bound synthesis of cell wall components. This impairs the ability of the hyphae to repair the lytic effect of the enzymes on the cell wall.

Sequential order of chitinase induction

Mycoparasitism, the multi step process in degradation of fugal cell wall is initiated by the extra-cellular secretion of hydrolytic enzymes. Among the various enzymes, chitinase isoforms play a vital role in cell wall lysis. Inbar and Chet (1995) demonstrated the formation of chitinase in *T. harzianum* by a lectin based physical interaction with the host pathogen. It was suggested to be the primary event of interaction and precede induction by chito oligomers. Chitinase (102 kDa) was induced by contact with host lectin whereas the induction of other chitinases requires the presence of living host. It concluded that the above enzymes might be responsible for the primary attack and induction of other chitinases (Inbar and Chet, 1995).

Spatial distribution of lytic enzymes

β-1,3 glucanase degrades the substrate β-1,3 glucan, which is the major component of fungal cell wall next to chitin. Physiological concentration of these fungi effectively inhibits the growth of many potential pathogenic fungi. Though this enzyme plays a major role in cell wall lysis, it is imperative to understand the *in vitro* production and sub-cellular localization of the constitutive expression of the enzyme, β-1,3 glucanase. Expression of β-1,3 glucanase by *T. harzianum* was observed in extracellular, periplasmic, cell wall bound internal fractions during submerged fermentation in the absence of lactose and glucose as carbon source. In submerged culturing of *T. harzianum* with lactose, β-1,3 glucanase was found in all extracellular, cell wall bound and in periplasmic fractions. The relative distribution of the enzyme in the cell is independent of the nature of carbon source and its concentration.

Chitinases are localized either outside or inside the cell and it is localized at sub cellular level in various fractions including cellwall, periplasmic space, vacuoles, and membranes (Manocha and Zhonghua, 1997). Extracellular chitinases are more important than the intracellular chitinases, since the extra cellular chitinases provide the first line of defense against plant pathogens (Graham and Sticklen, 1994).

Mycoparasitism

Mycoparasitism relies on the production of lytic enzymes by the mycoparasite for the degradation of cell walls of the host fungus. It involves chemotrophic growth of mycoparasite towards the host, recognition and attachment, secretion of lytic enzymes, penetration of host and lysis and killing of host (Chet, 1987). Isolates of *Trichoderma* spp. coiled around the hyphae of *R. solani*, which consequently lost their cell contents and collapsed (Chu and Wu, 1981). *Trichoderma* spp. attached to *S. rolfsii* or *R.solani* by hyphal coils, hooks, or appressoria. Lysed sites and penetration holes were found in hyphae of the pathogen following the removal of parasitic hyphae (Elad *et al.*, 1999). *Trichoderma* coiled over *Pythium* resulting in cell wall lysis, vacuolation and coagulation of protoplasm (Lifshitz *et al.*, 1986).

Toxin Degrading enzymes- α Glucosidase

It is an exocarbohydrase that catalyse the cleavage of α-residue from the non-reducing terminal of substrates to liberate α-glucose. Yamasaki *et al.* (1995) reported an α-glucosidase protein of 67 kDa with a pI value of 5.0 from *Trichoderma viride.* Though it is an important enzyme in food industry, its ability to inactivate the phytotoxin has been realized recently. Glucosidase from *T. viride*, TvMNT7 was able to inactivate RS toxin under *in vitro* conditions (Samiyappan *et al.*, 1997; Sriram *et al.*, 2000). The antifungal activity of the high molecular weight protein (110 kDa) was studied against inactivation of RS toxin as well as on the sclerotial germination and mycelial growth of *R. solani.* Enzyme assay and western blot analysis of the above protein indicated that it was α- glucosidase with pI value of 5.2 (Shanmugam *et al.*, 2001).

Indirect mechanism of action

Induced resistance

It occurs in a plant following the challenge inoculation of antagonist that triggers the expression of a set of genes which encode for pathogenesis – related proteins such as chitinases, β-1, 3-glucanases and thaumatin – like proteins with antifungal activity (Linthrost, 1991). Synthesis of phytoalexins may also occur. Application of *T. harzianum* T-39 suppressed the activity of *B. cinerea.* T-39 reduced the activity of pectolytic enzymes and increased the accumulation of pectic enzyme products. These sugars can elicit the host plant defence mechanisms, thus slowing the development of disease severity (Zimand *et al.*, 1996). Application of *T. harzianum* (Trichodex–T39) to roots or leaves controlled the disease caused by *B. cinerea* on leaves spatially separated from the site of application. It might be due to the induction of SAR (De Meyer *et al.*, 1998). Control of powdery mildew by T-39 (*T. harzianum*) may be due to induced resistance (Elad *et al.*, 1999). Growing of cucumbers in the presence of T-203, increased the growth rate of cucumber. T–203 penetrated the root cortical tissue and the cells of the root cortical tissue were strengthened in the area of penetration. The activities of both peroxidase and chitinase were increased in root and leaf tissues of the treated seedlings. It enlightens the role of SAR to be involved (Yedidia *et al.*, 1999).

Seed treatment of cotton with native strain of *T. virens* controlled the root rot of cotton incited by *R. solani.* Production of antibiotics such as gliotoxin and gliovirin is the prime reason for the management. But the mutant of the same native strain that lacks the ability to produce antibiotics was also found effective in the management of cotton root rot. The analysis of the root tissues challenged with both *T. virens* mutant and *R. solani* evidenced the presence of terpenoid phtoalexins like hemigossypol (39.78 mg/g), deoxyhemigossypol (22.62mg/g) and gossypol (94.84 mg/g) in the root tissues. The concentration of the same in *R. solani* treated plants was hemigossypol (4.0mg/g), deoxyhemigossypol (3.33mg/g) and gossypol (29.0 mg/g) in the root tissues. This mirrors the role of induced resistance by *T. virens* (Howell *et al.*, 2000).

Plant growth promotion

Growth promotion may occur either directly or indirectly. Promotion of growth may occur either through bio-control of deleterious pathogens (Baker, 1989) or through the production of plant harmones such as auxin, cytokinin or gibberellins (Arshad and Franken-berger, 1991). Baker *et al.* (1984) found that *T. harzianum* can also stimulate plant growth, including floricultural and horticultural plants (Chang *et al.*, 1986). Pepper seed germinated 2 days earlier in raw soils containing the fungus than in untreated controls. Flowering of periwinkle was accelerated, the number of blooms per plant on chrysanthemum was increased, and the height and weight of other plants were greater in either steamed or raw soil infested with *T. harzianum*. These responses have occurred consistently with population densities of *T. harzianum* higher than 10^8cfu g^{-1} of soil, either the bio-control agent was applied as conidial suspension or as mycelium and spore in a peat bran medium.

Growth stimulation by *T. harzianum* could be the result of production of plant hormones by antagonist, increased uptake of nutrients by plant, through nutrient mobilization and/or the control of one or more subclinical pathogens. *Pythium* spp. causes stunting plants and hinders the nutrient uptake of roots (Backer and Cook, 1988). Hence suppression of *Pythium* spp. could increase the plant growth (Salt, 1978). Increased growth by *Trichoderma* spp. was also induced by a diffusable growth-regulating factor produced by these organisms (Windham *et al.*, 1986).

Conclusion

Rhizobacteria and *Trichoderma* spp. are eco-friendly, alternative to chemical pesticides the use of which are regulated and sometimes forbidden. But, the market for rhizobacterial and *Trichoderma* inoculants is still expanding. Approximately twenty bacterial biocontrol products are being marketed. The discovery of many traits and genes that are involved in the beneficial effects of rhizobacteria and *Trichoderma* has resulted in better understanding of the performance of bio inoculants in the field and provides the opportunity to enhance the beneficial effects of biocontrol strains by genetic modification.

References

Arshad, M. and Franken-berger Jr., W. T. (1991). Microbial production of plant hormones. In " The rhizosphere and plant growth" (eds. D. l. Keister and P.B. Cregan,), pp 327- 324. Kluwer Academic Publisher, Dodrecht, The Netherlands.

Backer, J.O. and Cook, R.J. (1988). The role of siderophores in suppression of *Pythium* spp. and production of increased growth response of wheat by fluorescent pseudomonads. Phytopathology,**78**:778-782.

Baker, R. (1991). Induction of rhizosphere competence in the biocontrol fungus, *Trichoderma*. In: *Beltsville symposia in agricultural research. The rhizosphere and plant growth* (eds. D.L. Keister and P.B. Cregan), **14**: 221-228.

Baker, R., Elad. Y. and Chet, I. (1984). The controlled experiment in the scientific method with special emphasis on biological control. Phytopathology, **74**:1919-1021.

Baker, R.(1989). Improved *Trichoderma* spp. for promoting crop productivity. Trends in Biotechnology, **7**: 34-38.

Baldwin, J. E., Derome, A. E., Field, L., Gallagher, P. T., Taha, A. A., Thaller, V., Brewer, D. and Taylor, A. (1981). Biosynthesis of a cyclopentyl dienyl isonitrile acid in cultures of fungus *Trichoderma hamatum*. J. Chem.Soc. Comm., 1227-1229.

Bassler, B. L. (1999). How bacteria talk to each other: regulation of gene expression by quorum sensing. Current opinion in Microbiology, **2**: 582-587.

Benhamou, N. and Chet, I. (1993). Hyphal interactions between *Trichoderma harzianum* and *Rhizoctonia solani*: ultrastructure and gold cytochemistry of the mycoparasitic process. Phytopathology, **83**: 1062-1071.

Boda, B., Rebuffat, S., Elhajji, M. and Davoust, D. (1985). Structure of trichoarzianines A III C an antifungal peptide form *Trichoderma harzianum*. J. Am. Chem. Soc., **107**: 6011-6017.

Bruckner, H., Reinecke, C., Kripp, T. and Kieb, M. (1990). Screening isolation and sequence determination of a unique group of polypeptide antibiotics form filamentous fungi. Proclamation of International Mycological Congress. Regensburg, Federation Republic Germany. 224 p.

Carsolio, C., Gutierrez, A., Jimenez, B., Van Montagu, M. and Herrera-Estrella, A. (1994). Characterization of ech-42, a *Trichoderma harzianum* endochitinase gene expressed during mycoparasitism. Proceedings of National Academy of Sciences,USA, **91**:10903-10907.

Chang, Y.C., Chang, Y.C., Bakler, R., Kleifeld, O. and Chet, I. (1986). Increased growth of plants in the presence of the biological control agent *Trichoderma harzianum*. *Plant Disease,* **70**:145-148.

Cherif, M. and Benhamou, N. (1990). Cytochemical aspects of chitin breakdown during the parasitic action of a *Trichoderma* sp. on *Fusarium oxysporum f. sp. radicis lycopersici*. Phytopathology, **80**:1406-1414.

Chet, I. (1987). *Trichoderma* – application, mode of action, and potential as biocontrol agent of soil-borne plant pathogenic fungi, Innovative approaches in plant disease control (Ed. I. Chet,), John Wiley, New York, pp.137-160.

Chin A. Woeng, T.F.C., Bloemberg, G.V., Mulders, I., Dekkers, L. and Lugtenberg, B.J.J. (2000). Root colonization is essential for biocontrol of foot and root rot by Phenazine-1-carboxamide producing bacterium *Pseudomonas chlororaphis* PCL 1391. Molecular Plant Microbe Interactions, **13**:1340-1345.

Chu, F.F. and Wu, W.S. (1981). Biological and chemical control of *Rhizoctonia solani* by pea seed treatment. *Memoirs of the college of Agriculture.* National Taiwan University, **21**:19-28.

Claydon, N., Allan, M., Hanson, J. R. and Avent, A. (1987). Antifungal alkyl pyrones of *Trichoderma harzianum*. Transaction of British Mycological Society, **88**: 503-513.

Claydon, N., Hanson, J. R., Trunch, A. and Avent, A. G. (1991). Harzianolide, a butenolide metabolite from cultures of *Trichoderma harzianum*. Phytochemistry, **30**: 3802-3803.

Cuevas, V.C., Culiat, C.T., Manaligod, R.L. and Sajise, N.E. (1994). Induction of fungicide resistance for enhanced cellulase production of *Trichoderma harzianum* Rifai. Philippines Journal of Science, **123**: 177-183.

Dalsoglio, F.K., Bertagnolli, B.L., Sinclair, J.B. and Eastbrn, D.M. (1998). Production of chitinolytic enzymes and endoglucanases in the soybean rhizosphere in the presence of *Trichoderma harzianum and Rhizoctonia solani.* Biological Control, **12**:111-117.

De la Cruz, J., Hidalgo-Gallego, A., Lora, J. M., Benitez, T., Pintor-Toro, J. A. and Llobell, A. (1992). Isolation and characterization of three chitinases from *Trichoderma harzianum*. European Journal of Biochemistry, **206** : 859-867.

De Meyer, G., Birgirimana, J., Elad, Y. and Hofte, M. (1998). Induced systemic resistance of *Trichoderma harzianum* biocontrol of *Botrytis cinerea*. European Journal of Plant Pathology, **104**: 279-286.

Deane, E. E., Whipps, J. M., Lynch, J. M. and Peberdy, J. F. (1998). The purification and characterization of a *Trichoderma harzianum* exochitinase. Biochimica and Biophysica Acta, **1383**: 101-110.

DeMeyer, G., Capieau, K., Audenaert, K., Buchala, A., Metraux, J.P and Hofte, M. (1999). Nanogram amounts of salicylic acid produced by the rhizobacterium *Pseudomonas aeurogenosa* 7NSK2 activate the systemic acquired resistance pathway in bean. Molecular Plant Microbe Interactions, **12**:450-458.

Dennis, C. and Webster, J. (1971). Antagonistic properties of species group of *Trichoderma* production of non-volatile antibiotics. *Trans*action of British Mycological Society, **57**: 25-39.

Downing, K.J. and Thomson, J.A. (2000). Introduction of the *Serratia marcescens* chiA gene into an Endophytic *Pseudomonas fluorescens* for the biocontrol of phytopathogenic fungi. Canadian Journal of Microbiology, **46**; 363-369.

Downing, K.J., Leslie, G. and Thomson, J.A. (2000). Biocontrol of sugarcane borer *Eldana saccharina* by expression of the *Bacillus thuringiensis* cry1AC7 and *Serratia marcescens* chiA genes in sugarcane associated bacteria. Applied and Environmental Microbiology, **66**: 2804-2810.

Edenborough, M. S. and Hebert, R. B. (1988). Naturally occurring isocyanides. *Natural products* Reporter, **5**: 229-245.

El Hadjji, M., Rebuffat, S., Le Doan, T., Klein, G., Satre, M. and Bodo, B. (1989). Interaction of trichorzianines A and B with model membranes and with the amoeba Dictyostelium. Biochimica and Biophysica Acta, **978**: 91-104.

Elad, Y., David, D. R., Levi, T., Kapat, A., Kirshner, B., Gavrin, E. and Levine, A. (1999). *Trichoderma harzianum* T39 mechanisms of biocontrol of foliar pathogens. In: Modern fungicides and antifungal compounds II. (Ed. H. Lyr) Intercept. Ltd. Andover, Hampshire, U. K. pp- 459-467.

Elad, Y., Chet, I. and Henis, Y. (1982). Degradation of plant pathogenic fungi by *Trichoderma harzianum*. Canadian Journal of Microbiology, **28**: 719-725

Elad, Y., Chet, I. and Katan, P. (1980). *Trichoderma harzianum*. A biocontrol agent effective against *Sclerotium rolfsii* and *Rhizoctonia solani*. Phytopathology, **70**: 119-121.

Fray, R.G., Troup,J.P., Daykin, M., Wallace, A., Williams, P., Stewart, G.S.A.B. and Grierson, D. (1999). Plants genetically modified to produce N-acylhomoserine lactones communicate with bacteria. Nature Biotechnology, **17** : 1017-1020.

Geremia, R., Goldman, G.H., Jacobs, D., Ardiles, W., Vila, S.B., Van Montagu, M. and Herera-Estrella. (1993). Molecular characterization of the proteinase-encoding gene, Prb1, related to mycoparasitism by *Trichoderma harzianum*. Molecular Microbiology, **8**: 603-613.

Goldman, H.G., Hayes, C. and Harman, G.E. (1994). Molecular and cellular biology of biocontrol by *Trichoderma* spp. Trends in Biotechnology, **12**: 478-482.

Gotfredson, W. O. and Vangedal, S. (1965). Trichodermin, a new sesquiterpene antibiotic. Acta Chem. Scand., **19**: 1088-1102.

Graham , L.S. and Sticklen, M.B. (1994). Plant chitinases. Canadian Journal of Botany, **72**: 1057-1083.

Haran, S., Schickler, H., Oppenheim, A. and Chet, I. (1995). New components of the chitinolytic system of *Trichoderma harzianum*. Mycological Research, **99**: 441-446.

Harman, G. E., Hayes, C. K., Lorito, M., Broadway, R. M., DiPietro, A. and Tronsmo, A. (1993). Chitinolytic enzymes of *Trichoderma harzianum*: purification of chitobiosidase and endochitinase. Phytopathology, **83**: 313-318.

Howell, C. R. (1998). The role of antibiotics in biocontrol. In: *Trichoderma* and *Gliocladium* Vol.II.

(Eds: C. P. Kubicek and G. E. Houman) Taylor and Francin, London. p. 173-184

Howell, C. R., Hanson, L. E., Stipanovic, R. D. and Puckhaber, L. S. (2000). Induction of terpenoid synthesis in cotton roots and control of *Rhizoctonia solani* by seed treatment with *Trichoderma virens*. Phytopathology, **90**: 248-252.

Howell, C.R., Stipanovic, R.D. and Lumsden R.D. (1993). Antibiotic production by strains of *Glicaldium virens* and its relation to the biocontrol of cotton seedling diseases. Biocontrol Science and Technology, **3**: 435 – 441.

Inbar, J. and Chet, I. (1995). The role of recognition in the induction of specific chitinases during mycoparasitism by *Trichoderma harzianum*. Microbiology, **141**: 2823-2829.

Itoh, Y., Kodama, K., Furuya, K., Takahash, S., Heneishi, J., Takiguchi, Y. and Arai, M. (1980). A new sesquiterpene antibiotic heptelidic acid producing organisms. Fermentation, isolation and characterization. Journal of Antibiotic, **33**: 468-473.

Katan, J., Aharonnon, N., Cohen, E., Rubin, B. and Mathews, G. A. (2000). Biological control of foliar pathogens by means of *Trichoderma harzianum* and potential modes of action. Crop Protection, **19**: 709-714.

Kullnig, C. M., Krupica, T., Woo, S. L., Mach, R. L., Rey, M., Benitez, T., Lorito, M. and Kubicek, C. P. (2001). Confusion abounds aver identity of *Trichoderma* biocontrol isolates. Mycological Research (in press)

Lee, S.W. and Cooksey, D.A. (2000). Genes expressed in *Pseudomonas putida* during colonization of a plant pathogenic fungus Applied and Environmental Microbiology, **66**: 2764-2772.

Lifshitz, R., Windham, M.T. and Baker, R. (1986). Mechanism of biological control of pre-emergence damping-off by use of integrated chemical and biological controls. Phytopathology, **69**: 431-434.

Linthrost, H.J.M.(1991). Pathogenesis related proteins of plants. Critical Review in Plant Sciences, **10**: 123 – 150.

Lorito, M., Harman, G.E., Hayes, C.K., Broadway, R.M., Tronsmo, A., Woo, S.L. and Di-pietro, A. (1993). Chitinolytic enzymes produced by *Trichoderma harzianum*; Antifungal activity of purified endochitinase and chitobiosidase. Phytopathology, **83**: 302-307.

Lorito, M., Hayes, C.K., Di Pietro, A., Woo, S.L and Harman, G.E. (1994). Purification, characterization and synergistic activity of a glucan 1,3- â-glucosidase and an N-acetyl- â-glucosaminidase from *Trichoderma harzianum*. Phytopathology, **84**: 398-405.

Lorito, M., Mach, R. L., Sposato, P., Strauss, J., Peterbauer, C. K. and Kubicek, C. P. (1996). Mycoparasitic interaction relieves binding of Cre1 carbon catabolite repressor protein to promoter sequence of *ech*- 42 (endochitinase-encoding) gene of *Trichoderma harzianum*. Proceedings of National Academy of Sciences, USA, **93**: 14868-14872.

Mandels., M, Weber, J. and Parizek, R. (1971). Enhanced cellulase production by a mutant of *Trichoderma viride*. Applied Microbiology., **21**: 152-154.

Manocha, M.S. and Zhonghua, C. M. (1997). Immunochemical and cytochemical localization of chitinase and chitin in the infected hosts of a biotrophic mycoparasite, *Piptocephalis virginiana*. Mycologia, **89**: 185 – 194.

Merlier, A. M. O., Boire, J. M., Pons, J. B. and Renaud, M. C. (1984). European patent application EP 124388. Chemical Abstracts, **102**: 183-747.

Montenecourt, B.S. and Eveleigh, D.E. (1979). Production and characterization of high yielding cellulase mutants of *Trichoderma reesii*. Tech. Assoc. Pulp Paper Ind. Annu. Meet. pp. 101-108.

Nandakumar, R., Babu, S., Viswanathan, R., Raguchander, T. and Samiyappan, R. (2001). Induction of systemic resistance in rice against sheath blight disease by *Pseudomonas fluorescens*. Soil Biology and Biochemistry, **33**: 603-612.

Papavizas, G.C. (1985). *Trichoderma* and *Gliocladium*: Biology, ecology and potential for biocontrol.

Annual Review of Phytopathology, **23**: 23-54.

Pyke, T. H. and Dietz, A. (1966). U-21, 9263, a new antibiotic. Discovery and biological activity. Applied Microbiology, **14**: 506-510.

Raaijmakers, J.M. and Weller, D.M. (2001). Exploiting genotypic diversity of 2,4-diacetylphloroglucinol producing *Pseudomonas* spp.: Characterization of superior root colonizing *Pseudomonas fluorescens* strain Q8r 1-96 Applied and Environmental Microbiology **67**:2545-2554.

Rainey, P.B. (1999). Adaptation of *Pseudomonas fluorescens* to the plant rhizosphere. Environmental Microbiology, **1**: 243-257.

Ramamoorthy, V., Raguchander, T. and Samiyappan, R. (2002). Induction of defense-related proteins in tomato roots treated with *Pseudomonas fluorescens* Pf1 and *Fusarium oxysporum* f. sp. *lycopersici*. Plant and Soil, **239**: 55-68.

Ramamoorthy,V., Viswanathan,R., Raguchander,T., Prakasam, V. and Samiyappan, R. (2001). Induction of Systemic resistance by plant growth promoting rhizobacteria in crop plants against pests and diseases. Crop protection, **20**:1-11.

Rodriguez- Kabana, R. Kelley, W.D. and Curl, E.A. (1978). Proteolytic activity of *Trichoderma viride* with mixed culture in *Sclerotium rolfsii* in soil. Canadian Journal of Microbiology, **24**: 487-490.

Rousseau, A., Benhamou, N., Chet, I. and Piche, Y. (1996). Mycoparasitism of the extramatrical phase of *Glomus intraradices* by *Trichoderma harzianum*. Phytopathology, **86**: 434-443.

Salt, G.H. (1978). The increasing interest in minor pathogens. In: *Soil-borne plant pathogens. (*Eds. B. Schippers and W. Gams). Academic Press, New York. pp. 289-312.

Samiyappan, R. and Muthukrishnan, S. (1992). Improvement of biocontrol potential of fluorescent pseudomonads by Genetic Engineering. P.67 Proc. of 35th Annual West Central States Biochemistry Conference, Oct. 30-31, 1992. Kansas State University, Manhattan, Kansas 66506, USA.

Samiyappan, R., Raguchander, T., Sriram, S., Babu, S., Jaisankar, R., Xu, R., Muthukrishnan, S., Prakash, O. and Velazhahan, R. (1997). Enzymatic degradation of *Rhizoctonia* toxin involved in sheath blight disease of rice. General meeting of the International Program on Rice Biotechnology, Malacca, Malaysia, 15-19, September 1997. p. 230.

Samiyappan,R. Raguchander,T., Jayashree,K. and Nandakumar,R.(1999). Management of major rice diseases with fluorescent pseudomonads. In: Subramanian,M. (ed.), *Vistas of Rice*. Tamil Nadu Rice Research Institute, TNAU, Aduthurai, India. pp.111-121.

Schirmbock, M., Lorito, M., Wang, Y. L., Hayes, C. K., Arisan Atac, I., Scala, F., Harman, G. E. and Kubicek, C. P. (1994). Parallel formation and synergism of hydrolytic enzymes and peptaibol antibiotics: molecular mechanisms involved in the antagonistic action of *Trichoderma harzianum* against phytopathogenic fungi. Applied and Environmental Microbiology, **60**: 4364-4370.

Schnider, K.U., Seematter, A., Maurhofer, M., Blumer, C. Duffy, B., Gigot Bonnefoy, C. Reimmann, C., Notz, R., Defago, G., Haas, D. and Keel, C. (2000). Auto-induction of 2,4-diacetylphloroglucinol biosynthesis in the biocontrol agent *Pseudomonas fluorescens* CHAO and repression by the bacterial metabolites salicylate and pyoluteorin Journal of Bacteriology, **182**: 1215-1225.

Shanmugam, V., Sriram, S., Babu, S., Nandakumar, R., Raguchander, T., Balasubramanian, P. and Samiyappan, R. (2001). Purification and characterization of an extracellular a-glucosidase protein from *Trichoderma viride* which degrades a phytotoxin associated with sheath blight disease in rice. Journal of Applied Microbiology, **90**: 320-329.

Sivan, A. and Chet, I. (1989). The possible role of competition between *Trichoderma harzianum* and *Fusarium oxysporium* on rhizosphere colonization. Phytopathology, **79**: 198-203.

Sivasithamparam, K. and Ghisalberti, E. L. (1998). Secondary metabolism in *Trichoderma* and *Gliocladium*. In: Trichoderma and *Gliocladium* Vol.I. (Eds: C. P. Kubicek and G. E. Houman) Taylor and Francin, London. p. 139-191.

Sriram, S., Raguchander, T., Babu, S., Nandakumar, R., Shanmugam, V., Vidhyasekaran, P., Balasubramanaian, P. and Samiyappan, R. (2000). Inactivation of phytotoxin produced by the rice sheath blight pathogen *Rhizoctonia solani*. Canadian Journal of Microbiology, **46**, 520-524.

Tamimi, K. M. and Hutchinson, S. A. (1975). Difference between the biological effects of culture gases from several species of *Trichoderma*. Transaction of British Mycological Society, **69**: 455-463.

Tepiitski, M.,Robinson, J.B. and Bauer, W.D. (2000). Plants secrete substances that mimic bacterial N-acyl homoserine lactone signal activities and affect population density- dependent behaviors in associated bacteria. Molecular plant microbe- Interactions, **13**:637-648.

Tronsmo, A. Klemsdal, S.S., Hayes, C.K. Lorito, M. and Harman, G.E. (1993). The role of hydrolytic enzymes produced by *Trichoderma harzianum* in biological control of plant diseases, *Trichoderma reesei* cellulases and other hydrolases, Enzyme structure, Biochemistry, Genetic and Applications (Eds., P. Suominen and T. reinikainen), Foundation for Biotechnical and Industrial Fermentation Research, Helsinki, Finland. Volume 8. pp. 159-168.

Ulhoa, C. J. and Peberdy, J. F. (1991). Regulation of chitinase synthesis *in Trichoderma harzianum*. Journal of General Microbiology, **137** : 2163-2169.

Ulhoa, C.J. and Peberdy, J.F. (1992). Purification and some properties of the extracellular chitinase produced by *Trichoderma harzianum*. Enzyme and Microbial Technology, **14**: 236-240.

Van Loon, L.C., Bakker, P.A.H.M. and Pieterse, C.M.J. (1998). Systemic rsistance induced by rhizosphere bacteria. Annual Review of Phytopathology, **36**: 453- 483.

Van Wees,S,C.M., De Swart, E.A.M., Van Pelt, J.A., Van Loon, L.C. and Pieterse, C.M.J. (2000). Enhancement of induced disease resistance by simultaneous activation of salicylate and jasmonate dependent defense pathways in *Arabidopsis thaliana*. Proceedings of National Academy of Sciences,USA, **97**: 8711-8716.

Watts, R., Dahiya, J., Chaudary, K. and Tauro, P. (1988). Isolation and characterization of a new antifungal metabolite of *Trichoderma reesii*. Plant and Soil, **107**:81-84.

Wilhitte, S.E., Lumsden, R.D. and Skraney, D.C. (1994). Mutational analysis of gliotoxin production by the biological fungus *Glicladium virens* in relation to suppression of *Pythium* damping-off. Phytopathology, **84**:816-621.

Windham, M.T., Elad, Y. and Baker, R. (1986). A mechanism for increased plant growth induced by *Trichoderma* spp. Phytopathology, **76**:518-521.

Yamano, T., Hemmi, S, Yamamoto, I. and Tsubaki, K. (1970). Trichoviridin new antibiotic. Chemical Abstracts, **73**: 65093

Yamasaki, Y., Elbein, A. D. and Konno, H. (1995). Purification and properties of a-glucosidase from *Trichoderma viride*. *Bioscience,* Biotechnology and Biochemistry, **59**: 2181-2182.

Yedidia, I., Benhamou, N. and Chet, I. (1999). Induction of defense response in cucumber plants by the biocontrol agent *Trichoderma harzianum.* Applied Environmental Microbiology, **65**: 1061-1070.

Zimand,G., Elad, Y. and Chet, I.(1996). Effect of *Trichoderma harzianum* on *Botrytis cinerea* pathogenecity. Phytopathology, **86**: 1255 – 1260.

6

Heavy Metal Toxicitites and Resistance in Fungi: An Overview

R. Seeta Laxman

Division of Biochemical Sciences, National Chemical Laboratory, Pune 411 008

Recently lot of interest is generated on heavy metal pollution due to variety of man made activities, their impact on the environment and their remediation methods. Fungi are ubiquitous in nature and are known to produce useful secondary metabolities like enzymes, vitamins and antibiotics. More recently, fungal biomass from fermentation industry in the form of biosorbent granules is also being invesigated for metal removal form industrial effluents. Metals and organometallic compounds, e.g. biocides and wood preservatives are in use for number of years. Fungal strains isolated from metal contaminated sites are found to be more tolerant/ resistant than those isolated from garden soil. There are many mechanisms by which an organism can survive metal toxicity and understanding metal tolerance/resistance and their mechanisms will help identity potential isolates that may find application in metal bioremediation. This review mainly deals with metal toxicity, factors influencing toxicity and mechanisms of metal resistance.

Key words : Heavy metal, fungi, toxicity, tolerance, resistance

Fungi play an important role in decomposition and recycling of organic matter. A range of morphological forms from unicellular yeasts to highly branched filamentous forms including those forming fruit bodies such as mushrooms exist in nature. Fungi are ubiquitous in natural environments and important for industrial processes. They are the source of economically important substances such as ethanol, citric acid, antibiotics, polysaccharides, enzymes and vitamins.

Recent interest in the present state of the environment has resulted in research to evaluate the global impact of pollution on the biosphere. The level of metals in the environment has increased due to man made activities like minning, modern agricultural practices, extensive industrialization etc. Consequently, enormous amounts of metal containing wastewater and effluents are generated causing environmental pollution. These in turn affect the fauna and the various life forms. As the microorganisms inhabiting such habitats are constantly exposed to high metal concentrations, they have to adapt to these toxic metals in order to survive. Metals are present in various forms and levels in the environment and hence encountered in one form or the other by fungi during their life cycle

In recent times, increasing pollution of natural environment by toxic metals, metalloids, radionuclides and organometalliods has led to increase interest in metal tolerant/resistant fungi because of the ubiquitous and sometimes dominant presence of fungi in the metal polluted habitats. Many times metals are accumulated in fruit bodies of edible fungi. Apart from the amelioration of metal phytotoxicity by mycorrhizal fungi in reclamation of land, other application is the use of fungal biomass for detoxification of metal/ radionuclide containing industrial effluents (Maruthi Mohan, 1998a; Gray, 1998; Kapoor and Viraraghavan, 1995; Volesky, 2001).

Metal toxicity is the basis of many fungicidal preparations used to control plant pathogens and as wood preservatives and more recently this has led to interest of metal interactions in fungi. Though earlier interest was focussed to study and assess the extent of toxicity to metals, recent interest has shifted towards resistance and adaptation of fungi to toxic metals and evaluation of their potential for application in bioremediation of metals. Research has been particularly focused on physiological, biochemical and genetic explanations to resistance.

Metals are defined as electropositive elements forming hydroxides and whose salts exist as cations in solution. They are classified as hard and soft metals (Pearson, 1963). Metal ions such as Mg and Ca which are required in macro-quantities by microorganisms are hard acids, while Hg, Pb and Ag which are soft acids are not required for growth and are toxic to the organism. Trace elements such as Cr, Mn, Co, Fe, Ni, Cu and Zn are required only in minute quantities. Even trace metals present above a thresh-hold value exert toxic effects. Metals play important roles in cellular functions such as catalysis, osmotic responses, formation of charge and concentration gradients across membranes required for transport processes etc. Nearly two fifths of all proteins contain one or more metals and about one third of all enzymes require metal ions for catalysis. Deprivation of metals results in metal ion specific impairment of metabolism, growth and reproduction. Unlike organic pollutants, metal ions are indestructible and hence have cumulative effect on the biological systems. For e.g. Minamata and Itai-Itai diseases were found to be due to Hg and Cd poisoning respectively (Luckey and Venugopal, 1977).

Metal Toxicity

Toxic effects of metals are due to the strong coordinating ability of the metal (Ochiai, 1987). This can lead to: a) blocking of functional groups of important molecules such as proteins and enzymes or b) substitution of essential metals with toxic metals or c) conformational changes leading to denaturation of enzymes or disruption of membrane integrity etc. Since range of interactions can occur between fungi and metals, almost every aspect of their metabolism, growth and differentiation may be affected, depending upon organism, metal species, concentration and physicochemical factors (Gadd, 1986a, Gadd and White, 1989a). Thus toxic symptoms can vary widely between organisms, strains, the stage of growth and even different vegetative and reproductive forms of the same organism (Gadd and Mowll, 1985; Sabie and Gadd, 1990).

Amir and Pineau (1998) compared the effects of metals on the germination and growth of fungal isolates from metal contaminated soils. They studied the effects of Ni, Co, Fe, Mn and Mg upon spore germination and mycelial growth and found that nickel to be most toxic. Spore germination was more sensitive than mycelial growth. Heavy metal tolerance in fungi including *Penicillium funiculosum, Apergillus foetidus, Penicillium simplicissimum* to Ni, Co, Fe, Mn and Mg was studied by Valix *et al.* (2001). These strains were exposed up to 2000 ppm of heavy metals. They found that *Penicillium funiculosum* and *A.foetidus* were most tolerant to heavy metals while *P. simplicissimum* showed least tolerance. Levinskaite (1999) studied heavy metal resistance of micromycetes isolated from waste dump soil, unpolluted soil and chromium leather waste. The most resistant micromycetes to all metals tested were *Trichoderma viride* and *Exophiala jeanselmei* isolated from waste dump. Response of the other micromycetes to the metals altered individually, depending on the organism. It was found that the most resistant fungi belonged to the genus *Trichoderm*a. Sensitive micromycetes were mostly isolated from unpolluted soil, whereas metal tolerant fungi were mostly from the dump and chromated leather waste.

Blue mycelial formation was observed in *Cunninghamella blackesleeana* when grown in presence of copper. Copper at 9 mM concentration induced the formation of larger hyphae with rough granular surfaces, thicker cell walls and an abnormal and compressed cytoplasm in *C.blakesleeana* where decreased levels of Mg, Ca and increased levels of sulphur in the mycelia were also noticed. Cell walls prepared from copper toxic cultures of *C.blacksleeana* were shown to have enhanced metal binding capacity compared to that of the control cell walls (Venkateswarulu and Stotzky, 1989; Venkateswarulu *et.al.*, (1989). *N.crassa* when grown in presence of large amounts of Cu in media containing inorganic nitrate and phosphate as nitrogen and phosphorus sources respectively, develop blue colour due to high Cu content (12%) in the mycelia.

Such mycelia contained large amounts of chitosan, phosphate and amino groups with simultaneous decrease in their chitin content than when grown in absence of Cu. Supplementation of iron along with copper in the culture medium resulted in the disappearance of blue colour of the mycelia and restoration of normal growth and cell wall composition which was thought to be due to suppression of Cu uptake from the medium (Subramanyam *et. al*, 1983 a, b). Cu toxicity in *N.crassa* was also shown to result in deposition of glycogen granules with resultant ultrastructural changes. The walls of *N.crassa* grown in cadmium showed yellow colour and there was 2-3 fold increase in their protein content compared to controls.

For toxicity to be manifested, the contact between metal ion and cellular components is necessary. The initial site of contact is cell membrane and metal ions can cause irreversible damage to the cell membrane and integrity leading to changes in membrane permeability (Norris and Kelly, 1977; Cooney *et al.*, 1989). Passow and Rothstein (1959) showed that Hg ions caused extensive damage to the yeast cell membrane which was reflected in leakage of intracellular potassium ions as well as other constituents. Low concentrations of UO^{2+} ions have been shown to bind to the cell wall without themselves being transported inside and block the transport of sugar in yeast (Rothstein *et al.*, 1951).

Indirect mechanisms of metal toxicity may involve free radicals, which are deleterious to cells as they can take part in chain reactions which involve the breakdown of biological macromolecules. Aerobic organisms possess protective enzymes such as superoxide dismutases, which are metalloenzymes containing any one of the following metals, Mn, Fe or Cu/Zn (Galiazzo *et al.*, 1991), which eliminate the radicals produced by normal metabolism.

Horsfall (1956) has suggested that for a variety of fungi, the order of toxicity is Ag>Hg>Cu>Cr>Ni>Pb>Co>Cd>Zn>Ca as regards to spore germination. Metal toxicity can bring complex changes due to primary and secondary effects and cause characteristic morphological and visible changes. A chain of events can occur between the initial chemical effects and the final observed response and therefore, it is difficult to pin point one single key reaction responsible for the toxicity. For e.g. Ni toxicity in *N.crassa* is known to result in derangement of iron metabolism as well as carbohydrate metabolism (Maruthi Mohan and Sivarama Sastry, 1983a, 1984b). However, the extent of toxicity in the above metabolic pathways which results in overall growth inhibition is not clear. In *A.niger*, Co^{2+}, Ni^{2+} and Zn^{2+} toxicities impaired growth, acid formation and glucose utilization (Adiga *et.al.*, 1961; Seeta Laxman, 2000). In *S.cerevisiae*, Co toxicity inhibited glucose utilization (Veliky and Stefanec, 1964). Ni at 1 ppm inhibited sporulation of *A.niger*, *A.giganteus* and *A.gliocladium* sp. Higher concentrations (50 ppm) resulted in extensive branching and shortening of the hyphae of *Rhizopus stolonifer*. Ni^{2+} was also found to inhibit alcohol dehydrogenase (Fuhrmann and Rothstein, 1968).

Copper toxicity in *N.crassa* was enhanced on sole nitrate medium as compared to an ammonia nitrogen medium as Cu^{2+} inhibits nitrite reductase, thereby leading to nitrite accumulation which in turn results in toxicity (Venkateswarulu and Sivarama Sastry, 1979b). Nicholas and co workers showed that Cu toxicity in *N.crassa* results in depressed levels of nitrate reductase, nitrite reductase, cytochrome C oxidase and phosphatase activities which could be partially restored by excess molybdenum (Nicholas *et al.*, 1954; Nicholas and Commissiong, 1957). Glutamine synthetase activity was also inhibited in a non-competitive manner by copper in *N.crassa* (Srinivasa Rao *et al.*, 1984).

Cobalt toxicity in *N. crassa* was equivalent to 'conditional iron deficiency'. Catalase, peroxidase and succinic dehydrogenase were found to be depressed in cobalt toxicity in *N.crassa* (Sivarama Sastry *et al.*, 1962; Sivarama Sastry, 1963). The intracellular cobalt-iron antagonism in *N.crassa* was evident at the level of heme synthesis. Co inhibited the enzyme a-aminolevulinic acid dehydratase and in turn the synthesis of the porphyrin moiety essential for a number of enzyme catalyzing redox reactions. In cell free extracts, Co inhibited iron incorporation in to protoporphyrin when present in definite excess of iron (Padmanabhan and Sarma, 1966).

Chromium exists in both trivalent and hexavalent forms, latter being more toxic to a number of organisms. Venakata Ramana and Sivarama Sastry (1993; 1994) found that hexavalent Cr was more toxic than trivalent form in *N.crassa*. Cr^{6+} toxicity was found to be due to antagonism to sulfate uptake while Cr^{3+} toxicity was found to be due to conditional iron deficiency as indicated by a decrease in catalase, uricase and an increase in iron binding compound in to the medium (Venakta Ramana and Sivarama Sastry, 1994). In chromate resistant *N.crassa*, one of the two sulphate permeases responsible for lethal entry of Cr^{6+} in to the cells was shown to be absent (Marzluf, 1970). Hexavalent chromium was also more toxic than trivalent chromium to mycelial growth and spore germination of *Rhizopus stolonifer, Trichderma viride, Penicillium vermiculatum* and *Apergillus giganteus* (Babich *et al.*, 1982).

Organometallic compounds present in biocides are generally more toxic towards fungi than free metal ions. Toxicity varies with the number and the type of the organic groups (Blunden *et al.*, 1984). They disrupt the mitochondrial membranes and act on Cl^-/OH^- ionophores. Thus, they depolarize electrochemical gradients and consequently interfere with energy conservation (Cooney and Wuertz, 1989)

Tolerance/Resistance

Fungal survival in presence of toxic metals mainly depends on intrinsic, biochemical and structural properties, physiological and/or genetic adaptation including morphological changes of the organism and environmental modification such as metal speciation and availability of the metal.

Tolerance and resistance have been used interchangeably to define growth in presence of metal. An organism is defined to be tolerant if it can grow in presence high concentrations of a metal and toxic if the growth is inhibited. Gadd (1993) has defined tolerance as the ability of the organism to grow in presence of metals due to their inherent properties such as impermeable cell wall and/or environmental modification of toxicity. Intrinsic properties that can determine survival include possession of impermeable pigmented cell walls, extra cellular polysaccharide and metabolite excretion especially where it leads to detoxification of the metal species, by e.g. binding or precipitation (Gadd, 1990b). Resistance is defined as the ability of the organism to survive metal toxicity by means of mechanisms produced in direct response to metal such as synthesis of metallothioneins or γ–glutamyl peptides. Biological mechanisms implicated in fungal survival (as distinct from environmental modification of toxicity) include extra-cellular precipitation, complexation and crystallization, transformation of metal e.g. oxidation, reduction, methylation and dealkylation, decreased transport or impermeability, efflux, intracellular compartmentalization, precipitation and/or sequestration and is called resistance. An organism may employ one or more of these mechanisms to survive toxicity which can be influenced by the environmental factors. For e.g. metallothionein synthesis is a mechanism of Cu^{2+} resistance in *S.cerevisiae* yet Cu^{2+} binding or precipitation around the cell wall and intracellular transport are also components of the total cellular response (Gadd and White, 1989a). Maruthi Mohan (1998b) described resistance to denote the characteristic mutant derived from a sensitive wild type parent where the mutant strain has the ability to survive metal toxicity in comparison to the parent. Maruthi Mohan and his group have isolated a number of resistant strains to copper, nickhel and cobalt and used them very elegantly to elucidate the biochemical basis of toxicity/resistance in *Neurospora crassa*.

Factors Influencing Metal Toxicities

In fungi, metal effects can vary between organisms, strains, stage of growth and even different vegetative and reproductive forms of the same organism. (Gadd and Mowll, 1985; Sabie and Gadd,1990). A given organism may exhibit several mechanisms by which a potentially toxic metal species are detoxified both by metabolism dependent and independent and all affected by environmental factors. Several environmental factors can influence metal toxicities in microorganisms, both qualitatively as well as quantitatively (Collins and Stotzky, 1989). Physicochemical properties of a given environment or growth medium determines metal speciation, thereby influencing the biological availability of the metal and hence toxicity. Oxido-reduction potential (E_h), pH, presence of other cations and anions and particulate and soluble organic matter and clay minerals decrease/increase biological availability and hence influence the toxicity (Gadd and Griffiths, 1978). It is therefore clear that the concentration of free metal ion is important, although toxicity may still be exerted by many inorganic and organic metal complexes in the absence of free metal ions.

Effect of pH

Metal toxicity can be influenced by pH of the growth medium as metal speciation depends on pH. The pH also affects metal–fungus responses by its effects on metal speciation and mobility and indirectly by influencing other aspects of cell physiology and metabolism. Increasing the pH can result in the formation and precipitation of metal hydroxides or oxides. In aqueous solution, divalent metal cations form multiple hydroxylated species, some forming precipitates (Collins and Stotzky, 1989; Hughes and Poole, 1991). At higher pH values, some precipitates re-dissolve giving anionic hydroxometallate complexes, e.g. $Zn(OH)_4{}^{2-}$ (Hughes and Poole, 1991). Different hydroxylated forms possess differing toxicities while the pH values at which hydroxylated species form varies between different metals (Collins and Stotzky, 1989).

Decreasing the pH may increase the concentration of free metal ions in solution although, H^+ may compete with metal ions for cellular binding sites and reduce potential interactions with cells. External pH may also affect the speciation of metal binding ligands and therefore metal complexation. Many toxic interactions of heavy metals with fungal cells can be interpreted in terms of pH effects on metal speciation and availability (Newby and Gadd, 1986: Gadd, 1986a). Fungitoxicity of Cd may increase with increasing pH (pH 5-9), possibly as a result of formation of $CdOH^+$ which may enter cells more easily than Cd^{2+}, particularly as less H^+ will be present to compete with $CdOH^+$ at higher pH values. Cadmium is more toxic to *Alcalegenes faecalis, Bacillus cereus* and *Aspergillus niger* when the pH was increased from acidic to alkaline levels (Babich and Stotzky, 1977c). With certain other metals e.g. Ni, toxicity may decrease as the pH increases, possibly as a result of formation of hydroxylated species or pH effects on transport. A reduction in metal toxicity at low pH may result from H^+ competition and also pH effects on intracellular uptake. This has been described for *Penicillium ochro-chloron,* which is capable of growing in saturated $CuSO_4$ and high concentration of heavy metal salts.

With increasing pH, intracellular Cu accumulation dramatically increased in *Penicillium ochro-chloron* and at pH 6 and above, this fungus was sensitive to quite low concentrations of Cu (Gadd and White, 1985). Nickle was found to be less toxic to bacteria, Actinomyctes, yeast and filamentous fungi from 5.5 and above than at lower pH (Babich and Stotzky, 1982a; 1983). Babich and Stotzky (1981) found that response of different fungi to Mn toxicity varied with fungus to fungus at different pH values i.e. toxic to different fungi at different pH values. Increasing the pH from 3.5 to 4.7 caused an increase in toxicity of Cu^{2+} to the fungus *Aureobasidium pullulans* which was related to an increase in the uptake of Cu^{2+}. Hg^{2+} was more toxic to the spores and mycelium of *Fusarium lycopersici* under alkaline conditions (Horsfall, 1956). *Penicillium nigricans* was more tolerant to Cu^{2+}, Ni^{2+}, Co^{2+} and Cr^{3+} (even at concentrations as high as 4 $X10^{-2}$M) in acid media below pH 4.6 but failed to grow above pH 4.6 (Babich and Stotzky, 1982b). Effects of the above discussed nature are a reflection of the general feature that metal ion uptake tends to be higher around pH 5-7 and above than under acidic conditions, often due to either competition with protons or altered charge properties of cellwall/membrane.

Effect of Oxido-Reduction Potential of the Environment

The oxidation/reduction potential (E_h) of the environment can affect the availability of metal and therefore its toxicity. In a reducing environment (negative E_h), insoluble metal sulphides may form which exhibit little or no toxicity. In fact sulphide precipitation is a detoxification mechanism found in several fungi. The (E_h) may also effect speciation, e.g. chromium exists as Cr^{6+} or Cr^{3+} depending on the (E_h); Cr^{6+} exhibits greater toxicity towards hyphal growth and sporulation in several fungi. Hexavalent chromium Cr^{6+} was found to be more toxic than the trivalent Cr^{+3} to mycelial growth and spore germination of *Rhizopus stolonifer, Trichoderma viride, Penicillium vermiculatum* and *Aspergillus giganteus* (Babich *et al.*, 1982).

Effect of Media Constituents

The nature of the growth medium can sometimes profoundly influence metal toxicities. Where a metal has a role in, or affects a specific metabolic function, the composition of the medium determines the nature of the response to the metal ion in question under the prevailing conditions. For e.g., Mo^{+6} was toxic to *N.crassa* on a sulphate medium but not when methionine was provided as the sulphur source (Sivarama Sastry *et al.*, 1958) and suppression of copper toxicity in *N.crassa* by excess histidine in the medium (Subramanyam and Venkateswarulu, 1979a). Chromium toxicity in *Candida* sp. was modulated from Cr^{+6} resistance to Cr^{+6} hypersensitivity depending on the addition of methionine, cysteine or sulphate to the medium. If *Candida* sp. was grown in the presence of S-amino acids, especially methionine, it was more resistant than if the sulphur source was sulphate. When sulphate transport was enhanced by addition of djenkolic acid, *Candida* sp. became hypersensitive. However, *Rhodosporidium* sp. was always resistant to Cr^{+6} because sulphate transport was inefficient and it assimilated sulphur as S-amino acids (Pepi and Baldi, 1992). Cu was 15 fold more toxic to *N.crassa* on a sole nitrate medium in comparison to ammonium nitrogen medium due to inhibition of key enzyme nitrite reductase resulting in accumulation of highly toxic nitrite in to the medium. Cadmium was found to be more toxic to *N.crassa* on sole nitrate nitrogen medium than on ammonium nitrogen (Maruthi Mohan, 1998b).

Inorganic anions may affect toxicity by forming inorganic complexes by precipitation e.g. carbonates and phosphates (Gadd and Grifith, 1978). Increasing Cl^-, for example decreased Cd toxicity towards several filamentous fungi, possibly as a result of the formation of negative Cd-Cl coordination complexes (Babich and Stotzky, 1982b; Collins and Stotzky, 1989). Heavy metal toxicity towards fungi was also reduced in hard water, an effect attributed to CO_3^{2-} (Babich and Stotzky, 1981). Many heavy metal salts such as sulphides and phosphates are of low solubility, so readily precipitate out of solution. Silver chloride is highly insoluble and the concentration of free Ag in solution may be negligible in many cases, particularly in growing media (Hughes and Poole, 1991).

Dissolved and particulate organic matter in the environment/growth media can influence metal toxicity by complexation and binding, thereby generally reducing toxicity. Examples are natural and synthetic chelating agents, including plant, fungal and other microbial exudates and metabolites, humic acids and soil colloids and soluble organic chemicals and mixtures. It is commonly observed that metal toxicity towards fungi is reduced in complex media compared to simple defined media, such effects also being influenced by the pH. Media constituents like casamino acids, proteose, peptone and yeast extract can bind heavy metals and hence influence metal toxicity. Ramamoorthy and Kushner (1975) found that Cd bound to organic growth constituents in the order of: casaminoacids>proteose peptone>tryptone>yeast extract with no binding to peptone. Clay minerals can adsorb heavy metal cations and reduce their potential toxicity (Gadd and Griffiths, 1978; Babich and Stotzky, 1980). In general, minerals with high cation exchange capacity (CEC) are more effective in reducing toxicity than minerals with low CEC (Babich and Stotzky, 1977a; Bewley and Stotzky, 1983) For e.g. Montmorillonite offered better protection from Cd, Ni and Pb than kaolinite which had low CEC, such effects also being influenced by soil pH (Collins and Stotzky, 1989).

Metal Ion Antagonism

Cations can affect fungitoxcity of heavy metals by competing for binding sites on cell surfaces, including membranes, and transport systems. Metal ion antagonism i.e. influence of other metal ions present in the medium, plays a very important role in metal toxicity. The general mechanism underlying such interactions are: I) toxicity of one metal may cause conditional deficiency of another physiologically important metal ion. e.g. related toxicities of Co^{2+}, Ni^{2+} and Zn^{2+} in *N.crassa* (Maruthi Mohan, 1998 b). II) competitive interactions at the level of metal ion uptake may become operative when other metal ions are present in sizable amounts in the medium. Such interactions could be either generalized or due to specific metal ion antagonism. A metal ion that can powerfully suppress the uptake of a toxic metal ion will thereby counteract the toxicity in question, even though no other metabolic antagonism is involved. Typical example is the ability of excess Mg^{2+} ions to counteract the toxicities of a variety of toxic metal ions, a phenomenon first shown in bacteria by Abelson and Aldous (1950). Mg was found to reverse the toxic effects of Ni, Cd in *A.niger* (Seeta Laxman, 2000; Adiga *et al.*, 1961).

The underlying mechanisms of metal antagonisms were established in *N.crassa* in Co^{2+} and Ni^{2+} toxicities by the studies of Sivarama Sastry *et al.*, 1958; 1962). There is also specific antagonism, not at the uptake level but in metabolism. Healy *et al.* (1955) found that Co toxicity in *N.crassa* was equivalent in terms of metabolic consequences to iron deficiency, since there was a decrease in heme dependent enzymes; this effect but not that on growth was corrected by supplemented iron. Later on it was established that growth inhibition could also be competitively counteracted by more massive iron supplements (Sivarama Sastry *et al.*, 1962). Iron counteracted Cu toxicity in *N. crassa* in both nitrate as well as ammonium nitrogen media, but Mn counteracted only on ammonium nitrogen medium. Cu toxicity in *N.crassa* results in blockade of some of the metabolic steps in histidine biosynthesis (Subramanyam and Venkateswarulu, 1979a; Subramanyam *et al.*, 1983b). These effects were reversed by supplements of histidine. Analysis of free amino acid pools from mycelial extracts of copper toxic cultures of *N.crassa* showed significantly decreased levels of His, Arg and Tyr which were restored to normal levels by manganese (Subramanyam and Venkateswarulu, 1979b). Fe^{3+} reversed Co^{2+} and Ni^{2+} toxicities in *N.crassa* completely without having any influence on accumulation of these ions. Mg^{2+} counteraction was exclusively by suppression of toxic metal ion uptake similar to that observed in bacterial systems. The concentration of Co^{2+} and Ni^{2+} that become toxic have been shown to be function of Mg^{2+} level in the medium. Thus if Mg^{2+} concentration decreased, the toxic levels were much lower. Thus Mg^{2+} and Fe^{2+} are physiologically important ions which control heavy metal toxicities by two completely different mechanisms. The toxic effects of Co, Ni and Zn in *A.niger* could be reversed by excess of Mg^{2+} and Fe^{2+}. Competition between Co^{2+} and Ni^{2+} and Mg^{2+} in *N.crassa* is at cell surface. Fe^{3+} counteracts the adverse effects of these metals at intracellular sites. The nitrogen source influences the Co toxicity in *N.crassa* (Subramanyam and Sarma, 1968) where nitrate reductase increased while nitrite reductase decreased in cobalt toxicity. Cr^{6+} toxicity in *N.crassa* was found to be due to specific antagonism of sulfate uptake, methionine was effective in reversing toxicity, probably by repressing the synthesis of sulphate permeases responsible for chromate uptake. Calcium at a concentration of 0.5mM, completely removed the inhibitory effect of copper in *S.cerevisiae*. Mg exhibited a weak protective effect against the influence of Cu (Karamushka and Gadd, 1994).

Metal Resistance in Fungi

One of the strategies adapted by microorganisms to overcome toxicants in the environments is development of resistance. Microorganisms are known to adapt to elevated levels of toxic metals by evolving strategies to restrict the entry of metal ions in to the intracellular space thereby ensuring low in vivo concentrations. The first report of metal resistant fungus was *Penicillium glaucum* which was found to grow on carbonized wheat immersed in 2% $CuSO_4$.This organism was also able to grow in presence of other toxic metal such as Zn^{2+}, Ni^{2+}, Mn^{2+},Pb^{2+},Cd^{2+} and Ti^{2+}. Some of the first reports on metal resistant fungi were the species of *Aspergillus, Penicillium* growing on high concentrations of $CuSO_4$ and *Aconitium velatum* in media saturated with $CuSO_4$.

There is evidence for acquisition of heritable metal resistance by microorganisms. In case of bacteria, metal resistance is related to the acquisition of extrachromosomal DNA molecules i.e. plasmids (Silver, 1992). Plasmid carrying genes for resistance to arsenate, bismuth, cadmium, lead, zinc and mercury have been reported (Novick, 1967; Weiss *et al.*, 1977; Silver *et al.*, 1981). The resistance can be chromosomal or transposon coded and is either constitutive or inducible depending on the metal ion and/or the microorganism concerned.

Metal resistant organisms can also be obtained by adapting the organisms progressively on increasing concentrations of metal ions till ultimately adapted strain is obtained tolerant to higher concentration of toxic metal ion than parent strain. Many resistant strains of filamentous fungi have been isolated by gradual adaptation to increasing concentrations of metals such as copper and mercury. These include species of *Aspergillus, Penicillium, Sclerotina, Glomerella,*

Alternaria, Venturia and *Pyricularia.* They have also been isolated from seed treated with mercuric compounds. The predominant mercury tolerant species include *Pyrenophora avenae, Peniclillium crustosum, Cladosporium cladosporoides, Syncephalastrum racemosum* and *Elocladium atrum* (Gadd, 1993). But in all the above systems, the mechanism is not studied in detail. Metal resistant organisms can also be isolated by mutation using physical or chemical methods and selection on media containing toxic concentration of metals.

One of the major mechanisms of metal resistance is due to mutation of the membrane transport systems involved in accumulation of metal ions such that they abolish or greatly reduce entry of the metal ion into the cell. This mechanism is reported for copper, cobalt, nickel and zinc resistance in eukaryotic systems such as *Saccharomyces pombe* and *Neurospora crassa* (Maruthi Mohan, 1998 b).

The second mechanism is due to efflux. In certain microbial systems, As^{2+}, Sb^{3+} and Cd^{2+} are transported in to the cell by constitutively expressed cation transporters of broad specificity. e.g. magnesium transport system. In some of them, the resistance to toxic metal ions is found to involve a cellular exclusion mechanism (Cervantes and Guiterrez-Corona, 1994) based on active efflux as in some prokaryotes and possibly in *S.cerevisiae* (Wood and Wang, 1983; Nies, 1992).

In several organisms, metal ion resistance/detoxification is associated with the toxic ion induced synthesis of a variety of small molecular ligands such as metallothioneins, γ-glutamyl peptides, glutathione and extracellular ligands or by the enzymatic systems which oxidise or reduce metal ions to non-toxic species (Narender Reddy and Prasad, 1990; Gadd, 1993).

Copper resistant strains of *Rhizopus stolonifer, Cunninghamella blakesleeana* and *Mucor rouxii* bound significantly more copper in to the mycelial biomass than the corresponding parental strains (Garcia-Toledo *et al.*, 1985; Cervantes and Gutierrez-Corona, 1994). The strains of *R.stolonifer* and *C.blakesleeana* were found to be co-tolerant to Cd, Ni, Co and Pb. The copper resistant strain of *M.rouxii* cultured on presence of 1.6mM Cu^{2+} accumulated a 10 fold higher concentration of the metal in the cell wall fraction, and in a mixture of cytoplasm and membranes as compared to parental sensitive strain. Therefore, it was suggested that qualitative and quantitative changes in the cell wall components (chitin and chitosan) to be involved in copper detoxification (Cervantes and Gutierrez-Corona, 1994). Same authors found that a copper resistant strain of *T.viride* cultured in media containing toxic concentrations of Cu^{2+} developed a 5 fold thicker cell wall when compared to the parent wild type. Fifteen copper resistant mutants of *A.nidulans* were isolated which grew on higher concentrations of copper than wild type. All of them were found to result from a single gene mutation at two separate loci *CupA* and *CupB* (Phelan *et al.*, 1990).

Yeasts are extensively investigated organisms amongst the lower eukaryotes with respect to metal resistant mechanisms. A large number of resistant strains have been characterized and their mechanisms have been investigated in great detail both at molecular and biochemical level (Ashida *et al.*, 1963; Gadd, 1993; Cervantes and Gutierrez-Corona, 1994).

Metal toxicities/resistance in *Neurospora crassa* is extensively investigated and physiological, biochemical and genetic basis of metal resistance is well understood. A number of cobalt, nickel and zinc resistant strains of *N.crassa* are known at present. In several of these, metal ion uptake patterns are altered such that either there is a partial transport block, possibly due to altered specificities of the carrier systems which mediate uptake or a diminished binding of the toxic metal ions to the cell wall or even hyper-accumulation in some cases (Maruthi Mohan, 1998 b).

A cobalt resistant strain of *N.crassa,* Co^R was obtained by repeated subcultures on media containing high concentrations of Co and was also resistant to Ni^{2+} but not Zn^{2+}, and twice sensitive to Cu^{2+} in comparison to the parent strain (Venkateswarulu and Sivarama Sastry, 1973; 1979a). Mg^{2+} and iron reversed the toxicity in wild strain while in Co^R strain, only Mg^{2+} and not iron was able to reverse Co toxicity. The cobalt resistance in this strain was attributed to an altered iron utilization pathway unaffected by cobalt. Two other cobalt resistant mutant strains *cor-1* (10 fold resistant) and *cor-2* (20 fold resistant), also obtained by repeated subculturing on media containing cobalt, were also cross resistant to Ni but not to Cu or Zn (Maruthi Mohan, 1998 b). The primary mechanism for cobalt resistance was found to be due to reduced uptake or partial transport for cobalt ions. This was demonstrated to be due to a reduction in the metabolism independent cobalt binding to the cell walls of *cor* strains when compared to parent strains. Secondary mechanism was found to be due to increased efflux of Co^{2+} in the resistant strains. Third mechanism was found to be due to induction of unique cobalt binding protein (CBP) in *cor-2* strain.

Three nickel resistant strains of *N.crassa* Ni^{R1}, Ni^{R2} and Ni^{R3} were obtained by adaptation on nickel containing media (Maruthi Mohan and Sivarama Sastry, 1983a). All the strains were 4 fold resistant to Ni but differed in their cross resistance to Co and Cu. Ni^{R1}, Ni^{R2} were hyper-accumulators of Ni^{2+}, Co^{2+} and Zn^{2+} ions whereas Ni^{R3} accumulated much less than parent strain (Maruthi Mohan *et al.*, 1984 c). The hyper-accumulating strain Ni^{R2} was able to remove >90% nickel from 120 ppm solution while the parent was more effective in removing from 4 ppm nickel solution (Kumar *et al.*, 1992).

Cobalt transport in these strains was found to be non-saturable with respect to Co^{2+}. Ni^{R2} has highest rate of uptake followed by $Ni^{R1}$and Ni^{R3}. Cobalt uptake was both energy and temperature dependent and increased linearly with pH from 3 to 6. (Maruthi Mohan and Sivarama Sastry, 1984a). Copper toxicity in these strains showed that nitrate reductase levels were reduced in Ni^{R3} but not in $Ni^{R1}$and Ni^{R2}.

Three non-identical zinc resistant strains of *Neurospra crassa* were isolated out of which two strains *znr-1* and *znr-2* were isolated by adaptation on zinc containing media while *znr-3* was isolated by chemical mutagenesis followed by plating on zinc containing media (Maruthi Mohan, 1998 b). All of them were 2 to 3 fold resistant to zinc compared to wild strain and all of them had partial transport block for Zn^{2+} as a major mechanism of resistance.

A mercury resistant strain of *N.crassa* isolated by subjecting conidia to X-rays followed by plating on medium containing Hg was found to have an increased ability to generate methylmercury in presence of high concentrations of Hg (Landner, 1971). The methylation *in vivo* occurred by transfer of methyl group to mercury complexed to homocysteine. Methionine was found to inactivate the methylating enzyme.

Four cadmium resistant strains were isolated and characterized by Levine and Marzluf, (1989). Cadmium resistance in these mutants did not appear to result from increase in cytosolic, heat stable cadmium binding proteins since heat shock response was not induced in conidia by cadmium. Under various growth conditions, each of the mutants exhibited characteristic morphological alterations possibly involving thc cell wall or plasma membrane. Transport of cadmium in to the *N.crassa* cells was by more than a single uptake system; both energy dependent and independent uptake of cadmium has been observed. But, there was no difference in cadmium transport in mutants and wild type.

Marzluf (1970) isolated chromate resistant (CrO_4^{2-}), sulphate transport negative mutants of *N.crassa* with mutations at cys-13, cys-14 and cys-3 loci. The cys-13 mutant seems to be most resistant to chromate since the permease I coded by cys-13 was also responsible for the lethal entry of chromate (cited in Maruthi Mohan, 1998 b).

Metal Resistance Mechanisms

Extracellular precipitation

Many extracellular fungal products can complex or precipitate heavy metals. e.g. citric acid is an efficient metal chelator while oxalic acid interacts with metals and forms insoluble oxalate crystals around cell walls and in external medium (Murphy and Levy, 1983). Production of H_2S by some metal tolerant yeasts can precipitate metals and form metal sulphides in and around cell walls (Ashida *et al.*, 1963; Minney and Quirk, 1985) e.g. *S.cerevisiae* forms dark colour colonies due to formation of copper sulphide when grown on copper containing medium (Ashida *et al.*, 1963).

Siderophores

Iron is an important metal required by microorganisms and many filamentous fungi produce low molecular weight high affinity chelators called siderophores in iron limiting conditions (Winkelmann *et al.*, 1987, Winkelmann, 1992). Though they have high affinity for iron, they also bind other metals such as aluminium and gallium (Gascoyne *et al.*, 1991a, b). *Debaryomyces hansenii* produced riboflavin, or a related compound when grown in Fe-deficient media or in the presence of copper, cobalt and zinc.

Variety of fungi and yeasts secrete siderophores which remove iron from insoluble ferric forms and facilitate its transport in to cells (Raymond *et al.*, 1984). In *S.cerevisiae* Fe^{2+} is taken up faster than Fe^{3+} and uptake is strongly inhibited by Fe^{2+} trapping agent ferrozine. Iron is physiologically reduced by an iron induced trans-plasma membrane redox system and it seems that uptake of Fe^{2+} in *S.cerevisiae* is via a similar mechanism as for other divalent cations.

Metal binding to cell walls

The cell wall is the first site of interaction with metal species. Metabolism independent association of metal species to fungal walls may include ion exchange, adsorption, complexation, precipitation and crystallization (Mullen *et al*, 1992). Biosorption is a term, that is frequently used to describe non-directed physicochemical interactions between metal species and microbial biomass. The fungal cell wall thus has important protective properties and so may act as a barrier, controlling the uptake of solutes including potentially toxic metal species and also indirectly affects the intracellular ionic composition by restricting cellular water.

The cell wall is mainly composed of polysaccharides and is associated with protein, along with lipids and melanins. All filamentous fungi except Oomycetes contain chitin. Oomycetes possess cellulosic cell walls. Yeasts on the other hand consist of mannan-protein complex. The metal binding (biosorption) to fungal cell walls is a complex process involving different components and mechanisms and varies depending on wall structure and composition. The cell wall composition itself is affected by presence of toxic metals (Gadd and Griffith, 1980; Venkateswarulu and Stotzky, 1989; Venkateswarulu *et al.*, 1989). A variety of potential sites may be involved in metal sequestration including carboxyl, amine, hydroxyl, phosphate and sulphydryl groups although their relative significance is usually difficult to resolve (Strandberg *et. al.*, 1981). However, primary interaction probably involves binding to carboxyl and phosphate groups which may be enhanced by electrostatic attraction to negatively charged functional groups (Tobin *et al.*, 1990; Kapoor and Viraraghavan, 1997). Metabolism independent biosorption is usually rapid and unaffected in the temperature range of 4-30°C. In *Rhizopus arrhizus,* biosorption was related to ionic radius for. e.g. La^{3+}, Mn^{2+}, Cu^{2+}, Zn^{2+}, Cd^{2+}, Ba^{2+}, Hg^{2+}, Pb^{2+}, UO^{2-} and Ag^{+} but not Cr^{3+} or the alkali metal cations, Na^{+}, K^{+}, Cs^{+} and Rb^{+} which were taken up by the cell walls (Tobin *et al.*, 1984).

Melanins are important fungal pigments that enhance the survival of many species in response to environmental stress (Bell and Wheeler, 1986). Fungal melanins are located in and/or exterior to cell walls where they appear as electron dense deposits and granules on electron micrographs. These granules may be released in to the external medium and termed as extra-cellular melanins although this is generally of identical composition to wall associated melanin. Extra-cellular melanins are synthesized exterior to cells by secretion of phenol oxidases which oxidize phenolics or secretion of phenols that are subsequently oxidized. Fungal phenolic polymers and melanins contain phenolic units, peptides, carbohydrates, aliphatic hydrocarbons and fatty acids and therefore possess many potential metal binding sites. Oxygen containing groups in these substances including carboxyl, phenolic and alcoholic hydroxyl, carbonyl and methoxyl groups are particularly important in metal binding. The order of metal binding ability of fungal phenolic polymers and melanins followed the sequence Cu>Ca>Mg>Zn (Zunino and Martin, 1977).

A variety of melanins occur in fungi. Although it is apparent that many fungi can form extracelluar DOPA (3,4-dihydroxyphenylalanine)-melanin via tyrosinase-mediated oxidation of tyrosine, there is still no firm evidence for the existence of DOPA-melanin in fungal cell walls (Bell and Wheeler, 1986). Melanins in cell walls of basidiomycetes are derived from γ-glutaminyl -3,4-dihydroxybenzene (GDHB) or catechol. A variety of heavy metals can induce or accelerate melanin production in fungi and melanized cell forms, e.g. chlamydospores, can have high capacities to adsorb metals with virtually all the metal being located in the cell wall (Mowll and Gadd, 1984). Melanin from *Aureobasidium pullulans* can bind significant amounts of Cu^{2+} and Fe^{3+} (Gadd, 1984), as well as organometallic compounds e.g. tributyltin chloride (Gadd *et al.*, 1990).

Transport of Toxic Metal cations

The plasma and vacuolar membranes are the main transport membranes of fungi. Most of the work on metal transport in fungi has concerned K^+ and Ca^{2+}, largely because of their importance in fungal growth and differentiation. The transport of toxic metal is still poorly understood.

Transport systems in fungal cell membranes are usually classed as either carrier or channel systems. In the former, conformational changes in the transport protein are believed to result in alternate exposure of transport binding sites on each side of the membrane. Carriers include all metabolically coupled and H^+-gradient driven transport systems. Ion channels are a class of proteins that function as gated pores in the plasma membrane allowing the flow of ions down electrical and/or chemical gradients (Gustin *et al.*,1986).

In filamentous fungi and yeasts, energy dependent transport of many divalent cations has been demonstrated, the apparent affinity series being Mg^{2+}, Co^{2+}, $Zn^{2+}>Mn^{2+}>Ni^{2+}>Ca^{2+}>Sr^{2+}$. However, differences in accumulation may not be due to differences in affinities for the transport mechanisms as some reductions in the net rate of Ca^{2+} and Sr^{2+} uptake for e.g. may be due to increased leakage or efflux of these cations. In addition, depending on the metal and its concentration, there may be effects on the structural integrity of the membrane (Gadd and Mowll, 1983). The main primary transport systems which have been characterized in fungi derive energy from ATP hydrolysis and pump H^+ electrogenetically from the cytosol, creating a transmembrane electrochemical H^+ gradient, negative and alkaline inside. Such an electrochemical gradient has an electrical component, the membrane potential, and a chemical component, the pH gradient which are inter convertible and which can drive the transport of ionziable substances across membranes. Such secondary gradient–coupled transport systems exist for a variety of inorganic and organic solutes which are energized by coupling with passive reflux of H^+ proton and solute fluxes can be in the same (symport) or opposite (antiport) directions (Sanders, 1990). In fungi and yeast three main classes of H^+ pumping ATPases have been identified in mitochondrial, vacuolar and plasma membranes. The main function of mitochondrial ATPase is to synthesize ATP via mitochondial respiratory chain. Plasma membrane and vacuolar ATPases are associated with ion transport and compartmentalization.

Many divalent cations that are essential for growth and metabolism such as Mg^{2+}, Ca^{2+}, Cu^{2+}, Zn^{2+} and Mn^{2+} have to be accumulated from the environment. However, above certain concentrations they may be toxic and may cause impairment of cell metabolism, ultimately leading to cell death. Inessential metals may also be accumulated e.g.Cd^{2+} and Hg^{2+}. Potentially toxic metal ions e.g. Cu^{2+}, Co^{2+}, Ni^{2+}, Cd^{2+}, Mn^{2+}, mercurials and other organometallic compounds, can inhibit the yeast plamsa membrane ATPase by means of various binding interactions, both specific and non-specific. Such deleterious effects lead to a reduced ability in maintenance of the electrochemical gradients. For H^+ efflux, heavy metal inhibition generally increased with increasing metal concentration with a toxicity sequence of $Cd^{2+} > Cu^{2+} > Ni^{2+} > Zn^{2+} > Co^{2+}$, Mn^{2+}. These metals also inhibited K^+ uptake with a toxicity sequence of $Cd^{2+} > Cu^{2+} > Ni^{2+} > Co^{2+} > Mn^{2+}$, Zn^{2+}. K^+ uptake was considerably more affected than was H^+ efflux suggesting that the inhibition of K^+ uptake was not solely a result of inhibition of the H^+-ATPase and other factors such as increased membrane permeability to K^+ may also contribute (White and Gadd, 1987a, b). Uptake of divalent cations is inhibited or halted by metabolic inhibitors, low temperatures and absence of energy yielding substrates. Thus uptake depends on the metabolic state of the cell and may vary with different growth media and conditions. For maximal uptake rates, cells require adequate K^+ and phosphate (Fuhrmann and Rothstein, 1968) and the phosphate requirement may be indirect and related to the energetic state of the cell as shown for Mn^{2+} (Borst-Pauwels, 1981). Divalent cation transport is dependent on plasma membrane H^+-ATPase activity and a similar affinity sequence of divalent cations for stimulation of the ATPase for transport has been demonstrated (Fuhrmann, 1974). At low, non-toxic concentrations, it seems clear that high affinity uptake systems exist for, e.g.Mn^{2+} and Zn^{2+} in filamentous fungi (Startling and Ross, 1990).

For potentially toxic divalent metal cations, reduced transport in to yeast cells is a frequently observed mechanism of resistance, e.g. in metal-resistant strains of *S.cerevisiae.* (Joho *et al.*, 1991). However, Mn^{2+}-resistant strain accumulated more Mn^{2+} than the wild type strain, possibly by means of a more efficient internal sequestration system (Bianchi *et al.*, 1981). Those external factors that can reduce uptake of toxic cations may also result in protection, e.g., low external pH reduces the influx and toxicity of many metal ions (Gadd, 1986a; Gadd and White, 1989a), while protection of yeast cells by Ca^{2+} from Cd^{2+} toxicity was due to a reduction in Cd^{2+} uptake in the presence of Ca^{2+}. (Kessels *et al.*, 1985). Extrusion pumps have also been demonstrated for divalent cations, e.g.Ca^{2+} and Sr^{2+} (Nieuwenhuis *et al.*, 1981; Theuvent *et. al*, 1986). Earlier evidence indicated the existence of a Ca^{2+}/H^+ antiport on the plasma membrane although it is now proposed that Ca^{2+} efflux is via H^+/Ca^{2+}-ATPase with a stoichiometric ratio of at least $2H^+/Ca^{2+}$ in *Neurospora crassa* (Miller *et al.*, 1990).

Role of Metal Binding Proteins in Metal Resistance

It is evident that the metabolic consequence of metal ions in the environment are determined by the nature and magnitude of the metal ion uptake by the microorganism. Apart from the carrier systems for different metals, a number of specific metal binding molecules are important in the modulation of intracellular concentration of essential as well as toxic metals and affect metal toxicity and hence contribute the cellular resistance.

Metal binding proteins may achieve this by binding metal ions to cysteine thiolate groups as in metallothioneins (Hamer, 1986). Metallothioneins (MTs) are produced in response to metal toxicity. Animals synthesize MTs in response to copper, cadmium and zinc ion stress (Karin, 1985; Bremaner, 1987). Plants elaborate cysteine containing γ-glutamyl peptides which function in metal detoxification (Narender Reddy and Prasad, 1990; Robinson *et al.*, 1993). Some fungi elaborate both MT and γ-glutamyl peptides (Gadd, 1993; Cervantes and Guitrerrez-Corona, 1994). Both MTs and γ-glutamyl peptides are capable of sequestering a variety of metal ions. Generally only cadmium and copper are the prime inducers. These proteins were initially thought to be involved in homeostastis. However, a number of other functions were also proposed which include regulation of cellular metabolism, control of radiation and detoxification of free radicals, protection against ionizing radiation and detoxification of excess metal ions (Karin, 1985).

Polypeptides are named metallothioneins if they have following properties: low molecular mass, high metal content, high cysteine (Cys) content, lack aromatic amino acids and histidine, possess abundant Cys-X-Cys sequences (where X is an amino acid other than Cys), spectroscopic properties characteristic of metal thiolates and metal thiolate clusters. Metallothioneins are classified in to three classes by Rauser (1990).

Class I: Polypeptides with locations of cysteine closely related to those in equine renal metallothionein. Most vertebrate and some fungal metallothioneins, e.g. *Neurospora crassa* and *Agaricus bisporus* belong to this group (Lerch and Beltramini, 1983; Munger and Lerch, 1985).

Class II: Polypeptides with locations of cysteine only distantly related to those in equine renal metallothioneins. These are found in *S.cerevisiae* (Steffens, 1990).

Class III: Atypical metallothioneins, non-translationally synthesized metal thiolate polypeptides. Algae, plants and certain fungi such as *Schizosaccharomyces pombe* and *Candida glabrata* are examples of this class (Rauser, 1990). These occur as metal-binding complexes and behave like entities of relative molecular mass 10,000-30,000 in gel filtration (Rauser, 1990). However, these complexes are heterogeneous group of polypeptides for which no single name is suitable and called γ –glutamyl peptides (Winge *et al.*, 1989).

Metallothioneins

Metallothioneins are small cysteine rich polypeptides that can bind essential metals like Cu and Zn and inessential metals such as Cd. Copper resistance in *S.cerevisiae* is mediated by induction of a 6573 Da cysteine rich protein, copper metallothioneins, Cu-MT. This protein maintains low intracellular copper concentration. Though, this can also bind Cd and Zn *in vitro*, the protein is not induced by these ions and apparently does not protect against them (Butt, *et al.*, 1984 a, b; Winge *et al.*, 1985).

Copper resistant *S.cerevisiae* strains ($CUP1^r$) contain 2-14 or more copies of the *CUP1* locus, which are randomly repeated and may grow on media containing $\geq$ 2mM copper (Karin *et al*, 1984). Copper sensitive strains ($cup1^s$) do not grow at concentrations $\geq$ 0.15mM (Butt and Ecker, 1987). In *S.cervisiae*, copper resistance was shown to be due to a tandom amplification of *CUP1* gene based on unequally sister chromatid exchange (Fogel and Welch, 1982; Welch *et. al*, 1983) which in turn results in overproduction of Cu-MT leading to enhanced intracellular chelation capacity (Fogel *et .al*, 1983). Multi copy *CUP1* plasmids were also shown to enhance cadmium and copper resistance in yeast (Jayaprakash *et al.*, 1991). Continuous subculture of *S.cerevisiae* in the presence of increasing concentrations of copper can select hyper-resistant strains which appear disomic for chromosome VIII (White and Gadd, 1986). However, it should be noted that Cu-MT is not directly involved in Cu^{2+} uptake in *S.cerevisiae* (Lin and Kosman, 1990). The isolated protein from *S.cerevisiae* contains 8 mol copper [Cu^{+}] ligated to 12 cysteine per mol protein. (Winge *et al.*, 1985).

The Cu-MT gene of *Neurospora crassa* has been cloned and the nucleotide sequence determined. This gene codes for a 26 amino acid protein with molecular weight 2200 containing 7 different amino acids with 28% cysteine, 28% serine and 24% glycine. (Munger *et al.*, 1985; 1987). The structural gene encoding *N.crassa* Cu-MT has been cloned and transcriptional regulation studied (Munger *et al*, 1985; 1987). This is the smallest gene known to contain a single intron. The rapid decay of the induced mRNA to its initial value and the remarkable stability of the synthesized protein suggested an important role of Cu-MT in the defense mechanism of the fungus against this metal ion. It appears that copper is present as Cu^{1+} form and all the cysteines are ligated to form Cu complexes with 6 copper molecules per mole of protein (Lerch, 1980; Lerch and Beltramini, 1983). The large metal to protein binding ratio of

N.crassa Cu-MT makes it a good candidate for detoxification and also internal storage protein. Cu-MT also acts as a metal donor to the active sites of copper containing enzymes. Copper toxicity in *N.crassa* on a sole nitrate nitrogen medium results in blue mycelia (Subramaniam *et al.*, 1983 a, b). A 14 kDa protein was isolated from the copper toxic blue cell walls. The amino acid composition showed it to be devoid of cysteine (Suresh and Subramanyam, 1992). There was excess of phenolic compounds in the cell walls of blue mycelia and this could be the intrinsic tolerant mechanism towards copper toxicity in *N.crassa*.

A Cd-resistant strain of *S.cerevisiae* synthesized a cytoplasmic Cd-binding protein when grown in Cd-containing medium (Joho *et al.*, 1985a). This protein 9 KDa, exhibits the characteristics of a metallothionein, being rich in cysteine (18% mole) and having a high Cd content (63 µg Cd/mg protein). There is high similarity in amino acid composition between Cd binding protein and Cu-MT of *S.cerevisiae* (Inouhe *et al.*, 1989). Though Cd and Cu resistant mutants of *Apergillus nidulans* and Cd resistant mutants of *N.crassa* have been isolated and characterized, resistance mechanisms have not been elucidated. While metal-binding proteins or peptides are an obvious suggestion (Tomesett *et al.*, 1989), it seems that altered uptake and morphological changes may also be involved (Levine and Marzluf, 1989; Phelan *et al.*, 1990). Studies in *Candida glabrata* have revealed that two different types of MTs are produced in response to copper and cadmium, both of which have been implicated in the metal resistance (Mehra *et al.*, 1988). Copper resulted in the production of MT protein (class I) while Cd^{2+} induced $(\gamma\text{- glu-cys})_n$ peptide (class III). Copper resistant strains of *C.glabrata* obtained by repeated sub-culturing on copper containing media showed a tandem amplification of MT-II gene (30 copies) while MT-I gene remained unchanged (single copy). In *Saccharomyces pombe*, both copper and cadmium induced γ-glutamyl peptide synthesis was reported (Reese *et al.*, 1988). Constitutive expression of metallothionein was also shown to be responsible for resistance in a cadmium resistant strain of *S.cerevisiae* (Tohoyama *et al.*, 1992). There is very little genetic information available for other fungi.

Metal γ–glutamyl peptides

In addition to cysteine rich polypeptides of the metallothionein family, another group of small metal binding molecules i.e. cysteine containing γ-glutamyl peptides called phytochelatins are synthesized by fungi (Winge *et al.*, 1989; Rauser, 1990). They are some times called cadystins (Mutoh *et al*, 1991). These γ– glutamyl peptides are the main metal detoxification mechanism in plants and algae (Grill *et al.*, 1988) as well as a range of filamentous fungi and yeasts. They have the general formula (γ–Glu-Cys)n-Gly. Metal ions are bound within a metal-thiolate cluster composed of an oligomer of peptides (Winge *et al.*, 1989). Phytochelatins are not synthesized by *S.cerevisiae*. But *Schizosaccharomyces pombe* synthesizes at least seven different homologous phytochelatins in response to metals (Grill *et al.*, 1986).

There is similarity in structure between γ–glutamyl peptides and glutathione. In fact they can be viewed as linear polymers of the γ–glumayl-cysteine portion of glutathione. Linkage of glutathione metabolism with that of γ–glutamyl peptides has been kinetically demonstrated *in vivo* (Grill *et al.*, 1987) and also using mutants of *S.pombe* which lacked the glutathione biosynthetic enzymes, γ-glutamylcysteine synthetase and glutathione synthetase. These were unable to synthesize glutathione or γ–glutamyl peptides and were hypersensitive to Cd. (Mutoh and Hayashi, 1988). It is now known that γ–glutamyl peptides are synthesized by enzyme γ-glutamylcysteine dipeptidyl transpeptidase. This catalyses the transfer of the γ–glutamylcysteine dipeptide moiety to a growing chain of –(Glu-Cys)-Gly oligomers, thus synthesizing metal binding γ–glutamyl peptides (Grill *et al.*, 1989). Cd^{2+} is the best activator of the enzyme followed by Ag^{+}, Bi^{2+}, Pb^{2+}, Zn^{2+}, Cu^{2+}, Hg^{2+}, Au^{+} and it appears that the reaction product chelates the activating metal ions thus terminating the reaction (Grill *et al.*,1989).

The originally discovered two γ–glutamyl peptides in *S.pombe* , cadystin A and B (n=2 and 3 resp) were synthesized in response to Cd (Murasugi *et al.*, 1981), though other metals including Cu^{2+}, Pb^{2+} and Zn^{2+} and Ag^{+} were also effective (Grill *et al.*, 1986). However, the greatest amount of peptide synthesis occurs with Cd and seems larger peptides bind Cd more strongly than smaller ones (Hayashi *et al.*, 1986). When *S.pombe* was exposed to Cu^{2+}, Cu- γ -glutamyl peptide was synthesized similar in structure to the Cd- γ-glutamyl peptide (Reese *et al.*, 1988). However, only one Cu- γ -glutamyl peptide complex which did not contain labile sulphur was observed (Reese *et al.*, 1988). Cadmium induced but not Cu induced γ–glutamyl peptides contain labile sulphur. Cadmium stimulates the production of S^{2-} in *S.pombe* and *Candida glabrata* and a portion of this may be incorporated within the Cd- γ-gutamyl peptide complex (Mehra *et al.*, 1988). The sulphide content is dependent on the growth conditions and incubation time and concentrations of S^{2-} ranging from 0.1 to 1.5 mol/mol peptide have been shown. Incorporation of sulphide enhances the stability and metal content the peptide complex and this enhanced stability may increase the efficacy of the complex in maintaining low intracellular concentration of Cd^{2+} (Winge *et al.*, 1989) The extracellular and cell associated CdS colloid formed by interaction of Cd^{2+} with S^{2-} may also contribute to Cd^{2+} detoxification.

Synthesis of γ–glutamyl peptides is a predominant mechanism of Cd resistance in *S.pombe* although the genetics of Cd resistance in this fungus is poorly understood. *Candida glabrata* is unique in expressing both metallothioneins and γ–glutamyl peptide synthesis for metal detoxification in a metal specific manner (Mehra *et al.*, 1988; 1992a,b). It synthesizes two distinct MT polypeptides in response to copper which exhibit repeats of the Cys-X-Cys sequence common in mammalian MT and coordinates copper in thiolate clusters. The cysteine content of the two molecules is 30% (*S.cerevisiae* Cu-MT=22%; *N.crassa* Cu-MT=28%) with copper content being between 10-13 mol/mol protein. *C.glabrata* utilizes two types of regulatory metal binding molecules and affinity and/or coordination preference of the molecules may determine whether MT or γ–glutamyl peptides are synthesized. (Mehra *et al.*, 1992a,b).

Induction of γ–glutamyl peptides has so far been restricted to heavy metals and metalloids. However, other non-metallic inducing agents may induce their synthesis in yeasts. The antifungal agents, tetramethylthium disulphide (TMTD) or dimethyl dithiocarbamate (DMDTC), induce γ-glutamyl peptide synthesis to a lesser degree than $CdCl_2$ (but more than $ZnCl_2$ or $CuSO_4$) in *S.pombe*. Such results may indicate biochemical functions other than that of metal detoxification including intracellular metal transport of essential metal ion, Zn^{2+} and Cu^{2+}, detoxification of xenobiotics as scavengers of active oxygen (Mutoh *et al.*, 1991). Related phenomenon which have received relatively little attention are the possible roles of intracellular amino acids in metal resistance. Changes in amino acid pools of yeast can occur in response to exposure to Ni such that the formation of Ni-histidine complexes has been proposed as a survival mechanism (Joho *et al*, 1990). A cadmium binding peptide similar to phytochelatins of plants [γ-glutamyl cystinyl]$_n$ glycine was characterized from *N.crassa* grown on cadmium containing medium (Kneer *et al.*, 1992).

A cobalt resistant mutant *cor*-2 strain of *N.crassa* was characterized and was found to over produce a novel cobalt binding protein (CBP) (Maruthi Mohan, 1998b). This protein accounted for 12% of the total soluble protein and about 80% of total cobalt of the soluble fraction was located in CBP fraction of the resistant strain while the parent strain had 25% under similar conditions. The cobalt resistance in this mutant was attributed to this cobalt binding protein as it could immobilize the intracellular cobalt in to a non-toxic form. About 6 gramatoms of cobalt was bound per mole of protein. The amino acid composition revealed that it has 32% cysteine, 21% glycine like metallothioneins, but presence of aromatic amino acids tyrosine, tryptophan and acidic amino acids glutamic acid (18%) and aspartic acid (7%) make this a unique protein which is different form known metallothioneins. CBP was also found to be excreted in to growth medium in lag phase after mycelial

CBP production saturated. This CBP is specifically induced by cobalt and only binds to cobalt. Other metals like Ni, Cu, Zn, Mg or Fe neither induce nor were bound to this protein in vivo when incubated with cobalt.

Role of Small Molecules in Metal Resistance

Glutathione plays a critical role both in maintaining redox potential of the cell and in cellular detoxification mechanisms (Kaplowitz *et al.*, 1985; Meister and Anderson, 1983; 1988). Glutathione depletion from cells results in cytotoxic effects of sulphydral reactive alkalating agents, radiation thermal stress, reactive oxygen and metal ion stress (Meister and Anderson, 1983; 1988; Singhal *et al.*, 1987; Reese and Wagner, 1987). Glutathione appears to be the first molecule to be involved in detoxification of metal induced cytotoxicity before the induction of metallothionein (Hamer, 1986). How exactly glutathione protects against metal toxicity is unclear but presumably it is by chelation of metal ions. The possible involvement of glutathione in metallothioneine biosynthesis was shown in a cell wall deficient slime variant of *N.crassa* (Germann and Lerch, 1987). Zinc resistant strain of *N.crassa* was shown to have elevated levels of glutathione (oxidized+reduced) in zinc toxic conditions, when compared to the wild type (Maruthi Mohan, 1998 b).

Vacuolar Compartmentalization

The fungal vacuole is highly important organelle with functions including macromolecular degradation, storage of metabolites and cytosolic ion and pH homeostasis (Wiemken *et al.*,1979; Jones and Gadd, 1990). It is often considered similar to lysozyme as hydrolytic enzymes are contained in an acidic compartment although, it has greater similarity to vacuoles of plants (Klionsky *et al.*, 1990). The fungal vacuole has an important role in the regulation of the cytosolic concentrations of metal ions, both for essential metabolic functions and the detoxification of potentially-toxic metal ions. Many ions, including inorganic phosphate and monovalent and divalent cations are preferentially located in vacuoles (Ohsumi *et al.*, 1988). Localization of metal ions in the vacuole enables low cytosolic concentrations of, for example Ca^{2+}, to be maintained (Ohsumi and Anraku, 1983). Cytosolic concentrations of Mg^{2+} and Mn^{2+} may remain relatively constant even under considerable environmental perturbations (Lichko *et al.*, 1982). There is further evidence for compartmentalization of Zn^{2+}, Co^{2+} and iron in yeast vacuoles, which may contribute to tolerant behaviour (White and Gadd, 1986). Mutants defective in vacuolar function were sensitive to Ca^{2+} and other toxic metals (Kitamoto *et al.*, 1988). Polyphosphates, the only inorganic macromolecular anion found in the vacuole have an important role in maintaining ionic compartmentation and their biosynthesis accompanies vacuolar accumulation of Mg^{2+} or Mn^{2+} (Okorokov *et al.*, 1977). The significance of polyphosphate granules as a store for divalent cations has long been recognized (Doonan *et al.*, 1979).

Metal Transformations

Fungi as well as other microorganisms can effect chemical transformations of metals by e.g. oxidation, reduction, methylation and dealkylation (Gadd, 1992a) although in contrast to bacteria, detailed information is not available in several areas for fungi. Some enzymatic metal transformations may be involved in survival, since certain transformed metal species are less toxic/or more volatile than the original species. Plasmid mediated reduction of Hg^{2+} to Hg^0 is a well characterized bacterial metal resistance system and there is only a limited amount of evidence for analogous process in filamentous fungi and yeasts (Vonk and Sijpesteijn, 1973). Other reductions carried out by fungi include Ag^+ to Ag^0 which is deposited in and around cells (Kierans *et. al.*, 1991) and the reduction of Cu^{2+} to Cu^+ by cell wall-associated materials in *Debaryomyces hansenii* (Wakatsuki *et al.*, 1988, Wakatsuki *et al.*, 1991a,b). A copper-reducing enzyme, copper reductase, catalyzes Cu^{2+} reduction to Cu^+ with NADH or NADPH as electron donor (Wakatsuki *et al.*, 1991a,b). The location of copper reductase appears to be cell wall and this process may have a role in regulation of copper uptake (Wakatsuki *et al.*, 1991b). Fungal reductions of metalloids include reduction of tellurite to elemental tellurium which can appear as a black deposit located on the endoplasmic reticulum (Corfield and Smith, 1970; Smith, 1974) and the reduction of selenate to selenite to form amorphous selenium which imparts a red colour to fungal colonies owing to deposition, chiefly in walls and membranes (Konetzka, 1977). Reduction of hexavalent chromium to trivalent form by chromate resistant *Aspergillus* sp. is reported by Pakniker and Bhide (1993). The fungus was able to grow in media containing > 500 ppm chromium which was completely reduced in 6 days in a broad pH range of 4-9.

Methylation of Hg, and other metals and metalloids can be catalyzed by several fungi and may be viewed as a detoxification mechanism since methylated species are usually more volatile and may be lost from the environment (Landner, 1971; Yanni *et al.*, 1991, Gadd, 1992a, b). There is general agreement that methylation of Hg^{2+} involves methylcobalmin (Vitamin B_{12}) (Craig, 1986; Thayer, 1988). Biomethylation of As by transfer of carbonium ions (CH_3^+) from S-adenosylmethionine (SAM) has also been demonstrated (Thayer, 1984; 1989). Challenger (1945) first demonstrated that trimethylarsine could be produced by several fungi from arsenite. Biomethylation of Se appears to be by an analogous mechanism to that of arsenite, and several fungi can produce organoselenium compounds e.g.dimethylselenide from inorganic selenium ones (Chau and Wong, 1986). Certain fungi may be capable of production of volatile dimethyl telluride from tellurium salts (Konetzka, 1977; Chau and Wong, 1986). Apart from these examples, fungal involvement in the methylation of other metals and metalloids has not been examined in detail.

Organometallic compounds (which contain atleast one metal–carbon bond) arise in the environment naturally, by abiotic and biotic means and because of accidental or deliberate introduction. The latter mainly results from the use of organo-metallic and organo-metalloid compounds such as biocides e.g. organo-mercurials, organo-arsenials and organo-tins (Craig, 1986, Cooney and Wuertz, 1989). Degradation involves breaking of the metal-C bonds though detailed information on such processes is rather sparse for fungi. It should be noted that metal-C bonds can be disrupted by physico-chemical means although fungal activity may enhance the conditions necessary for abiotic attack by e.g. alteration of pH or metabolite excretion. Organo-tin degradation involves sequential removal of organic groups from the tin atom, a process generally resulting in decrease in toxicity (Blunden and Chapman, 1986). Degradation of tributyl-tin oxide and tribuyl-tin naphthenate used as wood preservatives can be achieved by fungal action, breakdown products including mono, di-butyltins (Barug, 1981). Organo-mercurials may be detoxified by organomercury lyase , the resulting Hg^{2+} being then reduced to Hg^0 by mercuric reductase (Tezuka and Takasaki, 1988), a similar mechanism to that found in bacteria (Gadd, 1992b). Trimethyl lead can be degraded by *Phaeolus schweinitzii*, a process with implications for fungal treatment of wastes (Macaskie and Dean, 1990b).

Studies so far have shown a wide variety of effects of toxic metal ions on fungi. The effects depend on the metal ion, media conditions and the organism being investigated. As can be seen, fungi manifest variety of mechanisms to tolerate/resist metal toxicity at different levels such as outside the organism, at the cell surface and inside the cell. The response varies with organisms and is influenced by a variety of environmental factors such as media composition, pH and oxidation-reduction potential etc. Similarly, the organism can survive by means of strong efflux mechanism or poor transport of the metal in to the cell. Once inside the cell, the metal can either be sequestered by metal binding proteins/peptides like metallothioneins or γ–glutamyl peptides or they can be transformed to less toxic forms by reduction/oxidation or methylation etc. However, mechanism of toxicity and site and enzymes/proteins inhibited are known only in very few cases. Once the mechanisms of toxicity become clear metal resistance can be studied by evaluating how the toxic sites evade the metal ion. Another approach is subcellular fractionation of toxic metal ions.

In such studies, the use of X-ray microprobes technique could greatly help in both location and quantification of the metal ions without disturbing the cellular structures. However, when metal ions are bound to specific proteins, selected protein fractionation procedures including metal chelate affinity chromatography can be used to quantitate the percent of the metal located in the soluble protein fraction. The use of extensively investigated fungi whose metal resistant mechanisms are well understood like *N.crassa* and *A.nidulans* will help in identification of metal resistant loci. Isolation of specific metal resistant genes can be undertaken with fungal transformation system by screening for metal resistant marker. The interaction of metal ions with specific transacting factors in the induction of specific genes (metallothioneins or others) and the regulation of its subsequent expression is only beginning to be understood in the yeast system and the same needs to be investigated in other filamentous fungi.

References

Abelson, P.H. and Aldous, E. (1950). Ion antagonisms in microorganisms : Influence of normal magnesium metabolism by nickel, cobalt, cadmium and manganese. J. Bacteriol., **60** : 401-404.

Adiga, P.R., Sivarama Sastry, K., Venkatasubramaniam, V. and Sarma, P.S. (1961). Interrelationships in trace element metabolism in *Aspergillus niger*. Biochem. J., **81** : 545-549.

Amir and Pineau (1998). Effects of metals on germination and growth of fungal isolates from New Caledonian ultramafic soils. Soil. Biol. Biochem., **30** : 2043-2054.

Ashida, J., Higashi, N. and Kikuchi, T. (1963). An electron microscopic study on copper precipitation by copper resistant yeast cells. Protoplasma., **57** : 27-32.

Babich, H. and Stotzky, G. (1977a). Reduction in the toxicity of cadmium to microorganisms by clay minerals. App. Environ. Microbiol., **33** : 696-703.

Babich, H. and Stotzky, G. (1977b). Effect of cadmium on fungi and on interactions between fungi and bacteria in soil: Influence of clay minerals and pH. Appl. Environ. Microbiol., **33** : 1059-1066.

Babich, H. and Stotzky, G. (1977c). Sensitivity of various bacteria including Actinomycetes and fungi to cadmium and influence of pH on sensitivity. Appl. Environ. Microbiol., **33** : 681-683.

Babich, H. and Stotzky, G. (1980). Environmental factors that influence the toxicity of heavy metal and gaseous pollutants to microorganisms. Critical Reviews in Microbiology, **8** : 99-145.

Babich, H. and Stotzky, G. (1981). Manganese toxicity to fungi: Influence of pH. Bull. Environ.Contam. Toxicol., **27** : 474-480.

Babich, H. and Stotzky, G. (1982a). Nickel toxicity to fungi: Influence of environmental factors. Ecotoxicol. Environ.Safety, **6** : 577-589.

Babich, H. and Stotzky, G.(1982b). Gaseous and heavy metal pollutants, In: Experimental Microbial Ecology Eds. Burns, R.G. and Slater, J.H., Blackwell, Oxford, England, 631.

Babich, H. and Stotzky, G. (1983). Temperature, pH, salinity, hardness and particulates mediate nickel toxicity to eubacteria, an actinomycete and yeast in lake stimulated estuarine and sea waters. Aquatic. Toxicol., **3** ; 195-198.

Babich, H., Schifferbauer, M.and Stotzky, G. (1982). Comparative toxicity of trivalent and hexavalent chromium to fungi. Bull. Environ. Contam. Toxicol., **28** : 452-455.

Barug, D. (1981). Microbial degradation of bis(tributyltin) oxide. Chemosphere, **10** : 1145-1154.

Bell, A. A. and Wheeler, M.H. (1986). Biosynthesis and functions of fungal melanins. Ann. Rev. Phytopathol., **24** : 411-451.

Bewley, R.J.F. and Stotzky, G. (1983). Effects of cadmium and zinc on the microbial activity in soil: Influence of clay minerals. Part 1: Metals added individually. Sci. Total Environ., **31** : 41-55.

Bianchi, M.E., Carbone, M.L., Lucchini, G. and Magni, G.E. (1981). Mutants resistant to manganese in *Saccharomyces cerevisiae.* Current Genetics., **4** : 215-220.

Blunden. S.J., Hobbs, L.A. and Smith, P.J. (1984). The environmental chemistry of organotin compounds. In : Bowen HJM ed. Environmental Biochemistry, London: Royal Society of Chemistry. 49-77.

Blunden, S.J. and Chapman, A. (1986). Organotin compounds in the environment. In: Craig P.J.Ed. Organometallic compounds in the environment. Harlow: Longman, 111-159.

Borst-Pauwels, G.W.F.H.(1981). Ion transport in yeast. Biochem. Biophys. Acta., **650** : 88-127.

Bremner, I. (1987). Nutritional and physiological significance of metallothioneins. In: Metallothioneins II. Eds. Kagi, J.H.R. and Kogima, Y., Birkhauser verlag, Basel, 81-107.

Butt, T.R. and Ecker, D.J. (1987). Yeast metallothioneins and applications in biotechnology. Microbiol. Reviews, **51** : 351-364

Butt, T.R., Sternberg, E., Herd, J. and Crooke, S.T. (1984). Cloning and expression of yeast *Saccharomyces cerevisiae* copper metallothionein gene. Gene, **27** : 23-33.

Butt, T.R., Sternberg, M., Gorman, J.A., Clark, P., Hamer., D.J, Rosenberg, M. and Crooke, S.T. (1984). Copper metallothioneins of yeast, structure of the gene and regulation of expression. PNAS (USA)., **81** : 3332-3337.

Cervantes, C and Gutierrez-Corona, F. (1994). Copper resistance in bacteria and fungi. FEMS Microbiol. Lett., **14** : 121-138.

Challenger, F. (1945). Biological methylation. Chem. Rev., **36** : 315-361.

Chau, Y.K. and Wong, P.T.S. (1986). Organic Group VI elements in the environment. In Craig, P.J. Ed. Organometallic compounds in the environment, Harlow: Longman, 254-278.

Collins, Y.E. and Stotzky, G. (1989). Factors affecting the toxicity of heavy metals to microbes In: Beveridge, T.J. Doyle, R. J. Eds. Metal ions and Bacteria, New York, Wiley, 31-90.

Cooney, J.J. and Wuertz, S. (1989). Toxic effects of tin compounds on microorganisms. J. Ind. Microbiol., **4** : 375-402.

Cooney, J.J. DeRome, L. Laurence, O.S. and Gadd, G.M. (1989). Effects of organotins and organoleads on yeasts. J. Ind. Microbiol., **4** : 276-288.

Corfield, P.S.and Smith, D.G. (1970). Endoplasmic reticulum as the site of potassium tellurite reduction in yeasts. J. Gen. Microbiol., **63** : 311-316.

Craig, P.J. (1986). Organometallic compounds in the environment. Harlow: Longman.

Donnan, B.D, Crang, R.E., Jensen, T.E. and Baxter, M. (1979). In situ X-ray energy dispersive microanalysis of polyphosphate bodies in *Aureobasidium pullulans.* J. Ultrastructure Research, **69** : 232-238.

Fogel, S. and Welch, J.W. (1982). Tandem gene amplification mediates copper resistance in yeast PNAS (USA)., **79** : 5342-5346.

Fogel, S., Welch, J.W. and Karin, M. (1983). Gene amplification in yeast: CUP1 copy number regulates copper resistance Curr. Genetics, **7**, 1-9.

Fuhrmann, G.F. (1974). Relation between divalent cation transport and divalent cation activated ATPase in yeast plasma membranes. Experientia, **30** : 686.

Fuhrmann, G.H. and Rothstein, A. (1968). The transport of zinc, cobalt and nickel in to yeast cells. Biochim. Biophys. Acta., **163** : 325-327.

Gadd, G.M. (1984). Effects of copper on *Aureobasidium pullulans* in solid medium, adaptation not necessary for tolerant behaviour. Trans. British. Mycol. Soc., **82** : 546-549.

Gadd, G.M. (1986a). The responses of fungi towards heavy metals. In Herbert, R.A., Codd, G.A. Eds. Microbes in Extreme Environments, London, Academic Press., 83-110.

Gadd, G.M. (1990b). Metal tolerance. In Edwards, C. Ed. Microbiology of Extreme environments, Milton Keynes, Open University Press, 178-210.

Gadd, G.M. (1992a). Microbial control of heavy metal pollution In Fry, J.C, Gadd, G.M, Herbert, R.A., Jones C.W., Watson Craik, I. Eds. Microbial control of environmental pollution, Cambridge, Cambridge University Press, 59-88.

Gadd, G.M. (1992b). Molecular biology and biotechnology of microbial interactions with organic and inorganic heavy metal compounds In Herbert, R.A., Sharp, R. J. Eds. Molecular biology and biotechnology of extremophiles. Glasgow, Blackie and Sons, 225-257.

Gadd, G.M. (1993). Interactions of fungi with toxic metals. New Phytol., **124** : 25-60.

Gadd, G.M. and Griffiths, A.J. (1978) Microorganisms and heavy metal toxicity Microbiol. Ecol., **4** : 303-317.

Gadd, G.M. and Griffiths, A.J. (1980). Effect of copper on morphology of *Aureobasidium pullulans.* Trans. British. Mycol. Soc., **74** : 387-392.

Gadd, G.M. and Mowll, J.L. (1983). The relationship between cadmium uptake, potassium release and viability in *Saccahromyces cerevisiae.* FEMS Microbiol. Lett., **16** : 45-48.

Gadd, G.M. and Mowll, J.L. (1985). Copper uptake by yeast like cells, hyphae and chlamydospores of *Aureobasidium pullulans* Exper. Mycol., **9** : 230-240.

Gadd, G.M. and White, C. (1985). Copper uptake by *Penicillium orchro-chloron*: influence of pH on toxicity and demonstration of energy dependent copper influx using protoplasts J. Gen. Microbiol., **131** : 1875-1879.

Gadd, G.M. and White, C. (1989a). Heavy metal and radionuclide accumulation and toxicity in fungi and yeasts. In : Poole, R. K .Gadd, G. M. Eds.. Metal microbe interactions. Oxford: IRL. Press, 19-38.

Gadd, G.M., Gray, D.J. and Newby, P.J.(1990). Role of melanin in fungal biosorption of trybutyltin chloride. Appl.Microbiol. Biotechnol., **34** : 116-121.

Galiazzo, F., Ciriolo, M.R., Carri, M.T., Civitareale, P., Marmocchi, L. Marmocchi, F. and Rotilio, G. (1991). Activation and induction by copper of Cu/Zn superoxide dismutase in *Saccharomyces cerevisiae*. Eur. J. Biochem., **196** : 545-549.

Garcia-Toledo, A., Babich, H. and Stotzky, G. (1985). Training of *Rhizopus stolonifer* and *Cunnighamella blakesleeana* to copper : co-tolerance to cadmium, cobalt, nickel and lead. Can. J. Microbiol., **31** : 485-492.

Gascoyne, D.J., Connor, J.A. and Bull, A.T. (1991a). Isolation of bacteria producing siderophores under alkaline conditions. Appl. Microbiol. Biotechnol., **36** : 130-135.

Gascoyne, D.J., Connor, J.A. and Bull, A.T. (1991b). Capacity of siderophores-producing alkalophilic bacteria to accumulate iron, gallium and aluminium. Appl. Microbiol. Biotechnol., **36** : 136-141.

Germann, U.A. and Lerch, K.(1987). Copper accumulation in the cell wall deficient slime variant of *Neurospora crassa*. Biochem. J., **245** : 479-484.

Gray, S.N. (1998). Fungi as potential bioremediation agents in soil contaminated with heavy or radioactive metals. Biochem. Soc. Trans., **26** : 666.

Grill, E., Winnacker, E.L. and Zenk, M.H. (1986). Synthesis of seven different homologous phytochelatins in metal-exposed *Schizosaccharomyces pombe* cells. FEBS Lett., **197** : 115-120.

Grill, E., Winnacker, E.L. and Zenk, M.H. (1987). Phytochelatins, a class of heavy metal binding peptides from plants, are functionally analogous to metallothioneins. PNAS (USA)., **84** : 439-443.

Grill, E., Thumann, J., Winnacker E.L. and Zenk, M.H. (1988). Induction of heavy metal binding phytochelatins by inoculation of cell cultures in standard media. Plant Cell Reports, **7** : 375-378.

Grill, E., Loffler, S., Winnacker, E-L. and Zenk, M.H. (1989). Phytochelatins, the heavy metal binding peptides of plants are synthesized from glutathione by a specific γ–glutamyl-cysteine dipeptidyl transpeptidase (phytochelatin synthase). PNAS (USA)., **86** : 6838-6842.

Gustin, M.C., Martinae, B., Saimi Y, Culberson, M.R. and Kung, C. (1986).. Ion channels in yeast. Science, **233** : 1195-1197.

Hamer, D.H. (1986). Metallothionein. Ann. Rev. Biochem., **55** : 913-951.

Hayashi, Y., Nakagawa, C.W. and Murasugi, A. (1986). Unique properties of Cd-binding peptides induced in fission yeast, *Schizosaccharomyces pombe*. Environ. Hlth. Perspectives., **65** : 13-19.

Healy,H.B., Cheng, S. and McElroy, W.D. (1955). Metal toxicity and iron deficiency effects on enzymes in *Neurospora crassa*. Arch. Biochem. Biophys., **54** : 206-214.

Horsfall, J.G. (1956). In Principles of Fungal Action. Waltham, Massachusetts : Chronica Botanica, Co.

Hughes, M.N. and Poole, R.K. (1991). Metal speciation and microbial growth-The hard (and soft). facts. J. Gen. Microbiol., **137** : 725-734.

Inouhe, M. Hiyama, M., Tohoyama, H., Joho, M. and Murayama, T. (1989). Cadmium binding protein in a cadmium resistant strain of *Saccharomyces cerevisiae*. Biochim. Biophys. Acta., **993** : 51-55.

Jayaprakash, A., Welch, J.W. and Fogel, S. (1991). Multi copy CUPI plasmids enhance cadmium and copper resistance in yeast. Mol. Gen. Genet., **226** : 363-368.

Joho, M., Imai, M. and Murayama, T (1985a). Different distribution of Cd^{2+} between Cd-sensitive and Cd-resistant strains of *Saccharomyces cerevisiae*. J. Gen. Microbiol., **131** : 53-56.

Joho, M, Inouhe, M. Tohoyama, H and Murayama, T. (1990). A possible role of histidine in nickel-resistant mechanisms of *Saccharomyces cerevisiae*. FEMS Microbiol. Lett., **66** : 333-338.

Joho, M., Tarumi, K., Inouhe, M., Tohoyama, H. and Murayama, T. (1991). Co^{2+} and Ni^{2+} resistance in *Saccharomyces cerevisiae* associated with a reduction in the accumulation of Mg^{2+}. Microbios, **67** : 177-186.

Jones, R.P. and Gadd, G.M. (1990). Ionic nutrition of yeast, the physiological mechanisms involved and applications for biotechnology Enz. Micrbiol. Technol., **12** : 402-418.

Kaplowitz,N., Aw, T and Ookhtens, M. (1985). The regulation of hepatic glutathione. Ann. Rev. Pharmacol. Toxicol., **25** : 715-744.

Kapoor, A. and Viraraghavan, T. (1995). Fungal biosorption-an alternative treatment option for heavy metal bearing waste water: a review. Bioresource. Technol., **53** : 195-206.

Kapoor, A. and Viraraghavan, T. (1997). Heavy metal biosorption sites in *A.niger*. Bioresource. Technol., **61** : 221-227.

Karamushka, V.I. and Gadd, G.M (1994). Influence of copper on proton efflux from *Saccharomyces cerevisiae* and the protective effect of calcium and magnesium. FEMS Microbiol. Lett., **122** : 33-38.

Karin, M. (1985). Metallothioneins : Proteins in search of function. Cell., **41** : 9-10.

Karin, M., Najarian, R., Haslinger, A., Valenzuela, P., Welch, J. and Fogel, S. (1984). Primary structure and transcription of an amplified genetic locus: The CUP1 locus of yeast PNAS (USA)., **81** : 337-341.

Kessels, B. G. F. Belde, P.J.M. and Borst-Pauwels, G.W.F.H. (1985). Protection of *Saccharomyces cerevisiae* against Cd^{2+} toxicity by Ca^{2+}. J. Gen. Microbiol., **131** : 2533-2537.

Kierans, M.,Staines, A.M., Bennet, H, and Gadd, G,M. (1991). Silver tolerance and accumulation in yeasts. Biology of Metals., **4** : 100-106.

Kitamoto, R., Yoshizawa, K, Ohsumi, Y and Anraku, Y. (1988). Mutants of *Saccharomyces cerevisiae* with defective vacuolar function J.Bacteriol., **170** : 2687-2691.

Klionsky, D.J. Herman, P.K. and Emr, S.D. (1990). The fungal vacuole: composition, function and biogenesis. Microbiol. Rev., **54** : 266-292.

Kneer, R., Kutchan, M.T., Hochberger, A. and Zenk, M.H. (1992). *Saccharomyces cerevisiae* and *Neurospora crassa* contain heavy metal sequestering phytochelatins. Arch. Microbiol., **157** : 305-310.

Konetzka, W.A. (1977). Microbiology of metal transformations. In: Weinberg E.D. Ed. Microorganisms and minerals, . New York: Marcel Dekker Inc., 317-342.x

Kumar, C.S., Sastry, K.S. and Maruthi Mohan, P. (1992). Use of wild type and nickel resistant *Neurospora crassa* for the removal of Ni^{2+} from aqueous medium. Biotechnol.Lett.14, 1109-1112.

Landner, L.. (1971). Biochemical model for the biological methylation of mercury suggested from methylation studies in vivo with *Neurospora crassa*. Nature, London, **230** : 452-453.

Lerch, K. (1980). Copper metallothionein, a copper binding protein from *Neurospora crassa*. Nature, London, **284** : 368-370.

Lerch, K. and Beltramini, M. (1983). *Neuropspora* copper metallothionein molecular structure and biological significance. Chemica. Scripta, **21** : 109-115.

Levine, W.B. and Marzluf, G.A. (1989). Isolation and characterization of cadmium resistant mutants of *Neurospora crassa*. Can. J. Microbiol., **35** : 359-365.

Levinskaite, L . (1999). Ekologija, **3** : 59-62.

Lichko, L.P. Okorokov, L.A. and Kulaev, I.S. (1982). Participation of vacuoles in regulation of levels of K^+, Mg^{2+} and orthophosphate ions in cytoplasm of the yeast *Saccahromyces carlsbergensis*, Arch. Microbiol., **132** : 289-293.

Lin, C.M. and Kosman, D. J. (1990). Copper uptake in wild type and copper metallothionein-deficient *Saccharomyces cerevisiae*. J. Biol. Chem., **265** : 9194-9200

Luckey, T.D. and Venugopal, B. (1977). Metal Toxicity in Mammals1, Plenum Press, New York, USA.

Macaskie, L.E. and Dean, A.C.R. (1990b). Trimethyl lead degradation by free and immobilized cells of an *Arthrobacter* sp. and by wood decay fungus *Phaeolus schweinitzii*. App. Microbiol. Biotechnol., **33** : 81-87.

Maruthi Mohan, P. (1998a). Biotechnology potential of fungi in metal bioremediation, Recent Trends in Mycoses, 165-169.

Maruthi Mohan, P. (1998b). Metal Toxicities and Resistance in Fungi. In: Microbes: For Health, Wealth and Sustainable Environment. Ed. A. Varma, Malhotra Publishing House, New Delhi, 433-465.

Maruthi Mohan, P. and Sivarama Sastry, K. (1983a). Interrelationships in trace element metabolism in metal toxicities in nickel resistant strains of *Neurospora crassa*. Biochem. J., **212** : 205-210

Maruthi Mohan, P. and Sivarama Sastry, K. (1984a). Cobalt transport in nickel resistant strains of *Neurospora crassa*. Current. Microbiol., **11** : 137-142.

Maruthi Mohan, P. and Sivarama Sastry, K. (1984b). Excretion of pyruvate in nickel toxicities in wild type and nickel resistant mutants of *Neurospora crassa*. J. Biosci., **6** : 283-288.

Maruthi Mohan, P., Prataparudra, M. and Sivaram Sastry, K. (1984c). Nickel transport in nickel resistant strains of *Neurospora crassa*. Curr. Microbiol., **10** : 125-128.

Marzluf, G.A. (1970). Genetic and metabolic controls for sulfate metabolism in *Neurospora crassa*. Isolation and study of chromate resistant and sulfate transport negative mutants. J. Bacteriol., **102** : 716-719.

Mehra, R. K. Tarbet, E.B., Gray, W.R and Winge, D.R. (1988). Metal specific synthesis of two metallothioneins and γ-glutamyl peptides in *Candida glabrata*. PNAS (USA)., **85** : 8815-8819.

Mehra, R.K., Thorvaldsen, J.L., Macreadie, I.G. and Winge, D.R. (1992a). Disruption analysis of metallothioneins encoding genes in *Candida glabrata*. Gene, **114** : 75-80.

Mehra, R.K. Thorvaldsen, J.L., Macreadie, I.G. and Winge, D.R. (1992b). Cloning system for *Candida glabrata* using elements from the metallothioneins IIa encoding gene that confer autonomous replication. Gene, **113** : 119-124.

Meister, A. and Anderson, M.E. (1983). Glutathione. Ann. Rev. Biochem., **52** : 711-760.

Meister, A. and Anderson, M.E. (1988). Glutathione metabolism and its selective modification. J. Biol. Chem., **263** : 17205-17208.

Miller, A.J. Vogg, G. and Sanders, D. (1990). Cytosolic calcium homeostasis in fungi: roles of plasma membrane transport and intracellular sequestration of calcium. PNAS (USA)., **87** : 9348-9352.

Minney, S.F. and Quirk, A.V. (1985). Growth and adaptation of *Saccharomyces cerevisiae* at different cadmium concentrations. Microbios., **42** : 37-44.

Mowll, J.L and Gadd, G.M. (1984).. Cadmium uptake by *Aureobasidium pullulans*. J. Gen. Microbiol., **130** : 279-284.

Mullen, M.D. Wolf, D. C. Beveridge, T.J. and Bailey, G.W. (1992). Sorption of heavy metals by the soil fungi *Aspergillus niger* and *Mucor rouxii*. Soil. Biol. Biochem., **24** : 129-135.

Munger, K. and Lerch, K. (1985). Copper metallothioneins from the fungus *Agaricus bisporus*: chemical and spectroscopic properties. Bichemistry **24**, 6751-6756.

Munger, K. Germann, N.A. and Lerch, K. (1985). Isolation and structural organization of the *Neurospora crassa* copper metallothionein gene. EMBO, **4** : 1459-1462.

Munger, K. Germann, N.A. and Lerch, K. (1987). *Neurospora crasssa* metallothioneins gene: regulation of expression and chromosomal location. J. Biol. Chem., **262** : 7363-7367.

Murasugi, A. Wada, C. and Hayashi, Y. (1981). Cadmium binding peptide induced in fission yeast *Schizosacchaomyces pombe*. J. Biochem. (Tokyo)., **90** : 1571-1574.

Murphy, R.J. and Levy, J.F. (1983). Production of copper oxalate by some copper tolerant fungi. Trans. Brit. Mycol. Soc., **81** : 165-168.

Mutoh, N. and Hayashi, Y. (1988). Isolation of mutants of *Schizosaccharomyces pombe* unable to synthesise cadystin, small cadmium binding peptides.Biochem. Biophys. Res. Comm., **151** : 30-39.

Mutoh, N. Kawabata, M. and Hayashi, Y. (1991). Tetramethylthiuram disulfide or dimethyldithiocarbamate induces the synthesis of cadystins, heavy metal chelating peptides, in *Schizosaccharomyces pombe*. Biochem. Biophys. Res. Comm., **176** : 1068-1073.

Narender Reddy, G. and Prasad, M.N.V. (1990). Heavy metal binding proteins/peptides: Occurrence, structure, synthesis and functions. A review. Environmental and Experimental Botany, **30** : 251-264.

Newby, P.H. and Gadd, G.M. (1986). Diffusion assays for determination of the effects of pH on the toxicity of heavy metal compounds towards filamentous fungi. Environ. Technol. Lett. 7, 391-396.

Nicholas, D.J.D. and Commissiong, K. (1957). Effects of molybdenum, copper and iron on some enzymes in *Neurospora crassa*. J. Gen. Microbiol., **17** : 699-707.

Nicholas, D.J.D., Nason, A. and McElroy, W.D. (1954). Molybdenum and nitrate reductase. Effect of molybdenum deficiency on *Neurospora* enzyme. J. Biol. Chem., **207** : 341-361.

Nies, D.H. (1992). Resistance to cadmium, cobalt and nickel in microbes. Plasmid, 27, 17-28.

Nieuwenhuis, B.J.W.M, Weijers, A.G.M. and Borst-Pauwels, G.W.F.H. (1981). Uptake and accumulation of $Mn^{2+}$and Sr^{2+} in *Saccharomyces cerevisiae*. Biochem. Biophys. Acta., **649** : 83-88.

Norris, P.R. and Kelly, D.P. (1977). Accumulation of cadmium and cobalt by *Saccharomyces cerevisiae*. J. Gen. Microbiol., **99** : 317-324.

Novick, R.P. (1967). Penicillinase plasmid of *Staphylococcus aureus*. Fed. Proc., **28** : 29-38.

Ochiai, E.I. (1987). General principles of biochemistry of the elements, New York, Plenum Press.

Ohsumi, Y. and Anraku, Y. (1983). Calcium transport driven by a proton motive force in vacuolar membrane vesicles of *Saccharomyces cerevisiae*. J. Biol. Chem., **41** :17-22.

Ohsumi, Y. Kitamoto, K. and Anraku, Y. (1988). Changes induced in the permeability barrier of the yeast plasma membrane by cupric ion. J. Bact., **170** : 2676-2682.

Okorokov, L.A., Lichko, L.P., Kodomtseva, V.M.Kholodenko, V.P., Titovsky, V.T., Lulaev, I.S. (1977). Energy dependent transport of manganese in to yeast cells and distribution of accumulated ions. Eur. J. Biochem., **75** : 373-377.

Padmanabhan, G. and Sharma, P.S. (1966). Cobalt toxicity and iron metabolism in *Neurospora crassa*. Biochem. J., **98** : 330-334.

Paknikar, K.M. and Bhide, J.V. (1993). Aerobic reduction and biosorption of chromium by a chromate resistant *Aspergillus* sp. In: Biohydrometallurgical Technologies, Ed. A.E. Torma, A.E., Apel, M.E. and Brierley, C.E., 237-244.

Passow, H. and Rothstein, A. (1959). Binding of mercury by the yeast cell in relation to the changes in permeability. J. Gen. Physiol., **43** : 621-633.

Pearson, R.G. (1963). Hard and soft acids and bases. J. Amer. Chem. Soc. 84, 3533-3539.

Pepi, M and Baldi, F. (1992).. Modulation of chromium (VI). toxicity by organic and inorganic sulfur species in yeasts from industrial wastes. Biometals, **5** : 179-185.

Phelan, A., Thurman, D.A. and Tomsett., A.B. (1990). The isolation and characterization of copper resistant mutants of *Aspergillus nidulans*. Curr. Microbiol., **21** : 255-260.

Ramamoorthy, S. and Kushner, D.J. (1975). Binding of mercuric and other heavy metals by microbial growth media. Microbiol. Ecol., **2** : 167-176.

Rauser, W.E. (1990). Phytochelatins. Ann. Rev. Biochem., **59** : 61-86.

Raymond, K.N., Muller, G. and Matzanke, B.F. (1984). Complexation of iron by siderophores, A review of their solution and structural chemistry and biological function. Topics. Curr. Chem., **123** : 49-102.

Reese, R.N. and Wagner, G.J. (1987). Effects of buthionine sulfximine on cadmium binding peptide levels in suspension culture of tobacco cells treated with cadmium, zinc or copper. Plant. Physiol., **84** : 574-577.

Reese, R.N., Mehra, R.K., Tarbet, E.B. and Winge, D.R. (1988). Studies on the γ-glutamyl Cu-binding peptide from *Scizosaccharomyces pombe*. J. Biol. Chem., **263** : 4186-4192.

Robinson, N.J., Tommey, A.M., Kuske, C. and Jackson, P.J. (1993). Plant metallothioneins. Biochem. J., **295** : 97-106.

Rothstein, A., Meier, R. and Hurwitz, L. (1951). The relationship of the cell surface to metabolism. J. Cell. Comp. Physiol., **37** : 57-62.

Sabie, F.T. and Gadd, G. M. (1990). The effect of zinc on the yeast mycelium transition of *Candida albicans* and examination of zinc uptake at different stages of growth. Mycol. Res., **94** : 952-958.

Sanders, D., (1990). Kinetic modeling of plant and fungal membrane transport systems. Ann. Rev. Plant. Physiol. Plant Mol. Biol., **41** : 77-107.

Seeta Laxman, R. (2000). Fungi toxicity of Metals *In* : Proc. Int. Symp. Processing of Chemical and Metallurgical Industries Wastes, Bhubaneshwar, 90-106.

Silver, S. (1992). Plasmid determined metal resistance mechanisms. Range and overview. Plasmid, **27** : 1-3.

Silver, S., Budd, K., Leathy, K.M., Sharo, W.V. and Hammond, D. (1981). Indusible plasmid determined resistance to arsenite and antimony (III). in *E. coli* and *Staphylococcus aureus*. J. Bacteriol., **146** : 983-996.

Singhal, R.K., Anderson, M.E. and Meister, A. (1987). Glutathione, first line of defence against cadmium toxicity FASEB. J., **1** : 220-223.

Sivarama Sastry, K. (1963). Interrelationships in trace element metabolism. In Adv. Biochem. Proc. First. Summer School in Biochemistry, Srinagar, India, 323-330.

Sivarama Sastry, K. Ramaiah, A. and Sarma, P.S. (1958). Influence of molybdenum toxicity on sulfur amino acid metabolism in *Neurospora crassa*. Biochem. Biophys. Acta., **30** : 438-442.

Sivarama Sastry, K. Adiga, P.R., Venkatsubramanium, V. and Sarma, P.S. (1962). Interrelationships in trace element metabolism in metal toxicities in *Neurospora crassa*. Biochem. J., **85** : 486-491.

Smith, D.G. (1974). Tellurite reduction in *Schizosaccharomyces pombe*. J. Gen. Microbiol., **83** : 389-392.

Srinivasa Rao, , Subramayam, C. and Venkateswarulu, G.(1984). Nitrogen metabolism in the blue mycelia of *Neurospora crassa* isolated from copper toxic cultures Curr. Microbiol., **10** : 79-84.

Startling, A.P. and Ross, I.S. (1990). Uptake of manganese by *Penicillium notatum*. Microbios., **63**: 93-100.

Steffens, J.C. (1990). The heavy metal-binding peptides of plants. Ann. Rev. Plant. Physiol. and Plant. Mol. Biol., **41** : 553-575.

Strandberg, G.W., Shumate, S.E. and Parrott, J.R. (1981). Microbial cells as biosorbents for heavy metals: accumulation of uranium by *Saccharomyces cerevisiae* and *Pseudomonas aeruginosa*. Appl. Environ. Microbiol., **41** : 237-245.

Subramanian, K.N., and Sarma, P.S. (1968). Effect of nitrogen source on cobalt toxicity in *Neurospora crassa*. Biochem. Biophys. Acta., **156** :199-202.

Subramanyam, C. and Venkateswarulu,G. (1979a). The effect of copper on histidine biosynthesis in *Neurospora crassa*. J. Biosci., **1** : 279-283.

Subramanyam, C. and Venkateswarulu, G. (1979b). Effect of copper and manganese on free amino acid contents of *Neurospora crassa*. Neurospora Newsletter, 26, 18.

Subramanyam, C., Venkateswarulu, G. and Rao, S.L.N. (1983a). Cell wall composition of *Neurospora crassa* under conditions of copper toxicity Appl. Environ. Microbiol., **46** : 585-590.

Subramanyam, C., Venkateswarulu, G. and Rao, S.L.N. (1983b). Accumulation of histidinol phosphate in copper toxic cultures of *Neurospora crassa*. Current. Sci., **52** : 879-880.

Suresh, K. and Subramanyam, C. (1992). Trypsin inhibitory activity in *Neurospora crassa*: A correlative study in relation to chitin synthesis in metal toxicity 61st Annual Meeting of Society of Biological Chemists, India, Hyderabad, Abstract 292.

Tezuka, T. and Takasaki, Y. (1988). Biodegradation of phenylmercuric acetate by organomercury resistant *Penicillium sp*. MR-2. Agric. Biol. Chem., **52** : 3183-3185.

Thayer, J.S. (1984). Organometallic compounds and living organisms. New York, Academic Press.

Thayer, J.S. (1988). Organometallic chemistry an overview. Weinheim: VCH, Verlagsgesellschaft.

Theuvenet, A.P.R., Nieuwenhuis, B.J.W.M, Van de Mortel, J. and Borst-Pauwels, G.W.F.H. (1986). Effect of ethidium bromide and DEAE-dextran on divalent cation accumulation in yeast. Evidence for an ion-exchange extrusion pump for divalent cations. Biochim. Biophys. Acta., **855** : 383-390.

Tobin, J.M., Cooper, D.G and Neufeld, R.J. (1984). Uptake of metal ions by *Rhizopus arrhizus* biomass. Appl. Environ. Microbiol., **47** : 821-824.

Tobin, J.M., Cooper, D.G. and Neufeld, R.J. (1990). Investigation of the mechanism of metal uptake by denatured *Rhizopus arrhizus* biomass. Enz. Microbiol. Technol., **12** : 591-595.

Tohoyama, H., Inagawa, A., Koike, H., Inouhe, M., Joho, M. and Murayama, T. (1992). Constitutive expression of the gene for metallothionein in a cadmium resistant yeast. FEMS Microbiol. Lett., **95** : 81-86.

Tomsett, A.B. Hodges, K.E., Cooley, R.N. and Thurman, D.A. (1989). Metal resistance in *Aspergillus nidulans* In: Metal ion Homeostasis- Molecular Biology and Chemistry, Eds. Hamer, D.H. and Winge, D.R., Alan, R. Liss, New York, 375-384.

Valix, M., Tang, J.Y and Malik, R. (2001). Heavy metal tolerance in fungi. Miner. Engg. 14, 499-505.

Veliky, E. and Stefanec, J. (1964). Der einflub von kobalt auf die mehrung von *Saccharomyces cerevisiae*. Naturwissenschften, **51** : 518-519.

Venakata Ramana, V. and Sivarama Sastry, K. (1993). Chromium-uric acid complex as growth substrates and inducers of uricase in *Neurospora crassa*. J. Inorg. Biochem., **50** : 107-117.

Venakata Ramana, V. and Sivarama Sastry, K. (1994). Chromium toxicity in *Neurospora crassa*. J. Inorg. Biochem., **56** : 87-95.

Venkateswarulu, G. and Sivarama Sastry, K. (1973). Inter relationship in trace element metabolism in metal toxicities in a cobalt resistant strain of *Neurospora crassa*. Biochem. J., **132** : 673-677.

Venkateswarulu, G. and Sivarama Sastry, K. (1979a). Cobalt transport in a cobalt resistant strain of *Neurospora crassa* . J. Biosci., **1** : 433-438.

Venkateswarulu, G. and Sivarama Sastry, K. (1979b). Studies on copper toxicity in *Neurospora crassa*. Indian. J. Biochem. Biphys., **16** : 84-87.

Venkateswarulu, G. and Stotzky, G. (1989). Binding of metals by cell walls of *Cunnighamella blakesleeana* grown in presence of copper and cobalt. Appl. Microbio.l. Biotechnol., **31** : 619-625.

Venkateswarulu, G. Yoder, M.J. and Stotzky, G. (1989). Morphological, ultrastructural and chemical changes induced in *Cunninghamella blakesleeana* by copper and cobalt. Appl. Microbiol. Biotechnol., **31** : 204-210.

Volesky, B. (2001). Detoxification of metal bearing effluents: biosorption for the next century. Hydrometallurgy, **59** : 203-216.

Vonk, J.W. and Sijpesteijn, A.K. (1973). Studies on the methylation of mercuric chloride by pure cultures of bacteria and fungi. Ann. van. Leewenhock., **39** : 505-513.

Wakatsuki, T. Iba, A. and Imahara, H. (1988). Copper reduction by yeast cell wall materials and its role on copper uptake in *Debaryomyces hansenii*. J. Ferment. Technol., **66** :257-265.

Wakatsuki, T. Hayakawa, S., Hatayama, T., Kitamura, T. and Imahara, H. (1991a). Solubilization and properties of copper reducing enzyme system from the yeast cell surface in *Debaryomyces hansenii*. J. Ferment. Bioengg., **72** : 79-86.

Wakatsuki, T. Hayakawa, S., Hatayama, T., Kitamura, T. and Imahara, H. (1991b). Purification and some properties of copper reductase from cell surface of *Debaryomyces hansenii*. J. Ferment. Bioengg., **72** :158-161.

Weiss, A.A., Murphy, S.D. and Silver, S. (1977). Mercury and organomercurial resistances determined by plasmids in *Staphylococcus aureus.* J. Bacteriol., **132** : 197-208.

Welch, J.W., Fogel, S., Catala, G. and Karin, M. (1983). Industrial yeasts display tandem gene iteration at the CUP 1 region Mol. Cell. Biol., **8** : 1353-1361.

White, C. and Gadd, G.M. (1986). Uptake and cellular distribution of copper, cobalt and cadmium in strains of *Saccharomyces cerevisiae* cultured on elevated concentrations of these metals. FEMS Microbiol. Ecol., **38** : 277-283.

White, C. and Gadd, G.M. (1987a). The uptake and cellular distribution of zinc in *Saccharomyces cerevisiae*. J. Gen. Microbiol., **133** : 727-737.

White, C. and Gadd, G.M. (1987b). Inhibition of H^+ ion efflux and K^+ uptake and induction of K^+ efflux in yeast by heavy metals. Toxicity Assessment, **2** : 437-444.

Wiemken, A., Schellenberg, M. and Urech, K. (1979). Vacuoles: the sole compartments of digestive enzymes in yeast (*Saccharomyces cerevisiae*).. Arch. Microbiol., **123** : 23-35.

Winge, D.R., Nielson, K.B., Gray and W.R., Hamer, D.H. (1985). Yeast metallothionein-sequence and metal binding properties. J. Biol. Chem., **260** : 14464-14470.

Winge, D.R., Reese, R.N., Mehra, R.K., Tarbet, E.B., Hughes, A.K. and Dameron, C.T. (1989). Structural aspects of metal γ–glutamyl peptides. In: Hamer, D.M., Winge, D.R. Eds. Metal ion homeostasis: molecular biology and chemistry. New York: Alan R. Liss Inc, 301-311.

Winkelmann, G. Van der Helm, D. and Neilands, J.B. (1987). Iron transport in microbes, plants and animals. Weinheim: VCR Verlagsgesellschaft.

Winkelmann, G.(1992). Structures and functions of fungal siderophores containing hydroxamate and complexone type iron binding ligands. Mycol. Res., **96** : 529-534.

Wood, J.M. and Wang, H.K. (1983). Microbial resistance to heavy metals. Environ. Sci. Technol., **17** : 582-590.

Yannai, S., Berdicevsky, I. and Duek, L. (1991). Transformations of inorganic mercury by *Candida albicans* and *Saccharomyces cerevisiae*. Appl. Environ. Microbiol., **57** : 245-247.

Zunino, H. and Martin, J.P. (1977). Metal binding organic macromolecules in soil. I. Hypothesis interpreting the role of soil organic matter in the translocation of metal ions from rocks to biological systems. Soil. Science., **123** : 65-76.

7

Modern Trends in Ergot Alkaloid Biosynthesis

P. Mehta * and A. Mehta

Department of Botany, Dr. H.S.Gour University, Saugor - 470 003

The ergot alkaloids are produced by a wide range of fungi primarily by species of *Claviceps*, a fungal parasite of grasses and cereals. The ergot alkaloids are considered as potential therapeutic agents in many important diseases. Ergot alkaloids are indole compounds formed from tryptophan, mevalonic acid and methionine. Chemically manipulated ergot derivaties possess altered pharmacologic effects. Saprophytic production of these alkaloids is more economical than parasitic cultivation on rye plants as in bioreactors the process is precisely controlled and secretion of alkaloids can take place on the same site. A few new natural alkaloids are known to have biogenetic importance. It is possible to understand the genetic control of alkaloid production and to find the genes responsible for specific alkaloids. Ergot alkaloid synthesis requires a change in differentiation which is considered to be a result of different expression of the genome under various conditions and at different times. Cultural conditions suitable for metabolism of primary substances are not suitable for alkaloid metabolism. The region of formation of ergot alkaloids is not generally the same as where alkaloids are found in increased concentrations. Ergot alkaloids are not waste physiological inert products of cell metabolism.

Key words : Ergot alkaloids, *Claviceps*, Chemistry, Biosynthesis, Clavines, Lysergic acid, Ergopeptines, Physiologial formation, Differentiation, Cultural condtions, Physiological role

The ergot alkaloids have many potential therapeutic uses some of which have been realized and others that have not survived the critical test of time. The ergot alkaloids have been of prime interest to pharmacologists due to their multiple actions which include effects on utrine and vascular muscle, and for their use as a tool for studying the mechanism of the sympathetic nervous system. Within the last few years newer aspects of the pharmacology of the ergot derivatives have been recognized (Rehacek and Mehta, 1993). One involves the action of certain ergot derivatives as inhibitors of prolactin secretion. In addition, new ergot compounds have been shown to produce stimulation of dopaminergic receptors and have been utilized effectively in the treatment of parkinsonism. This way, the newer ergot derivatives have been implicated as potential therapeutic agents in many

* Corresponding author E.mail: pradeepunisagar@hotmail.com

diseases (parkinsonism, acromegaly, amenorrhea-galactorrhea, suppression of postpartum lactation, treatment of breast cancer and possibly cancer of prostrate). Preparations containing ergot alkaloids or their semi-synthetic derivatives are also applied for the treatment of migrain (ergotamine), high blood pressure (dihydrocornine, dihydrokryptine). Ingestion of isoergine and lysergic acid diethylamide (LSD) are known to cause marked hallucinations.

Since the supply of natural grown ergots is not sufficient, the biotechnological production of ergot alkaloids has gained importance. The microbial technology of ergot alkaloids has two approaches: (i) the acquisition of knowledge for producing microorganism, and (ii) the development of large scale processes to exploit such knowledge. The current developments in biotechnology have opened up new perspectives in the production of ergot alkaloids. The characterization of enzymes catalysing metabolic conversions has been aided by the introduction of new separation techniques. In the last few years, progress has taken place in the description of enzymes involved in ergot alkaloid metabolism, followed by cloning efforts and the elucidation of primary structures. Structural genes and their control elements may then be manipulated to improve the production and properties of ergot alkaloids.

History

Ergot was formerly defined as the dried sclerotium of *Claviceps purpurea* (Fries) Tulasne developed on plants of rye, *Secale cereale* Linn. The ergot alkaloids are a group of pharamacologically active compounds produced primarily by species of *Claviceps*, a fungal pasrasite of grasses and cereals. The role of ergot (*Secale cornutum*), sclerotium of the pyrenomycete *Claviceps purpurea* (Fr.) Tul has undergone important changes from a dreaded toxic parasite on rye to an important source of biologically effective substances. In early days the medieval midwives used to collect the fungus from naturally infected plants and used it in the induction of child birth and in the control of postpartum bleeding. Many additional uses of the ergot alkaloids have been discovered since then. Toxic effects of ergot compounds have been recognized since the discovery that consumption of infected cereals infected by parasitic fungus, *Claviceps* could cause severe illness (ergotism). The disease is due to the effects of alkaloids produced by the fungus and contained in the ergot body. The ergot alkaloids can be used as drugs at low concentrations. It is of particular interest that even before the ergot was identified as a cause of ergotism, it was evidently used (ergometrine) in obstetrics, to enhance muscle contraction and was also used for abortions. Hence, *Claviceps purpurea* is apart from yeast, the first fungus which was biotechnologically exploited without its existence being known. However, because during these early days of medicinal use, neither its physiological action nor the appropriate dosage was understood, many deaths were brought about by the ergot extracts. At the end of 19th century the ergot was banned as a drug in Germany. Later, with increasing understanding of pharmacology, it was found that chemical alterations of various alkaloids improved their medical efficiency. Further, after a correlation between clinical indications and therapeutical dosage was developed, pure ergot alkaloids and their semisynthetic derivatives are being used extensively in medicine, at present.

The beginning of modern ergot research dates back to the isolation of the first crystalline alkaloid (ergotinine) in 1875, to the major pharmacological discoveries of ergotoxine in 1906 and to the discovery of first clinically useful compound, ergotamine, in 1918. At the beginning of fifties, the biosynthesis of ergot alkaloids received much attention, from a subject of fruitful speculation to the region of experimental research. Initially, chemical properties of the alkaloids were studied, while biologists studied the physiological aspects of the synthesis of alkaloids from a practical point of view, i.e. for both agronomists who were trying to obtain grain with lowest possible alkaloid content and for pharmacologists, who on the other hand were interested in high alkaloid yields. The dependence of pharmacological effectiveness on the alkaloid chemical structure, enabling preparation of substances for specific purposes was an important stimulus for the development of ergot alkaloid research.

Sources and Production of Ergot Alkaloids

In nature, the ergot alkaloids are formed primarily by the fungus *Claviceps*. More than 50 different alkaloids have been isolated from this ascomycetous genus which has about 50 species. All *Claviceps* species form ergot sclerotia on various plants. Atleast another 100 plants can be attacked by ergot to form the sphacelial but not the sclerotial stage of the fungus (Rehacek and Sajdl, 1990). Ergot alkaloids have also been found in organisms other than *Claviceps* (Table 1).

Table 1. Ergot alkaloids outside *Claviceps* (Rehacek and Sajdl, 1990)

	Organism	Alkaloid
Fungi imperfectii	*Aspergillus clavatus*	clavine type
	A. fumigatus	clavine type
	A.nidulans	clavine type
	Balansia sp.	clavine type
	Penicillium chermesinum	costaclavine
	P. concavorugulosum	chanoclavine-I

		rugulovasine A,B
	P. fumigatus	chanoclavine
		agroclavine
		elymoclavine
		festuclavine
		fumigaclavine A, B, C
	P. roqueforti	festuclavine
		isofumigaclavine A
		roquefortine A, B, C
	Sphacelia sorghi	dihydroergosine
	Hypomyces aurantius	ergokryptine
		ergocryptinine
		agroclavine
		elymoclavine
Phycomycetes	*Phycomyces nitens*	clavine type
	Rhizopus arrhizus	clavine type
Convolvulaceae	*Argyreia nervosa*	clavine type
		ergine
		ergometrine
		lysergic acid 2-hydroxyethylamide
	Ipomoea argyrophylla	ergosine
	I. hildebrandtii	cyclocalvine
		festuclavine
	I. ruberosa	ergine
		penniclavine
	I. violacea	chanoclavine
		cycloclavine
		elymoclavine
		lysergol
		ergine
		lysergic acid 2-hydroxyethylamide
		ergonovine
		ergosine
	Rivea corymbosa	clanoclavine
		cycloclavine
		elymoclavine
		lysergol
		ergine
		lysergic acid 2-hydroxyethylamide
		ergonovine
		ergosine
	Stictocardia sp.	clavine type

Table 2. Industrial submerged production (s) of ergot alkaloids (Rehacek, 1983)

Alkaloid	Company			
	A	B	C	D
Ergocornine	s	s	s	s
Ergocrystine	s	s	-	s
Ergokryptine	s	s	s	s
Ergotamine	-	s	s	-
Ergometrine	-	-	s	-
Lysergic acid	-	s	-	-
Paspalic acid	s	-	-	s

A Biochemie GmbH, Kundl/Tirol, Austria
B Fermitalia S.p.A., Milano, Italy
C Gedeon Richter, Budapest, Hungary
D Sandoz-Wander AG, Basal, Switzerland

Ergot alkaloids can be obtained in 3 basic ways: (i) by isolation from the sclerotia of *Claviceps purpurea* collected from the field (95% of the world production), (ii) by extraction of saprophytic cultures of other ascomycetes, especially of the *Claviceps* genus (Table 2), and (iii) by partial/ total synthesis. Sclerotia of *Claviceps purpurea* collected from rye may contain 0.8% peptide alkaloids on dry weight basis. This content would be the equivalent of about 500 mg / Lalkaloid level in a fermentation broth of 6% mycelial dry weight (Udwardy, 1980). Since the supply of naturally grown ergot is not sufficient as its field production is labour intensive and weather dependent, the biotechnological production of ergot alkaloids has gained importance. In some cases, it appears that the saprophytic production of ergot alkaloids is more economical than the parasitic cultivation on rye plants. In bioreactors, alkaloid yield may exceed 20% of the mycelial dry mass, while natural sclerotia yield is normally less than 1% or even less than 0.1% in some cases. Additionally, in bioreactors, the process is more precisely controlled, the cultivation is not limited to the specific season and the production and isolation of alkaloids can take place on the same site. Furthermore, it is possible to apply some manipulations which are not practicable under field conditions and above all clavines and simple lysergic acid derivatives are easier to manufacture than ergopeptines. Submerged large scale production of ergot alkaloids has been confined mainly to the commercially important simple derivatives of lysergic acid. Presently, the need for large quantities of some clavines increased the attention to be paid for the large scale fermentation process.

Chemistry

The ergot alkaloids constitute the largest known group of nitrogenous fungal metabolites. The common part of chemical structure of ergot alkaloids is the tetracyclic ergoling ring system (Fig. 1). Ergoline, a common part of ergot alkaloids is formed (Fig. 2) from tryptophan (Taber and Vining, 1959), mevalonic acid (Groger *et al.*, 1961) and methionine (Baxter *et al.*, 1961).

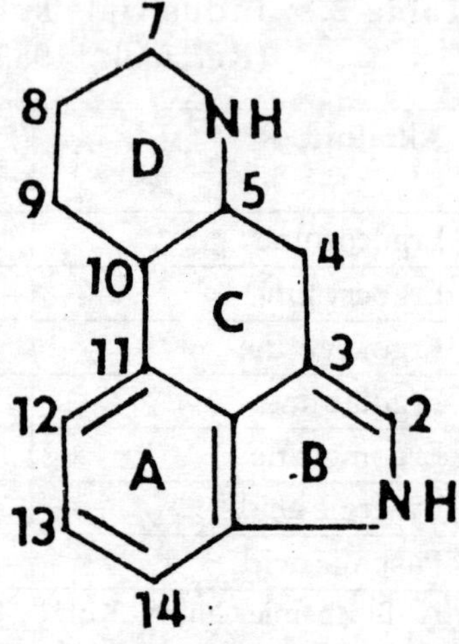

Fig. 1 Ergoline

Classification

The ergot alkaloids can be separated into four main structural groups: (i) clavine alkaloids, (ii) lysergic acids, (iii) simple lysergic acid amides and (iv) peptide alkaloids (Fig. 3).

The ergot alkaloids can be chemically manipulated to form derivatives that possess altered pharmacologic effects. Recently, a new group of alkaloids has been reported which interferes with DNA synthesis in mammalian cell (Eich *et al.*, 1981). Many laboratories have got success in isolating strains of *Claviceps purpurea* and their cultivation under submerged conditions to produce ergot peptide alkaloids (ergopeptine), simple derivatives of lysergic acid and clavine alkaloids (Vorisek and Rehacek, 1976).

The ergot alkaloids can be chemically manipulated to form derivatives that possess altered pharmacologic effects. Recently, a new group of alkaloids has been reported which interferes with DNA syntheis in mammalin cell (Eich *et al.*, 1981). Many laboratories have got success in isolating strains of *Claviceps purpurea* and their cultivation under submerged conditions to produce ergot peptide alkaloids (ergopeptine), simple derivatives of lysergic acid and clavine alkaloids (Vorisek and Rehacek, 1976).

1. *Clavines*

The clavines are (Fig. 4) substituted 6,8-dimethylergolines and include a few members, namely the chanoclavines, with a 6,7-seco D-ring. The degree of oxidation is a criterion for further differentiation in the group of clavine alkaloids. Some examples of clavine alkaloids are as follows:

1. Chanoclavine- I (secaclavine)
2. Chanoclavine-II
3. Isochanoclavine-I

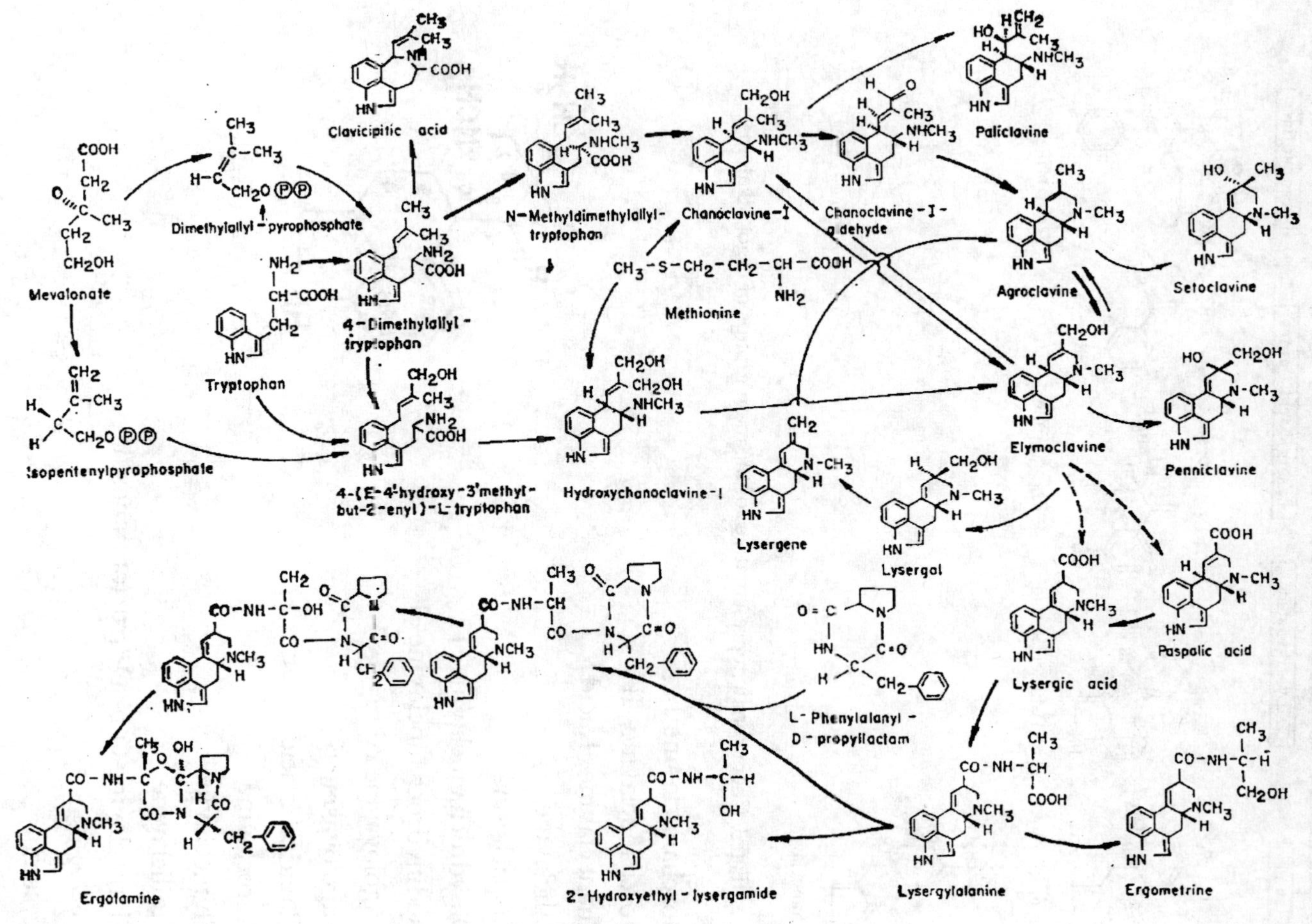

Fig. 2. Biosynthesis of clavine alkaloids

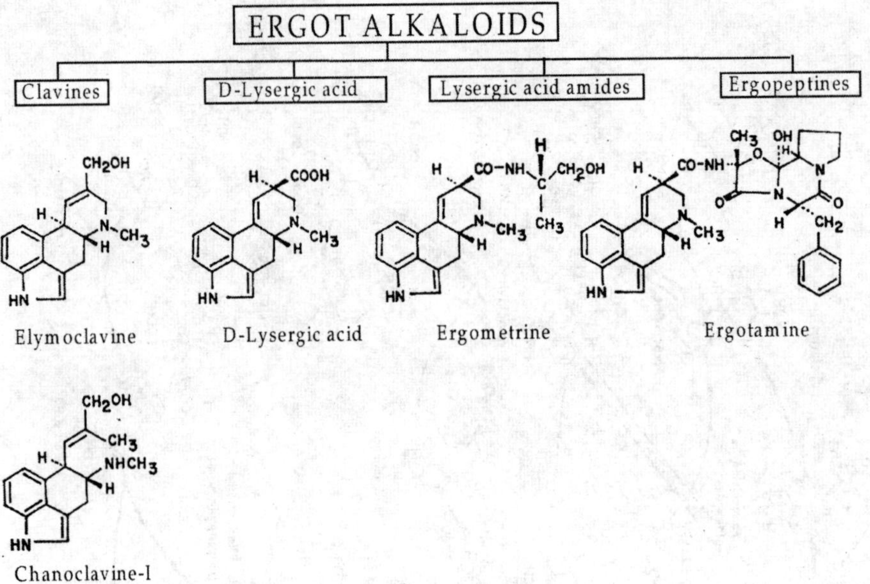

Fig. 3 Characterstic representatives of the main groups of ergot alkaloids

4. Norchanoclavine-I
5. Norchanoclavine- II
6. Chanoclavine- I acid
7. Paliclavine
8. Paspaclavine
9. Dihydrochanoclavine- I
10. Isodihydrochanoclavine- I
11. Rugolovasine A
12. Rugolovasine B
13. Clavicipitic acid
14. Agroclavine
15. Elymoclavine
16. Molliclavine
17. Elymoclavine-O-b- D- fructofuranoside
18. Lysergine
19. Lysergol
20. Isolysergol
21. Lysergene

Fig. 4. Clavine

22. Setoclavine
23. Isosetoclavine
24. Norsetoclavine
25. Penniclavine
26. Isopenniclavine
27. Festuclavine
28. Costaclavine
29. Pyroclavine
30. Dihydrolysergol- I
31. Dihydrosetoclavine
32. Fumigaclavine A
33. Fumigaclavine B
34. Roquefortine A (Isofumigaclavine A)
35. Roquefortine B (Isofumigaclavine B)
36. Cytoclavine
37. Aurantioclavine

2. *Lysergic acid and its simple derivatives*

The derivatives of lysergic acid are mostly amides, in which the amide part is a small peptide or a simple alkylamide. Non peptide amides of lysergic acid found in ergot are ergometrine, lysergic acid 2-hydroxyethylamide, lysergic acid amide (ergine) and paspalic acid (Fig. 5). Ergometrine (ergonovine and ergobasine) was also isolated from many laboratories. *Claviceps paspali* could also produce paspalic acid. Some examples of lysergic acid derivatives are:

1. Ergometrine (ergonovine, ergobasine)

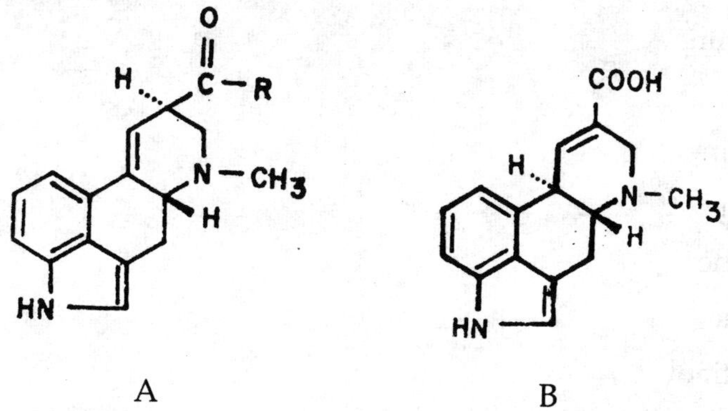

Fig. 5 Simple derivatives of lysergic acid (A) and paspalic acid (B)

2. Lysergic acid 2- hydroxyethylamide
3. Lysergic acid amide (ergine)
4. Lysergic acid

Name	R
Ergometrine (ergonovine, ergobasine)	$-HN-CH(CH_3)-CH_2OH$
Lysergic acid 2-hydroxyethylamide	$-HN-CH(CH_3)-OH$
Lysergic acid amide (ergine)	$-NH_2$
Lysergic acid	-OH

3. *Ergopeptines*

Ergotamine was the first alkaloid to be isolated from ergot crop and became commonly used compound for therapeutics. Ergotamine and other classical peptide ergot alkaloids (ergopeptines) are composed of lysergic acid and an L- proline containing complex tripeptide moiety. Ergopeptines are the only natural substances shown to have cyclol structure. Some examples are shown below:

1. Ergotamine
2. Ergosine
3. Ergovaline
4. Ergostine
5. Ergoptine
6. Ergonine
7. Ergobutine
8. Ergokrystine

9. α -Ergocryptine
10. α-Ergokryptine
11. Ergocornine
12. Ergobutyrine

New Alkaloids

Relatively few new natural alkaloids have been described recently (Table 3). Many of them have only biogenetic importance. Very recently, a fungal loline alkaloid which contributes to protection from aphids has been reported from a grass endophyte, *Neotyphodium uncinatum* (Wilkinson *et al.*, 2000).

Table-3. Some new natural ergot alkaloids (Rehacek and Sajdl, 1990) :

Alkaloid	Source
Chanoclavine- II	*Claviceps* sp.
Isochanoclavine- I	*Claviceps* sp.
Aurantioclavine- I	*Penicillium aurantiovirens*
Cycloclavine	*Ipomoea hildebrandtii*
	Aspergillus japanicum
Cividiclavine	*Penicillium citreoviride*
Dyhidrosetoclavine	*Claviceps paspali*
6- Norsetoclavine	*Claviceps* sp.
Elymoclavine- O- β -D fructofuranoside	*Claviceps* sp. SD 58
Epoxyagroclavine	*Penicillium corylophium*
Paliclavine	*C. paspali*
Paspaclavine	*C. paspali*
Clavicipitic acid	*C. fusiformis*
Ergoheptine	*Claviceps* sp.
Ergohexine	*Claviceps* sp.
Paspalic acid	*C. paspali*
Paspalicine	*C. paspali*
Paspaline	*C. paspali*
Ergocornom	*C. purpurea*
Ergobutine	*C. purpurea*
Ergobutyrine	*C. purpurea*
8- Hydroxyergotamine	*C.purpurea*
Ergosecaline	*C.purpurea*
Ergosine	*Ipomoea argyrophylla*
Dihydroergosine	*Sphacelia sorghi*
Ergostine	*C.purpurea*
Roquefortine A, B	*Penicillium roqueforti*
Rugulovasine A, B	*P.concavrugulosum*

Biosynthesis

Alkaloids can be formed into two ways, otherwise regulation of their formation is not possible. Ergot alkaloid is derived from tryptophan, mevalonic acid and methionine (Birch *et al.*, 1960 ; Groger *et al.*, 1960 ; Taylor and Ramsted, 1960). The treatment of these precursors depends on the appropriate genes being present in the cell. Ergot alkaloid biosynthesis genes are found to be present in *Claviceps* and *Neotyphodium* (Wang *et al.*, 1998). Tryptophan being the direct precursor of ergot alkaloids, is also a factor in the induction and depression of enzymes (Fig. 6.) catalysing alkaloid formation (Floss and Mothes, 1964 ; Vining, 1970 ; Krupinski *et al.*, 1976). The exogenous tryptophan consisting of depressed tryptophan synthesis removes inhibition of alkaloid production by phosphate (Robbers *et al.*, 1972). Tryptophan has a favourable effect on the biosynthesis of alkaloids as it decreases the content of cell protein (Rehacek and Malik, 1971). *Claviceps purpurea* has a gene which encodes dimethylallyltryptophan synthase for ergot alkaloid biosynthesis (Tsai *et al.*, 1995). The production of clavine alkaloids is normally accompanied by a decrease in the free energy and a tendency to achieve higher entropy (Rehacek *et al.*, 1973), while during biosynthesis of derivatives of lysergic acid, the free energy increases and entropy decreases. This free energy balance may show some relationship in explaining the frequent occurrence of clavine and lysergic acid derivatives and relatively rare occurrence of peptide alkaloids in *Claviceps*. Arntz and Tudzynski (1997) identified the genes which induce the alkaloid synthesis in *Claviceps* cultures. Ergopeptine biosynthesis well associates with *Claviceps purpurea* peptide synthetase gene (Annis and Panaccione, 1998). Tudzynski *et al.* (1999) gave an evidence for an ergot alkaloid gene which clusters in *C. purpurea.*

Clavine alkaloids lie at the beginning of the biogenetic order and classic ergot peptide alkaloids at the end. The oxidation order agroclavine – elymoclavine- lysergic acid is the main biogenetic pathway (Fig. 2). A variety of clavine alkaloids is formed due to action of various enzymes which direct elymoclavine and agroclavine from the main biosynthetic route. Such type of shunt activity is particularly evident in strains that lack the complete pathway and are unable to synthesize subsituted lysergic acid (Vining, 1980). The synthesis of the peptide alkaloids takes place by mediation of a multi-enzyme complex. The activating enzyme and the enzymes that catalyse synthesis of the peptide part can tolerate some change in the lysergic acid structure.

Physiological formation

Thorough research in the field of ergot alkaloid biosynthesis became possible only after the successful development of techniques for alklaloid

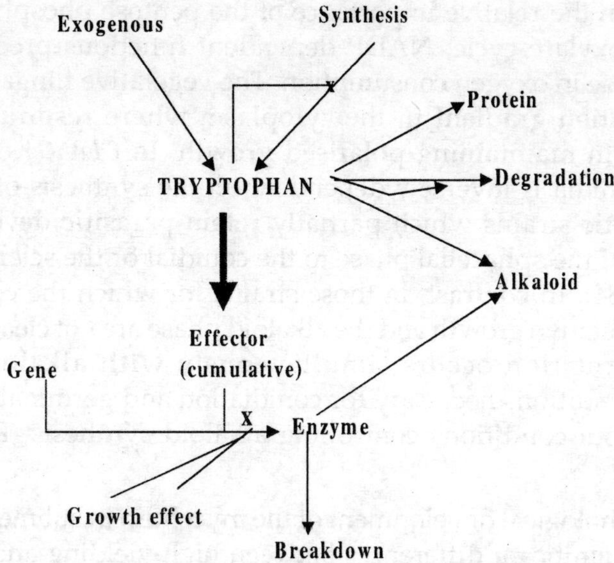

Fig. 6 Role of tryptophan in ergot alkaloid procduction (Rehacek, 1983)

production in saprophytic culture. Metabolic pathways involved in the synthesis have been studied in detail while physiological formation of ergot alkaloids have been worked out negligibly. The biosynthesis of ergot alkaloids is correlated with distinct genotypes (Esser and Tudzynski, 1979). Ergot alkaloids are produced under conditions of substrate limitation other than by carbon. Most of the enzymes and co-enzymes of basic metabolism and alkaloid biosynthesis are common. In submerged culture, alkaloid synthesis takes place under conditions of decreased proliferation and clearly defined degradation and resynthesis of nucleic acids and proteins (Erge *et al.*, 1972; Rehacek *et al.*, 1971, 1972). While Garre *et al.* (1998) isolated an extracellular catalase, Tenberge *et al.* (1999) could localize an extracellular β-1, 3 glucanase in *C. purpurea* during infection on rye. Recently, Sato *et al.* (2001) have studied the metabolic engineering of plant alkaloid biosynthesis. Jungehulsing and Tudzynski (1997) analysed the genetic diversity in *C. purpurea* using RAPD markers.

Differentiation

Ergot alkaloid synthesis requires a change in differentiation which is considered to be a result of different expression of the genome under various conditions and at different times (Turian, 1969). Morphological differentiation is preceded by synthesis of membrane, i.e. protein and lipids. Normally, the metabolism of the early sequences of fungal differentiation is characterized by an increase in the relative importance of the pentose phosphate route and that of the glyoxylate cycle, NADP dependent functions predominate and there is a decrease in oxygen consumption. The vegetative fungal cells initially form a polarisation gradient in the cytoplasm where respiration plays an important role in maintaining polarised growth. In *Claviceps purpurea*, the formation of conidia is inversely dependent on the synthesis of alkaloids for those saprophytic strains which partially retain parasitic development, i.e. differentiation of the sphacelial phase to the conidial or the sclerotial phase or both (Mehta,1984). In contrast, in those strains for which the conidial phase, the phase of vegetative growth and the alkaloid phase are not clearly separated, the conidia formation occurs simultaneously with alkaloid synthesis. Physiological conditions necessary for conidiation and germination of conidia are different from conditions controlling alkaloid synthesis (Rehacek *et al.*, 1974).

The morphological development of the mycelium in submerged *Claviceps* cultures show significant differences between high-yielding and degenerated cultures (Flieger *et al.*, 1982). Degenerated strains are no longer able to use large amounts of nutrients while the alkaloid yield of higher producing mutants could be further improved by supplying them with higher carbon and nitrogen concentrations (Mehta,1984). The character of fat hyphae and sclerotia like cells of submerged mycelium is reminiscent of the plectenchymatic structure of production sclerotia of parasitic cultures. Such sclerotia like cells influence the alkaloid production in *C. fusiformis* (Dickerson *et al.*, 1970).

Cultural condition

The aeration and agitation should meet the biochemical requirements. Cultural conditions suitable for metabolism of primary substances (Fig. 7) are not suitable for alkaloid metabolism (Taber and Vining, 1963). Submerged *Claviceps* cultures form alkaloids after using up the phosphate in the medium and accumulating polyols, lipids, polyphosphates and also inducers and precursors of alkaloids. Good alkaloid synthesis requires the presence of higher citrate level or some other intermediate in the Kreb's cycle in the cell. Cell differentiation and the onset of the alkaloid synthesis is governed both by phosphate limitation (Vining and Nair, 1966) and citrate utilization. The synthesis of alkaloids is unfavourably influenced by an increased phosphate concentration in a fermentation medium (Mehta, 1984; (Robbers *et al.*, 1972). High production strains are usually tolerant to higher phosphate concentrations (Arcamone *et al.*, 1970; Mehta, 1984). Phosphate plays a negative role in the induction of enzymes catalyzing the alkaloid synthesis (Krupinski *et al.*, 1976). Probably high phosphate levels seem to limit the synthesis of tryptophan. The rate of alkaloid synthesis and the activity of Kreb's

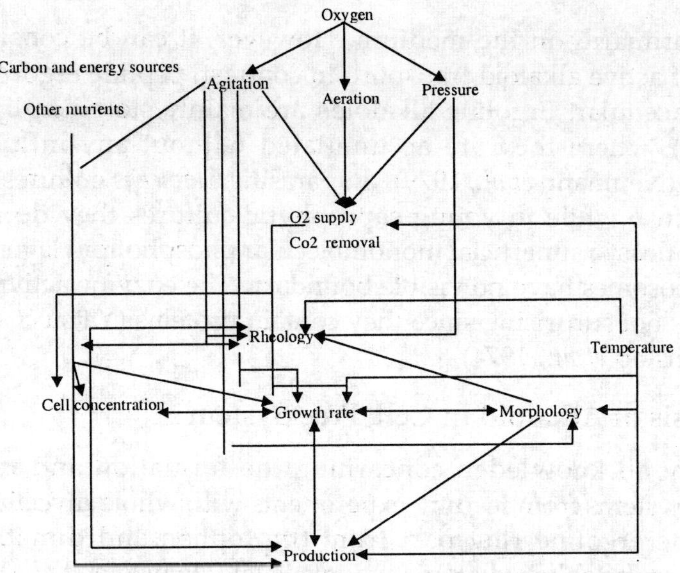

Fig. 7. Effect of cultivation conditions on the culture growth rate and metabolite production (Rehacek, 1983).

cycle were proved to be inversely proportional. The level of the Kreb's cycle enzymes decreased in favour of glyoxylate enzymes. The absence of citrate or another Kreb's cycle intermediate in the medium induced a specific complex of differentiation abberation which resulted in a 50% decrease of biomass, cessation of the alkaloid synthesis (agro-, elymo-, and chanoclavine) and medium acidification by lactate. Citrate is an intermediate which has an important role in metabolic regulation. It inhibits phosphofructokinase and stimulates glucose-6-phosphate dehydrogenase. It seems that the inability of submerged *Claviceps* cultures to grow on hexoses in absence of Kreb's cycle intermediates comes from the parasitic way of life of this pyrenomycete, where the level of citrate and malate in the phloem sap of the host plant is high. Therefore, in parasitic strains the impairment of mentioned regulatory relationships has not been excluded during the *Claviceps* evolution as a selection disadvantage.

Location of Biosynthesis

The region of formation of ergot alkaloids is not generally the same as where alkaloids are found in increased concentrations. In *Ipomoea* the alkaloids are formed in leaves while they are found accumulated in seeds which do not produce the alkaloids. The clavine alkaloids synthesis takes place in endoplasmic reticulum which later forms vacuoles containing lipoproteins (Vorisek and Rehacek, 1978). Clavines and simple derivatives of lysergic acid are found primarily in the medium. However, it can be considered as an indication of active alkaloid transport. In contrast, peptide ergot alkaloids are mostly intracellular. Ergoline alkaloids are mainly stored in lipid droplets (oleosomes), where they are accumulated without any influence on the metabolism (Neumann *et al.*, 1979). In parasitic *Claviceps* cultures cytoplasmic oleosomes fuse while in young saprophytic cultures they do not coalesce possessing atleast a superficial monomolecular phospholipid layer. In contrast, vacuolar oleosomes have no visible boundary. The enzyme activity inside the oleosomes is not surprising since they contain proteins (Yatsu *et al.*, 1971) and enzymes (Hislop *et al.*, 1974).

Biosynthesis of Alkaloid in Cell Free System

Mostly all knowledge concerning the formation and regulation of ergopeptines stems from feeding experiments with whole mycelium while the origin of the ergoline ring part from tryptophan and dimethylallylpyro phosphate was established at an enzymic level (Heinstein *et al.*, 1971 ; Groger and Sajdl, 1972; Sajdl and Rehacek, 1975). The cell free biosynthesis of agro-, elymo-, and chanoclavine-I and chanoclavine-II from tryptophan, iso pentyl pyrophosphate and methionine has already been reported (Cavender and Anderson, 1970). The cell free synthesis of ergot peptides from their constituent amino acids and D-lysergic acid has not yet been accomplished. To obtain such a suitable cell free system for this purpose might be use of protoplasts actively synthesizing ergopeptines. Like others, in *Claviceps*, the fusion of protoplast has a different effect (intra or interspecific). The intraspecific fusion originates heterokaryons able to complement both nutritional requirements and alkaloid synthesis while in the interspecific fusion aneuploids are formed which show nutritional requirement complementation only.

Physiological Role

Ergot alkaloids are not waste physiologically inert products of cell metabolism (Rehacek, 1972, 1974). Their biosynthesis plays a role in the physiology of the production organism by maintaing certain mechanisms of the metabolism and cell divisions in an operative state, especially under cultivation of the limited growth of the organism. Ergometrine added to a submerged culture of *Claviceps paspali* increases the alkaloid production by 120% (Rehacek, 1972 , 1974). Ergotamine and ergometrine reduce the intensity of conidia formation in a productive culture of *Claviceps purpurea* (Rehacek *et al.*, 1974). The effect of tryptophan on the enzyme system in the ergot alkaloid synthesis (Taber and Vining, 1963 ; Robbers *et al.*, 1972) suggests participation of the alkaloids in the feed-back inhibition and repression of the enzyme-forming system. The formation of clavine alkaloids is a part of endogenous metabolism releasing the energy required for maintenance for the integrity and viability of cells under conditions of limited growth (Rehacek, 1972, 1974). Clavine alkaloids probably may form a reservoir which may lead to formation of peptide alkaloids.

Conclusion

Ergot alkaloids biosynthesis lies at the level of genetic information apparatus and continues at the level of physiological realization. Genes involved in biosynthesis are found to be under rigorous metabolic control (Schmauder and Groger, 1976). Fungi and higher plants form ergolenes from the same precursors. Formation of ergot alkaloids is a result of changes in the physiological state than an indication of a new physiological state (Rehacek, 1972). Mechanism and enzymology of C-ring formation and the D-ring and of the peptide part are most important aspects of biosynthesis to be worked out. A cell free system for biosynthesis of the ergopeptines would greatly facilitate such studies. Main findings concerning the physiology of the alkaloid formation will be provided by the results of study of membrane processes in the productive organisms. Vacuoles are worth studying in this regard. Lipid deposit degradation in *Claviceps purpurea* has provided new data on the conversion of cell lipid reserves into an intravacuolar membrane complex. Determination of intracellular locus of ergopeptine synthesis and the problem of distribution of alkaloids and their transport at the molecular level are another areas of interest for future studies. Other important points to be examined (Mehta,1984) are : (i) shortening of the initial non-productive phase of alkaloid fermentation, (ii) immobilization of cells from the alkaloid producing phase on a suitable carrier, and (iii) preparation of alkaloids by co-fermentation. *Claviceps purpurea* mutants may produce broader spectrum and large amounts of some alkaloids of lower complexity, but for this we should know the genetic control of alkaloid production and to find the genes responsible for specific alkaloids. Any improvement of yield of alkaloids would need concerted breeding of the strains concerned. It requires a prerequisite knowledge of the life cycle and the ability to manipulate it in order to get controlled genetic recombination and the possibility of combining appropriate genes, obtained either by mutation or by selection from natural sources. The search for new natural alkaloids involves the testing of non-traditional organisms. Mutation of a gene can also lead to the formation of new compounds and to structural modifications inside the cell. The series of *Claviceps* ergot alkaloids can be widened by their semisynthetic analogues and thus can play an important role in fungal biotechnology (Mehta, 1984; Rehacek and Mehta, 1991, 1993; Lohmeyer and Tudzynski, 1997).

References

Annis,S.L. and Panaccione, D.G. (1998). Presence of peptide synthetase gene transcripts and accumulation of ergopeptines in *Claviceps purpurea* and *Neotyphodium coenophialum*. Can. J. Microbiol., **44** : 80-86.

Arcamone,F.,Cassinelli,G.,Feroni,F.,Penco,S.,Penella,P.andPol,C.(1970). Ergotamine production and metabolism of *C.purpurea* strain 275 FI in stirred fermentors. Can. J. Microbiol., **16**: 923-931.

Arntz, C. and Tudzynski,P. (1997). Identification of genes induced in alkaloid producing cultures of *Claviceps* sp. Curr. Genet., **31**: 357-360.

Baxter, R.M., Kandel, S. and Okany,A. (1961). Studies on the mode of incorporation of mevalonic acid into ergot alkaloids. Tetrahedron Lett. 596-600.

Brich, A.J., McLoughlin,B.J. and Smith,H. (1960). The biosynthesis of the ergot alkaloids. Tetrahedron Lett., **7**: 1-3.

Cavender, F.L. and Anderson, J.A. (1970). The cell free biosynthesis of clavine alkaloids. Biochem. Biophys. Acta, **208** : 345-348.

Dickerson,A.G., Mantle,P.G., and Szezyrbak, C.K. (1970). Autolysis of extracellular glucans produced *in vitro* by a strain of *Claviceps fusiformis*. J. Gen. Microbiol., **60** : 403-415.

Eich,E., Eichberg,D., and Muller, W.E.G. (1981). Clavines: New antibiotics with cytosatic activity. Biochem. Physiol., **33** : 523-526.

Erge,D., Wenzel,A. and Groger,D. (1972). Zur physiologie der alkaloidbuildung bei *Claviceps*-Arten. Biochem. Physiol. Pflanzen., **163** : 288-307.

Esser, K. and Tudzynski, P. (1979). Genetics of the ergot fungus *Claviceps paspali*. Theor. Appl. Genet., **53** : 145-149.

Fleiger,M., Sedmera,P., Vokoun,J., Ricicova,A.and Rehacek,Z. (1982). Separation of four isomers of lysergic acid hydroxyethylamide by liquid chromatography and their spectroscopic identification. J. Chromatogr., **236** : 453-459.

Floss,H.G. and Mothes,K. (1964). Uber den einfluss von tryptophan und analogenver-bindungen auf die biosyntheses von clavinealkaloiden in saprophytischer kulture. Arch. Mikcrobiol., **48**: 213-221.

Garre,V., Tenberge, K.B. and Eising, R. (1998). Secretion of a fungal extracellular catalase by *Claviceps purpurea* during infection on rye: Putative role in pathogenicity and suppression of host defense. Phytopath., **88** : 744-753.

Groger, D., Mothes,K., Simon,H., Floss,H.G. and Weygand, F. (1960). Uber den Einbau von mevalonsaure in das ergolinsystem der clavine-alkaloide. Z. Naturforsch., **15** B: 141-143.

Groger,D., Tyler, V.E.Jr. and Dusenberry, J.E. (1961). Investigations of the alkaloid of paspalum ergot. Lloydia, **24**: 97-102.

Groger,D. and Sajdl,P. (1972). Enzymatic conversion of chanoclavine-I. Pharmazie, **27**: 188.

Heinstein., P.F., Lee,S.L.and Floss,H.G. (1971). Isolation of DMAPP-trptophan dimethylallyl transferase from the ergot fungus (*Claviceps* sp.) Biochem. Biophs. Res., **44**: 1244-1251.

Hislop,E.C., Barnaby, V.M., Shellis,C. and Laborad, F. (1974). Localization of arabinofuranosidase and acid phosphate in mycelium of *Sclerotinia fructigena*. J. Gen. Microbiol., **81**: 79-99.

Jungehulsing, U. and Tudzynski, P. (1997). Analysis of genetic diversity in *Claviceps purpurea* by RAPD markers. Mycol. Res., **101**: 1-6.

Krupinski,V.M., Robbers,J.E. and Floss, H.G. (1976). Physiological studies of ergot. Induction of alkaloid synthesis by tryptophan at the enzymatic level. J. Bacteriol., **125**: 158-165.

Lohmeyer,M. and Tudzynski, P. (1997). Claviceps alkaloids. In: Fungal Biotechnology (Anke,T, ed), Chapman and Hall, Weinheim, pp. 173-185.

Mehta,P. (1984). Submerged fermentation of clavine alkaloids. Ph.D. Thesis. Czech Academy of Sciences, Praha, Czech Republic, pp. 202.

Neumann,D., Losecke,W., Maier,W. and Groger, D. (1979). Localization of alkaloids in sclerotia and suspension cultures of *Claviceps purpurea* (Fr.) Tul. Biochem. Physiol. Pfzen., **174**: 504-508.

Rehacek,Z and Malik, K.M. (1971). Cell pool tryptophan phases in ergot alkaloid fermentation. Folia Microbiol., **16**: 359-363.

Rehacek, Z. (1972). Physiology of formation of some antibiotics and ergot alkaloids. D.Sc. Thesis. Institute of Microbiology, Czech Academy of Sciences, Praha.

Rehacek, Z. (1974). Ergot alkaloids and some problems of the physiology of their formation. Zbl. Bakt. II. **129** : 20-49.

Rehacek,Z., Sajdl,P., Kozova,J., Malik,K.A. and Ricicova, A. (1971). Correlation of certain alternations in metabolic activity with alkaloid production by submerged *Claviceps*. Appl. Microbiol., **22**: 949-956.

Rehacek,Z., Sajdl,P., Kozova,J. and Ricicova, A. (1972). Physiological activities of ergoline alkaloids in submerged cultures of *Claviceps paspali* and *C.purpurea*. Folia Microbiol., **17**: 308-315.

Rehacek,Z., Sajdl,P. and Kremen,A. (1973). Thermodyanamic considerations in the formation of ergot alkaloids. Biotech. Bioeng., **15**: 207-211.

Rehacek,Z., Kozova,J., Sajdl,P and Vorisek,J. (1974). Physiology of conidial formation in submerged cultures of *C.purpurea* (Fr.) Tul. producing alkaloids. Can. J. Microbiol., **20**: 1323-1329.

Rehacek, Z. (1983). New trends in ergot biosynthesis. Process Biochemistry, 22-33.

Rehacek,Z. and Sajdl, P. (1990). Ergot alkaloids: Chemistry, Biological Effects, Biotechnology. Czech Academy of Sciences, Academia, Praha, Czech Republic,pp 376.

Rehacek,Z. and Mehta,P. (1991). Claviceps strains for saprophytic production of ergot alkaloids. The Genet. Engg. and Biotechnol., London, **116** : 19-21.

Rehacek,Z. and Mehta,P. (1993). Biological effects of ergot alkaloids. In: Fungal Ecology and Biotechnology (ed. Rai,B., Arora,D.K., Dubey, N.K. and Sharma,P.D.) Rastogi Publications, Meerut, India, pp. 275-287.

Robbers, J.E., Robertson,L.W., Horneman,K.M., Jindra,A. and Floss,H.G. (1972). Physiological studies on ergot. Further studies on the induction of alkaloid synthesis by tryptophan and its inhibition by phosphate. J. Bacteriol., **112**: 791-796.

Sajdl, P.and Rehacek,Z.(1975). Cyclization of chanoclavine-I by cell free preparations from saprophytic *Claviceps* strains. Folia Microbiol., **20** : 365-367

Sato, F., Hashimoto,T., Hachiya,A., Tamura,K., Choi,K.B., Morishige,T., Fugimoto, H. and Yamada, Y. (2001). Metabolic engineering of plant alkaloid biosynthesis. PANS., **98**: 367-372.

Schumauder,H .P. and Groger,D. (1976). Zur charakterisierung der anthranilat-synthetase bei *Claviceps*. Biochem. Physiol. Pflanzen., **129**: 471-486.

Taber,W.A. and Vining,L.C. (1959). Tryptophan as a precursor of ergot alkaloids. Chem. Ind. London, 1218-1219.

Taber,W.A. and Vining,L.C. (1963). Physiology of alkaloid production by *Claviceps purpurea* (Fr.) Tul. Correlation with changes in mycelial polyol, carbohydrate, lipid and phosphorous containing compunds. Can. J. Microbiol., **9**: 1-14.

Taylor,E.M. and Ramsted,E. (1960). Biosynthesis of lysergic acid in ergot. Nature, **188**: 494-495.

Tenberge,K.B., Brockmann,B. and Tudzynski, P. (1999). Immunogold localization of an extracellular β-1, 3-glucanase of the ergot fungus *Claviceps purpurea* during infection on rye. Mycol. Res., **103**: 1103-1118.

Tsai,H-F., Wang,H.,Gebler,J.C., Poulter,C.D. and Schardl, C.L. (1995). The *Claviceps purpurea* gene encoding dimethylallyltryptophan synthase, the committed step for ergot alkaloid biosynthesis. Biochem. Biophys. Res. Commu., **216** : 119-125.

Tudzynski,P., Holter,K., Correia,T., Arntz,C., Grammel,N and Keller, U. (1999). Evidence for an ergot alkaloid gene cluster in *Claviceps purpurea*. Molec. Genet., **261**: 133-141.

Turian, G. (1969). Differentiation fongique. Masson et Cie, Paris.

Udwardy.N.E. (1980). Consideration of the development of an ergot alkaloid fermentation process. Process Biochemem., **15**: 5-8.

Vining, L.C. (1970). Effect of tryptophan on alkaloid biosynthesis in culture of a *Claviceps* sp. Can. J. Microbiol., **16**: 473-480.

Vining,L.C. (1980). Conversion of alkaloid and nitrogenous xenobiotics. In: Economic Microbiology (Rose,A.H. eds), Academic Press, New York,pp. 523-573.

Vining,L.C. and Nair, P.M. (1966). Clavine alkaloid formation in submerged cultures of a *Claviceps* sp. Can. J. Microbiol., **12** : 915-931.

Vorisek, J. and Rehacek, Z. (1976). Origin of sclerotia like cells in submerged *Claviceps purpurea* producing clavine alkaloids. Arch. Microbiol., **107**: 321-327.

Vorisek,J. and Rehacek,Z. (1978). Fine structure localization of alkaloid synthesis in endoplasmic reticulum of submerged *C. purpurea*. Arch. Microbiol., **117**: 297-302.

Wang,J., Panaccione,D.G. and Schardl,C.L. (1998). Ergot alkaloid biosynthesis genes in *Claviceps* and *Neotyphodium* sp. Phytopath., **88** : S 133 (Abst).

Wilkinson,H.H., Siegel, M.R., Blankenship,J.D., Mallory, A.C., Bush,L.P. and Schardl,C.L. (2000). Contribution of fungal loline alkaloids to protection from aphids in a grass endophyte mutualism. Mol. Plant-Microbe Interactions, **13**: 1027-1033.

Yatsu,L.Y., Jacks,T.J. and Heusarling,T.P. (1971). Isolation of spherosomes/oleosomes from onion, cabbage and cotton seed tissues. Plant Physiol., **48**: 675-682.

8

Complexone Siderophores as Iron Transport Compounds of *Mucorales* (*Zygomycetes*)

Kintu Korat, B. P. Dave and H. C. Dube*

Department of Life Sciences, Bhavnagar University, Bhavnagar – 364 002, Gujarat

Twelve fungi belonging to *Mucorales* (class *Zygomycetes*) were examined for extracellular siderophore production. All except one, *Rhizopus* sp., produced siderophores as evidenced by positive CAS assay and CAS agar plate test. The eleven siderophore producing fungi produced complexone (= carboxylate) siderophores as confirmed by chemical test and bioassays using siderophore deficient strain *Morganella morganii* SBK3. *Rhizopus oryzae* was the highest siderophore producer (rhizoferrin equivalent) and *Syncephalastrum* sp. the least. Based on electrophoretic mobilities, siderophores fell into four groups. 1) acidic, 2) basic, 3) acidic, basic and 4) basic, neutral. All the fungi showed variation in the threshold concentration at which siderophore production was repressed. At one extreme was *Absidia ramosa* that did not produce siderophores at least concentration (0.5 μM) and at the other extreme was *Rhizopus nigricans* in which siderophore biosynthesis continued even at 21μM Fe (III). Rest of the fungi varied in the values between 1 – 18 μM Fe (III).

Key words: Siderophores, Complexones (= Carboxylates), Hydroxamates, Catecholates, Rhizoferrin

Siderophores of fungi have been known since 1952 when the first siderophore a ferrichrome, was detected from culture supernatants of *Ustilago sphaerogena* (Neilands, 1952). Since then, a variety of ascomycetes and basidiomycetes have been shown to produce siderophores (Winkelmann, 1992). Siderophores are extracellular, low molecular weight, iron-chelating compounds, that solubilize and transport iron into the microbial cell through the cognate transport system (Lankford, 1973). Siderophores of bacteria and fungi are **hydroxamates, catecholates,** or **mixed type,** depending on the groups that coordinate iron (Loper and Buyers, 1991). Hydroxamate siderophores, which are produced both by bacteria and fungi, coordinate iron with N-hydroxamic acid by amide bonds eg. ferrioxamines of bacteria and ferrichromes of fungi. Catecholates, produced only by bacteria, coordinate iron with catecholate hydroxy groups. **Enterobactin** a catecholate siderophore of *E. coli,*

*Corresponding Author E-mail : hcdube40@hotmail.com

is the strongest iron chelator known. Mixed siderophores contain both hydroxamate and catecholate siderophores, are exemplified by pyoverdines of fluorescent pseudomonads (yellow-green water soluble pigment that fluoresce under UV light; Meyer and Abdallah, 1978).

Until recently, it was assumed that fungal siderophores were exclusively hydroxamate type. A new citric acid containing siderophore, named **rhizoferrin** was discovered from *Rhizopus microsporus* var. *rhizopodiformis* (Thieken and Winkelmann, 1993). This belonged to the **complexone** (=carboxylate) type siderophores, as defined earlier for bacterial siderophores, **rhizobactin** isolated from rhizobia (Smith *et al.*, 1985) or **staphyloferrin** isolated from *Staphylococcus hyicus* (Miewes *et al.*, 1990). Rhizoferrin contains two citric acid residues, linked via amide bonds to **putrescine**. Putrescine is a well-known intermediate in fungal polyamine biosynthesis. Citric acid is generated in the citric acid cycle, which is also excreted by fungi under iron stress. Thus, both constituents are common products of primary metabolism of fungi, and their combination into one molecule seems to be an easy means to form a siderophore. The production of rhizoferrin has been observed in a variety of families (*Mucoraceae, Thamnidiaceae, Choanophoraceae* etc.) of the two orders *Mucorales* and *Entomophthorales*, confirming the view that rhizoferrin is, in fact, the principal siderophore of fungi belonging to the class *Zygomycotina* (Thieken and Winkelmann, 1992).

The present paper examines the occurrence of complexone (carboxylate) siderophores in *Mucorales*, their quantitative estimation, electrophoretic mobilities, and the threshold concentration of iron that represses siderophore production.

Methodology

Organisms: Twelve fungi belonging to class *Zygomyota* (all under *Mucorales*) used in this study were isolated in our laboratory. Their identification was confirmed by Dr. B.P.R. Vittal of Centre of Advanced Studies (CAS) in Botany, University of Madras, Chennai. All the organisms were maintained on Potato dextrose agar (PDA) slants and stored at 4° C until used.

Siderophore production

Glassware preparation : To minimize iron contamination, all glasswares were kept immersed in 6 N HCl overnight and rinsed with distilled water several times to remove traces of iron.

Siderophore production media : The Modified M9 medium (Shenker *et al.*, 1992) containing glucose (10 g), Na_2HPO_4 (7 g), KH_2PO_4 (3 g), NaCl (0.5 g),

NH_4Cl (1 g), $MgSO_4$, $7H_2O$ (0.25 g), $CaCl_2$ (0.015 g) were dissolved in 1 litre of distilled water and adjusted to pH 7.2. The medium was then supplemented with thiamine (0.0005 g) and $ZnCl_2$ (0.015 g), and rendered iron-free by adding 8-hydroxyquinoline dissolved in chloroform, as suggested by Messenger and Ratledge (1985). Aliquots of 50 ml of the media were dispensed in 500 ml conical flasks.

Detection of siderophores : 10^6 spores/ml of the 12 test fungi were used as inoculum. 15 days old cultures grown at 27°C were centrifuged and the cell - free supernatants were assayed for siderophore by standard assays as given below.

$FeCl_3$ test (Atkin, 1970): To 0.5 ml of cell-free culture supernatant, was added 0.5 ml of 2% aqueous $FeCl_3$ solution. Appearance of orange or red-brown colour indicated the presence of siderophores.

Chrome azurol sulfonate (CAS) assay (Schwyn and Neilands, 1987): To 0.5 ml of CAS solution was added 0.5 ml of cell – free culture supernatant. The change in colour of the blue dye to orange indicated the presence of siderophores. CAS assay solution was prepared as - 6 ml of hexadecyl trimethylammoniumbromide (10 mM HDTMA) solution was placed in a volumetric flask (100 ml) and diluted with water. A mixture of 1.5 mL iron (III) solution (1mM $FeCl_3$. 6 H_2O in 10 mM HCl) and 7.5 ml of 2mM aqueous CAS dye solution was slowly added with stirring. Anhydrous piperazine (4.307g) was dissolved in water and 6.25 ml of HCl (12M) was added to get a buffer solution (pH 5.6). It was added to above volumetric flask, made to 100 mL with distilled water to make CAS assay solution.

CAS agar plate test (Schwyn and Neilands, 1987): The test fungi were streaked on CAS agar plates and incubated for 48h at 30°C. Formation of orange zones around the colonies indicated the presence of siderophores. CAS agar plates were prepared by dissolving CAS (60.5 mg) in distilled water (50ml) and mixed with 10 ml of iron (III) solution (1mM $FeCl_3$. 6 H_2O in 10 mM HCl) with stirring. This solution was slowly added to 72.9 mg of HDTMA, dissolved in 40 ml of water. The resultant dark blue liquid was autoclaved at 15 lb psi. Also, a mixture of 750 ml H_2O; 100ml (10xMM9 salts);15 g agar; 30.2 g pipes (12 g of 50% w/w NaOH solution) was added. After cooling to 50°C, 30 ml of casamino acids (10%), as carbon source and supplements like vitamins and antibiotics (0.2%) sterililized by millipore filtration, were added to sterile CAS dye solution. The dye solution was finally added along the glass wall, with enough agitation to achieve mixing without generation of foam. Each plate contained 30 ml of blue agar.

Chemical characterization of siderophores

The methods most commonly used for detection of chemical nature include **chemical tests** and **bioassays.**

Chemical tests

Hydroxamates:

$FeCl_3$ test (Neilands, 1981): To 1 mL of culture supernatant was added 1-5 ml of freshly prepared aqueous 2% $FeCl_3$ solution. The formation of orange coloured ferric hydroxamate, showing maxima at 420 – 450 nm, indicated presence of hydroxamate siderophores.

Tetrazolium Salt test (Snow, 1954): To a pinch of tetrazolium salt, was added 1-2 drops of 2 N NaOH and 0.1 ml of the test culture-supernatant. Instant appearance of a deep-red colour indicated the hydroxamate nature of the siderophore.

Catecholates :

$FeCl_3$ test (Neilands, 1981): Formation of wine coloured complex on addition of 1 – 5 ml of freshly prepared 2% aqueous $FeCl_3$, that absorbed maximally at 495 nm, indicated the presence of catecholate siderophores.

Arnow's test (Arnow, 1937): To 1 ml of cell-free culture supernatant, was added 1 ml of 0.5 N HCl, 1 ml of nitrite molybdate reagent and 1 ml of 1 N NaOH, followed by distilled water to make the volumc to 5 ml. Absorbance was read at 500 nm, using 2, 3 dihydroxybenzoic acid as standard.

Carboxylates :

Vogel's chemical test (Vogel, 1987): To 3 drops of 2 N NaOH, was added 1 drop of phenolphthalein and then water. A light pink colour on addition of test sample, indicated carboxylate nature of siderophores.

Spectrophotometric assay (Shenker *et al.*, 1992): To 1 ml of cell – free culture supernatant, was added 1 ml of 250 μM $CuSO_4$ and 2 ml of acetate buffer (pH 4). A copper complex was formed with an absorption maximum between 190-280nm. There is no specific wavelength for absorption of copper complex and, thus, the entire wavelength (190-280 nm) was scanned to observe the peak of absorption of siderophore.

Bioassays:

Bioassays are more sensitive than the chemical assays for detection of siderophores. The indicator strains used were siderophore deficient strains:

Arthrobacter flavascens JG9 for hydroxamate, *Salmonella typhymurium* enb7 for catecholate, and *Morganella morganii* SBK3 for carboxylate siderophores. The indicator strains were iron - starved and seeded on AM medium (Haselwandter, 1998), and Luria Base (LB) agar medium (Thieken and Winkelmann, 1993). The culture – supernatants of test organisms were loaded on filter paper discs and deftly placed on surface of AM medium or LB medium. Formation of a growth zone around siderophore sample, after incubation, indicated the presence of hydroxamate, catecholate or carboxylate siderophores, depending on the indicator strain used.

Quantitative estimation of siderophores

Standard curves were prepared for absorbance (630nm) of rhizoferrin (standard for carboxylate siderophores) divided by absorbance (630nm) of reference solution (A_{630}/A_{ref}) as a function of siderophore concentrations (rhizoferrin 20-100 μM), that yielded a linear regression relationship. Reference solution contained all the ingredients except siderophore. To 1 ml of cell-free culture supernatant, was added 1 ml of CAS assay solution and absorbance noted at 630nm. The amount of siderophore produced was extrapolated from the standard curve.

Electrophoretic mobilities of siderophores

Electrophoresis was used to determine the net charge present on the siderophores as a function of pH. It was also used to find the pH suitable for driving siderophore into an organic solvent for purification (Schwyn and Neilands, 1987; Jalal and van der Helm, 1990). The net charge present on siderophore was examined by migration of siderophore towards cationic (acidic) or anionic (basic) electrode. Migration depends on net charge present on molecule and its molecular weight. The ionophoric mobilities were determined with a flat - bed device using a buffer at pH 5.6 (5.7 ml glacial acetic acid and 24.3 ml pyridine /L). 0.1 ml of siderophore solution adjusted to pH 5.6 was spotted on Whatman no. 3 paper. The electrophoresis was run at 30V/ cm for 1-2 h. The paper was dried at 110 ºC to remove all traces of pyridine and glacial acetic acid. It was then sprayed on both sides with CAS assay solution. After a few minutes, development of orange spots against light blue background, indicated location of migrated siderophores.

Regulation of siderophore production

Eleven siderophore - producing *Mucorales* were grown on Modified M9 medium supplemented with 0.5 μM to 27 μM Fe(III). After fifteen days of growth at 25° - 27°C, the mycelia were centrifuged out and the cell-free culture supernatants were examined by CAS assay for the presence of siderophores.

The siderophores were quantified by measuring A_{630}/A_{ref}.

Results and Discussion

The results (Table 1) clearly suggest that out of twelve *Mucorales*, all, except one (*Rhizopus* sp.), produced siderophores, as evidenced by positive reaction in CAS assay and CAS agar plate test. They showed negative reaction with $FeCl_3$ test due to methodological reasons as $FeCl_3$ complex formation was obstructed due to precipitation. The results indicate nonsuitability of $FeCl_3$ test for assaying mucoraceous siderophores. The results fall in line with those of Dave(1999), Vaidya (1999) and Vala (1999), who also found $FeCl_3$ test as unsuitable for assaying mucoraceous siderophores. The use of siderophores in iron uptake mechanisms under iron-deficient conditions is vindicated. The exceptional behaviour of *Rhizopus* sp., not utilizing siderophore for iron – uptake, cannot be explained. It might be that the siderophore is produced in feable quantity which is not detectable by available methods of assay.

Table 1 indicates that none of the eleven test fungi showed positive reaction with Tetrazolium Salt test and Arnow's test thus, ruling out hydroxamate and catecholate nature of their siderophores. $FeCl_3$ test (Neilands, 1981) could not be successfully performed for detection of hydroxamate and catecholate nature due to precipitation with $FeCl_3$. The results indicate positive reaction for carboxylate siderophores in Vogel's test and spectrophotometric assay(Shenker *et al.*, 1992) . In spectrophotometric assay, the siderophores showed characteristic maximum for carboxylate nature that varied in their absorbance, suggesting *Rhizopus oryzae* as highest siderophore producer and *Syncephalastrum* sp. the least. The organisms could be arranged in the following order according to relative amount of siderophores produced.

R. oryzae > *M. mucedo* > *R. microsporus* > *S. racemosum* > *R. rhizopodiformis* > *A ramosa* > *C. echinulata* > *C. elegans* > *Absidia* sp. > *R. nigricans* > *Syncephalastrum* sp.

Thus, all the eleven test fungi belonging to *Mucorales* produced exclusively carboxylate siderophores. Our results are in agreement with those of Winkelmann (1992), who proposed that fungi belonging to *Zygomycota,* which were previously not known to produce siderophores, produced a novel carboxylate (complexone) siderophore **rhizoferrin.** The first carboxylate siderophore identified was from *Rhizopus microsporus* var. *rhizopodiformis* (Thieken and Winkelmann, 1992).

Carboxylate nature of mucoraceous siderophores was furthur confirmed by bioassys using auxotrophs for siderophore production *Arthrobacter*

Table 1. Detection and chemical characterization of siderophores proudced by *Mucorales*.

Test Fungi (*Mucorales*)	$FeCl_3$ test	CAS assay	CAS agar plate test	Chemical Nature				
				Hydroxamate	Catecholate	Carboxylate		
				Tetrazolium salt test	Arrow's test	Chemical test	Spectrophotometric assay	
						Vogel test	x max (nm)	Absorbance (OD)
Absidia ramosa	-	+	+	-	-	+	247	1.97
Absidia sp.	-	+	+	-	-	+	229	1.51
Cunninghamella echinulata	-	+	+	-	-	+	264	1.83
C. elegans	-	+	+	-	-	+	264	1.80
Mucor mucedo	-	+	+	-	-	+	268	2.09
Rhizopus microsporus	-	+	+	-	-	+	239	2.05
R. nigricans	-	+	+	-	-	+	219	1.31
R. oryzae	-	+	+	-	-	+	269	2.51
R. rhizopodiformis	-	+	+	-	-	+	268	2.02
Rhizopus sp.	-	-	-	-	-	-	-	-
Syncephalastrum racemosum	-	+	+	-	-	+	220	2.03
Syncephalastrum sp.	-	+	+	-	-	+	240	0.40

flavascens JG9 for hydroxamates, *Salmonella typhimurium* enb7 for catecholates and *Morganella morganii* SBK3 for carboxylates. The results (Table 2) clearly indicate formation of growth zones around the introduced siderophores with *Morganella morganii,* confirming their carboxylate nature.

Table 2. Bioassays for detection of chemical nature of siderophores.

Test Fungi (*Mucorales*)	Hydroxamate *Arthrobacter flavascens* JG9	Catecholate *Salmonella typhimurium* enb7	Carboxylate *Morganella morganii* SBK3
Absidia ramosa	-	-	+
Absidia sp.	-	-	+
Cunninghamella echinulata	-	-	+
C. elegans	-	-	+
Mucor mucedo	-	-	+
Rhizopus microsporus	-	-	+
R. nigricans	-	-	+
R. oryzae	-	-	+
R. rhizpopodiformis	-	-	+
Syncephalastrum racemosum	-	-	+
Syncephalastrum sp.	-	-	+

Table 3. Amount of siderophore produced (rhizoferrin equivalent)

Test Fungi (*Mucorales*)	Siderophore production (x 10^2 µg/mL)
Absidia ramosa	16.6
Absidia sp.	13.4
Cunninghamella echinulata	16.3
C. elegans	15.8
Mucor mucedo	28.1
Rhizopus microsporus	24.2
R. nigricans	11.7
R. oryzae	29.3
R. rhizpopodiformis	18.4
Syncephalastrum racemosum	21.4
Syncephalastrum sp.	11.3

Among the eleven test isolates (Table 3), *R. oryzae* showed maximum siderophore production (29.3X10^2µg/ml), followed by *Mucor mucedo* (28.1x10^2 µg/ml) and *Syncephalastrum* sp. the least (11.3x10^2 µg/ml). Interestingly, the results show the same sequence as observed with absorbance values at max 190-280nm. The difference in the quantity of siderophore produced is a logical happening. There are reports of strain to strain variation as reported by Reigh and O' Connell (1988) and Roy *et al.* (1994). Our observations of *R. oryzae* and *M. mucedo* as relatively stronger producers of siderophores, finds support from earlier reports of Winkelmann (1992).

The results (Table 4) suggest that the siderophores fall into 4 groups. 1)

Table 4. Electrophoretic migration of siderophores.

Test Fungi (*Mucorales*)	Electrophoretic migration	Net charge
Absidia ramosa	Anion	Acidic
Absidia sp.	Anion	Acidic
Cunninghamella echinulata	Cation	Basic
C. elegans	Cation, No mobility	Basic, Neutral
Mucor mucedo	Cation	Basic
Rhizopus microsporus	Anion, cation	Acidic, Basic
R. nigricans	Anion	Acidic
R. oryzae	Cation	Basic
R. rhizpopodiformis	Cation	Basic
Syncephalastrum racemosum	Cation	Basic
Syncephalastrum sp.	Cation	Basic

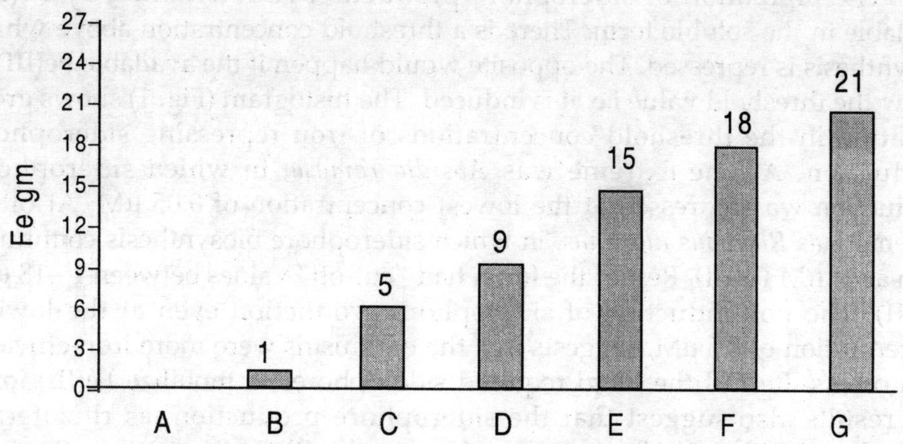

Fig. : 1 Repression of siderophore production at iron levels 0.5-27 μM/L
A: *A. ramosa*; B: *Absidia* sp., *C. elegans*; C: *M. mucedo*; D: *C. echinulata*, *Syncephalastrum racemosum*; E: *R. rizopodiformis*; F: *R. microsporus*, *R. oryzae*, Syncephalastrum sp., G: *R. nigricans*

acidic, 2) basic, 3) acidic, basic and 4) basic, neutral. Siderophores of *Mucorales* stand out distinct being invariably basic molecules; greater the basicity, greater is the ligand's affinity for iron (Crumbliss, 1990). The ferric complex of rhizoferrin is anionic throught the pH range prevailing in soils, and therefore, it is mobile in the soil (Shenker, 1996).

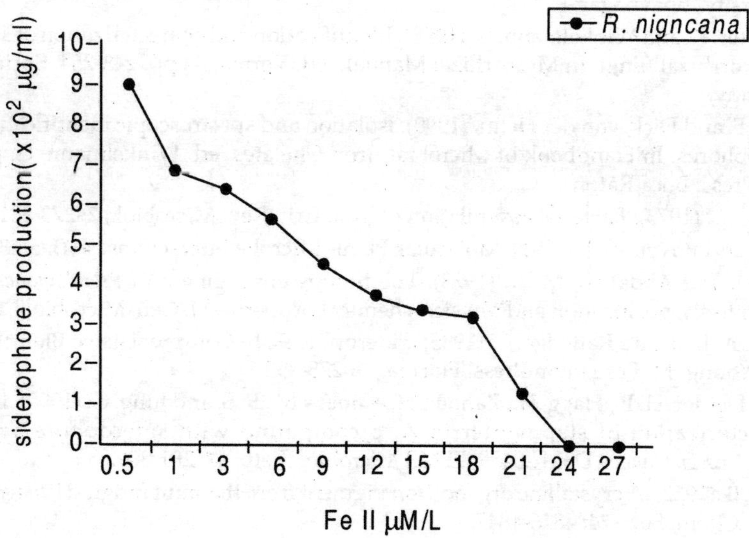

Fig. 2. Decrease in siderophore production-rhizoferrin equivalent x 10^2 μg/ml with increase iron (Fe III) concentration (0.5-27 μM/L)

The regulation of siderophore production is determined by Fe(III) available in the soluble form. There is a threshold concentration above which its synthesis is repressed. The opposite would happen if the available Fe(III) is below the threshold value i.e. it is induced. The histogram (Fig. 1) shows great variation in the threshold concentrations of iron repressing siderophore production. At one extreme was *Absidia ramosa,* in which siderophore production was repressed at the lowest concentration of 0.05 µM. At other extreme was *Rhizopus nigricans,* in which siderophore biosynthesis continued even at 21 µM Fe(III). Rest of the fungi had "cut-off" values between 1 - 18 µM Fe(III). The non-induction of siderophore production even at the lowest concentration of 0.5 µM, suggests that the organisms were more iron-efficient than others. Rest of the fungi required siderophores to mobilize Fe(III) ions. The results also suggest that the siderophore production, as rhizoferrin equivalent (x 10^2 µg/ml), gradually decreased with an increase in the iron concentration. This is explained by Fig. 2 for *R. nigricans*.

References

Arnow, L. E. (1937). Colorimetric determination of the components of 3,4-dihydroxyphenylalanine tyrosine mixtures. J. Biol. Chem., **118:** 531.

Atkin, C. L., Neilands, J. B. and Phaff, H. J. (1970). J. Bacteriol., **103:** 722-33.

Crumbliss, A. L. (1990). Aqueous solution equilibrium and kinetic studies of iron siderophores and model siderophore complexes. In Handbook of Microbial Iron Chelates. ed. Winkelmann, G. pp. 235-269. CRC Press. Boca Raton.

Dave, B. P. (1999). A Comparative study of microbial siderophores. Ph. D. Thesis. Bhavnagar University. Bhavnagar.

Haselwandter, K. and Winkelmann, G. (1998). Identification and charecterization of siderophores of mycorrhizal fungi. In Mycorrhizal Manual. ed. Verma, A. pp. 243-254. Springer Verlag, Germany.

Jalal, M. A. F. and Dick, van der Helm. (1990). Isolation and spectroscopic identification of fungal siderophores. In Handbook of Microbial Iron Chelates. ed. Winkelmann, G. pp. 235-269. CRC Press. Boca Raton.

Lankford, C. L. (1973). Bacterial assimilation of iron. Crit. Rev. Microbiol., **2:** 273-331.

Loper, J. E. and Buyers, S. J. (1991). Molecular Plant- Microbe Interactions., **4(1):** 5-13.

Mayer, J. M. and Abdallah, M. A. (1973). The fluorescent pigment of *Pseudomonas fluorescens*: biosynthesis, purification and physico-chemical properties. J. Gen. Microbiol., **107:** 319-328.

Messenger, A. J. M. and Ratledge, C. (1985). Siderophores. In Comprehensive Biotechnology. ed. Moo-Young, M. Pergamon Press, Florida., **3:** 275-295.

Miewes, J., Fielder, H. P., Hagg, H., Zahner, H., Konetschy, R. S. and Jung, G. (1990). Isolation and characterization of staphyloferrin A, a compound with siderophore activity from *Staphylococcus hyicus* DSM 20459. FEMS Microbiol. Lett., **67:** 201-206.

Neilands, J. B. (1952). A crystalline organo-iron pigment from the smut fungus *Ustilago sphaerogena* J. Am. Chem. Soc., **74:** 4846-4847.

Neilands, J. B. (1981). Microbial iron transport compounds (siderophores) as chelating agent. In Development of Iron Chelators for Clinical use. ed Martell, A. E., Anderson, W. J., Badman,

D. G. pp 13-31. Elsevier, North Holland, Amsterdam.

Reigh, G. and O' Connell, M. (1998). Siderophore production is strain specific in *Rhizobium*. In Nitrogen Fixation: Hundred Years After. ed. Bothe, H., De Bruijn, F. J., Newto, W. E. pp. 826-839. Gustav Fischer, New York.

Roy, N., Bhattacharya, P. and Chakrabartty, P. K. (1994). Iron acquisition during growth in an iron-deficient medium by *Rhizobium* sp. isolated from *Cicer arientium*. Microbiology., **140:** 2811-2820.

Schwyn, B. and Neilands, J. B. (1987). Universal chemical assay for the detection and determination of siderophores. Anal. Biochem., **160:** 47-56.

Shenker, M., Hadar, Y. and Chen, Y. (1996). Stability constants of the fungal siderophore rhizoferrin with various microelements and calcium. Soil. Sci. Soc. Am. J., **60:** 1140-1144.

Shenker, M., Oliver, I., Helmann, M., Hadar, Y. and Chen, Y. (1992). Utilization by tomatoes of iron mediated by a siderophore produced by *Rhizopus arrhizus*. J. Pl. Nutr., **15:** 2173-2182.

Smith, M. J., Shollery, J. N., Schwyn, B., Hoelden, I. and Neilands, J. B. (1985). Rhizobactin, a structurally novel siderophore from *Rhizobium meliloti*. J. Am. Chem. Soc., **107:** 1739-1743.

Snow, G. A. (1954). Mycobactin. A growth factor *Mycobacterium johnei*. II. Degradation and identification of fragments. J. Chem. Soc., 2588-2596.

Thieken, A. and Winkelmann, G. (1993). A novel bioassay for the detection of siderophores containing keto-hydroxy bidentate ligands. FEMS Microbiol. Lett., **111:** 281-286.

Thieken, A. and Winkelmann, G. (1992). Rhizoferrin: A complexone type siderophore of the *Mucorales* and *Entomophthorales* (*Zygomycetes*). FEMS Microbiol. Lett., **94:** 37-42.

Vaidya, S. Y. (1999). Microbial Pollution at Bhavnagar Coast. Ph. D. Thesis. Bhavnagar University. Bhavnagar.

Vala, A. K. (1999). Microbial Diversity at Bhavnagar Coast. Ph. D. Thesis. Bhavnagar University. Bhavnagar.

Vogel, A. I. (1987). Class reactions (reactions for functional groups). In Elementary Practical Organic Chemistry. Part 2. pp. 190-194. CBS Publishers.

Winkelmann, G. (1992). Structure and function of fungal siderophores containing hydroxamate and complexone type iron binding ligands. Mycol. Res., **96:** 529-534.

9

Fungal Cellulases for Novel Industrial Applications

Santosh Vyas, Anil Lachke and Absar Ahmad*
Division of Biochemical Sciences,
National Chemical Laboratory, Pune-411008

Cellulases produced by wide variety of fungi are industrially important hydrolytic enzymes. Generally, a typical fungal cellulolytic complex includes variety of hydrolytic and oxidative enzymes. Hydrolases such as endoglucanase (EC 3.2.1.4), cellobiohydrolase (CBH) (EC 3.2.1.91) and β-glucosidase (EC 3.2.1.21) are involved in degradation of crystalline cellulose to glucose. Oxidative enzymes, such as cellobiose dehydrogenase (EC 1.1.5.1), Cellobiose oxidase (EC 1.1.99.18) induces hydrolytic components of cellulase complex. In fungi cellulases occur in multiple forms and their multiplicity can be attributed to post-translational events such as glycosylation or proteolysis. fungal cellulases differ in their molecular characteristics as well as in their capacities to adsorb on the cellulose, substrate specificity and in their catalytic activities. The cellulases have wide range of applications. The main potential applications are in food, animal feed, textile, fuel, detergents and chemical industries, while the other ecofriendly applications of cellulases include those in environmental waste management and most important, in developing environmentally safe technologies for paper and pulp industries. In this article we will highlight some of the industrially important fungal cellulolytic systems.

Key-words : Cellulases, Characteristics, Applications, Detergents, Biodeinking, Textiles

Cellulose is the most abundant and renewable biopolymer on the Earth. An estimated synthesis rate of cellulose is approximately 4 X 10^{10} tonnes per year. Although it is the most promising renewable energy source to overcome our problems of energy resources, chemicals and food in the future, its enormous potential was recognized only after the cellulose degrading enzymes or cellulases were identified. The research activity on the cellulases started with the biodeterioration problem faced by US Army in South Pacific during World War II. It was found that the fungus is the causative agent for spoilage of tents, sandbags, soldier uniforms and other clothings. Several organizations within US Army had to set up laboratories to find an immediate solution to this problem. Later the fungus was identified as *Trichoderma viride*.

*Corresponding Author E-mail: aahmad786in@yahoo.com

It was very quickly recognized that cellulases, a group of hydrolytic enzymes could, certainly prove beneficial in converting cellulosic waste to glucose, soluble sugars and industrially important chemicals.

Cellulases have wide range of applications. The main potential applications are in food, animal feed, textile, fuel, detergents and chemical industries, while they are also important in medicinal/pharmaceutical industries, in genetic engineering research and in protoplasts production. The other ecofriendly applications of cellulases include those in environmental waste management and most important in developing environmentally safe technologies for paper and pulp industries. In this article we will highlight on some of the industrially important fungal cellulolytic systems. Particularly we have attempted to demonstrate the novel industrial applications of alkaline active fungal cellulases because of their increasing importance in paper and pulp industries as well as in the textile industries.

Fungal Cellulases

Cellulases are produced by wide variety of bacteria, fungi, actinomycetes, aerobes and anaerobes, mesophiles and thermophiles. However, fungi are the most studied organisms with respect to cellulose degradation. Generally, fungi hydrolyse cellulose to simple soluble sugars and utilize them as a carbon source in their metabolism. Some specialized group of fungal species e.g. *Pythium, Leptomytus* synthesize cellulases to help in extension of their cellulosic mycelial cell walls. Cellulases have been found to play key role in establishing host pathogen interactions in case of plant pathogens.

It has been observed that, fungal cellulases differ in their molecular characteristics such as molecular weight, amino acid composition and sequence, isoelectric point and carbohydrate content etc. They also differ in their capacities to adsorb on the cellulose, substrate specificity and even in their catalytic activities. Despite of such a variability, the general mechanism for cellulose hydrolysis, a commonly accepted one is shown fig 1.

Generally, a typical cellulolytic complex includes a variety of hydrolytic and oxidative enzymes. Hydrolytic enzymes such as endoglucanase [1,4-β-D glucan glucanohydrolase; (EC 3.2.1.4)] , cellobiohydrolase (CBH) [1,4-β-D glucan cellobiohydrolase ; (EC 3.2.1.91)], 1,4-β-D-glucan glucohydrolase (EC 3.2.1.74) and α-glucosidase [1,4-β-D glucoside glucanohydrolase, (EC 3.2.1.21)] are involved in degradation of crystalline cellulose to glucose. The enzyme action is generally initiated by random acting endoglucanases at amorphous regions within cellulose chain to produce cellooligosaccharides. The cellobiohydrolases then act sequentially by removing cellobiose units from

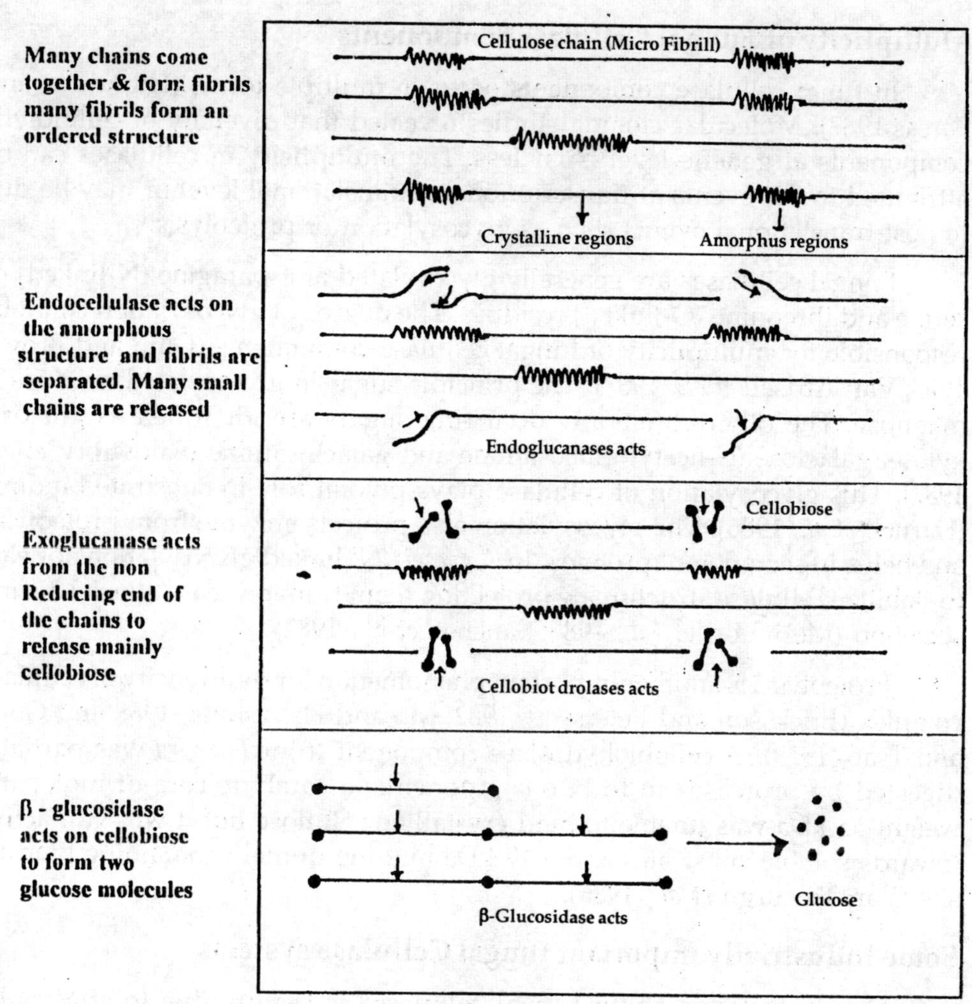

Fig. 1. Cellulases and their mode of action

non reducing ends of the cellulose chains. Finally, β–glucosidase completes the hydrolysis by degrading cellobiose to glucose (Eveleigh, 1987). Oxidative enzymes, such as cellobiose dehydrogenase (EC 1.1.5.1)(Westermark and Ericksson , 1974) ,Cellobiose oxidase (EC 1.1.99.18) (Morpeth,1985) have been characterized from cellulolytic fungus *Phanerochaete chrysosporium*. Oxidative enzymes have been found to induce hydrolytic components of cellualse complex. Lactonases [D- glucano- 1,5, lactonohydrolase (EC 3.1.1.17)] are responsible for hydrolysis of inhibitors of cellulase system (Bruchmann *et al.*, 1987).

Multiplicity of Fungal Cellulase components

In fungi cellulase components occur in multiple forms (Labudova and Farcas,1983). Molecular cloning studies revealed that diversity of cellulolytic components at genetic level is far less. The multiplicity in cellulases can be attributed to the events at transcriptional / translational level or may be due to post-translational events such as glycosylation or proteolysis.

Fungal cellulases are generally glycosylated at asparagine (N-linked) or serine and threonine (O-linked) residues. The degree of glycosylation could be responsible for multiplicity of fungal cellulase components (Gum and Brown 1977; Van Ardsell *et al.*, 1987). The principle sugar in glycosylated residues is mannose. The other commonly occurring sugars are identified as glucose, xylose, galactose, N-acetyl glucosamine and galactosamine (Salovuori *et al.*, 1987). This glycosylation of cellulase plays pivotal role in substrate binding. (Darnell *et al.*, 1986). The glycosylation also protects enzyme from proteolysis and helps in secretion of proteins. In *T. reesei* 'N' linked glycosylation appears to stabilize cellulase structure by protecting them from protease attacks during secretion (Merivuori *et al.*, 1985; Sanchez *et al.*, 1982).

Proteolysis is another important phenomenon for multiplicity of cellulase complex (Ericksson and Petersson, 1982; Mo and Hayashida, 1988 and Gong and Tsao, 1979). A cellobiohydrolase component from *T. reesei* was partially digested by proteases in to two components; a catalytic core of molecular weight 55 kDa was unable to bind crystalline cellulose but it was still active towards soluble substrates and a 10 kDa binding domain containing binding site (Van Tibeurgh *et al.*, 1986).

Some Industrially important fungal Cellulase systems

Our knowledge about fungal cellulases is mainly due to studies on hypersecretory fungi e.g. *Trichoderma reesei.* (Kubicek, 1992, Kubicek *et al.*, 1993), *Penicillium pinophilum* (Bhat *et al.*, 1989), *Penicillium funiculosum* (Wood and McCrae, 1982.), *Fusarium oxysporum* (Kumar *et al.*, 1991, Christakopaulos *et al.*, 1995a,b). *Aspergillus* niger (Okada, 1985, Hurst *et al.*,1977), *Sclerotium rolfsii* (Lachke and Deshpande,1988) and *Humicola* sp. (Hayashida *et al*, 1988; Schulien, 1997). However, other cellulolytic systems from, *Phanerochaete chrysosporium* (*S. pulverulentum* Ericksson, 1978),*Taleromyces emersonii* (McHale and Coughlan, 1980), and from anaerobic fungi belonging to genera *Neocallimastix, Cacomyces, Oprinomyces* (Wood, 1992) have also been very well characterized. These fungi generally produce all three major components of cellulolytic systems extracellularly for complete hydrolysis of crystalline cellulose to glucose; but the proportion of different cellulolytic components varies considerably. Here, we are shortly describing industrially exploited, extracellular cellulase systems from *Humicola insolens* and *Trichoderma reesei.*

Humicola insolens Cellulases

The thermophilic, micromycetes fungus *Humicola insolens* produces at least 7 different cellulases. As these cellulases are active in neutral to alkaline pH range (5-10) they have gained considerable commercial importance. They can be utilized for total biomass conversion as well as in other specialized commercial applications such as detergents, textile industry, and paper and pulp industries. (Schulein *et al.,* 1993) Out of seven different cellulases of *H. insolens,* five are endoglucanases while two belong to cellobiohydrolase class (Schulein, 1997). Based on the catalytic core sequence, they are classified under different families (Henrissat and Bairoch, 1993). The molecular properties of these different enzymes are compared in Table 1.

Table 1. Cellulases from *Humicola insolens* and their properties.

Enzyme	Family	MW KDa	pI	Presence of CBD
EGI	7	50	5.5	Absent
EG II	5	50	7.0	Present
EG III	12	26	5.2	Absent
EG V	45	43	5.2	Present
EG VI	6	43	5.0	Present
CBH I	7	72	4.5	Present
CBH II	6	65	Isozymes In the range of 4.6-5.2	Present

MW molecular weight; **pI** - isoelectric pH; **CBD** - Cellulose Binding Domain

These cellulases are also different in terms of their catalytic activities on different substrates. It was found that presence of cellulose binding domains (CBD) in cellulase lowers apparent Km. This was attributed to dispersing action of cellulose binding domain. Presence of CBD also reduces apparent Kcat values. (Schulein, 1997).

Trichoderma reesei Cellulases

The fungus *Trichoderma reesei* has been traditionally used by various enzyme manufacturers to produce fungal cellulases. It is one of the best-characterized fungal cellulase systems (Kubicek, 1992; Kubicek *et al.,* 1993). The fungus produces large amount of different cellulases that act synergistically for complete hydrolysis of crystalline cellulose to glucose. The fungus produces at least 3 different types of endoglucanases (EG I, EGII, EG III), 2 different cellobiohydrolases (CBH I, CBH II) and β–glucosidase (Shoemaker *et al.*, 1983, Teeri *et al.*, 1987) . Some of their molecular properties are highlighted in Table 2.

Table 2. Cellulases from *Trichoderma reesei* and their properties.

Enzyme	MW KDa	pI	Degree of Glycosylation %	References
EGI	54	4.7	4	Shoemaker *et al.*,1983 Bhikhabhai *et al.*, 1984
EG II	49.8	5.5	15	Saloheimo *et al.*, 1988
EG III	20-23.5s	7.0-7.7	12-15	Beldman *et al.*, 1985
CBH I	42-72	3.5-4.2	1.4-10.4	Bhikhabhai *et al.*, 1984 Bhikhabhai *et al.*, 1985
CBH II	50-58	5.0-6.3	8-18	Fagerstam and Peterson 1980.

MW molecular weight
pI - isoelectric pH

Applications of Cellulases

A cellulase complex containing all three major cellulolytic components are involved in the process of saccharification where agricultural waste residues can be converted to glucose, single cell proteins and biofuels like ethanol. During last 30 years of research this application has not yet been economically feasible due to cost of enzymes and lack of efficiency of cellulases. Recently, cellulases gained significant commercial importance due to their potential applications in food, animal feed, detergents, paper and pulp, and textile Industries.

Fungal Cellulases in Food Industry

Cellulases are used for various applications in food industries. (Beguin and Aubert, 1993; Coughlan, 1985a, 1985b; Mandels, 1985). 1) These enzymes facilitate the improvement in the extraction process of fruit juices and oils from seeds, 2) In the clarification of fruit juices, 3) For improving the soaking efficiencies and homogeneous water absorption of cereals, 4) In removal of external soybean coat in production of fermented soyabean foods, such as soysauce, 5) In the isolation of proteins from soybean and coconut, 6) for efficient isolation of starch from corn and sweet potato, 7) for the gelatinization of seaweeds to digest ball-milled lignocellulose which can be utilized as food additives.

Cellulases can also be used for (a) improving the nutritive quality of fermented foods, (b) improving rehydrability of dried vegetables and soup mixtures, (c) the production of cello-oligosaccharides, glucose and other soluble sugars from cellulosic waste and (d) for removal of cell wall which helps in releasing flavour enzymes, polysaccharides and proteins.

In brewery and wine industries, cellulases are used to hydrolyze α-1,3 and α-1,4 glucans, which are present in low-grade barley to help in filtration of beer. Cellulases are also used to increase the aroma in wines.

Fungal Cellulases in Detergents:

The enzymes used in detergents constitute major portion of industrial enzymes. Until now the use of enzymes in detergents was mainly limited to application of proteases, amylases, lipases as detergent additives. Recently, it was found that incorporation of cellulases as detergent components improves washing performance of detergents (Hoshiano *et al.*, 1997, Barbesgaard *et. al.*, 1984). It is known that washing of cotton containing fabrics by normal detergents generally causes a pronounced unpleasant harshness of the fabric. Along with dirt removal and colour clarification cellulase treatment also helps as fabric softener by reducing harshness of the cotton containing fabrics.

Several bacterial cellulases are available in the market that can be used for this purpose, but as the enzyme yields from bacteria are very low, their costs go up. Fungal cellulases can be more important from commercial point of view. Many types of cellulases derived from fungi have pH optimum around 5. As the pH increases above 7, their activity is greatly reduced. Usually the pH of the detergent solution is 7.0-10.5. Fortunately, certain fungal cellulases produced from strains of *Humicola* sp , *Fusarium* sp. , *Myceliophthora* sp. and *Cephalosporium* sp. (Kang *et al.*,1995) are active at a pH range mentioned earlier. Various proteases are also incorporated in the commercial detergents. Resistance to action of proteases is another desirable criterion for the enzyme to be used in detergents. The mechanism of cellulase action in detergent is different than other detergent enzymes that act directly on the soil and dirt in the cloths. Endoglucanase component of cellulolytic complex acts at the amorphous regions of fabric. This results in loosening of fibrilar structure of the cotton cloths, which in turn helps in removal of soil and dirt from these regions. The monocomponent enzyme formulations of fungal endoglucanases are preferred, as they are required in low quantities (Schulein *et al.*, 1997 b). These monocomponent cellulases do not show action over crystalline regions of cotton fibers hence cloths are not damaged even after repeated washing cycles.

Fungal Cellulase Enzymes in Textile Applications:

Cellulase offers an environment friendly alternative to the harsh chemicals being used for fabric finishing in textile industry. The successful application of cellulases in textile industry started with the process called 'Biostoning'. It was first introduced in Europe in 1989 and USA in 1990. For biostoning, the cellulase enzyme is used in place of pumice stone to produce characteristic abraded, faded appearance of jeans. In this process the enzyme cuts bond between cotton fibres and insoluble, surface adhering particles of indigo pigment to produce the necessary faded effect. The enzyme process is more attractive as it overcomes several limitations of traditional process. Even small amount of enzyme dose replaces several kilogrammes of stones in the washing machines that indirectly helps in higher processing capacities as well as high productivity. The enzyme treatment is fast and also avoids damage due to mechanical action of stones both on the garments and washing machines.

In the new fashion trends Denim jeans with worn-out looks are very much popular. Considering great potential in the business, R&D departments of various enzyme-manufacturing firms are actively assessing the various cellulase preparations. Several commercial formulations are available in the market and few are listed in table 3. According to Cavaco Paulo (1998),

Table 3. Commercial Cellulase Formulations for Textile Industry

Product	Manufacturer	Application	Remark
IndiAge® 2XL	Genencore	Low contrast Denim finishing	Conventional cellulase
IndiAge® Super L	Genencore	High- contrast Denim finishing	Engineered cellulase component, low operating temperature and neutral pH.
Denilite™	Novo Nordisk	Denim bleaching	Low contrast finishing
DeniMax®	Novo Nordisk	Denim finishing	Monocomponent cellulase High Contrast finishing
STONAG FP3	Dyadic International	Denim bleaching Denim finishing	Operative at neutral to alkaline pH.
ROCKSOFT™ BTU	Dyadic International	Polishing of cotton and lyocell fabrics	pH self adjusted Temperature 50-60° C
IndiAge® RFW	Genencore	Biofinishing of cotton, linen and rayon clothes.	Engineered cellulase component,
Cellusoft®	Novo Nordisk	For Biopolishing of cotton and lyocell and their blends	Monocomponent cellulase, pH 5.0,active in the presence of detergents.

cellulases rich in endoglucanase components are important in the process of 'Biostoning'. Avicel adsorbable endoglucanases are suitable for this application. It was found that cellulases with hydrophobic amino acid domains on the surface binds indigo dye and act as emulsifier helping them to separate from cellulose fibres (Gusakov *et al.*, 2000).

The 'Biostoning' process is very much sensitive to the effect of pH. Generally conventional acid cellulases works best around pH 5. If the process is carried out at pH 5 back staining problem appears. That means blue indigo dye released by enzyme action gets attached back on the fibre surface. This results in loss of contrast and also slight reddening of the original shade. However, this problem can be avoided if the process is carried around neutral pH. For this reason neutral cellulases operating best around neutral pH are in great demands for ' Biostoning '.

Biopolishing with fungal cellulases :

'Biopolishing' is the registered trademark of the Novo-Nordisk Industries for the process of cellulase treatments to cotton fibres and their blends. However, the technology began in 1988 in Japan and then it was introduced in Europe in 1993. It is routenly used by Germans and Italians to polish textiles for luxurious designer cloths. Cellulases have also proven useful in polishing of jute and its different blends.

A hairy ball of fuzz protruding out of yarn is called as pill in textile trade. These pills present serious quality problem because they result in unattractive knotty fabric appearance. Cellualse treatments cleave the ends of the fibres protruding out from the fabric. Biopolishing with cellulases have following additional benefits.1) It improves pilling resistance of cloths for long durations. 2) The fabrics show clean fuzz free appearance. 3) Biopolished textiles show improved drapability and softness.

Biopolishing of cotton clothes is generally carried out after bleaching steps since fabric is clean and more accessible to cellulase for the enzymatic action. During the process pH is maintained around 4.5 to 5.5 with temperature between 40-55°C. The choice of enzyme formulation containing multicomponent, or monocomponent cellulase enzyme depends on the particular type of fabrics or kind of machinery used. The process is generally carried out in the presence of anionic or cationic detergents and different fabric softeners are also used during the process. The susceptibility of enzymes to these various agents is important criterion for selection of enzymes for the process.

Cellulases in Paper and Pulp Industries :

Paper manufacturing is one of the largest industries in the world. Fungal cellulases are proven useful in various ecofriendly applications in paper and pulp industry (Kent Kirk and Jeffries, 1996)

The mechanical process to improve paper strength by enhancing fibrillation and inter fibre bonding is called as beating or refining of pulp. Cellulases can be used for 'Biobeating' or Pulp refining. The first process of refining paper fibrils with cellulases from white rot fungus *Trametes suaveolens* was patented in 1968 by Yerkes. The process was successful in reducing beating time. Presently various commercial formulations such as Pergalase, Maxazyme, Celluclast 1.5L are available in market for this application.

Enhancement of drainage rates with cellulase enzymes can be another success story of cellulases in paper industry. Drainage rate determines the loss of water during paper production. Low drainage rate severely affects machine speed that ultimately results in high-energy consumption with low-productivity. This problem is more critical when secondary fibres are used for paper production. Secondary fibre pulp with the commercial cellualse preparation can dramatically enhance drainage rates of recycled fibers (Pommier *et al.*, 1990). Commercial cellulase preparations such as Maxazyme 2000, Cellulase 250 P and SP 249 are available in the market for this application.

'Biodeinking' with Cellulases :

The demand for paper consumption is increasing all over the world. Any modern office produces waste paper that includes laser printed papers and photocopier papers. Toners and laser printing inks are synthetic nylon or polyester based polymers with embedded carbon black that do not disperse readily. Toners are difficult to disperse than conventional oil based inks. Toners remain as large, flat rigid particles that separate poorly, from fibres during flotation and washing stages. Costly dewatering and dispersion steps followed by additional flotation and washing are required to improve the deinking of laser printed paper. Here is the real challenge for paper technologists because none of the conventional deinking methods can remove these inks from paper. Moreover, it is the most valuable grade paper available in market. In the enzymatic deinking of mixed office waste paper cellulases play a key role in releasing thermally fused toners from fibre surfaces (Elgir *et al.*, 2000, Sreenath *et al.*,1996, Morbak *et al.*, 1998 , Sanciolo *et al.*, 2000). The process of removing ink and other contaminants from recycled fiber is termed as deinking.

The most favoured fungal sources for research and production of cellulases are *Aspergillus niger, Trichoderma viridae (T.reesi), Penicillium* sp.,

Fusarium sp. *and Humicola* sp. Each of these microorganisms can produce different kind of cellulase system that differs in their mode of action as well as properties like activity towards crystalline cellulose, activity and stability in acid or alkaline pH. Different kinds of cellulases work best under different operational conditions. Acid cellulases that hydrolyze cellulose in acidic pH range of 3-6 can be used in acid sized (sulphite paper) paper. Alkaline cellulases that are effective between pH 8-10 are more useful since alkaline sized papers are more acceptable owing to their high fibre flexibility. Calcium carbonate is often used in the paper as a filler and brightner. It dissociates from fibres in alkaline environments. This also leaves fibres with alkaline pH hence alkaline active and stable cellulases are necessary to process them for deinking. This automatically saves additional acid treatment required for acid cellulases to work. However, microbial alkaline cellulases are very rare. A few *Bacillus* sp. produce cellulases that are active in alkaline pH. Unfortunately bacterial cellulases lack complete cellulase complex. Thermal stability of enzyme is an additional important prerequisite for enzyme complex to be used for development of a commercial process. These all requirements led to shift research focus on screening of alkalophilic fungi that are capable of producing thermostable cellulases and are operative in a pH range of 8-10. Novozyme 342 from Novo Nordisk industries, Denmark is the commercially available enzyme preparations for deinking process. It is obtained on a large scale from micromycetes fungus *Humicola insolense.* This enzyme complex is active in a broad pH range of 5-10.

The Mechanism of Enzymatic Deinking

Understanding the behavior of the enzymes towards fibre is essential to minimize the related negative effects on the strength of the paper as well as quality. The role of cellulases in deinking is still not clear. Jeffries *et al.* (1994) suggested that the enzymes having higher filter paper degrading activities are effective in deinking. However, Qin *et al.* (1998, 2000) have suggested that the filter paper degrading activity of the cellulase complex had detrimental effect on paper strength. According to their studies when cellulases with different cellobiohydrolases and endoglucanases activities from *Aspergillus* L22 and *Trichoderma pseudokoningii* S28 were used they have different effects on the fibre properties. During deinking process individual endoglucanase treatment does not result in the decrease of average fibre length or coarseness, which is important to modify the pulp fiber. The synergistic effect of the endoglucanases and cellobiohydrolases could lead to the cross breakage, and this is probably the major cause for the decrease in the paper strength. Gubitz *et al.* (1998a,b) treated laser-printed wastepaper individually and with combinations of purified endoglucanases from *Gloeophyllum sepiarium* and *Gloeophyllum trabeum* and found that pure endoglucanases were responsible for most of the success in deinking operations. Although all enzyme treatments increased freeness and slightly decreased intrinsic fibre strength, the strength of the handsheet's was maintained or marginally improved. The nature of the individual endoglucanase plays an important role in this process. According to their observations different deinking efficiencies and strength properties were measured with both the endoglucanases, those are closely related enzymes in terms of molecular properties and substrate specificities. Recently in our studies we have found that the alkaline active cellulases from alkalotolerant *Fusarium* sp. are useful for deinking of MOW paper. The viscometric data showed that low molecular weight endoglucanase from alkalotolerant *Fusarium* sp. is more random in its action and has more access to the surfaces of cellulose fibers. The randomly acting endoglucanases are responsible for overall deinking process.

In the flotation deinking process the pulp is diluted to 1% consistency. Now the air bubbles are injected through the pulp from the bottom. The ink particles that are more hydrophobic get attached to bubbles that rise to the top creating froth layer. This layer is swept off by skimming action. Here the stability of air bubble is the most important factor. Generally ink particles are 40-50 μ in size. Additional chemicals termed as 'collectors' are needed if the ink chemicals are smaller in size. All the ink particles can't be removed in a single run. So the process is repeated couple of times. The mechanism of enzymatic deinking suggested that the enzyme molecules reduce the hydrodynamic drag that in turn increases filtration and flotation rates. In flotation process, the air bubble rising to the top carries ink particles along with cellulose fibres. These fibres exert negative effects on flotation efficiencies. Cellulases simply 'shaves' fibres from the ink particles where fibres will settle down and ink particles will adhere to the air bubble in much better way experiencing much less drag from interaction with water. Thus, the ink particles can be carried to the top of the tank much easily. In short, the cellulase enzyme complex reduces the hydrodynamic drag produced due to the attachment of fibres to the floating air bubbles. The enzyme complex can also reduce specific surface area of fibres, thereby lowering the interaction with contaminants. The enzyme activity enhances the draining resulting in enhancing filtration or flotation process. Once the flotation cycles are over, the pulp passes through different micro screens (0.15 mm) under pressure and the accepted fibres undergo subsequent washing steps. This results in removal of stickies and other contaminants.

Advantages of Enzymatic Deinking

The deinking with enzymes works best with chemically pulped fibre as compared to mechanically pulped fibre. This is because lot of lignin is associated with mechanically pulped fibres. As compared to water deinked pulps the enzyme releases upto 95% of toner particles from recycled fibres. The properties of deinked pulp such as tensile index, burst index, tear index are comparable to pulps from conventional deinking procedures. The enzymatic process preserves fiber integrity and also shows improved pulp brightness. Effluent samples from the deinking plants indicated lower oxygen demands as well as less toxicity than that of traditional processes.

The enzymatic" biodeinking" process requires fewer amounts of chemicals. This is a double benefit as it saves cost of waste treatment along with saving cost of chemicals required for the process. It also avoids landfill disposal costs. Enzyme deinked pulp shows high drainage rates resulting in faster machine speed and less energy cost.

The technology for paper and pulp manufacturing is changing rapidly. This industry is becoming a major prospective market for the enzyme industry. However, enzymatic deinking is relatively a new process. Many of the paper and pulp industries all over the world are aware of this technology. They have even developed some unique pilot plants to explore the ecofriendly enzymatic technology of the future. The technologists from Forest Products Laboratory (Madison, WI, USA) have prepared some cost analysis of enzymatic deinking. It is approximately US $ 1.60 to 2.40 per 1000 Kg of the recovered paper. In some cases, depending upon the quality of the waste paper, the expenses can be as high as US $ 18 per 1000 Kg of the raw materials. In comparison, conventional deinking, chemical cost is between US $ 3.30 to 8.00 when sodium hydroxide (1%) is used alone or in combination with hydrogen peroxide. Considering many factors the cost of conventional deinking process and proposed enzymatic process will not differ much. Thus the Biodeinking with cellulases can fulfill the demands for better quality papers in ecofriendly way. Indeed it is the novel ecofriendly way of recycling the wastepaper.

Miscellaneous

Cellulases have also found useful in other various applications. These miscellaneous applications of cellulases are described in table 4.

Table 4. Miscellaneous Applications of cellulase enzymes

Area	Applications
Animal Feed	Improving digestibility of usual forage. Processing of wood and wood product as feed
Pharmaceuticals	Digestive aids Elimination of unwanted fibers (phytobezoars)
Pollution control	Cleaning of solid waste in municipal dumps. Digestion of excreta in septic tanks and drainage
Fermentation	Providing substrate for the synthesis of methane, ethanol, glycerol, citric acid, lactic acid, vitamins, antibiotics, single cell proteins and in breweries.
Flavors	Cellulases treatments reduces maturation period of tobacco, tea leaves.
Research	To dissolve plant cell walls and to obtain protoplasts from plant cells.

Concluding remarks

Cellulases is a group of enzymes that hydrolyze α 1- 4 - glucosidic linkage of cellulose. Considerable research has been carried out on fungal cellulases during last three decades with special attention to their molecular properties and catalytic properties. More recently the possibilities of their use in specialized applications like detergents, textile industry and paper and pulp industries have been assessed in different laboratories. Extracellular alkalistable cellulases from fungal sources are commercially successful because of the excellent qualities and productivities. The other enzyme systems from *Trichoderma* sp., *Aspergillus* sp., *Penicillium* sp., Fusarium sp. and *Humicola* sp. are found to be suitable to a certain extent in this respect.

The search for novel cellulases from extremophilic microorganisms and their applications in ecofriendly industrial processes will be the major driving forces for cellualse research in years to come.

Acknowledgement

S.V. is thankful to the Council of Scientific and Industrial Research (CSIR), Government of India, for financial assistance.

References

Barbesgaard, P.O, Jenson, G.W. and Holm, P. (1984). Detergent cellulases, U.S.patent No. 4,435,307. Nov Industri A/S Denmark.

Beguin, P. and Aubert, J.P. (1993). The biological degradation of cellulose, FEMS Microbiol. Rev., **13**: 25-58.

Beldman, G., Searle-Van Leewen, M.F., Rombouts, F.M. and Voragen, F.G.J. (1985). The cellulase of *Trichoderma viride* Purification characterization and comparison of all detectable endoglucanases , exoglucanases and α-glucosidases. Eur. J. Biochem., **146** :301-308.

Bhat, K.M., McCrae, S.I. and Wood, T.M. (1989). The endo-(1-4)-α-D-glucanase system of *Penicillium pinophilum* cellulase: isolation, purification and characterization of five major endoglucanase components, Carbohydr. Res., **190** : 279-29.

Bhikhabhai, R. Johansson, G. and Petterson, G. (1984). Isolation of cellulolytic enzymes from *Trichoderma reesei* QM 9414, J. Appl. Biochem, **222** : 729-36.

Bhikhabhai, R., Johansson, G. and Petterson, G. (1985). Cellobiohydrolase from *Trichoderma reesei.* Int. J. Peptide protein Res., **25** : 368- 75.

Bruchmann, E. E., Schach, H.S. and Graf, H. (1987). Role and properties of lactonase in cellulase system, Biotech. Appl. Biochem., **9** : 146-59.

Cavaco-Paulo, A., Cortez, J. and Almeida, L. (1998). The effect of cellulase treatment in textile washing processes, J. Soci. Dyers and Colour., **113** : (7/8) 218-222.

Christakopoulos, P., Kekos, D., Macris, B.J., Claeyssens, M. and Bhat, M.K. (1995a). Purification and Characterization of less randomly acting endo -1,4-α-D-glucanase rom culture filtrates of *Fusarium oxysporum*, Arch. Biochem. Biophys., **316** : 428-33.

Christakopoulos, P., Kekos, D., Macris, B.J., Claeyssens, M. and Bhat, M.K. (1995b). Purification and mode of action of low molecular mass endo -1,4-α-D-glucanase from *Fusarium oxysporum*, J. Biotechnol, **39** : 85-93.

Coughlan, M.P. (1985 a). Cellulases production properties and applications. Biochem. Soc.Trans, **13** : 405-406.

Coughlan, M.P. (1985 b). The properties of fungal and bacterial cellulases with comment on their production and application, In: Biotechnology and Genetic Engineering Reviews, Vol **3** (Russell, G.E., editor) Interscience : 39-109 .

Darnell, J., Ladisch, H. and Baltimore, D. (1986). "Molecular Cell Biology" Scientific American, NewYork.

Elegir, G., Panizza, E. and Canetti, M. (2000). Neutral enzyme-assisted deinking of xerographic office waste with a cellulase/amylase mixture. Tappi J., **83**:11:71.

Errickson, K.E. (1978). Enzyme mechanisms involved in cellulose hydrolysis by the rot fungus *Sporotrichum pulverulentum*. Biotechnol. Bioeng., **70**: 317-332.

Errickson, K.E. and Peterson, B. (1982). Purification and partial characterization of two acid proteases from the white rot fungus *Sporotrichum Pulverulentum* Eur. J. Biochem., **124**: 635-42.

Eveleigh, D.E. (1987). Cellulase : a perspective phil., Trans. R. Soc. Lond., **321**:435-7.

Fagerstam, L.G. and Petterson, L.G. (1980). 1,4-α-glucan cellobiohydrolases of *Trichoderma reesei* QM 9414. A newtype of cellulolytic synergism , FEBS lett., **119**: 97-100.

Gong, C.S. and Tsao, G.T. (1979). Cellulase and biosynthesis regulation , Ann. Rep.Ferment.Proc., **3**:111-40.

Gubitz, G.M., Mansfield, S.D., Bohm, D. and Saddler, J.N. (1998 a). Effect of endoglucanases and hemicellulases in magnetic and flotation deinking of xerographic and laser- printed papers, J. Biotechnol., **65**: 209-215.

Gubitz, G.M., Mansfield, S.D., Bohm D. and Saddler, J.N. (1998 b). Effectiveness of two endoglucanases from *Gloeophyllum* species in deinking mixed office waste paper, Int. Conf Biotechnol Pulp Pulp Ind, 47th, **C**: 135-138.

Gum, E.K. and Brown, R.D. Jr. (1977). Comparison of four purified extracellular 1-4,α-D glucan cellobiohydrolase from *Trichoderma viride.* Biochim. Biophys.Acta, **492:** 225-31.

Gusakov, A.V., Sinitsyn, A. P., Berlin A.G., Markov, A. V. and Ankudimova N. V. (2000). Suraface hydrophobic amino acid Resiues in Cellulase molecules as a structural factor responsible for their high Denim washing performance, Enzyme Microb. Technol., **27**: 664-671.

Hayashida, S., Ohta, K. and Mo, K. (1988). Cellulases from *Humicola Insolens* , *Humicola grisea* , Methods In Enzymology, **160**: 323-332.

Henrissat, A. and Bairoch, B. (1993). New families in the classification of glycosyl hydolase based on amino acid sequence similarity, Biochem. J., **293**: 781-788.

Hoshino, E. and Susumo, I. (1997). *Enzymes in detergency*, Marcel Dekker Inc., **9** : 149

Hurst,P.L.,Nielsen, J., Sullivan, P. A. and Shepherd, M.G. (1977). Purification and Properties of cellulase from *Aspergillus niger* Biochem. J.,**165**:33-41.

Jeffries, T. W., Klungness, J. H., Marguerite, S. and Cropsey, K. R. (1994). Comparison of enzyme enhanced with conventional deinking of xerographic and laser-printed paper. Tappi J, **77**,4: 173-179.

Kang M. K. and Rhee, Y. H. (1995). Carboxymethyl cellulases active and stable at alkaline pH from alkalophilic *Cephalosporium* sp.RYM-202, Biotech Lett., **175**: 507- 12.

Kent Kirk, T. and Jeffries, T.W. (1996), Role of microbial enzymes in pulp and Paper Processing , ACS symposium Series . *655*, Ch., **1**:1-13.

Kubicek, C.P. (1992). The cellulase proteins of *T. reesei* : the structure ,multiplicity, mode of action and regulation formation, Adv. Biochem. Eng., **45** : 1-27.

Kubicek, C.P., Messner, R., Gruber, F., Mach, R.L. and Kubicek -Pranz, E.M. (1993). The *Trichoderma reesei* Cellulase regulatory puzzle from the interior of a secretory fungus. Enzyme microb.Technol., **15**: 90-99.

Kumar, P.K.R., Singh, A. and Schgerl, K. (1991). Formation of acetic acid from cellulosic substrates, Appl. Microb. Biotechnol., **34:** 570-572.

Labudova, I. and Farkas, V. (1983). Multiple enzyme forms in the cellulase system of *Trichoderma reesei* during its growth on cellulose, Biochim.Biophys.Acta,**744** :135-40.

Lachke, A.H. and Deshpande, M.V. (1988). *Sclerotium rofsii*. Status in cellulase research . FEMS Microbiology Reviews., **54**: 177-184.

Mandels, M. (1985). Application of cellulases, Biochem Soc.Trans., **13**: 414-415.

McHale, A. and Coughlan, M.P. (1980). Synergistic hydrolysis of cellulose by scomponents of extracellular cellulase system of *Taleromyces emersonii*, FEBS Lett. **117** : 319-322.

Merivuori, H., Seiglar K.M., Sands J.A. and Montenecourt, B.S. (1985). Biochem Soc.Trans., **13**: 411.

Mo, K. and Hayashida, S. (1988), Synergistic hydrolysis of cellulose by components of extracellular cellulase system of *Taleromyces emersonii* Agric. Biol.Chem., **52** :1683-8.

Morpeth, F.F. (1985). Some properties of cellobiose oxidase from white rot fungus *Sporotrichum pulverulentum* , Biochem.J., **228**: 557-64.

Morkbak, A. L. and Zimmermann, W. (1998). Deinking of recovered mixed office paper by treatment with a single endo-glucanase I from *Humicola insolens* and a neutral surfactant. Prog. Pap. Recycl., **73**: 33-37.

Okada, G. (1985). Purification and properties of cellulase from *Aspergillus niger* Agric. Biol. Chem., **49**: 1257

Pommier, J.C., Goma, G., Fuentes, J.L. and Rousset, C. (1990). Tappi J., **7312**: 197.

Qin, M., Gao, P., Qu, Y., Fu, Y., Shao, Z. and Quan, W. (1998). Physical characteristics of enzymically modified fibers from old newsprint , Emerging Technol. Pulping Papermaking Fast-Grow. Wood, Proc. Int. Symp., 462-73.

Qin, M., Gao, P., Qu, Y., Fu, Y., Shao, Z. and Quan, W. (2000). Enzymatic deinking of old newsprint, Zhongguo Zaozhi (Chinese) **194** : 7-13.

Saloheimo, M., Lehtovaara, P. Pentilla, M., Teeri, T.T., Stahlberg, J., Johansson, G., Petterson, G., Claeyssens, M., Tomme, P. and Knowles, J.K.K (1988). EGIII , a new endoglucnase from *T. reesei* : The characterization of both gene and enzyme .Gene, **63**, 11-21.

Salovuori, J. Macarow, M. Rauvala, H., Knowles, J.K.C. and Kaarianen, L. (1987). Biotechnology **5**: 152-55.

Sanchez, A., Villanueva, J.R. and Villa, T.G. (1982). Effect of Tunicamycin on Exo-1-3-α-D.glucanase synthesis and secretion by cells an protoplasts of *Saccharomyces cerevisiae*, J. Gen. Microbiology, **128** : 3051-3055.

Sanciolo, P., Warnock, H., Harding, I., Forbes, L. and Lonergan, G. (2000). Microscopy study: liberation of ink from fibers during enzymic deinking of mixed office papers, Prog. Pap. Recycl., **93** : 22-30.

Schulein, M. (1997). Enzymatic properties of cellulases from *Humicola insolens* , J. Biotechnol., **57** : 71-81.

Schulein, M., Tikhimirov, D.F., and Charlotte, Schou. (1993). *Humicola Insolens* alkaline cellulases. Proc. Of second Tricel Symposium on *Trichoderma Reesei* cellulases and other Hydrolases, **8**: 109-116.

Schulein, M. and Levering, K. (1997). Fungal cellulases composition containing alkaline- CMC-Endoglucanase And Essentially No cellobiohydrolase. United State patent No. 5691178.

Shoemaker, S., Watt, K., Tsitovsky, G. and Cox, R.(1983). Characterization and properties of cellulases purified from *Trichoderma reesei* strain L27. Bio.Technol., **1**:687-90.

Sreenath, H.K., Yang, V.W., Burdsall, H. and Jeffries, T.W. (1996). Toner removal by alkaline active cellulases from desert Basidiomycetes, Enzymes for pulp and Paper processing, *American* Chemical Society. Ch., **21** : 207-19.

Teeri, T.T., Lehtovaara, P., Kaupinnen, S., Salovuori, I. and Knowles, J.K. C. (1987), Homologous domains in Trichoderma reesei cellulolytic enzymes : gene seqence and expression of cellobiohydrolase II , Gene, **51** : 43-52.

Van Arsdell, J.N. Kwock, S., Schweickart, V., Ladner, M.B., Gelfand, D.H. and Innis, M.A. (1987). Cloning and characterization and expression in *Saccharomyces cerevisiae* of endogucanase I from *Trichoderma reesei* , Bio. Technol., **5**: 60-64.

Van Tibeurgh, H., Tomme, P., Claeyssens, M., Bikabhai, R. and Pettersson, G. (1986). Limited proteolysis of the cellobiohydrolase I from *Trichoderma reesei*: separation of functional domains. FEBS Let., **204**: 223-7.

Westermark, U. and Errickson , K -E. (1974). Celllobiose : Quinone oxido reductase- a new wood -degrading enzymes from white rot fungi. Acta. Chem. Scand., **28** : 209-14.

Wood, T.M. and McCrae, S.I. (1982). Purification and some properties of a (1,4)-α-D-glucan glucohydrolase associated with the fungus *P. funiculosum*, Carbohydr. Res., **110** : 291-303.

Wood, T.M. (1992). Microbial enzymes involved in degradation of cellulose component of plant cell wall, In Rowett Research Institute Annual Report, pp 10-24.

Yerkes, W.D. (1968). United States patent No : 3,406,089.

10

Aflatoxin : Contamination, Toxicosis and Preventive Measures

R.J. Verma*

Department of Zoology, University School of Sciences, Gujarat University, Ahmedabad –380 009

Aflatoxins are highly substituted coumarin derivatives produced by *Aspergillus flavus* and *A. parasiticus* group of fungi. Although natural occurrence of aflatoxin in various food products have been reported from various countries and on vast array of crops, comparatively higher concentrations of aflatoxins were recorded from tropical and subtropical countries where environmental conditions are more congenial for mouldy growth and toxin production. Ingestion of aflatoxin-contaminated food/feed stuffs cause serious pathological condition called aflatoxicosis. Aflatoxin is well knwon hepatotoxic, hepatocarcinogenic, teratogenic and mutagenic agent. Various factors influencing aflatoxin toxicity and preventive measures to reduce toxicity have been discussed in this review.

Key words : Aflatoxin, aflatoxicosis, contamination, preventive measures.

Among the micro-organisms, fungi are more commonly associated with most of the cereals, oil seeds, dry fruits and spices. The growth of fungi on various food/feed stuffs not only decreases their nutritive value and taste but also add toxins.

That fungi produce toxic substances was suspected ever since some mushrooms were found to be poisonous in the early days of mycology. Ergotism or St. Anthony's fire caused by *Claviceps purpurea* was well known to the ancient Chinese, Greeks, Romans and Arabs; and was an epidemic scourge in Europe from the middle ages until the nineteenth century. Outbreak of alimentary toxic aleukia in USSR in 1930 was found to be causally associated with consumption of wheat grains infested by fusarium moulds; years later the fungus was shown to produce trichothecenes. Acute cardiac beri-beri outbreaks in 1952 and 1955 in Tokyo and in 1971 in Korea were strongly suspected to be due to the use of mouldy rice.

Aflatoxin research had a dramatic beginning. In 1960 one hundred thousand young turkeys in various poultry farms in England died in

*Crresponding Author E-mail: zooldeptgu@satyam.net.in

opisthotonous following earlier signs of anorexia, lethargy and muscular weakness. At post-mortem, subcutaneous haemorrhage and pale colour liver with extensive necrosis and biliary proliferation were recorded (Sargeant *et al.*, 1961). Concurrent with loss of turkey poults, large-scale losses of ducklings, partridges and pheasants poults were also reported (Asplin and Carnaghan, 1961). Investigations revealed that the common ingredient of feed in various farms were incorporation of imported Brazilian peanut meal.

Occurrence of this incidence attracted many groups of scientists to work together for isolation and characterization of the toxic compound present in the Brazilian peanut meal. Hartley *et al.* (1963) isolated 4 fluorescent compounds on thin layer chromatograms using methanol: chloroform (90:2) and designated them as aflatoxin B1, B2, G1 and G2 in order of decreasing Rf. The infrared and UV absorption spectra were similar indicating related structures; B2 and G2 were established as dihydro-derivatives of B1 and G1.

This incidence lead world over major research efforts aimed at the identification of the toxins produced by moulds growing on foods and feeds, quantification of these toxins and an evaluation of biological effects on human beings and animals. The fungi that are known to produce toxic chemicals are found in almost all major taxonomic groups. Toxic chemicals produced by fungi are collectively called as mycotoxins (*mykes*= fungi; *toxicum*= poison) and the disease associated with ingestion of mycotoxin contaminated food/ feed stuffs as mycotoxicosis. Mycotoxins, a group of secondary toxic fungal metabolites are relatively low molecular weight organic compounds characterized by their diversity, their frequent specificity with regard to the taxonomy of the producing organisms and their production during the stationary phase of batch cultures.

For farm animals or human populations whose diet largely contain plant products, mycotoxins may be produced directly by growth of moulds on animal feed or human food. Illnesses caused by consumption of such contaminated food/feed stuffs are conveniently referred to as primary mycotoxicosis. It is also possible that toxicosis may occur as a result of ingestion of animal products such as milk or meat which have not themselves been contaminated by mould growth. Illnesses arising from such sources are referred as secondary mycotoxicosis.

Amongst all the mycotoxins so far identified, aflatoxins have received maximum attention because of its toxicological characteristics and also because of the fact that they are most frequent contaminants of a variety of foods consumed by human beings. The name of aflatoxin is derived from the fungus *Aspergillus flavus* (*A. flavus* toxin).

Aflatoxins (B1, B2, G1, G2) are highly substituted coumarin derivatives containing a fused dihydrofurofuran moiety. AFB1 and AFB2 were named because of their strong blue fluorescence under UV light, whereas AFG1 and AFG2 fluoresced greenish yellow. The B-toxins were characterized by the fusion of a cyclopentenone ring to the lactone ring of the coumarin structure, while G-toxins contained an additional fused lactone ring. AFB1 and to a lesser extent AFG1 are responsible for the biological potency of aflatoxin-contaminated meals and crude fractions derived from toxigenic *Aspergillus flavus* cultures. These two toxins possessed an unsaturated bond at the 2,3 position (the 8, 9 position according to IUPAC nomenclature) on the terminal furan ring. AFB2 and AFG2 are essentially biologically inactive unless these agents are first metabolically oxidized to AFB1 and AFG1 *in vivo*.

Allcroft and Carnaghan (1963) found that cows when fed with toxic groundnut meal, excreted a toxic factor in the milk which had the same biological effect on 1-day old ducklings as had aflatoxin B1. De Iongh *et al.* (1964) showed by TLC that this toxic factor designated as 'milk toxin' was a blue fluorescent compound with an Rf lower than that of aflatoxin B1. Holzapfel *et al.* (1966) isolated two components from the urine of sheep fed with mixed aflatoxins, which were similar to the 'milk toxin' and designated them as M1 and M2.

Aflatoxins are soluble in methanol, chloroform and other organic solvents but are only sparingly soluble in water (10 to 30 μg/ml). These toxins strongly absorb UV light (362 nm) with extinction coefficients in methanol or ethanol varying from 17 000 for AFG2 up to 24 000 for AFB2. Though aflatoxins are quite stable in foods and feeds, they are rapidly inactivated by extremes of pH (less than 3 or more than 10), oxidizing agents or exposure to UV light in the presence of oxygen.

Aspergillus flavus and *A. parasiticus* are important toxigenic moulds producing aflatoxins. During early period of recognition, it was considered as a part of storage flora of inadequate post-harvest storage and poor storage conditions during distribution of commodities such as peanuts. Ciegler (1975) has reported that pH, moisture and temperature are three important factors that govern the production. However, it has now become increasingly recognized that the contamination of some commodities especially peanuts and maize with aflatoxin is a far more complex phenomenon which may involve infection and aflatoxin production in the field. Although contamination of crop in field is undoubtedly influenced by insect damage, *A. flavus* is also known to infect intact kernels of maize by colonizing and growing down the external silk (Bilgrami *et al.*, 1980; Sinha and Ranjan, 1989).

All *Aspergillus flavus* group of fungi are not toxigenic (Raper and Fennell, 1965). Morphologically it is not possible to distinguish toxigenic and non-toxigenic strains. Genetical factors might be responsible for variations in aflatoxin production by different strain (Maggon *et al.*, 1969; Ciegler, 1975). Heathcote and Hibbert (1978) assigned the interconversion of aflatoxin B1 into different types of aflatoxins by toxigenic strains. However, there are ways of biochemical differentiation between toxigenic and non-toxigenic strains; the former produce more of acetaldehyde which acts as a precursor for aflatoxin production.

The quantity of aflatoxins produced by isolates of *A. flavus* vary widely with the nature of the strain (Diener and Davis, 1966). Toxigenic isolates of *A. flavus* varied also in their capacity to elaborate different components of aflatoxins. However, the frequency of *A. flavus* isolates to produce aflatoxins other than aflatoxin B1 was comparatively low. Some isolates of *A. flavus* produced only the B-compound; some only the G; but the majority produced both B and G compounds. None of the toxigenic isolates produced B2, G1 and G2 in absence of B1 (Lillehoj *et al.*, 1977). Most isolates of *A. parasiticus* produced all four aflatoxins.

The fungal association with food and feed materials varied with the nature of the species and substrates on which they grew. Reports from India and abroad suggest that the incidence of fungal infestation depends on the composition of the substrates (Hitokoto *et al.*, 1984; Bilgrami, 1984). Factors like variations in climatic conditions and place of collections modulate the pattern of fungal incidence in different samples. Oil seeds and their products are comparatively higher risk substrates.

Extensive surveys have been made all over the world with regard to aflatoxin contamination of cereals, oil seeds, pulses, dry fruits, nuts, spices (Dwarkanath *et al.*, 1969; Bean *et al.*, 1972; Scott, 1973; Hesseltine, 1974; Nagarajan *et al.*, 1974; Diener, 1976; Basappa *et al.*, 1977; Singh, 1984), fleshy fruits, vegetables, milk, poultry and cattle feeds (Katoch *et al.*, 1975; Neelkantan *et al.*, 1978; Jagdish Kumar and Veernarayan Gowda, 1979; Nusrath and Nahdi, 1983; Bilgrami, 1984; Bilgrami and Sinha, 1987) and oil cakes (Kolhe *et al.*, 1994; Wagle, 1970; Sahay and Ranjan, 1987). Reddy *et al.* (1983) reported various mycotoxin contaminations in mixed food from Andhra Pradesh and detected highest concentration of aflatoxin B1 among all the mycotoxins studied.

Eight peanut varieties viz., Dhokya Red, Dhokya White, Ghungari Simple, G-2, G-51, HB-11, G-97 and Akola Farm were screened for aflatoxin B1 contamination. Maximum amount (144 ppb) was present in Dhokya Red seeds followed by HB-11 and G-2. The Ghungari Simple seeds contained least amount of AFB1 contamination (Verma *et al.*, 1996a). Among different oil seeds aflatoxin contamination was highest in peanut seeds followed by cotton, sesamum and castor (Kolhe *et al.*, 1994). Aflatoxin content was highest in peanut cakes than in castor, cotton or sesamum. Analysis of oil samples revealed that aflatoxin contamination was highest in peanut followed by sesamum, cotton and castor (Verma *et al.*, 1997).

Although natural occurrence of aflatoxin in agricultural products has been reported from many countries and on vast array of crops, the levels of aflatoxins detected vary greatly from area to area. In general, higher quantities of aflatoxins were recorded in commodities from tropical and subtropical countries where environmental conditions are more congenial for moldy growth and toxin formation. In a survey of peanut products in North America, 19% of 1416 samples examined were contaminated with an average level of 1 μg/kg, whereas, in Thailand 49% of 216 samples contained AFB1 at an average level of 424 μg/kg (Shank *et al.*, 1972). As much as 260 ug aflatoxin/kg was found in the sample of oat in Sweden shown contaminated with *A. flavus*. Very severe contamination may sometimes occur. In parts of India 100% of maize samples have been found contaminated with aflatoxin in the range of 6250-15600 μg/kg (Krishnamachari *et al.*, 1975).

Verma *et al.* (1995) analysed various samples of betel nuts, tobacco, coriander pulses, pan masala and pan masala with tobacco. Amongst these, maximum contamination was in betel nuts followed by tobacco and pan masala. Heroin may be contaminated with aflatoxin. Two samples contained 507 and 1346 pg/g of aflatoxin B1 in heroin. Intravenous heroin users are probably injecting aflatoxin B1 into their systemic circulation and thereby suppress cell-mediated immunity (Lamplugh *et al.*, 1988; Hendrickse and Maxwell, 1988).

It appears that biological risk of exposure to aflatoxin is much lesser in technologically developed countries than in developing ones. Reduced exposure to aflatoxin is technically and economically feasible through the prevention of contamination of food products or raw materials by rapid post-harvest drying of crops and controlled storage conditions.

Contamination of finished foods/feeds with moulds invariably leads to the exclusion of that product from the food chain in developed economies. However, in most developing countries this is generally not possible and mouldy foods such as cereals are often a regular part of the daily diet. In parts of Africa where liver cancer is extremely prevalent, regular occurrence of aflatoxins and fusarial toxins in the cereal foods have been recognized.

Undoubtedly stricter adherence to good food and feed hygiene would greatly reduce mycotoxin presence.

Incidence of Aflatoxicosis and Pathological Symptoms

In 1960, large-scale death of young turkeys, ducklings, patridges and pheasant poults in various poultry farms in Britain was directly correlated with ingestion of aflatoxin-contaminated Brazilian peanut meal.

In controlled feeding experiments where cattle consumed feed containing 2000 ppb of AFB1, the liver exhibited bile duct proliferation, an increase in connective tissue and degradation of hepatocytes (Allcroft and Lewis, 1963). One of the earliest outbreaks of aflatoxicosis in cattle from India was recorded by Sastry *et al.* (1965). This disease was characterised by loss of appetite, diarrhoea, dullness, ascites, emaciation, anaemia and icterus and affected 24 Murrah buffaloes. It was speculated to be due to feeding peanut cakes to the animals. An outbreak of aflatoxicosis in dairy cattle involving death of 58 cross-bred cattle was reported by Gopal *et al.* (1968). The symptoms of affected cattle include anorexia, apathy, depression, corneal opacity, ascites, diarrhoea and occasional nervous symptoms like excitement and walking in circles. Peanut cake was major component in the feed mix and aflatoxin ranging between 1.1 to 25 ppm was found in the mixture. Gopal *et al.* (1969) reported an outbreak of aflatoxicosis in Karnataka which involved death of 2219 chicks with symptoms of sudden anorexia, loss of weight, staggering gait and convulsive movements. Subacute and chronic cases showed enteritis, ascites and histopathological changes typical of aflatoxicosis such as yellow liver, fatty changes and biliary epithelial hyperplasia.

In an outbreak at a private farm in Kullu valley 4000 rabbits died. In 15 autopsies carried out on the animals died by chronic toxicity, the characteristic changes in the liver were marked proliferation of connective tissue and hyperplasia of bile duct (Mehrotra and Khanna, 1973). Manickam *et al.* (1985) reported an incident where nine buffaloes were affected and one of them died. The affected animals were fed with sorgham stalk for about 4 months as the only source of feed and they showed dyspnoea, uncoordinated movements, sudden fall and inability to get up without assistance and a loss of sensory perception in the hindquarters.

During 1974 an outbreak of hepatitis occurred among tribal areas extending over 200 villages of Banaswada district of Rajasthan and Panchmahal district of Gujarat. The disease was characterized by jaundice, rapidly developing ascites and portal hypertension. About 500 persons predominantly male (sex ratio 2:1) in the age group of 5-14 and above 30 years were affected and more than 100 persons died of the disease. Maize, the staple food of these people in the area was found to be contaminated with *A. flavus* and it showed aflatoxin level ranging between 6500 to 15600 μg/kg. In autopsied liver bile duct proliferation and multinucleate giant cells were noticed (Krishnamachari *et al.*, 1975).

Indian childhood cirrhosis (ICC) mainly confined to Indian subcontinent has been attributed to aflatoxin consumption (Amla *et al.*, 1974). A similar disease affected 3 children in Uganda whose diet consisted mainly of cassava, beans, fish and meat (Serck-Hanssen, 1970). The oldest, a 15 year old boy died 6 days after onset of signs including oedema of the extremities, vomiting and abdominal pain. The cassava from the house- hold contained 1.7 mg aflatoxin/kg.

In Thailand cases of a different syndrome in children were associated with acute aflatoxin consumption. Cases of encephalopathy and fatty degeneration of viscera (EFDV) were reported in North-Eastern Thailand where there was a high incidence of aflatoxin in the food (Olson *et al.*, 1971). An acute paediatric syndrome with a high mortality rate known as Reye's syndrome has been associated with aflatoxin poisoning (Reye *et al.*, 1963). Death usually occurs within 72 hours with symptoms of encephalopathy and fatty degeneration of viscera. Aflatoxin B1, G1 and B2 were detected in the liver and other vital organs of the children in Thailand where the disease was endemic (Olson *et al.*, 1971).

The incidence of kwashiorkor was found to be greater in warm humid climates during the rainy season where the staple foods are starchy and low in protein eg. plantain, rice, maize (Hendrickse, 1984). Such countries include Uganda, parts of Zaire, Southern India, Mexico, Central America, parts of Brazil and South Africa. These starchy foods provide excellent substrates for aflatoxin production. Aflatoxins have been detected in foods in Sudan both from market and from homes of the children examined (Hendrickse *et al.*, 1982) in addition to human breast milk in the same area (Hendrickse, 1984).

An outbreak of food poisoning resulting in 13 deaths in children occurred in Malaysia during the Chinese festival of the Nine Emperor Gods in 1988. The food poisoning was attributable to aflatoxins and boric acid. Clinical features included an initial Reye-like syndrome with vomiting, fever, diarrhoea, abdominal pain, anorexia, seizures and coma. Patients died in acute hepatic and renal failure. High levels of aflatoxins B1, B2, G1, M1, M2 and aflatoxicol were found in various organs (Chao *et al.*, 1991).

Factors Influencing Aflatoxicosis

Poultry animals are acutely sensitive to AFB1. Young turkeys and ducklings are more susceptible to aflatoxin than are chickens. Guinea pigs and rabbits are among the more sensitive rodents to the acute toxicity of AFB1. The rats exhibit moderate susceptibility to acute toxicity of aflatoxins. A species classified as being resistant to the acute toxicity of aflatoxins is the mouse. Non-human primates are sensitive to the acute toxic effects of AFB1 (Wong and Hsieh, 1980). The type of hepatic lesions varies with the species of monkey. Centrilobular necrosis with biliary proliferation and fibrosis were characteristic for cynomologous and African green monkeys (Cuthbertson *et al.*, 1967) rather than the focal or periportal necrosis as in rhesus monkey.

Results of a number of studies indicate that in comparison with males, female rats are more resistant to both acute toxic and carcinogenic effects of aflatoxins (Butler *et al.*, 1969; Ward *et al.*, 1975). Male rats (Charles River CD) fed for more than one year on a diet containing AFB1 at 0.2 mg/kg and diethylstilbestrol at 4 mg/kg developed liver tumours (8/40) than those fed the same aflatoxin diet without diethylstilbestrol (25/35) indicating an important sex hormone effect. Studies from our laboratory also revealed that although both testosterone and estradiol can retard aflatoxin-induced hemolysis, comparatively the latter was found to be more potent (Raval and Verma, 1993).

Young animals are comparatively more susceptible to aflatoxin than the adults. Feeding protein deficient diets to rhesus monkeys enhanced characteristic aflatoxin-induced liver injuries. Animals developed fatty liver, bile duct proliferation and fibrosis and died with gastrointestinal haemorrhage (Madhavan *et al.*, 1965). Weanling rats fed with a diet containing less protein (4% casein) and 50 μg aflatoxin/animal/day died on 18/19th days with extensive liver damage (Madhavan and Gopalan, 1975). Newberne and Wogan (1968) observed that rats fed a diet containing 9% protein developed higher incidence of liver tumours (11/15) in a shorter period of time (8 months) than rats fed a diet containing 22% protein (incidence 7/14 after 10 months). Both groups of rats were given total dose of 375 μg of AFB1/animal by gastric intubation over 3 weeks at the beginning of the experiments.

Hendrickse (1985) reported that high protein intake protects while a deficient protein intake increases the susceptibility to aflatoxins. Many geographic region endemic to the possibility of mycotoxins in the food chain are also regions of dietary protein deficiency.

Although AFB1 is known primarily as a hepatocarcinogen, it also induces colon cancers in rats fed on diets deficient in vitamin A. In numerous experimental studies vitamin A has been shown to prevent, delay or retard carcinogenesis induced by a variety of chemicals (Sporn and Newton, 1981; Moon *et al.*, 1983) and consequently vitamin A has been regarded as a potential inhibitors of human cancer. Newberne (1987) has reported that colon DNA in vitamin A deficient rats binds more aflatoxin than does DNA from normal, supplemented rats and that this binding can be returned to levels equivalent to those of controls by supplementation with vitamin A for as short a period as 48 hours.

Less glutathione (GSH) was recorded in the liver of vitamin A deficient rats compared to those of control rats; this was accompanied by more aflatoxin-GSH conjugate in the liver of deficient animals accounting in part for the decreased unconjugated GSH. The decreased GSH would not permit normal levels of other GSH-conjugates, and in this manner may have contributed to the greater toxicity of AFB1 in deficient rats.

Microsomes catalysed reactions of AFB1 leading to its activation and DNA adduct formation have been found to be inhibited under in vitro conditions by vitamin A compounds (Bhattacharya *et al.*, 1984; Firozi *et al.*, 1987).

Riboflavin (vitamin B2) is an essential cofactor for a number of enzymes, including some involved in the detoxification of certain carcinogens like the carcinogenic azo dyes. Adequate levels of riboflavin and protein have dramatically reduced the carcinogenicity of certain of these compounds in animal models and in human studies. Because riboflavin plays a key role in essential metabolising systems, it is important in maintaining optimum tissue and blood levels of this vitamin to ensure proper function of enzyme systems concerned with respiratory energy generation and biochemical detoxification. Aflatoxin reduced riboflavin levels in blood an effect enhanced by irradiation.

Pyridoxine deficiency enhances aflatoxin-induced carcinogenesis (Foy *et al.*, 1966). Aflatoxins are less toxic to animals fed near toxic doses of selenium. Toxicity is also modified by the methionine, choline and fat contents of diets (WHO, 1979).

The folic acid coenzymes are specifically concerned with biochemical reactions involving the transfer and utilization of the single carbon (C1) moiety. In those anaemia which were due to simple folic acid deficiency or due to deficiency of other factors (as is the case in uncomplicated pernicious anaemia), the haematological response to folic acid was not permanent and the neurologic symptoms in pernicious anaemia (combined system disease) remained unchanged. It is now apparent that while folic acid derivatives do have an effect on haematopoiesis, other factors are also necessary for the complete development of the blood cells.

Concurrent administration of vitamin C readily decreases the extent of cytogenetic damages induced by aflatoxins. Similar effects of vitamin C have also been observed in the in vitro systems (Bhattacharya *et al.*, 1987). Though the exact mechanism of vitamin C mediated bioantimutagenicity is not known; it is suggested that this vitamin perhaps prevents the toxin molecule to bind with DNA, the latter being the main mechanism to cause DNA strand breakage or to interfere with its (DNA) normal replication process (Jacobsen *et al.*, 1987).

Vitamin E caused an increase in the frequency of abnormal cells, which was not significant. Vitamin E administration minimizes the toxin caused abnormalities and the concurrent variant of the administration appears to be most effective. The precise protective mechanism of vitamin E against the toxic effects of aflatoxin is still not known. However, it can be said that like vitamin C, vitamin E also prevents the formation of guanine-toxin adduct and thereby removes any blocks in DNA replication.

Preventive Measures

Fungal growth especially of *Aspergillus flavus* and *A. parasiticus* on food/feed stuffs not only decreases its nutritive value but also add aflatoxin (secondary toxic fungal metabolites released on substratum as exotoxin). Since consumption of aflatoxin contaminated food/feed stuffs causes primary and secondary aflatoxicosis, there is a need to develop effective control measures to reduce or prevent fungal infestation and toxin production. Various methods tested for minimizing aflatoxin contamination are as follows:

The use of resistant varieties of seeds towards aflatoxigenic fungi and careful drying and storing procedures can reduce fungal infestation thus diminish food contamination by aflatoxin.

Visual inspection for mould growth before processing can serve as an initial screening technique but toxin-producing fungi can be present without detectable aflatoxins and vice-versa. Because aflatoxin distribution in a contaminated, unprocessed commodity is uneven, adequate sampling is essential for effective maintaining.

The toxin is generally concentrated in a small proportion of seeds that are often different in colour. Segregation of discoloured seeds by sorting can significantly reduce the aflatoxin levels in some crops such as groundnuts.

The toxin is not eliminated from food- stuffs or animal feeds by ordinary cooking or processing practices and since pre- and post-harvest procedures do not ensure total protection from aflatoxin contamination, techniques for decontamination have been developed. As aflatoxins can be chemically degraded in vitro, several oxidizing agents, alkalies, calcium hydroxide, formaldehyde (Piva *et al.*, 1985), hydrogen peroxides and ammonia (Bagley, 1979) are currently used for chemical decontamination of animal feeds.

Visualizing the seriousness of aflatoxin problem in human and animal health, attempts were made to control fungal growth and toxin production by using certain chemicals such fungicides (Buchanan and Shepherd, 1981; Ramadevi and Polasa, 1985), pesticides (Draughan, 1983; Draughan and Ayers, 1981), herbicides (Bean and Southall, 1983), plant extracts (Singh, 1983), antibiotics, phenolics (Singh, 1983) essential oils (Garg and Dengre, 1983), propionic acid and butyric acid (Ghosh and Haggblom, 1985), lauric acid derivatives (Ramadevi and Polasa, 1984), volatile compounds (Batt *et al.*, 1983; Reddy *et al.*, 1983), food preservatives (Lee *et al.*, 1986) and spices (Hitokoto *et al.*, 1980). All these substances reduce fungal growth and toxin production.

Aflatoxin contamination of various food/feed stuffs occurs due to infestation, growth and aflatoxin production by *Aspergillus flavus* and *A. parasiticus*. We made an attempt to evaluate effect of incubation period on mycelial growth and aflatoxin production by *A. parasiticus* in SMKY liquid medium at 28±2°C. Data revealed a time-dependent increase in dry mycelial mass and concentration of total, as well as individual aflatoxins with a maximum on 10 days. Although *A. parasiticus* produced all four types of aflatoxins, but initially only B1 and G1 were produced. It is also evident from the data that aflatoxins were mostly produced at the onset of the stationary phase.

To study effect of micronutrients (vitamin A, riboflavin, folic acid, ascorbic acid, lactic acid and citric acid) on growth and aflatoxin production, similar experiments were done as mentioned above. On addition of vitamin A (500-1500 IU/ml) a concentration dependent decline in growth and aflatoxin production was recorded. Complete inhibition of growth was observed on 1500 IU/ml and higher concentrations (Verma *et al.*, 1996 b). On addition of riboflavin (0.5 to 10 mg/ml) both growth and aflatoxin production were inhibited. At 10 mg/ml concentration the decrease in aflatoxin production was 90% though growth was only inhibited by 56% (Verma *et al.*, 1996 a)

In presence of folic acid (0.5 to 10 mg/ml) growth and aflatoxin production were inhibited. Maximum inhibition (approx. 90%) was recorded when 10 mg/ml folic acid was added into incubation medium (Verma *et al.*, 1996 b). Addition of ascorbic acid (0.5 to 10 mg/ml) decreased mycelial growth and aflatoxin production. Though mycelial growth was decreased by 30% at 10 mg/ml concentration, decrease in aflatoxin production was about 90%

(Verma *et al.*, 1996 b). A concentration-dependent decrease in mycelial growth and aflatoxin production was recorded by adding lactic acid (0.2 to 1 mg/ml) in the incubation medium. Total inhibition of mycelial growth and toxin production was observed at 1 mg/ml concentration. Addition of citric acid (0.5 to 10 mg/ml) in the medium, a concentration-dependent decrease in mycelial growth and aflatoxin production was observed which was maximum (90%) at 10 mg/ml concentration (Verma *et al.*, 1994).

In all the micronutrients studied the decrease in aflatoxin B1 and G1 production was comparatively more than aflatoxin B2 and G2. Lactic acid and vitamin A proved to be more potent causing total inhibition of growth and toxin production. These micronutrients are not harmful to human beings and animals, rather are beneficial. In addition, being a part of natural food products, incidence of fungal infestation as well as incidence of aflatoxicosis is found to be less. Outbreaks of aflatoxicosis are more pronounced in countries where malnutrition is more. People taking diet deficient in proteins or vitamins are more prone to aflatoxicosis than those who are better nourished.

The human diet is an unavoidable source of exposure not only to carcinogens (Ames, 1983) but also to a variety of compounds which can inhibit experimental carcinogenesis (Fiala *et al.*, 1985). Bhattacharya *et al.* (1984) reported an inhibition of in vitro DNA-AFB1 adduct formation in the presence of selenium, riboflavin, butylated hydroxyanisole (BHA), butylated hydroxytoluene (BHT), retinal and retinal derivatives. Other studies note that vitamin E (Draper and Bird, 1984), vitamin B6 and vitamin C (Shamberger, 1984) may be possible antimutagenic agents. Rosin and Stitch (1983) observed a suppression of AFB1 metabolites when AFB1 was incubated with chlorogenic acid and a liver S-9 metabolizing system.

Compounds in food that are capable of altering the metabolism of carcinogens such as AFB1 may play a vital role in the prevention and inhibition of carcinogen modification. Wheelar *et al.* (1987) have evaluated effects of several naturally occurring food components (selenium, vitamin E, B6 and C, pyridoxine) on the in vitro metabolism of AFB1 by liver microsomal enzyme metabolising system with varying concentrations of each nutrient. All these nutrients either inhibited or reduced the metabolism of AFB1.

The induction of phase II enzymes such as the glutathione-S-transferases by ethoxyquin or other dietary antioxidant is a prominent biochemical effect of antioxidant treatment in rat. After one day on a semi-purified diet supplemented with 0.4% ethoxyquin, the specific activities of glutathione-S-transferases were significantly elevated about 1.5 fold. Maximal induction of 4 to 5 fold was observed after 1 week on the antioxidant diet and persisted throughout the feeding period. Removal of ethoxyquin from the diet resulted in a rapid diminution of glutathione-S-transferase activities such that basal activities were reached within 10 days. Rats fed on ethoxyquin at levels as low as 0.05% showed significant elevations of enzyme activities after 2 weeks of diet.

These findings indicate the transient nature of dietary alteration that could have significant impact on the short-term initiation of phase of carcinogenesis.

In summary the present review clearly indicates that aflatoxin pose serious health hazard in animals and human beings when consumed. Therefore following components can be proposed for proper aflatoxin control programme:

a. Education to farmers to improve the method of storage

b. Surveillance of food stuffs and animal feeds for the presence of aflatoxins

c. Application of appropriate food processing technology to separate the contaminated food components from the clean ones, and

d. Consumption of balanced diet containing sufficient concentrations of antioxidants.

References

Allcroft, R. and Carnaghan, R.B.A. (1963). Groundnut toxicity: an examination for toxin in human food products from animals fed toxic groundnut meal. Vet. Rec., **75**: 259-263.

Allcroft, R. and Lewis, G. (1965). Groundnut toxicity in cattle. Experimental poisoning of calves and a report on clinical affects in older cattle. Vet. Rec., **75** : 487-493.

Ames, B.W. (1983). Dietary carcinogenesis and anticarcinogens. Science, **221**: 1256-1264.

Amla, I., Sreenivasamurthy, V., Jayaraj, A.P. and Parpia, H.A.B. (1974). Aflatoxin in Indian childhood cirrhosis- A review. J.Trop. Pediat. Environ. Child Health, **20** : 729-731.

Asplin, F.D. and Carnaghan, R.B.A. (1961). The toxicity of certain groundnut meals for poultry with special reference to their effect on ducklings and chickens. Vet. Rec., **73**: 1215-1219.

Bagley, E.B. (1979). Decontamination of corn containing aflatoxins by treatment with ammonia. J. Amer. Oil Chem. Soc., **56** : 808-811.

Basappa, S.C., Sreenivasamurthy, V. and Rajalakshmi, D. (1977). Analysis of aflatoxin in groundnut and its products. Indian J. Technol., **15** : 311-313.

Batt, C., Solberg, M. and Ceponis, M. (1983). Effects of volatile components of carrot seed oil on growth and aflatoxin production by *A. parasiticus*. J. Food Sci., **48**: 762-764.

Bean, C.H. and Southhall, A. (1983). Effect of pyridazinone herbicides on growth and aflatoxin release by *Aspergillus flavus* and *Aspergillus parasiticus*. Appl. Environ. Microbiol., **46**: 503-505.

Bean, G.A., Schillinger, J.A. and Klarman, W.L. (1972). Occurrence of aflatoxins and aflatoxin producing strains of *Aspergillus* sp in soyabeans. Appl. Microbiol., **24** : 437-439.

Bhattacharya, R.K., Firozi, P.F. and Aboobaker, V.S. (1984). Factors modulating the formation of DNA adduct by aflatoxin B1 *in vitro*. Carcinogenesis., **5** : 1359-1362.

Bhattacharya, R.K., Francis, A.R. and Shetty, T.K. (1987). Modifying role of dietary factors on the mutagenicity of aflatoxin B1 *in vitro* effects of vitamins. Mutat. Res., **188** : 121-128.

Bilgrami, K.S. (1984). Mycotoxins in food. J. Indian Bot.Soc., **63** : 109-120.

Bilgrami, K.S., Prasad, T., Misra, R.S. and Sinha, K.K. (1980). Aflatoxin production by isolates of Aspergillus flavus associated with grains in standing maize crop. Biol. Bull. India, **2** : 32-33.

Bilgrami, K.S. and Sinha, K.K. (1987). Aflatoxin in India. In: Aflatoxins in maize, Zuber MS, L9illehoj EB, Renfro BL (eds) Proceedings of the workshop (IMMYT, Mexico, DF) pp. 349-358.

Buchanan, R.L. and Shepherd, A.J. (1981). Inhibition of *Aspergillus parasiticus* by thymol. J. Food Sci., **4**: 976-979.

Butler, W.H., Greenblatt, M. and Lijinsky, W. (1969). Carcinogenesis in rats by aflatoxins B1, G1 and B2. Cancer Res., **29** : 2206-2211.

Chao, T.C., Maxwell, S.M. and Wong, ST (1991). An outbreak of aflatoxicosis and boric acid poisoning in Malaysia: a clinico-pathologic study. J. Pathol., **164** : 225-233.

Ciegler, A. (1975). Mycotoxins: Occurrence, chemistry and biological activity. Lloydia, **38** : 21-35.

Cuthbertson, W.F.J., Larursen, A.C. and Pratt, D.A.H. (1967). Effect of groundnut meal containing aflatoxin on cynomologous monkeys. Brit. J. Nutr., **21** : 893.

De, Iongh, H., Vles, R.O., van and Pelt, J.G. (1964). Milk of mammals fed on aflatoxin-containing diet. Nature, **202** : 466-467.

Diener, U.L. (1976). Environmental factors influencing mycotoxin formation and contamination of food. Proc. Amer. Phytopath. Soc., **3**: 126-139.

Diener, U.L. and Davis, N.D. (1966). Aflatoxin production by isolates of Aspergillus flavus. Phytopathology, **56**: 1390-1393.

Draper, H.H. and Bird, R.P. (1984). Antioxidants and cancer. J. Agri. Food Chem., **32**: 433.

Draughan, F.A. (1983). Suppression of aflatoxin production with pesticides. South Coop. Ser. Bull., **279** : 81-86.

Draughan, F.A. and Ayers, J.C. (1981). Inhibition of aflatoxin production by selected insecticides. Appl. Environ. Microbiol., **41**: 972-976.

Dwarkanath, C.T., Sreenivasamurthy, V. and Parpia, H.A.B. (1969). Aflatoxin in Indian peanut oil. J. Food Sci. Technol., **6**: 107-109.

Fiala, E.S., Reddy, B.S. and Weisburger, J.H. (1985). Naturally occurring anticarcinogenic substances in food stuffs. Ann. Rev. Nutr., **5** : 295-321.

Firozi, P.F., Aboobaker, V.S. and Bhattacharya, R.K. (1987). Action of vitamin A on DNA adduct formation by aflatoxin B1 *in vitro*. Carcinogenesis, **5** : 1359-1362.

Foy, H., Gilman, T., Kondi, A. and Preston, J.K. (1966). Hepatic injuries in riboflavin and pyridoxine deficient baboons- Possible relations to aflatoxin-induced hepatic cirrhosis and carcinoma in Africans. Nature, **212** : 150-153.

Garg, S.C. and Dengre, S.L. (1983). Antifungal activity of the essential oil of Tugetes erecta. Riechst Aromen Kosmet., **33**: 218.

Ghosh, J. and Haggblom, P. (1985). Effect of sublethal concentration of propionic acid or butyric acid on growth and aflatoxin production by Aspergillus flavus. Int. J. Food Microbiol., **2**: 323-330

Gopal, T., Zaki, S., Narayanaswamy, M. and Preml ta, S. (1968). Aflatoxicosis in dairy cattle. Indian Vet. J., **45** : 707-712.

Gopal, T., Zaki, S., Narayanaswamy, M. and Premlata, S. (1969). Aflatoxicosis in fowls. Indian Vet. J., **46** : 348-349.

Hartley, R.D., Nesbitt, B.F. and O'Kelly, J. (1963). Toxic metabolites of Aspergillus flavus., Nature **198** : 1056-1058.

Heathcote, J.G. and Hibbert, J.R. (1978). Aflatoxins: Chemical and biological aspects. Elsevier Scientific Pub. Co., Amsterdam.

Hendrickse, R.G. (1984). The influence of aflatoxins on child health in the tropics with particular reference to kwashiorkor. Trans Royal Soc.Trop. Med. Hyg., **78** : 427-435.

Hendrickse, R.G. (1985). Aflatoxins and child health in tropics. Chronicle, **15** : 138-156.

Hendrickse, R.G. and Maxwell SM (1988). Heroin addict, AIDS and aflatoxin. Br. Med. J., **296**: 1257.

Hendrickse, R.G., Coulter, J.B.S., Lamplugh, S.M., Macfarlane, S.B.J., Williams. T.E., Omer, M.I.A. and Suliman, G.I. (1982). Aflatoxins and kwashiorkor: a study in Sudanese. Brit. Med. J., **288** : 843-846.

Hesseltine, C.W. (1974). Natural occurrence of mycotoxins in cereals. Mycopath. Mycol. Appl., **53**: 141-153.

Hitokoto, H., Morozumi, S., Wakike, T., Salkai, S. and Kurta, H. (1980). Inhibnitory effect of spices on growth and toxin production of toxigenic fungi. Appl. Environ. Microbiol., **39**: 818-822.

Holzapfel, C.W., Steyn, P.S. and Purchase, I.F.H. (1966). Isolation and structure of aflatoxins M1 and M2. Tetrahedron Letters, **25** : 2799.

Jacobsen, J.F., Refolo, L.M., Conley, M.P., Sambamurti, K. and Humayun, M. (1987). DNA replication blocking properties of adducts formed by aflatoxin B1-2,3-dichloride and aflatoxin B1-2,3-oxide. Mutat. Res., **179** : 89-101.

Jagdish, Kumar, B., Veeranarayana and Gowda, D.K. (1979). Some observations of aflatoxin levels in livestock fed. Indian Vet. J., **56** : 497-500.

Katoch, R.E., Bhowmic, K.B.D. and Katoch, B.S. (1975). Preliminary studies on mycoflora of poultry feed and litter. Indian Vet. J., **52** : 759-762.

Kolhe, A.S., Verma, R.J. and Dube, H.C. (1994). Aflatoxin contamination in oil cakes. Indian Phytopathology, **47** : 270-274.

Krishnamachari, K.A.V.R., Bhat, R.V., Nagarajan, V. and Tilak, T.B.G. (1975). Hepatitis due to aflatoxicosis: An outbreak in western India. Lancet, **I**: 1061-1063.

Lamplugh, S.M., Hendrickse, R.G., Apeagyei, F. and Muanmut, D.D. (1988). Aflatoxins in breast milk, neonatal cord blood and serum of pregnant women. Br. Med. J., **296**: 968.

Lee, S.J., Hanna, M.A. and Bullerman, L.B. (1986). Carbon dioxide and aflatoxin production in high moisture corn treated with potassium sorbate. Cereal Chem., **63**: 82-85.

Lillehoj, E.B., Fennell, D.I. and Kwolek, W.F. (1977). Aflatoxin and Aspergillus flavus occurrence in 1975 corn harvest from a limited region of Iowa. Cereal Chem., **54** : 366-372.

Madhavan, T.V. and Gopalan, C. (1975). Effect of dietary protein on aflatoxin-induced liver injury in weanling rat. Arch. Pathol., **80**: 123-126.

Madhavan, T.V., Tulpule, P.G. and Gopalan, C. (1965). Aflatoxin- induced hepatic fibrosis in rhesus monkeys. Arch. Pathol., **79**: 466-469.

Maggon, K.K., Vishwanathan, L., Venkitasubramanian, T.A. and Mukherji, K.G. (1969). Aflatoxin production by some Indian strains of Aspergillus flavus link ex. Frie. J. Gen. Microbiology, **59**: 119-124.

Manickam, A., Palaniswami, A. and Neelakantan, S. (1985). Mycotoxicosis in buffaloes. Indian Vet. J., **62** : 711-714.

Mehrotra, M.L. and Khanna, R.S. (1973). Aflatoxicosis in Angora rabbits. Indian Vet. J., **50**: 620-622.

Moon, R.C., McCormick, D.L. and Mehta, R.G. (1983). Chemoprevention of animal tumours by retinoids. In: Modulation and Mediation of Cancer by Vitamin A. Meyskens FL, Prasad KN (eds) Karyer, Basel. pp. 47-57.

Nagarajan, V., Bhat, R.V. and Tulpule, P.G. (1974). Aflatoxin production in sunflower (*Helianthus annus*) seed varieties. Curr. Sci., **43**: 603-604.

Neelakantan, S., Swaminathan, R., Balasubramanian, T., Balasaraswathi, R., Indira and Jasmine, G. (1978). Aflatoxin in commercial poultry feeds. Indian Poult. Gaz., **62**: 40.

Newberne, P.M. (1987). Interaction of nutrients and other factors with mycotoxins. In: Mycotoxins in Food. Krogh P (ed.) Academic Press, London. P. 191.

Newberne, P.M. and Wogan, G.N. (1968). Sequential morphologic changes in aflatoxin B1 carcinogenesis in the rat. Cancer Res., **28**: 770-781.

Nusrath, M. and Nahdi, S. (1983). Occurrence of aflatoxins and other mycotoxins in cattle feeds from Hyderabad. Indian Phytopathology, **36** : 106-109.

Olson, L.C., Bourgeois, C.H., Cotton, R.B., Harikul, S., Grossman, R.A. and Smith, T.J. (1971) Encephalopathy and fatty degeneration of the viscera in North-Eastern Thailand: Clinical syndrome and epidemiology. Pediatrics, **47**: 707-716.

Piva, G., Pietri, A. and Carini, E. (1985). Detoxification of peanut meal contaminated with aflatoxin B1 using calcium hydroxide and paraformaldehyde and aflatoxin M1 content in milk. Zoetec Nutr. Anim., **11** : 303-310.

Ramadevi, G. and Polasa H (1985). Diacetyl- A growth and toxin inhibitor of *Aspergillus*. J. Stored Prod. Res., **21**: 195-197.

Ramadevi, G. and Polasa, H. (1984). Inhibition of growth and toxin production by lauric acid derivatives in *Aspergillus* species. Curr. Sci., **53** : 811-813.

Raper, K.B. and Fennell, D.I. (1965). The Genus Aspergillus. The Williams and Wilkins Co., Baltimore, Maryland. P. 686.

Raval, P.J. and Verma, R.J. (1993). Steroid hormones retard aflatoxin-induced cytotoxicity. Indian J. Expt. Biol., **31** : 780-781.

Reddy, A.S., Reddy, S.M. and Girisham, S. (1983). Influence of volatile compounds on aflatoxin production by Aspergillus flavus. In: Mycotoxins in Food and Feed. Bilgrami KS, Prasad T, Sinha KK (eds) Allied Press, Bhagalpur. P. 289-292.

Reye, R.D.K., Morgan, G. and Baral, J. (1963). Encephalopathy and fatty degeneration of the viscera: A disease entity in childhood. Lancet, **2** : 749-752.

Rosin, M.P. and Stich, H.F. (1983). The identification of antigenotoxic and anticarcinogenic agents in food. In: Diet, Nutrition and Cancer: From Basic Research to Policy Implications Roe DE (ed) Loss, New York, pp. 141-154.

Sahay, S.S. and Ranjan, K.S. (1989). Assessment of natural occurrence of aflatoxin in oil cakes. Biol Bull. India, **1** : 51-53.

Sargeant, K., Sheridan, A., O'Kelly, J. and Carnaghan, R.B.A. (1961). Toxicity associated with certain samples of groundnuts. Nature, **192** : 1096-1097.

Sastry, G.A., Narayana, J.V., Rao, P.R., Christopher, K.J. and Hill, K.R. (1965). A report on the groundnut toxicity in Murrah buffaloes in Andhra Pradesh. Indian Vet. J., **42** : 79-82.

Scott, P.M. (1973). Mycotoxins in stored grains, feeds and other cereal products. In: Grain storage: Part of a System, Sinha RN, Muir WE (eds) Avi. Pub.Co. Inc., Westport, Cunn.

Shamberger, R.J. (1984). Nutrition and Cancer. Plenum Press, New York.

Serck-Hanssen, A. (1970). Aflatoxin induced fatal hepatitis, A case report from Uganda. Arch. Environ. Health, **20** : 729-731.

Shank, R.C., Wogan, G.N. and Gibson, J.B. (1972). Dietary aflatoxins and human liver cancer I. Toxigenic molds in foods and foodstuffs of tropical south east Asia. Food Cosmet. Toxicol., **10**: 51.

Singh, A. (1984). Survey and study of the mycotoxins associated with dry fruits and spices. Ph.D. thesis, Bhagalpur University, Bhagalpur.

Singh, P. (1983). Control of aflatoxin through natural plant products. Proc. Symp. Mycotoxin in food and feed. Bilgrami KS, Prasad T, Sinha KK (eds) Allied Press, Bhagalpur. Pp. 307-315.

Sinha, A.K. and Ranjan, K.S. (1989) BGY fluorescence and aflatoxin contamination in insect damaged maize crop. Indian Phytopathol., **42** : 514-518.

Sporn, M.B. and Newton, D.L. (1981). Retinoids and chemoprevention of cancer. In: Inhibition of tumour induction and development. Zedeck MS, Lipkin M (eds) Plenum, New York. PP. 71-100.

Stoloff, L. (1982). Carcinogens and mutagens in the environments, In:Vol. I, Food products. Stitzh HF (ed.) pp. 98-120. CRC Press, Boca Raton, Florida.

Verma, R.J., Mehta, D.N., Raval, P.J. and Dube, H.C. (1994). Effect of citric and lactic acids on mycelial growth and aflatoxin production. In: DAE symposium on Stress and Adaptive responses in Biological systems, MSU, Baroda. pp. 360-364.

Verma, R.J., Kolhe, A.S. and Dube, H.C. (1995). Aflatoxin contamination in chewing products. Proc. Nat. Acad. Sci. India, **65(B)II** : 167-170.

Verma, R.J., Kolhe, A.S. and Dube, H.C. (1996 a). Aflatoxin contamination in different varieties of peanuts. Indian Phytopathology, **49**: 62-66.

Verma, R.J., Raval, P.J., Mehta, D.N. and Dube, H.C. (1996 b). Inhibitory effect of vitamin A and C on mycelial growth and aflatoxin production by Aspergillus parasiticus. Indian Phytopathology, **49**: 98-100.

Verma, R.J., Raval, P.J., Mehta, D.N. and Dube, H.C. (1996 c). Riboflavin and folic acid prevents aflatoxin production. Indian Phytopathology, **49** : 101-103.

Verma, R.J., Mehta, D.N., Raval, P.J. and Dube, H.C. (1997). Aflatoxin contamination in oil seeds, oil cakes and oil samples. Indian Phytopathology, **50** : 77-82.

Wagle, N.G. (1970). Detection and estimation of aflatoxin in groundnuts and groundnut products. Indian Standard Inst. Bull., **22** : 299-301.

Ward, J.M., Sontag, J.M., Weisburger, E.K. and Brown, C.A. (1975). J. Natl. Cancer Inst., **55**: 107.

Wheelar, J.L., Harrison, M.A. and Koehler, P.E. (1987). In vitro metabolism of aflatoxin B1 with microsomal enzymes in the presence of selected nutrients. J. Food Sci., **52** : 1432-1433.

Wong, Z.A. and Hsieh D.P.H (1980). The comparative metabolism and toxicokinetics of aflatoxin B1 in monkey, rat and mouse. Toxicol. Appl. Pharmacol., **55** : 115-125.

World Health Organization (1979). Environmental Health Criteria, 11: Mycotoxins, pp. 68-71, World Health Organization, Geneva.

Zuber, M.S. (1977). Influence of plant genetics on toxin production in corn. In: Mycotoxins in Human and Animal Health. Rodricks JV, Hesseltine CW, Mehlman MA (eds) Pathotox, Park Forest South, Illinois USA. P. 173-179.

11

Diversity and Biotechnological Applications of Indian Species of *Phoma*

M K Rai

Department of Biotechnology, Amravati University, Amravati-444602, Maharashtra

The genus *Phoma* is widely distributed in nature and is characterized by the formation of single-celled, hyaline pycnidiospores borne inside pycnidia. Speciation in this genus has been solely based on host alone. However, on the basis of morphological and cultural studies, this problem has been solved. Integrated approach for speciation seems to be a solution to this problem. The selected species of *Phoma* has been described. There is need of use of molecular markers for differentiation of species in the genus. The pigment producer strains of *Phoma* may be improved by using molecular techniques for over production in order to utilize the fullest potential of the species. The biotechnological applications of the genus *Phoma* particiuarly in the field of production of antibiotics, agrochemicals and herbicides, anthroquinones, etc. are discussed.

Key words : *Phoma*, Sphaeropsidales, Diversity, Biotechnology, Molecular markers.

The genus *Phoma* belongs to order Sphaeropsidales of Deuteromycotina and the species have been reported from wide variety of hosts/ substrates, particularly from plant and soil (Table 1). They have been recovered from aquatic and aerial (Rai, 1981), marine (Sugano *et al.*, 1991) and entomopathogenic environment (Narendra and Rao, 1974) and also found to cause diseases in human beings (Shukla *et al.*, 1984; Baker *et al.*, 1987; Rai, 1989; Zaitz *et al.*, 1997). This genus can be easily differentiated from related *Phyllosticta,* by formation of relatively smaller pycnidiospores. In addition, the later produces a hyaline sheath around the pycnidiospores. A special appendage is also produced by the pycnidiospores, which can be visible in aqueous condition. The genus is characterised by single-celled, hyaline pycnidiospores borne inside a fruiting body referred to as pycnidium which varies from globose, sub-globose to coalescing forms.

E-mail : mkrai123@rediffmail.com; pmkrai@hotmail.com

The existing Indian species of *Phoma* have been recognised on the basis of host alone, and thus the importance of host specificity for the taxonomy of *Phoma* has been much emphasized and often overestimated. The assumption that each host genus or species was colonized by a specialized *Phoma* species prompted many mycologists to ignore morphological characters when erecting new *Phoma* species. Usually, a morphological species may attack various host plants; for example, *P. exigua* has been reported by investigators on different hosts (Bhowmik and Singh, 1976; Raut, 1977; Kanaujia, 1978; Kamal and Singh, 1979; Wadje and Deshpande, 1981; Rao and Thirumalachar, 1981). As a matter of fact, the criterion of identification should be in such a way that it should be possible to identify a *Phoma* species in case host identification is difficult, particularly when floral parts are lacking, or when the fungus is grown on artificial media.

In the past, several new species have been erected from India giving sole importance to host (Tandon and Bilgrami, 1960; Bilgrami, 1963; Agarwal and Sahni, 1964; Dutta and Ghosh, 1965; Chandra and Tandon, 1965, 1966; Hasija, 1966; Pavgi and Singh, 1966; Shreemali, 1972, 1973, 1978; Jamaluddin *et al.*, 1975; Maiti *et al.*, 1978; Rai and Misra, 1981; Agrawal and Misra, 1981). However, in nineties erection of species of *Phoma* was minimized (Borborua, 1990; Firdousi *et al.*, 1990; Chile, 1992; Mittal, 1992).

The present review is focused on the diversity of the species of *Phoma* and their biotechnological importance in the field of antibiotics and also for production of anthraquinone pigments, which have use in textile industries.

Diversity of Indian *Phoma*

Indian species of *Phoma* were studied in detail in culture and, on the basis of morphological and cultural characters 20 broad morphological groups were formed (Rai, 1985, 1986a, 1986b, 1987, 1989, 2000; Rai and Rajak, 1982a, 1982b, 1983a, 1983b, 1985, 1986a, 1986b, 1993; Rajak and Rai, 1982a, 1982b, 1983a, 1983b, 1984a, 1984b, 1985; Rai *et al.*, 1985). These groups include *P. medicaginis* Malbr. and Roum. var. *pinodella* (Jones) Boerema, *P. medicaginis* var. *medicaginis* Malbr. and Roum., *P. pomorum* Thum., *P. herbarum* Westend., *P. exigua* var. *exigua* Desm., *P. tropica* Schneider and Boerema, *P. glomerata* (Corda) Wollenweber Hochapfel, *P. sorghina* (Sacc.) Boerema, *P. multirostrata* (Mathur *et al.*) Dorenbosch and Boerema, *P. capitulum* Pawar *et al.*, *P. betae* Frank, *P. jolyana* Pirozynski and Morgan-jones, *P. fimeti* Brun., *P. chrysanthemicola* Hollos., *P. complanata* (Tode ex Fr.) Desm., *P. destructiva* Plowr., *P. eupyrena* Sacc., and *P. arachidicola* Marasas *et al.*

Table 1: *Phoma* species reported from India

Species	Reference
P. allahabadense Tand and Bilgra.	Tandon and Bilgrami (1960)
P. anaphalidis Weh.	Wehmeyer (1964)
P. arcuata (Cke) Sacc.	Singh and Gupta (1977)
P. arenarie Weh.	Wehmeyer (1964)
P. artabotrii Sharma	Sharma (1971)
P. asparagi Sacc.	Kheswalla (1936)
P. aspidioticola Naren and Rao	Narendra and Rao (1974)
P. baldratti Sacc.	Varma (1975)
P. balliensis Srivastava	Srivastava (1979 a)
P. betae (Oudem.) Frank	Kisan Singh *et al.* (1973) Shreemali (1973), Singh *et al.* (1973)
P. boerhaaviae Shreemali	Shreemali (1972)
P. borassi Srivastava	Srivastava (1979 b)
P. buddleiae Cooks	Bagchee and Singh (1960)
P. cajani (Rangel) Khune and Kapoor	Khune and Kapoor (1981)
P. campanulata Lall and Yadav	Lall and Yadav (1964)
P. capparidina Pass.	Kamal and Singh (1979)
P. canscorae Rabench	Rabenhorst (1878)
P. caperatae Venoux	Shreemali (1973)
P. capitula Pawar *et al.*	Singh and Atheya (1961), Rao (1966); Pawar *et al.* (1967)
P. capci Magnanghi	Varma (1975), Sydow and Butler (1916)
P. casuarinae Tassi	Sydow and Butler (1916
P. chenopodii Pavgi and Singh	Pavgi and Singh (1966)
P. chrysanthemicola Holls.	Srivastava (1953)
P. comosa Pavgi and Gupta	Pavgi and Gupta (1967)
P. convolvulii Weh.	Wehmeyer (1964)
P. crotalariae Weh.	Wehmeyer (1964)
P. cucurbitacearum (Fr.) Sacc.	Rao (1964, 1966)
P. desmonchi Rabench.	Rabenhorst (1878)
P. destructiva Plower	Grover (1965), Shreemali (1973), Vyas and Panwar (1976)
P. dracaenae Pavgi and Upadhyay	Pavgi and Upadhyay (1966)
P. ehretiae Chandra and Tandon	Chandra and Tandon (1966)
P. emblicae Jamal. *et al.*	Jamaluddin *et al.* (1975)
P. euphorbiae Sacc.	Rangaswami *et al.* (1970), Seshadri *et al.* (1972)
P. exigua Desm.	Tiwari and Agrawal (1972), Singh and Agarwal

	(1973), Bhowmik and Singh (1976), Raut (1977), Kamal and Singh (1979), Kanaujia (1979); Wadje and Deshpande (1981,86), Rai and Rajak (1986b); Mishra (1987)
P. *eupyrena* Sacc.	Gangwane and Deshpande (1973, 74), Khanna and Chandra (1977)
P. *fimeti* Burm.	Manoharachary and Rama Rao (1975)
P. *foevata* (Lev.) Sacc.	Lloyd (1904-1919)
P. *fourcroyae* Thuem.	Sydow and Butler (1916)
P. *gardenicola* Mhalsker	Varma (1975)
P. *garflorida* Chandra	Chandra and Tandon (1965), and Tandon Singh (1981)
P. *geophila* Saxena *et al.*	Saxena *et al.* (1969)
P. *glomerata* (Corda) Wr and Hochapf.	Goyal *et al.* (1971), Padma and Mukeji (1972), Ayachi (1972), Putto and Sohi (1975), Varma (1975), Mathur and Prakash (1976), Pathak and Chauhan (1976), Mukeshwar *et al.* (1976), Prakash (1978), Kamal and Singh (1979), Maiti *et al.* (1979), Agarwal and Bakshi (1981), Kanaujia and Rajkishore (1981)
P. *graminicola* Sacc.	Wehmeyer (1964)
P. *herbarum* West.	Singh and Singh (1965), Chiplonkar (1970), Gangwane and Deshpande (1973, 74), Varma (1975), Vijaya Kumar and Rao (1976), Khanna and Chandra (1977), Srivastava (1979), Rajak and Rai (1984), Rai (1987), Chile *et al.* (1992)
P. *hibernica* Grimes	Saksena (1955), Mukerji (1966), Saxena *et al.* (1969), Ranga Rao and Mukerji (1972), Krishnamurthy and Verma (1974)
P. *indianensis* Deshp. and Mantri	Agarwal *et al.* (1972)
P. *insidiosa* Tassi	Verma and Khan (1965), Nema *et al.* (1971)
P. *ipomoeae* Singh and Kamal	Singh and Kamal (1979)
P. *ixorae* Hasija	Hasija (1966)
P. *jatropae* Shreemali	Shreemali (1978)
P. *jasminicola* Agarwal and Sahni	Agarwal and Sahni (1965)
P. *jolyana* Pirozy and Morg.	Purohit *et al.* (1973), Khare and Dhingra (1975), Sridhar and Ullasa (1978), Sohi and Prasher (1981)
P. *justiciae* Patch.	Singh and Agarwal (1973)
P. *khasianum* var. *chalterjeanum* Sen Gupta	Pandey *et al.* (1981)
P. *keckii* (Thuem.) Sacc.	Thüemen (1880)
P. *kissarensis* Gupta and Madan	Gupta and Madan (1977)
P. *lantanae* Verwood and Duplessis	Singh and Agarwal (1974)
P. *levelillei* Boerema and Bollen	Sharma *et al.* (1990)
P. *liliana* Chandra and Tandon	Chandra and Tandon (1965)

P. limitata (Peck.) Boerema	Sud and Agarwal (1972)
P. lingam (Tode ex. Fr.) Desm.	Munjal and Gill (1961), Rao (1966), Varma (1975)
P. lucknowensis Agrawal and Mishra	Agrawal and Mishra (1981)
P. macrophoma Mclp.	Uppal *et al.* (1935)
P. macrostoma Mont.	Jamaluddin (1977b), Sharma (1979), Kamal and Singh (1979), Chauhan and Pandey, (1991)
P. macrostoma var. *incolorata* Horne, Boerema and Dorenbosch	Rajak (1981)
P. mali Schulz. and Sacc.	Agarwal and Sharma (1968)
P. medicaginis Malbr.	Haware and Nene (1981), Rai and Rajak (1982a), Bisht and Nath (1991)
P. meghalyansis Srivastava and Das	Srivastava and Das (1980)
P. mentzeliae Dutta and Ghosh	Dutta and Ghosh (1965), Chauhan and Pandey (1991)
P. microspora Bala. and Narayna	Balasubramanian and Narayanaswamy (1981)
P. minutella Sacc. and Perzig.	Kamal and Singh (1979)
P. minutispora Mathur	Mathur (1967)
P. muehlenbeckiae Cooke and Massee	Varma (1975)
P. multirostrata Mathur *et al.*	Srivastava *et al.* (1984), Rai and Rajak (1986a), Philip *et al.* (1994)
P. multispora Pawar *et al.*	Pawar and Thirumalachar (1966), Pawar *et al.* (1967)
P. musicola Speg.	Sohi *et al.* (1972)
P. nainiense Bilgrami	Bilgrami (1963)
P. nebulosa (Pers. Fr.) Berk.	Singh (1984)
P. nivea (Syd.) Sriv. and Bharg.	Srivastava and Bhargava (1979)
P. nocturni Sohi *et al.*	Sohi *et al.* (1972)
P. nux-vomica Gupta	Gupta (1974)
P. oryzae Cooke and Massee	Chowdhury (1947)
P. ostiolata Pawar *et al.*	Pawar *et al.* (1967)
P. ostiolata var. brunnea Pawar *et al.*	Pawar *et al.* (1967)
P. palmarum (Cooke) Sacc.	Uppal *et al.* (1935)
P. pannioa Petr.	Wehmeyer (1964)
P. papayae Sohi *et al.*	Sohi *et al.* (1972)
P. pardanthi Died.	Sydow and Butler (1916)
P. pedicularis var *minor* Weh.	Wehmeyer (1964)
P. pentatropidis Singh and Kamal	Kamal and Singh (1979)
P. pimprina Mathur *et al.*	Mathur *et al.* (1959)
P. piperis-betle Maiti *et al.*	Maiti *et al.* (1978)
P. polyanthis Died.	Sydow *et al.* (1916)
P. pomorum Thuem.	Khanna and Chandra (1975) Rai (1986b)

P. prunicola (Oper.) Wr. Hochapf	Varma (1975)
P. psidii Henn	Srivastava *et al.* (1964) Varma (1975), Rao (1966)
P. prunicae Srivastava *et al.*	Srivastava *et al.* (1966)
P. putaminum Speg.	Kamal *et al.* (1980), Goyal and Pathak (1983)
P. rabiei (Pass) Khune and Kapoor	Khune and Kapoor (1981)
P. rhododendri Cooke	Varma (1975)
P. sabdariffae Sacc.	Chowdhury (1955), Anonymous (1962)
P. saccharina Syd.	Sydow and Butler (1916) Jain *et al.* (1960)
P. salicina West.	Sydow and Butler (1916)
P. sirodesmii Chona and Munjal	Chona and Munjal (1955)
P. sorghina (Sacc.) Boerema	Mathur and Prakash (1976), Mishra and Shivpuri (1978), Kamal *et al.* (1981), Agarwal and Bakshi (1981), Saxena *et al.* (1981), Rai (1989)
P. terrestre Saksena *et al.*	Saksena *et al.* (1966)
P. terricola Agni	Agnihotrudu (1961)
P. theicola Petch.	Chaudhuri *et al.* (1937)
P. thesii Weh.	Wehmeyer (1964)
P. tridacis Weh.	Wehmeyer (1964)
P. triticum Muller	Ayachi (1972)
P. tropica Sch. and Boerema	Agarwal (1977), Srivastava (1978), Srivastava and Sinha (1979), Kamal and Singh (1979)
Pandey *et al.* (1980)	Khune and Kapoor (1981), Barde (1988), Bisht (1989), Chakrabarty *et al.* (1989)
P. typharum Sacc.	Ponappa (1968)
P. usthrina Rai and Mishra	Rai and Mishra (1981)
P. vasicae Shreemali	Shreemali (1972)
P. vignae Henn.	Rao (1964)
P. vitis Bon.	Goyal *et al.* (1971)

Boerema and his co-workers (1968) studied some of the Indian species and found that several newly described ones are similar to already known species of *Phoma* (Table 2).

Integrated approach of speciation

The erection of new species of *Phoma* based on the host and sometimes on shape and size of pycnidia and pycnidiospores, may be misleading particularly when it is difficult to identify the host or when the substrate is other than plant/ animal. Taking these into consideration, there has been an awakening among the fungal taxonomists in eighties and ninenties concerning erection of new species. The best way was to study the so-called new species in pure culture and an integrated approach based on cultural,

Table 2. Morphologically identical isolates of some Indian *Phoma* species studied on different culture media

1. *P. capitulum* Pawar *et al.*
   = *P. ostiolata* Pawar *et al.*
   = *P. ostiolata* var. *brunnea* Pawar *et al.*
2. *P. herbarum* West.
   = *P. hibernica* Grimes
3. *P. macrostoma* var. *macrostoma*
   = *P. limitata* (Peck) Boerema
   = *P. mororum* Sacc.
   = *P. salicinia* Westd.
4. *P. medicaginis* Malbr. and Roum.
   = *P. jatropae* Shreemali
5. *P. minutispora* Mathur
   = *P. oryzae* Cooke and Massee
6. *P. multirostrata* (Mathur and al.) Dorenbosch and Boerema
   = *P. ushthrina* Rai and Misra
   6.1. *P. multirostrata* var. *macrospora*
   = *P. lucknowensis* Saksena *et al.*
   = *P. liliana* Chandra and Tandon
   = *P. terrestris (= P. terrestre)* Saksena *et al.*
   = *P. ehretiae* Chandra and Tandon
   6.2. *P. multirostrata* var. *microspora* (Allescher) Boerema
   = *P. microspora* Balasubramanian and Narayanaswami
   = *P. lucknowensis* Agarwal and Misra
7. *Phoma pomorum* Thum
   = *P. prunicola* (Sacc.) Wr. and Hochapf.
8. *P. sorghina* (Sacc.) Tassi
   = *P. aspidioticola* Narendra and Rao
   = *P. insidiosa* Tassi
   = *P. indiansis* Deshpande and Mantri

morphological, physiological, determination of ontogeny, biochemical evaluation or in ambiguous cases molecular approaches was applied for identification and differentiation of a particular species (Rai, 1981 and Montel *et al.* 1991).

Rai (1981) studied Indian species of *Phoma* in pure culture on malt-agar, oat-agar and rice-agar, and proposed a key to the identification and differentiation of different species of *Phoma.*

Table 3. Genera included in *Phoma* after detailed morphological and cultural studies *in vitro*.

***Phoma* species**	**Earlier name**
P. arachidis-hypogeae Rao	*Phyllosticta arachidis-hypogeae*
P. jolyana Piro and Morgan- Jones	*Peyronellaea nainensis*
P. sorghina (Sacc.) Tassi	*Peyronellaea indianensis*

Selected Indian species of *Phoma* based on cultural characteristics:

P. arachidicola Marasas *et al.*

Colonies grey to dull-yellow, closely appressed at the margin, attaining a diam. of 5-5.5 cm on oat-agar in 14 days; pycnidia dark-brown, abundant, globose to sub-globose to flask-shaped, pseudoparenchymatous, ostiolate, 100-250 µm; conidia hyaline, l-celled, mostly ovoid, few ellipsoid, ends obtuse, 4.8-8 X 2.5-3.9 µm; chlamydospores produced after 2 weeks, single or in groups, dark-brown, 2-to 3-celled, globose to sub-globose, thick-walled, 7-15 µm.

Formation of chlamydospores single or in groups is a typical character of the present fungus.

P. betae Frank.

Colonies greenish-grey to black, attaining similar diameter on oat-agar and malt-agar in 7-days; pycnidia black, abundant, globose to sub-globose, pseudoparenchymatous, ostiolate, 74-150 µm; conidia hyaline, 1-celled, mostly ovoid on rice-agar and malt agar, ellipsoid on oat-agar, relatively broader, 7 X 3.2-4 µm; chlamydospores in chains, grey, relatively larger-celled.

The present species is characterized by relatively broad conidia (Boerema and Dorenbosch, 1973).

P. capitulum Pawar *et al.*

Colonies with cottony aerial mycelium, oval, attaining a diameter of 5 cm on oat-agar in 7-days; pycnidia 50-105 X 50-80 µm, usually in clusters of up to 20 specimens in a row, borne on radiating, dense, blackish hyphal strands, globose with 1-3 ostioles on a short neck, citrine to honey when young then olivaceous to olivaceous-black, smooth, glabrous, with thin walls, in concentric zones, or sometimes in clusters, after 7-14 days, mostly on agar, but also partly or entirely in the agar; exudate grey-saffron; conidiogenous cells 3-7 X 4-7 µm, phialidic, globose, thin-walled; conidia 3.2-4.4 X 2-3 µm, average 3.8-2.6 µm, broadly and shortly ellipsoidal with 1 or 2 guttules; NaOH spot test negative.

It is halophilic species and occurs in forest soil, saline soil and marine environment.

***P. chrysanthemicola* Hollows**

Colonies dark-grey with compact aerial mycelium, attaining a diameter of 3-3.5 cm on oat-agar in 7-days; pycnidia black, globose to sub-globose, compound, parenchymatous, ostiolate, 88-250 µm; conidia hyaline, l-celled, mostly ovoid on rice-agar and malt-agar, ellipsoid on oat-agar, 4-7.3 X 2.5-3.4 µm; orange-red discoloration of the medium appears.

An important diagnostic character of this fungus is the occurrence of dark-brown to black pseudosclerotial, irregular masses (Dorenbosch, 1970).

***P. complanata* (Tode ex Fr.) Desm.**

Colonies cottony with much aerial mycelium, attaining a diameter of 6.7 cm on oat-agar in 7-days; pycnidia thick-walled, dark-yellow to black, globose, parenchymatous, 74-150 µm; conidia hyaline, 1- to 2-celled, 3.9-11 X 2-3.2 µm.

A typical identifying character of the fungus is formation of massive and thick-walled mature pycnidia (Boerema, 1976).

***P. destructiva* Plowr.**

Colonies dark-green with much aerial mycelium, attaining a diameter of 5-6 cm on oat-agar in 7 days; pycnidia black, globose to sub-globose, pseudoparenchymatous, ostiolate, 78-155 µm; conidia hyaline, 1-celled, mostly ellipsoid on oat-agar, ovoid on rice-agar and malt-agar, 4.5-7.5 X 3.2-3.5 µm; chlamydospores in chains, single-celled, grey, non-guttulated or less guttulated.

***P. eupyrena* Sacc.**

Colonies green with compact aerial mycelium, attaining a diameter of 4.2-4.8 cm on oat-agar in 7-days; pycnidia abundant, black, sub-globose to globose, parenchymatous, 74-160 µm; conidia hyaline, 1-celled, mostly ellipsoid, few ovoid, guttulated, 3-7-6.2 X 2-3.5 µm; typically small, black chlamydospores produced in chains.

Formation of typically small chlamydospores in chains is stable character of this species (Dorenbosch, 1970). The chlamydospores of this fungus differ from *P. medicaginis* var. *pinodella* in their size. The former are relatively small in size and less guttulated.

P. exigua var. exigua Desm.

Colonies grey to black with irregularly scalloped margins, attaining a diameter of 6-7 cm on oat-agar in 7-days; pycnidia black, abundant, sub-globose to globose, some coalesce to form irregular fructifications, ostiolate, parenchymatous, 62-290 μm; conidia hyaline, 1-2- celled, ovoid on malt-agar and rice-agar, ellipsoid on oat-agar, 3.7-11.1 X 2-3.7 μm.

In spot test, the greenish-blue discolouration of the agar medium on application of a drop of NaOH takes place, which is an important diagnostic character. This greenish colour gradually becomes reddish.

***P. exigua* var. *inoxidabilis* Boerema and Veig**

Colonies black with irregularly scalloped margins, attaining a diameter of 7 cm on oat agar in 7 days; pycnidia black, abundant, sub-globose to globose to flask-shaped, some coalesce to form irregular pycnidia, parenchymatous, ostiolate 67-286 μm; conidia hyaline, 1-celled, mostly ellipsoid, few ovoid, 3.9-7.4 X 2.5-3.5 μm; lack of pigment production in this species is an important criterion for differentiation with *P. exigua* var. *exigua.*

***P. fimeti* Brun.**

Colonies ashy to green with compact aerial mycelium, attaining a diameter of 1.7-3 cm on oat agar in 7 days; pycnidia black, globose to subglobose, pseudoparenchymatous, ostiolate, 74-160 μm; conidia hyaline 1-celled, mostly ovoid, few ellipsoid, ends obtuse, 2.9-5.2 X 2-3.2 μm; yellow discolouration of the medium on malt-agar appears.

P. fimeti is known for its slow rate of growth, and thus can easily be differentiated from other species.

***P. glomerata* (Corda) Wollenw. Hochapf.**

Colonies dull, dark-yellow, flat with little aerial mycelium, in sectors, some with woolly aerial mycelium, others with pycnidial types, attaining a diameter of 4.5-7 cm on oat-agar in 7 days; pycnidia dark-yellow to black, abundantly produced, in concentric rings, globose to subglobose, some cylindrical, parenchymatous, ostiolate, 74-160 μm; conidia hyaline 1-celled, mostly ellipsoid, some ovoid, 4-8.3 X 2.5-3.8 μm; NaOH spot-test- the medium turns tea-brown; Chlamydospores and dictyochlamydospores (*Alternaria*-like) produced in chains.

P. herbarum Westend.

Colonies grey to green, massive mycelial mat (mycelial type), uniform rate of growth, attaining a diameter of 4-5 cm in 7 days on malt-agar and oat-agar, in sectors; pycnidia grey to black, sub-globose to globose, usually parenchymatous, sometimes with short neck, ostiolate, 74-150 µm; conidia hyaline, 1-celled, mostly ovoid on malt-agar, ellipsoid on oat-agar, 3.7-8.1 X 2-3.2 µm; red pigments produced, which on application of NaoH changes blue.

Sometimes pycnidial type strains of *P. herbarum* also occur, characterized by the production of abundant pycnidia and sparse mycelium. This creates confusion with *P. glomerata*. The latter can be differentiated from the former by the presence of chlamydospores and dictyochlamydospores.

At the 8th International Botanical Congress at Paris (1954) *P. herbarum* was selected as the type species of the "form genus" *Phoma* (Boerema, 1964, 1970).

P. jolyana Pirozynski and Morgan-Jones

Colonies greyish-black, attaining a diameter of 4-5 cm on oat-agar in 7 days; pycnidia black, sub-globose to globose, compound, parenchymatous, ostiolate, 88-290 µm; conidia hyaline, 1-celled, ellipsoid on oat agar, ovoid on rice-agar and malt-agar, few kidney-shaped, 3.7-8.5 X 2-3.8 µm; chdamydospores and dictyochlamydospores in chains, abundantly produced on malt-agar, multicellular and laterally arranged.

A typical character of this species is chlamydospores in chains and lateral arrangement of dictyochlamydospores.

P. macrostoma var. *macrostoma* Mont.

Colonies pinkish-red with much aerial mycelium, attaining a diameter of 7-7.5 cm on oat-agar in 7 days; pycnidia black, globose to sub-globose, pyriform, parenchymatous, ostiolate, ostiole upto 48 µm wide, 80-200 µm; conidia hyaline, 1-celled, some 2-celled or 3-celled mostly ellipsoid, a few ovoid, a few kidney-shaped, 7.5-14.2 X 2.5-3.7 µm.

The variety *macrostoma* can be easily distinguished from another variety, viz., *incolorata* by lack of reddish-pigments in the hyphae (Boerema and Dorenbosch, 1970).

P. macrostoma var. *incolorata* (Horne.) Boerema and Dorenbosch

Colonies black, attaining a diameter of 7-7.5 cm on oat-agar in 7 days;

pycnidia greyish-black, sub-globose to globose, parenchymatous, ostiolate, 110-210 µm; conidia hyaline, 1-2-celled, mostly ellipsoid, few ovoid, 4-11.9 X 2-3.5 µm; reddish pigments absent in hyphae.

This variety of *P. macrostoma* can be differentiated from variety *macrostoma* of the same species by absence of red pigments (Boerema and Dorenbosch, 1970).

P. medicaginis Malbr. and Roum. var. *medicaginis* (Jones) Boerema

Colonies grey to black, flat, attaining a diameter of 5-6.2 cm on oat agar in 7 days; pycnidia grey to black, scattered, some in radial rows, globose to sub-globose, parenchymatous, 80-175 µm; conidia hyaline, 1-celled, mostly ellipsoid, few ovoid, 4-8.7 X 2.5-3.3 µm; chlamydospores abundantly produced, grey to black, in chains, sub-globose to globose, highly guttulated.

Formation of crystals on malt-agar is one of the most important characteristic features of this fungus. The production of these dendric crystals differ from strain to strain. These can only be produced after some weeks of growth. Another important and stable character of this species is production of chlamydospores.

P. medicaginis* var. *pinodella Malbr. and Roum.

Colonies buff, flat, attaining a diameter of 4-4.8 cm on oat-agar in 7 days; pycnidia black, globose to sub-globose, parenchymatous, ostiolate, 85-215 µm; conidia hyaline, 1-celled, mostly ellipsoid, some ovoid, 3.7-9.6 X 2-2.5 µm.

P. medicaginis var. *pinodella* can be differentiated from var. *medicaginis* by different growth pattern, absence of chlamydospores and few crystals, scattered, and only in cultures over 1-month old.

P. multirostrata (Mathur *et al.*) Dorenbosch and Boerema

Colonies grey to black, flat with little aerial mycelial mat, attaining a diameter of 7-7.2 cm on oat-agar in 7 days; pycnidia in concentric rings, abundant, globose to sub-globose, parenchymatous, ostiolate, sometimes 1 to 2 ostioles in a pycnidium, 80-160 µm; conidia hyaline, 1-celled, ovoid on malt-agar and rice-agar ellipsoid on oat-agar with obtuse ends, sometimes one end acute and other end obtuse, guttulated 4-7.2 X 2.2-3.1 µm; rosy-buff conidial slime produced.

The pycnidia of this fungus are typically parenchymatous and ostiolate, which seems to be stable and diagnostic characters of this species.

P. pomorum Thum.

Colonies greyish-black, with much aerial mycelium, attaining a diameter of 6-6.5 cm on oat-agar in 7 days; pycnidia black, sub-globose to globose, coalesce to form irregular fructifications with many ostioles, parenchymatous 73-225 µm; conidia hyaline, 1-celled, mostly ovoid, few ellipsoid, end obtuse, 4-7.4 X 2-3.7 µm; abundant production of single-celled chlamydospores in chains in combination with dictyochlamydospores.

It can be distinguished from *P. glomerata* especially by its possession of chains of single chlamydospores as well as dictyochlamydospores.

P. sorghina (Sacc.) Boerema, Doren. and V. Kest. Comb. Nov.

Colonies pink with much aerial mycelium, attaining a diameter of 5.5-7.2 cm on oat-agar in 7-days; pycnidia grey, globose, sub-globose to flask-shaped, pseudoparenchymatous, ostiolate, 75-155 µm; conidia hyaline, 1-celled, mostly ellipsoid, few ovoid, 3.9-7.1 X 2-2.7 µm; yellow discolouration of the medium in acidic condition below the colony; chlamydospores abundant, in chains, mostly globose.

A typical diagnostic character of this species is production of chlamydospores and dictyochlamydospores in chains, and red pigment in hyphae.

P. tropica Schneider and Boerema

Colonies olivaceous to greasy with much aerial mycelium, attaining a diameter of 4.5-6.2 cm on oat-agar in 7 days; pycnidia dark-brown, sub-globose to globose to flask-shaped, some coalesce to form irregular fruiting bodies, pseudoparenchymatous, ostiolate, 88-300 µm; conidia hyaline, 1-celled, mostly ellipsoid, some ovoid, 3.5-6.8 X 2.2-2.6 µm.

The pycnidia of this fungus are typically flask-like, pseudoparenchymatous, and ostiolate.

Biotechnolological applications

Antibiotic production by Phoma

Some species of the genus *Phoma* are known to produce biologically active fungal metabolites (Pearce 1997; Singh *et al.*, 1997; Baxter *et al.*, 1998). Among these, squalestatin production by *Phoma* sp. is worthy of note. A *Phoma* sp. (IMI 332962) known to produce the pharmaceutically active metabolites squalestatin-1 (S1) and squalestatin-2 (S2) (Baxter *et al.*, 1998).

Table 4. Antibiotic production by *Phoma*

S. No.	*Phoma* sp.	Active chemical/ product	Reference
1	*Phoma exigua* var. *exigua*	Antibiotic E (antibacterial and antifungal) Cytochalasin B	Boerema and Howeler, 1967
2	*Phoma* sp.	Equisetin (anti HIV)	Dombrowski *et al*., 1999
3	Marine *Phoma* sp.	Phomactin A	Sugano *et al*., 1991
4	*P. exigua* var. *heteromorpha*	Cytochalasin F	Capasso *et al*., 1991
5	*Phoma* sp.	Squalestatin (antiinfective agents)	Dawson *et al*., 1992 Sidebottom *et al*., 1992; Pearce , 1997
6	*Phoma* sp.	Squalestatin 1, 2 (S1, S2)	Baxter *et al*., 1992
7	*Phoma* sp.	Antiviral agent (against HIV) Equisetin	Hazuda *et al*., 1999
8	*Phoma* sp.	Antiviral agent (against HIV) Equisetin and Phomasetin	Singh *et al*., 1998
9	*Phoma* sp.	Antitumour (Fusidienol A)	Singh *et al*., 1997
10	*Phoma* sp.	Antibiotic, antitumour, RKS-1778	Kakeya *et al*., 1997
11	*Phoma* sp.	Pesticide (Octahydronaphthol derivative MK8383, Patent)	Wakui *et al*., 1999
12	*Phoma* sp.	Antibiotic activities	Sponga *et al*. 1999

There are some antitumour agents produced by *Phoma* spp. (Singh *et al*., 1997) particularly Fusidienol-A, which is the second member of the fusidienol family of inhibitors to possess a novel tricyclic oxygen-containing heterocycle with a 7/6/6 ring system. Fusidienol-A was isolated from an unidentified species of fungal genus *Phoma* (MF 6118) collected from a vegetation sample. Singh *et al*. (1998) reported equisetin and phomasetin production by species of *Phoma*, which are useful against AIDS. Dombrowski *et al*. (1999) isolated equisetin derivatives from *Fusarium* or *Phoma* sp. (MF6070). This showed HIV virus integrase inhibitor activity. Due to its potential, the fungus can be used against AIDS. Sponga *et al*. (1999) reported antibiotic activities in micro-organisms isolated from marine sediments. These included besides *Phoma*, *Acremonium*, *Alternaria*, *Aspergillus*, *Cephalosporium*, *Chaetomium*, *Cladosporium*, *Geotrichum*, *Fusarium*, *Gliomastix*, *Humicola*, *Paecilomyces*, *Penicillium*, *Pestalotia*, *Plectosphaerella*, *Scopulariopsis*, *Stachybotrys*, and *Trichoderma* . The fungi exhibited antimycotic activities against *Enterococcus faecium*, *E. coli*, and *Candida albicans*.

Agrochemicals and herbicides

Octahydronaphthol derivative MK8383 of *Phoma* sp. or its salt contains potent pesticidal activity against some plant pathogenic fungi (Wakui *et al.*, 1999). Pandey (2000) evaluated the mycoherbicidal potential of indigenous fungal pathogens against *Lantana camara* L., which is an obnoxious weed of India. There are about 40 fungi islolated from different parts of Madhya Pradesh. Maximum efficacy was shown by *Alternaria alternata* followed by *Phoma multirostrata*. The later fungus is thermophillic and originally reported from India by Mathur and Thirumalachar (1959). It is widely distributed in subtropical , tropical regions and warm green houses.

Eco-friendly anthraquinones

Earlier all colorants were obtained from natural sources such as: plants, lichens, insects or shellfish (Hobson *et al.*, 1997). Anthraquinones were predominant among these as examplified by red dyes recovered from the insect *Kermocococcus ibicish* (Thompson, 1971). The natural dyes are better than the man-made dyes in fastness and brilliance (Taylor, 1986). The synthetic anthraquinone dyes are the second most important group of organic colorants listed in the colour index and tend to predominate in the violet, blue and green hue sectors (Shore, 1990).

The chemical synthesis of anthraquinones require the use of strong acids at high temperature and heavy metal catalysts, as a consequence of which environmentally hazardous effluents and byproducts are produced. With increasing awareness of the environment degradation by industry, the disposal of industrial effluent is becoming more costly and strictly regulated. A return to dyes extracted from molluscs, insects, fungi or plants grown in their environment is not the intention. These methods are not commercially practical and can in many ways be regarded as environmentally unfriendly and totally impractical. For example 150 000 insects reared on 0.16 hectares of cactus plants are required to produce 1 Kg of cochineal dye. In order to obtain the dye, the insects are put in bags and dried slowly by long exposure to the sun. However, a microbial source cultivated in bioreactors is by no means out of question.

There are many fungi which produce anthraquinones as a secondary metabolite. Fungal anthraquinones as polyketide-derived secondary metabolite occur widely in many genera of fungi. Compared with the commercially available hydroxyanthraquinones most possess an adiitional methyl substitution in position three, e.g. emodin and this allows a study of the effect of such a group on the dyeing properties of dyestuffs derived from them. A fungal anthraquinone cynodontin (1,4,5,8-tetra hydroxy-3-methylanthraquinone) was produced in sufficient purity to allow it to be transformed using a simple chemical step to a dye product and this was compared with a comercially available close analogue (Hobson *et al.*, 1997).

Phoma exigua Desm. produces pigments. Bick and Rhee (1966) have reported that *P. exigua* var. *foveata* contains many anthraquinone pigments, such as, pachybasin, chrysophanol, emodin and phomarin. In acid condition, this complex of pigments becomes yellow and in alkaline conditions red. This character is based on ammonia test described by Logan and Khan (1969). On malt-agar, *P. exigua* var. *foveata* gives pinkish colour after exposure to ammonia. This is due to reaction of diffusible anthraquinones and their reaction with ammonia. In old cultures, anthraquinone pigments crystallize out as yellow-green crystals. Tichelaar (1974) found that the fungicide thiophanate-methyl accelerates and increases the crystallization process of the pigments.

Both *P. exigua* var *exigua* and *P. exigua* var. *inoxydabilis* produce cytochalasin B, which are also known as "phomine" (Bousquet and Barbier, 1972; Scott *et al.*, 1975)

Some isolates of *P. exigua* var. *foveata* produce antibiotic substance E similar to isolates of the ubiquitous *P. exigua* var. *exigua* (Boerema and Howeler, 1967). It is a colourless substance and can easily be demonstrated in cultures by sodium hydroxide test. On application of a drop of sodium hydroxide at the margin of colonies on malt agar oxidation takes place and pigment alpha converts into pigment beta. Pigment alpha is red-purple at pH <10.5 and blue-green at pH >12.5. Pigment beta is yellow at ph <3.5 and red at pH >5.5. Anthraquinone pigments are also produced by some other *Phoma* species.

Besides anthraquinones some species of ***Phoma*** produce phomalgin A (Pedras *et al.*, 1995). The potential of these pigments can be utilized commercially.

Need for screening and strain improvement

There is need for screening of *Phoma* and *Phoma*-like fungi for production of antibiotics, agrochemicals, herbicides and also for production of anthraquinones. Not only this, other active chemicals present in such fungi should also be isolated. Indeed, it should be an objective of biotechnologists to improve strains of *Phoma* in order to produce antibiotics, herbicides and anthraquinones more efficiently.

Conclusion

The taxonomy of *Phoma* has been based on host alone for a long period. Consequently, a large number of species have been erected. However, this problem has been largely solved by Boerema and his co-workers. The use of integrated approach for identification of different species of *Phoma* seems to be more fruitful, and has been widely realized by the mycologists. Host alongwith the characters stable on different culture media (rate of growth, presence or absence of chlamydospores, size of pycnidia and pycnidiospores, type of pycnidial wall, texture, presence or absence of ostioles, guttulation and septation in conidia, pigmentation and crystallization, etc.) should be used for identification and differentiation of species of *Phoma*. Use of molecular markers for the ambiguous species of *Phoma* should be used in order to solve the inherent problem of identification and differentiation which have been traditionally based on morphological characters. Application of PCR (Polymerase Chain Reaction) technology may prove to be a boon in solving this problem of species differentiation.

In the present biotechnological era, application of *Phoma* can be utilized for the production of antibiotics, agrochemicals and biocides. The species of *Phoma* should be screened to search for useful secondary metabolites. Furthermore, pigment producing species in general, and anthraquinone producing species in particular can be utilized in textile industry as natural and eco-friendly coloring agents. The existing pigment forming species/ strains can be improved by the help of molecular techniques in order to develop over producer strains of secondary metabolites.

Acknowledgments

The author is grateful to Dr G H Boerema, former Head, and Dr Gruyter, Director, Plant Protection Institute, Wageningen, The Netherlands, for supply of literature and help in various ways and to Professor R.C. Rajak, Department of Bio-Sciences, RDVV, Jabalpur, M.P. for guidance and encouragement.

REFERENCES

Agarwal, G.P. (1977). *Phoma putaminum* Speg.- A new record of Indian soil. Indian J Microbiol, **17**: 35

Agarwal, G.P. and Sahni, V.P. (1964). Fungi causing plant diseases at Jabalpur (M.P.). Mycopath et Mycol Appl, **22**(2): 245-247.

Agarwal, G.P. and Sahni, V.P. (1965). Fungi causing plant diseases at Jabalpur (M.P.)-XI. Mycopath et Mycol Appl, **27**: 136-143.

Agrawal, S.C. and Misra, J.K. (1981). *Phoma lucknowensis* sp. Nov. From Indian alkaline soils. Curr Sci, **50**(13): 600.

Agarwal, R.K. and Sharma, V.C. (1968). Storage rot disease of apple. Indian Phytopath, **2**: 294-298.

Agarwal, V. K., Mathur, S.B. and Neergaard, P. (1972). Some aspect of seed health testing with respect to seed borne fungi of rice, wheat, black gram, green gram and Soyabean growth in India. Indian Phytopath, **25** : 91-100.

Agarwal, S. and Bakshi, S. (1981). New leguminous substrate records for some fungi in India. Geo-bios, **8** : 192.

Agnihothrudu, V. (1961). Rhizosphere mycoflora of tea (Camellia sinensis (L.) O. Kuntze) in relation to the root-rot caused by Ustulina zonata (Lev.) Sacc. Soil Sci, **91**: 133-137.

Anonymous (1962). *Report of Indian Agricultural Research Inst.*, Division of Mycology and Plant Pathology, New-Delhi. Agric Res, **2** : 8.

Ayachi, S. (1972). *Studies on the mycoflora of wheat grains*, Ph D thesis, University of Jabalpur, M.P. (unpublished)

Bagchee, K.D. and Sind, V. (1960). Indian Forest Records. New series, **1**(10): 199-348.

Baker, J.G., Salkin, I.F., Forgacs, P., Haines, J.H. and Kemna, M.E. (1987). First report of subcutaneous phaeohyphomycosis of the foot caused by *Phoma minutella.*, J. Cli Microbiol, **25**-120:2395-2397.

Balasubramanian, P. and Narayanaswamy, P. (1981). A note on a new blight disease of groundnut caused by *Phoma microspora sp.* Nov. In Tamil Nadu. Indian Phytopath, **33**: 133-136.

Barde, A.K. (1988). New host records of fungi of Balaghat from India. Indian J Mycol Pl Path, **18**(2): 217-218.

Baxter, A., Fitzgerald, B.J., Hutson, J.L., McCarthy, A.D., Motteram, J.M., Ross, B.C., Sapra, M., Snowden, M.A., Watson, N.S. (1992). Squalestatin 1, a potent inhibitor of squalene synthase, which lowers serum cholesterol *in vivo*. J. Biol. Chem., **267** (17): 11705-11708

Baxter, C.J., Magan, N., Lane, B., Wildman, H.G. (1998). Influence of water activity and temperature on in vitro growth of surface cultures of a Phoma sp. And production of the pharmaceutical metabolites, squalestatin S1 and S2. Applied Microbiol. Biotechnologl., **49(3)**:328-332.

Bhowmik, T.P. and Singh, A. (1976). Occurrence of *Phoma exigua Desm.* On sunflower. Curr Sci, **45**: 776-777.

Bick, I.R.C. and Rhee, C. (1966). Anthraquinone pigments from *Phoma foveata* Foister. Biochemical Journal, **98**:112-116.

Bilgrami, K.S. (1963). Association of a new species of *Phoma* with *Pleuospora herbarum (Pers.) Rahb.* Curr Sci, **32**: 174-175.

Bisht, I.S. (1989). Leaf-spot of Soyabean and Kuhi caused by *Phoma tropica*- a new record from India. Indian Phytopath, **42**(3): 480.

Bisht, I.S. and Ram Nath (1991. New host records from India. Indian Phytopath, **44**(3): 199.

Boerema, G.H. (1964). *Phoma herbarum Westend.*, The type species of the form genus *Phoma Sacc.* Persoonia, **3** : 9-16.

Boerema, G.H. (1969). The use of the term forma specialis for *Phoma*-like fungi. Trans Br Mycol Soc., **52**: 509-513.

Boerema, G.H. (1970). Additional notes on *Phoma herbarum*. Persoonia, **6**: 15-18.

Boerema, G.H. (1976). The *Phoma* species studied in culture by R W G Dennis. Trans Br Mycol Soc, **67**(2): 288-319.

Boerema, G.H. and Dorenbosch, M.M.J. (1968). Some *Phoma* species recently described from marine soils in India. Trans. Br. Mycol. Soc., **51**: 145-169.

Boerema, G.H. and Dorenbosch, M.M.J. (1970). On *Phoma macrostoma Mont.* A ubiquitous species on woody plants. Persoonia, **6**: 49-58.

Boerema, G.H. and M.M.J., Dorenbosch (1973). The *Phoma* and *Ascochyta* Species described by Wollenweber and hochapfel in their study on fruit rotting. Studies in Mycology, **3**: 1-50.

Boerema, G.H. and L.H. Howeler (1967). *Phoma* species from *Viburnum*. Trans. Br. mycol. Soc., **63** (1):109-114.

Borborua, A. (1990). New potato diseases from Assam. Indian J Mycol Pl Pathol, **20** (2): 162-163.

Bousquet, J.F. and Barbier, M. (1972). Sur l'activite' phytotoxique de trois souches de *Phoma exigua et al.* presence de la cytochalasine B (ou phomine) dans leur milieu de culture. Phytopathologische Zeitschrift, **75**;365-367.

Capasso, R., Evidente, A. and Vurro, M. (1991). Cytochalasins from *Phoma exigua* var. *heteromorpha*. Phytochemistry, **30**(12):3945-3950.

Chakrabarty, P.K., Shyam, K.R. and Bharadwaj, S.S. (1989). Some new records of microorganisms associated with curd-rot of cauliflower. Indian J Mycol Pl Pathol, **19**(1): 118-120.

Chandra, S. and Tandon, R.N. (1965). Three new folicolous fungi. Curr Sci, **34**: 257-259.

Chandra, S. and Tandon, R.N. (1965). Two new leaf spot fungi. Curr Sci, **34** : 365-366.

Chandra, S. and Tandon, R.N. (1966). Three new leaf infecting fungi from Allahabad. Mycopath et Mycol, Appl **29**: 273-276.

Chaudhuri, H., Kapur, V.S., Bhatia, K.L. and Anand, J.S. (1937). Disease of tea busch in Kangra valley (Punjab). J Agri Sci, **7** : 565-573.

Chauhan, S. and Pandey, B.N. (1991). *Cercospora populina* and *Phoma macrostoma* causing foliage diseases of *Populus deltoides*. Indian Phytopath, **44**(4): 549-551.

Chile A, Khare, V. Kachhwaha, M. Mehta, A. and Mehta, P. (1992). A new *Phoma* rot of Dolichos lablab, Indian Phytopath, **45** (3): 387.

Chiplonkar, A. (1970). Fungi imperfecti (Sphaeropsidales) from Maharastra, India. Mycopath et Mycol Appl, **42**: 161-164.

Chona, B.L. and Munjal, R.L. (1955). Notes on miscellaneous Indian fungi. II. Indian Phytopath, **8**: 184-198.

Chowdhury, S. (1947). A Gibberella blight of rye hitherto unrecorded from India. Curr Sci, **16**: 152-153.

Chowdhury, S. (1955). Notes on fungi of Assam-I Lloydia **18**: 82-87.

Dawson, M.J., Farthing, J.E., Marshall, P.S., Middleton, R.F., O'Neill, M.J., Shuttleworth, A., Stylli, C, Tait, R..M, Taylor, P.M., Wildman H.G., Buss, A.D., Langley, D. and Hayes, M.V. (1992). The squalestatins, novel inhibitors of squalene synthase produced by a species of *Phoma* I. Taxonomy, fermentation, isolation, physico-chemical properties and biological activity. J. Antibiot, **45**: 639-647.

Dombrowski, A.W., Hastings, J.C., Polishook, J.D. and Singh, S.B. (1999). New equisetin derivatives are HIV virus integrase-inhibitors. Biotechnological abstract.

Dorenbosch, M.M.J. (1970). Key to nine ubiquitous soil-borne *Phoma*-like fungi. Persoonia **6**: 1-14.

Dutta, B.G. and Ghosh, G.R. (1965). Soil fungi from Orissa (India)-IV. Soil fungi of paddy fields. Mycopath et Mycol Appl, **25**: 316-322.

Firdousi, S.A., Rai, A.N. and Vyas, K.M. (1990). A new leaf-spot disease of *Eleodendron glaucum* from India. Indian Phytopath, **43**(1): 122.

Gangawane, L.V. and Deshpande, K.B. (1973-74). Studies on rhizosphere mycoflora of groundnut-IV. A list of fungi isolated from rhizosphere, rhizoplane and soil. Sydowia, **27**: 312-316.

Goyal, J.P., Desai, B.G. Bhatnagar, L.G. and Pathak, V.N. (1971). Fungal collection from Rajasthan state of India. Sydowia, **25**: 172-175.

Goyal Rita and Pathak, N.C. (1983). A new host of *Phoma putaminum* from India. Indian Phytopath, **13**(1): 107.

Grover, R.K. (1965). *Phoma* rot of tomato. Indian Phytopath, **18**: 388-389.

Gupta, P.C. (1974). Parasitic fungi on medicinal plants from India -II. Some pycnidial and perithecial forms. Mycopath et Mycol Appl, **54** : 127-130.

Gupta, P.C. and Madan, R.L. (1977). Fruit rot diseases of Ber (*Zizyphus mauritiana*) L. From Haryana. Indian Phytopath, **30**:554-555.

Hasija, S.K. (1966). Additions to the fungi of Jabalpur, M.P.-V. Mycopath et Mycol Appl, **28**: 33-41.

Hazuda, D., Blau, CU, Felock, P., Hastings, J., Pramanik, B., Wolfe, A., Bushman, F., Farnet, C., Goetz, M., Williams, M., Silverman, K., Lingham, and R., Singh, S. (1999). Isolation and characterization of novel human immunodeficiency virus integrase inhibitors from fungal metabolites. Antiviral Chem Chemother, **10** (2): 63-70.

Haware, M.P. and Y.L., Nene (1981). *Phoma blight*- a new disease of chickpea. Pl Dis, **65**: 282.

Hobson, D.K., Edwards, R.L., and Wales, D.S. (1997). Cynodontin: a secondary metabolite and dyestuff intermediate. J. Chem. Tech. Biotechnol., **70**:343-348.

Jain A.C., Nikam, B.G., Kulkarni, S.N. and Sharma, O.P. (1960). Fungi of Gwalior and Indore region. The Vikram, **4**: 181-187.

Jamaluddin (1977 a). Two new diseases of grape berries. Curr Sci, **46**: 320.

Jamaluddin (1977 b). A new leaf-spot of Agathis caused by *Phoma*. Proc Nat Acad Sci, India, **47** (B): 71-72.

Jamaluddin, Tandon, M.P., and Tandon, R.N. (1975). A fruit rot of Aonla (*Phyllanthus emblica* L.) caused by *Phoma sp.* Proc Nat Acad Sci India, **45** (B): 75-76.

Kakeya, H., Morishita, M., Onozawa, C, Usami, R, Horikoshi, K, Kimura, K, Yoshihama and M, Osada, H (1997) RKS-1778, its preparation, a cell cycle inhibitor and an antitumour agent produced by *Phoma* sp. J. Nat. Prod., **60**(7): 669-672.

Kamal and Singh, R.P. (1979). Fungi of Gorakhpur VII. Indian J Mycol Pl Pathol , **9**: 170-172.

Kamal, Singh, R.P. and Bhargava, K.S. (1978). Sour rot of lemon new to India. Indian Phytopath, **31** : 76-77.

Kamal, Singh, R.P. and Kumar, P. (1981). Fungi of Gorakhpur XXVIII. Indian J Mycol Pl Pathol, **8** (2) : 35-39.

Kamal, Kumar, P. and Singh, R.P. (1980). Fungi of Gorakhpur XVI. Indian Phytopath, **33**: 551-556.

Kanaujia, R.S. (1979). Notes on a new fungal disease of *Alocacia indica*. Indian Phytopath, **32**: 463-464.

Kanaujia, R.S. and Raj Kishore (1981). Annonated list of fungi of Faizabad V. Indian J Mycol Pl Pathol, *II*: 292-293.

Khanna, P.K. and Chandra, S. (1975). Some new market disease of vegetables. Curr. Sci., **44**:390.

Khanna, P.K. and Chandra, S. (1977). Some new leaf-spot diseases.I. Proc nat Acad Sc, India, **47B**: 118-124.

Kheshwalla, K.F. (1936). A *Phoma* disease of asparagus. Indian J Agric Soc, **6**: 800-802.

Kisan Singh, Sen, C. Srivastava, S.N. and Agnihotri, V.P. (1973). Two new disease of sugar beet from India. Curr Sci, **42**: 69.

Khune, N.N., and Kapoor, J.N., (1981). A new disease of pigeon pea. Indian Phytopath, **34**(2): 258-260.

Krishnamurthy, V., Goswami, B.K. and Gupta, J.N. (1972). *Helminthosporium rostratum* causing leaf-spot of cucurbitaceous host, Muskmelon (*Cucumis melo*). Indian Phytopath, **25**: 302-303.

Lall, S.P. and Yadav, A.S. (1964). A preliminary list of microfungi associated with the decaying stem of *Triticum vulgare* L. and *Andropogon sorghum* l. Indian Phytopath, **17**: 208-211.

Laxminarayana, P. and Reddy, S.M. (1979). Some post harvest diseases of brinjal in India. Indian J Mycol Pl Pathol, **9**: 214-219.

Lloyd, C.G. (1904-1919). *Mycological letters* 1-69 each separately paged, Cincinnati, Ohio.

Logan, C. and Khan, A.A. (1969). Comparative studies of *Phoma* sp. associated with potato gangrene in Northern Iyreland. Trans Br mycol Soc, **52** : 9-17.

Logan, C. and O'Neill, R. (1970). Production of an antibiotic by *Phoma exigua.* Trans Br mycol Soc, **55**: 67-75.

Maiti, S., Dinesh, C., Khatua, and Sen, Chitreshwar (1978). A new leaf-spot disease of betel vine. Indian J Mycol Pl Pathol, **8**(2): 217.

Maiti, S., Sahambi, H.S. and Jana, B.K. (1979). A new purple leaf-spot of *Hippeastrum from India.* Indian Phytopath, **32***:* 120-121.

Mishra, D. (1987). A new leaf-spot disease of *Chenopodium album* caused by *Phoma exigua*. Indian J Mycol Pl Pathol, **17**(3): 353.

Mishra, A. and Shivpuri, A. (1978). Observations on disease problems of certain fruit plants in Rajasthan. Indian J Mycol Pl Pathol, **8**: 83.

Manoharachary, C.H. and Rao, P., Rama (1975). Two new additions to the Indian soil fungi. Indian Phytopath, **28**: 427-428.

Mathur, P.N. (1967). *Studies on members of Sphaeropsidales among Indian soil fungi and morphological and cultural studies on some members of plectascales from soils in India*, Ph.D. Thesis, University of Agra, Agra, U.P. State, India.

Mathur, P.N. and Thirumalachar, M .J (1959). Studies on some Indian soil fungi.I. Some new or noteworthy Sphaeropsidales. Sydowia, **13** : 143-147.

Mathur, R.S and Prakash, J. (1976). New records of fungi isolated from seed and leaf spots of some pasture plants in India. Indian Phytopath, **29**: 54-55.

Mittal, R.K. (1992). Two new diseases of *Vigna mungo* and *V. radiata.* Indian Phytopath, **45**(3): 387.

Montel, E., Bridge, P.D. and Sutton, B.C. (1991). An integrated approach to *Phoma* systematics. Mycopathologia, **115** (2):89-103.

Mukerji, K .G. (1966). Ecological studies on the micro-organic population of the usar soils. Mycopath et Mycol appl, **29**: 339-349.

Munjal, R.L. and Gill, H.S. (1961). Occurrence of backleg disease of Cabbage in India. Indian Phytopath, **14** : 16-19.

Narendra, D.V. and Rao, V.G. (1974). A new entomogenous species of *Phoma.* Mycopath et Mycol Appl, **54** : 135-140.

Nema, K. G., Dave, G. S. and Khosla, H.K. (1971). A new glμme blotch of wheat. Pl Dis Reptr, **55** : 95.

Padma, R and Mukerji, K.G. (1972). Fungi in the root region of *Rauwolfia serpentina* and *R. canescens.* Ind Phytopath, **25**: *104-107.*

Pandey, R.S., Khati, D.V.S. , Shukla, D.N. and Bhargava, S.N. (1980). A new *Phoma* rot of bael fruit. Indian Phytopath, **33** : 319.

Pandey S, Srivastava, S.N. and Misra , S.R. (1981). A new seedling blight of *Solanum khasianum* in India. Curr Sci, **50**: 141.

Pandey, S.K. (2000). Mycoherbicidal potential of some fungal pathogens for the management of *Lantana camara* L. , Ph.D. Thesis (unpublished) submitted to R.D. University, Jabalpur, M.P., India.

Pathak, P.D. and Chauhan, R.K.S. (1976). Two new leaf-spot diseases of *Crotalaria juncea* L. Caused by *Pleospora infectoria* Fuckel and *Phoma glomerata* Corda. Curr Sci , **45**: 706.

Pavgi, M.S. and Singh, U.P. (1966). Parasitic fungi from north India. VII. *Mycopath* Mycol Appl, **30** : 261-270.

Pavgi, M.S. and Gupta, P.C. (1967). Some folicolous fungi on pineapple from India. Sydowia, **21**: 96-99.

Pavgi, M.S. and Singh, U.P. (1966). Parasitic fungi from North India-VII. Mycopath et Mycol Appl, **30**: 260-271.

Pavgi, M.S. and Upadhyay, H.P. (1966). Parasitic fungi from North India-VI. Mycopath et Mycol Appl, **30**: 257-260.

Pawar, V.H., Mathur, P.N., and Thirumalachar, M.J. (1967). Species of *Phoma* isolated from marine soil in India. Trans Br Mycol Soc, **50**: 259-265.

Pawar, V.H. and Thirumalachar, M.J. (1966). Studies on Haplophilic soil fungi from Bombay. Nova Hedwigia, **12** : 497-508.

Pearce, C. (1997). Biologically active fungal metabolites. Adv. Appl. Microbiol., **44**: 1-80.

Pedras, M. Soledade, C., Janet, L., Taylor and victor M. Morales (1995). Phomaligin A and other yellow pigments in *Phoma lingam* and *P. wasabiae*. Phytochemistry, **38** (5):1215-1222.

Philip Tomy, Govindaiah and Saramangala, H.S. (1994). Two new host records in mulberry. Indian Phytopath, **47**(2) : 214.

Ponappa, K.M. (1968). Some interesting fungi-III. On Typha. Proc Ind Acad Sci *B*, **68**: 157-180.

Prakash, Om (1978). A new mango disease and its cure. Indian J Mycol Pl Pathol, **8**: 32-33.

Purohit D K, Panwar, K.S. and Vyas, N.L. (19730. Sphaeropsidales new to India. Indian Phytopath, **26**: 591-592.

Putto, B.L. and Sohi, H.S. (1975). Seed mycoflora of egg plant. Ind J Mycol Pl Pathol, **5**: 111-113.

Rabenhorst, L. (1878). Fungi *europai exsiccati*. Hedw **17** : 31, 44-47, 59-63, 88-90, 171-176.

Rai, J.N. and J.K. Misra (1981). A new species of *Phoma* from Indian alkaline soil. Curr Sci , **50**: 377.

Rai, M.K. (1981). Studies on some Indian Sphaeropsidales with special reference to *Phoma* and related fungi. Ph.D. Thesis, University of Jabalpur, Jabalpur, Madhya Pradesh, India.

Rai, M.K. (1985). Taxonomic studies of species of *Phoma* isolated from air. J Econ Tax Bot, **7**(3): 645-647.

Rai, M.K. (1986a). Two new diseases of *Albizzia lebbek* and *Jasminum sambac*. Acta Bot Indica, **14**: 238-239.

Rai, M.K. (1986b). A new report of *Phoma pomorum* Thum from India. Geo-bios, **5**: 157.

Rai, M.K. (1987). Afirst report of *Phoma herbarum* from tank water of Pune. Geo-bios, **6**: 66-67.

Rai, M.K. (1989). *Phoma sorghina* infection in human beings. Mycopath, **105**: 167-170.

Rai, M.K. (2000). *Phoma* research in India: A review, In: Integrated Management of Plant Resources (Eds: M K Rai, Ajit Varma and R C Rajak), Scientific Publisher, Jodhpur, Rajasthan, India

Rai, M.K. and Rajak, R.C. (1982a). A new leaf-spot disease of *Ailanthus excelsa*. Curr Sci , **51**(2): 98-99.

Rai, M.K. and Rajak, R.C. (1982b) A report of leaf-spot disease of *Holoptelea integrefolia* Planch caused by *Phoma sorghina*. Indian J Mycol Pl Pathol, **12**(2): 342.

Rai, M.K. and Rajak, R.C. (1983a). Studies on two species of *Phoma* from aquatic environment. Geo-bios New Reports, **2**(28): 135.

Rai, M.K. and Rajak, R.C. (1983b). A first report of two species of *Phoma* from tank water. Geo-bios, **2**: 136-137.

Rai, M.K. and Rajak, R.C. (1984). *Phoma jatropae*, a synonym of *Phoma medicaginis var. medicaginis.* J Econ Tax Bot, **5**(2) : 395-396.

Rai, M.K. and Rajak, RC. (1986a). *Phoma multirostrata* on three new hosts. J Econ Tax Bot, **8**(1): 229-230.

Rai, M.K. and Rajak, R.C. (1986b). A new disease of *Citrus medica* and the identity of its causal organism *Phoma exigua* var. *foevata* (Foister) Boerema. Indian J Mycol Pl Pathol, **17**(3): 320-321.

Rai, M.K. and Rajak, R.C. (1993). Distinguishing characteristics of selected *Phoma* species. Mycotaxon, **XL VIII**: 389-414.

Rajak, R.C. (1981). New Sphaeropsidales from India. Indian J Mycol Pl Pathol, **11**: 73-79.

Rajak, R.C. and Rai, M.K. (1982a). Species of *Phoma* from legumes. Indian Phytopath, **35**(4): 609-611.

Rajak, R.C. and Rai, M.K. (1982b). Effect of different colours of light and temperature on the morphology of three species of *Phoma in vivo*. Jabalpur Univ Journal, **1**: 1-6.

Rajak, R.C. and Rai, M.K. (1983a). Effect of different factors on the morphology and cultural characters of 18 species and 5 varieties of *Phoma* .I. Effect of different media. Bibl. Mycol, **91**: 301-317.

Rajak, R.C. and Rai, M.K. (1983b). Cholesterol content in twenty species of *Phoma*. Nat Acad Sci Letters, **6** (4): 113-114.

Rajak, R.C. and Rai, M.K. (1984a). Effect of different factors on the morphology and cultural characters of *Phoma*. II. Effect of different hydrogen-ion-concentrations. Nova Hedwigia, **40**: 299-311.

Rajak, R.C. and Rai, M.K. (1984b). A new leaf-spot disease of *Jasminum pubescence* caused by *Phoma herbarum*. Ind J Mycol Pl Pathol , **14**(2): 173.

Rajak, R.C. and Rai, M.K. (1985). A key to the identification of species of *Phoma* in pure culture. J Econ Tax Bot , **7**(3): 588-590.

Ranga, Rao, V. and Mukerji, K.G. (1972). Studies on charcoal rot disease of *Abelmoschus esculentus*, fungal flora in the root zone of healthy and infected plants. Ann Inst Pasteur, **122**: 81-90.

Rangaswami, G., Seshadri, V.S. and Lucky, K.A., Channamma (1970). Fungi of south India, University of Agriculture sciences, Bangalore pp 193.

Rao, V.G. (1964). Some of new market and storage diseases of fruits and vegetables in Bombay, Maharastra. Mycopath et Mycol Appl , **23** : 297-310.

Rao, V.G. (1964). Some new records of market and storage diseases of vegetables in Bombay, Maharastra. Mycopath et Mycol Appl, **27**: 49-58.

Rao, V.G. (1966). An amount of the market and storage disease of fruits and vegetables in Bombay, Maharastra. Mycopath et Mycol Appl, **28**: 165-176.

Rao, S. and U., Thirumalachar (1981). *Phoma exigua* infecting brinjal leaves. Indian Phytopath., **34**:37.

Raut, J.G. (1977). *Phoma* leaf blight of sunflower. Indian Phytopath, **30**: 269-270.

Rollo, F., Salvi, R. and Torchia, P. (1990). Highly sensitive and fast detection of *Phoma tracheiphila* by Polymerase chain reaction. Appl. Microbiol. Biotechnol, **32**(5): 572-576.

Saxena, A.K., Saksena, S.B. and Narang, V. (1981). Diseases of ornamental plants at Gwalior. Indian Phytopath, **34**: 118.

Saxena, A.S., Mukerji, K.G. and Agarwala, M.K. (1969). Spread of fungal spores causing allergic disease. Aspects of Allergy. Appl Immun, **2**: 175-180.

Saksena, R.K., Krishna, Nand and Sarbhoy, A.K. (1966). *Phoma terrestris* sp. Nov. From an Indian soil. Mycopath et Mycol Appl, **29** : 85-86.

Saksena, S.B. (1955). Ecological factors governing the distribution of soil microfungi in some forest soil of Sagar. J Ind Bot Soc., **35** : 262-298.

Scott, P.M., Harwig, J, Chen, Y.K. and Kennedy, B.P.C. (1975). Cytochalasins A and B from strains of *Phoma exigua* var. *exigua* and formation of cytochalasin B in potato gangrene. Journal of general Microbiology, **87** : 177-180.

Seshadri, V.S., Lucy, A.K. , Channamma and Rangaswami, G. (1972). A few new records of fungi from India. Mycopath et Mycol Appl, **25** : 246-252.

Shankhapal, K.V. and Hatwalne, V.G. (1974). Sour rot of Nagpur oranges. Indian Phytopath, **27**: 445-446.

Sharma, K.D., Bhattacharya, Sravani and Bhadauria, Seema (1990). Two new records of fungi from India. Indian Phytopath, **43**(1): 122-123.

Sharma, N.D. (1979). Some records of fungi from India. Indian Phytopath, **32** : 511-512.

Shore, J., (1990). Colorants and auxiliaries. Society of Dyers and Colourists, England.

Shreemali, J.L. (1972). Two new pathogenic fungi causing diseases on indian medicinal plants. Ind J Mycol Pl Pathol, **2** : 84-85.

Shreemali, J.L. (1973). Sphaeropsidean fungi obtained from Indian phylloclades. Indian Phytopath, **26** : 56-62.

Shreemali, J.L. (1978). Two new species of *Phoma* from India. Indian J Mycol Pl Pathol, **8**: 220-221.

Shrivastava, S.N.S. (1953). On the occurrence of *Phoma chrysanthemicola* Hollos on *Chrysanthemum sp.* Curr Sci, **22** : 216.

Shukla, N.P., Rajak, R.K. Agarwal, G.P. and Gupta, D.K. (1984). *Phoma minutispora* as a human pathogen. Mykosen, **27** : 255-258.

Sidebottom, PJ, Highcock, RM, Lane, S.J., Procopion, PA and Watson, NS (1992). The squalestatins, novel inhibitors of squalene synthase produced by a species of *Phoma* II. Structural elucidation. J. Antibiot., **45: 648-658.**

Sidhu, J.S. (1977). Studies on leaf-spot disease of ber (*Zizyphus mauriana* Lam.), M. Sc. Thesis, PAU, Ludhiana.

Singh, A.K. (1984). *Phoma nebulosa* in river Ganges. Indian Phytopath, **37**(4): 741-742.

Singh, B. and Atheya, S.C. (1961). A new fruit rot disease of *Capsicum annum L.* Sci and Cul, **27**: 447.

Singh, I.D. and Gupta, R.D. (1977). Two new leaf-spot diseases. Curr Sci, **46** : 539.

Singh, I.N. (1981). Some new host records from India. Indian Phytopath, **34** (2): 233-234.

Singh, P.V. and Singh, R.S. (1965). *Phoma herbarum* Westend on Murraya exotica L. Ind J Microbiol, **5**: 2.

Singh, S. and Kamal (1978). Fungi of Gorakhpur-VI. Indian Phytopath, **31**: 315-317.

Singh, S.B., Ball, R.G., Zink, D.L., Monaghan, R.L., Polishook, J.D., Sanchez, M. Pelaez, F., Silverman, K.C. and Lingham, R.B. (1997). Fusidienol - A: a novel Ras farnesyl-protein – transferase inhibitor from *Phoma* sp. J. Org. Chem., **62**(21) : 7485-7488.

Singh, S.B., Zink, D.L., Goetz, M.A., Dombrowski, A.W., Polishook, J.D. and Hazuda, D.J.S. (1998). Equisetin and a novel opposite stereochemical homolog phomasetin, two fungal metabolites as inhibitors of HIV-1 integrase, Biotechnology abs No. 98—06361.

Singh, S.M. (1974). Studies on some Sphaeropsidales with special reference to amerosporous forms. Ph D Thesis, University of Jabalpur, M.P. state, India.

Singh, S.M. and Agarwal, G.P. (1973). On some folicolous Sphaeropsidales. Proc Nat Acad Sci, India, **43 B**: 73-80.

Singh, S.M., and Agarwal, G.P. (1974). Some Sphaeropsidales new to India. Indian Phytopath, **27** : 244-246.

Sohi, H.S. and Prasher, T.B. (1981). A new leaf-spot disease of *Bilbergia nutans* H. Wendl. Caused by *Phoma jolyana* Piroz. And Morg. Curr Sci, **50**: 324-325.

Sohi, H.S., S.L., Sharma, Nayak , S.K. and K Shyam (1972). New records of fungi from Himachal Pradesh. Indian Phytopath, **25**: 439-441.

Sponga, F., Cavaletti, L., Lazzarini, A., Borghi, A., Ciciliato, I., Losi, D. and Marinelli, F. (1999). Biodiversity and potentials of marine-derived microorganisms. J. Biotechnol, **70** (1-3):65-69.

Sridhar, T.S. and Ullasa, B.A. (1978) Leaf-blight of guava- a new record. Curr Sci, **47**: 442.

Srivastava, R.C. (1979a). Two new species of *Phoma* from India. Mykosen, **22** : 184-186.

Srivastava, R.C. (1979b). Fungi causing plant diseases at Jaunpur-II. Indian Phytopath, **32**: 289-290.

Srivastava, R.C., and Das, J. Sandhya (1980). *Phoma meghalyansis* Srivastava et Das sp. Nov. From India. Arch Protistenk, **123** : 324-325.

Srivastava, R.C. and Sinha, A.B. (1979). Three new host records from India. Curr Sci , **48**: 352.

Srivastava, S.L., Bisht, Suman and Singh, Jasveer (1984). New fungi from phyllosphere. Indian Phytopath, **37**(4): 741.

Srivastava, Y.N. (1978). Leaf-spot fungi from India. Indian Phytopath, **31**: 549.

Srivastava, Y.N. and Bhargava, K.S. (1979). Three fungal host records from India. Indian Phytopath, **21** : 114-116.

Sud, V.K. and Agarwala, R.K. (1972). New records of pome and stone fruits tree diseases from Himachal Pradesh-II. Indian Phytopath, **25**: 580-582.

Sugano, M., Sato, A., Lijima, Y., Oshima, T., Furuya, K. Kuwano, H., Hata, T. and Hanzawa, H. (1991). Phomactin A: a novel PAF antagonist from a marine fungus *Phoma* sp. J. Am. Chem. Soc., **113**(14): 5463-5464.

Sydow H and E J Butler (1916). Fungi indiae orientalist Pars V. Ann Mycol, 14: 177-220.

Tandon, R.N. and Bilgrami, K.S. (1960). A new host for *Phyllosticta cycadina Pass Sci and Cult*, **26** : 277-278.

Taylor, G.W. (1986). Natural dyes in textile applications: Rev. Prog. Coloration, **16** :53-61.

Thompson, R.H. (1971). *Naturally occurring quinones*, 2nd edn., Academic press, London, New York.

Thuemen, F.V. (1880). *Fungorum novorum exoticorum* Decas altera. 36-38.

Tichelaar, G.M. (1974). The use of thiophanate-methyl for distinguishing between the two *Phoma* varieties causing gangrene of potatoes. Netherlands J of Plant Pathology, **80**: 169-70.

Tiwari, D.P. and Agarwal, P.D. (1972). *Phoma exigua* from grassland of Jabalpur. Curr Sci, **41** : 77.

Uppal, B.N., Patel, M.K. and Kamat, M.N. (1935). The fungi of Bombay, **VIII**: 1-56.

Varma, U. (1975). Taxonomic and nutritional study on certain Sphaeropsidales. Ph.D. Thesis, Bhagalpur University, pp 17-59.

Verma, V.S. and Khan, A.M.(1965). Fungi associated with sorghum seeds. Mycopath et Mycol Appl, **27** : 314-320.

Vijaya Kumar, C.S.K. and Rao, A.S. (1976). A report of leaf spot disease on some vegetable, fodder and ornamental plants. Curr Sci, **45** : 309-310.

Vyas, N.L. and Panwar, K.S. (1976). Some new post harvest disease of fruits. Indian Phytopath, **29**: 94-96.

Wadje, S.S. and Deshpande, K.S. (1981). A new leaf-spot of cotton caused by *Phoma exigua*. Indian Phytopath, **33** : 126-127.

Wadje, S.S. and Deshpande, K.S. (1986). Phylloplane microflora of cotton: interaction between *Phoma exigua* and other fungi. Indian Bot Reptr, **5**(2): 177-178.

Wakui, F., Mikoshiba, H. and Tachibana, K. (1999). Light-stable agrochemical microbicide preparation, *biotechnological abs* No. 99-06952.

Wehmeyer, L.E. (1964). Some fungi imperfecti from Punjab and Kashmir. Mycologia, **55**: 313-335.

Zaitz , C., Heins, E. M. , de Freitas, R. S., Arriagada, G.L., Ruiz, L., Totoli, S.A., Marques, A.C., G.G. Rezze, Muller, H., Valente, N.S. and Lacaz, C.D. (1997). Subcutaneous phaeohyphomycosis caused by *Phoma cava*. Report of a case and review of the literature. Rev Inst Med Trop Sao Paulo., **39**(1):43-48.

12

Ecological and Functional Diversity of Thermophilic Fungi

Seema Rawat and B.N.Johri*

Department of Microbiology
G.B. Pant University of Agriculture and Technology
Pantnagar-263145, U.S. Nagar, Uttaranchal

The main area of interest in thermophilic fungi has in the past been restricted to their functional behaviour. The genetic diversity although an important aspect, has not been of much interest among the workers. Till now, there is just a horizontal advancement in this direction. The present panorama needs to be extended towards vertical dimensions. The biopotentiality of thermophilic fungi has been widely exploited in various industrial sectors, but their full biotechnological potential remains to be tapped. This chapter is devoted to examine the structural diversity of thermophiolic fungi in soil, differential type of composts, wood chip piles, coal spoils tips, stored grains ,other thermophilic sites and certain non-thermophyllic environments. The genetic and functional attributes of the diversity of these fungi have also been discussed.

Key Words: Ecological diversity, functional diversity, thermophilic fungi

The first thermophilic fungus, *Mucor pusillius* was discovered by Lindt over a century ago (1886). However, the real early beginnings are attributed to Miehei (1905) who made pioneering investigations of self-heating hay and discovered two important thermophilic fungi, *Malbranchea pulchella* var. *sulfurea* and *Thermoascus aurantiacus*. His book "Die Selbsterhitzung des Heus" published in 1907 is a distinguished landmark in the development of this subject. However, even after this monumental work, a wide lacuna remained since few mycologists devoted themselves to study of this group. These few workers, too were more interested in their functional potential rather than the detailed morphology, diversity and taxonomy. A comprehensive account of the systematic morphology and general biology of thermophilic fungi came into picture with the monograph of Cooney and Emerson (1964) which served as the major basis of understanding of thermophilic fungi. Cooney and Emerson (1964) described thermophilism as

* Corresponding Author E-mail: bhavdishnjohri@rediffmail.com

the inability of a fungal species to grow below 20°C as the real hallmark of thermophilism rather than their ability to grow above 50°C. The inherent ability of this group to occupy a temperature niche that would preclude most other fungi resulting in their ubiquitous distribution, is a point of special interest.

The community structure of this small group of Eumycota is very heterogeneous exhibiting a wide structural distributional pattern. Their population is found both in natural and man-made habitats and though limited to a handful species, encompasses various taxanomic groups. Interestingly, these fungi have been recovered not only from hot climatic regions but also from cold and stressed environments (Ellis, 1980a, b; Mouchacca *et al.*, 1995). Their ubiquitous presence worldwide raises many questions regarding their survival and dispersal. The presence of self-heating masses of organic debris all around the globe lends sustainability to this community. These developments have now been compiled in a book by Johri *et al.* (1999) to highlight particularly their biotechnological potential.

It is really interesting to explore out their geographically diverse reservoir not only from the viewpoint of having a fair picture of species spectrum but also for understanding their functionality in a given habitat. It would also provide information regarding the starting point of their dispersal. The survival and dominance exhibited by thermophilic fungi in a wide array of habitats is attributed to their inherently strong extracellular enzymatic machinery, release of volatile sporostatic and fungistatic compounds and colonization potential (Satyanarayana and Johri, 1981). Available information suggests that these species can exert both positive and negative interactions with other microflora in a habitat.

Thermophilic fungi are not only structurally diverse but also genetically and functionally distinct. Early studies of community structure of thermophilic fungi were based on their recovery on specific media and propagule counts, which revealed their abundance and successional pattern. However, due to limited nature of such studies, a substantial gene pool could not be established. Moreover, these studies would only provide an insight into their structural and functional diversity. The use of molecular tools in recent years has however added considerable information to extend our knowledge of this interesting group. The availability of immuno- and molecular probe of various kinds opens newer possibilities, which would permit retrieval of information concerning the role of thermophilic fungi in colonization, succession and *in situ* functioning in various habitats. Use of immunoprobe in the study and distribution of *Thermomyces lanuginosus* has already been documented by Dewey *et al.* (1989) and Rajasekaran and Maheswari (1993). Thus, a beginning has already been made to add to the dimension of our knowledge, which extends to their genetic diversity. Detailed study of thermophilic fungal diversity would help in their fullest exploitation in industries.

The present chapter is devoted to the structural, genetic and functional diversity of thermophilic fungi. While information on some aspects is still scanty due to only horizontal advancements in that area but an effort to compile all such information has been made. However, readers may find a bias towards the diversity of thermophilic mycoflora of compost because they are primarily compost inhabitants, and secondly because our own active engagement in ecology during the last years has been with mushroom compost system.

Structural Diversity

Thermophilic fungi, an ecologically diverse group, shows no geographical restrictions in their distributional pattern. They have been isolated form an array of natural and man-made habitats. The most significant natural habitats are different soil types and places where self-heating of plant materials result in high temperatures (Hemida, 1992; Mouchacca *et al.*, 1995; Straatsma *et al.*, 1994). Besides, a diverse population has been recovered from man-made habitats such as cooling towers, effluent from nuclear power reactors and ducts employed for thermal insulation (Johri and Satyanarayana, 1986). They may occupy these habitats either as the resting propagalues or as active mycelia (Cooney and Emerson, 1964). Their ubiquity suggests wide dissemination machinery through vegetative or reproductive propagules to places of man-made activities (Johri *et al.*, 1999).

A high temperature alongwith high moisture content, which significantly influences the thermogenic environment, are considered to be two decisive factors in their distribution (Festenstein, 1966; Bilai, 1984). In view of this, tropics have been looked upon as natural abode of these fungi. Tropical sites do favour their colonization as a consequence of elevated temperature and more competitive environment. However, it was therefore, surprising to observe that their population in tropical soil was not particularly different from that in temperate soil. It is possible that the inoculum density in tropical soils and plant material is, however, greater and that the variation in temperature results in changed fungal spectrum (Chang and Hudson, 1967; Satyanarayana, 1978). Interestingly, some non-thermogenic environments like coal and moist soils in Australia, Antarctic and sub-Antarctic soils (Ellis, 1980 a, b) and aquatic sediments where the bottom temperature never exceeds 6°C or 7°C, (Tubaki *et al.*, 1974) have also been found to harbour thermophilic fungi.

The distributional structure of these fungi in self-heated habitats appears to be regulated, atleast in part, through antibiosis and self-inhibition, which otherwise also plays a decisive role in fungal ecology. Compost and soil are the two important ecological niche for thermophilic mycoflora which present a fierce competition between the colonizing microflora (Subrahmanyam, 1999).

The structural diversity of this heterogeneous fungal community in some important habitats is discussed below.

Soils

It is an excellent habitat for thermophilic fungi. They have been isolated from sun-heated mud (Tansey and Brock, 1973), desert soils (Hemida, 1992, Mouchacca *et al.*, 1995), coalmine soils (Johri and Thakre, 1975) and mangrove soils (Jaitly and Rai, 1982). Thermophilic growth in soil is made possible by solar heating of surface layers of soil (i.e., insolation). The top soil in tropical, sub-tropical and temperate regions under the influence of solar radiation gets heated to 50-70°C facilitating growth and multiplication of theromophilic organisms (Vertaja, 1949).

It had long been a matter of debate whether they are active inhabitants of soil or not. Early workers such as Miehei (1907) and Noack (1912) had considered soil as a possible natural habitat, particularly in the tropics where solar heating of the surface could provide temperatures conducive to growth of thermophilic fungi. Most workers have considered their ubiquitous presence in soil to be a result of their growth *in situ*. Soil being a heterogeneous system makes phenomonology of microbial growth in it, too complex to derive a conclusion based on soil plating which detects presence of only viable propagules (Maheshwari *et al.*, 1987). An indirect approach to determine whether they are active inhabitants of soil or not lies in the comparison of their geographical distribution in soil. If the species diversity and concentration of propagules of thermophilic fungi is uniformly higher in subtropical and tropical regions compared to temperate regions, then it would suggest that soil is a natural habitat of thermophilic fungi. However, Hedger (1975) found a comparable number of propagules in samples from Java and U.K. Soils in tropical countries therefore, do not seem to have a higher population of thermophilic fungi than those of temperate countries. Thus, although widespread, they are ordinarily not an active component of soil microflora.

Maheshwari *et al.* (1987) studied the diversity of thermophilic fungi in Indian soils. The dominant fungi were, *Chaetomuim thermophile* var. *coprophile*, *Humicola grisea*, *Melanocarpus albomyces*, *Sporotrichum thermophile*, *Thermoascus aurantiacus* and *Torula thermophila*. Their study revealed that India did not harbour an exceptionally rich thermophilic microbiota (less than 50% of the known species). Soils from fewer localities in U.K. or USA yield as many or more species of thermophilic fungi. The presence of thermophilic fungi in temperate (Hudson, 1974), sub-tropical and tropical countries (Thakur, 1977;

Sandhu and Singh, 1981) suggest that their widespread presence in soil is a consequence of dissemination of spores from foci of growth in self-heating masses of organic materials which occur world-wide because of human practices (Cooney and Emerson, 1964; Tansey and Brock, 1978). The dissemination theory also explains that thermophilic fungi are found in soils in climatic regions where (soil) temperatures are generally suboptimal for their growth. *Thermomyces lanuginosus*, most thermophilic of all fungi, is the most common soil fungus in cool climates (Sandhu and Singh, 1981; Maheshwari *et al.*, 1987). Ellis and Keane (1981) found the widest range of species of thermophilic fungi in soils from cool and moist climatic zones in Australia. This is most likely because in these regions agriculture and composting practices are prevalent resulting in high output of thermophilic propagules into air. An extensive flora of obligate thermophiles has been reported from certain volcanic areas of the Ural Mountains (Loginova *et al.*, 1962). Apinis (1963) observed a comparable frequency of occurrence of *Aspergillus fumigatus, Absidia ramosa, Byssochlamys nivea* and *Sporotrichum thermophile* in alluvial soils near Nottingham. The cultivated and desert soils exposed continuously to cement dust particles were found to be dominated by *Aspergillus fumigatus, Chaetomium thermophile, Malbranchea pulchella* var. *sulfurea, Myriococcus albomyces, Rhizomucor pusillus* and *Torula thermophila* (Hemida, 1992).

Abdel Rahman *et al.* (1992), while studying the effect of different sources of urea on ureolytic activity and diversity of fungi in soil, found a majority to be mesophilic, except *Aspergillus fumigatus, Aspergillus terreus* and *Rhizomucor pusillus.* Chen Kuei *et al.* (1996) added two new thermophilic fungi to the known list, *Thermoascus taitungiacus* sp. nov. with a *Paecilomyces* anamorph and *Thielavia pingturgia* sp. nov., isolated from field soil and sugarcane soils, respectively in Taiwan.

The variation in the abundance of individual species depends upon the type of soil, depth, season of the year, concentration of organic materials and isolation procedure employed (Subrahmanyam, 1999). In alluvial soils, the thermophilous microorganisms are more frequent in top soils (25 cm); highest frequency of thermophilous fungi has been reported in the turf layer consisting of rich root system and/or plant residues made of root system and /or plant residues made of wood or grass litter (Apinis, 1972; Evans, 1971 a, b; Gochenaur, 1975). Soil temperature, moisture, pH and nutritional status determine the development and distribution of thermophilic microbiota. Tansey and Jack (1976, 1977) observed a decrease in the occurrence of thermophilic fungi with increase in shade; maximum frequency of occurrence was observed in sun-heated soil and minimum in tree-shaded areas. The diversity and richness was found to decrease with increase in depth of the soil. In summers, considerable number of thermophilic fungi has been recorded in sun-heated mud of ponds and lakes (Subrahmanyam, 1999).

Soil is a complex habitat with fiercely competitive environment. High temperature conditions, secretion of antimicrobial substances (Somkuti and Greenberg, 1978; Subrahmanyam, 1982) and volatile sporostatic compounds (Satyanarayana and Johri, 1981) favour abundance and richness in soil. But in widely fluctuating temperature regimes in soil, thermophiles appear to be outcompeted by mesophiles as the respiratory rate of latter is independent of temperature change (Rajasekaran and Maheshwari, 1993). The pigmented form of thermophiles exhibit better tolerance to unfavourable conditions than hyaline forms (Subrahmanyam, 1975). At low temperature conditions they remain viable with low metabolic activity. Some species are known to remain in viable state in grass tissues during winter months as clustered mycelia (Apinis, 1972).

Besides, soil plating methods, immunoprobe and molecular probe have been used to study the diversity of thermophilic fungi in soil. Rajasekaran and Maheshwari (1993) used an antibody probe to study the richness of *Thermomyces lanuginosa* in soil.

Compost

It is the most important habitat, which serves as a rich reservoir of thermophilic fungi. The concentration of thermophilic propagules per g material is approximately 10^6 times higher in composts than in soil (Maheshwari *et al.*, 1987). Thermophilic fungi, therefore, are considered primarily as compost fungi.

Composting represents an astonishing example of thermogenic, solid state fermentation process resulting from a succession of microbial populations. It begins with mesophilic bacteria, actinomycetes and fungi as primary colonizers and thermophilic microflora as the climax community. Most currently known thermophilic fungi were originally isolated from hay and compost. Their prevalence in compost is explained on the basis of the prolonged elevated temperatures in such substrate (20-80^{0}C), humidity (70-80%), aerobic conditions (0-21% O_2), and supply of carbon and nitrogen (10:1) which favour the development of thermophilic microflora.

Wheat Straw Compost

In wheat straw compost, the initial high population of mesophilic fungi results in peak heating in the central region of the pile wherein temperature rises rapidly and reaches a plateau around 50^{0}C. Thermophilic fungi rapidly develop replacing the mesophilic population and persist until the compost cools down (Subrahmanyam, 1999).

Chang and Hudson (1967) studied the diversity of mycoflora of wheat straw compost. *Absidia ramosa, Aspergillus fumigatus, Chaetomium thermophile, Humicola grisea* var. *thermoidea, H. insolens, H. lanuginosa, Malbranchae pulchella* var. *sulfurea, Mucor pusillus* and *Stilbella thermophila* occurred abundantly in the compost. The mesophilic fungi were soon killed off by the high temperature developed during the first few days; *A. fumigatus* and *M. pusillus* were also suppressed by the high temperature. *Mucor pusillus* did not recur even when the temperature of the compost later become suitable presumably due to lack of easily available carbon sources in the later stages. In contrast *A. fumigatus* persisted, with its ability to utilize cellulose and hemicellulose. *Humicola lanuginosa* developed early and persisted throughout composting due to its ability to lead commensal life with other forms (Hedger and Hudson, 1974). Besides, this eukaryotic thermophile can tolerate wide temperature fluctuations on either side of optima and elaborates a variety of hydrolytic enzymes that help in continued presence. *Chaetomium thermophile, Humicola insolens, H. lanuginosa* and *Talaromyces dupontii* develop abundantly in the plateau period and rapidly utilise cellulose and hemicellulose. *Sporotrichum thermophile* alongwith some mesophilic fungi succeed when the temperature of the compost dropped. However, Moubasher *et al.* (1982) reported that thermophilic fungi were not present at peak high temperature in wheat and broadbean straw composts. When it cooled down to 51.5^{0}C, *Myriococcum albomyces, Penicillium dupontii* and *Sporotrichum thermophile* were found in abundance.

Paddy Straw Compost

In paddy straw compost, *Chaetomium thermophile, Humicola* spp. and *Sporotrichum thermophile,* are abundantly present. (Satyanarayana, 1978). Satyanarayana and Johri (1984) observed that the colonizing ability of thermophilic fungi on paddy straw was directly proportional to the inoculum concentration. For example, colonization by *Humicola lanuginosa, Sporotrichum thermophile* and *Torula thermophila* increased with higher inoculum dose. During peak heating period, only a few thermophilic fungal propagules were present exhibiting high rate of respiration.

Miscanthus Straw Compost

Klamer *et al.* (2001) studied the succession of mycoflora during eight months of composting of Miscanthus straw and pig slurry in well insulated containers. Before peak heating, *Aspergillus fumigatus* and *Rhizomucor pusillus* were dominant. Forms developing after peak heating could be divided into two groups: those appearing from day 15 to 27, and others developing from day 50 to 225. The first group was dominated by, *Paecilomyces variotii, Scytalidium thermophilum and Thermomcyces lanuginosus*, and the second by, *Acremonuim* spp. and *Thermomcyces lanuginosus.* The Brillouin diversity index changed with temperature; diversity was high before peak heating, low during elevated temperature, and increased again during the third phase of composting. Temperature was the main controlling parameter that changed fungal community during the first month of composting. Based on principal component analysis (PCA), it could be concluded that when the temperature reached the ambient level, only minor change in fungal community was detected although some changes in certain species were still discernible.

Municipal Compost

Municipal wastes generally contain substrates rich in lignohemicellulose, among other substrates. Thermophilic fungi play a significant role in the conversion of these materials into farmyard manure or single cell protein (Stutzenberger *et al.*, 1970; Barnes *et al.*, 1972; Kane and Mullins, 1973; Eriksson and Larson, 1975). This aroused interest in studying the diversity of mycoflora in municipal compost. The unique ability of some species to degrade plastic substances envisaged special interest in their study (Eggins and Mills, 1971; Brown *et al.*, 1974).

Subrahamanyam *et al.* (1977) isolated a new genus *Thermomucor* besides other fungi from municipal waste compost in India. *Thermoascus aurantiacus* and *Myceliophthora thermophila* were the two major thermophilic fungi recovered from municipal waste compost (Sen *et al.*, 1980; Subrahmanyam, 1980). Kane and Muller (1973) reported *Thermoascus aurantiacus* and other thermophilic fungi at the beginning of the digestion period, which persisted throughout the period of compositing. These workers were of the opinion that high temperature, acidity and anaerobic conditions could limit fungal growth in the interior of the compost, and thus restrict the role of these moulds. They recorded the presence of thermophilic fungi at all times during composting of municipal refuse and did not find any apparent succession within the species.

Mushroom Compost

It is a rich reservoir of microbial types exhibiting an interesting successional pattern of mesophilic and thermophilic communities. The thermophilic fungal community is an important extremely diverse component of compost biota making up the largest percentage of microbial biomass.

The dominance of microbial crop changes with the two phases of composting in phase I composting (20-70^{0}C) due to uncontrolled self-heating in windrows for about 1-2 weeks, the outer cooler region of the stacks harbours mesophilic micoflora while in the centre where temperatures can reach upto 70^{0}C or higher, anaerobic bacteria predominate. The intervening zone with temperature range of 50-65^{0}C favours growth of aerobic thermophilic bacteria, fungi and actinomycetes (Johri and Rajni, 1999). After this period, the compost is subjected to highly aerobic thermophilic microbial phase II (pasteurization) which runs at temperatures of 40-60^{0}C for 6 days. In the beginning of phase II composting, thermophilic fungi and actimomycetes extensively colonize the plant matter until the temperature reaches 60^{0}C as an outcome of slow peak heating for about two days. High microbial activity and substrate compaction are responsible for elevated temperature of the composting material. Rise in temperature slows down the microbial activity and consequent decomposition rates. At high temperature, most of the pathogenic and non- pathogenic microorganisms except spores of actinomycetes and thermophilic fungi such as *Scytalidium thermophilum*, are killed (Straatsma *et al.*, 1991).

Thermophilic fungi grow extensively during the last phase of composting process from spores that survive the pasteurization temperatures (Straatsma *et. al.*, 1989). Thus thermophilic mycoflora dominates the scene at later stages of composting and contributes significantly towards the quality of composts. However, their presence throughout the course of composting (Table 1) is

Table 1. Numbers of thermophilic fungi in mushroom compost at various stages of its preparation.

Source of fungi	Log_{10} $CFUg^{-1}$	*S.thermophilum frequency*		*S. thermophilum* recovery (%)[a,b]	
Wheat straw	3.4	Infrequent	6	ND	
Drainage from compost	3.9	Infrequent	2	ND	
Horse droppings	2.3	Infrequent	3	ND	
Ingredient mixtures, start self-heating	3.3	Half	26	28	3
End phase I	2.4	Abundant	12	3	8
End phase II	6.5	Exclusive	5	90	54

[a] Standard deviations of CFU counts and of recovery are 1.00 and 17.3, respectively.

Compared with data of ingredient: mixture, CFU counts and recovery were significantly lower at the end of phase I and higher at the end of phase II.

[b] Counts of fungal recovery from washed particles are restricted to *S. thermophilum*

ND, not determined

Source: Straatsma *et al.* (1994)

largely responsible for maintenance of biological equillibrium that ultimately leads to unique selectivity wherein *Agaricus bisporus* multiplies without competition. Fergus (1964) described *Humicola grisea* var. *thermoidea, H. insolens* as dominant species among eight thermophilic and thermotolerant moulds isolated during phase II of peak heating viz., *Chaetomium thermophilum, Humicola grisea* var. *thermoidea, H. insolens, H. lanuginoa, Mucor pusillus, Stilbella thermophilum and Talaromyces dupontii.*

The pace of fungal succession is at least partly a result of the ecophysiological conditions in compost (Satyanarayana *et al.*, 1992). The pioneer thermophilic mycoflora of compost comprises of fast growing and rapidly sporulating fungi such as *Aspergillus fumigatus* and *Rhizomucor* spp. with a pH optima below 7.0 and temperature optima of about 40^{0}C. When self heating and ammonification starts and pH reaches 9.0, these pioneer species disappear and pave way for *Talaromyces thermophilus* and *Thermomyces lanuginosus* which exhibit high thermal death point and pH tolerance, thus dominating the scene during massive heat production. At the end of the composting process, about 50-70% of the compost biomass is constituted by thermophilic fungi (Sparling *et al.*, 1982; Weigant, 1992). While most species are eliminated, *Scytalidium thermophilum* appears as near exclusive species after phase II composting and constitutes the climax species in mushroom compost (Straatsma *et al.*, 1994). Actinomycetes and bacteria, however, appear to play a decisive role in successful colonization by *Scytalidium thermophilum.* This thermophilic fungus promotes growth of *Agaricus bisporus* by positively influencing the surface growth rate of mushroom mycelium and by exerting inhibitory influence on the growth of pathogenic microorganisms.

The microbial biomass content of the composting material also changes with microbial activity during the composting sequence (Derix *et al.*, 1990) The actinomycetes and fungal mycelium comprises of a much higher proportion of the total biomass than bacteria although it is difficult to define the specifics of the domineering species. Straatsma *et al.* (1994) reported the biomass ratio of 1.8:1 of fungi to prokaryotes after phase II composting while Weigant (1992) reported an estimated ratio of 0.9:1 and 2.3:1 for the conventional and experimental composts, respectively. Alexander (1997) reported a ratio of 1:30: > 10^3 between the counts of *Scytalidium thermophilum, Aspergillus fumigatus,* Zygomycetous forms and prokaryotes in compost.

Thermophilic fungi of the *Torula-Humicola* complex are a dominant component of microbial community during phase II. They are inherently close partners in the degradation processes and provide selectivity to the compost (Straatsma *et al.*, 1989; Op-Den Camp *et al.*, 1990). Various groups are actively engaged in studying the ecology of mushroom compost. Among these,

Straatsma's group in Holland, has laid greater emphasis on population dynamics of *Scytalidium thermophilium* for better compost management. Straatsma *et al.* (1994) isolated twenty two species of *thermophilic* fungi from mushroom compost, of which, nine viz., *Chaetomium thermophilum, Malbranchea sulfurea, Myriococcum thermophilum, Scytalidium thermophilum, Stilbella thermophila , Thielavia terrestris* and two unidentified Basidiomycetes were promotory for mycelial growth of *Agaricus bisporus.*

In our group, the community structure of thermophilic fungi in mushroom compost has been studied for sometime now and few leads are available. Rajni *et al.* (1998) observed nearly similar microbial distribution pattern in compost as corroborated by Straatsma *et al.* (1991) with predominance of *Scytalidium thermophilum* although the inputs in the European and Indian composts are substantially different. These workers also observed morphological variability of this fungus in compost with differences in their growth rate and promotion of mycelial growth. In 20d schedule of compost preparation, *S. thermophilum* was detected from first turning (i.e., after 5th day of composting) till fifth turning (i.e. 17th day) whereas species of *Paecilomyces* was present only during the last turning i.e., 20th day. Two isolates of *Malbranchea cinnamonea* were recorded between the second and fifth turning stage (8th to 17th day). The presence of *Paecilomyces* sp. and *Malbranchea cinnamonea,* which are normally slow growers, provides an opportunity to evaluate their influence on *in situ* mycelial extension of *Agaricus bisporous.* After 24 d, the population of *S. thermophilum* was 10^8 propagules g^{-1} of compost.

The cultivation of mushroom under controlled conditions in a growth room changes the quality and quantity of mycoflora in the compost. However, Klamer *et al.* (1998) reported *Aspergillus fumigatus* and *Rhizomucor pusillus* as predominant species before peak heating and *Paecilomyces variotii, Scytalidium thermophilum* and *Thermomyces lanuginosa* as dominant forms after peak heating. In casing soil the total thermophilic and thermotolerant population was observed to decrease with the time of sampling, maximum counts at day zero (0.08 x 10^7 $cfug^{-1}$) were represented by *Aspergillus fumigatus, Malbranchea cinnamonea, Malbranchea* sp., *Sporotrichum thermophile* and mycelia sterilia (Rajni *et al.*, 1998). This population remained constant until 24 h of formalin treatment. After 48 h, *Malbranchea cinnamonea* disappeared. After first week, *Aspergillus fumigatus* and *Rhizomucor* sp. were obseved but after 3rd and 4th week, species of *Rhizomucor* only were isolated along with mycelia sterilia. *Rhizomucor* sp. dominated from second week after formalin treatment.

Wood Chip Piles

Industrial wood chip piles are an interesting example of self-heating plant matter. Such piles often spontaneously heat which aggravates deterioration and frequently leads to spontaneous ignition. The temperature in wood chip piles generally reaches 60⁰C but occasionally can rise to ignition within a few weeks of storage (Hajny *et al.*, 1967). Thus, it constitutes an excellent ecological niche for the development and growth of thermophilic fungi. The rise in temperature is contributed significantly by microbial metabolism and respiration of wood parenchyma cells (Springer *et al.*, 1971; Feist *et al.*, 1971, 73). The microbial succession begins at the centre of pile and builds up a large population of thermophilic fungi which can cause considerable loss due either to discolouration or reduction in the quality of wood chips or losses on account of spontaneous ignition (Greaves, 1975; Bergman and Nilsson, 1981; Ogbonna and Pugh, 1983).

Tansey (1971) carried out detailed studies of the diversity of thermophilic fungi in wood chips collected at Fiberboard Corporation San Joaquin Caiton Plant in Antioch, California. *Thermoascus aurantiacus* was abundant in fresh chip samples whereas *Chaetomium thermophile* var. *dissitum, Humicola grisea* var. *thermoidea* and *Thermoascus aurantiacus* were isolated from unheated chip samples. *Humicola lanuginosa, Talaromyces emersonii* and *Thermoascus aurantiacus* were abundant in heated samples while *Humicola lanuginosa, Sporotrichum thermophile* and *Talaromyces emersonii* were abundant on charred samples. *Chaetomium thermophile* var. *dissitum* and *Thermoascus aurantiacus* were also common in charred samples. Auer *et al.* (1988) observed *Aspergillus* sp., *Rhizomucor* sp. *Sporotrichum* sp. and *Thermoascus aurantiacus* as dominant fungi on piles of *Eucalyptus* chips in Sao Paulo.

Coal Spoils Tips

It provides an environment suitable for the establishment of a variety of thermophilic organisms. The greater plant turnover, extreme heat and lack of suitable soil crumbs, favour thermophilic microorganisms in coal spoils (Evans 1971a, b). However, with gradual accumulation of humus, temperature decreases and this promotes development of thermotolerant and mesophilic forms. The survival of obligate thermophilic microflora at this stage is usually dependent on the production of overwintering structures. *Thermoascus aurantiacus* is abundant in coal spoil tips; sexual spores resistant to unfavourable conditions determine success in survival and distribution. *Aspergillus fumigatus, Mucor pusillus and Thermomyces lanuginosus* are dominant forms not only in this habitat but also in other habitats like senescent grass leaves (Apinis and Chesters, 1964), sewage (Chang and Hudson, 1967; Klopotek, 1972) and peat and bog soils (Kuster and Locci, 1964). The population of thermophiles in coal spoil tips is similar to that in natural grasslands but is markedly lower than the density known in composts, manure and sewage (Evans, 1971b).

The unique thermophilous species present in this habitat are, *Acrophylospora fusispora, Calcariosporium thermophile, Mortierella wolfii, Penicillium argillaceum, Sphaerospora saccata* and *Talaromyces laycillanum* (Evans, 1971b).

In Indian coal mine soils, the development and abundance of thermophilous fungi was found to vary with the plant cover. The coal mine soils of Chhindwara district of Madhya Pradesh, harboured fourteen thermophilic and thermotolerant fungi belonging to Phycomycetes, Ascomycetes and Deuteromycetes with an overall frequency percentage of 7.5, 9.1 and 16.6, respectively (Johri and Thakre, 1975). The most prominent fungi in unamended soils were, *Absidia corymbifera, Rhizopus microsporus, Rhizopus rhizopodiformis* and *Torula thermophila*. Addition of glucose to the soils resulted in recovery of, *Achaetomium macrosporum, Aspergillus fumigatus, Sporotrichum sp., Thermoascus aurantiacus, Thermomyces lanuginosus* and *Thielavia minor*. The regular appearance of slow growing *Sporotrichum* sp. in all soil samples appear to have resulted from the slow pace of natural microbial competition during the period of incubation. The observed low density was on account of closeness to colliery areas, devoid of plant cover.

Stored Grains

Stored grains constitute another excellent substrate for colonization by thermophilic fungi (Cooney and Emerson, 1964). Suitable temperature for microbial growth is a consequence of metabolic activity of mesophilic microorganisms and insects. Available food and moisture in the grains provide an ideal habitat for growth and multiplication of fungi (Flannigan and Sellars, 1972).

Apinis (1963) reported thermophilic fungi on the aerial parts of the standing crops while Clarke *et al.* (1967) reported their presence in freshly harvested grains. However, kernels of barley and wheat collected before harvest were devoid of these organism (Flannigan, 1974). The air, dust and debris, under storage conditions, increase the inoculum concentration and temperature that facilitates the development of thermotolerant fungi such as *Absidia corymbifera* and *Aspergillus fumigatus*. The metabolic activity of fungi raises the temperature and obligate thermophiles such as *Humicola lanuginosa* appear on stored grains (Clarke *et al.*, 1969).

Wareing (1997) studied the diversity of thermotolerant and thermophilic fungi in shipments of food-aid grains and large bag stacks of maize stored in Sub-Saharan Africa. The dominant fungi were, *Aspergillus candidus, A. flavus, A. fumigatus, Paecilomyces variotii, Rhizomucor pusillus, Thermoascus aurantiacus, T. crustaceus* and *Thermomyces lanuginosus*. *Aspergillus* sp. was the primary colonizer. The change in succession pattern was dependent on change in temperature. *Paecilomyces variotii* could not grow above 50^{0}C whereas *Thermoascus crustaceus* and *Thermoascus aurantiacus* were dominant at 55^{0}C and 60^{0}C, respectively.

Moubasher *et al.* (1980) studied the effect of temperature and moisture content on seed borne fungal diversity of the Egyptian peanuts. *Aspergillus fumigatus* was not only extremely dominant but was the main component of the total fungal flora. *Mucor pusillus, Humicola grisea* var. *thermoidea, H. insolens, H. lanuginosa* and *Sporotrichum thermophile* were the other dominant species. The total count of thermophilic fungi observed at 45^{0}C significantly increased with change in moisture content to 21%.

Aspergillus fumigatus and *Rhizomucor pusillus* were the most common thermophilic fungi in hazlenut and walnut seeds at 45^{0}C. *Humicola grisea* var. *thermoidea* and *Thermoascus aurantiacus* were rarely found in walnuts (Abdel Hafez and Saber, 1993). The frequency of occurrence of thermophilous fungal species on processed oat samples stored in cans, cardboard boxes and plastic bags was in the order, *Aspergillus fumigatus*> *Rhizopus microsporus*>*Rhizopus oryzae* > *Paecilomyces* sp. (Nogueira *et al.*, 1996).

Other Thermogenic Sites

A number of other thermogenic sites such as hay, nesting materials, dung of herbivorous animals, volcanic regions, hot springs and thermal effluents of nuclear power station and industries have also been examined for thermophilic fungal diversity.

Malbranchea pulchella var. *sulfurea, Mucor pusillus, Humicola lanuginosa, H. stellata* and *Thermoascus aurantiacus* were reported to predominantly colonize self-heated hay (Cooney and Emerson, 1964). *Chaetomium thermophile* var. *dissitum, Myriococcum albomyces* and *Torula thermophila* were dominant species in nesting materials of birds (Cooney and Emerson, 1964). This association is traced to the presence of excreta in the nesting materials wherein elevated temperatures develop during the incubation of eggs or rearing of the young. Cooney and Emerson (1964) have in particular drawn attention to the nests of a group of birds, the Megapodiidae (or incubator birds) as possible natural site of the occurrence of thermophilic fungi. These birds are native of Australia and the islands of Southwestern Pacific and appear to have probably existed for over 50-60 million years. The birds incubate their eggs inside large mounds (compost piles) which they construct out of forest litter wherein the microbial activity warms them upto 33-50^{0}C for several months (Seymour and Bradford, 1992). Thermophilic fungi probably have evolved from their mesophilic counterpart in such warm, humid and aerobic environment however scientific information supporting such a possibility is missing.

The dung of herbivorous animals, from rabbits to elephant, is another niche for thermophilic fungi. *Chaetomium thermophile* var. *coprophile*, *Humicola lanuginosa*, *Mucor pusillus*, *M. miehei* and *Torula thermophila* were reported from horse dung; *Chaetomium thermophile* var. *coprophile* and *Humicola insolens* in cow dung; *C. thermophile* var. *coprophile* and *Malbranchea pulchella* var. *sulfurea* in goat dung, and *Humicola insolens* and *Humicola lanuginosa* were dominant species in sheep dung. *Humicola grisea* var. *thermoidea* was a dominant species of elephant dung while *Sporotrichum* sp. was an inhabitant of mixed dung (Cooney and Emerson, 1964). Hasenekoglu and Yesilyurt (1996) studied the thermophilic and thermotolerant microfungal flora of dung in sheep and cow stables in some districts and villages of Erzurum province, Turkey. Eight thermophilic and five thermotolerant taxa were isolated from 300 dung samples in which *Aspergillus fumigatus* was the dominant.

Rippon *et al.* (1980) studied the diversity of thermophilic fungi in samples of water, foam, microbial mat, soil and air obtained from areas associated with the cooling canal of a nuclear power station. All sites were thermally enriched from which 53 thermophilic and thermotolerant taxa of fungi were recovered. *Aspergillus fumigatus* was dominant in these samples whereas *Rhizomucor pusillus* and *R. miehei* were particularly abundant in heated water and adjacent soils. Foam acted as a concentrator of spores so that many colony forming units of fungi were often found in such samples. *Chaetomium thermophile*, *Mucor pusillus*, *Humicola grisea*,. *H. insolens*, *H. lanuginosa* and *Scytalidium thermophilum* were the dominant mycoflora recovered from the samples collected from the industrial thermogenic sites of Uttaranchal, India (Gupta, 2000).

Non-Thermogenic Environments

Propagules of thermophilic fungi have been recorded in non-thermogenic environments like red wood, cooling towers, floor dust, dust on sparingly used books in air-conditioned libraries, Arctic and Antarctic soils and Himalyan soils (Abdel Malleket *et al.*, 1988; Ellis, 1980 b; Sandhu and Singh, 1981; Subrahamanyam, 1978). Tubaki *et al.* (1974) reported thermophilic fungi in sediments of rivers where the bottom temperature never exceeded 7^0C. Authors concluded that such species were perhaps carried away from the surrounding terrestrial habitats. Abdel Mallek *et al.* (1988) reported thermophilic form alongwith mesophilic and keratinolytic fungi in floor dust samples collected from students residential halls of Assiut University, Egypt. *Aspergillus fumigatus* was the commonest thermotolerant species, recovered with high frequency.

Thakre *et al.*(1991) isolated fungi responsible for biodeterioration of books and journals in Indian libraries. The predominant thermophilic and thermotolerant fungi, found in dusts on sparingly used books ultimately leading to biodeterioration were, *Aspergillus fumigatus, Chaetomium thermophile, Sporotrichum thermophile* and *Thermoascus aurantiacus*. The mycological analysis of sedimented dust samples from oases of western desert Egypt revealed presence of, *Chaetomium thermophilum, Rhizomucor pusillus, Talaromyces thermophilum* and *Thermomyces lanuginosus* as dominant thermophilic mycoflora (Abdel Hafez *et al.*, 1992).

Genetic Diversity

The genetic diversity of thermophilic fungi has not been widely studied. It is however attracting attention in order to assign correct taxonomic status based on the traditional and molecular tools. The study of genomic DNA/RNA composition, gene structure mainly with a view to examine intron architecture/ organization and also regulatory sequences and "motifs" related to gene expression, is in progress.

de Bertoldi *et al.* (1972,1973) observed interesting variations in GC content among strains of *Humicola grisea* var. *thermoidea* and *Thermomyces lanuginosa*. The mole % GC content of species of thermophilic *Humicola* varies between 34.8-55.5 and in *Thermomyces,* between 30.0- 40.2. Although, *Humicola brevis* and *Torula terresteris* possess almost similar GC content (30.6 and 30.4 %) but these two organisms are not related which may be due to difference in their base sequences (Meyer and Phaff, 1969). de Berloldi *et al.* (1973) attributed the discrepancies between the traditional criterion of species determination, DNA composition of *Humicola* and related fungal genera, to the absence of sexual cycle. Rodrigues *et al.*, (1991) suggest that high mutation rates can cause variability and genomic differences in *Humicola* sp. which are further accentuated by polynucleated characteristic of this species.

The mutagenic mechanisms could have evolved in favour of GC formation in thermophiles (Wildeman, 1988). This assumption is based on the fact that non-inducible (or low-repressed) genes of thermophilic fungi show different pattern. A comparison of nucleotide sequence of 5.8 S rRNA should be the central focus of studies on genetic diversity of thermophilic fungi. The nucleotide sequence of *Thermomyces lanuginosus* is closely related to other fungi, but exhibits high overall homology with other 5.8 S rRNA (Wildeman and Nazar, 1981). The 5 S rRNA sequence of *T. lanuginosus* exhibits a high degree of similarity with other eukaryotic 5 S rRNA (Wildeman and Nazar, 1982). This ubiquitous thermophile shows app. 75% base equivalence with that of

Neurospora. The 24 S rRNA sequence was also highly homologous to sequences of other fungal rRNAs (Walker *et al.*, 1990). Nazar *et al.* (1987) analysed rDNA from *Thermomyces lanuginosus* and found that 18-25 S rRNA intergenic region was significantly shorter then other organisms and distinct in residue distribution, when compared with other known fungal internally transcribed spacer (ITS) region sequences.

An important feature observed in some thermophilic moulds is lack of introns in asp A of *Rhizomucor miehei* and *R. pusillus*; trp C of mesophilic, *Aspergillus niger* exhibits this property (Gurr *et al.*, 1987). However, the positional conservation is not well observed in fungal cellulase genes. A comparison between cellobiohydrolase genes of *Humicola grisea* with other fungi displayed different positional pattern for their respective genes (Azaevedo *et al.*, 1999). Azaevedo *et al.* (1990) have reported a unique intron for the cbh 1.1 in *Humicola grisea* cellulase genes. These workers also observed 4 regions in the catalytic domain of the putative cbh 1.2 gene harbouring the consensus sequence for splicing and generation of the four introns.

The genetic variation exhibited by *Torula-Humicola* complex has drawn wide attention. Araujo *et al.* (1983) have distinguished homokaryotic strains of *Humicola grisea* var. *thermoidea* in two groups displaying uniform DNA profile reflecting heterogenicity in the wild genome using RAPD analysis. Straatsma and Samson (1993) studied the genetic diversity among *Scytalidium thermophilum* isolates using RAPD analysis. These strains exhibited a distinct pattern of amplified DNA bands.

The RAPD analysis and sequence analysis of ITS region of rDNA has been found to show wide genetic variation in *Torula- Humicola* complex (Lyons *et al.*, 2000). RAPD analysis of 34 geographically diverse isolates revealed two distinct groups showing differences in the banding pattern. An examination of the genetic distance matrix indicated differences between isolates from *Scytalidium thermophilum* cultural types 1 and 2. The sequence analysis of ITS 1, 5.8 S and ITS 2 region of rDNA suggested high homology between the isolates with minor sequence variation. This could be correlated with differences between the isolates based on morphology and thermogravimetric characteristic (Straatsma and Samson, 1993; Lyons and Sharma, 1998). The genomic DNA variations thus facilitate differentiation of sub-groups within the species. These sub-groups do not appear to be dependent on their geographical origin. The RAPD groupings mirror closely the morphological and thermogravimetric data for *S. thermophilum* isolates and provides further evidence of the variation, which exists between the species complex. Isolates of *S. thermophilum* bear close similarity with those of *Humicola* grisea var. *thermoidea* and *Humicola insolens*. In our lab, Rajni (1999) studied the genetic diversity between the black and white strains of *Scytalidium thermophilum* by RAPD analysis. A distinct band pattern was exhibited by these isolates. The protein profiling of *Torula-Humicola* complex exhibited *Scytalidium thermophilum* to be more closely related to *Humicola grisea* var. *thermoidea* while the latter was more related to *H. insolens* than *H. lanuginosa*; *H. insolens* and *H. lanuginosa* were more distantly related. Currently this work is being expanded to study the genetic diversity among various thermophilic isolates of compost.

Functional Diversity

The early description of thermophilic fungi from hay and retting guayule infused interest in unravelling their *in situ* functionality especially pertaining to extracellular enzymatic machinery viz., polysaccharases, proteases and lipases (Johri *et al.*, 1999). The wide enzymatic potential exhibited by these fungi with different phsico-chemical characteristics helps in the functioning of each component under different biophysical conditions. The theory of natural selection suggests that an organism capable of producing one or more thermostable enzyme would be better equipped to function at elevated temperature depending on the metabolic importance of the particular enzyme.

The predominance of thermophilic fungal community in the two major habitats viz., soil and compost is attributed to their avidity for degrading various components of complex organic matter by extracellular enzymes, fungistatic action, production of antimicrobials, volatile sporostatic products, and phenomenon of self-inhibition. These features give them a competitive edge over others and help in establishing positive and negative interactions within a given ecosystem. The wide exploration of the enzymatic machinery of individual microbial component and their *in situ* enzymatic action in their own niche would help in greater understanding of the relationship between structural and functional diversity.

Several thermophilic fungal species such as *Humicola lanuginosa*. *Scytalidium thermophilum* and *Sporotrichum thermophile* secrete amylase and lipase that would extend the lytic action beyond the hemicellulose component (Johri *et al.*, 1991; Rawat, 1998; Khokar, 1999). *Aspergillus fumigatus*, *Chaetomium thermophilum*, *Humicola grisea* var. *thermoidea*, *H. insolens* and *Torula thermophila* are known to grow rapidly by virtue of their strong cellulolytic and xylanolytic enzyme system (Rana *et al.*, 1996). Dusterhoft *et al.* (1997) have reported that species of *Humicola* degrade cellulose and hemicellulose of paddy straw by producing cellulases and glucosidases.

The polysaccharolytic activity of thermophilic mycoflora plays a major role in their successful colonization in compost. The loss of cellulose and lignocellulose and increase in protein content during the composting period reflects on the increased polysaccharolytic activity of fungal biomass resulting increased level of reducing sugars (Iiyama *et al.*, 1996). Thermophilic fungal community constitutes a dominant population of the functional niche occupied by them in the compost. Each crop of microflora utilises the available substrate as also the cellular components of its predecessors for growth and sustenance.

The increase in cellulolytic and amylolytic activity during composting is a reflection of change in the population and community structure of the resident microflora. The enzymatic diversity of compost microbiota is known to result in specificity and successional change besides providing a niche to various species to survive in the absence of simple sugars. Stanek (1969) observed stimulation of growth of mushroom mycelium by cellulase decomposing microflora mainly, actinomycetes and fungi. It is, however, difficult to draw a conclusive relationship between restricted cellulolysis and growth promotion of *Agaricus bisporus* since species such as *Aspergillus fumigatus* and *Corynascus thermophilum* are cellulolytic but not growth promotry whereas the reverse is true for *Chaetomium thermophilum* and *Sporotrichum thermophilum* (Straatsma *et al.*, 1994, 1995). Thus, growth promotry species were cellulolytic but not the pioneers of compost biota. This influence is exerted by the climax species of mushroom compost, *Scytalidium thermophilum,* due perhaps to production of a complete complement of enzyme machinery (Rawat, 1998; Khokar, 1999; Tiwari, 2000).

More detailed, *in situ* enzymatic investigations would probably give us a better understanding of relationship between structural and functional diversity of thermphilic fungal community. Rajni (1999) observed that during the 24 day composting sequence of button mushroom, the level of organic carbon decreased from 18.12 to 10.57% while that of cellulose and lignocellulose from 32.3-23.0% and 52.4 – 43.1%, respectively. An increase of 77.5% and 86% in protein and reducing sugar level was observed. The level of amylase endo- cellulase and exo-cellulase also changed. Based on the regression analysis between the population of *Scytaliduim thermophilum* and chemical parameters, a 92.5% change in lignocellulose was found as a result of rise in population of *S. thermophilum* in compost. A change of 96%, 83% and 88% in nitrogen, organic carbon and cellulose respectively was due to population change in *S. thermophilum.*

An increase in cellulase activity and decrease in laccase activity was observed after the addition of casing soil to the surface of compost colonized by *Agaricus bisporus* (Gillman *et al.*, 1994). The ureolytic activity of *Aspergillus fumigatus* in non-amended soil and that of *A. fumigatus*, *A. terreus* and *Rhizomucor pusillus* in urine and human excreta amended soil, was found to aid in successful colonization (Abdel Rahman *et al.*, 1992). The successful secondary colonization by *Aspergillus fumigatus*, *Chaetomium cellulolyticum*, *C. thermophile* var. *coprophile*, *Humicola insolens*, *H. lanuginosa*, *Mucor pusillus*, *M. miehei*, *Myceliophthora thermophilum*, *Rhizopus rhizopodifornis*, *Sporotrichum thermophile*, *Thermoascus aurantiacus* and *Thielavia terrestris* was due to their ability to produce cellulases and hemicellulases (Jain, 1999).

The thermophilic fungi produce not only wide range of extracellular enzymes but also exhibit enzyme multiplicity which helps them function efficiently under different eco-physiological conditions. Multiplicity of cellulases and xylanases has been widely reported in *Chaetomium thermophile* var. *coprophile*, *Humicola grisea* var. *thermoidea*, *Melanocarpus albomyces*, *Talaromyces emersonii* and *Thermoascus aurantiacus* (Ganju *et al.*, 1989; Thakur *et al.*, 1992; Tuohy *et al.*, 1993; Kalogeris *et al.*, 1998; Prabhu and Maheshwari, 1999).

Conclusion

The study of structural diversity complemented with genetic and functional diversity of thermophilic fungi would give us a total picture of species spectrum of its community alongwith their functional potential. The structural distribution of this community has not been widely examined across the globe. There has not been a concerted effort to explore ' hot spots', variety of habitats available in hot and humid topics in search of novel isolates. The survival strategies and competitive behaviour of this community in their niche is very little known. The non-availability of suitable techniques during the last decades had precluded the investigations of *"in situ"* behaviour of microorganisms in their natural habitat, but, with the advent of immuno- and molecular-probes the knowledge in this area has undergone a sea change. The main area of interest in thermophilic fungi has in the past been restricted to their functional behaviour. The genetic diversity, although, an important aspect has not been of much interest among the workers. Till now, there is just a horizontal advancement in this direction. The present panorama needs to be extended towards vertical dimensions. The biopotentiality of thermophilic fungi has been widely exploited in various industrial sectors, but their full biotechnological potential remains to be tapped. The use of molecular technologies in this arena has to be extended. Their predominance in compost is based on their strong enzymatic potential which necessitates search for more potent isolates whose preinoculation in mushroom compost would provide selectivity to compost besides increasing productivity of mushroom. Thus, new dimensions to our existing knowledge still need to be added which is likely to be instrumental in throwing light on their ecological adaptive behaviour and exploitation of their degradation capacity of plant material.

References

Abdel-Hafez, A.I.I. and Saber, S.M. (1993). Mycoflor and mycotoxin of hazelnut (*Corylus avellana* L.) and walnut (*Juglans regia* L.) seeds in Egypt. Mykoflora and mykotoxine der haselnuss- (*Corylus avellana* L.) and walnuss (*Juglans regia* L.) fruchte in Agypten. Zentralblatt-fur-Microbiologie, **148**: 137-147.

AbdelHafez, S.I.I., Moharram, A.M., Ismail, M.A. Razak, A.A. and Ghanem, E.H. (1992). Mycoflora of sedimented dust in the oases of western desert Egypt. Proc. second Regional Mycological Conference, Cairo, Egypt, 7-10 Oct. 1992, 73-87.

AbdelMallek, A.Y., Bagy, M.M.K. and Moharram, A.M. (1988). Fungi of the floor dust in students residential halls of Assiut University, Egypt. Egypt. J. Botany., **31**: 69-80.

Abdel-Rahman, T.M.A., Ali, M.I.A., Salama, A.A.M., Abo-Ellil, A.H. and El-Selhdar, A.A.H. (1992). Ureolytic activity and soil mycoflora as affected by amendment with different urea sources. Egypt. J. Microbiol., **27**: 143-155.

Alexander, M. (1997). Introduction to soil microbiology. Wiley, New York. 467 p.

Al-Fassi, F.A., Malibari, A.A. and Moustafa, M. A. (1994). Physiological studies on ten thermophilic and thermotolerant fungi isolated from different locations in the western region of Saudi Arabia. Arab Gulf J. Scientific Res., **12**: 321-340.

Apinis, A.E. (1963). Occurrence of thermophilous microfungi in certain alluvial soils near Nottingham. Nova Hedgw. 5: 57-78.

Apinis, A.E. (1963). Thermophilic fungi of coastal grasslands. In: Soil organisms. (Eds: J. Doeksrn and J. Van der Drift). North Holland Publications, Amsterdam, 127-938.

Apinis, A.E. (1972). Thermophilous fungi in certain grassland. Mycopathol. Mycol. App., **48**: 63-74.

Apinis, A.E. and Chesters,C.G.C. (1964). Ascomycetes of salt marhes and salt dunes. Trans Br. Mycol. Soc., **47**: 419-435.

Araujo, E.F., Barros, E.G., Galdas, R.A. and Silva, D.O. (1983). Beta-glucosidase activity of a thermophilic cellulolytic fungus, *Humicola* sp. Biotechnol. Lett., **5**: 781-784.

Auer, C.G., Krunger, T.L. and Barrichelo, L.E.G. (1988). Thermophilic fungi in self-heating piles of *Eucalyptus* chips. Instituto-de-Pesquisas-e-Estudos-Florestais., No. **38**, 28-321.

Azevedo, M.O., Felipe, M.S.S. and Satyanarayana, T. (1999). Molecular and general genetics. In: Thermophilic moulds in Biotechnology (Eds: B.N. Johri, T. Satyanarayana and J. Olsen). Kluwer Academic Publishers, Netherlands, 317-342.

Azevedo, M.O., Felipe, M.S.S., Alstolfi-Filho, S. and Radford, A. (1990). Cloning, sequencing and homologies of the cbh-1 (exoglucanase) gene of *Humicola grisea* var. *thermoidea*. J. Gen. Microbiol., **136**: 2569-2576.

Barnes, T.G., Eggins, H.O.W. and Smith, E.L. (1972). Preliminary stages in the development of a process of microbial upgrading of waste papers. Int. Biodeterio. Bull., **8**: 112-116.

Bergman, O. and Nilsson, T. (1981). Studies on outside storage of saw mill chips. Inst. Virkeslara Sukogshogasskolan, Stockholm Res. Note. R 71, pp. 43.

Bilai, T. (1984). Thermophilic micromycetes species from mushroom cellar composts. Microbiol. Zh., **46**: 35-38.

Breton, A. and Zwaenepoel, P. (1991). Succession of moist hay mycoflora during storage. Can. J. Microbiol., **37**: 248-251.

Brown, B.S., Mills, J. and Hulse, J.M. (1974). Chemical and biological degradation of waste plastics. Nature, **250**: 161-163.

Chang, Y. and Hudson, H.J. (1967). Fungi of what straw compost. I. Ecological studies. Trans. Br. Mycol. Soc., **54**: 649-666.

Chen-Kuei, Yu, Zueiching, C., Chen, K.Y. and Chen, Z.C. (1996). *Thielavia pingtungia* sp. nov. a thermophilic ascomycete from Taiwan. Mycotoxon., **60**: 241-247.

Chen-Kuei, Yu., Chen-Zueiching, Chen, K.Y. and Chen, Z.C. (1996). A new species of *Thermoascus* with a *Paecilomyces anamorph* and other thermophilic *Thermoascus* sp. from Taiwan. Mycotaxon., **60**: 225-240.

Clarke, J.H., Hil, S.T., Niles, E.V. and Howard, M.A.R. (1969). Ecology of microflora of moist barley in sealed silos on farms. Pest Infest. Res., **1968** : 17.

Clarke, J.H., Niles, E.V. and Hill, S.T. (1967). Ecology of the microflora of moist barley in sealed silos on farms. Pest. Infest. Res.,**1966**: 14 -16.

Cooney, D.C. and Emerson, R. (1964). Thermophilic fungi: An account of their biology, activities and classification (Eds: D. C. Cooney and R. Emerson). Freman and Co. San Francisco Pub, 180p.

de Bertoldi, M., Lepidi, A.A. and Nuti, M.P. (1972). Classification of the genus *Humicola* Traeen: I. Preliminary reports and investigations. Mycopathol. Mycol. Appl., **46**: 289-304.

de Bertoldi, M., Lepidi, A.A. and Nuti, M.P. (1973). Significance of DNA base composition in classification of *Humicola* and related genera. Trans. Br. Mycol. Soc., **60**: 77-85.

Derix, P.J.L., Op Den Camp, H.J.M., Vander Drift, C., van Griensven, L.J.L.D. and Vogels, G.D. (1990). Biomass and biological activity during the production of compost used as a substrate in mushroom cultivation. Appl. Environ. Microbiol., **56**: 3029-3034.

Dewey, F.M., MacDonald, M.M. and Phillips, S.I. (1989). Development of monoclonal antibody-ELISA DOT BLOT and DISPTICK immunoasays for *Humicola lanuginosa* in rice. J. Gen. Microbiol., **135**: 361-374.

Dusterhoft, E.M., Linssen, V.A.J.M., Vorgen, A.G.J. and Beldman, G. (1997). Purificaton, characterization and properties of two xylanases from *Humicola insolens.* Enzyme Microbial Technol., **20**: 437-445.

Eggins, H.O.W. and Mills, J. (1971). *Talaromyces emersonii,* a possible biodeteriogen. Int Biodet. Bull., **7**: 105-108.

Ellis, D.H. (1980 a). Thermophilic fungi isolated from a heated aquatic habitat. Mycologia, **72**: 1030-1033.

Ellis, D.H. (1980 b). Thermophilous fungi isolated from some Antarctic and sub-Antarctic soil. Mycologia, **72**: 1033-1036.

Ellis, D.H., and Keane, P.J. (1981). Thermophilic fungi isolated from some Australian soils. Aust. J. Bot., **29**: 689-704.

Eriksson, K.E. and Larsson, K. (1975). Fermentation of waste mechanical fibres from newsprint mill by the fungus *Sporotrichun pulverulentum.* Biotechnol. Bioeng., **17**: 327-348.

Evans, H.C. (1971 a). Thermophilic fungi of coal spoil tips I, Taxonomy. Trans. Br. Mycol. Soc., **57**: 241-254.

Evans, H.C. (1971 b). Thermophilic fungi of coal spoil tips II, Occurrence and temperature relations. Trans Br. Mycol. Soc., **57**: 255-266.

Evans, H.C. (1972). Thermophilus fungi isolated from the air. Trans. Brit. Mycol. Soc. **59**: 516-519.

Feist, W.C., Springer, E.L. and Hajny, G.J. (1973). Effect of storing wood chips at elevated temperatures. TAPPI J., **56**: 1-95

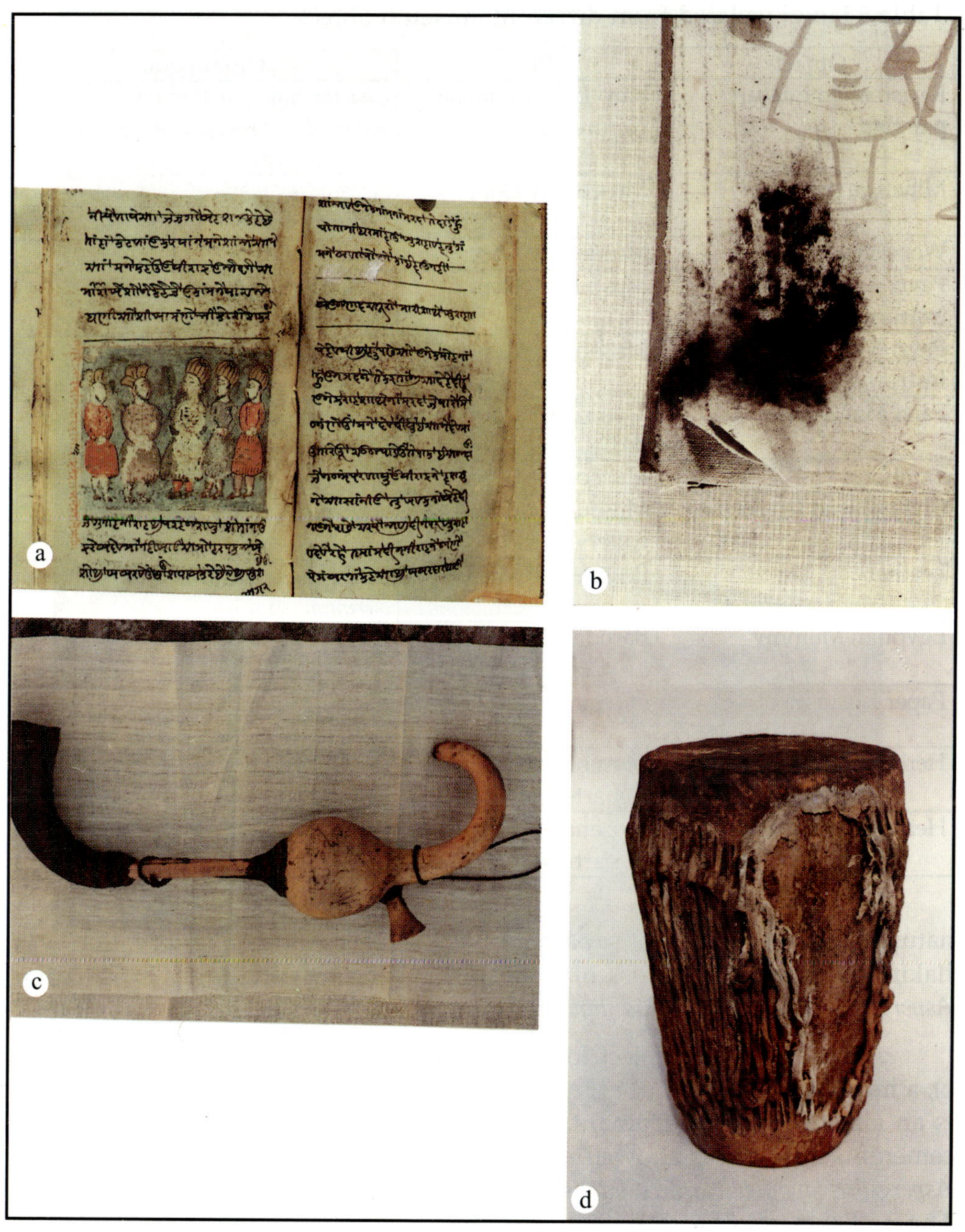

Plate 1

(a) A Parsi manuscript showing fungal infection (b) A photograph showing black growth due to *Aspergillus niger*. (c) Musical instrument Parvi (d) Tambura showing deteriorated strings

Plate 4

(a) Damaged Herbarium sheet of *Marchantia.* (b) Black coloured growth on *Lycopodium.* (c) Herbarium sheet of *Lycopodium* (d) Damaged leaf of *Ficus.*

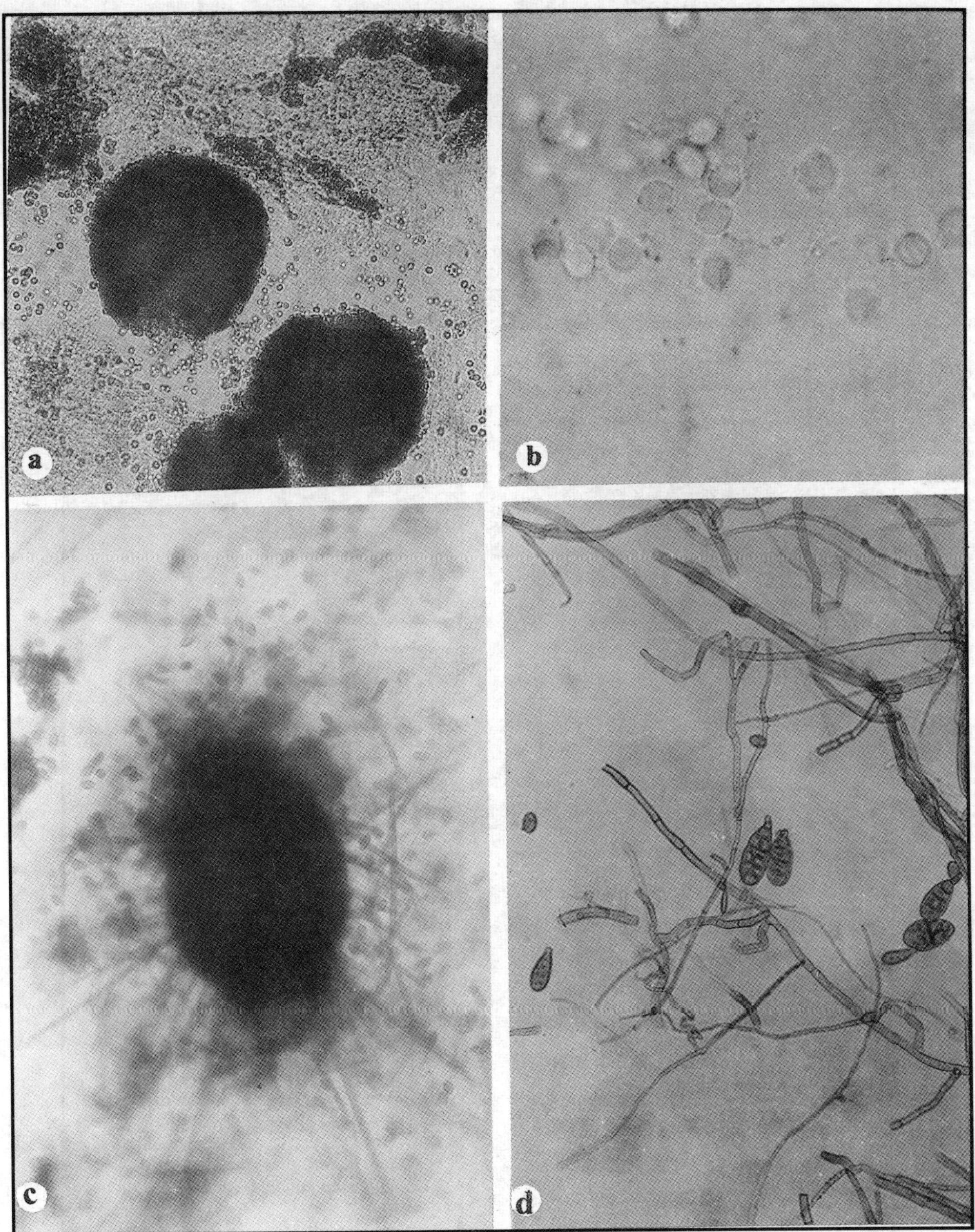

Plate 5

(a) Cliestothecia and ascospore of *E. Nivea* X94 (b) Photomicrograph showing ascospore of *E. nivea,* a fungal biodeteriogen X940 (c) Perithecium of *Chaetomium atrabrunneum* with straight hairs and oval ascospores X126 (d) Branched septate mycelium and conidia on *Alternaria alternata.*

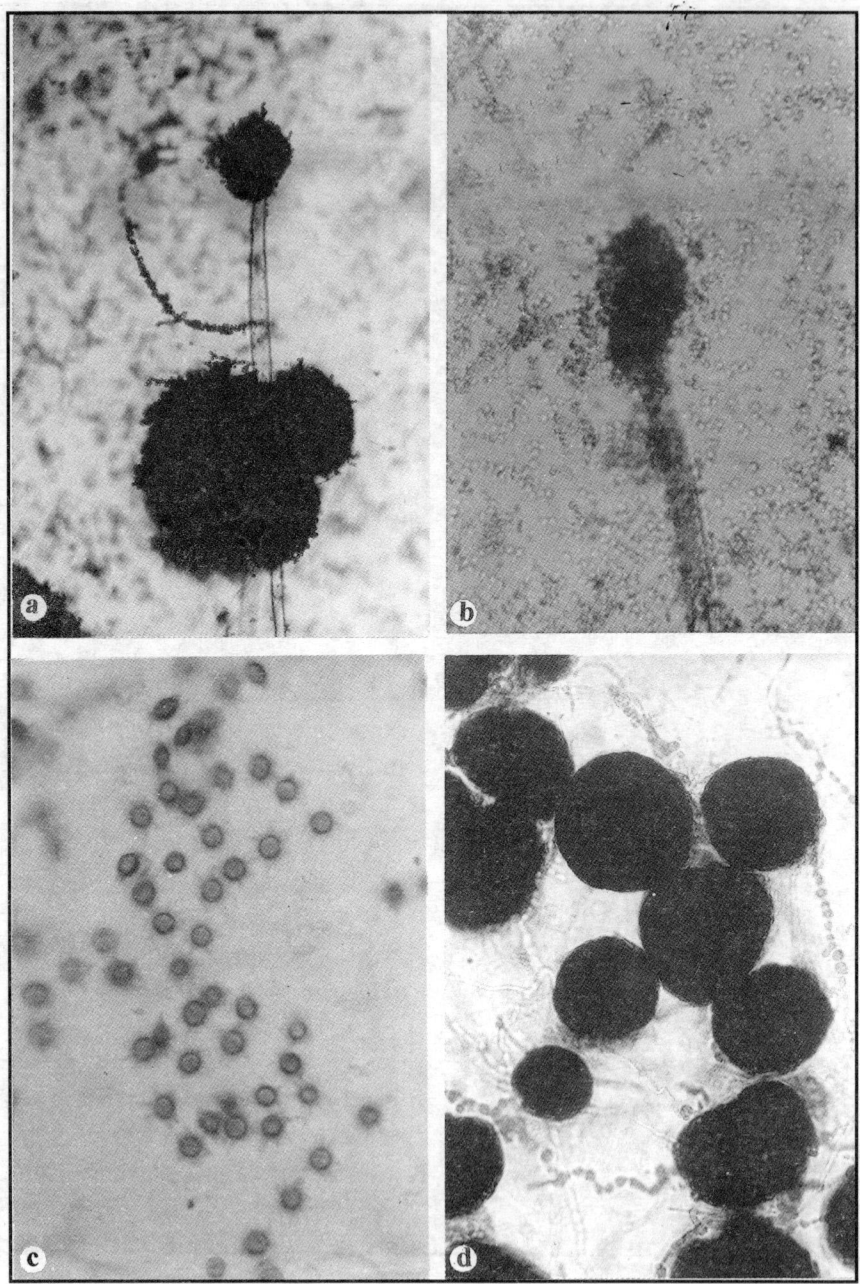

Plate 6

(a) Conidiophore and conidial heads of *Aspergillus niger* X 126 (b) Conidial head of *Aspergillus flavus* X 126 (c) Photomicrograph showing ascospores of *E. nidulans* X940 (d) Dark spherical cliestothecia of *Monascus ruber* isolated from indoor air of Egyptian gallery Fungal Biodeterioration of Museum objects in Baroda showed the presence of *Curvularia pallesance* and *Aspergillus niger*, water mark can be clearly seen on this herbarium (Plate 4 fig a). Herbarium sheet of *Lycopodium cerrnum* shows fungal growth of *Cladosporium herbarum* (Plate 4 fig b). Other two herbaria showed the presence of *A. niger* (Plate 6 fig. c, d).

Methods to Control Biodeterioration

Chemicals such as Para dichloro benzene, Thymol, Methyl bromide, Ethylene oxide are used to control fungal infection. Ethylene oxide has been banned in many countries due to its carcinogenic nature. Thymol, which is even today used to control fungal growth on museum, thereby causing artwork framed in them to adhere to the glass. Also reported is the softening of oil paints and yellowing of paper.

Some of the natural plant products are screened for their fungicidal and fungistatic properties are, *Allium satiuvm* (Kshemakalyani *et al.*, 1990; Arya *et al.*, 1995; Abdul Hafez and El-said 1997), *Eugenia caryophyllus* (Chinguang *et al.*, 1995), *Eucalyptus citrodora* (Baruah *et al.*, 1996), *Curcuma longa* (Saju *et al.*, 1998) etc.

Shah and Arya (2001) found oils of cashew nut, neem, eucalyptus and Godavaj powder (Sweet flag) effective against four fungal biodeteriogens. Aqueous leaf extracts of a large number of plant have been tried (Sekhawat and Prasad 1971;) Khanna and Chandra 1972; Dixit *et al.*, 1983; Arya and Mathew, 1990; Arya *et al.*, 1995). Dry rhizome powder has volatile oil with insecticidal properties. Its use is common in Oriental Institute of Baroda. It was found antifungal too. *Eucalyptus* leaves may be effective due to presence of Cineole in them (Arya, 1988). It has been observed that the activity of extracts may be lost due to autoclaving (Lokesha *et al.*, 1986).

Conclusion

The need of the hour is to make periodic survey of museum objects to find out any damage caused by microbes, to isolate the fungi, study the detailed behaviour of fungi and suggest suitable control measures. These methods need not be just effective but meet the other criteria like preserving aesthetic quality, economically viable, eco-friendly, broad, spectrum, with no side effects so as to enable us to preserve our works of art for posterity.

Acknowledgements

I thank Prof. G. P. Senan, Head, Department of Botany, M. S. University of Baroda, Vadodara for providing necessary laboratory facilities. I am grateful to the Director, Agharkar Research Institute, Pune, and the Head, Division of Plant Pathology, IARI, New Delhi, for confirming the identity of the fungi. Thanks are due to U.G.C. for awarding a Project on 'Eco-friendly approach to control fungal biodeteriogens in Museums of Gujarat'. I also record thank to Dr. S.K. Bhowmik, Ex. Director Baroda Museum and Picture Gallery for his encouragement during this study.

References

Abdul Hafex, S.I.I. and El-said, A.H.M. (1997). Effect of garlic onion and sodium benzoate on the mycoflora of pepper, Cinnamon and Rosemary in Egypt. Int. Biodet. and Biod., **39**: 100-101

Arya, A. (1988). Control of *Phomopsis* fruit rots by leaf extracts of certain medicinal plants. In: (Kaushik M.P. ed) Indigenous Medicinal Plant Symp. Pub. by Today & Tomorrow's Printers and Publishers New Delhi 41-46 pp.

Arya, A., Chauhan, R. and Arya, C. (1995) Effect of allicin and extract of garlic and *Bignonia* on two fungi. Indian J. Mycol. Pathol., **25** (3) 316-318.

Arya, A. and Mathew, D.S. (1990). Control of chiku fruit rots by leaf extracts of certain medicinal plants. Res. J. Pl. Env., **6 (1)**: 31-33.

Baruah, P., Sharma, R.K., Singh, R.S. and Ghosh, A. C. (1996). Fungicidal activity of some naturally occurring essential oils against *Fusarium moniliforme*. J. of Essential Oil Research, **8 (4)**: 411-412.

Chile, S. (1993). Study of airborne fungal spores at Jabalpur with special reference to allergenicity. Ph. D. Thesis, R.D. University, Jabalpur.

Chingduang, S., Manoch, L., Tantayaporn, S., Vong Kaluang, C. and Siriacha, P. (1995). Fungal deterioration of book in libraries In: Preprints of International Conference on Biodeterioration of Cultural Property, Bangkok, Thailand pp. 472-480.

Davies, R.R. and Smith, L.P. (1973). Aspects of human allergy and fungal spores. Clinical Allergy, **3**:263-267.

Dixit, S. N., Dube, N.K. and Tripathi, N.N. (1983). Fungitoxic essential oils vis-a-vis disease control. In: Recent Advances in Plant Pathol. (Eds.) Husain A., Singh K., Singh B. P. and Agnihotri V. P. Print House (India) Lucknow pp. 248-252.

Foster, R.A. (1986). The larger provincial museums and art galleries, In: Manual of curatorship, Butterworths, London, 66-75.

Gregory, P.H. (1973). The Microbiology of the Atmosphere, Leonord Hill Books, Aylesbury, England 1973 2nd end. pp. 377.

ICOM, (1990). ICOM statutes code of Professional ethics, Paris pp. 3.

Khanna, K.K. and Chandra, S. (1972). Antifungal activities in some plant extracts., Proc. Nat. Acad. Sci. India, **42 (B)**: 300-302.

Kshemakalyani, S.B., Telore, R., Madhavi, B. and Patel, G. S. (1990). The effect of Allicin and extracts of garlic on *A. flavus* and *A. parasiticus*. Indian J. of Mycol. and Pl. Pathol. p **20(5)**:247-248.

Lakshmikant and Mathur, S. (1991). Biodeterioration of cultural property by cellulolytic fungi. In: Biodet. of Culture Prop. (Eds. Agrawal O. P. and Dhawans S.) Macmillan India Ltd. New Delhi: 375-386.

Lokesha, S., Vasantha, K. and Shetty, H.S. (1986). Effect of plant extracts on growth and sporulation of *Aspergillus flavus*. Plant Dis. Res. 1 (1-2): 7981c. f.: Rev. Pl. Path. 67 (19880:4864).

Mishra, G. and Mishra, K.B. (1992). Aeromycoflora of a mica mine. In : National symp. on Management of microbes in Service of mankind, Allahabad Univ., 19-21.

Nair, S.M. (1972). In: Convservation in Tropics. Proc. of AST-Pacific Seminars on con. of cul. properties (Ed. Agrawal, O. P.) Rome 150-158.

Nair, S. M. (1980), Insect pest of textile materials in museum. In : Conservation in Tropics. (Ed Agarwal O. P.) pp 150.

Nigam, R. K. and Pathak, N. C. (1991). Environmental mycology of a tannery and a textile unit at Kanpur: Allergical aspects. In: Biodet. of Cultural Prop. (Eds. Agarwal, O. P. and Dhawan, S.) Macmillan India Ltd. New Delhi: 375-386.

Riecelli, A. (1999), Fungal growth on samples of paper: Inhibition by new antifungals., Restaurator, 20**(2)**: 97-107.

Saju, K. A., Venugopal, M. N. and Mathew, M.J. (1998). Antifungal and insect-repellent activities of essential oil of turmeric (*Curcuma longa* L.). Curr. Sci. **75** (7) 660-662.

Sekhawat, P. S. and Prasad, R. (1971). Antifungal activities of some plant extracts on Inhibition of spore germination. Indian Phytopath. **24**: 800-802.

Shah, N.R. and Arya, Arun (2001). Studies on some fungal biodeteriogens. Pub. by Bhartiya Kala Prakashan, New Delhi, 212.

Spieksma, F. T. M. (1980). Importance of aerobiological studies for the prediction of human diseases. In: Aerobiological Studies for the prediction of human diseases Aerobiol. Symp. Munich : 307-315.

Tilak, S. T., Sharma, B.R.N., Sengupta, S.R. and Kulkarni, R. L. (1970). The deterioration of Ajanta and Ellora paintings, In Con. of Cul. Prop. In India Proceedings of V Seminar, New Delhi 77-82.

Verma, K. S. and George, A.M. (1997). Aeromycoflora of Jabalpur. Res. J. chem. and Env., **1**: 29-31.

Wadhwani, K., Srivastava, A. K., Jamil, Z., Rizvi, H.I., Mishra, A. K. and Negi H.(1986). Jour. Rec. Adv. Appl. Sci. 1:8-12.

Walston, S. (1976). Proceedings of the ICCM National conference, Canberra 145.

14

Keratinophilic Fungi from Soils of Hills and their Keratinolytic Ability

Divya Bhadauria and R.K.S. Kushwaha*

Department of Botany, Christ Church College, Kanpur 208.001

One hundred ninety four isolates of ten keratinophilic fungal species were isolated from the hills of Assam, Chandigarh, Kathmandu, Kullu, Manali, Mussoorie, Nainital, Sagar, Shimla and Uttarkashi for the first time. Out of 194 isolates 155 isolates were tested for their keratinolytic ability. All of them were found positive.

Key words : Keratinophilic fungi, hills, keratinolytic ability.

Keratinophilic fungi represent an important component of soil microflora where they decompose the highly resistant keratin, a proteinaceous substrate. This unique group may exist and proliferate actively in soil under favourable conditions, particularly in places where they can utilize various forms of keratin. Frequency of occurrence of keratinophilic fungi in the soil is influenced by a number of biotic and abiotic factors (Otcenasek, 1978). Soil that are rich in keratinous materials are most conducive for the growth and occurrence of keratinophilic fungi (Otcenasek, 1978; Mercantini *et al.*, 1980, 1986).

Among the multitude of fungal species known to colonize keratin some are able to parasitize man and animals causing disease (Gentles, 1962; Singh, 1969; Hubalek and Hornick, 1977).The keratinophilic fungi are responsible for the break down of any keratin containing wastes such as hair, fur and feather. These fungi therefore play a significant role in the break down of soil debris. The genus *Chrysosporium* is composed of a large group of species frequently isolated from soil (Oorschot, 1980; Sigler *et al.*, 1986; Cano and Guarro, 1994; Kushwaha, 2000). The majority of them are anamorphs of dermatophytes and have keratinophilic properties. The identification of these species is simple when the teleomorph is present. The frequency of occurrence of geophilic dermatophytes and related keratinophilic fungi in a particular area depends on the frequency of the animals visited (Pugh and Mathison, 1962; Garg, 1966; Randhawa and Sandhu, 1965; Sur and Ghosh, 1980; Deshmukh and Agrawal, 1983 a, b). Soil is recognized as the basic substrate for keratinophilic fungi. The studies from several countries reported their

E-mail : *rks_kushwaha@yahoo.co.in

occurrence in soil. Sufficient data has accumulated in the past few years on the ecology and distribution of such fungi from several parts of the world. The limitations of the study of keratinophilic fungal flora also emphasizes the need of the study of affinity of these fungi to the various types of keratin substrates and unusual habitats.

The words 'keratinolytic', 'keratinophilic' and 'dermatophytes' vary greatly. These words will be used in the following senses : **keratinolytic** - a fungus able to digest keratinized substrate completely; **keratinophilic** - a fungus may or may not be able to digest keratinized substrate completely while inhabiting it; **dermatophyte** - a fungus belonging to one of the genera *Keratinomyces, Microsporum, Trichophyton* or *Epidermophyton,* whether pathogenic to man or animal or not. All dermatophytes are keratinolytic.

The first report of distibution of keratinophilic fungi in Indian soils can be traced back to 1955 when Dey and Kakoti reported the occurrence of *Microsporum gypseum* from soil samples of Dibrugarh, Assam. Later on, more reports of isolation of fungi of the same nature were presented (Shrivastava and Kushwaha, 1983; Nigam and Kushwaha, 1984, 1985, 1986, 1989 a, b, 1990; Deshmukh, 1985; Jain *et al.*, 1985; Kushwaha *et al.*, 1985; Dixit and Kushwaha, 1990; Awasthi and Gotewal, 1990). In general the ecological habitats mapped out and reported to be rich in keratinophilic fungi are the soils from densely populated areas such as public gardens, play grounds, poultry farms, cattle sheds, places inhabited by animals (Pugh and Mathison, 1962; Randhawa and Sandhu, 1965; Garg, 1966; Sur and Ghosh, 1980; Deshmukh and Agrawal, 1983 a, b) and decomposed organic matter. Little attention on systemic survey of soil inhabiting dermatophytes and related keratinophilic fungi in unusual habitats of India and abroad has been paid. Caretta and Piontelli (1977) commented that the isolation of keratinophilic fungi from habitats with non - keratinous substrates may represent only a means of dispersal of these fungi. They concluded that the type of keratin is not the main factor responsible for the distribution of different keratinophilic fungi but represent a special survival habitat. Dermatophytes and other species are able to attack native keratin by penetration (English, 1963) and enzymatic activity (Vanbreuseghem, 1952; Mathison, 1965). The process of colonizing keratin may differ in various soils but in general saprophytes with a high competitive ability are finally replaced by keratinophilic fungi.

During the present study, 106 soil samples were collected from different habitats of hills. Soil samples were taken from superficial soil layer depth not exceeding 2 - 3 inches with sterilized plastic spoon in sterilized polythene bags and brought to the laboratory. In the laboratory, the soil samples were placed at room temperature if not processed promptly. During the collection of soil samples, preference was given to the areas which were freqently inhabitated by human and animals. Main attention was drawn to garden and park soils. Protein release was studied by using method of Nigam and Kushwaha (1992).

Keratinophilic fungi from different habitats

Keratinophilic fungi and related dermatophytes isolated from various habitats were studied. During the present study 106 soil samples collected from different habitats of various localities yielded 194 keratinophilic fungi belonging to 6 genera namely *Acremonium* (49 isolates), *Chrysosporium* (97 isolates), *Ctenomyces* (3 isolates), *Malbranchea* (35 isolates), *Microsporum* (7 isolates) and *Paecilomyces* (3 isolates). Isolates mentioned here showed some differences in morphological characters. Hundred per cent samples were found positive for keratinophilic fungi (Table 1 and Fig. 1). Garden and park soil yielded the maximum number of these fungi, while minimum was recorded in field soil. The per cent distribution and relative density of each fungus in different habitats were given in Table 2 and Fig. 2 respectively. From 37 samples of garden soils, 68 keratinophilic fungi were isolated, while 46, 36, 29 and 15 keratinophilic fungi were recorded from 26 samples of park soil, 17 samples of picnic spot soil, 17 samples of religious places soil and 9 samples of field soil respectively. Out of 68 keratinophilic fungi of garden soil, 15 isolates of *Acremonium* sp., 11 isolates of *Chrysosporium indicum*, 1 isolate of *C. keratinophilum*, 10 isolates of *C. queenslandicum*, 14 isolates of *C. tropicum*, 12 isolates of *Malbranchea gypsea*, 2 isolates of *Microsporum gypseum* and 3 isolates of *Paecilomyces fusisporus* were observed. Among 46 fungi isolated from park soil, 11 isolates of *Acremonium* sp., 8 isolates of *C. indicum*, 1 isolate of *C. keratinophilum*, 7 isolates each of *C. queenslandicum* and *C. tropicum*, 3 isolates of *Ctenomyces serratus*, 6 isolates of *Malbranchea gypsea*, 3 isolates of *Microsporum gypseum* were observed. Among 36 fungi isolated from picnic spot soil, 12 isolates of *Acremonium* sp., 3 isolates of *C. indicum*, 1 isolate of *C. keratinophilum*, 7 isolates each of *C. queenslandicum*, 3 isolates of *C. tropicum*, 1 isolates of *C. xerophilum*, 8 isolates of *Malbranchea gypsea*, 1 isolate of *Microsporum gypseum* were observed. Among 29 fungi isolated from religious places soil, 10 isolates of *Acremonium* sp., 1 isolate each of *C. indicum* and *C. keratinophilum*, 6 isolates each of *C. queenslandicum*, 5 isolates of *C. tropicum*, 6 isolates of *Malbranchea gypsea* were observed. Among 15 fungi isolated from field soil, 1 isolate of *Acremonium* sp., 3 isolates each of *C. indicum*, *C. queenslandicum* and *C. tropicum*, 4 isolates of *Malbranchea gypsea*, 1 isolate of *Microsporum gypseum* were observed during present study.

Studies of these keratinophilic fungi revealed 100 per cent samples to be

Table 1. Frequency (per cent) of keratinophilic fungi from soil of different localities

S. No.	Fungus	Localities										Total
		A	B	C	D	E	F	G	H	I	J	
	No. of samples	14	29	13	8	7	6	7	6	6	10	106
	No. of samples positive	14	29	13	8	7	6	7	6	6	10	106
	% samples positive	100	100	100	100	100	100	100	100	100	100	100
1	*Acremonium sp.*	64.28 (9)	65.51(19)	69.23 (9)	25.00 (2)	14.28 (1)	-	-	-	33.33 (2)	70.00 (7)	46.22(49)
2	*Chrysosporium indicum*	21.42 (3)	10.34 (3)	15.38 (2)	37.50 (3)	14.28 (1)	66.66 (4)	42.85 (3)	50.00 (3)	33.33 (2)	20.00 (2)	24.52(26)
3	*Chrysosporium keratinophilum*	-	3.44 (1)	7.69 (5)	-	14.28 (1)	-	-	-	-	10.00 (1)	3.77 (4)
4	*Chrysosporium tropicum*	14.28 (2)	37.93(11)	46.15 (6)	37.50 (3)	28.57 (2)	16.66 (1)	42.85 (3)	33.33 (2)	16.66 (1)	30.00 (3)	32.07(34)
5	*Chrysosporium tropicum*	35.71 (5)	31.03 (9)	15.38 (2)	25.00 (2)	57.14 (4)	33.33 (2)	14.28 (1)	16.66 (1)	33.33 (2)	40.00 (4)	30.18(32)
6	*Chrysosporium xerophilum*	-	-	7.69 (1)	-	-	-	-	-	-	-	0.94 (1)
7	*Ctenomyces serratus*	-	-	-	-	-	-	14.28 (1)	-	33.33 (2)	-	2.83 (3)
8	*Malbranchea sp.*	42.85 (6)	37.93(11)	53.84 (7)	25.00 (2)	-	16.66 (1)	-	66.66 (4)	16.66 (1)	30.00 (3)	33.01(35)
9	*Microsporum gypseum*	-	3.44 (1)	-	25.00 (2)	-	16.66 (1)	28.57 (2)	-	16.66 (1)	-	6.60 (7)
10	*Paecilomyces sp.*	-	-	-	-	-	50.00 (3)	-	-	-	-	2.83 (3)
	No. of species isolated	5	7	7	6	5	6	5	4	7	6	58
	No. of appearances	(25)	(55)	(28)	(14)	(9)	(12)	(10)	(10)	(11)	(20)	(194)

A = Shimla, B = Manali, C = Kullu, D = Uttarkashi, E = Nainital, F = Assam, G = Kathmandu, H = Sagar, I = Mussoorie, J = Chandigarh.

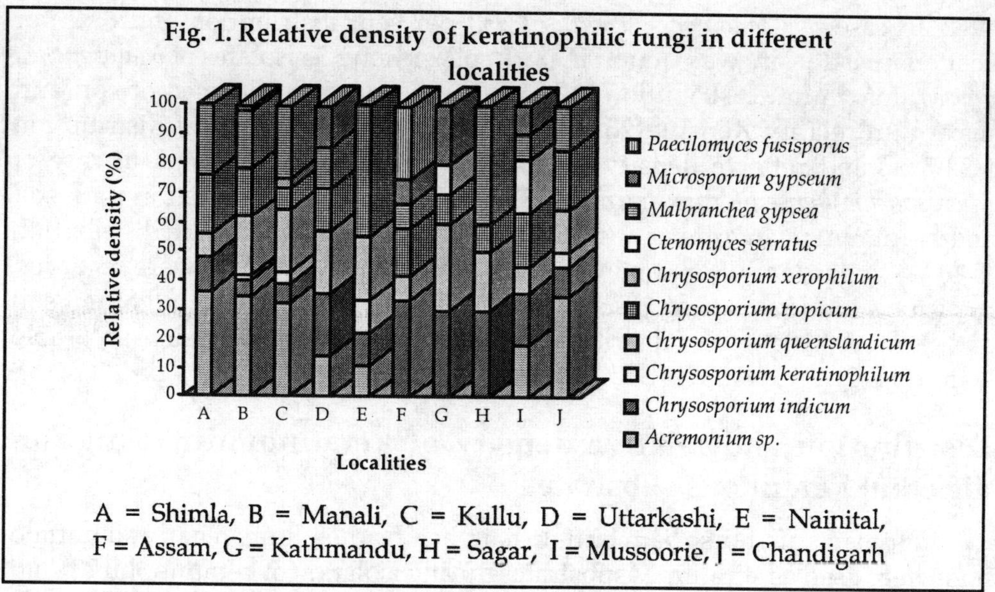

Fig. 1. Relative density of keratinophilic fungi in different localities

A = Shimla, B = Manali, C = Kullu, D = Uttarkashi, E = Nainital, F = Assam, G = Kathmandu, H = Sagar, I = Mussoorie, J = Chandigarh

positive for them. Among these fungi *Chrysosporium* represented 96 isolates, while *Acremonium* sp. represented 49 isolates and *Malbranchea* represented 36 isolates was next to it. *Chrysosporium* was recorded most frequently in all the habitats. It showed highest frequency as compared to other fungal genera. Among *Chrysosporium* isolates the maximum distribution (41.17 %) of *C. queenslandicum* was recorded in picnic spot soil. It was followed by *C. tropicum* (37.83 %) in was garden soil, while minimum distribution (2.70 %) of *C. keratinophilum* in garden soil. *C. keratinophilum* was not isolated in field soil. The relative density of genus *Chrysosporium* recorded in various habitats was as follows : garden (52.94 %), park (50.00 %), religious places soil (44.82 %), field (60.00 %) and picnic spot (41.66 %). Maximum (20.58 %) relative density of *C. tropicum* was in garden soil, while in garden soil minimum (1.47 %) relative density was observed in case of *C. keratinophilum*. In case of park soil maximum relative density of *C. indicum* (17.39 %) was noted, in religious places soil maximum relative density (20.68 %) was observed in case of *C. queenslandicum.* But in case of field soil all the fungi showed similar relative density and in picnic spot soil maximum relative density (19.44 %) was recorded in case of *C. queenslandicum*. Minimum relative density in park (2.17 %), in religious places soil (3.44 %), and in field (2.77 %) was recorded in case of *C. queenslandicum.* Among 49 isolates of *Acremonium* sp., maximum distribution (70.58 %) was recorded in picnic spot soil, while minimum (11.11 %) in field soil. The relative density of *Acremonium* sp. recorded in various habitats was as follows : garden (6.66 %), park (23.91 %), religious places soil (34.48 %), field (6.66 %) and picnic spot (33.33 %).

In case of *Ctenomyces serratus* 11.53 per cent distribution and 6.52 per cent relative density were found in park soil. Among 36 isolates of *Malbranchea gypsea,* maximum distribution (47.05 %) was found in picnic spot soil and maximum relative density (26.66 %) in field soil. Its minimum distribution (23.07 %) and relative density (13.04 %) was found in park soil. In case of *Microsporum gypseum* maximum (11.53 %) distribution was noted in park soil and maximum relative density (6.66 %) in field soil, while minimum distribution (5.40 %) in garden soil and minimum relative density (2.77 %) were recorded in picnic spot. *Paecilomyces fusisporus* (3 isolates) only observed in garden soil and showed 8.10 per cent distribution and 4.41 per cent relative density (Table 2 and Fig. 2).

Distribution and relative density of keratinophilic fungi on different keratinous substrates

Human hair, horse hair, buffalo hair, hen feather, human nail and buffalo horn were used as keratinous substrates during isolation of keratinophilic fungi from the soil. The results were shown in Table 3 and Fig. 3. Horse hair yielded maximum number of fungi followed by human hair, hen feather, human nail, buffalo horn and buffalo hair. All the isolated fungal species appeared on horse hair except *C. keratinophilum* and *C. xerophilum.* On human hair, all the isolated fungi observed except *C. xerophilum* and *Paecilomyces fusisporus.*

Table 2. Frequency (per cent) of keratinophilic fungi in different habitats

S. No.	Fungus	Habitats					
		Garden	Park	Religious Places	Field	Picnic Spot	Total
	No. of samples	37	26	17	9	17	106
	No. of samples positive	37	26	17	9	17	106
	% samples positive	100	100	100	100	100	100
1	*Acremonium* sp.	40.54	42.3	58.82	11.11	70.58	46.22
2	*Chrysosporium indicum*	29.72	30.76	5.88	33.33	17.64	24.52
3	*Chrysosporium keratinophilum*	2.70	3.84	5.88	-	5.88	3.77
4	*Chrysosporium queenslandicum*	27.02	26.92	35.29	33.33	41.17	31.13
5	*Chrysosporium tropicum*	37.83	26.92	29.41	33.33	17.64	30.18
6	*Chrysosporium xerophilum*	-	-	-	-	5.88	0.94
7	*Ctenomyces serratus*	-	11.53	-	-	-	2.83
8	*Malbranchea gypsea*	32.43	23.07	35.29	44.44	47.05	33.96
9	*Microsporum gypseum*	5.40	11.53	-	11.11	5.88	6.60
10	*Paecilomyces fusisporus*	8.10	-	-	-	-	2.83

Fig. 2. Relative density of keratinophilic fungi on different keratinous substrates

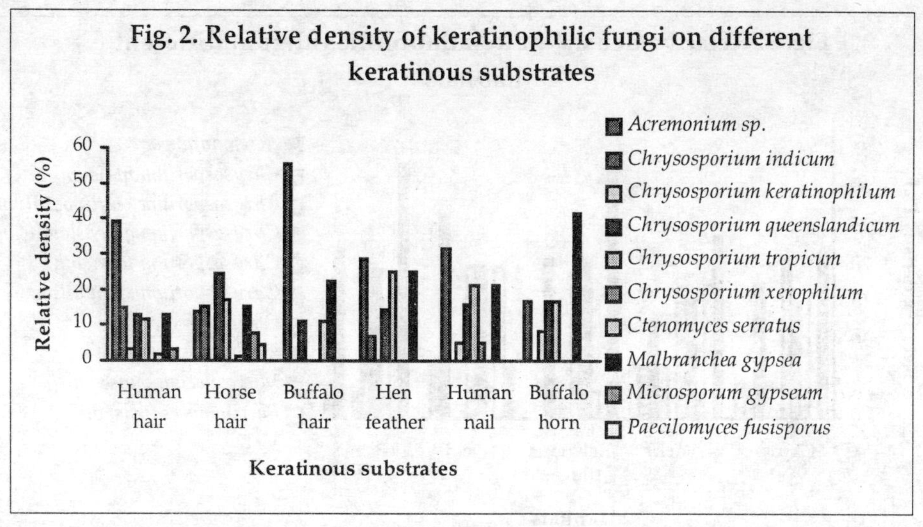

Acremonium sp. was recorded most frequently on all the keratinous substrates except buffalo hair. It showed 77.41, 23.68, 50.00, 60.00 and 33.33 per cent distribution on human hair, horse hair, hen feather, human nail and buffalo horn, respectively. *C. queenslandicum* and *Malbranchea gypsea* were recorded most frequently on all the keratinous substrates. *C. queenslandicum* showed 25.80, 42.10, 20.00, 25.00, 30.00 and 33.33 per cent distribution on human hair, horse hair, buffalo hair, hen feather, human nail and buffalo horn respectively. *C. xerophilum* was isolated on human nail only and *Paecilomyces fusisporus* was isolated on horse hair only.

Table 3. Frequency (per cent) of keratinophilic fungi on different keratinous substrates

S. No.	Fungus	Keratinous substrates					
		Human hair	Horse hair	Buffalo hair	Hen feather	Human nail	Buffalo horn
1	*Acremonium* sp.	77.41	23.68	-	50.00	60.00	33.33
2	*Chrysosporium indicum*	29.03	26.31	100.00	12.50	-	-
3	*Chrysosporium keratinophilum*	6.45	-	-	-	10.00	16.66
4	*Chrysosporium queenslandicum*	25.8	42.1	20.00	25.00	30.00	33.33
5	*Chrysosporium tropicum*	22.58	28.94	-	43.75	40.00	33.33
6	*Chrysosporium xerophilum*	-	-	-	-	10.00	-
7	*Ctenomyces serratus*	3.22	2.63	20.00	-	-	-
8	*Malbranchea gypsea*	25.8	26.31	40.00	43.75	40.00	83.33
9	*Microsporum gypseum*	6.45	13.15	-	-	-	-
10	*Paecilomyces fusisporus*	-	7.89	-	-	-	-

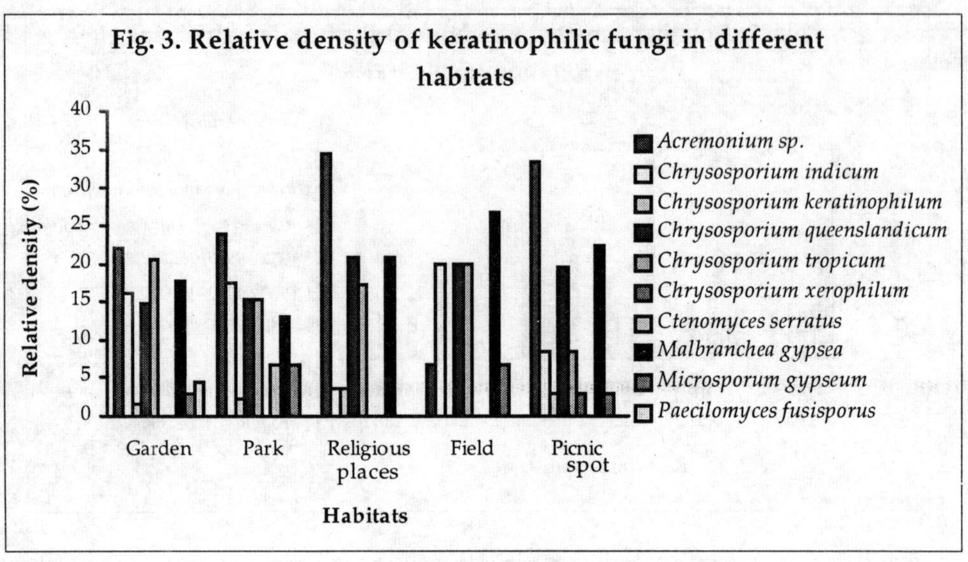

The relative density of C. *queenslandicum* as recorded from various keratinous baits is as follows : human hair (13.11 %), horse hair (24.61 %), buffalo hair (11.11 %), hen feather (14.28 %), human nail (15.78 %) and buffalo horn (16.66 %). The relative density of *Malbranchea gypsea* was 13.11, 15.38, 22.22, 25.00, 21.05 and 41.66 per cent on human hair, horse hair, buffalo hair, hen feather, human nail and buffalo horn, respectively. *Acremonium* sp. was observed from all the substrates except buffalo hair and its relative density was as follows : 39.34, 13.84, 28.57, 31.57 and 16.66 per cent from human hair, horse hair, hen feather, human nail and buffalo horn, respectively.

The result provided information on the nature of the distribution of keratinophilic fungi in unexplored habitats of India. The perusual of data revealed that some species of *Chrysosporium* were frequent in their distribution. During this study *C. indicum, C. queenslandicum* and *C. tropicum* were isolated from all the habitats. The data provide the nature of distribution of genus *Acremonium* sp., which was isolated from all the habitats in the present study. Parihar (1998) isolated *Acremonium* sp. from pot soil, garden soil and zoological park soil. Nigam and Kushwaha (1989 a) reported this genus from house dust. Gupta (1994) isolated it from garden and forest soil. Saksena (1993) isolated *Acremonium* sp. from zoological park also. Katiyar (1998) found *A. implicatum* from lake. Among genus *Chrysosporium, C. indicum* was found to be the most common keratinophilic fungus and it was isolated from all the habitats and all the keratinous substrates except human nail and buffalo horn. Gupta (1994) isolated it from the soil of crop field, garden and zoological park. Saksena (1993) isolated *C. indicum* from forest soil and Parihar (1998) isolated it from pot soil and garden soil. *C. indicum* was earlier reported from plains and hills of India. However, soil of plains confirmed the greater preponderance of this fungus in usual habitats than on the hills (Garg, 1966). It is found that the men and animals dwelling areas and soil rich in organic and keratin substances favoured the distribution of *C. indicum*. Isolation of this fungus is in accordance with previous reports from India (Padhye *et al.*, 1966; Sur and Ghosh, 1980; Verma *et al.*, 1982; Deshmukh and Agrawal, 1983 a, b). This species was most dominant in Egypt and Chilean Ander (Abdel - Fattah *et al.*, 1982) and has been reported to occur in soils of Galapagos islands (Ajello and Padhye, 1974), Rumania (Evolceanui and Alteras, 1966) and Mozambique (Oorschot, 1980).

C. keratinophilum was isolated from all habitats except field soil. Its isolation from India has been reported by several workers (Garg, 1966; Jain, 1983; Deshmukh and Agrawal, 1985; Jain *et al.*, 1985). This fungus was also isolated from house dust of Kanpur (Nigam and Kushwaha, 1986). Parihar (1998) reported this fungus from pot soil and garden soil also. *C. queenslandicum* was isolated from all the habitats. *C. queenslandicum* was, however, restricted in distribution. Although, its occurrence in house dust of India confirms its survivability in nutritionally less soils (Nigam and Kushwaha, 1986). *C. tropicum* was also isolated from all the habitats. It was also isolated from zoological park soil and pot soil (Parihar, 1998). Saksena (1993) also isolated it from all habitats except zoological park. Randhawa and Sandhu (1965) reported it from cattle field in Punjab, while Garg (1966) reported its isolation from garden, cultivated field and river sand. Cattle field soils have been earlier described as suitable habitats for the occurrence of *C. tropicum* (Deshmukh and Agrawal, 1983 a; Singh and Agrawal, 1983). Other reportes have confirmed its isolation from poultry farm, crop field and forest soil (Kushwaha and Agrawal, 1976; Jain, 1983). Nigam and Kushwaha (1986) isolated *C. tropicum* from house dust soil. This fungus was also isolated from different parts of the world such as Europe, New Guinea, Soloman islands, United states, Belgium, Italy (Carmichael, 1962; Ajello *et al.*, 1965 a, b) indicating its cosmopolitan distribution. *C. tropicum* has also been isolated from lava bed soil of Galapagos islands and organic matter rich soil of Easter islands (Ajello and Alpert, 1972; Ajello and Padhye, 1974). Hubalek and Hornick (1977) established the infection potential of *C. keratinophilum* and *C. tropicum* in white mouse. Saidi *et al.* (1994) studied pathogenicity of *C. tropicum* and stated that this fungus was probable cause of a dermatomycosis in chicken. Vroey (1976) isolated it from skin lesions in humans. *C. xerophilum* was isolated from picnic spot soil only. It was first isolated from spoiled high moisture prunes. Katiyar (1998) reported this fungus from river above, lake above, lake below, pond above and canal below.

High percentage of occurrence of various species of *Chrysosporium* in different habitats may be due to the fact that the soil of these usual habitats were rich in organic, inorganic and keratin substrates and ability to digest these substrates easily by the species of *Chrysosporium* has been confirmed (Carmichael, 1962; English, 1965, 1969). Nigam (1987) while studying the ecology of *Chrysosporium* elaborated its competitive survivability with other fungi present in the environment. She also found *Chrysosporium* to be widely distributed. *Myceliophthora* anamorph of *Ct. serratus* was isolated from park soil baited with human hair, horse hair and buffalo hair. It was isolated frequently from zoological park soil baited with human hair, human nail, cow hair and horse hair. Parihar (1998) reported it from pot soil baited with human hair only. Saksena (1993) isolated *My. vellera* from forest soil. Gupta (1994) also isolated it from zoological park soil. From college campus soil (Nigam and Kushwaha, 1990) and from hospital waste (Dixit, 1991), it was also reported. *Malbranchea* was isolated from all habitats and on all keratinous baits. This fungus appears to be common in its distribution in India (Kushwaha and Agrawal, 1976; Jain and Agrawal, 1979; Jain, 1983; Dixit, 1991; Saksena. 1993; Gupta, 1994; Parihar, 1998) and in abroad (Rebell and Taplin, 1970; Orr and Kuehn, 1972; Ajello and Padhye, 1974; Sigler and Carmichael, 1976). *Microsporum gypseum* was isolated from all habitats except religious places soil. This fungus was recorded from pot soil and zoological park soil also (Parihar, 1998). *M. gypseum* was first reported by Dey and Kakoti (1955) from an animal house in Assam. Randhawa and Sandhu (1965) found it to be the second most common species in Indian soils. Nigam and Kushwaha (1986) found it in house dust. Dixit (1991) also isolated it from all the habitats, which he studied. Desai *et al.* (1959) reported its distribution in swimming pool, while Padhye (1961) isolated it from barn yard, chicken house, dairy farm where the soil was presumably intermixed with keratin substrates showing high probability of occurrence of *M. gypseum*. Jain (1983) also reported the colonization of peacock feathers burried in soil of poultry farm and cattle field. These are evidences for the saprophytic existance of this fungus, which disclose the wide spread occurrence of *M. gypseum* in garden soil and farm soil in Britain (Stockdale, 1958). This fungus have also been encountered in one or more countries of each continent (Ajello *et al.*, 1965 a, b; Frey, 1965; Rogers, 1971; Muhammad and Lalji, 1978; Mercantini *et al.*; 1980; Abdel - Fattal *et al.*, 1982; Ulfig and Korcz, 1983; Calvo *et al.*, 1984). Punsola and Guarro (1984) reported this fungus from cats, dogs and rodents. In human this fungus causes *Tinea corporis* or *Tinea capitis*, was reported by Hayashi and Toshitani (1983). *Paecilomyces fusisporus* was recorded from garden soil baited with horse hair only. This fungus was also isolated from forest litter, crop field, animal house floor sweeping soil, on horn and hair, keratinous baits

(Kushwaha, 1976). This fungus was also isolated from soil of Patharia forest (Saksena, 1955). *P. varioti* was isolated from jute fibre and fabric and also from wax paper. Some keratinophilic fungi were reported from Nigerian soil samples (Ogbonna and Pugh, 1987).

Among 194 isolates, 155 isolates were tested for protein releasing ability. Protein released in to the medium was measured after 30 days of incubation period. The fungi tested here were *Acremonium* sp. (49 isolates), *Chrysosporium indicum* (13 isolates), *C. keratinophilum* (4 isolates), *C. queenslandicum* (27 isolates), *C. tropicum* (27 isolates), *C. xerophilum* (1 isolate), *Malbranchea gypsea* (31 isolates) and *Microsporum gypseum* (3 isolates). Among these 155 isolates, the highest protein was (433.99 μg/ml) released by *C. indicum* 8. Out of 49 isolates of *Acremonium* sp. highest protein (410.33 μg/ml) was released by *Acremonium* sp. 37 and the lower values of protein (208.99 μg/ml) was released by *Acremonium* sp. 38. In case of 13 isolates of *C. indicum*, highest protein (433.99 μg/ml) was released by *C. indicum* 8, while minimum protein (3.15.32 μg/ml) was released by *C. indicum* 11. Four isolates of *C. keratinophilum* released 358.66, 340.65, 369.99 and 325.65 μg/ml protein. The highest protein (371.66 μg/ml) and the lower values of protein (200.99 μg/ml) were recorded by *C. queenslandicum* 21 and *C. queenslandicum* 23. Among 27 isolates of *C. tropicum* the highest protein (385.99 μg/ml) and the lowest protein (223.66 μg/ml) were observed by *C. tropicum* 11 and *C. tropicum* 16. *C. xerophilum* released 226.32 μg/ml protein, while out of 31 isolates of *Malbranchea gypsea*, highest 269.66 μg/ml protein was released by *M. gypsea* 17. Three isolates of *Microsporum gypseum* released 377.65, 352.66 and 359.33 μg/ml protein from feathers (Table 4).

The results presented here indicate that the fungi which grow on feathers released high amount of protein in the culture filtrate as also indicated by Kushwaha (1995). Some strains which are of common occurrence in soil are found to degrade feathers and released large amount of protein. Nigam and Kushwaha (1989 c) observed the decomposition of feathers and hairs by *C. tropicum*. Paotetti *et al.* (1975) studied the microbiological production of protein hydrolyzates from feather keratins. Preferential utilization of feather keratin by keratinophilic fungi was also studied by Kushwaha (1976) and Parihar and Kushwaha (1997). Rapid degradation of feathers by certain keratinophilic fungi in soil may lead to recycling of keratinous waste.

All the fungi discussed above are keratinophilic in nature and a few of them are well known pathogens causing skin diseases of men and animals. If these etiologic agents are present in an active state even for a short period, they can serve as potential foci of infection for man and animals inhabiting these localities. Their occurrence in the soil is of epidemiological importance.

Table 4. The protein released (μg/ml) from bird feathers during the growth of keratinophilic fungi

S. No.	Fungi tested	Fungus + Keratin control	Total protein
1	*Acremonium* sp. 1	159.66 ± 18.08	290.32
2	*Acremonium* sp. 2	161.33 ± 19.56	288.99
3	*Acremonium* sp. 3	193.99 ± 27.06	302.65
4	*Acremonium* sp. 4	176.99 ± 28.27	315.65
5	*Acremonium* sp. 5	176.66 ± 37.94	265.99
6	*Acremonium* sp. 6	170.99 ± 24.64	273.32
7	*Acremonium* sp. 7	164.66 ± 20.60	244.99
8	*Acremonium* sp. 8	162.33 ± 17.56	267.33
9	*Acremonium* sp. 9	193.33 ± 37.07	268.66
10	*Acremonium* sp. 10	221.99 ± 36.81	295.99
11	*Acremonium* sp. 11	185.66 ± 38.74	281.32
12	*Acremonium* sp. 12	201.66 ± 36.11	298.66
13	*Acremonium* sp. 13	233.33 ± 45.69	332.33
14	*Acremonium* sp. 14	216.33 ± 35.63	298.33
15	*Acremonium* sp. 15	214.99 ± 20.07	316.65
16	*Acremonium* sp. 16	210.99 ± 26.82	303.32
17	*Acremonium* sp. 17	171.66 ± 31.02	299.32
18	*Acremonium* sp. 18	181.33 ± 35.56	275.99
19	*Acremonium* sp. 19	193.66 ± 35.69	247.66
20	*Acremonium* sp. 20	206.99 ± 26.96	314.32
21	*Acremonium* sp. 21	248.99 ± 41.24	386.32
22	*Acremonium* sp. 22	187.33 ± 25.44	265.33
23	*Acremonium* sp. 23	178.33 ± 22.12	253.99
24	*Acremonium* sp. 24	198.66 ± 31.30	261.32
25	*Acremonium* sp. 25	184.33 ± 27.37	321.33
26	*Acremonium* sp. 26	218.66 ± 21.06	322.66
27	*Acremonium* sp. 27	183.33 ± 26.56	260.33
28	*Acremonium* sp. 28	183.66 ± 29.06	238.32
29	*Acremonium* sp. 29	185.99 ± 28.61	261.32
30	*Acremonium* sp. 30	192.66 ± 43.32	266.32
31	*Acremonium* sp. 31	176.66 ± 26.01	214.32
32	*Acremonium* sp. 32	161.33 ± 21.75	258.66
33	*Acremonium* sp. 33	160.33 ± 18.29	261.99
34	*Acremonium* sp. 34	180.99 ± 17.71	250.99
35	*Acremonium* sp. 35	177.66 ± 33.94	278.99

36	*Acremonium* sp. 36	163.66 ± 20.34	265.66
37	*Acremonium* sp. 37	281.33 ± 57.06	410.33
38	*Acremonium* sp. 38	160.99 ± 19.07	208.99
39	*Acremonium* sp. 39	244.33 ± 30.08	368.99
40	*Acremonium* sp. 40	184.33 ± 33.08	310.66
41	*Acremonium* sp. 41	237.99 ± 31.53	332.99
42	*Acremonium* sp. 42	179.66 ± 33.12	303.66
43	*Acremonium* sp. 43	175.66 ± 32.06	263.66
44	*Acremonium* sp. 44	229.66 ± 38.74	361.99
45	*Acremonium* sp. 45	217.99 ± 30.35	317.32
46	*Acremonium* sp. 46	210.33 ± 29.08	309.66
47	*Acremonium* sp. 47	202.66 ± 31.83	299.99
48	*Acremonium* sp. 48	200.66 ± 21.49	285.32
49	*Acremonium* sp. 49	195.33 ± 37.22	272.33
50	*Chrysosporium indicum* 1	214.66 ± 39.23	363.32
51	*Chrysosporium indicum* 2	200.33 ± 37.22	325.66
52	*Chrysosporium indicum* 3	206.66 ± 28.78	352.66
53	*Chrysosporium indicum* 4	216.66 ± 35.33	369.32
54	*Chrysosporium indicum* 5	225.66 ± 58.70	363.32
55	*Chrysosporium indicum* 6	211.33 ± 27.37	370.66
56	*Chrysosporium indicum* 7	240.66 ± 32.72	385.66
57	*Chrysosporium indicum* 8	252.99 ± 30.21	433.99
58	*Chrysosporium indicum* 9	236.33 ± 35.63	366.66
59	*Chrysosporium indicum* 10	220.66 ± 38.29	355.32
60	*Chrysosporium indicum* 11	203.66 ± 36.56	315.32
61	*Chrysosporium indicum* 12	210.33 ± 41.11	322.66
62	*Chrysosporium indicum* 13	211.66 ± 27.25	335.32
63	*Chrysosporium keratinophilum* 1	192.33 ± 31.69	358.66
64	*Chrysosporium keratinophilum* 2	177.99 ± 34.03	340.65
65	*Chrysosporium keratinophilum* 3	201.33 ± 32.08	369.99
66	*Chrysosporium keratinophilum* 4	201.99 ± 32.19	325.65
67	*Chrysosporium queenslandicum* 1	165.99 ± 23.36	264.32
68	*Chrysosporium queenslandicum* 2	175.33 ± 29.56	260.66
69	*Chrysosporium queenslandicum* 3	193.33 ± 34.80	326.99
70	*Chrysosporium queenslandicum* 4	171.33 ± 33.90	236.33
71	*Chrysosporium queenslandicum* 5	161.66 ± 21.59	285.32
72	*Chrysosporium queenslandicum* 6	176.66 ± 33.21	294.66
73	*Chrysosporium queenslandicum* 7	184.99 ± 24.19	333.65
74	*Chrysosporium queenslandicum* 8	161.66 ± 20.07	322.99
75	*Chrysosporium queenslandicum* 9	167.66 ± 22.06	317.32

76	*Chrysosporium queenslandicum* 10	171.99 ± 28.15	320.65
77	*Chrysosporium queenslandicum* 11	200.99 ± 30.13	336.99
78	*Chrysosporium queenslandicum* 12	193.99 ± 72.27	331.65
79	*Chrysosporium queenslandicum* 13	195.33 ± 42.84	342.99
80	*Chrysosporium queenslandicum* 14	199.99 ± 33.59	332.65
81	*Chrysosporium queenslandicum* 15	197.33 ± 27.91	310.99
82	*Chrysosporium queenslandicum* 16	166.99 ± 22.06	277.99
83	*Chrysosporium queenslandicum* 17	182.33 ± 36.85	335.99
84	*Chrysosporium queenslandicum* 18	186.33 ± 28.68	289.99
85	*Chrysosporium queenslandicum* 19	208.99 ± 25.29	353.32
86	*Chrysosporium queenslandicum* 20	165.66 ± 21.49	287.66
87	*Chrysosporium queenslandicum* 21	211.66 ± 24.06	371.66
88	*Chrysosporium queenslandicum* 22	199.33 ± 32.65	311.33
89	*Chrysosporium queenslandicum* 23	177.99 ± 34.17	200.99
90	*Chrysosporium queenslandicum* 24	176.66 ± 27.62	285.99
91	*Chrysosporium queenslandicum* 25	208.33 ± 30.98	319.99
92	*Chrysosporium queenslandicum* 26	185.99 ± 41.56	302.99
93	*Chrysosporium queenslandicum* 27	189.33 ± 27.00	322.99
94	*Chrysosporium tropicum* 1	165.66 ± 24.70	317.99
95	*Chrysosporium tropicum* 2	177.33 ± 29.08	287.99
96	*Chrysosporium tropicum* 3	171.33 ± 26.09	279.33
97	*Chrysosporium tropicum* 4	181.99 ± 42.66	298.99
98	*Chrysosporium tropicum* 5	181.66 ± 27.62	330.32
99	*Chrysosporium tropicum* 6	177.33 ± 20.56	279.66
100	*Chrysosporium tropicum* 7	188.66 ± 28.90	314.99
101	*Chrysosporium tropicum* 8	164.33 ± 22.64	267.33
102	*Chrysosporium tropicum* 9	185.66 ± 40.27	321.32
103	*Chrysosporium tropicum* 10	206.33 ± 36.23	343.99
104	*Chrysosporium tropicum* 11	242.99 ± 18.08	385.99
105	*Chrysosporium tropicum* 12	240.66 ± 32.57	373.32
106	*Chrysosporium tropicum* 13	199.33 ± 27.37	322.33
107	*Chrysosporium tropicum* 14	196.33 ± 26.40	329.33
108	*Chrysosporium tropicum* 15	198.66 ± 31.60	342.99
109	*Chrysosporium tropicum* 16	160.66 ± 21.69	223.66
110	*Chrysosporium tropicum* 17	183.99 ± 29.61	301.32
111	*Chrysosporium tropicum* 18	225.33 ± 28.68	348.66
112	*Chrysosporium tropicum* 19	163.33 ± 20.02	267.99
113	*Chrysosporium tropicum* 20	171.33 ± 25.56	277.66
114	*Chrysosporium tropicum* 21	176.33 ± 29.32	294.33
115	*Chrysosporium tropicum* 22	169.99 ± 31.86	237.65

116	*Chrysosporium tropicum* 23	207.33 ± 29.32	342.33
117	*Chrysosporium tropicum* 24	195.99 ± 31.57	303.65
118	*Chrysosporium tropicum* 25	202.99 ± 36.29	318.65
119	*Chrysosporium tropicum* 26	210.33 ± 26.09	314.33
120	*Chrysosporium tropicum* 27	208.99 ± 23.58	321.99
121	*Chrysosporium xerophilum*	160.66 ± 19.07	226.32
122	*Malbranchea gypsea* 1	168.66 ± 24.64	212.99
123	*Malbranchea gypsea* 2	165.33 ± 21.85	205.33
124	*Malbranchea gypsea* 3	163.66 ± 20.72	194.66
125	*Malbranchea gypsea* 4	167.99 ± 22.24	242.65
126	*Malbranchea gypsea* 5	163.66 ± 19.07	232.32
127	*Malbranchea gypsea* 6	165.99 ± 22.24	218.99
128	*Malbranchea gypsea* 7	164.33 ± 20.91	219.33
129	*Malbranchea gypsea* 8	163.66 ± 23.36	225.99
130	*Malbranchea gypsea* 9	157.66 ± 18.08	187.99
131	*Malbranchea gypsea* 10	165.66 ± 21.69	225.99
132	*Malbranchea gypsea* 11	165.99 ± 21.29	207.99
133	*Malbranchea gypsea* 12	176.99 ± 27.06	206.32
134	*Malbranchea gypsea* 13	165.33 ± 23.48	194.66
135	*Malbranchea gypsea* 14	174.33 ± 33.38	207.33
136	*Malbranchea gypsea* 15	181.66 ± 34.56	202.32
137	*Malbranchea gypsea* 16	164.99 ± 21.49	186.99
138	*Malbranchea gypsea* 17	189.66 ± 35.06	269.66
139	*Malbranchea gypsea* 18	165.66 ± 21.69	199.32
140	*Malbranchea gypsea* 19	168.66 ± 22.24	197.99
141	*Malbranchea gypsea* 20	172.99 ± 25.06	196.99
142	*Malbranchea gypsea* 21	169.33 ± 20.02	218.99
143	*Malbranchea gypsea* 22	165.33 ± 23.48	226.33
144	*Malbranchea gypsea* 23	165.99 ± 21.59	186.99
145	*Malbranchea gypsea* 24	166.99 ± 22.06	188.65
146	*Malbranchea gypsea* 25	168.33 ± 23.11	200.99
147	*Malbranchea gypsea* 26	164.66 ± 19.77	188.32
148	*Malbranchea gypsea* 27	169.99 ± 23.79	200.32
149	*Malbranchea gypsea* 28	169.99 ± 20.72	215.99
150	*Malbranchea gypsea* 29	169.99 ± 18.08	195.65
151	*Malbranchea gypsea* 30	170.99 ± 28.49	193.65
152	*Malbranchea gypsea* 31	165.66 ± 21.69	197.99
153	*Microsporum gypseum* 1	208.99 ± 30.17	377.65
154	*Microsporum gypseum* 2	230.66 ± 32.09	352.66
155	*Microsporum gypseum* 3	214.33 ± 36.07	359.33

The data are mean of three replications; ± indicates standard deviation.

Observations from above study reveals that examined samples of all habitats studied, found to be inhabited by non pathogenic as well as pathogenic keratinophilic fungi.

Acknowledgements

This work was carried out under All India Co-ordinated Project on Taxonomy sponsored by Ministry of Environment and Forests. We are thankful to Dr. Pervez E. Deen, Principal for providing research facilities.

References

Abdel - Fattah, H.M.; Moubasher, A.H. and Maghazy, S.M. (1982). Keratinophilic fungi in Egyptian soils. Mycopathologia, **79** : 49 - 53.

Ajello, L. and Alpert, E.M. (1972). Survey of Easter Island soils for keratinophilic fungi. Mycologia, **64** : 161 - 166.

Ajello, L. and Padhye, A.A. (1974). Keratinophilic fungi of the Galapagos Island. Mykosen, **17** : 239 - 243.

Ajello, L.; Varsavsky, E. and Delvingt, W. (1965 a). Keratinophilic fungi from Belgian soils. Trans. Brit. Mycol. Soc., **48** : 417 - 421.

Ajello, L.; Varsavsky, E.; Sotgiu, G.; Mazzoni, A. and Mantovani, A. (1965 b). Survey of soils for human pathogenic fungi from Emilia - Romanyna region of Italy. Mycopath. Mycol. et. Appl., **26** : 65 - 70.

Awasthi, P.B. and Gotewal, S. (1990). Prevelance of keratinophilic fungi in certain soils of Rohilkhand division in relation to Edaphic factors conditions. Proc. Natn. Acad. Sci., **76** : 37.

Calvo, A.; Vidal, M. and Guarro, J. (1984). Keratinophilic fungi from urban soils of Barcelona, Spain. Mycopathologia, **85** : 145 - 147.

Cano, J. and Guarro, J. (1994). Studies on keratinophilic fungi. III. *Chrysosporium siglerae* sp. nov. Mycotaxon, **51** : 75 - 79.

Caretta, G. and Piontelli, E. (1977). Isolation of keratinophilic fungi from soil in Pavia, Italy. Sabouraudia, **13** : 33 - 37.

Carmichael, J.W. (1962). *Chrysosporium* and some other aleurosporic hyphomycetes. Can. J. Bot., **40** : 1137 - 1173.

Desai, S.; Marquis, L. and Bhatt, M.L.A. (1959). Mycologic study on common superficial mycoses. Bull. Cal. Sch. Trop. Med., **7** : 37.

Deshmukh, S.K. (1985). Isolation of dermatophytes and other keratinophilic fungi from soil of Mussoorie, India. Mykosen, **28** : 98 - 101.

Deshmukh, S.K. and Agrawal, S.C. (1983 a). Prevalence of dermatophytes and other keratinophilic fungi in soils of Madhya Pradesh (India). Mykosen, **26** : 574 - 577.

Deshmukh, S.K. and Agrawal, S.C. (1983 b). Isolation of keratinophilic fungi from coastal habitats of Goa (India). Kavaka, **11** : 53 - 54.

Deshmukh, S.K. and Agrawal, S.C. (1985). Degradation of human hair by some dermatophytes and other keratinophilic fungi. Mykosen, **28** : 463 - 466.

Dey, N.C. and Kakoti, L.M. (1955). *Microsporum gypseum* in India. J. Indian Med. Assoc., **25** : 160 - 164.

Dixit, A.K. (1991). Occurrence and antagonistic studies of keratinophilic fungi with special reference to *Microsporum* species. Ph.D. Thesis, Kanpur University, Kanpur pp. 350.

Dixit, A.K. and Kushwaha, R.K.S. (1990). Keratinophilic fungi of Andman Islands. India. Ind. J. Microbiol., **30** : 349 - 350.

English, M.P. (1963). The saprophytic growth of keratinophilic fungi on keratin. Sabouraudia, **2** : 115 - 130.

English, M.P. (1965). The saprophytic growth of non keratinophilic fungi on keratinized substrate and a comparision with keratinophilic fungi. Trans. Brit. Mycol. Soc., **48** : 219 - 235.

English, M.P. (1969). Destruction of hair by *Chrysosporium keratinophilum*. Trans. Brit. Mycol. Soc., **66** : 357 - 358.

Evolceanui, R. and Alteras, J. (1966). The occurrence of *Trichophyton indicum* Randhawa and Sandhu, 1963 its immunobiological properties. (Preliminary report). Mycopath. et. Mycol. Appl., **28** : 57 - 62.

Frey, D. (1965). Isolation of keratinophilic and other fungi from soil collected in Australia new Guinea. Mycologia, **57** : 202 - 215.

Garg, A.K. (1966). Isolation of dermatophytes and other keratinophilic fungi from soil in India. Sabouraudia, **4** : 259 - 264.

Gentles, J.C. (1962). Keratinophilic fungi and diseases caused by them. In G.Dalldorf, Fungi and Furgous Diseases, Chpter II, C. Thomas, Springfield, I II.

Gupta, M. (1994). Decomposition of hair and feathers by keratinophilic fungi. Ph.D. Thesis. Kanpur University, Kanpur pp. 201.

Hayashi, N. and Toshitani, S. (1983). Human infection with *Microsporum gypseum* in Japan. Mykosen, **26** : 527 - 530.

Hubalek, Z. and Hornick, M. (1977). Experimental infection of white mouse with *Chrysosporium* and *Paecilomyces*. Mycopathologia, **62** : 173 - 178.

Jain, P.C. (1983). Prevalance of keratinophilic fungi in soil of Madhya Pradesh and Uttar Pradesh, India. Geobios, **2** : 10 - 13.

Jain, P.C. and Agrawal, S.C. (1979). Some additions to Indian *Malbranchea*. Kavaka, **7** : 69 - 72.

Jain, Madhu; Shukla, P.K. and Shrivastava, O.P. (1985). Keratinophilic fungi and dermatophytes in Lucknow soil and their global distribution. Mykosen, **28** : 98 - 101.

Katiyar, S. (1998). The keratinomycetes from bottom sediments and their hair invasion ability. Ph.D. Thesis, CSJM University, Kanpur pp. 209.

Kushwaha, R.K.S. (1976). Studies on keratinophilic fungi from soils. Ph.D. Thesis. Saugar University, India pp. 259.

Kushwaha, R.K.S. (1995). Biodeterioration of feathers by keratinomycetes isolated from a museum of Spain. 3rd International Conference on Biodeterioration of Cultural Property. 289 - 302.

Kushwaha, R.K.S. (2000). The genus *Chrysosporium*, its physiology and biotechnological potential. In Biology of dermatophytes and other keratinophilic fungi. Eds. R.K.S. Kushwaha & J. Guarro. Revista Iberoamericana de Micologia, Spain, 66 - 76.

Kushwaha, R.K.S. and Agrawal, S.C. (1976). Some keratinophilic fungi and related dermatophytes from soil. Proc. Ind. Natl. Sci. Acad., **42** : 102 - 110.

Kushwaha, R.K.S.; Srivastava, J.N. and Nigam, N. (1985). Dermatophytes and keratinophilic fungi in soil of India. IX ISHAM Congress, Georgia. R. **3 - 3**.

Mathison, G.E. (1965). An autoradiographic technique for the detection of the cystine and peptides containing cystine or its derivatives in keratin digestion. Mycopath. Mycol. Appl., **27** : 225 - 232.

Mercantini, R.; Marsella, R.; Caprilli, F. and Dovgiallo, G. (1980). Isolation of dermatophytes and correlated species from the soil of public gardens and parks in Rome. Sabouraudia, **18** : 123 - 128.

Mercantini, R.; Marsella, R.; Lambiase, L. and Belardi, M. (1986). Isolation of keratinophilic fungi from floors in Rome (Italy) kindergarten and secondary schools. Mycopathologia, **94** : 109 - 115.

Muhammad, S.I. and Lalji, N. (1978). The distribution of geophilic dermatophytes in Kenyan soil. Mycopathologia, **63** : 95 - 97.

Nigam, N. (1987). Studies on *Chrysosporium* and allied fungi from soil. Ph.D. Thesis, Kanpur University Kanpur pp. 188.

Nigam, N. and Kushwaha, R.K.S. (1984). Keratinophilic fungi and related dermatophytes from Kanpur. Proc. 7th Ind. Bot. Soc., 18.

Nigam, N. and Kushwaha, R.K.S. (1985). Eight new keratinophilic fungal records from India. Proc. Natn. Acad. Sci., India. 55th Ann. Sess., 26.

Nigam, N. and Kushwaha, R.K.S. (1986). Keratinophilic fungi in house dust of Kanpur. Proc. Natn. Acad. Sci., India. 56th Sess. 36 - 37.

Nigam, N. and Kushwaha, R.K.S. (1989 a). Some new reports on keratinophilic fungi. Curr. Sci., **58** : 1374.

Nigam, N. and Kushwaha, R.K.S. (1989 b). Occurrence of non - keratinophilic fungi on keratin. Trans. Mycol. Soc., Japan, **30** : 1 - 8.

Nigam, N. and Kushwaha, R.K.S. (1989 c). Decomposition of feathers and hairs by keratinophilic fungi. Ind. J. Microbiol., **29** : 241 - 244.

Nigam, N. and Kushwaha, R.K.S. (1990). Occurrence of keratinophilic fungi with special reference to *Chrysosporium* species in Indian soil. Sydowia, **42** : 200 - 208.

Nigam, N. and Kushwaha, R.K.S. (1992). Biodegradation of wool by *Chrysosporium keratinophilum* acting singly or in combination with other fungi. Trans. Mycol. Soc. Japan, **33** : 481 - 486.

Ogbonna, C.I.C. and Pugh, G.J.F. (1987). Keratinophilic fungi from Nigerian soil. Mycopathologia, **99** : 115 - 118.

Oorschot, van, C.A.N. (1980). A revision of *Chrysosporium* and allied genera. Stud. Mycol., **20** : 1 - 89.

Orr, G.F. and Kuehn, H.H. (1972). Notes on Gymnoascaceae II. Some Gymnoascaceae and keratinophilic fungi from Utah. Mycologia, **64** : 55 - 72.

Otcenasek, M. (1978). Ecology of the dermatophytes. Mycopathologia, **65** : 67 - 72.

Padhye, A.A. (1961). *Microsporum gypseum* from soil in Poona. Prof. Agharkar Commemoration Volume 117 - 119.

Padhye, A.A.; Mishra, S.P. and Thirumalachar, M.J. (1966). Occurrence of soil inhabiting dermatophytes and other keratinophilic fungi from soil in Poona. Hindustan Antibiotic Bull., **9** : 90 - 93.

Paotetti, C.; Sili, C. and Balloni, W. (1975). Preliminary experiments on the microbiological production of protein hydrolyzates from feather keratins. Agric. Ital., **75** : 227 - 239.

Parihar, Pramila (1998). Exploitation of non pathogenic and pathogenic keratinophilic fungi for their potential to degrade keratin. Ph.D. Thesis, CSJM University, Kanpur pp. 195.

Parihar, Pramila and Kushwaha, R.K.S. (1997). Keratin decomposition in forest soil in a successive manner. National Conference on Environmental Biopollution. 25p.

Pugh, G.J.F. and Mathison, G.E. (1962). Studies on fungi in coastal soils III. An ecological survey of keratinophilic fungi. Trans. Brit. Mycol. Soc., **45** : 567 - 572.

Punsola, L. and Guarro, J. (1984). Distribution of mating types of the *"Microsporum gypseum* complex" in Spanish soils. Mykosen, **27** : 191 - 193.

Randhawa, H.S. and Sandhu, R.S. (1965). Survey of soil inhabiting dermatophytes and related keratinophilic fungi of India. Sabouraudia, **4** : 71 - 79.

Rebell, G. and Taplin, D. (1970). Dermatophytes, their recognition and identification. U. Miami press. Coral Gables, Florida.

Rogers, A.L. (1971). Isolation of keratinophilic fungi from soil in the vicinity of Bogota, Colombia. Mycopath. et. Mycol. Appl., **44** : 261 - 264.

Saidi, S.A.; Bhatt, S.; Richard, J.L.; Sikdar, A. and Ghosh, G.R. (1994). *Chrysosporium tropicum* as a probable cause of mycosis of poultry in India. Mycopathologia, **125** : 143 - 147.

Saksena, S.B. (1955). Taxonomic studies of the soil microfungi of some forest soils of Sagar. Saugar Univ., J. **1** : 30 - 31.

Saksena, V. (1993). Studies on anamorphs and teleomorphs of *Arthroderma* and related fungi. Ph.D. Thesis, Kanpur University, Kanpur pp. 263.

Shrivastava, J.N. and Kushwaha, R.K.S. (1983). Dermatophytes from Indian soil. III Internatn. Mycol. Cong., 638.

Sigler, L. and Carmichael, J.W. (1976). Taxonomy of *Malbranchea* and some other hyphomycetes with arthroconidia. Mycotaxon, **4** : 349 - 388.

Sigler, L.; Guarro, J. and Punsola, L. (1986). New keratinophilic species of *Chrysosporium*. Can. J. Bot., **64** : 1212 - 1215.

Singh, M.P. (1969). Mycotic dermatitis in camels. Ind. Vet. Jour., **46** : 854 - 856.

Singh, K.V. and Agrawal, S.C. (1983). Keratinophilic fungi from some substrates in central India. Geobios, **10** : 165 - 171.

Stockdale, P.M. (1958). Occurrence of *Microsporum gypseum*, *Keratinomyces ajelloi* and *Trichophyton terrestre* in some British soils. Nature London, **182** : 1754 - 1758.

Sur, B. and Ghosh, G.R. (1980). Keratinophilic fungi from Orissa, India, I. Isolation from soils. Sabouraudia, **18** : 269 - 274.

Ulfig, K. and Korcz, M. (1983). Isolation of keratinophilic fungi from sewage sludge. Sabouraudia, **21** : 247 - 250.

Vanbreuseghem, R. (1952). Keratin digestion by dermatophytes : a specific diagnostic method. Mycologia, **44** : 176 - 182.

Verma, H.N.; Sinha, B.K. and Das, U.L. (1982). Isolation of keratinophilic fungi from soil in Bihar (India). Mykosen, **25** : 449 - 452.

Vroey, C. de (1976). Sur quelques Ascomycetes isoles de lesions cutanees chez I'homme. Bull. Soc. Fr. Mycol. Med., **5** : 161 - 163.

15

Diversity of Rust Fungi in Oudh Region of Uttar Pradesh

R.S.Kanaujia and C.K. Yadav

Department of Botany, K.S.S. Post Graduate College, Ayodhya 224123, Faizabad (U.P.)

In all 207 rust species belonging to 15 genera related to 3 major families were recorded. Melampsoraceae having 4 genera with 16 species on 20 hosts, Pucciniaceae with 8 genera and 166 species from 22 hosts, and 3 form genera having 25 species from 31 hosts were identified. The host from 32 families of flowering plants were found parasitized by different rust species. Thirteen rust genera represented by 127 species were collected from 150 hosts of dicots and 7 genera with 86 species were recorded from 137 monocots; grasses being the main hosts. Poaceae widley followed by leguminosae domianted the spectrum of diversity. Poaceae alone contributed 85 hosts for 68 rust species. Amongst the total collection, 25 including 19 species of *Puccinia*, 4 of *Uromyces* and one each of *Aecidium* and *Uredo* warranted attraction. Of the above, *Puccinia cenchri, P. dietellii, P. fragosoana, P. leptochloae-unifloare, P. magnusiana* and *P. suveolens* were new to India, whereas *Puccinia cenchri* on *Cenchrus setigerus* Vahl., *P. dietelii* on *Chloris barbata* Sw. and *P. letpochloae-uniflorae* on *Leptochloa panicea* (Retz.) Ohwi. constituted the new host records.

Key-words: Diversity, rust fungi, host, Uttar Pradesh, India

Most of the fungi are innocuous, many are beneficial and thousands of others are known to be pathogens causing diseases to plants. These are omnipresent owing to their tremendous and untiring capacity of adoptation to various climatic conditions, ability to grow on surprisingly wide range of substrates, production of variety of vegetative propagules with long perennating ability and their unending and frequent mutability which perhaps not only overnumber them other groups of plants but also produces a remarkable quality of diversity. These phytopathogenic fungi had much economic significance as it is evident by the fact that there would hardly be a plant or animal which has not been fallen prey to the mercilesness of one or the other fungi. Not only this, the hyperparasitic and mycoparasitic forms do not spare the members of their own race. Associated with the various hosts these develop such qualities which enable to set them well on a particular host of their liking and this provides a sound background for great diversity among them. Among various pathogenic forms of fungi like smuts, mildews, wilts, rots and blights, rusts occupy the most important place owing to their occurrence on almost all the cereal crops which constitute more than 90 per cent of our usual diet. The ravages of rust are known wherever plants are cultivated. These include nearly all the groups of economic plants such as cereals, grasses, forage crops, garden fruits, orchard plants, green house and garden ornamentals and trees of the forests.

About 7000 or more species are strict parasites of conifers and flowering plants. These are not only specific to hosts but in many cases specialized to biological races, which are confined to a certain species of a host genus or even to a special agricultural varieties or strains of a host species. This ultimately established in them a great structural, physiological and to some extent genetical diversity. A lot of work has ben added to the diversity of rusts in various regions of the world (Butler, 1914; Butler and Bisby, 1931; Barclay, 1989-1890; Ahmad, 1981, 1989; Ahmad *et al.*, 1969; Arther, 1907; Bhagyanarayana and Ravinder, 1988; Berkeley, 1936; Bilgrami *et al.*, 1979, 1981; Cummins, 1971; Dalela and Sinha, 1964; Egan, 1981; Fischer, 1953; Flor, 1953; Goswami and Singh, 1973; Joshi, 1978; Mckenzie, 1910; Mehta, 1940, 1952; Prasad, 1951; Stakman, 1914; Vasudeva, 1960; Yadav, 1953; Zhunag, 1989), however, the Eastern Uttar Pradesh, adds some sporadic reports (Kanaujia and Yadav, 1993; Yadav, 1993;Kanaujia and Raj Kishore, 1981; Kanaujia and Singh, 1977, 1978) and needs thorough investigation. Keeping this in mind a work on diversity of rusts in this region was undertaken in 1990. The present paper embodies the findings of the investigation obtained so far. The further work along these lines in progress.

As a whole, 207 rust species falling to 15 rust genera were recorded, off which 25 were found to be very interesting. Fifteen genera represented 3 major rust families, viz., Melampsoraceae (4 genera), Pucciniaceae (8 genera) and form genera (3 form genera) were recorded. Melampsoraceae had 4 genera comprising of 16 species on 20 hosts, Pucciniaceae had 8 genera comprising of 166 species on 22 hosts and Form Genera represented 25 species on 31 hosts (Kanaujia and Yadav, 1993).

Out of 207 species of rust fungi recorded during systematic survey of rust fungi in Faizabad, Pucciniaceae alone contributed 122 rust species. The number of hosts parasitized was also the maximum in this case. A total of 32 families. (25 of Dicotyledons and 7 of Monocotyledons) and 1 of Gymnosperm were found to be parasitized by rusts. Out of 32 families of flowering plants and Coniferaceae, Graminae widely followed by Leguminosae dominated, so far as the number of rust species concerned. 85 hosts from Graminae harboured 68 rust fungi (Kanaujia and Yadav, 1993).

Twenty five very interesting rust fungi viz., *Aecidium crini* Kalchber., *Puccinia barbeyi* (Roum.) Magn., *P. cenchri* Diet. & Holw., *P. fragosoana* Beltr., *P. heterospora* Berk & curt., *P. leptochloae* Art. & Fromme., *P. leptochloae-*

uniflorae Cumm., *P. magnusiana* Koern., *P. phragmitis* (Schum) Koern., *P. romagnoliana* Maire & Sacc., *P. stenotaphri* Cumm., *P. suaveolens* (Link.) Rostrup., *Puccinia* species. 1,2,3,4,5,7,8 and 9 , *Uromyces capitatus* Syd., Uromyces species- *2,3,4 and Uredo zizyphi* pat., which were found to be new host fungus association for India and/ or world are being described here in this paper. Few species have been identified upto generic level.

1. ***Aecidium crini*** Kalchbr.

(Butler, E.J.& G.R. Bisby 1931. The Fungi of India Imp. Counc. Agr. Res., India Sci. Monogr. XVVIII, 237 pp)

The disease started as a small water soaked spot of light yellow colour on abaxial surface of leaf during extreme humid days of continuous rains. The mycelium develops below epidermis rapidly. The mycelium is cylindrical, profusely branched, septate, dikaryotic. It flourished between the intercelluar spaces from upper to lower epidermis. The mesophyllous cells soon collapsed leaving the ample space for the mycelium, fully developed spots are upto 6 cm in diameter. These showed the light coloured concentric rings, soon after the chains of spores are visible at the surface of the epidermis and the whole surface becomes dark yellow coloured at places the secretions oozed out as fine drops. Aecia mostly hypophyllous on the abaxial leaf surface, concave, upto 200 µ in diameter, peridia lacirate, wall regularly polyhedral (mostly tetrahedral), laterally compressed, very prominent forming peridium, Aeciospores sessile, in chains, spherical, single celled,double walled, wall smooth, dikaryotic, bright golden yellow when seen under microscope, spore 12 - 25 x 12-20 µ in diameter, spherical globose to ellepsoidal, wall smooth, uredospores and teleutospores were not found (Figure-1).

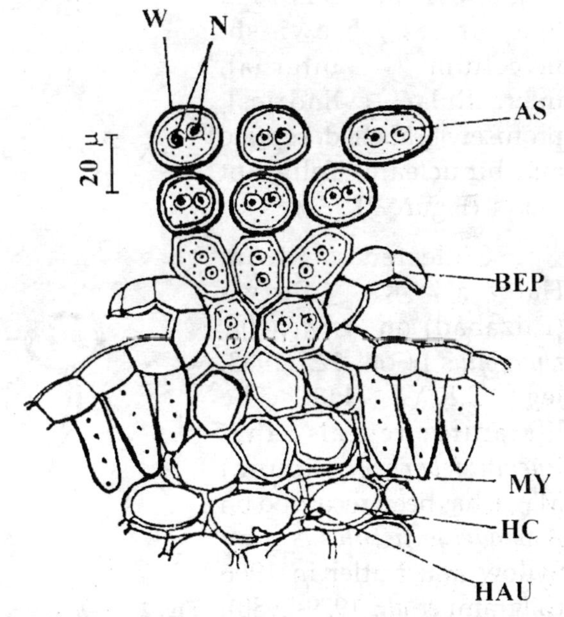

Fig. 1. ***Aecidium crini*** Kalchbr. on ***Crinum maculatum*** L. showing aecial stage: (W-Wall, N-Nuclei, BEP-Broken epidermis, AC-Aeciospore, MY-Mycelium HC-Host cell and HAU- Haustoria)

Collected on *Crinum maculatum* L. from

Professors colony, station road, Ayodhya on 20.7.1993, leg. C.K.Y. –85. The fungus is known on *Crinum asiaticum L., C. latifolium* and *Crinum* species (Bilgrami *et al.,* 1979, 1961). This has not been known on *Crinum maculatum* L from India so far, therefore, it is described here as a new fungus host association for country.

2. *Puccinia barbeyi* (Roum). Magn.

(Sydow &Butler, Fungi Indiae Orientalis, part, I, Ann. Mycol. 425, 1906).

The leaves become covered with numerous uredosori on the tips, loose their green colour, drop and dry up, spots reddish brown and very minute; uredosori on tips in the leaves, at first as little yellowish points which turn into reddish brown pustules, uredospores small, densely produced, 16-22x 16-22μ, single celled, binucleate, wall 1.5 –2 μ, light yellowish in colour, echinulate, mostly obovoid or globose; pedicels 31 μ long or less, brownish; mycelium internal, intercellular, cylindrical, profusely branched, septate and binucleate; Telia not found (Figure-2).

Collected from Halkara Ka Purwa (Faizabad) on *Asphodelus tenuifolius* L. on. 12.2.1993, leg. C.K.Y. –111. The literature reveals that *Puccinia barbeyi* (Roum.) Magn. has been recorded on *Asphodelus fistulosus* by Sydow and Butler in 1906 (Bilgrami *et al.,* 1979-1980). So far, it has not been found on *Asphodelus tenuifolius* L., therefore, it is described here as a new fungus host association for country.

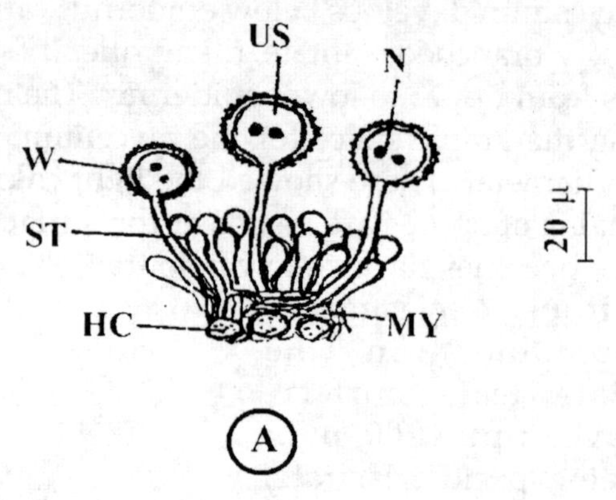

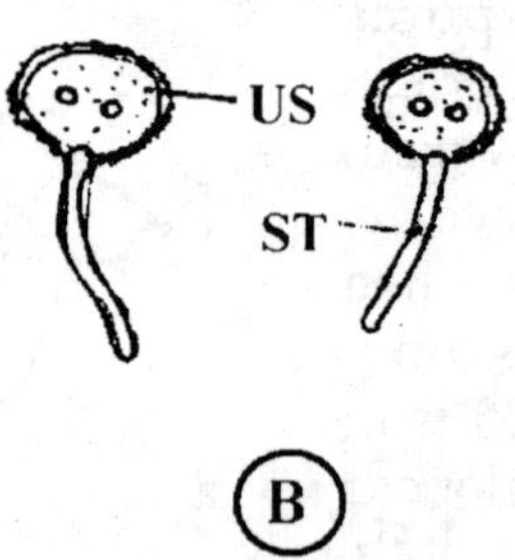

Fig. 2. ***Puccinia barbeyi*** (Roum) Magn. from ***Asphodelus tenuifolius*** L. showing uredinial (A) stage and uredospore (B) (HC-Host cell, ST-Stalk, W-wall, US-Uredospore, N-Nucleus, My-Mycelium).

3. *Puccinia cenchri* Diet. & Holw.

(Holway, Bot. Gaz. 24: 28:, 1997)

Rust sori produced on every part of the plant, viz., stem, leaves and leaf sheaths except inflorescence, spots pin head size to 1mm. usually spherical, Some times in rows, rust appears from mid September to mid December, light yellow to reddish brown in colour. Aecia not found; uredinia amphigenous or mainly on adaxial leaf surface, cinnamon brown; Spore 25-3 x 19-31 µ thick, prominently echinulate, cinnamon brown; germpores 2 or some times 3, equatorial; mycelium profusely branched , light in colour, in intercelluar spaces of hypodermis and mesophyll below epidemis; Telia on abaxial leaf surface. lesser than uredinia, covered by epidermis, blackish brown inconspicuous, spores 37-51 x 17-24 µ, mostly oblong or clavate, wall 1.5 µ thick at side, 3.5 µ apically, golden smooth, pedicel colourless to light yellowish, thin walled, upto 78 µ long (Figure 3).

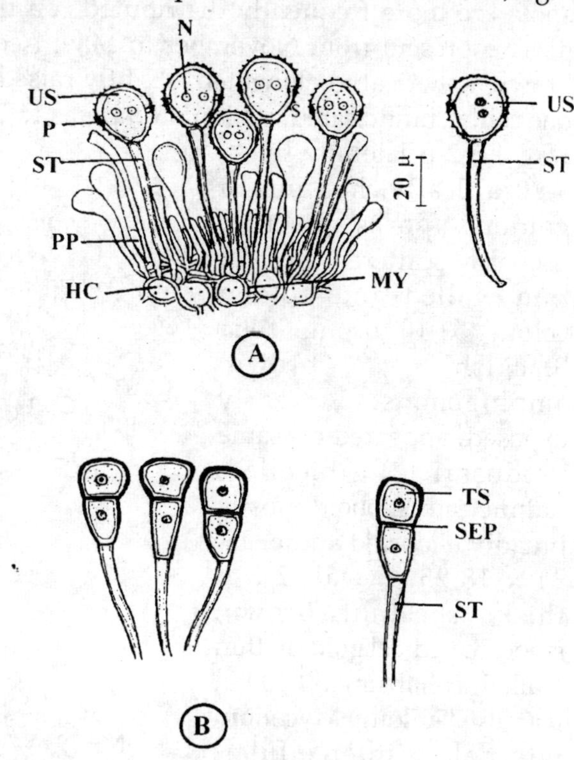

Fig. 3. ***Puccinia cenchri*** Diet. & Holw. on ***Cenchrus setigerus*** Vahl. showing Uredinial (A) and Telial (B) stages: (HC-Host cell, PP- Paraphysis, ST-Stalk, P-Pore, US-Uredospore, N-Nucleus, MY-Mycelium, TS-Teleutospore and SEP-Septum).

Collected on *Cenchrus setigerus* Vahl from various places of Faizabad on 10.10.1991, leg. C.K.Y.278. Earlier, *Puccinia cenchri* Diet. & Holw. has not been reported from India. It has been reported on *Cenchrus ciliaris* Fig. & De Not from Central Africa. Its type specimen was reported by E.W.D. Holway on *Cenchrus multiflours* from Guadalajara of Mexico on October 12, 1896. Earlier, it was known as *Uredo Cenchricola* Holway. The fungus was collected widely and abundantly in whole of the district Faizabad, matches in greater details to *Puccinia cenchri* Diet. & Holow. It is a new fungus species for India and new fungus host association for the country and the world.

4. *Puccinia dietelii* Sacc. & Syd.

(Saccardo, Sylloge Fungorum, **14**:358, 1899.)

Rust sori yellow or pale yellowish when young and dark brown when mature, spherical to elongated, some times arranged in rows, generally on both the surfaces of the leaves, spots 0.5 mm to 1 cm in diameter; number of spots are more frequently distributed on the dorsal side than the ventral; disease present from November to May; Aecia not obtained; Uredosori pale brown, spherical to elongated, slightly raised, uredospore densely produced, one celled, binucleated, echinulate, spores 16-28 x 16-25 µ ellipsoid or obovoid, wall 1.5-2 µ laterally thick, 3-10 apically, hyaline to golden, pores obscure, probably scattered, pedicel thin walled and golden colour, 53-10 µ long. Telia blackish brown, amphigenous, early exposed, appeared on same uredosori, bicelled, uninucleate, globoid mostly broadly ellepsoid spore, 25-34 x 18-25 µ, wall 2-3 µ thick, chestnut brown, smooth, pedicel golden, thin walled, collapsing 50-100 µ, usually 75µ long. Mycelium internal intercellular, cylindrical, profusely branched, binucleate and septate (Figure 4).

Collected on *Chloris barbata* Sw. from Asharfabad (Faizabad) on 17.11.1991, leg. CKY-264. This species has great resemblance with *Puccinia dietelii* Sacc. & Syd. Collected by Bertholomen on *Chloris verticillata* Rooks from rooks country, Kansal.

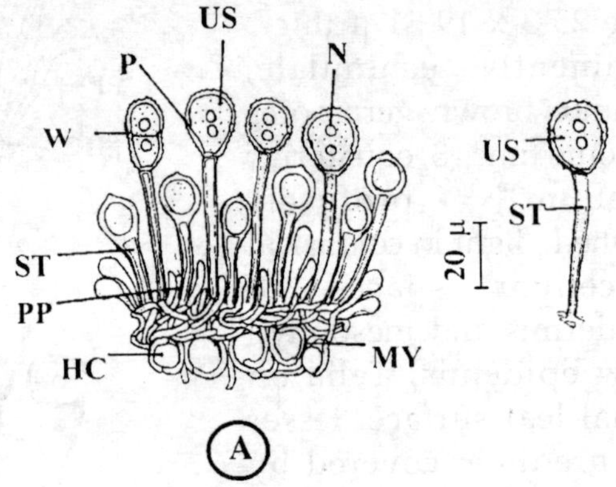

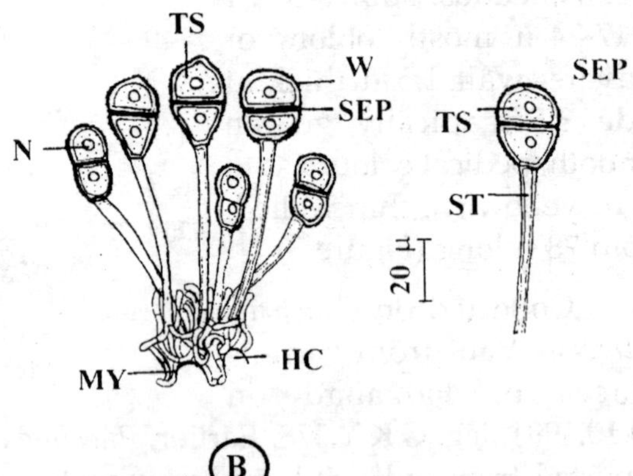

Fig. 4. ***Puccinia dietelii*** Sacc. & Syd. from ***Chloris barbata*** SW. - showing uredinial (A) and Telial (B) Stages : (HC-Host cell, PP- Paraphysis, ST-Salk, W-Wall, P-Pore, US-Uredospore, N-Nucleus, MY-Mycelium, TS-Teleutospore and SEP-Septum).

It has also been found on *Chloris* species, *Dactyloctenium aegyptiacum* (L.) Beauv. From India this has not been reported so far and is, therefore, reported for the first time. It also causes a new fungus host association for India and the world.

5. *Puccinia fragosoana* Beltram.

(Beltram Mem. Roy. Soc. Espn. Hist. Nat. **50** : 249,1921).

Rust sori reddish brown to dark brown, spherical to slightly elongated, generally on both the surfaces of the leaf, spots pin head size when young and 0.5-1.5 μ when mature number of spots more frequently distributed on adaxial leaf side than abaxial one, present throughout the year but more frequent from January to March. Aecia not traced; uredosori reddish brown, spherical to elongated, slightly raised, wide open when matured; uredospores densely produced, one celled, binucleate, spores 28-44 x 28-31 μ, variable but mostly obovoid, spherical to slightly ellepsoid, wall echinulate, uniformly thickened upto 2μ, germ pores 4 or 5 , pedicel light yellow, usually up to 75μ long; telia blackish brown exposed, compact, spore 24-34 x 14-23μ. Mostly ellipsoid, wall uniformly 2-2.5μ thick at side and mostly ellepsoid, wall uniformly 2-2.5 μ thick at side and slightly thicker apically, chestnut brown, smooth; pedicel mostly thin walled, collapsing or not mostly brownish, upto 45μ long, usually brown near the spores, mycelium internal, intercellur, cylindrical, profusely branched, binucleate and septate (Figure 5).

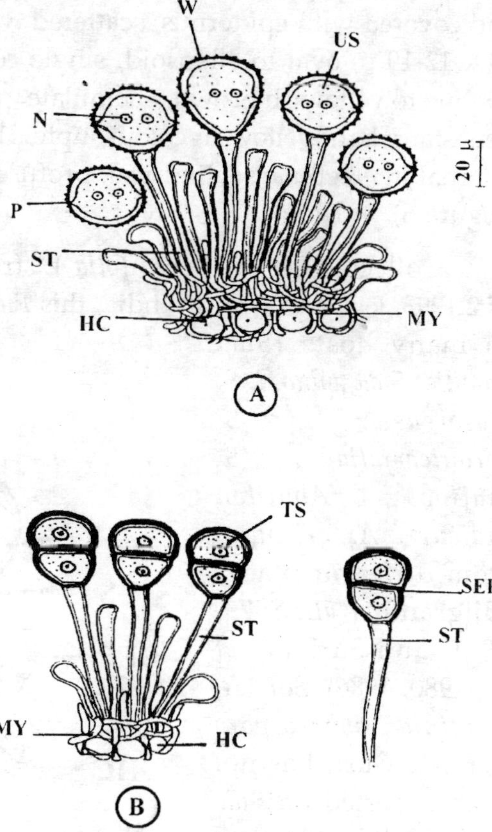

Fig. 5. ***Puccinia fragosoana*** Beltr. showing Uredinial (A) and Telial (B) stages on ***Imperata cylindrica*** (Linn.) Beauv.: (HC-Host cell, ST-Stalk, P-Pore, MY-Mycelium, TS-Teleutospore and SEP-Septum).

This species was collected on *Imperata cylindrica* (Linn.) Beauv. from Raghunath Das Ki Chhawani, Ayodhya, present throughout the year, leg. CKY-265. This species in greater details matches with *Puccinia fragosoana* Beltr. on same hosts collected from Spain by Beltran on *Schizachyrium sanguineum* (Retz) Atst. It is also reported on *Imperata cylindrica* (Linn.) Beauv. from Sierra, leone and South Africa. Earlier this species of *Puccinia* has not been reported from India. It is, therefore, a new fungus host association for country.

6. ***Puccinia heterospora*** Berk & Curt.

(Sydow H. and E.J.Butler, 1906, Fungi Indiae Orientalis Part. I Ann. Mycol. 4:421-445.)

Pustules appeared mostly on stem and branches, uredosori very small and covered with epidermis, scattered, yellowish brown in colour, spores 12-19 x 12-19 μ, oval to ellepsoid, single celled binucleate, wall 1.5 –3 μ thick, hyaline to yellowish brown, ehcinulate, pores 4-6 scattered, spores pedicillate, persistant, light yellow in colour, upto 31μ long; telia not obtained; mycelium internal intercellular, cylindrical, profusely branched, binucleate and septate (Figure 6).

Collected on *Sida cordifolia* L. from Hanumangharhi, Ayodhya on 15.2.1993, leg. CKY-148. In India, this *Puccicinia* species has been reported on so many hosts (*Sida humilis, Sida spinosa, S. mysorensis, S. veronicaefolia, S. glutinosa, Abutilon indicum, A. crispum*) from different places (Bilgrami *et al.*, 1979, 1981, and Sarbhoy *et al.*, 1980, 1986). Sor far, *Puccinia heterospora* Berk. & Curt. has not been reported on *Sida cordifolia*. It is, therefore, described here as a new host fungus association for country.

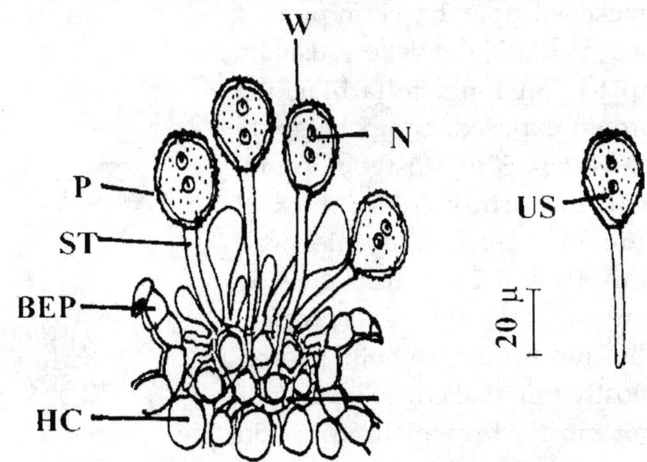

Fig. 6. ***Puccinia heterospora*** Berk. and Curt. from ***Sida cordifolia*** L. showing uredinial stage (HC-Host cell, BEP-Broken epidermis, ST-Stalk, P-Pore, W-Wall, N-Nucleus and US-Uredospore).

7. ***Puccinia leptochloae*** Arth. & Fromme

(Arth. & fromme, Torreya, **15**:263, 1915).

The pustules are formed on the abaxial leaf surface and on leaf sheaths, scattered as well as arranged in streaks, light yellow to brown, raised, upto 0.5-1 mm long, aecia not traced; uredosori light yellow to reddish brown, on abaxial leaf sides, uredospores densely produced, 19-25 x 19-22 μ, globoid or obovoid, single celled binucleate, wall 1.5-2.5 μ, golden or cinnamon brown, pores 4-6 scattered, pedicels elongated thick walled; telia blackish brown in colour, on abaxial leaf surfaces densely produced and not widely opend; teleutospores 25-34 x 17-24μ, broadly ellipsoid, bicelled, uninucleate, wall 2.5-4 μ thick at sides and 4-7 μ apically, dark chestnut colour, pedicels thick walled, usually not collapsing, golden brown attaining a length of 53μ; mycelium internal, Intercelluair or intracellular, cylindrical, profusely branched, binucleate and septate (Figure-7).

This specimen was collected on *Leptochloa filiformis* (Lam.) Beauv. from Professors colony, Station Road, Ayodhya on 5.5.1993, leg CKY-268. This species of *Puccinia* collected from Professors Colony, Station Road, Aydohya matches in major detalis to *Puccinia leptochloae* Arth & Fromme on *Leptochloa filiformis* (Lem.) Beauv. from southern U.S.A. to Guatemala Puerto Rico southward to Argentina. It produced aecia on *Telinum pamiculatum* by inoculation (Cummins, **Mycologia, 55**;73-78, 1963). This species of *Puccinia* on

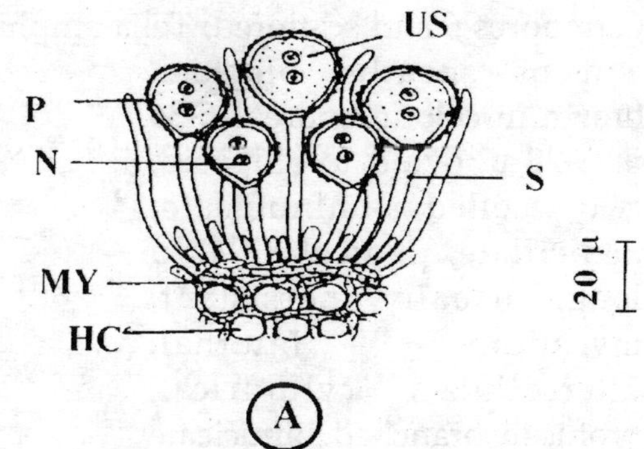

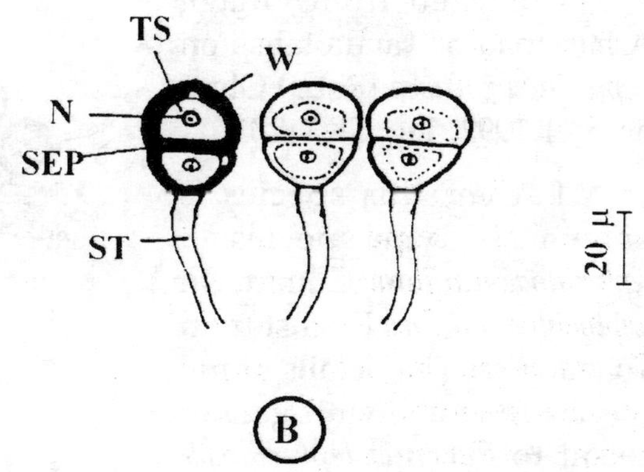

Fig. 7. ***Puccinia leptochloae*** Arth. & Fromme from ***Leptochloa filiformis*** (Lam.) Beauv. showing Uredinial (A) and Telial (B) stages: (HC-Host cell, My-Mycelium, N-Nucleus, P-Pore, US-Uredospore, ST-Stalk, SEP-Septum, W-Wall and TS-Teleutospore).

Leptochloa filiformis is, therefore, recognized as *Puccinia leptochloae* Arth. & Fromme, Earlier, this has not been reported from India, it is, therefore, new fungus host association for country.

8. ***Puccinia leptochloae-uniflorae*** Cumm.

(Cummins, G.B., **The Rust Fungi**, Springer verlag, Berlin, 1971).

Rust sori occurred on both the surfaces of the leaves, spots pinhead size to 1.5 mm diameter, raised, spherical to elongated, yellowish in colour, 4-7 mm, dark brown, present in streaks and distributed sporadically on the leaves at maturity; uredinia mostly on adaxial leaf surface, orange when fresh, colourless and dry when old; spores 19-28 x 19-22 μ, mostly broadly ellepsoid or broadly obovoid, wall 2-2.5 μ thick colourless or pale yellowish, echinulated, germpores 6 and scattered; Telia amphigenous mostly on abaxial leaf surface and on sheaths, blackish brown in colour, spores 25-34 x 21-28 μ, ovoid to ellipsoid, two celled uninucleate, pedicillate, pedicel 37-47 μ long, usually persistant; mycelium internal, intercellular, cylindrical, profusely branched, binucleate and septate (Figure-8).

Collected from village Asharfabad of Tanda tahsil on *Leptochloa panicea* (Retz.) Ohwi on 5.11.1992, leg CKY-203.

Earlier, this species is known as type species *P. leptochloae-uniflorae* Cumm. On *Leptochloa uniflora* Hochst from Tonganyica. The details of my species resemble with greater extent to *Puccinia leptochloae-uniflorae* Cumm. This has not been reported from India so far, it is, therefore, reported as new host fungus association for country and for the world.

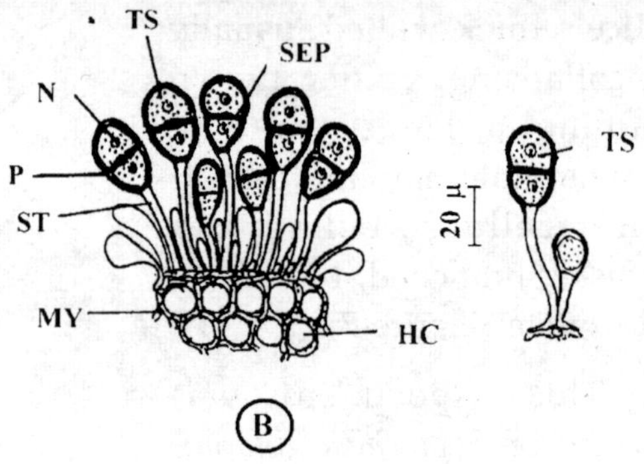

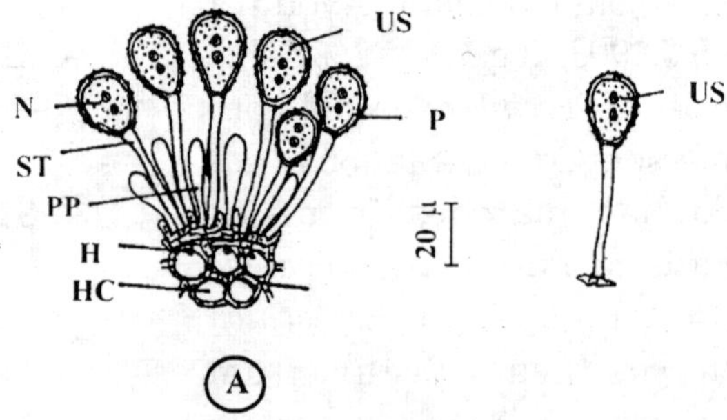

Fig. 8. ***Puccinia leptochloae-uniflorae*** Cumm. from ***Leptochloa Panicea*** (Retz.) owhi. showing Telial (A) and Uredinial (B) stages. (My-Mycelium, ST-Stalk, P-pore, N-nucleus, TS-Teleutospore, SEP- Septum, HC-Host cell, PP-Paraphysis, US-Uredospore and H-Haustorium).

9. *Puccinia magnusiana* Koern.

(Koernicke; Hedwigia, **15**:179, 1876).

Rust sori appeared generally on every part of the plant, viz, stem, leaves, leaf sheaths and inflorescence; spots 1 mm to 15 mm diameter, spherical to irregular shape and yellowish red to dark brown in colour, on abaxial leaf surfaces but in severe cases appeared on both surfcaes of leaf; aecia not found; uredinia amphigenous, mostly on adaxial leaf surface, yellowish brown in colour, pulvinate, with clavate or clavoid capitate hyaline or yellowish paraphyses whose wall is 1.5-4 μ thick apically, becoming thinner below; uredospore 19-41 x 13-25 μ, mostly ellepsoid, oblong-ellepsoid, or oval, wall 1.5-3 μ thick, hyaline to yellowish brown, echinulate pores obscure, 8-10, scattered; spores pedicillate, persistant, light yellow in colour, 69-94 μ long and 5-6 μ wide; telia amphigenous and on leaf and leaf sheaths, early exposed in large blackish brown confluent group upto 3 cm long, compact; spores 28-53 x 13-22 μ, variable, mostly clavate or oblong clavate, sometimes ellipsoid, the apex mostly or narrowly rounded, wall 1.5-4 μ thick, deep golden or chestnut brown, smooth, pedicels thick walled or not, only partially collapsing, hyaline to brownish, persistent, from 47-135 μ long; mycelium internal, interecellular, intracellular in post pathogenic stage cylindrical, profusely branched, binucleate, and septate (Figure 9).

Collected on *Arundo donax* L. from Avadh university campus, Faizabad on 11.11992, leg. CKY-230. *Puccinia magnusiana* Koern. has been found parasitizing different hosts like species of *Phragmitis* from different parts of the globe. It has also been found on *Arundo donax* L., however, it has not been recorded from India. This constitutes its first report as new fungus and new host record for world.

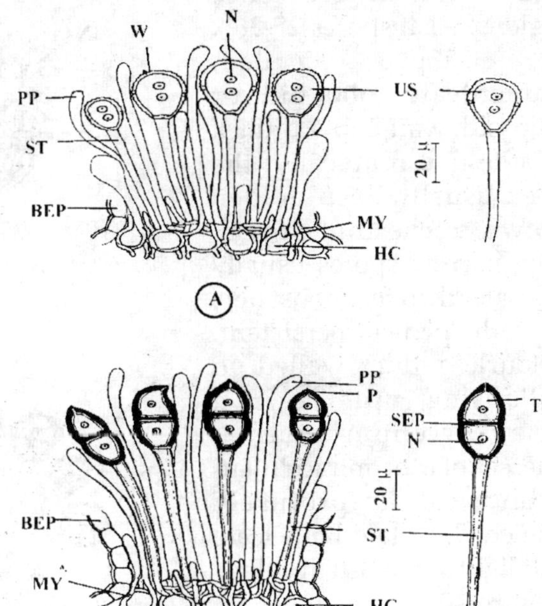

Fig. 9- ***Puccinia magnusiana*** Koern. showing Uredinial (A) and Telial (B) stages on ***Arundo donax* L.:** (BEP-Broken epidermis, ST-Stalk, PP-Paraphysis, W-Wall, N-Nucleus, US-Uredospore, MY-Mycelium, HC-Host cell, SEP-Septum, P-Pore and TS-Teleutospore).

10. *Puccinia phragmitis* (Schum.) Koern.

(Sydow & Butler, Fungi Indiae Orientalis)

Pats II. Ann. Mycol . 5: 485-515, 1907.

The pustules appeared on the leaves, stems and branches; sori are small rounded, pin head to 1 mm diameter scattered on the lower surface of the leaf and finally spread over both the surfaces. Sometimes remain scattered on leaf surface and form circular ring like patches, the sori are powdery, brown to dark tan brown in appearance; Aecia not obtained; uredinia raised to circular ringoid, present on both the surfaces of the leaf, brownish in colour, rings upto 7 mm in diameter; spores 19-28 x 19μ, ellepsoid or obovoid, single celled binucleate, wall 3-5 μ thick at side, 4-6 μ apically, yellow to goden brown, echinulate, germpores 3-6 equatorial, pedicel persistent, colourless, thick walled, not collapsing, upto 76μ long; telia dark tan brown at maturity, mostly on both the sides but frequently distributed on adaxial side of the leaves, also frequent on stem, branches and and inflorescence regions, telial spores 25-36 x 22 -28μ, bicelled, uninculeate, obovoid or ellipsoid, wall 2 to 4 μ thick at side, 5-8 μ apically, the apex usually deep golden brown to chestnut brown, long narrow spores usually are paler than the robust one, smooth pedicel persistent, colourless, thick walled no collapsing, upto 36-76 μ long. Mycelium internal, intercellular or intracellular, cylindrical profusely branched, septate, binucleate and light yellowish in colour (Figure-10).

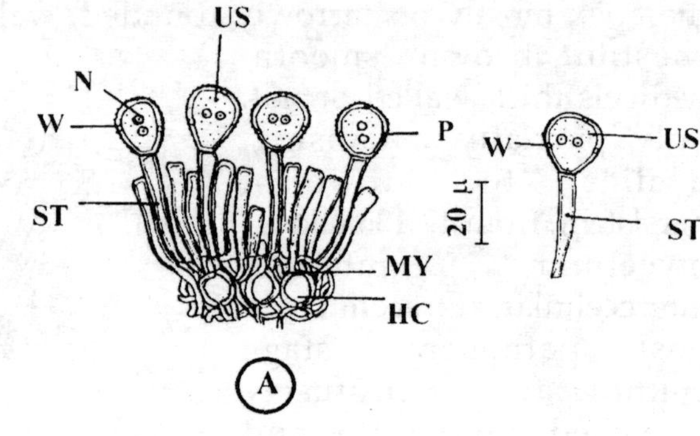

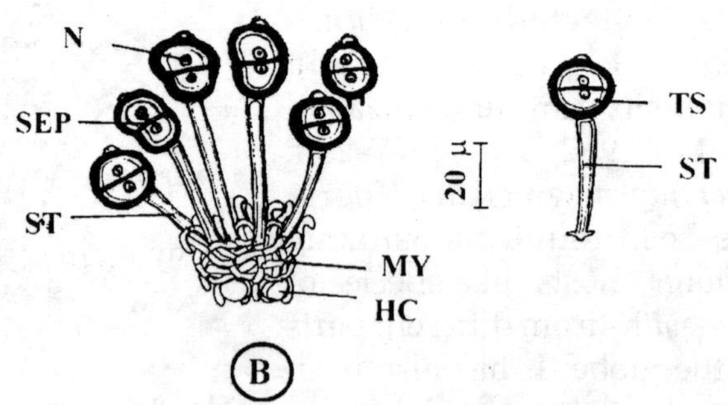

Fig. 10- ***Puccinia phragmitis*** (Schum.) Koern. showing Uredinial (A) and Telial (B) stages on ***Rumes vesicarius*** Linn.: (ST-Stalk, W-Wall, N-Nucleus, US-Uredospore, P-Pore, MY-Mycelium, HC-Host cell and SEP-Septum).

Collected on Rume vesicarius Linn.) from Professors Colony, Station Road, Ayodhya on 19.3.1992, leg. CKY-180. This species showed the resemblace to *Puccinia phragmitis* (Schum) Koern. In India this species has been reported on *Phragmites karka* from Waldaiar, Tamil Nadu and on Rume species from Himalaya, North Deharadun, U.P. (Bilgrami *et al.*, 1979, 1981). So For, *Puccinia phragmitis* has not been recorded on *Phragmites communis* L. This species does not match with *Puccinia invenusta on Phragmites communis* reported by Kanaujia in 1997 from Faizabad. It is, therefore, described here as new fungus host association for India.

11. *Puccinia romagnoliana* Maire & Sacc.

(Sacc. XVII: 374:1905)

The sori appeared on the adaxial side of the leaves and leaf sheaths The sori 1 mm to 5 mm in diameter, elongated to spherical, arranged in a row and dark brown in colour; uredosori raised, circular to elongated, present on adaxial leaf surface, yellowish brown, paraphyses clavate or capitate, hyaline to yellowish; spores 16-31 x 16-25 µ, mostly ellepsoid, oblong ellepsoid or oval, wall 2-3 µ thick, yellowish brown, echinulate, pores obscure, 8-10 scattered, pedicels yellowish, thin walled upto 60 µ long; telia blackish to brown on adaxial leaf surface below epidermis; spore 31-56 x 13-19µ, elongated, two celled, uninuclete, wall uniformly 1-5 µ thick or very slightly thicker apically, golden brown or chestnut brown, smooth, pedicel yellowish, thin walled , persistent,upto 62 µ long, 3-6 µ wide, mycelium, internal intracellular or intercellular, profusely branched, clyindrical, binucleate and septate (Figure-11).

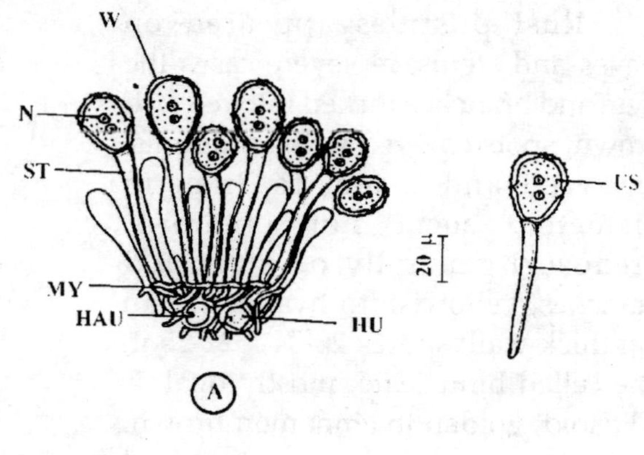

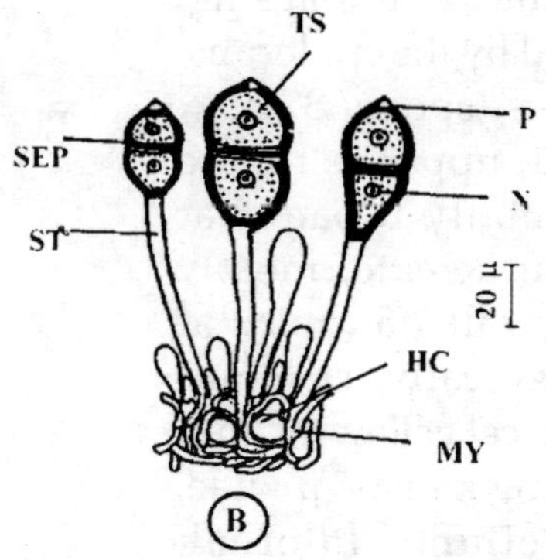

Fig. 11- ***Puccinia romagnoliana*** Maire. & Sacc. showing Uredinial (A) and Telial (B) stages on ***Cyperus longus*** L.W.: (HC-Host cell, HAU-Haustorium, MY-Mycelium, ST-Stalk, N-Nucleus, W-Wall, SEP-Septum, TS-Teleutospore and P-Pore).

Collected from Naya Ghat, Ayodhya on *Cyperus longus* L. on 15.4.1993, leg. CKY-237. This species has also been found on *Cyperus kyllingia* Endl. On 15.5.1992, leg. CKY 224. Telia were not recorded on *C. kyllingia.* In India, this species has been reported from so many hosts (*Cyperus rotundus* L., *Cyperus tagetum C. compressus, C.arenarius, C. tuberosus, C. capitatus* Retz., *C.iria*, C *compositus* L. and *C. esculentus* L.) from different places (Bilgrami *et al.*, 1979, 1981; Sarbhoy *et al.*, 1980, 1986). So far, *Puccinia romagnoliana* has not been reported on *Cyperus longus* L. It is, therefore, described here as a new fungus host association for the country.

12. *Puccinia stenotaphri* Cumm.

(Cummins, G.B., 1990, Bull. Torrey Bot. Club, 87:40)

Rust pustules appeared on leaves and stems, in severe cases the stem and branches turned into reddish brown, spots covered with epidermis, scattered and 1 mm to 3mm in diameter; aecia not obtained, uredosori generally on stem and branches, yellowish to hyaline, 1.5 to 3 µ thick wall; spores 23-34 x 22-25 µ, one celled binucleate, mostly oval or ellipsoid, golden to cinnamon brown, echinulate, pores 4, equatorial, pedicel verey short, some times upto 16µ long; telia tan red, covered by the epidermis, without paraphyses, spores 37-53 x 16-25 µ, two celled, upper cell more elongated, abnormally broader at upper region, uninucleate, mostly angularly clavate, wall 1.5µ thick at side and 2,5-5 µ apically, chestnut brown, smooth, pedicel yellowish, thin walled,upto 37 µ long and 4-6 µ wide, persistent; mycelium internal, intercellular, profusely branched, cylindrical, binucleate and septate (Figure-12).

Collected on *Sesbania grandiflora* Pers. from Avadh University Campus,

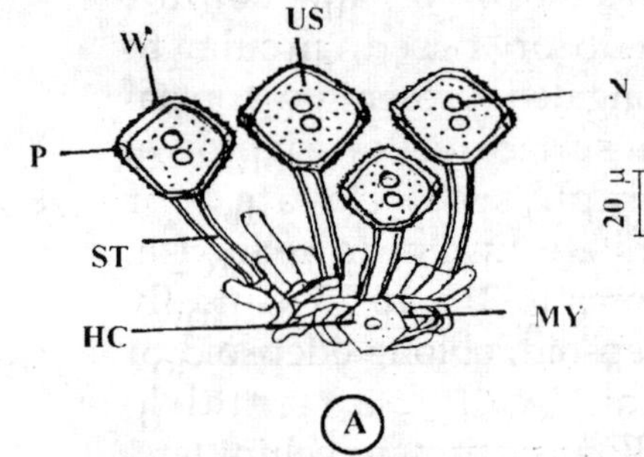

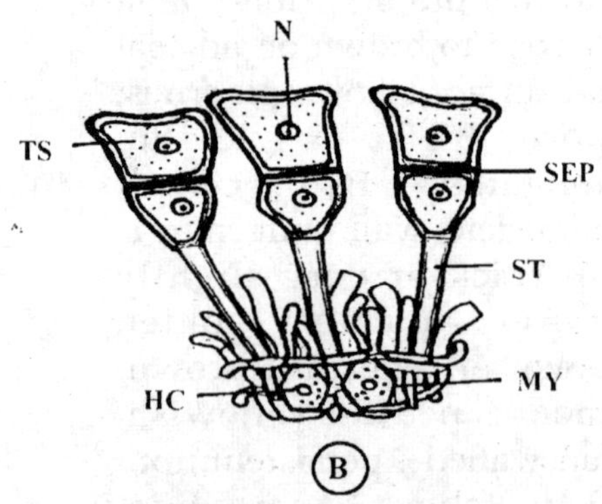

Fig. 12- ***Puccinia stenotaphri*** Cumm. from ***Sesbania grandiflora*** Pers. showiing Uredinial (A) and Telial (B) stages.: (HC-Host cell, ST-Stalk, P-Pore, W-Wall, US-Uredospore, N-Nucleus, MY-Mycelium, TS-Teleutospore, and SEP-Septum).

Faizabad on 11.1.1992, leg, CKY-32. This *Puccinia* species in greater details matches with *Puccinia stenotaphri* Cumm. which was reported by Cummins on *Pennisetum hordeoides* (Lam.) Steud *P setosum* (Swartz.) L. Rich., *Stenotaphrum dimidiatum* (L) Brongn., *S secundatum* (Walt.) O. ktze, *Stereochlaena cameronii* (Stapf.) Pilger. In India it is found on *Pennisetum typhoides* (Burm. f.) Stapf. & Hubb. (Sethi, 1969). So far, it has not been found on *Sesbania grandiflora*, it is , therefore, described here as a new fungus host association for country.

13 *Puccinia suaveloens* (Link) Rostrup.

(Sacc. VII, 633, 1988; Sydow and Butler, Fungi Indiae Orientalis, part. VI, Ann. Mycol., 10 : 242-280.)

Rust pustules appeared on leaves, stems and inflorescence, pustules 1 mm to 10 mm long, spheroid to ellipsoid, raised, yellowish to dark brown in colour, in severe cases the whole plants including the stem,leaves and the inflorescence become stunted and much branched as a result of pathogenesis; Aecia not found; uredinia amphigenous or mostly on adaxial leaf surface, cinnamon brown; spores 19-28 x 19 – 28 µ, broadly ellipsoid or obovoid, wall 2.5 –3.5 µ thick, golden or reddish brown, echinulate; pores 8-12, generally scattered, pedicels upto 62µ long and 4.5µ wide; telia like the uredinia, spores 44-49 x 28-31 µ, variable but mostly golden brown, finally echinulate, verrucose, germpore mostly apical in upper cell, midway to hilum in lower cell, pedicel fragile and always broken near the spore, upto 65-70 µ long; mycelium internal, intercellular, cylindrical, profusely branched and septate (Figure-13).

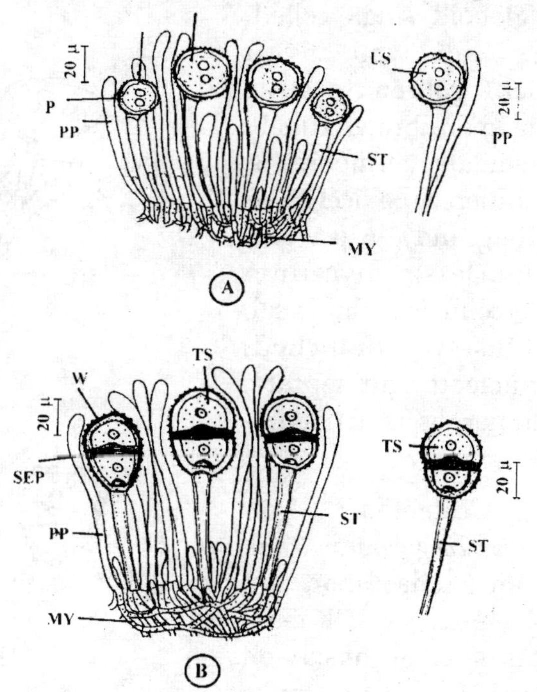

Fig. 13- *Puccinia suaveolens* (Link) Rostrup. form *Cirsium arvens* (Linn.) Scop. showing uredinial (A) and Telial (B) stages: (PP-Paraphysis, P-Pore, W-Wall, US-Uredospore, ST-Stalk, MY-Mycelium, SEP-Septum, and TS-Teleutospore.)

Recorded on *Cirsium*

arvense (Linn.) Scop, collected from Raghunath Das Ki Chhavani, Ayodhya (Faizabad) on 10.1.1992, leg., CKY-54.

The fungus though reported on *Cirsium argyracanthis* from Kashmir and on *C. arvense* from Pusa, Bihar (Butler and Bisby revised by Vasudeva, 1960, p 111). The author's fungus has echinulations on teleutospores in addition to the species reported on *Cirsium arvense* (Linn). It is, therefore, described here as a new var. nov. of *Puccinia suaveolens* (Link.) Rostrup var. nov. This, however, needs further confirmation.

14. *Puccinia* species-1

Rust pustules produced on abaxial leaf surface, spots pointed to pin head size, spherical reddish brown in colour, scattered on the leaves; uredinia produced mostly on the abaxial side of the leaf, brownish in colour, spores 25-36 x 22-28 µ, broadly ellepsoid, broadly ovovoid or globoid, single celled, binucleate, wall 2.5-3 µ thick, golden to near cinnamon brown, finely echinulate, germpores 3-6, scattered, pedicel 25-47 µ long and 3-6 µ wide, colourless; mycelium intercellular, cylindrical, profusely branched, binucleate and septate; telia not observed (Figure-14).

Collected on *Anthistiria gigantean* Vahl. From Asharfabad on 2.11.1991, leg. CKY-35. This species has been identified only upto generic level. It does not match with any of the existing species on the grasses.

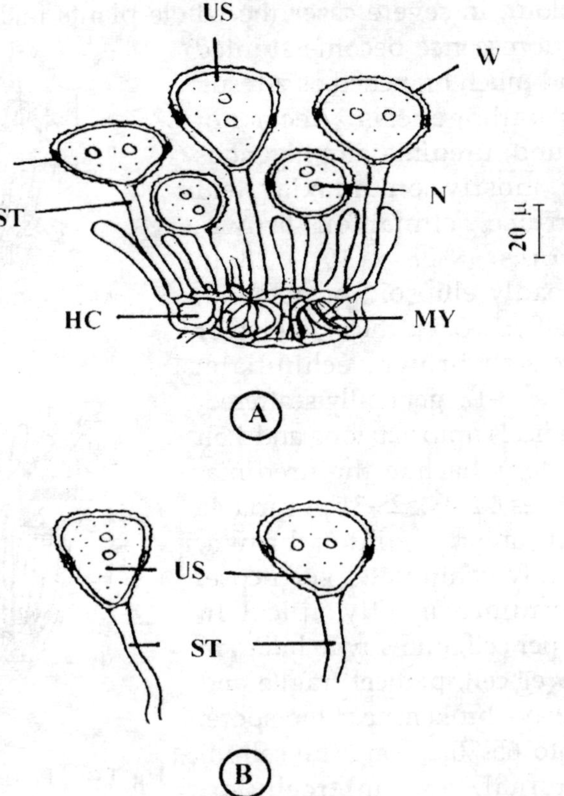

Fig. 14- *Puccinia* species-1 from *Anthistiria gigantea* Vahl. showing uredinial stage (A) amd Uredospore (B): (HC-Host cell, St-Stalk, P-Pore, US-Uredospore, W-Wall, N-Nucelus and MY-Mycelium).

15. *Puccinia* species-2.

Rust pustules appeared on adaxial surface of leaves and spikes, spots 1 mm ot 10 mm long, circular, spherical, elongated and slightly raised, reddish brown to dark brown in colour, pustules first appeared in August and most favoured in the month of September-October; uredosori reddish brown, 1 mm to 10 mm long, spherical to elongated and wide open, spores 12-22 x 9-16 µ, mostly obovoid, globoid or ellipsoid, germpores 4, wall 2 µ thick, echinulate, pedicel present, 37-64µ long and 3-6 µ wide, mycelium internal, intercellulr, profusely, branched, cylindrical, septate and binucleate; telia not observed (Figure-15).

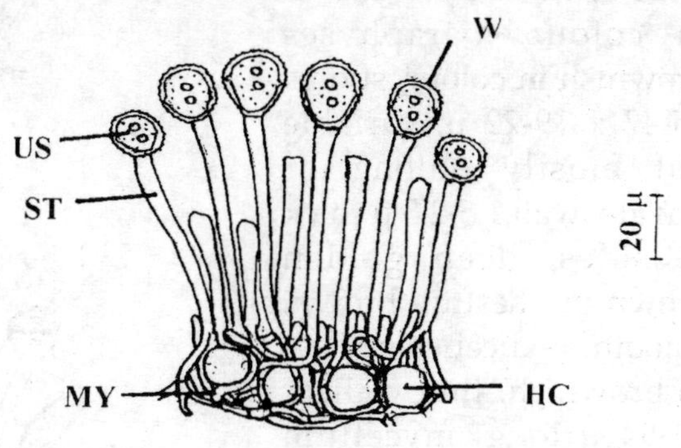

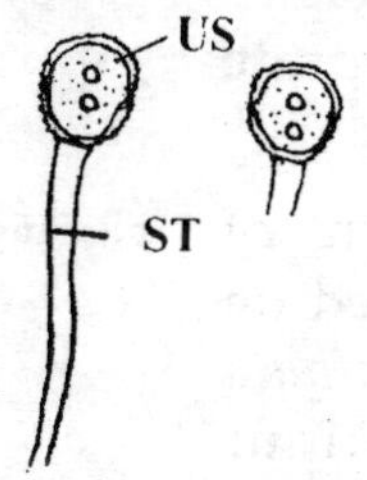

Fig. 15- ***Puccinia*** species- 2 on ***Cyperus flavidus*** Retz. showing uredinial stage: (My-Mycelium, ST-Stalk, US-Uredospore, W-Wall, HC-Host cell).

Collected on *Cyperus flavidus* Retz. from various places of the district Faizabad during August to November, 1991-1992; leg. CKY-34. This Puccinia species-2 does not match with any existing species on *Cyperus*. The further work is in progress. This for time being has been described as new host fungus association for the country.

16. *Puccinia* species-4

Pustules appeared on adaxial leaf surface, spots 0.5 mm to 15 mm, arranged in a long streak, spherical to elongated, raised, yellowish brown to dark blackish brown in colour, aecia not found; uredinia amphigenous or mostly on adaxial surface of the leaf in linear series or chlorotic streaks, pointed to 0.5 mm, mostly spherical, bright orange yellow in colour with peripheral clyindrical or mostly saccate paraphyses that collapse radially, wall colourles, 0.5 µ thick, spores 19-31 µ x 16-19 µ, mostly ellepsoid or obovoid, wall 1.5-2 µ thick, pale yellowis or nearly colourless colsely echinulate, germ pores 10-14, scattered pedicels upto 75µ long, telia mostly on abaxial leaf surface and sheath is in linear series, wide open blackish brown in colour, paraphyses brownish in colour, spores 25-47 x 19-22 µ, variable but mostly oblong or clavate, wall 1.5-2.5 µ thick at sides, deep golden brown or chestnut brown, smooth, pedicel colourless to brownish, thin walled, 62-94 µ long, mycelium internal, intercellular, intracellular, cylindrical, profusely branched, binucleate and septate (Figure 16).

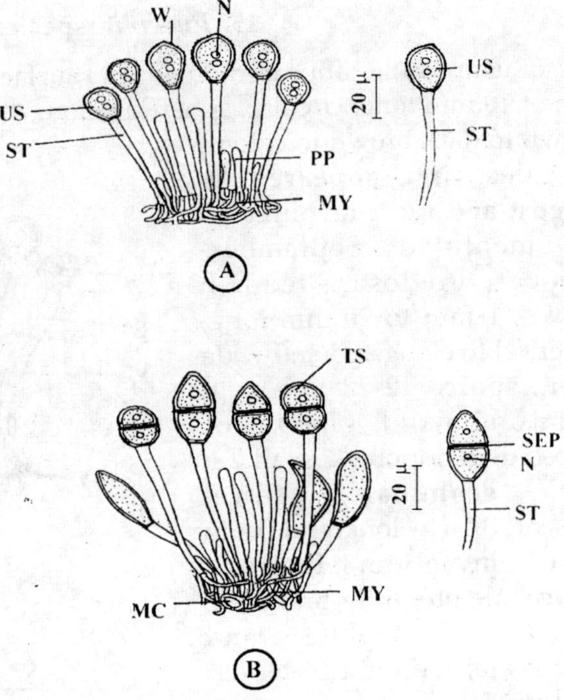

Fig. 16- ***Puccinia*** species-4 showing uredinial (A) and Telial (B) stages on ***Desmostachya bipinnata*** Stapf. ST-Stalk, US-Uredospore, W-Wal, N-Nucleus, PP-Paraphysis, MY-Mycelium, TS-Teleutospore, HC-Host cell, SEP-Septum).

This specimen of *Puccinia* was collected on *Desmostachya bipinnata* Stapf. from Sadipur (Faizabad) on 11.12.1992, leg, CKY-73. This species has been identified only upto generic level. This does not match with any existing species of *Puccinia* on grasses.

17. *Puccinia* species-5

The pustules appeared as black powdery mass on the abaxial surface of the leaves, scattered in the form of patches, 1 mm to 5mm in diameter;uredosori dark brown, densely produced spores 19-28 x 19-28 µ, spherical to obovoid, single celled, binucleate, wall 4-6 µ, smooth, dark black in colour, pores 3-6, scattered, pedicel colourless, upto 70µ long, mycelium internal, intercellular, cylindrical, profusely branched, binucleate and septate; telia not found (Figure-17).

Collected on, *Leptochloa filiformis* (Lam.) Beauv. from Professor's Colony, Station Road, Ayodhya (Faizabad) µ on 11.4.1992, leg. CKY-52. So far, a species of Puccinia has not been reported on *Leptochloa filiformis* Beauv. from India, it is, therefore, reported here as a new host fungus association for country.

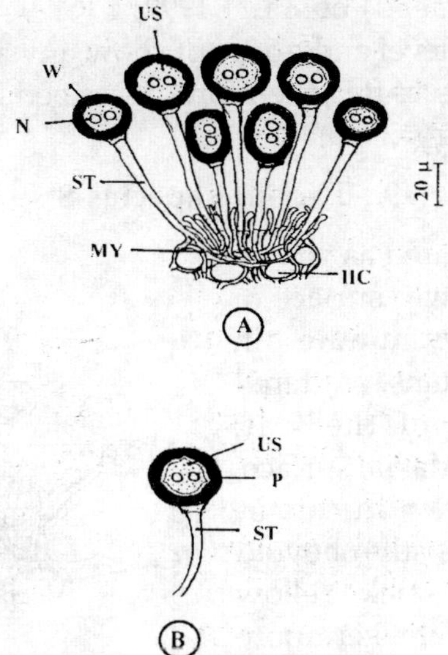

Fig. 17 ***Puccinia*** species-5 from ***Leptochola filiformis*** (Lam.) Beauv. showing uredinial stage (A) and Uredospoe (B): (MY-Mycelium, ST-Stalk, N-Nucleus, Wall, US-Uredospore, HC-Host cell and P-Pore).

18. *Puccinia* species-7

The pustules appeared on both the sides of the leaves, spherical to elongated. spots pin head size to 1mm, scattered on whole leaf, arranged in streaks upto 10 mm long; sori yellowish coloured when young and red brown at maturity, number of spots more frequently distributd on abaxial than the adaxial surface of leaf, frequently found from August to May; uredosori yellowish to reddish brown, spherical to slightly raised, wide open when matured, spores densely produced 25-31 μ x 22-28μ, mostly broadly oval or obovoid, wall 1.5-2.5 μ thick, golden or pale cinnamon brown, echinulate, germpores 4, rarely 3 equatorial; pedicels profusely branched, binucleate and septate (Figure- 18).

Collected on *Oropetium thomaeum* (Linn). Trin. from K.S.S. P.G. College,

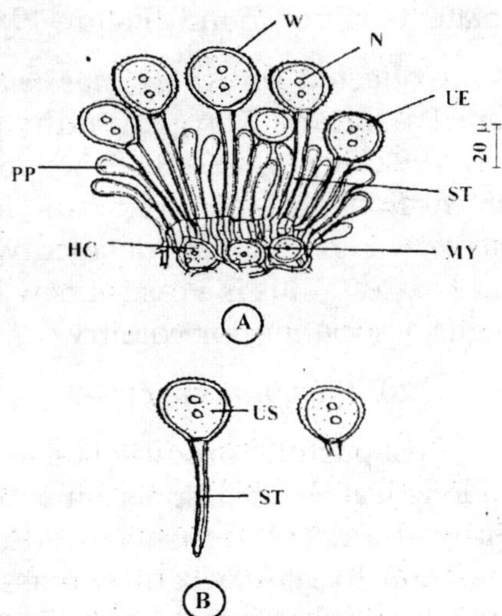

Fig. 18- ***Puccinia*** species-7 showing uredinial stage (A) and uredospore (B) on ***Oropetium thomaeum*** (Linn.) Trin. (HC-Host cell, PP-Paraphysis, W-Wall, NUCLEUS, US-Uredospores, ST-Stalk and MY-Mycelium).

Campus, Ayodhya (Faizabad) on 15.11.1992; CKY-94. This species has been idenified only to generic level and does not show any affinity with any known species of *Puccinia* on grasses It is, therefore, described here as new host fungus association for the country.

19. *Puccinia* species-8

Rust pustules appeared as yellow powdery mass on the lower surface of the leaves, pustules very minute pin head size, mostly near veins, scattered on the whole surface of the leaf; uredosori mostly on the adaxial surface, golden to yellowish brown in colour, spore 12-22 x 12-16 μ, mostly obovoid or ellepsoid, wall 1-1.5 μ thick. Yellow to pale brownish, echinulate, germpore 3-4, scattered,; pedicillate, pedicel colourless, mostly broken near the spore, 33-47 μ long, and 2-3 μ wide; mycelium internal, intercellular, cylindrical profusely branched, binucleate and septate; telia not found (Figure-19).

Collected on *Ruta graveolens* L. from Parikrama Marg, Ayodhya on 11.12.1992; leg. CKY-88. This species has not been reported on *Ruta graveolens* L. so far. The detailed work is in progress. This is a case of new host fungus association for country.

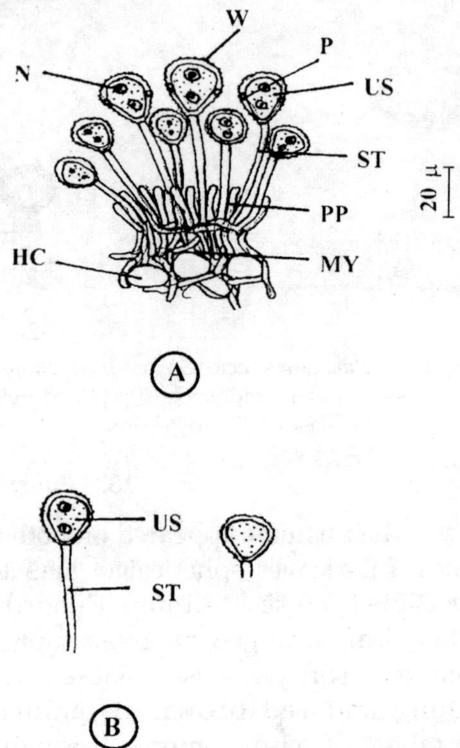

Fig. 19- ***Puccinia*** species-8 on ***Ruta graveloens*** L. showing uredinial stage (A) and Uredospore (B): (HC-Host cell, N-Nucleus, W-Wall, P-Pore, US-Uredospore, ST-Stalk, PP-Paraphyses and MY-Mycelium).

20. *Puccinia* species-9

Rust pustules produced on leaves and long leaf sheaths, spots 2 mm to 15 mm in diameter, densely produced yellowish to redish brown mass on abaxial surface of leaf and leaf sheaths, also arranged in rows, in severe cases the sori appeared on both the surfaces rust present on this host throughout the year, heavy infection appeared in the month of August to September and January to February,only uredinia were obtained on the host, uredosori present on both the surfaces of the leaves but very common on adaxial surface, spores are yellowish brown to reddish brown in colour, spore 25-62 x 19-35μ, broadly ellesoid, broadly obovoid or globoid, wall upto 5μ thick, pores 4-5 scattered echinulate, pedicel present but broken near the spore wall thin and uniform, pedicel 16-56 μ long and 6-9 μ wide, colourles to light yellow. Mycelium internal, intercellular, cylindrical, profusely branched, septate and binucleate; telia not obtained (Figure-20).

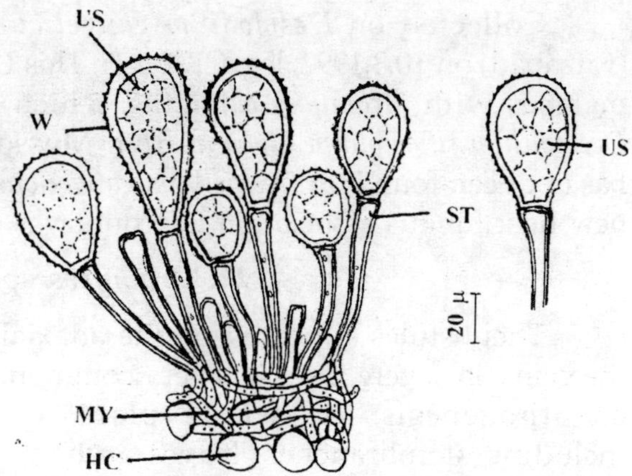

Fig. 20- *Puccinia* species-9 form ***Saccharum*** species showing uredinial stage (HS-Host cell, MY-Mycelium W-Wall, US-Uredospore, and ST-Stalk).

Collected on ***Saccharum*** species (locally known as **Ramsar**) from Asharfabad throughout the year. Leg. CKY-9. This species of Puccinia (Species-9) does not match with any existing species of ***Puccinia*** on ***Saccharum.*** The uedospore size is usually much larger than any of the known species of Puccinia on this host. It is for time being, has been narrated here as a new fungus host association for India.

21. *Uromyces capitatus* Syd.

(Sydow, H., Milton. J.H. and R.N. Tadon, Fungi Indiae-III, Mycol, **35**:222-243, 1937)

Pustules appeared on the abaxial side of the leaves, The sori were light brown to golden brown in colour, uredosori were densely produced, 22-31 x 12-16 μ, single celled, binucleate, 1.5-2.5 μ thick, yellowish or dull brown, echinulate, pores 3-4, equatorial, pedicels yellow to brownish, thick walled upto 47μ long, mycelium internal, intercellular, cylindrical profusely branched, binucleate and septate; telia not found (Figure-21).

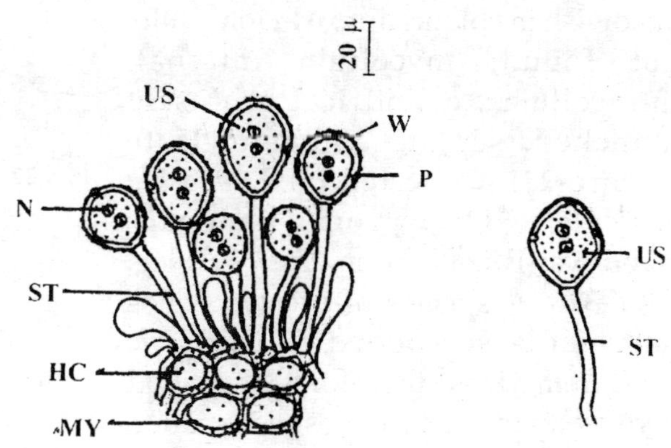

Fig. 21- ***Uromyces Capitatus*** Syd. from ***Desmodium gangeticum*** DC. showing uredosorus and uredospore (MY-Mycelium, HC-Host cell, ST-stalk, N-Nucleus, US-Uredospore, P-Pore, and W-Wall).

Collected on *Desmodium gangeticum* DC from Halkara Ka Purwa (Faizabad) on 10.3.1992, leg, CKY-295, This Uromyces species in greater details matches with *Uromyces capitatus*, which has been reported in India on *Desmodium tiliaefolium* G. Don from Mussoorie and Nainital, U.P. So far, it has not been found on *Desmodium gangeticum* it is, therefore, regarded here as new host fungus association for country.

22. *Uromyces* species-2

Thepustules appeared on the adaxial side of the leaves, mostly neary the veins, in severe cases it is very common on stem and branches. As a result of pthogenesis the whole plant including stembranches, leaves and inflorescence become deformed, hypertrophied and much branched. The pustules ranging from 1 mm to 4 mm and yellowish to reddish brown in colour. Uredosori yellowish, scattered or circular and coalesce to form irregular patches on adaxial side of the leaves, raised; spore 28-44 x 19-28 μ, mostly ellepsoid or broadly ellesoid, single celled and binucleate, wall 2-4 μ thick, hyaline or verey aple brownish, echinulate, pores verey obsure, scattered, pedicel thick walled, light yellowish in colour, upto 64μ long; telia not found; mycelium internal, intercellular, cylindrical profusely branched. septate and binucleate (Figure-22). Collected on *Euphorbia geniculata* Ort, from Ranopali, Ayodhya (Faizabad) on 25.2.1993, leg. CKY-326. A species of *uromyces* has not sofar been reported on *Euphorbia genicultata,* it is, therefore, described here as a new fungus host association for country.

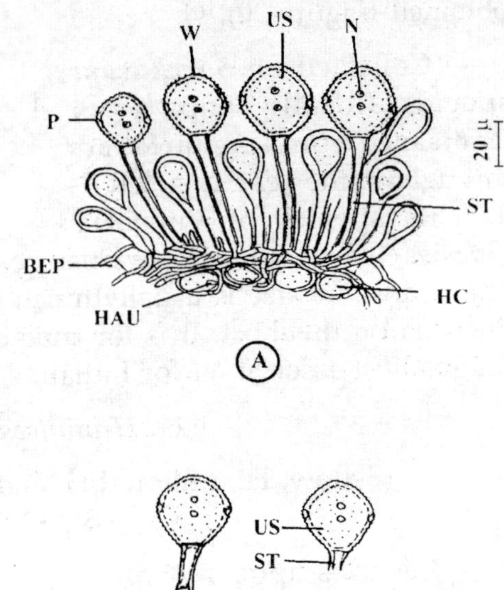

Fig. 22- ***Uromyces*** **species-2 from *Euphorbia geniculata* DC.** showing uredinial (A) stage and uredospore (B): (HAU-Haustoria, BEP-Broken epidermis, P-Pore, W-Wall, US-Uredospore, N-Nucleus, ST-Stalk and HC-Host cell).

23. *Uromyces* species-3

Uredinia not found; teleutosori appeared on abaxial side of the leaves, reddish brown, spores 28-34 x 16-19 μ, dark blackish brown in colour,

thick, yellowish, echinulate, pores single celled uninucleate, ovate or elliptical, mostly elongated and narrower at apx, wall 2-4μ thick, smooth pedicel colourless, thick walled, not collapsing, upto atleast 37μ long (Figure-23).

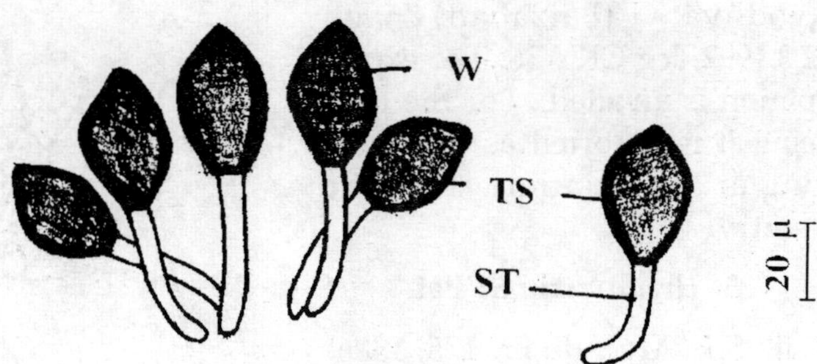

Fig. 23. ***Uromyces*** species-3 from ***Sonchus arvensis*** Linn. Showing Teleutospore: (W-Wall, ST-Stalk and TS-Teleutospore).

Collected on Sonchus arvensis Linn. from Professors Colony Station Road, Ayodhya (Faizabad), on 18.2.1993, leg. CKY-327.

24. *Uromyces* species-4

Rust pustules appeared on chlorotic regions of abaxial leaf surfaces, spots very minute, light yellow to red brown in colour; uredosori on abaxial leaf surface, yellowish; uedospores 16-19 x 12-16 μ, single celled binucleate, globoid or obovoid, wall 1,5-2 μ thick, yellowish, echinulate, pores 3-4 scattered, pedicels colourless, upto 16 μ long and 3 μwide; telia on abxial side of the leaves, dark brown to black in colour, spores 25-31 x 19-22μ, single celled and uninucleate; ellepsoid to oblong ellepsoid, wall 2-4 μ thick, tan black smooth; pedicels colourless, upto 36 μ long; mycelium internal, intercellular, cylindrical, profusely branched, binucleate and septate (Figure-24).

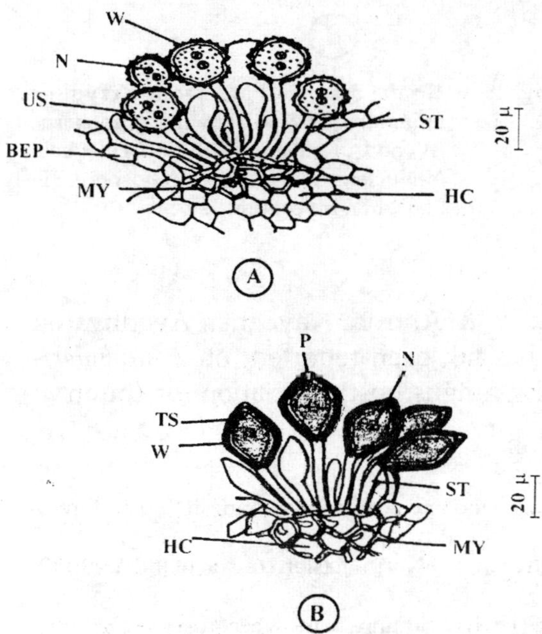

Fig. 24- ***Uromyces*** species-4 from ***Ageratum conyzoides*** showing uredinial (A) and telial (B) stages: (MY-Mycelium, BEP-Broken epidermis, US-Uredospore, N-Nucleus, W-Wall, ST-Stalk, HC-Host cell, TS-Teleutospore, P-Pore).

Collected on *Ageratum conyzoides* from Professors Colony, Station Road,

Ayodhya (Faizabad) on 12.3.1992, leg CKY-36. The expert opinion is awaited., For the time being it is reported as new host fungus association for the country.

25. *Uredo zizyphi* Pat.

(Bull. Soc. Myc. de Fr. 135, 1886)

The rust pustules appeared on adaxial surface of the leaves, rusty brown in colour, pin head to 2 mm in diameter, circular to elongated, at severe stage the spots are frequently distributed on both the surfaces of the leaves, pustules appeared mostly near the mid rib and vein; uredospores 12-43x 12-22μ, single celled, binucleate, ellepsoid or broadly ellepsoid, wal 2-3 μ thick, pores 3-4 equatorial, yellowish brown, pedicel smooth, thin walled upto 34 μ long; mycelium internal, intercellular, cylindricl profusely brached, binucleate and septate (Figure 25).

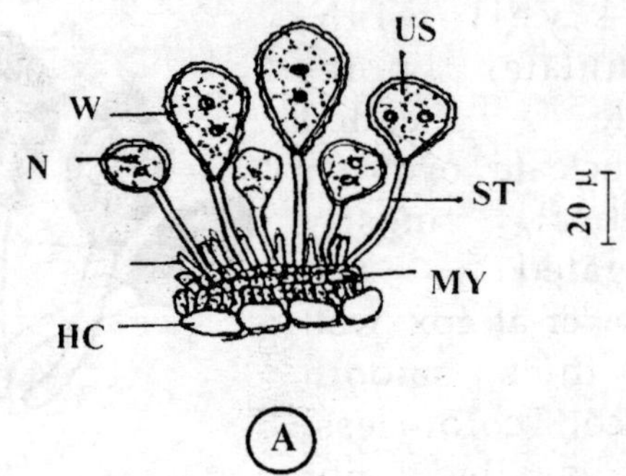

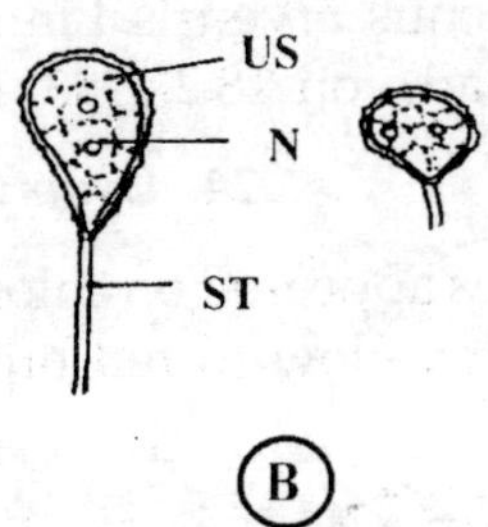

Fig. 25- *Uredo zizyphi* Pat. from ***Zizyphus numularis*** W.& A.Showing uredosorus (A) and uredospore (B): (HC-Host cell, N-Nucleus, W-Wall, US-Uredospore, ST-Stalk, MY-Mycelium).

Collected on *Zizyphus numularis* W. & A. from Nayaghat, Ayodhya on 10.3.1993, leg. CKY-215. This species has not been reported on *Z. numularis* W. & A. in India. It is, therefore, a new fungus host association for country.

References

Ahmad, G.U. (1981). Studies on rusts from Nagaland and Arunanchal Pradesh. Indian Phyto-pathol, 34:240-241

Ahmad, G.U. (1989). A taxonomic survey of rust fungi of North- Eastern region of India. 42:149-150.

Ahmad, S.T., Singh, Sheodhan and Mishra, D.P. (1969). Additions to the wheat rust races in India. Races 14 and 38 of *Puccinia striiformis* identified during 1965. **Ibid**, 22:522-524.

Arthur, J.C. (1907). Cultures of Uredineae in 1906. J.Mycol., **13:** 189-205.

Bayanarayana, G and Ravinder, E.J. (1988) Notes of two Indian species of *Ravenalia*. Trans. Brit. Mycol. Soc., **90**:477-479.

Barclay, A. (1889-1890). Descriptive list of Uredineae occurring in the neighbourhood of Shimla (Western Himalayas).1. J. Asiatic Soc. Bengal, **56**: 350-375, 1987; III, **Ibid, 59**:75-112.

Berkeley, M.J. (1936) Fungi. IN. English Flora. J.E. Smith (ed.).

Bilgrami, K.S., Jamaluddin and Rizvi, M.A. (1979). Fungi of India Part- I. List and References. (Today and Tomorrow Printer and Publishers, New Delhi,. pp.467)

Bilgrami, K.S., Jamaluddin and Rizvi, M.A. (1981). Fungi of India Part-II, Host Index and Add enda, pp 140 + 128, Ibid, New Delhi.

Butler, E.J. (1914). Notes on some rusts in India. Ann. Mycol., **12**: 76-32.

Butler, E.J. and Bisby, G.R. (1931). Fungi of India Sci. Monogr.-1, Imp. Counc. Agr. Res., India, Calcutta.

Cummins, G.B. (1971). The Rust Fungi of Cereals, Grasses and Bamboos. Springer-Verlag Berlin, Heidelberg. New York, pp. 569.

Cummins, G.B. (1983). IIlustrated Genera of Rust Fungi (Rev.Ed.) Amer. Phytopath. Soc. Meinneapolis, USA.

Daleala, G. and. Sinha, S. (1964). Further studies of biometry of uredospores: *Puccinia* **Proc. Acad. Sci.**, India, Sec. B, **7**: 245-250.

Fischer, G.W. (1953). Some of the 125 rusts of grasses. In : *Plant Diseases* the year book of Agric., p. 276.

Flor, H.H. (1953). Epidemiology of flax rust in North- Central States. Phytopathology, 43:624-628.

Goswami, R.N. and Singh, K.L. (1973). Uredinales of North-East . Ind. Phytopath. **26**: 310-314.

Joshi, L.M. (1976). Recent contribution towards epidemiology of wheat rust in India. *Ibid,* **29** (1): 1-16

Kanaujia, R.S.and Singh, C.S. (1977). Some new host records of parsitic fungi from. Faizabad.-II, **Ibid, 30**: 115.

Kanaujia, R.S. and Singh, C.S. (1978). Some new host records of parasitic fungi. I. *Ibid,* **31**: 218-219.

Kanaujia, R.S. and Rajkishore (1981). Annonated list of fungi of Faizabad. Indian J. Mycol. *Pl. Pathol.,* **11**; 292-293.

Kanaujia, R.S. and Yadav, C.K. (1993). Rust fungal flora of Faizabad.Abstr, Zonal IPS Meeting at Gorakhpur on 18-19 Dec. 1993.

McKenzie, E.H.C. (1981). New Zealand rust fungi: Additions and corrections. Newzealand J. Bot., **19** (2): 227-232.

Mehta, K.C. (1940). Further studies on cereal rusts in India.I. *Sc.* Monogr. Council Agric. Res., Indian Indian Phytopathol, **14**:224 pp.

Mehta, K.C. and Prasad, R. (1951). *Further studies on cereal rusts in India.* II. I.C.A.R. Monogr., New DELHI. Rust of wild grasses. Curr. Sci., **20**:243.

Sarbhoy, A.K., Varshney, J.L. and Agrawal, D.K. (1980). Fungi of India-1971-1976. Navyug Trader, New Delhi.

Sarbhoy, A.K., Agrawal, D.K. and Varshney, J.L. (1986). Fungi of India, pp 273, S.K. Dutta Associated Publishing Company, New Delhi.

Sethe, A.V. (1969) A new rust reocerd of *Pennisetum typhoides* Stapf & Hubb. Mycologia, **61**:198-199.

Stakmen, E.C. (1914) A study in cereal rusts. Physiologic forms. Minn. Agric. Exp. Stat. Bull. p.138.

Sydow, H. and Butler, E.J. (1906). Fungi Indiae Orientalis Part. Ann. Mycol., **4** : 425-455.

Vasudeva, R.S. (1960) *The Fungi Of India.* Rev. by E.J. Butler and G.R. Bisby I.C.A.R., New Delhi.

Yadav, A.S. (1953) A contribution to the knowlege of Uredineae of Bihar Indian Phytopathology, **6**:83-91.

Yadav, C.K. (1993) Sudies of Rust Fungi of Faizabad With Special reference to Cereal Rust., PH. D. Thesis, Avadh University, Faizabad

Zhuang, J.V. (1989) Species of *Puccinia* on Liliflorae in China. Mycosystema, **2** : 175-198.

16

Diversity and Systematics of Agaricales of Central India

R.C. Rajak, Deepak Rahi, Kamlesh Shukla and A.K. Pandey

Department of Biological Science,
R.D. University, Jabalpur- 482001 (M.P.)

Central India with its rich forest wealth harbours a number of micro- and macroscopic fungi. This chapter deals with diversity, systematics and qualitative composition of Agaricales from Central India which enriches our knowledge in fungal biodiversity.

Key-words: Diversity, systematics, Agaricales, Central India

Diversity or Biological diversity is the wealth of life on earth, the millions of plants, animals and microorganisms, the gene they contain and the intricate ecosystem they help build in to the living environment. All these species used directly or indirectly for food and agriculture , feed for domesticated animals and also provide essential raw materials and services for life support such as fiber, fuel, fertilizer and pharmaceuticals. Thus, biodiversity makes the ecosystem stable, functional and environmentally sustainable.

Fungi are vital to the function and maintenance of earth's ecosystem of biosphere. As major contributors in biogeochemical cycles, they perform unique and indispensable activities in circulation of matters in the world on which all other organisms, including humans, depend. It has been reported that less than 3% of the world's fungi have yet been described and it is not improbable that their real number on earth exceeds even that of insects (Hawksworth, 1991). That total number of fungi described so far, half may fall into the category of larger fungi, mainly belonging to basidiomycetes.

Agaricales- Status of Systematics: International Scenario

In past, the systematists were unable to recognize fungi as a special group of plants and named them in arbitrary manner. Before Linnaeus nomenclature, mushrooms were placed under generic name Fungus and the name *Agaricus* was used by various systematists for polypores and some lamellate bracket fungi. Linnaeus (1753) in his *Species Plantarum* placed fungi alongwith higher plants and took up the name Agarics in modern sense for gill bearing hymenomycetes and Boletes for fleshy,pileate porate fungi.

The word 'Agaric' has been derived from the Greek word 'Agaricon', meaning the gilled fungi and bolete (bolos-a globular mass, a king of mushroom) constitute Agaricales (hymenomycetes). Agaricales is characterized by hymeniun covered initially by volva or annulus and exposed at maturity (Rea, 1922) or fleshy fruit body consisting of monomitic inflated hyphae with transverse nuclear spindle in the basidium (Webster, 1970) and excluding Aphyllophorales and Gasteromycetes. The limits of Agaricales have been well defined by Singer (1986) . It comprises of 3 suborders, 17 families, 230 genera and approximately 5000 species. Some more species have also been added to the above list of species of Agarics by latter workers.

According to latest International Code of Botanical Nomenclature (Greuter *et al.*, 1988), all publications in Agaricales are considered valid from Linnaeus (1753) onwards. The workers previous to Linnaeus (1753) were aware of the Agarics. They grouped all the lamellated forms in genus Fungus. Friesian (1821-1832) system of classification of Agaricales remained predominant in the nineteenth century and first quarter of twentieth century. In his three volume work of "Systema Mycologicum"he grouped Agarics under Hymenini of class hymenomycetes. He divided Hymenini into two sub orders viz., Clavati and Pileati. The Agarics alongwith *Polypores*, *Thelephores* and *Hydnum* were classified under sub-order Pileati. On the basis of the spore colour he segregated the large genus *Agaricus* into 5 series:-

Spore white	Series I	:	Leucospori
Spore rosy red	Series II	:	Hyporphodic
Spore rusty red	Series III	:	Dermini
Spore purplish black or dark in colour	Series IV	:	Pratellae
Spore black	Series V	:	Coprinarii

Although Friesian system of classification was purely artificial, it has been proved useful towards a natural classification with regard to the Agarics. Many earlier workers contributed significantly to Agaricales. Ainsworth (1971) recognized 220 genera, 3250 species and only two families in Agaricales. Smith (1973), Schaefer (1975a), Pegler and Young (1979)also made good contributions in Agaricales. Hawksworth *et al.* (1983) recognised 230 genera and 400 species under Agaricales in the Dictionary of fungi. From these account it appears that the various treatments suggested for Agaricales is still controversial. It is likely that the system would gradually stabilize in coming years.

National Scenario

The systematics of Indian Agarics dates back to the period of British regime in India. The first record of Agaricales from India was of *Fomes pectinatus* by Klotzsh (1833) which was identified later on as *Polyporus pectinatus* of family Polyporaceae in 1839 followed by *Trogia montagnei* (Montagne, 1842) from Nilgiri hills (T.N.) and *Lentinus javanicus* (Leveille, 1844) from Bombay (Maharashtra).

Thus, upto the mid of nineteenth century major additions of Agarics from India were made by imperial mycologists. Later on individual and isolated reports have been published on the systematics of Agarics by Indian workers. (Bose 1919, 1920b, 1921, 1949; Natarajan, 1975, 1977b, 1978a,b, 1979; Rawla *et al.*, 1982, 1983, 1987; Saini *et al.*, 1982, 1988, 1991; Manjula 1983; Lakhanpal *et al.*, 1987, 1988). Bilgrami *et al.* (1991) listed 896 species of Agarics from India to which a few more species have been added by the latter workers.

Agaricales of Central India

Agarics are an important component of the forest ecosystem occupying diverse ecological niches. They appear mostly during the Monsoon season and post Monsoon season in the months of October to January. Mushrooms prefer to grow in a variety of habitats, some in open while the others require sheltered shady grounds. Some grow under or on certain kinds of trees, while others are found in mixed woods. Many of them form ectomycorrhizal relationship with forest plants. Mushrooms are mostly saprophytic growing on the ground, decaying leaves, bark, wood, manure etc. Some mushrooms are pathogenic and attack the roots and trunks of trees causing diseases. They also fasten the decomposition process by growing on dead plants due to their unique enzymatic weapons especially lignolytic and cellulolytic enzymes thus adding to the fertility of the soil and recycling of elements.

In recent years there has been an increasing interest in the Agaricales due to various reasons. Unfortunately many parts of the country including central India are still under explored. Keeping this in view, an attempt has been made to explore the lesser known, unexplored Agarics of central India. Extensive,systematic and periodical survyes of various regions especially tribal regions were made to recover the Agaric flora. Consequently it was observed that the tropical forests of central India abounds enormous Agarics and many of them are on the verge of extinction due to habitat destruction by various degrading activities of human beings. The prominent families of the order Agaricales encountered in this central part of India have been described (Rahi, 2001; Shukla, 2001).

Agaricaceae Fr.

Out of 25 genera reported from all over the world (Singer 1986) , Manjula (1983) listed 11 from India. This family is represented in this region by three important genera: *Agaricus, Lepiota* and *Leucoagaricus.*

Agaricus L.ex. Fr.

The taxonomy of the genus has extensively been discussed by Singer (1986). The colour and homogeniety of the spore wall and the lack of clamp connections in the cortical layers define this genus completely. Singer (1986) reported 35 species from all over the world. Manjula (1983) reported 19 Indian species of *Agaricus* while Bilgrami *et al.* (1991) included 24 species. Besides, the genus has been also studied by various other Indian workers including Vrinda *et al.* (1997), Pahil *et al.* (1999) and Saini *et al.* (1997).

In our studies five species of the genus viz., *A campestris, A. basianulosus. A placomyces, A. bisporus* and *A. bitorquis,* have been encountered. All the species reported here are terrestrial, growing on humus rich soil in solitary to gregarious manner. *A. campestris* is characterized by globose to broadly campanulate pileus with ovoid spores while *A. basianulosus* has white glabrous pileus with conspicuos clyindrical stipe. *A. placomyces* possesses pileus with sooty brown squamules, while two spored basidia are the main characteristic of *A. bisporus. A. bitorquis* is characterized by convex pileus with slight umblicate centre, four spored basidia and subglobose spores. It was observed that the species recovered possess slight variations in morphological features from species reported by earlier workers which might be due to differences in ecological conditions. Only few species are slightly poisonous at times, others edible.

Lepiota (Pers. ex.) S.F. Gray

The taxonomy of the genus has been extensively reviewed by Singer (1986) who reported 37 species. The genus has also been studied by various workrs including Banerjee (1947) and Atri *et al.* (1996, 2000). None of the species of *Lepiota* is used for foods nor sold in markets. However, there are atleast three poisonous species which may be confused with edible ones. The genus is characterized by non-repent, non filamentous hyphae, with or without clamp connection, pluteoid habit, annulate veil and scales on the surface of pileus, spores pseudoamyloid,usually grow on the soil and on various plant debris, living or dead.

In central India this genus is represented by five species viz., *L. longicauda, L. mimica, L. procera, L. subclypeolaria* and *L. sistrata. Lepiota longicauda* is characterized by subglobular spore shape, 6-8 µm in diameter, white to cream coloured umbonate pileus while *L.mimica* possesses a strong unbonate pileus. Brown coloured pileus with white lamellae and spore dimension of 1.6-2.2x3.3-4.4 µm is the characteristic feature of *L. procera* while *L. subclypeolaria* is characterized by 7 cm elongated stipe length. *L. sistrata* possesses white, convex pileus with pinkish scales and its diameter is not more than 2cm. These species slightly differed in their colour and morphological features with species described by earlier workers which might be due to some geographical variations. None of the reported species possess edibility.

Leucoagaricus (Locqueir) Sing.

The genus was studied by various workers in India. Manjula (1983) included four species while Bilgrami *et al.* (1991) listed only *L. exoriatus.* The genus is characterized by pure white spore print, smooth or ornamented spore wall, always metachromatic in cresyl blue, hyphae without clamp connection, surface of the pileus scaly, fibrilose or pubescent, volva none. Usually grow on soil, saw dust, dung or on decayed wood. The genus is intermediate between *Macrolepiota* and *Leucocoprinus.* The absenceof clamp connections in the trama of pileus and stipe makes it possible to distinguish it from *Macrolepiota.* The distinction of *Leucoagaricus* from *Leucocoprinus* appears to be simple because of the long pectinate pileus of the latter.

During our studies, one species *Leucoagaricus fuligineus* was found growing on forest litter in solitary to gregarious manner with characteristic greyish brown, fibrillose squamules on pileus surface. The species has also been reported from Nilgiri by Manjula (1980). The macroscopic and microscopic features of the species resemble well with *L. fuligineus* as described by Pegler (1977) except slight variations in pileal colour.

Amanitaceae Roze

The family is represented by more than hundred species of the genus *Amanita* world over (Singer, 1986) and in India it is represented by 28 species.

Amanita Pers. ex. Hooker

Lakhanpal (1997) recovered 28 species from North India. Manjula (1983) included 9 species and Bilgrami *et al.* (1991)listed only 5 species from India. The genus is characterized by sulcate pileal margin, inamyloid, smooth spores, alternating lamellae, lamellulae free, spore print white, basidia voluminous, cystidia absent, rarely present, hymenophoral trama bilateral, stipe central, volvate, content fleshy, hyphae with or without clamp connections. It appears on the ground, very rarely on other substrates, usually in the forests, almost consmopolitan in distribution.

During our studies seven species of *Amanita* viz, *A. caesaria, A. fulva, A. gemmata, A.pachycolea, A. pantherina, A. vaginata* and *A. verna* were collected. Out of theses only *A. caesaria* was found to possess edibillity. The species possess conspicous yellow-orangish pileus with well developed volva and yellowish lamellae. *A. fulva* was characterized by reddish chocolate brown to greyish brown pileus, covered with white to yellowish white patches with white annulus and scaly volva. *A. pachycolea* has dark brown to yellowish brown pileus, covered with white to creamy warts. *A. pantherina* is characterized by smoky brown to brown pileus with whitish warts, white annulus and smooth volva and *A. vaginata* has a dull grey to smoky grey pileus while white to cream coloured pileus with saccate, white, cup shaped volva are the main characteristics of *A. verna.* All the mentioned species were found to grow terrestrial in solitary to scattered manner.

Cantharellaceae. J. Schroet

Donk (1964) recognised four genera in this family viz, *Cantharellus, Goosensia, Pseudocraterellus* and *Craterellus,* whereas Corner (1966) in his monograph on Cantharelloid fungi included *Arrhenia, Cantharellula, Cantharellus, Caripia, Craterallus, Dichantharellus, Geopetalum, Gloecantharellus, Gomphus, Goosensia, Hygrophoropsis, Leptoglossum Mycenella, Nyctalis, Pseudocraterellus, Pterygellus Rinbachia* and *Trogia* in family Cantharellaceae. Singer (1986) treated *Cantharellus* and *Craterallus,* the two traditional genera of the family as 'Genera excludenda' ET in 'Cretaesedes', that is genera which belong somewhere else but have been confused with Agaricales. In central India the family is represented by the genus *Cantharellus* only.

Cantharellus Fries

The genus *Cantharellus* has been extensively studied by many workers including Manjula (1983), Purkayastha and Chandra (1985). Bilgrami *et al* (1991) listed four species of *Cantharellus* from India. Lakhanpal (1997) reported three Indian species. The genus is characterized by thin to fleshy basidiocarps, white or coloured pileus, lamellae decurrent, sometimes anastomosing folds, often forked dichotomously, whitish or brightly coloured or greyish to violaceous, stipe solid, spores usually white, mostly humicolous, some ectomycorrhizal, almost cosmopolitan in distribution.

During our investigations *Cantharellus minor* has been collected. The species possesses excellent edibility among local tribals and inhabitants of the region and is characterized by 1.0-2.5 cm pileus diameter, thin context, solid stipe, hollowing with age, and spores having dimensions of 7.0-10.0 x .4.0-7.0 µm. The species was found involved in ectomycorrhizal association with Bamboo plants (*Dendrocalamus strictus*).

Coprinaceae Roze

The family is represented by seven important genera. In central India it is represented by the genus *Coprinus* only.

Coprinus (Pers.ex.) S.F. Gray

The taxonomy of the genus has been well discussed by Singer (1986), who reported 70 species from all over the world. The best and modern treatment of the genus is found in the keys of European species particulary Kuhner & Romagnesi, Lange (1952) which clearly defined the generic concept of the fungus. Several are harvested by amateurs every year as long as they are very young and are considered as first class edible mushrooms. About twenty one species of the genus have been recorded frequently from different parts of India (Banerjee, 1947; Lange and Smith, 1953; Ghosh *et al.*, 1974). The genus is characterized by conical or campanulate pileus when young, expanding on maturity, margin deeply plicate, spores reticulate hymenium consisting of isolated basidia, spore print black. Usually found growing on dung, soil, peat, around the living trunk or on dead wood.

During our investigations, two species viz, *Coprinus comatus* and *C. niveus* have been recovered from central India. *C. comatus* is characterized by terrestrial habitat, white, pinkish or brown with recurved scales, large pileus with 6-15 cm length and spore dimensions of 10-14x6-8 µm *C. niveus* is characterized by spores with dimensions of 15-19x11-13 µm. These species were found growing abundantly during rainy season on dead organic matter and on forest litter.

Pluteaceae Roze

The family is represented by three important genera viz. *Volvariella, Chamaeota* and *Pluteus*. Of these, the genus *Volvariella* represent the family in central India.

Volvariella Speg.

Singer (1986) described 22 species from all over the world. The genus was also studied by many workers including Manjula (1983) who reported 13 Indian species and Bilgrami *et al.* (1991) listed 15 species. Atri *et al.* (1996)

described only one species from Punjab. The genus is of much economic importance as *Agaricus* and is characterized by pluteoid habit of the carpophore, inverse hymenophoral trama, pink ot brownish pink spore print, basidia normal, cystidia often present, stipe central, always with a distinct membranous volva at the base, hyphae without clamp connections. Usually found on soil and on wood, dung or on straw.

In Central India the genus *Volvariella* is represented by three species viz, *V. volvacea, V. media and V. speciosa. V. volvacea* is characterized by dull brown, snuff brown to date brown or cream pileus with spore size of 7-10 x 5-6 µm.*V. media* possess whitish pileus having 3-5cm. diameter and white volva. White to pale grey pileus with diameter of 7-15cm and membranous volva are the important characteristic features of *V. speciosa.* All these species were found growing solitary to scattered on grassy land as well as on Sal (*Shorea robusta*)litter during rainy season. Out of these, only *v. volvacea* was found edible and quite popular among tribals of this region.

Polyporaceae (Fr.) Fr.

The traditional genera of the family are *Pleurotus, Lentinus* and *Panus.* These genera have been shown to be artifical in their classical limits. During the last decades, several mycologists have contributed to their dismemberment,leaving a comparatively, small nucleus of species within each of the old genera. However, the delimitation of the three genera from each other and the choice of nomenclatory type in each of the various generic taxa has been any thing but unanimous. Two approaches were possible, both anatomical, either the presence or absence of metuloids could be made with the major distinguishing character between *Panus* and *Pleurotus* with *Lentinus* *m*aintained according to the traditional macroscopical differences, or the delimitation. After checking on all the key species and a large number of additional types, the decision was made in favour of the second principle. In central India we collected only two genera viz. *Lentinus* and *Pleurotus.*

Lentinus Fr.

Singer (1986) reported 14 species of *Lentinus* from all over the world. Besides, the genus was also studied by various workers including Roy and Samajpati (1978), Manjula (1983) who incorporated 17 Indian species, Bilgrami *et al.* (1991) listed 33 species from India. The genus is characterized by pure white spore print, pileal surface either scaly or even, fibrillose, pubescent or glabrous, hyphae without clamp connections, grows on soil, saw dust and decayed wood. Almost cosmopolitan in distribution.

During our investigation one important species, *Lentinus cladopus* has been recovered from central India. The species possess excellent edibility among tribals of the region and found growing on dead log during rainy season. The species is characterized by slender basidiocarp, depressed pileus 2-5 cm diameter, stipe often bifurcated at the base and spores with dimension of 6.5-8.5 x 3.7-4.5 µm.

Pleurotus (Fr.) Quel

The taxonomy of the genus *Pleurotus* has been studied by Singer (1986) who reported 38 species from all over the world. Manjula (1983) listed 12 Indian species of *Pleurotus* while Bilgrami *et al.* (1991) included 21 species from India. Besides, the genus has been frequently recorded from various regions of India by many other workers. It is characterized by pleurotoid habit, vividly coloured pileus, lamellate hymenophore, hymenophoral trama irregular, spore print white or cream coloured, basidia normal, cheilocystidia usually present, stipe present, more rarely absent, hyphal system mono, di or lamphimitic growing usually on wood. During our studies three important species of *Pleurotus* viz, *P. ostreatus*, *P. sajor caju* and *P. florida* have been recovered. It was observed that of these three species, *P. ostreatus* possess purplish tint on pileus with spores having dimension of 10-13x4.5-5.0 µm. *P. sajor caju* was characterized by densely crowded lamellae and fusoid to ventricose cheilocystidia. *P. florida* possess close lamellae and clavate cheilocystidia. Sporophores of all the three species was generally found growing in clusters or in gregarious manner on the logs of Sal tree (*Shorea robusta*). However, the species differed slightly in their shape with described species.

Russulaceae Roze.

Singer (1986) retained the family Russulaceae in Agaricales and included two genera, *Russula* and *Lactarius.* Pegler and Young (1979) on the basis of ultrastructural investigations recognised Russulales as an autonomous order, with two families, Russulaceae and Etasmomycetae. In the family Russulaceae they included three more genera in addition to *Lactarius* and *Russula.* The conventinal two generic concept is being followed here. The family is represented by two genera, *Russula* and *Lactarius* in central India.

Russula Pers. ex. S.F. Gray

The taxonomy of the genus has been studied by Singer (1986). He reported 317 species from all over the world while Atri and Saini (1989) included 337 species. Manjula (1983) and Bilgrami *et al.* (1991) listed 13 Indian species. Atri *et al.* (1994) in "Indian Russulaceae Roze-A check list" listed 67 species of *Russula* from India. The genus is characterized by vividly coloured pileus, heteromerous trama, spores with amylaceous ornamentations, stipe fleshy, hymenophore lamellate, lamellulae present, free to decurrent, hymenophoral trama subregular to regular or intermixed in *Russula* mostly intermixed with numerous sphaerocysts. Spore print white-cream, stipe usually central, rarely on living or dead plant material, mostly obligatory ectomycorrhizal, fewer species facultative ectomycorrhizal or non mycorrhizal. During our investigations 22 species of *Russula* have been collected from central India. These are *R. albida, R. breviceps, P. cyanoxantha, R. decolorans, R. delica, R. densifolia, R. emetica, R. farinipes, R. foetens, R. illota, R. laurocerasi, R. lepida, R. lilacea, R.rosacea, R. rosea, R. rubsescens, R. rubicunda, R. sanguinea. R. sororia, R. sub-foetens, R. violeipes and R. virescens. R. albida* is characterized by pileus having creamish to white peripheri with yellowish centre and hollow stipe. *R. breviceps* possess white pileus with yellowish patches, context discolouring to reddish with crowded lamellae. Violet to purplish colour pileus with white flesh, white lamellae, and white to cream stipe are characteristic features of *R. cyanoxantha.* In case of *R. decolorans* pileus is of different colour mixed with red and buff yellow and stipe creamish or white. *R delica* is characterized by unchanging colour of the context, white to cream pileus, close and thick lamellae. *R. densifolia* is recognised by yellowish brownish to creamish grey pileus and solid stipe. *R. emetica* is conspicuous with pink to blood bright pileus and acrid taste. *R. farinipes* is characterized by unchanging flesh and close lamellae. Yellow pileus, yellowish lamellae and white to yellowish stipe are the important characteristics of *R. foetens* and *R. illota* possess red to cinnamon red pileus and mild taste. *R. laurocerasi* is characterized by brownish yellow pileus, white to yellowish lamellae and cream to yellowish brown stipe. Red to pale red pileus and mild taste are the characteristic feature of *R. lepida. R. lilacea* possess pinkish to dark greyish red pileus with white to yellowish lamellae. *R. rosacea* is characterized by pileus having the red centre, fading to pink in old age with white flesh. Pink to pale dark red pileus with reddish flesh and creamy white to yellowish lamellae are the important features of *R. rosea. R. rubescense* possess red to creamish-pale pileus and pinkish white pileus. *R. rubicunda* is characterized by pink pileus with white patches while *R. sanguinea* possess pur white to cream stipe with pinkish tint and scarlet or dark greyish red pileal surface. Greyish brown to hazel brown, usually paler or straw coloured pileal surface, cream to yellowish cream lamellae and white coloured stipe are the characteristic features of *R. sororia. R. sub- foetens* is characterized by ochre to yellowish or straw coloured pileus, dark yellow to yellowish white lamellae with yellowish cream or brown stipe. *R. violeipes* possess dark violet to greyish, creamish violet pileus with white to cream flesh. *R. virescens* is characterized by discolouring flesh with crowded lamellae. Out of these 22 species *R. cyanoxantha, R. laurocerasi, R. sanguinea, R. lepida, R. rosea, R. lilacea, R. brevicepa* and *R. delica* were found o possess excellent edibility among local tribals of the region. It was observed that all the species were involved in ectomycorrhizal association with Sal tree (*Shorea robusta*). All the species found actively growing on forest litter in solitary, gregarious to scattered manner during monsoon period. The characters of all the species described above match with that of species described by earlier workers. However, there are some differences in morphological features, which might be due to variations in ecological conditions.

Lactarius (D.C. ex) S.F. Gray

Singer (1986) reported 122 species of *Lactarius* from all over world. The genus has been studied and reported by many workers from India including Manjula (1983) who listed 11 Indian species while Bilgrami *et al.* (1991) included 14 species in Fungi of India. Atri *et al.* (1994) in "Russulaceae Roze-A check list" included 48 species of *Lactarius.* The genus is characterized by fleshy pileus and stipe, often vividly coloured, hymenophore lamellate with alternating, sub-decurrent to decurrent lamellae, basidia normal, cheilocystidia sometimes present, hymenophoral trama sub-regular to sub-irregular and intermixed with laticeferous hyphae and in very few cases also with shaerocyst, spore print white to cream, stipe central, context with latex, either forming droplets of latex which may be watery or milky or merely moistened. Usually on the grounds in woods, usually froming mycorrhiza, more rarely on the decayed woods, practically cosmopolitan.

During our investigations a total of seven species of *Lactarius* have been reported. These are *L. angustifolius, L. controversus, L. deliciosus var. deliciosus, L. piperatus var. piperatus, L. stramineous, L. volemus var. flavus* and *L. corrugis. Lactarius angustifolius* is characterized by feebly corrugated, yellowish brown pileus upto 6 cm in diameter and *L. deliciosus var deliciosus* is recognized by pileus having the diameter of 5.5 cm, convex to depressed in shape and enrolled margin. Cellular pileal cuticle is the important characteristic feature of *L. piperatus var. piperatus* . *L. stramineous* is characterized by having the spores with dimension of 6-14 x 6-8 µm, white pileus with depressed centre and white to cream lamellae. *L. volemus* var. *flavus* possess greyish white with yellowsih. toned pileus and latex staining the flesh brown. All these species were found edible among tribals and local inhabitants of the region and were involved in ectomycorrhizal association with Sal tree (*Shorea robusta*). Characters of all these species matche well with those species described earlier except some slight variations in morphological features.

Tricholomataceae Roze.

Tricholomataceae is one of the largest family of Agaricales, cosmopolitan in distribution occupying a variety of ecological niches. Out of 98 genera reported all over the world, forty one are recorded from India (Manjula, 1983). In central India the family is represented by three important genera viz. *Clitocybe, Termitomyces* and *Tricholoma.*

Clitocybe Kummer

The taxonomy of the genus has been well studied by Singer (1996) who reported 115 species from all over the world. Manjula (1983) listed 4 Indian species while Bilgrami *et al.* (1991) included 14 species of *Clitocybe.* Some species of the genus are edible others are poisonous causing diseases with characteristic Muscarine syndrome. The genus is characterized by strictly clitocyboid carpophores often depressed or umblicate in the centre of the pileus with lamellae deeply decurrent or adnate, pileus dry or hygrophanous, lamellae thin, concolorous with pileus, spore print pure white, spores smooth, always inamyloid, basidia normal, hymenophoral trama regular, context of the pileus leathery, stipe central, eccentric, solid or hollow, grow on soil humus and on rotten woods. Nearly cosmopolitan, from the tropics to the arctic.

During our studies three species of *Clitocybe* viz., *C. dealbata, C. infundibuliformis* and *C. pumila* have been colected from central India. *C. dealbata* is characterized by plano-convex pileus of 2-4 cm diameter, 3 cm long stipe with spore size of 4x3-3.5 µm .Pileus diameter of 2.7cm with characteristic depressed to infundibuliformpleus shape. 6 cm long stipe and spores of size 6-8x3-4 µm are the important characteristic features of *C. infundibulifomis. C. pumila* is characterized by convex pileus, slightly depressed centre 1-2 cm in diameter, 1 cm long stipe and spores having size of 4-3 µm. Out of these *C. infundibuliformis* possess excellent edibility among tribals of the region.

Termitomyces Heim

Thirteen species of the genus have been recorded globally (Singer, 1986). From India eleven species have been listed (Bilgrami *et al.,* 1991). The genus has also been studied by Atri *et al.* (1995). According to Heim who considered *Podabrella,* a subgenus of *Termitomyces,* the two genera are undoubtdely closely related. However, In *Podabrella* the primordia does not develop with the termite nests and the carpophores are devoid of constant and true pseudorrhiza and veil, the epicutis of the pileus is always and in all parts a cutis rather than a trichodermal palisade or a hymeniform structure. Most species of the genus are highly valued edible mushrooms. They are linked with the economical importance of the termites because of their role in the biology of many termite species, usually distributed in tropics of Asia, Africa and South Pacific. The genus is characterized by pluteoid habit of the carpophores with fleshy consistency, prominent and sharphy differentiated umbo, stipe central, spore print pink to cream, lamellae free to adnexed but decurrent tooth, stipe with a pseudorrhiza and with simple or double veil epicutis consisting of filamentous, hyaline hyphae, hymenophoral trama bilateral, spores imamyloid, basidia normal, clamp connections absent.

During our studies a total of five species have been collected and reported here from central India. They are *T. clypeatus, T. heimii, T. microcarpus, T. rabuorii* and *T . radicatus. Termitomyces clypeatus* is characterized by blackish brown strongly spiniform perforatorium with dark brown pileal surface. *T. heimii* possess thick white pileus with greyish centre, having a diameter of more than 10 cm and woolly annulus. *T. microcarpus* is recognized by numerous cystidial elements while chestnut brown pileal surface with obtusely conical perforatorium are the main characteristics of *T. rabuorii. Termitomyces radicatus* is separated from other species by absence of any of the cystidial elements and presence of pseudorrhiza.

Tricholoma (Fr.) staude

Singer (1986) reported 88 species from all over world. Manjula (1983) listed 3 species of *Tricholoma* while Bilgrami *et al* (1991) included 9 Indian species. Some species of *Tricholoma* are important edible mushrooms. It is also claimed that it has antitumour activity while few species are deadly poisonous containing muscarine and epimuscarine. The genus generally has tricholomatoid habit, pileus viscid, glabrous, or fibrillose, naked to squammose, spore print pure white, basidia normal, cystidia usually none, hymenophoral trama regular to sub-regular, stipe central, context of pileus fleshy, mild, mostly grow on earth in woods, open fields, prairies or in deep moss. During this study one species, *T. georgii* has been collected and recorded here from central India. The species was found growing solitary to gregarious manner in bamboo forest during rainy season. The specis is characterized by white to cream coloured pileus with a diameter of 2-8 cm., sphercial to conical with slight umbo, stipe 6-8 cm long and spore size ranging from 6-9x5-7µm. The collection was found typical of *T. georgii* as described by Fries, however, it differ in not having the hygrophanous consistency.

Conclusions

After the convention on bilogical diversity, drawn up in 1992, a remarkable resurgence in interest in the variety of life with which we share the earth, is seen. Fungi in general and higher fungi in particular are peculiar eukaryotic organisms exhibiting remarkable diversity in their reproductive stages. The diverse biogeographical regions of central India are rich in macroflora which becomes evidenced by the present studies. Evaluating diversity of Agaricales in this region and knowledge about their implication in ecosystem and for the wealth of mankind is a formidable task. Moreover, during the last few years a spectacular phase of growth in the demand for collection of fungal germplasm is witnessed as a direct result of increased interest in biotechnological processes and they are being recognized as a major microbiological resource. The present work may fill the gap in the systematic understanding of Agaricales of central India and also help in improving the comprehensive Agaricales map of India.

Acknowledgement

Authors are thankful to Department of Biotechnology, New Delhi for financial assistance to carry out the present work.

References

Ainsworth, G.C. (1971). Ainsworth and Bisby's Dicitionary of Fungi CMI.Kew, surrey England.

Atri, N.S., Saini, M.K. and Saini, S.S. (1994). Indian Russulaceae roze- A check list, In : current Researches in Plant sciences, (eds T.A.Sharma, S.S.Saini, M.L.Trivedi, & M. Sharma), Published by Bishen Singh Mahendra Pal Singh, Dehradun, India.

Atri, N.S., Saini, S.S. and Kaur, G. (1995). Taxonomic studies on the North Indian Agarics: the genus *Termitomyces* Heim. Mushroom Res., **4**:7-10.

Atri, N.S., Saini, S.S. and Kaur, G. (1996). Three species of Agarics from Patiala. Mushroom Res., **5**:77-80.

Atri, N.S., Saini, S.S. and Kaur, A. (2000). Taxonomical studies on Agarics from Punjab the genus *Lepiota* (Pers. ex Fr.) Gray. Mushroom Res., **9**(2): 71-77.

Atri, N.S. and Saini, S.S. (1989). Family Russulaceae Roze- A Review. In: Plant science Research in India (eds M.C. Trivedi, B.S. Gill & S.S. Saini). Today & Tomorrow's Printers, Delhi, PP. 115-128.

Banerjee, S.N. (1947). Fungus flora of Calcutta and suburbs. I Bull. Bot. Soc., Bengal, **1**:3754.

Bilgrami, K.S. Jamaludin and Rizwi, M.A. (1991). Fungi of India Part I lists & Refrences. Todays & Tomorrow's Printers and Publishers, New Delhi.

Bose, S. R. (1919). Description of Fungi in Bengal-I: Proc Indian Asoc. Cult Sci., **4**:109-114.

Bose, S.R. (1920b). Description of Fungi from Bengal. Proc. Sci Convention Indian Assoc. *Cult. Sci* for the year 1918 PP. 136-143.

Bose, S.R. (1921). Possibilities of mushroom industry in India by cultivation. Agaric. J. India, **16**:643-647.

Bose, S.R. (1949). Horse-hair fungus from Bengal *Proc. India Sci Cong.* Sec. VI, **56**:123.

Corner, E.J.H. (1966). A monograph of Cantharelloid fungi. Ann. Bot. Mem., **2**:1-255.

Donk, M.A. (1964). A conspectus of the families of Aphlyllophorales Persoonia, **3**:199-324.

Fries, E.M. (1821-1832) Systema Mycologicum. Vols. I-III. Gryphiswald.

Ghosh, R.N., Pathak, N.C. and Singh, M.S. (1974). Studies on Indian Agaricales II Proc. Natl. Acad. Sci. Ind., **44** : 125-128.

Greuter, W., Buret, H.M., Chalonea, W.G., Demoulin, V., Grolle, R., Hawksworth, D.L., Nicolson, O.H., Silva, P.C., Staefleu, F.A., Voss, E.T. and Mcneill, J. (1988). International code of Botanical Congres Berlin, July-August, 1987. Regnum vegetable 118 Konigstein.

Hawksworth, D.L., Sutton, B.C. and Ainsworth, G.C. (1983). Dictionary of the fungi, CMI Kew, Surrey, England.

Hawksworth, D.L. (1991). Fungal dimension of biodiversity. Mycol. Res., 95: 641-655.

Klotzsch, J.F. (1833). Fungi exotici collectini bus britannorum. Linnaea, 7:478-490.

Kuhner, R. and Romagnesi , H. (1953). Flore analytique des chanpignos superieurs (*Agaricus, Boletus, Cantharellus*), I-XIV, 1-556, Masson & Lie, Paris.

Lakhanpal, T.N., Bhat, R.P. and Kamaraja, K. (1987). *Lactarias sanguiflus* Fr. An edible mushroom new to India. Curr. Sci, **56**(3):148.

Lakhanpal, T.N., Sharma, R. and Kumar, A. (1988). *Boletus edulis* Fr. Bull. An edible mushroom new to India. Curr. Sci, 57 (11): 611.

Lakhanpal, T.N. (1997). Diversity of Mushroom Mycoflora in the North- West Himalaya. In: Recent Researches in ecology, Environment and Pollution 10 :35-68 (eds. S.C. Sati, J. Saxena and R.C. Dubey). Today & Tomorrow's Printers & Publishers, New Delhi-5 (India) .

Lange, M. (1952) Species concept in the genus. *Coprinus-Dansk.* Bot. Ark., **14**(6):1:164)

Lange, M. and Smith, A.N. (1953). The *Coprinus ephemerus* group. Mycologia, **57**:747-780.

Leveille, J.H. (1844). Fungi, in voyage dans I' Inde per victor Jacquemont pendant les annees 1829-1832, Tome 4:179.

Linnaeus, C. (1753). Species Plantarum, 2.561-1200 Indexes Holmaiae.

Manjula, B. (1983). A revised list of the agaricoid and boletoid basidiomycetes from India and Nepal. Proc. Ind. Acad. Sci., **92**:81-213.

Montagne, J.F.C. (1842). Cryptogamae Nilgherensis. Ann. Sci, Nat. Ser II, **18**:12-23.

Natarajan, K. (1975). South Indian Agaricales. I. Termitomyces Kavaka, **3**:63-66.

Natarajan, K. (1977a). South Indian Agaricales. II. Mycologia, **69**:185-189.

Natarajan, K. (1977b). South Indian Agaricales. III. Kavaka, **5**:35-39.

Natarajan, K. (1978). South Indian Agaricales, VI. Kavaka, **6**:65-70.

Natarajan, K. (1978a). South Indian Agariclaes. IV: A new species of *Psathyrella.* Mycologia, 70:1259-1261.

Natarajan, K. (1978b). South Indian Agaricales, V. Kavaka, **6**:65-70.

Natarajan, K. (1979). South Indian Agaricales. VI. Termitomyces heimii Mycologia, **71**:853-855.

Pahll, V.S., Smith, J.F. and Ellioff, T.J. (1999). Cytological studies of homokaryotic and dikarynotic stocks of tropical *Agaricus bitorquis* isolates. Mushroom Res.,.8(1): 1-70.

Pegler, D.N. (1977). A preliminary Agaric flora of East Africa. *Kew Bull.* additional series VII. Royal Botanic gardens Kew.

Pegler, D.N. and Young, T.W.K. (1979). The Gasteroid Russulales; Trans Br. Mycol. Soc., **72**:353-388.

Purkayastha, R.P. and Chandra, A. (1985). Manual of Indian edible mushrooms. Today & Tomorrow's Printer & Publishers, New Delhi.

Rahi, Deepak Kumar (2001). Studies on the Edible Tribal Mushrooms of Madhya Pradesh and Developoment of Technology for Large Scale Production. Ph. D. thesis unpublished, R.D. University, Jabalpur.

Rawla, G.S. Sarwal, B.M. and Arya, S. (1982). Agarics new to India. I. Nova Hedwigia **36**: 433-444.

Rawla, G.S., Arya, S. and Sarwal, B.M. (1983). Specis of *Termitomyces* Heim from Chandigarh, India. Bibliotheca Mycologica, **91**:13-21.

Rawla, G.S., Locquin, M.V. and Sarwal, B.M. (1987). Three new species of *Tricholoma* (Fr.) Quel. In Proc. International Conferences on Science and Cultivation Technology of edible fungi (eds. T.N. Kaul & B.M. Kapur) PP. 383-388. CSIR Jammu Tawi.

Rea, C. (1922). British Basidiomycetes, XI Cambridge University Press. U.K.

Roy, A. and Samajpati, N. (1978). Agaricales of West Bengal –II *Ind.* J. Mush, 4(2):17-23.

Saini, S.S., Atri, N.S. and Singer, R. (1982). North Indian Agaricales. II Sydowia, 35:236-241.

Saini, S.S., Atri, N.S. and Bhupal, M.S. (1988). North-Indian Agaricales. Indian Phytopath, 41(3):662-665.

Saini, S.S., Atri, N.S. and Gupta, A.K. (1991). Additional Studies on North-West Indian Agarics. In: Advances in Mushroom Science (National Symposium on Mushrooms) (ed M.C. Nair), pp 6. College of Agriculture, Vellayai Thriuvanthapuram.

Saini, S.S., Atri, N.S. and Gupta, A.K. (1997). Studies on the genus *Agaricus* Fr. the subgenus *Agaricus* section *Sanguinolenti* Schaeff. Et Moller from north west India. Mushroom Res., **6**(2):53-58.

Schaefer, R.L. (1975). The major groups of Basidiomycetes. Mycologia, **67**:1-18.

Shukla, Kamlesh Kumar (2001). Studies on Agaricales of Madhya Pradesh with Special Reference to Cultivation of Tribal Edible Mushrooms. Ph. D. thesis unpublished, R.D. University, Jabalpur.

Singer, R. (1986). The Agaricales in the modern Taxonomy, 4th edn. Seven koeltz Scientific Books, Koenigastein, Germany.

Smith, A.H. (1973). Agaricales and related Gasteromycetes. In ' The fungi': An advanced treatise Vol. IV B. 421-446 (eds. G.C. Ainsworth, A.S.Sussman & F.K. Sparrow), Academic Press, New York and London.

Vrinda, K.B., Pradeep, C.K., Mathew, S. and Abraham, T.K. (1997). Agaricales from Western Ghats- V. Mushroom Res., **6**(1):7-10.

Webster, J. (1970). Introduction to fungi Cambridge University Press. Cambridge.

17

Biodiversity of Acellular Slime Fungi (Myxomycetes) in the Royal Kingdom of Bhutan

Rajesh Sharma[1] and T.N. Lakhanpal[2]

[1]*Department of Botany, DAV College, Chandigarh*

[2]*Department of Biosciences, H.P. University, Shimla - 171005, H.P.*

Extensive survey of Royal Kingdom of Bhutan has resulted in the discovery of 85 acellular slime molds. Among these eight are new taxa and all are new to Bhutan. Futher the biodversity is comparable to Eastern Himalayas.

Key-Words: Biodiversity, slime molds, Bhutan

The myxomycetes are an interesting group of organisms, which are marked by two equally predominant phases in their life cycle. The first phase of their life cycle is animal-like while the second stage is plant-like in nature. This dual behaviour is the basis of controversy regarding the animal or plant-like affinities and the changes in the classification and placement of these organisms time and again.

For many years, the myxomycetes have been largely collected and described from N.W. Himalayas by Thind and his students. The first attempt to describe myxomycetes from eastern India was made by Lodhi (1934), Bruhl and Gupta (1927), and Agnihothrudu (1958, 1959). Systematic study of myxomycetes from Eastern Himalayas was initiated by Prof. K.S. Thind and his students. Extensive surveys, of the eastern Himalayas, including the Royal Kingdom Bhutan were undertaken by the senior author, under the guidance of Prof. Thind from 1978-1981. This paper lists all the species collected and described from the Royal Kingdom of Bhutan.

Vegetational Account of Bhutan

The territory of Bhutan includes many river valleys, the most important of them being the Manas Valley. The mountain ranges stretch mostly north south. In general, it may be said that Bhutan forms a transition between the floras of Sikkim and West China. The forests of Bhutan can be classified into the following types.

1. Tropical and Subtropical Forests:

The outer spurs receive very heavy rainfall and are as a consequence densely forested with tropical and subtropical genera. The density of vegetation in the valleys, however, depends on the topography and the amount of rainfall received. The moisture-laden winds reach the interior or valleys, which are disposed in north-south direction and such valleys, which receive comparatively less rainfall, have bare slopes and support moist deciduous forests. *Shorea* and *Pinus* may or not be present. *Shorea* when present, dominates the vegetation and occurs along with *Bauhinia, Careya arborea, Dillenia, Melastoma, Syzygium* and others.

In the tropical evergreen forests along the river banks, the dominant genera are *Dallenia, Duabanga, Hydnocarpus* and *Talauma,* with numerous climbers, epiphytes and orchids. In the *Quercus-Rhododendron-Schima* forests, the dominant oak is *Quercus griffithii* associated with *Castanopsis* and *Engelharditia. Pinus roxburghii* reaches a lower limit at about 600m. On the dry slopes of the lower valley *Euphorbia royleana* is commonly met with.

2. Temperate Forests:

In the interior, wherever there is sufficient moisture, a temperate flora, in which *Rhododendron, Acer* and *Betula* are prominent, is developed. In some ranges, the forest is dominated by magnolias, oaks, *Strobilanthes* and others. In the inner valleys are also found the characteristic temperate conifer forests, consisting of *Abies spectabilis* (*A. densa*), *Larix griffithiana, Picea spinulosa, Pinus* spp. and *Tsuga dumosa. Larix griffithiana* may occur in associatation with an undergrowth of rhododendrons.

3. Alpine Forests:

In the alpine zone, the vegetation varies between the moist alpine type on the south and the dry steppe type on the northern aspects. *Primula sikkimensis* is abundant in alpine zone.

Places of Collections:

1.	Chimakothi	2037m	(Tsimalakha, Chukha Dam site, Bunakha, 2000-2500m). Temperate forests. Pure angiospermous forests at Tsimalakha and Chukha, mixed at Bunakha.

2.	Thimphu	2275m	(Nawephu, Namseling, Begana, Chankaphug, Honsu, Yusipang, Dochula, Motithang, Fajudin 2200-3100m). Predominantly mixed temperate forests.
3.	Wangdi	1250m	(Punakha, 25km from Wangdi towards Tongsa, 15km from Wangdi towards Dochula (Mindleygang, 1250-2250m). Mixed forest are 15 km, temperatre angiospermous forests towards Tongsa.
4.	Paro	2095m	(D'Dzong, Chailesa, 2050-2700m). Temperate forests. Patches of pure confiers, but predominantly mixed forests.
5.	Ha Dzong	2690m	(9-10km from 11a on Thimphu road, Kana, 2600-300m).

The diferent locations in Bhutan were visited during August to October, 1980-81. The Collections are made of the fructifications along with a part of substratum and then transferred to appropriate cardboard boxes. The data pertaining to habitat, colour, locality, type of forest, date of collection and collections number allotted is recorded in the notebook at the spot. Examination of the fresh specimen is done in water mount. The dried specimens, are first treated with 70% alcohol, and then revived in 2-3% KOH and mounted in glycerine and after treatment with alochol, these are placed in lactophenol. All the collections have been made by the senior author, the identification has been done jointly. All the specimens are deposited in the herbarium of Botany Dept., Punjab Univ. Chandigarh, India (PAN). Collections have been also submited to Dr. (Mrs.) N.E. Nannenga Bremekamp. The Netherlands, and Dr. Marie L. Farr. USDA, Beltsville, Maryland, USA.

The classification followed is of Alexopoulos (1973) and Farr (1976) Identification are based on the monographs of Martin and Alexopoulos, (1969), Lister and Lister (1911, 1925), Thind (1977) and Lakhanpal and Mukerji (1981).

Sub class-I Ceratiomyxomycetidae Martin

Ceratiomyxa fruticulosa (Mull.) Macbr. N. Am. Slime-Moulds 18, 1899.

Collections Examined : PAN 15603, 15611, 15647, 15658, 15675.

Habitat : Dead and decaying wood and dead *Pinus* cone.

Sub class - II Myxogasromycetidae Martin emend. Ross Martin, in Ainsworth, Dict. Fungi ed. 5. 497. 1961, emed. Ross, Mycologia **65** : 483. 1973.

Order-Liceales Macrbride N.A. Slime-Moulds, ed. 2. 237. 1922.

Family-Liceaceae Rost. Versuch 4. 1873 (as Tribus).

Licea castanea G. Lister Jour. Bot. **49 :** 61. 1911
Collections Examined : PAN 15884
Habitat : Living bark of *Pinus* sp.

Family-Reticulariaceae Rost. Versuch 6. 1873 (as Tribus).
Tubifera dimorphotheea Nann. Brem &Loerak, K. Ned, Akad. Wet. Proc. C. 84, 1981, 233.
Collection Examined : PAN 15449.
Habitat: Dead and decaying stumps.

Tubifera ferruginosa (Batsch) J.F. Gmel. Syst. Nat. **2:** 1472. 1791
Collections Examined: PAN 15492, 15497, 15502, 15508, 15509,
Habitat: Dead and decaying wood, dead wood covered with mosses and living tree trunk.

Dictydiaethalium plumbeum (Schum.) Rost in Lister, Mycet. 157. 1894.
Collections Examined: PAN 15582
Habitat Dead log.

Lycogala epidendrum (L.) Fries Syst. Myc. **3:** 80. 1829
Collections Examined: PAN 15550, 15563, 15564
Habitat: Dead and decaying wood covered with mosses and burnt stump.

Lycogala exiguum Morgan Jour. Cinc. Soc. Nat. Hist. **15:** 134. 1893.
Collections Examined: PAN 15425, 15437, 15519, 15567, 15585
Habitat: Dead and decaying wood and burnt stump.

Enteridium intermedium (Nann-Brem.) Farr, var. nov.
Collections Examined: PAN 15506, 15516, 15517
Habitat: Dead stump.

Family-Cribrariaceae Rost. Versuch 5. 1873 (as Tribus).
Cribraria argillacea (Pers.) Pers. Neues Mag. Bot. **1:** 91. 1794
Collections Examined: PAN 15623, 15653, 15656, 15657, 15673.
Habitat: Dead and decaying wood.

Cribraria costata Dhillon et. Nann. Brem. K. Ned. Akad. Wet. Proc. C. **81**: 141. 1978.
Collections Examined: PAN 15116
Habitat: Dead wood.

Cribraria intricata Schrad. Nov. Gen. P1. 7. 1797
Collections Examined: PAN 15742, 15690
Habitat: Dead wood.

Cribraria languescens Rex Proc. Acad. Phila. **43:** 394. 1891
Collections Examined: PAN 15290
Habitat: Dead stump.

Cribraria macrocarpa Schrad. Nov. Gen. P1, 8. 1797
Collections Examined: PAN 15452, 15444, 15498, 15501, 15512.
Habitat: Dead and decaying wood.

Cribraria splendes (Schrad.) Pers. Syn. Fung. 191. 1801
Collections Examined: PAN 15489
Habitat: Dead wood.

Cribraria tenella Schrad. Nov. Gen. P1. 6. 1797.
Collections Examined: PAN 15451.
Habitat: Dead and decaying wood.

Dictydium cancellatum (Batsch) Macbr. N. Am. Slome-Moulds 172. 1899
Collections Examined: PAN 15507, 15518, 15527, 15533, 15583.
Habitat: Dead and decaying wood.

Dictydium mirabile (Rost.) Meylan Bull. Soc. Vaud. Sci. Nat. **57**: 305. 1931.
Collections Examined: PAN 15433, 15453, 15454, 15607, 15629, 15642.
Habitat: Dead and decaying wood and dead stump covered with mosses.
Order-Trichiales Macbride N.A. Slime-Moulds, ed. 2. 237. 1922

Arcyria cinerea (Bull.) Syn. Fung. 184. 1801
Collections Examined: PAN 15531, 1559, 15572, 15588, 15628.

Arcyria denudata (L.) Wettst. Verh. Zool.-Bot. Ges. Wien **35**: Abh. 535. 1886
Collections Examined: PAN 15649, 15730.
Habitat: Dead wood.

Arcyria gulielmae Nann.-Brem. K. Ned. Akad. Wet. Proc. C, **74**: 352-365. 1971
Collections Examined: PAN 15579
Habitat: Dead wood.

Arcyria manga Rex var. *rosea* Rex Proc. Phila. Acad. **45**: 365. 1893
Collections Examined: PAN 15524
Habitat: Dead log.

Arcyria stipata (Schw.) A. Lister Mycet. 189, 1894
Collections Examined: PAN 15581
Habitat: Decayed log.

Arcyria sp. nov.
Collections Examined: PAN 15558
Habitat: Dead and decaying stump.

Metatrchia vesparium (Batsch) Nann Brem. K. Ned. Akad. Wet. C. **69**: 146. 1966.
Collections Examined: PAN 15467, 15469
Habitat: Dead and decaying wood

Hemitrichia calyculata (Speg.) Farr Mycologia **66**: 887. 1974
Collections Examined: PAN 15474, 15593
Habitat: Dead and decaying wood and dead stump covered with mosses.

Hemitrichia clavata (Pers.) Rost. in Fuckel, Jahrb. Nass. Ver. Nat. **27-28**: 75. 1873
Collections Examined: PAN 15457, 15486
Habitat: Dead and decaying wood.

Hemitrichia imperialis G. Lister Trans. Brit. Mycol. Soc. **14**: 226. 1929
Collections Examined: PAN 15465, 15543
Habitat: Dead stump.

Hemitrichia serpula (Scop.) Rost. in Lister, Mycet. 179. 1894
Collections Examined: PAN 15434, 15476, 15576, 15604
Habitat: Dead and decaying wood and twigs.

Trichia botrytis (J.F. Gmel.) Pers. Neues Mag. Bot. **1**: 89. 1794.
Collections Examined: PAN 341
Habitat: Bark of a tree covered with mosses.

Trichia decipiens (Pers.) Macbr. N. Am. Slime-Moulds 218. 1899
Collections Examined: PAN 15424, 15430, 15431, 15438, 15440
Habitat: Dead and decaying wood.

Trichia favoginea (Batsch) Pers. Neues Mag. Bot. **1**: 90. 1794
Collections Examined: PAN 15470, 15569, 15570, 15589, 15723
Habitat: Dead and decaying wood, bark stump, mosses.

Trichia floriformis (Schw.) G. Lister Jour. Bot. **57**: 110. 1919
Collections Examined: PAN 15471, 15473, 15520, 15546, 15554
Habitat: Dead and decaying wood at the base of angiospermic tree trunk covered with mosses.

Trichia scabra Rost. Mon. 258. 1875
Collections Examined: PAN 15442
Habitat: Dead log.

Trichia varia (Pers.) Pers. Neues Mag. Bot. **1**: 90. 1794
Collections Examined: PAN 15434, 15439, 15460, 15463, 15464
Habitat: Dead and decaying wood and burnt wood.

Order-Physarales Macrbride N.Am. Slime-Moulds, ed. 2. 237. 1922
Family-Physaraceae Rost. Versuch 9. 1873 (as Tribus).

Leocarpus fragilis (Dicks.) Rost. Mon. 132. 1874
Collections Examined: PAN 15715
Habitat: Angiospermic tree trunk, dead stump, *Pinus* needles and dead leaves.

Badhamia macrocarpa (Ces.) Rost. Mon. 143. 1874.
Collections Examined: PAN 15561
Habitat: Dead wood.

Fuligo septica (L.) Wiggers Prim. F1. Holsat. 112. 1780
Collections Examined: PAN 15436, 15466, 15475, 15553, 15556
Habitat: Dead wood, wood covered with mosses, leaves and debris and living *Pinus* tree trunk.

Craterium leucocephalum (Pers.) Ditmar in Sturm, Deuts. F1. Pilze **1**: 21. 1813.
Collections Examined: PAN 15526, 15667, 15668, 15721
Habitat: Dead and living leaves.

Craterium minutum (Leers) Fries Sys. Myc. **3**: 151. 1829
Collections Examined: PAN 15526, 15667, 15668, 15721
Habitat: Dead wood, twig, leaves, *Pinus* needles, dead fern and mosses.

Physarum bivalve Pers. Ann. Bot. Usteri **15**: 5. 1795
Collections Examined: PAN 15681
Habitat: Dead twigs, angiospermic leaves and fern fronds.

Physarum compressum Alb. & Schw. Consp. Fung. 97. 1805
Collections Examined: PAN 15
Habitat : Dead wood, twigs and leaves.

Arcyria stipata G. Lister Jour. Bot. **59**: 252. 1921
Collections Examined: PAN 15551, 15648
Habitat: Dead wood, twig and leaves.

Physarum conglometratum (Fries) Rost. Mon. 108. 1874. Not. *P. conglomeratum* Massee, 1892.
Collections Examined: PAN 15494
Habitat: Dead angispermic twigs and leaves.

Physarum javanicum Racib Hedwigia **37**: 53. 1898
Collections Examined: PAN 15482
Habitat: Dead angiospermic log.

Physarum luteolum Peck Ann. Rep. N.Y. State Mus. **30**: 50. 1878
Collections Examined: PAN 15652
Habitat: Dead angiospermic leaves and *Pinus* needles.

Physarum melleum (Berk. & Br.) Massee Mon. 278. 1892
Collections Examined: PAN 15740,
Habitat: Dead wood, twig, leaves and sack.

Physarum nutans Pers. Am. Bot. Usteri **15**: 6. 1795
Collections Examined: PAN 15591, 15595, 15596, 15676, 15679
Habitat: Dead and decaying wood.

Physarum psittacinum Ditmar in Saturm, Deuts, F1. Pilze **1**: 125. 1817
Collections Examined: PAN 15432, 15433, 15606, 15618, 15698.
Habitat: Dead and decaying wood and dead *Pinus* stump.

Physarum pusillum (Berk. & Curt.) G. Lister Mycet. Ed. 2. 64. 1911
Collections Examined: PAN 15671
Habitat: Dead and decaying wood.

Physarum stellatum (Massee) Martin Mycologia **39**: 461. 1947
Collections Examined: PAN 15677, 15732.
Habitat: Dead and decaying wood and *Pinus* stump.

Physarum straminipes A. Lister Jour. Bot. **36**: 163. 1898
Collections Examined: PAN 15503
Habitat: Dead mosses growing on a decaying log.

Physarum sulphureum Alb. & Schw. Consp. Fung. 93. 1805
Collections Examined: PAN 15683, 16695
Habitat: Dead leaves.

Physarum tenerum Rex Proc. Acad. Phila. **42**: 192. 1890
Collections Examined: PAN 15597
Habitat: Dead and decaying wood.

Physarum viride (Bull.) Pers. Ann. Bot. Usteri **15**: 6. 1795
Collections Examined: PAN 15426, 15428, 15496, 15540, 15587
Habitat: Dead and decaying wood, burnt stump, angiospermic twigs and dead *Pinus* cone.

Physarum sp. nov.
Collections Examined: PAN 15477, 15481
Habitat: Dead log.

Family-Didymiaceae Rost. Versuch 12, 1873 (as Tribus)

Diachea leucopodia (Bull.) Rost. Mon. 190. 1874
Collections Examined: PAN 15447, 15488
Habitat: Dead wood, twigs, grasses and leaves.

Diderma badhamioid Thind The Myxomycetes of India. I.C.A.R., pp. 315-316. 1977
Collections Examined: PAN 15538
Habitat: On mosses growing on decayed stump and bark of tree.

Diderma subfloriformis F. Candoussau et Nann-Brem. Cryptog. Mycol. Tome **1**: 201-204.1980.
Collections Examined: PAN 15568
Habitat: Decaying log.

Didymium melanospermum (Pers.) Macbr. N. Am. Slime-Moulds 88. 1899
Collections Examined: PAN 15500, 15510, 15511
Habitat: Dead and decaying log covered with mosses.

Didymium nigripes (Link) Fries Syst. Myc. **3**: 119. 1829
Collections Examined: PAN 15650, 15682, 15703
Habitat: Dead angiospermic leaves, twigs, *Pinus* needles ang grass.

Didymium squamulosum (Alb. & Schw.) Fries Symb. Gast. 19. 1818
Collections Examined: PAN 15651
Habitat: Dead angiospermic leaves and plant debris.

Didymium sp. nov.
Collections Examined: PAN 15575
Habitat: Dead angiospermic leaves.

Lepidoderma tigrinum (Schrad.) Rost. in Fuckel, Jahrb. Nass. Ver. nat. **27-28**: 73. 1873
Collections Examined: PAN 15490, 15515, 15536, 15537, 15545
Habitat: Dead and decaying wood and dead wood covered with mosses.

Lepidoderma tigrinum (Schrad.) Rost. var. nov.
Collections Examined: PAN 15455
Habitat: Dead and decaying angiospermic log.

Lepidodermopsis martinii Lakhanpal Norw. J. Bot. **25**: 196. 1978
Collections Examined: PAN 15456, 15531, 15414, 15541, 15664
Habitat: Dead and decaying wood, twigs, leaves and *Pinus* needles.

Sub class-Stemonitomycetidiae Ross Mycologia **65**: 483. 1973

Enerthenema papillatum (Pers.) Rost. Mon. App. 28. 1876
Collections Examined: PAN 15739
Habitat: Dead *Pinus* stump.

Stemonitis aequalis (Peck) Massee var. *microspora* Nann.-Brem. & Y. Yamam. K. Ned. Akad. Wet. Proc. C. **86**: 237-239. 1983.
Collections Examined: PAN 15562
Habitat: Decaying log.

Stemonitis axifera (Bull.) Macbr. N. Am. Slime-Moulds 120. 1889
Collections Examined: PAN 15427, 15450, 15584, 15590, 15605
Habitat: Dead and decaying wood, burnt *Pinus* stumps and angiospermic leaf.

Stemonitis farrensis Lakhanpal & Mukerji Acta Bot. Indica **5**: 60. 1977
Collections Examined: PAN 15573
Habitat: Dead wood and burnt wood.

Stemonitis flavogenita Jahn Verh. Bot. Ver. Brand. **45**: 165. 1904
Collections Examined: PAN 15443, 15488, 15525, 15704
Habitat: Dead and decaying wood and living grass.

Stemonitis fusca Roth Mag. Bot. Romer & Usteri **1** (2): 26. 1787
Collections Examined: PAN 15600, 15666
Habitat: Dead and decaying wood.

Stemonitis nigrescens Rex Proc. Acad. Phila. **43**: 392. 1891
Collections Examined: PAN 15586, 15612
Habitat: Dead and decaying wood and burnt stump.

Stemonitis pallida Wingate in Macbr., N. Am. Slime-Mouls 123. 1899
Collections Examined: PAN 15441
Habitat: Dead and decaying wood and burnt stump.

Stemonitis smithii Macbr. Bull. Nat. hist. Univ. Iowa **2**: 381. 1893
Collections Examined: PAN 15429, 15445, 15487, 15529, 15712
Habitat: Dead and decaying wood and burnt stump.

Stemonitis splendes Rost. Mon. 195. 1874
Collections Examined: PAN 15594, 15728
Habitat: Dead wood.

Stemonitis sp. nov.
Collections Examined: PAN 15480, 15478,m 15479, 15493
Habitat: Dead stump.

Comatricha acanthodes Alexop. Mycologia **50**: 55. 1958
Collections Examined: PAN 15885
Habitat: Living bark of *Pinus* tree.

Comatricha nigra (Pera.) Schoret Krypt.-F1. Schles. **3** (1): 118. 1885
Collections Examined: PAN 15505, 15599
Habitat: Dead and decaying wood.

Comatricha parvispora Dhillon and Nann.-Brem. K. Nad. Wet. Proc. C. **80**: 260. 1977
Collections Examined: PAN 15622
Habitat: Dead and decaying wood.

Comatricha pulchella (C. bb.) Rost. Mon. App. 27. 1876
Collections Examined: PAN 15446
Habitat: Dead wood, twigs and leaves.

Lamproderma arcyrionema Rost. Mon. 208. 1874
Collections Examined: PAN 15620, 15626, 15720, 15737
Habitat: Dead and decaying wood, stump covered with mosses and burnt *Pinus* stump.

Lamproderma columbinum (Pers.) Rost. in Fuckel, Jahrb. Nass. Ver. nat. **27-28**: 69. 1873
Collections Examined: PAN 15542
Habitat: Dead log.

Lamproderma columbinum (Pers.) Rost. form nov.
Collections Examined: PAN 15547
Habitat: Dead stump covered with mosses.

Stemonitis rhizoidepes Nann.-Brem., sharma & Thind, n. sp.
Collections Examined: PAN 13467, 13469, 15534, 15491
Habitat: Stump of trees.

Conclusion

Eighty five taxa of acellular slime molds have been recorded in this paper from the Royal Kingdom of Bhutan. This is the first survey of slime molds from this Himalayan Kingdom. The analysis of taxa shares a similarity of species between those listed here and those recorded from other parts of Eastern Himalyas. All the species are new record from Bhutan. Among these eight taxa are new to science.

References

Agnihothrudu, V. (1958). Notes on Fungi from North-East India II. An underscribed Myxomycete from Assam. J. Indian Bot. Soc., **37**: 499-503.

Agnihothrudu, V. (1959). Notes on Fungi from North-East India IV. Myxomycetes. J. Indian Bot. Soc., **38**: 418-451.

Agnihothrudu, V. (1959). Notes on Fungi from North-East India IV. Myxomycete. J. Indian Bot. Soc., **38**: 4453-491.

Alexopoules, C.J. (1973). Myxomycetes. The Fungi, An Advanced Treatise (eds. G.C. Ainsworth, F.K. Sparrow and A.S. Sussman), IV-B Chapter III. 39-60. Academic Press, New York.

Bruhl, P. and Gupta, J.S. (1927). Indian Slime-Fungi (Myxomycetes or Mycetozoa). First contribution. Jour. Dept. Sci. Univ. Calcutta **8**: 101-122.

Farr, M.L. (1976). Flora Neotropica. Mon. No. 16. 1-304. New York.

Lakhanpal, T.N. and Kumerji, K.G. (1981). Indian Myxomycetes. pp 1530. J. Cramer, Germany.

Lister, A. (1911). A monograph of the Mycetozoa (ed. 2, revised by G. Lister). 1-302. British Museum (Natural History), London.

Lister, A. (1925). A monograph of the Mycetozoa (ed. 3, revised by G. Lister). 1-296. British Museum (Natural History), London.

Lodhi, S.A. (1934). Indian Slime-Moulds (Myxomycetes). Punjab Univ. Pub. Lahore (now in Pakistan)

Martin, G.W. and Alexopoulos, C.J. (1969). The Myxomycetes. 561 pp. Univ. of lowa Press. lowa, USA.

Thind, K.S. (1977). The Myxomycetes of India. ICAR, New Delhi.

18

Diversity of Fleshy Fungi in Eastern Uttar Pradesh

C.S. Singh and A.B. Prasad

Department of Botany, K.S.S.PG. College, Ayodhya- 224 123, U.P.

A total of 274 species of fleshy fungi were recorded from Eastern Uttar Pradesh. Agaricales were predominantly occurring forms followed by Aphyllophorales, the species of Sphaeriales, Lycoperdales, Pezizals were represented by their moderate numbers while Auriculariales, Dacrymycetales, Helotiales, Nidulariales, Sclerodermatales, Phallales, Tremellales and Tuberales were very few in respect to their comparative numbers. Of the species recorded, 20 were excellently edible, 72 edible, 104 inedible, 14 medicore, 18 poisonous and and 46 others whose edibility was not established. Amongst the species recorded during two years of observation (1997-88) and 1998-99) 18, viz., *Amanita umbrinolutea, Colcybe gambosa, Cystolepiota bucknalii, Lactarius glaucescens, L. piperatus, L. uvidus, L. vallereus, Lepiota acutisquamosa, L. alba, L. felina, Lepiota sp., Megacollybia platyphylla, Pholiota lenta, Russula mairei, R. mustellina, R. rubra, Vovariella fibrillose* and *Marasmius brunuclus* were reported for the first time from India.

Key words: Diversity, fleshy fungi, Uttar Pradesh, India

India is a rich treasure of natural resources and her glorius hertiage preserves the original knowledge about plants since vedic times. The Ayurveda, being subsidiary of Vedas, is an authoritative source of multidisciplinary knowledge including the potentials of plants belonging to various kinds and varieties. A closer look on literature indicates that indigenous materia medica is dominated by the drugs of botanic origin, representing major portions of phanerogamic plants. However, different shcools representing epochs did play a little attention to cryptogamic plants including mushrooms for the human welfare.

Mushrooms are represented by gill fungi (agarics), porefungi (boleti) and more precisely the earth stars (Geastrum), Stinkhorns (Palloides), buffballs (Lycoperdon) and truffles (Tubers). Many of them are edible, famour for their palatability and delicacies fetching high market prices. Ofthese, more are collected from nature and comparatively their lesser number is cultivated under defined conditions whose cultivation technology is widely documented.

Although, the widly occurring edible mushrooms are many in number but only a few of them are traditionally consumed by the local inhabitants because of their unawareness in respcet of their edible nature. Amongh such mushrooms, some are specifically mycorrhizal which grow in association of roots of forest trees and grasses, and other non-mycorrhizal which thrive independently.

Non-mycorrhizal mushrooms include gill fungi with long and short stalks, non-gill members, tooth and coral fungi, false morels and cauliflower fungi (Lakhanpal, 1994). From the above mentioned forms, edible ones are yet to be exploited commercially for cultivation. Many mushroom species grow wild in association of tree roots, represent a very remarkable relationship between feeder roots of plant and beneficial fungi which live on root surface or inside it. This relationship of fungi with roots of higher plants is a hindrance to achieve indoor cultivation of such useful fungi (Lakhanpal et al.,1991). These fungi are certainly mycorrhizal in nature which can only be cultivated outdoors with specific forest trees. Such cultivation will need to be encouraged in area where deforestation is rapidly progressing (Rambelli, 1983). Studies of these fungi in respect of their shelflife and preservation may benefit the people involved in such work.

The role of mycorrhiza in mineral cycling, improving the quality of planting stock in nursery and to obtain good survival of seedlings and growth after transplanting has been widely recognised. Mycorrhizal technology, therefore, has assumed greater relevance in forest production of tropical pines, conifers, eucalyptus, poplars and Oaks. In addition to multiple benefits to plants through mycorrhiza in forest species, the potential of certain fungi to produce edible fruit bodies, while in mycorrhizal association is of greater importance and practical application of which can lead to successful afforestation, additional timber and food production.

Exploitation of mycorrhizal fungi for the dual benefit of symbiotic association with tree species and fruit body production has been attempted with great success only with species of *Tuber* and *Tricholoma* amongst fungi which have been established as mycorrhizal farmers (Jeffries, 1987.)

The technology practicising in France and other European Countries convincingly establishes that the dual purpose inoculation of samplings with appropriate ectomycorrhizal fungi is practicable. Although many mycorrhizal fungi have been identified as edible, there have been no considerable attempt to exploit these macromycetous fungi for mushroom production in India. Singh (2001) has reported 28 species of ectomycorrhizal fungi from Tikari forest of Gonda which belonged to both basidiomycetes and ascomycetes. The basidiomycetous forms were species of *Amanita, Boletus, Cortinarius, Suillus, Laccaria, Inocybe, Lactarium, Russula, Tricholoma, Thelephora, Scleroderma* and

Pisolithus. Certain such ascomycetous members included *Tuber, Helvella* and *Morchella* etc.

Besides edible species some are poisonous, a few of them are dangerous and detrimental to human health while others are used as medicines and they are operative upon the body on the principle 'the use of poison' as antidote for poison' (Yang and Jong, 1987). A few edible mushrooms have also gained importance in modern medicine for their various pharmacological values. The prominent medicinal ingredients of mushrooms are mostly polysaccharides which are different in their composition, mode of connections, degree of ramification and polymerization etc., yet they all are immunostimulants with no poison. All of them have effect of strengthening health and increase immunity.

Mushrooms have been extensively studied in western countries, while India in this respect is relatively less explored. Earlier contribution to mushrooms in India were made by foreigners, mostly from the places of tourist interests. Lt. Col. Kirtikar was first to record some mushroom from Calcutta in 1918.

A survey of the literature on Indian Agarics reveals that Natrajan and his colleagues in South India, Sathe and co-workers in South West India, Kapoor and his students around Delhi, Purkayastha and Chandra (1985) in West Bengal and Lakhanpal *et al.* (1986) in North Western Himalayas explored the mushroom flora of respective areas.

A few reports by Walting and Gregory (1980) from Kashmir, a revision of Berkeley's work from eastern Himalayas by Horak (1980) and the Himalayan stretch in India with its abundant mycoflora was explored by Saini and Atri (1981, 1982a, 1982b); Shad (1985, 1989); Rawla and Sarwal (1983). In H.P. University, Shimla, a systematic work on mushroom and Toadstools of North Western Himalayas was started in 1976 and carried out by Abraham et al., (1980); Hongo (1965); Kaul and Kachroo (1974); and Kaul *et al.* (1978) Where 6500 collections of mushrooms and Toadstools belonging to 223 species, 59 genera and 15 families of Agaricales were made. Studies on fleshy fungi of Garhwal Himalayas was made by Bhatt et al., when they reported four species of genus Russula (*R. albida, R. amoenolens, R. brevipes var. acrior, R. dissimulans*) in 1995; six of *Amanita* (*A. albocreata, A. muscariavar regalis, A. pachycolea, A. peckiana, A. phalloides and A. virosa*) in 1999 and nine species of *Lactarius* (*L. indigo var indigo, L. zonarius var reparius, A. piperatus var glaucescence, L. subdulcis, L. camphoratus, L. fuliginosus, L. piperatus var piperatus, L. Sanguifluus and L. subpurpurescens*) in 2000 when their availability was authenticated.

Very little informations are available about the wild edible fungi of

Madhya Pradesh where sporadic attempts have been made by Dadwal *et al.* (1989); Rai (1992) Harsh *et al.* (1993) and Rai and Sharma (1996). They concerned primarly with the edible fungi, particularly in respect of their habitats, time of collection and possible marketing potentials.

Although, Assam is very rich in fungal flora growing in shady forest floors as well as on the cultivated land due to congenial climatic conditions but till recently very few species of mushrooms have been reported from this region. Kalitha and Coworkers have reported twenty different fungal flora including some popular species of *Agaricus, Lepiota, Pleurotus,* and *Termitomyces,* Gogoi *et al.* (2000) on the other hand, have listed seventeen mushroom species from stateof Assam which included the species of *Cantharellus, Chlorophyllum, Coprinus, Hygrophorus, Lactarius, Lepiota, Lycoperdon, Marasmius, Pholiota, Pleurotus, Termitomyces* and *Tricholoma.*

Studies on Agarics from wild in Punjab was carried out by Atri *et al.* (2000) when six taxa of genus *Lepiota* viz; *L. oregonensis var macrospora, L. parvannulata, L. luteophylla, L. josserandii, L. thierssii and L. luteocephala* were reported.

Four Agaric species viz., *Agaricus caribeus, A. endoxanthus, Lepiota pyrrhaes* and *Leucoagaricus hortensis* were recorded from Western Ghats region of Kerala (Vrinda *et al.*, 1999) and a large bolete (*Phlebonus portentosus*) was reported new to India from this region (Vrinda *et al.*, 2000).

Objectives

North estern part of India with its varied topography, diversified vegetation and climatological fluctuations can be successfully exploited for the growth of macrofungi in wild habitat. The need of today is to explore lesser known species to enrich our dish and improve the existing strains through selection, breeding and by use of mutagenic agents. In spite of the fact that this region is rich in the resources of edible fungi, no planned efforts have been made so far to collect and conserve their germ plasm from the forests and hilly areas. A fragmentary report of six important mushroom species has been made only by Singh and Prasad (1989). The tarai belt of North eastern Himalayas, particularly is famous for bumper production of rice crop leaving its straw in abundance constituting huge bulk of lignocellulose which is utilized by higher fungi as their food.

There is vast cope of study of edible fungi which grow wild in forests, orchards and grass lands of this region. This part of state alone is glorified by the forests and grass lands which come under foot hills of Nepal. These forests and grasslands should be explored for the edible fungi growing there which will be in benefit of people of these regions in some way or other. There is also a vast scope for new taxa and for the availability of better strains whose germplasm cane be collected and used in breeding experiments to improve the variety of mushroom. In view of above, present study has been carried out which is based on observation of mushroom growing in wild habitats during 1997-98 and 1998-99 in Faizabad and adjacent tarai belt of Gonda district.

Methodology

Present study was carried out from June to October during 1997-98 and 1998-99. The specimens were collected from the grasslands, woodlands, deciduous forests, different orchards, road side trees, ruined buildings, huts, old mud flats and from starw and dung heaps.

Full bloomed and complete sporocarps of fresh specimens were collected from their natural habitats. They were then kept in sterilized polythene packets separately and sealed with stapples. These were then brought to the laboratory for examination. Date of collection, locality, nature of substrate, odour, colour of fresh specimen and the nature of latex, if present, were noted. Their ecological habitats and occurrence in solitory or groups were also recorded. For preparation of spore prints, stipe of the fruitbody was cut and cap was set on a piece of paper (for white-spore fungi black paper was used and half white and half black for any other mushroom) turning the gill downwards. A drop of water was mounted on the cap in order to minimise the drying of tissue. The material was then placed in a container and incubated for sometimes (2-24 hrs, depending on the nature of fruit body) at 20-25^{0}C. Lower temperature is generally preferred in order to minimise bacterial contamination. Finally, the cap was removed carefully and characteristics of spore-print were noted. The slides of the section of gill were prepared and specimen were preserved in a mixture of distilled water, alcohol and formalin in ratio of 70: 25:5.

Histochemical tests of basidiocarp was performed with various chemicals which produced specific colour reaction on contact with various structures of basidiocarp. The macro chemical reagents used in this study were Melzer's reagent, Ferric Chloride and Ferrous Sulphate.

Observations

Number of Species

During two years of observation greater number of species were recorded during 1997-98 and comparatively their lesser number in 1998-99. The species

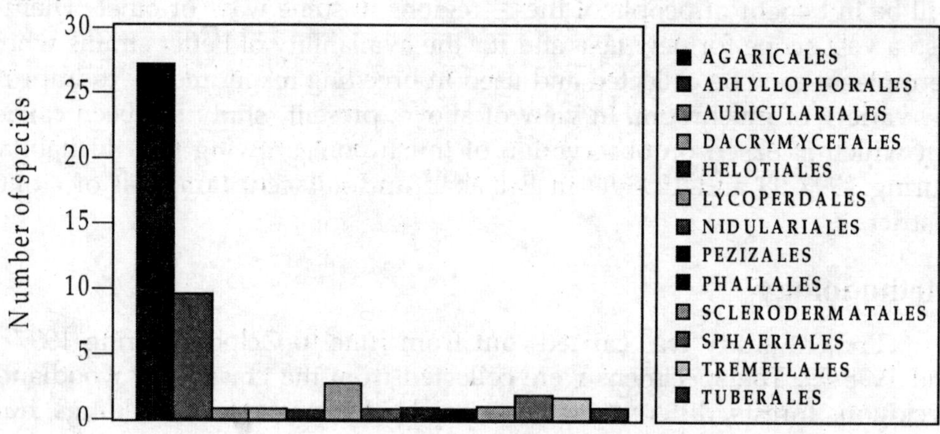

Fig. 1. Fleshy fungi recorded in 1997-99

were greater in number in month of september during 1997-98 and in August in 1998-99. The lesser number of species was noticed in July and October in both the years of observation. Amongst species recorded during the observation, Agaricales were predominantly occuring forms while Auriculariales, Dacrymycetales, Helotiales, Nidulariales, Sclerodermatales, Phallales, Tremellales and Tuberales were represented by their very fewer number. Aphyllophorales were next to Agaricales while Sphaeriales, Lycoperdales, Pezizales were represented by their moderate numbers (Fig.1).

Amongst different families of agaricales, tricholomataceae was predominantly occurring followed by Russulaceae, Coprinaceae, Agaricaceae and Cortinariaceae. The Strophariaceae, Amanitaceae, Pluteaceae, boletaceae,entolomataceae were comparatively fewer in their numbers while Bolbitiaceae, Hygrophoraceae, Crepidotaceae and Paxillaceae were represented by their 4,3,2 and 1 species, respectively (Fig.2).

Among different familes of order Aphyllophorales, Polyporaceae was dominant in its occurrence followed by Clavariaceae, Stereaceae and Thelephoraceae which were represented by their moderate numbers. The Ganodermataceae, Hydnaceae, Schizophyllaceae and Aurisclepiaceae were represented by their 2,2,2, and 1 species, respectively (Fig.3).

Species of different ecological groups

The species recorded during observation were found growing on different types of substrates and accordingly they were grouped into: (i) species growing on fresh plant residues, (ii) species growing on partially decomposed plant residues, (iii) species growing on very decomposed plant

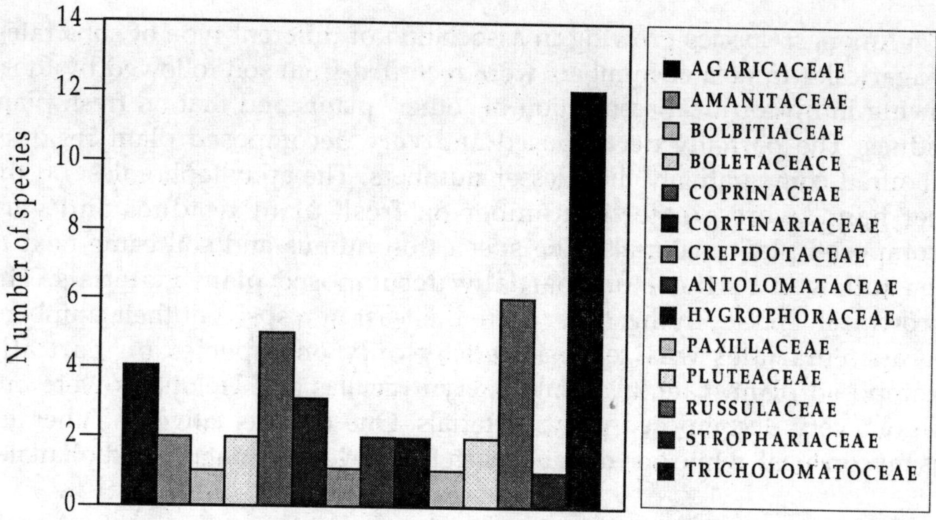

Fig. 2. Families of Agaricales reported

materials, (iv) species growing on soil or humus, and (v) Mycorrhizal species, which grow in association with root of plants.

Greater number of species was recorded from soil followed by mycorrhizal forms, the fungi occurring on very decomposed plant materials exhibited their moderate number followed by those on fresh plant residues. The partially decomposed plant materials harboured the smallest number of species.

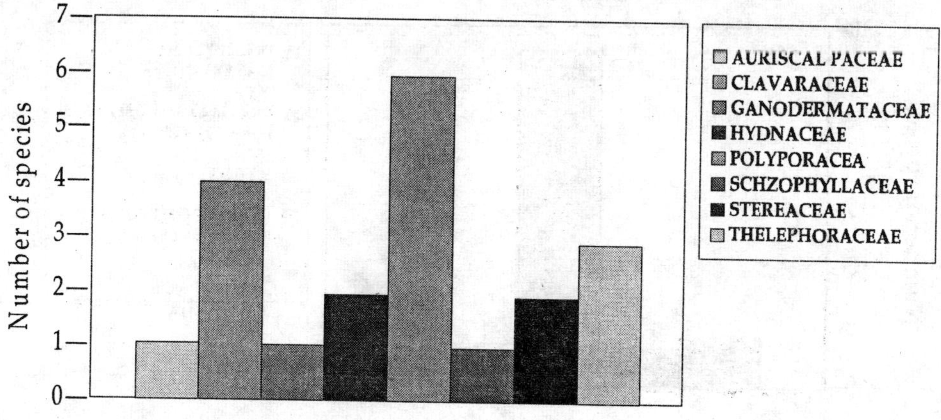

Fig. 3. Families of Aphyllophorales reported

Amongst species growing in association of different type of substrates, the agaricales in greater numbers were recorded from soil followed by those growing in mycorrhizal association of other plants and that on fresh plant residues. The partially decomposed and very decomposed plant residues harboured comparatively their lesser numbers. The aphyllophorales, on the other hand, were greater in number on fresh plant residues and very decomposed plant materials. The species on humus and soil came next to them followed by those on partially decomposed plant materials. The mycorrhizal species in this order were the least in respect of their numbers. The auriculariales was represented by only one species on partially decomposed plant materials while dacrymycetales and Helotiales were one each on very decomposed plant materials. One representative of Tuberales was mycorrhizal while one member each from sclerodermatales and phallales

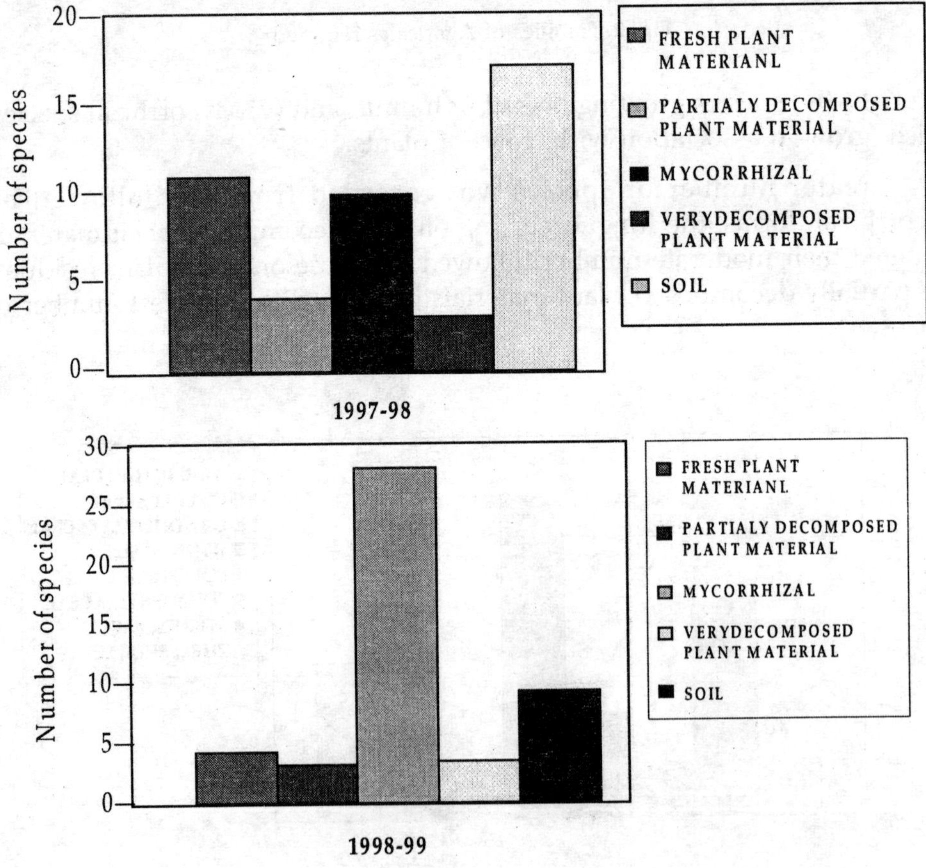

Fig. 4. Species of different Ecological group

were recorded from soil. Tremellales were represented by two species on very decomposed plant materials while lycoperdales were six in number recorded from soil. Nidulariales was one each on very decomposed plant materials and soil while pezizales was represented by four species from soil only. Three species of sphaeriales were recorded on very decomposed plant materials and five from soil.

The mushrooms recorded on fresh plant materials included *Coprinus radians, Craterellus cornucopoides, Crepidotus variabilis, Galerina mutabilis, Ganoderma applanatum, Laccaria lacata, Lentinus squarrulosus, Oudmansiella radicata, Pholiota destruens, Schizophyllum thelephoraterestris, Sterium hirsutum, Thelephora palmata, Xeramphelina campanella, Heterobasidium annosum and species of Flammulina, Phellinus, Pleurotus* and *Polyporus.*

The species growing on partially decomposed plant materials were: *Armillaria tabescens, Auricularia mesentrica, Gerronema Chrysophyllum Hericum erinaceum, Lentinus, Cladopus, Pholiota aurivella, Pluteus pallitus, Polyporus betulinus, Schizophyllum commune, Spongipellis pachydon, Thelephora terrestris* and *Tremetes versicolor.*

The mycorrhizal forms included the species of *Amanita, Boletus, Clavaria, Inocybe, Lactarius, Russula, Tricholoma* and *Cantherellus cibarius Cystoderma amianthianum, Hebeloma Crustuliniforme, Leccinum carpinum, Leucocortinarius bulgiber, Paxillus involutus, Elaphomyces granulatus* and *Entoloma sinuatum.*

The species growing on very decomposed plant materials were: *Calocera viscosa, Cyathus, striatus, Daedalia quercina, Daldinia concentrica, Fomes fomentarius Helvellallacunosa, Hydnum coralloides, Laccaria amythystes, Lacrymaria velutina, Lentinellus ochleatus, Lenzites betulinus, Oudmansiella sp., Sclerotinia fructicola, Sparasis crispa, Tremella reticulata, Trametes americana* and species of *Mycena, Pleurotus, Marasmius, Stereum* and *Xylaria.*

Species harbouring soil comprised *Agrocybe dura, Armillaria mellea, Asterospora lycoperoides, Bolbitius vitellinus, Boletus impolitus, Calocybe gambosa, Cordiceps militaris, Cryptotrama chrysopeplum, Cystolepiota bucknellii, Gyromita seculenta, Hygrophorus chrysodon, Leccinum crocipodium, Lycoperdon gigantia, Megacollybia platyphylla, Microglossum virdae, Morchella esculenta, Naucoria semiorbicularis omphalina epichysium, Phallus impudicus,Pheocollybia festiva, Teberia conspersa, Verpa vohemica* and species of *Agaricus, Bovista, Clavaria, Clitocybe, Collybia, Coprinus, Cortinarius, Geastrum, Hygrocybe, Hypholoma, Inocybe, Lactarius, Lepiota, Lepista, Marasmius, Micromphale, Paneolus, Pluteus, Psathyrellus, Russula, Scleroderma, Termitomyces, Tricholoma* and *Volvariella.*

Species of different utility group:

On the basis of utility of species in different forms these were grouped into: (1) Excellently edible, (ii) edible, (iii) Inedible, (iv) poisonous, and (v) mediocre species.

A total of 274 mushroom species recorded during study comprised 20 species of excellent edibility. These were larger in size and fleshy in texture with possibility of high nutritive value, excellent taste and aroma. These were *Agaricus augustus, A. bitorquis, A . compestris, A. macrosporus, Agrocybe dura, Armillaria mellea, Calocyble gambosa, Coprinus comatus, Flammulina velutipes,*

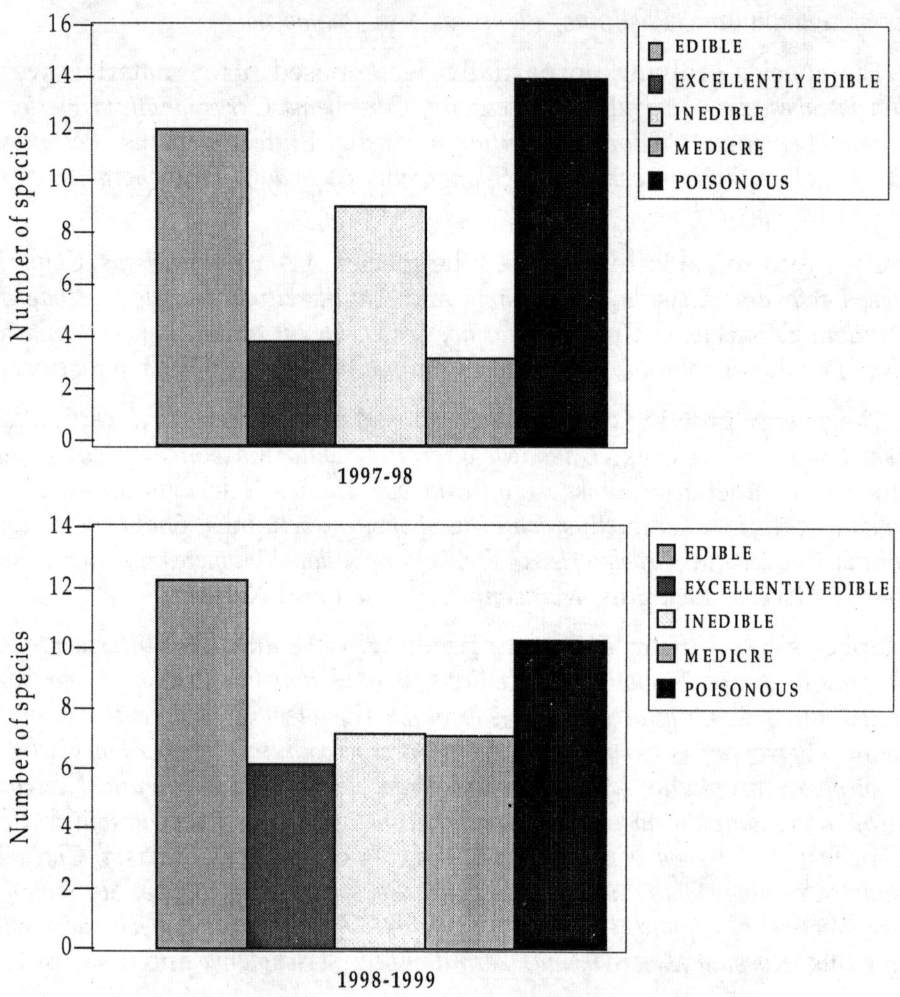

Fig. 5. Species of different utility groups

Lepiota procera, Lycoperdon giganita, Megacollybia platuphylla, Pleurotus eous, P. eryngii, P. ostreatus, P. pulmonaris, Russula cynoxantha, Stropharia ferri, Volvariella fibrillosa and *V. indica.*

The edible species came next to the excellently edible category and they have been considered good in respect of their edibility, however, they have not been considered suitable for commercial cultivation. These species were 72 in number and comprised various species of *Amanita, Boletus, Clavaria, Clitocybe, Collybia, Coprinus, Entoloma, Geastrum, Lactarius, Lepiota, Marasmius, Micromphale, Mycena, Panaeolus, Polyporus, Russula, Stereum, Trameta, Collybia, Coprinus, Cortinarius, Laccaria, Lepiota, Lepista, Pholiota, Pleurotus, Psathyrella, Termitomyces* and 45 other mushroom speices.

The inedible species were not cosidered utilizeable in form of food because they are ephemeral and decompose soon after their blooming, others are very hard and stiff in texture and remaining are too small to be used as food and these are considered useless. Such species are: *Marasmius brunneolus, Amanita vaginata, Asterophora lycoperoides, Auricularia mesentrica, Bolbitius eleuriales, B. vitelinus, Boletus aereus, B. elegans, Tricholoma albobrunneus, Tuberia conspersa* and *Verpa vohemica.*

The mediocre species are those whose medicinal properties have already been explored. This category of species included: *Amanita fulva, Boletus pseudoscaber, Bovista nigrascens, B. plumbea, Collybia butricia, C. radicata, Cystoderma amianthinum, Hygrocybe brevespora, Leccinum caprinum, Oudmansiella radicata, Pluteus cervinus, Russula delica, R. lepida, Volvariella bomacina.*

The poisonous species were *Agaricus xanthodermus, Amanita pantherina, A. phalloides, A. verna, A. virosa, Clitocybe dealbata, Entoloma sinuata, Hebeloma crustuliniformae, Hygrocybe nigrascesn Inocybe asterospora, I. fastigiata, I. geophylla, I. napipes, Lactarius fuliginosus, Lepiota cristata, paneolus atra,. Russula emetica* and *Tricholoma vergatum.*

Description of Some Important taxa:

During the survey of fleshy fungi of Faizabad and Gonda districts in the year 1997-98 and 1998-99, 274 species were collected of which following eighteen species were considered new to India. The specimens were deposited in Mushroom museum of Botany Department, K.S. Saket Post Graduate College, Ayodhya (Faizabad) on their respective accession numbers.

Amanita umbrinolutea Secr. (Fig. 6)

Habitat: Mycorrhizal; *Date of collection:* 18.8.1997; *Locality:* Ranopali, Ayodhya

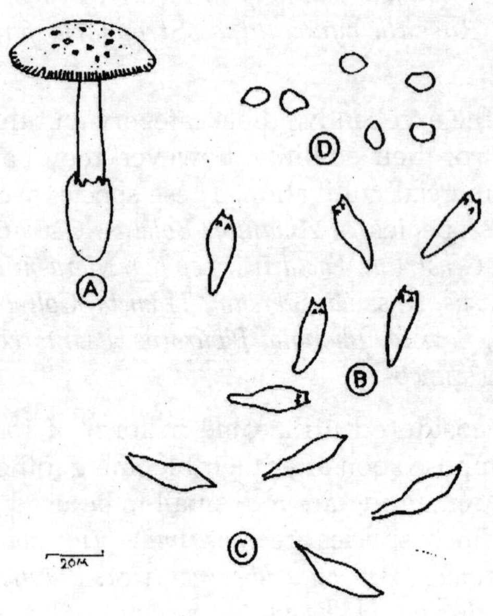

Fig. 6. *Amanita umbrinolutea* Secr. A. Basidiocarp; B. Basidia; C. Cystidia; D. Basidiospores

(Faizabad); *Pileus:* 6-11 cm in diameter, conical than expanded, colour ochre grey, veilar remain present on pileus surface, non-hygrophanous, margin striate, flesh white, soft and odourless; *Stipe:* 9-10 x 0.8-1.5 cm in size, colour light ochre grey, narrow at the top, hollow when mature, velum absent volva thin, membranous sheath like, white flesh, fragile, odourless; *Gill:* Crowded, free, long, white, equal, hymenophoral trama irregular, dermatocystidia 30-35 x 11-14 µ in size; *Basidium:* 25-30 x 10-11 µ in size, tetrasterigmatic; *Spore print:* White; *Spore:* White, smooth, elliptical, non-amyloid and 9-11 x 7-9 µ in size; *Odour:* Odourless; *Edibility:* Edible

Calocybe gambosa (Fr.) Sing. ex Donk. (Fig. 7)

Habitat: Saprophyte, growing in grass land, solitory; *Date of collection:* 5.9.1998; *Locality:* Takpura, Faizabad; *Pileus:* 6-10 cm in diameter, hemispherical, convex bright pale colour, glabrous, margin incurved, cuticle easily dettachable, smooth, cracked at maturity, non-hygrophanous, flesh white and about 1 cm thick, agreeable odour; *Stipe:* 7-9 x 1.5 cm in size, central, cylindrical, base bulbous, white, surface dry, fibrillose, solid, flesh compact, thick, white, annulus and ulva absent; *Gill:* adnate, very

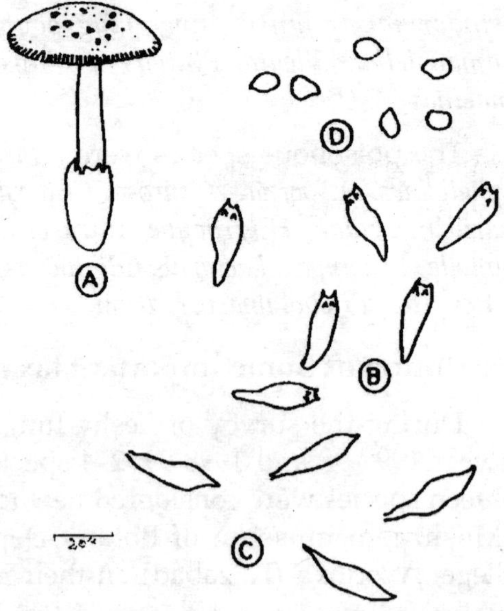

Fig. 7. *Calocybe gambosa* (Fr.) Sing. ex Donks : A. Basidiocarp; B. Basidia; C. Cystidia; D. Basidiosproes.

crowded, creamy white, emarginate, separable, unequal, hymenophoral trama regular, denmatocystidia 30-32 x 10-13 μ in size; *Basidium:* Clavate, tetrasterigmatic and 20-232 x 10-12 μ in size; *Spore-print:* White colour; *Spore:* White, smooth, elliptical, small non, amyloid and 6-7 x 4-4.5 μ in size; *Odour:* Pleasant; ***Edibillity:*** Excellently edible.

Cystolepiota bucknallii (Berk. S. Br.) Sing. & Clemencon (Fig. 8)

Habitat: Saprophyte, growing in garden soil; *Date of collection:* 10.8.1997; *Locality:* Tulsingar, Ayodhya (Faizabad); *Pileus:* 4-8 cm in diameter, hemisphaerical with a slight central hump, glabrous, margin incurved and minutely toothed, cuticle grey in colour and easily dettachable, non- hygrophanous, flesh brown, thick, unpleasent smell; *Stipe:* 6-8 x 0.5-1.2 cm in size, grey in colour, central , dark and narrow towards base, smooth, solid, annulus present, flesh compact brown with disagreeable smell; *Gill:* Moderately crowded, free, brown in colour, soft, thin, hymenophoral trama regular, cheilocysti dia clavate and 28-38 x 11-14 μ in size; *Basidium:* Globose, tetrasterigmatic and 21-24 x 10-14 μ in size; *Spore print:* Brown in colour; *Spore:* Hyaline, small cynophilic, smooth, elliptical, inamyloid and 6-7 x 4-5 μ in size; *Odour:* Disagreeable smell; *Edibility:* Not knwon.

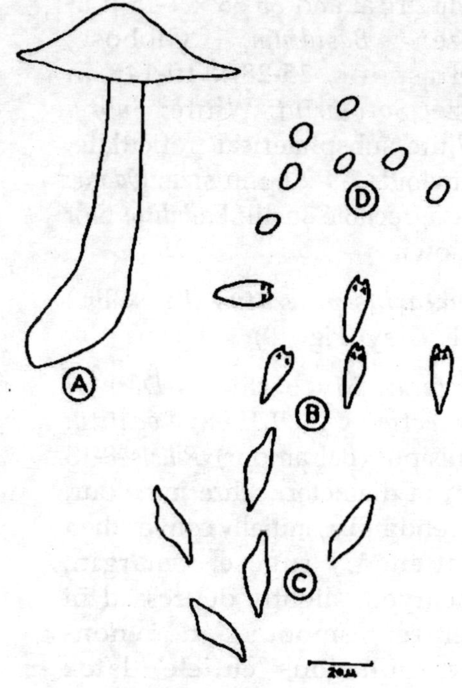

Fig. 8. *Cystolepiota bucknallii* (Berk & Br.) Singh. clemencon : A. Basidiocarp; B. Basidia; C. Cystidia; D. Basidiospores

Lactarius glaucescens Gressel (Fig. 9)

Habitat: Teak stands of Jarwa forest; *Date of collection:* 7.8.1997; *Locality:* Jarva forest (Balrampur); *Pileus:* 7-13 cm in dia meter, white, initially convex then funnel shaped, surface smooth, dry, margin incurved, flesh white and hard with disagreeable smell, latex white; *Stipe:* 3-6 x 1.0 –2.0 cm in size, white, cylindrical, equal in thickness, flesh creamy white turning bluish green when cut, odourless; *Gill:* Moderately crowded, very decurrent, creamy, white in colour, hymenophoral trama irregular with sphaerocystidia, pleurocystidia cylindrical and 35-38 x 1-15 μ in size; *Basidium:* Globose, tetrasporic, 25-28 x 10-14μ in size; *Sore-print:* White; *Sopre:* White subsphaerical, reticulate, amyloid, 7-9 x 6μ in size; *Odour:* Disagreeable smell; *Edibility:* Not known

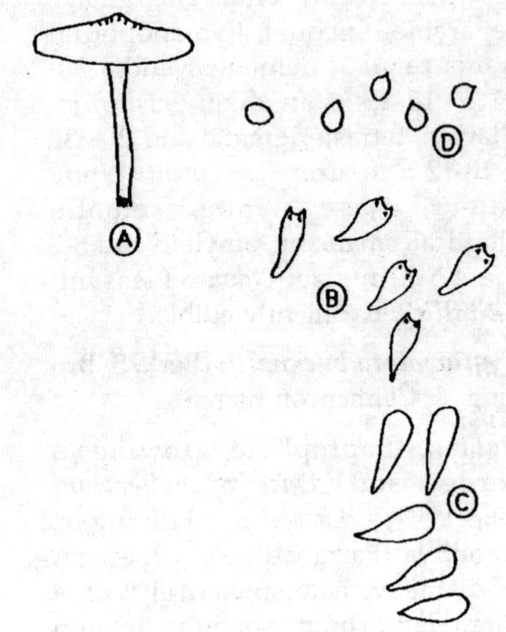

Fig. 9. *Lactarius glaucescens* Crossl. : A. Basidiocarp; B. Basidia; C. Cystidia; D. Basidiospores.

Lactarius piperatus (L. ex Fr.) S.F. Gray (Fig. 10)

Habitat: Mycorrhizal; *Date of collection :* 20.9.1998; *Locality:* Tulsipur (Balrampur); *Pileus:* 8-13 cm in diameter, ochre in colour, infundibular, initially convex then flattened, rugose, margin, incurved, slightly depressed in centre, smooth and non-hygrophanous cuticle, latex white, unchanged on exposure; *Stipe:* 5-8 cm in size, cap colour, cylindrical, purinose, hollow when mature, annulus and Volva absent, flesh white, hard, odourless; *Gill:* Creamy white, crowded, subdecurrent, forked, unequal, trama heteromerous with shpaerocystidia, pleurocystidia pyriform with 30-35 x 10-13 μ in size; *Basidium:* 24-27 x 8-11μ in size, globose and quadristerigmatic; *Spore-print:* Creamy white; *Spore:* White, well ornamented, rough, subsphaerical and 7-8 x 5-7μ in size and amyloid; *Odour:* Unpleasant; *Edibility:* Not Known.

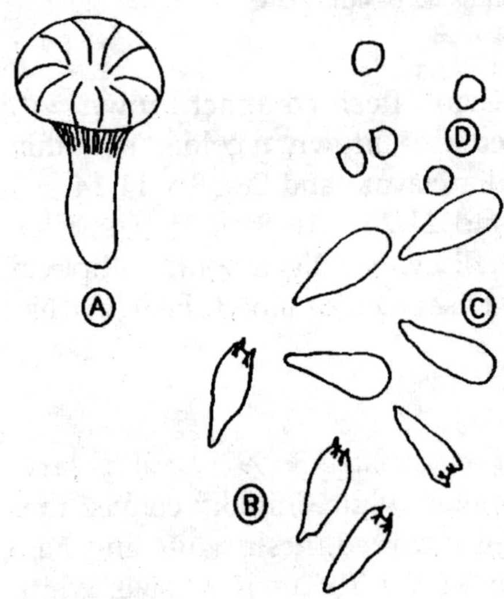

Fig. 10. *Lactarius piperatus* (L. ex Fr.) S. F. Gray: A. Basidiocarp; B. Basidia; C. Cystidia; D. Basidiospores.

Lactarius uvidus (Fr. ex. Fr.) Fr. (Fig. 11)

Habitat: Damp soil of mixed forest; *Date of collection:* 5.10.1998; *Locality:* Tikari forest (Gonda); *Pileus:* 8-13 cm in

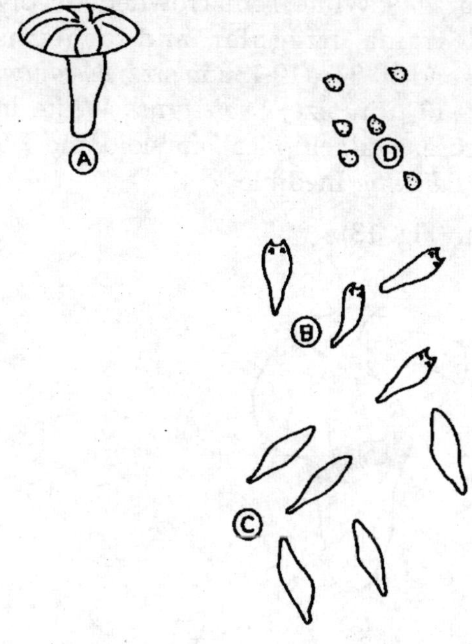

Fig. 11. *Lactarius uvidus* (Fr. ex Fr.) Fr. : A. Basidiocarp; B. Basidia; C. Cystidia; D. Basidiospores.

diameter, colour brownish grey, convex, centrally depressed, margin incurved and cuticle thin, non-hygrophanous, flesh thin and turns lilac when exposed in air, latex white turning lilac when exposed in air; *Stipe:* 4-7 x 0.8- 1.2 cm in size, colour brownish grey, thickness equal, rugose, hard, annulus and volva absent, odourless, on bruising turns lilac; *Gill:* Crowded, semidecurrent, white, on maturity turning plae brown in colour, unequal, hymenophoral trama heteromerous with sphaerocystidia, pleurocystidia lecythi form and 30-35 x 12-15µ in size; *Basidium:* Quadristerigmatic, globose and 22-25 x 12-14µin size; *Spore-print:* Ochre in colour; *Spore:* Ochre in colour, ellipsoidal, ornamented, amyloid and 10-12 x 5-8µ in size; *Odour:* Unpleasant; *Edibility:* Not known.

Lactarius vellereus (Fr.) Fr. (Fig. 12)

Habitat: Mixed forest, sandy soil, scattered; *Date of collection:* 5.10.1998; *Lacality:* Tikari foest (Gonda); *Pileus:* 10-15 cm in diameter, colour white, very depressed in centre, infundibuliform, surface velvety, smooth, glabrous, non-hygrophanus, margin incurved and irregular, flesh white and thick latex white on exposure turns light grey; *Stipe:* 5-8 x 2-2.5 cm in size, white, soft, smooth, fibrous, cylindrical, hollow, narrow at the base and thick at the top, flesh

Fig. 12. *Lactarius vellereus* (Fr.) Fr. : A. Basidiocarp, B. Basidia; C. Cystidia; D. Basidiospores.

white and turns ochre when exposed in air; *Gill:* White, semi crowded, deeply decurrent, unequal, hymenophoral trama irregular and contains sphaerocystdia, pleurocystidia lenceolate and 30-32 x 10-13µ in size; *Basidium:* Clavate, tetrasterigmatic and 27-30 x 10-13 µ in size; *Spore-print:* White in colour; *Spore:* White, ornamented, verrucose, oval, elliptical, amyloid and 7-9 x 6-7µ in size; *Odour:* Unpleasant smell; *Edibility:* Inedible

Lepiota acutesquamosa (weinm.) kumm. (Fig. 13)

Habitat: Saprophyte, growing in garden soil; *Date of collection:* 10.10.1997; *Locality:* Darshan Nagar (Faizabad); *Pileus:* 8-13 cm is diameter, pluteoid, hemisphaerical, conical, brown colour, cuticle rough and contains spiny dark brown scale, non-hygrophanous, dettachable, margin curved with veilar remain, flesh white strong with unpleasant smell; *Stipe:* 7-12 x 1.0-1.2 cm in size, brown in colour, cylindrical with enlarged base, fragile, velum present, flesh white, strong with unpleasant smell; *Gill:* White, free, crowded, forked, unequal, hymenophoral trama regular, cheilocystidia lecythiform and 28-32 x 10-13 µ in size; *Basidium:* Globose, tetrasterigmatic and 23-26 x 9-11µ in size; *Spore-print:* White in colour; *Spore:* White, elliptical, pseudoamyloid, 6-8 x 3.5-5µ in size; *Odour:* Disagreeable smell; *Edibility:* Inedible

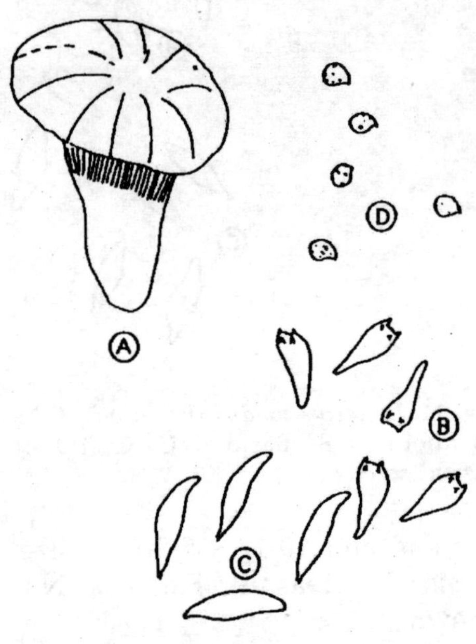

Fig. 13. *Lepiotal acutesquamosa* (Weinm.) Kummer : A. Basidicarp; B. Basidia, C. Cystida; D. Basidiospores.

Lepiota alba (Bres) Sacc. (Fig. 14)

Habitat: Pluteoid, scattered, growing in grassland; *Date of collection:* 7.8.1997; *Locallity:* Tulsinagar, Ayodhya (Faizabad); *Pileus:* 5-8 cm in diameter, pale brown in colour, campanulate and finally expanded, cuticle smooth, dry, few small scales scattered on the surface, slightly dark in centre, margin slightly incurved and splitted at few places, flesh thin and brown; *Stipe:* 6-8 x 0.7-1.1 cm in size, cylindrical, equal in thickness, cap colour, veil present, fragile, flesh thick brown; *Gill:* Pale brown, not crowded, unequal, free, hymenophoral trama regular, cheilocystidia clavate and 26-29 x 10-13µ in size; *Basidium:* Globose, tetrasterigmatic, 20-23 x 10-11µ in size; *Spore-print:* Pale brown in colour; *Spore:* Oval, smooth, pale brown, pseudoamyloid and 7-9 x 5-7µ in size; *Odour:* Unpleasant; *Edibility:* Not unknown

Lepiota felina (Pers ex Fr.) Karst. (Sensu Quel.) (Fig. 15)

Habitat: Grass land, pluteoid, solitory; *Date of collection:* 9.8.1997; *Locality:* Katra (Gonda); *Pileus:* Colour pale brown, 5-8 cm in diameter, campanulate and finally flat, margin incurved with veil ar remain, cuticle rough and coveredwith circularly arranged fibrillose scale, dry, flesh thin, brown in colour; *Stipe:* 4-6 x 0.6 -0.9 cm in size, covered with cottony brown fibre below the veilar ring, base slightly enlarged and distinguishable, cylindrical, fragile, flesh white or brown in colour; *Gill:* Brown in colour, moderately crowded, free, unequal, hymenophoral trama regular, cheilocystdia lecythiform and 27-31 x 11-14µ in size; *Basidium:* Clavate, tetrasterigmatic and 22-25 x 10-11µ in size; *Spore-print:* Light brown; *Spore:* Brown, oval, smooth, pseudoamyloid and 5-6 x 4-5µ in size; *Odour:* Unpleasant; *Edibility:* Inedible

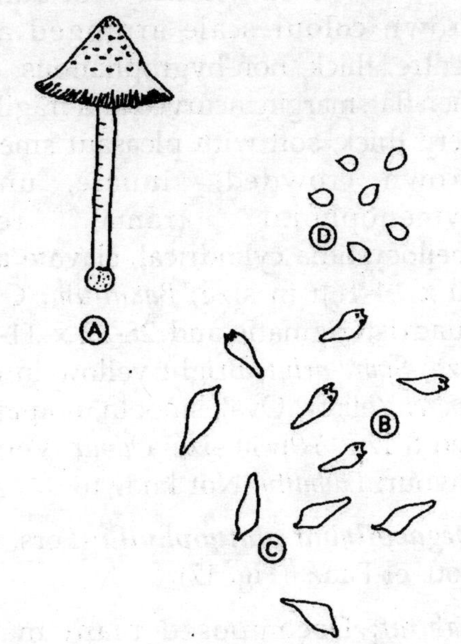

Fig. 14. *Lepiota alba* (Bers.) Sacc. : A. Basidiocarp; B. Basidia; C. Cystidia; D. Basidiosproes.

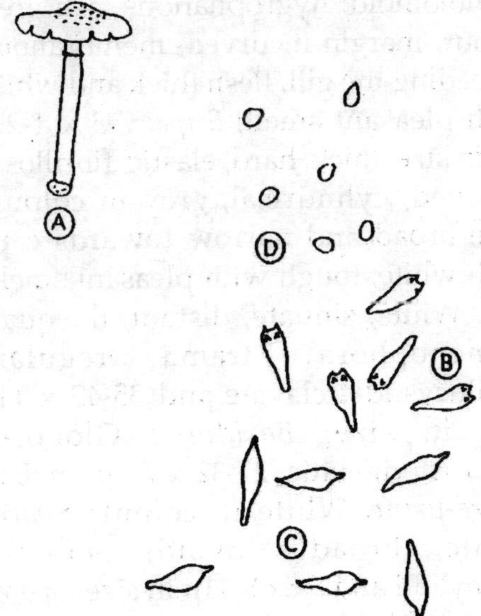

Fig. 15. *Lepiota felina* (Fers. ex Fr;) Karst. (Sensuuel.): A. Basidiocarp; B. Basidia; C. Cystidia; D. Basidiospores.

Lepiota sp (Fig. 16)

Habitat: Road side, in sandy soil, scattered; *Date of collection:* 29.8.1998; *Locality:* Ahmadpur (Faizabad); *Pileus:* 15-20 in diameter, agaricoid, brick brown in colour, cuticle with dark brick brown colour scale arranged around centre, thick, non-hygrophanous, convex then flat, margin incurved and fragile, flesh very thick, soft with pleasant smell; *Gill:* Brown crowded, sinuate, unequal, hymenophoral trama regular, cheilocystidia cylindrical, clavate and 35-40 x 14-16μ in size; *Basidium:* Clavate, quadristerigmatic and 28-32 x 11-14μ in size; *Spore-print:* Bright yellow in colour; *Spore:* Yellow, Oval, smooth, non-amyloid and 8-12 x 6-9μ in size; *Odour:* Very good flavour; *Edibility:* Not known

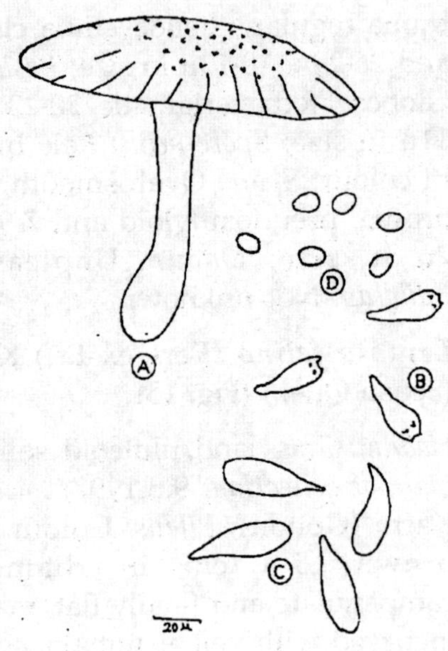

Fig. 16. *Lepiota (Sp.)* : A. Basidicarp; B. Basidia; C. Cystidia; D. Basidiospres.

Megacollybia platyophylla (Fers. ex Fr.) Kotl. et Pauz. (Fig. 17)

Habitat: Decomposed plant material, solitory; *Date of collection:* 10.9.1997; *Locality:* Katra (Gonda); *Pileus:* 10-16 cm in diameter, umbonate,. convex then flat, tricholomoid, hygrophanous, ash grey colour, margin incurved, membranous exceeding the gill, flesh thick and white with pleasant smell; *Stipe:* 7-3 x 1-2.5 cm in size, thick, hard, elastic, fibrillose, grooved, cylindrical, grey in colour, base broad and narrow towards cap, flesh white, tough with pleasant smell; *Gill:* White, sinuate, distant, unequal, hymenophoral trama regular, cheilocystidia clavate and 35-42 x 11-13μ in size; *Basidium:* Globose, quadristerigmatic, 27-32 x 7-9μ in size; *Spore-print:* White in colour; *Spore:* White, broadly ovoid, smooth, inamyloid and 6-8x 8-11μ in size; *Spore-print:* White in colour; *Spoore:* White, broadly ovoid, smooth, inamyloid and 6-8 x 8-11μ in size; *Odour:* Pleasant smell; *Edibility:* Edible

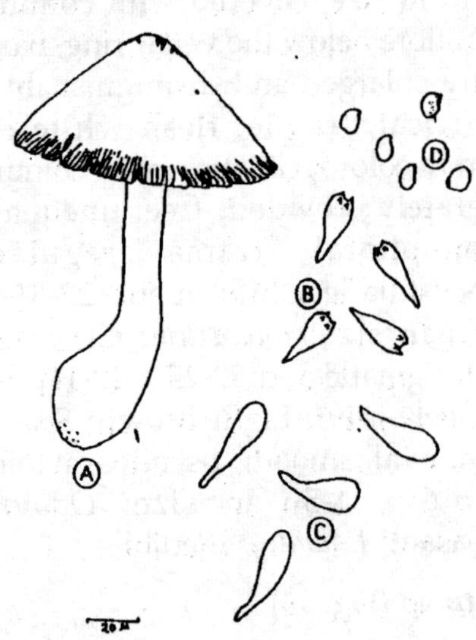

Fig. 17. *Megacollybia platyophylla* (Pers. ex Fr.) Kotl. et Pauz : A. Basidiocarp; B. Basidia; C. Cystidia; D. Basidiospores.

Pholiota lenta (Frs.) Singer (Fig. 18)

Habitat: Garden soil; *Date of collection:* 16.8.1997; *Locality:* Ranopali, Ayodhya (Faizabad); *Pileus:* 6-8cm in diameter, white finally olive yellow in colour, pluteoid, convex finally flat, slightly dark colour in centre, glabrous, hygrophanous, cuticle dettachable, margin incurved, flesh brown, thin with unpleasant smell; *Stipe:* 7-9 x 1-1.4 cm in size, brown colour, covered with brown cottony scales, bulbous base, cylindrical,dry velum short lived, volva absent, slightly curved at the base, flesh white with unpleasant smell; *Gill:* Crowded, adnate, cap colour, unequal, hymenophoral tramaregular, thin walled, pleurocystidia present 30-36 x 12-15µ in size; *Basidium:* Clavate, quadristerigmatic, 28-31 x 10-12µ in size; *Spore-print:* Rusty brown; *Spore:* Rusty brown in colour, smooth, ovoid and 6-8 x 4-6µ in size; *Odour:* Unpleasant smell; *Edibility:* Not known

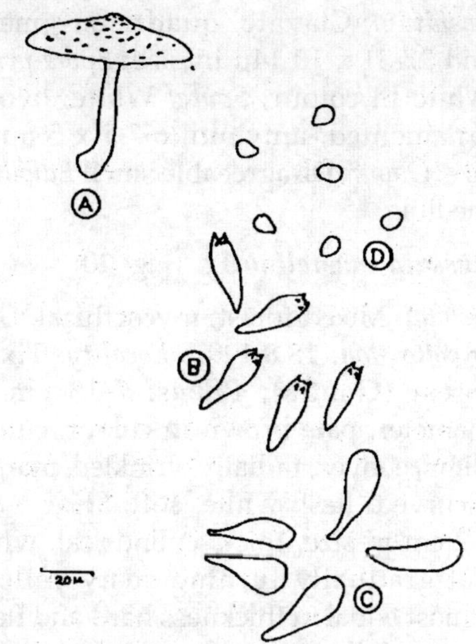

Fig. 18. *Pholiota lenta* (Fr.) Sing A. Basidiocarp; B. Basidia; C. Cystidia; D. Basidiospores.

Russula mairei Singer (Fig. 19)

Habitat: Mixed forest, mycorrhizal; *Date of collection:* 18.91998; *Locality:* Jarwa forest (Balrampur); *Pileus:* 6-8 cm in diameter, blood red colour, very convex than flat, slightly depressed in the centre, margin incurved, irregular, slightly viscous, velvety, flesh white and soft, unpleasant smell; *Stipe:* 3-5 x 1-2cm in size, white slightly reddish yellow, cylindrical, hard, rugose, flesh cream colour, thick and soft with unpleasant smell; *Gill:* White, crowded, unequal, forked,obtuse, intervenose, hymenophoral trama irregular with sphaerocystidia, dermatocystidia

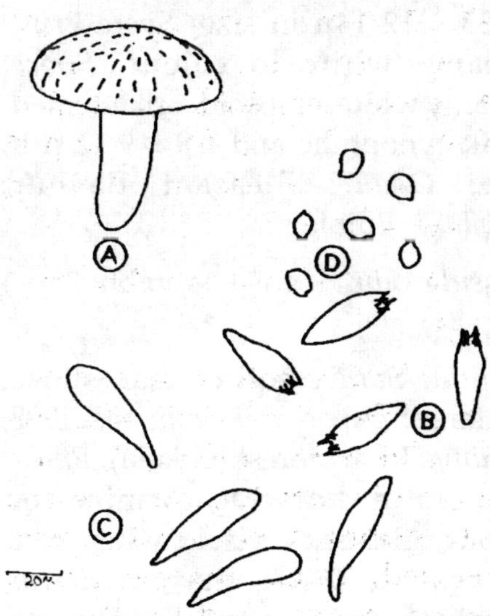

Fig. 19. *Russula mairei*. Sing : A. Basidiocarp; B. Basidia; C. Cystidia; D. Basidiospores.

cylindrical and 35-40 x 11-14 μ in size; *Basidium:* Clavate, quadristerigmatic and 27-31 x 10-14μ in size; *Spore-print:* White in colour; *Spore:* White, ovoid, ornamented, amyloid, 6-7.5 x 5-6μ in size; *Odour:* Disagreeable smell *Edibility:* Inedible

Russula mustelina Fr. (Fig. 20)

Habitat: Mixed forest, mycorrhizal; *Date of collection:* 18.8.1997; *Locality:* Tikari forest (Gonda); *Pileus:* 7-13 cm in diameter, pale brown in colour, cuticle slimy, convex, radially wrinkled, margin incurved, flesh white, soft; *Stipe:* 5-8 x 1-2 cm in size, thick, cylindrical, white but gradually turning rusty yellow, almost equal in thickness hard and flesh white; *Gill:* Adnexed, crowded, thick, white and finally straw colour, flesh tough and white, hymenophoral trama irregular, dematocystidia cylindrical measuring 36-43 x 14-16μ in size; *Basidium:* Clavate, tetrasterigmatic, 31-33 x 12-14μ in size; *Spore-Print:* Creamy white in colour; *Spore:* Creamy white, ellipsoid, ornamented, weak cynophilic and 7-9 x 9-12 μ in size; *Odour:* Pleasant flavour; *Edibility:* Edible

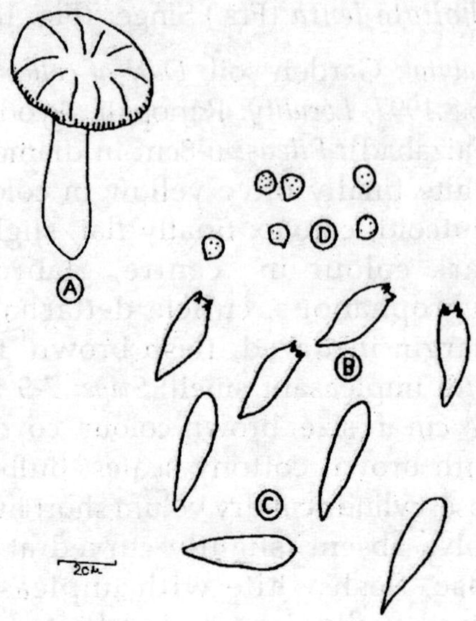

Fig. 20. *Russula mustelina* FR. : A. Basidiocarp; B. Basidia; C. Cystidia; D. Basidispores.

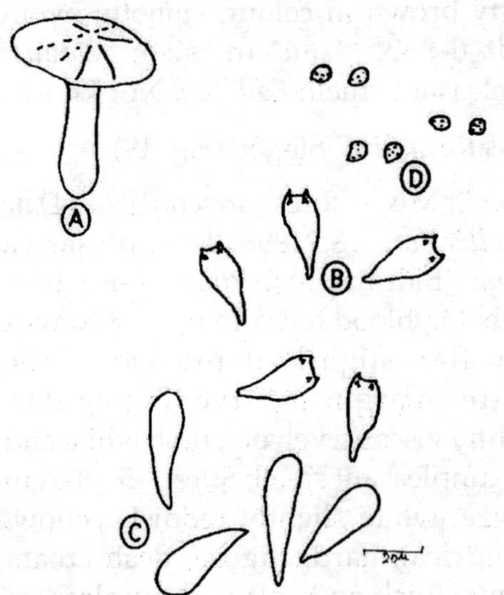

Fig. 21. *Russula rubra.* (Fr. Sensu Romagnesi) : A. Basidiocarp; B. Basidia; C. Cystidia; D. Basidiospores.

Russula rubra (Sensu. krombh. Bers.) (Fig. 21)

Habitat: Sandy soil of teak stand, scattered; *Date of collection:* 10.8.1998; *Locality:* Tikari forest (Gonda); *Pileus:* 5-10 cm in diameter, carmine red colour, glabrous, viscid when wet, depressed, thick, margin acute, incurved, convex and finally flat, hollow in centre, flesh white and thick; *Stipe:* 5-10 x 1.0- 1.5 cm in size, slighly pink in colour, cylindrical, lowerhalf thick, rugose, velum and volva absent, flesh white; *Gill:* Adnexed, crowded, lightpale in colour, unequal, forked, intervenose, hymernophoral trama irregular with sphaerocystidia, dematocystidia clavate 40-42 x 10-13μ in size; *Basidium:* Clavate, tetrasterigmatic, 30-34 x 10-13μ in size; *Spore print:* Pale brown in colour; *Spore:* Pale brown in colour, elliptical, oval, ornamented wall, amyloid, 6-7.5 x 5-6μ in size; *Odour:* Fruity odour with bitter taste; *Edibility:* Not known;

Volvariella fibrillosa (Bers. in Rick) Sing. (Fig. 22)

Habitat: Grassland, Scattered, Centrally stipitate; *Date of collection:* 7.10.1998; *Locality:* Ahmadpur (Faizabad); *Pileus:* 6-13 cm in diameter, brown colour, initially globose, cuticle made up of radially arranged hyphae, non-hygrophanous, margin incurved, fleshy thick, white with pleasant odour; *Stipe:* 7-12 x 0.8-1.6 cm in size, narrow towards top and thick at the base, basal bulbous part covered with volva, cylindrical brown colour cuticle, flesh white with pleasant odour; *Gill:* Free, very crowded, brown colour, uneuqual, delicate, hymenophoral trama inverse, cheilocystidia clavate and 26-30 x 12-14μ in size; *Basidum:* Globose, tetrasterigmatic, 20-24 x 10-13 μ in size; *Spore print:* Brownish pink; *Spore:* Brownish pink in colour, oval, smooth, cynophilic, inamyloid, 7-9 x 6-7 μ in size; *Odour:* Pleasant odour; *Edibility:* Excellently edible.

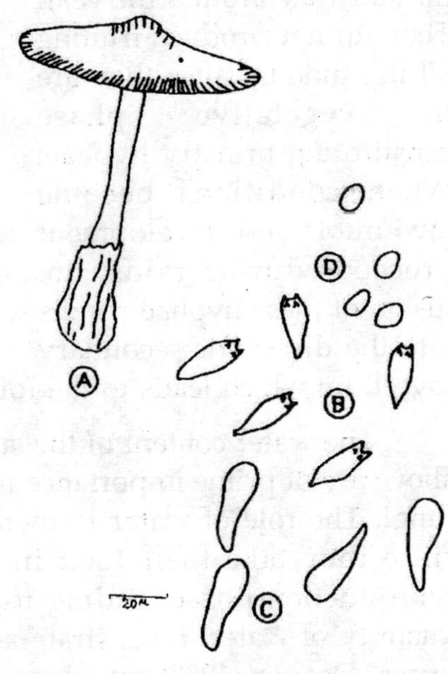

Fig. 22. *Volvariella fibrillosa* (Bres. in Rick) Sing. : A. Basidiocarp; B. Basidia; C. Cystidia; D. Basidiospores.

Marasmius brunneolus (Berk. & Br.) Pelger (Fig. 23)

Habitat: Saprophyte, growing on rotting leaves of bamboo; *Date of collection:* 20.8.1998; *Locality:* Harkara-ka-Purwa, Ayodhya (Faizabad); *Pileus:* 1-1.6 cm in diameter, convex, smooth, thin, pseudoamyloid with pink striation and incurved margin; *Stipe:* 3-6 cm long, black, hairy, thin, central pseudoamyloid; *Gill:* Adnate, fairly distant, thin, brown, unequal with entire margin, regular hymenophoral trama, lecithiform cystidia 30-36 x 11-14μ in size; *Basidium:* Tetrasporic, 21-34 x 9-13μ in size with 2-2.5μ long sterigmata; *Spore-print:* Brick brown in colour; *Spore:* Ellipsoidal, brown, smooth, inamyloid, 6-10 x 3-

4.5μ in size; *Odour:* Garlic like; *Edibility:* Inedible

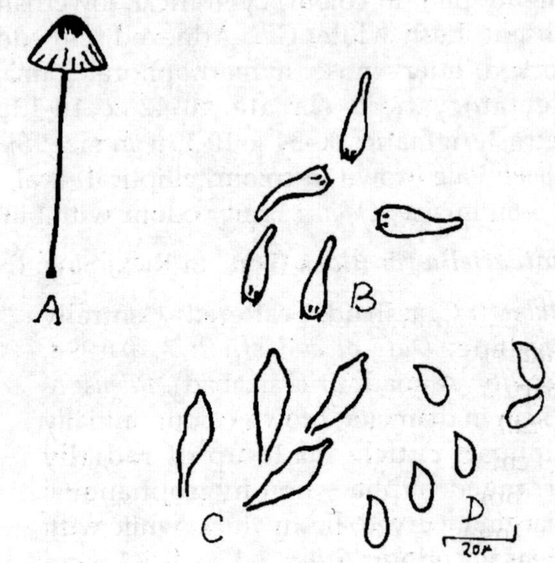

Fig. 23. *Marasmius brunneolus* (Brek., & Br.) Pezler : A. Basidiocarp; B. Basidia; C. Cystidia; D. Basidispores.

Discussion

The basidiocarp of mushroom species are observed only during rainy season inspite of the fact that these are in association with the substrate around the year. They do not produce fruiting all the time because they are in vegetative phase constituting primary hyphae, When conditions become favourable and development proceeds during rains, the fusion of these hyphae results into the dicaryotic secondary mycelium which leads to the formation of basidiocarp.

The water content of the substrate and relative humidity of atmosphere above are of prime importance in controlling the gorwth and reproduction of fungi. The role of water in metabolism of fungi is of increased importance since they take their food in form of solution. In larger fungi, sexual reproduction phase leading to fructification, in general, requires greater quantity of water in substrate and higher relative humidity than vegetative phase. Deacon (1980) mentioned that 70% relative humidity is probably the lowest limit for initiating fruit body formation and 80% during their further development. The atmospheric humidity above 90% is detrimental to good fruiting of mushroom species.

Temperature also has profound effect on growth and reproduction of fungi, the temperature between 25-35^{0}C or below it, in general is suitable for fruit body formation in mushroom species, however, there is a great variation in temperature requirement of individual species (Zaadrazil, 1975).

The greater relative humidity, more rainfall and moderate temperature range probably accounted for more fructification and hence appearance of greater number of species in the month of August and September. Their complete absence during April, May and June may be due to high temperature, low relative humidity of atmosphere and no rainfall rendering soil to its dryness, preventing fruit body formation in these fungi. However, in monthof November, December, January and February these species were not recorded because of low temperatue range and low relative humidity of atmosphere and possibly little or no rainfall in these months leading to sterile nature of mycelium in these forms. The more number of species during 1997-98 than in 1998-99 was due to obvious reasons (Fig 1 and 2).

Bels (1976) categorization of substrate for growth of different groups of mushroom species has a direct bearing on cultivation practices of various such species. Full information on phanerogamic plants in vicinity of these fungi is also important since these provide nutrition by their dead remains that sustain fungal flora. Symbiotic forms are known as mycorrhizal species which grow in association of plant root. The relation of mycoflora with the substrate and their preference for different habitats has profound effect on the appearance and distribution of species.

Very little work has been done on ecology of macro fungi due to their ephemeral nature and uncertainty of appearance, however, it is the first step towards the domestication of such fungi under artificial conditions. Ramsbottom (1953) is of opinion that there is abundant scope for field observation which may help in unrevelling some of the problems in mushroom cultivation.

The development of fruit bodies of larger fungi requires large amout of nutrients and they appear only after a fairly extensive vegetative mycelium has developed. Many of such fungi are characterized by production of enzymes capable of degrading cellulose, hemicellulose and lignin and thus are able to grow on agricultural waste materials. Species of *Pleurotus* are characterized by the rapidity of mycelial growth which facilitates speedy penetration in the substratum by the fungus, simplifying the cultivation as a whole. The growth of mushroom is directly dependent on nutrient present in substratum and their availability to fungus.

References

Abraham, S.P., Kachroo, J.L. and Kaul, T.N. (1980). Fleshy fungi from Gulmarg forests, India, Kavak, **8** : 29-40.

Atri, N.S., Saini, S.S. and Kaur, Amanjeet. (2000). Taxonomical studies on Agarics from Punjab: the genus *Lepiota* (Pers. ex Fr.) Gray. Mushroom Research. **9**(2): 71-77.

Bels, P.J. (1976). In the Biology and cultivation of Edible Mushroom (Eds. S.T. Chang and W.A. Hayes). Academic Press, New York.

Bhatt, R.P., Bhatt, V.K. and Gaur, R.D. (1995). Fleshy fungi of Garhwal Himalayas: The genus *Russula*. Indian Phytopath, **48**(4) : 402-411.

Bhatt, V.K. Bhatt, R.P. Gaur, R.D. and Singh, M.D. (1999): Mushrooms of Garhwal Himalayas: The genus *Amanita* Pers,. ex Hooker. Mushroom Research **8**(2): 1-8.

Bhatt, V.K., Bhatt, R.P. and Gaur, R.D. (2000). Mushrooms of the Garhwal Himalayas: the germs *Lactarius*. pers. ex S.F. Gray. Mushroom Research, **9**(1): 11-18.

Dadwal, V.S., Soni, V.K. and Jamalluddin. (1989). J. Trop. Forest., **5**(1): 85-88.

Deacon, J.W. (1980). Introduction of Modern Mycology. Blackwell scientific publication London.

Gogoi, R. Majumdar, D. and Puzari, K.C. (2000). New additions to mushroom flora of Assam. Mushroom Research, **9**(1): 5.

Harsh, N.S., Rai,. B.K. and Ayachi, S.S. (1993). J. Trop. Forest, **9**(3): 270-279.

Hongo, T. (1965). Fungi of Hindukush collected by Mr. K. Honda, Acta Phytotax, Geobot., **21**:117-118.

Horak, E. (1980). Indian Boletales and Agaricales, Revisions and new taxa, Sydowia, **33**:88-110.

Jeffries, P. (1987). Use of mycorrhiza in Agriculture. Critical Reviews in Biotechnology, **5**:319-357.

Kaul, T.N. and Kachroo, J.L. (1974). Common edible mushrooms of Jummu and Kashmir. Indian J. Mush., 1:517-529.

Kaul, T.N. Kachroo, J.L. and Raina, A. (1978). Common edible mushroom of Jammu and Kashmir. Indian J. Mush., **1**:29-31.

Lakhanpal, T.N. (1994). Prospects of mushrooms from the Biotechnology. p. 15-22.

Lakhanpal T.N., Sharma, R. and Singh, L. (1988b). Systematics and ectomycorrhizal relationship of some mushrooms in N.W. Himalayas. In: Mycorrhiza for Green Asia (eds.) Mahadevan, A., Raman, N. and Natarajan, K. Madras Univ. p. 66-69.

Lakhanpal, T.N., Shad, O.S. and Sagar. A. (1991). Mycorrhizae: A possible deterrent in artificial cultivation of morels. Accepted for publication in Curr. Sci. Banglore (Ref. N. 776).

Purkayastha, R.P. and Chandra, A. (1985). Manual of Indian Edible Mushroom Ist (eds) Today & tomorrow's Printers and Publishers, New Delhi.

Rai, B.K. (1992). PH. D. Thesis, R.D. University, Jabalpur, India.

Rai, B.K. and Sharma, N.D. (1996). Studies on *Termitomyces heimii:* A staple mycofood from mandla. Vasundhara, **1**: 91-93.

Rambelli, A. (1983). Manual of Mushroom cultivation. FAO Rome, p. 65.

Ramsbottom, J. (1953). Mushroom and Toadstool: A study of the activity of fungi. 306 pp. Collins, London.

Rawla, G.S. and Sarwal, B.M. (1983). Taxonomic studies on Indian Agarics-I, Russulacea Bibliotheca Mycologica, **91**: 23-46.

Saini, S.S.and Atri, N.S. (1981). *Russula foetens* (Pers.) Fr. A new record for India. Curr. Sci., **50**: 460-461.

Saini, S.S. and Atri, N.S. (1982a). North Indian Agaricales-I Indian Phytopath., **35**: 365-372.

Saini, S.S. and Atri N.S. (1982b). North Indian Agaricales-III Proc. Indian Nat. Sci. Acad., **48**: 452-457.

Shad, O.S. (1985): Studies on some edible fungi of Himachal Pradesh. M. Phil. Dissertation, H.P. University, Shimla.

Shad, O.S. (1989). Biological studies on Morchella species (Morels) of Himachal Himalayas, Ph.D. thesis, H.P.Unversity, Shimla.

Singh, C.S. (2001). Mycorrhizal mushrooms from Tikari forest of Gonda vasundhara (In Press).

Singh,. C.S. (1989). Proc. Ind. Sci. Congr. Assoc., No. 38, Prasad, A.B.

Vrinda, K.B. Pradeep C.K., Mathew, S., and Abraham, T.K. (1999). Agaricales from Western Ghats VII Mush. Reseach, **8**(2): 9-120.

Watling, R. and Gregovy, N.M. (1980). Larger fungi from Kashmir. Nova hedwigia., **32**: 493-564.

Yaug, C.Y. and Jong, S.C. (1987). Medical Mushrooms in China Mushroom Sci., **12**: 631-643.

Zadrazil, F. (1975) Z. Pflanzonen Bodenkde, 263.

19

Taxonomy and Parasitism of *Alternaria* spp. associated with Cucurbitaceous Vegetables

Udit Narain, Mukesh Srivastava and Pinky Rani

Department of Plant Pathology
C. S. Azad University of Agriculture and Technology, Kanpur - 208002

Family cucurbitaceae constitutes the major source of fruit vegetables in India. The leaf spots, blight and rotting of the fruits in the field and storage are the major disease problems of these vegetables. Three species of the genus *Alternaria* viz., *A. alternata*, *A. cucurmerina* and *A. tenuissima* were found associated with cucurbitaceous vegetables. Studies on symptomatology and etiology were made for the identification of disease and associated causative fungal species. Sixteen plant species of cucurbits were found to be parasitized by the species of *Alternaria*. *A. cucumerina* appeared to be widespread causing leaf spots and blight of fourteen hosts and *A. alternata* was associated with nine cucurbits, while *A. tenuissima* was confined to only four cucurbitaceous plants. The morphological characters of all the three *Alternaria* species associated with these vegetables have been described for their differentiation and identification. A very simple and feasible key has been framed for the ready identification of *Alternaria* spp.

Key Words : ***Alternaria alternata*, *A. cucurmerian*, *A. tenuissima*, *parasitism*, *taxonomy*.**

The members of the family Cucurbitaceae are the major source of vegetables in India and constitute an important part of the daily diet. These vegetables are mostly grown throughout the year in India in all the seasons of *Rabi*, *Kharif* and *Zaid*. Family Cucurbitaceae is represented by 90 genera and 750 species, which are especially used for their edible fruits (Chakravarty, 1959). Some of the important cultivated plants are *Citrullus lanatus* (water melon), *Cucumis melo* (musk melon), *Cucumis sativus* (cucumber), *Lagenaria siceraria* (bottle gourd), *Cucurbita maxima* (winter squash), *Momordica charantia* (bitter gourd), *Trichosanthes dioica* (pointed gourd), *Coccinia indica* (scarlet gourd), *Citrullus vulgaris* var. *fistulosus* (round gourd), *Cucurbita pepo* (winter squash), *Trichosanthes anguina* (snake gourd) and *Luffa acutangula* (ridge gourd).

These vegetables are the good source of income to the growers and play an important role in human nutrition. Regular use of these vegetables provides

E-mail : mukeshcsau@redifmail.com; mukeshcsauknp@yahoo.com

Table 1 : Nutritive value of different cucurbitaceous vegetables

Cucurbits	Protein	Carbohydrate	Mineral	Fat	Moisture	Vitamin A	Vitamin B
Cucumber	0.4g	2.5g	0.3g	0.1g	96.3%	-	7.0mg
Musk melon	0.6g	5.4g	0.6g	0.1g	92.8%	450 I.U	32mg
Water melon	0.2g	3.3g	0.3g	0.2g	95.8%	-	1.0mg
Bottle gourd	0.2g	2.5g	0.3g	0.1g	96.1%	-	6.0mg
Pumpkin	1.4g	4.6g	0.5g	0.1g	92.6%	84 I.U	2.0mg
Bitter gourd	1.6g	4.2g	0.8g	0.2g	92.4%	210 I.U	8mg
Ridge gourd	0.5g	3.4g	0.3g	0.1g	95.2%	56 I.U	5mg
Pointed gourd	2.0g	2.2g	0.5g	0.3g	92.0%	255 I.U	29mg

the most essential and protective substances, such as protein, carbohydrate, minerals, salts and vitamins. The nutritive value as per 100 gm of edible portion in different cucurbits is given in Table 1 (Gopalan *et al.*, 1982).

In addition to their significant role in the human dietary, the medicinal and therapeutic values of cucurbits are also very important.

The production of these crops must be increased to meet the demand of vegetables of our country. Taking into the consideration the rapid growth of population and alarming situation of vegetables and nutritive deficiency in the country, there is a great need of co-ordinated efforts by the scientists for improving these important vegetable crops in average productivity.

A critical review of the reasons for low yield reveals that diseases are one of the major limiting factors in the successful cultivation of these crops. A few important diseases, which take heavy toll every year, are leaf blight, powdery and downy mildews, *Alternaria* leaf spots, seedling blight, wilt, root knot, bacterial wilt and mosaic (Muncie, 1947; Singh, 1995). Out of these maladies, Alternaria leaf spot caused by *Alternaira cucumerina* (Ellis & Ev.) Elliot is assuming considerable importance now a days. The disease was previously confined to the states of Punjab (Chahal *et al.*, 1970) and Rajasthan (Khandelwal and Prasada, 1970) but it was reported from Uttar Pradesh also (Narain and Prasad, 1981; Ahmad and Narain, 2000; Narain and Srivastava, 2000). Apart from the association of *A. cucumerina* with the Cucurbitaceous hosts, two more species of the genus *Alternaria* viz. *A. alternata* and *A. tenuissima* have also been found to cause leaf spots and blight (Sharma and Bhargava, 1977; Narain *et al.*, 1985, 1997).

It has been estimated that upto 51% loss is caused in severe outbreaks of fruit rot of summer squash caused by *A. alternata* in Kulu valley in Himachal Pradesh. In Rajasthan, the maximum loss in watermelon and pumpkin has been recorded up to 88.3% and 80.0%, respectively (Bhargava and Singh,

1985). In a survey report, heavy losses in different crops viz. water melon (80.7%), pumpkin (60.8%), ridge gourd (22.2%) and bitter gourd (29.0%) have been recorded in different districts of Rajasthan like Jaipur, Ajmer, Tonk, Udaipur and Chittorgarh (Prasada *et al.*, 1975).

Some of the important diseases of Cucurbitaceous crops caused by *Alternaria* species reported form India are given in Table 2.

Symptomatology of Disease

Leclerg (1931) described Alternaria leaf spot of cucurbits caused by *A. cucumerina*. The spots are mostly produced on watermelon and melon. The spots begin as yellow brown flecks, 0.5mm diameter, usually on the upper surface of the crown leaves. They have a light green halo with concentric rings more common on the upper surface of the leaf. Later, the lesions become covered with the dark olive green conidial masses.

Chahal *et al.* (1970) observed the leaf spot disease of cucurbits caused by *A. cucumerina* in Punjab. The spots are circular to elliptical and olivaceous in

Table: 2- *Alternaria* species associated with cucurbits in India.

Sl. No.	Disease	Host	Cause organism	Reported by
1.	Leaf blight	Water melon	*Alternaria alternata*	Narain *et al.* (1985)
2.	Leaf spot	Water melon	*Alternaria cucumerina*	Singh *et al.* (1975)
3.	Fruit rot	Water melon	*Alternaria cucumerina*	Mathur and Shekhawat (1992)
4.	Fruit rot	Bottle gourd	*Alternaria alternata*	Singh and Chohan (1980)
5.	Fruit rot	Bottle gourd	*Alternaria cucumerina*	Narain and Srivastava (2000)
6.	Leaf blight	Bottle gourd	*Alternaria cucumerina*	Chahal *et al.* (1970)
7.	Leaf spot	Cucumber	*Alternaria alternata*	Laxminarayanan and Reddy (1976)
8.	Leaf blight	Cucumber	*Alternaria cucumerina*	Chahal *et al.* (1970)
9.	Fruit rot	Cucumber	*Alternaria tenuissima*	Narain and Srivastava (2000)
10.	Leaf blight	Pumpkin	*Alternaria cucumerina*	Chahal *et al.* (1970)
11.	Leaf blight	Scarlet gourd	*Alternaria cucumerina*	Narian and Prasad (1981)
12.	Leaf blight	Bitter gourd	*Alternaria cucumerina*	Ahmad and Narain (2000a)
13.	Fruit rot	Bitter gourd	*Alternaria cucumerina*	Narain and Srivastava (2000)
14.	Fruit rot	Round gourd	*Alternaria alternata*	Narain and Srivastava (2000)
15.	Fruit rot	Pointed gourd	*Alternaria cucumerina*	Narain and Srivastava (2000)
16.	Fruit rot	Sponge gourd	*Alternaria tenuissima*	Narain and Srivastava (2000)
17.	Fruit rot	Bitter gourd	*Alternaria tenuissima*	Sharma and Bhargava (1977)
18.	Fruit rot	Pointed gourd	*Alternaria alternata*	Sahu Kritigyan and Singh (1980)

colour, which vary from 3-10 mm in size. In case of severe infection, several lesions coalesce to cover large area and cause death of the leaves. The young and actively developing leaves are almost free from the disease or may have very small necrotic areas. The severely infected leaves show burning effect, which ultimately dry up and may fall prematurely (Fig.3).

Narain and Prasad (1981) described Alternaria leaf spot of scarlet gourd caused by *Alternaria cucumerina*. The disease appears in the form of numerous, minute spots scattered on the leaf surfaces. The spots later enlarge, become nearly circular, 3-12mm in diameter, brown to dark brown in colour with distinct zonations and profuse sporulation of the fungus (Fig. 5).

Khandelwal and Prasada (1970) reported Alternaria leaf spot as a serious disease of cucurbits in Rajasthan. In the initial stage, the lesion starts as small circular, reddish brown spots, which gradually increase in size in concentric manners. Lesions often coalesce to form large necrotic areas (Fig. 6).

The disease is more serious in the month of July and August on Cucurbitaceous hosts and in general symptoms initially appear as small brown spots on the margin of the leaves (Fig.7,8&9), which later increase rapidly inwards and become circular to irregular in shape and turn light to dark brown in colour with slight distinct or indistinct zonations (Fig. 1&3). Two or more such spots may coalesce to form bigger spots. In well-advanced stage, some of the spots are covered with dark olive green mass of the fungal spores (Fig. 2&4).

Due to the prevalence of high humidity at ground level, the older leaves get infected and more than 50% areas are covered by the necrotic portion produced by the fungus. The lesions also appear on stems, fruits and petioles in favourable conditions. Majority of fruits lye on the ground and they get infected readily in presence of the moisture (Narain and Srivastava, 2000). On cucumber fruits, the lesions are sunken and mostly observed at the curvature points of fruit (Fig. 10) whereas, scattered, light brown lesions of different sizes appear with profose sporulation on the surface of sponge gourd fruit due to *A. alternata* (Fig. 11). Dark brown lesions appear mostly at the distant end of pointed gourd (Fig. 12) and bottle gourd (Fig. 14). Lesions on bitter gourd are irregular in appearance with profuse sproulation of *A. cucumerina,* which later leads rotting of the fruits (Fig. 13).

Ahmad and Narain (2000 b) reported that leaf spot of bitter gourd due to *A. cucumerina* appeared in the first fortnight of July and maximum disease intensity was noticed within the temperature range of 25^0 to 28^0C and average relative humidity more than 80%. Rainfall was greatly responsible for the severity of infection and disease development. The disease intensity was minimum in the month of October when there was very little or no rainfall, low atmospheric temperature and less humidity.

Taxonomy and Morphological Characters of the Pathogens

The genus *Alternaria* is one of the most commonly encountered fungi of the Dictyosporae (Gr. dictiyon= net and spore = seed, spore). The conidia are rather large and multicellular with both transverse and longitudinal or oblique septa (muriform). They are usually borne in chains (acropetal). Conidia are formed in long chain (*A. alternata*) or in short chains (*A. tenuissima*) or may be formed singly (solitary) on the apex of the conidiophores or seldom a secondary conidium is also produced as in *Alternaria cucumerina* (Neergaard, 1945; Joly, 1964; Ellis, 1971and 1976). All these species are represented by the multitude of isolates ranging from saprophytes to strong parasites having polyphagous nature.

Joly (1959) described the morphological variation of species in the genus *Alternaria* and made the monographic study of the genus *Alternaria* and proposed simple key for determination of the most common species dividing them in three sections (Joly, 1964). Nishihara (1968) for the first time reported *Alternaria cucumerina* from Japan. The fungus caused leaf spots of *Cucurbita pepo.* Ellis (1971 and 1976) has described the characters of the most common species of *Alternaira* in his monograph of Hyphomycetes. The distinguishing characters of *Alternaria* spp. from India have also been described by Subramanian (1971).

The casual organism of blight of watermelon was identified in India as *Alternaria cucumerina* (E. & E.) Elliot by Khandelwal and Prasada (1970) and also Ibrahim (1975a) identified its causal organism as *A. cucumerina* from Egypt.

Later on, the leaf blight, leaf spot, veinal necrosis and fruit rot of cucurbits caused by *A. cucumerina, A. alternata* and *A. tenuisima* were reported from Punjab (Chahal *et al.*, 1970) Rajasthan (Khandelwal and Prasada, 1970), Andhra Pradesh (Laxminarayanan and Reddy, 1976) and Uttar Pradesh (Narain and Prasad, 1981; Ahmad and Narain, 2000a; Narain and Srivastava, 2000).

The diagnostic characters of all the three *Alternaria* species associated with different Cucurbitaceous hosts are described below :

1. ***Alternaria alternata*** (Fr.) Keissler (*Beih. Bot. Zbl.*, **29** : 434, 1912) :

Colonies moderately fast growing, usually black or olivaceous black, sporulation abundant; **mycelium** septate, branched, olive buff, 2.5-9.5 mm in width, sometimes swollen (12.5-23.5 mm) to from chain of chlamydospores;

Symptoms of disease on fruits (Fig. 10 to 14)

Fig. 10: Cucumber; ig. 11: Sponge gourd; Fig. 12: Pointed gourd; Fig. 13: Bitter gourd; Fig. 14: Bottle gourd

conidiohpores arising singly or in groups, usually simple, septate, straight or variously curved, geniculate, pale to mid olivaceous brown, smooth and 24.5-68.5 mm in length and 3.50-7.25 mm in width; **conidia** formed in long and often branched chains, obclavate, obpyriform, ovoid or ellipsoidal, dark brown, smooth or verruculose, 8.50-53.50 x 5.30-21.50 mm in size ; **beaks** often short, conical or cylindric but never equaling length of conidium, usually lighter in colour, 1.0-45.5 mm in length and 1.5-6.5 mm in width and 0-3 septate (Fig.15).

2. ***Alternaria tenuissima*** (Kunze ex Pers.) Wilts. (*Trans. Br. mycol. Soc.* **18**:157, 1933) :

Colonies growing faster, centre sometimes raised and deep grayish; **mycelium** septate, branched, hyaline at first, later olive buff, 2.8-10.7 mm in width, sometimes swollen (15.0-21.5 mm) to form chains of chlamydospores ; **conidiophores** produced either as side branches from cells of hyphae or terminally to hyphae, also may arise from chlamydospores, more or less cylindrical, simple or rarely branched, septate (2-5 cross septa), sometimes geniculate also with scars, dark brown, smooth measuring 19.60-75.0µm in length and 2.70-6.20 µm in width; **conidia** produced on short chains (3-6), conical to oval, dark olive buff, smooth, tapering gradually into beaks, generally with 3-9 cross and 0-8 longitudinal or oblique septa, slightly or not constricted at septa, measuring 12.0-65.0 mm in length and 7.0-15.5 mm in width ; **beaks** usually smaller to body in culture but in nature upto same length of spore body, sometimes forming a terminal swelling with a prominent scar, 3.50-70.0mm in length and 3.50-5.50µm in width and with 0-3 cross septa (Fig. 16 a,b,c&d).

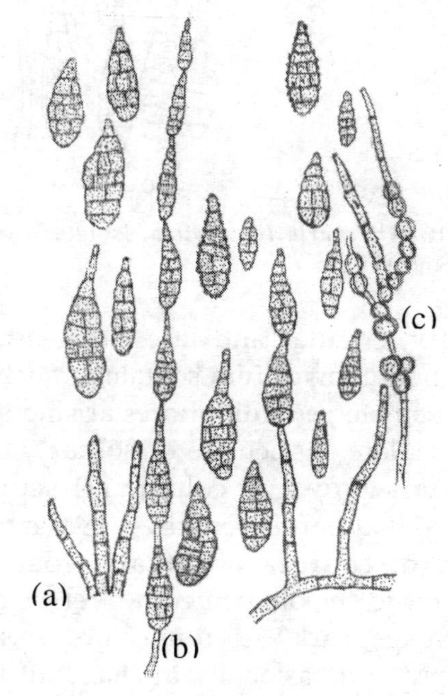

Fig. 15. *Alternaria alternata* (a) conidiophores (b) conidial chain (c) chlamydospores

3. ***Alternaria cucumerina*** (Ellis & Everh.) Elliott (*Am. J. Bot.* **4** : 472, 1917) :

Colonies fast growing, amphigenous, usually in the beginning dull white,

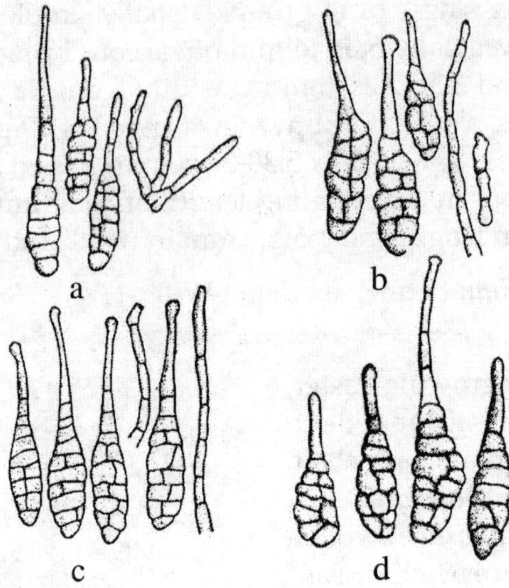

Fig.16. ***Alternaria tenuissima.*** Isolates from (a) Phoot, (b) Long gourd, (c) Bitter gourd (d) Snake gourd

fluffy, circular and later turn into dark olivaceous green, sporulation abundant, **mycelium** septate, branched, dark olivaceous in colour,4.0-6.0 mm in diameter; **conidiophores** arising singly or in groups, septate, erect, simple, geniculate, fasciculate or solitary, light to dark brown in colour, 1-4 septate, 42.5-110.0 mm in length, 6.0-9.0 mm in width; **conidia** obclavate, rostrate, cylindric, smooth, sometimes verruculose with age, dark brown in colour, formed singly or occasionally in chains of two, 54.3-80.2 mm in length, 17.6-27.2 mm in width with 6-10 transverse septa and 2-12 longitudinal or oblique septa; **beaks** usually pale brown, septate, not branched, long, cylindrical, 25.2-70.0μm in length and 1.0-2.5μm in width with 0-5 cross septa (Fig.17).

The comparative morphological differences of all the three species of *Alternaria* parasitic on cucurbits or given in Table 4.

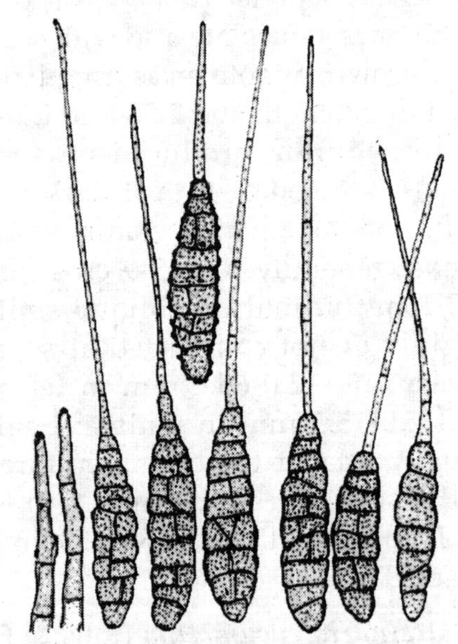

Fig. 17. *Alternaria cucumerina*

Table 4. Comparative differentiation in three *Alternaria* spp. occurring on cucurbitaceous hosts

Morphological characters	*A. alternata*	*A. tenuissima*	*A. cucurmerina*
Conidiophores	Olive brown, erect, simple, septate	Dark brown, erect simple or branched, septate	Dark brown, erect, simple, septate
Length	24.5-68.5μm	19.60-75.0μm	42.5-110.0μm
Width	3.50-7.25μm	2.70-6.20μm	6.0-9.0μm
Conidia			
Conidia in chain	Up to 20 or even more	3-6	Single (rearely 2)
Shape	Polymorphic	Conical to oval	Oblavate rostrate,
Colour	Dark brown	Dark olive buff	Olive brown
Cross septa	3-10	3-9	6-10
Long. septa	0-8	0-8	2-12
Length	8.50-53.50μm	12.0-65.0μm	54.3-80.2μm
Width	5.30-21.50μm	7.00-15.50μm	17.6-27.2μm
Beak			
Length	1.0-45.5μm	3.50-70.0μm	25.2-75.1μm
Width	1.5-6.5μm	3.50-5.50μm	1.0-2.5μm
Cross septa	0-3	0-3	0-5

Key to *Alternaria* spp. occurring on cucurbitaceous hosts

Conidia formed in long and often branched chains consisting up to 20 or even more

Conidia usually polymorphic, often with short conical or cylindrical beaks .... *A. alternata*

Conidia formed in short chains (3-6)

Conidia conical to oval with many cross and longitudinal septa, sometime swollen terminally and may be of up to the same length of spore body..................... *A. tenuissima*

Conidia formed singly or rarely up to two

Conidia obclavate, reostrate, comparatively larger with many cross and longitudinal septa and with long filiform beaks.................... *A. cucumerina*

Host Range

Jackson (1958) in host range studies of *A. cucumerina* included watermelon, muskmelon, cucumber, 3 squash spp. bergerkin and several native plants while. Khandelwal and Prasada (1970) for *A. cucumerina* included watermelon, muskmelon, cucumber and guar. Chahal *et al.* (1970) also reported *A. cucumerina* on musk melon, cucumber, water melon, bottle gourd and vegetable marrow whereas Cirsan (1976) reported isolations of *Alternaria* from 18 vegetable species.

The three species of genus *Alternaria* viz., *A. alternata, A. tenuissima* and *A. cucurmerina* were found to be associated with sixteen Cucurbitaceous vegetable hosts. *A. alternata* was found to parasitise nine species and *A. tenuissima* was confined to only four vegetables. *A. cucurmerina* was wide spread causing most serious disease and occurring on fourteen cucurbitaceous hosts (Table 3). Under natural conditions, the disease intensity varied from 30-60%

Table: 3 - *Alternaria* spp. associated with cucurbitaceous hosts

Alternaria cucumerina	
Citrullus lanatus	Water melon
C. vulgaris var. fistulosus	Round gourd
Coccinia indica	Scarlet gourd
Cucumis melo	Musk melon
C. melo var. *momordica*	Phoot
C. melo var. *utilissimus*	Long gourd
Cucumis sativus	Cucumber
Cucurbita maxima	Winter squash
Cucurbita pepo	Summer squash
Lagenaria siceraria	Bottle gourd
Luffa acutangula	Ridge gourd
Momordica charantia	Bitter gourd
Momordica dioica	Small bitter gourd
Trichosanthes dioica	Pointed gourd
Alternaria alternata	
Citrullus lanatus	Water melon
Coccinia indica	Scarlet gourd
Cucurbita maxima	Winter squash
Cucumis sativus	Cucumber
Lagenaria siceraria	Bottle gourd
Luffa acutangula	Ridge gourd
Luffa cylindrica	Sponge gourd
Momordica charantia	Bitter gourd
Trichosanthes dioica	Pointed gourd
Alternaria tenuissima	
Citrullus melo var. *momordica*	Phoot
C. melo var. *utilissimus*	Long gourd
Momordica charantia	Bitter gourd
Trichosanthes anguina	Snake gourd

Physiology and Survival of the Pathogens

Temperature is known to have profound effect on the growth of microorganisms. Neergaard (1945) recorded that the optimum temperature for the growth of *A. alternata* was 25°C, whereas Ashour and El-Kadi (1959), Verma (1969) and Singh *et al.* (1980) observed that optimum temperature for the growth and sporulations of *A. alternata* was 25°C. Khandelwal and Prasada (1970) reported that *A. cucumerina* grew at temperature range of 10-35°C with optimum temperature for growth and sporulations being 25-28°C and no growth occurred at lower temperature. Evans *et al.* (1992) found that *A. cucumerina* could grow at a temperature range of 15-21°C.

Hydrogen ion concentration (pH) has a pronounced effect on the growth and sporulation of fungi. Ashour and El-Kadi (1959) observed that the best growth and sporulation of *Alternaria alternata* occurred at pH level of 6.0 but Verma (1969) reported that optimum pH for the growth and sporulation of this fungus was 6.6. Khandelwal and Prasada (1970) reported *A. cucumerina* to grow at pH range of 3.0 to 9.0 but optimum pH for the growth and sporulation was 6.0.

Ibrahim *et al.* (1975 b) studied the survival of *A. cucumerina,* the causal organism of leaf spot disease of cucurbits and reported that conidia failed to germinate after storage for 4 months at room temperature, while mycelial thread in dry water melon leaf tissues retained its viability after 2 years. Mycelial fragments were pathogenic to water melon leaves, causing spots, similar to those induced by conidial suspension. Mycelium in plant debris supplied inoculum for primary infection, while conidia caused further disease spread. The fungus was not internally seed borne.

Disease Management

Varietal resistance

Musk melon verities UFG 508, G 509, G510, G511 and G515 have been found to be resistant to *A. cucumerina* (Halsey, 1980). Thomas *et al.* (1990) found *Cucumis melo* vark Mr-1 to be resistant to *A. cucumerina.* Boyhan and Nortan (1992) found that the musk melon variety AC 82-37-2 was resistant to Alternaria blight caused by *A. cucumarina.*

The reaction of water melon varieties, Durgapura, Meetha, AY, WHY and WHY-4, was highly susceptible, RW –177-3, RW–1, RW –187-2 and Milan as moderately susceptible, while Sel-1 and Sugarbaby were resistant against Alternaria leaf spot (Mathur and Shekhawat *et al.,* 1992). Chhipa *et al.* (1995) reported that water melon varieties WHY–3, WHY–4, Honeydew, Instant,

RW–177-3, S7-3-21 were resistant to Alternaria blight due to *A. cucumerina*.

Katiyar *et al.* (2001a) screened 14 germplasm of bottle gourd under artificial conditions of inoculation for the source of resistance in bottle gourd against *A. cucumerina* in which none was found immune (free). However, three varieties/cultures viz. Azad Harit, 7002 and 7003 proved resistant and rest were rated as moderately resistant or susceptible to *A. cucumerina*.

Use of chemicals

The endeavours made by several workers to evaluate the efficacy of chemicals in inhibition of the growth of *Alternaria* spp. *in vitro* gave promising results.

Sahni and Singh (1967) tested fungicides in laboratory and found that Arasan and Thiram completely inhibited the growth of *A. alternata*.

Ibrahim *et al.* (1975) tested seven fungicides against the growth of *A. cucurmerina in vitro* and under field conditions and found that adequate control was obtained by Daconil 2787, Difolatan 4-F, Dithane M–45 and Difolatan 80 W.P.

Crisan (1976) reported that Vitavax (0.2%) was effective against all the isolates of *A. alternata* and *A. cucurmerina in vitro*. Out of ten fungicides screened in laboratory conditions against *Alternaira alternata*, Dithane M-45, Duter and Dithane Z-78 were found to be very effective as they caused total inhibition of the vegetative growth of the fungus (Ahmad and Narain *et al.*, 1998).

Ahmad and Narain (1998) carried out the field experiment on the control of Alternaria leaf spot of bitter gourd caused by *A. cucurmerina*, using eight spray fungicides. Indofil M-45 was observed to be most effective in reducing the disease intensity followed by Indofil Z–78, Kavach and Blue copper –50, which were statistically superior also amongst all the other fungicides tested.

Katiyar *et al.* (2001b) obtained best control of Alternaria leaf spot of bottle gourd caused by *A.cucurmernaria* with the application of Indofil M-45 (0.2%) and Chlorothalonil (0.2%) at 15 days interval.

References

Ahmad, Shahid and Narain, U. (1998). Fungicidal management of Alternaria leaf spot of bitter gourd. Ann. Pl. Protec. Sci,. **6** : 95-96.

Ahmad, Shahid, and Narain, U. (2000a). A new host record of *Alternaria cucumerina* on bottle gourd. Indian Phytopath., **53** : 234.

Ahmad, Shahid and Narain, U. (2000b). Effect of temperature, R.H. and rainfall on development of leaf spot of bitter gourd. Ann. Pl. Protec. Sci., **8** : 114-115.

Ashour, W.E. and El Kadi, M. (1956). Cultural studies of *Fusarium semitectum, Alternaria tenuis* and *Rhizoctonia solani* causing damping off of tomato seedlings. A. in Shams. Sci. Bull., **3** : 57-68.

Bhargava, A.K. and Singh, R.D. (1985). Comparative study of Alternaria blight, losses and causal organism of cucurbits in Rajasthan. Indian J. Mycol. Pl. Pathol., **15** (2): 150-154.

Boyhan, G.E. and Norton, J.D. (1992). Inheritance of resistance of Alternaria leaf blight in musk melons. Hort. Science., **27** : 1114-1115.

Chahal, D.S.; Chohan, J.S. and Sidhu, G.S. (1970). Alternaria leaf spot of cucurbits in Punjab. Indian Phytopath., **23** : 580-581.

Chakravarty, N.L. (1959). Monograph on Indian Cucurbitaceae (taxonomy and distribution). Botanical Survey of India, **17** (1) : 6-7.

Chhippa, H.P.; J.P. Muthra and Shivpuri, Asha (1995). Present status of cucurbit diseases in Rajasthan. Indian J. Mycol. Pl. Pathol., **25** : 90-91.

Crisan, A. (1976). Aspects of the spread, biology and control of *Alternaria* species parasitic on vegetables in Cluj. Napoca district. Cuntributii Botanice. pp. 11-12 (*R.P.P.*, **56** : 1046, 1977).

Ellis, M.B. (1971). *Dematiaceous Hyphomycetes,* C.M.I., Kew, England, 608 pp.

Ellis, M.B. (1976). *More Dematiaceous Hyphomycetes,* C.M.I., Kew, England, 411-427 pp.

Evans, K.J., Nyquist, W.E. and Latin, R.X. (1992). A model based on temperature and leaf wetness duration for establishment of Alternaria leaf blight of musk melon. Phytopathology., **82** : 890 –895.

Gopalan, C. ; Rama Sastri, B. V. and Balasubramanian, S.C. (1982). Nutritive value of Indian foods. Indian Council of Medical Research, National Institute of Nutrition, Hyderabad.

Halsey, L.H. (1980). U.F.G. 508, G 509, G 510, G 511 and G 515 musk melon breeding lines. Hort. Science, **15** : 538.

Ibrahim, A.N.; Abdul- Hak, T.M., Fadl, F.A. and Mahrous, M.M. (1975a). Certain morphological and pathological characters of *Alternaria cucumerina,* cause of water melon leaf spot disease in Egypt. *Acta Phytopathologica Academiae Scientirum Hungaricae,* **10** : (3/4) : 301-307.

Ibrahim, A.N.; Abdul- Hak, T.M., Fadl, F.A. and Mahrous, M.M. (1975 b). Survival of *Alternaria cucumerina,* the causal organism of leaf spot disease of cucurbits. *Acta. Phytopathologica Academiae Scientirum Hungaricae,* **10** : (3/4) : 309-313.

Ibrahim, A.N.; Fadl, F.A. and Mahrous, M.M. (1975). Effect of different fungicides on the severity of water melon leaf spot disease. Egyptian J. Phytopath., **7** : 49-57.

Jackson, C.R. (1958). Symptoms and host parasite relation on Alternaria leaf spot disease of cucurbits. Phytopathology , **4**: 181-187.

Joly, P. (1959). Morphological variations and the idea of species in the genus Alternaria. Bull. Soc. Mycol. Fr., **75** : 149-158.

Joly, P. (1964). *Le Genre Alternaria.* Editions Paul Lachevalier, Paris, 150 pp.

Katiyar, Archana ; Kant, Sandhya, Chauhan, S.S. and Alka (2001a). Sources of resistance in bottle gourd to Alternaria leaf spot. Ann. Pl. Protec. Sci., **9** (1) : 155-157.

Katiyar, Archana ; Kant, Sandhya, Chauhan, S.S. and Alka (2001b). Chemical control of Alternaria leaf spot of bottle gourd. Ann. Pl. Protec. Sci., **9** (2) : 339-341.

Khandelwal, G.L. and Prasada, R. (1970). Taxonomy, physiology and control of Alternaria leaf spot of watermelon (*Citrulus lanatus*). Indian Phytopath., **23** : 81-84.

Laxminarayanan, P. and Reddy, S.M. (1976). Post harvest disease of some Cucurbitaceous vegetables from Andhra Pradesh. Indian Phytopath., **29**: 57-59.

Leclerg, E.L. (1931). The relation of leaf blight to sun scald of honey dew melon. Phytopathology., **21** : 97.

Mathur, K. and Shekhawat, K.S. (1992). Fruit rot of watermelon. Indian J. Mycol. Pl. Pathol., **22** : 80.

Muncie, J.H. (1947). Cucumber diseases. Canner Civ., **16** : 12-13.

Narain, A.; Swain, N.C., Sahoo, K.C., Dash, S.K. and Shukla, V.D. (1985). A new leaf blight and fruit rot of water melon. Indian Phytopath., **38** : 149-150.

Narain, U.; Katiyar, Archana and Chauhan, S.S. (1997). Detection and diagnosis of *Alternaria* spp. associated with Cucurbitaceous vegetables. Paper presented in National Symposium on Emerging Issues in Plant Pathology held at R.D. University, Jabalpur on 16-18 Feb. 1997, p. 83.

Narain, U. and Prasad, R. (1981). Alternaria leaf spot of *Kundru* from India. Plant Sci., **13** : 96.

Narian, U. and Srivastava, M. (2000). Detection and diagnosis of *Alternaria* spp. associated with fruits of Cucurbitaceous vegetables. *Proc. Indian Phytopath. Soc. Golden Jubilee* Vol., **2** : 718-719.

Neergaard, P. (1945). *Danish species of Altenaria* and *Stemphylium*. Oxford University Press, London, 560 pp.

Nishihara, N. (1968). Leaf spot of *Cucurbita pepo* caused by *Alternaria cucumerina. Ann. Phytopath. Soc. Japan*, **34** : 152-155.

Prassada, R., Khandelwal, G.L. and Jain, J.P. (1975). Epidemiology, forecasting and control of Alternaria blight of cucurbits. Report of S.K.N. Coll. Agric. Jobner, Rajasthan, 301-308 pp.

Sahni, M.L. and Singh, R.P. (1967). Bioassay of tetramethyl thiuramisulphide fungicides. Indian Phtopath., **20** : 71-73.

Sahu Kritagyan, S.P. and Singh, S.P. (1980). Fruit rot of pointed gourd (T*richosanthes dioica*) in Bihar. Indian Phytopath., **33** : 308-309.

Sharma, N. and Bhargava, K.S. (1977). Fruit rot bitter gourd. Indian Phytopath., **30** : 557-558.

Singh, C.S., Kanaujia, R.S. and Chaudahry, R.L. (1975). Paracitic fungi of Faizabad. Botanique, **6** : 151-154.

Singh R.S. (1995). *Diseases of vegetable crops*. Oxford & I.B.H. Pub. Co., New Delhi, 346 pp.

Singh, R.S. and Chohan, J.S. (1980). Fungal fruit rot of bottle gourd in North India. Indian Phytopath., **33** : 598-599.

Subramanian, C.V. (1971). *Hyphomycetes*. I.C.A.R. Publication, New Delhi.

Thomas, C.E.; Mecreight, J.D. and Jourdain, E.L. (1990). Inheritance of resistance of *Cucumis melo* line MR-1 to *Alternaria cucumarina*. Plant Disease., **74** (1): 868-870.

Verma, V. (1996). Effect of temperature and hydrogen-ion concentration on three pathogenic fungi. Sydowia, **23** : 164-168.

20

The Diseases in Forest Nurseries Caused by *Macrophomina phaseolina* and Their Management

Jamaluddin, V. S. Dadwal and K.K. Soni

Tropical Forest Research Institute, P.O. RFRC
Jabalpur- 482 021

Macrophomina phaseolina **(Tassi) Goid. with its sclerotial stage (*****Rhizoctonia bataticola*** **(Taub). Butler) caused severe root rot in seedlings of** ***Pinus caribaea*** **Morelet,** ***Eucalyptus*** **sp.,** ***Casuarina equisetifolia*** **L. Beefweed and** ***Sesbania grandiflora*** **Pers. The diseases were consistently recorded in the local forest nurseries and also in pine nurseries at Amarkantak and Jagdalpur, M.P. A perusal of literature indicated that there is no record of** ***M. phaseolina*** **on** ***P. caribaea, Eucalyptus*** **and** ***Sesbania grandiflora*** **from India. In the present chapter, an account of diseases comprising symptomatology, cultural characters, pathogenicity, mode of entry and disease development, histopathology and epidemiology as well as management of diseases in forest nurseries is presented.**

Key-words : Disease, symptoms, forest nursery, fungi, ***Macrophomina phaseolina***

Macrophomina phaseolina (Tassi) Goid. with its sclerotial phase (*Rhizoctonia bataticola* (Taub.) Butler) caused severe root rot in the seedlings of *Pinus caribaea* Morelet., *Eucalyptus* sp., *Casuarina equisetifolia* L. Beefweed and *Sesbania grandiflora* Pers. The diseases were significantly more important and caused severe losses to the seedlings in different forest nurseries as well as in plantations in M.P. *M. phaseolina* is a common soil borne plant pathogen which commonly occurred in all types of soil. There were few important reports regarding this disease in forest nurseries (Qureshi, 1956; Hodges, 1962; Smith and Bega, 1964, 1966; Seymour, 1969; Chellman, 1971; Rowan, 1971 Seymour and Cordell, 1979, Soni *et al.*, 1985). Jamaluddin *et al.* (1982) reported *M. phaseolina* on *P. caribaea* first time from India. A perusal of literature indicated that there is no report of *M.phaeseolina* on *Eucalyptus* sp. and *S. grandiflora* from India, so far. The present chapter incorporates the different pathological aspects of the diseases in four important forestry species.

Disease Occurrence

The disease was consistently recorded from nurseries and plantations of *Pinus caribaea* at Amarkantak, Jagdalpur and Jabalpur. In December 1979, a severe incidence of charcoal root rot of *P. caribaea* was observed in pine nurseries and plantations at Jagdalpur which caused 20 to 30 per cent and 10 to 20 per cent mortality, respectively. In tube seedings of *Eucalyptus* of 15 days to 5 month sold the disease was observed but the percentage of occurrence was less than the disease in nursery beds at Pariyat nursery, Jabalpur. One year old *Casuarina* plants also exhibited the charcoal root rot symptoms at TFRI, Jabalpur during 1981. Two to nine months old seedlings of *Sesbania* were found to be susceptible with *M. phaseolina.* The occurrence of disease was more in nursery beds as compared to tube seedlings in Barha nursery and Pariyat nursery, Jabalpur.

Symptoms

Disease symptoms in *P. caribaea*

The young plants of varying age were susceptible to this infection. The symptoms of the disease started from the top towards the base. The needle turned yellow, then brown from the top down ward. The terminals of the seedlings eventually wilted and bent over (Plate I Fig. 2). The young plants in plantations became dead as early as 30 to 45 days from the initiation of the disease symptoms. The dried needles exhibited brushy appearance. The rot occurred on both tap and lateral roots which appeared deep rusty brown. Black dot like sclerotia of the fungus were observed after removing the bark of the root (Plate I Fig. 3).

These sclerotia loosened the contact between bark and the wood. The sclerotia were more in the root but less abundant in the basal part of the stem. Pycnidial stage of the fungus also appeared in the dead seedlings at the advance stage of infection.

Disease symptoms in *Eucalyptus* sp.

Initially leaves started yellowing which dried up and then turned brown. The crown bent from the top downward. The leaves shrivelled and dried up. Finally the seedlings resulted to drying and death (Plate I Fig. 7). Brown to black minute dots of pycnidia and sclerotia appeared on the dead parts of the seedlings giving charred appearance due to death of cortical tissues. The basal leaves first exhibited pycnidia formation which developed successively upwards.

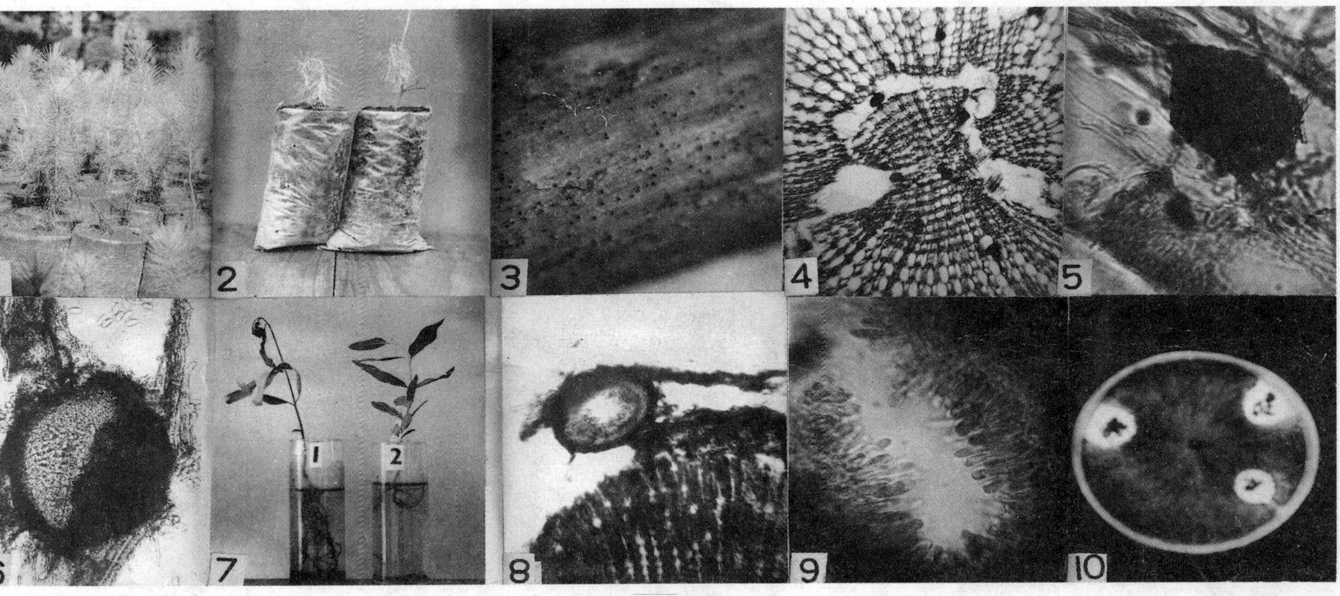

PLATES - I

Fig. 1. Healthy seedlings of *P. caribaea*; Fig. 2. Diseased seedlings of *P. caribaea*; Fig. 3. Exposed root of *P. caribaea* showing sclerotia under stereobinocular microscope x 20; Fig. 4. T.S. of root showing sclerotia in the stelar region x 150; Fig. 5. T.L.S. of P. caribaea root showing sclerotia in the medullary rays x 600; Fig. 6. T.S. of stem showing pycnidia and conidia in P. caribaea x 100; Fig. 7. i. Diseased seedling of *Eucalyptus* ii. healthy seedling of *Eucalyptus*; Fig. 8. T.S. of *Eucalyptus* root showing pycnidia in the cortical region x 150; Fig. 9. Pycnidia and conidia of *M. phaseolina* in *Eucalyptus* x 600; Fig. 10. Isolate of *M. phaseclina* from *P. caribaea* showing zone of inhibition with *P. sclerotioium*.

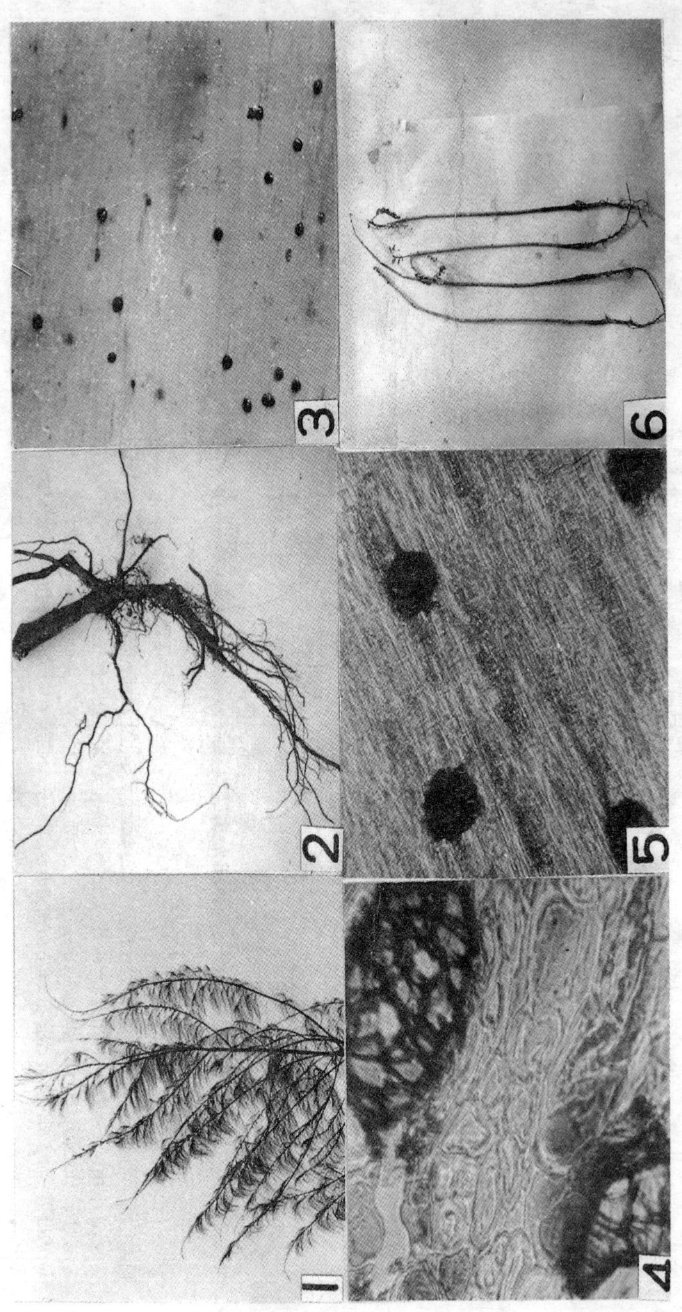

PLATE II

Fig. 1. Diseased plant of *Casuarina equisetifolia;* Fig.2. Diseased root of *C. equisetifolia;* Fig 3. Exposed root of *Casuarina* showing sclerotia with connecting threads of mycelium under Stereobinocular microscope x 20; Fig. 4. Vertical section of bark of *Casuarina* showing sclerotia x 600; Fig. 5. T.L.S. *Casuarina* root showng sclerotia in the stelar region; Fig. 6. Diseased seedling of *Sesbania grandiflora.*

Symptoms in *Casuarina* seedlings

The symptoms of the disease started from, the top to downwards (Plate II Fig. 1 and 2). The rot occurred on both tap and lateral roots. Black sclerotia of the fungus mostly developed under the surface of bark which resulted in the spiliting of bark. The sclerotia were common on the root and collar region of the stem (Plate II Fig. 3) Finally the seedlings became dried and dead.

Symptoms in *Sesbania grandiflora*

The seedlings of *Sesbania grandiflora* exhibited black necrotic scars on the basal portion of the stem near the soil level. Ultimately yellowing in the foliage and defoliation in the seedlings occurred. The necrotic lesions increased gradually and the sclerotia of the fungus developed in the seedlings above the ground level (Plate II Fig.6). The root system of the seedlings gradually decayed due to successive growth of the thick walled, brown, septate, branched mycelium under the surface of bark resulting in drying and death of the seedlings. The pycnidial stage of the fungus could not be detected in the diseased samples.

Cultural Variation of the Isolates

The pathogen was successfully isolated on P.D.A. medium by placing the bits of infected root or from the soil following the procedure described by M Cain and Smith (1972). Growth was fast and covered the petridish within 5 to 7 days at 27 to 35^{0}C. The mycelium of *P. caribaea* isolate produced abundant cottony and effused growth as compared to other isolates (Table 1).

Pathogenicity

The pathogenicity of the organism was confirmed in the seedings by different methods as mentioned below:

1. The potting mixture was infested with organism and kept for few days (5 days) and then the seedlings were transplanted.
2. The seedlings in polythene bags were drenched with the sclerotial suspension of the fungus in sterilized distilled water.
3. The root system of the seedlings was immersed in the sclerotial suspension of the fungus for 48 hours and then transplanted in the polythene bags.

It was observed that the seedlings which were kept in sclerotial suspension of the pathogen exhibited the symptoms of the disease within 10 days and the mortality was 60 percent. In the remaining two methods in

Table 1. Cultural characters of the isolates of *M. phaseolina*

M. phaseolina on *P. caribaea*	*M. phaseolina* on *Eucalyptus* sp.	*M. phaseolina* on *C. equisetifolia*	*M. phaseolina* on *S. grandiflora*.
Colony-appeared dull, cottony, effused turned dark brown.	Appeared brown, fast growing.	Appeared dull, cottony with superficial growth of hyphae.	Appeared brown effused cottony, hyphae superficial.
Hyphae-abundant, septate, dark brown, upto 7.5 µ broad.	Septate, branched upto 5.0 µ broad.	Septate, branched, upto 5.0 µ broad.	Septate, branched, upto 5.0 µ broad.
Sclerotia-brown, to black, compact, globose 50.-75.0 x 25.0-50.0 µ in size.	Developed superficially giving black appearance to the media; black, compact thick walled, globose to variable 63.0-300. Ox 52.5-136.5 µ in size.	Compact, dark brown to black, thick walled, globose, 80.0-200.0 µ in diameter.	Abundant in growing culture, black, compact, thickwalled, 60.0-90.0 µ in diameter.
Pycnidia-subcarbonaceous, globose, thick walled, 200.0-400.0 x 200.0-320.0 µ in size; conidia hyaline, unicellular, thickwalled, smooth, oval to oblong, 20.0-30.0 x 12.5-15.0 µ in size.	Subcarbonaceous; globose, thick walled, with small ostiole, 94.5-472.5 x 94.5-315.0 µ in size; conidia hyaline, unicellular, thin walled, ovoid to fusoid, 16.25-25.0 x 5.0-7.5 µ in size.	Absent in the host.	Not observed.

which the seedlings separately in polythene bags and soil were treated with the sclerotial suspension of the fungus. The disease was mainfested after 30 to 40 days of inoculation and percentage of infection ranged between 30 to 40 percent. The seedling s inoculated with the fungus exhibited the same symptoms as observed under natural conditions. The younger seedlings of 2 to 3 months age were easily got infected as compared with that of older ones i.e. more than six months old.

Mode of Penetration and Development of Symptoms

The infection by the pathogen is substantially influenced by root exudates. The different isolates germinated well in the crude root exudates of *P. caribaea*, *Eucalyptus* sp., *Casuarina equisetifolia* and *Sesbania grandiflora*. The activity of *M. phaseolina* was summarized by Dhingra and Sinclair (1978) who mentioned that various compunds stimulated sclerotial germination. The entry by the pathogen could be initiated directly through the cuticle and epidermis, with or without complex organised infection cushions or infection pegs. The entry could also be made through wounds and natural openings. The isolates from *P. caribaea* and *Casuarina equisetifolia* varied considerably in their mode of penetration as compared to *Eucalyptus* or *Sesbania* isolates as they infected the seedlings within 10 to 15 days while the other two isolates taken 15 to 30 days for developing the infection. The infection also varied with the isolates used and the plants species involved. Reddish brown, water soaked lesions appeared on the stem just at the ground level. The pathogen spreaded both upward and downwards in the roots. The dead tissue was converted with abundant black sclerotia appearing as an ashy grey covering like a soot. Along with stem rot and wilt, root were also attacked. Usually the tap root turned black and later became rotten, shredded and covered with sclerotia.

Histopathology of Diseases

The studies concerned with pathological anatomy were conducted by obtaining the thin sections (T.S.; TLS & RLS) of root and basal part of the stem of *Pinus caribaea* and also by obtaining simple T.S. of other broad leaved seedlings under study. The hyphae of the fungus attached to the root bark. Some times developed infection cushions. The fungus invaded the wood intercellularly, later intracelluarly and that was followed by the formation of sclerotia (Plate I, Fig.4) . The sections (T.S., TLS and RLS) clearly indicated that the sclerotical accumulation in the medullary rays (Plate I, 4 and 5). At some places in the wood compact mass of resistant hyphae accumulated in the form of sclerotia which disorganised the surrounding cells by forming the ward cavities in the wood. The affected trachieds and the medullary rays exhibited brown colour. Pycnidial stage of the fungus also developed in the under ground and basal parts of the stem of seedlings (Plate I, Fig. 6).

The large size of sclerotia and pycnidia the apparent disruption of host-cell walls by sclerotia and appearance of sclerotia in the basal part of the stem suggested that the sclerotia in tissues were indicative of host cell death.

In *Eucalyptus* seedlings formation of pycnidia alongwith the characteristic condition was observed after death of parenchymatous cells (Plate I Fig.9) . Sclerotia of pathogen also developed in the medullary region of the root and stem.

Usually the sclerotia were detected under the bark of lateral as well as tap roots in the seedlings of *Casuarina*. These sclerotia were connected with the thick walled brown septate threads of mycelium and formed chains under the surface of roots (Plate 2 Fig. 4). The formation of sclerotia was less in stelar region as compared to extra-stelar region (Plate 2 Fig. 5).

In *Sesbania grandiflora* sclerotial and mycelial stage were commonly observed on the basal portion of stem. The root system of S. grandiflora exhibited black charred symptoms. The mycelial growth was poor in the stelar region as compared to extrastelar region. Mainly parenchymatous tissues exhibited the sign of disorganisation by the pathogen which resulted drying and death of the seedlings. The pycnidial stage of the pathogen was not detected.

Epidemiology of the Diseases

Macrophomina phaseolina is a well known soil-borne pathogen but later disseminated through air and multiplied rapidly at appropriate soil moisture and high temperature. The fungus persisted in the soil for long periods as an actively growing mycelium or as dormant sclerotia. The seedlings were more rapidly and severely infected at 29^{0}C to 35^{0}C than at lower temperature i.e. below 25^{0}C. *M. phaseloina* was a poor competitor in the soil and existed primarily as sclerotia. Warm and dry conditions were favourable for development of the charcoal root rot it caused. Such conditions were not much suitable for the growth and development of other soil-borne fungi. Thus the saprophytic survival of *M. phaseolina* was more and sufficient amount of inoculum potential was produced. Due to regular watering the pH of the potting mixture shifted towards neutrality or alkalinity which favoured the infection. Acidic soil reduced the activity of the organism.

Management of the Diseases in Nursery

The diseases appeared in nurseries as well as in plantation during the hot and dry summer months. The disease was very severe and produced high inoculum density which might affected the future health of the nurseries crop and plantations. The saprophytic stage (Sclerotial stage) of this fungus was very resistant to the physical factors of the soil. *M. phaseolina* developed rapidly at appropriate soil moisture and high temperature. The fumigation of potting mixture with 30 percent formalin or methyl bromide controlled the inocula mixed with it. Our preliminary trials indicated that certain fungicides including Brassicol (Penta Chloro nitro benzene), Dithane M-45 (Zinc ion and manganese ethylene bis dithio carbamate) or thiride (Tetra methyl thiram disulfide) at 0.3 percent as well as bavistin (Methyl 2 benzimidazole carbamate) 0.2 percent may be applied by soil drenching in *Pinus caribaea* (Jamaluddin *et al.* 1982). In case of *Eucalyptus, Casuarina* and *Sesbania* seedlings bavistin 0.2 percent and dithane M-45 0.2 percent have been found suitable for controlling the disease. No phytotoxic symptoms due to the fungicides could be observed at 0.2 percent or 0.3 percent cocentration. Certain antagonistic fungi were isolated from the soil of pine nursery and rhizosphere., *Penicillium sclerotiorum* and *Eupenicilium shearai* considerably checked the growth of organism in vitro. Only *M. phaseolina*, isolated from *P. caribaea* developed 5 mm broad zone of inhibition with *P. sclerotiorum* (Plate I Fig.10) and *Eupenicillium shearai*, but *Eucalyptus* isolate produced 5 mm zone of inhibition with *P. sclerotiorum* only. The remaining two isolates from *Casuarina equisetifolia* and *Sesbania grandiflora* failed to exhibit any zone of inhibition with the antagonistic fungi. *Trichoderma harzianum* and *T. pseudokoninghii* were also tried but they failed to exhibit any zone of inhibition with different isolates of M. phaseolina. The disease percentage could be reduced to 30 or 40 percent if a suspension of *P. sclerotiorum* or *E. shearai* is applied by soil drenching to the seedlings of *Pinus caribaea* and *Eucalyptus*. The disease could be managed by appropriate moisture content in the soil, pH should be between 6.4 to 6.7 and less organic matter content in the potting mixture.

References

Chellman, C. W. (1971). Insects diseases and other problems of Floridas trees. Flo. Dept. Agric. and Consumer Serv. Bull No. 195, 156 pp.

Dhingra, O.D. and Sinclari, J.B. (1978). Biology and pathology of *Macrophomina phaseolina*. *Univ. Fee* de Vieosa, Brazil 166 p.

Hodges, C.S. (1962). Black root rot of Pine seedlings. Phytopathology, **52**: 210-219.

Jamaluddin, V.S. Dadwal and K.K. Soni (1982). Studies on charcoal root rot of *Pinus caribaea* Indian Forester, **108** : 618-622.

Mc Cain, A.H. and Smith, R.S. (1972). Quantitative assay of *Macrophomina phaseolina* from soil. Phytopathology, **62**: 1098.

Qureshi, I.M. (1956). Mortality of *Casuarina equisetifloia* in plantations in Bombay State. *Proc. 8th Silvic Conf.* Dehra Dun 1951: Pt **2**, 329-330.

Rowan, S.J. (1971). Soil fertilizations, fumigation and temperature effect serverity of black root rot of slash pine. Phytopathology, **61**: 184-187.

Seymour, C.P. (1969). Charcoal root rot of nursery grown pines in florida. Phytopathology, **59**: 89-92.

Seymour, C.P. and Cordell, C.E. (1979). Control of charcoal root rot with methyl bromide in forest nurseries. South. Jour. App. For., **3**: 104-108.

Smith, R.S. Jr. and Bega, R.V. (1964). *Macrophomina phaseloina* in the forest tree nurseries of California. Plant Dis. Rept., **48**: 206.

Smith, R.S. Jr. and Bega, R.V. (1966). Root disease control by fumigation in forest nurseries. Plant Dis. Reptr., **50**: 245-248.

Smith, W.H. (1969). Germination of *Macrophomina phaseolina* as affected by *Pinus lambertiana* root exudates. Can. J. Microbiol., **15**: 1387-1391.

Soni, K.K., Dadwal, V.S. and Jamaluddin (1985). Charcoal root and stem rot of *Eucalyptus*. Euro Jour. of Forest Pathol., **15** : 397-401.

21

New Leads in Biology and Control of *Sclerotium rolfsii*

B.K. Sarma, D.P. Singh and U.P. Singh*

*Department of Mycology and Plant Pathology,
Institute of Agricultural Sciences, Banaras Hindu University,
Varanasi-221005*

Recent developments in the study of biology and control of the soilborne pathogen-*Sclerotium rolfsii,* are discussed in this chapter. The differentiating factors (morphologial and biochemical) of the pathogen from two closely related species of the genus *Sclerotium,* viz., *S. delphinii* and *S. coffeicola,* have been described. The role of extracellular as well as constitutive enzymes in *S. rolfsii* pathogensis and factors affecting production of such enzymes are also discussed in details. Influence of oxalic acid on phenol oxidizing enzymes in the hosts and relationship between lipid peroxidation and melanin formation during sclerotial biogenesis and their maturation, and the role of some antioxidant growth factors are described. Factors affecting teleomorph formation in *S. rolfsii* are explained. Leaf spot causing *S. rolfsii* in relation to infection, symptom production and environmental factors conducive to disease development are discused. The chapter also deals with variability in the pathogen. Recent approaches for the control of *S. rolfsii* through biocontrol agents and transgenic plants have been described. Biological control with *Trichoderma harzianum, Gliocladium virens* and plant growth-promoting rhizobacteria has been discussed with particular emphasis on the molecular aspects of the interaction between the pathogen and the biocontrol agents, *T. harzianum* and *G. virens.* Similarly, the use of transgenic plants transformed with a chitinase gene for the control of the pathogen has also been dealt with.

Key words : *Sclerotium rolfsii,* biology, control

Sclerotium rolfsii Sacc. (Teleomorph: *Athelia rolfsii* (Curzi) Tu & Kimbrough) is one of the most important and destructive fungal pathogen of many mono- and dicotyledonous plants encompassing more than 500 host species and occurs in all parts of the world causing collar rot, southern blight, root rot and leaf spot diseases. It grows as white thread like mycelium with tan to dark brown sclerotia produced in the colony. It is soil-borne usually causes infection at the collar region of the plants resulting into wilting. The pathogen does not produce inoculum for secondary infection in the neighbouring plants in the same season (Fry, 1982), and it becomes severe only when large amount of initial inoculum is present in the soil. Root diseases

* Corresponding Author E-mail: ups@banaras.ernet.in

caused by *S. rolfsii* generally become severe in moist soil at warm temperature (Cook and Papendick, 1972). Due to its wide host range, crop rotation is not an adaptable cultural practice, which is usually recommended for other soil-borne diseases. Sclerotia formed on undecomposed plant tissues in the field are capable of initiating infection and serve as the primary inoculum of disease in the field.

Biology

The form genus *Sclerotium* was described by Saccardo (Saccardo, 1913) to include fungi which formed differentiated sclerotia from sterile mycelia. Several diverse fungi were included in this genus, and all were characterized by the production of small, tan to dark brown or black spherical sclerotia internally differentiated into a rind, cortex and medulla (Punja and Rahe, 1992). Several plant pathogenic fungi with this sclerotial type and with no known sexual or conidial states were placed in the genus *Sclerotium*. However, as the teleomorph was discovered, the affinities of these fungi to the Basidiomycotina (Punja, 1988) or Ascomycotina (Krause and Webster, 1972) were established. *S. rolfsii* Sacc., is one of the best known members of this genus, which is a widespread destructive plant pathogen (Aycock, 1966; Punja, 1985). It differs from two other related species, viz., *S. coffeicola* and *S. delphinii*, in growth, morphology and physiology. Maximum sclerotial production in *S. coffeicola, S. rolfsii* and *S. delphinii* was achieved on potato dextrose agar (PDA) medium at incubation temperatures of 30, 35 and 20°C, respectively. Under the same temperature ranges the total sclerotia produced by the three species were 7, 1043, and 47 per Petri dish and the average sclerotial diameter was 5.1, 0.6 and 3.5 mm for *S. coffeicola, S. rolfsii* and *S. delphinii*, respectively. The cortical layer is 15-20 cell layer-thick in *S. coffeicola* as compared to 4-10 in the other two species. Histochemical staining of sclerotia shows maximum carbohydrates and metachromatic granules in medulla, followed by cortex and mostly confined in the cytoplasm. Proteins are also distributed in both the cortex and mostly confined in the cytoplasm. Proteins are also distributed in both the cortex and medulla of *S. rolfsii* and *S. delphinii*, but occurred only in the medulla of *S. coffeicola* in higher concentration than the other two species. Phenolic compounds are found in rind cells of all sclerotia, with *S. coffeicola* containing the least amount, and are localized in the walls. All the three species produce oxalic acid, pectinase and polyphenoloxidase in culture and production is maximum in *S. rolfsii* followed by *S. delphinii* and *S. coffeicola*. These observations show a close similarity among the three *Sclerotium* species. Thus, the species can be differentiated by colony morphology, size of sclerotia, histochemical staining, composition of sclerotia, and growth response to temperature (Punja and Damiani, 1996).

S. rolfsii has an extensive host range, partly due to its prolific growth rate and production of oxalic acid and cell wall degrading enzymes (Punja, 1985). There are several cell wall degrading enzymes reported to be secreted by the pathogen (Punja, 1985). Bhosale and Shewale (1995) reported production of extracellular and constitutive enzyme β-galactosidase by the pathogen. Enzyme production is enhanced in the presence of galactose, raffinose, cellobios, sucrose, xylose, maltose, cellulose and pectin. However, cellulose and diammonium hydrogen phosphate are found to be the best sources. Surfactants such as, Sag, Paraffin oil, Tween 20 and Tween 80 increase enzyme production. Maximum production of enzyme is observed at pH 2.7 and at 60°C. Role of oxalic acid, released by the pathogen during its pathogenesis, is thought to be vital in lowering the pH of the host tissues. Moreover, according to Ferrar and Walker (1993), oxalic acid secreted by the pathogen affected the activity of O-diphenol oxidase by oxalic acid. In contrast, in *P. vulgaris* pods the secreted oxalic acid reduces pH to a level where O-diphenol oxidase is inactive. Thus, Ferrar and Walker (1993) postulated that this action of oxalic acid enhances the success of the pathogen by suppressing the host defence mechanisms. Similarly, Sachslehner and Haltrich (1999) isolated and purified a thermostable acidic endo-β-1, 4-D-mannase from the fungus. This enzyme is formed even under non-induced and derepressed conditions, i. e., after depletion of glucose used as the only carbohydrate substrate for its formation. The enzyme is especially stable in the acidic region with an exceptional half-life of 41 days at pH 4.5 and 50°C. Thus, the reason for its wide host range and high degree of pathogenicity could be understood from these results.

Since conidial production is absent in *S. rolfsii,* resting sclerotia serves as the primary inoculum for the pathogen. Biogenesis of sclerotia in this fungus is associated with lipid peroxidation. Georgiou (1997) observed a direct relationship between number of sclerotia produced by the fungus and lipid peroxidation levels in the colonies. His observation included that lipid peroxides of membranous and cytoplasmic origin occurred in sclerotial exudate. He also suggested their lipid peroxidation may be due to oxidative stress caused by growth condition of the mycelia. According to Ellil (1999), melanin pigments constitute 13.9% of the chemical composition of sclerotial walls of *S. rolfsii* and is associated with proteins, reducing sugars and amino acids. The lipid and ash contents in the sclerotial walls are double than those in the hyphal walls of the fungus. Increasing age of culture and maturation of the sclerotia are always accompanied by increased lipid peroxides and melanin pigments. Such behaviour may indicate that lipid peroxidation and melanin formations are operating parallel during sclerotial biogenesis and maturation. These two processes depend on the theory of oxidative stress, as affected by growth conditions. Both processes could be stopped or sharply retarded when subjected to some antioxidant growth factors such as vitamins (ascorbic acid), micro-elements (selenium), and sulfhydryl compounds (glutathione). A clear relationship among oxidative stress, myceliogenic germination and lytic activity via melanin production is observed. This finding appears promising in applying a new control measure against diseases caused by sclerotium producing fungi without using traditional toxic fungicides.

Teleomorph

The first description of teleomorph of *S. rolfsii* was made by Goto (1930) based on his studies on the perfect stage in culture. The teleomorph was provisionally placed in *Corticium centrifugus* (Lev.) Bres. Curzi (1931), however, designated the teleomorph as a new species, *Corticium rolfsii* Curzi. The species was transferred later to *Pellicularia* by West (1947) and to *Botryobasidium* by Venkatarayan (1950). Talbot (1973) suggested that the basidial stage belonged to the genus *Althelia.* Tu and Kimbrough (1978) gave the specific epithet as *Athelia rolfsii* (Curzi) Tu and Kimbrough in their study of the perfect stages of the *Rhizoctonia-Sclerotium* complex. Basidial hymenia of *A. rolfsii* have been reported infrequently in nature (Ghosh and George, 1955; Kulkarni and Ahmed, 1967a, West, 1947) but can be induced in culture (Ahmed and Kulkarni, 1966; Ahmed *et al.*, 1965, 1967, 1968; Goto, 1930; Kulkarni and Ahmed, 1965, 1966, 1967b; Misra and Haque, 1960; Mundkur, 1934; Punja *et al.*, 1982). However, in all these reports, the time required for basidiocarp induction was from 2 weeks to several months. Since the teleomorphs of most known isolates of *S. rolfsii* have not been induced or discovered, it is uncertain whether they all belong to *A. rolfsii.* Because of their widespread occurrence and economic importance, more rapid and dependable methods of basidial induction in such isolates are needed. Singh *et al.* (1996) were able to develop basidial stage on potato dextrose agar (PDA) medium supplemented with fresh rhizome powder of *Cyperus rotundus* within 5-7 days. This medium will help in the study of biology of the fungus encompassing sclerotial to basidial stage. Athelial stage could also be induced by incorporation of only *C. rotundus* rhizome meal in distilled water (Singh *et al.,* 1996). Similarly, methanolic extract of *C. rotundus* rhizomes also induces athelial stage but in comparison to the fresh rhizome it takes more than 10 days to induce it. Another report by Prithiviraj *et al.* (2000) showed that athelial stage induction in *S. rolfsii* on the same *C. rotundus* rhizome meal-incorporated medium is dependent on the bivalent cation calcium (Ca^{2+}) and when it was chelated with ethylenediamine tetra-acetic acid (EDTA), induction of athelial stage was hampered. Moreover, they have also reported that athelial stage is sensitive to white fluorescent light though other lights, viz., blue, green and red had no effect on athelial stage induction. Maximum athelial stage was produced under continuous darkness. Similarly, athelial stage is temperature-dependent as beyond 30ºC its formation is prevented. Different pH levels (2-8) do not affect athelial stage induction. Prithiviraj *et al.* (2000) also demonstrated that athelial stage induction is negatively correlated with sclerotium formation by the fungus. The more the athelial stage, the less will be the sclerotia.

Infection

S. rolfsii is commonly known to infect the host plants at the collar region and causes wilting of the entire plant. However, there are certain reports (Singh and Pavgi, 1965) of spotted leaf rot disease in a number of taxonomically unrelated plant species belonging to cultivated, ornamentals and several common weeds. The infection usually starts as a chlorotic translucent spot on a leaf above the ground, which becomes necrotic at its center and expands to 5 to 10 mm in diameter. Only a few leaves on a plant are infected. The tissues show a wet rot that develops fast in humid weather but is arrested with a dry spell. Concentric zones like a target board invariably appears on the necrotic tissues. There is no defoliation but occasionally shot hole is discernible. Rarely two or more spots develop closely which coalesce later. In a few cases, root and stem rots are also observed. Abundant sclerotia are produced on the rotting stems. However, in all the cases, a tiny sclerotium develops in the center of the spots, as a rule. It appears as a dull whitish hyphal knot initially, turning hard circular pinkish to brown resembling a mustard seed.

Variability

Variability in this fungus has been studied by several workers (Harlton *et al.*, 1995; Nalim *et al.*, 1995; Okabe *et al.*, 1998; Okabe *et al.*, 2000; Prithiviraj *et al.*, 1996). Harlton *et al.* (1995) observed that within a mycelial compatibility group (MCG), isolates were often from the same host and geographical area; however, widely diverse isolates were also grouped within the same MCG. Variation in nuclear rDNA [internal transcribed spacer (ITS) regions] was also observed using the restriction fragment length polymorphism (RFLP) technique. There was more than one subspecific grouping in the isolates of the fungus and some of these groupings were correlated with their MCG. However, isolates within a MCG also showed different ITS-RFLP patterns and certain patterns were also dispersed among different MCGs. similarly, Okabe *et al.* (1998) observed variations in the southern blight fungal isolates in Japan. On the basis of ITS-RFLP analysis of nuclear rDNA, they formed five groups and morphological differences were observed in sclerotia of various groups. Out of five, three groups were identified as *S. rolfsii* and the two resembled *S. delphinii* in RFLP patterns with large sclerotia requiring relatively low optimal growth temperature (28ºC). Sclerotial size of the latter is variable and temperature dependent often becoming indistinguishable from those of *S. rolfsii* at high temperatures. Hyphal anastomosis (imperfect fusion) was observed among different ITS-RFLP groups and different isolates belonging to the same groups. These results indicate that population of this fungus in Japan consists of several different subgroups, although morphological differences are not always evident. Similarly, variablity in *S. rolfsii* solates of soybean in India was studied by Ansari and Agnihotri (2000). They observed morphological, physiological and pathological variations among the isolates. Basad on mycelial compatibility, they found 12 races of the pathogen. Oxalic acid production by the fungus also varied in the isolates and the oxalic acid production was correlated with pathogenicity of the fungus in soybean.

Scanning Electron Microscopy (SEM) and isozyme studies conducted by Prithiviraj *et al.* (1996) revealed structural and biochemical variations between two types (large and small) of segregated sclerotia of a single *S. rolfsii* isolate. They observed that the small sclerotia are round to spherical and brown with smooth surface, the large sclerotia are typically brown, irregular in shape with highly pitted surface. They conclude that the size difference between the two types of sclerotia is basically due to the medullary region where the hyphae were loosely interwoven specially in large sclerotia. The isozyme pattern of GPI-1 (glucose phosphate isomerase-1) of both sclerotial types was the same but significant differences were observed in the relative mobility of PEP-1 (peptidase-1). However, these two sclerotial types are not genetically stable. Since this isolate of *S. rolfsii* is capable of producing teleomorph, genetic segregation in the isolate is thought to be due to meiosis rather than mutation or entry of foreign gene through hyphal anastomosis.

Control

Despite continuous efforts over the past one century since the report by Rolfs in 1892 of *S. rolfsii* on tomato to manage this problem (Aycock, 1966), it continues to plague growers and cause considerable economic loss. Control efforts have often met with limited success partly due to the extensive host range, prolific growth rate and ability to produce large number of sclerotia that may remain viable in soil for several years. Furthermore, control measures effective for a particular crop in an area may not be adaptable elsewhere due to regulatory or economic constraints. There are reports from India as well as outside that the host range of the pathogen is increasing and thereby causing concern for its successful control in the fields where it poses great threat to the growers. Some of the newly recorded hosts of collar rot causing *S. rolfsii* are:

Ficus auriculata (Mohan *et al.*, 2000), *Crossandra infundibuliformis* (Naidu, 2000), *Mentha citrata* and *M. spicata* (Singh *et al.*, 1999), *Iris xiphium* (Madhu Meeta and Madhu Bhatia, 1992), *Eustoma grandiflorum* (McGrover *et al.*, 2000), *Solanum pseudocapsicum* (Garibaldi *et al.*, 2000), etc. There are also newly recorded hosts showing symptoms of leaf spot rot causes by *S. rolfsii* such as *Dioscorea* species (Amusa, 1999) and *Piper nigrum* (Larez *et al.*, 1996), etc.

Various control measures of *S. rolfsii* have been described in literature including use of synthetic fungicides. Numerous fungicides were reported to control the pathogen successfully but major limitations to the widespread use of fungicides to control *S. rolfsii* are the requirement of large amounts of the chemicals as the inoculum is soil-borne. Moreover, the efficacy of a particular compound may vary depending on the crop and results are not consistent from one growing season to another. Looking into the environmental hazard posed by the fungicides, more ecologically sound and feasible control measures include use of plant products and biocontrol agents. Transgenic plants also became a primary field of research for the control of this pathogen. Prithiviraj *et al.* (1996) showed that methanol extract of *Aegle marmelos* inhibited sclerotial and mycelial growth of *S. rolfsii*. Remirez-Chavez *et al.* (2000) showed fungicidal activity of crude extract of *Heliopsis longipes* against *S. rolfsii* at 15-20 μl/ml. Similarly, Patnaik *et al.* (1996) showed the antifungal property of some essential oils of plants *in vitro* against *S. rolfsii*. However, none of them has shown promise under field conditions.

Biocontrol agents (BCA) including fungi and bacteria are reported to protect crops from *S. rolfsii* infection. Among the fungal bioagents, *Trichoderma* and *Gliocladium* speices are the major antagonistic fungi which are ubiquitous in natural soil (Papavizas, 1985). Antibiosis and mycoparasitism by these fungi are the major characteristics through which they reduce the inoculum of *S. rolfsii* (Elad *et al.*, 1982; Mukherjee *et al.*, 1995; Upadhay and Mukhopadhyay, 1986). During antibiosis, both volatile and nonvolatile secondary metabolites are implicated in restricting the vegetative growth of pathogenic fungi (Claydon *et al.*, 1987; Dennis and Webster, 1971a, b; Upadhaya and Mukhopadhyay, 1986). *Trichoderma* spp. attack the pathogen by excreting lytic enzymes including β-1,3-glucanase(s), proteinase(s) and chitinase(s), enabling them to degrade host cell walls and thus reduce inoculum potential (Chet *et al.*, 1993; Geremia *et al.*, 1993; Goldman *et al.*, 1994; Lorito *et al.*, 1993). The chitinolytic system of *T. harzianum* strain TM was found to have six distinct chitinolytic enzymes: two were identified as β-1, 4-N acetylglucosaminidases that hydrolyze chitin to N-acetylglucosamine monomers in an exotype fashion (CHIT 102 and CHIT 73) (Haran *et al.*, 1995). Elad *et al.* (1982) found that *Trichoderma* isolates produce chitinases and glucanases when grown on live mycelium of *S. rolfsii* in soil. The differential expression of the chitinolytic system of *Trichoderma* during mycoparasitic interactions has only recenly been studied. Inbar and Chet (1995) used direct confrontation between *Trichoderma* and *S. rolfsii* to identify specific chitinases expressed during parasitism. They detected a change in the chitinolytic enzyme profile during the interaction between the fungi. Before coming into contact with each other, both fungi contained a protein with constitutive β-1, 4-*N*-acetylglucosaminidase activity. After contact, the chitinolytic activity of *S. rolfsii* disappeared, whereas that of the *Trichoderma N*-acetylglucosaminidase (CHIT 102) greatly increased. As the interaction proceeded, the activity of CHIT 102 diminished concomitantly with the appearance of CHIT 73.

Benhamou and Chet (1996) studied ultrastructural and cytochemical aspects of the interaction between *S. rolfsii* and *T. jarzianum* through scanning and transmission electron microscopy to assess the potential role of enzymatic hydrolysis in the antagonistic process. SEM investigations revealed that hyphae of *T. harzianum* grew abundantly on the sclerotial surface and formed a dense lunched mycelium that appeared to establish contact with the outer host cells through a thin mucilage. Observations of cross-sections of parasitized sclerotia by light microscopy showed that hyphae of the antagonist multiplied on the sclerotial surface and displayed the ability to penetrate the rind. Growth of the antagonist in the rind layer was mainly intracellular and host-wall penetration was achieved by means of constricted hyphae. Ultrastuctural observations showed that *Trichoderma* growth and development coincided with extensive host cell alterations, such as retraction and aggregation of the cytoplasm and vacuole breakdown. In the invaded outer rind cells, host cell walls apparently were not altered, as judged by their preserved structure. In contrast, cell breakdown due to host cell-wall disruption was observed more frequently in inner rind cells adjacent to medullary cells. Ingress of *T. harziamum* hyphae in the medulla was characterized mainly by a change in the mode of growth from intra-to intercellular. *Trichoderma* hyphae did not penetrate the medullary cells, although the latter showed pronounced alterations, such as cytoplasm disorganization and aggregation. The use of wheat germ agglutin/ovomucoid-gold complex for localization of chitin monomers resulted in regular labeling of both host and antagonist cell walls, even when sclerotia were massively colonized. Chitinolytic degradation at a distance from the site of *Trichoderma* penetration was never observed. There was no indication of cell-wall disorganization in either the host or the antagonist, as shown by the regular distribution of the labeling, even in zones of penetration by constircted *Trichoderma* hyphae. In the medulla, gold labeling was regularly distributed over the thick host cell walls, even when the medullary cells showed obvious signs of disorganization. When ultrathin sections of parasitized sclerotia were incubated with gold complexed β-1, 3-glucanase for localization of β-1, 3-glucans, a regular distribution of gold particles was observed over the walls of both outer and inner rind cells, even when these exhibited a disorganized cytoplasm that was reduced to a few remnants of aggregated material. Incubation with gold-complexed lipoprotein lipase yielded a pattern of labeling similar to that obtained with gold-complexed β-1, 3-glucanase. Gold particles were evenly distributed over the host cells walls in the rind layer.

Similarly, numerous reports exist in literature citing effective biocontrol of *S. rolfsii* by *Gliocladium virens in vitro* as well as under field conditions (Pierto *et al.*, 1993; Lumsden *et al.*, 1992; Mukherjee *et al.*, 1995; Wilhite and Straney *et al.*, 1996). The main mode of action of the BCA is parasitism as well as antibiosis. The BCA colonize sclerotia of *S. rolfsii* immediately when it comes in contact with sclerotia and penetrate to the inner layers of sclerotia where it sporulates (Mukherjee *et al.*, 1995). Similarly, the BCA produce chitinolytic enzymes that include chitin 1,4-β- chitobiosidase and glucan *N*-acetyl-β-D-glucosaminidase, and glucan 1,3-β-glucosidase, known to be deletorious for *S. rolfsii* (Pierto *et al.*, 1993). Similarly, the BCA also produces an antibiotic, gliotoxin which is inhibitory to *S. rolfsii* at 50 μg/ml. Several compounds like some mixtures of fatty acids, viridin, dimethyl gliotoxin, viridiol, mixture of phenolic acids like ferulic acid are also detected in the culture filtrate of BCA. But inhibitory effect of all the compounds is not the same and depends on the interacting pathogen and the BCA isolate (Lumsden *et al.*, 1992). Although, gliotoxin is very effective against *S. rolfsii* many a times the BCA fails to control the pathogen significantly. One of the basic reasons for such a behaviour of BCA is attributed to its timing of gliotoxin production. A relatively short period of biocontrol activity limits the use of this bicontrol agent in certain applications. *G. virens* synthesizes gliotoxin only within a short 16-h period during replicative growth. An apparent lack of gliotoxin production in the later growth phases was due to the cessation of synthesis rather than an increase in gliotoxin catabolism. Medium-transfer experiments indicated that cessation of gliotoxin synthesis could not be explained by gliotoxin feedback inhibition, a diffusible inhibitor, or changing the nutritional status of the medium over a 2-h response time. These results demonstrate that the regulation of gliotoxin biosynthesis is a major determinant in the kinetics of gliotoxin appearance and focuses the need for further study on the regulation of gene expression for greater efficacy of the BCA (Wilhite and Straney, 1996).

Plant growth-promoting rhizobacteria (PGPR) are rhizosphere inhabiting bacteria that protect the plants from various pathogens through various mechanisms, viz., induced systemic resistance in the host (Maurhofer *et al.*, 1994; van Peer *et al.*, 1991), antibiotic production (Thomashaw and Weller, 1988; Maurhofer *et al.*, 1995), growth promotion of the host (Schippers *et al.*, 1987), competition for nutrients with the pathogens (Duijff *et al.*, 1993); Leeman *et al.*, 1996), etc. *Pseudomonas fluorescens* is one that has been reported to protect hosts against pathogen attack through these mechanisms (Dauijff *et al.*, 1993; van Peer *et al.* 1991; Keel *et al.*, 1992). Antagonistic effect of *P. fluorescens* against *S. rolfsii* has been reported by Rangeshwaran and Prasad (2000) under dual culture. They have also reported that the BCA is also effective under field conditions as 60-63% *S. rolfsii* infection-free sunflower plants were obtained in comparison to control following treatment with BCA. Similarly, in another report, Thiribhuvanamala *et al.*, (1999) also reported disease suppression in tomato upto 80% following treatment with *P. fluorescens* under pot culture. In another report, Patil *et al.* (1998) showed that *P. fluorescens* strain FDP-15 significantly improved groundnut seed germination, nodulation, dry matter production, pod yield and protected seedlings from *S. rolfsii* infection.

Use of transgenic plants for the control of *S. rolfsii* has also been reported by Punja and Raharjo (1996). They used three lines of cucumber cv. Endeavor, each transformed with a chitinase gene originating from *Petunia* (acidic), tobacco (basic) or *Phaseolus vulgaris* (basic) using *Agrobacterium tumefaciens* and compared with non-tyransgenic plants from response to inoculation with *S. rolfsii.* In growth chamber studies using whole plants and *in vitro* inoculation with detached leaves, no differences in disease development (rate and final levels) were detected between transgenic and non-transgenic plants. Carrot cultivars Nanco and Golden State transformed with 2 chitinase genes (from *Petunia* and tobacco) were evaluated for response to inoculation with *S. rolfsii.* A detached petiole inoculation method was used to compare transgenic and non-transgenic plants. The rate and final extent of lesion development after 7 days were significantly (P = 0.01) lower in transgenic plants expressing the tobacco (basic) chitinase gene upon inoculation with *S. rolfsii* but not in plants expressing *Petunia* (acidic) chitinase gene. It is concluded that the efficacy of chitinase gene transformation as a strategy for enhancing disease tolerance in plants can be influenced by the plant species used, the type of chitinase protein expressed and the characteristics of the fungal pathogen.

References

Ahmed, L. and Kulkarni, N.B. (1966). Studies on the basidial formation by *Sclerotium rolfsii* Sacc. II. A new medium for the basidial development by *S. rolfsii* isolate from *Colocasia.* Mycopath Mycol Appl, **29**: 331-332.

Ahmed, L., Kulkarni, N.B., and Moore, B.B. (1965). Studies on the basidial formation by *Sclerotium rolfsii* Sacc. XI. Attempts made to induce the basidial stage on culture media. Sydowia, **19** : 259-267.

Ahmed, L., Kulkarni, N.B., and Patil, B.C. (1967). Studies on the basidial formation by *Sclerotium rolfsii* Sacc. VI. A note on the additional medium promoting basidial stage by *Colocasia* isolate of *S. rolfsii.* Sci Cult, **33**: 26-27.

Ahmed, L., Kulkarni, N.B. and Patil, B.C. (1968). Studies on the basidial formation by *Sclerotium rolfsii* Sacc. VIII. Abortive basidial formation on media. Mycopathol Mycol Appl, **34**: 86-192.

Amusa, N.A. 1999. Concentric leaf spot of yam (*Dioscorea* spp.) in South-Western Nigeria. Mycopathologia, **148**: 33-36.

Ansari, M.M., and Agnihotri, S.K. (2000). Morphological, physiological and pathological variations among *Sclerotium rolfsii* isolates of soybean. Indian Phytopath., **53**: 65-67.

Aycock, R. (1966). Stem rot and other diseases caused by *Sclerotium rolfsii.* N.C. Agric. Exp. Stn. Tech. Bull., **174**: 202 pp.

Benhamou, N. and Chet, I. (1996). Parasitism of sclerotia of *Sclerotium rolfsii* by *Trichoderma harzianum:* ultrastructural and cytochemical aspects of the interaction. Phytopathology, **86**: 405-416.

Bhosale, J.H. and Shewale, J.G. (1995). Aryl beta-galactosidase from *Sclerotium rolfsii:* physiological and biochemical studies. Hindustan Antibiotics Bulletin, **37** : 25-36.

Chet, I., Barak, Z. and Oppenheim, A. (1993). Genetic engineering of microorganisms for improved biocontrol activity. In: Biotechnological Prospects for Plant Disease Control. I. Chet, ed. Wiley-Liss, New York, pp. 211-235.

Claydon, N., Allan, M and Hanson, J.R., Avent., A.G. (1987). Antifungal alkyl pyrones of *Trichoderma harzianum.* Trans Brit Mycol Soc., **88**: 503-513.

Cook, R.J. and Panedick, R.I. (1972). Influence of water potential of soils and plants on root diseases. Annu Rev Phytopathol, **10**: 349-374.

Curzi, M. 1931. Studi su lo *"Sclerotium rolfsii"*. Boll Staz Patol Veg Roma, Ser 2, **11**: 306-373.

Dennis, C. and Webster, J. (1971a). Antagonistic properties of species-groups of *Trichodermka.* 1. Production of non-volatile antibodies. Trans Brit Mycol Soc., **57**: 25-39.

Dennis, C. and Webster, J. (1971b). Antagonistic properties of species-groups of *Trichoderma.* 1. Production of non-volatile antibodies. Trans Brit Mycol Soc., **57**: 41-48.

Duijff, B.J., Meijer, J.W., Bakker, P.A.H.M. and Schippers, B. (1993). Siderophore-mediated competition for iron and induced resistance in the suppression of Fusarium wilt of carnation by fluorescent *Pseudomonas* spp. Netherlands J. P1. Pathol., **99**: 277-289.

Elad, Y., Chet, I. and Henis, Y. (1982). Degradation of plant pathogenic fungi by *Trichoderma harzianum.* Can. J. Microbiol, **28**: 719-725.

Ellil, A.H.A.A. (1999). Oxidative stress in relation to lipid peroxidation, sclerotial development and melanin production by *Sclerotium rolfsii.* J. Phytopathol. **147**: 561-566.

Ferrar, P.H. and Walker, J.R.L. (1993). O-Diphenol oxidase inhibition - an additional role for oxalic acid in the phytopathogenic arsenal of *Sclerotinia sclerotiorum* and *Sclerotium rolfsii.* Physiol. Mol. Pl. Pathol., **43**: 415-422.

Fry, W.E. (1982). Principles of Plant Disease Management, Academic Press, New York.

Garibaldi, A., Minuto A. and Gullino, M.L. (2000). First report of *Sclerotium rolfsii* on Jerusalem Cherry (*Solanum pseudocapsicum*) in Europe. Plant Dis., **84**: 1048.

Georgiou, C.D. (1997). Liqid peroxidation in *Sclerotium rolfsii:* a new look into the mechanism of sclerotial biogenesis in fungi. Mycol. Res., **101**: 460-464.

Geremia, R., Goldman, G., Jacobs, D., Ardiles, W., Vila, S., van Montagu, M. and Herrera-Estrella, A. (1993). Molecular characterization of proteinase-encoding gene, *prb* 1, related to mycoparasitism by *Trichoderma harzianum.* Mol. Microbiol., **8**: 603-613.

Ghosh, T. and George, K.V. (1955). *Sclerotium rolfsii* Sacc. on juice and its perfect stage. Indian J. Agric. Sci., **25**: 171-173.

Goldman, H.G., Hayes, C. and Harman, G.E. (1994). Molecular and cellular biology of biocontrol by *Trichoderma* spp. Trends Biotechnol., **12**: 478-482.

Goto, K. (1930). On the perfect stage of *Sclerotium rolfsii* sacc. produced on culture media. J Soc Trop Agr, **2**: 165-175.

Haran, S., Schickler, H., Oppenheim, A. and Chet, I. (1995). New components of the chitnolytic system of *Trichoderma harzianum*. Mycol. Res., **99**: 441-446.

Harlton, C.E., Levesque, C.A. and Punja, Z.K. (1995). Genetic diversity in *Sclerotium* (*Athelia*) *rolfsii* and related species. Phytopathology, **85**: 1269-1281.

Inbar, J. and Chet, I. (1995). the role of recognition in the induction of specific chitinase during mycoparasitism by *Trichoderma harzianum*. Microbiology., **141**: 2823-2829.

Krause, R.A. and Webster, R.K. (1972). The morphology, taxonomy and sexuality of the rice rot fungus, *Magnaporthe salvinii* (*Leptosphaeria salvinii*). Mycologia, 64: 103-114.

Keel, C., Schnider, U., Maurhofer, M., Voisard, C., Burger, D., Hass, D. and Defago G. (1992). Suppression of root diseases by *Pseudomonas fluorescens* CHA0: Importance of the bacterial secondary metabolite 2,4-diacetylphloroglucinol. Mol Plant Microbe Interact, **5**: 4-13.

Kulkarni, N.B. and Ahmed, L. (1966). Studies on the basidial formation by *Sclerotium rolfsii* Sacc. v. Basidial stage of *S. rolfsii* isolates from potato on a new medium containing organic nitrogen compound. Sydowia, **19**: 162-164.

Kulkarni, N.B. and Ahmed, L. (1967). Studies on basidial formation by *Sclerotium rolfsii* Sacc. VII. A modified medium inducing basidial stage by wheat isolate of *S. rolfsii.* Sci. Cult., **33** : 127-128.

Kulkarni, N.B. and Ahmed, O. (1965). Studies on the basidial formation by *Sclerotium rolfsii* Sacc. IVa. Development of basidial stage of *S. rolfsii* isolates from *Colocasia* and groundnut on a new medium. Sydowia, **19** : 159-161.

Kulkarni, N.B. and Ahmed, O. (1967a). Studies on the basidial formation by *Sclerotium rolfsii* Sacc. I. Basidial formation on artificially inoculated host. Sci. & Cult., **33**: 73-75.

Kulkarni, N.B. and Ahmed, O. (1967b). Studies on the basidial formation by *Sclerotium rolfsii* Sacc. VII. A modified medium inducing basidial stage by wheat isolate of *S. rolfsii.* Sci. & Cult., **33**: 127-128.

Larex, C.R., Laynez, J.A. and Marin, de C.N. (1996). Zonate leaf spot of pepper (*Piper nigrum*) caused by *Sclerotium rolfsii* in Caripito, Monagas State, Venezuela Fitopatologia Venezolana, **9**: 10-12.

Leeman, M., den Ouden, F.M., van Pelt, J.A., Dirkx, F.P.M., Stejil, H., Bakker, P.A.H.M. and Schippers, B. (1996). Iron availability affects induction of systemic resistance to Fusarium wilt of radish by *Pseudomonas fluorescens.* Phytopathology, **86**: 149-155.

Lorito, M., Harman, G.E., Hayes, C.K., Broadway, R.M., Tronsmo, A., Woo, S.L. and di Pietro, A. (1993). Chitinilytic enzymes produced by *Trichoderma harzianum:* antifungal activity of purified endochitinase and chitobiase. Phytopathology **83**: 398-405.

Lumsden, R.D., Ridout, C.J., Vendemia, M.E., Harrison, D.J., Waters, R.M. and Walter, J.F. (1992). Charactirization of major secondary metabolities produced in soilless mix by a formulated strain of the biocontrol fungus *Gliocladium virens.* Can. J. Microbiol., **38**: 1274-1280.

Madhu, Meeta and Madhu, Bhatia. (1992). Some new host records on ornamental plants. Indian Phytopathol., **45**: 279-280.

Maurhofer, M., Hase C., Meuwly, P., Metraux, J.P. and Defago, G. (1994). Induction of systemic resistance of tobacco to tobacco necrosis virus by the root-colonizing *Pseudomonas fluorescens* strain CHA0: Influence of the *gacA* gene and of pyoverdine production. Phytopathology, **84**: 139-146.

Maurhofer, M., Keel, C., Haas, D. and Defago, G. (1995). Influence of plant species on disease suppression by *Pseudomonas fluorescens* CHA0 with enhanced antibiotic production. Plant Pathol, **44**: 44-50.

McGovern, R.J., Bouzar, H. and Harbaugh, B.K. (2000). Stem blight of *Eustoma grandiflorum* caused by *Sclerotium rolfsii*. Pl. Dis., **84**: 490.

Mishra, A.P. and Haque, S.Q. (1960). Perfect stage of *Sclerotium rolfsii* sacc. Nature (London), **186**: 567.

Mohan, L., Paranidharan, V. and Prema, S. (2000). New diseases of timal fig (*Ficus auriculata*) in India. Indian Phytopathol, **53** : 496.

Mundkur, B.B. (1934). Perfect stage of *Sclerotrium rolfsii* Sacc. in culture. Indian J Agric Sci, **4**: 779-782.

Mukherjee, P.K., Mukhopadhyay, A.N., Sarmah, D.K. and Sgresthan, S.M. (1995). Comparative antagonistic properties of *Gliocladium virens* and *Trichoderma harzianum* on *Sclerotium rolfsii* and *Rhizoctonia solani*-its relevence to understanding the mechanisms of biocontrol. J Phytopathol, **143** : 275-279.

Nadiu, P.H. (2000). Crossandra- a new host record for *Sclerotium rolfsii*. Indian Phytopathol, **53**: 496-497.

Nalim, F.A., Starr, J.L., Woodard, K.E., Segner, S. and Keller, N.P. (1995). Mycelial compatibility groups in Texas peanut field populations of *Sclerotium rolfsii*. Phytopathology, **85**: 1507-1512.

Okabe, I., Morikawa C., Matsumoto N. and Yokoyama, K. (1998). Variation in *Sclerotium rolfsii* isolates in Japan. Mycoscience **39**: 399-407.

Okabe, I., Morikawa C. and Matsumoto, N. (2000). Variation in southern blight fungus in Japan detected by ITS-RELP analysis. Japan International Research Center for Agricultural Sciences, **34**: 93-97.

Papavizas, G.C. (1985). *Trichoderma* and *Gliocladium:* Biology, ecology and the potential for biocontrol. Ann. Rev. Phytopathol, **23** : 93-121.

Patil, R.P., Jagadeesh, K.S., Krishnaraj, P.U. and Kulkarni, J.H. (1998). Bacterization of groundnut with *Pseudomonas fluorescens* for the control of collar rot caused by *Sclerotium rolfsii* Sacc. Karnataka J. Agri. Sci., **11**: 423-425.

Pattnaik, S., Subramanyam, V.R. and Kole, C. (1996). Antobacterial and antifungal activity of ten essential oils *in vitro*. Microbios, **86**: 237-246.

Pierto, A. di, Lorito, M., Hayes, C.K., Broadway, R.M. and Harman, G.E. (1993). Endochitinase from *Gliocladium virens:* isolation, characterization and synergistic antifungal activity in combination with gliotoxin. Phytopathology, **83**: 308-313.

Prithiviraj, B., Khiste, S., Ram, D. and Singh, U.P. (1996). Effect of methonal extract of *Aegle marmelos* leaves on mycelial growth and sclerotium formation in *Sclerotium rolfsii*. Intern J Pharmacog, 34: 148-150.

Prithiviraj, B., Sarma, B.K., Srivastava, J.S. and Singh U.P. (2000). Effect of $Ca2^+$, $Mg2^+$, light, temperature and pH on athelial stage formation of *Sclerotium rolfsii* Sacc. J P1 Dis Prot, **107**: 274-278.

Prithiviraj, B., Singh, U.P., Kobayashi, K., Sugawara, K., Nishimura, R. and Sakari-Wada, A. (1996). Variability in sclerotial morphology in *Sclerotium rolfsii:* SEM and isozyme studies. J P1 Dis Prot, **103**: 495-501.

Punja, Z.K. (1985). Biology, ecology and control of *Sclerotium rolfsii*. Ann Rev Phytopathol, **23**; 97-127.

Punja, Z.K. (1988). *Sclerotium* (*Athelia*) *rolfsii,* a pathogen of many plant species. Pp. 523-534. In: Genetics of plant pathogenic fungi. Vol. 6. Ed., G.S. Sidhu. Academic Press, London.

Punja, Z.K. and Damiani, A. (1996). Comparative growth, morphology and physiology of three *Sclerotium* species. Mycologia, **88** : 694-706.

Punja, Z.K., Grogan, R.G. and Adam, J.V. (1982). Influence of nutrition, environment and the isolate on basidiocarp formation, development and structure in *Athelia* (*Sclerotium*) *rolfsii*. Mycologia, **74**: 917-926.

Punja, Z.K. and Raharjo, S.H.T. (1996). Response of transgenic cucumber and carrot plants expressing different chitinase enzymes to inoculation with fungal pathogens. Plant Di., **80**: 999-215.

Punja, Z.K. and Rahe, J.E. (1992). *Sclerotium*. Pp. 166-170. In: Methods for research on soilborne phytopathogenic fungi. Eds., L.L. Singleton, J.D. Mihail and C.M. Rush. APS Press, St. Paul.

Ramirez-Chaver, E., Lucas-Valdez, L., Virgen-Calleros, G. and Molina-Torres, J. (2000). Fungicidal activity of afinnin and *Heliopsis longipes* crude extract on two *Sclerotium* species. Agrociencia, **34** : 207-215.

Rangeshwaran, R. and Prasad, R.D. (2000). Biological control of *Sclerotium rolfsii* rot of sunflower. Indian Phytopath, **53** : 444-449.

Saccardo, P.A. (1913). *Sclerotium rolfsii.* Sylloge Fungorum XXII. Pavia, Italy. 1500 pp.

Sachslehner, A. and Haltrich, D. (1999). Purification and some properties of a thermostable acidic endo-o-1, 4-d-mannase from *Sclerotium* (*Athelia*) *rolfsii.* FEMS Microbiol. Lett., **177**: 47-55.

Schippers, B., Bakker, A.W. and Bakker, P.A.H.M., (1987). Interactions of deletorious and beneficial rhizosphere microorganisms and the effect of cropping practices. Annu Rev Phytopathol, **25**: 339-358.

Singh, A., Chaturvedi, C., Kalra, A. and Singh, H.B. (1999). Two species of *Mentha* as new hosts of *Corticium rolfsii.* Bulletin OEPP, **29**: 211.

Singh, U.P. and Pavgi, M.S. (1965). Spooted leaf rot of plants - a new sclerotial disease. P1 Dis Reptr, **49** : 58-59.

Singh, U.P., Prithiviraj, B., Khiste, S., Dalai, S. and Singh, A. (1996). Induction of sexual reproduction in *Sclerotium rolfsii, Ustilago cynodontis* and *Cintractia limitata* by *Cyperus rotundus* rhizomes. Can J Bot, **74** : 803-806.

Talbot, P.H.B. (1973). Aphyllophorales I: general characteristics: thelephoroid and cupuloid families. In : the Fungi. Vol. IVB. Ends., GC Ainsworth, FK Sparrow and AS Sussamn. Academic Press, New York, pp. 327-349.

Thiribhuvanamala, G., Rajeswari, E. and Duraiswamy, S. (1999). Biological control of stem rot of tomato caused by *Sclerotium rolfsii* Sacc. Madras Agril J, **86** : 30-33.

Thomashaw, L.S. and Weller, D.M. (1988). Role of phenazine antibiotic from *Pseudomonas fluorescens* in biological control of *Gaeumannomyces* graminis var. *tritici.* J Bacteriol, **170**: 3499-3508.

Tu, C.C. and Kimbrough, J.W. (1978). Systematics and phylogeny of fungi in the *Rhizoctonia* complex. Bot Gaz, **139**: 454-466.

Upadhaya, J.P. and Mukhopadhyaya, A.N. (1986). Biological control of *Sclerotium rolfsii* by *Trichoderma harzianum* in sugarbeet. Trop Pest Management, **32** : 215-220.

Van Peer, R., Niemann, G.J. and Schippers, B. (1991). Induced resistance and phytoalexin accumulation in biological control of Fusarium wilt of carnation by *Pseudomonas* sp. strain WVS417r. Phytopathology, **81** : 728-734.

Venkatarayan, S.V. (1950). Notes on some species of *Corticium* and *Pellicularia.* Indian Phytopath., **3**: 81-86.

West, E. (1947). *Sclerotium rolfsii* Sacc. and its perfect stage on climbing fig. Phytopathology, **37**: 67-69.

Wilhite, S.E. and Strancey, D.C. (1996). Timing of gliotoxin biosynthesis in the fungal biological control agent *Gliocladium virens* (*Trichoderma virens*). Appl Microbiol Biotechnol, **45**: 513-518.

22

Piriformospora indica: A Novel Fungus

Shailendra Kumar[1], Satyendra K Garg[1*] and Ajit Varma[2]

[1]*Department of Microbiology, Dr. Ram Manohar Lohia Avadh University Faizabad-224001*

[2]*School of Life Sciences, Jawaharlal Nehru University, New Delhi-110067*

The symbiotic fungus *Piriformospora indica* belonging to Hymenomycetes (Basidiomycota) was discovered from north-west desert of Rajasthan, India. The fungus has functional similarities with arbuscular mycorrhizal fungus, which colonizes roots of various plants except crucifers and myc⁻ mutants of soybean and pea. It exhibits phytopromotional effect after colonization. The fungus is proved as a phosphorus transporter from substrate to the plant. The fungus has been reported fungicide resistant against ethylene bisdithiocarbamate, methyl benzimidazole-2-yl carbamate (MBC) and dimethyl 4,4′-O-phenylene bis (3-thioallophanate). It has biocontrol properties against various plant pathogenic fungi.

Key words : ***Piriformospora indica*, arbuscular mycorrhizal fungus, biological control agent, phosphorus transporter, bisdithiocarbamate, thiophenate, micropropagation.**

The soil microflora promote plant growth by providing necessary nutrients as a result of their colonization of the rhizosphere (*Azotobacter*, *Azospirillum*, *Bacillus* and cyanobacteria) or by symbiotic association (*Rhizobium*, Mycorrhizal fungi and *Frankia*). They may also regulate cyclic processes in the ecosystem such as decomposition of organic matter, nitrogen fixation, phosphorus solubilization, secretion of growth promoting substances and by providing protection to the plants against pathogens (Mukerji *et al*., 1998).

Mycorrhizal symbiosis is association of certain groups of soil fungi with the plant root system (Sahay *et al*., 1998). Benefits of the association include increased crop yield, protection against root pathogens, increased tolerance to environmental stress and perhaps, most important, a reduction in the use of chemical fertilizers (Linderman, 1994). The mycorrhiza-plant partnership is the basic and integral part of plant survival and growth (Varma and Schuepp, 1995). Inoculum production of mycorrhizal fungi for application to the plant is very different problem (Varma, 1995a). Being obligate symbionts they do not grow in the absence of their host.

*Corresponding Author E-mail: rrshailendra@hotmail.com

Piriformospora indica is a symbiotic fungus, discovered by Varma *et al.*, (1998) from sand-dunes in Thar desert of Rajasthan, north-west India. It resembles the arbuscular mucorrhizal fungus (AMF) with respect to morphological and ultrastructural features, but belongs to Hymenomycetes (Basidiomycota). *P. indica* drastically improves plant growth and overall biomass and can be easily cultivated on variety of synthetic media. Hyphae colonize the root and show inter- and intracellular structures (vesicle and hyphal coils). Chlamydospores are formed both inside the root tissues and extremely into the rhizosphere (Varma *et al.*, 1998, 2001). The properties of the fungus have been patented (Varma, A. and Franken, P., European Patent Office, Muenchen, Germany. patent No. 97121440.8-2105, Nov. 1998). The culture has been deposited Braunchweig, Germany (DMS No. 11827).

P. indica has a wide host range- monocots and dicots including legumes and orchids (Blechert *et al.*, 1999; Varma *et al.*, 1999; Varma *et al.*, 2001). Tissue culture raised plantlets treated with *P. indica* (biologically hardened) show 90-100% survival on transfer of plantlets to the pots and later to the field, while untreated plantlets exhibit low survival ranging from 10 to 60% (Sudha, 1998; Sahay, 1999). The properties of the fungus can be exploited for studying mycorrhizal associations.

Occurrence

P. indica has been reported from the rhizosphere of woody shrubs *Prosopis juliflora* and *Zizyphus nummularia* in western Rajasthan deserts. Several glomaceous fungi were screened (Neeraj *et al.*, 1991) and later some spores were allowed to germinate on minimal agar medium (Varma, 1995b). A pure culture was obtained which was later named *P. indica*. A similar fungus is reported from New South Wales, Australia. Comparison of 18S rDNA sequence studies reveal the presence of similar fungus *Leptochloa fusca* and *Oryza minuta* in rhizosphere soils of Pakistan and Phillipines, respectively (Varma *et al.*, 2001).

Morphological Features and Taxonomic Status

Morphological features vary on different media. Incubation under shaking conditions suppress the growth in minimal medium (MM1) broth cultures, but not on any other substrate. The fungal biomass is considerably enhanced under shaking conditions when grown in *Aspergillus* medium (Kaefer medium, Kaefer, 1977). On Moser B medium, the colonies appear compact, wrinkled with furrows and constricted. Mycelium produces fine zonation and a great amount of white aerial hyphae (Verma *et al.*, 1998; Varma *et al.*, 2001). The axenic culture of *P. indica* on agar surface and in culture broth are depicted in Plate I.

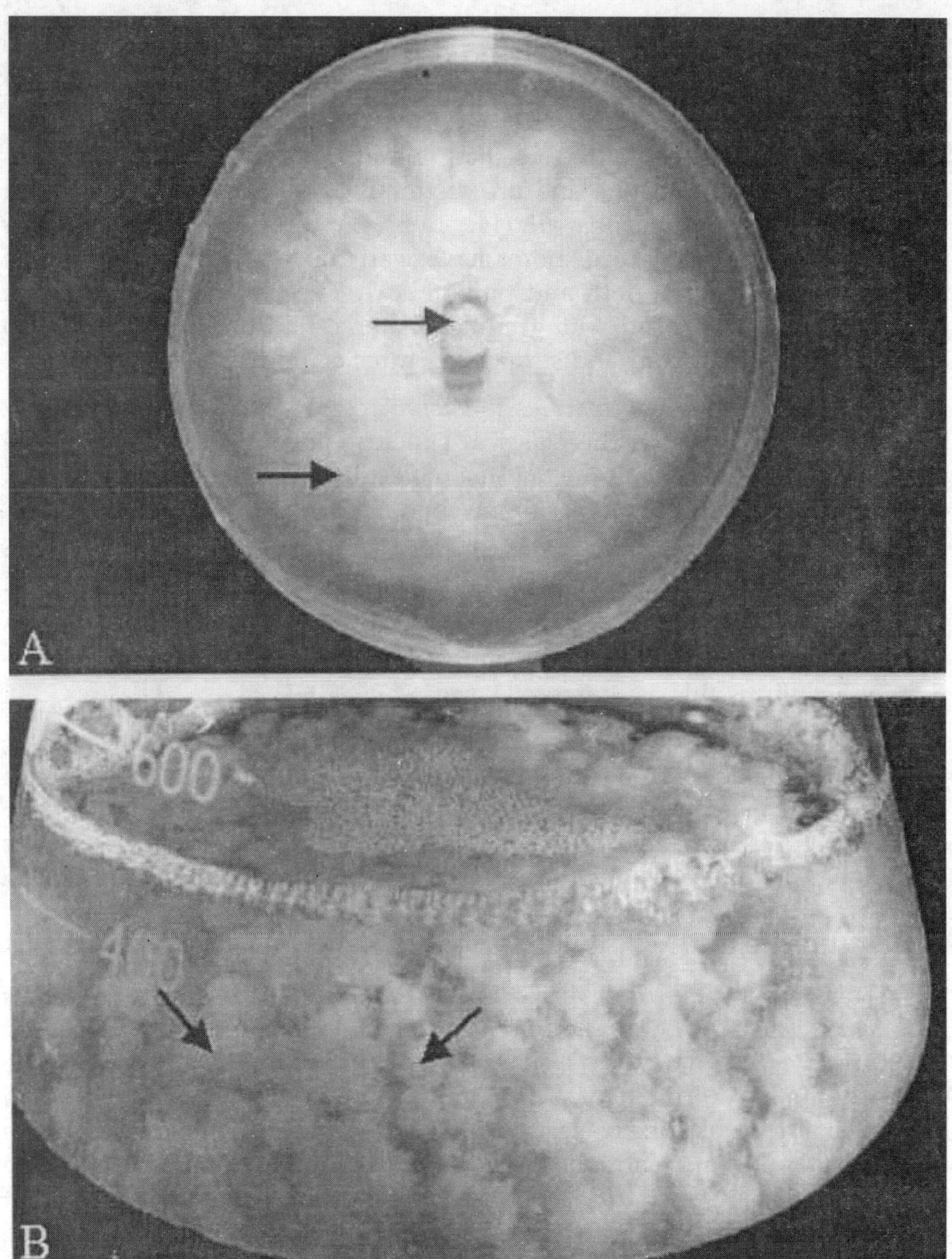

Plate I: (A) Fungus growing on Kaefer agar in Petridish (B) Axenic culture of *Piriformospora indica* in Kaefer broth

The fungal hyphae are highly interwoven and overlapped with each other giving the appearance of simple interwined cord. Young mycelia are white and almost hyaline, but incospicuous zonations are recorded in other cultures. Hyphae are thin walled ranging from 0.7 to 3.5 μm in diameter. Hyphae show anastomosis and are irregularly septate. The single septum or compartments could contain more than one nucleus. The chlamydospores appear singly or in clusters and are distinctive due to their pear-shaped structure. The chlamydospores are (14-) 16-25(-33) μm in length and (-9) 10-17(-20) μm in width. Mature spores have walls up to 1.5 μm thick, which appear low layered, smooth and pale yellow. The chlamydospores have cytoplasm densely packed with granular material and usually contain 8-25 nuclei. Clamp connections and sexual structures could not be observed.

There are only a few features to characterize *P. indica* morphologically to group it according to the classical species concept. According to the ultrastructure of the septal pore and the molecular data, the fungus is placed within the Hymenomycetes (Basidiomycota).

Molecular sequence analysis of partial 18s rDNA sequence (525 nucleotide position) placed *P. indica* within the Basidiomycota close to the *Rhizocotonia solani* group (Ceratobasidiales) (Verma *et al.*, 1998). A maximum likelihood analysis on almost complete 18s rDNA sequences (1550 nucleotide position) confirms this assignment (Varma *et al.*, 2001).

There is no existing genus, which covers all the characters of the new fungus. The members of Basidiomycota which may be related according to the molecular or ultrastructural data are not known to be able to produce chlamydospores or have a different way of life. Therefore, a new genus is erected and the fungus is designated *Piriformospora indica* (Verma *et al.*, 1998).

Ultrastructure of the Fungal Hyphae

Transmission Electron Microscopy (TEM) reveals the structural details of *P. indica*. The cell wall of the fungus is very thin and shows multi-layered structures. Septa consist of dolipores with continuous parenthosomes. The dolipores are very prominent with a multilayered cross-wall and a median swelling mainly consisting of electron-transparent material. In median sections of the septal pores, the parenthosomes are always straight and have the same diameter as the corresponding dolipore. Parenthosomes are flat discs without any detectable perforation (Verma *et al.*, 1998). The parenthosomes are in contact with the endoplasmic reticulum (ER) membrane, which are mostly found near the dolipore. These features formed the basis for the systematic position within the Hymenomycetes (Varma *et al.*, 2001).

Root Colonization Studies

Fungus *P. indica* colonizes plants of a wide host range including *Zea mays, Nicotiana tabacum, Petroselinum crispum, Populas tremula, Setaria italica, Oryza sativa, Pisum sativum, Glycine max, Cicer arietinum, Solanum melongena, Artemisia annua, Bacopa monniera, Azadirachta indica, Spilanthes calva, Withania somnifera, Dalbergia sisso* and *Terminalia arjuna*. The symbiont also colonizes terrestrial orchids viz., *Dactylorhiza purpurella, D. maculata* and bryophyte *Funaria hygrometrica* (Rai *et al.*, 2001; Varma *et al.*, 2001). Interestingly it does not colonize members of Cruciferae and several myc^- mutants of soybean and pea. Colonization results in extramatrical, inter- and intracellular hyphae and pear-shaped chlamydospores.

The hyphae first colonize the surface by producing appressoria-like structures. Subsequently, they enter and traverse through the cortical cells and produce hyphal coils and vesicles. Like arbuscular mycorrhizal (AM) fungi, hyphae never invade the endodermis and the aerial portion. Protein profile of the root shows significant qualitative changes but protein profile of leaf shows least qualitative but some quantitative change. Root colonization by the fungus has been studied either in pot culture or root organ culture or by both the techniques. Plate II, shows the root colonization by *P. indica* studied under light microscope.

Phyto-promotional Effects

Fungal hyphae have been reported to improve the growth and overall biomass production of various plants. *P. indica* colonization shows phyto-promotional effects to the plants like *Z. mays, N. tabacum, P. crispum, P. tremula, B. monniera, A. indica, S. calva, W. somnifera, D. sisso, S. italica, P. sativum, G. max, C. arietinum, S. melongena* and *A. annua*. The roots are highly colonized and the plants highly promoted by interacting with *P. indica* (Sudha, 1998; Sahay, 1999; Sharma, 2000; Kumar, 2000; Rai *et al.*, 2001; Varma *et al.*, 2001).

It has also been observed that not only the fungal hyphae but also the **culture filtrate** containing fungal exudates exerts positive effect on plant growth. An increase in root length, shoot length and plant biomass in *P. indica*-treated host plants, maize, *Bacopa monniera, Nicotiana tabacum, W. somnifera, S. calva, D. sisso* and *Solanum melongena* has been reported (Rai *et al.*, 2001; Varma *et al.*, 2001).

Phosphorus Transport

Plants take phosphorus from soil either directly by roots or indirectly

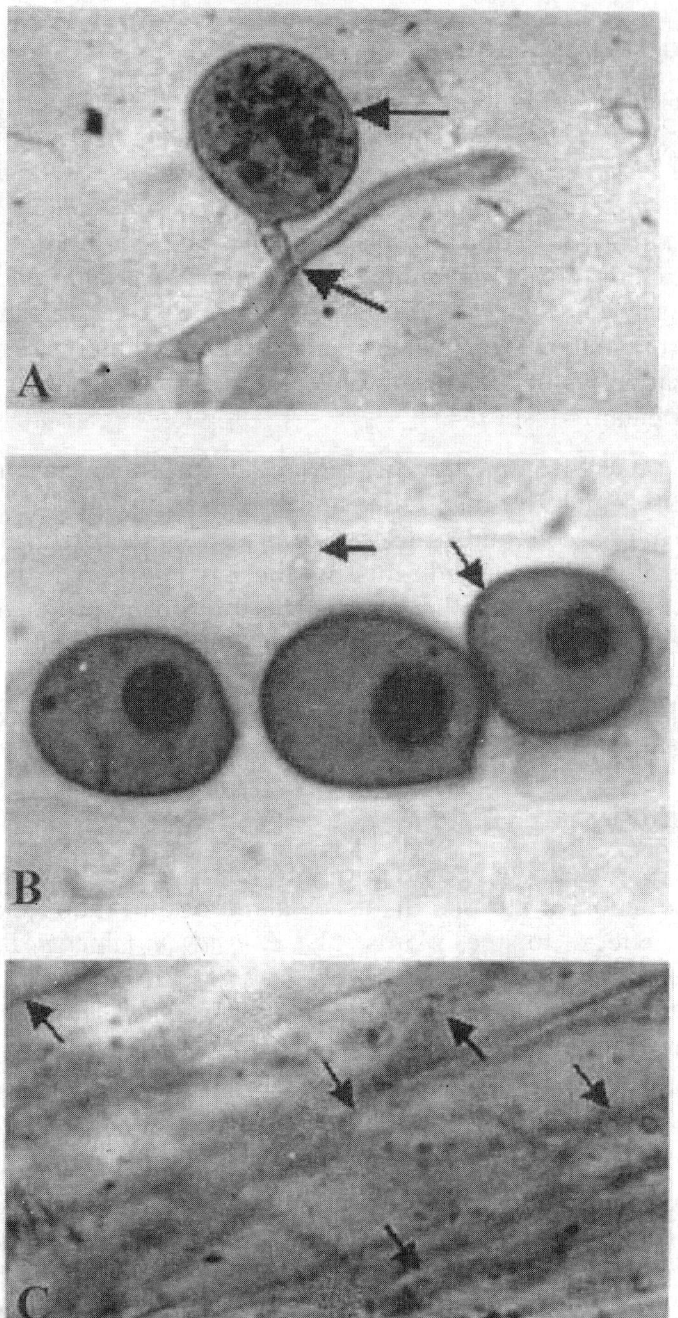

Plate II: Roots of *Bacopa monniera* stained with trypan blue showing colonization of root cells with *Piriformospora indica* observed under light microscope (1000 X)

(A) Fungal spore with hypha (B) Intracellular spores (C) Inter-intracellular hyphae

through fungal association which may increase phosphorus accumulation beyond the limits, allowing mycorrhized plants ultimately to out-perform non-mycorrhized plants (Koide, 1991). Transportation of radiolabelled phosphorus (P^{32}) by the fungus *P. indica* to host plant has been proved by experiments (Sudha, 1998), which suggest that *P. indica* is important for phosphorus acquisition by the root especially in the arid and semi arid regions (Varma *et al.*, 2001).

Serological Characterization

Antibodies, the main tool of immunochemical technique, can be used to characterize extra-and intracellular surfaces and to quantify cellular constituents. Immunochemistry provides elegant tool for these tasks (Goebel *et al.*, 1998; Hahn *et al.*, 1998). Immunological studies of the fungus indicate that *P. indica* is closely related with *Glomus mosseae*. The antiserum raised against *P. indica* recognizes *Glomus mosseae, Scutellospora gilmorei* and *Gigaspora gigantea*. This serological similarity of *P. indica* with AMF is strengthened by the physiological property of *P. indica* to colonize the roots of diverse hosts (Sharma, 2000).

Immunofluorescence tests indicate the cross-reactivity of *P. indica* and thereby an affinity with the spores of *Glomus intraradices* (Varma *et al*, 2001). The Western blot analysis shows that the antiserum raised against *P. indica* could strongly recognize four protein bands of *G. mosseae* antigens. The high molecular weight of these protein bands raise the possibility that the serological homologies between these two fungi is largely associated with the glycosylated components of the proteins (Sharma, 2000).

Applications of the Fungus

P. indica is a culturable mycorrhiza-like fungus, having various applications and can be employed as biological hardening tool for micropropagated plantlets. It acts as biocontrol agent against pathogenic fungi. It has been studied that various fungicides have least or no effect on fungus growth and colonization of plant root by *P. indica*.

Biological hardening tool

The mortality rate and 'transient transplant shock' on transfer of plantlets from the tissue culture laboratory to the field is very high. The application of AMF as a tool for biological hardening has solved a part of the problem (Varma and Schuepp, 1994). *P. indica* is proved as potent biological hardening tool for tissue culture raised micropropagated plants. The survival rate of *P. indica* treated plants is as high as >94%. Morphological observations of the inoculated plants show a better revival and regeneration capacity than untreated control (Sahay and Varma, 1999). Root colonization and survival rate of the micropropagated plants has been studied with *Nicotiana tabacum*, *Bacopa monniera*, *Solanum melongena*, etc (Sudha, 1998; Sahay, 1999; Sudha *et al.*, 2002).

Biocontrol agent against pathogenic fungi

Arbuscular mycorrhizal fungi are major components of rhizosphere and reduce soil-borne diseases and/or the severity of the diseases caused by phytopathogens (Dehne, 1982). Like other mycorrhizal fungi, *P. indica* also has activity against phytopathogens. Growth of *Gaeumannomyces graminis* (seed pathogen) is completely arrested when co-inoculated on solidified nutrient medium. Similar results have been reported when *P. indica* is allowed to interact with *Aspergillus sydowii* (Varma *et al.*, 2001).

Compatibility with fungicides

Several fungicides are in regular practice to protect agricultural crops and forestry plants from pathogenic fungi. The mycorrhizal fungus can be affected by the use of fungicides. Fungicides either have no negative effects (Jalali and Domsch, 1975) or even positive effects were reported with triadimefon on mycorrhiza development (Dehne, 1985; 1986). *P. indica* has been reported to be resistant to certain fungicides used in agriculture fields. Ethylene bisdithiocarbamate shows negative effect on *P. indica* at very high concentration (200 ppm in culture broth while normal recommended dose is 10 ppm in soil). Whereas, benomyl fungicides show different effects on the fungus; methyl benzimidazole-2-yl carbamate (MBC) is neutral while dimethyl 4,4′-ortho-phenylene bis (3-thioallophanate) inhibits the growth at higher concentration. Plant root colonization by *P. indica* is not affected by the application of the fungicides used (Kumar, 2000; Kumar and Varma, 2002).

Immobilization of fungal spores/hyphae

Nelson and Griffin (1916) proposed the concept of immobilized enzymes, attached to solid surface. The term immobilization may be interpreted as the presence of microorganisms or individual cells either on the surface or within the particles. Immobilization of ecto-mycorrhizal fungal mycelia in sodium/calcium alginate beads has been found an efficient tool for the production of inoculants of mycorrhizal fungi (Le Tacon *et al.*, 1985; Mauperia *et al.*, 1987). Immobilization of hypahe and spores of *P. indica* in alginate beads has been done (Varma *et al.*, 2001). Spores and hyphae of *P. indica* retain viability and germination capacity even after two months (Roy, 1999). Immobilized inocula was stored at ambient temperature for more than two months without loss of viability. This gives the conclusion that immobilized fungus may be stored and transported at ambient temperature without loss of viability.

Conclusion

Piriformospora indica can be used as a model AMF as this is a cultivable fungus on synthetic media with many functional similarities with AM fungi such as, broad and diverse host spectrum, hyphae extramatrical, inter- and intracellular hyphae never invade the endodermis, chlamydospores in extraradically and within cortical tissues, sexual stages not seen, positive phytopromotional effects on tested hosts (monocots, dicots including legumes), phosphorus mobilizer and transporter, tool for biological hardening of micropropagated plantlets, potent biological control agent against root pathogens and resistant to several commonly used fungicides.

References

Blechert, O., Kost, G., Hassel, A., Rexer, R.H. and Varma, A. (1999). First remarks on the symbiotic interactions between *Piriformospora indica* and terrestrial orchids. In: Mycorrhizae 2nd Edition. Varma, A. and Hock, B. (eds), Springer-Verlag, Germany, pp 683-688.

Dehne, H.W. (1982). Interaction between vesicular arbuscular mycorrhizal fungi and plant pathogens. Phytopathol, **72**: 1115-1119.

Dehne, H.W. (1985). Influence of pesticides on the development of vesicular arbuscular mycorrhizae. Colloq *INRA*, **31**: 255-263.

Dehne, H.W. (1986) Untersuchungen zum Einflub von Pflanzenbehandlungsmittlen auf das Afutreten der VA-Mycorrhiza. Meded Fac Landbouww Rijksuniv. Genet., **51**: 465-475

Goebel, C., Hahn, A., Giersch, T. and Hock, B. (1998). Monoclonal antibodies for the identification of arbuscular mycorrhizal fungi. In: Mycorrhiza Manual. Varma, A. (ed). Springer-Verlag Berlin Heidelberg New York, pp. 271-187.

Hahn, A., Goebel, C. and Hock, B. (1998). Polyclonal antibodies for the identification of AM fungi. In: Mycorrhiza Manual. Varma, A. (ed). Springer-Verlag Berlin Heidelberg New York, pp. 255-270.

Jalali, B.L. and Domsch, K.M. (1975). Effects of systematic fungicides on the development of mycorrhiza. In: Endomycorrhizae. Sanders, F.E., Mosse, B. and Tinker, P.B. (eds). Acad. Press, New York, pp:619-626.

Kaefer, E. (1977). Meiotic and mitotic recombination in *Aspergillus* and its chromosomal aberration. Adv Genet, **19**: 33-131.

Koide, R.T. (1991). Nutrient supply, nutrient demand and plant response to mycorrhizal infection. New Phytol, **117**:365-386.

Kumar, S. (2000). Characterization of *Piriformospora indica*: a mycorrhiza-like fungus. M.Phil. Dissertation submitted to the Jawaharlal Nehru University, New Delhi, India.

Kumar, S. and Varma, A. (2002). Influence of Commonly used Fungicides on *Piriformospora indica in vitro* and *in vivo*. Mycorrhiza (communicated).

Le Tacon, F., Jung, G., Mugnier, J., Michelot, P. and Mauperia, C. (1985). Efficacy of a forest nursery of an ectomycorrhizal inoculum produced in a fermentor and entrapped in polymeric gels. Can J Bot, **63**: 1664-1668.

Linderman, R.G. (1994). Mycorrhiza for plant health. In: Symposium series. Fleger, F.L. and Linderman, R.G. (eds), APS Press, St Paul, Minnesota, pp 1-25.

Mauperia, C., Mortier, F., Garbaye, J., Le Tacon, F., and Carr, G. (1987). Viability of an ectomycorrhizal inoculum produced in a liquid medium and entrapped in calcium alginate gels. Can J Bot, **65** : 2326-2329.

Mukerji, K.G., Mandeep, and Varma, A. (1998). Mycorrhizosphere microorganisms: screening and evolution. In: Mycorrhiza Manual. Varma, A. (ed). Springer-Verlag Berlin Heidelberg New York, pp. 85-97.

Neeraj, Shankar, A., Mathew, J. and Varma, A.K. (1991). Occurrence of VA mycorrhizae within the Indian semi-arid soils. Biol Fert Soils, **11**:140-144.

Nelson, J.M. and Griffin, E.G. (1916). Adsorption of invertase. J Am Chem Soc., **38**:1109-1111.

Rai, M., Acharya, D., Singh, A. and Varma, A. (2001). Positive growth responses of the medicinal plants *Spilanthes calva* and *Withania somnifera* to inoculation by *Piriformospora indica* in a field trial. Mycorrhiza, **11**:123-128.

Roy, A (1999). Immobilization of *Piriformospora indica*. M.Phil. Dissertation submitted to the Jawaharlal Nehru University, New Delhi, India.

Sahay, N.S. (1999). Interaction of *Piriformospora indica* with tissue culture raised plant. Ph.D. Thesis submitted to the Jawaharlal Nehru University, New Delhi, India.

Sahay, N.S. and Varma, A. (1999). *Piriformospora indica*: a new biological hardening tool for micropropagated plants. *FEMS* Microbiol Lett, **181**: 297-302.

Sahay, N.S., Sudha, Singh, A. and Varma, A. (1998). Trends in endomycorrhizal research. Indian J Exptl. Biol, **36**:1067-1086.

Sharma, J. (2000). Characterization of arbuscular mycorrhizal fungi using immunological techniques. Ph.D. Thesis submitted to the CSJM University, Kanpur, India.

Sudha (1998). *In vitro* study of endosymbionts associated with tissue culture-raised medicinal plants. Ph.D. Thesis submitted to the Jamia Hamdard University, New Delhi.

Sudha, Srivastava, P.S. and Varma, A. (2002). Biological hardening of micropropagated *Bacopa monniera* (Linn.) by mycorrhizas. Plant and Soil (In press).

Varma, A. (1995a). ecophysiology of arbuscular mucorrhizal fungi. In: Mycorrhizae 2nd Edition, Varma, A. and Hock, B. (eds), Springer-Verlag, Germany, pp 561-591.

Varma, A. (1995b). Arbuscular mycorrhizal fungi: the state of art. CRC Crit Rev Biotechnol, **15**:179-199.

Varma, A. and Schuepp, H. (1994). Infectivity and effectiveness of *Glomus intraradices* on micropropagated plants. Mycorrhiza, **5**: 29-37.

Varma, A. and Schuepp, H. (1995). Mycorrhiza: their application in micropropagated plants. *CRC* Crit Rev Biotechnol, **15**: 313-328.

Varma, A., Singh, A., Sudha, Sahay, N., Sharma, J., Roy, A., Kumari, M., Rana, D., Thakran, S., Deka., D., Bharti, K., Franken, P., Hurek, T., Blechert, O., Rexer, K.H., Kost, G., Hahn, A., Hock, B., Maier, W., Walter, M ., Strack, D. and Kranner, I. (2001). *Piriformospora indica*: a cultivable mycorrhiza-like endosymbiotic fungus. In: Mycota IX, Springer-Verlag Series, pp 123-150.

Varma, A., Verma, S., Sudha, Sahay, N., Britta, B. and Franken, P. (1999). *Piriformospora indica*–a cultivable plant growth promoting root endophyte with similarities to arbuscular mycorrhizal fungi. Appl Env Microbiol, **65**: 2741-2744.

Verma, S., Varma, A., Rexer, K.H., Hassel, A., Kost, G., Sarbhoy, A., Bisen, P., Butehorn, B. and Franken, P. (1998). *Piriformospora indica*, gen. et sp. nov., a new root-colonizing fungus. Mycologia, **90**: 896-903.

23

Piriformospora indica and Plant Growth Promoting Rhizobacteria: An Appraisal

Rajani Malla[1], Anjana Singh[1], Md. Zeyaullah[1], Vikas Yadav[1], Suniti, Anuprita Verma[1], Ajit Varma[1*] and Mahendra Rai[2]

[1]*School of Life Sciences, Jawaharlal Nehru University, New Delhi-1100067*
[2]*Department of Biotechnology, Amravati University, Maharastra-444602*

Most of the plant roots are surrounded by bacteria (rhizosphere, rhizoplane or mycorrhizosphere). Among them the beneficial microorganisms provide a wide range of vital services in natural ecosystems. These services include nutrient uptake, nitrogen fixation, decomposition of organic residues, promotion of plant growth, protection of plants against root pathogens and formation of soil structure. In this chapter, authors have presented an overall view of mycorrhizal fungi, Plant Growth Promoting Rhizobacteria (PGPRs) and Mycorrhization Helper Bacteria (MHBs). For clarity of the presentation, an effort was made to present some recent unpublished work carried out in the author's laboratory, where the newly discovered fungus *Piriformospora indica* was studied to ascertain the interaction between *Pseudomonas fluorescens, Azospirillum* sp, *Bacillus coagulans* and *P. indica* under *in vitro* condition.

Key words : Mycorrhizae, Mycorrhiza Like Fungus, Plant Growth promoting Rhizobacteria (PGPRs), Mycorrhization Helper Bacteria (MHBs), Bacteria Like Organisms (BLOs).

There is wide spread agreement among plant and soil scientists that mycorrhiza are beneficial to the growth and health of plant and soil. Mycorrhizae provide an intimate link between the soil environment and the functional nutrients absorbing system of the plant. Although most research observations on mycorrhizae have been concerned with nutrient uptake, awareness of their potential importance in many diverse aspects of a plant ability to grow and survive in natural and man-altered environments is important aspect. The modification of plant roots by symbiotic fungi into the distinct structural characteristics of mycorrhizas result in an unique and intriguing component of the rhizosphere. Mycorrhizal associations involve 3-way interactions between host plants, mutualistic fungus and soil factors (Declerck *et al.*, 2000; Franken and Requena 2001; Morton and Bruns 2000; Morton and Redecker 2001; Schußler *et al.*, 2001).

*Corresponding Author E-mail : ajitvarma73@hotmail.com

Types of Associations-

I. Vesicular-arbuscular mycorrhizae (VAM or AM)

II. Ectomycorrhizae (ECM)

III. Ectendo-, arbutoid-and monotropoid mycorrhizal associations

IV. Orchid mycorrhizas

V. Ericoid mycorrhizas

VI. The Australian lily *Thysanotus*

These six groupings are tools for examining the broad understanding structure and function. All of these groupings do share the commonality of a host plant having a fungal associate in intimate, mutualistic symbiosis where both partners are living and exchange of substance can occur in both directions, i.e., a biotrophic association (Varma, 1995). These mycorrhizas have been found to influence function in the rhizosphere e.g., surface area for nutrient absorption, microbe colonization and source of carbon for microbes, physical contact with the soil and protection from pathogens. They are strictly dependent on nutrition of their host plant, whereas, others like ericoid fungi can be easily grown in pure culture (Varma and Bonfante, 1994).

The ectomycorrhizae (ECM) are found almost exclusively on woody perennials. They form sheath of fungal mycelium that covers the absorbing root. The presence of "Hartig's net" is a key diagnostic feature of this association (Agerer, 1999). This net of hyphae extends into the cortex but is confined to the intercellular spaces and never penetrates the wall into individual cortical cells (intracellular) although root cap cells may be invaded (Kottke and Oberwinkler, 1986). The fungus never penetrates beyond the epidermis or epiblemma. The ericaceous mycorrhizae occur only in the order Ericales, which may be ericoid, arbutoid and monotropoid generally related to the type of root system of the host. The ericoid mycorrhizae are endomycorrhizae in the general sense. Since the fungal symbiont penetrates the cortical cell walls and invaginates the plasmalemma (as do all endomycorrhizal types). Infected cells appear to be fully packed with fungal hyphae. The arbutoid mycorrhizae have certain characteristics found in both ECM and other endomycorrhizae. Intracellular penetration of cortical cells can occur, as can the formation of a sheath and a "Hertig 'snet" is present. Although little experimental work has specifically examined the function of arbutoid mycorrhizae and were functionally comparable to ECM. The monotropoid mycorrhizae are observed on the achlorophyllous plants in the family monotropaceae. These mycorrhizae are very similar to the ECM and form a distinct sheath and "Hartig's net". However, they exhibit a distinctive type of intracellular penetration in cortical cells that is unlike other endomycorrhizal types (Duddridge and Read, 1982). The ect-endomycorrhizae form a "Hartig's" net in the cortex of the root but develop little or no sheath. These fungi are most likely to be the imperfect stage of Ascomycetes. The orchid mycorrhizae are some what unusual in the sense that the "roles" are reserved in this symbiosis at least in the early stages of the Orchid's life cycle where the plant is not autotrophic. The fungal symbiont appears to provide carbon to the host plant. The fungal association is of the endomycorrhizal type where the fungus penetrates the cell wall and invaginates the plasmalemma and forms hyphal coils, within the cell. The term Arbuscular Mycorrhizas (AM) refers to the presence of intracellular structures- vesicles and arbuscules that form in the root during various phases of development. These mycorrhizae are most commonly reported group since they occur on a vast taxonomic range of plants, both herbaceous and woody. The AM fungi are in great contrast to the ECM in both anatomical features and morphology. There is no formation of a sheath of mycelium and more importantly, intracellular penetration of cortical cells takes place. Internally arbuscules, vesicles or coils can form within individual cortical cells. Generally, AM causes few changes in the root morphology, but the physiology of the host plant changes significantly. For example tissue concentrations of growth regulating compounds and other chemical constituents change, photosynthetic rates increase. Since AM fungi are major components of the rhizosphere, it is logical that they could effect the incidence and severity of root disease. Much of the literature suggests that the AM fungi reduce soil borne disease or the effects of disease caused by fungal pathogens. In addition, mycorrhizal plants have greater tolerance to toxic plants, to root pathogens, drought, high soil temperature, saline soil, adverse soil pH and to transplant shock than non-mycorrhizal plants (Bagyaraj and Varma, 1995). Spores of many fungi have the capacity to survive indefinitely even under extreme ecological stresses (Varma, 1991). The potential of AM fungi as "biofertilizer and bioprotector" to enhance crop production is well recognised, but not well exploited because of the current agronomic practices, with their implications for environment (Harley, 1991).

Functions of Mycorrhizal Fungi

Mycorrhizal fungus having unique carbon strategies can efficiently couple soil mineralization and nutrient uptake by the plant roots. In many cases the mycorrhizal system actually "bridges" functionally across the rhizosphere, and provides an organic link between the root and the bulk soil. Mycorrhizal plants have an enhanced capability over non-mycorrhizal plants in acquiring nutrients from soil, particularly where the nutrient is poorly soluble and less in concentration. This efficiency in nutrient uptake, primarily through increased absorbing surface area, is likely to be one of the most important features of mycorrhizae.

The agricultural significance of arbuscular mycorrhizae lie principally in its ability to assist the plant to absorb inorganic phosphorous (Pi) from the surrounding soil. Phosphate ions are not very mobile in soil and a phosphate depletion zone often .develops around roots when phosphate is scarce. Mycorrhizal hyphae are able to extend from the mycelium within the root to beyond the depletion zone and translocate phosphate directly to the host. Plants with restricted root systems having short stubby roots with few root hairs are generally not very efficient in extracting soil phosphate and tend to benefit considerably from establishment of arbuscular mycorrhizae. The mycorrhizal association often causes a dramatic improvement in phosphate deficient soil. Absorption of trace element such as zinc and copper can also be enhanced. There is evidence that mycorrhizal plant can contain higher concentration of growth hormones than their non-mycorrhizal equivalent.

For many land plants mycorrhizal association are one way of generating adequate phosphorus absorption from reserves in the soil. The central role of the fungal symbiont in the provision of greater efficiency of phosphate absorption has been shown both *in vitro* and *in vivo* (Buecking and Heyser, 2001; Harrison *et al.*, 1995; Hodge *et al.*, 2001; Rausch *et al.*, 2001). Modification of phosphate uptake properties of root systems arising from mycorrhizal infection, consequently, depend largely upon development of extramatrical hyphae in soil and then absorption of phosphorus, translocation of phosphorus through hyphae and mycelial development within root tissues and transfer of phosphorus from fungus to root cells (Joner and Johansen, 2000).

Mycorrhizae readily absorb soluble phosphate from the labile pool of phosphorus in soils but their role in the removal or breakdown of insoluble, condensed or complex forms of phosphate has still not been clearly elucidated.

Despite many beneficial functions of arbuscular mycorrhizal fungi their effective use for biotechnological application remain incomplete. The bottleneck is the absence of axenic culture as these fungi are obligate phyto-symbiont.

Piriformospora indica- Fungus of the Millenium

An axenically cultivable mycorrhiza-like-fungus, has been described by Varma (Jawaharlal Nehru University, New Delhi) and his collaborators (Verma *et al.*, 1998). It was originally found along with arbuscular mycorrhizal spores from the rhizosphere soil samples of spineless cacti and *Cenchrus* sp (desert grass) growing in North-West Rajasthan. The fungus was named *Piriformospora indica* based on its characteristic pear-shaped chlamydospores (Fig.1). According to the analysis of 18s rDNA and 28s rDNA the ultrastructure of the septal pore, its phylogenetic relationship within the Hymenomycetes (Basidiomycota) and closest to *Rhizoctonia* group and *Sebacina vermifera* i.e. the fungi (Verma *et al.,* 1998; 2001). *P. indica* tremendously improves the growth and overall biomass production of diverse hosts, including legumes, medicinal and economically important. Pronounced growth promotional effect was also seen with terrestrial orchids. The fungus promises to be a potential orchid mycorrhizal fungus (Singh and Varma, 2000). This fungus is able to colonize a variety of host plants resulting in the improvement of growths compared as non-colonized control plants (Rai *et al.,* 2001; Varma *et al.,* 1999). This fungus resembles AM fungi in several functional and physiological characteristics (Singh *et al,* 2000; Varma *et al.,* 1999, 2001). It improves the growth and biomass of wide host range and is an efficient phosphate solubilizer and transporter (Sudha *et al* 1998; Varma *et al.,* 2001). Like AM fungi, *P.indica* does not colonize the roots of members of Crucifereae or the Myc- mutants of soybean (*Glycine max*), pea (*Pisum sativum*) (Singh, 2001). More than 90% of the micropropagated plantlets of tested hosts treated with the fungus survived on transfer from laboratory to open field conditions (Sahay and Varma, 1999, 2000). The fungus also protects the plantlets from some potent root pathogens (Varma *et al.,* 2001). The fungus also provides protection when inoculated to the tissue culture raised plantlets by overcoming the 'transient transplant shock' on transfer to the field rendering almost 100% survival. The fungus forms inter- and intracellular hyphae in the root cortex, often differentiating into dense hyphal coils (arbuscule-like structures), spore and vesicle-like structures. Like-AM fungi, hyphae multiply within the host cortical tissues and never traverse through the endodermis. Likewise, they also do not invade the aerial portion of the plant (stem and leaves). Interestingly, the host spectrum of *P. indica* is very much alike AMF.

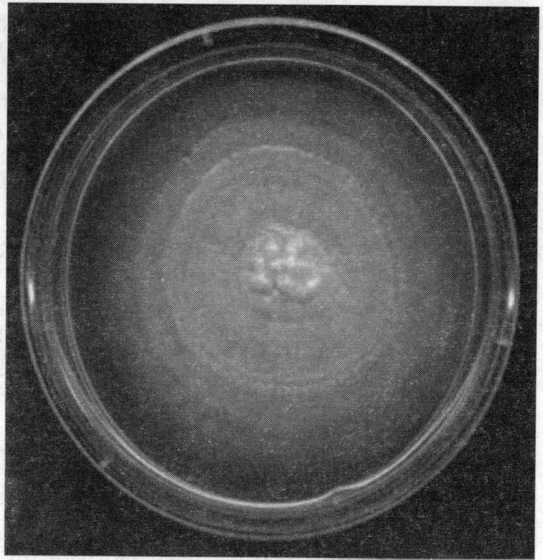

Fig. 1. A typical growth of *Piriformospora indica* on Kaefer medium (1977). Inoculum was incubated for 7 days at 28 °C.

P. indica colonized the roots of host plants as diverse as *Zea mays* L., *Nicotiana attenuata* L., *Petroselinum crispum* L., *Populus tremula* L., *Setaria italica* L., *Oryza sativa* L., *Sorghum vulgare, Glycine max* (L.) Merr., *Cicer arietinum* L., *Solanum melongena* L., *Artemissia annua* L., *Pisum sativum* and *Bacopa monniera* (L.) Wett (Varma *et al.*, 1999, 2001). Also like AMF, *P. indica* did not colonize the members of *Brassicaceae* and the myc⁻ mutants of *Glycine max* and *Pisum sativum.* Only exception was *Arabidiopsis thaliana,* where *P. indica* interacted vigorously. The molecular signals responsible for the interaction and symbiosis is currently under investigation. This fungus could be mass multiplied on cheap and simplified medium. Spores were found both extramatrical and intracellular in the root system. The fungus has great potential for application in agro-forestry, flori-horticulture, arboriculture, viticulture and especially for better establishment of tissue culture raised plants- much needed for the application in plant industry. This would open up numerous opportunities for the optimization of plant productivity in both managed and natural ecosystems, while minimizing risks of environmental damage.

Phosphorous is an essential mineral nutrient for the plant growth. Plants acquire this mineral from the environment either directly by their roots or indirectly from mycorrhizal fungi, which form inter-or intracellular symbiotic association with the roots. AM fungal colonization may increase the rate of phosphorus accumulation beyond the limits, which can be currently utilized, thus reducing the rate of phosphorus utilization efficiency. Such momentary "luxury consumption" of phosphorus may, however, serve a storage function and be utilized subsequently, allowing mycorrhizal plants ultimately to out perform non-mycorrhized plants. Recent radioactive P transport experiments suggested that *P. indica* is important for P acquisition by the root especially in the arid and semi arid regions where phosphorus deficiency is well known.

Acid phosphatases have been observed to be active in *P. indica* mycelium. The fungus could utilize a variety of inorganic and organic phosphate sources, which is in accordance with the broad range of substrates utilized by the acid phosphatases of many fungi. Besides this, phosphate starvation of *P. indica* led to an overall 27% increase in the intracellular acid phosphatase activity. This increase was probably due to the appearance of a P-repressible isoform of acid phosphatase in addition to the constitutive one observed in the enzyme staining of the native polyacrylamide gels. The significance of these enzymes in the phosphate transport needs to be further substantiated by the studies on the plant roots colonized with *P. indica.*

Molecular sequence data are useful for systematics, where morphological characters, like the taxonomically decisive sexual states, are missing. A neighbour- joining analysis on comparisons of partial 18S rDNA sequences

(525 nucleotide position) placed *P. indica* within the Basidiomycota close to the *Rhizoctonia solani* group (*Ceratobasidiales*). A maximum likelihood analysis on almost complete 18S rDNA sequences (1550 nucleotide positions) confirmed this assignment. Similar results were obtained by distance and parsimony methods. A comprehensive phylogenetic analysis of *Rhizoctonia* using sequences from mitochondrial and nuclear rDNA on more representatives should provide an insight into the evolution of this important group and clearly the evolutionary relationship of *P. indica* and *Rhizoctonia* within the Hymenomycetes. A 28S rDNA was analyzed and no change was found with respect to the taxonomic status (Varma *et al.*, 2001,2002).

The properties of the fungus, *Piriformospora indica*, have been patented (Varma, A. and Franken, P., 1997, European Patent Office, Muenchen, Germany. Patent No. 97121440.8-2105, Nov. 1998). The culture has been deposited at Braunsweich, Germany (DMS No.11827).

Rhizobacteria

Bacteria have great potential for biocontrol, plant growth promotion, formation of soil aggregates, and hence in the stabilization of soil structure, and bioremediation of xenobiotics. The establishment of beneficial bacteria on root systems of plants via seed bacterization has long been of major interest to agricultural researchers. This is due to the secretion of nutrients into the surrounding environment. This environment or the volume of soil that is influenced biologically and biochemically by the living root, is known as the rhizosphere. The term "rhizosphere" was introduced by Hiltner (1904) but has since been modified and redefined. Simply it is the volume of the soil influenced by the root. Although the extent may vary with soil type, plant species, age and many other factors. It is usually assumed to extend from the root surface (rhizoplane) out in to the soil for up to a few millimeters, or possibly a few centimeters in extreme cases in some desert and sand dune plants (Varma, 1995). The cortex may be invaded by apparently harmless bacteria and fungi (Greaves and Darbyshire, 1972) cells of the cortex and epidermis may die or be damaged while the root as a whole remains alive and healthy. The rhizosphere can therefore be considered as a microbial continuum stretching out in to the soil from the root endodermis (Old and Nicholson, 1978.) It is frequently divided in to ectorhizosphere (i.e. the soil region) and the endorrhizosphere, which includes the rhizoplane and the root that are invaded by microbes. The ectorhizosphere can extend a substantial distance from the root with the development of mycorrhizal fungal associations and is sometimes termed the mycorrhizosphere. The symbiotic establishment of mycorrhizal fungi on plant is affected in various ways by the other microorganisms of the rhizosphere, which are rhizospheric bacteria. The rhizobacteria once considered as passive by standards of the root environment, are now known to affect plant health, development and environment adoption, both beneficially and detrimentally. Biological and metabolic activities of this root soil interface are among the most critical factors contributing to the support of life on our planet. The microbial population, which is directly or indirectly related to plant growth and health, are the Plant Growth Promoting Rhizobacteria (PGPRs). These bacteria inhabit the rhizosphere of plant and promote the plant growth. The most common PGPRs are the strains of *Pseudomonas, Azospirillum, Rhizobium, Bradyrhizobium, Azotobacter* and *Bacillus*. These microorganisms play a vital role in mobilization of different nutrients from organic and inorganic non-available form to available forms and *vice versa*. It is well documented that mycorrhizal fungi improve plant growth and enhance microbial activity in the rhizosphere of mycorrhizal plants. It has been found that some bacteria present in the mycorrhizosphere can modify or alter the establishment of mycorrhizal symbiosis. The endomycorrhizae, *Glomus intraradices* can negatively influence the growth and survival of *P. fluorescence* DF5. The mechanism behind the effect of *G. intraradices* on *P. fluorescence* DF57 did not appear to require cell to cell contact (Ravnskov *et al.*, 1999). "Rhizosphere-Mycorrhizosphere" system can be tailored to help plants to establish and survive in nutrient deficient or degraded habitats during periods of stress (Smith and Bowen, 1979; Varma, 1998, 1999).

Effects of AM fungi on other microbial groups including nitrogen fixing bacteria, Actinomycetes and morphological and physiological groups of bacteria varied with each fungal species. In soil, many microorganisms occur in close proximity and they interact in a unique way. The microorganisms which include fungi, Actinomycetes, bacteria, viz., plant growth promoting rhizobacteria, soil pseudomonads, nitrogen fixing bacteria, phosphorus solubilizing bacteria and others affect mycorrhizal development on plant roots and the survival of mycorrhizal fungi in one way or the other. The possible mechanism by which PGPRs aid plant growth include suppression of root pathogens through chelation of iron or production of antibiotics (Burr and Caeser, 1984; Weller, 1985), fixation of nitrogen (Chanway and Holl, 1991), and production of plant hormones (Holl *et al.*, 1988). PGPRs are synergistic with mycorrhiza in stimulating plant growth (Meyer and Linderman, 1986).

Some of these bacteria which constantly promote mycorrhizal development, are mycorrhization helper bacteria (MHBs). MHBs can improve the success of ectomycorrhizal inoculation of planting stocks with fungi selected for their outstanding growth stimulation after outplanting. There are a number of fundamental question which remained unanswered about MHBs and suggested some investigation priorities in this new field of mycorrhiza research. From the germination of propagules in the soil to the achievement of the true symbiosis, the fungus has to grow in the soil and colonize the root surface before reaching receptive entry points. During this free-living stage of its development, it interacts with rhizospheric bacterial population (Bowen 1980; Marschner 1995; Rambell, 1973).

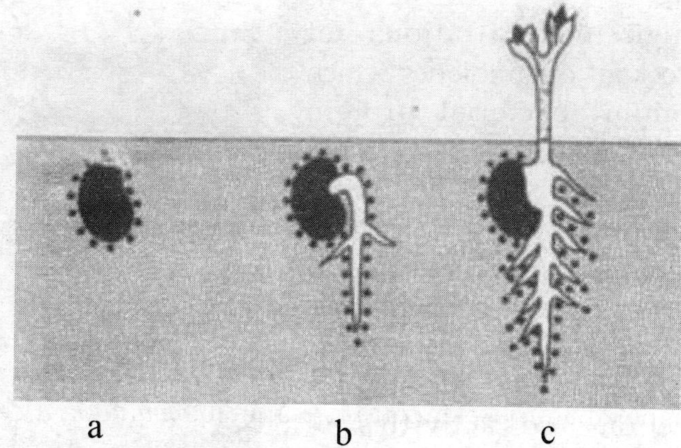

Fig. 2. Root colonization. Bacterial inocula on seeds planted into field soils multiply in the seed zone in response to seed exudation before germination. Root colonizers transfer from the seed zone to the developing root where they multiply and persist through the growing season. c.¦. (Kloepper 1989). a, seed placed for germination; b, newly developed radical; c, plumule and radicle developed. Note the colonization of the entire young seedling by the rhizospheric bacteria.

Although most of the interactions are competitive, some can be beneficial to the mycorrhizal infection process. MHBs are bacteria associated with mycorrhizal roots and mycorrhizal fungi which selectively promotes the establishment of mycorrhizal symbiosis (Garbaye, 1994).

Rhizobacteria are those members of the total rhizosphere bacteria that are able to colonize roots. Root colonization is defined as the bacterial capacity to multiply and keep pace with the growing root in field soil. From the practical point of view, it is important that colonization can follow as a result of inoculation (Fig. 2). Methods for monitoring bacterial root colonization, involving antibiotic resistant strains, have allowed detailed ecological analyses of bacterial root colonization under field conditions. The impact of rhizobacteria on plant growth and health may be neutral, deleterious or beneficial.

Endomycotic bacteria within fungus hyphae

From a semiaquatic location in the Thar desert near Jaisalmer, Rajasthan, India, in course of characterization of the fungal microflora in the rhizosphere of a sedge, an easily culturable coenocytic fungus was isolated which showed an interesting phenotype in carrot- root- organ cultures: in the presence of high concentrations of cocktail of antibiotics which inhibit bacterial protein biosynthesis, roots were not colonized, but without these antibiotics or in presence of ampicillin, an inhibitor of bacterial cell wall synthesis, colonization was pronounced. Modes of action and activities of antibiotics suggested that bacteria effect the interaction of the fungus with the root. Transmission electron microscopy (TEM) images of vital fungal hyphae showed that the cytoplasm harbored bacteria –like organisms (BLOs) (Fig.3a-c). These BLOs were sensitive to antibiotics as these cells were on treatment lysed within the hyphae. Interestingly, fungal hyphae devoid of BLOs had failed to grow on defined medium (Fig.3d).

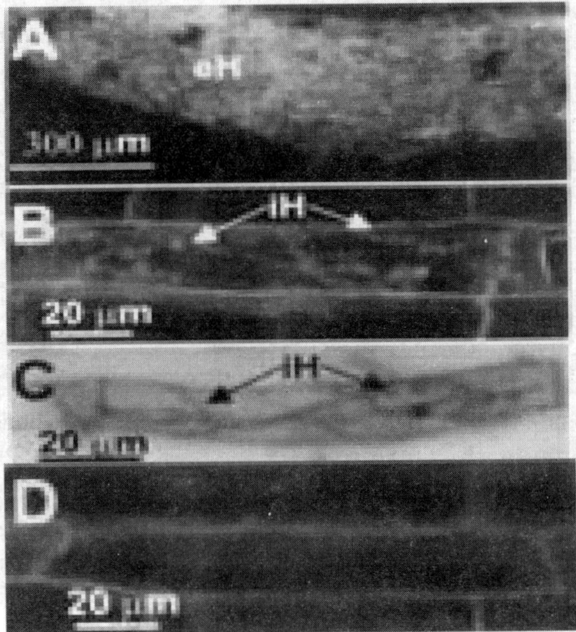

Fig. 3. Effect of pre-incubation with antibiotics (cocktail) on the pattern of colonization of carrot-root-organ-cultures by the fungal isolate. A, B, and C, fluorescent microscopy of inoculated carrot root organ cultures stained with the Live/Dead BacLight Bacteria Viability Kit (Molecular probes, Europe), Confocal RGB image. C, bright field illumination, root were stained with trypane blue; e (i) H, extracellular and intracellular hyphae. (Courtesy T. Hurek, Germany).

Piriformospora indica and their Field Applications

A pronounced growth response following inoculation with *P. indica* was observed in both *Spilanthes calva* and *Withania somnifera*. The plants treated with *P. indica* were superior in development to control plants. A significant increase in shoot length was observed in inoculated plants. Microscopic examination of stained root samples revealed a high colonization by *P. indica* of over 62 and 73 per cent root length in *S. calva* and *W. somnifera*, respectively. The basal stem and leaf areas of treated plants were also enhanced. Interestingly, some large, 2.3 (±0.06) x 1.54 (±0.38) cm^2. Kidney-shaped inflorescence was observed on inoculated *S. calva* plants together with normal, round inflorescence. These kidney-shaped inflorescence were never observed in untreated control plants. The lengths of the inflorescence and the number of flowers on inoculated *S. calva* plants were also increased relative to controls. Similarly, the number of flowers on the inoculated plants of *W. somnifera* was higher than on controls. For both medicinal plant species, seed counts were higher for treated than for control plants. (Rai *et al.*, 2001,2002).

The over all root biomass of the inoculated plants was higher than that of the corresponding controls. The fresh and dry weights of both underground and aboveground parts of *S. calva* and *W. somnifera* inoculated plants were higher than the controls. The net primary productivity of the inoculated *S. calva* and *W. somnifera* plants was 0.06 and 0.23 g/plant/day, respectively. These values were higher than those of control plants (0.02 and 0.12 g/plant/day, respectively).

MD is used as an index to compare receptivity of different plant species to AM fungi. This can also be used for other endophytes, such as *P. indica*. In the present study, the term ED is used instead of MD, as the test organism does not develop a typical mycorrhizal association. The ED value was 211.13 for *S. calva* and 671.90 for *W. somnifera*. These data suggest that *P. indica* has greater influence on the growth of *W. somnifera* than that of *S. calva*. The MD of a host plant can be altered by factors such as soil-type, soil P content, mycorrhizal species. The more intense root proliferation in treated plants observed here may be due to the synthesis of as yet unidentified extracellular phytohormones by *P. indica*.

We have investigated the influence of *P. indica* on active principle of *S. calva*, a plant of high commercial value. It was that both inoculated *P. indica* and uninoculated plants of S. *calva* exhibited antifungal activity against two human pathogenic fungi, viz., *Fusarium oxysporum* (opportunistic human pathogen) and *Trichophytum mentagrophytes* (potential human pathogen). We also report a significant antimycotic efficacy in aqueous and petroleum ether extracts of *S. calva*. It was also found that petroleum ether extract of *S. calva* was more effective than aqueous extract in inoculated as well as uninoculated plants. The MIC of *F. oxysporum* was recorded as 125 µl/ml when treated with aqueous extract of inoculated plants, whereas 250 µl/ml aqueous extract of inoculated plant was required to check the growth of *T. mentagrophytes*. This suggests that *F. oxysporum* was found to be more sensitive than *T. mentagrophytes*. The value of MIC for *F. oxysporum* was very high (1000 µl/ml) when aqueous extract of uninoculated was tested. Moreover, aqueous extract of uninoculated plant of *S. calva* failed to check the growth of *T. mentagrophytes* which is potential fungal pathogen causing skin and nail infection in human beings.

A remarkable antifungal activity of petroleum ether extract of inoculated plants of *S. calva* was recorded. An amount of 62.5 µl/ml extract was sufficient to inhibit the growth of 1 ml spore suspension of *F. oxysporum*, whereas in extract of uninoculated plants the MIC value reached to 500 µl/ml of spores. Similarly, for *T. mentagrophytes* 125 µl extract of inoculated plants was enough to arrest the growth, while MIC value was quite high (1000 µl/ml) in case of petroleum ether extract of uninoculated plants. The antimycotic activity of extract of inoculated plants of *S. calva* was significantly higher as compared to uninoculated plants. Furthermore, extract of petroleum ether was found to be more active than aqueous extract, which indicates that there must be enhanced quantity of Spilanthol in inoculated plants. Chemical analysis of the roots of the plant revealed the increase in Spilanthol content. The safed musli (*Chlorophytum tuberosum*) also showed a positive response in the field trials with AM fungi and those treated with *P. indica*.

Neem plants (*Azadirachta indica*) were treated with two arbuscular fungi, *Scutellospora gilmorei* and *Glomus mosseae* and *P. indica*. The treatments were conducted both under natural (unautoclaved) and sterile (autoclaved) potted soils. The parameters selected were plant height and total biomass of the host plant. Neem plants inoculated with *P. indica* were found to be drastically improved with respect to morphological and physiological features in contrast to those inoculated with *G. mosseae* and *Sc. gilmorei* and untreated controls. The fresh and dry weight of the shoots of *P. indica* plants showed 28.14 and 46.25 per cent increase over the control, respectively. Whereas, *G. mosseae* treatment resulted in only 14.07 and 21.25 per cent increase in the fresh and dry weight. *Sc. gilmorei* recorded an increase of 17.59 and 23.75 per cent over the control, respectively. The fresh and dry weight of the roots of *P. indica* treated plants showed 31,03 and 66.66 per cent increase over the control, respectively. Whereas, *G. mosseae* treatment resulted in only 10.34 and 20.51 per cent increase in the fresh and dry weight. *Sc. gilmorei* showed an increase of 14.94 and 33.33 per cent increase over the control, respectively. Interestingly, the 7 days old culture filtrate of *P. indica* recorded an increase in plant growth and over all development over the control, however, the impact steadily slowed down over a period of one year.

The total population of bacteria in the "Mycorrhizosphere" soil was greater than those reported from non-mycorrhized roots. This was true for all the seasons studied. Invariably, higher counts were recorded from the roots of *Sc. gilmorei* and *P. indica* treated and control plants during winter period recorded lower counts. Interestingly, in summer season counts were also recorded for the plants receiving fungus *G. mosseae*. The scenario was considerably changed in spring. All the treatments had almost identical number of counts except *G. mosseae* where the bacterial counts were low. In monsoon, low bacterial colonies were observed, in absence of any inoculum as well as when *G. mosseae* was used as a fungal inoculum in comparison to other treatments.

Micropropagation of *Bacopa monniera*, an important medicinal plant has been achieved in *in vitro* culture technique. Biological hardening, during the transplantation of micropropagated plantlets from laboratory to field was successful after inoculation with one of the six AM fungi viz., *G. mosseae*, *G. fasciculatum*, *G. etunicatum*, *Sc. gilmorei*, and *Sc. calospora*. Control plants showed least survival (12%), in contrast, the rate of plantlets survival on inoculation with *Sc. gilmorei* was 93% . The rate of plantlet survival also varies among the species of *Scutellospora* (*Sc. gilmorei* 93, *Sc. margarita* 23, and *Sc. calospora* 23 %, respectively) and *Glomus* (*G. etunicatum* 87, and *G. mosseae* 69%, respectively). *Sc. gilmorei* proved best for the plantlet survival and maximum percent root colonization under present experimental conditions. Percent root colonization is not directly related to the plantlet survival as shown by *Sc. margarita* and *Sc. calospora*, *G. etunicatum* supported higher biomass production. For the higher survival of micro propagated plantlets physical hardening is not the only factor, since after the transfer of the plantlets from culture vials to the pot and finally to the field conditions; they are not only exposed to different physical and chemical environmental conditions but they also exposed to different biological communities in the rhizosphere "Mycorrhiozosphere". Biological hardening of the plantlets by the beneficial microbes at the weaning stage gives a pre-conditioning to the

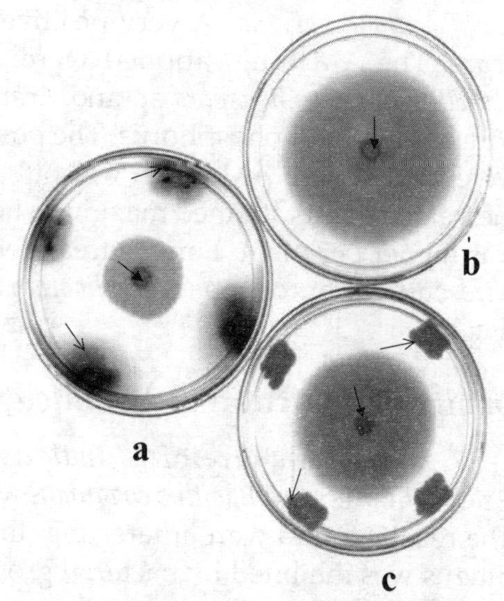

Fig. 4. Approximately 4.0 mm circular agar disc infested with hyphae and spores was placed in the center of the petriplates containing solidified agar medium. Bacterial inocula were placed on four sides of the petriplates. Care was taken to keep a distance of about 3.0 cm between the two microorganisms. The petriplates were incubated at 28 °C for 7 days. a, interaction of *P. indica* with *Pseudomonas fluorescens*; b, fungus alone; c, with *Azospirillum*. Solid arrows (⟶) indicate the fungal growth and open arrows (⟶) indicate the bacterial smear.

plantlets and hence increase the rate of survival. The soil microflora was also enriched to a great extent. This resulted in better soil binding and soil aggregation. This increase in the diverse microflora have enhanced the soil fertility (Kumari, 2002).

A similar field trial was done for another medicinal plant *Adhtoda vasica* with *P. indica*. The fungus treated plantlets showed a very positive response towards fungus and growth throughout the development. An identical phytopromotional effect was seen for another medicinal cum ornamental plant, *Tagetes erecta* plant in pots.

The timber cum fuel plants tested were *Tectona grandis, Delbergia sisso* and *Terminalia arjuna*. A very positive response was recorded in the last two years. The AM fungi applied were: *G. mosseae, G. aggregatum, Glomus* sp, *Scutellospora* sp, *Gigaspora* sp and *Acaulospora* sp. Trials with teak were tested with and without phosphorus. The positive response was seen while the plants were treated with AM Fungi without phosphatic compound. In *Delbergia sisso* the treated plants attained maximum height, about six feet. Number of branches was higher per plant. Fungus treatment promoted the biomass production. *In vitro* cultivation of tobacco (*Nicotiana attenuata* showed a positive response (Fig. 4).

In vitro Interaction of *P. indica* with Selected Rhizobacteria

A growth study of *P. indica* with PGPRs *Pseudomonas fluorescence, Azospirillum* sp and *Bacillus coagulans* was made on a defined solidified medium. The results found were interesting. In the case of *Pseudomonas fluorescence* the fungus was inhibited by bacterial growth (Fig 5a). The fungal growth without bacteria is given in (Fig 5b). In contrast, *Azospirillum* showed no inhibition of the fungus (Fig.5c). Interestingly, *P. fluorescence* culture produced a bulk amount of extracellular deep bluegreen and greenish yellow coloured pigment. This compound seems to be pyocianin and

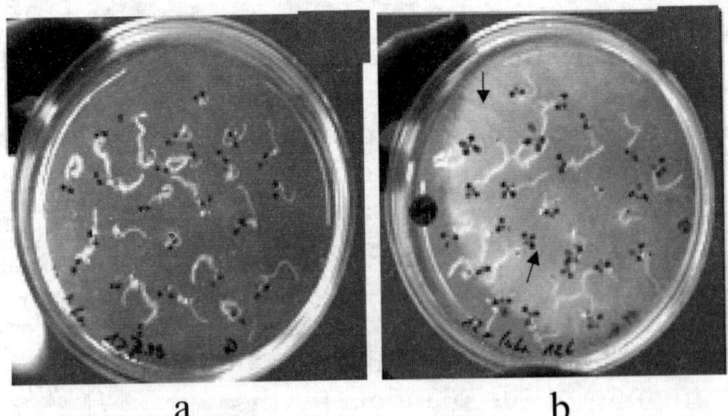

Fig. 5. Three days old young tobacco seedlings were placed on a solidified minimal agar medium where fungus was allowed of grow earlier. Petriplates were incubated for 7 days at 25°C. a, control (no fungus); b, treated with *P. indica*.

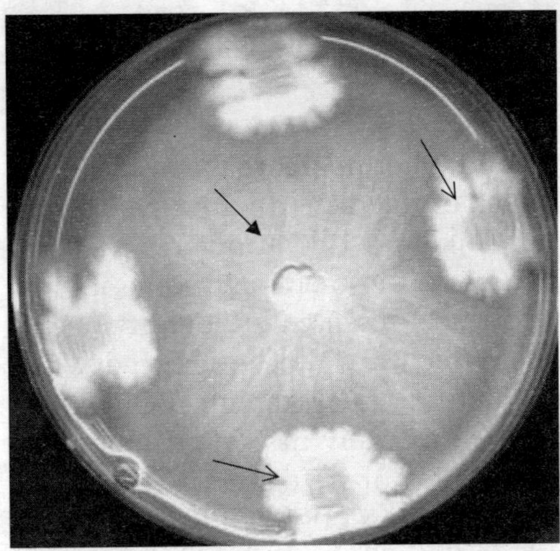

Fig.6. Interaction of *P. indica* with *Bacillus coagulance*. Both the organisms were placed on the solidified agar media at the same time. Petriplates were incubated for 7 days at 25°C.

pyoverdin or fluorescain. In another experiment, the culture filtrate was removed from the broth (after 72 hr incubation at 28°C) and passed through bacterial filter (Millipore, Millex-GV, 0.22mm Filter Unit, Made in France). About 100 µl of colored culture filtrate was placed in wells at four corner of plate. The fungus was placed in wells the center as a disc (4mm in diameter) containing hyphae and spores. Incubation was allowed in dark for 5 days at 28°C. Bacterial culture filtrate did not inhibit the fungus growth (Fig 6). *Bacillus coagulans* showed a positive interaction with P. indica (Fig.7). Both the interacting microorganisms profusely grew and attained normal development.

The exact nature of the blue coloration needs to be established. It seems that the interaction with the fungus produced some inhibitory compounds, the exact chemical identification needs to be done. Interaction experiments between these rhizobacteria with *P. indica* is being followed under *in vitro* and *in vivo* conditions. It is obvious that there is a strong interaction between the eucaryote and prokaryotes at the molecular level. Further studies would enable to understand the

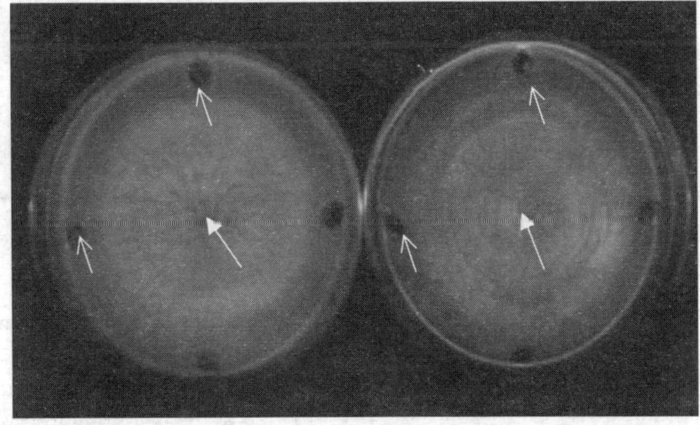

Fig.7. *Pseudomonas fluorescens* was allowed to grow in a broth culture for three days. The culture filtrate was taken out and passed through a disposable bacterial membrane sterile filter. 100 ml of aliquot was placed on the four sides of the petriplates maintaining a uniform distance of about 3.0 cm from the fungus inoculum (left). Control contained an equal quantity of the sterile bacterial medium (right). Petriplates were incubated for 7 days at 25 °C.

mechanism and the molecular basis of this strong "cross talk".

Conclusion

Among the wide variety of bacteria and other microorganisms living in the rhizosphere, it is probable that all types of interaction with mycorrhizal symbiosis occur from inhibition to stimulation and with various mechanisms involved. Nevertheless, as shown by evidence of experiments and some applied results, it gives a new dimension to mycorrhizal research by identifying the category of rhizosphere microorganisms, which can not be ignored when studying mycorrhizal symbiosis in their natural environment. PGPRs and MHBs (Mycorrhization Helper Bacteria) are especially important whenever the competitiveness and population dynamics of mycorrhizal fungi are considered. More research is needed before the concept of PGPRs and MHBs can be generalized. Fundamental research on the cellular and molecular basis of the interactions might contribute to the understanding of beneficial association between plants and fungi.

Acknowledgement

The authors are thankful to Department of Biotechnology, Science and Technology, Council of Scientific and Industrial Research, University Grants Commission and Indo-German Joint Cooperation for partial financial assistance. PGPRs were obtained from Division of Microbiology, Indian Agricultural Research Institute, New Delhi and Professor D J Bagyaraj, University of Agricultural Sciences, Bangalore.

References

Agerer, R. (1999). Anatomical characteristics of identified Ectomycorrhizas: An attempts toward a natural classification. A. varma, B.Hock (Eds.). Mycorrhiza 2nd Ed. Springer-Verlag Berlin Heidelberg, pp 633-682.

Bagyaraj, D.J. and Varma, A.K. (1995). Interaction between VA mycorrhizal fungi and plants, and their importance in sustainable agriculture in arid and semi arid tropics. Adv. Microbiol. Ecol., **14**: 119-142.

Bowen, G.D. (1980). Misconceptions, concepts and approaches in rhizosphere biology. Contemporary Microbial Ecology. London Academic Press, 283-304.

Buecking, H. and Heyser, W. (2001). Microautoradiographic localization of phosphate and carbohydrates in mycorrhizal roots of *Populus tremula* x *Populus alba* and the implications for transfer processes in ectomycorrhizal associations. Tree Physiology, **21**: 101-107.

Burr, T. J. and Caeser, A. (1984). Beneficial plant bacteria. Critical Reviews in Plant Sciences, **2**: 1-20

Chanway, C.P. and Holl, F.D. (1991). Biomass increased and associative nitrogen fixation of mycorrhizal *Pinus contorta* seedlings inoculated with a plant growth promoting *Bacillus* strain. Canadian Journal of Botany, **69**: 507-511.

Declerck, S., Cranenbrouck, Dalpe, Y., Seguin, S., Grandmougin-Fergani, A., Fontaine, J. and Sancholle, M. (2000). Glomus proliferum sp nov: a description based on morphological, biochemical, molecular and monoxenic cultivation data. Mycologia, **92**: 1178-1187

Duddridge, J. and Read, D. J. (1982). An ultrastructural analysis of the development of mycorrhizas in *Rhododendron ponticum*. Can. J. Bot., **60**: 2345-2356

Franken, P. and Requena, N. (2001). Analysis of gene expression in arbuscular mycorrhiza: new approaches and challenges., New Phytologist, **150**: 431-439

Garbaye, J. (1994). Helper bacteria; a new dimension to the mycorrhizal symbiosis. New Phytol., **128**: 197-210

Greaves, M. P. and Darbyshire, J. F. (1972). The ultrastrusture of the mucilagenous layer on plant's roots. Soil. Biochem., **4**: 443-449

Harley, J. L. (1991). Mycorrhizae: State of art. In: Norris J. R. Read D. J. and Varma A. K. (eds.), Methods in Microbiology., **23**: 1-17. Academic Press, New York

Harrison, M. J. and Buuren, L. van M. (1995). A phosphate tronsporter from the mycorrhizal fungus Glomus versiforme. Nature, **378**: 26-29.

Hiltner, L. (1904). Uber neuere Erfahrungen und Problem auf dem Gebeit der Bodenbakteriologie und unter besonderer Berucksichtigung der Grundungung und Brache. Arb Dtsch. Landwirt. Ges., **98**: 119-140.

Hodge, A., Campbell, D. C. and Fitter, A. H. (2001). An arbuscular mycorrhizal fungus accelerates decomposition and acquires nitrogen directly from organic material. Nature, **413**: 297-301.

Holl, F.B., Chanway Turkington , C.P. and Radley, R.A. (1988). Response of crested wheatgrass *Agropyron cristatum* L. Perennial ryegrass *Lolium perenne* and white clover *Trifolium repens* L. to inoculation with *Bacillus polymixa*. Soil Biology and Biochemistry, **20**: 19-24.

Joner, E.J. and Johansen, A. (2000). Phosphatase activity of external hyphae of two arbuscular mycorrhizal fungi. Mycol. Res., **104**: 81-86.

Kaefer, E. (1977). Meiotic and mitotic recombination in *Aspergillus* and its chromosomal aberrations. Advances Genet., **19**: 33-131.

Kloepper, J.W. Lifshitz and Zablotowicz, R.M. (1989). Free living bacterial inocula for enhancing crop productivity.

Kottke, I. and Oberwinkler, F. (1986). Mycorrhiza of forest trees- Structures and function. Trees, **1**: 1-24.

Kumari, M. (2002). Mycorrhiza for better establishment of neem plants (biological hardening) and enhancement of therapeutic properties. Ph. D. thesis. B. R. Ambedkar Bihar University, Muzaffarpur..

Marschner, H. (1995). The soil- root interface (Rhizosphere) in relation to mineral nutrition. *in* Mineral Nutrition of Higher Plants, 2nd edition, Acad. Press, London. pp. 537-595.

Meyer, J. R. and Linderman, R. G. (1986). Response of subterranean clover to dual inoculation with vesicular Arbuscular mycorrhizal fungi a plant growth promoting bacterium, Pseudomonas putida. Soil Biology and Biochemistry, **18**: 185-190.

Morton, J. B. and Bruns, T. D. (2000). Ancestral lineages of arbuscular mycorrhizal fungi (Glomales). Mol Phylogenet Evol., **14**: 276-284.

Morton, J.B. and Redecker, D. (2001). Two new families of Glomales, Archeosporaceae and Paraglomaceae with two new genera *Archeospora* and *Paraglomus* based on concordant molecular and morphological characters. Mycologia, **93**: 181-195.

Old, K.M. and Nicholson, T.H. (1978). The root cortex as part of a microbial continuum. In Microbial Ecology (m.w. Loutit and J.A.R. Miles,eds.) Springer-Verlag Berlin pp: 291-294.

Ravnskov, S., Nybroe, O. and Jakobsen, I. (1999). Influence of an arbuscular mycorrhizal fungus on *Pseudomonas fluorescens* DF 57 in rhizosphere and hyphosphere soil. New Phytol., **142**: 113-122.

Rai, M.K., Singh, A., Arya, D. and Varma, A. (2001). Positive growth responses of *Withania somnifera* and *Spilanthes calva* were inoculated with *Piriformospora indica* in field. Mycorrhiza, Germany, **11**: 123-128.

Rai, M.K., Varma, A. and Pandey, A.K. (2002). Enhancement of antimycotic potential in *Spilanthes calva* after inoculation of *Piriformospora indica*, a new growth promotor, Mycoses, Germany (in press).

Rambelli, A. (1973). The rhizosphere of mycorrhizae. Ectomycorrhizae, their ecology and physiology. New York: Academic Press, 299- 349.

Rausch, C., Daram, P., Brunner, S., Jansa, J., Lalol, M., Leggewie, G., Amrhein, N. and Buchert, M. (2001). A phosphate transporter expressed in arbuscle-containing cells in potato. Nature, **414**: 462-466.

Sahay, N.S. and Varma, A. (1999). *Piriformospora indica*: a new biological hardening tool for micropropagated plants. FEMS Microbiol Lett., **181**: 297-302.

Sahay, N.S. and Varma, A. (2000). Biological approach towards increasing the survival rates of micropropagated plants. Curr Sci., **78**: 126-129.

Schußler, A., Schwarzott, D. and Walker, C. (2001). A new fungal phylum, the *Glomeromycota*: phylogeny and evolution. Mycological Research, **105**: 1413-1421.

Singh, A. (2001). Intergeneric hybridization between a yeast strain and a plant growth –promoting endophytic fungus by isolation, fusion and regenaration of somatic protoplasts. Ph D thesis, Jawaharlal Nehru University, New Delhi.

Singh, A., Sharma, J., Rexer, K-H and Varma, A. (2000). Plant productivity determinants beyond minerals, water and light. *Piriformospora indica*, a revolutionary plant promoting fungus. Current Science, **79**: 101-106.

Singh, A and Varma, A. (2000). Orchidaceous mycorrhizal fungi. In: Mukherji KG (eds) Mycorrhizal fungi, Kluwer academic Press, Amsterdam pp. 265-288.

Smith, S.E. and Bowen, G.D. (1979). Soil temperature, mycorrhizal infection and nodulation in *Medicago truncatula* and *Trifolium truncatula*. Soil Biol. Biochem., **11**: 469-473.

Sudha, Hurek, T. and Varma, A. (1998). Active translocation of phosphate (P32) to rice and carrot by *Piriformospora indica*. In: Ahonen- Jonnarth U, Danell E, Fransson P, Karen O, Lindahl B. Rangel I. Finlay R. (eds) Second International congress on Mycorrhiza, Uppsala, Sweden July 5-10.

Varma, A. (1991). History of mycorrhizal research in India current status. Indian J. Microbiol, **31**: 323-332.

Varma, A. (1995). Ecophysiology and Application of Arbuscular Mycorrhizal Fungi in Arid Soils. Mycorrhiza. (ed) A Varma and B Hock. Springer-Verlag, Berlin, Heidelberg, New York pp., 561-591.

Varma, A. (1998). Mycorrhizae: Their role in sustaining wasteland. In: Problems of Wasteland development and Role of Microbes (ed Mishra A), AMIFEM Publications pp. 161-181.

Varma, A. (1999). Hydrolytic Enzymes from Arbuscular Mycorrhizae: the current Status. Mycorrhiza. (ed) A Varma and B.Hock Springer- Verlag, Berlin, Heidelberg, New York pp 373-389.

Varma, A. and Bonfante, P. (1994). Utilization of cell wall-related carbohydrates by ericoid mycorrhizal endophytes. Symbiosis, **16**: 301-313.

Varma, A., Singh, A., Sudha, Sahay, N.S., Sharma, J., Roy, A., Kumari, M., Rana, D., Thakran, S., Deka, D., Bharti, K., Hurek, T., Blechert, O., K-H, Rexer, Kost, G., Hahn, A., Maier, W., Walter, M., Strack, D. and Kranner, I. (2001). *Piriformospora indica*: An axenically culturable mycorrhiza like endosymbiotic fungus. In: Hock B (ed) Mycota IX, Springer-Verlag, Germany pp. 123-150.

Varma, A., Singh, A., Sudha, Sahay, N.S., Kumari, M., Bharati, K., Sarbhoy, A.K., Maier, W., Walter, M. H., Strack, D., Franken, P., Singh, A., Malla, R. and Hurek, T. (2002). *Piriformospora indica*: A plant stimulator and pathogen inhibitor arbuscular mycorrhiza-like fungus. In: Microorganisms in Bioremediation, (eds) DK Markandey, Capital Book Company, New-Delhi, India pp. 71-89.

Varma, A., Verma, S., Sudha, Sahay, N,. Britta, B. and Franken, P. (1999). *Piriformospora indica*- a cultivable plant growth promoting root endophyte with similarities to arbuscular mycorrhizal fungi. Applied and environmental microbiology USA, **65**: 2741-2744.

Verma, S., Varma, A., Rexer, K-H, Kost, G., Sarbhoy, A., Bisen, P., Buetehorn, B. and Franken, P. (1998). *Piriformospora indica*, gen. et sp. nov., a new root colonizing fungus. Mycologia, **95** : 896-903.

Weller, D.M. (1985). Application of fluorescent Pseudomonads to control root diseases. pp 137 140 in C.A. Parker, A.D. Rovira, K. T.moore and P.T.W. Wong (eds.) Ecology and management of soilborne plant pathogens. Proceedings of section 5 of the fourth international congress of plant pathology. The American Phytopathological Society, St. Paul., USA.

24

The Genus *Ravenelia* Berk. (Uredinales) from Andhra Pradesh

G. Bagyanarayana*, P. Ramesh and M.Raju

Department of Botany, Osmania University, Hyderabad-500 007, A.P.

A mycofloristic study of *Ravenelia* species infecting different fabaceous hosts from Andrha Pradesh is made. *Ravenelia delonicii* is described as a new species. *R. odorotissima* is recorded for the second time from India. *R. acaciae-senegalae, R. fragrans* var. *evernia, R. parasnathii, R. tandonii* are new records to South India. *Acacia trita* is the new host record for *R. parasnathii.*

Key words : ***Ravenelia, Fabaceae, Taxonomy, Andhra Pradesh***

With 170 species *Ravenelia* Berk. is the third largest genus of Uredinales and is known to infect the fabaceous hosts. *Ravenelia* is characteristic in the possession of telial heads with pendent, hyaline cysts on a compound pedicel. A systematic study of 13 *Ravenelia* species infecting different Fabaceous hosts from Andhra Pradesh is made. *Ravenelia delonicii* is described as a new taxon, *R.odorotissima* is being recorded second time from India. *R.acaciae-senegalae, R.fragrans* var. *evernia, R.parasnathii, R.tandonii* are new records to South India. *Acacia trita is the new* record for *R.parasnathii.*

The rust infected specimens were collected from different places of Andhra Pradesh. Spore scrapings were treated with saturated chloral hydrate solution then with water and ultimately mounted in lactophenol. The identification of the rust taxa was done by referring to various standard manuals (Sydow, 1915; Arthur, 1934; Cummins, 1978; Cummins and Hiratsuka, 1983). Fungi of India manuals (Butler and Bisby rev. by Vasudeva, 1960; Rangaswamy *et al.*, 1970; Mukerji and Jayanti Bhasin, 1986; Sarbhoy *et al.* 1986) were regularly referred to ascertain the distribution of concerned rust taxa.

*Corresponding Author e-mail : gbagyan@rediffmail.com

Ravenelia Berkeley, Gard. Chron. 132. 1853

Spermagonia conical, mostly subcuticular, indeterminate, (type 7) or rarely subepidermal, determinate (type 5).

Aecia subepidermal or subcuticular, erumpent, mostly uredinoid with pedicillate spores or in a few species aecidioid with catenulate spores; aeciospores ovid or spherical, verrucose or echinulate.

Uredinia mostly subepidermal but sometimes subcuticular, erumpent and pulverulent, paraphysate or aparaphysate, paraphyses peripheral or intermingled with spores; urediniospores globoid, ellipsoid or oblong, contents hyaline or coloured, wall echinulate or verrucose, germpores variously arranged.

Telia subepidermal or sometimes subcuticular, erumpent, spores typically 1-celled in a few species, 2-celled by horizontal septum, strongly adherent laterally in discoid heads, pigmented, surface smooth or sculptured, subtended by colourless hygroscopic cysts, pedicels compound, germpore 1 in each cell, basidium external.

Lectotype: *Ravenelia glandulosa* Berk. and Curt. on *Tephrosia*

Ravenelia is the third largest genus of Uredinales with 170 species, having tropical and subtropical distribution. The genus is confined to the hosts of *Fabaceae*. Most *Ravenelia* species perhaps all, reported on Euphorbiaceae have been assigned to a separate genus *Kernkampella*. The telial heads with pendent hyaline cysts born on a compound pedicel are very characteristic of the genus, *Ravenelia*. The genera *Haploravenelia* and *Pleoravenelia* are based on one celled or two celled teliospore characters. Long described *Neoravenelia* on *Prosopis juliflora* on the basis of caeomoid aecium in contrast to aecidioid and peridiate ones present in *Haploravenelia*. The genera *Cystomyces* and *Spumula* are separated from *Ravenelia* on account of the presence of simple pedicels. *Sphaerophragmium* lacks characteristic cysts of *Ravenelia*. Thirumalachar and Mundkur (1949) proposed 3 sections under *Ravenelia*. 1. Two celled teliospore heads in the head *Pleoravenelia*. 2. All spores one celled in teliospore head, but with aecidioid aecia *Haploravenelia*. 3. All spores one celled in teliospore head but with caeomoid aecia *Neoravenelia*.

Key to the species of *Ravenelia*.

On Caesalpinaceae

1. Urediniospore germpores obscure ...*R. delonicii*

1. Urediniospore germpores scattered ..*R. cassiaecola* var. *berkeleyi*

On Fabaceae (S.S.)

1. Urediniospore germpores equatorial ..2
1. Urediniospore germpores scattered ..*R. ornata*
2. Teliospore heads upto 100 μm, surface slightly tuberculate ..*R. stictica*
2. Teliospore heads upto 80 μm, surface prominently appendiculate ..*R. hobsoni*

On Mimosoideae

1. Urediniospore germpores scattered ..2
1. Urediniospore germpores equatorial or not known ..3
2. Teliospore surface smooth ..*R. fragrans var. evernia*
2. Teliospore surface spikata ..*R. evansii*
3. Urediniospores length upto 18 μm ..*R. sessilis*
3. Urediniospores length more than 18 μm ..4
4. Teliospore surface smooth ..*R. odorotissima*
4. Teliospore surface ornamented ..5
5. Teliospore surface papillate ..6
5. Teliospore surface verrucose or warty ..7
6. Teliospores minutely papillate ..*R. japonica*
6. Teliospores prominently papillate ..*R. parasnathii*
7. Teliospore surface verrucose ..*R.acaciae-senegalae*
7. Teliospore surface warty ..*R. tandonii*

***Ravenelia delonicii* Bagyanarayana, John Ravinder and Ramesh sp. nov. (Fig.1)**

Spermagoniis et aeciis ignota. Urediniis hypophylliis, aparaphysata, circularis, pallide brunnea, subepidermalis, erumpentis, pulverulentis; urediniosporiis 19.5-23.4 x 11.7-15.6 μm, ovatis vel elliposoidiis, membrana 1.6 μm crassa, echinulatis.

Teliis hypophyllis, aggregatis, subepidermalis, erumpentis, pulverulentis, brunneis; teliosporiis compoundiis, 66.3 117 μm diam; papillatiis, hyalino; teliosporiis individual is 15.6-23.4 x 11.7-19.5 μm, cystibus minuta, pedicillatis, pedicello compoundis, hyalino.

Holotypus : In foliis vivis *Delonix elata* (L.) Gamble, (Caesalpinaceae), Sri Anantagiri Hill, Vikarabad (A.P.), E.J. Ravinder, Dec., 1992. O.U., Myc. Herb. Ured. No. 187 & HCIO.

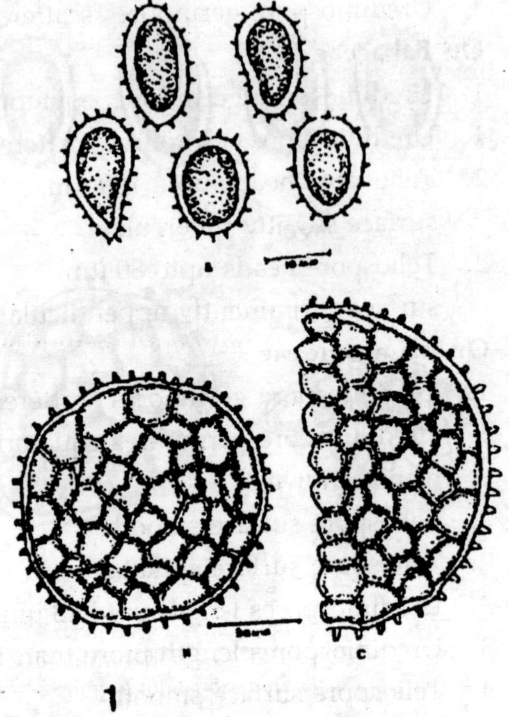

Fig 1. *Ravenelia delonicii* sp. nov. a. Urediniospores B, Teliospore C. Part of teliospore enlarged

Spermogonia and aecia not present. Uredinia hypophyllous, associated with the telia, aparaphysate 0.5 - 1 mm diam., circular, light brown, subepidermal, erumpent, pulverulent; urediniospores 19.5 - 23.4 x 11.7 - 15.6 µm, ovate or ellipsoid, wall 1.6 um thick, echinulate.

Telia hypophyllous, closely aggregated, subepidermal, erumpent, pulverulent, brown 0.5 - 1 mm diam., teliospore heads deep brown, 66.3 - 117 µm, papillate, papilla minute, conical, hyaline; each teliospore 15.6 - 23.4 x 11.7 - 19.5 µm, cysts minute, pedicillate, pedicel compound and hyaline.

Although Caesalpinaceae family has several genera, *Cassia* and *Caesalpinia* are the two genera, which are predominately parasitized by the rust genus *Ravenelia*. Of the several *Ravenelia* species reported on these genera, the present rust is comparable with *R. bella* (on *Cassia emarginata*) and *R. cassiaecola* (on *Cassia nictitans*), since both have aparaphysate uredinia. *R. bella*, differs from *R. delonicii* in having urediniospores with prominently scattered germpores and larger teliospore heads. Wheras *R. cassiaecola* differs in having smaller urediniospores with conspicuous germpores, and smaller teliospore heads. Hence the present rust taxon *R. delonicii* is a distinct species. Apart from these *Delonix* is a new host genus for *Ravenelia*.

Ravenelia odorotissima Tyagi and Prasad. Sci & Cult. 44 (6) : 268 - 269. 1978.(Fig.2)

Spermogonia, aecia and uredinia not known. Telia epiphyllous, round, black, 0.5 - 1mm. diam, subepidermal, erumpent, pulverulent, aparaphystate;

telial heads discoid, yellowish brown, spherical, margin wavy, 56 - 89 μm in diam, 8 - 16 cells per head. The spores arranged in 2 : 1 ratio of central cells to the peripheral cells, central cells 16 - 19 μm across, cysts equal to the number of marginal spores, free and pendulous, pedicillate, pedicel hyaline and compound.

Host : On the living leaves of *Albizia odorotissimai* Benth. (Mimosoideae), Sri Tirumala Hills, Chittoor dist, A.P. G.Bagyanarayana, O.U.Myc. Herb. Ured. No.186.

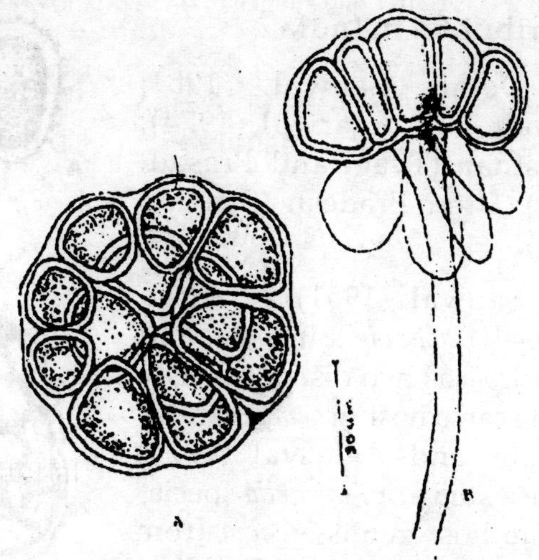

Fig 2. *Ravenelia odorotissima* Tyagi & Prasad A. Surface view of a teliospore B. Vertical Section of teliospore with cysts

Distribution in India:

Rajasthan (Tyagi and Prasad, 1978)

R. odorotissima was erected by Tyagi and Prasad (1978). Earlier on *Albizia odorotissima, R. japonica* Diet. and *R. sessilis* have been reported. *R. odorotissima* differs from *R. japonica* in the characters of the cyst apparatus and telial head. *R. japonica* has spirally oriented teliospores in the head, with the number always more than 25 cells per head, while in *R. odorotissima* the number never exceeds 16 cells per head. *R. sessilis* has prominently appendiculate teliospores while those of *R. odorotissima* are smooth.

***Ravenelia acaciae-senegalae* Sanwal. Sydowia 5: 412. 1951. (Fig.3)**

=*Ravenelia acacicola* Sanwal. Sydowia 5: 412 - 414. 1951.

Spermogonia and aecia not known. Uredinia amphigenous, mostly epiphyllous, scattered or aggregated, 1 to 1.5 μm diam; subepidermal, erumpent, pulverulent, urediniospores 19.5 - 31.2 x 15.6 - 19.5 μm, wall 1.6 μm, thick, echinulate, germpores 4, equatorial.

Telia amphigenous, mostly epiphyllous, scattered or aggregated, 1 - 1.5 mm diam., subepidermal, erumpent, pulverulent, teliospore heads 85.8 - 97.5 um, each teliospore 15.6 - 19.5 μm; cysts hyaline, wall 3.2 μm thick.

Host : On the living leaves of *Acacia catechu* (Linn.) Willd. (Mimosoideae). Sri Tirumala Hills, Chitoor Dist. A.P., G.Bagyanarayana, O.U. Myc. Herb. ured No 187.

Distribution in India

Delhi (Sanwal, 1951; Kapoor and Agarwal, 1974); Rajasthan (Tyagi and Prasad, 1972); Uttar Pradesh (Sanwal, 1951).

Sanwal (1951) reported *Revenelia acaciae-senegalae* and *R.acacicola* as two distinct species on the same host *Acacia senegalae*. Kapoor and Agarwal (1974) made a study of *Ravenelia* species on the host genus *Acacia* from India. They examined both the type specimens of *R. acaciae--senegalae* and *R.acacicola.* As they could not find the variations as reported by Sanwal, they treated *R. acacicola* as a synonym, of *R. acaciae - senegalae.*

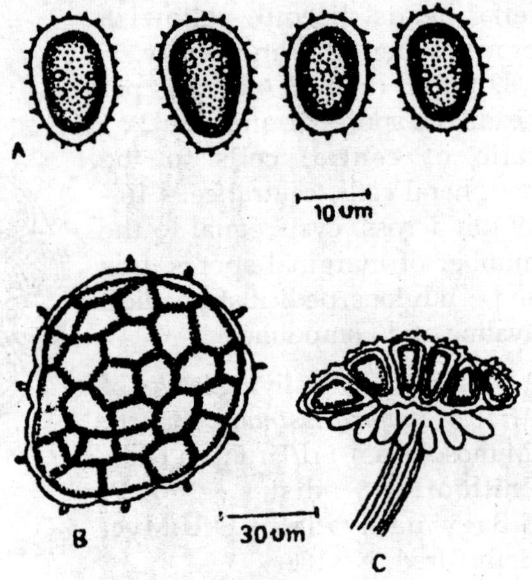

Fig 3. *Ravenelia acociae-senegalae* Sanwal A. Urediniospores B. Teliospore, C. Vertical section of teliospore with cysts

Ravenelia fragrans Long var. *evernia* (Syd) Baxt. Mycologia 57: 7. 1965.(Fig.4)

= *R. fragrans* Long. Both Gaz. 35: 123. 1903.

= *Ravenelia evernia* Syd. Ann. Mycol 31: 87. 1933.

= *Ravenelia dysocarpae* Long & Godd., Mycol 31: 670. 1939.

Spermogonia and aecia not known. Uredinia amphigenous, dense, cinnamon brown, 0.2 - 0.4 µm diam., subepidermal, erumpent, pulverulent, urediniospores 19.5 - 23.4 x 15.6 - 19.5 µm, wall 1.6 µm thick, prominently echinulate, germpores 4 -6 , scattered, paraphysate, paraphyses 46.8 µm long.

Telia amphigenous, dense, blackish brown, 0.2-0.6 mm diam., subepidermal, erumpent, pulverulent, teliospore heads discoid, 54.6 - 78 x 42.9 74.1 µm, each telisopore cell 1.5 - 15.6 µm. smooth, cysts hyaline pedicillate, pedicel 58.5 µm long.

Host: On the living leaves of *Mimosa* sp (Mimosoideae). Sri Tirumala Hills, Chitoor Dist, A.P. 10th Feb., 1994. G. Bagyanarayana, O.U. MYC. Herb. Ured. No.245.

Distribution in India

Bihar (Yadav and Thirumalachar, 1995) Madhya Pradesh (Sydow and Mitter, 1933; Mishra, 1969; Mundkur and Thirumalachar, 1946.); Maharashtra (Mundkur and Thirumalachar, 1946); Rajasthan (Tyagi and Prasad, 1972); Uttar Pradesh (Sydow and Mitter, 1933; Mundkur and Thirumalachar, 1946).

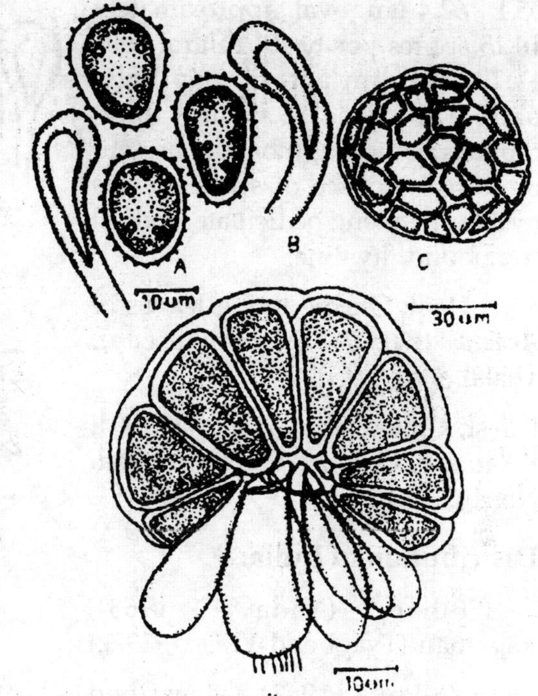

Fig 4. *Ravenelia fragrance* Long var. evenia (Syd.) Baxt. A. Urediniospores B. Paraphyses, C. Teliospore D. Vertical section of teliospore with cysts

R. fragrans was first reported by Long (1903) on *Mimosa fragrans*. Sydow *et al.* (1933) reported *R.evernia* as a new species on *Mimosa* sp. Later, Long and Goodding (1939) reported another new species *R.dysocarpae* on *Mimosa dysocarpae* Benth. Mundkur and Thirumalachar (1946) collected a rust on *Mimosa rubicaulis* L. and reported it as *R.evernia* Syd. They also examined the type of *R. dysocarpae* and considered it as a synonym of *R. evernia*. Baxter (1965) critically examined the type material of *R. evernia* and *R. fragrans* and recognised *R. evernia* as a variety of *R. fragrans* raher than a seperate species. *R. fragrans* var. *evernia* is being reported for the first time from South India.

***Ravenelia parasnathii* Yadav Ind. Phytopath.** 16: 146. 1963. (Fig.5)

= *Ravenelia acacieae - caesiae* Tyagi & Prasad. Science & Culture 44: 271: 1978.

= *Ravenelia pennatae* Durrieu, Cryptogamie Mycologie 1: 64. 1980.

Spermogonia and aecia not known. Uredinia amphigenous, dense, closely aggregated, 0.2 mm diam. subepidermal, erumpent, pulverulent, urediniospores 19.5 - 23.4 x 11.7 - 15.6 µm ovoid to elliposid, wall 1.6 µm echinulate.

Telia amphigenous, dense, closely aggregated, subcuticular, erumpent, pulverulent, blackish brown, 0.5 - 1mm diam., teliospore heads 58.5 - 78 x

35.1 - 62.4 µm, oval, approximately 10-15 spores per head, central cells 11.7 - 15.6 µm across, papillate, papilla hyaline, 9.6 µm long, capitate, numerous tubercles present on dorsal surface of spores; cysts hyaline, pendent; pedicillate, pedicel compound, hyaline.

Host: On the living leaves of *Acacia trita* L (Mimosodieae). Thalakona

Forest, Chitoor dist, A.P., Jan 1994, P.Ramesh, O.U. MYC. Herb. Ured. No. 167.

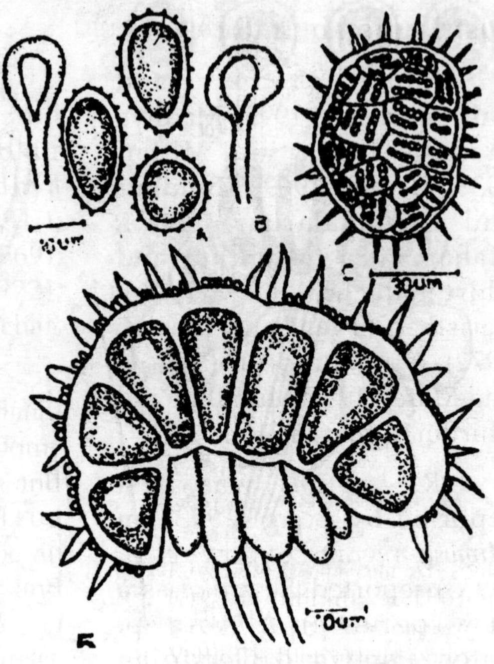

Fig 5. *Ravenelia parasnathii* Yadava A. Urediniospore B. Paraphyses, C. Teliospore, D. Vertical Section of telisopore with cysts

Distribution in India:

Bihar (Yadav, 1963); Rajasthan (Tyagi and Prasad, 1972)

Yadav (1963) described *Ravenelia parasnathii* as a new species on *Acacia canescens* from Bihar. Tyagi and Prasad (1978) described *Ravenelia acaciae - caesiae* on *Acacia* and Durrieu (1980) described *Ravenelia pennatae* on *Acacia* sp. Recently, Mordue (1988) made a critical study of these three species and treated *Ravenelia acaciae - caesiae* and *Ravenelia pennatae* as the synonyms of *Ravenelia parasnathii. Acacia trita* is the new host species record for *R. parasnathii* and this rust fungus is being reported for the first time from South India.

***Ravenelia tandonii* Syd.** Ann. Mycol. 35: 227. 1937. (Fig.6)

Spermogonia and aecia not known. Uredinia epiphyllous, rarely amphigenous, scattered, minute, subepidermal, erumpent, pulverulent, cinnamon brown, 0.3 mm diam., urediniospores 11.7 - 31.2 x 15.6 - 19.5 µm, elliposid or ovoid, wall 1.6 µm thick echinulate, germpores 4, equatorial.

Telia amphigenous, dense, closely aggregated, subcuticular, erumpent, pulverulent, black, 0.3 mm diam., teliospore heads 78 - 93.6 µm, 6 - 8 cells across, teliospores single celled, 15.6 - 23.4 x 11.7 - 15.6 µm, wall 1.6 µm thick, dark cinnamon brown, wall of peripheral spores have warty ornamentation, cysts hyaline, pendent, persistent, pedicel compound, hyaline or light brown.

Host: On the living leaves of *Acacia sundra* (Roxb.) Willd., (Mimosoideae). Horseley Hills, Chitoor Dist,A.P., P.Ramesh, O.U.MYC.Herb.Ured.No.161.

Distribution in India:

Bihar (Yadav, 1963); Maharashtra (Chavan and Bakare, 1974; Date, 1979); Rajasthan (Tyagi and Prasad, 1972); Uttar Pradesh (Sydow, 1913; Sydow, Mitter and Tandon, 1937)

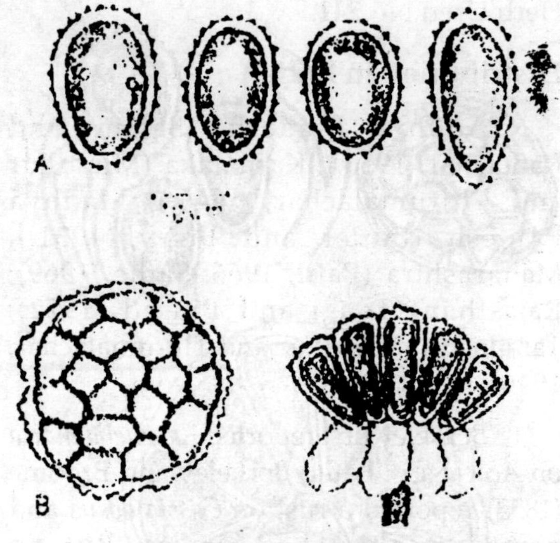

Fig 6. *Ravenelia tandonii* Syd. A. Urediniospore, B. Teliospore, C. Vertical section of teliospore with cysts

Ravenelia tandonii was orignially described by Sydow *et al.* (1937) on *Acacia catechu* Willd from Uttar Pradesh. Apart from U.P. this rust is reported from Bihar, Rajasthan and Maharashtra. For the first time this rust is being reported from South India. *R. tandonii* has ellipsoid to ovoid or slightly curved urediniospores, with equatorial germ pores. The teliospore heads show warty outer wall.

***Ravenelia cassiaecola* Atk. var. *berkeleyi* (Mundk. & Tghirum.) Cumm. & Baxter. Mem. N. Y. Both Gard. 28: 38 - 39. 1976.(Fig.7)**

= *Ravenelia berkeleyi* Mundk. & Thirum. Imp. Mycol. Inst. Kew: Mycol. Papers 16: 19 - 20. 1946.

= *Ravenelia indica* Berkeley, in Gardeners Chron. P. 132. 1853.

Spermogonia and aecia not seen. Uredinia amphigenous, mostly hypophyllous, scattered, circularly deposited, subcuticular, erumpent, pulverulent, brown, 0.4 - 0.7 mm diam., 15.6 - 19.56 x 11.7 - 15.6 µm, globose, subglobose or broadly elliposid, golden brown, wall 1.6 µm, echinulate, germpores 5 -6 scattered.

Telia amphigenous, usually hypophyllous, blackish brown, 0.4 - 1mm diam., subcuticular, erumpent, pulverulent, teliospore heads discoid, 50.7 - 7.12 um, each teliospore 15.6 - 19.5 um, chest nut brown, smooth, cysts hyaline, pendent, pedicel compound, 50.7 - 175.5 µm long.

Host: On the living leaves and stems of *Cassia absus* L. (Caesalpinaceae). Thalakona Forest, Chitoor Dist,A.P. 26th Jan., 1994, P. Ramesh, O.U.MYC.

Herb.Ured.No.241.

Distribution in India:

Andhra Pradesh (Salam and Ramachar, 1956.); Karnataka (Mundkur and Thirumalachar, 1946.); Madhya Pradesh (Butler and Bisby, 1931.); Maharashtra (Patil, 1966; Sathe, 1969); Rajasthan (Tyagi and Prasad, 1972); Tamilnadu (Mundkur and Thirumalachar, 1946)

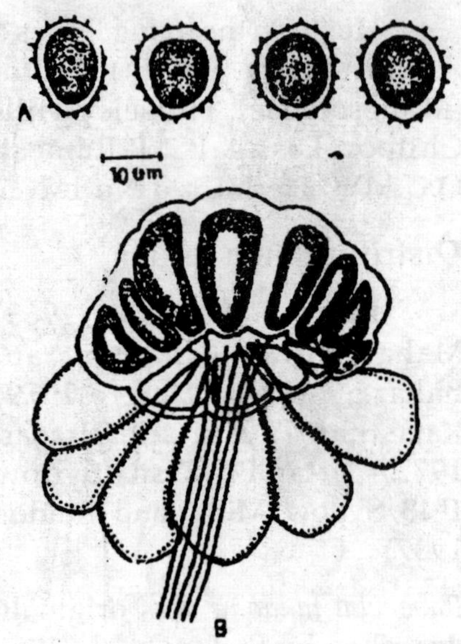

Fig 7. *Ravenelia cassiaecola* Atk. var. *berkeleyi* (Mundk. & Thirum.) Baxter. A. Urediniospore, B. Vertical section of teliospore with cysts

Berkeley first reported *Ravenelia indica* on *Acacia* sps. Later, Berkeley and Broome (1853) reported a rust on *Cassia absus* and named it also as *R. indica*. But on verification this host turned out to be an *Albizia* sps but not *Cassia*. Mundkur and Thirumalachar (1946) created *R.berkeleyi* as a new taxon for a rust infecting *Cassia*. Prior to Mundkur and Thirumalachar Atkinson (1891) reported a rust on Cassia as *R.cassiaecola*. However, Cummins and Baxter (1976) during their study of the genus *Ravenelia* found *R. berkeleyi* to be very much similar to that of *R. cassiaecola* except for minor morphological differences. Therefore, they treated *R. berkeleyi* as a var. of *R. cassiaecola*.

Ravenelia evansii **H & P.Sydow. Ann. Mycol 10: 440, 1912(Fig.8)**

= *Ravenelia acaciae - arabicae* Mundkur. & Thirum. Mycological Papers 16: 17. 1946.

Spermogonia minute, usually associated with aècia, fructicolous, pale brown with dark coloured exudate at the ostiole, subcuticular, conical, determinate in growth, without peridium, 40 - 55 µm high 50 -90 µm wide, flexuous hyphae and periphyses absent; spermatiophores aseptate, hyaline, smooth, thin walled, arranged in palisade like hymenium.

Aecia present on pods, dense in clusters, creamy to pale yellow, aecial cups subepidermal, erumpent, pulverulent; aeciospores catenulate, subglobose to elliposid, often angular 15.6 - 23.4 x 11.7 - 19.5 µm, wall 1.6 µm, verrucose.

Uredinia amphigenous, 0.4 - 0.7 mm diam., subepidermal, erumpent, pulverulent, hyaline to light brown; urediniospores 19.5 - 27.3 x 15.6 19.5 µm,

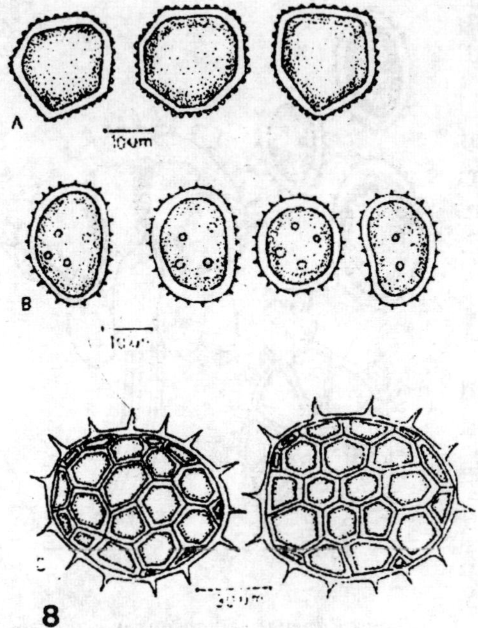

Fig 8. *Ravenelia evansii* H& P. Sydow, A. Aeciospore B., Urediniospore, C. Teliospores

globose to ovate, wall 1.6 µm, echinulate, germpores 3 - 4, scattered.

Telia amphigenous, scattered, erumpent, pulverulent, black, circular, 0.4-0.9 mm diam., teliospore heads discoid, 85.8 - 97.5 µm, light chestnut brown, wall 3.2 µm, each teliospore 19.5 - 23.4 µm long, appendages spikate, 2 to 9.6 µm long, cysts hyaline and pendent, pedicillate, pedicel hyaline.

Host: On the living leaves and pods of *Acacia arabica* Willd. (Mimosoideae). Thalakona Forest, Chitorr Dist. A.P. 24th Jan, 1995, G. Bagyanarayana, O.U.MYC. Herb.Ured.No.243.

Distribution in India:

Andhra Pradesh (Bagyanarayana and John Ravinder, 1988); Karnataka (Mundkur and Thirumalachar, 1946; Rangaswami *et al.* 1970; Kapoor and Agarwal 1974); Taminadu (Mundkur and Thirumalachar, 1946)

Sydow and Sydow (1912) first reported *Ravenelia evansii* on *Acacia robusta* Burch from Pretoria, South Africa. From India Mundkur and Thirumalachar (1946) reported *Ravenelia acaciae - arabicae* on Acacia *arabica* Willd. However, Kapoor and Agarwal (1974) treated *R. acaciae - arabicae* as a synonym of *R. evansii.* Sydow's (1912) original description of *R. evansii* pertained to telial stage only. Later, Doidge (1939) reported the uredinial stage. Recently, Bagyanarayana and John Ravinder (1988) gave a detailed description of the spermogonial and aecial stages of this rust fungus.

***Ravenelia hobsoni* Cooke. Journ. Roy Micr. Soc. III : 386. 1880.(Fig.9)**

Spermogonia and aecia not known. Uredinia hypophyllous. dense, closely aggregated, subepidermal, erumpent, pulverulent, black. 0.5 - 1 mm diam., urediniospores 19.5 - 23.4 x 7.8 - 11.7 µm., globoid or ovoid, cinnamon brown, wall 1.6 - 3.2 µm thick, echinulate.

Telia hypophyllous, dense, brown, subcuticular, erumpent, pulverulent, 0.5 - 1mm diam., chestnut brown; teliospore head 58.5 - 78 µm diam., each teliospore 19.5 - 31.2 x 11.7 19.5 µm, appendiculate, appendages unbranched, 5 - 13.5 µm, wall 1.6 µm thick.

Host: On the living leaves of *Pongamia glabra* Went (Fabaceae). Sri Tirumala Hills, ChittorDist,A.P., 10th Jan. 1994, G.Bagyanarayana, O.U. MYC. Herb. Ured. No 165.

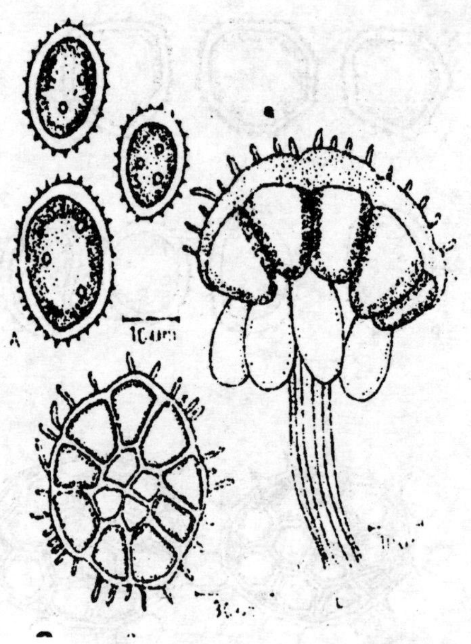

Fig 9. *Ravenelia hobsoni* Cooke A. Urediniospores, B. Teliospore, C. Vertical section of teliospore with cysts

Distribution in India:

Andhra Pradesh (Salam and Ramachar, 1956); Bihar (Gupta and Verma, 1987); Madhya Pradesh (Jain *et al.*, 1966); Maharashtra (Patil, 1966; Sathe, 1969; Chavan and Patil, 1972); Rajashtan (Tyagi and Prasad, 1972); Tamilnadu (Ramakrishnan and Subramanian, 1952); Uttar Pradesh (Pavgi and Singh, 1969).

R. hobsoni was first reported by Cooke (1880) on *Pongamia glabra* Went. Long (1903) splitted the genus *Ravenelia* into *Pleoravenelia, Haploravenelia* and *Neoravenelia.* This was done on the basis of septation of central teliospore cells and presence or absence of peridium in the aecial stage. Pavgi and Singh (1969) reported the spermagonia of *R.hobsoni* for the first time from Varanasi (U.P.). Later Sathe (1969) reported the occurrence of aecia and he described them as uredinoid in nature and subcuticular in position. But Chavan and Patil (1972) stated that they are subepidermal, not subcuticular as stated by Sathe (1969). They reported the *Pongamia* rust as *Haploravenelia hobsoni* (Cooke) Ito.

Ravenelia japonica **Diet. and Syd. Hedwigia 37 : 216. 1898.(Fig.10)**

Spermogonia and aecia not present. Uredinia amphigenous, mostly epiphyllous, scattered or circularly aggregated, subcuticular, erumpent, and pulverulent, 0.5 - 1 µm diam. cinnamon brown, aparaphysate, urediniospores 18.6 - 22 x 15.5 - 19 µm, ovoid or elliposid, wall 2 - 2.5 µm thick, minutely echinulate, germpores 3-4, equatorial.

Telia epiphyllous, caulicolous, and fructicolous, black, minute, 0.3-0.4 mm diam, circular, subcuticular, erumpent; teliospore heads dark brown to black, 78 - 117 µm diam, central cells (-15.5) 18.6 - 31 µm across, peripheral cells ornamented with minute papillae, cysts hyaline and pendent; pedicel compound, hyaline.

Host: On the living leaves of *Albizia odorotissima* Benth, (Mimosoideae). Sri Tirumala Hills, Chittoor Dist, A.P., 26th Feb, 1994, G.Bagyanarayana, O.U.MYC. Herb. Ured. No.164.

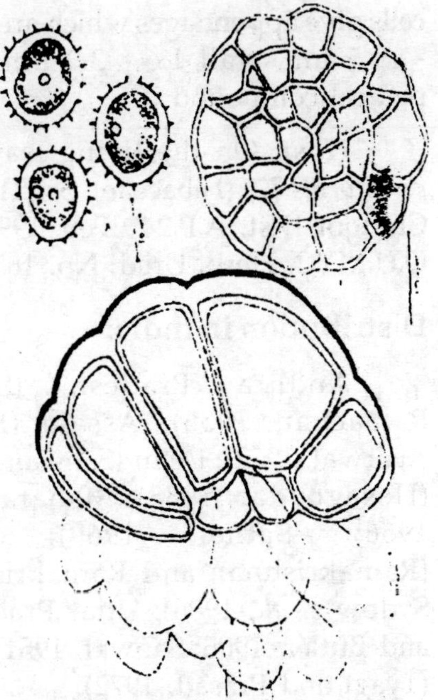

Fig 10. *Ravenelia japonica Diet. & Syd.* A. Urediniospores B. Teliospore, C. Vertical section of telisopore with cysts

Distribution in India:

Andhra Pradesh (Salam and Ramachar, 1956); Maharastra (Sathe, 1969); Uttar Pradesh (Sydow, 1913; Sydow, Mitter and Tandon,1 937; Kapoor and Agarwal, 1972); Rajasthan (Tyagi and Prasad, 1978)

Ravenelia japonica was originally described from Japan parasitizing *Albizia julibrissan* by Dietel and Sydow. *Ravenelia japonica* differs from *R.odorotissima* in the characteristcs of teliospore head and cysts. In *R.japonica* cysts are equal to the number of teliospores in the head, but in *R. odorotissima* they are in 2 : 1 ratio.

Six different speices of *Ravenelia* are reported from India on genus *Albizia* (Mukerji and Jayanthi Bhasin, 1986). As per revisionary studies done by Kapoor and Agarwal (1972) only two *Ravenelia* species viz., *R. sessilis* and *R. japonica* occur on *Albizia* in India. Kapoor and Agarwal based on their studies found out that the report of occurrence of *R. albiziae-amarae* and *R. clemensae* on *Albizia* was because of misdetermination made by the reporting authors. In addition *R. indica* was made a synonym of *R. sessilis*.

***Ravenelia ornata* Syd. Annal, Mycol. 4: 437. 1906. (Fig.11)**

Spermogonia and aecia not known. Uredinia hypophyllous, scattered, subepidermal, erumpent, pulverulent, 0.5 - 0.7 mm diam., brown, urediniospores 15.6 - 23.4 x 15.6 - 19.5 µm, globose or ovid, wall 1.6 µm, finely echinulate, germpores 4-5, scattered; paraphysate, paraphyses capitate, apically brown, stalk hyaline, 39 - 58 µm long.

Telia hypophyllous, black, subcuticular, circular, minute, erumpent, pulverulent 0.5 mm diam., teliospore heads 77.5 - 93 x 31 - 62 µm in diam., globose, deep brown, each teliospore 23.4 - 31.2 x 11.7 - 19.5 µm, peripheral

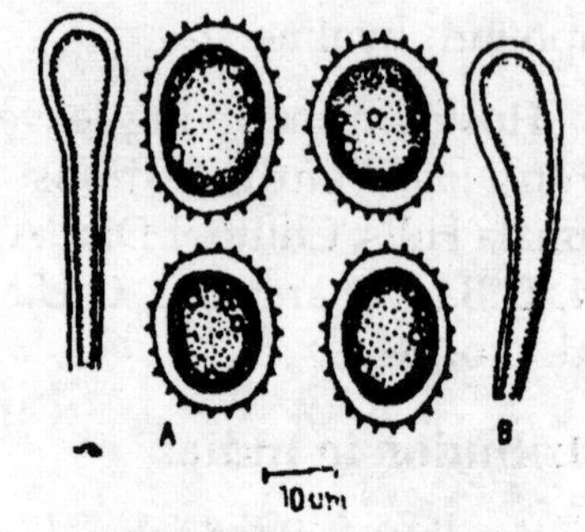

cells give appendages which are capitate, 11.7 - 19.5 µm, wall 1.6 - 3.2 µm. pedicillate, pedicel compound.

Host: On the living leaves of *Abrus precatorius* L., (Fabaceae). Sri Tirumala Hills, Chittoor Dist, A.P.24th Feb. 1994, P.Ramesh, O.U.MYC. Herb. Ured, No. 165.

Distribution in India:

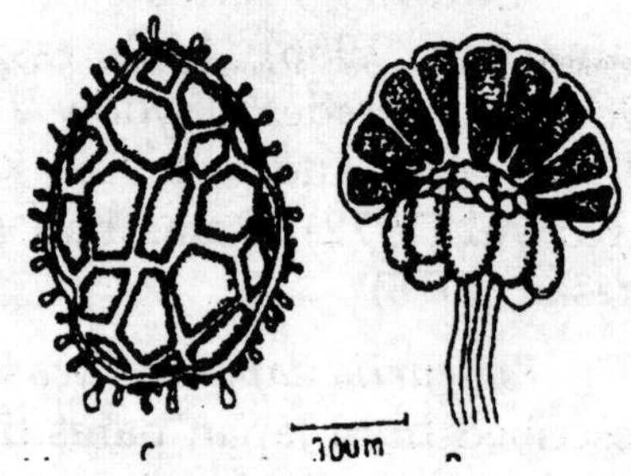

Fig 11. *Ravenelia oranta* Syd. A. Urediniospores, B. Teliospore, C. Vertical section of telisopore with cysts

Andhra Pradesh (Salam and Ramachar, 1956); Assam (Kapoor and Agarwal, 1972); Delhi (Sanwal, 1951); Kerala (Hosagoudar, 1988); Maharashtra (Patil, 1966; Sathe, 1969); Tamilnadu (Ramakrishnan and Ramakrishnan, 1948; Sydow *et al.*, 1906); Uttar Pradesh (Sydow and Butler, 1906; Sanwal, 1951); Rajuasthan (Tyagi and Prasad, 1972).

Ravenelia ornata was first reported by Sydow (1906) on *Abrus pulchellus* from Dehradun (U.P). Later, Petch (1909) reported *R. ornata* on *A. precatorius* from Ceylon. From India Sanwal (1951) reported *R. ornata* on *Abrus precatorius.* Sanwal (1951) considered this rust as a *Pleoravenelia* sp, but Kapoor and Agarwal (1972) treated this as *R. ornata* only. Patil (1966) reported for the first time the spermogonial stage of *R.ornata.*

***Ravenlia sessilis* Berk. In Journ. Linn. Soc Bot. 14: 93. 1875.(Fig.12)**

= *R. indica* Berk. in Gardeners Chron. P. 132. 1853.

=*R. albiziae-amarae* Bacc. Ann. Bot. Roma.14: 129. 1917.

Spermogonia and aecia not present. Uredinia amphigenous, dense, subcuticular, erumpent, pulverulent, cinnamon brown, 0-5 - 0.8 mm diam., urediniospores 15.6 - 23.4 x 11.7 - 15.6 µm, wall 1.6 µm, echinulate, germpores 4, equatorial.

Telia amphigenous, dense, closely aggregated, subcuticular, erumpent, pulverulent, black, 0.8 mm diam., teliospore heads 78 - 101.4 x 78 - 83.6 µm, each teliospore 15.6 - 23.4 x 11.7 - 19.5 µm, wall 3.2 µm, chestnut brown, papillate, papilla hyaline, 1- 1.6 µm long, cysts hyaline, pendent, pedicel compound, hyaline.

Host: On the living leaves of *Albizia amara* (Roxb.) Bovi. (Mimosoideae). Thalakona Forest, Chitoor dist. A.P. 27th Jan, 1994, P.Ramesh, O.U. MYC.Herb. Ured. No.168.

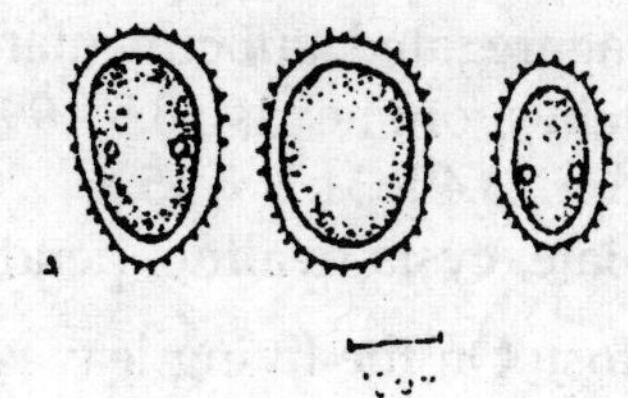

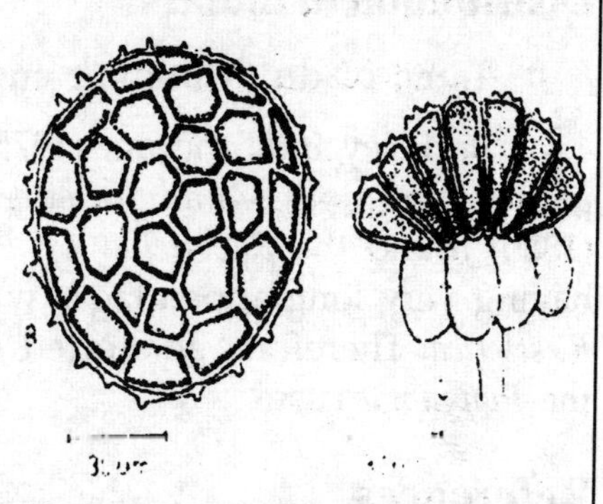

Fig 12. *Ravenelia sessills* Berk A. Urediniospores B. Teliospore, C. Vertical section of telisopore with cysts

Distribution in India:

Andhra Pradesh (Salam and Ramachar, 1956); Arunachal Pradesh (Roy, 1965); Bihar (Wilson, 1963; Yadav 1963; Sydow and Butler, 1906); Kerala (Kapoor and Agarwal, 1972); Madhya Pradesh (Sydow and Butler, 1901; Jain *et al.*, 1966); Maharashtra (Patil, 1966; Sathe, 1965); Punjab (Sydow and Butler, 1906); Rajasthan (Tyagi and Prasad, 1972); Uttar Pradesh (Sydow and Butler, 1906); West Bengal (Sydow and Butler, 1906)

Ravenelia sessilis is a widely distributed rust fungus from India. The echinulate uredospores possess equatorial germpores. The discoid teliospore heads have minute papillae.

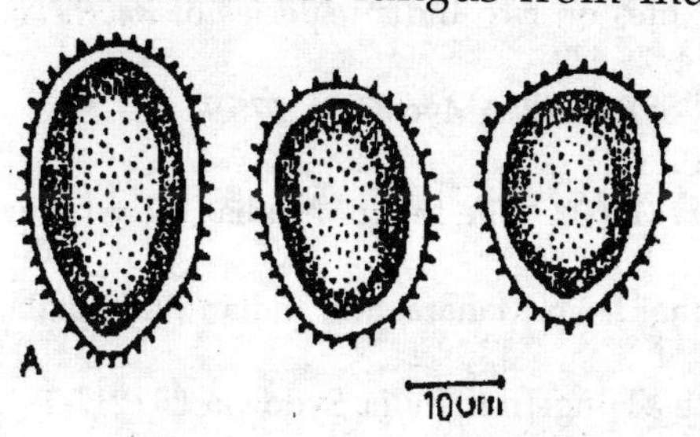

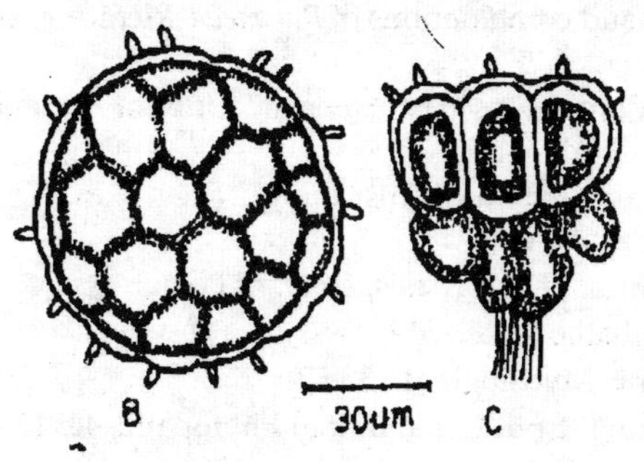

Fig 13. *Ravenelia stictica* Berk A. Urediniospores B. Teliospore, C. Vertical section of telisopore with cysts

Ravenelia stictica **Berkeley & Broom. J. Linn. Soc Bot. 14: 93. 1875.(Fig.13)**

=*R. munduleae* P. Henn. Ann Roy. Inst. Roma 6. 86. 1896.

Spermagonia and aecia not known. Uredinia hypophyllous, dense, closely aggregated, subcuticular, erumpent, pulverulent, brown, 0.5 mm diam., urediniospores 15.6 - 27.3 x 15.6 - 19.5 μm, globose or ovoid, wall 1.6 μm, cinnamon brown, minutely echinulate, germpores 2, equatorial, indistinct.

Telia hypophyllous, dense,

closely aggregated, subcuticular, erumpent, pulverulent, dark brown, 0.6 mm diam., teliospore heads 58.5 - 99.5 x 62.4 - 85.8 µm, discoid, deep brown, each teliospore 23.4 - 31.2 x 15.6 - 23.4 µm, appendiculate, appendages hyaline, tuberculate, cysts hyaline, pendent, pedicillate, pedicel hyaline, compound.

Host: On the living leaves of *Mundulea* sp. (Fabaceae).

Distribution in India:

Tamil Nadu (Mundkur and Thirumalachar, 1946)

Berkeley and Brooome (1975) described *Ravenelia sticitica* as a new speices parasitizing both *Mundulea suberosa* and *Pongamia glabra*. However, Cook (1880) found the *Ravenelia* on *Pongamia glabra* to be different, since it was having very long appendages when compared to the minute appendages of *R. stictica*. Therefore, he erected *R. hobsoni* as a new species to accommodate the *Pongamia* rust.

References

Arthur, J.C. (1934). "Manual of the rusts in United States and Canada". Purdue Research Foundation, Lafayette, U.S.A.

*Atikinson, 1891. Bot. Gaz., **16**:313.

Bagyanarayana, G. and Ravinder, E.John (1988). Notes on two Indian species of *Ravenelia*. Transactions of British Mycological Society, **90** : 477 - 479.

Baxter, J.W (1965) Studies of North American species of *Ravenelia*. Mycologia, **57**:79.

*Berkeley and Broome (1853). Gardeners Chronicle.P.132.

Butler, E.J. and Bisby, G.R. revised by R.S. Vasudeva (1960). "The fungi of India:, New Delhi, ICAR.

Chavan, P.B. and Bakare, U.B (1974). Some rust fungi from Maharashtra Indian Phytopath., **27**:206.

Chavan, P.B. and Patil, S.K. (1972). Studies in some Rust Fungi from India. Sydowia, **26** : 217-281.

Cooke, M.C. (1880). Fungi of India. Grev, **8** : 93 - 96.

Cummins, G.B. and Baxter, J.W. (1976). New species and combinations of *Ravenelia*. Memories of the New York Botanical Garden, **28**: 38-40.

Cummins, G.B. (1978). "Rust Fungi on Legumes and Composites in N. America". Univ. of Arizona Press. Tucson, Arizona, U.S.A.

Cummins, G.B. and Hiratsuka, Y. (1983). "Illustrated Genera of Rust Fungi "(Rev. Edn.) The American Phytopathological Soc. St. Paul, Minnesota, U.S.A.

Date, K.G. (1979) Studies on Rust Fungi of Maharashtra. Ph. D. Thesis, Univ. of Poona.

Doidge E.M. (1939) South African Rust Fungi-III. Bothalia, **3** : 487-512.

Durrieu, G. (1980). Uredinales du Nepal. Cryptogamie Mycologie, **1** : 33-68.

Gupta, B.K. and Verma, R.P. (1987). New hosts of fungi from India. Indian Phytopath., **40**: 123-124.

Hosagoudar, V.B. 1988. Uredinales of Kerala. J. Ecom. Tax. Bot., **12**: 265-272.

Jain A.C., Nikam, B.G., Kulkerni, S.N. and Sharma, O.P. (1966). Fungi of Gwalior and Indore region. The Vikram, **4**: 181-187.

Kapoor, J.N. and Agarwal, D.K. (1972). Indian species of *Ravenelia* on *Abrus* and *Albizzia*. Indian. Phytopath., **25** : 551-554.

Kapoor, J.N. and Agarwal, D.K. (1974). Indian species of *Ravenelia* on *Acacia*. Indian Phytopath.,**27** : 666-669.

Long, W.H. (1903). The Ravenelias of United States and Mexico. Bot. Gaz., **35**: 111-133.

Long, W.H. and Goodding (1939). *Ravenelia dysocarpae*. Mycologia, **31**: 670.

Misra, R.P. (1969). The Uredineae of Jabalpur (M.P)-II Proc. Bihar Acad. Agric. Sci., **17**:96.

Mordue, J.E.M. (1988). A reassessment of *R. parasnathi*, it's morphology, synonymy and host range. Trans, Br. Mycol. Soc., **90**(3) : 19.

Mukerji, K. G. and Bhasin, Jayanthi. (1986). Plant Diseases of India. Tata McGraw Hill, New Delhi.

Mundkur, B.B. and Thirumalachar, M.J. (1946). Revisions and additions to Indian fungi - I. Mycol. Pap. Imp. Mycol Inst., **16** : 1 - 27.

Patil, S.D. (1966). The genus *Ravenelia* Berk. M.V.M Patrika, **1**: 52 - 55

Pavgi, M.S. and Singh, U.P. (1969). Morphology of pycnia of some *Ravenelia* species. Mycologia, **61** : 826-830.

*Petch, T (1909) Ann. Roy. Bot. Garden. Peradeniya, **4** : 302.

Ramakrishnan, T.S. and Ramakrishnan, K. . (1948). Additions to Fungi of Mardras -IV. Proc. Indian Acad. Sci., **278** : 34-36.

Ramakrishnan, K. and Subramanian, C.V. (1952). The fungi of India - A second supplement. J. Madras Univ., **21B** : 303 - 305.

Rangaswamy, G. Seshadri, V.S. and Lucy, K.A. and Channamma, K.A. (1970). "Fungi of South India", University of Agricultural Sciences. Bangalore

Roy, A.K. (1965). Additions to the fungus flora of Assam -I Indian Phytopath, **18**: 327-355.

Salam, M.A. and Ramachar, P.. (1956). Additions to our knowledge of the rusts of Hyderabad - II., J. Indian Bot. Soc., **35**: 152-157.

Sanwal, B.D. (1951). On some new and noteworthy *Ravenelias* from India- I. Sydowia, **5** : 384 - 387 and 412 - 417.

Sarbhoy A.K., Varshney J.L and Agarwal, D.K (1986). "Fungi of India" 1977-91. New Delhi.

Sarbhoy, A.K., D.K. Agarwal and J.L. Varashney (1996). *:Fungi of India"* 1982-92. CBS Publishers and Distributors, New Delhi. 350 pp

Sathe Arun, V. (1969). Studies in some Indian rusts with special reference to life-cycles, sex behaviour and evolution. Bull Bot. Survey India 11: 169

Sydow H and Butler E.J(1901). Fungi Indiae Orientalis. Pars I. Ann. Mycol., **4**: 424-445.

Sydow, H. and P. (1912). Novae Fungorum species-I. Ann.Mycol., **10**: 405-410.

Sydow, P. and H. (1915). "Monograhia Uredinearum". Vol III. pp.257.

Sydow, H. & P. and Butler, E.J. (1906). Fungi Indiae orientalis. Pars I, Ann. Mycol., **4** : 423-445.

Shdow, H & P. and J.H. Mitter. (1933). Fungi Indici, I. Ann. Mycol., **31** : 84 - 97.

Sydow, H. & P., Mitter J.H. and Tandon, R.N. . (1937). Fungi Indici III. Ann. Mycol., **35** : 222 - 243.

Tyagi, R.N.S. and Prasad, N. (1972). The monographic studies on genus *Ravenelia* occurring in Rajasthan. Indian J. Mycol. and Plant Pathol., **2**: 108-130.

Tyagi, R.N.S. and Prasad, N. (1978). Some new Ravenelias from Rajasthan. Science and Culture, **44**: 268.

Yadav, A.S. (1963). Contributions to the knowledge of Uredineae of Bihar-III. Ind. Phytopath., **16** : 138 - 142.

Yadav A.S and Thirumalachar, M.J. (1955). A contribution to the knowledge of Uredineae of Bihar-I. Indian. Phytopath., **8**: 143-149.

25

Indian Mucorales : An Appraisal

A.K. Sarbhoy*

Division of Plant Pathology, IARI, New Dehli-110012

Forty genera of Mucorales comprising of more than 175 are know to occur till 2001 from India. These are placed in thirteen families. Three new genera *Saksenea, Dichotomocladium* and *Apophysomyces* are added from India apart from 75 different new species and varieties were recognised.

Key words : India, Mucorales

The origin of the concept that the mucors (infecting Bread fungi) constitute an order is not readily determined. It is true that Fries had erected an order for the mucors but in that certain genera which he included are obviously not Phycomycetes and furthermore, the group was placed under Hyphomycetes. The begining of the real understanding of the Mucorales dates from 1873 when van Tieghem and Le Monnier and later van Tieghem envisioned these as belonging in a family Mucoraceae. This manner of treatment was essentially the same as used by Berlese and Detoni in Saccardo with the exception that the subfamily Chaetocladiacae had been added. Schroter treated the mucors as a suborder and later Fischer gave us apparently a modern concept in his order Mucorinae, which embodied five families. Subsequent workers such as Zycha, Hesseltine and Benjamin have closely followed this system and all have recognized the group as constituting an order.

The order is distinguished from the related order Entomophthorales by the nature of the zygospore wall, which is said to be formed from the modification of the gametangial wall in the Mucorales but not in the Entomophthorales. From a practical point of view this is not of much help. Actually in the Entomophthorales conidia are present which often are discharged as conidia though later they may function as sporangia. They are often found on insects or other animals. In the Mucoarles sporangia are the rule and, when conidia are formed they are never forcibly discharged, never catenulate and are always borne directly on vesicles or on phialides upon swollen vesicle-like structures except possibily certain forms as in Mortierellaceae where one spored sporangia are known.

*Corresponding Author E-mail : sarbhoy@yahoo.co.in

The mycelium of the mucors is usually much more extensive than that of Entomophthorales. A third related order, Zoopagales, is distinguished because of parasitic habit of its members on soil and aquatic animals with reporoduction by means of conidia which may be in chains. They are attached to the host by haustoria-like organs.

Hesseltine (1995) recognized nine families in this order and 42 genera out of 57 common generic names recorded so far. The rest of them he placed under doubtful list. In 1960, Benjamin created a new group "Merosporangiferous Mucorales" in which he placed some of these most fascinated mucors showing mero sporangiferous like structures.

Mycologists use many types of characters to contribute to their evolutionary studies, including morphology, anatomy ultrastructural features, biochemistry, nucleic acid sequences, and various other attributes. A number of biochemical characters have been important in showing similarties. between Fungi and *Stramenopila.* Chemical techniques such as chromatography and protein electrophoresis have been useful in the comparision of fungal pigments and isoenzymes. The early use of chemical spot tests in the study of lichenized fungi has developed into every day use of chromatography to determine the presence of secondary metabolites in the group. Most recently, molecular techniques have come to mycology and provide a wealth of characters (Bruns *et al.*, 1991, 1993; Hibbet, 1992a; Kohn, 1992).

Many mycologists look upon Zygomycetes as a group characterised primarily by Mucorales, this obscures a broader view involving a radiation of at least three evolutionary lines (R.A .Humber, personal communication, 1994).

1. Many species have extensive mycelia and a high level of asexual reproductive structure differentiation. Sexual reproduction in this group is characterised by out crossing (heterothallism) determined by elaborate hormonal control. This lineage is exemplified by Mucorales.
2. Homothallic species that have relatively undifferentiated asexual reproductive structures and limited somatic growth are found in Entomophthorales.
3. Heterothallic species with elaborate asexual reporduction and parasitic habit are found in Zoopagales.

Additionally, Endogonales and Glomales appear to comprise a fourth evolutionary line of ancient origin.

Work on Mucorlaes in India started forty years ago although the first report of a definite number of this order was made by Currey in 1873 on flowers of *Hibiscus rosasinensis* from Calcutta. He named it as *Choanephora cunninghamia* which was later relegated to *Choanephora infundibulifera. C. simsonii* on flowers of *Zinnia elegans, Cunninghamella elegans* from the soils of Pusa were also reported by the turn of the century.

In sixities Saksena and his students studied the various aspects of Mucorales from soil and excreta of animals in India. Sarbhoy (1967) listed twenty five genera comprising of one hundred and thirty five species while Tandon (1968) listed nineteen genera and sixty seven species. However, now forty genera of Mucorales comprising of more than one hundred and seventy species are knwon to occur in this sub-continent from thirteen families. The author is of the opinion that a proper survey of these fungi are yet to be made in India and many more species and genera will be added in the near future.

The family can be distinguished by multispored sporangia, columella always present, sporangiospores liberated by breaking up the sporangial wall, zygospores rough, suspensors not tong like. It comprises of eighteen genera, out of which fourteen are reported from India. The genera like *Mucor, Absidia, Gongronella, Circinella, Pilobolus, Rhizopus, Phycomyces* and *Zygorhynchus* are most common in occurrence.

The cell wall of Mucoraceae is composed of chitosan and other polysaccharides viz., glucosamine, galactose, polyglucu-zonides, polyphosphatases, proteins, lipids, purines, pyrimidines, magnesium and calcium (Bartinic-Garcia, 1968;Kreger, 1964; Bartinici-Garcia and Reyes, 1968; James *et al.*, 1928).

Mucor

More than six hundred names have been proposed in the genus *Mucor* and most of these have no connection with the above genus but owing to inadequate description and data they are still to be found listed under this genus (Sarbhoy, 1967). This multiplicity of the names is the cause of much confusion and make the identification difficult of the species, on the basis of shape, size of sporangiospores and colour of the culture. The sections are Sphaerosporus, Rammanianus, Fragilis, Hiemalis, Racemosus, Flavus and Mucedo. Hesseltine (1954) created two more sections, Genevensis, in which he placed all the homothallic species forming zygospores and Macromucor, the sporangia are more than 500 µm in diameter with sporangiospores usually upto 10 cm in height.

Sarbhoy (1967) revised the Sphareosporus section of the *Mucor* and also proposed a key for identification on the basis of smooth and echinulate sporangiospores and recognised nine species and variety. This group was later studied by Mehrotra and Mehrotra (1974).

The other sections of *Mucors* also need a thorough revision in the light of recent advances. Although Schipper (1975, 1978, 1989) worked on Circinelloides and Racemosus group of *Mucor* alongwith several unrelated species which are placed in other groups under standard conditions. This has not been possible so far due to non-ability of voucher specimens and isolates from different sources and the existing descriptions are also grossly inadequate.

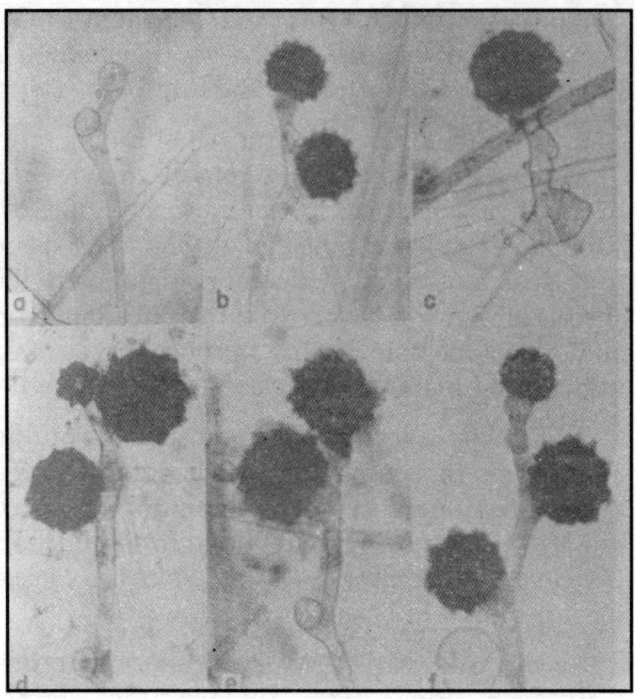

Fig. 1. *Mucor azygospora* Mehrotra, B.S. and Usha Baijal

Fifty seven different species of *Mucor* have been recorded from India *Mucor ardhlaengiktus* Mehrotra and Mehrotra (1973) and *M. thermophilus* Sharma and Sarbhoy (1993) have been added recently from India.

Absidia, Mycocladus, Chlamydoabsidia and *Gongronella*

These four genera are closely related to each other and originally placed in the genus *Absidia* like forms in which the zygospores were not surrounded by appendages as in other species of *Absidia* where it has finger like projections. Hesseltine and Ellis (1964) recongnised *Absidia* and *Gongronella.* He separated *Absidia* into three series on the basis of distinct shape of the sporangiospores i.e. cylindrical, globose, ovoid and cruneate. So far twenty five different species of *Absidia* are recorded from the world but only fifteen from India. Prakash and Sarbhoy (1993) added *Absidia californica* Ellis and Hesseltine and *A. cylindrospora* var. *nigria* Hesseltine and Ellis and *A. cylindrospora* Hesseltine and Ellis from compost soils of Delhi.

The genus *Gongronella* can be distinguished from *Absidia* on the basis of strictly globose apophysis which consequently shows a constriction between the apophysis and attachment of sporangial wall. All the species possess *Mucor* like zygospores without appendaged suspensors. Vanova (1968, 1969) also contributed and described a few species of *Absidia* from Czechoslavakia.

Rhizomucor

There are three genetically distinct thermophilic species of Mucor which are reintroduced as *Rhizomucor* i.e. *R. pusillus, R. miehei* and *R.. tauricus*. (Schipper, 1978). Formation of stolons and rhizoids are morphologically different in these species. Sporangiospores are branched and each branch bearing multispored terminal sporangium. Zygospore globose covered with blunt projection.

Rhizopus

It is differentiated on the basis of stolons and sporangiospores which arise above the rhizoids. Thirteen species have been recognised on the basis of biochemical and morphological characters (Inui and Izuka, 1965) of which eleven species have been described from India. Schipper and Stalpers (1984) maintained in *R. stolonifer* group only two species with several varieties and other taxa are treated as synonyms *R. oryzae*. Prakash and Sarbhoy (1993) have added *Rhizopus microsporus* var. *Pseudochinensis* combination nov. from compost soils of Delhi. Some serious efforts are needed to revise the genus in India.

Circinella

The genus *Circinella* differs from *Mucor* in having sporangiospores with circinate branches terminated into a globose sporangia with persistent wall. Schroeter (1886) recognized *Circinella* as subgenus of *Mucor* but this view was not adopted (Hesseltine, 1954 and Zycha *et al.*, 1969). There are only eight species recognised in this genus out of which six are recorded from India. Patil and Kale (1981) recorded a new species of *C. indica* from soils of Pune, M.S. but it is not validly published.

Pilobolus, Pilaria and *Utharomyces*

Pilobolus comprises of twenty one species and *Pilaria* of three species. Ten species of *Pilobolus* and *Pilaria anamola* have been described from India (Nand and Mehrotra, 1968). *Utharomyces* has two species, one of which *U. indica* has been reported from peacock dung from India (Sarbhoy *et al.*, 1974) All attempts have failed to culture this species. This species was relegated by Kirk *et al.* (1980) into *Utharomyces epallocaudus* because under SEM the sporangiospores of *U.indica* are echinulate.

Phycomyces

It has globose sporangia containing many sporangiospores and typically *Mucor* like columella. It differs from *Mucor* in having tong shaped suspensors of the zygospores and these also possess dichotomously branched dark species or protuberances which have some resemblance to *Absidia* in not having apophysate sporangia and tong shaped zygospores scattered through the aerial portion of the colony. Three species are recognised of which two are reported from India. Zygospores are formed at 15 to 20 ^{0}C in corn steep liquor medium (Hesseltine and Ellis, 1966).

Zygorhynchus

Zygorhynchus as soil inhabitant is closely related to *Mucor* like sporangium with many sporangiospores. It differs from *Mucor* in having homothallic zygospores, these are borne between suspensors of unequal size and shape. *Mucor* is typically heterothallic but in these the suspensors are of equal size and usually produced near the surface of substrate. *Zygorhynchus* has six species and one variety. The species considerably differ in their nutritional requirements which can be used for specific differentiation. *Zygorhynchus psychrophiles* is the species found in Finland which grows well at 0 to 15^0 C.

Pirella

The genus *Pirella* is characterised by the presence of two types of sporangia borne on the same sporangiosphore. The terminal ones are globose and later are of different shape and borne circinately. Only one species of *Pirella reniformis* is recorded from soils. Zygospore of this genus is of *Mucor* type. Under different environmental and nutritional conditions the main morphological characters remain unchanged.

Actinomucor and *Dicarnophora* are also reported from India. *Spinellus, Syzygites, Parasitella* have not been reported from India so far.

Cuninghamellaceae

It is characterised by branched or unbranched conidiophore, bearing only conidia on terminal vesicle, sporangia and sporangiola never formed. Conidia borne singly, hyaline and echninulate. Zygospores *Mucor* like with later suspensor verrucose. Baijal and Mehrotra (1980) in a reassessment studies of Indian Cuninnghamellas recognised five species and their varieties out of the seven known from the world. *C. blakesleeana* var. *verticillata* (Paine) Baijal and Mehrotra. *C. echinulata* var. *indica* Baijal and Mehrotra and *C. indica*

Baijal and Mehrotra. *Thamnocephalis ovalispora* and *Mycotypha microspora* were isolated from dung from India.

Mortierellaceae

The family represented by *Mortierella, Haplosporangium, Dissophora, Gamsiella and Aquamortierella* were not included in the family by Gams (1969) for unknown reasons. Nine sections consisting of *Isabellina, Simplex, Alpina, Schumuckeri, Polycephala, Actinomortierella, Hygrophila, Stylospora* and *Spinosa* were recognised. Linnemann (1966) recognised eleven sections whereas Gams (1969 , 1976, 1978) recognised only ten. About ninety species were recognised on the basis of morphological characters of sporangiophores, spores, sporangiola, stylospores and chlamydospores in these sections. In India, only fifteeen species and a variety have been recorded. Heterothallism is pronounced in several species and a few species of the genus are known to be homothallic and form zygospores with a thick hyphal coat (Zycha, 1935). Attempts to germinate zygospores have failed so far. The genus *Haplosporangium* which is characterised by one or raely two spored, sporangiola and sporangiophores meso or acrotonically branched, has so far not recorded from India. Linnemann (Zycha *et al.*, 1969)included *Echniosporangium* in Mortierellaceae but not *Aquamortierella* as did Hesseltine and Ellis (1973) *Echnisporangium* was placed in Sakesenaeceae by Ellis and Hesseltine (1974).

Choanephoraceae

Three genera *Blakeslea, Gilbertella* and *Choanephora* are recognised. *Blakeslea* can be distinguished from *Choanephora* by the presence of few spored sporangioferous heads and the absence of conidia. The presence of monsporus sporangiola in different species of *Blakeslea* indicated evolutionary trend from multispored sporangia (via *Mucor, Absidia* and *Rhizopus*) to single spored sporangiola. Six species of *Chaonephora* and the three species of *Blakeslea* are recorded from India.

The monotypic genus *Gilbertella* is characterized by heterothallic zygospores, possessing a *Mucor* like sexual stage and a *Choanephora* like asexual stage. From its life cycle it is concluded that *Gilbertella* is a form intermediate between Mucoraceae and Choanephoraceae (Hesseltine, 1960). A new variety of the above genus has been recorded.

Radiomycetaceae

It consists of two genera *Radiomyces* and *Hesseltinella* Both the genera were originally placed in Thamnidiaceae or Choanephoraceae. Embree (1959) suggested a new family for *Radiomyces* due to similarity in character with Mucoraceae and Thaminidaceae which was then named Radiomycetaeceae (Ellis and Hesseltine, 1974). *Hesseltinella* has a single stalked, many spored sporangiola whereas *Radiomyces* has many single or multispored stalked sporangioles borne on secondary vesicle. *Radiomyces* is homothallic whereas *Hesseltinella* presumably is heterothallic. No. record of these two genera have so far been made from India.

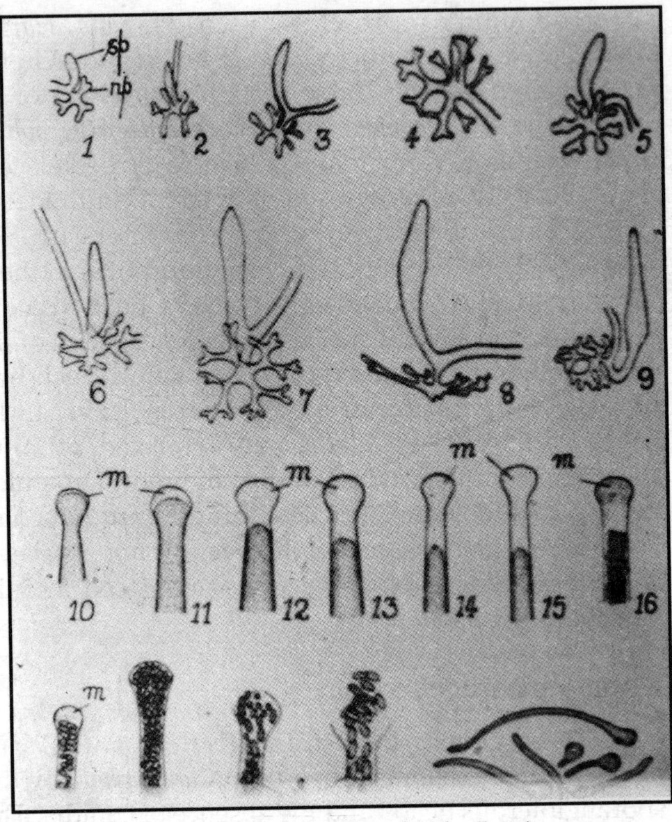

Fig. 2. *Mucor baineri* Mehrotra, B.S. and Usha Baijal

Sakseneaceae

It contains two monotypic genera, *Sakesnaea* and *Echinosporangium*, former is named after late, Prof. Ram Kumar Saksena under whose guidance the work on Mucorales was initiated in India. This is admitted as an unusual family. Zygospores are unknown in the both the genera. The genus *Sakenaea* was recorded from soils of India.

Thamnidiaceae

It is recognised by stalked sporangioles borne laterally on sporangiospores bearing columellate sporangia, usually without sterile spines, sporangiolar branches often arising in whorls formed below the terminal sporangium. Sporangiola globose to pyriform, one to few spores often without a well developed columella. Sporangiospores hyaline, smooth walled, zygospores formed between two unadorned suspensors. These are usually saprophytic and often prefer low temperatures. In the family seven genera

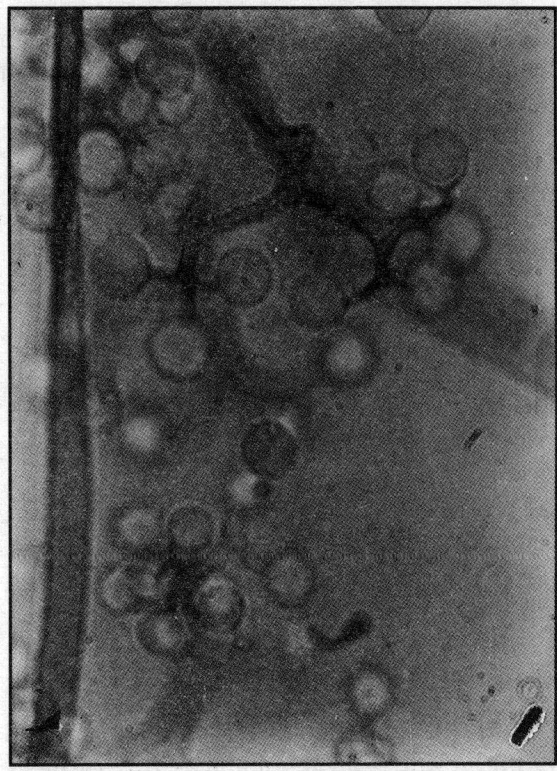

Fig. 3. *Dichotomocladium hesseltinei=Chaetocladium hesseltinei* (Mehrotra, B.S. and A.K. Sarbhoy, *Mycologia*, 1960) Benny and Benjamin, *Aliso*, 8: 391, 1976.

and twenty species have so far been recorded. The most common genera are *Dichotomo cladium, Helicostylum, Chaetocladium, Thamnidium, Chaetostylum and Thamnostylum,* Upadhyay (1973) made a comprehensive study of several taxa classified in *Helicostylum* and suggested that the genus is comprised of species having different relationship and accordingly recongnised only *Helicostylum* and *Thamnostylum.* In *Helicostylum* sporangium is without an apophysis, forming globose sporangiola borne on straight or recurved branches and lacking stolons and rhizoids whereas in *Thamnostylum* the rhizoid and stolons, apophysate sporangia are present. The pyriform sporangiola borne on circinate branches. He recognised three species in each genera, *Helicostylum cordense* and two species of *Thamnostylum* have been recorded from India.

Cokeromyces, Hesseltinella and *Radiomyces* are now placed in the family Radiomycetaceae instead in Thamnidiaceae (Upadhyay, 1970) on the basis of sporangiola borne on secondary vesicle and zygosporic characters.

Benny and Benjamin (1976) created several new genera *Dichotomocladium, Ellisomyces, Fennellomyces* and *Zychae* after critically going through this family. Many species placed in synonymy and new combination were made.

Merosporangiferous Mucorales

The merosporangia may be borne, usually in groups, on the terminal enlargements of simple or branched sporophores as in Syncephalastraceae and many species of Piptocephalidiaceae or highly modified branches of fertile hyphae, in Dimargaritaceae and Kickxellaceae (Benjamin, 1959). The number of spores poduced in each merosporangium varies with species to species, it usually posseses ten to fifteeen spores or it may be reduced two to one.

Syncephalastraceae

Syncephalastrum, the monotypic genus of the family Syncephalastraceae which has cosomopolitan in origin and recorded invariably from soils (Sarbhoy, 1967 and Sarbhoy *et al.,* 1974).

Piptocephaldiaceae

The family Piptocephalidiceae is very well represented in India, eleven species and a variety of *Syncephalis* (Mehrotra, 1959; Mehortra and Prasad, 1964, 1965, 1967 and 1969; Sarbhoy and Singh, 1976) and five species and a variety of *Piptocephalis* have so far been recorded from soil and excreta (Mukerji, 1968; Sarbhoy, 1967; Sabhoy *et al.,* 1974). Zygospore is a useful taxonomic criteria for identification in *Pipocephalis* (Leadbeater and Mercer, 1956 and 1957a and b), although it is reported in only nine species out of the thirty described so far.

Dimargaritaceae

Dimargaritaceae has four important genera i.e. *Dimargis, Dispira, Splinalia* and *Tieghemiomyces* but only *Dimargis* is recorded from India (Mehrotra and Baijal, 1963, 1968). Zygospore like bodies are recorded in all the members of the family but their true nature has been demonstrated in some species only.

Kickxellaceae

In Kickxellaceae, the most common genera are *Coemansia* and *Linderina* (Linder, 1943), which have only one spored sporangia in sporocladia as the characteristic feature for identification. It comprises of eight genera and twenty one species. Nine species of *Coemansia* were isolated from soil of India (Kakkar and Mehrtora, 1970; Sarbhoy and Singh,1981). *Lindernia* is also reported from India (Benjamin, 1958).

As one focusses his attention on this long history of four decades on the Taxonomy of Zygomycetes; it is evident that the group has developed more interest during the last fifty years. However, more concerted attempts should be made to sudy the "merosporangiferous, Mucorales" of India because some of the genera i.e. *Spinalia, Tieghemiomyces, Corkeromyces, Dispira* etc. are yet to be recorded. This will give us a wide range of overlapping characters used which may help us in resovling certain confusion due to incomplete information.

Mating reaction

The Mucorales are characterised by the mating of two morphologically identical gametangia, termed as progametangia. Blakeslee (1904) demonstrated the phenomenon of sexuality in this order. He encountered heterothallism, a condition in which the mycelia from a single sporangium could not mate and required two compatible strains for zygospore formation. These compatible strains were designated plus (+) and minus (-) in his opinion. Mucorales also showed homothallism in which the mycelia drawn from a single spore could form zygospores. Burgeff (1924) implicated a chemical which is responsible for this mating. The (+) and (-) strains *Mucor mucedo,* even when separated by a membrane, could induce one another to form zygospore. Satina and Blakeslee (1926) noticed that the (-) strains produced twice the amount of carotene as did (+) strains. Goodwin and Griffths (1952) observed this in *Phycomyces blakesleanus* and *P.nitens.* Chodat and Schooper contradicated this observation, in *Mucor hiemalis* and found that (+) strains contained more B-carotene than the (-) strains.

Benjamin (1959) outlined the physiological and morphological changes in zygospore formation. Accordingly, when two compatible hyphae come in contact and are stimulated to produce branches which grows towards each other. These branches, or progametangia, remain in contact at their apices, swell somewhat, becoming more or less clavate and delimit single gamatengia terminally by the formation of transverse walls. The remaining portion of the original progametangia is now known as the suspensor.The walls separating the two gametangia dissolve and form one large common multinucleate cell. The young zygosporangium is rather thin walled, enlarged rapidly and is thickened by the deposition of several concentric layers of wall material. In many species the outer layer may became irregularly thickened so that the zygosprangium at maturity, is conspicuosuly roughened with more or less elevated tooth like projection. The phenomena trisporic acid induces zygophores development in zygomycetes. The mycoparsite interaction of *Parasitellia parasitica* with its host *Absidia glauca* has some unusual features. The mycoparasite recognizes only complementary mating types of the host. Like in a mating reaction, the (+) type of *P. parasitica* infects exculsively the (-) type of *A. glauca* (Burgeff, 1924). An additional argument for physiological relations between mating and infection stems from the analysis of mating deficient mutants of *A. glauca,* some of which can no longer be infected with *P.parasitica* (Wostemeyer 1995 unpublished). The mating type dependency of infection led (Vetter *et al.,* 1994) them to the assumption that sexual phenomena system itself plays a role in host recognition. In order to gain access to this phenomenon at the molecular level; Vetter *et al.* (1994) isolated cDNA clones, derived from mRNAs which are specifically regulated by trisporic acid. Two clones hybridize with two size classes of mRNA, one of which is regulated in the Northern blots, where as the second one seems to be transcribed constitutively. In Southern blots with genomic DNA, two of the three DNAs hybridize with repetitiveDNA, the third one has at least two chromosomal copies. This was confirmed by earlier workers (Sutter, 1970; van den Ende *et al.*, 1972; Prakash and Sarbhoy, 1992). Raper (1952) defined the above chemical as a substance produced in one portion of the organism and transported by another means including diffusion to other portion of the same species responses. These hormones/chemicals are classified on the basis of their specific mode of actions into "erotactins", where movement are induced like spore attractants. Erotofins are those which induce the directional growth of the sexual structures in response to chemotrophline agents while "erogens" are those which control the induction and differentiation of sexual structures (Machlis, 1972). The mucoralean hormone is, therefore an erogen. Since the process of mating may involve the movement of mycelia over considerable distances, this movement was known as "zygotropin". Austin (1969) and his coworkers, isolated a hormone from *M. mucedo* and *B. trispora,* which was identical to trisporic acid. Although the production of zygospores is not clearly understood from a genetical point of view, zygospores are an excellent means of confirming the phylogenetic relationship between two strains. However, an incomplete mating reaction does not indicate a separate genetic line. This may be due to lack of suitable temperature, nutrient factor or compatible strains and age of culture (Schippr, 1978; Baijal and Mehrotra, 1980; Prakash and and Sarbhoy, 1994), though similarity in morphology of zygospores is related to phylogenetic entity. Conclusion ferreted out from the studies that the efficiency of zygospore formation depends upon the mating partners as coroborated by Blakeslee *et al.*,(1927) and Hesseltine and Anderson (1956). Schipper (1978) was of the opinion that in some *Mucor* species a fewer number of zygospores formed by using standard (+) and (-) strains. Few of the strains of *Rhizopus* which did not form zygospores even with the designated (+) and (-)strains are considered as neutral.

The use of scanning electron and transmission electron micrographs for exploring the zygospore morphology was introduced by Hawker and Gooday (1968) and later by Hawker and Beckett (1971). Schipper *et al.*(1975) studied in detail the zygosporic ornamentation in *Mucor* and *Zygorynchus.* Sarbhoy (1983) studied the zygospore ornamentation in different species of the genus *Absidia* in detail by light as well as scanning microscopic studies. Some of these characters were used in delimiting the species into proper taxonomic position.

The first description of the zygosporic stage of R.. *stolonifer* was given by de Bary and Woronin (1886) in cultures on fleshy fruits. Although many parthenospores and incomplete conjugation were also observed. Callen (1940) obtained"hybrid zygospores" when he contrasted the type strain of *R.sexualis* with (+) and (-) strains of *R. stolonifer* which were latter called as induced parthenospores by Schipper (1978). Hawker and Syrop (1973), Harris and Dennis (1980) and Schipper (1984) conducted temperature and relative humidity studies on the initiation of zygospores. *Rhizopus stolonifer* group possesses homothallic and heterothallic strains and within group morphological differences were observed. The ornamentation of the outer zygospore wall differs from tht of the *R. micosporus* and *R. oryzae* groups. Homothalllic strains were not observed in *R. oryzae* group (Schipper, 1984). Prakash and Sarbhoy (1992) were succesful in inducing the zygospores initiation in diferent species in *Rhizopus* with the help of retinol (Vitamin A) a precrusor of trisporic acid. The carotenes are also directly involved in the sexual procees in Mucorales (Austin *et al.*, 1969; 1970). They reported that trisporic acids, sex hormones in the Mucorales are oxidative products of β-carotene. Briboch *et al.* (1974) confirmed biosynthetic route of β-carotene to trisporic acid in *Blakeslee trispora.* Sporopollemin, identified in the zygospore wall of *Mucor mucedo* (Gooday *et al.*, 1973) is an oxgygenated polymer of carotenoids (Brooks and Shaw, 1968). The sporepollemin plays a role for increased carotenogenesis observed during sexual reporduction of many members of Mucorales (Gooday *et al.*, 1973). Fuch and Gooday (1978) reported the presence of sporopollemin in *Phycomyces blakesleanus*. Mathur and Sarbhoy (1994) reported that four to five fold increase of β-carotene production in intraspecific matings of *M. hiemalis F. hiemalis* and *Phycomyces blakesleanus* over that produced by either strain alone. The plus strains of the same species were better elaborators of β-carotene than (-) strains.

Nuclear Pattern

Mucoraceous nuclei are very notorious and small and present difficulties in the study of divisions through the usual cytological techniques (Cutter, 1942, 1946). The nuclear developmental patterns of Mucorales can be grouped into *Mucor, Rhizopus,* and *Sporidinia* types. In *Mucor* pattern, all the functional zygospore nucleus undergo karyogamy followed by immediate reduction in chromosome number prior to the onset of dormancy. Segregation of sex is complete in heterothallic species of meiosis. There is no clear cut fusion or reduction in azygous variety i.e. *Zygorynchus vuillemini* var. *agamous.* The *Rhizopus* pattern might have arisen from *Mucor* pattern by the restriction of karyogamy to certain favoured nuclear and deferement of meiotic process until the time of germination. The condition further led to *Phycomyces* pattern, where association of nuclei occurs early in the formation of zygote, the fusion being delayed until the time of germinatioin. This appears more specialised because all the fusion nuclei do not under go meiosis in the first generation but certain nuclei may not persist throughout the life cycle without karyogamy. The persistence of unfused nuclei correlated with only partial meiosis and possible degeneration of fusion nuclei may indicate the origin of *Sporidinia* pattern in which karyogamy is apparently obsolete and unfused nuclei are regularly present throughout the life cycle. Cutter (1942) observed delicate, long thread like mitochondaria in the developing sporangia of *Mucor genevensis* in the regions of active cytoplasmic movement. As soon as the cytoplasmic activity ceases there mitochondaria were observed in granular forms only. After a period of rest the zygospore may germinate by the formation of small germ tube which terminates in a germ sporangium (Cutter, 1942). More work is needed on the germination of zygospores of Mucorales so that a clear cut picture will emerge which will throw some light on the germination pattern of zygospores although Buckley and Sommer (1966) did some work on *Rhizopus stolonifer*.

Physiology

It is well known that all the natural substrates are not equally suitable for the growth of fungi. Many of the complex are insoluble or colloidally soluble viz. cellulose, starch and proteins. Before such compounds can be utilized, they must be changed into low molecular weight compounds which are soluble in water. This accomplished by enzymes. In nature, the fungi rarely come across as simple sugars. Schroter and Muller (1966) studied forty six strains of *Absidia corymbifera,* twenty four strains of *Mucor pusillus* and four of *M. miehei,* on the basis of biochemical tests (assimilation of 41 carbon sources, 3 nitrogen sources and vitamin dependence) and concluded that strains *A. ramosa* and *A. italania* were one single species. In *M. pusillus* and *M. miehei,* which are closely related morphogically and biochemical patterns are found to be similar Sarbhoy (1963) reported that the presence and absence of certain amino acids in several species of *Zygoryhnchus, Absidia, Mucor* and *Mortierella* can be taken as good subsidiary characters for distinguishing these genera and their species. It has also been observed that under different environmental and nutritional conditions, the principal morphological characters between the two species of *Pirella* remain unchanged (Mehrotra and Singh, 1970).

Mucorales are capable of synthesizing a large proportion of oil of low free fatty acids. Mehrotra and Nand (1971) were of the opinion that the fatty acids of Mucorales can be exploited as potential source of oil in industries. Gunasekaran *et al.* (1972) investigated the changes of lipid components during the germination the total lipids decreased then gradually increased during the later periods. The content of neutral lipids also decreased during germination, whereas the content of polar lipids increased. Mathur and Sarbhoy (1996) reported potential strains for the production of β-carotene and lipid identified.

The quantum of work on nutrition of Mucorales is not sufficient enough to reach any decisive conclusion. It is therefore, essential that several isolates from varied ecological conditions of the same species be studied with respect to their nutritional needs before these criteria can be of some use in classification of fungi. The aspects which need to be exploited to study the potential strains of lipid, carotene fermentation processes and other secondary metabolites which are useful for the mankind.

References

Austin, J. J., Bu' Lock, J. D. and Gooday, C. W. (1969). Trisporic acid biosynthesis and carotenogensis in *Blakeslea trispora*. Biochem.J., **113**:34.

Austin, J. J., Bu' Lock, J. D. and Drake, D. (1970). The biosynthesis of trisporic acids from β-carotene via retinol and trisporal. Experientia., **26**: 348-349.

Baijal, U. and Mehrotra, B.S. (1980). The genus *Cunninghamella*-a reassessment. *Sydowia.*, **33**: 1-13.

Benjamin, R.K. (1958). The merosporangiferous Mucorales. Aliso., **4**:321-433.

Benjamin, R.K. (1960). Two new members of the mucorales. Aliso., **4**:523-530.

Blakeslee, A. F. (1904). Sexual reporduction in Mucorinae. Proc. Am. Acad. Arts.Sci., **40**:205-319.

Blakeslee, A. F., Cartledge, J. L., Welch, D. S. and Bergner, A. D. (1927). Sexual dimorphism in Mucorales I. Intraspecific reactions. Bot. Gaz., **84**:27-50.

Brooks, C. J. W. and Shaw, G. (1968). Chemical structure of the exine of pollen walls and a new function for carotenoids in nature. Nature., **219**:532-533.

Buckely, P. M. and Sommer, N. F. (1966). Electron micrographs of germinating sporangiospores of *Rhizopus stolonifer*. Phytopath., **56**:872.

Bure geff, H. (1924). Untersuchungen ubersexualitat und parasitisumus bei Mucorinee I. *Bot. Abhamdl.* Goebel, **14**:1-135.

Callen, E.O. (1940) The morphology, cytology and sexuality of the homothallic *Rhizopus sexualis* (Smith) Callen. An. Bot., **4**: 791-818.

Currey, F. (1873) On a new genus in order Mucedines. J. Linn. Soc. Bot., **13**: 333, 518.

Cutter, V. M. Jr. (1942). Nuclear behaviour in the Mucorales-The Mucor pattern. Bull. Torrey. Bot. Club., **69**: 480-508.

Cutter, V. M. (1946). The genus *Cunninghamella* (Mucorales). Farlowia., **2**:321-343.

De Bary, A. and Woronin, M. (1866). Beitrage zur morphologie and physiologie der pilze II. Frankfurt am main.

Ellis, J. J. and Hesseltine, C. W. (1974) . Two new families of Mucorales. Mycologia., **54**:87-95.

Embree, R. W. (1959). Radiomyces, a new members of the Mucorales. Am. J. Bot., **4**:25-30.

Gams, W. (1973) Gliederuŋgs prinzipien in der Gattug *Mortierella*-Nova Hedwigia., **18**: 30-43.

Gooday, G. W. (1973). Differentiation in the Mucorales. Symp. soc. Gen. Microbiol., **23**: 269-294.

Gooday, G. W., Fawcett, P., Green, D. and Shaw, G. (1973). The formations of fungal sporollenin in the zygospore wall of *Mucor mucedoa:* role for the sexual carotegenesis in Muorales. J.Gen. Microbiol., **74**: 233-239.

Goodwin, T. W. and Griffiths, L.A. (1952). Studies in carotengenesis 5. Carotene production by various mutants of *Phycomyces blakesleanum* and *P. nitens.* Biochem. *J.*, **52**: 499-501.

Gunasekaran, M., Weber, D. J. and Hess, W. M. (1972). Changes lipid composition during spore germination of *Rhizopus arrhizus.* Trans. Br. Mycol. Soc., **52**:241-248.

Fuch, B. and Gooday, G. W. (1978). Sporopollenin in *Phycomyces blakesleanus* spore carotenoid polymer. Trans. Br. Mycol. Sco., **70**: 307-309.

Harris, J.E. and Dennis, C. (1980). Distribution of *Mucor piriformis, Rhizopus stolonlifer* in relation to their spoilage of strawberries. Trans. Br. Mycol. Soc., **75**:445-450.

Hawker, L.E. and Beckett, A. (1971). Fine structure and development of the zygospore of *Rhizopus sexualis* (Smith) callen. Trans. R. Phil. Soc. London, B., **263**:71-100.

Hawker, L.E. and Gooday, M.A. (1968). Development of the zygospore wall in *Rhizopus sexualis* (smith) callen. J.Gen. Microbiol., **54**:13-20.

Hawker, L.E. and Syrop, M. (1973). The effect of low temperature on ultrastructure of hyphae and young zygospores of *Rhizopus sexualis* (Smith) Callen. Protoplasma., **78**:57-68.

Hesseltin, C.W. (1954). The section genevensis of the genus *Mucor*. Mycologia. **46**: 358-366.

Hesseltine, C. W. (1955). Genera of mucorales with note on their synonymy. Mycologia., **47**: 344-363.

Hesseltine, C. W. (1961). Carotenoids in the fungi mucorales. *Tech. Bull.* U.S. Dept. Agr. No. 1245.

Hesseltine, C.W. and Ellis, J. J. (1964). The genus *Absidia: Gangonella* and cylindrical spored species of *Absidia.* Mycologia., **46**:568-601.

Hesseltine, C. W. and Ellis, J. J. (1966). Species of *Absidia* with ovoid sporangiospores. Mycologia, **48:** 761-785.

Hesseltine, C. W. and Ellis, J. J. (1973). In: The Fungi, ed. G.C. Anisworth, F.K. Sparrow, A.S. Sussman, 4 A. New York; Academic Press, pp. 187-217.

Hesseltine, C. W. and Anderson, R.F. (1956). The genus *Thamnidium* and a study of the formation of its zygospores. Am. J. Bot., **43**:696-703.

Leadbeater, G. and Mercer, C. (1956). Zygospores in *Piptocephalis cylindrospora* Bain. Trans. Br. Mycol. Soc., **39:** 17-20.

Leadbeater, G. and Mercer, C. (1957a). Zygospores in *Piptocephalis.* Trans. Br. Mycol. Soc., **40**:109-116.

Leadbeater, G. and Mercer, C. (1957b). *Piptocephalis virginiana* sp. Trans. Br. Bycol. Soc., **40**:461-471.

Linder, D. H. (1943). The genera *Kirkxella, Martensella, Coemansia.* Farlowia, **1**:49-77.

Machlis, L. (1972). The coming of age of sex hormones in plants. Mycologia., **64**:235-247.

Mathur, N. and Sarbhoy, A. K. (1994). Studies on Mucoraceous fungi-Taxonomy, lipid production with special reference to carotene. Ph. D. Thesis, IARI, New Delhi.

Mehrotra, B. S. and Baijal, U. (1963). A new species of *Dimargaris* from India. Can.J. Bot., **41**: 499-502.

Mehrotra, B. S. and Baijal, U. (1968). Is. *Blakselea* a valid genus? Jour. Elisha Mitchell. Sci. Soc., **84**:207-210.

Mehrotra, B.S. and Kakkar, R.K. (1970). *Coemansia aciculifera* Linder var. *suhagensis.* var. nov. from India. Sydowia, **24**:39-41.

Mehrotra, B. S. and Prasad, R. (1964). *Syncephalis depressa* from India. Mycologia., **56**:905-908.

Mehrotra, B. S. and Prasad, R. (196). Species of *Syncephalis* from India. Sydowia., **19**:112-116.

Mehrotra, B.S. and Prasad, R. (1967). Species of *Syncephalis* from India. Sydowia., **21**:125-129.

Mehrotra, B. S. and Prasad, R. (1969). An interesting new speceis of *Syncephalis* from India. Sydowia., **23:**92-94.

Mukerji, K. G. (1968). Fungi of Delhi: IV.A new species of *Piptocephalis* from India. Mycologia., **60:**326-330.

Nand, K. and Mehrotra, B. S. (1968). Species of *Pilobolus* and *Pilaria* from India. Sydowia, **22:**29-306.

Prakash, R. and Sarbhoy, A.K. (1993). *Mucor thermophilus* spec. nov. and other Mucorales from India. Zentrabl. Mikrobiol., **148:**531-534.

Raper, J. R. (1952). Chemical regulation of sexual process in the Thallophytes. Bot. Rev., **18:**447-545.

Sarbhoy, A. K. (1963). Nutritional studies on six member of Mucorales. III. Utilization of different mixtures of aminoacids. Llyodia., **26:**236-242.

Sarbhoy, A. K., Gridharilal and Varshney, J. L. (1974). Fungi of India. Supplement. IV. Navyug, Traders Publisher. 184 pp.

Sarbhoy, A. K. and Singh, S. N. (1981). Two new records of the genus *Coemansia* from India. *Zbl. bakt.* II. Abt. **136:**83-85.

Satina, S. and Blakseslee, A. F. (1926). Studies on biochemical differences between (+) and (-) sexes in *Mucors*-2. A preliminary report on the manoilov reaction and other tests. Proc. Nat. Acad. Sci., **12:**191-196.

Schipper, M. A. A. (1975). On *Mucor mucedo, M. flavus* and related species. Stud. Mycol. **10:** 1-33.

Schipper, M. A. A. (1978). On certain species on *Mucor* with a key to all accepted species. 2. On the genera *Rhizomucor* and *Parasitella*, Stud. Mycol., **17:** 1-71.

Schipper, M. A. A. (1984). A revision of the genus *Rhizopus* I. The *R. stolonifer* group and *R.oryzae.* Stud. Mycol., **25:**1-19.

Schipper, M. A. A. and Stalpers, J. A. (1984). A revision of the genus *Rhizopus* II. The *Rhizopus microsporus* group. Stud. Mycol., **25:**20-34.

Schipper, M. A. A. , Samson, R. A. and Stalpers, J. A. (1975). Zygospore ornamentation in the genera *Mucor* and *Zygorynchus.* Persoonia., **8:**321-328.

Schroeter, J. (1886). Mucorinae. In cohn, Kryptogamen Flora Von Schlesien, **3:**198-217.

Sutter, R. P. (1970). Effect of light on β-carotene accumulation in *Blakeslea trispora.* J. Gen Microbiol. **64:**215-221.

Tandon, R. N. (1968). Mucorales of India. ICAR Publication pp.117+12. Appendix II.

Upadhaya, H. P. (1973). *Helicostylum* and *Thamnnostylum* (Mucorales). Mycologia, **65:**733-735.

Van der Ende ***et al.*** (1972). Cited: Characterization of specific cDNA clones of the zygomycete. *Parasitella parasitics,* derived from mRNAs which are regulated by the pheromone trisporic acid. Microbiol. Res., **149:**17-22.

Vanova, M. (1968). Contribution to the taxonomy of the genus *Absidia* (Mucorales) I. *Absidia macrospora* sp. nov. Ceska Mykologici, **22:**296-300.

Vanova, M. (1969). Contribution to the taxonomy of the genus *Absidia* (Mucorales) II. A reapprisal of the characters. Ceska Mykologici. **23:**187-189.

Vant ieghem, P. (1876). Troisieme memoire sur les Mucorines. *Ann. Sci. Nat. Ser.* IV. Botan., **4:**312-398.

Vetter, M., Wostemeyer, J. and Burtmester, A. (1994). Cited: Characterization of specific cDNA clones of the zygomycete *Parasitella parasitica,* derived from mRNAs which are regulated by the pheromone trisporic acid. *Microbiol. Res.,* **149:**17-2.

Zycha, H. (1935). Pilze II. Mucorineae, in *Kryptogamenflora der Mark branderburg*. pp. 264. Leipzing.

Zycha, H., Sepmann, R. and Linnemann, G. (1969). Mucorales. Verlag von J. Cramer. pp. 355.

Mehrotra, B. S. and Prasad, R. (1969). An important new species of [illegible] from India. Sydowia 23:[illegible].

[illegible] (19[illegible]). [illegible] A new species of [illegible] from India. [illegible]

[illegible] and Mehrotra, B. S. (1968). [illegible] from India. [illegible]

[illegible] and [illegible] (19[illegible]). [illegible] new and other [illegible] from [illegible].

[illegible] (19[illegible]). Chemical [illegible] in the [illegible]

[illegible]

[illegible] and [illegible] (19[illegible]). [illegible] of India. [illegible]

[illegible]

[illegible]

[illegible] (19[illegible]). [illegible]

[illegible]

[illegible] and [illegible] (19[illegible]). [illegible]

[illegible] (19[illegible]). [illegible]

[illegible] (19[illegible]). [illegible]

[illegible] (19[illegible]). [illegible] Mycologia [illegible]

[illegible]

[illegible] (19[illegible]). [illegible]

[illegible] (19[illegible]). [illegible]

[illegible]

[illegible] and [illegible] (19[illegible]). [illegible]

[illegible] (19[illegible]). [illegible]

26

Loculoascomycetes from India

Ch. Ramesh

Dept. of Botany, Karnatak University, Dharwad-580 003

Fungi are eukaryotic microorganisms which have now been given the rank of separate kingdom. The Ascomycotina is the largest and complex group of fungi. Due to diversity in pattern morphology of the ascomata and highly heterogenous nature, these fungi have posed a great problem to the taxonomists. The Loculoascomycetes comprise a group segregated from the Euascomycetous forms on the basis of their bitunicate asci and ascostromatic ascomata. Loculoascomycetes appear as saprotrophs, paratrophs or hyperparasites of superficial fungi and insects, as internal parasites fruiting on green leaves and stems, as parasites fruiting on dead leaves and stems or as saprotrophs on dead leaves, herbaceous stems, wood, dung and plant debris. Solutions of residual taxonomic problems will definitely unravel the status presently known members of this ecological group of fungi. Such will first ensure a sound restructuring process of available information on Loculoascomycetes taxa. Taxonomic stability will also pave the way for valid new discoveries of interest.

Key words: Loculoascomycetes, occurrence, taxonomy, characteristics

Fungi are eukaryotic heterotrophic microorganisms which have their own status. They differ from all other groups of organisms on this planet on account of their peculiar life style, physiology, biochemistry and exclusively absorptive mode of nutrition. They are known to colonize, multiply, sporulate and live in or on water, soil, leaf surface, litter, dead organic matter, living plants and animal parts, keratin substances, fruits, insects, submerged, skeletonized leaves and other organic substrates.

Hawksworth (1974), Booth (1971 a, b), Dennis (1978) and Sutton (1980) gave detailed account of techniques for collection, culture and preservation of micro fungi. Kendrick *et al.* (1979) worked for methods of establishing a connection between anamorph and telomorph.

The Ascomycotina is the largest and complex group amongst fungi. Due to diversity in pattern morphology of the ascoma and highly heterogenous nature, these fungi have posed a great problems to the taxonomists. Lindau (1897) was the first to attempt a classification of Ascomycetes, chiefly based on gross morphological characteristics. While proposing a classification he recognized three types of ascocarps, namely 1) Cleistothecium in which the ascocarp is completely closed and asci and ascospores are released by the disintegration of ascocarp wall. 2) Perithecium-flask shaped, with true ostioles for the discharge of asci and ascospores and 3) Apothecium or disc shaped ascocarp in which the asci are arranged in parallel rows on the exposed surface. These ascoma types formed the main basis of three series within the Euascomycetes.

A great deal of work has since been done on the developmental pattern of centrum, cytology and sexuality, nature of ascus wall and manner of dehiscence and origin and nature of interthecial threads, which provide criteria of fundamental importance in the proper understanding and rearrangement of this highly heterogenous group of fungi. On this basis two distinct series are recognized namely the 'Ascoloculares' and 'Ascohymeniales' as proposed by Nannfeldt (1932), pseudothecium and perithecium of Miller (1949), series of 'bitunicatae' and 'unitunicatae' of Luttrell (1951) and Protunicatae and Eutunicatae of Gaumann (1964) based on the fundamental characters as stated above. Various investigators have contributed a great deal to these fundamental aspects, which eventually helped in formulating of phylogenetic system of classification in place of the artificial grouping, thus shifting the emphasis in the classification of these fungi from the outward form and morphology to the more fundamental criteria derived from a study of ascocarp "centrum" as originally defined by Miller (1949).

The Ascomycetes are divided into three subclasses, Hemiascomycetes, Euascomycetes and Loculoascomycetes. The Loculoascomycetes are comprised of groups segregated from the Euascomycetes primarily on the basis of their bitunicate asci and ascostromatic ascocarps.

Loculoascomycetes appears superficial epiphytes, parasites or hyperparasites of fungi and insects, as internal parasites on green leaves and stems, as parasites fruiting on dead leaves and stems or as saprotrophs on dead leaves, herbaceous stems, wood, dung and plant debris.

General Characteristics

A) Bitunicate Asci:

The primary character of the Loculoascomycetes is presence of bitunicate ascus. The ascus wall consists of two layers which are separable in the normal course of ascospore discharge as a thin, inextensible outer layer or ectoascus, and a thick extensible inner layer or endoascus. The very young ascus is thin walled. During most of its development, however, the endoascus is conspicuously thickened, especially towards the apex where it is indented or partially penetrated by a tubular channel forming a sub apical chamber into which the protoplast extends. This marks the pore through which the ascospores are later ejected. Although the endoascus becomes thinner as the ascus matures, the sub apical chamber characteristically reminds visible as an indentation in the thicker apex. As the ascus swells prior to dehiscence the entire wall appears uniformly thin.

At dehiscence the bitunicate structures becomes apparent. The ectoascus splits at the apex. In some species it ruptures circumstantially throwing off the apex as a thimble shaped cap. The endoascus expands as a tubular extension which doubles or triples the original length of the ascus. The margin of the ruptured ectoascus is visible as a faint line around the ascus. The margin of the ruptured ectoascus is visible as a faint line around the ascus below a slight or definite bulge in the freed endoascus. The wall of the expanded endoascus is uniformly thin. The ascospores move up in the endoascus. One advances into the elastic pore in the tip and hangs there briefly at its widen diameter. Spores constricted at the septa are trapped momentarily at a constriction. As the first spore is shot away the second spore moves into the pore and, as pressure rebuilt is in turn shot away, the ascus tip recoiling slightly at each discharge. The whole component of spores may be ejected in rapid succession or there may be longer pauses between spores or between groups of spores. After all spores are discharged pressure is rapidly released and the ascus shrinks. The endoascus swells until only a narrow lumen remains, and the sub apical chamber is again evident. The ectoascus is now readily apparent as a wrinkled collar constructing the endoascus.

In many Loculoascomycetes, such as *Sporormia* spp. the entire process of ascus dehiscence and ascospore discharge occurs normally in detached asci or in sections of ascocarps. In water mounts, it may be observed easily under the microscope.

In *Botryoshpacria* spp., the endoascus begins to swell and dissolve as soon as it emerges from the ectoascus, and the ascospores have often been described as oozing from the ascus in a gelatinous sheath. In such species, placing single ascocarps or a bit of substrate bearing ascocarps on squares of moistened filter paper in the lid of petri dishes and inverting them over an agar plate will demonstrate that ascospores are forcibly discharged. The irregular pattern of groups of 8 ascospores on the agar resulting from the discharge of spores single and with greater or lesser intervals between discharges is presumptive evidence that the ascus is bitunicate. Close groupings of 8 ascospores, on contact, is characteristic of unitunicate asci, in which spore discharge usually is simultaneous. Even in unitunicate asci with narrow pores, such as those of the Clavicipitaceae which discharge their spores successively, the spores follow one another quickly and smoothly without a pause and readjustment of the ascus between spore discharges typical of Bitunicate asci. The spores consequently follow more merely similar trajectories and usually are more closely grouped on the agar. Spore patterns may be easily recognized if the plates are examined after only a few asci have discharged. On a thin clear agar, spores may be observed by inverting the plates on the microscope stage and focusing on the surface of the agar through the bottom of the dish. The target area should be marked on the bottom of the dish so that the lid bearing the ascospores can be rotated.

Ascospore discharge often may be observed directly under the microscope or stereomicroscope by mounting moistened whole ascospores or substrate bearing ascospores on dry slide or at the edge of the moistened squares of filter paper so that the ostioles or hymenial surfaces can be focused on in profile. If a number of mature asci are present, discharge of a spore may be observed during the short period before the mounted ascospores become too dry.

A characteristic structure of most bitunicate asci is the 'Nasse' (Chaedfaud, 1960) formed in the peripheral layer of the cytoplasm against the walls of the subapical chamber. This usually has the form of four vertically arranged refringent rods on staining with cotton blue. The rods may branch and anastomose to form a network. In a few genera of Loculoascomycetes such as *Pyrenophora* and *Leptosphaerulina* the sub apical chamber of the ascus has a flange like extension into the lateral walls of the endoascus and some stages of ascus development is characteristically hat shaped or trilobate in optical longitudinal section. In *Pyrenophora* and *Wettsteinina,* a non-amyloid annulus staining with blue-black inks and congo red is differentiated within the endoascus wall surrounding the sub-optical chamber (Errikson, 1967). A refractive apical apparatus is present also in *Didymoshaeria rhizosphere* (Kohlmeyer and Kohlmeyer, 1967) and in *Otthia prumi* (Munk, 1957). Generally the presence of more complex apical structures such as the refractrive chitinoid annulus of the Diaporthaceae and Sordariaceae the double amyloid annulus of the Xylariaceae and the cap like apical plug of the Clavicipitaceae or a positive reaction to iodine are characters that indicated exclusion from the Loculoascomycetes. The completely structured asci *Melogramma spiniferum* and of *Helicascus,* which in other characters seen related to Euascomycetes, however, have been described as bitunicate (Doguet, 1960; Kohlmeyer, 1969a). The bitunicate structure of the ascus is the most useful single characters in placing a fungus in the Loculoascomycetes. Structures of mature and immature asci alone, even in long dried herbarium specimens is usually sufficient for identification.

B) Ascostromatic Ascomata:

Historically the Loculoascomycetes (Luttrell, 1955) first as the Dothidiineae of Theissen and Sydow (1918) and later as the Ascoloculares of Nannfeldt (1932) have been distinguished from Euascomycetes primarily on the basis of the ascostromatic nature of the ascospore. In Euascomycetes the ascogonia, whether free or in a stroma, become surrounded by an envelope of hyphae which develops into the peridium of the ascospore. In Loculoascomycetes the ascogonia appears within a stromatic initial which develops directly into the ascospore and the asci lie in unwalled locules in the stroma. The ascostromatic nature of the ascoma is most clearly apparent, and earlier recognized in the Myriangiaceae in which the asci are individually distributed in the stroma. In the Dothideaceae, the multiloculate stroma contains groups of asci in locules which obviously are not separated from the surrounding stroma or may be reduced to a thin wall and globose or flask shaped, solitary, pseudothecium which is indistinguishable from the solitary perithecium of the Pyrenomycetes. Tissue types occasionally are helpfull to differentiate since there is a tendency among Loculoascomycetes to produce parenchymatous stromal tissues. This is expressed in the vertically oriented rows of angular cells in the ascostromata of many of the Dothideaceae and Leptopeltidaceae and in the origin of the ascocarp initial . Through three-dimensional division of a single cell or a few cells in fungi such as *Sporormia.*

Although the ascostromatic ascocarp is the fundamental feature of the Loculoascomycetes, it is not an exclusive character. A meristogenous origin of the ascocarp similar to that in *Sporormia* occur in some fungi which are generally classified in the Euascomycetes (Barr, 1956).

C) Diagnostic Types of Internal Structure:

Although no single type of ascoma structure is an index to all Loculoascomycetes, the structural types found in Loculoascomycetes generally are distinct from those characterizing the various groups of the Euascomycetes.Three major types exhibited in the Myringiales, Dothideales and Pleosporales distinguish the majority of the Loculoascomycetes.

In the Myringiales, the asci are scattered individually and usually at several levels in relatively unaltered stromal tissue. The ascostromata are crustose, pulvinate, discoid or subglobose and asci are distributed throughout the stroma or in restricted fertile regions. The asci are globose and thick walled. They expand individually through the soft end or crumbling stromal tissue to discharge their spores above the surface.

In the Dothidcales the ovate to short cylindrical asci typically are in fascicles in small perithecoid locules in uni or multi-loculate ascostromata. They push up as a group into disintegrated centrum parenchyma and are paraphysate. The locules open at the apex through ostiloes which are lysigenous or less commonly schizogenous and periphysate. The periphyses may be abundant and in some species such as *Scirrhia acicola, Mycosphaerella ascophylla* (Kohlmeyer and Kohlmeyer, 1968), *Metacapnodium juniperi* (Corlett, 1970) and *Pleurisperma dahlbergiae* (Srivanesan, 1970 a), periphyses like hyphae projecting inwarded from the laterals of the pseudothecium occupy most of the space above the developing fascicle of asci. The perithecial centrum of the Diaporthales (Euascomycetes) is similarly parenchymatous, but the more numerous, unitunicate asci have refractive annulus in their apices and become detached to fill the perithecial cavity.

In the Pleosporales, the clavate to long cylindrical asci typically occur in broad basal layers in large perithecoid locules in uni or less commonly, multiloculate ascostromata. The ostioles are lysigenous or commonly schizogenous and periphysate. The asci are interspersed with pseudoparaphyses. Even at maturity the attachment of pseudoparaphyses at the apex of the locule can be established and this will distinguish them from the paraphyses of the Euascomycetes. In some of the smaller pleosporaceous fungi the pseudoparaphyses are sparse and the asci relatively few.

A fourth structural type is represented by the Patellariaceae in which the ascocarps are essentially identical with apothecia. In *Rhytidhystenium rufulum* ascognonia arise in a stroma. A palisade of vertically oriented paraphysoidal hyphae attached at the both base and apex develops across the upper part of the stroma, and asci grow up among the paraphysodial hyphae. The overlying stromal tissue splits and as the ascoma expands to its mature apothecial shape, the apical ends of the paraphysoidal hyphae separate from overlying stromal tissue. Their freed tips regenerate and produce numerous short branches with swollen tips which form an apothecium over the asci. At maturity, there is little to distinguish these paraphysoidal hyphae from the paraphyses of Discomycetes except their tendency to produce lateral branches at various points among their length. In other possible related, forms such as Pseudoscypha (Reid and Pirozynski, 1966), the paraphysoidal hyphae arise from the surface of a flat basal stroma or subiculum. Their tips are free from the beginning and at present there is no basis for distinguishing them from paraphyses. Presence of the bitunicate ascus is the primary reason for including these fungi in the Loculoascomycetes and ascus structure itself has been inadequately studied in this group.

D) Septate Ascospores:

Unicellular ascospores occur in relatively few genera of Loculoascomycetes. If the ascospores are unicellular, the odds are perhaps nine to one that the genus belongs in the Euascomycetes and even higher if the spores are globose or provided with germ slits or poses or conspicuously ornamented. None of the Loculoascomycetes has allantoid spores. If the ascospores are single septate, the odds are approximately two to one that the genus is in the Loculoascomycetes. If the ascospores are phragmosporous or dictyosporous the odds are nine to one or higher that the genus is in the Loculoascomycetes.

Classification

In 1897, Lindau established the Order Dothideales for Ascomycetes in which the fruit bodies are without peridia, formed in a stroma type of structure that had long been recognized in *Dothidea* and similar fungi. The nucleus of the Loculoascomycetes appeared in 1907 when Hohnel expanded Dothideales to include the Myriangiaceae, Dothidiaceae and Pseudosphaeriaceae and recognized the existence of simple perithecoid forms for which he created the Pseudosphaeriaceae. Theissen and Sydows (1918) erected Dothidiineae in 1918 to comprise the Myriangiales, Dothideales and Pseudosphaeriales as an extension of Hohnel's concept which Hohnel himself opposed. Theissen (1916, 1919) attempted to relate the three orders of Dothidiineae on the basis of the common character of forming asci in mono-ascus locules. Subsequent development studies have not supported Theissen's concept but his ideas like those of Hohnel were of significance to ascomata. Nannfedt (1932) further expanded the Dothidiineae to include the Hemisphaeriales and the Hysteriales and proposed the name Ascoloculares for this group, which he recognized as a series phylogenetically distinct from Plectomycetes, Pyrenomyctes and Discomycetes. Nannfeldt noted additional characters of the Ascoloculares with the thick walled, bitunicate asci and the tendency to form the parenchymatous tissue and dictyosporous ascospores and conidia. Luttrell (1955) gave this group the rank of subclass as the Loculoascomycetes, transposing the syllables of Ascoloculares to provide a name corresponding in form to those of other subclass of Ascomycetes. The concept of Loculoascomycetes, therefore, represents a current opinion. The validity of this concept is being challenged, most significantly by Chadefaud (1960) and subjected for continuing critical examination.

Historically the Loculoascomycetes (Luttrell, 1955), first as the Dothidiineae of Theisen and Sydow (1918) and later as the Ascoloculares of

Nannfeldt (1932) have been distinguished from Euascomycetes primarily on the basis of the ascostromatic nature of the ascoma. It is obvious from as examination of compound ascoma of forms such as *Dothidea*, in which the asci occur in clusters in numerous unwalled locules in a large stroma that ascostromatic ascomata exit.

Because of intergradations even within species, it is also obvious that the ascostroma may be reduced to a uniloculate perithecoid structures. The difficulty is in distinguishing simple pseudothecia from simple perithecia. Miller (1928, 1949) and Nannfeldt (1932) attempted to distinguish the pseudothecia of Loculoascomycetes from the Euascomycetes on the basis of ontogeny. In pseudothecilaes forms the ascogonia appear within a stromatic initial which develops directly into the ascocarp and asci which lie in unwalled locules in the stroma.

In euthecial forms, the ascognoium, whether free of within a stroma, becomes surrounded by an envelope of hyphae which develops into the peridium of the ascoma. Even if this ontogentic definition of a pseudothecium is accepted as a generally useful one, the formation of ascostromatic ascomata is not an absolute index to the Loculoascomycetes. Although the ascomata of the Coronophoraceae appeared to be ascostromatic, Nannfeldt (1932) correctly excluded this family from the Loculoascomycetes and ascostromatic ascocaprs have been found in other groups of Euascomycetes such as the Plectascales (Barr, 1956) and the Chaetomiaceae (Whiteside, 1962). On the other hand ascocarps of a few fungi such as *Cochliobolus* (Shoemaker, 1935) which are generally included in the Loculoascomycetes don't fall within Miller's (1928, 1949) ontogenetic definition of an ascostroma.

Theissen (1916, 1919) attempted to define the Loculoascomycetes in terms of common developmental patterns, characterized by the formation of asci in Monascus locules, but failed. However, it is necessary to guard against the tendency to define structural types in terms of the group and the group in terms of structural types- for example to define pseudoparaphyses as vertical hyphae occurring in the centrum of Loculoascomycetes and Loculoascomycetous fungi with pseudoparaphysis.

Although the morphology of the basidium has been made the primary basis of classification in the Basidiomycets, ascus morphology has been employed hardly at all in the classification of the Ascomycetes except for the separation of Discomycetes. Asci are divided into two major types the bitunicate type in which the ascus is surrounded by two distinct walls and the unitunicate type in which there is a single wall.

The primary character of the Loculoascomycetes is the bitunicate ascus.

The bitunicate ascus is the most useful criterian of the Loculoascomycetes because it is clearly apparent in simple squash mounts of mature ascomata. It is also most consistent character of the Loculoascomycetes. Munk (1962) warned that any group set up on the basis of a single character must atuomatically prove to be uniform for this character.

The term "bitunicate" was introduced and defined by Luttrell (1951). The bitunicate ascus is surrounded by two distinct walls, a thin inextensible outer wall and a thick extensible inner wall. The outer wall ruptures at the apex, and the inner wall expands to form a long cylindrical sac. Typically the ascospores are ejaculated successively through an elastic pore in the apex of the expanded inner wall.

The bitunicate ascus was first reported by Currey (1856) in *Pleospora herbarum* and by Pringsheim (1858) in *Pleospora scirpicola* and was fully described by De Barry (1887) who summarized the observation that had been made at the time. Good descriptions have been given by Hodgetts (1917) of *Leptoshaeria acuta* and by Hoggan (1927) of *Plowrightia ribesia.* Luttrell's definition was based on his own experiments. However, this definition needs further examination in relation to contemporary information. Various workers have examined the development and morphology of the bitunicate ascoma. Electron microscopic studies revealed that the walls i.e. ectotunica and endotunica are composed of more than one layer, and therefore the bitunicate ascus may be considered to be multi-layered. However, the unique behaviour of the two walls is confirmed and explained by their characteristic arrangements of wall elements. Therefore, the term bitunicate doesn't seem to be inaccurate. The significance of the bitunicate ascus became apparent after the Loculoascomycetes had been assembled on the basis of other criteria. It has been suggested that the thickened wall of the bitunicate ascus is a response to development within a stroma. However, there is no evidence that the delicate internal tissue of pseudothecia offer greater resistance to expansion of asci than do those of euthecia or that wall thickness is related to ability of asci to penetrate tissue. On the contrary, cells capable of expanding and exerting pressure are commonly thin walled. Although the bitunicate ascus seems to be the most fundamental character of the Loculoasceomycetes, its critical value remains open to question. In an apparently Lolculoascomycetous genus *Cochliobolus,* the asci seem to be unitunicate. On the other hand bitunicate asci occur in the Patellariaceae and there is as yet no indication of how the ascocarps in this family are to be distinguished from the apothecia of many operculate Discomycetes that also develop from stromatic initials . Attempts to arrive at an exclusive definition of the Loculoascomycetes have proved disappointing. No criterion that has been suggested is without exception. In the Ascomycetes it is necessary to define groups in terms of the complex characters exhibited by typical members and to acknowledge to existence of forms of more or less uncertain relationship, which can be appended only doubtfully to any group.

The Loculoascomycetes are resistant to organization into orders showing patterns of correlated characters. The order Myriangiales centered on forms such as *Myriangium* and *Elsinoe* with pulvinate ascostromata containing globose asci regularly distributed in monoascus locules persisted through all classification. Theissen and Sydow (1918) placed pluriloculate forms in the Dothideales and unilocular forms in the Pseudosphaeriales. Luttrell (1951, 1955) recognized the Dothideales with asci having a paraphysate clusters and the pleosporales for those with asci among pseudoparaphyses. Muller and Von Arx (1962) retained the pseudosphaeriales for forms with perithecooid pseudothecia and erected the Dothiorales for these with apothecioid pseudothecia. As per the first half of last century any discussion of the classification of the pyrenomycetes had to proceed from Lindau's systemand further based on monumental revision of Loculoascomycetes and Pyrenomycetes (Muller and Von Arx, 1962; Von Arx and Muller, 1954, 1975; Barr, 1976b, c, 1979).

The composition and relationship of the Loculoascomycetes are shown diagrammatically in Figs. 1 and 2 (Luttrell, 1955). Luttrell (1951) has given the reasons for considering the Loculoascomycetes distinct from the true pyrenomycetes. It is possible that ascostromatic forms could be derived from true pyrenomycetes at many points by reduction of perithecia and develop within a stroma and this process probably has occurred for example, in the Coryneliales and Cornomphorales. However in order to derive the species in Loculoascomycetess from the pyrenomycetes by this process it is necessary to assume that the bitunicate ascus also developed at numerous lines and coincidentally with the loss of perithecial walls. This seems too great coincidence to be credible more so that the species with bitunicate asci represent a separate monphyletic line. This conclusion was expressed by Luttrell when the created this subclass.

Objections to this taxonomic hypothesis was noted previously by Luttrell (1951). Now it seems less serious. These were the apparent lack of correlation between the production of bitunicate asci and ascostromatic ascocarps in several species and groups. It was the occurrence of bitunicate asci in forms considered to be Discomycetes such as Hysteriaceae, e.g. *Keithia juniperi* Miller and species of *Lecanidion* and *Tryblidiella*. Luttrell (1953) studied and showed that ascoma in *Glonium stellatum* and presumably in other members of the

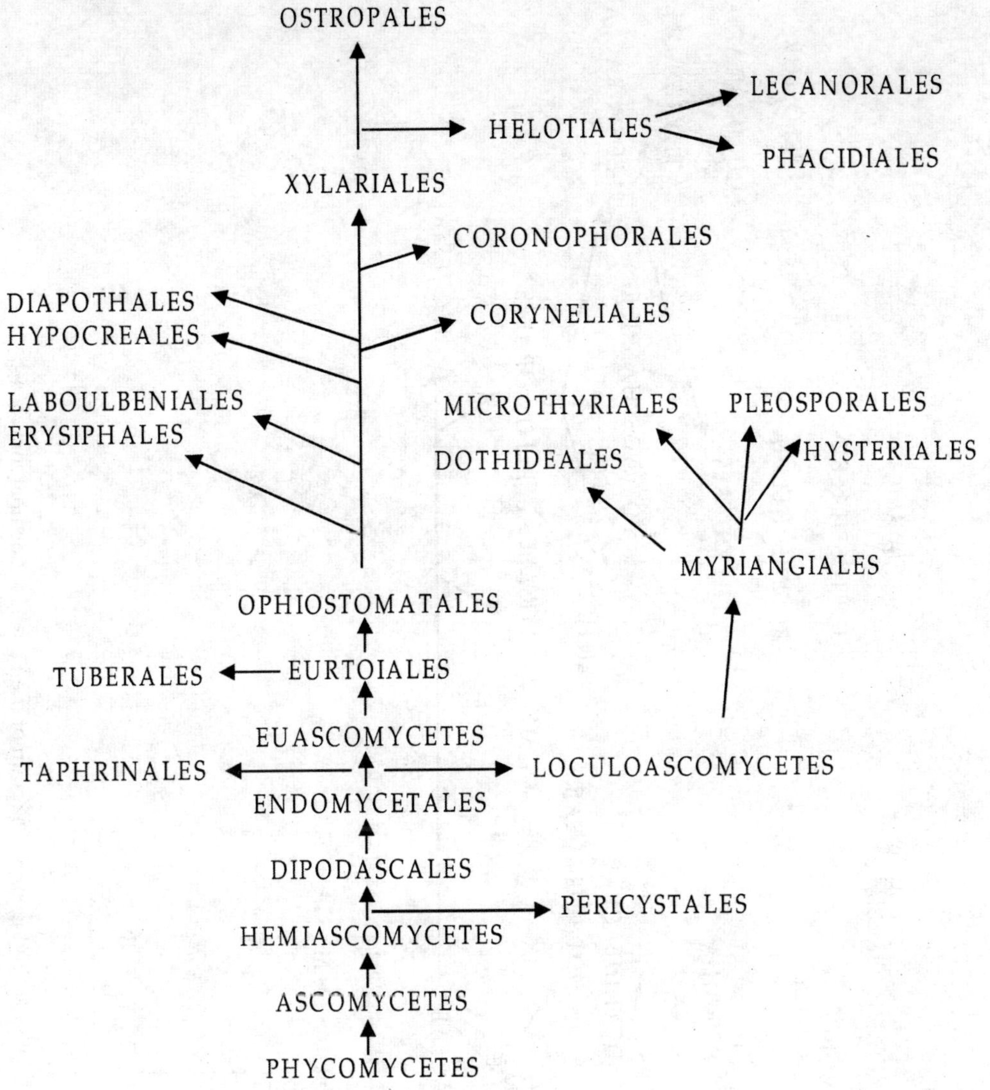

Fig. 1. Phylogenetic Diagram of the Asocmycetes

Hysteriaceae is an ascostromata. Pantidou and Korf (1954) found that the formation of an ascostromatic ascocarp was correlated with the production of bitunicate asci in *Keithia juniperi* this species doesn't belong to the Discomycetes genus *Keithia* but is synonymous with *Coccodothis sphaeroidea* The and Syd., a species which has been included among the ascostromatic forms in the Pseudosphaeriales. Pantidou and Korf's report is of special importance, since in this case the presence of the bitunicate ascus served as a

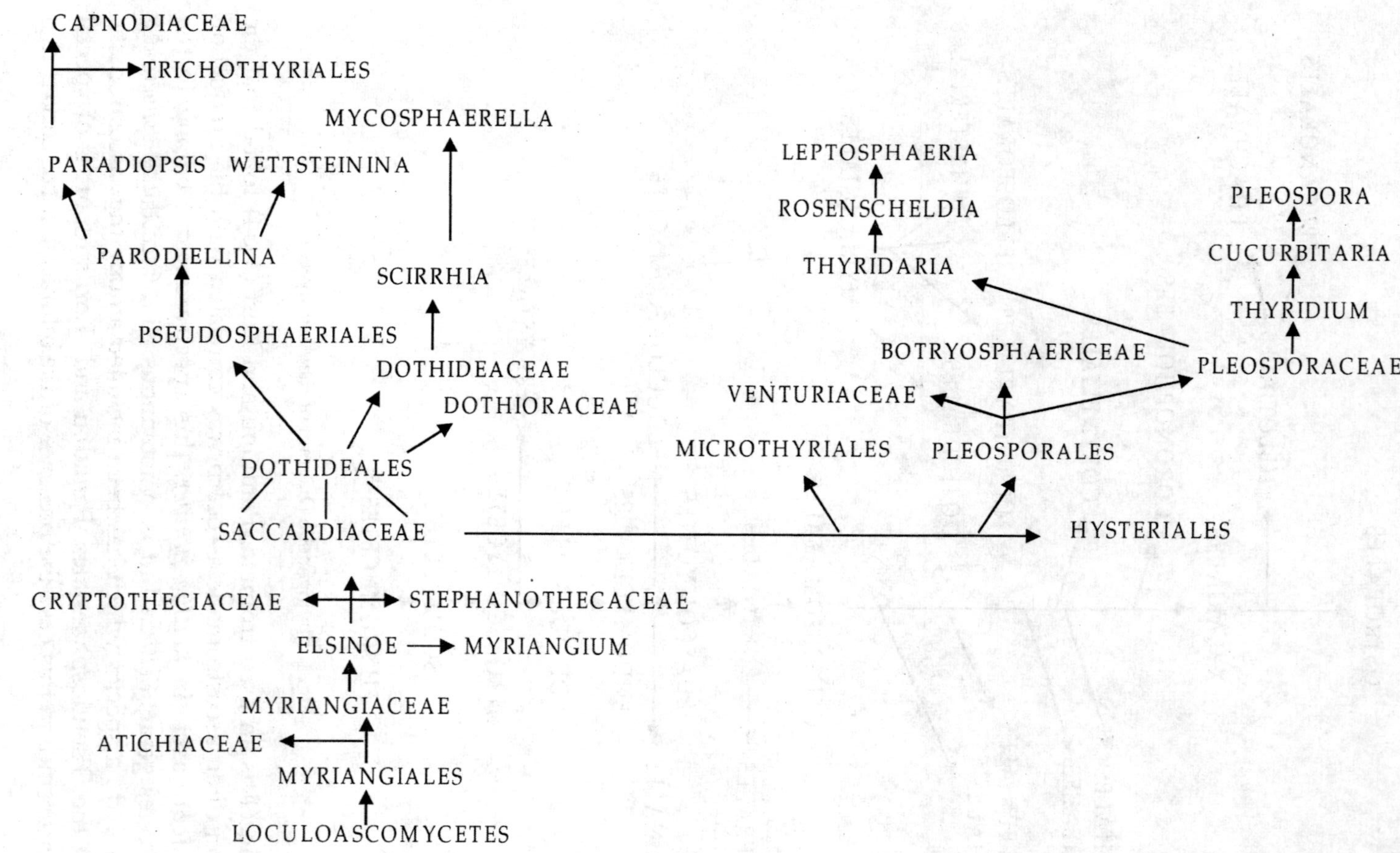

Fig. 2. Phylogenetic Diagram of the Sub-class Loculoascomycetes

valid indication of the correct taxonomic position of a fungus which otherwise had previously been misinterpreted. The transfers of *Coccodothis* back and forth from the Patellariaceae to the Pseudosphaeriales (Muller and Sanwal, 1954; Petrak, 1954) emphasized the importance of developmental studies to determine the nature of the ascoma in species of these genera. The formation of ascostromatic ascomata in species with unitunicae asci was unique particularly those in the Coryneliales and Cornomophoales. This was clearly indicated in *Caliciopsis pinea* of Coryneliaceae.

As mentioned previously, it is quite possible that an ascostroma could be produced by reduction of the perithecial walls in stromatic pyrenomycetes. This process apparently has occurred in the Coryneliales and Coronophorales. The members of these orders have arrived at an ascostromatic structure resembling that of the Loculoascomycetes but by a different path being derived from perithecial forms in the pyrenomycetes. They are definitely excluded from the Loculascomycetes by their unitunicate asci. It is significant that without mentioning the occurrence of unitunicate asci as the reason, perhap Sintuitively excluded the coronophorales from the true Pyrenomycetes by different authors.

The ascomycotina fungi belonging to the Loculoascomycetes have been arranged in four orders (Nannfeldt, 1932), five by Luttrell (1951, 195, 1973) three by Von Arx and Muller (1975). Barr (1976a) divided the class into eight orders (1977 unpublished) and prepared a set of mimeographed keys to orders and recognized families of Loculoascomycetes.

The classification followed by Luttrell (1973) was based on morphological data along with developmental studies, whereas Muller and von Arx's family keys were based on ecological characters. We feel it evident that morphological data must be fully utilized in a satisfactory system of classification.

The Loculoascomycetes probably arose from forms in the Hemiascomycetes (Fig. 1), although none in the later group are known at present gives any indication of the origin of the bitunicate ascus. Forms with bitunicate asucs developed a tendency to produce their ascogonia within a stroma. The ascogenous hyphae penetrated the stroma and formed ovoid asci in individual locules scattered at different levels in the undifferentiated stroma. The species presently known, which most closely approach the primordial forms are found among the lower Myringiales in *Elsinoe* and *Atichia* (Fig. 2). The genus *Myriangium* represents a specialized side branches from the *Elsinoe* type in which the stroma becomes more massive and differentiated into sterile and fertile regions.

About 1764 species of Loculoascomycetous fungi are known from terrestrial habitat throughout the world (Von Arx and Muller, 1975) but in contrast to this, only 51 species have been described from oceans and estuaries (Kohlmeyer and Kohlmeyer, 1979) nearly about 630 species are reported from India (Bilgrami *et al.*, 1979, 1981, Ramesh, 1985). The marine Loculoascomycetes are well represented in intertidal habitats (Kohlmeyer, 1969b) and a number of them have been collected on substrates continuously submerged in the sea (Jones, 1968; Bryne and Jones, 1972). The majority are the members of the Pleosporaceae and belong to the well known terrestrial genera, *Pleospora, Leptosphaeria, Sphaerulina* and *Didymella*. Other genera including *Paraliomyces, Kymadiscus, Manglicola* and *Pontoreia* appear to have only marine species.

Most of the fungi in this group are represented in the terrestrial environment and new genera and species are constantly added to the list. As contrast to the voluminous work carried out in the field of Ascomycetes especially Plectomycetes and Pyrenomycetes by galaxy of eminent mycologists (Bilgrami *et al.*, 1979, 1981), little attention has been given to the study of sub class Loculoascomycetes. In India some reports have been published viz. Wehmeyer (1961) reported the presence of Loculoascomycetes from Punjab and Kashmir. Sivanesan *et al.* (1974) described a new genus *Bitrimonospora* with *B. indica* as type species from Marathwada region (Maharashtra). Sivanesan *et al.* (1976) described another new genus *Thaxteriellopris* with *T. lignicola* as type species from Mount-Abu (Rajasthan). Anahosur (1969, 1970) and Dharne and Muller (1969) reported some Ascomycetes from Maharashtra. Srinivasulu (1972) described a new genus *Tilakiella* from Nasik. Patil and Ramesh (1983, 1984, 1986, 1987) and Ramesh (1986, 1987a, b, c, d, e, 1988a, b, c, 1990a, b, c, 1992, 1993a, b, c, d, e) reported Loculoascomycetes from Maharashtra .

Studies on marine Loculoascomycete were undertaken from Maharashtra Borse (1983) and Ramesh (1985) and Tamil Nadu and Kerala by Raghu Kumar (1973).

The coastline of India including those of Andaman and Nicobar and the Lakshadweep islands is over 7,000 kms out of which above 560 kms, occupied by the Konkan region, lies between the Sahyadris and the Arabian sea. It forms the western boundaries of the Maharashtra state. Mangrove forest with varied flora are found in coastal muddy creeks of Konkan. Most of the fungi were found on intertidal wood and mangrove plant parts in marine habitats.

Kohlmeyer and Kohlmeyer (1979) recognized 149 Ascomycetes with

28 doubtufl species. Some new genera and new species of Ascomycetes have recently been described. The Loculoascomycetes are well represented in intertidal habitats and have 51 marine representatives (Kohlmeyer, 1969a; Kohlmeyer and Kohlmeyer, 1979).

The majority are members of the Pleosporaceae and belong to well known terrestrial genera like *Didymella, Didymosphaera, Leptosphaeria, Mycosphaerella* and *Pleospora.* Other genera including *Halolthia, Helicascus, Magnicola, Paraliomyces, Pontoporeia* and *Thalossoascus,* appear to have only marine species. Table 2 provides a list of marine Loculoascomycetes occurring in these localities.

Conclusion

This chapter deals with the details of Loculoascomycetous fungi along the coast of Maharashtra. Table 1 provides a list of different Loculoascomycetous fungi occurring in various localities.

In all, 153 speccies belonging to 59 genera of Loculoascomycetous fungi have been isolated during the investigations as shown below.

Orders	Genera	Species
Myriangiales	2	2
Dothideales	9	14
Pleosporales	38	122
Hysteriales	8	13
Microthyriales	2	2
Total	**59**	**153**

Table 2 gives an overall picture of this work in relation to the work carried out in India.

Out of 153 species collected and described, 30 are new taxa, 2 are new genera and many of these are being reported for the first time from Maharashtra.

1. Establishment-of new taxa

Vestergrenia ixorae Ramesh, *Mycosphaerella, kankeshwarensis* Ramesh, *M. magniferae* Ramesh, *Bagnisiella jawaharensis* Ramesh, *Parodiella eujeniae* Ramesh, *Didymella thevetiae* Ramesh, *Botryosphaeria alibagensis* Ramesh, *B. purandharensis* Ramesh, *Sporormia millingtonii* Ramesh, *Didymosphaeria mallotii*

Table No. 1: Loculoascomycetous fungi from India and Maharashtra

Genus	No. of Spp. Reported in this work	New to Science	New Reports for India	New Reports for Maharashtra
(1)	(2)	(3)	(4)	(5)
Myriangium	1	-	-	1
Elsinoe	2	-	-	2
Chaetothyrim	2	-	1	1
Pilgeriella	1	-	1	-
Leptoshaerulina	2	-	1	1
Vestergrenia	1	1	-	-
Auerswaldia	1	-	1	-
Mycosphaerella	3	2	-	1
Sphaerulina	1	-	-	1
Bagnisiella	2	1	1	-
Dothiora	1	-	1	-
Apiosporina	1	-	1	-
Parodiella	4	1	3	-
Platychora	1	-	1	-
Didymella	3	1	2	-
Teratospohaeria	1	-	1	-
Pyrenostigme	1	-	1	-
Botryosphaeria	8	2	6	-
Lophiotrema	1	-	1	-
Sporormia	3	1	1	1
Didymosphaeria	11	3	6	2
Otthia	1	1	-	-
Ophiobolus	1	4	-	1
Acanthophiobolus	1	1	-	-
Cochliobolus	2	1	1	-
Leptospora	2	1	1	-
Thaxteria	1	-	1	-
Thaxteriella	2	1	-	1
Thaxteriellopsis	1	-	1	-
Herpotrichia	2	-	2	-
Herpotrichiella	1	-	1	-
Massarina	4	1	3	-
Massaria	4	1	3	-
Thyridaria	1	-	1	-
Passeriniella	1	-	1	-

(1)	(2)	(3)	(4)	(5)
Leptosphaeria	20	3	17	-
Keissleriella	5	-	5	-
Trematosphaeria	10	1	9	-
Melanomma	4	1	3	-
Berlesiella	1	-	1	-
Pleospora	16	1	15	-
Comoclathris	1	1	-	-
Dictyodothis	1	1	-	-
Teichospora	1	-	1	-
Pleomassaria	1	1	-	-
Bitrimonospora	1	1	-	-
Quasiconcha	1	1	-	-
Baerlagellopsis	1	1	-	-
Pyrenophoropsis	1	1	-	-
Glonium	2	-	2	-
Gloniella	3	-	3	-
Glyphium	1	-	1	-
Lophium	1	-	1	-
Chaetoscutula	1	-	1	-
Pseudoscypha	1	-	1	-
Lecanidion	2	-	1	1
Tryblidiella	1	-	1	-
Moeszopeltis	1	-	1	-
Leptopeltopsis	1	-	1	-
Total genera = 59	**153**	**32**	**108**	**13**

Ramesh, *D. shirkolaensis* Ramesh, *D. vaitarnensis* Ramesh, *Otthia buteae* Ramesh, *Acanthophiobolus anogeissi* Ramesh, *Cochliobolus buteae* Ramesh, *Leptospora elaeodedri* Ramesh, *Thaxteriella garugii* Ramesh, *Massarina lantanae* Ramesh, *Massaria symloc* Ramesh, *Leptosphaeria garciniae* Ramesh, *L.kilabandarensis* Ramesh, *L. vasaii* Ramesh, *Trematoshpaeria bambusae* Ramesh, *Melanomma thespesiae* Ramesh, *Pleopsora purnadharensis* Ramesh, *Comoclathris bouganvilli* Ramesh, *Dictyodothis jawaharensis* Ramesh, *Pleomassaria phoenicsis* Ramesh, *Btrimonospora adananthii* Ramesh, *Quasiconcha matheranensis* Ramesh.

The two new genera *Pyrenophoropsis* and *Boerlagellopsis.*

2. New reports from India.

The following species are being reported for the first time from the country.

Table No. 2 : Occurrence of marine Loculoascomycetes from India and Maharashtra

Name of the Fungus	1	2	3	4	5	6	7	8	9	10	11	12	13	14	15	16	17	18
Mycosphaerella avicenniae	-	+	-	-	-	+	-	-	-	+	+	-	-	+	-	-	-	-
Dactylospora heliotriphuys	+	+	-	+	-	-	-	-	-	-	-	-	-	-	-	-	-	-
Didymosphaeria enalia	+	+	-	+	-	+	+	+	+	-	+	+	+	+	+	+	+	+
Helicascus kanalosnus	-	+	-	-	-	-	-	-	-	-	+	+	-	-	-	+	-	-
Leptosphaeria albubunctata	-	+	-	-	-	-	-	-	+	-	-	-	-	-	-	-	-	-
L. australiensis	-	-	-	-	-	-	-	-	-	-	+	-	-	-	-	-	-	-
L. bombayensis	-	-	+	-	-	-	-	-	-	-	-	-	-	+	-	-	+	-
L. marina	-	+	-	+	-	-	+	+	+	-	-	+	-	-	-	+	-	-
L.neomaritima	-	-	-	-	-	-	-	-	-	-	-	-	-	-	-	+	-	-
L. obiones	-	-	-	+	-	-	-	-	+	-	-	-	-	-	-	-	-	-
L. rhizophorae	-	+	-	+	-	+	+	+	-	-	+	+	-	-	-	+	-	-
Pleospora avicenniae	-	+	-	-	+	-	-	+	-	-	-	+	-	+	+	-	+	-
P. malvenensis	-	-	-	-	-	-	-	-	-	-	+	-	-	-	+	-	+	-
P.pelagica	-	+	-	+	-	-	-	-	-	-	-	-	-	-	-	-	-	-
P.rhizophorae	-	-	-	+	-	-	-	-	+	-	-	-	-	-	-	-	-	-
P. sonnertiae	-	+	-	+	-	-	-	+	-	+	+	-	-	-	+	+	-	+

Note: + = Present; - = Absent

1. Alibag, 2., Bankot 3. Bombay, 4. Dabhol, 5. Dahanu, 6. Devgarh, 7. Ganapatipule 8.Ilarnai, 9. Jaigarh, 10. Mahim, 11. Malvan, 12.Ratnagiri, 13. Redi,14. Revanda, 15. Revas, 16. Shriwardhan, 17. Vasai., 18. Vengurla.

Chaetothyrium syzygii, Pilgeriella periosporoides, Leptosphaerulina americana, Auerswaldia winteri, Sphaerulina myriadea, Bagnisiella bacteridia, Apiosporina collinissi, Parodiella caespitosa, P. reticulata, P. saccharidicola, Platychora ulmi, Didymella barbieri, D. olendri, Pyrenostigme spiarune, Teratosphaeria dispersa, Botryosphaeria arctostaphyli, B. berengeriana, B. fistulae, B. synconophila, Lophiotrema oenothera, Sporormia ticinensis, Didymosphaeria anaxea, D. brunneola, D. endoleucae, D. sellae, D. winteri, Ophiobolous herpotrichus, Cochliobolus sporoboli, Leptospora rubella, Thaxteria didyma, Thaxteriellopsis bambusicola, Herpotrichia juniper, H.macrotrichia, Herpotrichiella moravica, Massarina albizziae, M. chrysopogonis, M.spartii, Massaria occulata, M. pupula, Thyridaria incrustans, Passeriniella discors, Leptosphaeria arvensis, L. baldingerae, L. epicalmia, L. fusispora, L. gallicola, L. haematites, L. iridigena, L. junicola. L. medicaginis, L.niessleana, L. patellaeformis, L.planiuscula, L. praetermissa, L. recutita, L. salsolae, L. suffulta, L. tenuis, Keissieriella culmifida, K. linearis, K. ocellata, K. taminensis, Trematosphaeria callisperma, T. confertula, T. melina, T. nuclearia, T. olearum, T. passerinii, T. pertusa, T. schweinitizii, T. seminude, Melanomma holmskjoeldii, M. sporadicum, M. sylvanum , Berlesiella setosa,

Pleospora agaves, P. amelanchiers, P. coloradensis, P. eccoronis, P. equiseti, P. gigaspora, P. juglandina, P. laricinia, P. lutea, P. obtuse, P. prtoteisopora, P. pyrenaica, P. rainierensis, P. rudis, P. vitalbae, Teichospora oleicola, Glonium lineare, G. subtectum, Gloniella byssiseda, G. pusilla, G. scortechiniana, Pseudoscypha abietis, Lecanidion pusillum, L. clavisporum.

3. New reports from Maharashtra

The species reported in the present work are *Myrinagium duriae, Elsinoe bitancourtiana, Chaetothyrium quaraniticum, Mycosphaerella palmicola, Botryosphaeria ingiae,Sporormia minima, Didymosphaeria ingniaria, Ophiobolus herpotrichus, O. cajani, Leptosphaeria culmifraga, L. eustoma, L. coniothyrim, Melanomma citricola, Pleospora scrophulariae, P. njegustensis, Thazteriella silentvalliensis, Lecanidion coorgicii.*

References

Anahosur, K.H. (1969). Some Ascomycetes from Maharashtra (India). Sydowia, **23**:95-99.

Anahosur, K.H. (1970). Some Ascomycetes from Maharashtra (India). Sydowia, **23**:63-68.

Barr, M.E. (1956). The development of the Ascocarp in *Phaeotrichum hystericinum*. Can. J. Bot., **34:**563-568.

Barr, M.E. (1968). The Venturiaceae in North America. Can. J.Bot.,**46:**799-864.

Barr, M.E. (1976a). Pespectives in the Ascomycotina Mem. Newyork. Bot. Gard., **28:**1-8.

Barr, M.E. (1976b). The genus *ostreichnion.* Mycotaxon, **3**(1): 81-88.

Barr, M.E. (1976c) .*Loculohypoxylon* a new genus. Mycotaxon, **3**(3): 325-329.

Barr. M.E. (1979) .A classification of Loculoascomycetes. Mycologia, **71**:935-937.

Bstista, A.C. (1959). Monografia dos fungos Micropelteaceae. univ. Recife Inst. Mycol. Publ. **56:**1-519.

Batista A.C. and Cifferri, R. (1962). The Chaetothyriales. Sydowia, **3:**1-129.

Batista A.C.and Cifferri R. (1963). *Capnodiales Sacardoa,* **2:**1-196.

Bilgrami M.S., Jamaluddin and Rizwi M.A. (1979). Fungi of India Part-i (List and references) **467** pp., Today and Tomorrow's printers and Publishers, New Delhi.

Bilgrami M.S., Jamaluddin and Rizwi M.A. (1981). Fungi of India Part-II (List and references) **268** pp., Today and Tomorrow's printers and Publishers, New Delhi.

Booth, C. (1971a). Introduction to general Methods. In: Methods in Microbiology. Ed. Booth, C. Acad. Press Newyork.

Booth, C. (1971b). Fungal culture media. In: Methods in Microbiology. Ed. Booth, C. Acad. Press Newyork.

Borse, B.D. (1983). Marine fungi from Maharashtra. Ph.D. thesis of Poona University, Poona. 178pp.

Bryne, P.P. and Jones, E.B.G. (1972). Second Int. Symposium on marine mycology. Bremerhaven Sept. 1972.

Chadefaud, M. (1960). Les vegetaux on vascularis in M Chadefaud and L. emberger. Traitede botanique Systematique Tomel. Masson and Cie, Paris. 1018pp.

Corlett, M. (1970). Ascocarp development of two species of Sooty molds. Can.J. Bot.,**48**:991-995.

Currey, F. (1856). On the reproductive organs of certain fungi, with some remarks on germination. Jour. Microscop. Sci., **4**:198.

DeBary, A. (1887). Comparative morphology and biology of the fungi, Mycetozoa and bacteria. Clarendon Press, Oxford.

Dennis, R.W.G. (1978). British Ascomycetes. 3rd Ed.Revised-XXVI+1-58 pp. J. Cramer, Vaduz.

Dharne, C.G. and Muller, E. (1969). Notes on some interesting Ascomycetous fungi from India. Sydowia, **23**:77-80.

Doguest, G. (1960). Etude due *Melogramma spiniferum* de Not. Pyrenomycete ascohymenie annaelasce. bitunique Rev. Mycol. (Paris), **25:** 15-26.

Erikson, O. (1967). On graminicolous Pyrenomycetesfrom Fennoscandia I. Dictyosporous species Ark. Bot., **(2) 6:** 339-380.

Gaumann, E. (1964). Die pilze-Zauel. 541s, Basel.

Hawksworth, D.L. (1964). Mycologists handbook. Common wealth mycological Institute, Kew.

Hedjarouda, G.A. (1968). Etudes taxonomiques sur les *Phaeasphaeria* Miyake et leurs formes voisines (Ascomycetes). Sydowia, **22:** 57-107.

Hodgetts, W.J. (1917). On the forcible discharge of spores of *Leptoshaeria acuta*. New Phytol., **16:**139-146.

Hoggan, I. (1927). The parasitism of *Plowrightia ribsia* on the currant. Trans. Bri. Mycol. Soc., **12:**27-43.

Jones, E.B.G. (1968). The distribution of marine fungi on wood submerged in the sea. In "Biodeterioration of Materials" (A.H. Walters and J.J.Elphick,eds.) pp. **460-485.** Elsevier. Amsterdam.

Kendrick, B., Samuels, G.J., Webster, J. and Luttrell, E.S. (1979). Techniques for establishing connections between anamorph and telemorph. In: The whole fungus Vol-II pp. **635-651** (ed. B.Kendrick) Museums of Canada, Canada.

Kohlmeyer, J. (1969a). Ecological notes on fungi in mangrove forests. Trans. Brit. Mycol. Soc., **53:**237-250.

Kohlmeyer, J. (1969b). Marine fungi of Hawaii including the new genus Heliascus. Can. J.Bot., **47:** 1469-1487.

Kohlymeyer, J. and Kohlmeyer, E. (1967). "Iconoes fungorum maris" Vol.l.Crarner, Lehre.

Kohlmeyer, J. and Kohlmeyer, E. (1968). "Iconoes fungorum maris" Vol.l.Crarner, Lehre.

Kohlmeyer, J. and Kohlmeyer, E. (1979)Marine mycology, The higher Fungi, pp.698, Academic press Newyork.

Lindau, G. (1897) Pyrenomycetineae In Engler and Prantl. P. 321-491, Die Naturlichen pflanz enfamillien wihelm Engelmann, Leipzing.

Luttrell, E.S.(1951). Taxonomy of Pyrenomycetes. Univ. Mo. Stud., **24**(3) 120pp.

Luttrell, E.S.(1953). Development of the ascocarp in *Glonium stellatum*. Am. J. Bot., **40:**626-633.

Luttrell, E.S. (1955). The ascostromatic ascomycetes. Mycologia, **47:**511-532.

Luttrell, E.S.(1973). Loculoascomycetes in Ainsworth, G.C. *et al.* The fungi IV. A.135-219.

Miller, J.H. (1928). Biologic studies in the Sphaeriales I. Mycologia, **20:** 187-213, 305-339.

Miller, J.H. (1949). A revision of the classification of the ascomycetes with special emphasis on the pyrenomycetes. Mycologia, **41:**99-127.

Mirza, F. (1968). Taxonomic investigations on the ascomycetous genus Cucurbitaria S.F. Grey. Nova Hedwigia, **16:**161-213.

Muller, E. and Sanwal, B. D. (1954). Uber die Gattugen Microcyclus Sacc: Coccoidella V.Hohn. Coccodothis Thesis. et. Syd. and Coccodothella Thesis et Syd. Sydowia, **8:**231-244.

Muller, E. and Von Arx, J.A. (1962). Die Guttungen der didymospoen pyrenomyceten, beitrage zur kryptogamen flora der schweiz, vol.II part 2,Buchler and Co. Bern. 922 pp.

Munk, A. (1957). Danish Pyrenomycetes. Dansk Botanisk. *Arkiv.* Vol.I.Crarner, Lehre., **17:** 491pp.

Munk, A. (1962). An approach to an analysis of taxonomic method with main references to higher fungi. Taxon, **11:**185-190.

Nannfeldt, J.A. (1932). Studien uber die Morphologie und systematic der nichtlichenisierten ubioereykateb, Discomycten. Nova- Acta. Rigiae Soc. Sci. upsalielsis, Ser. IV. **8**(2):368pp.

Pantidou, M.E. and Korf, R.P. (1954). A revision of the genus *Keithia.* Mycologia, **46:**386-388.

Patil, S.D. and Ramesh, Ch. (1983). Loculoascomycetes from Maharashtra. Life Sciences Advances ,**2:**215-217.

Patil, S.D. and Ramesh, Ch. (1984). Some *Pleospora species* from Maharashtra. Indian Bot. Reptr., **3**(1):80.

Patil, S.D. and Ramesh, Ch. (1986). Some Loculoascomycetes from Maharashtra. Ind. Bot. Reptr., **5**(1): 27-31.

Patil, S.D. and Ramesh, Ch. (1987). Notes on Some Fungi of Pleosporaceae (Loculoascomycetes) from Maharashtra (India). Trans. Myco. Sco. Japan., **22**(3) 229-236.

Petrak, F. (1954). Ergebisse einer Revision der Grund typen verschiedener Gattungen der Ascomyzeten and Fungi imperfectii. Sydowia, **8:** 287-303.

Pringshiem, N. (1858). Uber das austreten der sporen von *Sphaeria scirpi* Aus. Ihren chlauchen. Jahrb. Wiss. Bot., 1:189-192.

Raghukumar, S. (1973). Marine Lignicolous fungi from India. Kavaka, **1:**73-85.

Ramesh, Ch. (1985). Studies in the Ascomycetes of Maharashtra with special emphasis on Loculoascomycetes. Ph.D. thesis, University of Poona, Poona.

Ramesh, Ch. (1986). Some Loculoascomycetes From Maharashtra, Genus-*Mycosphaerella*, Curr. Sci, **55**(7):369-370.

Ramesh, Ch. (1987a). Some Loculoascomycetes From Maharashtra. Acta Botanica. Indica., **15**(1):89.

Ramesh, Ch. (1987b) The Genus *Sporormia* From Maharashtra, India. *Bot.Reptr.*, **5**(2):148.

Ramesh, Ch. (1987c) Loculoascomycetes From Maharashtra, Genus-*Bagnisiella* India. *Bot. Reptr.*, **5**(2):203.

Ramesh, Ch. (1987d), Frequency of occurrence of Loculoascomycetous Fungi From Maharashtra.Proc. Nat. Symp. Fungal. Ecol., 79-85.

Ramesh, Ch. (1987e). Three Loculoascomycetes New to India. Acta. Botanica. India, **15**(1):106.

Ramesh, Ch. (1988a). Loculoascomycetes From Maharashtra. Ind.J. Bot., **12**(1): 78-80.

Ramesh, Ch. (1988b). Loculoascomycetes From Maharashtra. (India). Ind. Bot.Reptr., 7 (1&2):54.

Ramesh, Ch. (1988c). A New Species of *Vestergrenia, V. Ixorae* From Maharasthra. Ind. Bot. Reptr., 7(1&2):105.

Ramesh, Ch. (1990a). Some Loculoascomycetes From Maharashtra. J.Ind. Bot.Soc., **69**:165.

Ramesh, Ch. (1990b). Some Loculoascomycetes From India. Genus-*Botryosphaeria*. Indian. Phytopath., **40**(3):295-300.

Ramesh, Ch. (1990c). Studies on Loculoascomycetes, Genus-*Didymoshpaeria*. Indian Phytopath., **43**(3):361-367.

Ramesh, Ch. (1992). The Genus-*Massarina* From Maharashtra, (India). Ind.Bot.Reptr., **11**:1-5.

Ramesh, Ch. (1993a). The Genus-*Melanomma* From Maharashtra, Asian. J.pit. Sciences., **5**(1):33-37.

Ramesh, Ch. (1993b). The Genus-*Trematosphaeria* From Maharashtra. Ind.J.Bot., **17**:56-58.

Ramesh, Ch. (1993c). Some Loculoascomycetes From Maharashtra, Genus-*Keisseriella*. Ind. Bot. Reptr., **22**:23-27.

Ramesh, Ch. (1993d). The Genus *Pleospora* From India. J.Kar.Univ. (Sci), **37**:17-32.

Ramesh, Ch. (1993e). The genus *Leptosphaeria* From India.J. Kar.Univ(Sci), **37**:45-61.

Reid, J. and Pirozynski, K.A. (1966). A new Loculoascmycetes on *Abies balsamed* (Linn) Mill. Can. J. Bot., **44**:351-354.

Shoemaker, R.A.(1935). Biology, cytology and taxonomy of *Cochliobolus sativus*. Can.J. Bot., **33**:562-576.

Sivanesan, A.(1970a). *Plurisperma dalbergiae* gen. et. sp. nov. Trans. Brit. Mycol. Soc., **54**:495-496.

Sivanesan, A., Talade, U.K. and Tilak, S.T. (1974). *Bitrimonosperma indica gen.et.sp.nov.* A new Loculoascomycete from India. Trans. Brit. Mycol. Soc., **63**(3):595-596.

Sivanesan, A., Panwar, K.S. and Kaur, S.J. (1976). Thaxteriellopsis lignicola gen.et.sp.nov. A new Loculoascomycete from India. Kavaka, **4**:39-42.

Srinivasulu, B.V.(1972), *Tilakiella* a new genus from India. Indian Phytopathol., **25**(2):291-293.

Sutton, B.C.(1980). The Coelomycetes. Common Wealth Mycological Institute, Kew.

Theissen, F. (1916). Mycologische. Abbandlangen. *Verhandl.* Zool.Bot. Gest. Wien., **66:**296-400.

Theissen, F. (1919). Neue. orignal-untersuchungen. *Verhandl.* Zool.Bot.Ges.Wien, **69:**1-24.

Theissen, F. and Sydow, H. (1918). Vorent wurfe zu den Pseudosphaeriales. Ann. Mycol., **16:**1-34.

Von Arx, J. and Muller, E. (1975). A reevaluation of the bitunicate Ascomycetes with keys to families and genera. Stud. Mycol., **9**:1-159.

Whiteside, W.C. (1962). Morphological studies in the Chaetomiaceae III. Mycologia, **54:**611-620.

27

Biodiversity of Coelomycetous Fungi from Tamil Nadu

J. Muthumary

Centre for Advanced Studies in Botany, University of Madras, Guindy Campus Chennai-600 025

A brief review on research work in Coelomycetes is given including the contributions made by Indian mycologists. The various characteristics of this group of fungi is outlined with few examples. Data relating to about 138 species in 62 genera of Coelomycetes collected from various places in Tamil Nadu is included. For easy identification of the various species, details about the conidiomata, conidiogenous cells and conidia for each species included are listed. The cultural characters of some selected Coelomycetes are discussed.

Key words : Taxonomy, Coelomycetes, conidiomata, conidiogenous cells, conidia, cultural characters, species from Tamil Nadu.

Coelomycetes are anamorphic fungi in which conidia are produced within a cavity lined by fungal or host tissue or sometimes by these two tissues. The conidia are produced within pycnidial, acervular, stromatic, cupulate and pycnothyrial conidiomata. These groups of fungi are known mainly from temperate and tropical regions, which are capable of surviving in a wide variety of ecological situations. They have been reported from leaf litter and other organic debris, soil, water, air and also as parasites on many plant parts causing diseases. Many species are pathogenic to economically important crops and ornamental plants causing lesions, blights, anthracnose, cankers and diebacks. Many Coelomycetes are also responsible for rots in fruits and seeds. Recently many of these fungi have been reported to be present as endophytes in a variety of plant species.

These groups of fungi were recognized as early as 1796. Great progress has been achieved by research efforts of many mycologists in these fungi (Sutton, 1961,1962,1963, 1967a and b, 1968, 1969, 1971, 1972a and b, 1973, 1974, 1975, 1977, 1980; Punithalingam, 1970, 1974,1979,1989; Nag Raj and Kendrick, 1971, 1972 a and b, 1978; Nag Raj and Morgan Jones, 1973, Nag Raj, 1974, 1975a and b, 1977, 1978, 1981, 1983, 1985a-d, 1986a-c, 1993). Studies on Indian Coelomycetes had been pursued by Subramanian and Reddy (1974) and followed by Muthumary and co-workers (Masilamani and

Muthumary 1994, 1995, 1996; Muthumary, 1981, 1986a-c, 1987a-c, 1988, 1993; Muthumary and Sutton, 1986; Muthumary and Vanaja, 1986; Muthumary *et al.*, 1986, 1993; Muthumary and Masilamani, 1989; Muthumary and Murugan, 2000). The primary aim of this chapter is to present data relating to about 138 species in 62 genera of Coelomycetes collected from various parts in Tamil Nadu and to provide an easy and practical approach to the identification of these fungi. Details on conidiomata, conidiogenous cells, conidium development and conidial morphology in each species is given.

A survey was conducted in Tamil Nadu and about 300 coelomycetous fungi were recorded from these areas.

Methodology

Coelomycetes grow on dead or live stems, branches, twigs, leaves, barks and fruits. They normally cause necrotic lesions or cankers. When looking for Coelomycetes on these substrates the presence of conidiomata should be confirmed by using a hand lens. The collection data should be recorded along with the name of the host as it is essential for identification of the species. The specimen should be air-dried to be stored in the herbarium. Proper microscopic examination of the material in the laboratory is essential because characters such as the type of conidiomata, the relationship of the conidiomata to the substrate, the conidiogenous layers, the tissues comprising the conidiomatal wall layers and the ostiole are important for the correct identification of the species. Apart from these characteristics, conidium development, conidial morphology, paraphyses and behaviour in artificial culture are also taken into consideration for proper identification.

Taxonomic Characters in Coelomycetes

In Coelomycetes the conidium is formed in specialized structures such as acervuli or pycnidia. It may not be possible to simply classify these conidiomata into two broad categories because of their morphological diversity. However, the five morphological categories - acervuli, pycnidia, stromata, cupulate and pycnothyrioid pycnidia recognized by Von Höhnl (1923) have been widely accepted.

Acervuli are subcuticular, inter epidermal, sub epidermal or sub peridermal in the host and the enclosing fungal layer is absent. The conidiogenous cells arise from the upper layers of the fungal stroma lining the cavity of the acervulus [Fig. 2 (1-2)]. The overlying host tissues are ruptured irregularly by the developing mass of conidia. Often septate, brown, thick-walled setae are seen randomly distributed throughout the conidiomata while

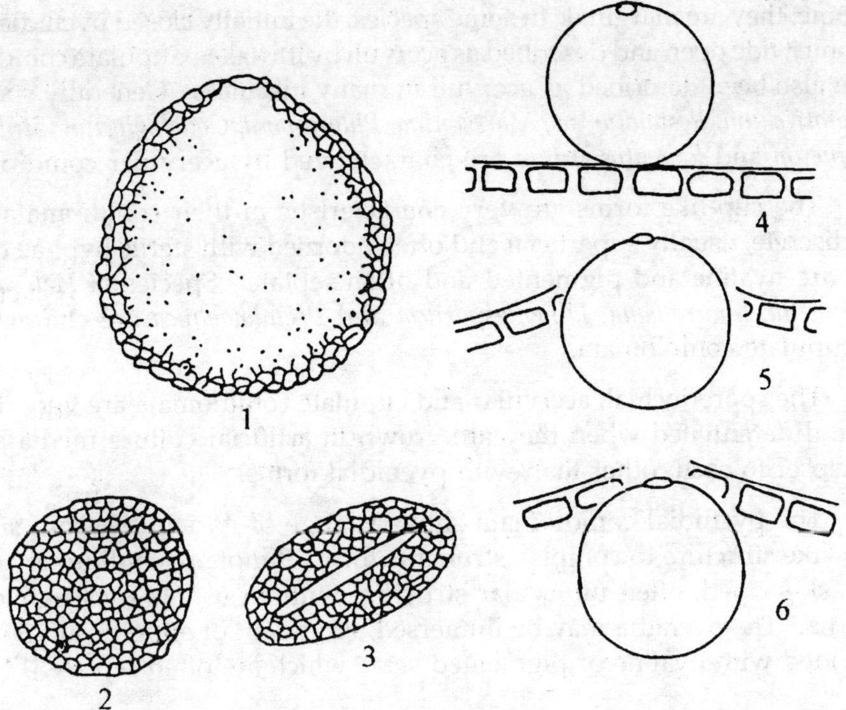

Fig. 1. 1. Vertical section of pycnidium with central circular ostiole, 2. Surface view of a pycnidium with central circular ostiole, 3. Surface view of a pycnidium showing dehiscence by a longitudinal raphe, 4. Superficial-pycnidium, 5. Semi-immersed pycnidium, 6. Immersed pycnidium

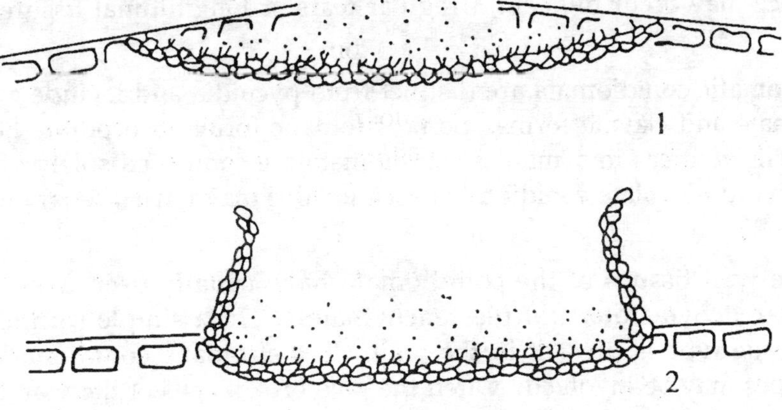

Fig. 2. 1. Vertical section of an acervulus, 2. Vertical section of a cupulate conidioma

in some, they are marginal. In some species, the initially closed pycnidia finally become wide open and described as acervuli by mistake. Cupulate conidiomata have also been described as acervuli in many instances. Generally species of *Colletotrichum, Pestalotiopsis, Marssonina, Phloeospora, Colletogloeum, Stilbospora, Coryneum* and *Seimatosporium* are characterized by acervular conidiomata.

The cup-like forms are very characteristic in their conidiomata which are discrete, usually superficial and often adorned with sterile hyphae or setae that are hyaline and pigmented and often septate. Species of *Heteropatella, Diarimella, Stauronema, Dinemasporium* and *Pseudolachnea* are characterized by cupulate conidiomata.

The sporodochial, acervular and cupulate conidiomata are very difficult to be differentiated when they are grown in artificial culture media as they are closer to each other than with pycnidial forms.

The pycnidial conidiomata show a range of variability from simple - flask like structure to complex stromatic form. Simple pycnidium is globose to flask shaped, often unilocular structure with an ostiole for the release of conidia. The pycnidia may be immersed, erumpent or superficial. Some are provided with hyaline or pigmented setae which are often septate [Fig. 1 (1-6)].

The pycnothyrial conidiomata are distinct from pycnidial or stromatic categories. They are generally flat, hemispherical or irregular. Often they are superficial or subcuticular. They are composed of an upper shield, sometimes supported by a central column of cells. The conidiogenous cells are seen only at the base of the conidioma. Sometimes there is a regular ostiole. Otherwise dehiscence may occur by a slit, irregular tears or longitudinal fissure [Fig. 4 (1-7)].

Stromatic conidiomata are distinct from pycnidia and include a variety of pulvinate and clavate forms. Some stromatic forms incorporate host and fungal cells whereas in some the conidiomata are composed solely of fungal tissue. In multiloculate conidiomata each loculus may have a separate ostiole [Fig. 3(1-4)].

The wall tissues of the conidiomata have usually been described as pseudoparenchymatous and plectenchymatous. In a simple pycnidium, a single tissue type is present in the wall. In a stromatic conidioma several tissue types may be involved. Often the pale brown cells of the wall become darker near the ostiole. In some pycnidia, the wall is several cells thick (*Phyllosticta*) and is composed of outer textura of dark brown, thick walled cells and inner textura prismatica of hyaline, thin walled cells. Very complex wall tissues are present in some genera eg. *Cornicularielia*. In the more common

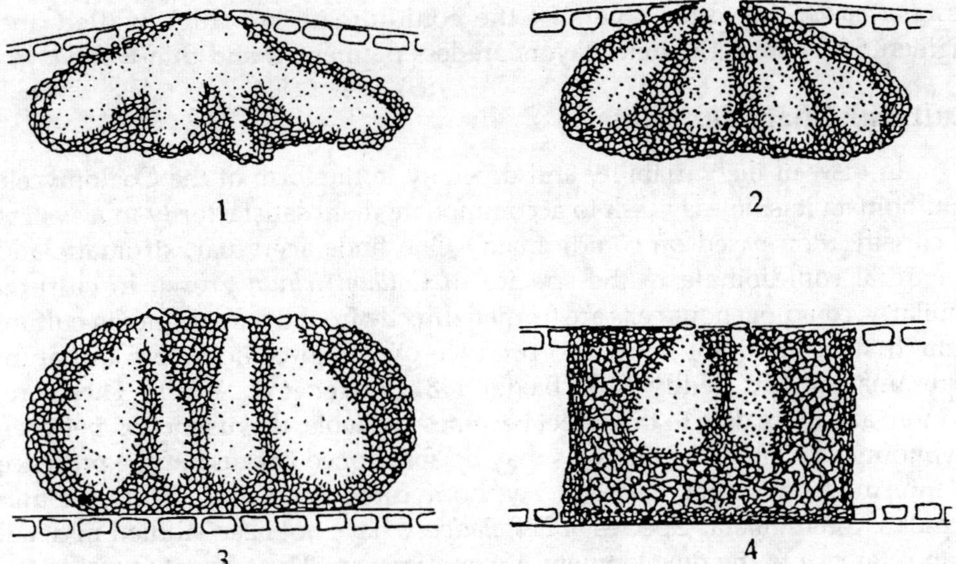

Fig. 3. 1. Convolute and immersed stroma in vertical section, 2. Multilocular and immersed stroma in vertical section, 3. Multilocular and superficial stroma in vertical section, 4. Pseudostromatic conidiomata in vertical section

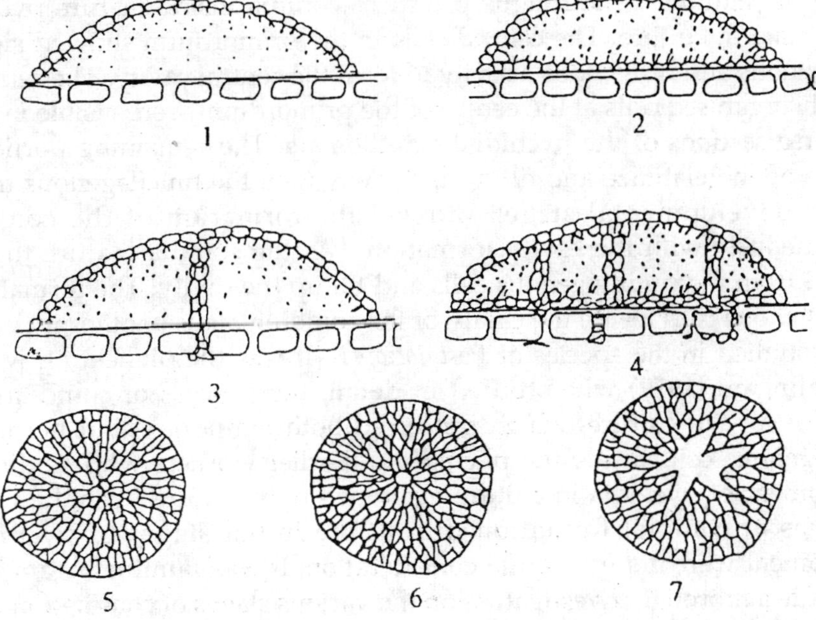

Fig. 4. 1. Pycnothyrium with only the upper wall, 2. Pycnothyrium with lower and upper walls, 3. Pycnothyrium showing a central supporting column, 4. Multiloculate pycnothyrium with several supporting columns, 5. Pycnothyrial dehiscence from the margin, 6. Pycnothyrial dehiscence by central ostiole, 7. Pycnothyrial dehiscence by irregular fissures

Pestalotiopsis, the outer layers of the conidiomata are thick walled and pigmented whereas the inner layers are less pigmented and thin-walled.

Cultural Characteristics

In view of the variability and diversity in the form of the Coelomycete conidiomata it is no easy task to accommodate them satisfactorily in a system of classification based on conidiomata. One finds acervular, stromatic and pycnidial conidiomata in the species of *Colletotrichum* grown in culture. Similarly, conidiogenous cells are formed directly from the mycelium in culture though such fungi are known to produce pycnidial conidiomata in nature (Subramanian and Reddy 1974; Baxter 1981; Baxter *et al.*, 1985). Therefore, as more and more data to this effect becomes available, the distinction between Hyphomycetes and Coelomycetes may be abandoned because of the presence of intermediary stages between Hyphomycetes, acervular, stromatic and cupulate conidiomata. Species of *Pestalotiopsis* have not been studied in detail with reference to the development of conidiomata. The present study in the author's laboratory reports for the first time the various stages of development of conidiomata in species of *Pestalotiopsis* with interesting observations. The developmental morphology of the conidiomata in these species has shown interesting features. The fungus produces conidia inside mature pycnidial conidiomata in culture. The central cells in the primordium showed signs of schizogenous and lysigenous activity to form the central cavity. The remnants of the disorganised cells at the centre of the primordium were visible in many transverse sections of the pycnidial conidiomata. The remaining portions of the cells often gelatinize and fill up the cavity with the mucilagenous matrix. Further developmental stages showed the formation of the conidium simultaneously with the cavity formation. As the conidia mature they are released from the conidiogenous cells and fill up the cavity. The formation of conidiogenous cells inside the cavity of the conidiomata is another interesting feature studied in the species of *Pestalotiopsis.* It was the pioneering work of Punithalingam (1979) who studied in detail, the process of conidiation in *Ascochyta* spp. in culture, which produced both temporary and permanent conidiogenous cells inside the pycnidium. Earlier workers who studied the development of pycnidia in culture examined only mature pycnidia but the processes of conidial formation during the initial stages of pycnidium development were not given due consideration. It was Punithalingam (1979) who made a thorough investigation on the various stages of conidiation inside a pycnidial conidioma. According to him, the cells lining the pycnidial cavity produce conidia initially only to get sloughed off as soon as their function is over. Once a layer of conidiogenous cells lining the pycnidial cavity is completely dislodged, a new layer of cells lining the cavity, functions as conidiogenous cells. Thus, the processes of conidiation and dislodgement are repeated continuously. These are the two main processess by which the pycnidial cavity gradually increases in size and becomes filled with conidia and sloughed off conidiogenous cells. This is another way adopted by nature to produce greater number of conidia inside a pycnidium where the space is very limited. During the development, as the primordium increases in size, the central cavity is to be formed because the primordium is a compactly packed pseudoparenchymatous tissue. Nag Raj (1981) suggested that the slime, which originates through lysis occurring during cavity formation, play an important role in conidium dispersal. The presence of mucilagenous matrix during the formation of the pycnidial conidiomata was reported in certain other pycnidial Coelomycetes also (Masilamani and Muthumary, 1994, 1995, 1996). The pioneering work of Punithalingam (1979) who studied in detail the process of conidiation in *Ascochyta* species in culture revealed a similar process in this fungus.

The production of temporary and permanent conidiogenous cells in *Ascochyta* is construed as a general feature of *Ascochyta* species in culture (Punithalingam, 1979). Similarly, the dual conidiation process was also observed in species of *Pestalotiopsis* in the present investigation. The dual conidiation process was observed in species of *Phyllosticta caryotae* Shen in culture (Murugan 1999). Punithalingam (1979) reported that some of the observations made with *Ascochyta* in culture were in conformity with the interpretations of Archer (1926) during his investigation on cavity formation and ostiole development in some members of Sphaeropsidales. This suggests that probably these fungi which produce pycnidial conidiomata in culture follow some specific regularity in their conidiomatal development. It is not possible to conclude that these stages of developmental processes occur during the formation of pycnidia in nature.

Conidium Ontogeny

The significance of conidiogenesis in classification of conidial fungi gained recognition only after the publication of Hughes (1953) system. The concepts of Hughes have been extended to many Coelomycetes in the following years and new genera of Coelomycetes were established based on conidium ontogeny. According to Sutton (1980), the application of conidium ontogeny to the taxonomy of Coelomycetes is somewhat more difficult because in Coelomycetes conidiogenous structures are much smaller and consequently more difficult to interpret.

Not much information is available regarding the electron microscopic details on conidium ontogeny in Coelomycetes. Similarly, time lapse photomicrography, which is readily applied in the study of Hyphomycetes has not been used at all in the Coelomycetes. Therefore, basic problems of conidium ontogeny in many Coelomycetous fungi are yet to be solved.

The available data on the conidium ontogeny in Coelomycetes show that phialides and proliferating phialides are very common in many Coelomycetous forms followed by blastic-single and blastic-sympodial ontogeny. Blastic-annellidic and polyblastic ontogeny is less common especially in pycnidial forms.

Conidia

The conidium in Coelomycetous fungi exhibits diversity in form, from amerospores to helicospores and the complex form of conidium as seen in *Geastrumia*. Sutton (1973) recognized two types of septa in Coelomycetes: the eusepta are present usually in common genera and the distosepta are present in genera like *Coryneum, Massariothea* and *Pestalotiopsis*. Concolorous pigmented conidia are common whereas the versicoloured conidium where the end cells remain hyaline and the median cells show variable degrees of pigmentations is present in species of *Pestalotiopsis*. In many taxa with hyaline conidia, the walls are usually smooth, and develop pigmentation and ornamentation of the walls when they mature. Often the conidium in many Coelomycetes bear appendages which are described variously as pedicels, setae and cilia. Appendages are common in Coelomycetes which are adapted to special ecological niches such as marine habitats, aquatic environment and also terrestrial conditions. According to Nag Raj (1993) nearly 40% of Coelomycetes possess appendages in their conidia.

The appendages may be cellular (*Bartalinia lateripes*) or extracellular (*Phyllosticta, Urohendersonia*) (Nag Raj, 1993). Based on cytological studies Punithalingam (1989) listed five major types of appendages 1) enucleate mucilaginous 2) enucleate, tubular and filiform 3) enucleate, tubular 4) nucleate tubular and 5) nucleate with apical cell / appendages. Both cellular and extra cellular appendages vary in the developmental processes. Lot of information with reference to the cytochemistry of the extra cellular appendages is very essential for a better understanding of these novel structures. A comprehensive treatment of appendaged Coelomycetes by Nag Raj (1993) provides very valuable information about these appendages.

The primary concern of this chapter is to present data relating to 135 species in 62 genera of Coelomycetes and to provide practical approach to the identification of these fungi collected in Tamil Nadu. Diagnostic characters of this group of fungi like conidial type, conidiogenous cells and conidial

Table 1 : Coelomycetes from Tamil Nadu with taxonomic characters

S.No	Fungus Name	Conidiomata	Conidiogenous Cells	Conidia
1.	*Amerosporiopsis gaubae* Petrak.	Pycnidial unilocular	Phialidic	Amerosporous, hyaline
2.	*Amerosporium concinnum* Patrak	Pycnidial, unilocular	"	"
3.	*Aplosporella beaumontiana* Ahmad	Stromatic	"	Amerosporous, brown
4.	*Aplosporella* sp. (1)	Pycnidial, Eustromatic	Holoblastic	Amerosporous, brown
5.	*Aschersonia aleyrodis* Webber	Pycnidial, Eustromatic	"	Amerosporous, hyaline
6.	*Ascochyta avenae* (Petrak) Sprague & Johnson	Pycnidial, unilocular	Phialidic	Didymosporous, hyaline
7.	*Ascochyta caricae* Pat.	"	"	"
8.	*Ascochyta festucae* Punithalingam	"	"	"
9.	*Ascochyta graminicola* Sacc.	"	"	"
10.	*Ascochyta imperfecta* Pk.	"	"	"
11.	*Ascochyta lycopersici* (Plour.) Brun.	"	"	"
12.	*Ascochyta pisi* Lib.	"	"	"
13.	*Ascochyta sorghi* Sacc.	"	"	"
14.	*Ascochyta* sp. (2)	"	"	Didymosporous, hyaline
15.	*Asteroma alneum* (Pers. ex Fr.) Sutton	Acervular	"	Amerosporous, hyaline
16.	*Asteroma padi* Dc. Ex Fr.	"	Phialidic	Amerosporous, hyaline
17.	*Bartalinia robillardoides* Tassi.	Pycnidial, unilocular	"	Amerosporous, hyaline

18.	*Bartalinia* sp.(4)	Pycnidial, unilocular	Phialidic	Amerosporous, hyaline with 2-3 apical appendages
19.	*Botryodiplodia theobromae* Pat.	Pycnidial	Holoblastic	Didymospore, brown with longitudinal striation
20.	*Brycekendrickia indica* Nag Raj	Eustromatic	Phialidic	Amerosporous, hyaline
21.	*Camarosporellum nervisequum* (Tassi) Tassi	Pycnidial, unilocular	Holoblastic, annellidic	Dictyosporous, brown
22.	*Camarosporium quaternatum* (Hazsl.) Schulz.	"	Holoblastic, annellidic	Dictyosporous, brown.
23.	*Camarosporium* sp. (3)	"	Annellidic	Dictyosporous, brown
24.	*Ceuthospora lauri* (Grev.) Grev.	Eustromatic	Phialidic	Amerosporous,hyaline
25.	*Chaetoconis polygoni* (Ell. & Ev.) dements	Pycnidial, unilocular	"	Didymosporous, apical cell with a single appendage.
26.	*Chaetomella acutiseta* Sutton & Sarbhoy.	Cupulate, stromatic, unilocular.	"	Amerosporous, hyaline
27.	*Chaetomella raphigera* Swift	"	"	"
28.	*Ciliochorella mangiferae* Syd. Apud Syd. & Mitter	Stromatic	"	Phragmosporous, end cells with appendages
29.	*Coleophoma crateriformis* (Cour. & Mont.) Hohn.	Pycnidial, unilocular	"	Amerosporous, hyaline
30.	*Coleophoma cylindrospora* (Desm.) Hohn.	"	"	Amerosporous, hyaline

31.	*Coleophoma empetri* (Rostr.) Petrak	"	"	"
32.	*Coleophoma* sp. (1)	Pycnidial	"	"
33.	*Colletotrichum capsici* (Syd.) Butl. & Bisby	Acervular	Phialidic	"
34.	*Colletotrichum caudatum* (Sacc.) Pk.	"	Phialidic	"
35.	*Colletotrichum dematium* (Pers. ex Fr.) Grove	"	"	"
36.	*Colletotrichum dracaenae-fragrentis* (Mori.) Petrak & Sydow	"	"	"
37.	*Colletotrichum gloeosporoides* (Benz.) Sacc.	"	"	"
38.	*Colletotrichum gomphrenae* Srivatsawa	"	"	"
39.	*Colletotrichum graminicola* (Ces.) Wilson	"	"	"
40.	*Colletotrichum lindemuthianum* (Sacc. & Magn.) Br. & Cav.	"	"	"
41.	*Colletotrichum sp.* (5)	Acervular	"	"
42.	*Colletotrichum suscum* Laub.	"	"	"
43.	*Coniella castaneicola* (Ell. & Ev.) Sutton	Pycnidial, unilocular	Phialidic	Amerosporous, hyaline
44.	*Coniella fragariae* (Oud.) Sutton	"	"	Amerosporous, hyaline to pale brown
45.	*Coniothyrium fuckelii* Sacc.	Pycnidial, unilocular	Holoblastic, annellidic	Didymosporous brown
46.	*Conostroma quercicola* Muthumary	Stromatic	"	Amerosporous,hyaline
47.	*Coryneum indicum* Sutton & Rizwi	Acervular	Annellidic	Didymosporous, constricted at the septum,guttulate hyaline to pale brown

48.	*Coryneum* sp. (1)	Acervular	Annellidic	Didymosporous, constricted at the septum,guttulate hyaline to pale brown
49.	*Cryptosporiopsis coryli* (Pk.) Sutton	Acervular	Holoblastic	Amerosporous, hyaline
50.	*Cryptosporiopsis fasiculata* (Toge ex Tul.) Patrak	"	"	"
51.	*Cryptosporiopsis malicorticis* (Cordil.) Nannfeldt	"	"	"
52.	*Cryptosporiopsis nobilis* (Sacc.) Pirozynski & Morgan-Jones	"	"	"
53.	*Cryptosporiopsis* sp. (1)	Acervular,	Holoblastic	Amerosporous, hyaline
54.	*Cylindrogloeum trillii* (Ell. & Ev.) Arx	Stromatic,	Phialidic	Amerosporous,hyaline
55.	*Cytospora chrysosperma* Pers. ex Fr.	Stromatic	Phialidic	Amerosporous,hyaline
56.	*Diarimella setulosa* Sutton	Cupulate	Phialidic	"
57.	*Dinemasporium aberrans* Sutton	"	"	Amerosporous,hyaline with two setulae
58.	*Dinemasporium graminum* (Bark.) Lev.	Cupulate	Phialidic	Amerosporous,hyaline end cells with appendages
59.	*Dinemasporium strigosum* (Pers. ex Fr.). Saccardo	Cupulate	Phialidic	Amerosporous,hyaline, end cells with appendages
60.	*Diplodia punctata* Lev.	Pycnidial, unilocular	Monoblastic	Dictyosporous, brown
61.	*Diplodia* sp. (3)	Pycnidial, unilocular	Monoblastic	Dictyosporous, brown
62.	*Diplodia sycina* Mont. Saccardo	Pycnidial, unilocular	Holoblastic	Didymosporous, brown

63.	*Discosia strobilina* Lib.	Stromatic dimidiate	Holoblastic	
64.	*Discosia strobilina* Lib.	Dimidiate stromatic multilocular	Holoblastic	Phragmosporous,hyaline dorsiventral filiform setula formed at the concave side of the conidium
65.	*Discosia vagans* de Not.	"	"	"
66.	*Dislogloeum veronicae* (Lib.) Petrak	Acervular,	Phialidic	Amerosporous, pale brown
67.	*Hainesia lythri* (Desm.) Hohn.	Cupulate	"	Amerosporous, hyaline
68.	*Harknessia* sp. (1)	Stromatic, unilocular	Holoblastic	Amerosporous, brown with single basal appendage
69.	*Harknessia thujina* Ell. & Ev.	Stromatic,	"	"
70.	*Kendrickomyces indicus* Sutton, Rao & Mhaskar.	Stromatic, multilocular	Phialidic	Amerosporous, hyaline
71.	*Macrophomina phaseolina* (Tassi.) Goid.	Pycnidial, unilocular	Phialidic	Amerosporous hyaline with apical mucilagenous appendage
72.	*Marssonina rosae* (Lib.) Died.	Acervular, unilocular	Annellidic	Didymosporous, constricted at the septum, guttulate.
73.	*Melanconium* sp. (1)	Stromatic	"	Amerosporous,brown
74.	*Microsphaeropsis centaureae* Morgan-Jones	Pycnidial, unilocular	Phialidic	Amerosporous, brown verruculose.
75.	*Mirimyces pulcher* Nag Raj	Acervular	Phialidic	Amerosporous, hyaline
76.	*Monostichella indica* Sutton	Pycnidial	"	Amerosporous, hyaline 5 celled
77.	*Mycotribulus mirabilis* Nag Raj & Kendrick	Pycnidial unilocular	Holoblastic	Amerosporous with 3 apical appendages and a single basal appendage, hyaline.

78.	*Pestalotiopsis carissae* Guba	Acervular,	Holoblastic, annellidic	5 celled Phragmosporous, 3 apical appendages and single basal appendage concolorous
79.	*Pestalotiopsis cruenta* (Sydow)Stey.	"	"	5 celled Phragmosporous, 3 apical appendages and single basal appendage versicolored
80.	*Pestalotiopsis disseminata*(Thuem.)Stey.	"	"	5 celled Phragmosporous, 3 apical appendages and single basal appendage concolorous
81.	*Pestalotiopsis elasticola* P. Henn.,	"	"	5 celled Phragmosporous, 3 apical appendages and single basal appendage versicolorsed
82.	*Pestalotiopsis funerea*(Desm.)Stey.	"	"	5 celled Phragmosporous, 3 -5 apical appendages and single basal appendage concolorous
83.	*Pestalotiopsis gossypii* Hori.	"	"	Phragmosporous, 3 -4 apical appendages and single basal appendage concolorous
84.	*Pestalotiopsis guepini* (Desm.) Stey	"	"	"
85.	*Pestalotiopsis ixorae* Rangel	"	"	5 celled Phragmosporous, 3 - apical appendages and single basal appendage versicolored
86.	*Pestalotiopsis javanica* Guba.	"	"	Phragmosporous, 3 - apical appendages and single basal appendage concolorous

87.	*Pestalotiopsis mangiferae*(Hennings)Stey.	"	"	5 celled Phragmosporous, 3 apical appendages and single basal appendage versicolored
88.	*Pestalotiopsis palmarum* (Cooke) Stey.	"	"	"
89.	*Pestalotiopsis phoenicis* Vize	"	"	3 apical appendages and single basal appendage concolorous
90.	*Pestalotiopsis* sp. (13)	Acervular	Holoblastic, annellidic	Phragmosporous, 3 apical appendages and single basal appendage
91.	*Pestalotiopsis sydowiana* Bres.,	"	"	5 celled Phragmosporous, 3 - apical appendages and single basal appendage versicolored
92.	*Pestalotiopsis theae* Sawada	Acervular	Holoblastic, annellidic	5 celled Phragmosporous, 3 - apical appendages and single basal appendage concolourous
93.	*Pestalotiopsis versicolor* (Speg.) Stey.	"	"	5 celled Phragmosporous, 3 apical appendages and single basal appendage, versicolored
94.	*Pestalozziella parva* Nag Raj	"	Holoblastic, sympodial	Amerosporous with dichotomously branched apical appendages, hyaline.
95.	*Phaeocytostroma S*acchari (Ell. & Ev.) Sutton.	Stromatic, unilocular	Phialidic	Amerosporous, hyaline
96.	*Phoma* sp. (9)	Pycnidial unilocular	"	Aseptate,or occasionally with single septate,hyaline
97.	*Phomopsis abdita* (Sacc.) Trav.	Eustromatic	"	Amerosporous, hyaline
98.	*Phomopsis archeri* Sutton	"	"	"
99.	*Phomopsis dalbergiae* Sahni	"	"	"

100.	*Phomopsis dracaenae* Sahni	"	"	"
101.	*Phomopsis eugeniae* Punithalingam	"	"	"
102.	*Phomopsis lantanae* (Costa & Camara) Sutton	"	"	"
103.	*Phomopsis mangiferae* Ahmad	"	"	"
104.	*Phomopsis obscurans* (Ell. & Ev.) Sutton	"	"	"
105.	*Phomopsis* sp. (9)	"	"	"
106.	*Phomopsis terminaliae* (P.Henn.) Sutton	"	"	"
107.	*Phyllosticta bokensis* (P.Henn.) Van der Aa,	Pycnidial unilocular	Holoblastic	Amerosporous, hyaline with mucilaginous apical appendage
108.	*Phyllosticta colocasiicola* Hohnel.	"	"	"
109.	*Phyllosticta murrayicola* Van der Aa	"	"	"
110.	*Phyllosticta* sp. (9)	"	"	"
111.	*Phyllosticta spinarum* (Diedicke) Nag Raj & Morelet	"	"	"
112.	*Phyllosticta* state of *Guignardia mangiferae* Roy.	"	"	Amerosporous, hyaline with mucilaginous apical appendage
113.	*Phyllosticta toxica* Ellis & Martin	"	"	"
114.	*Pilidium acerinum* Kunze apud Kunze & Schmidt	Stromatic	Phialidic	Amerosporous, hyaline
115.	*Pseudolachnea longiciliata* (Hino & Katumoto) Sutton	Cupulate to stromatic	"	Phragmosporous hyaline
116.	*Pyrenochaeta ilicis* Wilson.	Pycnidial	Phialidic	Amerosporous, hyaline

117.	*Pyrenochaeta* sp. (1)	"	Phialidic	Amerosporous, hyaline
118.	*Robillarda depazeoides* (W. & C.) Sacc.	"	"	Didymosporous,hyaline with 2-3 apical appendage .
119.	*Seimatosporium rosae* Corda.	Acervular	Holoblastic, Annellidic	Phragmosporous, hyaline
120.	*Seiridium anceps* (Sacc.) Sutton.	"	Holoblastic, annellidic	Phragmosporous, dark brown, distoseptate, except the apical and basal cell, apical cell with single appendage
121.	*Selenophoma* sp. (1)	Stromatic	Holoblastic	Amerosporous, hyaline
122.	*Septoria achyranthis* Chona and Munjal	Pycnidial	Sympodial, blastic	Scolecosporous, hyaline, septate.
123.	*Septoria apicola Speg.*	"	"	"
124.	*Septoria chrysanthemella* Sacc.	"	"	"
125.	*Septoria* sp. (1)	Pycnidial	"	
126.	*Septoria villarsiae* Desm.	"	"	"
127.	*Sphaerellopsis filum* (Biv-Bern Ex Fr.) Sutton	Stromatic	"	Dictyosporous, hyaline
128.	*Sphaeropsis tumefaciens* Hedges	Pycnidial, unilocular	Holoblastic	Amerosporous, hyaline
129.	*Stagnospora brideliae* Thirum & Narasimhan	"	Holoblastic, annellidic	Phragmosporous, pale brown
130.	*Tetranacrium gramineum* Hudson & Sutton	Stromatic	holoblastic	Phragmosporous, hyaline
131.	*Tiarosporella graminis* (Pirozynski & Shoemaker) Nag Raj	Pycnidial, unilocular	"	Amerosporous, hyaline with funnel shaped, mucoid appendage at the apex

132.	*Tiarosporella madreeya* (Subram & Ramakr.) Nag Raj	"	"	Amerosporous, hyaline with conical mucilaginous appendage at the apex
133.	*Truncatella* sp. (1)	Acervular	Holoblastic, annellidic	Phragmosporous, brown
134.	*Urohendersonia pongamiae* Nag Raj & Ponnappa	Pycnidial, unilocular	"	Phragmosporous, brown with mucilaginous apical appendage
135.	*Waydora typica* (Rodway) Sutton	Eustromatic, multilocular	Phialidic	Amerosporous, hyaline

morphology are included for all the species in Table-1. A full description for each species is not possible, as the space is limited. To illustrate the type of conidioma seen in this group of fungi, the diagrammatic representation followed by Kirk *et al.* (2001) is given (Fig.1-4).

References

Archer, W.A. (1926). Morphological characters of some Sphaeropsidales in culture. Ann. Mycol., **24**: 1-84.

Baxter, A.P. (1981). A study of the morphology and taxonomy of certain South African species of *Colletotrichum* Corda. M.Sc. Thesis, University of Pretoria, Pretoria, South Africa.

Baxter, A.P., Van Der Westhuizen, G.C.A. and Eicker., A. (1985). A review of literature on the taxonomy, morphology and biology of the fungal genus *Colletotrichum*. Phytophylactica, **17**: 15-18.

Hughes, S.J. (1953). Conidia, conidiophores and classification. Can. J. Bot., **31**: 577-659.

Kirk, P.M., Cannon, P.F., David, J.C. and Stalpers, J.A. (2001). Dictionary of the Fungi, 9th Edition, CABI pub. UK.

Masilamani, S. and Muthumary, J. (1994). Development of Conidiomata in *Ciliochorella mangiferae* (Coelomycetes). Mycol. Res. **98**(8): 857-861.

Masilamani, S. and Muthumary, J. (1995). Pycnidium ontogeny in *Coleophoma cylindrospora.* Mycol. Res. **99**(6): 693-696.

Masilamani, S. and J. Muthumary. (1996). Development of conidiomata in *Botryodiplodia theobromae.* Mycol. Res., **100**(11): 1383-1387.

Muthumary, J. (1981). Studies on Coelomycetes: Contribution to a monographic account of *Septoria* species in India, Doctoral Thesis, Univ. of Madras.

Muthumary, J. (1986a). *Heteropatella indica* Muthumary sp. nov. from India. Curr. Sci., **55**(5): 250-251.

Muthumary, J. (1986b). *Pilidium acerinum* Kunze A new generic record for India. Curr. Sci., **55**(16): 791-792.

Muthumary, J. (1986c). *Conostroma quercicola* sp. nov. from India. Curr. Sci., **55**(21): 1081-1082.

Muthumary, J. (1987a). Observations on conidiogenesis in *Entomosporium mespili* (Dc ex Duby) Sacc. Curr. Sci. **56**(4): 195-196.

Muthumary, J. (1987b). *Pseudolachnea longicilliata* (Hino and Katumotto) Sutton - A new record for India. Curr. Sci., **56**(7): 306-307.

Muthumary, J. (1987c). Notes on *Cytosphaera mangiferae* Died. Curr. Sci., **56**(9): 426-427.

Muthumary, J. (1988). *Scyphospora phyllostachydis* Kantschaveli A new generic record for India. Curr. Sci., **57**(11): 621-622.

Muthumary, J. (1993). The gelatinous covering of a *Phyllosticta* conidium - SEM study. Mycologist., **7**(1) : 16.

Muthumary, J. and Sutton, B.C. (1986). *Coryneum quercinum* sp. nov. on *Quercus alba* from India. Trans. Br. Mycol. Soc., **86**(3): 512-515.

Muthumary, J. and Vanaja, R. (1986). Development of conidiomata in *Coniella fragariae*. Trans. Br. Mycol. Soc., **87**(1): 109-113.

Muthumary, J., Abbas, S.Q. and Sutton, B.C. (1986). A reassessment of *Monochaetia terminalia.* Trans. Br. Mycol. Soc., **87**(1): 103-108.

Muthumary, J. and Masilamani, S. (1989). Addition to the Coelomycetes of India. J. Indian. Bot. Soc., **68**(1): 131-133.

Muthumary, J., Jayachandra, J.A. and Preetha, M. Bhagavathy (1993). Development of conidiomata in the *Phyllosticta* State of *Guignardia mangiferae* Roy and observation on the fine structure of the conidium. Mycotaxon., **XLVII**: 147-155.

Murugan, M. and Muthumary, J. (2001). Developmental morphology and ultrastructure of *Pestalotiopsis maculans*. Mycotaxon, **LXXIX**, 455 - 465.

Nag Raj, T.R. (1974). Icones generum coelomycetum. VI. Biology series #13, University of Waterloo, Waterloo, Canada. 42 pp.

Nag Raj, T.R. (1975a). Genera Coelomycetum. XI. Hyalotia, Hyalotiella and Hyalotiopsis. Can. J. Bot., **53**: 1615-1624.

Nag Raj, T.R. (1975b). Genera Coelomycetum. XII. Tracyella and Amerodiscosiella. Can. J. Bot., **53**: 2435-2442.

Nag Raj, T.R. (1977). *Psilonia, Acanthotheciella* and *Kazulia* gen. nov. Can. J. Bot., **55**: 1599-1622.

Nag Raj, T.R. (1978). Genera Coelomycetum. XVI. *Allelochaeta, Basilocula, Ceuthosira, Microgloeum, Neobarclaya, Polynema, Pycnidiochaeta* and *Xenodomus*. Can. J. Bot., **56**: 686-707.

Nag Raj, T.R. (1981). Coelomycete Systematics. In Biology of conidial Fungi I (ed. G.T. Cole and B. Kendrick). Pp. 43-79. Academic Press. New York.

Nag Raj, T.R. (1983). Genera coelomycetum. XXI. *Strasseria*, and two new anamorph-genera: *Apostrasseria* and *Nothostrasseria*. Can. J. Bot., **61**: 1-30.

Nag Raj, T.R. (1985a). Redisposals and redescriptions in the *Monochaetia-Seiridium, Pestalotia-Pestalotiopsis* complexes. I. The correct name for the type species of *Pestalotiopsis*. Mycotaxon, **22**: 43-51.

Nag Raj, T.R. (1985b). Redisposals and redescriptions in the *Monochaetia-Seiridium, Pestalotia-Pestalotiopsis* complexes. II. *Pestalotiopsis besseyii* comb. nov. Mycotaxon, **22**: 52-63.

Nag Raj, T.R. (1985c). Redisposals and redescriptions in the *Monochaetia-Seiridium, Pestalotia-Pestalotiopsis* complexes. III. *Monochaetia ilicina* comb. nov. Mycotaxon, **22**: 64-70.

Nag Raj, T.R. (1985d). Redisposals and redescriptions in the *Monochaetia-Seiridium, Pestalotia-Pestalotiopsis* complexes. IV. The true identity of *Monochaetia miersi*. Mycotaxon, **22**: 71-75.

Nag Raj, T.R. (1986a). Redisposals and redescriptions in the *Monochaetia-Seiridium, Pestalotia-Pestalotiopsis* complexes. V. *Monochaetia alnea* and *Monochaetia berberidicola*. Mycotaxon, **26**: 187-198.

Nag Raj, T.R. (1986b). Redisposals and redescriptions in the *Monochaetia-Seiridium, Pestalotia-Pestalotiopsis* complexes. VI. *Pestalotia decolorata* and *Pestalotia heteromorpha*. Mycotaxon, **26**: 199-210.

Nag Raj, T.R. (1986c). Redisposals and redescriptions in the *Monochaetia-Seiridium, Pestalotia-Pestalotiopsis* complexes. VII. *Pestalotia citrina, P. maura* and *Pestalotiopsis uvicola*. Mycotaxon, **26**: 211-222.

Nag Raj, T.R. (1993). Coelomycetous anamorphs with appendage-bearing conidia. Mycologue publications 331 Daleview PI. Waterloo, Ontario, Canada N2L 5M5.

Nag Raj, T.R. and Kendrick, B. (1971). Genera Coelomycetarum. I. *Urohendersonia*. Can. J. Bot., **49**: 1852-1862.

Nag Raj, T.R. and Kendrick, B. (1972a). Genera Coelomycetarum, II *Doliomyces*. Can. J. Bot., **50**: 45-48.

Nag Raj, T.R. and Kendrick, B. (1972b). Genera Coelomycetarum, III *Pestalozziella*. Can. J. Bot., **50**: 607-617.

Nag Raj, T.R. and Kendrick, B.(1978). Genera Coelomycetarum. XV. *Belaina, Belainopsis* and *Crucellisporium*. Can. J. Bot., **56**: 708-714.

Nag Raj, T.R. and Morgan-Jones, G. (1973). Genera Coelomycetarum. VIII. *Rhabdogloeopsis* Petrak and *Rhabdogloeum* Sydow. Can. J. Bot., **51**: 565-569.

Punithalingam, E. (1970). Studies on Sphaeropsidales in culture. Mycol. Pap., **119**: 1-24.

Punithalingam, E. (1974). Studies on Sphaeropsidales in culture. II. Mycol. Pap., **136**: 1-64.

Punithalingam, E. (1979). Graminicolous *Ascochyta* species. Commonwealth Mycological Institute. Mycol. Pap. No.142.

Punithalingam, E. (1989). Techniques for staining fungal nuclei and appendages. Journ. Linn. Soc. Lond., **99**: 19-32.

Subramanian, C.V. and Reddy, K.R.C. (1974). The genus *Discosia* I. Taxonomy. Kavaka, **2**: 57-89.

Sutton, B.C. (1961). Coelomycetes I.Mycol. Pap., **80**: 1-16.

Sutton, B.C. (1962). *Colletotrichum dematium* (Pers. ex Fr.) Grove and *C. trichellum* (Fr. Ex Fr.) Duke. Trans. Br. Mycol. Soc., **45**: 222-232.

Sutton, B.C. (1963). Coelomycetes II. *Neobarclaya, Mycohypallage, Bleptosporium* and *Cryptostictis*. Mycol. Pap., **88**: 1-50.

Sutton, B.C. (1967a). Two new genera of the Sphearopsidales and their relationships with *Diachorella, Strasseria* and *Plagiorhabdus*. Can. J. Bot., **45**: 1249-1263.

Sutton, B.C. (1967b). *Libertina*, a synonym of *Phomopsis*. Trans. Br. Mycol. Soc., **50**: 355-358.

Sutton, B.C. (1968). *Kellermania* and its generic segregates. Can. J. Bot., **46**: 181-196.

Sutton, B.C. (1969). Type studies in *Coniella, Anthasthoopa* and *Cyclodomella*. Can. J. Bot., **47**: 603-608.

Sutton, B.C. (1971). Coelomycetes IV. The genus *Harknessia* and similar fungi on *Eucalyptus*. Mycol. Pap., **123**: 1-46.

Sutton, B.C. (1972a). *Wakefieldia punctata* Amaud and *Blennoria buxi* Fr. Sensu Amaud Trans. Br. Mycol. Soc., **59**: 285-294.

Sutton, B.C. (1972b). Nomenclature of *Ceuthospora, Pyrenophora* and *Blennoria* (Fungi). Taxon, **21**: 319-326.

Sutton, B.C. (1973). Coelomycetes. Ch. 11 pp. 513-582, In The Fungi, Volume IV A, Eds.: Ainsworth, G.C., Sparrow, F.K. and Sussman, A.S., Academic Press, New York, U.S.A. 621 pp.

Sutton, B.C. (1974). Miscellaneous Coelomycetes on *Eucalyptus*. Nova Hedwigia, **25**: 161-172.

Sutton, B.C. (1975). Coelomycetes V. *Coryneum*. Mycol. Pap., **138**: 1-224.

Sutton, B.C. (1977). Coelomycetes VI. Nomenclature of generic names proposed for Coelomycetes. Mycol. Pap., **141**: 1-253.

Sutton, B.C. (1980). The Coelomycetes Fungi imperfecti with pycnidia, acervuli and stromata. Commonwealth Mycological Institute, England. 696 pp.

Von Höhnel, F. (1923). System der Fungi Imperfecti Fuckel. Mykol. Unters., **1**(3) : 301-369.

28

Vermiculariopsiella Bender : Present Status of Species Diversity

T.S. Keshavaprasad, Maria D'souza and D. J., Bhat
Department of Botany, Goa University, Goa-403 206, India

Genus *Vermiculariopsiella* Bender is briefly reviewed in this paper. Three new species, *V. elegans*, *V. parva* and *V. indica*, producing solitary, cylindrical, straight or slightly curved, smooth, aseptate, thick-walled, colourless, phialoconidia with rounded ends at the tip of conidiogenous cells arranged in palisade manner in setose sporodochium, recovered as litter and endophytic fungi from various plant species in India, are added. A taxonomic key is provided to the species in the genus.

Key Words: Biodiversity, tropical fungi, taxonomy.

Hohnel (1918) transferred *Excipula immersa* Desm. into *Vermiculariopsis* v. Hohnel, as *V. immersa* (Desm.) Hohnel. The new genus being a later homonym of *Vermiculariopsis* Torrend, Bender (1932) proposed *Vermiculariopsiella* Bender for *Excipula immersa* Desm. as *Vermiculariopsiella immersa* (Desm.)Bender. The original diagnosis of the genus included species with one-celled conidia. Considering the greater pertinence of this feature at the specific rather than generic level, Pasqualitti and Zucconi (1992) suggested inclusion of species also with septate conidia in the genus. This provided room for retension of *V. falcata* Nawawi *et al.* (1990) characterized by the presence of 3-septate conidia in the genus.

While adding two more new species, *V. falcata* Nawawi, Kuthubutheen and Sutton and *V. parvula* Nawawi Kuthubutheen and Sutton, Nawawi *et al.* (1990) revised the genus *Vermiculariopsiella* with three new combinations and provided a key for then recognised species. *V. immersa* var. *ramosa* Sutton was elevated to the species level as *V. ramosa* (Sutton) Nawawi Kuthubutheen and Sutton, *Oramasia hirsuta* Urries var. *cubensis* Castaneda and *Gyrothrix cornuta* V. Rao & de Hoog were recognized as *V. cubensis* (Castaneda) Nawawi Kuthubutheen and Sutton and *V. cornuta* (V. Rao & de Hoog) Nawawi Kuthubutheen and Sutton, respectively. Pasqualitti and Zucconi (1992) added *V. arcicula* Pasqualitti and Zucconi into the genus.

Till date, *Vermiculariopsiella* accommodates 7 species. Of these, 3 species, viz., *V. cornuta*, *V. cubensis* and *V. ramosa* possess branched setae. The other 4 species produce unbranched setae and narrow conidia with less than 4 µm width. Of these, *V. falcata* produces falcate, 3-sepate hyaline conidia of 36-47 x 1.5-2 µm. *V. immersa* and *V. parvula* produce aseptate hyaline conidia of 13-23 x 1.5-2.5 µm and 8-13 x 2-2.5 µm, respectively with rounded base and pointed, slightly curved apex. The apex of conidiogenous cell is recurved in the former while it is narrow, long and straight in the latter. *V. arcicula* produces aseptate hyaline fusiform conidia of 15-19.5 x 2.5-4 µm.

During studies on the taxonomy and diversity of microfungi, three new hyphomycetous fungi belonging to *Vermiculariopsiella* were recovered as endophytes from *Saraca asoca* (Roxb.) De Wilde from forests of Western Ghats in Goa State and *Pimenta dioica* (L.) Merr. from a mixed plantation in north Malabar region of Kerala State, India. Two of these were also isolated from leaf litter of the respective plants. The fungi are described below.

Fresh leaves of *Saraca asoca* and *Pimenta dioica* were thoroughly washed and subjected to a 'three-step-isolation' process of Fisher and Petrini (1986) for isolation of endophytic fungi. Decaying leaves of the plants were thoroughly washed in deionized tap water and subjected to (i) 'moist chamber incubation' (Hawksworth, 1974) and (ii) 'particle plating technique' (Bills and Polishook, 1994) for recovery of litter fungi. Conidiophores with a crown of conidia appeared on the surface of the twigs from 4th day and on culture media after two weeks of incubation.

Taxonomic Part

Vermiculariopsiella elegans sp.nov. (Fig.1)

Coloniae irregularise, rhizoidacea ad marginis, nigra ad centrum et hyalinae ad ambitum, cum fructificatum mucosum, inverto incouloris vel pallide brunneus, 5.4 cm in diam. in 7 dies. *Mycelium* partim superficiale, ex hyphis laevis, tenuitunicatus, septatis, ramosis, pallide brunneus, 3.5 µm lat., compositum. *Sporodochia* disseminata, dissita, punctiformia, circumscripta, irregulariter circularia, nonnumquam coalescentia, conidis agglutinatis cremea, 80-130 µm. *Setae* numerosae, laeves, 1-3-septata, ereta, curvata vel modice flexuosa, simplex, 103-165 µm longa, 6-7 µm lat. ad basim, 5-5.5 µm ad supra et 3.5 im ad apicem, prope basim crassae, crassitunicatae, brunneae, gradatim tenuiores tunicatae et pallidiores ad angustiorem et leviter rotundatum apicem. *Conidiophora* laevia, septata, ramosa, atrobrunnea ad nigra, crassitunicata, 8-12 x 3-5 µm. *Cellulae conidiogenae* monophialidicae, intergrata vel discretae, 15-30 µm long, 4-5 µm lat. ad basim, 4.8-6.2 µm lat. ad supra et 1.3 µm lat. ad apicem, experse collarettis. *Conidia* solitaria, cylindricale, aseptata, levia, hyalina 20-27 x 6-8 µm.

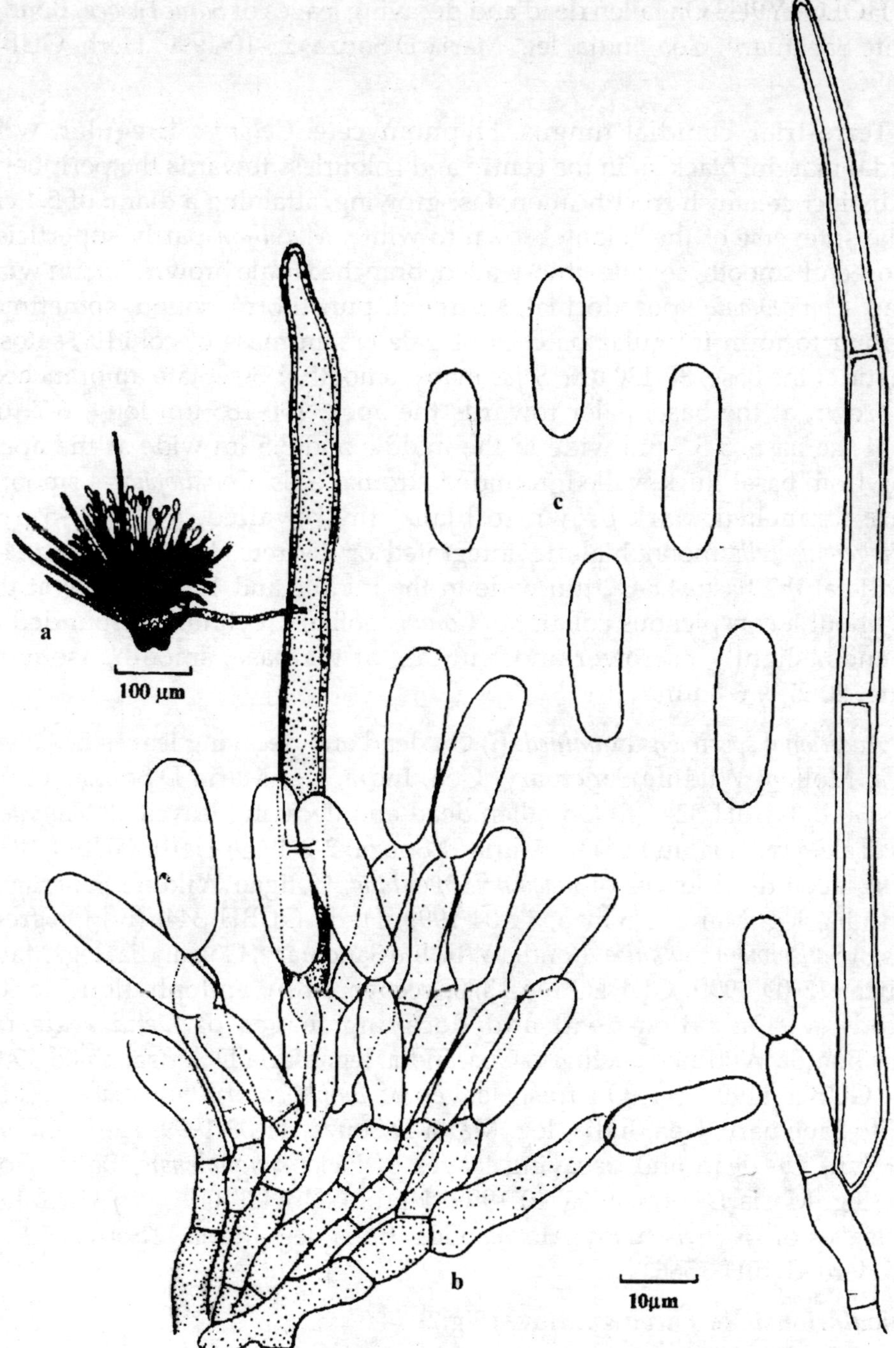

Fig.1. *V. elegans* sp.nov: a. Habit, b. A sporodochium with setae, c. Conidia.

HOLOTYPUS On fallen dead and decaying leaves of *Saraca asoca*, Bondla Wildlife Sanctuary, Goa, India, leg. Maria D'Souza, 25-10-1999, Herb. GUBH No.1496.

Terrestrial, conidial fungus, Hyphomycete. *Colonies* irregular, with rhizoidal margin, blackish in the centre and colourless towards the periphery, with distinct creamish fructification, fast growing, attaining a diam. of 5.4 cm in 7 days; reverse of the colony brown to white. *Mycelium* partly superficial, composed of smooth, septate, thin-walled, branched, pale brown, 3.5 µm wide hyphae. *Conidiomata* sporodochial, scattered, punctiform, round, sometimes coalescing to form irregular patches of pale cream mass of conidia, setose, stromatic at the base, 80-130 µm. *Setae* many, smooth, 1-3-septate, unbranched, dark brown at the base, paler towards the apex, 100-165 µm long, 6-7 µm wide at the base, 5-5.5 µm wide in the middle and 3.5 ìm wide at the apex, arising from basal thick-walled, rounded stromal cells. *Conidiophores* smooth, septate, branched, dark brown to black, thick-walled, 8-12 x 3-5 µm. *Conidiogenous cells* monophialidic, integrated or discrete, 15-30 µm long, 4-5 µm wide at the base, 4.8-6.2 µm wide in the middle and 1.5 µm wide at the tip, without a conspicuous collarette. *Conidia* solitary, cylindrical rounded at both ends, slightly narrower and truncate at the base, smooth, aseptate, hyaline, 20-27 x 6-8 µm.

Additional specimens examined: (i) On dead and decaying leaves of *Careya arborea*, Mollem Wildlife Sanctuary, Goa, India, leg. Maria D'Souza, 11-03-1999, Herb. GUBH 324. (ii) On fallen dead and decaying leaves of *Mangifera indica*, Taleigao, Goa, India, leg. Maria D'Souza, 2-7-1999, Herb. GUBH 1038. (iii) On fallen dead leaves of *Flacourtia montana,* Cotigao Wildlife Sanctuary, Goa, India, leg. Maria D'Souza, 11-04-1999, Herb. GUBH 344. (iv) On fresh leaves of *Sanseiviera zeylanica,* Bondla Wildlife Sanctuary, Goa, India, leg. Maria D'Souza, 15-09-1999, GUFCC No.1370; recovered by endophytic isolation method. (v) On fallen dead and decaying leaves of *Dendrocalamus strictus,*Bondla Wildlife Sanctuary, Goa, India, leg. Maria D'Souza, 15-08-1999, Herb. GUBH 1190. (vi) On fresh leaves of *Bambusa arundinaceae,* Bondla Wildlife Sanctuary, Goa, India, leg. Maria D'Souza, 15-08-1999, Herb. GUBH 1195.. (vii) On dead and decaying leaves of *Ficus benghalensis,* Baga, Goa, India, leg. Maria D' Souza, 11-10-1999, Herb. GUBH 1405. (viii) On fallen dead leaves of *Helictris ixora*y Alorna, Goa, India, leg. Maria D'Souza, 19-11-1999, Herb GUBH 1566.

Vermiculariopsiella parva **sp. nov. (Fig.2)**

Coloniae irregularise, levia, subhyalina ad centrum et hyalina ad ambitum, cum fructificatum mucosum, aurantium, disseminatum, 6 cm diam. in MEA

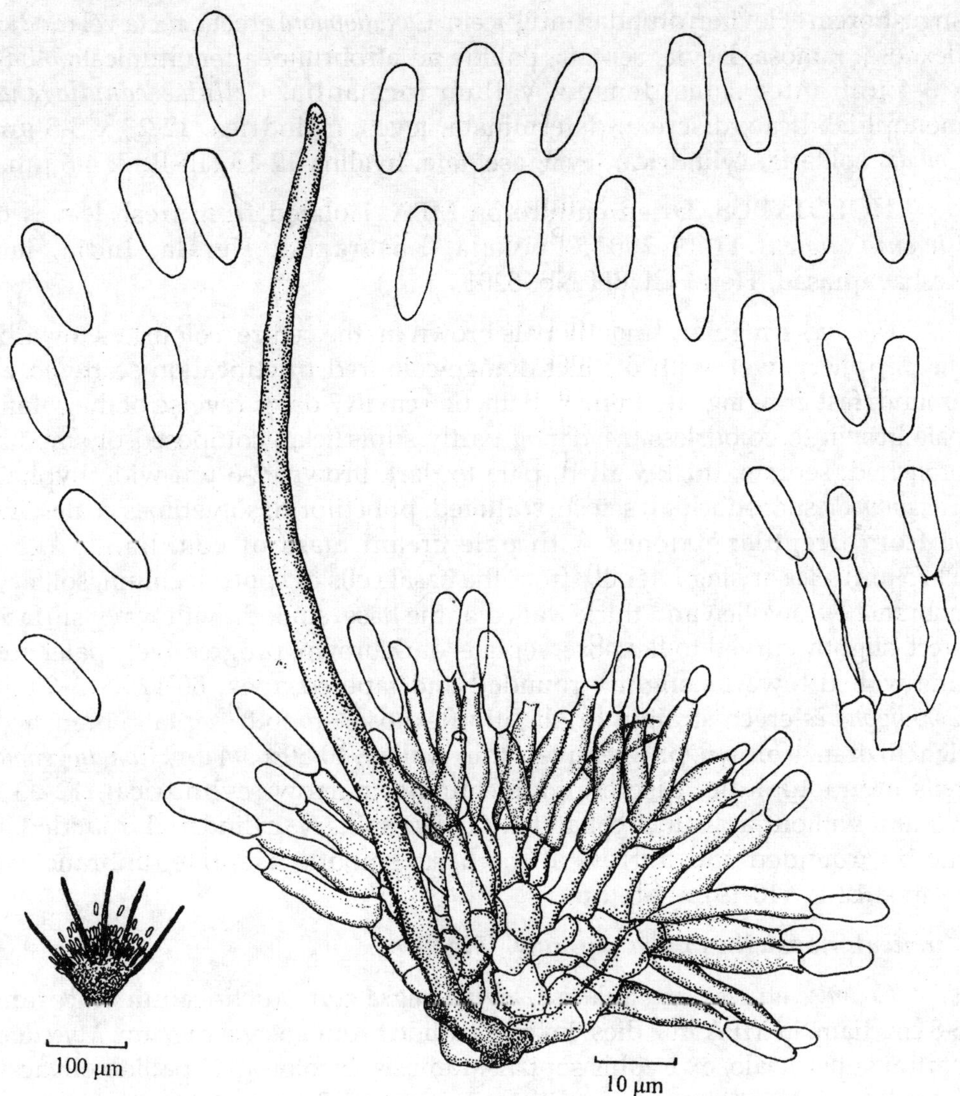

Fig.2. *V. parva* sp.nov: a. Habit, b. A sporodochium with a seta, c. Condiogenous cells with conidia, d. Conidia.

in 7 dies, inverto pallide brunneus vel. Incouloris. *Mycelium* partim superficiale, ex hyphis septatis, ramosis, crassitunicatus, incoloris vel pallide olivaceis, 2-6 μm lat. compositum. *Conidiomata* sporodochia, setosa, dissita, punctiformia, circumscripta, irregulariter circularia, nonnumquam coalescentia, conidis agglutinatis cremea, 25-125 x 55-75 μm. *Setae* numerosa, ereta, curva vel modice flexuosa, simplex, septata, laeves, 80-125 x 3-6 μm, prope basim crassa, crassitunicata, brunneae, gradatim tenuiores tunicatae et pallidiores ad angustiorem et leviterrotundatum apicem. *Conidiophora* erecta, recta vel modice flexuosa, ramosa, laevia, septata, pallide ad atrobrunnea tenuitunicata, 40-80 x 3-4 µm, inter setas densum vallum formantia. *Cellulae conidiogenae* monophialidicae, discreta, determinata, leves, cylindrica, 12-23 x 3-5 µm. *Conidia* solitaria, cylindrica, levia, aseptata, hyalina 12-15 (10-19) x 4-6 µm.

HOLOTYPUS, Dried culture on MEA, isolated from fresh leaves of *Pimenta dioica*, 11-11-2001, Pookala, Kasaragod, Kerala, India, leg. Keshavaprasad, Herb. GUBH No. 5201

Colonies irregular, smooth, pale brown in the centre, colourless towards the periphery, wet, with distinct orange-coloured fructification scattered all around, fast growing, attaining a diam. of 6 cm in 7 days; reverse of the colony pale brown to colourless. *Mycelium* partly superficial, composed of smooth, branched, septate, thick-walled, pale to dark brown, 2-6 µm wide hyphae. *Conidiomata* sporodochial, setose, scattered, punctiform, sometimes coalescing to from irregular patches with pale cream mass of conidia, 25-125 x 55-75 µm; *setae* arising laterally from the basal cells of sporodochium, solitary, unbranched, swollen and thick-walled at the base, smooth with wavy surface, erect, slightly curved to flexuose, septate, dark brown, progressively paler and thin-walled towards slightly rounded and tapered apex, 80-125 x 3-6 µm. *Conidiophores* erect, straight or slightly flexuose, smooth, septate, branched, light to orange brown, progressively thin-walled, 40-80 x 3-4 µm. *Conidiogenous cells* monophialidic, discrete, determinate, narrow, cylindrical, 12-23 x 3-5 µm, without a distinct collarette. *Conidia* solitary, cylindrical, rounded at the tip, rounded and narrow at the base, smooth, aseptate, unbranched, hyaline,12-15 (10-19) x 4-6 µm.

***Vermiculariopsiella indica* sp. nov. (Fig.3)**

Coloniae effusae, irregularise, leviae, nigra cum fructificatum mucosum, 6-8 cm diam. in MEA in 7 dies, inverto pallide brunneus vel nigrum. *Mycelium* partim superficiale, ex hyphis septatis, ramosis, incoloris vel pallide olivaceis 2 ìm lat., et hyphis atrofucus, crassitunicatus, 3-6 µm lat., compositum. *Conidiomata* sporodochia, setosa, disseminata, punctiformia, circumscripta, irregulariter circularia, nonnumquam coalescentia, conidis agglutinatis cremea, 120-150 x 100-200 µm. *Setae* numerosa, erecta, curva vel modice flexuosa, simplex, septata, leves, 125-300 x 4-6 µm prope basim crassae, crassitunicata, brunnea, gradatim tenuiores tunicatae et pallidiores ad angustiorem et leviterrotundatum apicem. *Conidiophora* erecta, recta vel modice flexuosa, laevia, septata, ramosa, pallide brunnea, tenuitunicata, 60-100 x 6-10 µm longa. *Cellulae conidiogenae* monophialidicae, integrata vel discreta, determinata, cylindricales, laeves, tenuitunicata, hyalina, expers collerettis, 15-26 x

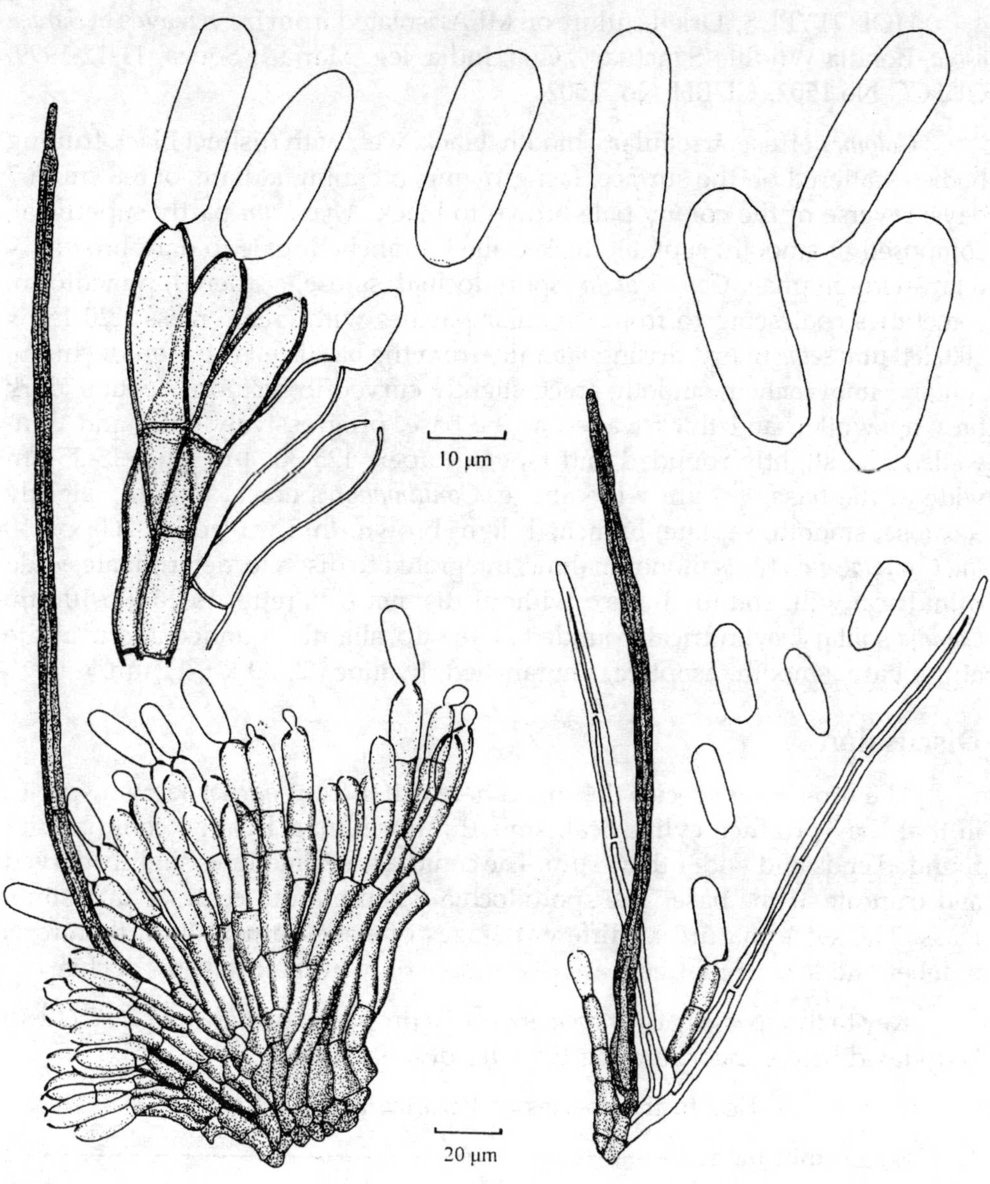

Fig.3. *V. indica* sp. nov: a. A sporodochium with a seta, b. Setae with condiophores, c. Condiogenous cells with conidia, d. Conidia.

6-10 μm. *Conidia* solitaria, cylindrica, levia, aseptata, tenuitunicata, hyalina 22-30 x 8-11 μm.

HOLOTYPUS, Dried cutlure on MEA, isolated from fresh leaves of *Saraca asoca*, Bondla Wildlife Sanctuary, Goa, India, leg. Maria D'Souza, 11-12-1999, GUFCC No.1502, GUBH No. 1502.

Colonies effuse, irregular, smooth, black, wet, with distinct black fruiting bodies scattered on the surface, fast growing, attaining a diam. of 6-8 cm in 7 days; reverse of the colony pale brown to black. *Mycelium* partly superficial, composed of smooth, septate, thick-walled, branched, pale to dark brown, 2-6 μm wide hyphae. *Conidiomata* sporodochial, setose, scattered, punctiform, sometimes coalescing to from irregular patches with cream mass, 120-150 x 100-200 μm; *setae* many, arising laterally from the basal cells of sporodochium, solitary, unbranched, smooth, erect, slightly curved to flexuose, septate, dark brown, swollen and thick-walled at the base, progressively paler and thin-walled and slightly rounded and tapered apex, 125-300 μm long, 15-25 μm wide at the base, 4-6 μm wide above. *Conidiophores* erect, straight, slightly flexuose, smooth, septate, branched, light brown, thin-walled, 60-100 x 6-10 ìm. *Conidiogenous cells* monophialidic, integrated to discrete, determinate, wide cylindrical with rounded apex, without distinct collarette, 15-26 x 6-10 μm. *Conidia* solitary, cylindrical, rounded at the tip, slightly rounded and truncate at the base, smooth, aseptate, unbranched, hyaline, 22-30 x 8-11 μm.

Discussion

The three new species described here differ from earlier known species in that they produce cylindrical, smooth, unicellular, hyaline conidia with rounded ends and wider than 4 μm. The conidia are sometimes slightly curved and truncate at the base. The sporodochia when mature coalesce into slimy mass. The setae mature at different stages of development and, therefore, number and size of the setae are discernible only in mature sporodochia.

Key to the species of *Vermiculariopsiella* proposed by Nawawi *et al.* (1990) is updated below incorporating the later described taxa.

Key to the species of ***Vermiculariopsiella***

1. Setae unbranched -- 2
1. Setae branched -- 6
2. Conida cylindrical with rounded ends ---------------------------------- 3
2. Condia with pointed apex -- 4
3. Conidia 20-25 x 6-8 ìm ---------------------------------- ***V. elegans* sp.nov.**

3. Conidia 12-15 (10-19) x 4-6 ìm------------------------------ *V.parva* **sp.nov.**

3. Conidia 22-30 x 8-11 ìm---------------------------------- *V. indica* **sp. nov.**

4. Conidia falcate, apex curved and pointed----------------------- *V. falcata*

4. Conidia fusiform, slightly curved, 15-19.5 ìm long------------*V. arcicula*

4. Conidia cylindrical, straight, apex acute, base rounded------------------5

5. Conidia 14-25 ìm long; conidiogenous cell apex recurved - *V. immersa*

5. Conidia 8-13 ìm long; conidiogenous cell apex narrow, straight ------- --- *V. parvula*

6. Setae mostly once dichotomously branched ------------------- *V. ramosa*

6. Setae thrice dichotomously branched---------------------------- *V. cornuta*

6. Setae with short primary and secondary branches, terminal cells with hair-like appendages -- *V. cubensis*

Acknowledgement

This work is supported by a research grant to Dr. D. J. Bhat from the Ministries of Science and Technology and Environment & Forests, Government of India, New Delhi. We are indebted to Prof. Roger D. Goos, University of Rhode Islands, for kindly reviewing the manuscript.

References

Bender, H.B. (1932). The genera of Fungi Imperfecti. Mycologia, **14:** 410-412.

Bills, G.F. and Polishook, J.D. (1994). Abundance and diversity of Microfungi in leaf litter of a lowland rain forest in Costa Rica. Mycologia, **86:** 187-198.

Hawksworth, D.L. (1974). *Mycologist's Handbook.* Commonwealth Mycological Institute. Kew, U.K.

Hohnel, F.Von. (1918). Dritte vorläufige Mitteilung mycologischer Ergebnisse (Nr.201-304). Ber.dtsch. Bot.Ges., **36:** 309-317.

Nawawi, A., Kuthubutheen, A.J. and Sutton, B.C. (1990). New species and combinations in *Vermiculariopsiella* (Hyphomycetes). Mycotaxon, **37:** 173-182.

Pasqualetti, M. and Zucconi, L. (1992). *Vermiculariopsiella arcicula*, a new dematiaceous hyphomycete from Sardinia, Italy. Mycotaxon, **43:** 1-7.

Petrini, O. (1991). Fungal endophytes of tree leaves. In: *Microbial ecology of leaves.* (Eds. J.H. Andrews and S.S. Hirano). Springer, New york. pp. 179-197.

29

The Genus *Phaeoramularia* Muntañola in Indian Sub-continent

Sanjay K. Singh

Mycology and Plant Pathology Group, Plant Sciences Division, Agharkar Research Institute, Pune-411 004

A monographic review of the cercosporoid genus *Phaeoramularia* and its 107 valid species hitherto reported is discussed. Taxonomic descriptions and other details of 31 species from Indian sub-continent, especially from India, Nepal and Pakistan have been given in brief. The species concept, economic significance and the status of *Phaeoramularia* in Indian sub-continent have been discussed. The delimitation and other taxonomic problem that existed, with the related genera of cercosporoid fungi are discussed. In addition to the morpho-taxonomy, the possibilities of using molecular tools in taxonomy of *Phaeoramularia* have been also suggested.

Key words: Biodiversity, Cercosporoid fungi, *Phaeoramularia*, Indian sub-continent.

Phaeoramularia Munt. is an important genus of cercosporoid hyphomycetes due to its ability to cause diseases in ornamental, fruit plants and forests. The species of *Phaeoramularia* colonize the leaves and rarely the fruits. The majority have inter-tropical distribution, whereas a few venture into North Temperate Zone. The present communication includes a review of the hitherto described species of *Phaeoramularia* and brief taxonomic details of species published from Indian sub-continent and evolution of the 'species concept' in this genus.

Historical Review

Muntañola (1960) established the genus *Phaeoramularia* to accommodate two brown-spored species of *Cercospora* Fres. bearing thickened conidial scars and internal vegetative mycelium. The type species, *P. gomphrenicola* (Speg.) Munt., a foliicolous fungus occurring on *Gomphrena glauca* (Spreng.) Moq., was originally named as *Cercospora gomphrenicola* Speg. (Spegazzini, 1882). All the four species were segregated on the basis of production of distinct leaf lesions and thickened conidial scars on the conidiophores, similar to some

e-mail: singhsksingh@rediffmail.com

other genera in the *Cercospora*-complex. They can be distinguished by their habit, branched conidiophores emerging through stomata, and conidia produced in chains, which are sometimes branched. Muntañola had also given the culture characteristics of the type species (Deighton, 1979). Sutton (1970) added *Phaeoramularia maculicola* (Rom. & Sacc.) Sutton (=*Torula maculicola* Rom. and Sacc.) on *Populus* sp. from Canada followed by Marasas and Bredel (1974) and Matsushima (1975), who added new species *Phaeoramularia kellermaniana* Marasas and Bredel *and P. hachijoensis* Matsushima, collected from South-Africa and Japan, respectively, on dead stem of *Triticum aestivum* and *Cinnamomum japonicum*.

Von Arx (1974) placed *Phaeoramularia* Munt., *Mycovellosiella* Rangel and *Stenella* Sydow in synonymy with *Cladosporium* Link *ex* Fries on the basis of their similarities in axenic cultures. However, Deighton (1979) and subsequent workers did not find this tenable. *Phaeoramularia* remained, however, a relatively small genus for well over a decade after its establishment, until 21 species were included (Ellis 1976; Deighton, 1976, 1979). However, all but one was proposed as new combination (Deighton, 1976) in *Cercospora*. A plethora of taxa have been added in this genus and *Phaeoramularia* contains a total of 107 species to this date.

Liu and Guo (1982) described nine species from China, of which four *viz*., *Phaeoramularia acericola, P. euphorbiae, P. helianthi* and *P. markhamiae* on *Acer truncatum, Euphorbia lathyris, Helianthus debilis* and *Markhemia caudafilina*, respectively were described as new and *P. barringtoniae, P. eupatorii-odorati, P. fusimaculens, P. panicillata* and *P. trilobi* as comb. nov.

Phaeoramularia weigelae, P. pruni and *P. montana* were described as new species and comb. nov., respectively on *Weigelia japonica, Prunus* sp., and *Chamaenerion* from China (Guo, 1986, 1987, 1989). Subsequently, Guo (1993) added *Phaeoramularia schisandrae* from *Schisandra chinensis* .

Phaeoramularia tabebuiae Muchovej and Ferreira and *P. pachirae* (Chupp and Muller) Muchovej and Ferreira were reported from Brazil on *Tabebuia serratifolia* and *Bombax insigne*, respectively (Muchovej and Ferreira, 1981, 1991). *Phaeoramularia tabebuia* was cultured on PDA.

Phaeoramularia isotomae and *P. acanthicola* were described by Deighton (1986, 1987) on *Isotoma longiflora* and *Acanthus arboreus*, from Trinidad, (West Indies) and Uganda, respectively in which *P. acanthicola* (Hansford) Deighton (=*Cercospora acanthicola* Hansford),was described as comb. nov., while the former was described as new. Moreover, *Phaeoramularia angolensis* a potent leaf and fruit disease causing agent was reported from Africa and Asia (Yemen) as a new combination [=*Cercospora angolensis* Carvalho & Mendes] by Kirk

(1986) primarily on *Citrus* species.

More recently, Braun (1992) reported *Phaeoramularia minitissima* (Desm.) Braun (=*Selenosporium minitissima* Des.) and *P. punctiformis* (Schlechtendal) Braun [=*Fusidium punctiformae* Schlechtendal] from Germany. Similarly, *Phaeoramularia hoenelii* Petzoldt nom. nov. [Bas. *Fusicladium heterosporum* v. Hohn.) was described by Petzoldt (Braun,1992) collected on *Epilobium parviflorum* from Australia (non *P. heterospora* Ell. and Galloway).

Shin and Braun (1993) reported *P. valerinicola* from Korea, on *Valeriana fauriei* and Braun and Rogerson (1993) described *P. coalescenc*, a common leaf pathogen, as new combination (=*Cercospora coalescence* Davis) on species of *Ribes*. *Phaeoramularia collomiae* Braun and Rogerson was reported to infect *Collomia linearis* in USA. Recently, Morgan-Jones (1997) has published a comprehensive note on taxonomic status of these species. In addition, Braun (1993) made significant contributions to the genus by recombining many species from other generic segregates such as *Phaeoramularia scandicearum* (Magnus) Braun [=*Cercospora scandicearum* Magnus], *P. momordicae* (Heald and Wolf.) Braun, (=*Ramularia momordicae* Heald and Wolf.) Braun, *P. viburni* (Ell. and Everh.) Braun (=*Ramularia viburni* Ell. and Everh.) and *P. actaeae* (Ell. and Halw.) Braun (=*Ramularia actae* Ell. and Halw) on the hosts *Torolis japonica*, *Momordica balsamina*, *Viburnum lentago* and *Actaea alba*, respectively. Later on Braun (1995) described *P. ahmadii* (Petr.) Braun as comb. nov. (=*C. ahmadii* Petr.) on *Marsdenia roylei*. Garcia and Pons transferred *Cercospora alabamensis* Atk. to *Phaeoramularia. alabamensis* (Atk.) Garcia and Pons found in Alabama, USA (Garcia *et al.*,1996).

Phaeoramularia chuppii (Viegas) Inácio and Dianes was described as comb. nov. (=*Cercospora chuppii* Viegas) on *Ormosia arborea* Harms from Brazil (Inácio *et al.*, 1996). Similarly, Inácio and Dianese (in Crous *et al.*, 1997) transferred *Cercospora pyrostegiae* Viegas into *Phaeoramularia* Munt. as *P. pyrostegiae* (Viegas) Inácio and Dianes causing leaf spots on *Pyrostegia venusta* in Sao Paulo, Campinas (Crous *et al.*, 1997). Further, *Phaeoramularia heliotropi* (Ell. and Everh.) Pons and Sutton (=*Cercospora heliotropii* Ell. and Everh.) was reported on two species of *Heliotropium* (Pons and Sutton, 1996). Shin and Braun (1996) reported *P. weigelicola* (syn. *Ramularia weigelicola* Speg.) as causal agent of *Weigelia* sp. but, this species is different from already known *Phaeoramularia weigeliae* Guo & Liu, reported on the same host genus. Guo (1996) reported *Phaeoramularia vexans* (Massd.) Guo (=*Cercospora vexans* Massd.) on *Frageria vesca* from Italy.

Braun and Melnik (1997) reported *P. levieri* (Magnus) Braun, comb. nov. (*Fusicladium levieri* Magnus) from Georgia and Japan. Crous and Sutton (1997) reported *P. digitaria* as a new species on *Digitaria diagonalis*. Crous *et al*., (1999a) transferred 3 species of *Cercospora* occurring on *Croton* spp., namely, *C. crotonis* Ell. & Everh., *C. crotonis-oligandri* Yen and Gilles and *C. maritima* Tracy and Earle to *Phaeoramularia* Munt.

Braun (1999) proposed *Phaeoramularia gilbertii* (Speg.) Braun (=*Cercospora gilbertii* Speg.) on *Iresine celosiodes* from Uruguay and *P. occulata* (Ell. and Kellerm.) Braun (=*Cercospora occulata* Ell. and Kelerm.) on *Vernonia baldwini* from Kansas. *Phaeoramularia destruens* Anderson *et al*., occurring on *Dichondra* sp. from Argentina, was reported as new by Anderson *et al*., (1998) from Brazil. Similarly, *P. adenocalymmae* (Mull. and Chupp) Crous and Camara was proposed as comb. nov. of *Cercospora* sp. (Crous and Camara, 1998) from Brazil. Crous (1998) reported *Phaeoramularia eucalyptorum* as cause of leaf spot disease of *Eucalyptus saligena* in S. Africa. Braun (1998) added *P. cephalanthi* (Ell. & Kellerm.) Braun (=*Cercospora cephalenthi* Ell. and Kellerm.) as comb. nov.

Phaeoramularia malpighiae Braun and Muchacca was described from Tahiti, French Polynesia, on leaves of *Malpighia punicifolia* (Braun *et al*., 1999). *Phaeoramularia. saururi* (Ell. and Everh.) Crous and El-gholl was reported as comb. nov. by Crous *et al.* (1999 b) on species of *Saururus* from Louisiana , U.S.A.

Generic features of *Phaeoramularia*

Habit: foliicolous. *Infection spots* slight discolouration to necrotic. Caespituli punctiform, effuse, velvety. Mycelium internal. *Stroma* absent to well developed, substomatal to intra-epidermal or erumpent. Conidiophores in small to large fascicles, rarely solitary or in sporodochial conidiomata, unbranched to branched, pigmented, non-geniculate to geniculate, transversely septate. Conidiogenous cells integrated, terminal to intercalary, or conidiophores reduced to a single conidiogenous cell, polyblastic, sympodial, cicatrized, conidial scars slightly thickened to thickened and more or less darkened. Conidia catenate in unbranched to branched chains, (rarely solitary), variable in shape, amerosporous to scolecosporous, (0)-1 to pleuri-transversely septate, hilum thickened or somewhat darkened. Conidial secession schizolytic.

Type species: *Phaeoramularia gomphrenicola* (Speg.) Munt., **Lilloa**, 30: 165-232, 1960. = *Cercospora gomphrenicola* Speg., **An. Soc. cient argenti**. 13: 11-35, 1882. [Ellis (1971, 1976), Deighton (1979), von Arx (1983), Pons & Sutton (1996), Braun (1996, 1997), Singh *et al.* (1995)].

Species concept

The species concept in cercosporoid fungi is a combination of morphological features and biological criteria. Morphologically discernible differences are the general fundamentals for the delimitation of the species. Morphologically indistinguishable species on host plants belonging to the same genus or species are treated as single species. The polyphagus species in this group of fungi are seldom encountered. Species of *Phaeoramularia* are rather host specific with each being restricted to a single host species or genus. The exceptions are listed in Table 1.

Table 1 : List of *Phaeoramularia* species with multiple host range

Species	Host species
Phaeoramularia angolensis	*Citrus sinensis, C. aurantium*
P. antipus	*Lonicera flava, L. sempervirens, L. glaucenscens, L. dioca, L. hirsuta, L. sullivantii, P. coalescens, Ribes bracteosum, R. inerme*
P. dioscoreae	*Dioscorea bulbifera, D. japonica, D. subclavatum*
P. fusimaculens	*Echinochloa crusgalli, Panicum miliaceaum, Setaria plicata*
P. hoehnelii	*Epilobium parviflorum, E. hirsutum*
P. marmorata	*Rhus coriaria, R. glutinosa, R. villosa,*
P. occidentalis	*Cassia occidentalis, C. corymbosa, C. merilandica*
P. oldenlandiae	*Borreria sinensis, Oldenlandia sp.*
P. periclymeni	*Lonicera periclymenum L. implexa, L. albiflor*
P. plucheae	*Pluchea odorata, P. purpurescens*
P. rhois	*Rhus petitiana, R. villosa*
P. scandicearum	*Torilis japonicus, Cherophylum femulum, C. hirsutum*
P. sudanensis	*Clematis longicauda, Clematis sp.*

Economic Significance

Phaeoramularia angolensis is of specific importance, since it causes diseases of leaf and fruit on *Citrus* plants over a wide geographical area, and poses a threat to the global Citri-culture. Diseased fruits become juiceless and unsuitable for consumption. Yield losses range from 50-100 %. The disease was reported for the first time in 1952 from Mozambique (Southern Africa).

In addition, many species are found on forest plants. The infection by *Phaeoramularia* species causes loss of chlorophyll, leading to reduction in photosynthetic area of the plants and its biological activity. According to an estimate, their average infection, on susceptible plants often accounts for as much as 60 % reduction in net photosynthesis and overall growth. Though the forests are one of the largest revenue earners and controller of environmental quality, the colossal loss caused by foliicolous fungi has not attracted the attention of professional biologists.

Our knowledge about the range of distribution and full spectrum of biodiversity and its taxonomy is prerequisite to have a thorough understanding of the role of foliicolous fungi in the forest ecosystems. Hsieh and Goh (1993) have estimated that one species of *Cercospora sensu lato* occurs over every 10th species of Taiwanese vascular plants and that the cercosporoid fungi alone may be a group of 25,000 species word-wide.

A few species of cercosporoid fungi are reported to be 'hyperparasites' on other fungi and colonize lichens, as well. A few have also been found to be potential biocontrol agents of weeds. Among these some species in the *Cercospora* complex have been of special interest; *Cercospora piaropi* Tharp and *C. romanii* Conway as control agents of *Eichhornia crassipes* (Mart.) Solms. (Julien, 1992*)*, *Mycovellosiella lantanae* (Chupp) Deighton against *Lantana camara* (Deighton, 1974; Freeman, 1981), *Phaeoramularia* sp. for *Ageratina adenophora* (=*Eupatorium adenophorum*) (Julien l.c.), *Mycovellosiella perfoliata* (Ell. & Everh.) Munt.-Cvtk as classical biological control agents for the tropical weed *Chromolaena odorata* (Barreto and Evans, 1994). Recently, Barreto *et al.* (1999) identified *Cercospora pistiae* on *Pistia stratioites*, an aquatic weed. It is a damaging pathogen but prospects of its use as a biocontrol are not considered to be very promising due to its slow growth and poor *in vitro* sporulation. The tropical weed *Heliotropium indicum* and *H. europium* are controlled by some species of *Cercospora* (Pons and Sutton, 1996). In addition, some of the *Cercospora* species produce 'cercosporin' a toxin acting as photosensitizer on the host plants (Daub and Ehrenshaft, 2000)

The genus *Phaeoramularia* in Indian sub-continent

Survey of the literature indicates that in the Indian sub-continent *Phaeoramularia* species are described from India and Nepal. Recently, one species (*P. ahmadii*) has been reported as new combination from Pakistan (Islamabad). Out of total 38 *Phaeoramularia* spp. so far described from the Indian sub-continent, 31 have been reported from India, 6 from Nepal and 1 from Pakistan. It is noteworthy to know that, Dublish and Singh (1977) reported *Phaeoramularia pergulariae* as a new species from the Indian sub-continent for the first time. Deighton (in Ellis, 1976) has made significant contribution on the taxonomy of *Phaeoramularia* based on examination of authentic type specimens of the Indian species described under *Cercospora*, which resulted in transfer of five species to *Phaeoramularia*. Thirumalachar, Vasudeva, Ramakrishanan, Kamat and co-workers and recently, Kamal and co-workers have made significant contributions to Hyphomycetes in general and cercosporoid fungi in particular. This communication details the descriptions of the 31 *Phaeoramularia* species described from Indian sub-continent (except those given in Ellis, 1976; Deighton, 1979).

Phaeoramularia adenostemae **Kamal, B.K. Gupta and C. Gupta (1990);** Indian Phytopath. 43: 143-149

Leaf lesions amphigenous, circular to irregular, necrotic, light brown. Caespituli mostly hypophyllous, effuse, velutinous, olivaceous brown. Mycelium internal. Stromata substomatal, pseudoparenchymatous, pale olivaceous, 20.8-26 µm wide. Conidiophores macronematous, fasciculate (ca. 5-12), unbranched, septate, smooth-walled, sinuous over the more distal part, olivaceous brown, 41.4-91 x 2.6 -5.2 µm. Conidiogenous cells integrated, terminal to intercalary, sympodial, polyblastic, cicatrized, scars thickened. Conidia acropleurogenous, catenate, in unbranched chains, 2-5 septate, smooth-walled, obclavatocylindric, pale olivaceous brown, 18.2-68 x 2.6-3.8 µm.

On leaves of *Adenostema* sp. (Asteraceae), India, IMI 290960 holotype.

Phaeoramularia adenostemmatis **Verma & Kamal (Budathoki *et al.*, 1989);** Indian Phytopath. 42: 53-60

Leaf lesions hypogenous, ill defined or none. Mycelium internal. Stromata substomatal, pseudoparenchymatous, pale olivaceous, 12.5 x 18.5 µm. (10.5-25 x 15.5 µm). Conidiophores arising from stromata, semi-macronematous, fasciculate (ca.4-10), unbranched, smooth-walled, pale olivaceous, 34.5-78.5 x 2-4.5 µm. Conidiogenous cells integrated, terminal, polyblastic, sympodial, cicatrized, scars thickened (2.5-3 µm wide). Conidia acropleurogenous, solitary to catenate, mostly in unbranched chains, septate, smooth-walled, olivaceous, 18-45.5 x 3-4.5 µm.

On leaves of *Adenostemma viscosa* (Asteraceae), Nepal, IMI 303484 isotype.

Phaeoramularia ahmadii **(Petr.) Braun (1995);** Mycotaxon 55: 236 = *Cercospora ahmadii* Petrak, Sydowia, **9**: 549 (1955)

Leaf lesions amphigenous, almost absent to angular, often vein limited, 1-6 mm diam. Caespituli hypophyllous, punctiform, dark. Mycelium internal. *Stromata* olivaceous brown, 15-50 µm diam. Conidiophores arising from stromata, fasciculate, 0-1 (-3) septate, smooth-walled, subcylindrical to geniculate, sinuous, olivaceous, 20-40 (-65) X 3-6.5 µm. Conidiogenous cells integrated, cicatrized, scars thickened and darkened. Conidia in short chains,

0-5 septate, subcylindrical (-obclavate), pale olivaceous, hila thickened and darkened, 10-50 (-70) X 3-6.5 µm.

On leaves of *Marsdenia roylei* (Asclepiadaceae), Pakistan, GZU 2544 holotype.

***Phaeoramularia asclepiadacearum* Gadpandey, Dwivedi and Rai (1998);** Indian Phytopath. 51: 36-42

Leaf lesions amphigenous, necrotic. Caespituli hypophyllous. Mycelium internal. Stromata substomatal, pseudoparenchymatous, light to mid olivaceous, 11-39 µm diam. Conidiophores arising from stromata, macronematous, profusely branched, 6- septate, smooth-walled, light to mid olivaceous, 7-96 x 2.5-5 µm. Conidiogenous cells integrated, terminal, scars indistinct, sometimes denticulate. Conidia catenate in branched chains, 7-septate, clavate to obclavate, or obclavatocylindric, base obconicotruncate, light olivaceous, hila slightly thickened, 14-71.5-x 3.5-7 µm.

On leaves of unidentified host (Asclepiadaceae), India, IMI 356512 holotype.

***Phaeoramularia bauhiniae* Singh, Bhalla and Kamal (1999);** Mycol. Res. 103: 271-274 [Fig. 1]

Leaf lesions amphigenous, irregular, punctiform, dark brown to blackish. Caespituli hypogenous, spreading on lower leaf surface, dark brown. Mycelium internal. Stromata substomatal, pseudoparenchymatous, olivaceous brown, 14 X 38µm. Conidiophores arising from stromata, micronematous to semi-macronematous, fasciculate (ca.3-10), unbranched, 0-2 septate, smooth-walled, straight to flexuous, cylindrical, pale olivaceous, 5-24.5 x 2.5-4.5µm. Conidiogenous cells integrated, terminal, rarely polyblastic, sympodial, scars thickened. Conidia acrogenous, solitary to catenate in unbranched chains, 0-5 septate, smooth-walled, light brown, cylindrical, apex subobtuse to rounded, base obconicotruncate, hila conspicuously thickened, 6.5-37.5 x 2-4µm.

On leaves of *Bauhinia bahli* (Caesalpiniaceae), India, HCIO 42566 holotype.

***Phaeoramularia braunii* Singh, A. Singh and Kamal (1997);** Mycol. Res. 101: 530-534 [Fig. 2]

Leaf lesions hypogenous, rectangular to irregular, golden brown. Caespituli hypophyllous, discrete, slightly raised and closely arranged in irregular spots, punctiform, pale brown. Mycellium internal. Stromata subepidermal, pseudoparenchymatous to prosenchymatous, olivaceous brown, 18-42 µm wide. Conidiophores arising from stromata, semi-

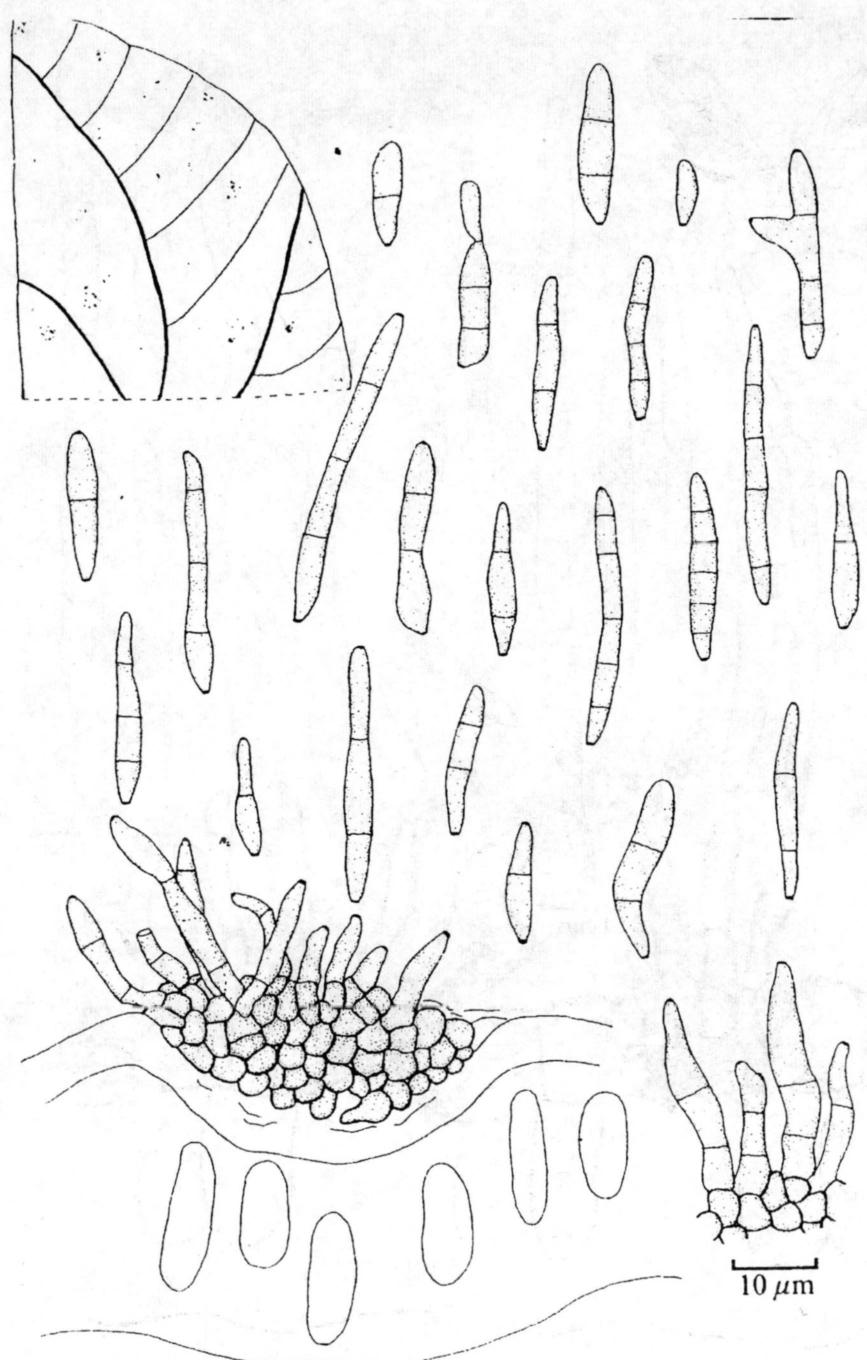

Fig. 1. *Phaeoramularia bauhinia* Singh, Bhalla and Kamal

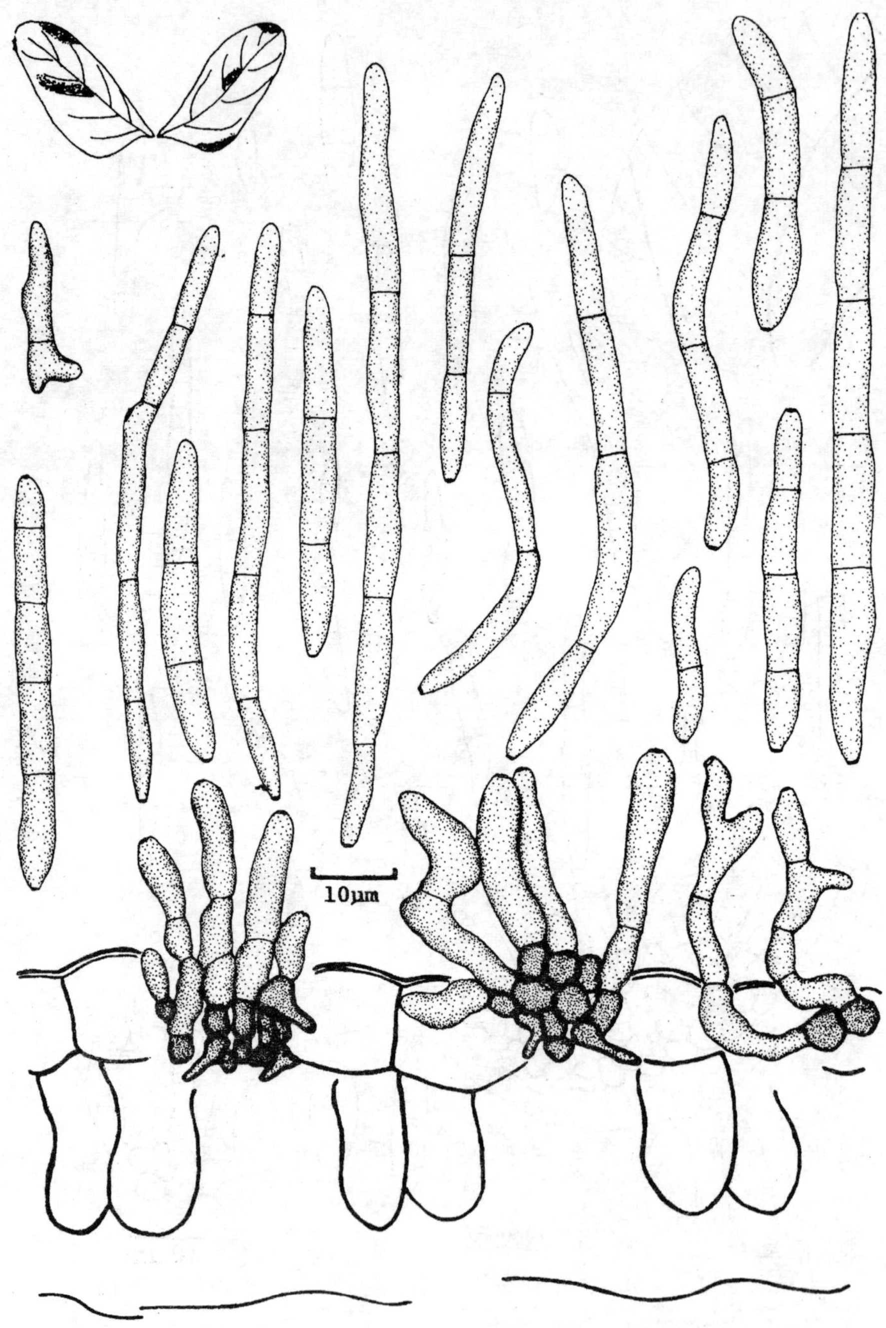

Fig. 2. *Phaeoramularia brauni* Singh, S.K. and A. Kamal

macronematous to macronematous, fasciculate (ca.3-8), branched, (0-) 1-2 septate, smooth-walled, cylindrical, rarely geniculate, pale olivaceous 21-62 x 3.5-6.25 µm. Conidiogenous cells integrated, terminal, sympodial, polyblastic, cicatrized, scars thickened. Conidia acropleurogenous, almost sickle-shaped, solitary to catenate, rarely in branched chains, (0-) 1-5 septate, smooth-walled, cylindrical to slightly obclavate, apex subobtuse to rounded, base obconical, subhyaline, hila thickened, 12-49 x 2.5-5.5µm.

On leaves of *Chloroxylum* sp.(Rhamnaceae), India, HCIO 42045 holotype.

Phaeoramularia buteae **Kamal, B. K . Gupta and C. Gupta (1990);** Indian Phytopath. 43: 143-149

Leaf lesions amphigenous, irregular, coalescing, necrotic, light to dark brown. Colonies hypophyllous, effuse, brown. Mycelium internal. Stromata partly immersed, partly superficial, pseudoparenchymatous, mid to dark olivaceous. Conidiophores arising from stromata, caespitose, macronematous, unbranched, septate, straight to flexuous, smooth-walled, light olivaceous, 38.5-109.4 x 3-4.8 µm. Conidiogenous cells integrated, terminal, polyblastic, sympodial, cicatrized, scars indistinct. Conidia acropleurogenous, solitary to catenate, 1-5 septate, subhyaline, 20-70.5 x 2-4 µm.

On leaves of *Butea frondosa* (Fabaceae), India, IMI 237155 holotype.

Phaeoramularia caesalpiniacearum **Gadpandey, Diwedi & Rai (1998);** Indian Phytopath. 51: 36-42

Leaf lesions amphigenous. Caespituli amphiphyllous, effuse. *Mycelium* internal. Stromata substomatal, pseudoparenchymatous, mid to dark olive, 6.5-32 µm. diam. Conidiphores arising from stromata, macronematous, 7-septate, unbranched, smooth-walled, light to mid-olivaceous, 14-123 x 3-5.5 µm. Conidiogenous cell integrated, terminal, sympodial, polyblastic, cicatrized, scars thickened. Conidia catenate in branched chains, 17-septate, cylindrical to obclavatocylindric, base obconicotruncate, protuberant, subhyaline to light olivaceous, hila slightly thickened, 5-125 x 2-3.5 µm.

On leaves of *Cassia tora* (Caesalpiniaceae), India, IMI 355881 holotype.

Phaeoramularia calotropidicola **Singh and Chaudhary (Singh *et al.*, 1996);** Mycotaxon 58: 137-145 [Fig. 3]

Leaf lesions amphigenous, at first yellowish-brown, later on dark blackish brown. Caespituli hypophyllous, effuse, dark brown. Mycelium internal. Stromata substomatal, pseudoparenchymatous, pale brown, 9-10 ìm wide. Conidiophores arising from stromata, semi macronematous, fasciculate (ca.

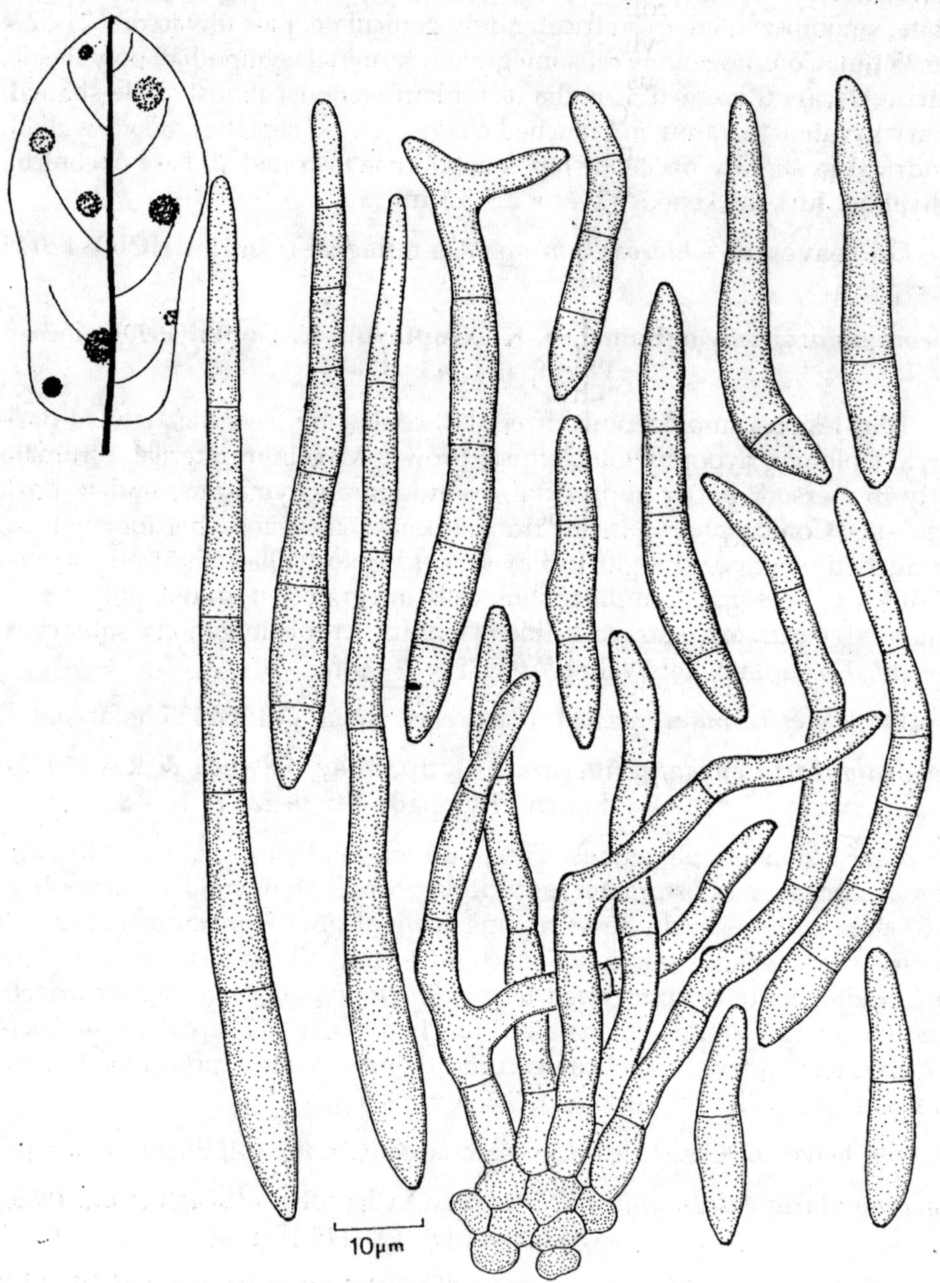

Fig. 3. *Phaeoramularia calotropidicola* Singh and Chaudhary

4-5), branched, 2-6 septate, smooth-walled, cylindrical, pale olivaceous brown, 64-116 x 2.5-5.5µm. Conidiogenous cells integrated, terminal to intercalary, polyblastic, sympodial, cicatrized, scars thickened (1µm diam.). Conidia acropleurogenous, solitary to catenate in rarely branched chains, 1-6 septate, smooth-walled, subcylindrical to narrowly obclavate, subhyaline to pale olivaceous brown, attenuating towards each ends, hila thickened, 32-146 x 2.5-7 µm.

On leaves of *Calotropis gigantea* (Asclepiadaceae), Nepal, HCIO 41713 holotype.

Phaeoramularia calotropidis **(Ell. and Everh.) Kamal *et al.*, (1990):** Mycol. Res. 94: 714-717.

= *Cercospora calotrpidis* Ell. and Everh., Rept. Mo Bot. Gdn. 9: 120 (1898)

Leaf lesions hypogenous, circular to irregular, coalescing to become large, black. Caespituli effuse, hypophyllous, brownish black to balck. Mycelium internal. Stromata substomatal, prosenchymatous, dark olivaceous, 18 µm wide. Conidiophores arising from stromata, caespitose, largely fasciculate, macronematous, mononematous, unbranched, septate, smooth-walled, straight to flexuous, sometimes geniculate, pale olivaceous, 20.5-87.5 x 3.5-7.5µm. Conidiogenous cells integrated, terminal, sympodial, polyblastic, cicatrized, scars thickened. Conidia acropleurogenous, catenate in unbranched chains, 1-4 septate, smooth-walled, slightly curved, obclavatocylindric, light olivaceous, 25-87.5 x 3.5-7.5µm.

On leaves of *Calotropis procera* (Asclepiadaceae), India, IMI 314694 holotype.

Phaeoramularia clematidis **Singh and Chaudhary (Singh *et al.*, 1995);** Mycotaxon 54: 57-66 [Fig. 4]

Leaf lesions amphigenous, usually small, angular to irregular, dark brown to blackish. Caespituli hypophyllous, shortly floccose, pale brown. *Mycelium* internal. *Stromata* substomatal, pseudoparenchymatous, brown, 17-28 µm wide. Conidiophores arising from stromata, semi- macronematous to macronematous, fasciculate (ca. 5-8), 1-3 septate, smooth-walled, sometimes slightly geniculate, cylindrical, pale brown to pale olivaceous brown, 20-48.5 x 2-4 µm. Conidiogenous cells integrated, terminal, polyblastic, sympodial, cicatrized, scars thickened (1 µm in diam.). Conidia acropleurogenous, mostly solitary to catenate, 1-2 septate, smooth-walled, subcylindrical, attenuating toward each ends, hila conspicuous, subhyaline to very pale brown, 14-38 x 2.5-5µm.

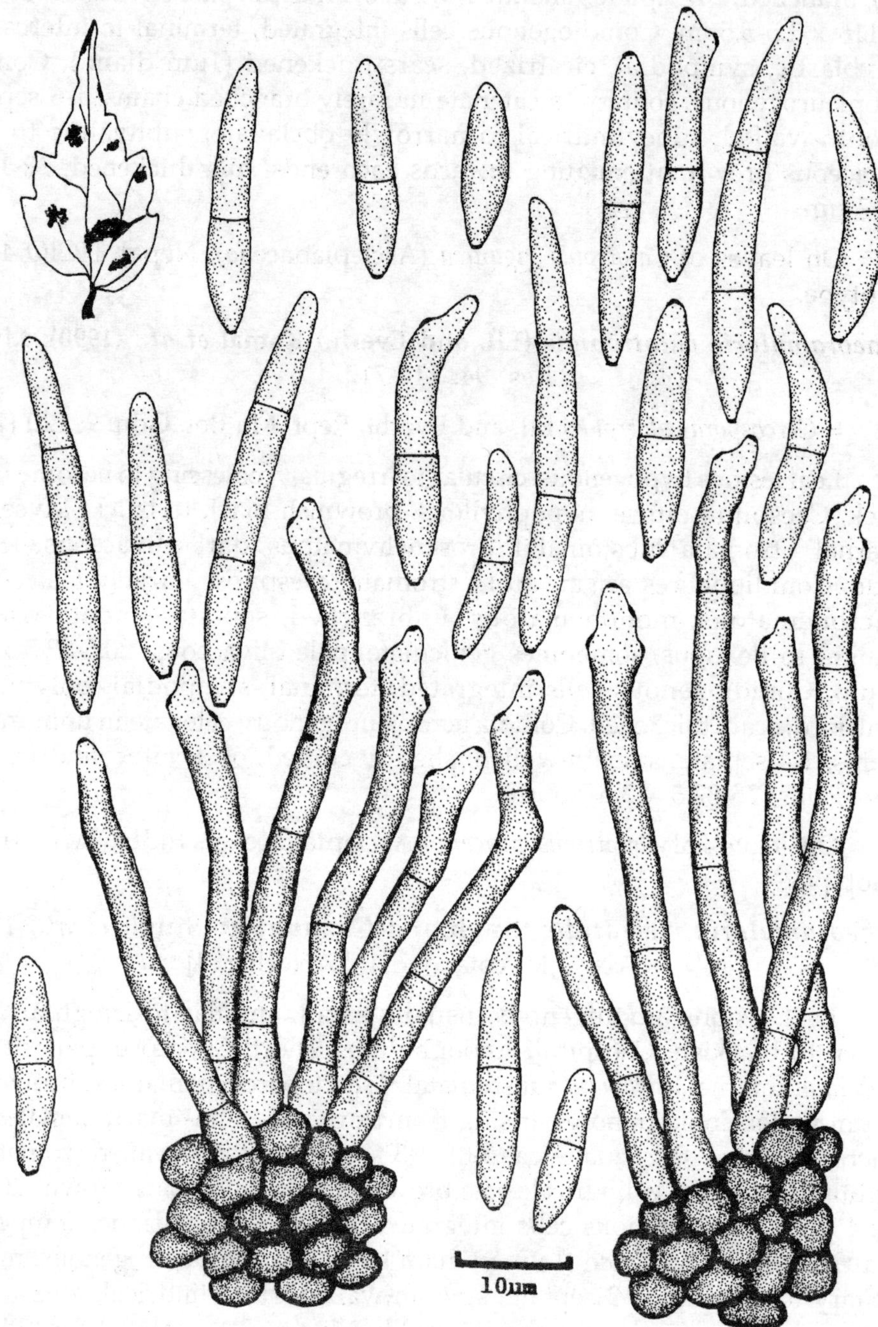

Fig. 4. *Phaeoramularia clematidis* Singh and Chaudhary

On leaves of *Clematis gouriana* (Ranunculaceae), Nepal, HCIO 41486 holotype.

***Phaeoramularia cordiae* Kumar and Kamal (1980);** Nat. Acad. Sci. Lett. India 3: 109-110

Leaf lesions indistinct. Caespituli hypophyllous, effuse, pale olivaceous. Mycelium internal. Stromata substomatal, pseudoparenchymatous, 18-28 µm. wide. Conidiophores arising from stromata, macronematous, fasciculate, unbranched, 1-2 septate, smooth-walled, cylindrical, subhyalin, 24-50 x 4-6 µm. Conidiogenous cells integrated, terminal, sympodial, polyblastic, cicatrized, scars thickened. Conidia acropleurogenous, solitary to catenate, in branched chains, 6-septate, smooth-walled, subcylindrical to cylindrical, base obconicotruncate, apex subobtuse to rounded, hila thickened, subhyaline, 30-132 x 4-6.5 µm.

On leaves of *Cordia dichotoma* (Boraginaceae), India, GPU/SK-77 isotype.

***Phaeoramularia cyperi* Gupta, Padhi and Chowdhry (1981);** Curr. Sci. 50: 140

Leaf lesions amphigenous, circular to elongated, deep brown, 0.5–2.0 mm diam. Caespituli hypophyllous, effuse, brown. Mycelium internal. Stromata pseudoparenchymatous, mid-brown 15–18 µm diam. Conidiophores arising from stromata, macronematous, fasciculate (ca. 7-9), caespitose, rarely branched, 3-10 septate, smooth-walled, olivaceous brown, 45-90 x 3.5-6 (-7) µm. Conidiogenous *cells* integrated, terminal, polyblastic, cylindrical, apex rounded, cicatrized, scars thickened. *Conidia* acropleurogenous, solitary to catenate in branched chains, 0-1 septate, smooth-walled, cylindrical, ellipsoidal or fusiform with rounded ends, pale olivaceous brown, 4 – 22 x 2 – 5 µm.

On leaves of *Cyperus alternifolius* (Poaceae), India, HCIO 32895 holotype.

***Phaeoramularia ficina* Singh & Chaudhary (Singh *et al.*, 1996);** Mycotaxon 58: 137-145 [Fig. 5]

Leaf lesions hypogenous, vein limited, appearing at first as ill-defined, irregularly shaped, pale cinnamon-brown blotches, later spread to cover large area of the leaf. *Caespituli* hypophyllous, effuse, pale brown. *Mycelium* internal. *Stromata* substomatal, pseudoparenchymatous, feebly developed, pale brown, 14-16 µm wide. *Conidiophores* arising from stromata, macronematous, fasciculate (ca.3-5), rarely branched, 1-3 septate, smooth-walled, more often, flexuous, cylindrical, pale brown, 13.5-50 x 2.5-4.5 µm. *Conidiogenous cells* integrated, terminal, sympodial, polyblastic, cicatrized, scars thickened (1µm in diam.). *Conidia* acropleurogenous, solitary to catenate, rarely in branched chains, 1-14 septate, smooth-walled, subcylindrical to narrowly obclavate,

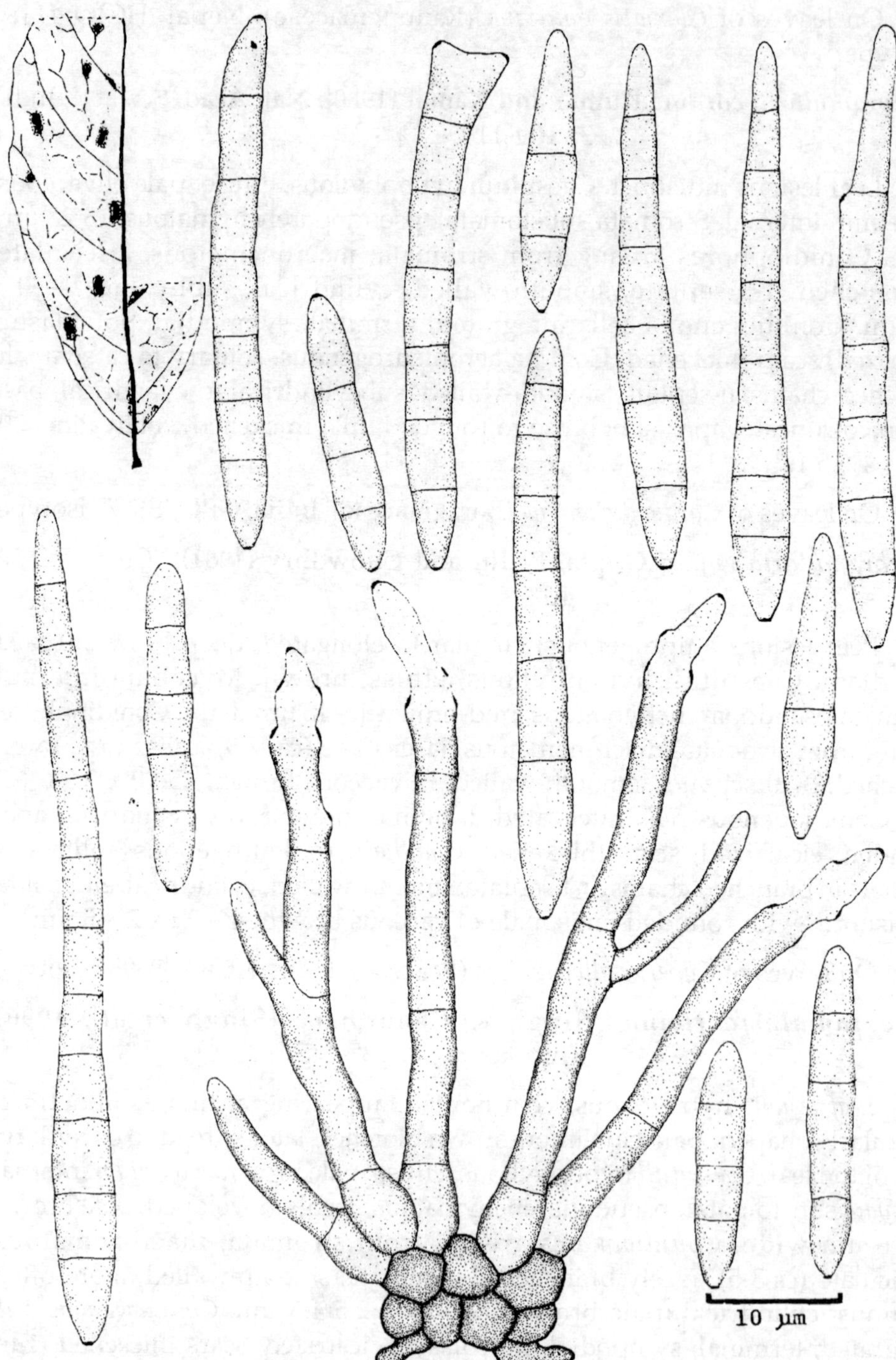

Fig. 5. *Phaeoramularia ficina* Singh and Chaudhary

apex obtuse, subhyaline to pale olivaceous, (3-8) septate, hila thickened, 18-63 x 1.5-4.5µm.

On leaves of *Ficus heterophylla* (Moraceae), India, HCIO 30857 holotype.

***Phaeoramularia hyptidigena* Kamal, Moses & Chaudhary (1990);** Mycol. Res. 94: 714-717.

Leaf lesions amphigenous, irregular, coalescing to cover a major portion of the leaf surface, necrotic, reddish brown. *Caespituli* hypogenous, effuse, greyish brown. *Mycelium* internal. *Stromata* partly immersed, partly superficial, erumpent, pseudoparenchymatous, mid-dark olivaceous, 19-28 µm diam. *Conidiophores* arising from stromata, macronematous, fasciculate (ca.7-16), unbranched, 1-3 septate, smooth-walled, light brown, 21.5-60 x 3.5-5µm. *Conidiogenous cells* integrated, terminal to intercalary, sympodial, cicatrized, scars thickened. *Conidia* acropleurogenous, catenate in branched chains, 1-8 septate, smooth-walled, cylindrical, subhyaline, 1.5-78 x 2.5-5.5µm.

On leaves of *Hyptis suaveolens* (Lamiaceae), India, IMI 337031 holotype.

***Phaeoramularia indica* Singh, Chaudhary, Meenu (1997)**; Mycol. Res. 101: 863-866. [Fig. 6]

Leaf lesions amphigenous, up to 10 mm diam. Caespituli hypophyllous, light grey. Mycelium internal. Stromata substomatal, sub cuticular to superficial, pseudoparenchymatous, brown, 10-15µm diam. Conidiophores arising from stromata, fasciculate, branched, 0-4 septate, smooth-walled, light brown, cylindrical to subcylindrical, geniculate, 5-44.5 x 2.5-5.5 µm. Conidiogenous cells integrated, terminal to intercalary, sympodial, polyblastic (mostly bi-blastic), paler near the conidiogenous region, cicatrized, scars thickened and darkened. Conidia acropleurogenous, solitary to catenate in branched chains, 1-7 septate, smooth-walled, straight to curved, subcylindrical to obclavate, slightly narrower towards the base, apex subobtuse to rounded, light brown, hila thickened, 9.5-136 x 3 -5.5 µm.

On leaves of *Acalypha indica* (Euphorbiaceae), India, IMI 366370 holotype.

***Phaeoramularia legrelii* Rao and Yadav (1991);** Crypt. Bot. 2: 188-191.

Leaf lesions hypogenous, necrotic, orbicular to irregular, scattered throughout on dorsal leaf surface. Caespituli hypophyllous. Mycelium internal. Stromata pseudoparenchymatous, dark coloured, feebly developed. Conidiophores arising from stromata, fasciculate, unbranched, 1-3 septate, cylindrical upto 20 µm long. Conidiogenous cells integrated, terminal, cicatrized, scars thickened and slightly darkened. Conidia acrogenous,

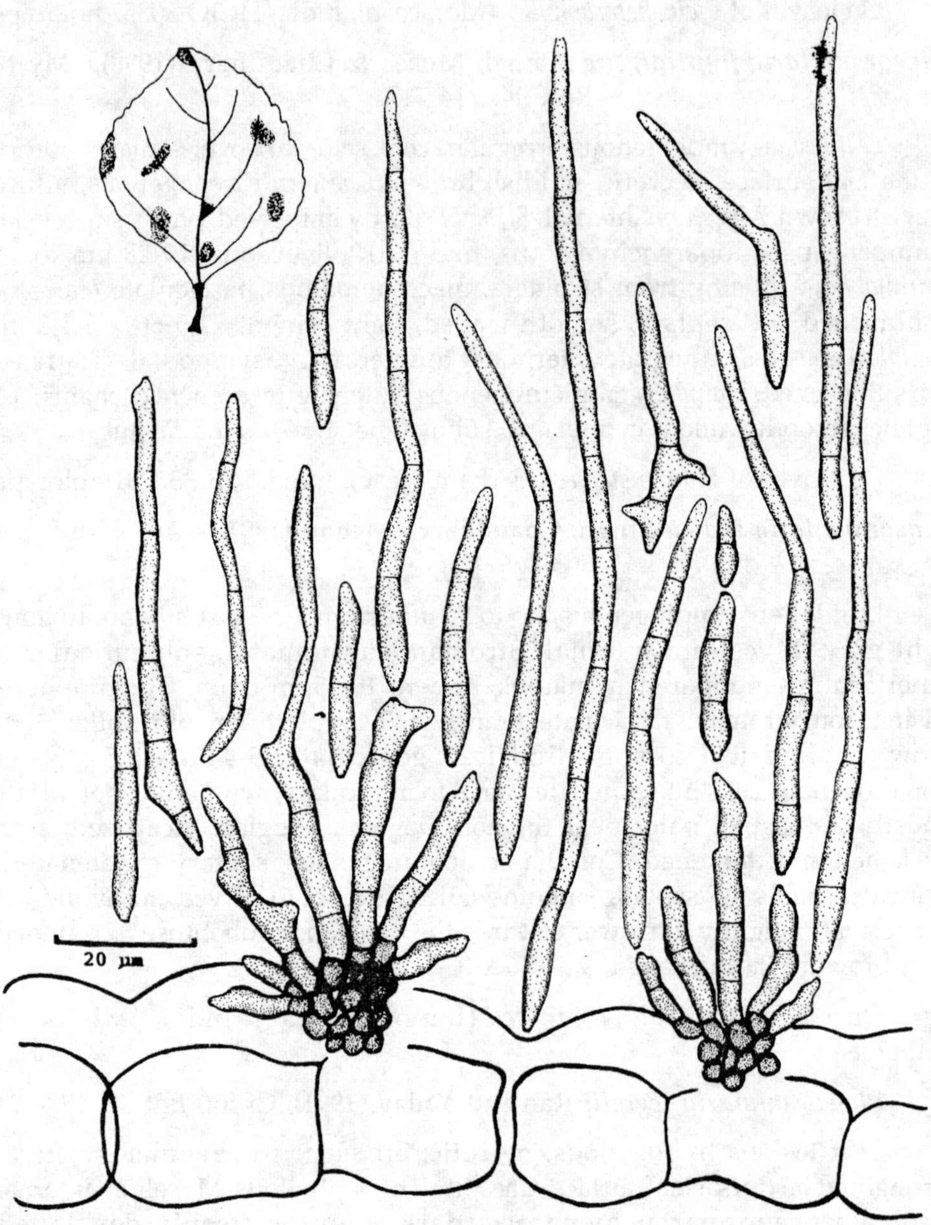

Fig. 6. *Phaeoramularia indica* Singh, Chaudhary and Meenu

catenate, 1-3 septate, smooth-walled, base coniform, subhyaline, 26.5-53 x 3.5-5 µm.

On leaves of *Punica granatum* var. *legrellei* (Punicaceae), India, AMH 4146 holotype.

***Phaeoramularia marsdeniae* Singh, Bhalla and Kamal (1999);** Mycol. Res. 103: 271-274 [Fig. 7]

Leaf lesions amphigenous, rectangular to irregular, spread over whole leaf surface, later becomes vein-limited, dark brown to blackish, 1-10 mm wide. Caespituli hypogenous, effuse dark brown. Mycellium internal. Stromata substomatal, pseudoparenchymatous, olivaceous to dark brown, 12.5 x 34 µm. Conidiophores arising from stromata, semi-macronematous to macronematous, fasciculate (ca.3-6), rarely branched, 2-4 septate, smooth-walled, sometimes geniculate, constricted at the septa, olivaceous brown, 50-91.5 x 4-8µm. Conidiogenous *cells* integrated, terminal to intercalary, polyblastic, sympodial, scars thickened. Conidia acropleurogenous, solitary to catenate in simple to branched chains, 1-6 septate, smooth-walled, cylindrical to obclavate, sometimes constricted at the septa, apex obtuse to rounded, base slightly obconicotruncate, pale brown to olivaceous brown, hila conspicuously thickened, 22.5-74.5 x 4.5-8.5 µm.

On leaves of *Marsdenia verticulata* (Asclepiadaceae), India, HCIO 42219 holotype.

***Phaeoramularia moracearum* Singh, Chaudhary and Meenu (1997);** Mycol. Res. 101: 863-866 [Fig. 8]

Leaf lesions amphigenous, angular, often vein limited, greyish-white. Caespituli hypophyllous, punctiform, greyish-white. Mycelium internal. Stromata subepidermal, prosenchymatous, elongated to ovoid, mid-brown, cells 1.5-4.0 µm wide. Conidiophores arising from stromata, macronematous, fasciculate (ca. 6-8), unbranched, septate, smooth-walled, geniculate, 42.5-54 x 2.5-5.5 µm. Conidiogenous cells integrated, terminal, polyblastic, sympodial, slightly paler in conidiogenous region, cicatrized, scars thickened. Conidia acropleurogenous, mostly solitary to catenate, 0-4 septate, apex subobtuse to blunt, sometimes slightly attenuating at both ends, subhyaline to pale brown, hila thickened, 10-98.5 x 2-4.5µm.

On leaves of *Ficus religiosa* (Moraceae), India, IMI 366371 holotype

***Phaeoramularia nepalensis* Singh and Chaudhary (Singh *et al.*,1995);** Mycotaxon 54: 57-66 [Fig. 9]

Leaf lesions amphigenous, sometimes coalescing and becoming quite

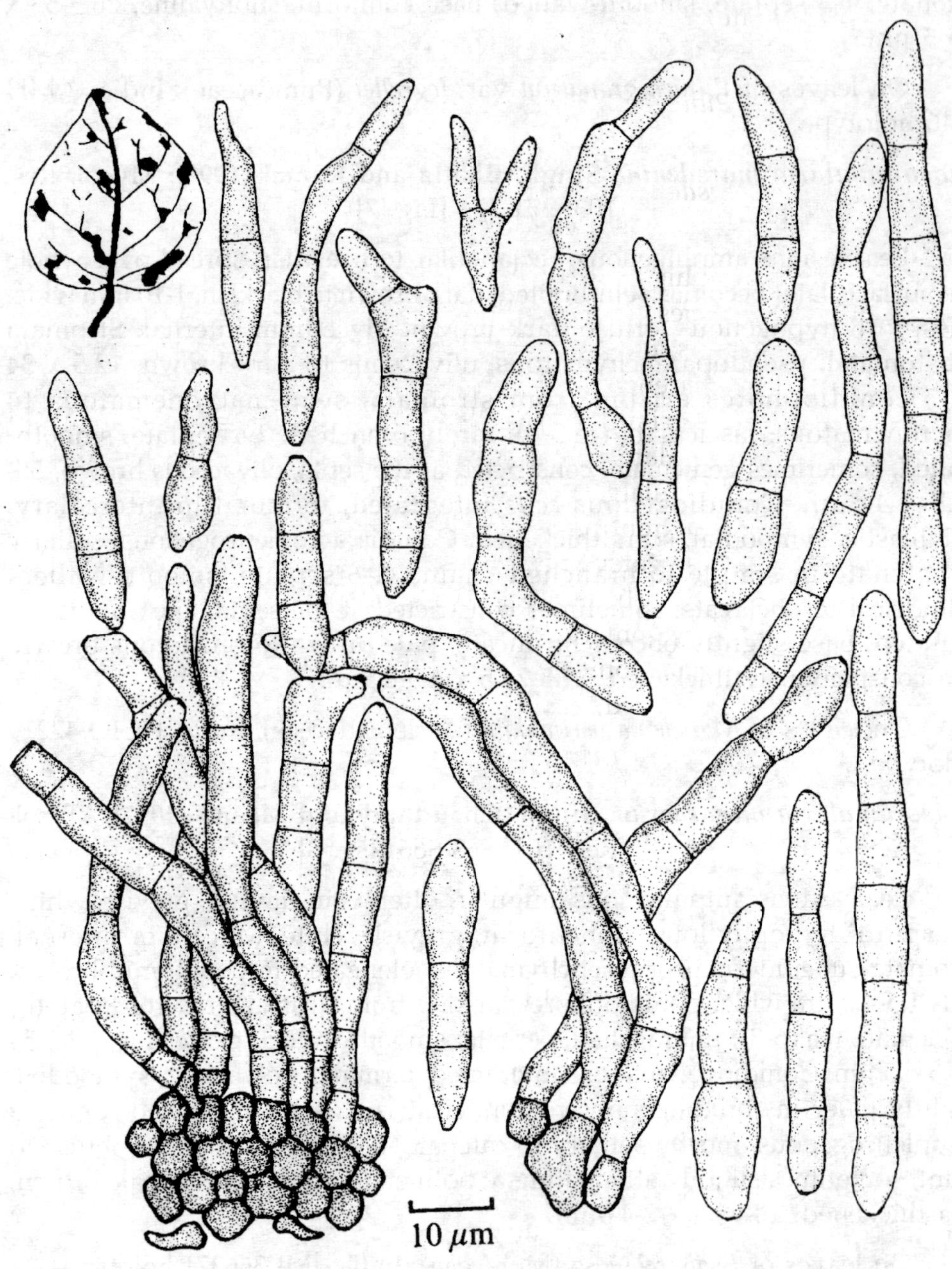

Fig. 7. *Phaeoramularia marsdeniae* Singh, Bhalla and Kamal

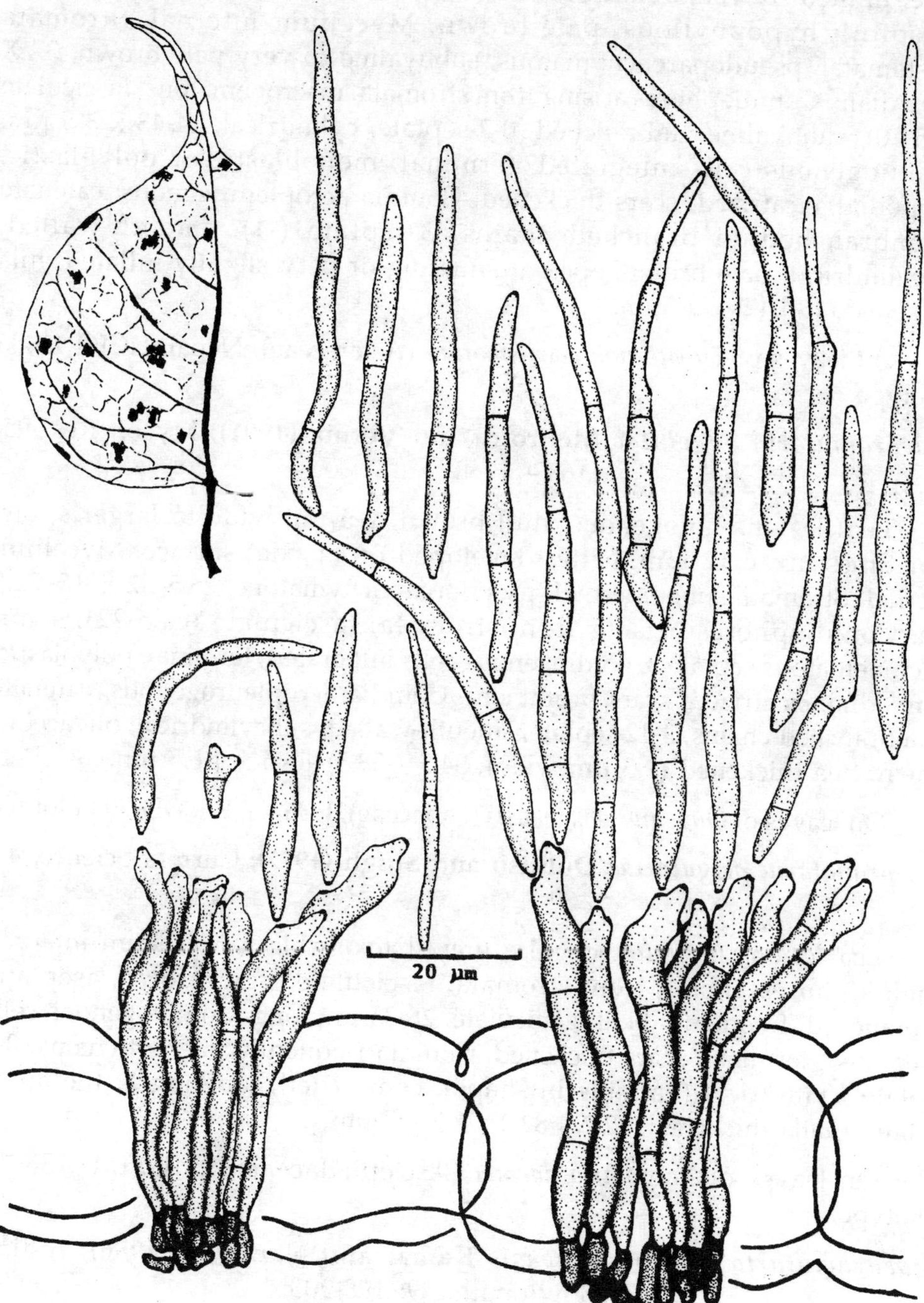

Fig. 8. *Phaeoramularia moracearum* Singh, Chaudhary and Meenu

large, limited by veins, mid brown, often with a darker, narrow margin. Caespituli hypophyllous, pale brown. Mycelium internal. Stromata substomatal, pseudoparenchymatous, subhyaline to very pale brown, 37 X 52µm diam. Conidiophores arising from stromata, macronematous, fasciculate (ca.5-10), subhyaline, unbranched, 0-2-septate, cylindrical, 14-48 x 3-5 µm. Conidiogenous cells integrated, terminal, monoblastic or polyblastic, sympodial, cicatrized, scars thickened. Conidia acropleurogenous, catenate in unbranched to branched chains, 3-septate (-1), smooth-walled, subcylindrical, pale brown, apex attenuating or very slightly inflated, hila thickened, 18-64.5-x 2-5µm.

On leaves of *Eupatorium adenophorum* (Asteraceae), Nepal, HCIO 41486 holotype.

Phaeoramularia ougeiniae **Mehrotra and Verma (1991);** Mycol. Res. 95: 1163 – 1168 .

Leaf lesions hologenous, dull brown, 2-5 µm wide to larger spots, sometimes up to 12 mm, better developed on abaxial surface. Mycelium internal. Stromata subepidermal, pseudoparenchymatous, 26.5–32 x 8.5–21.5 µm. Conidiophores arising from stromata, fasciculate (ca.5-22), semi-macronematous septate. Conidiogenous cells integrated, terminal, polyblastic, sympodial, cicatrized, scars prominent. Conidia acropleurogenous, catenate in unbranched chains, 3-12 septate, smooth-walled, subcylindrical, olivaceous brown, hila thickened (2-3 µm wide), 34.5–(-58.5)–85.5 x 4-6.5 µm.

On leaves of *Ougeinia oojeinensis* (Fabaceae), India, F.R.I. DD-50 holotype

Phaeoranularia pergulariae **Dublish and Singh (1977);** Current Science, 43: 269

Leaf lesions angular to circular, greyish to oily black. Mycelium internal. Conidiophores arising from stromata, fasciculate, light brown, aseptate, unbranched, frequently bent, geniculate, 28-35 µm long. Conidiogenous cells integrated, terminal, scars thickened. Conidia frequently in short chains, 0-1 septate, cylindrical (variable in shape), constricted, tapering at the apex, coloured, hila thickened, 12.2 – 82.2 x 3.5 – 7 µm.

On leaves of *Pergularia daemia* (Asclepiadaceae), India, IMI 188632 holotype.

Phaeoramularia pithoragarhensis **Kamal and Naraiyan (1986);** Indian Phytopath. 39: 198-203.

Leaf lesions amphigenous, brown, sub spherical, upto 1.5 cm in diam. Caespituli hypophyllous, effuse, brown. Mycelium internal. *Stromata* erumpent,

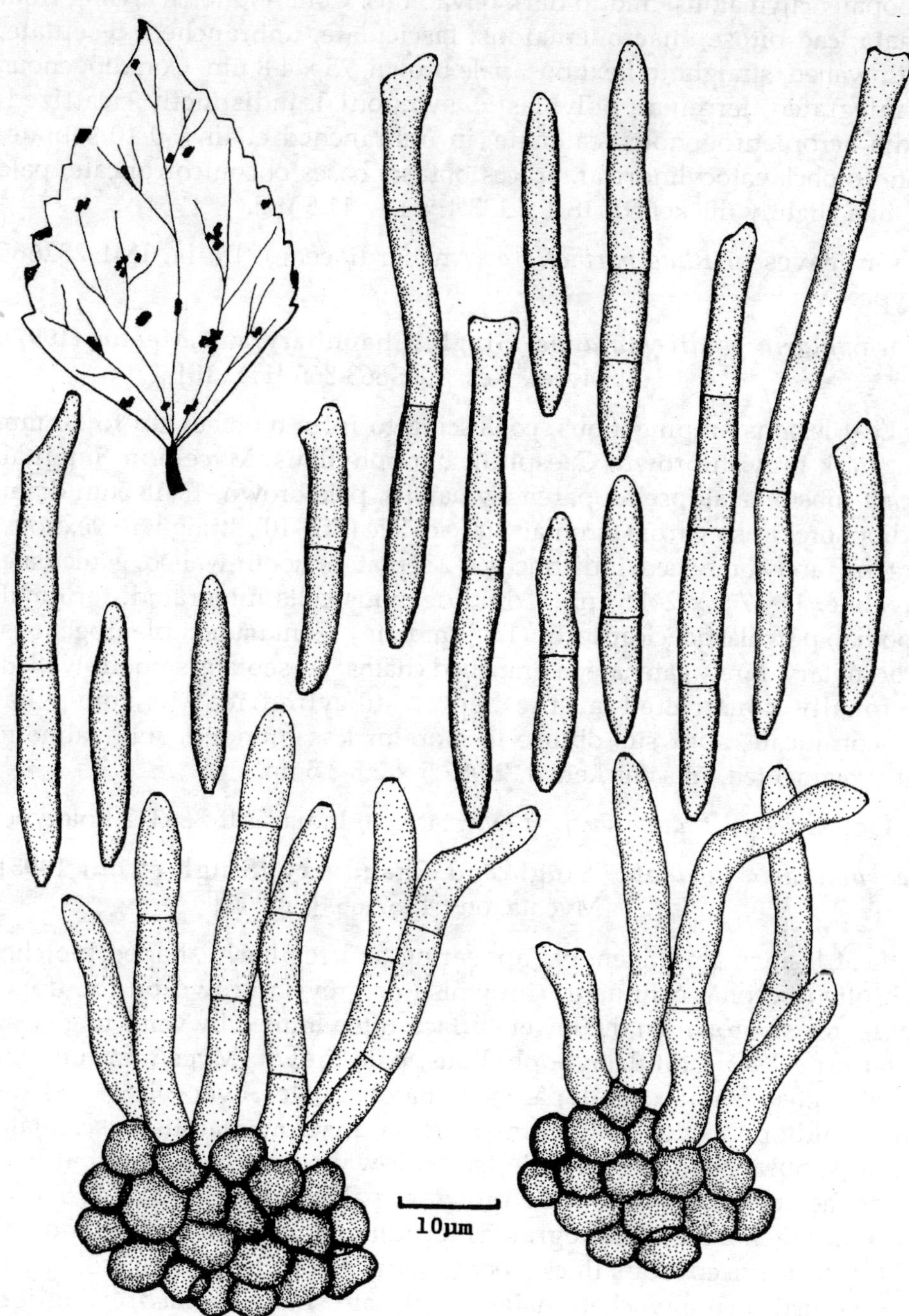

Fig. 9. *Phaeoramularia nepalensis* Singh and Chaudhary

pseudoparenchymatous, mid to dark olivaceous. Conidiophores arising from stromata, caespitose, macronematous, fasciculate, unbranched, 5-septate, smooth-walled, straight to flexuous, pale brown, 75 x 4-8 µm. Conidiogenous cells integrated, terminal, polyblastic, sympodial, indistinctly cicatrized. Conidia acropleurogenous, catenate, in unbranched chains, 0-10 septate, clavate to obclavatocylindrical, apices obtuse, bases obconicotruncate, pale olive, hila slightly thickened, 18.4 x 133.4 x 4.4– 11.5 µm.

On leaves of *Rhus parvioflora* (Anacardiaceae), India, IMI 282665 holotype.

Phaeoramularia psidii-guajavae **Singh, Chaudhary and Meenu (1997);** Mycol. Res. 101: 863-866 [Fig. 10]

Leaf lesions amphigenous, coalescing to form blotches up to 20 mm diam., dark greyish-brown. Caespituli hypophyllous. Mycellium internal. *Stromata* substomatal, pseudoparenchymatous, pale brown, 15-18.5µm diam. Conidiophores arising from stromata, fasciculate (ca.5-10), straight to flexuous, divergent, rarely branched, indistinctly 0-2 septate, smooth-walled, geniculate, subhyaline, 11-27.5 x 2-4.5 µm. Conidiogenous cells integrated, terminal, sympodial, polyblastic, cicatrized, (1.5 µm diam). Conidia acropleurogenous, mostly solitary, rarely catenate in branched chains, 1-4 septate, smooth-walled, occasionally constricted at the septa, subcylindrical to somewhat obconicotruncate, apex sub-obtuse to more or less rounded, subhyaline to slightly pigmented, hila thickened, 20-87.5 x 2.5-3.5µm.

On leaves of *Psidium guajava* (Myrtaceae), India, IMI 366372 holotype.

Phaeoramularia pyricola **Singh and Chaudhary (Singh *et al.*, 1995);** Mycotaxon 54: 57-66. [Fig. 11]

Leaf lesions amphigenous, appearing as irregularly shaped blotches which often become confluent, brownish or greyish brown on the upper surface, greyish brown on the lower surface, often limited by veins to give an angular aspect. Caespituli hypophyllous, pale brown. Mycelium internal. Stromata substomatal, pseudoparenchymatous, pale olivaceous, 31-34 µm diam. Conidiophores arising from stromata, macronematous, fasciculate, flexuous, unbranched, 1-3 septate, smooth-walled, often somewhat bent towards the apex, geniculate, pale brown to pale olivaceous, 24.5-46.5 x 3-6 µm. Conidiogenous cells integrated, terminal to intercalary, sympodial, polyblastic, cicatrized, scars thickened. Conidia acropleurogenous, solitary to catenate, mostly in unbranched chains, 0-5-septate, smooth-walled, cylindrical, subhyaline, attenuating toward each end, hila thickened, 15-64 x 2.5-5.5 µm.

On leaves of *Pyrus communis* (Rosaceae), Nepal, HCIO 41489 holotype .

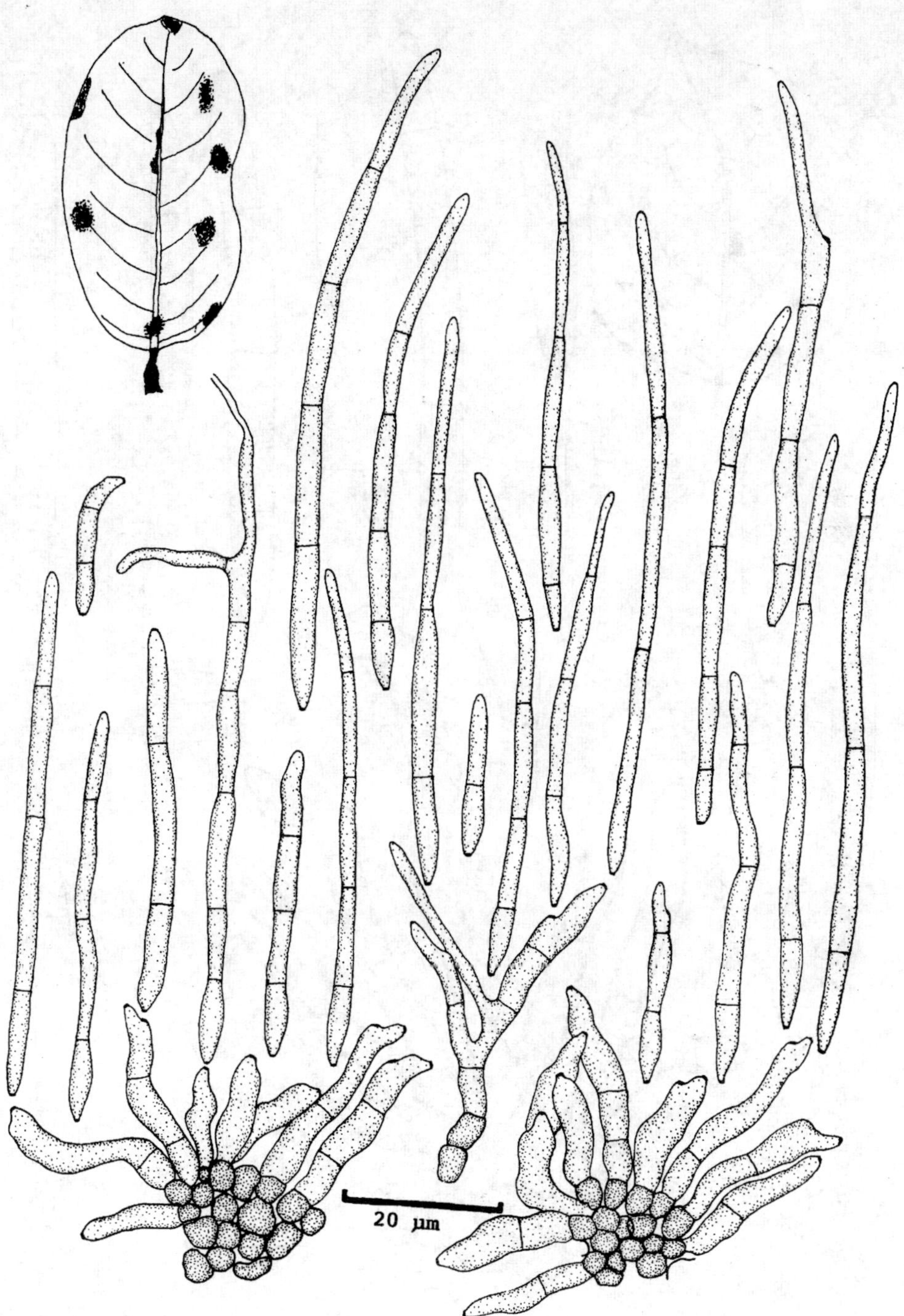

Fig. 10. *Phaeoramularia psidii-guajavae* Singh, Chaudhary and Meenu

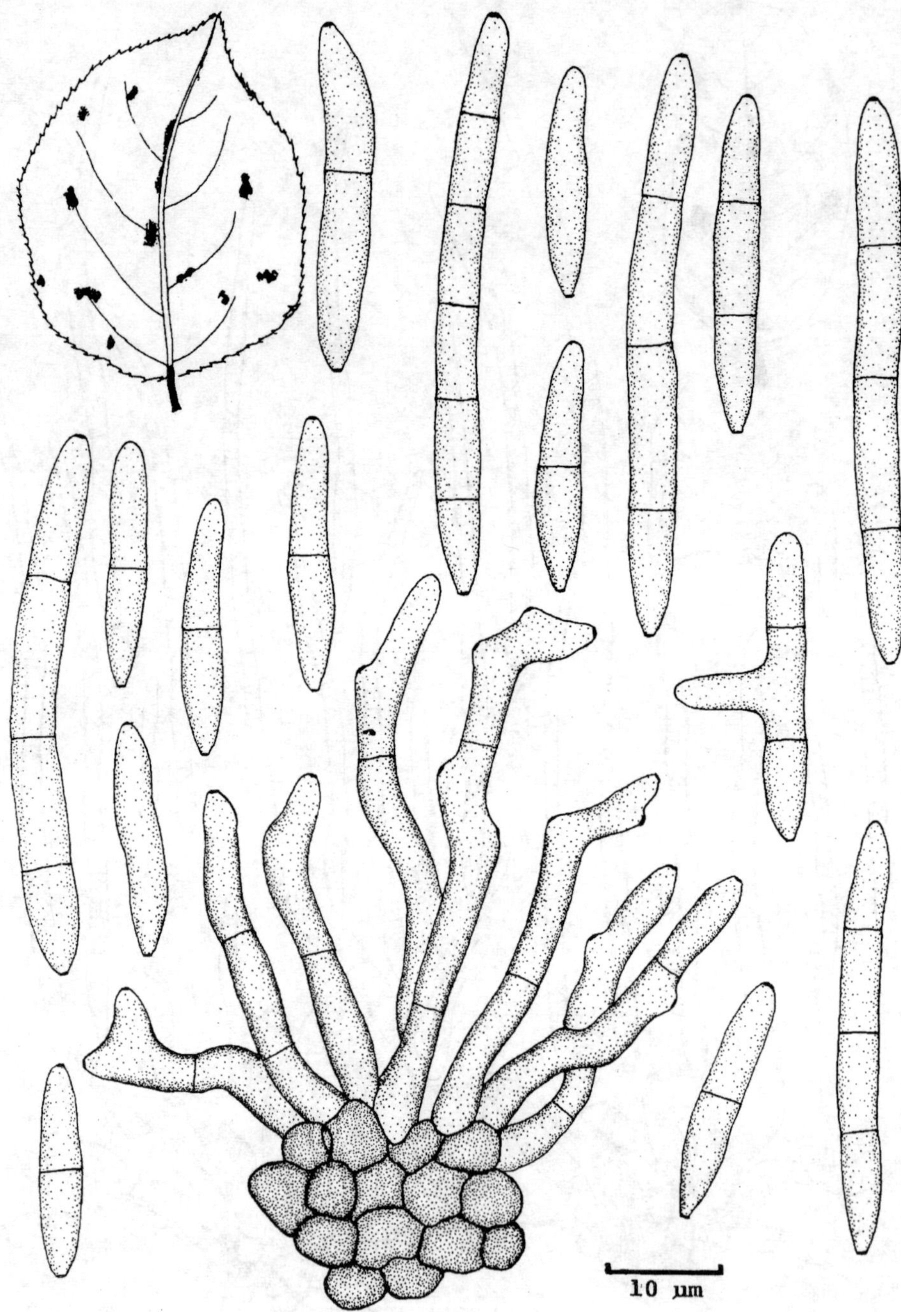

Fig. 11. *Phaeoramularia pyricola* Singh and Chaudhary

Phaeoramularia pyrigena **Singh and Chaudhary (Singh *et al.*, 1996);** Mycotaxon 56: 137-145. [Fig. 12]

Leaf lesions amphigenous, vein limited, small, at first pale grayish-brown, later becoming dark blackish-brown, up to 4 µm wide. Caespituli hypophyllous, discrete, olivaceous to brown. Mycelium internal. Stromata substomatal, pseudoparenchymatous, pale brown, 18-28 µm wide. Conidiophores arising from stromata, macronematous, fasciculate, rarely branched, 1-3 septate, smooth-walled, cylindrical, geniculate, pale brown to olivaceous, 31-52 x 3-6 µm. Conidiogenous *cells* integrated, terminal to intercalary, polyblastic, cicatrized, scars thickened. Conidia acropleurogenous, straight or very slightly curved, solitary to catenate in unbranched to branched chains, 0-2 septate, smooth-walled, cylindrical, pale olivaceous brown, hila slightly thickened, 10-31 x 1.5-4 µm.

On leaves of *Pyrus pashia* (Rosaceae), Nepal, HCIO 41490 holotype.

Phaeoramularia tephrosiilcola **Singh, Bhalla and Kamal (1999);** Mycol. Res. 103: 271-274. [Fig. 13]

Leaf lesions amphigenous, rectangular to irregular, primarily light brown later becoming brown to dark tiny spots up to 1-1.5 µm wide. Caespituli hypogenous, discrete, grayish brown. Mycelium internal. Stromata substomatal or subepidermal, pseudoparenchymatous, olivaceous brown, 14 x 19 µm. Conidiophores arising from stromata, semi-macronematous to macronematous, fasciculate (ca. 2-6), unbranched to rarely branched, 0-2 septate, smooth-walled, geniculate, olivaceous brown, 16.5-45 x 4-5µm. Conidiogenous cells integrated, terminal, polyblastic, sympodial, scars thickened. Conidia acropleurogenous, solitary to catenate in simple to rarely branched chains, straight to curved, subcylindrical to obclavate, smooth-walled, subhyaline to pale brown, 1-6 transversely septate, constricted at septa, apex sub acute to rounded, base narrowly obclavate, hila moderately thickened, 16-41 x 3.5-4.5µm.

On leaves of *Tephrosia hirta* (Fabaceae), India, HCIO 42110 holotype.

Phaeoramularia tinosporae **Kamal, Moses and Chaudhary (1990);** Mycol. Res. 94: 714-717.

Leaf lesions amphigenous, circular to irregular, brown to brownish black. Mycelium internal. Stromata immersed, prosenchymatous, 16 µm diam. Conidiophores arising from stromata, macronematous, fasciculate (ca. 4-11), branched, septate, straight to flexuous, olivaceous brown, 77.5-168 x 3-5.5µm. Conidiogenous cells integrated, terminal to intercalary, sympodial, cicatrized, scars thickened and darkened. Conidia acropleurogenous, catenate, 2-6

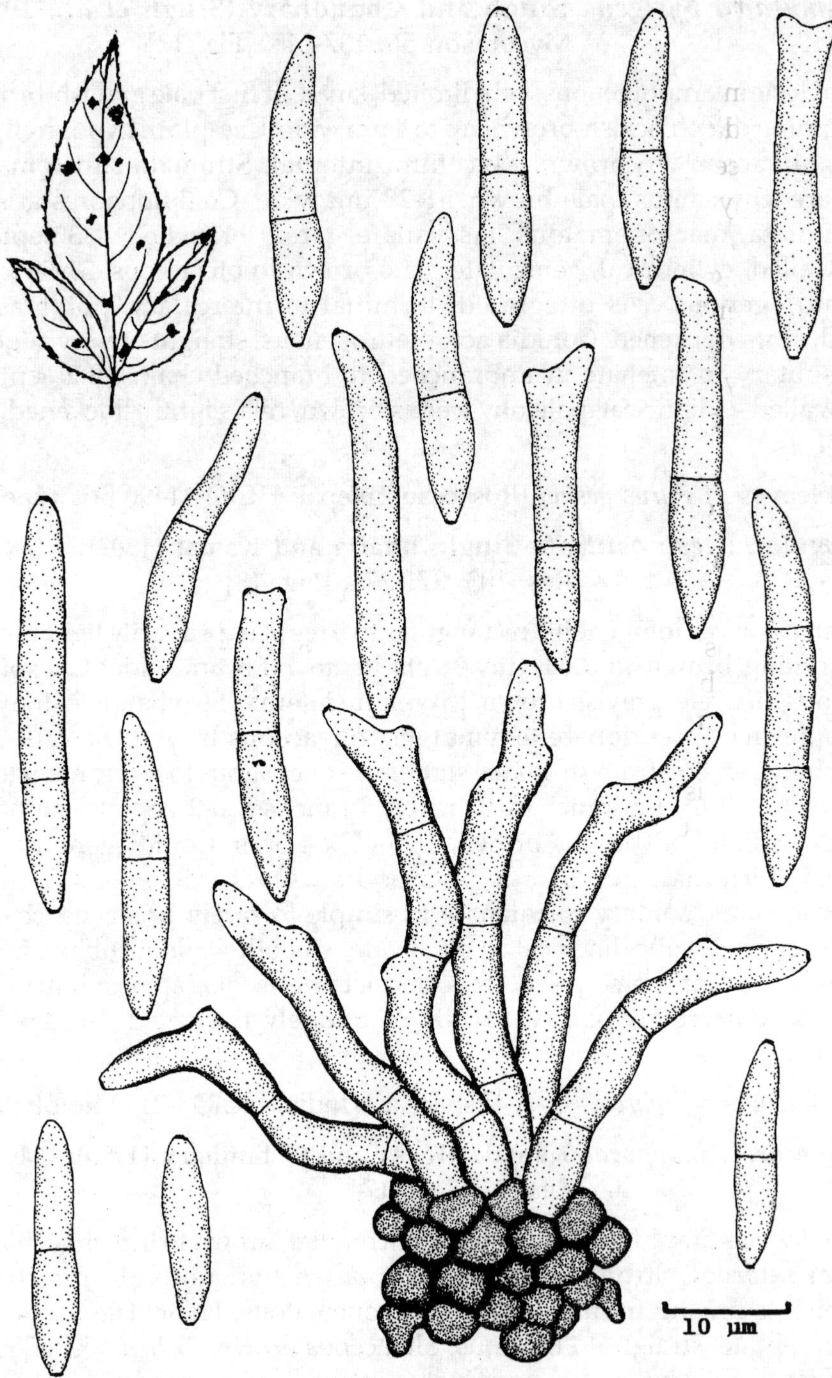

Fig. 12. *Phaeoramularia pyrigena* Singh and Chaudhary

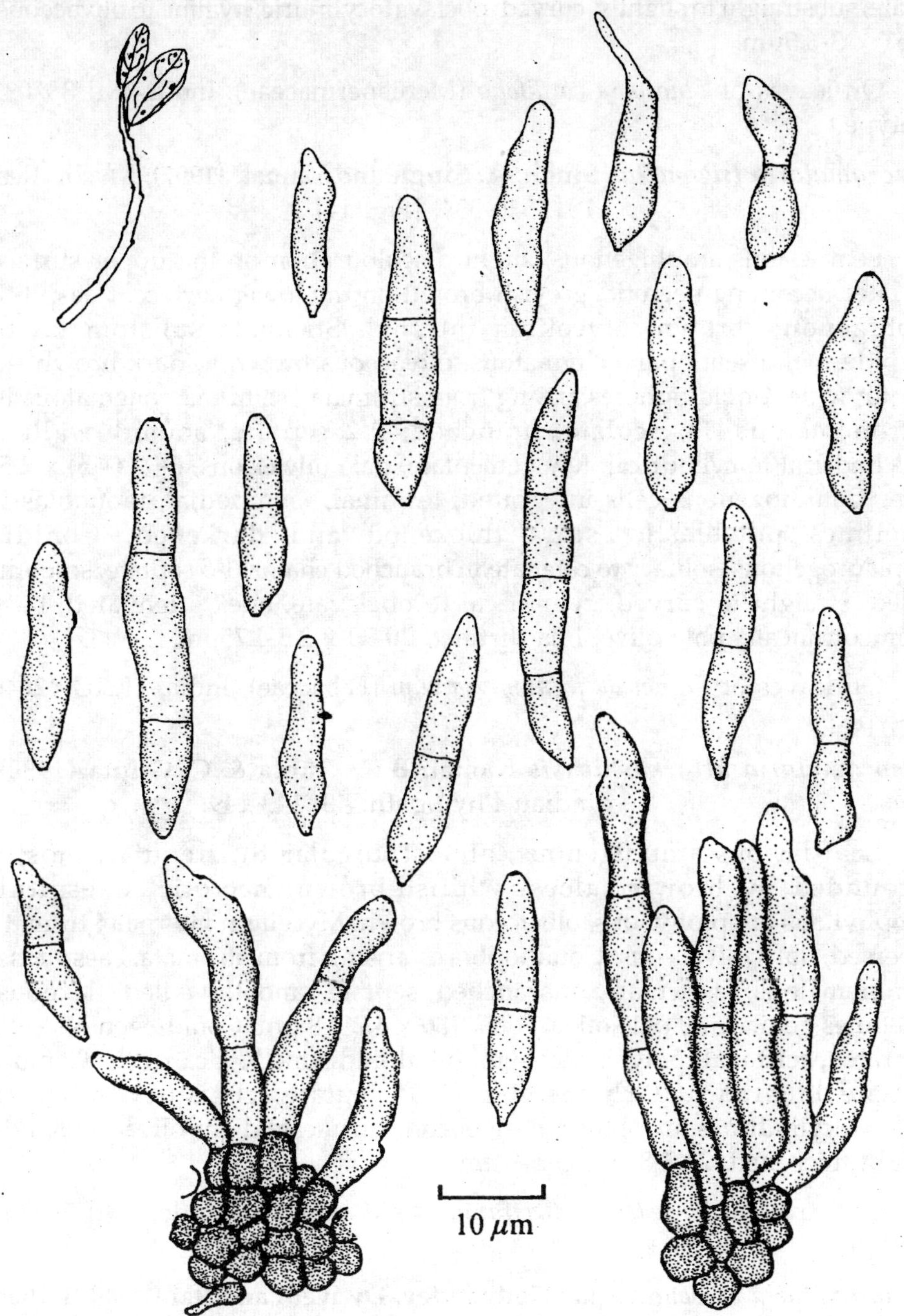

Fig. 13. *Phaeoramularia tephrossicola* Singh, Bhalla and Kamal

septate, substraight to slightly curved, obclavatocylindric, hyaline to olivaceous, 0.5-51 x 3-6.5µm.

On leaves of *Tinospora cordifolia* (Menispermaceae), India, IMI 337032 holotype.

Phaeoramularia trigonellae **Singh, A. Singh and Kamal, (1997);** Mycol. Res. 101: 530-534. [Fig. 14]

Leaf lesions amphigenous, slight discolouration on the upper surface and later becoming necrotic, greyish brown on the lower surface. Caespituli amphigenous, brown. Mycelium internal. Stromata substomatal to subepidermal, pseudoparenchymatous, olivaceous brown to dark brown, 15 x 16 µm wide. Conidiophores arising from stromata, semi-macronematous to macronematous, fasciculate, branched, 0-2 septate, smooth-walled, subcylindrical to cylindrical, few geniculate, pale olivaceous, 6-32 (-45) x 2.5-5 µm. Conidiogenous cells integrated, terminal, sympodial, monoblastic sometimes polyblastic, scars thickened and darkened. Conidia acropleurogenous, solitary to catenate in branched chains, 1-5 septate, smooth-walled, straight to curved, cylindrical to obclavate, apex attenuated, base obconicotruncate, pale olive, hila distinct, 20-90 x 2.5-4.75µm.

On leaves of *Trigonella foenum-graecum* (Fabaceae), India, HCIO 42046 holotype.

Phaeoramularia uttarkashiensis **Kamal, B.K. Gupta & C. Gupta (1990);** Indian Phytopath. 43: 143-149.

Leaf lesions amphigenous, almost circular to irregular, mostly surrounded by brown haloes, whitish-brown, necrotic. Caespituli hypophyllous or epiphyllous, olivaceous brown. Mycelium internal. Stromata immersed, dark olivaceous. Conidiophores arising from stromata, caespitose, macronematous, fasciculate, unbranched, septate, smooth-walled, flexuous, sometimes geniculate, pale olivaceous, 180 x 4.2-7.8 µm. Conidiogenous cells integrated, terminal, polyblastic, sympodial, indistinctly cicatrized. Conidia catenate in branched chains, 1-6 (3-5) septate, clavatocylindric to obclavatocylindric, apex obtuse, base obconicotruncate, light olivaceous, hila slightly thickened, 24.3-81.6 x 3-5.4 µm.

On leaves of *Pogostemon plectranthoides* (Lamiaceae), India, IMI 283306 holotype.

Phaeoramularia verbenacearum **Gadpandey, Dwivedi and Rai (1998)** Indian Phytopath. 51: 36-42.

Leaf lesions amphigenous, predominantly epigenous, irregular, necrotic.

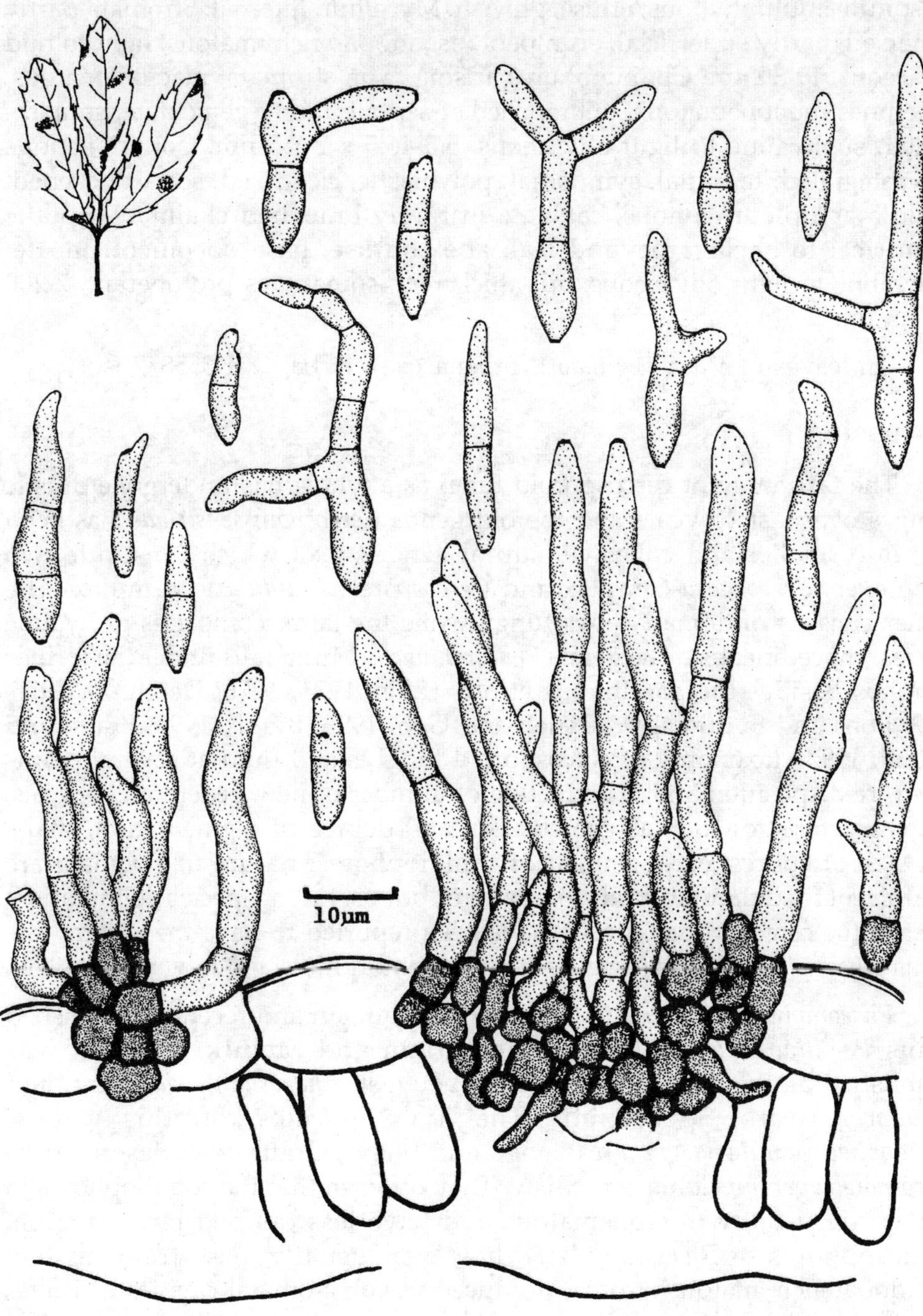

Fig. 14. *Phaeoramularia trigonellae* Singh S.K. and Kamal

Caespituli amphiphyllous, effuse, greyish. Mycelium internal. Stromata partly immersed, partly superficial, erumpent, pseudoparenchymatous, light to mid olivaceous, 19-39 µm. Conidiophores arising from stromata, macronematous, sometimes mononematous, unbranched, 4-septate rarely geniculate, smooth-walled, subhyaline to light olivaceous, 5.5-46.5 x 2-3.5 µm. Conidiogenous cells integrated, terminal, sympodial, polyblastic, cicatrized, scars thickened. Conidia acropleurogenous, catenate in rarely branched chains, 9-septate, cylindrical to obclavatocylindrical, apex obtuse, base obconicotruncate, subhyaline to light olivaceous, hila thickened, sometimes protuberant, 7-200 x 2-4.5 µm.

On leaves of *Vitex negundo* (Verbenaceae), India, IMI 355877 isotype.

Discussion

The taxonomy of cercosporoid fungi as a whole has undergone drastic changes over last 40 years and the old genus *Cercospora sensu lato* has been split into smaller and cohesive natural generic taxa which constitute two complexes, *Cercospora*-complex and *Cercosporella-Ramularia* complex. The former consists of dematiaceous fungi, while the latter comprises of hyaline species (mucedineous). The genus *Phaeoramularia* Munt. falls under the former complex. Besides the continuous efforts (Ellis, 1971, 1976; Deighton, 1976, 1979; Pons and Sutton, 1996; Hsieh and Goh, 1993; Braun, 1996; Braun and Melnik, 1997), taxonomy of cercosporoid fungi is still fluid at several points due to re-disposition and introduction of number of new genera and species. The characteristics such as geniculation and degree of pigmentation, shape and size of spores, arrangement of conidiophores, nature of stromata are insufficient for delimitation at generic level, but useful for species delimitation. Most of the cercosporoid species have been reported to have the ascomycete genus *Mycosphaerella* Johanson as their teleomorphic stage (Sivanesan 1984).

Phaeoramularia is close to *Cladosporium* Link: Fr. and recently established genus *Pseudophaeoramularia* Braun (Braun and Melnik, 1997). It was established as the intermediate genus between *Phaeoramularia* Munt. and *Pseudocercospora* Spegazzini. The type species of this genus, *Pseudophaeoramularia geranii* (Cooke and Shaw) Braun was described as *Cercosporella geranii* Cooke and Shaw. They differ in that the conidiophores in former are usually micronematous and have less defined fertile region. Conidiophores of *Phaeoramularia*, however, usually arise from distinct pseudoparenchymatous stroma, produced in sub-stomatal cavities of leaves and, as a result, are arranged in dense fascicles in contrast to the solitary type in *Cladosporium*. Although, conidium production, conidiophore and conidial scar characteristics are similar in the two genera, the conidia of *Phaeoramularia* are much longer and slender. *Pseudophaeoramularia* is distinct in having catenate or disarticulate conidia and sub-conspicuous, slightly darkened and refractive conidial scars (Braun and Melnik, 1997).

The delimitation of the genus *Phaeoramularia* Munt. has been problematic as is evident from the fact that, 45 of the 107 currently recognized species, were included in *Cercospora* Fres. and others were placed in diverse genera such as as *Fusicladium* Link, *Passalora* Fr., *Ragnhildiana* Solheim, (=*Mycovellosiella* Deighton), *Ramularia* Unger, *Selanosporium* Corda (=*Fusarium* Link ex Fr.), *Septocylindrium* Bon., *Septosporium* Corda, *Torula* Pers. ex Fr., *Ramularia* Unger, *Cercosporella* Sacc. and *Pseudocercospora* Speg. Since 1960, additional 48 new species have been described. Most species of *Phaeoramularia* are distinguished by a combination of morphological distinctions.

However, some of the morphological features seem to be affected by environmental conditions and therefore distinctive criteria need to be verified with the help of molecular techniques. Recently, Stewart *et al.* (1999) and Crous *et al.* (2000) used molecular techniques to find out the phylogenetic relationships among the anamorphic cercosporoid genera with *Mycosphaerella* Johanson, as their teleomorph.

The perusal of literature on the genus *Phaeoramularia* in Indian sub-continent shows that of the 48 new species described under this genus, around 60% are from India. The majority of these have been described from a very small geographical region of North eastern U.P. and adjacent areas. This leaves a large scope for exploration of the remaining regions of India for new taxa of *Phaeoramularia*. In addition to the opportunities in morpho-taxonomy, the work on use of molecular tools in taxonomy of *Phaeoramularia* species needs to be undertaken.

Acknowledgements

This paper is based on a work carried out under the guidance of Professor Kamal at Gorakhpur University. He has been a source of inspiration in the development of my understanding of fungal taxonomy. I thank Director, Agharkar Research Institute, Pune, for providing necessary facilities.

References

Anderson, F., Delhey, R. and Braun. U. (1998). A new species of *Phaeoramularia* (Hyphomycete) causing leaf spots on *Dichondra* sp. (Convolvulaceae). Mycotaxon, **67**: 491-494.

Arx, J.A. von. (1974). The genera of fungi sporulating in pure culture. 2nd ed. J.Cramer, Vaduz.

Barreto, R.W. and Evans, H.C. (1994). The mycobiota of the weed *Chromolaena odorata* in Southern Brazil with particular reference to fungal pathogens for biological control. Mycol. Res., **98**: 1107-1116.

Barreto, R.W., Evans, H.C. and Hanada, R.E. (1999). First record of *Cercospora pistiae* causing leaf spot of water lettuce (*Pistia stratioites*) in Brazil, with particular reference to weed control. Mycopathologia, **144**: 81-85.

Braun, U. (1992). Taxonomic notes on some species of the *Cercospora*-complex. Nova Hedwigia, **55**:211-221.

Braun, U. (1993). Taxonomic notes on some species of the *Cercospora* – complex. Mycotaxon, **48**: 275-298.

Braun, U. (1995). Miscellaneous notes on phytopathogenic hyphomycetes. II. Mycotaxon, **55**: 223-242.

Braun, U. (1996). Taxonomic notes on some species of the *Cercospora*-complex. IV. Sydowia, **48**: 205-217.

Braun, U. (1998). A monograph of *Cercosporella, Ramularia* and allied genera (Phytopathogenic hyphomycetes). Vol.2. IHW-Verlag, Eching, p.298.

Braun, U. (1999). Taxonomic notes on some species of the *Cercospora* complex. V. Schlechtendalia, **2**: 1-28.

Braun, U., Mouchacca, J. and Mckenzie, E.H.C. (1999). Cercosporoid hyphomycetes from New Caledonia and some other South Pacific islands. N Z J. Botany, **37**: 297-327.

Braun, U. and Rogerson, C.T. (1993). Phytoparasitic hyphomycetes from Utah. Mycotaxon, **46**: 263-274.

Braun, U. and Melnik, V.A. (1997). Cercosporoid fungi from Russia and Adjacent countries. Proc. Komarov Bot. Institute, **20**: 1-130

Braun, U. (1999). Taxonomic notes on some species of the *Cercospora* -complex. V. Schlechtendalia, **2**: 1-28.

Budathoki, U., Verma, R.K. and Kamal. (1989). New species of *Phaeoramularia, Pseudocercospora* and *Stenella* from Nepal. Indian Phytopath., **41**: 53-60.

Crous, P.W. (1998). *Mycospherella* spp. and their anamorphs associated with leaf spot diseases of *Eucalyptus*. Mycologia Memoire, **21**: 1-170.

Crous, P.W., Aptroot, A., Kang, J.C., Braun, U. and Wingfield, M.J. (2000).The genus *Mycospherella* and its anamorphs. In: Molecules, morphology and classification towards monophyletic genera in the ascomycetes. (K.A. Seifert, W.Games, P.W. Crous and G.J. Samuels, eds.). Studies in Mycology, **45**: 107-121.

Crous, P.W. and Sutton, B.C. (1997). New cercosporoid fungi fungi from Southern Africa. S African J. Botany, **63**: 282.

Crous, P.W., Acelino, C. and Barreto, R.W. (1997). Cercosporoid fungi from Brazil 1. Mycotaxon, **64**: 405-430.

Crous, P.W. and Camara, M.P.S. (1998). Cercosporoid fungi from Brazil 2. Mycotaxon, **68**: 299-310.

Crous, P.W., El-Gholl, N.E., Walker, S.E. and Schubert, T.S. (1999b). Angular leaf spot disease of *Saururus* caused by *Phaeoramularia saururi* comb. nov. Mycotaxon, **72**: 7-8.

Crous, P.W., Braun, U. and Alfenas, A.C. (1999a). Cercosporoid fungi from Brazil 3. Mycotaxon, **72**: 171-176.

Daub, M.E. and Ehrenshaft, M. (2000). The photoactivated *Cercospora* toxin: contributions to Plant Disease and fundamental Biology. Ann. Rev. Phytopathol., **38**: 461-490.

Deighton, F.C. (1974). Studies on *Cercospora* and allied genera. V. *Mycovellosiella* Rangel and a new species of *Ramulariaopsis*. Mycol. Pap., **137**, 1-73.

Deighton, F.C. (1976). Brown leaf mould of *Capsicum* caused by *Phaeoramularia capsicicola*. Trans. Br. Mycol. Soc., 67: 140-142.

Deighton, F.C. (1979). Studies on *Cercospora* and allied genera. VII. New species and redispositions. Mycol. Pap., **144**: 1-56.

Deighton, F.C. (1986). Misidentification of *Cercospora effusa*. Trans. Br. Mycol. Soc., **86**: 637-641.

Deighton, F.C. (1987). New species of *Pseudocercospora* and *Mycovellosiella*, and new combinations into *Pseudocercospora* and *Phaeoramularia*. Trans. Br. Mycol. Soc., **88**: 365-391.

Dublish, P.K. and Singh, P.N. (1977). A new species of *Phaeoramularia* from India. Sci. and Cult., **43**: 269.

Ellis, M.B. (1971). Dematiaceous Hyphomycetes. CMI, Kew, England.

Ellis, M.B. (1976). More Dematiaceous Hyphomycetes. CMI, Kew, England.

Freeman, T.E. (1981). Use of conidial fungi in Biological control. In biology of conidial fungi 2 (G.T.Cole & Bryce Kendrick, eds.), pp. 143-165. Academic Press, New York, U.S.A.

Fresenius, G. (1863). Beitrage zur Mikologie. Frankfurt.

Garcia, C.E., Pons, N. and de Rojas, B. (1996). Fitopatologia Venezuelana, **9**: 31.

Gadpandey, K.K., Dwivedi, D.K., and Rai, A.N. (1998). Three new species of *Phaeoramularia* from India. Indian Phytopath., **51**: 36-42.

Guo, Y. L. (1986). Studies on Hyphomycetes of Shennongjia. II. *Phaeoramularia* and *Pseudocercospora*. Acta Mycol. Sinica Suppl., **I**: 342-347.

Guo, Y.L. (1987). Studies on Hyphomycetes of Shennongjia. IV. *Cercospora, Cercosporidium* and *Phaeoramularia*. Acta Mycol. Sinica, **6**: 225-228.

Guo, Y.L. (1989). Foliicolous hyphomycetes of Shennongjia,Hubei Province. Fungi and Lichens of Shennongjia, 341-385.

Guo, Y.L. (1993). *Cercospora* and allied genera. Mycosystema, **6**: 67.

Guo, Y.L. (1996). *Cercospora* and allied genera and species record from the Qinling Mountains. Mycosystema, **9**: 93.

Gupta, D., Padhi, B. and Chaudhary, P.N. (1981). A new species of *Phaeoramularia* from India. Curr. Sci., **50**: 140.

Hsieh, W. H. and Goh, T. K. (1993). *Cercospora* and similar fungi from Taiwan. Maw Chang Book Company, Taipei.

Inácio, C.A., Ferlanetto, C., Hernandez-Gutierrez, A. and Dianese, J.C. (1996). Some *Cercospora* species originally described by Ahmesh Pinto Viegas. Fitopatologia Brasileira, **21**: 405-409.

Julien, M.H.(1992). Biological control of weeds. A world catalogue and their target weeds. CAB International, U.K.

Kamal, Moses, A.S. and Chaudhary, R. (1990). Two new species and a new combination in *Phaeoramularia* in Uttar Pradesh, India. Mycol. Res., **94**: 714-717.

Kamal and Narayan, P. (1986). Fungi from hilly tracks of Uttar Pradesh I. Indian Phytopath., **39**: 198-203.

Kamal, Gupta, B.K. and Gupta, C. (1990). Three new species of *Phaeoramularia* from Uttar Pradesh. Indian Phytopath., **43**: 143-149.

Kirk, P.M. (1986). *Phaeoramularia angolensis* (CMI Description of Pathogenic Fungi and Bacteria No. 843). Mycopathologia, **94**: 177-178.

Kumar, P. and Kamal. (1980). A new species of *Phaeoramularia* from India. Nat. Acad. Sci. Lett. India, **3**: 109-110.

Liu, X.J. and Guo, Y.L. (1982). Studies on some species of the genus *Phaeoramularia* in China . Acta Phytopath. Sinica, **12**: 1-45.

Marasas, W.F.O. and Bredell, I.H. (1974). A new species of *Phaeoramularia*. Bothalia, **11**: 217-220.

Matsushima, T. (1975). Icones microfungorum a Matsushima lectorum. Kobe. 209 pp.

Mehrotra, M.D. and Verma, R.K. (1991). Some new hyphomycetes associated with leaf spots of trees in India. Mycol. Res., **95**: 1163-1168.

Morgan-Jones, G. 1997. Notes on Hyphomycetes. LXXIII. Redescription of *Phaeoramularia coalescens, Phaeoramularia eupatorii-odorati* and *Phaeoramularia pruni*. Mycotaxon, **61**: 363-373.

Morris, M. J. (1989). Host specificity studies of a leaf spot fungus, *Phaeoramulaia* sp. for the biological control of croton weed (*Ageratina adenophora*) in South Africa. Phytophylactica., **21**: 281-283.

Muchovej, J.J. and Ferreira, F.A. (1981). A new species of *Phaeoramularia* from Brazil. Mycologia, **73**: 345-347.

Muchovej, J.J. and Ferreira, F.A 1991. A new combination for *Cercospora pachirae*. Mycol. Res., **95**: 767-768.

Muntañola, M. (1960). Algunos Hyphomycetes criticos, notas y descripciones. Lilloa, **30**: 165-232.

Patil, M.S. and Thite, A.N. (1997). Fungal Flora of Radhanagari (Kolhapur). J. Shivaji Univ. (Sci.), **17**: 149-162.

Pons, N. and Sutton, B.C. (1996). *Cercospora* and similar fungi on *Heliotropium* weeds. Mycol. Res., **100**: 815-820.

Rao, V. G. and Yadav, B. R. D. (1991). Some hitherto undescribed fungi infecting ornamentals. Crypt. Bot., **2**: 188-191.

Shin, H.D. and Braun, U. (1993). Notes on Korean Cercosporae and allied genera (I). Mycotaxon, **49**: 351-362.

Shin, H.D. and Braun, U. (1996). Notes on Korean Cercosporae and allied genera (II). Mycotaxon, **58**: 157-166.

Sivanesan, A. (1984). The bi-tunicate ascomycetes and their anamorphs. International Books and Periodical Supl. Ser., New Delhi pp.701.

Singh, S.K., Singh, A. and Kamal. (1997). Additions to *Phaeoramularia* and *Stenella* from the Indian sub-continent. Mycol. Res., **101**: 530-534.

Singh, S.K., Bhalla, K. and Kamal. (1999). New pathogenic *Phaeoramularia* spp. on forest plants. Mycol. Res., **103**: 271-274.

Singh, S.K., Chaudhary, R.K. and Meenu. 1997. Some new species of *Phaeoramularia* causing foliar spots in North-Eastern U.P., India. Mycol. Res., **101**: 863-866.

Singh, S.K., Chaudhary, R.K. and Morgan Jones, G. (1995). Notes on Hyphomycetes. LXVI. Three new species of *Phaeoramularia* from Nepal. Mycotaxon, **54**: 57-66.

Singh, S.K., Chaudhary, R.K. and Morgan Jones, G. (1996). Notes on Hyphomycetes. LXXII. Further novel leaf spots inducing species of *Phaeoramularia* from the Indian sub-continent. Mycotaxon, **58**, 137-145.

Spegazzini, C. (1882). Fungi argentini. Pugillus quartus. An. Soc. cient argent., **13**: 11-35.

Stewart, E. L., Liu, Z. Crous, P.W. and Szabo, L. (1999). Phylogenetic relationships among some cercosporoid anamorphs of *Mycospherella* based on r-DNA sequence analysis. Mycol. Res., **103**: 1491-1499.

Sutton, B.C. (1970). Forest microfungi. IV. A leaf spot of *Populus* caused by *Cladosporium subsessile*. Can. J. Bot., **48**: 471-477.

30

New Insights in Species Diversity in Genus *Gonatophragmium* Deight.

M.S. Tripathi and V. Tripathi

Mycology Laboratory, Deptt. of Botany, DDU Gorakhpur University, Gorakhpur-273 009

The present chapter is devoted to the description and illustration of novel taxa of *Gonatophragmium* Deighton viz., *G. semi-lunatum, G. daedalacanthi, G. justiciae, G. jarwaensis, G. lamiacearum, G. pogostemonidis, G. deightonii, G. kushinagarensis, G. moracearum* collected on *Pentatropis spiralis* (Asclepiadaceae), *Daedalacanthus purpurascens, Justicia quinquangularis* (Acanthaceae), *Anisomeles ovata, Colebrookia oppositifolia, Pogostemon lanceolatus,* (Lamiaceae), *Ficus hispida, Ficus glomerata* (Moraceae) respectively and amendation of *G. kuanense* Rai collected on *Miliusa tomentosa* (Annonaceae) from the forest flora of North-Eastern U.P. in Indian subcontinent.

Key words : Foliicolous, hyphomycete, morphotaxonomy, novel taxa, *Gonatophragmium*.

Deighton, (1969) established the foliicolous hyphomycete genus *Gonatophragmium* with *G. mori* (Sawada) Deighton as the type species. The generic circumscription was based on a collection made on *Morus* species (Moraceae)occurring widely in tropics. The genus was characterized by producing eye-catching amphigenous zonate leaf lesions having fluffy deposition on the abaxial surface with distinctly nodose, solitary conidiophores producing multiseptate phragmoconidia on denticles emerging from integrated conidiogenous cells along the terminus usually in bunch. The genus remained monotypic untill Mulder, (1973) added one more species, *G. mangiferae* Mulder on *Mangifera indica* (Anacardiaceae) again from a tropical country, Burma. After a gap of over two decades a third species, *G. kuanense* Rai collected on *Miliusa tomentosa* (Annonaceae)was described by Rai (1996) from India.

North-Eastern Uttar Pradesh has been under intensive and extensive investigation of foliicolous fungi for last more than twenty years and as such attempts have been made to look into different foliicolous forms to study the host range of a particular fungus species and taxonomic characterization of hitherto undescribed taxa. It is with this view point that an extensive survey of different forest areas of this region in Indian subcontinent was made during close of the present century and millennium as well for this fungus genus.

The result was that not only different plant genera of Moraceae but of other families also were found infested with the *Gonatophragmium.*

Upon critical examination it was found that entirely different specific entities are parasitizing different genera of Angiospermic families in different localities. An important distinguishing feature that escaped the close observation of earlier mycologists is the presence of septal bands corresponding to each conidial septum, especially in the novel fungal taxon collected on the same host species on which the type species was reported. Certain other fungal entities showed such features. Therefore, this new morphotaxonomic feature can be used as a limiting character in describing some novel fungal taxa, if not all.

This paper, therefore, is devoted to *Gonatophragmium* along with redescription of the type species.

Observations and Descriptions of Novel Taxa

Collections made from different forest reservoirs of North- Eastern U.P. were processed into the laboratory. Scrap mounts and hand cut sections of the infected leaf samples were prepared in water, lactophenol, cotton-blue mixture and lactofuchsin so as to observe clear morphotaxonomic diagnostic features. Camera lucida drawings were made and descriptions as per current international standards were executed. Pertinent literatures including monographs, research papers from CAB IMI, Kew, England in determining the taxonomic status of the fungi concerned were consulted. In addition, to justify novel identity of the fungi, their descriptions were compared with the earlier described allied forms.

G. semi-lunatum sp. nov. (Fig.1):

Maculae amphigenae, singulae vel caespitosae et irregulares, necroticae, brunneae, atro marginatae, in aetate auctae et areas magnas tegentes, plerumque zonatae, variabilissima. **Coloniae** hypophyllae, effusae, brunneae, velutinae. **Mycelium** superficiales et parte immersae, ramosae, pallide brunneae, laevia, septatae, 3.0 μm diam. **Stromata** nulla. **Conidiophora** solitaria, lateralibus vel terminalibus, micronematosa vel semimacronematosa, mononematosa, bilaterales ramosa, septata, erecta vel suberecta et fere repentia, flexuosa, nodosa fere septata, intumescentiis lateralibus vel terminalibus cellulae conidiogenae, pallide brunneae, variabilissima, 20.0-159.5x3.0-3.5 μm. **Cellulae conidiogenae** integratae terminales vel intercalares, monoblasticae vel polyblasticae, in intumescentiis dispositae, denticulatae, denticula 1-8, cicatricosa, cicatrices conspicue. **Conidia** solitaria, holoblastica, sicca,

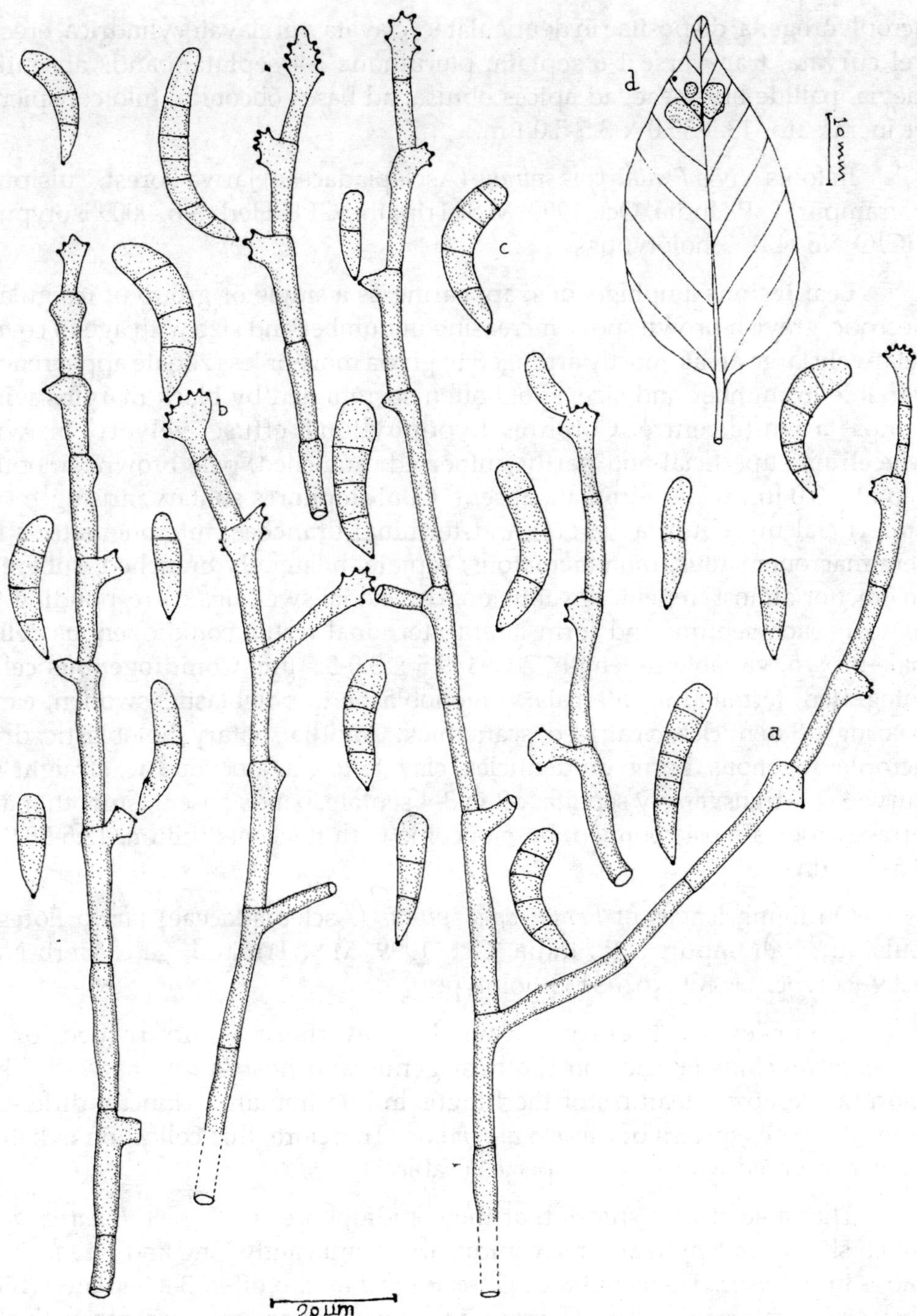

Fig. 1. *Gonatophragmium semi-lunatum sp. nov.* (a) Conidiophores. (b) Conidiogenous cells. (c) Conidia (d) Infection spots. Bars. a-c = 20 µm. d = 10 mm.

acropleurogena, dispositae in denticulatae, clavata vel clavatocylindrica, erecta vel curvata, transverse 1-6 septata, plerumque 3-4 septata, bands absentis, laevia, pallide brunneae, ad apices obtusa, ad bases obconica, hilo conspicuo et incrassato, 13.5-35.0 x 3.5-5.0 μm.

In foliis vivis *Pentatropis spiralis* (Asclepiadaceae) Jarwa Forest, Tulsipur, Balrampur, U.P., India, Oct. 1999, M.S. Tripathi, GPU Herb No. 9009 isotypus, HCIO No. 43133 holotypus.

Leaf lesions amphigenous, appearing as a single or group of irregular, necrotic, greyish-brown spots, increasing in number and size with age to cover relatively large areas, mostly arranged to give a more or less zonate appearence, variable in number and size, spots often surrounded by black margins with almost a white centre. **Colonies** hypophyllous, effuse, velvety, brown. **Mycelium** superficial and partly immersed, branched, pale brown, smooth, septate, 3.0 μm wide. **Stromata** absent. **Conidiophores** solitary, arising from superficial mycelium as lateral and terminal branches, micronematous to semimacronematous, mononematous, septate, bilaterally branched, suberect to erect or almost repent, flexuous nodose, nodal swellings corresponding to almost each septum and form lateral/terminal fertile conidiogenous cells, pale-brown, variable in length, 20.0-159.5 x 3.0-3.5 μm. **Conidiogenous cells** integrated, terminal to intercalary, monoblastic to polyblastic, swollen, each bearing 1-8 denticles, cicatrized, scars thick. **Conidia** solitary, holoblastic, dry, acropleurogenous, borne on denticles, clavate to clavatocylindric, straight to curved, 1-6 transversely septata, often 3-4 septate, bands absent, smooth, pale brown, apex almost rounded, base obconic with thickened hilum, 13.5-35.0 x 3.5-5.0 μm.

On living leaves of *Pentatropis spiralis* (Asclepiadaceae) Jarwa Forest, Tulsipur, Balrampur, U.P., India, Oct. 1999, M.S. Tripathi, GPU Herb No. 9009 isotype, HCIO No. 43133 holotype.

A survey of literature reveals that there is no record of a *Gonatophragmium* species on the host genus and host family as well. The morphotaxonomic features of the fungus in question are distinctly different from the type species of *Gonatophragmium*. Therefore, this collection is liable to be compared with its type species (Table 1).

The table clearly shows that the conidiophores of *G. semi-lunatum* are much shorter in length and the conidia are significantly long and wider than those in *G. mori*. The conidia of *G. semi-lunatum* are often 3-4 septate while they are 3 septate in *G. mori*. The conidiogenous cells are having comparatively larger number of denticles. The conidia are more curved to appear lunar in shape hence, the species name.

Table 1 : Comparision between *G. semi-lunatum sp. nov.* and *G. mori* Deighton

Fungus spp.	Conidiophores	Conidia
G. mori Deighton.	branched with slightly lateral or terminal nodose swellings, denticulate, pale brown, 500x3.0-4.5 µm.	clavate to cylindric, often slightly curved with slightly thickened scar, pale brown, 1-7 septate, often 3 septate, 13.0-26.0x 4.5 µm.
G.semi-lunatum sp. nov.	branched bilaterally with lateral or terminal nodose swellings, cicatrized denticles with thick and dark scars, denticles 1-8, pale brown, 20.0-159.5 x 3.0-3.5 µm.	clavate to clavatocylindric, more curved or lunar in shape with thickened dark scar, pale brown 1-6 septate, often 3-4 septate, 13.5-35.0 x 3.5-5.0 µm.

Description and illusration of this collection as taxon of separate species rank therefore, is warranted.

G. daedalacanthi sp. nov. (Fig. 2)

Maculae amphigenae, singulae vel caespitosae et fere circulares vel irregulares, necroticae, brunneae, atro-marginatae, in aetate auctae et areas magnas tegentes, plerumque zonatae, variabilissima. **Coloniae** hypophyllae, effusae, brunneae, velutinae. **Mycelium** superficiales et parte immersae, profuse ramosae, pallide brunneae, laevia, septatae, 4.0 µm diam. **Stromata** absentis. **Conidiophora** solitaria, lateralibus vel terminalibus, micronematosa vel semimacronematosa, mononematosa, unilaterales ramosa, septata, erecta vel suberecta et fere repentia, flexuosa, nodosa fere septata, intumescentiis lateralibus vel terminalibus cellulae conidiogenae, pallide brunneae, variabilissima, 35.0-91.0 x 3.5-4.0 µm. **Cellulae conidiogenae** integratae, intercalares vel terminales, polyblasticae, raro monoblasticae in intumescentiis dispositae, denticulatae, denticula 1-6, cicatricosa, cicatrices conspicue. **Conidia** solitaria, holoblastica, sicca, acropleurogena dispositae in denticulatae, fere clavata vel raro clavatocylindrica, curvata, transverse 1-3 septata, plerumque 3 septata, bands fere septata, pallide brunneae, ad apicem obtusa, ad basim obconica, hilo conspicuo et incrassato, 12.5-23.5 x 3.0 –5.0 µm.

In foliis vivis *Daedalacanthus purpurascens* (Acanthaceae), Jarwa Forest, Tulsipur, Balrampur, U.P., India, Oct. 1999, M.S. Tripathi, GPU Herb No. 9008 isotypus, HCIO No. 43132 holotypus.

Leaf lesions amphigenous, small, single, greyish-brown in young stage, later forming large, sometimes in groups, greyish-white in colour with age which ultimately form zonate rings, greyish-white centre surrounded by dark

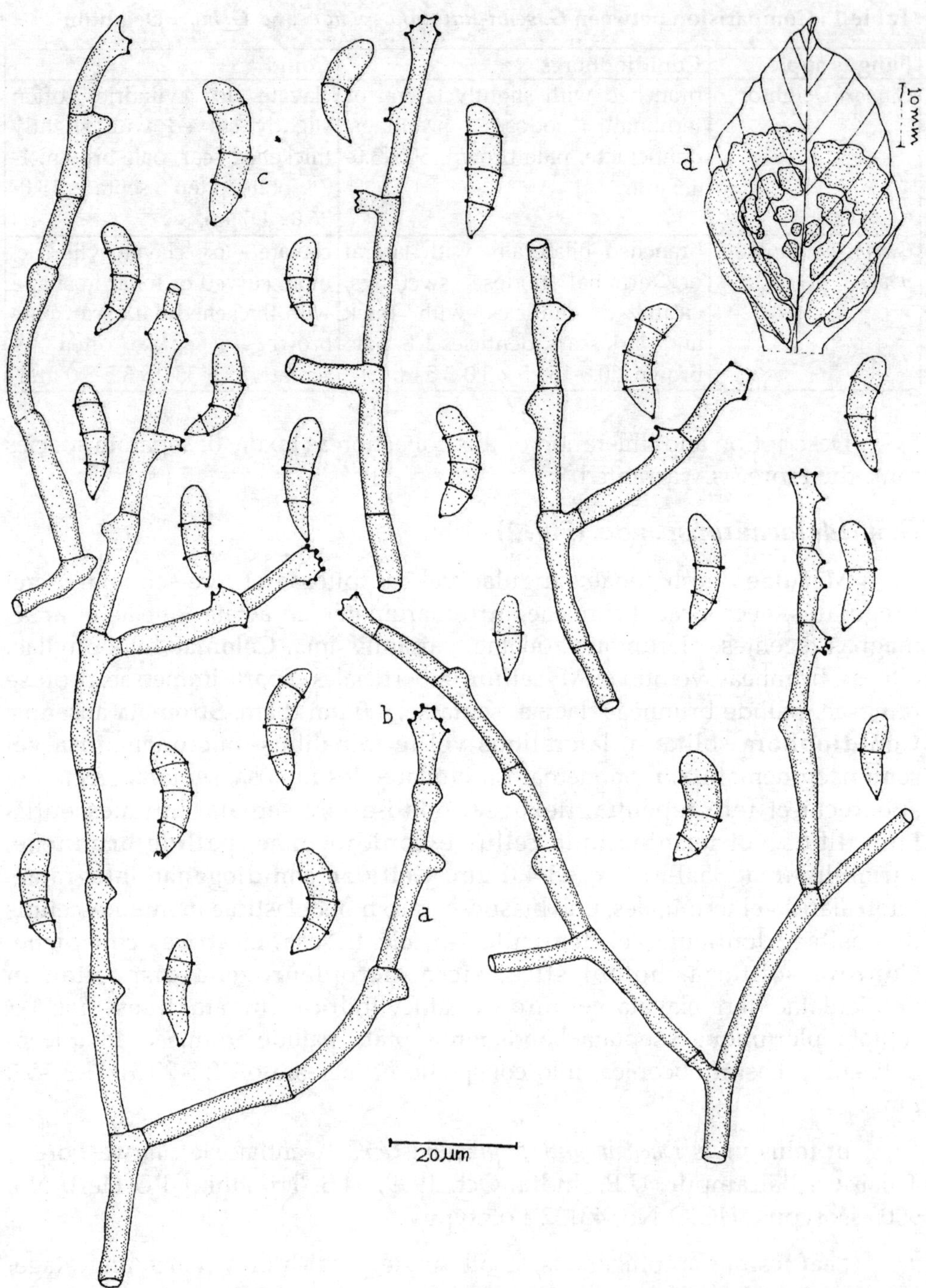

Fig. 2. *Gonatophragmium daedalacanthi sp. nov.* (a) Conidiophores (b) Conidiogenous cells. (c) Conidia. (d) Infection spots. Bars. a-c = 20 µm. d = 10 mm.

brown to somewhat blackish margins which increase in number with the age, sometimes single spot cover the whole leaf, variable in size, almost circular to irregular. **Colonies** hypophyllous, greyish-brown appear as fluffy deposition between the zonate rings, effuse. **Mycelium** internal and external, external mycelium comprising superficial, profusely branched, subhyaline to light brown, smooth walled, septate, 4.0 μm wide. **Stromata** absent. **Conidiophores** solitary, arising from external mycelium as lateral or terminal branches, micronematous to semimacronematous, mononematous, septate, branched unilaterally suberect to erect or almost repent, nodose, nodal swelling appear corresponding to almost each septum and form the fertile conidiogenous region terminally or laterally, nodes bilateral, pale brown, variable in length, 35.0-91.0 x 3.5-4.0 μm. **Conidiogenous cells** integrated, denticulate, 1-6 denticles, terminal to intercalary, polyblastic, rarely monoblastic, swollen, cicatrized, bearing thickened conidial scars. **Conidia** solitary, holoblastic, dry, acropleurogenous, borne on denticles, almost clavate to rarely clavatocylindric, curved/sickle-shaped, 1-3 transversely septate, often 3 septate, bands on each septum, pale brown, apex almost rounded, base obconic, hilum thickened, 12.5-23.5 x 3.0-5.0 μm.

On living leaves of *Daedalacanthus purpurascens* (Acanthaceae), Jarwa Forest, Tulsipur, Balrampur, U.P., India, Oct. 1999, M.S. Tripathi, GPU Herb No. 9008 isotype, HCIO No. 43132 holotype.

A thorough survey of literature on the taxonomy of *Gonatophragmium* reveals that a species of the same has not yet been reported ony any host genus of Acanthaceae. However, another novel taxon of species rank viz., *G. justiciae sp. nov.* collected on different host but same family, is also being described in the present paper. To justify its novelty the present collection is being compared with the type species of the fungus in question and its allied novel taxon. The comparison is given in Table 2.

It is evident from the data presented in table 2 that *G. daedalacanthi sp.nov.* is different from its type species and the allied novel taxon in having significantly smaller conidiophores. The conidia of the present collection are also smaller but slightly wider than the other two allied species. Conidial septation is limited to three only in the present collection while it is seven in the type species. The conidia are never aseptate in *G. daedalacanthi sp. nov.* but in *G. justiciae sp. nov.* aseptate conidia are also found. Denticles in each conidiogenous cell of the present collection is maximum as compared to the rest allied taxa.

All these morphotaxonomic features of the collected fungus merit its description and illustration as a taxon of a separate species rank.

G. justiciae sp.nov. **(Fig.3)**

Maculae amphigenae, singulae, et irregulares necroticae, brunnae in aetate auctae et areas magnas tegentes, plus minusve zonatae. **Coloniae** hypophyllae, effusae,pallide brunneae,velutinae. **Mycelium** superficialibus, parte immersae, profuse ramosae, pallide brunneae, laevia, septate, 2.5-3.5 µm diam. ***Stromata*** nulla. **Conidiophora** solitaria, micronematosa to semi macronematosa, mononematosa, bilaterales ramosa, septata suberecta, fere repentia, flexuosa, laevia, nodosa fere septata, intumescentiis lateralibus vel terminalibus denticulas conoides ferentibus praedita, pallide brunneae, variabilissima, 123.5-37.5 x 2.5-3.5 µm. **Cellulae conidiogenae** in conidiophores incorporatae, terminales vel intercalares, mono-deinde polyblasticae, in intumescentiis dispositae, denticulatae, denticula 1-5, plerumque 3-4 denticula, cicatricosa, cicatrices conspicue. **Conidia** solitaria, holoblastica, sicca, plerumque 3 septata, laevia, pallide brunneae, ad apicem fere circulares, ad basim obconica, hilo conspicuo et incrassato, dispositae in denticulate, bands fere septata, 13.0-24.5 x 3.0-4.5 µm.

In foliis vivis ***Justicia quinquangularis*** (Acanthaceae) Jarwa Forest, Tulsipur, Balrampur, U.P., India, Kamal and M.S. Tripathi, Oct.1999, GPU Herb No. 9011/99 isotypus, HCIO No. 43135 holotypus.

Leaf lesions amphigenous, appearing as a single, irregular, necrotic brown spot with greyish white surrounding, increasing in size with age to cover large area, more or less zonate. **Colonies** hypophyllous, brown, velvety, effuse. **Mycelium of hyphae** external and partially internal, profusely branched, pale brown, smooth, septate, 2.5-3.5 µm. **Stromata** absent, **Conidiophores** solitary, micronematous to semimacronematous, mononematous, septate, branched bilaterally, suberect almost repent, flexuous, smooth, nodose at almost each septum, bearing lateral and terminal conidiogenous cells, pale brown, variable in length, 37.5-123.5 x 2.5-3.5 µm. **Conidiogenous cells** integrated, lateral to terminal, monoblastic to polyblastic, swollen, each bearing 1-5 denticles, generally 3-4 denticles cicatrized, scars thick. **Conidia** solitary, holoblastic, dry, acropleurogenous, clavate to clavatocylindric, relatively more curved, 0-3 septate, often 3 septate, smooth, pale brown, apex almost rounded, base obconic, with thickened hilum, borne on denticles, bands on each septum 13.0-24.5 x 3.0-4.5 µm.

On living leaves of ***Justicia quinquangularis*** (Acanthaceae), Jarwa forest Balrampur, Tulsipur, U.P., India, Kamal and M.S. Tripathi; Oct. 1999, GPU Herb No. 9011/99 isotype, HCIO No. 43135 holotype.

No species of *Gonatophragmium* has been described so far on the host genus *Justicia quinquangularis* (Acanthaceae) as evident from literature survey.

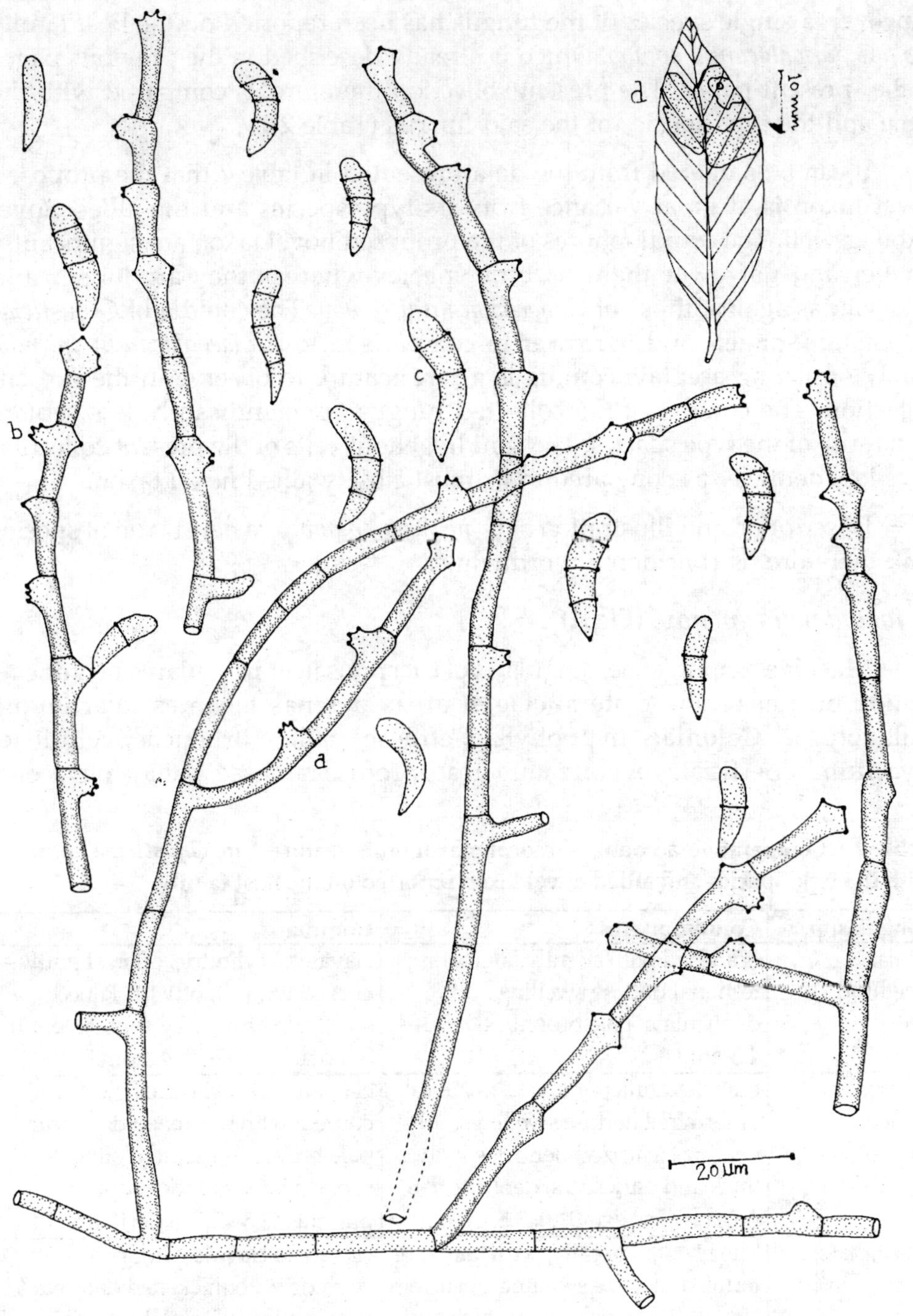

Fig. 3. *Gonatophragmium justiciae sp. nov.* (a) Conidiophores. (b) Conidiogenous cells. (c) Conidia. (d) Infection spots. Bars a-c = 20 µm. d = 10 mm.

However, a single species of the fungus has been reported on the host family viz., *G. daedalacanthi sp.nov.* which is already described in the previous pages of the present paper. The present collection therefore, is compared with the same and the type species of the said fungus (Table 2).

It can be gathered from the data presented in table 2 that the proposed novel taxon is at great variance from its type species and the allied novel taxon as well. The conidiophores of the proposed novel taxon are significantly smaller and narrower than the type species whereas they are longer and narrower as against those of *G. daedalacanthi sp.nov.* The conidia of *G. justiciae sp. nov.* are longer and narrower in comparison to *G. daedalacanthi sp. nov.* Rarely occurring aseptate conidia is a rare feature to observe in the present collection. The conidia of the collected fungus are significantly less septate than those of the type species. The conidiogenous cells of the present collection have less denticles as compared to its most closely allied novel taxon.

Description and illustration of *G. justiciae sp.nov.* as a new taxon of species rank therefore, is considered worthwhile.

G. jarwaensis sp. nov. (Fig.4)

Maculae amphigenae, singulae vel caespitosae et irregulares necroticae, pallide brunneae, in aetate auctae et areas magnas tegentes, plerumque multizonatae. **Coloniae** hypophyllae, effusae, pallide brunneae, velutinae. **Mycelium** superficiales et parte immersae, profuse ramosae, pallide olivaceae,

Table-2 : Comparative account of morphotaxonomic features of *G.justiciae sp. nov.* with the type species and allied novel taxon reported on the host family.

Fungus spp.	Conidiophores	Conidia
G. mori Deighton	branched with slightly lateral or terminal nodose swellings, denticulate, pale brown, 500x3.0-4.5 μm.	clavate to cylindric, often slightly curved with slightly thickened scar, pale brown, 1-7 septate, often 3 septate 13.0-26.0x4.5 μm.
G.daedalacanthi sp. nov.	branched unilaterally, with lateral or terminal nodose swellings, bearing cicatrized denticles with thick and dark scars, denticles 1-6, pale brown, 35.0-91.0x3.5-4.0 μm.	clavate, rarely clavatocylindrical, curved, with thickened dark scar, pale brown, 1-3 septate often 3 septate, bands at each septum present, 12.5-23.5x3.0-5.0 μm.
G.justiciae sp. nov.	branched bilaterally, with lateral or terminal nodose swellings, bearing cicatrized denticles with thick and dark scars, denticles 1-5, pale brown, 37.5-123.5x2.5-3.5 μm.	clavate to clavatocylindrical, curved, with thickened dark scar, pale brown, 0-3 septate, often 3 septate, bands at each septum present, 13.0-24.5x3.0-4.5 μm.

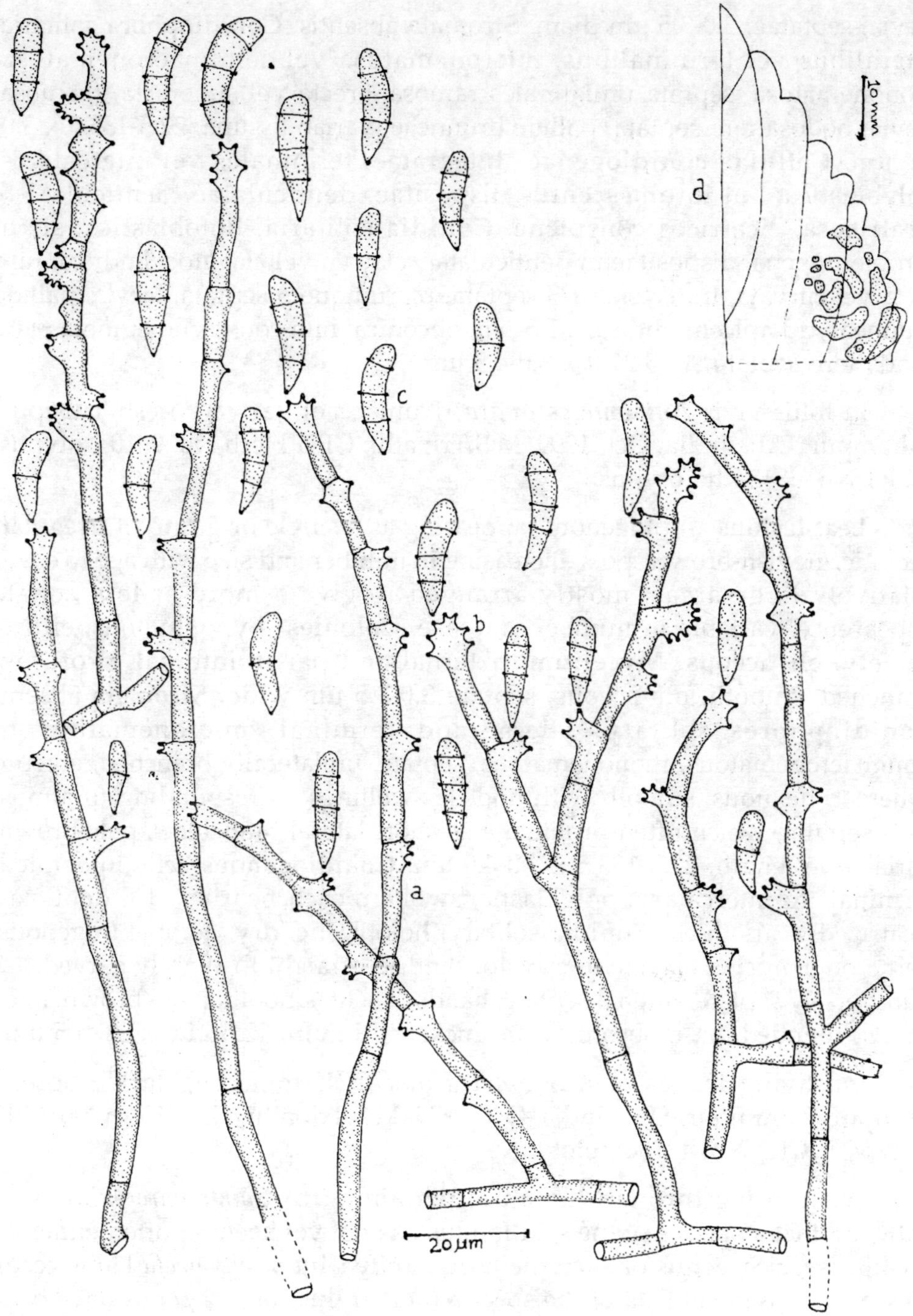

Fig. 4. *Gonatophragmium jarwaensis sp. nov.* (a) Conidiophores. (b) Conidiogenous cells. (c) Conidia. (d) Infection spots. Bars. a-c = 20 µm. d = 10 mm.

laevia, septatae, 3.0-3.5 µm diam. **Stromata** absentis. **Conidiophora** solitaria, lateralibus vel terminalibus, micronematosa vel semimacronematosa, mononematosa, septata unilaterales ramosa, erecta vel suberecta, flexuosa, laevia, nodosa fere septata, pallide brunneae, variabilissima, 22.5-133.5 x 3.0-4.0 µm. **Cellulae conidiogenae** integratae, terminales vel intercalares, polyblasticae, in intumescentiis dispositae, denticulatae, denticula 1-8, cicatricosa, cicatrices conspicue. **Conidia** solitaria, holoblastica, sicca, acropleurogena, dispositae in denticulatae, clavata vel clavatocylindrica, raro recta vel curvata, transverse 1-3 septata, plerumque 3 septata, laevia, pallide brunneae, ad apicem obtusa, ad basim obconica, hilo conspicuo et incrassato, bands fere septata, 12.5-21.0 x 3.0-4.5 µm.

In foliis vivis ***Anisomeles ovata*** (Lamiaceae), Jarwa Forest, Tulsipur, Balrampur, U.P., India, Oct. 1999, M.S.Tripathi, GPU Herb No. 9010 isotypus, HCIO No. 43134 holotypus.

Leaf lesions amphigenous, appearing as a single or group of irregular, necrotic, greyish-brown spots, increasing in number and size with age to cover relatively large areas, mostly arranged to give a more or less zonate appearence, variable in number and size. **Colonies** hypophyllous, effuse, velvety, olivaceous. **Mycelium** external and partly internal, profusely branched, smooth, olivaceous, septate 3.0-3.5 µm wide. **Stromata** absent. **Conidiophores** solitary, lateral and terminal, micronematous to semimacronematous, mononematous, septate, unilaterally branched, erect to suberect, flexuous, smooth with nodose swellings corresponding to almost each septum, which often proliferate as short lateral branchlets, pale-brown, variable in length, 22.5-133.5 x 3.0-4.0 µm. **Conidiogenous cells** integrated, terminal to intercalary, polyblastic, swollen, each bearing 1-8 denticles, cicatrized, scars thick. **Conidia** solitary, holoblastic, dry, acropleurogenous, borne on denticles, clavate to clavatocylindric, straight to slightly curved, 1-3 transversely septate, often 3 septate, bands absent, smooth, pale- brown, apex almost rounded, base obconic with thickened hilum, 12.5-21.0 x 3.0-4.5 µm.

On living leaves of ***Anisomeles ovata*** (Lamiaceae), Jarwa Forest, Tulsipur, Balrampur, U.P., India, Oct. 1999, M.S.Tripathi, GPU Herb No. 9010 isotype, HCIO No. 43134 holotype.

After going through literature pertaining to *Gonatophragmium* it is gathered that a species of the said fungus has not yet been reported either on the host species, genus or even the host family (*Anisomeles ovata*-Lamiaceae). However, two novel taxa of the species rank of this fungus genus have been collected on the host family during the course of present investigation viz., *G. lamiacearum sp. nov.* and *G. pogostemonidis sp.nov.* and they are being described in the present paper. The present collection is compared with the type species of the same fungus and its allied taxa to confirm its novelty in Table 3.

It can be gathered from the data presented in the table that the present collection is entirely different from the type species and its allied taxa. The conidiophores and conidia of the present collection are markedly smaller than that of *G. mori* Deighton. The conidiophores of *G.jarwaensis sp.nov.* are longest as compared to its allied taxa but the conidia of the same are narrowest and smallest. Number of denticles per conidiogenous cell are highest in the present collection. Bands at each conidial septum are markedly present in *G. jarwaensis sp. nov.* while they are absent in all the other allied taxa and the type species as well.

All these morphotaxonomic features of the present collection prove its novelty as a separate taxon of species rank.

G. lamiacearum sp. nov. (Fig.5)

Maculae amphigenae, singulae vel caespitosae, fere circulares, necroticae, brunnae, atro-marginate, in aetate auctae et areas magnas tegents, plerumque multi-zonatae. **Coloniae** hypophyllae, effusae, pallide brunneae, velutinae. **Mycelium** superficialibus et parte immersae, profuse ramosae, pallide brunneae, laevia, septatae, 4-5µm diam. **Stromata** absentis. **Conidiophora** solitaria, micronematosa vel macronematosa,mononematosa, bilaterales ramosa, septata, erecta vel suberecta, fere repentia, flexuosa, laevia, nodosa fere septata, intumescentiis intercalaribus vel terminalibus, cellulae conidiogenae ferentibus praedita, pallide brunneae, variabilissima, 50.0-102.0 x 3.0-4.0 um. **Cellulae conidiogenae** integratae, terminales vel intercalares (laterales), monoblastic vel raro-polyblasticae, in imtumescentiis dispositae, denticulatae, denticula 1-5, plerumque 4 denticula, cicatricosa, cicatrices conspicue. **Conidia** solitaria, holoblastica, sicca, acropleurogena, clavata vel clavatocylindrica, recta vel curvata, transverse 1-6 septata, plerumque 4-6 septata, laevia, pallide bruneae, ad apices obtusa vel fere circulares, ad basim obconica, hilo conspicuo et incrassato, dispositae in denticulate bands absentis, 12.5-32.5 x 3.0-5.0 µm.

In foliis vivis ***Colebrookia oppositifolia*** (Lamiaceae) Duma Forest, Nichlaul, Maharajganj, U.P., India, Kamal and M.S. Tripathi, Sept. 1999, GPU Herb No. 9003/99 isotypus, HCIO No. 43120 holotypus.

Leaf lesion amphigenous, appearing as a single or group of almost circular, necrotic, dark brown spots with black margins, increasing in number and size with age, cover large area, sometimes many distinct zonate rings. **Colonies** hypophyllous, pale brown, velvety, effuse. **Mycelium of hyphae**

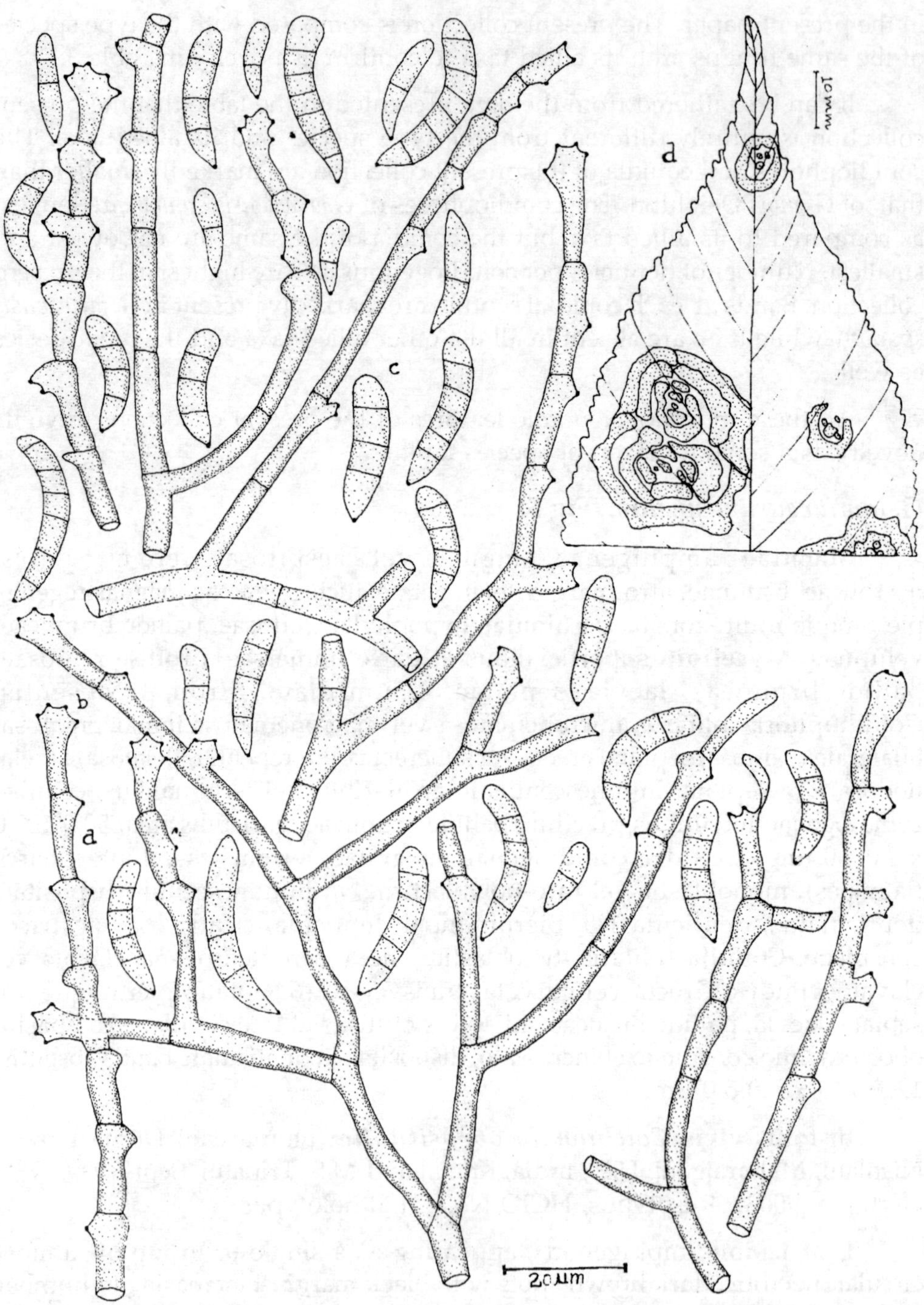

Fig. 5. *Gonatophragmium lamiacearum sp. nov.* (a) Conidiophores. (b) Conidiogenous cells. (c) Conidia. (d) Infection spots. Bars. a-c = 20 mm. d = 10 mm.

superficial, partly immersed, profusely branched, pale brown, smooth, septate, 4.0-5.0 µm wide. **Stromata** absent. **Conidiophores** solitary, micronematous to semimacronematous, mononematous, septate, branched bilaterally, erect to suberect or repent, flexuous, smooth, nodose at almost each septum, bearing intercalary (lateral)and terminal conidiogenous cells, which rarely proliferate as short lateral branchlets, pale brown variable in length, 50.0-102.0 x 3.0-4.0 µm. **Conidiogenous cells** integrated, intercalary to terminal, monoblastic to polyblastic, swollen, each bearing 1-5 denticles, generally 4 denticles, cicatrized, scars thick. **Conidia** solitary, holoblastic, dry, acropleurogenous, clavate to clavatocylindric, substraight to curved, 1-6 transversely septate, often 4-6 septate, smooth, pale brown, apex obtuse to almost rounded, base obconic, with thickened hilum, borne on denticles, bands absentis, 12.5-32.5 x 3.0-5.0 µm.

On living leaves of ***Colebrookia oppositifolia*** (Lamiaceae) Duma Forest Nichlaul, U.P., India, Kamal and M.S.Tripathi, Oct. 1999. GPU Herb No. 9003 isotype, HCIO No. 43120 holotype.

No *Gonatophragmium* species has so far been reported on the host species or on the host genus (*Colebrookia oppositifolia*-Lamiaceae) as indicated by the survey of literature. Therefore, the present collection is being compared with its allied taxa viz., *G. jarawaensis sp. nov.* and *G. pogostemonidis sp. nov.* collected on the same host family and described in the present paper and with the type species of the fungus genus in question. This comparative account is given in Table 3.

The table shows remarkable differences in the morphotaxonomic features of the present collection as against its allied taxa and the type species. *G. lamiacearum sp. nov.* is different from its allied taxa in having shortest conidiophores. The conidia of the novel taxon are largest in size than those of *G. jarwaensis sp. nov.* and *G. pogostemonidis sp. nov.* The conidia of the present collection are more septate than its allied taxa and denticles in each conidiogenous cells are more in number than in *G. pogostemonidis sp. nov.* while less in number than in *G. jarwaensis sp. nov.* Conidial bands are absent. The conidiophores and conidia of the present collection are significantly smaller than its type species, *G.mori* Deighton.

All these morphotaxonomic features of the present collection warrant its description and illustration as a taxon of separate species rank.

G. pogostemonidis sp. nov. (Fig.6)

Maculae amphigenae, singulae vel caespitosae, et irregulares vel raro circulares, necroticae, atrobrunnae, atro-marginate, in aetate auctae et areas magnas tegents, zonatae. **Coloniae** hypophyllae, effusae, brunneae, velutinae. **Mycelium** superficialibus et parte immerse, profuse ramosae, pallide

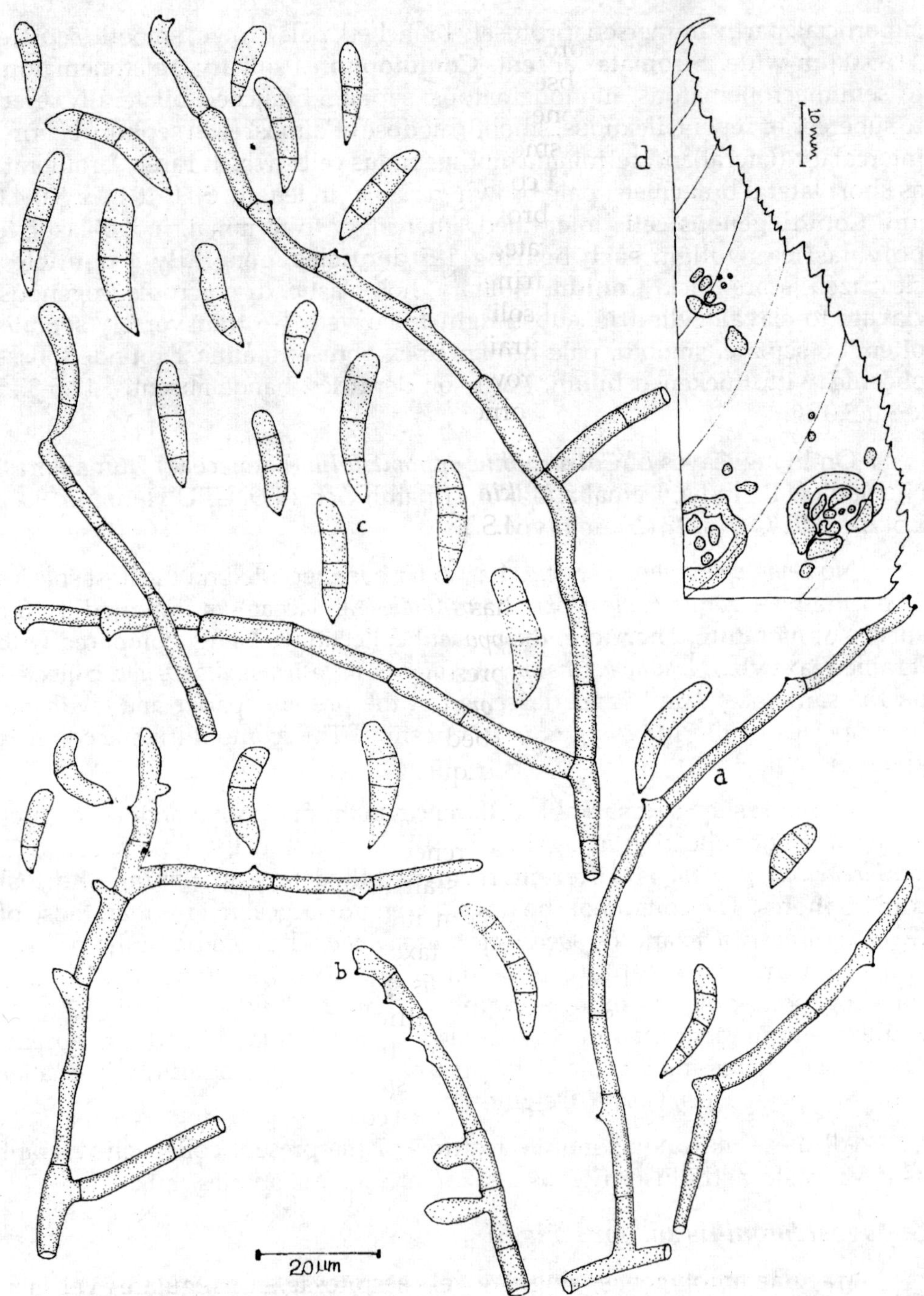

Fig. 6. *Gonatophragmium pogostemonidis sp. nov.* (a) Conidiophores. (b) Conidiogenous cells. (c) Conidia. (d) Infection spots. Bars. a-c = 20 µm. d = 10 mm.

Table 3 : Comparative account of morphotaxonomic features of *G. jarwaensis sp. nov.* with its type species and allied taxa.

Fungus spp.	Conidiophores	Conidia
G. mori Deighton	branched with slightly lateral or terminal nodose swellings, denticulate, pale brown, 500x3.0-4.5 µm.	clavate to cylindric, often slightly curved with slightly thickned scar, pale brown, 1-7 septate, often 3- septate. 13.0-26.0x4.5 µm.
G. lamiaceraum sp. nov.	branched bilaterally, with intercalary or terminal nodose swellings, bearing cicatrized denticles with thick and dark scars, denticles 1-5, pale brown, 50.0-102.0x3.0-4.0 µm.	clavate to clavato cylindrical, substraight to curved, with thickened dark scar, pale brown, 1-6 septate, often 4-6 septate bands at each septum absent, 12.5-32.5x3.0-5.0 µm.
G. pogostemonidis sp. nov.	branched bilaterally, with intercalary or terminal nodose swellings, bearing cicatrized denticles with thick and dark scars, denticles 1-2, pale brown, 26.0-122.5x2.5-3.5 µm.	clavate to clavatocylindrical, slightly curved with thickened dark scar, pale brown, 1-3 septate, often 3 septate, bands at each septum absent, 11.5-24.0x3.0-5.0 µm.
G. jarwaensis sp. nov.	branched unilaterally, with intercalary or terminal nodose swellings, bearing cicatrized denticles with thick and dark scars, denticles 1-8, pale brown, 22.5-133.5x3.0-4.0 µm.	clavate to clavato-cylindrical, substraight to slightly curved, with thickened dark scar, pale brown, 1-3 septate, often 3 septate, bands at each septum present, 12.5-21.0x3.0-4.5 µm.

brunneae, laevia, septata, 3.0-3.5 µm diam. **Stromata** absentis. **Conidiophora** solitaria, micronematosa vel macronematosa, mononematosa, bilaterales ramosa, septata, erecta vel suberecta, fere repentia, flexuosa, laevia, nodosa fere septata, intumescentiis intercalaribus vel terminalibus, cellulae conidiogenae ferentibus praedita, pallide brunneae, variabilissima, 26.0-122.5 x 2.5-3.5 µm. **Cellulae conidiogenae** integratae, terminales vel intercalares, monoblastic vel raro-polyblasticae, in intumescentiis dispositae, denticulatae, denticula 1-2, cicatricosa, cicatrices conspicue. **Conidia** solitaria, holoblastica, sicca, acropleurogena, clavata vel clavatocylindrica, curvata, transverse 1-3 septata, plerumque 3 septata, laevia, pallide brunneae, ad apicem obtusa, ad basim obconica, hilo conspicuo et incrassato, dispositae in denticulate, bands absent, 11.5-24.0x 3.0-5.0 µm.

In foliis vivis ***Pogostemon lanceolatus*** (Lamiaceae) Duma Forest, Nichlaul Maharajganj, U.P., India, Kamal and M.S.Tripathi. Sept. 1999, GPU Herb No. 9002 isotypus, HCIO No. 43119 holotypus.

Leaf lesions amphigenous, appearing as a single or group of irregular, necrotic, dark brown spots with black margins, increasing in number and size with age, cover large leaf areas, zonate. **Colonies** hypophyllous, brown, velvety, effuse. **Mycelium of hyphae** superficial,partially immersed, profusely branched, pale brown, smooth, septate, 3.0-3.5 µm wide. **Stromata** absent. **Conidiophores** solitary, micronematous to semimacronematous, mononematous, septate, branched bilaterally, suberect, almost repent, flexuous, smooth, nodose at almost each septum, bearing intercalary and terminal conidiogenous cells, pale brown, variable in length, 26.0-122.5 x 2.5-3.5 µm. **Conidiogenous cells** integrated, intercalary to terminal, monoblastic, rarely polyblastic, swollen, each bearing 1-2 denticles, cicatrized, scars thick. **Conidia** solitary, holoblastic, dry, acropleurogenous, clavate to clavatocylindric, slightly curved, 1-3 transversely septate often 3 septate, smooth, pale brown, apex obtuse, base obconic,with thickened hilum, borne on denticles, bands absent, 11.5-24.0 x 3.0-5.0 µm.

On living leaves of ***Pogostemon lanceolatus*** (Lamiaceae), Duma Forest, Nichlaul, Maharajganj, U.P., India, Kamal and M.S. Tripathi; Sept. 1999. GPU Herb No. 9002 isotype, HCIO No. 43119 holotype.

A survey of literature reveals that a *Gonatophragmium* species has not previously been collected either on the host species or on the host genus (*Pogostemon lanceolatus-* Lamiaceae). The present collection is being compared with its type species and the allied taxa viz., *G. jarwaensis sp. nov.* and *G. lamiacearum sp. nov.* reported on Lamiaceae in the present chapter. Their comparative accounts are given in Table 3.

The table presented is indicative of the fact that *G. pogostemonidis sp.nov.* is distinctly different from the type species as well as the allied taxa of its own. The conidiophores of the present collection are longer and narrower than *G. lamiacearum sp. nov.* but shorter and narrower than *G. jarwaensis sp.nov.* The conidia of the novel taxon are least septate than its type species and smaller than *G. lamiacearum sp.nov.* but larger than *G.jarwaensis sp. nov.* Denticles in conidiogenous cells of the novel taxon are least as compared to its allied taxa. Conidial bands are absent. The conidiophores and conidia of the present collection are significantly smaller than the type species, *G. mori* Deighton.

It is with this conclusion that the present collection has been treated here as a novel taxon of distinctly different species rank.

G. deightonii sp.nov. (Fig.7)

Maculae amphigenae, singulae vel caespitosae, fere circulares, necroticae, atrobrunnae, atro-marginate, in aetate auctae et areas magnas tegents, plerumque multizonatae. **Coloniae** hypophyllae, effusae, pallide brunneae, velutinae. **Mycelium** superficialibus et parte immersae, profuse ramosae, pallide brunneae, laevia, septatae, 3.5-4.0 µm wide. **Stromata** absentis.

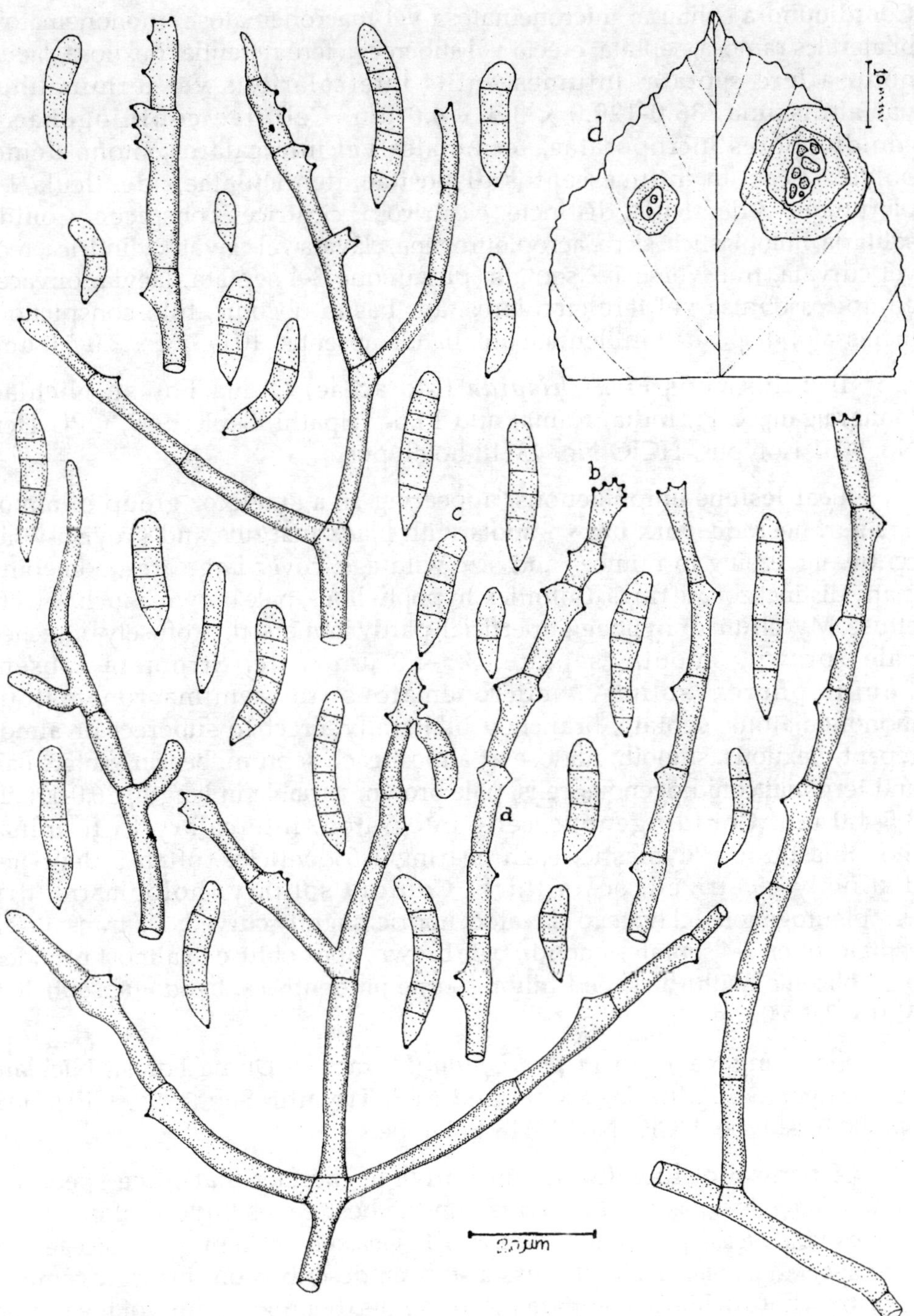

Fig. 7. *Gonatophragmium deigthonii sp. nov.* (a) Conidiophores. (b) Conidiogenous cells. (c) Conidia. (d) Infection spots. Bars. a-c = 20 µm. d = 10 mm.

Conidiophora solitaria, micronematosa vel macronematosa, mononematosa, bilaterales ramosa, septata, erecta vel suberecta, fere repentia, flexuosa, laevia, nodosa fere septata, intumescentiis intercalaribus vel terminalibus, variabilissima, 36.0-120.0 x 3.5 x 4.0 μm. **Cellulae conidiogenae** in conidiophores incroporatae, terminales vel intercalares, mono-deinde polyblasticae, in intumescentiis dispositae, denticulatae, denticula 1-6, plerumque 3 denticula, distincte, cicatricosa, cicatrices conspicue. **Conidia** solitaria, holoblastica, sicca, acropleurogena, clavata vel clavatocylindrica, recta vel curvata, transverse 1-7 septata, plerumque 3-4 septata, laevia, olivacea, ad apices obtusa vel fere circulares, ad basim obconic, hilo conspicuo et incrassato, dispositae in denticulate, bands absentis, 10.0-31.0 x 3.0-5.0 μm.

In foliis vivis ***Ficus hispida*** (Moraceae) Duma Forest, Nichlaul Maharajganj, U.P., India, Kamal and M.S. Tripathi, Sept. 1999, GPU Herb No. 9001 isotypus, HCIO No. 43118 holotypus.

Leaf lesions amphigenous, appearing as a single or group of almost circular, necrotic, dark brown spots with black margins and greyish white centre, increasing in number and size with age, cover large area, sometimes many distinct zonate rings. **Colonies** hypophyllous, pale brown, much velvety, effuse. **Mycelium of hyphae** superficial, partly immersed, profusely branched, pale brown, smooth, septate, 3.5-4.0 μm wide. **Stromata** absent. **Conidiophores** solitary, micronematous to semimacronematous, mononematous, septate, branched bilaterally, erect to suberect or almost repent, flexuous, smooth, nodose at almost each septum, bearing intercalary and terminal conidiogenous cells, pale brown, variable in length, 36.0-120.0 x 3.5-4.0 μm. **Conidiogenous cells** integrated, intercalary to terminal, monoblastic to polyblastic, each bearing 1-6 denticles, often 3 denticles, distinctly cicatrized, scars thick. **Conidia** solitary, holoblastic, dry, acropleurogenous, clavate to clavatocylindric, slightly curved, 1-7 transversely septate, often 3-4 septate, smooth, pale brown, apex obtuse to almost rounded, base obconic, with thickened hilum, borne on denticles, bands absent, 10.0-31.0 x 3.0-5.0 μm.

On living leaves of ***Ficus hispida*** (Moraceae) Duma Forest, Nichlaul, Maharajganj, U.P., India, Kamal and M.S. Tripathi, Sept.1999, GPU Herb No. 9001 isotype, HCIO No. 43118 holotype.

A perusal of literature is indicative of the fact that three species of *Gonatophragmium* has been described on the host genus but on different host species viz., *Ficus* sp., *Ficus hispida* and *F. glomerata* of family Moraceae. The type species, *G. mori* Deighton has also been described on this host genus. It appears whorthwhile to compare the morphotaxonomic features of the present collection with those of the type species and other two fungal species collected on the same host genus of Moraceae viz., *G. kushinagarensis sp. nov.* and *G.*

moracearum sp. nov. which are being described in the present paper. The comparison is presented in Table 4.

The comparative account of Table 4 shows that the present collection is different from its type species in having significantly smaller and narrower conidiophores. It is also different from the allied taxa, *G. kushinagarensis sp. nov.* and *G. moracearum sp. nov.* in having smaller and narrower conidiophores on one hand and larger and wider conidiophores on the other. The conidia of *G. deightonii sp. nov.* are longer, wider and more septate than the type species and the allied taxa as well. Bands at each septum of conidia are absent in sharp contrast to its allied taxa.

G. kushinagarensis sp. nov. (Fig.8)

Maculae amphigenae, singulae vel caespitosae, fere circulares, necroticae, atrobrunnae, atro-marginate, in aetate auctae et areas magnas tegents, multizonatae, variabilissima. **Coloniae** hypophyllae, effusae, brunneae, velutinae. **Mycelium** superficialibus et parte immersae, ramosae, pallide

Table 4 : Comparative account of morphotaxonomic features of *G. deightonii sp. nov.* with its allied taxa.

Fungus spp.	Conidiophores	Conidia
G.mori Deighton	branched with slightly lateral or terminal nodose swellings, denticulate, pale brown, 500x3.0-4.5 μm.	clavate to cylindric, often slightly curved with slightly thickened scar, pale brown, 1-7 septate often 3 septate, 13.0-26.0x4.5 μm.
G. kushinagarensis sp. nov.	branched unilaterally, with intercalary or terminal nodose swellings, bearing cicatrized denticles with thick and dark scars, denticles 1-7, pale brown, 32.5-135.5x3.0-3.5 μm.	clavate to clavato-cylindrical, substraight to curved, with thickened dark scar, pale brown, 1-3 septate, often 3 septate, bands at each septum present, 10.5-22.5 x 3.0-4.5 μm.
G. moracearum sp. nov.	branched unilaterally, with intercalary or terminal nodose swellings, bearing cicatrized denticles with thick and dark scars, denticles 1-4, pale brown, 41.0-107.0 x 2.5-3.5 μm.	clavatocylindrical, substraight to curved, with thickened dark scar pale brown, 1-6 septate, often 3 septate, bands at each septum present 15.5-26.0 x 2.5-3.5 μm.
G.deightonii sp.nov.	branched bilaterally, with intercalary or terminal nodose swellings, bearing cicatrized denticles with thick and dark scars, denticles 1-6, pale brown, 36.0-120.0x3.5-4.0 μm.	clavate to clavatocylindrical, slightly curved, with thickened dark scar, pale brown, 1-7 septate, often 3-4 septate, bands at each septum absent, 10.0-31.0x3.0-5.0 μm.

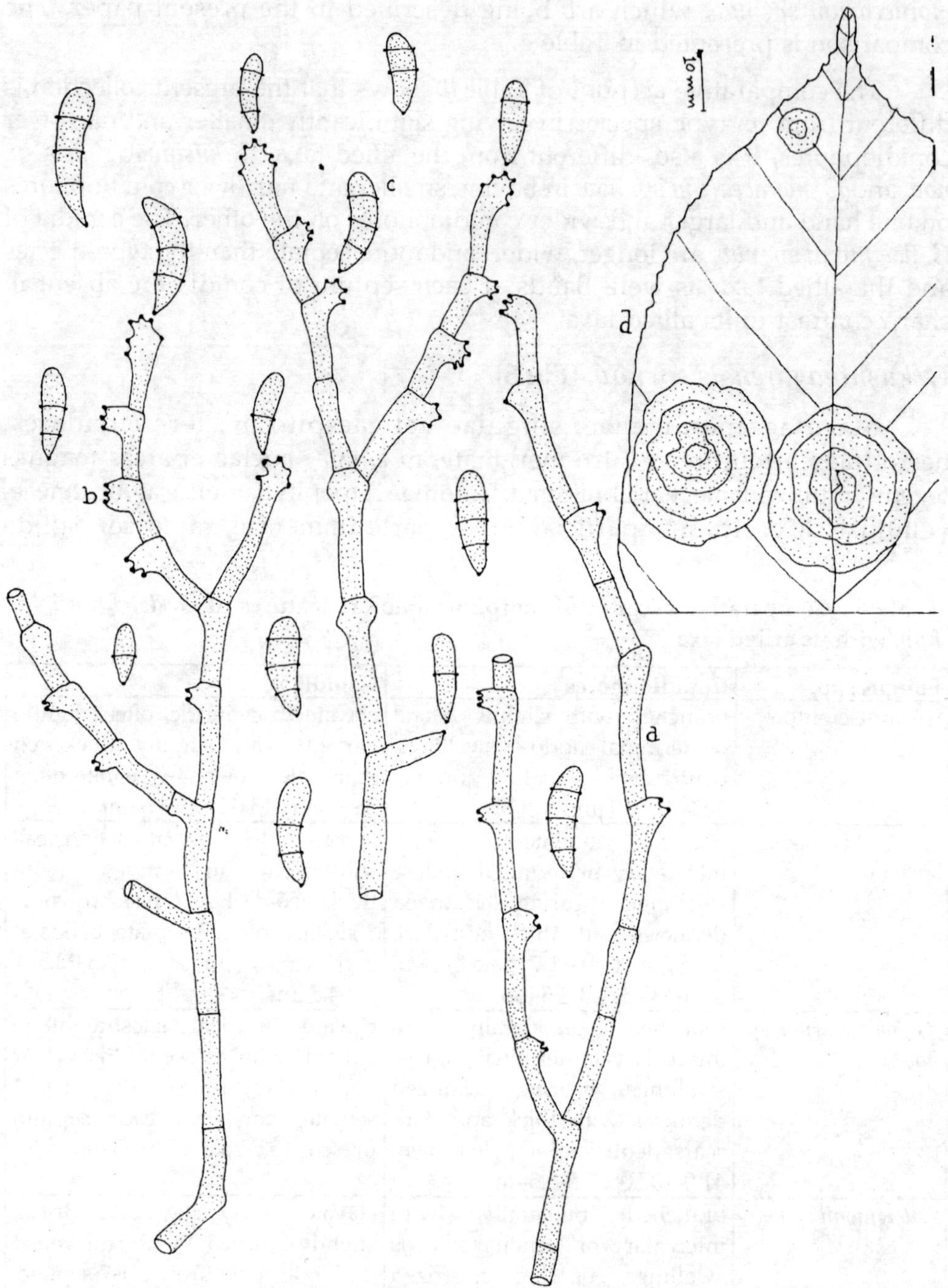

Fig. 8. *Gonatophragmium kushinagarensis sp. nov.* (a) Conidiophores. (b) Conidiogenous cells. (c) Conidia. (d) Infection spots. Bars. a-c - 20 µm. d = 10 mm.

brunneae, laevia, septatae , 3.0-3.5 µm diam. **Stromata** absentis. **Conidiophora** solitaria, micronematosa vel macronematosa, mononematosa, unilaterales ramosa, septata, erecta vel suberecta, fere repentia, flexuosa, laevia, nodosa fere septata, intumescentiis intercalaribus vel terminalibus variabilissima, 32.5-135.5 x 3.0-3.5 µm. **Cellulae conidiogenae** in conidiophoris incorporate, terminales vel intercalares, mono-deinde, polyblasticae, in intumescentiis dispositae, denticulatae, denticula 1-7, plerumque 3 denticula, distincte cicatricosa, cicatrices conspicue. **Conidia** solitaria, holoblastica, sicca, acropleurogena, clavata vel clavatocylindrica, suberecta vel curvata, transverse 1-3 septata, plerumque 3 septata, laevia, olivacea, ad apicem obtusa vel fere circulares, ad basim obconica, hilo conspicuo et incrassato, dispositae in denticulate, bands et fere septata, 10.5-22.5 x 3.0-4.5 µm.

In foliis vivis ***Ficus hispida*** (Moraceace), Kushinagar, U.P., India, Nov. 1999, M.S. Tripathi, GPU Herb No. 9013 isotypus, HCIO No. 43137 holotypus.

Leaf lesions amphigenous, appearing as a single or group of almost circular, necrotic, dark brown spots with black margins and greyish white centre, increasing in number and size with age, cover large leaf areas, sometimes many distinct zonate rings. **Colonies** hypophyllous, effuse, much velvety, brown. **Mycelium of hyphae** superficial, partly immersed, profusely branched, pale brown, smooth, micronematous to semimacronematous, mononematous, septate, unilaterally branched, erect to suberect or almost repent, flexuous, smooth, nodose at almost each septum, bearing intercalary and terminal conidiogenous cells, pale brown, variable in length, 32.5-135.5 x 3.0-3.5 µm. **Conidiogenous cells** integrated, intercalary to terminal, monoblastic to polyblastic, swollen, each bearing 1-7 denticles, often 3 denticles, distinctly cicatrized, scars thickened. **Conidia** solitary, holoblastic, dry, acropleurogenous, clavate to clavatocylindric, substraight to almost curved, 1-3 transversely septate, often 3 septate, smooth, pale brown, apex obtuse to almost rounded, base obconic with thickened hilum, borne on denticles, bands present, 10.5-22.5 x 3.0-4.5 µm.

On living leaves of ***Ficus hispida*** (Moraceae), Kushinagar, U.P., India, M.S.Tripathi, Nov. 1999, GPU Herb No. 9013 isotype, HCIO No. 43137 holotype.

Earlier literature on the taxonomy of this fungus speaks of the fact that three different species of *Gonatophragmium* viz. *G. mori* Deighton, *G. deightonii sp. nov.* and *G. moracearum sp. nov.* have so far been described on the same host species, *F. hispida*. Except for *G.mori* Deighton the type species of the fungus genus in question, others are described currently in the present chapter as novel taxons of species rank. The morphotaxonomic features of *G.*

kushinagarensis sp. nov. are compared with allied taxa to confirm its novelty (Table 4).

As gathered from the data, *G. kushinagarensis sp.nov.* is found distinctly different in having much smaller and narrower conidiophores than *G.mori* Deighton. However, the conidiophores of the present collection are longer but narrower than those of *G. deightonii sp. nov.* and *G. moracearum sp. nov.* The conidia of present fungus species are smallest, least septate and less wider in comparison to its allied taxa. Septal bands are present corresponding to each conidial septa. Number of denticles per conidiogenous cells are maximum in the present collection.

Therefore, to accommodate the present collection its discription and illustration as taxon of novel species rank is considered worthwhile.

***G. moracearum sp. nov.* (Fig.9)**

Maculae amphigenae, singulae, fere circulares, necroticae, brunneae, atro-marginatae, in aetate auctae et areas magnas tegentes, zonatae. **Coloniae** hypophyllae, effusae, brunneae, velutinae. **Mycelium** superficiales et parte immersae, profuse ramosae, pallide brunneae, septatae, 3.0-3.5 µm diam. **Stromata** absentis. **Conidiophora** solitaria, laterales vel terminales, micronematosa vel semimacronematosa, mononematosa, septata, unilaterales ramosa, erecta vel suberecta repentia, flexuosa, laevia, nodosa fere septata, intumescentiis intercalaribus vel terminalibus denticulas conideas ferentibus praedita, pallide brunneae, in magnitudinae variabilissima, 41.0 – 107.0 x 2.5-3.5 µm. **Cellulae conidiogenae** integratae, terminales vel intercalares, monoblasticae vel polyblasticae, in intumescentiis dispositae, denticulatae, denticula 1-4, cicatricosa, cicatrices distincte. **Conidia** solitaria, holoblastica, sicca, acropleurogena, dispositae in denticulate, clavatocylindrica, raro sub erecta vel fere curvata, transverse 1-6 septata, plerumque 3 septata, bands fere septata, laevia, pallide brunneae, ad apices circulares, ad bases obconica, hilo conspicuo et incrassato, 15.5 – 26.0 x 2.5-3.5 µm.

In foliis vivis ***Ficus glomerata*** (Moraceae), Jarwa Forest, Tulsipur, Balrampur, U.P., India, Oct. 1999, M.S. Tripathi, GPU Herb No. 9012 isotypus, HCIO No. 43136 holotypus.

Leaf lesions amphigenous, appearing as a single almost circular, necrotic brown spots with black margins and greyish white centre, increasing in size with age to cover large areas, zonate. **Colonies** hypophyllous, effuse, velvety, dull brown, smooth, septate, 3.0-3.5 µm wide. **Stromata** absent. **Conidiophores** solitary, lateral and terminal, micronematous to semimacronematous, mononematous, septate, unilaterally branched, erect to suberect, flexuous,

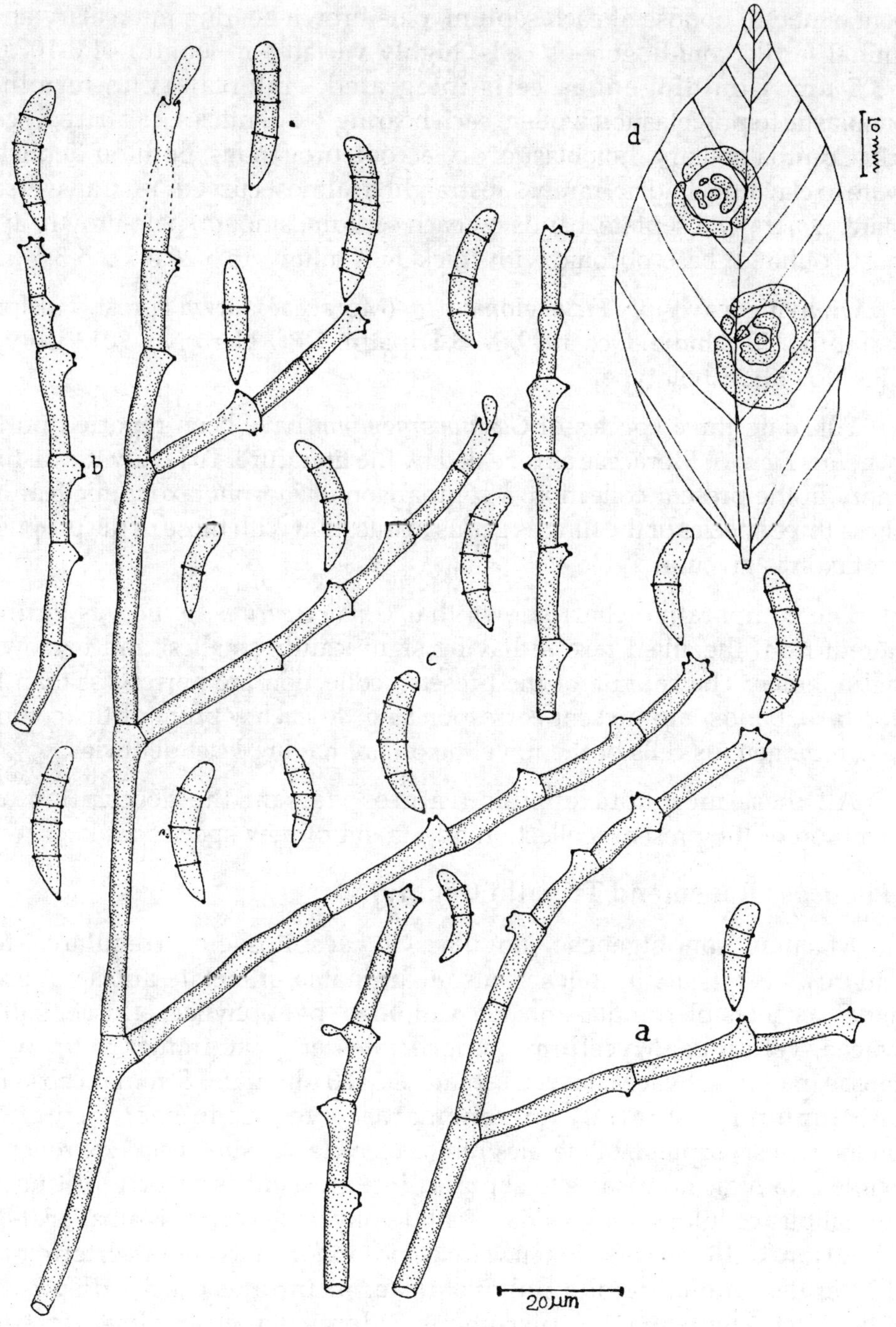

Fig. 9. *Gonatophragmium moracarum sp. nov.* (a) Conidiophores. (b) Conidiogenous cells .(c) Conidia. (d) Infection spots. Bars. a-c = 20 µm. d = 10 mm.

repent, smooth, nodose at each septum, pale-brown bearing intercalary and/ terminal fertile conidiogenous cells, highly variable in length, 41.0-107.0 x 2.5-3.5 µm. **Conidiogenous cells** integrated, intercalary to terminal, monoblastic to polyblastic, swollen, each bearing 1-4 denticles, cicatrized, scars thick. **Conidia** solitary, holoblastic, dry acropleurogenous, born on denticles, clavate to clavatocylindric, rarely substraight to almost curved, 1-6 transversely septate, generally 3 septate, bands on each septum, smooth, pale-brown, apex almost rounded, base obconic with thickened hilum, 15.5-26.0 x 2.5-3.5 µm.

On living leaves of ***Ficus glomerata*** (Moraceae) Jarwa Forest, Tulsipur, Balrampur, U.P., India, Oct. 1999, M.S.Tripathi, GPU Herb No. 9012 isotype, HCIO No. 43136 holotype.

Till date three species of *Gonatophragmium* have been reported on the host genus *Ficus* of Moraceae as revealed by the literature. To justify the distinct identity of the present collection, a comparison of morphotaxonomic features of these three species of the fungus genus in question with those of the proposed new taxon is given in Table 4.

The comparative chart shows that *G.moracearum sp. nov.* is entirely different from the allied taxa in having significantly smallest and narrowest conidiophores. The conidia of the present collection are narrowest than the allied taxa. Bands are present corresponding to each septum of the conidia. The conidiogenous cells of the novel taxon are having least denticles.

All these morphotaxonomic features warrant the description and illustration of the present collection as a taxon of new species rank.

G. kuanense Rai emend Tripathi (Fig. 10)

Maculae amphigeneae, singulae vel caespitosae et irregulares, fere circulares, necroticae, brunneae, atro-marginatae, in aetate auctae et areas magnas tegentes plerumque zonatae. **Coloniae** hypophyllae, effusae, pallide olivacea, velutinae. **Mycelium** superficiales et parte immersae, profuse ramosae, pallide olivacea, laevia, septate, 2.5-3.0 µm diam. **Stromata** absentis. **Conidiophora** solitaria, micronematosa vel semimacronematosa, mononematosa, septata, bilaterales ramosa, erecta vel suberecta fere repentia, flexuosa, laevia, nodosa fere septata, intumescentiis intercalaribus vel terminalibus cellulae conidiogenae, pallide olivacea, variabilissima, 15.0-89.0 x 2.5-3.0 µm. **Cellulae conidiogenae** in conidiophoris incorporatae, terminales vel laterales, mono-deinde polyblasticae, in intumescentiis dispositae, denticulatae, denticula 1-7, plerumque 3 denticula, cicatricosa, cicatrices conspicue. **Conidia** solitaria, holoblastica, sicca, acropleurogena, dispositae in denticulate, clavatocylindrica, transverse 1-3 septata, plerumque 3 septata,

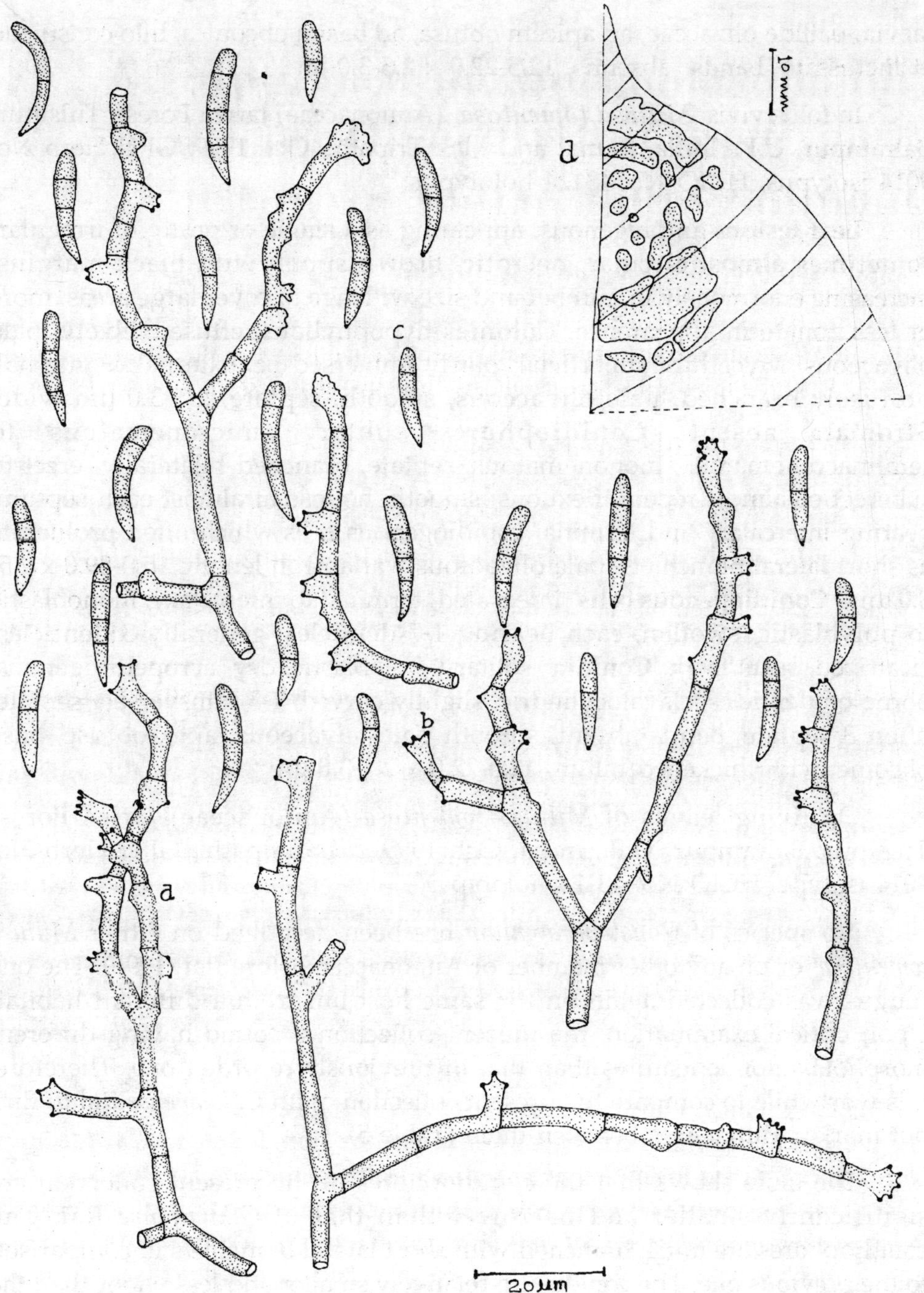

Fig. 10. *Gonatophragmium kuanense* Rai emend Tripathi. (a) Conidiophores. (b) Conidiogenous cells. (c) Conidia. (d) Infection spots. Bars. a-c = 20 µm. d = 10 mm.

laevia, pallide olivaceae, ad apicem obtusa, ad basim obconica, hilo conspicuo et incrassato, bands absentis, 12.5-22.0 x 2.5-3.0 µm.

In foliis vivis ***Miliusa tomentosa*** (Annonaceae) Jarwa Forest, Tulsipur, Balrampur, U.P., India, Kamal and M.S. Tripathi, Oct. 1999, GPU Herb No. 9014 isotypus, HCIO No. 43124 holotypus.

Leaf lesions amphigenous, appearing as a single or group of irregular, sometimes almost circular, necrotic, brown spots with black margins, increasing enormously in number and size with age to cover large areas, more or less zonate in appearence. **Colonies** hypophyllous, effuse, velvety, pale olivaceous. **Mycelium** superficial, partly immersed i.e. external & internal, profusely branched, pale olivaceous, smooth, septate, 2.5-3.0 µm wide. **Stromata** absent. **Conidiophores** solitary, micronematous to semimacronematous, mononematous, septate, branched bilaterally, erect to suberect or almost repent, flexuous, smooth, nodose at almost each septum, bearing intercalary and terminal conidiogenous cells which often proliferate as short lateral branchlets, pale olivaceous, variable in length, 15.0-89.0 x 2.5-3.0 µm. **Conidiogenous cells** integrated, terminal to intercalary, monoblastic to polyblastic, swollen, each bearing 1-7 denticles, generally 3 denticles, cicatrized, scars thick. **Conidia** solitary, holoblastic, dry, acropelurogenous, borne on denticles, clavatocylindric, slightly curved, 1-3 transversely septate, often 3 septate, bands absent, smooth pale olivaceous, apex obtuse, base obconic, with thickened hilum, 12.5-22.0 x 2.5-3.0 µm.

On living leaves of *Miliusa tomentosa* (Annonaceae) Jarwa Forest Tulsipur, Balrampur, U.P., India, Oct. 1999, M.S.Tripathi, GPU Herb No. 9014 isotype, HCIO No. 43124 holotype.

No species of *Gonatophragmium* has been described on either *Miliusa tomentosa* or on any other member of Annonaceae before Rai (1996). The one fungus was collected again on the same host but from a different habitat. Upon critical examination, the present collection is found having different morphotaxonomic features than that of previously recorded one. Therefore, it is wortwhile to compare the present collection with *G. kuanense* Rai to find out marked differences between them (Table 5).

The table shows that the conidiophores of the present collection are significantly smaller and narrower than that of *G.kuanense* Rai. The conidiophores are much branched with short lateral branchlets in comparison to the previous one. The conidia are relatively smaller and less wider than the earlier reported species. The conidiogenous cells are having larger number of denticles in the present collection. All these features are in sharp contrast with *G. kuanense* Rai. Keeping in view the gross morphological features of the

Table-5 : Comparative account of morphotaxonomic features of *G.kuanense* Rai and *G.kuanense* Tripathi.

Fungus spp.	Conidiophores	Conidia
G. kuanense Rai	branched with lateral or terminal nodose swellings, bearing cicatrized conoid denticles with thick & dark scars, brown, 41.0-184.0x 3.0-4.0 µm.	clavate to clavatocylindrical or fusiform smooth with thickened dark scar, brown, 1-3 septate, 14.0-25.5 x 2.5-4.0 µm.
G. kuanense Rai emend Tripathi	branched bilaterally, intercalary, lateral or terminal nodose swellings, lateral nodose swellings proliferate as short lateral branchlets, cicatrized denticles with thick and dark scars, denticles 1-7, pale brown, 15.0-89.0x2.5-3.0 µm.	clavatocylindrical, slightly curved with thickened dark scar, pale brown, 1-3 septate, often 3 septate, 12.5-22.0x 2.5-3.0 µm.

fungus in question, occurring on the same host species, it appears that inadequate attention was given by the earlier author while describing *G. kuanense* Rai.

Acknowledgements

Authors wish to thank the Head, Deptt. of Botany, DDU Gorakhpur University, Gorakhpur for providing laboratory and library facilities. Thanks are also due to the Curator, H.C.I.O., New Delhi for providing accession number of the type specimens.

References

Deighton, F.C. (1969). Mycological Paper, **117**:13-30.

Ellis, M.B. (1971). Dematiaceous Hyphomycetes, CMI, Kew, England.

Ellis, M.B. (1976). More Dematiaceous Hyphomycetes, CMI, Kew, England.

Mulder, J.L. (1973). Trans. Br. Mycol. Soc., **60** : 160-162.

Rai, A.N. (1996). A new species of *Gonatophragmium* from the forest flora of the Indian sub-continent. Mycol. Res., **100** (10): 1263-1264.

Tripathi, M.S. (2000). Studies on Taxonomy of Foliicolous Fungi Causing Tar-spots and related symptoms in Indian subcontinent. Ph. D. Thesis, DDU Gorakhpur University, Gorakhpour.

31

Current Status of Species Diversity in *Stenella* Sydow in Indian Sub-Continent

Sapna Chaudhary and R.K. Chaudhary

Mycology Laboratory, Department of Botany
D.D.U. Gorakhpur University, Gorakhpur-273009, U.P.

This chapter deals with up to date account of hitherto described species of Stenella from Indian sub-continent. In addition we described the illustrations and descriptions of seven novel taxa of species rank of *Stenella* Sydow viz., *S. caseariicola, S. cassigena, S. ehertigena, S. euphorbiicola, S. glycosmiidis, S. michelina* and *S. smilacismacrophyllae* collected on *Casearia tomentosa, Cassia fistula, Ehretia laevis, Antidesma ghissembila, Glycosmis pentaphylla, Michelia champaca* and *Smilax macrophylla,* respectively.

Key-words : Fungi, hyphomycetes, *Stenella,* novel taxa, diversity, Indian sub-continent

Genus *Stenella* was established by Sydow in 1930 with *Stenella araguata* Syd. as the type species. This genus and its type species has been redescribed and illustrated by Ellis (1971). Deighton (1979) transferred *Biharia vangueriae* to *Stenella vangueriae. Stenella* has some relationship with *Cercospora, Cercosporidium, Cladosporium, Mycovellosiella* and it has been confused with these species. Deighton (1979) was of the opinion that *Stenella* has close similarities with *Mycovellosiella* and it is also closely related to the members of dematiaceous hyphomycetes especially *Cercospora* and *Cladosporium.* Deighton (1979) emphasized the diagnostic characteristics of the genus *Stenella* to distinguish it from *Mycovellosiella. Stenella* differs from *Mycovellosiella* in having rough walled mycelial hyphae and usually narrow rod-like conidia which are in most cases catenulate and verruculose, brown, subcylindric to narrowly obclavatocylindric and more than 2 septate. The conidial scars are thickened but lie flat against the spore wall and are not raised.

Separation of *Stenella* from *Cercospora, Mycovellosiella* and *Cladosporium* has been effected by Deighton (1971, 1979); Mulder (1975,1982); Singh and Kamal (1978); Yen *et al.* (1982). Later more species were recombined and shifted to *Stenella* from other genera by De Hoog *et al.* (1983); Sivanesan (1984); Goh and Hsieh (1987 d); Walker and White (1991); Braun (1995,1996); Kamal and Naraiyan (1986); Sutton and Crous (1977); Mc Kenziae (1982).

A total of 28 species have been recombined from other generic seggregates of Cercosporoid fungi to form genus *Stenella*. Out of these 28 species, 15 species were recombined from *Cercospora*, 1 from *Chaeotrichum*, 2 from *Cladosporium*, 1 from *Biharia*, 1 from *Eriocercospora*, 2 from *Heterosporium*, 2 from *Ramularia*, 1 from *Veronaea*, 2 from *Napicladium* and 1 from *Pseudocercospora*.

Genus *Stenella* is of wide spread occurrence parasitizing various species of flowering plants and ferns. In addition few saprotrophic species found on dead plant parts (De Hoog *et al.*, 1983; Matsushima, 1983, 1985, 1989; Castenada, 1985, 1986; Castenada and Kendrick, 1919; Yip, 1989). *Stenella* Syd. is mainly tropical and subtropical fungi (Deighton, 1971, 1979; Ellis, 1971, 1976; Mulder, 1975, 1982, 1989; De Hoog *et al.*, 1983; Priest, 1991. *S. lythri* (Westened) Mulder is however an exception and it is confined to temperate regions of Europe, including Britain.

Overall literature on *Stenella* reveals that 119 species have been described so far from across the world, of which 69 reported from Indian subcontinent and collected on 34 different host families of flowering plants barring one on Pteridophyte family.

Sarbajna and Chattopadhaya (1991) described *S. plumeriae* on *Plumeria acutifolia* (Apocynaceae). Two species of *Stenella* viz., *S. alocasiae* and *S. colocasiae* on *Alocasia indica and Colocasia antiquorum* (Araceae) respectively have been described by Sarbjna and Chattopadhyay (1991).

Chaudhary *et al.* (2001) described *S. caryotae-urensis* on *Caryota urens* (Arecaceae). Four species of *Stenella* have been described on host family Asclepiadaceae. viz., *S. cynanchi* Yen *et al.* (1982) on *Cynanchum callitata, S. ceropegiae* Patil and Sawant (1991) on *Ceropegia bulbosa*; Srivastava *et al.* (1994, 1995) described *S. telosmae* and *S. pentatropidis* on the living leaves *Telosma pallida* and *Pentatropis spirale* respectively.

Only one species each of *Stenella* is found to be reported on each families, Burseraceae, Boraginaceae and Berberidaceae, viz., *S. garugae on Garuga pinnata* Yen *et al.* (1982) (Burseraceae), *S.myxa* on *Cordia myxa* (Sydow) Yen *et al.* (1982) (Boraginaceae) which is a new recombination made from *Cercospora myxa*, while *S. mahoniae* Verma *et al.* (1989) was collected on *Mahonia nepalensis* (Berberidaceae).

Two species of *Stenella* are described on Bignoniaceae viz., *S. oroxylicola* Yen *et al.* (1982) on *Oroxylum indicum* and *S. heterophragmae* Misra *et al.* (1997) on *Heterophragma* sp. Kamal *et al.* (1980) described *S. capparidis* on *Capparis sepiaria* (Caesalpiniaceae). Mulder (1982) recombined *Cercospora capparidicola* into *S. capparidicola*. Further Singh *et al.* described *Capparis zeylanica* on different hosts.

Four species have been described till now on family Caesalpiniaceae. These are *S. chandleri* (Hansf.) Singh and Kamal (1979) (a new recombination from *Cercospora chandleri).* Abbasi and Shukla (1980) reported *S. cassiae* on *Cassia fistula*. Later in 1999, Misra *et al.* reported *S. cassicola* on host *Cassia fistula*.

Three species of *Stenella* have been described on family Celastraceae. viz. *S. celastrae* Rai and Kamal (1990) on *Celastrus paniculata, S. hippocratiace* Srivastava *et al.*(1995) on *Hippocratia arborea* and *S. elaeodendri* Kamal *et al.* (1980) on *Elaeodendron roxburghii*. However, only one species of *Stenella* each on families, Dioscoreaceae and Dipterocarpaceae is described as *S. dioscoreicola* Yen *et al.* (1982) on *Dioscorea* sp. and *S. shoreae* Khan & Kamal (1995) on *Shorea robusta* respectively.

Three species of *Stenella* have been reported so far on family Euphorbiaceae viz., *S. brideliicola* Srivastava *et al.* (1994) collected on *Bridelia stipularis* and *S. gorakhpurensis* (Kamal and Kumar) Hoog *et al.* (1983) which was a recombination made from *Veronea gorakhpurensis*. Recently, Chaudhary *et al.* (2001) described *S. bischofiae- javanicae* on *Bischofia javanica* in the same family.

A good number of *Stenella* species have been reported on family Fabaceae, two of these are recombination to *Stenella* from other genera. Deighton (1971) placed *Cercospora canavaliae* H. and P. Syd (1914) in *Stenella* as *S. canvaliae*. Kamal and Narayan (1987) placed *Eriocercospora moghaniae* Singh under *Stenella* as *S. moghaniae*. This species was reported on *Moghania prostrata*. Other species described on Fabaceae are *S. crotalariicola* Chaudhary *et al.* (1991) on *Crotalaria* sp; *S. fabacearum* Srivastava *et al.* (1997) on *Butea parviflora* and recently Chaudhary *et al.* (2001) described two more species, *S. browniaeicola* on *Brownea hybrida* and *S. millettiae* on *Milletia ovalifolia*. Only one species viz. *S. cinnamomi* on *Cinnamomum sulphureum* has been reported on host family Lauraceae.

Two species of form genus *Stenella* have been described on family Liliaceae viz., *S. smilacis* Kumar *et al.* (1980) and *S. smilacicola* Shaw and Alcorn (1993) both on *Similax macrophylla*. One species each have been reported from families Magnoliaceae and Malvaceae. These are *S. schizandrae* Pavgi and Singh (1971) on *Schizandra grandiflora* and *S.kydiae* Singh and Kamal (1987) on *Kydia calycina*, respectively.

Three species of *Stenella* viz., *S. anamirtae* Srivastava *et al.* (1995), *S. tiliacorae* Sarbjna (1990) and *S. stephaniae* Yen *et al.* (1982) have been reported respectively on hosts *Anamirta paniculata, Tiliacora racemosa* and *Stephania hernandifolia* of family Menispermaceae. One species each of *Stenella* each on families Moraceae and Myrtaceae, respectively have been reported. These are *S. ficina* Kamal *et al.* (1981) on *Ficus benghalensis* and *S.eugeniicola* Misra *et al.* (1981) on *Eugenia heyniana.*

One species reported on family Myrsinaceae, is *S. embeliae* Rajak (1981) collected on *Embelia robusta.* However in 1989, Rai and Kamal described *S. embeliae* on *Embelia tegeriam- cottam* but being homonym of *S. embeliae* Rajak (1981) it was considered invalid. Only one species of *Stenella* is reported on families Meliaceae and Rosaceae. Yen *et al.* (1982) reported *S. cedrelae* on *Cedrela toona* (Meliaceae) while Khan *et al.* (1991) reported *S. prinsepiae* on *Prinsepia utilis* (Rosaceae). *S. sonapathrienis* Khan *et al.* (1994) was reported on host *Ventilago madraspatna* (Rhamnaceae). Two species viz., *S. rhododendriicola* Misra *et al.* (1991) on *Rhododendron campanulatum* and *S. rhododendri* Verma and Kamal (1987) have been reported on family Rhododendraceae.

Largest number of species of *Stenella* have been reported on family Rubiaceae. These are *S. plectroniae* Ponnappa (1968) on *Plectronia parviflora, S. vangueriae* (Thirum and Mishra) Deighton (1979) (a new combination from *Biharia vanguriae* on host *Vangueria* sp.). Further Yen *et al.* described 3 species of *Stenella* on Rubiaceae : *S. canthii* on *Canthium dedymum, S. coffeae* on *Coffea benghalensis* and *S. xeromphigena* on *Xeromphis uliginosa* (1982a, 1982b). Three more species of *Stenella* viz., *S. naucleae* Das (1990), *S. meynae-laxiflorae* Srivastava *et al.* (1995) and *S. hyptiantherae* Singh *et al.* (1997) have been reported on hosts *Nauclea excelsa, Meyna laxiflora* and *Hyptianthera stricta* respectively.

A new recombination *Cercospora aeglicola* Das (1958) and *S. aegle* by Prasad (1967) as *S. aeglicola* (Das) McKenziae was proposed by McKenziae (1982) on *Aegle marmelos* of family Rutaceae. Another species reported on same family is *S. murrayae* Khan *et al.* (1995) on *Murraya paniculata.* One species of *Stenella* is reported in the families viz., Santalaceae, Schizaeceae, Thymeliaceae and Verbenaceae. Chaudhay *et al.* (1996) described *S. osyrina* on *Osyris wightiana* (Santalaceae). *S. lygodii* Sarbajna (1990) on *Lygodium* sp. is the only species of *Stenella* reported on a non-angiospermic host. In 1991, Walker recombined *Heterosporium wikstromiae* (Petek) as *S. wikstromiae.* The last family on which *Stenella* is reported is Verbenaceae. Misra *et al.* (1997) reported *S. lantanae* on *Lantana indica.*

A thorough survey of forest areas, grasslands, agricultural lands etc. was made to study foliicolous plant pathogens in their natural habitat. The infected leaf samples were collected and herberia thereof were prepared.

Microscopic slides of these specimens were prepared by scrapping method, using razor blade, in a drop of lactophenol-cotton blue. Camera lucida drawings of the same were made.

To justify the distinct identity of novel taxon, morphotaxonomic features of these were compared with those of morphotaxonomically allied taxa. In case when no species was found to be described on host genus, host species or host family than the novel species was compared with type species, *S. araguata*. For detalied study relevant lietrature such as monographs, journals, reviews, various research papers published upto recent date have been looked into.

Observations and Descripitoins

(a) Table-1 : List of *Stenella* Species Reported from Indian Subcontinent

S. No.	Fungus Spp.	Family and Host (Apocynaceae)	Place of Collection
1	*S.plumeriae* Sarbajna & Chattopadh (1991)	*Plumeria acutifolia*	W.Bengal, India
2	*S. ichnocarpicola* Singh *et al.* (2001)	*Ichnocarpi frutescentis*	U.P, India
		Annonaceae	
3.	*S. polyallthiae* Chaudhary *et al.* (2002)	*Polyalthia-suberosa*	U.P., India
		Araceae	
4	*S. alocasiae* Sarbajna & Chattopadh (1991)	*Alocasia indica*	W.Bengal, India
5	*S. colocasiae* Sarbajna & Chattopadh (1991)	*Colocasia antiquorum*	W.Bengal, India
		(Arecaceae)	
6	*S. cynanchi* Yen *et al.* (1982)	*Caryota urens*	U.P. India
		(Asclepiadaceae)	
7	*S.cynanchi* Yen *et al.* (1982)	*Cynanchum callilulu*	W.Bengal, India
8	*S.ceropegiae* Patil & Sawant (1991)	*Ceropegia bulbosa*	Maharashtra, India
9	*S. pentatropidis* Srivast. *et al.* (1995)	*Pentatropis spirale*	U.P. India
10	*S. telosmae* Srivast. *et al.* (1994)	*Telosma pallida*	U.P. India
		(Burseraceae)	
11	*S. garugae* Yen *et al.* (1982)	*Garuga pinnata*	W. Bengal, India
		(Boraginaceae)	
12	*S. myxa* (Sydow) Yen *et al.* (1982)	*Cordia myxa*	W. Bengal, India
13	*S.mahoniae* Verma *et al.* (1989)	*Mahonia nepalensis*	Nepal
		(Bignoniaceae)	
14	*S. heterophragmae* Misra *et al.* (1997)	*Heterophragma* sp.	U.P. India
15	*S. oroxylicola* Yen *et al.* (1982)	*Oroxylum indicum*	W.Bengal, India

		(Caesalpiniaceae)	
16	*S. chandleri* (Hansf) Singh & Kamal (1979)	*Casia fistula*	U.P. India
17	*S. cassiae* Abbasi & Shukla 1980 (*S. cassiae* kamal *et al.*(1980)	*Cassia fistula*	U.P. India
18	*S. cassicola* Misra *et al.* (1999)	*Cassia fistula*	U.P. India
19	*S. capparidis* Kamal *et al.* (1980)	*Capparis sepiaria*	U.P. India
20	*S. capparidicola* (Hansf & Thirum) Mulder (1982)	*Capparis sepiaria*	Mysore, India
21	*S. cassiae-torae* Singh *et al.* (2001)	*Cassia torae*	U.P., India
		(Celastraceae)	
22	*S. celastrae* Rai & Kamal (1990)	*Celastrus paniculata*	U.P. India
23	*S.hippocratiae* Srivast *et al.* (1995)	*Hippocratia arborea*	U.P. India
24	*S. elaeodendri* Kamal *et al.* (1980)	*Elaeodendron roxburghii*	U.P. India
		(Dioscoreaceae)	
25	*S. dioscoreicola* Yen *et al.* (1982)	*Dioscorea sp.*	W. Bengal, India
		(Dipterocarpaceae)	
26	*S. shoreae* Khan & Kamal (1995)	*Shorea robusta*	U.P. India
		(Euphorbiaceae)	
27	*S. bridelicola* Srivast. *et al.* (1994)	*Bridelia stipularis*	U.P. India
28	*S. bischofiae-javanicae* Chaudhary *et al.* (2001)	*Bischolfia-javanica*	U.P. India
29	*S. gorakhpurensis* (Kamal & Kumar)Hoog *et al.* (1983)	*Glochidion multiloculare*	U.P. India
		(Fabaceae)	
30	*S. browniaeicola* Chaudhary *et al.* (2001)	*Brownea bydria*	U.P. India
31	*S. millettiae* Chaudhary *et al.* (2001)	*Millettia ovalifolia*	U.P. India
32	*S. canavaliae* (H &psyed) Deighton (1971)	*Canavalia* sp.	India
33	*S. moghaniae* (Singh) Kamal & Narayan (1987)	*Moghania prostrata*	Uttaranchal, India
34	*S. crotalariicola* Chaudhary *et al.* (1994)	*Crotalaria* sp.	U.P., India
35	*S. fabacearum* Srivast *et al.* (1994)	*Dolichos* sp.	U.P., India
36	*S. buteae* Misra *et al.* (1997)	*Butea parviflora*	U.P., India
37	*S. cinnamoni* Hosagoudar & Braun (1996)	*Cinnamomum sulphureum*	Tamil Nadu, India
		(Lamiaceae)	
38	*S. lamiacearum* Chaudhary *et al.* (2002)	*Teucrium* sp.	U.P., India

		(Liliaceae)	
39	*S. smilacis* Kamal *et al.* (1980)	*Smilax macrophylla*	U.P., India
40	*S. smilacicola* Shaw & Alcorn (1993)	*Smilax macrophylla*	Bihar, India
		(Magnoliaceae)	
41	*S. schizandrae* Pavgi & Singh (1971)	*Schizandra grandiflora*	W. Bengal, India
		(Malvaceae)	
42	*S. kydiae* Singh & Kamal (1987)	*Kydia calycine*	U.P., India
		(Menispermaceae)	
43	*S. anamirtae* Srivast *et al.* (1995)	*Anamirta paniculata*	U.P., India
44	*S. tiliacorae* Sarbajna (1990)	*Tiliacora racemosa*	W. Bengal, India
45	*S. stephaniae* Yen *et al.* (1982)	*Stephania hernandifolia*	W. Bengal, India
		(Moraceae)	
46	*S. ficina* Kamal *et al.* (1981)	*Ficus benghalensis*	U.P., India
		(Myrtaceae)	
47	*S. eugeniicola* Misra *el al.* (1999)	*Eugenia heynianu*	M.P., India
		(Myrsinaceae)	
48	*S. embeliae* Rajak (1981) ≈ (*S. embeliae* Rai Kamal) (1989)	*Embelia robusta*	M.P., India
		(Meliaceae)	
49	*S. cedrelae* Yen *et al.* (1982)	*Cedrela toona*	W. Bengal, India
		(Passifloraceae)	
50	*S. mirzapurensis* Singh *et al.* (2001)	*Parsiflora foetida*	U.P., India
		(Rosaceae)	
51	*S. prinsepiae* Khan *et al.* (1991)	*Prinsepia utilis*	Nepal
		(Rhamanaceae)	
52	*S. sonapathriensis* Khan & Kamal (1984)	*Ventilago madraspatna*	U.P., India
		(Rhododendraceae)	
53	*S. rhododendricola* Misra *et al.* (1999)	*Rhododendron campanulatum*	Nepal
54	*S. rhododendri* verma & Kamal (1987)	*Rhododendron campanulatum*	Uttaranchal, India
		(Rubiaceae)	
55	*S. plectroniae* Ponnappa (1968)	*Plectronia parviflora*	Bangalore, India
56	*S.vangueriae* (Thirum & Misra) Deighton	*Vangueria* sp.	India
57	*S. naucleae* Das	*Nauclea excelsa*	W. Bengal, India
58	*S. meynae-laxiflorae* Srivast *et al.* (1995)	*Meyna laxiflora*	U.P., India
59	*S. hyptiantherae* Singh *et al.* (1997)	*Hyptianthera stricta*	M.P., India

60	*S. canthi* Yen *et al.* (1982a)	*Canthium dedymum*	W. Bengal, India
61	*S. coffeae* Yen *et al.* (1982 b)	*Coffea bengalensis*	W. Bengal, India
62	*S.xeromphigena* Yen *et al.* (1982b)	*Xeromphis uliginosa*	W. Bengal, India
63	*S. rubiacola* Chaudhary *et al.* (2002)	*Wedlandia tinctoria*	U.P., India
		(Rutaceae)	
64	*S. agelicola* (Das) Mckenziae Comb nov. (1982) *S.aegle* S.S. Prasad 1968.	*Aegle marmelos*	Bihar, India
65	*S. murrayae* Khan *et al.* (1994)	*Murraya paniculata*	U.P., India
		(Santalaceae)	
66	*S. osyrina* Chaudhary *et al.* (1996)	*Osuris wightiana*	W. Bengal, India
		(Schizaeaceae)	
67	*S. lygodii* Sarbajna (1990)	*Lygodium* sp.	W. Bengal, India
		(Thymeliaceae)	
68	*S. wikstromiae* (Petch) Walker (1991)	*Wikstroemia indica*	Sri Lanka
		(Verbenaceae)	
69	*S. lantanae* Misra *et al.* (1997)	*Lantana indica*	India

(b) Generic circumscripitions and novel taxa proposed

Genus *Stenella* can be distinguished from other dematiaceous hyphomycetes on the basis of the characters given by Mulder (1975, 1982); Deighton (1979) and De Hoog *et al.* (1983).

1. It produces necrotic blotches on host plant parts.
2. It produces superficial mycelium which is usually verruculose.
3. Conidiophores are macronematous arising in fascicles from substomatal/ superficial stromata and / singly as lateral or terminal branches of superficial hyphae.
4. Conidiogenous cells are polyblastic, integrate, cicatrized with conspicuous more or less thickened conidial scars.
5. Conidia cylindrical, subcylindrical, obconicotruncate, multiseptate, verruculose.

The descriptions and illustrations of the novel taxa alongwith the comparative chart are given hereunder :

Stenella casaieriicola *sp. nov.* (Fig.1)

Maculae amphigenae, necroticae, sub-orbiculares vel orbiculares, coalescentes irregulariter magne et interdum tagentes per totam folii, infernae brunneae, supernae griseum **Coloniae** hypophyllae, effusae, griseum fuscae. **Mycelium** ex hyphis externum vel internum; hyphis externis ramosis, septatis

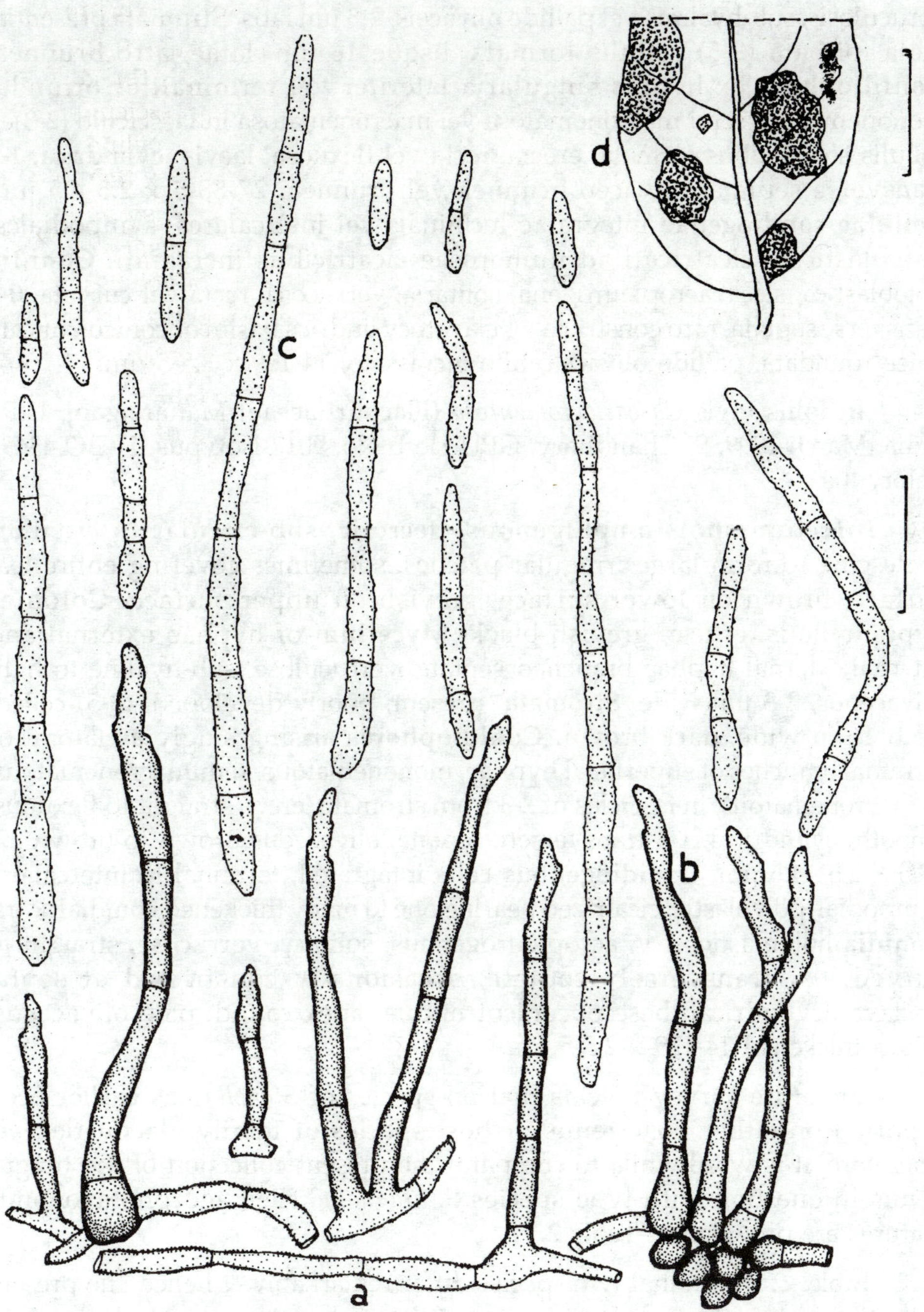

Fig. 1 : *Stenella casieriicola sp. nov.* (a) Superficial mycelium with conidiophores, (b) Conidiophores with few stromatal cells, (c) Conidida, (d) Leaf spots; Bars : a-c = 20µm, d = 20 mm.

veruculosis, sub-hyalinis vel pallide olivaceis, 2-3 µm latis. **Stromata** presentia, mala formata (3-5) cellulis formata, usque 16 um. latae, atro brunnea. **Conidiophora** ex hyphis singularia lateriter vel terminaliter oriunda, mononematosa, semi macronematosa vel macronematosa in fassciculo (2-3)ex cellulis stromatibus formata, erecta, recta vel flexuosa, laevia, cylindrica, 1-4 transverse septata, olivaceo brunnea vel brunnea, 27-88-5 x 2.5-4.5 µm. **Cellulae condiogenae** integratae terminals vel intercalares, sympodiales, polyblasticae, cicatricati ad numerosae cicatricibus incrassati. **Conidia** holoblastica, sicca, acropleurogena, solitaria, verrucosa, recta vel curvata, 0-7 transverse septata, raro constricta, obclavatocy lindrica basim obconicotruncata apice roundata, pallide olivacea, hilo incrassato, 14-129 x 2.5-5 µm.

In foliis vivis *Casearia tomentosa* (Flacourtiaceae) Maharajganj, U.P., India, March 1999; S. Chaudhary, GPU Herb No. 9012 isotypus, HCIO 43698 holotypus.

Infection spots amphigenous, necrotic, sub-circular to circular, coalescing to form large irregular patches, sometimes covering entire leaf surface, brown on lower surface, greyish on upper surface. **Colonies** hypophyllous, effuse, greyish black. **Mycelium** of hyphae external and internal;external hyphae branched septate, verruculose, sub-hyaline to light olivaceous, 2-3 µm wide. **Stromata** present, poorly developed, (3-5) celled, upto 16 in wide, dark brown. **Conidiophores** arising singly as lateral or terminal branches of superficial hyphae, mononematous, semimacronematous, to macronematous, in fascicles of 2-3 from stromata. erect, straight to flexuous, smooth, cylindrical, 1-4 transversely septate, olivaceous brown to brown 27-88.5 x 2.5 –4.5 µm. **Conidiogenous cells** integrated, terminal to intercalary, sympodial, polyblastic, cicatrized bearing one to many thickened conidial scars. **Conidia** holoblastic, dry, acropleurogenous, solitary, verrucose, straight to curved, 0-7 transversely septate, occasionally constricted at septa, obclavatocylindrical, base obconicotruncate, apex round, pale olivaceous, hilum thickened 14-129 x 2.5-5 µm.

Literature survey reveals that no species of *Stenella* has earlier been reported on either host genus or host species of family Flacourtiaceae. Therefore, it is worthwhile to compare the present collection of the fungus genus in question with type species *S. araguata*. Their morphotaxonomic features are provided in Table 2.

Table –2 shows that two species differ remarkably & hence, the present collection is described as novel species of *Stenella*.

Stenella cassigena *sp. nov.* (*Fig.2*)

Maculae amphigenae, effusae, irregulariter extensae per totam folii.

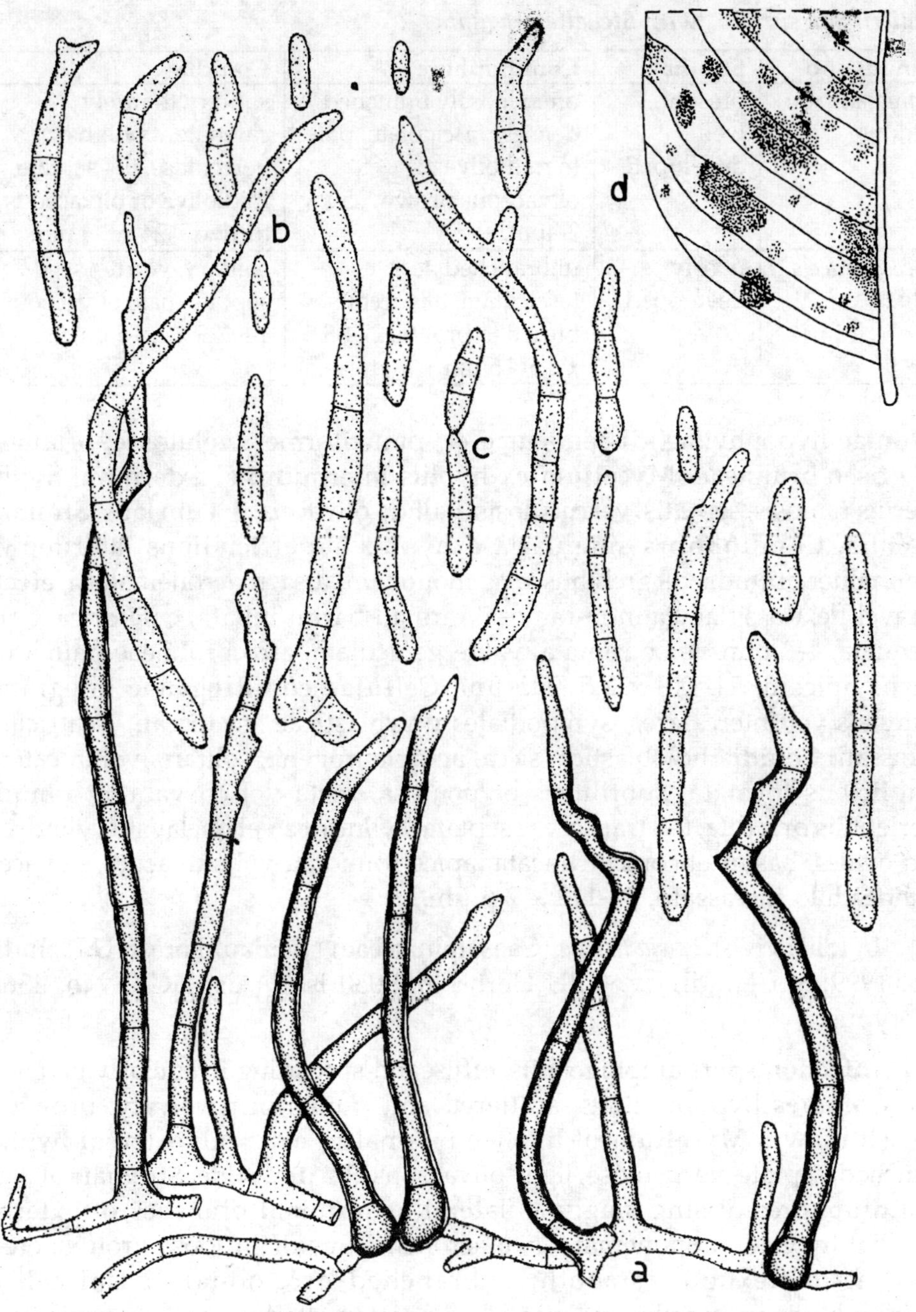

Fig. 2 : *Stenella cassigena sp. nov.* (a) Superficial mycelium with conidiophores, (b) Conidia, (c) Branched conidia, (d) Leaf spots; bars : a-c = 20 μm, d = 20 mm

Table 2 : Comparative account of morphotaxonomic features of *Stenella casaieriicola sp. nov.*, with *Stenella araguata*

Fungus spp.	Stromata	Conidiophore	Conidia
S.araguata Sydow	present, well developed	often loosely branched, densely fasciculate, pale to mid olive or olivaceous brown, 65 x 2-4μm.	solitary to rarely catenate, smooth or verrculose, 0-4 septate, mid olive or olivaceous brown 7-24 x 2-4 μm.
S. casaieriicola sp.nov.	poorly developed	unbranched, less fasciculate, olivaceous brown to brown, 27-88.5 x 2.5-4.5 μm.	solitary, verrucose, 0-7 septate, pale olivaceous, 14-129 x 2.5-5 μm.

Coloniae hypophyllae, disseminatae et punctiformes, velutosae, brunneae vel griseo brunneae. **Mycelium** ex hyphis internum vel externum; hyphis externis ramosis, septatis, verruculosis, pallide olivaceis, 3-4 μm latis. **Stromata** absentia. **Condiophora** singularia exhyphis superficialibus lateriter vel terminaliter oriunda vel grebibus laxa, mononematosa, macronematosa, erecta, rectavel flexuosa, laevia, non-ramosa, raro ad basim inflata, cylindrica, apice attenuata, 1-8 transverse septata, valde geniculata, atro brunneae palliodora versus apicem, 21-148 x 2.5 –5.5 μm. **Cellulae conidiogenae** integratae, terminals vel intercalares, sympodiales polyblasticae, cicatricati, cicatricibus incrassati, **Conidia** holoblastica, sicca, acropleurogena, solitaria vel in catenis simplicibus formata, subtiliter verruculosa, recta vel curvata, et ramulis laterietibus praedita, 0-8 transervse septata, cylindrica vel obclavato cylindrica, raro ovales, basim obconicotruncata, apice roundata vel sub-acuta, olivaceo-brunnea,hilo incrassato, 18-112 x 2-6 μm.

In foliis vivis *Cassia fistula* (Caesalpiniaceae) Nichlaul Forest, U.P., India, Feb. 1999; S. Chaudhary, GPU Herb No. 9030 isotypus, HCIO No. 43680 holotypus.

Infection spots amphigenous, effuseand spreading irregularly on entire leaf. **Colonies** hypophyllous, scattered and punctiform velvety, brown to greyish brown. **Mycelium** of hyphae internal to external; external hyphae branched, septate, veruculose, light olivaceous, 3-4 μm. wide. **Stromata** absent. **Conidiophores** arising singly as lateral or terminal branches of external mycelial hyphae, mononematous, macronematous or in loose groups, erect, straight to flexuous, smooth, unbranched occasionaly basal cell of conidiophores is swollen cylindrical and attenuated apex, 1-8 transversely septate, strongly geniculate, dark brown but paler near apex 21-148 x 2.5 –5.5 μm. **Conidiogenous cells** integrate, terminal to intercalary, sympodial, polyblastic, cicatrized bearing thickened conidial scars. **Conidia** holoblastic, dry acropleurogenous solitary or catenate in unbranched chain, finely verruculose, straight to curved, rarely showing protuberances laterally, 0-8 transversely septate, cylindrical to obclavatocylindrical, base obconicotruncate, rarely oval, apex round to sub acute, olivaceous brown, hilum thickened, 18-112 x 2-6 µm.

Three species of *Stenella* viz. *S. cassiae* Abbasi and Shukla (1980), *S. chandleri* (Hansf) Singh and Kamal (1978) and *S. cassicola* Misra *et al.* (1999) have already been collected on the same host genus, *Cassia fistuala* as evident from literature survey. However the present collection is compared with all the three species of *Stenella* described earlier. Their comparative account being presented in Table 3.

From Table-3 it is evident that the present collection is distinctly different from its allied species and hence this proposal.

Stenella ehretigena *sp. nov.* (Fig.3)

Maculae amphigenae coalescentes extendents omnes per totam folii, supernae pallide brunneae, infernae brunneae. **Coloniae** hypophyllae, effusae,

Table 3 : Comparative account of morphotaxonomic features of *Stenella cassigena sp. nov.* with its allied species.

Fungus spp.	Conidiophore	Conidia
S. cassicola Misra *et al.*	arising singly, straight to slightly curved, geniculate unbranched, brown, 33-90 x 1.5-4 µm.	rarely catenate in unbranched chain, strainght to curved, finely verruculose, 2-9 septate sub-hyaline to olivaceous brown, 19-51 x1.5 –4.5 µm.
S. cassiae Abbasi & Shukla	arising singly or sometimes in fascicles, straight to flexuous, geniculate, unbranched, brown, 85.5-150 x 4.5-7 µm.	solitary, straight to curved, thick walled, verruculose, 1-22 septate, brown, 12-153 x 3.5-8 µm.
S.chandleri (Hansf) Singh & Kamal	unbranched, olivaceous brown, 28-120 x 3-6.5 µm.	catenate in unbranched chain, echinulate, 2-6 septate, subhyaline to pale olive, 120-55.5 x 2.5-4 µm.
S. cassigena sp. nov.	arising singly or in loose fascicles of 2-3, straight o highly flexuous, strongly geniculate, dark brown, 21-148 x 2.5-5.5 µm.	solitary to catenate, straight to strongly curved, finely verruculose, 0-8 septate, olivaceous brown, 18-112 x 2-6 µm.

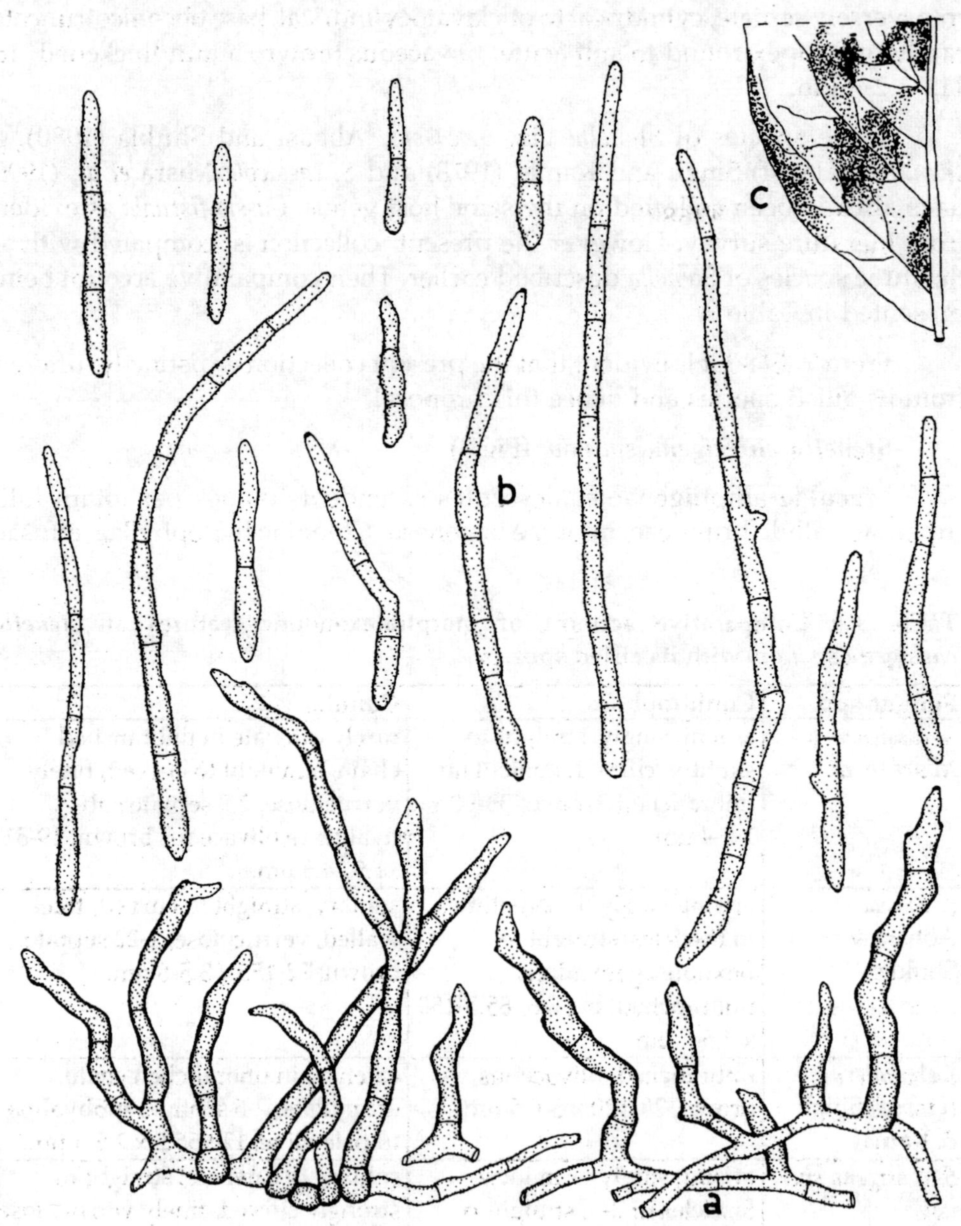

Fig. 3 : *Stenella ehretigena sp. nov.* (a) Superficial mycelium with conidiophores, (b) Conidia, (c) Leaf spots; Bars : a-b = 20 µm, c = 20 mm.

bruneae. **Mycelium** ex hyphis internum vel plerumque externum; hyphis externis ramosis, septatis veruculosis, sub-hyalinis, usque 2µm. latis, **Stromata** absentia. **Conidiophora** singularia vel in gregibus laxa ex hyphis superficialibus lateriter vel terminaliter oriunda, mononematosa, macronematosa, sub-erecta vel erecta , recta vel flexuosa, laevia, non-ramosa vel ramosa, cylindrica, 1-6 transverse septata, geniculata, atro olivacea, 18-92 x 2-4.5 µm. **Cellulae conidiogenae** integratae, terminales vel intercalares, sympodials, polyblasticae, cicatricati, ad numerosae cicatricibus incrassati, **Conidia** sicca, holoblastica, acropleurgena, simplicia, solitaria vel raro in catenis simplicibus formata subtiliter verruculosa, recta vel plerumque curvata, 1-8 transverse septata, obclavatocylindrica, basim obconicotruncata, apicem roundata, raro cicatricata, sub-hyalina vel pallide olivacea, 21.5-129 x 2-4 µm.

In foliis vivis *Eheretia laevis* (Boraginaceae) Nepal, Jan. 1999 S. Chaudhary, GPU Herb No. 9010 isotypus, HCIO No. 43086 holotypus.

Infection spots amphigenous, coalescing, extending to cover entire surface of leaf, light brown on upper surface, brown on lower surface. **Colonies** hypophyllous, effuse, brown, **Mycelium** of hyphae internal to mostly external; external hyphae branched, septate, verruculose, sub-hyaline upto 2 µm. wide. **Stormata** absent. **Conidiophores** arising singly or in loose groups as lateral or terminal branches of external hyphae, mononematous, macronomatous, sub-erect to erect, strainght to flexuous, smooth walled, unbranched or branched, cylindrical, 1-6 trans versely septate, geniculate, dark olivaceous, 1892 x 2-4.5 µm. **Conidiogenous cells** integrated, terminal to intercalary, sympodial, polyblastic , cicatrized berning one to many thickened conidial scars. **Conidia** dry, holobalstic, acropleurogenous, simple, solitary to rarely catenate in unbranched chain, finely verruculose, straight to mostly curved. 1-8 transversely septate, obclavatocylindrical, base obconicotruncate apexround, rarely cicatrized, sub-hyaline to light olivaceous 21.5-129 x 2-4 µm.

Earlier literature on the taxonomy of this fungus speaks of the fact that no species of *Stenella* has earlier been reported on host genus *Eheretia*. However, one species of the said fungus, *S. myxa* Yen *et al.* is reported on a different host genus *Cordia myxa* of same host family. Therefore, the morphotaxonomic features of *S. eheretigena sp. nov.* are compared with that of *S. myxa* to confirm its novelty. This comparison is given in table 4.

From this Table, it is obvious that the descriptions and illustrations of *S. ehertigena* sp. nov., therefore is considered worthwhile.

Stenella euphorbiicola sp. nov. (Fig. 4)

Maculae amphigenae, sub-orbiculares vel orbiculares, deinde coalescentes et irregulariter per totam folii, pallide brunneae. **Coloniae**

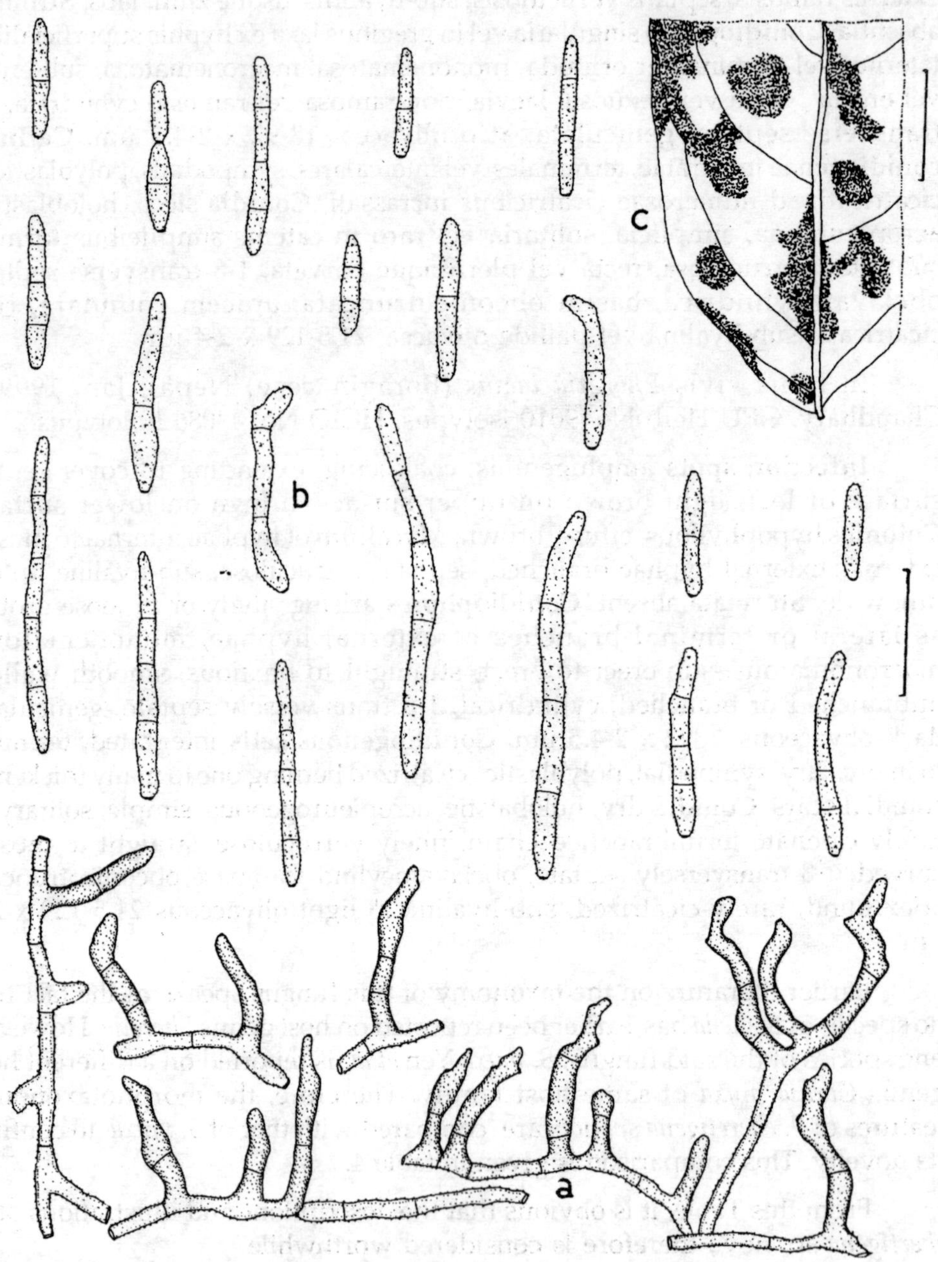

Fig. 4 : *Stenella euphorbicola sp. nov.* (a) Superficial mycelium with conidiophores, (b) Conidia, (c) Leaf spots; Bars : a-b = 20 µm, c = 20 mm.

Table 4 : Comparative account of morphotaxonomic features of *Stenella eheretigena sp. nov.* and *stenella myxa.*

Fungus spp.	Conidiophore	Conidia
S.myxa Yen *et al.*	solitary, simple, continuous or 1-2 septate (rarely 3) sub-geniculate towards apex, olivaceous to pale olivaceous, 13-26 x 5-6 µm.	cylindric, 2-14 septate, olivaceous brown to pale brown olivaceous, base attenuate and truncate, 24-10 x 5-6 µm.
S. ehertigena sp. nov.	solitary or in loose groups from hyphae, less geniculate, unbranched to rarely branched, 1-6 septate, dark olivaceous, 18-92 x 2-4.5 µm.	solitary to rarely catenate cylindrical to obclavatocylindrical, 1-8 septate, subhyaline to light olivaceous, base obconicotruncate, 21.5-129 x 2-4 µm.

hyphophyllae, effusae, atro bruneae. **Mycelium** ex hyphis internum vel externum; hyphis externis, romosis, septatis, subtiliter verruculosis, repentibus, pallide olivaceis, 2-2.5 µm latis. **Stromata** absentia. **Conidiophora** singularia ex hyphis lateriter vel terminaliter oriunda mononematosa, semimacronematosa vel macronematosa, sub-erecta vel erecta, recta vel flexuosa, laevia 1-3 transverse septata, sub-cylindrica vel cylindrica pallide olivacea, 9-34.5 x 2-4 µm. **Celulae conidiogenae** integratae, terminales vel intercalares, sympodiales, polyblasticae, cicatricati and numerosae cicatricibus incrassati. **Conidia** sicca, holoblastica, acropleurogena, solitaria vel in catenis simplicibus formata, subtiliter verruculosa, recta vel leniter curvata, 1-4 transverse septata, obclavatocylindrica, basim obconicotruncata, apice sub-acuta vel roundata, pallide olivacea, 16.5-61 x 2.5-3.5 µm.

In foliis vivis *Antidesma ghissembila* (Euphorbiaceae), Maharajganj, U.P., India, Feb. 1999, S. Chaudhary, GPU Herb No. 9015 isotypus, HCIO No. 43695 holotypus.

Infection spots amphigenous, sub-circular, later coalescing to form irregular patches on entire leaf, light brown. **Colonies** hypophyllous, effuse, dark brown. **Mycelium** of hyphae internal and external; external hyphae branched, septate, finely verruculose, repent, light olivaceous, 2-2.5 µm. wide. **Stromata** absent. **Conidiophores** arising singly as lateral or terminal branches of superficial mycelial hyphae, mononematous, semimacronematous to macronematous, sub-erect to erect, straight to flexuous, smooth, 1-3 transversely septate sub cylindrical to cylindrical, pale olivaceous, 9-34 x 2-4 µm. **Conidiogenous cells** integrated, terminal to intercalary, sympodial, polyblastic, cicatrized bearing one to few conidial scars. **Conidia** holoblastic, dry, acropleurogenous, solitary to catenate in unbranched chain, finely verruculose straight to slightly curved, 1-4 transversely septate, obclavatocylindrical, base obconicotruncate to conicotruncate, apex subacute to round, light olivaceous 16.5-61 to 2.5-3.5 µm.

Till date four species of *Stenella* viz., *S. bischofiae-javanicae* Chaudhary *et al.* (2001), *S. brideliicola* Srivastava *et al.* (1994), *S. elaeodendri* Kamal *et al.* (1980) and *S. gorakhpurensis* (Kamal and Kumar) Hoog *et al.* (1983) have earlier been reported on different host genera but on same host family (Euphorbiaceae) as revealed by the literature survey. To justify the distinct identity of the present collection, a comparison of morphotaxonomic features of these four species of the fungus in question with those of the proposed new taxon is given in Table. 5.

As seen in Table-5, *S. euphorbicola* is entirely different from its allied species.

***Stenella glycosae Chaudhary, S & R.K. sp. nov.* (Fig.5)**

Maculae amphigenae parvae primo orbiculares, diende irregulariter extensae, supernae, fuscae, infernae pallide brunneae. **Coloniae** hypophyllae, effusae, brunneae. **Mycelium** ex hyphis internum vel plerumque externum; hyphis externis septatis, ramosi, verruculosis, pallide olivaceis, 2-3 µm. latis. **Stromata** absentia. **Conidiophora** singularia ex hyphis superficialibus lateriter vel terminaliter oriunda, mononematosa, macronematosa, erecta, recta vel flexuosa, laevia, non-ramosa, 0-3 transverse septata, cylindrica, olivaceo

Table 5 : Comparative account of morphotaxonomic features of *Stenella euphorbicola sp. nov.* with allied species.

Fungus spp.	Conidiophore	Conidia
S. bischofiae-javanicae Chaudhary *et al.*	simple, branched to unbranched, 2-8 septate, olivaceous, 22-55 x 3-5 µm. superficially arising condiophores, 1-2 septate, 14-18.5 x 2.4 µm.	solitary to rarely catenate, verruculose, 2-6 septate, plae olivaceous, 21-58 x 2-5 µm.
S. bridelicola Srivastava *et al.*	unbranched, 90-310 x 5-7.5 µm. dark brown.	verruculose, 1-8 septate, 20-104 x 4-7 µm.
S. elaeodendri Kamal *et al.*	often loosely branched, thick, olivaceous brown, 126 x 1.5-6 µm.	smooth, 1-17 septate, 45-132 x 3.5-7
S. gorakhpurensis (Kamal & Kumar) de Hoog	unbranched, bearing only terminal conidia, 120 x 3.9 µm.	smooth , 1-3 septate, 21.6 x 4.3 µm.
S.euphorbiicola sp. nov.	unbranched, 1-3 spetate, pale olivaceous, 9-34 x2-4 µm.	finely verruculose, 1-4 septate, 16.5-61 x 2.5-3.5 µm.

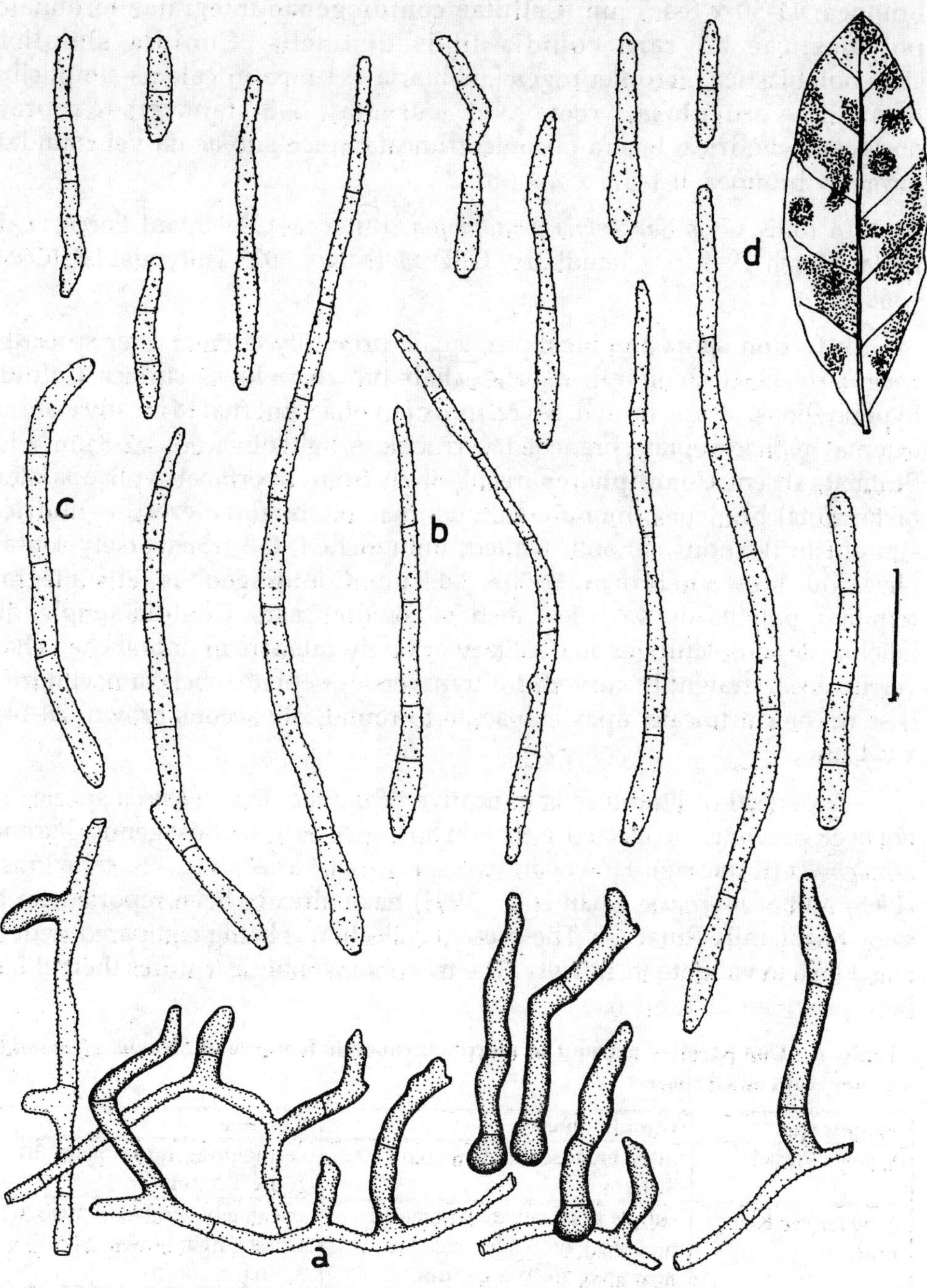

Fig.5 : *Stenella glycosmidis sp. nov.* (a) Superficial mycelium with conidiophores (b) Conidia, (c) Branched conidia, (d) Leaf spots ; Bars : a-c = 20 μm, d = 20 mm.

brunnea, 11-50 x 3-4.5 µm. **Cellulae conidiogenae** integratae terminales, polyblasticae ad raro conidia libus distinctis. **Conidia** simplicia, sicca,holoblastica, acropleurogena, solitaria vel raro in catenis simplicibus formata, veruculosa, recta vel curvata, 1-8 tansverse septata, obclavatocylindrica, basim obconicotruncata apice sub-acuta vel roundata, olivaceo- brunnea, 0-141.5 x 2-4 µm.

In foliis vivis *Glycosmis pentaphylla* (Rutaceae), Nichlaul Forest, U.P., India, March 1999, S. Chaudhary, GPU Herb No. 9023 isotypus, HCIO No. 43687.

Infection spots amphigenous, small, primarily circular later spreading irregularly, blackish on upper surface, light brown on lower surface. **Colonies** hypophyllous, effuse, brown. **Mycelium** of hyphae internal to mostly external; external hyphae, septate, branched, verruculose, light olivaceous, 2-3 µm. wide. **Stromata** absent. **Conidiphores** arising singly from superficial hyphae as lateral or terminal branches, mononematous, macronematous, erect, cylindrical, straight to flexuous, smooth walled, unbranched, 0-3 transversely septate, olivaceous brown to brown, 11-50 x 3-4.5 µm. **Conidiogenous cells** integrate, terminal, polyblastic, with few distinct conidial scars. **Conidia** simple, dry, holoblastic acropleurogenous, solitary or rarely catenate in unbranched chain, verruculose, straight to curved, 1-8 transversely septate, obclavatocylindrical, base obconicotruncate, apex sub-acute to round, olivaceous brown, 20-141.5 x 2-4 µm.

A perusal of literature is indicative of the fact that *Stenella* species has not been previously collected either on host species or on host genus *Glycosmis pentaphylla* (Rutaceae). However, two species of *Stenella* viz., *S. aegle* Prasad (1968) and *S. murrayae* Khan *et al.* (1994) have already been reported on the same host family Rutaceae. The present collection is being compared with the allied taxa to validate its novelty. The morphotaxonomic features thereof have been provided in Table 6.

Table 6 : Comparative account of morphotaxonomic features of *Stenella glycosmidis sp. nov.* with allied species

Fungus spp.	Conidiophore	Conidia
S. aegle Prasad	often branched, brown upto 120 x 3-8 µm.	verruculose, 1-13 septate, 30-160 x 3-5 µm.
S. murrayae Khan *et al.*	simple to mostly dichotomously branched, brown but pale brown near apex 30-95 x 4-6 µm.	serrulately veruculose upto 3-15 septate, light brown, 24-230 x 3.5-6 µm.
S. glycosmiidis sp. nov.	unbranched, 0-3 septate olivaceous brown to brown, 11-50 x 3-4.5 µm.	verruculose, 1-8 septate, olivaceous brown, 20-141.5 x 2-4 µm.

The data given in Table 6 indicates that the collection in question differs entirely from its allied taxa. Therefore, it is described as new species.

Stenella michelina *sp. nov.* (Fig. 6)

Maculae amphigenae, irregulares, coalescentes extensae per totam folii, infernae griseo-brunneae, supernae fuscobrunneae. **Coloniae** hypophyllae, effusae, cum mass crassa conidiorum pulvacearum, griseo-brunneae. **Mycelium** ex hyphis internum; hyphis externis ramosis, septatis, verruculosis, modice vel pallide olivaceis, 2.5-3.5 µm latis. **Stromata** absentia. **Conidiphora** singularia ex hyphis superficialibus terminaliter vel lateriter oriunda mononematosa, semimacronematosa, vel macronematosa, cylindrica, laevia, erecta, recta vel flexuosa, non-ramosa vel raro-ramosa, 2-4 transverse septata, brunnea, palliodora verses apicem 35-87 x 2-4.5 µm. **Cellulae conidiogenae** integratae, terminales vel intercalares, sympodiales, polyblasticae, cicatricati ad numerosae cicatricibus incrassati. **Conidia** holoblastica, sicca, acropleurogena, solitaria, subtiliterverruculosa, recta vel leniter curvata, 0-10 transverse septata, cylindrica vel obclavatocylindrica, raro ovales, basim obconicontruncata, apice subacuta vel roundata, modice vel pallide brunnea, hilo incrassato, 11-98 x 2-3.5 µm.

In foliis vivis *Michelia champaca* (Magnoliaceae), Nepal, June, 1998, S. Chaudhary, GPU Herb No. 9031 isotypus, HCIO 43791 holotypus.

Infection spots amphigenous, irregular, coalescing to spread on entire leaf surface, greyish brown on lower surface blackish brown on upper surface. **Colonies** hypophyllous, effuse, bearing thick mass of powdery conidia, greyish brown. **Mycelium** of hyphae internal and external; external hyphae branched, septate, veruculose, mid to plae olivaceous, 2.5-3.5 µm. wide. **Stromata** absent. **Conidiophores** arising singly as terminal or lateral branches of superficial hyphae, mononematours, semi-macronematous to macronematous, cylindrical, smooth, erect, straight to flexuous, unbranched or occasionally branched, 2-4 transversely septate, brown but paler near apex, 35-87 x 2-4.5 µm. **Conidiogenous cells** integrated, terminal to intercalary, sympodial, polyblastic, cicatrized bearing one to many conidial scars. **Conidia** holoblastic, dry, acropleurogenous solitary, finely veruculose, straight to slightly curved, 0-10 transversely septate, cylindrical, obclavatocylindrical, rarely oval, base obconicotruncate, apex sup-acute to round, mid to plae brown, hilum thickened, 11-98 x 2-3.5 µm.

After going through literature pertaining to *Stenella* it is gathered that a single species of the said fungus, *S. schizandrae* Pavgi and Singh (1971) has been reported earlier on host family Magnoliaceae. However, one species, *S. michelina* sp. nov. has been collected on different host of the same family during

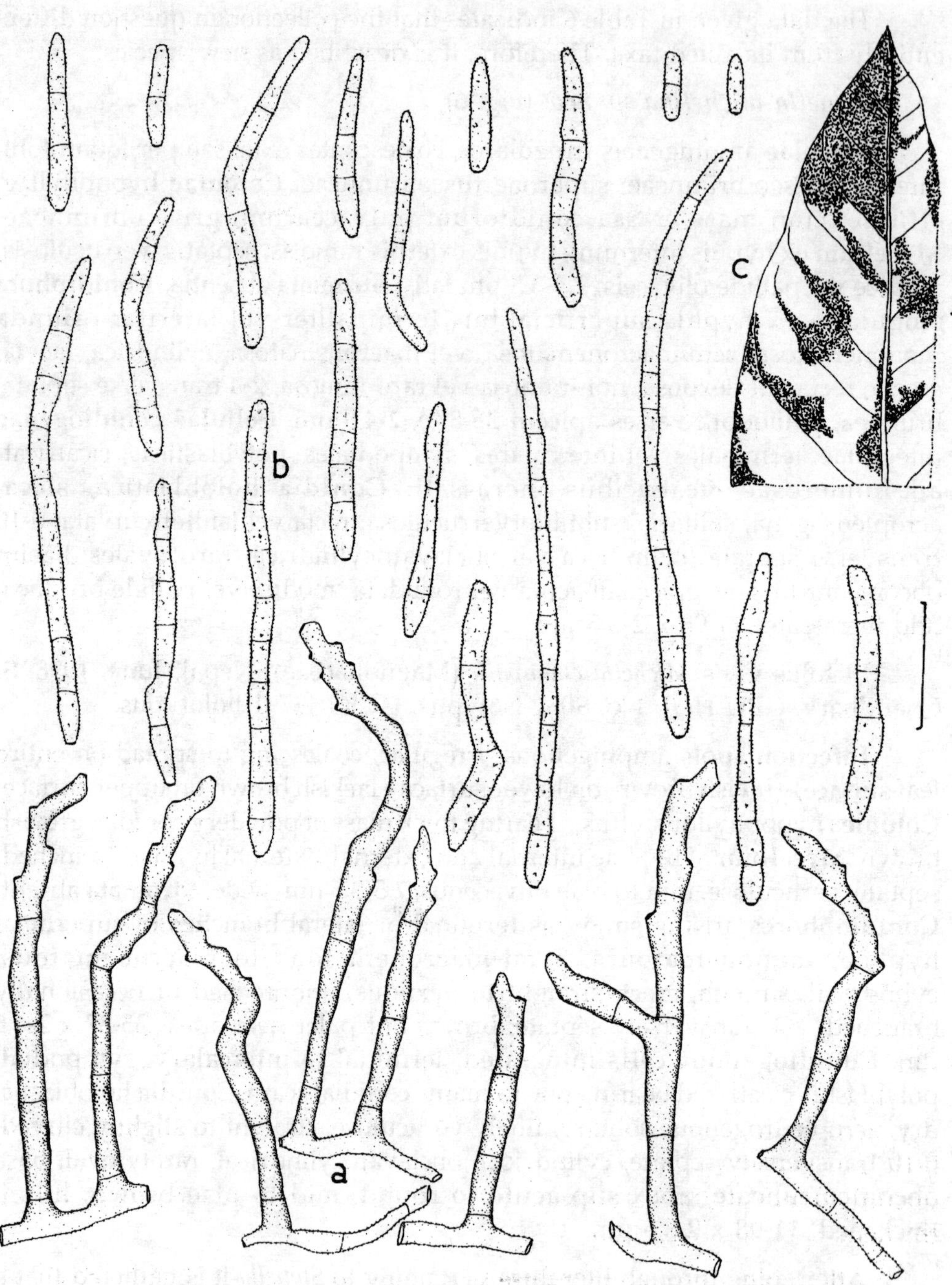

Fig. 6 : *Stenella michelina sp. nov.* (a) Superficial mycelium with conidiophores (b) Conidia, (c) Leaf spots; Bars : a-b = 20 mm, c = 20 mm.

the course of present investigation. It is compared with the allied taxon to confirm its novetly, the morphotaxonomic features there of being provided in Table 7.

Table 7 : Comparative account of morphotaxonomic features of *Stenella michelina sp. nov.* and *Stenella schizandrae*

Fungus spp.	Conidiophore	Conidia
S. schizandrae Pavgi & Singh	generally solitary, straight, simple, mid to dark brown, geniculate, upto 100x3-4 µm. 5-7 apex.	solitary to catenate, usually flexuous, often strongly curved, 1-24 septate, pale to dark brown, brown, verrucose, 25-170 x 4-6 µm.
S. michelina sp. nov.	solitary, occasionally branched, erect to curved or flexuous, brown to pale brown, 35-87 x 2-4.5 µm.	solitary, finely verruculose, mostly straight, 0-10 septate, mid to pale brown, 21-98 x 2-3.5 µm.

As indicated from Table-7, our collection is distinctly differnt from the allied taxa and warrants its description and illustration as a taxon of separate species rank.

Stenella smilacis- macrophyllae sp.nov. (Fig.7)

Maculae amphigenae, raro sub-orbiculares vel plerumque orbiculares, fuscae ad centrum margine fusco-brunneae. **Coloniae** hyphophyllae, atro brunneae vel fusco brunneae. **Mycelium** ex hyphis internum vel plerumque externum; hyphis externis septatis, ramosis, verruculosis per stromatibus emergentis, olivaceo brunneis, 2-3 µm. latis. **Stromata** raro cellulis formata. **Conidiophora** ex hyphis superficialibus lateriter vel terminaliter oriunda et singularia per stromatibus emergentis, mononematosa, semimacronematosa vel macronematosa, recta vel flexuosa, laevia, non ramosa vel raro-ramosa, 1-4 transverse septata, sub-cylindrica vel clyindrica, brunnea vel atro brunnea, 12-84 x 3-4.5 µm. **Celulae conidiogenae** integratae terminales vel intercalares, polyblasticae ad conidia libus distinctis. **Conidia** holoblastica, sicca, acropleurogena, solitaria, verrucosa, parietibus crassis,recta vel curvata, 2-6 transverse septata, raro constricta ad septata, cylindrica, basim roundata vel leniter attenuata, apicem roundata, atro brunnea,hilo incrassato, 15-46 x 4-6.5 µm.

In foliis vivis *Smilax macrophylla* (Lilliaceae) Nichlaul Forest, U.P. India, Sept. 1997, S. Chaudhary, GPU Herb No. 9021 isotypus, HCIO No. 43689 holotypus.

Infection spots amphigenous, rarely sub-circular to mostly circualr, black in centre while dark brown in margine. **Colonies** hypophyllous, dark brown to blackish brown. Mycelium of hyphae internal to mostly external;

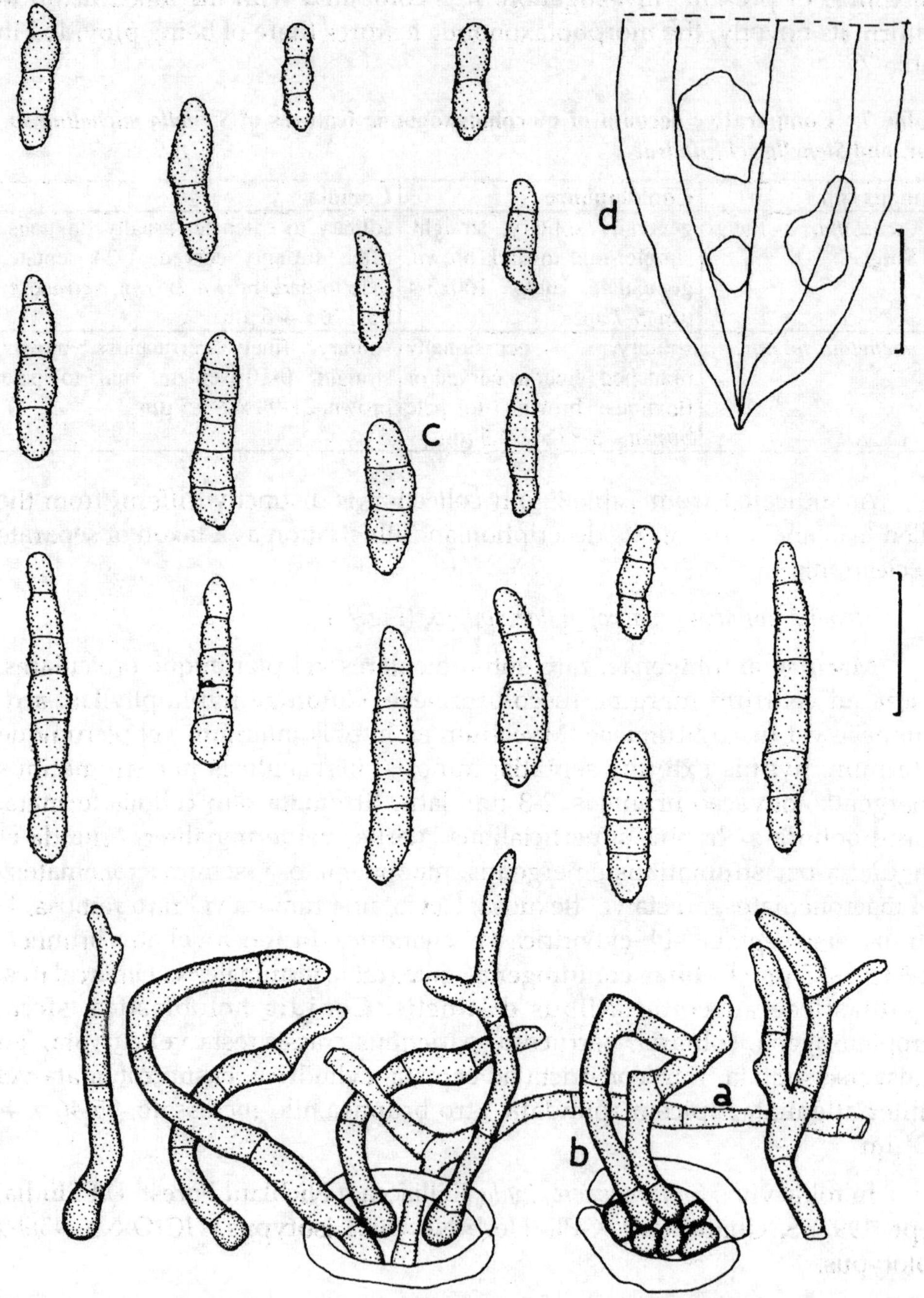

Fig. 7 : *Stenella smilacis macrophyllae sp. nov.* (a) Superficial mycelium (b) Conidiophores with few stromatal cells, (c) Conidia, (d) Leaf spots ; Bars : a-c = 20 μm, d = 20 mm.

external hyphae septate, branched, verruculose coming out through sub-stomatal cavity, olivaceous brown, 2-3 µm wide. **Stromata** present, reduced to few cells. **Conidiophores** arising singly as lateral or terminal branches from external hyphae or through sub-stomatal cavity, mononematous, semi-macronematous to macronematous, straight to flexuous, sub-cylindrical to cylindrical, smooth walled, unbranched or rarely branched, 1-4 transversely septate, brownto dark brown, 12-84 x 3-4.5 µm. **Conidiogenous cells** integrated, terminal to intercalary, polyblastic with few distinct conidial scars. **Conidia** holoblastic, dry, acropleurogenous, solitary, verrucose, thick walled, straight to curved, 2-6 transversely septate occasionally constricted at septa, cylindrical, base round to slightly attenuated, apex round dark brown, hilum thickened, 15-46 x 4-6.5 µm.

Literature survey pertaining to the morphotaxonomy of the fungus genus in question reveals that two species of *Stenella* viz., *S. smilacis* Kumar *et al.* (1980) and *s. smilacicola* Shaw and Alcorn (1993) have already been reported on host genus *Smilax*. The proposed novel taxon *S.smilacis- macrophyllae* sp. *nov.* is found to be morphotaxonomically different from the allied taxa. The comparative account is given in Table 8.

Table 8 : Comparative account of morphotaxonomic features of *Stenella smilacis-macrophyllae sp. nov.* with allied taxa

Fungus spp.	Stromata	Condiophore	Conidia
S. smilacicola Shaw & Alcorn	present	arising in fascicles from stroma, pale yellowish brown, 131-210 µm.	obclavate to linear, 1-14 septate, often constricted at septa, pale brown, 16.4-105 x 3.3-6.6 µm.
S.smilacis Kumar *et al.*	absent	arising usually singly, terminally or laterally from hyphae, brown, 61-165 x upto 3.25 µm.	cylindric, 0-4 septate, sub-hyaline, finely verruculose, 11-21.5 x 2.8-3 µm.
S.smilacis macrophyllae sp.nov.	present, reduced, sub-stomatal	solitary or in loose groups, brown to dark brown, 12-84 x 3-4.5 µm.	cylindrical, verrucose, thick walled, 2-6 septate, dark brown, 15-46 x 4-6.5 µm.

It is gathered from the comparative acount given above that this collection deserves treatment as hitherto undescribed taxon of species rank.

Acknowledgements

The authors are thankful to Head Department of Botany DDU Gorakhpur University, Gorakhpur, for providing library and laboratory facilities. Authors are also thankful to Prof. Kamal for providing helpful suggestions in determination of species.

References

Abbasi, P. and Shukla, D.N. (1980). A new species of *Stenella* from India, Curr. Sci., **49 (2):** 71.

Chaudhary, S. Anuradha and Tripathi, M.S. (2002). Some Hitherto undescribed species of genus *Stenella* from U.P., India. Jour., Basic, Appl. Mycol., **1.**

Chaudhary, R.K., Singh, S.K. and Morgan-Jones, J. (1996). Notes on Hyphomycetes LXXI. New spcies of *Stenella, Stenellopsis and Tretospora* from Nepal. Mycotaxon, **57** :201-209

Chaudhary, R. Gupta, C. and Kamal (1991). New sp. of *Heteroconium, Pseudocercospora and Stenella* from India, Mycol. Res., **95 (a):** 1079

Chaudhary, S., Sharma, N. and Kamal (2002). Three new species of Stenella. Indian Photopath, 55 (1) : 57-60.

Castaneda, Ruiz R.F. and Kendrik, W.B. (1991). University of Waterloo Biology Series, **35** : 113-116.

Castaneda, Ruiz, R.F. (1985). *Deuteromycotina de cuba Hyphomycetes (III) (Cuba) Institute de Investigaciones Fundamentales en Agricultura Tropical –"Alejandro de Humboldt"-A.C.C.:* 70 pp.

Castaneda, Ruiz R.F. (1996). *Fungi Cubense, Institute de Investigaciones Fundamentales en Agricultura Tropical" Alejandro de Humboldt"*:15

Das, A.K. (1990). A new specis of *Stenella* Sydow from West Bengal, Journal of Mycol. Path. Res., **28 (2):** 148

Deighton; F.C. (1979). Studies on *Cercospora* and allied genera VII. New sp. and redispositions Mycol., pp. **144**:53

Deighton, F.C (1971). Brown leaf mould of *Canavalia* caused by *Stenella canavaliae* comb. nov. Trans. Br. Mycol. Soc., **56. (3)**: 412

Ellis, M.B. (1976). More Dematiaceous Hyphomycetes. CMI, Kew, England (1976)

Hosagoudar, V.B. and Braun, U (1976). Two new Indian Hyphomycetes. Indian, Phytopath., **48 (3)**: 260-262; 1995 (1996)

Hoog, G.S. de, Rahman and Boukhout (1983). *Ramichloridiu, Veronaea & Stenella.* Generic delinitations, new combinations & two new species Trans. Br. Mycol. Soc., **81 (3)** : 487

Khan, M.K., Khan, M.S. and Kamal (1994). New sp of *Pseudocercospora and Stenella* from India. Mycol. Res., **98 (11)** : 1279

Khan, K.K., Budathoki, U. and Kamal (1991) New foliicolous hyphomycete from Kathmandu Valley. Indian Phytopath., **44 (1)**: 24

Khan, M.K. and Kamal (1995). Notes on Hyphomycetes LXIV. New species of *Mycovellosiella, Phaeoisariopsis, Sirosporium and Stenella.* Mycotaxon., **54**:34

Kamal, Kumar,P and Rai, B (1981). A new interesting hyphomycete from India. Indian J. Mycol. Pl. Path II. **1** : 144

Kamal and Naraiyan, P (1987). Fungi from hilly tracts of U.P.I Indian Phytopath., **39 (2)**: 201

Kamal, Rai, B. and Kumar, P. (1980a). A new species of *Stenella* from India, Curr. Sci., **49** : 234-235.

Kamal, Singh, R.P. and Kumar, P. (1980). Fungi of Gorkhpour XVIII. Sydowia., **33**: 162-165

Kumar, P, Shukla, D.N. and Kamal (1980). A new sp. of *Stenella* from India. Curr. Sci., **49 (6)**: 234

Kranz (1968). New Hyphomyceten aus guinea Sydowia., **20:** 211-217

Misra, S., Srivastava, A.K. and Kamal (1999). Further additions to *Stenella* from India & Nepal Mycol. Res., 103 **(3)**: 268-270

Misra, S. Srivastava, N and Srivastava A.K. (1997) Mycol. Res., **101**: 278-280

Matsuhina, T. (1983). Matsushima Mycol Memoirs., **3**:1-89

Matsushina, T. (1985). Matsushima Mycol Memoirs., **4**:18-19

Matsustina, T. (1987). Matsushima Mycol Memoirs,. **5**:30

Mulder, J.L. (1975). Notes on *Stenella.* Trans. Br. Mycol. Soc., **65**:517.

Mulder, J.L (1982). New species and combination in *Stenella.* Trans. Br. Mycol. Soc., **79 (3)** : 476

McKenziae, E.H.C. (1982). New hyphomycete on monocotyledon. N.Z. Jr. Bot., **20 (3)**:252

Patil, M.S. and Sawant (1991). Studies on Hyphomycetes 1. Indian Phytopath., **44 (1)**:18

Pavgi and Singh, (1971). Parasitic fungi from North India IX. Sydowia, **24(1-6)** :117

Prasad, S.S. (1968). A new sp. of *Stenella* on leaves of *Aegle marmelos.* Indian Phytopath., **20 (3)** : 253

Ponnappa, K.M. (1968). Some interesting fungi II. *Cercospora hygrophilae sp. nov.* and *Stenella plectroniae* sp. nov. Indian Acad Sci., **B. 67**:33

Rajak, R.C.(181) A new sp. of *Stenella* from India, Acta Bot. Indica, **9(1)**:132

Rai; A.N. and Kamal (1989). A new *Stenella* species from India. Mycol. Res., **93(3)**:398

Rai, A.N. and Kamal (1990). A new sp. of *Stenella* from India. Indian Phytopath., **43(2)**:214

Sarbajna, K.K. Chattopadh. (1991). New *Stenella* sp. from India. Mycopath. Res., **29(1)**:31-35 .

Sarbajna, K.K. (1990). New foliicolous hyphomycetes from India Jr. Economic and Taxonomic Botany, **14(2)** : 489-291

Singh, S.K., Singh A. and Kamal (1997). Additons to *Phaeoramularia and Stenella* from the Indian sub-continent. Mycol. Res., **101(5)**: 530-534

Singh, S and Kamal (1979). Fungi of Gorakhpur, VI Indian Phytopath., **31(3)**:316

Singh, A.K. and Kamal (1987). Fungi from hilly tracts of U.P.1. Indian Phytopath., **39(2)** :274.

Singh, S.K., Bhalla, K. and Bhat, D.J. (2001). Four new foliicolous hyphomycetes from Vindhya Hills, India. J. Mycol. Pl. Pathol, **31**(1) : 277-286.

Sydow, H. (1930). Fungi Venezuelani Annals. Mycologici. **28**: 29-224

Shaw D.E. and Alcorn, J.L. (1993). Australian systematic Botany. **6(3)**:274

Srivastava, K., Srivastava, A.K. and Kamal (1994). New sp. of *Stenella* from India. Mycol. Res., **98(5)**: 516-518

Srivastava, K. Srivastava, A.K. and Kamal (1995). Aditions to *Stenella* from India. Mycol. Res. **99(2)** : 233-236

Verma, R.K. and Kamal (1987). New species of *Pseudocercospora, Phaeoisariopsis, Sarcinella & Stenella* from U.P. India Trans. br. Mycol. Soc., **89(1)**: 67-72

Verma, R.K., Budathoki U and Kamal (1989). New species of *Phaeoramularia, Pseudocercospora and Stenella* from Nepal. Indian Phytopath., **42(1)**:58

Walker, J (1991). Species of *Phaeoisariopsis and Stenella* on *Wikstroemia* (Thymeliaceae). Mycol. Res., **95(8)**: 1011

Yen J.M., Kar, A.K. and Das, B.K. (1982a). Studies on Hyphomycetes from West Bengal, India, I, *Cercospora* and allied genera of genera of West Bengal 1. Mycotaxon., **16(1)**:51-55.

Yen, J.M., Kar, A.K. and Das, B.K. (1982b). Studies on Hyphomycetes from West Bengal, India II. *Cerocospora* and allied genera of West Bengal 2. Myctoaxon., **16(1)** :70-76

Yen, J.M. Kar, A.K. and Das , B.K. (1982c). Studies on Hyphomycetes from West Bengal, India. *Cercospora* and allied genera of West Bengal 3. Mycotaxon. **16(1)**: 90-92

Yip, H.Y. (1989). *Stenella alyxiae* sp. nov. the anamorph of *Mycospharella.* Mycol. Res., **92(3)**: 373-375.

32

An Appraisal of Morphotaxonomic Species Diversity in *Corynespora* Güssow in Indian Sub-Continent

Nidhi Sharma[1] and Sanjay Srivastava[2]

[1]*Mycology Laboratory, Department of Botany*
D.D.U. Gorakhpur University, Gorakhpur 273009 (U.P.)
[2]*Department of Botany, H.C.P.G. College, Varanasi (U.P.)*

The Indian subcontinent is known to embody a very rich diversity of fungi including the foliicolous hyphomycete genus *Corynespora* on account of turnover of green vegetation all the year round in one or the other part combined with diverse set of climatic conditions. In this connection nine new species and one new variety have been described here. These are *C. buchananiae*, *C. crotalariicola*, *C. jabalpurensis*, *C. luffae-cylindricae*, *C. meliacearum*, *C. melongenae*, *C. nana* var. *bifurcata*, *C. oleacearum*, *C. robusta* and *C. xanthiigena* on the living leaves of *Buchanania lanjan* (Anacardiaceae), *Crotalaria sericea* (Fabaceae), *Dodonaea viscosa* (Sapindaceae), *Luffa cylindrica* (Cucurbitaceae), *Azadirachta indica* (Meliaceae), *Solanum melongena* (Solanaceae), *Lantana camara* (Verbenaceae), *Nyctanthes arbor-tristis* (Oleaceae), *Justicia betonica* (Acanthaceae) and *Xanthium strumarium* (Asteraceae). An account of *Corynespora* species described so far from Indian subcontinent is also provided. The perusal of literature on morphotaxonomic diversity of the genus *Corynespora* is suggestive of the fact that Indian mycologists have described maximum number of species, in Indian subcontinent. Our survey and observations during the past couple of years indicates that the end is still not in sight and a complete spectrum of species diversity in this foliicolous form genus warrants a thorough survey of the country in particular and Indian subcontinent in general.

Key words : Biodiversity, foliicolous fungi, *Corynespora*, novel taxa

The genus *Corynespora* was established by Güssow in the year 1906 while studying a glass house disease of cucumbers. This fungus was earlier described by Cooke (1896) as *Cercospora melonis* on melon. Güssow proposed it as a new genus on the ground that the conidia are formed in chain and there is a hyaline isthmus connecting conidiophore and conidium and called it *Corynespora mazaei* ignoring the earlier epithet proposed by Cooke. Lindau (1910), treated the two species as one under the name *Corynespora melonis* (Cooke) Lindau.

In the late autumn of every year the cowpea (*Vigna sinensis*) growing in the valley of Yangtze River, China, were attacked by a serious disease. This disease was caused by a fungus which produced large slender and pale olivaceous brown conidia with thick exospore, the colour of its conidia intensified with age, turning brown. The same fungus was later observed on soybean (*Glycine max*) and was identified by Tai (1936) and Teng (1939) as *Cercospora vignicola.* Kawamura (1931) first described *Cercospora vignicola* on cowpea in Japan and was later identified by Liu (1948) as *Helminthosporium vignae.*

Wei (1950) reinvestigated the genus and accommodated *Cercospora vignicola* Kawamura (1931) and *Helminthosporium cassiicola* Berk. & Curt. (in Berkeley, 1869) with *Corynespora cassiicola* as a type species. The type species was characterised by non stromatic conidiophores terminally proliferating through the scars of fallen conidia reaching a length upto 800 µm and conidiogenous cells bearing catenate, distoseptate conidia with hyaline exospore and conspicuous hilum.

The genus *Corynespora* contains 79 species described from all over the world, most of which are reported to be parasites on plant leaves, but some species have also been reported to survive saprophytically on plant parts including leaves, barks, wood etc. Of the total 26 species of *Corynespora* reported from the Indian subcontinent, 23 were described de novo while the rest 3 were originally placed in genera like *Phaeoisariopsis, Dendrographium* and *Helminthosporium* and were later recombined in *Corynespora* due to obvious reasons.

The first contribution made from India was by Solheim (1933) who described *Corynespora colebrookiae* on *Colebrookia oppositifolia* from the specimen collected from Maharashtra. Ellis (1961) described *C. siwalika* on *Helicteres isora,* which was a recombination of *Helminthosporium siwalikum* Subramanian (1956). In the same year, Rao (1956) described *C. filicola* on *Ficus hispida* from Warangal, A.P.

Subsequently a new variety of *C. gigaspora* (*C. gigaspora* var. *microspora*) and a new species *C. indica* were described by Munjal and Gill (1962) on unidentified twigs from Dehradun.

Ellis (1976) recombined *Dendrographium kamatii* Rao Vasant as *Corynespora kamatii;* the fungus being collected from South India on dead stem of a species of *Vitis.*

C. longispora was described by Saikia and Sarbhoy on dead herbaceous stems from Assam, India in 1981. Again in 1990 Deighton recombined *Phaeoisariopsis inornata* Jong and Morris as *C. inornata* collected on dead branches from Chakrata, India.

A major contribution was made by Meenu *et al.* (1997, 1998a, 1998b) who described nine species of *Corynespora. C. alstoni* on *Alstonia scholaris* collected from Butwal, Nepal, *C. euphorbiacearum* on *Manihot esculentus* from Gorakhpur, India and *C. woodfordiana* on *Woodfordia fruiticosa* from Naraingarh, Nepal was reported by Meenu, Singh and Singh (1997). Meenu and Kamal (1998) described three species from Gorakhpur viz., *C. nana* on *Lantana indica, C. rosacearum* on *Eriobotrya japonica* and *C. morindae-tinctoriae* on *Morinda tinctoria. C. cucurbitaecola, C. jasminiicola C. trictoides* were described by Meenu, Kharwar and Bhartiya (1998) on the hosts *Coccinia grandis* from Kathmandu, Nepal, *Jasminum arborescence* from Tansen, Nepal and *Triumphetta rhomboidea* again from Kathmandu, Nepal respectively.

Singh *et al.* (2000a) added three more species to *Corynespora* from India viz., *C. ulmacearum* on *Trema orientalis* from Gorakhpur, *C. schleichericola* on *Schleichera trijuga* from Siddharthnagar and *C. pedaliacearum* on *Sesamum indicum* from Gorakhpur.

In the same year, again three species were reported by Singh *et al.* (2000b), these are *C. elephantopi, C. lepidagathii-hyalinae* and *C. melanthesae* on the hosts *Elephantopus scaber* from Gorakhpur, *Lepidagathis hyalina* from Siddharthnagar and *Melanthesa rhomboides* from Mahrajganj.

Sharma *et al.* (2002) reported 3 more species of *Corynespora* viz., *C. trematicola* on *Trema orientalis, C. viticola* on *Cayratia carnosa* from Gorakhpur and *C. solanii* on *Solanum indicum* from Nichlaul forest.

In this paper 9 further new species and 1 new variety occurring on living leaves of different hosts belonging to 10 different families are described and illustrated. These are *C. buchananiae, C. crotalariicola, C. jabalpurensis, C. luffae-cylindricae, C. meliacearum, C.melongenae, C.nana* var. *bifurcata, C. oleacearum, C. robusta* and *C. xanthiigena* in addition to an account of hitherto validly described new species of *Corynespora* from Indian subcontinent.

Experimental Requirements

The infected leaf samples with distinct symptoms were taken to the laboratory in separate polythene bags. The specimens were dried by the method used for phanerogamic plant parts; care being taken to keep the symptoms intact and clear. Microscopic slides were prepared in lactophenol and cotton blue mixture (which served as stain and mountant both) from scrappings taken from infected leaf portions. The slides were examined and Camera Lucida drawing prepared under suitable eye piece (10X, 15X) and objective (10X, 40X 45X) combinations so as to show that the most diagnostic morphotaxonomic features can be made out easily. The taxonomic determinations were made with the help of most recent and pertinent literatures and expertise of the resident mycologists. The species are described as new to science on the basis of comparison made with their allied taxa which may be described on the same host or host family and if a species of the genus in question (*Corynespora*) has not been reported on either of these then it is compared with the type species, *Corynespora cassiicola*. The morphotaxonomic features taken for comparison included presence or absence of stromata, nature of vegetative hyphae, length of conidiophores and conidia, branching of conidiophores, number of proliferations in conidiophores, shape of conidia, presence of proliferations in conidia, number of pseudosepta and thickening of hila. As per the international rule, Latin diagnoses have been provided for English descriptions of novel taxa. Nomenclature of the novel taxa has been done according to rules of International Code of Botanical Nomenclature.

Morpho Taxonomy

a) An account of *Corynespora* species validly described from the Indian subcontinent till now along with the collection details (Table-1).

b) Generic circumscription and novel taxa described.

The genus *Corynespora,* in the light of present knowledge, is characterized as follows:

"Parasitic or secondary parasitic; mostly causing necrotic symptoms; mycelium endophytic or occasionally superficial, conidiophores arise perpendicularly to the surface of substratum from hyphae emerging through the epidermis, or occasionally from aerial hyphae with or without a bulbous base, slightly or sometimes in tufts, proliferating terminally through the scar of the previous conidium or sometimes through the injured conidiophore, bearing a single conidium or conidial chain at the apex; conidia terminal, often connected with the conidiophore with a hyaline isthmus, under certain conditions, forming short chains in acropetal succession, pyriform, obclavate or more rarely cylindrical, multiseptate, pale brown becoming dark with age, with a thin, coloured pellicle, a thick colourless spore and a prominent, dark annular basal hilum."

Description and illustrations of the new species of *Corynespora* are as follows:

Corynespora buchananiae Sharma and Srivastava sp. nov. (Fig.1)

Maculae amphigenae, irregulares, atro brunneae ad supernae, brunneae ad infernae. *Coloniae* hypophyllae, effusae, griseae. *Mycelium* ex hyphis internum, ramosis, laevis, subhyalinae vel pallide olivaceae. *Stromata* absentia.

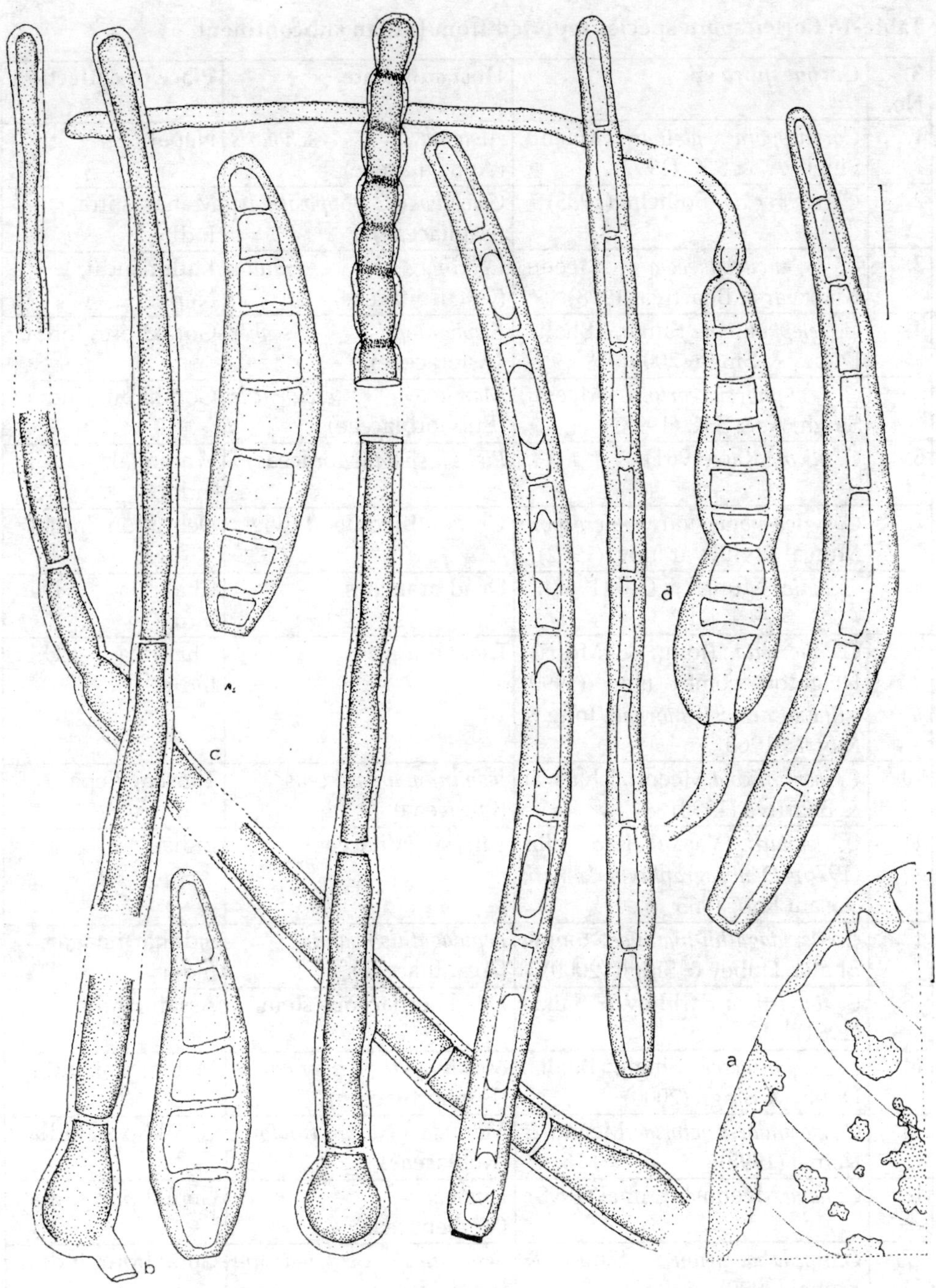

Fig.1: *Corynespora buchananiae* sp. nov. (a) Leaf spots, (b) Mycelium, (c) Conidiophores, (d) Conidia; Bars: a=20mm, b-d 20 μm.

Table-1 : *Corynespora* species reported from Indian subcontinent

S. No.	***Corynespora* sp.**	**Host substrate**	**Place of collection**
1.	*Corynespora alstoni* Meenu, Singh, A. & S.K. (1997)	*Alstonia scholaris* (Apocynaceae)	Nepal
2.	*C. colebrookiae* Solheim (1933)	*Colebrookia oppositifolia* (Lamiaceae)	Maharashtra, India
3.	*C. cucurbitaecola* Meenu, Kharwar & Bhartiya (1998)	*Coccinia grandis* (Cucurbitaceae)	Kathmandu, Nepal
4.	*C. elephantopi* Singh, Bhalla, Dubey & Singh (2000)	*Elephantopus scaber* (Asteraceae).	Gorakhpur, India
5.	*C. euphorbiacearum* Meenu, Singh, A. & S.K. (1997)	*Manihot esculentus* (Euphorbiaceae)	Gorakhpur, India
6.	*C. filicola* Rao (1961)	*Ficus hispida* (Moraceae)	Warangal, A.P., India
7.	*C. gigaspora* var. *microspora* Munjal & Gill var. nov. (1962)	Unidentified dead twigs	Dehradun, India
8.	*C. indica* Munjal & Gill (1962)	Dead branches	Chakrata, U.P., India
9.	*C. inornata* (Jong & Morris) Deighton comb. nov. (1990) =*Phaeoisariopsis inornata* Jong & Morris, 1968	Dead branches	Chakrata, U.P., India
10.	*C. jasminiicola* Meenu, Kharwar & Bhartiya (1998)	*Jasminum arborescens* (Oleaceae)	Tansen, Nepal
11.	C. *kamatii* (Vasant Rao) Ellis (1976)=*Dendrographium kamatii* Vasant Rao, 1963	*Vitis* sp. (Vitaceae).	India.
12.	*C. lepidagathii-hyalinae* Singh, Bhalla, Dubey & Singh (2000).	*Lepidagathis hyalina* (Acanthaceae)	Siddharthnagar, India
13.	*C. longispora* Sarbhoy & Saikia (1981)	Dead herbaceous stem	Assam, India
14.	*C. melanthesae* Singh, Bhalla, Dubey & Singh (2000)	*Melanthesa rhamnoides* (Euphorbiaceae)	Mahrajganj, India
15.	*C. morindae-tinctoriae* Meenu & Kamal (1998)	*Morinda tinctoria* (Rubiaceae)	Gorakhpur, India
16.	*C. nana* Meenu & Kamal (1998)	*Lantana indica* (Verbenaceae)	Gorakhpur, India
17.	*C. pedaliacearum* Singh & Kamal (2000)	*Sesamum indicum* (Pedaliaceae)	Gorakhpur, India
18.	*C. rosacearum* Meenu & Kamal (1998)	*Eriobotrya japonica* (Rosaceae)	Gorakhpur, India

19.	*C. schleichericola* Singh, A. and S.K. & Kamal (2000)	*Schleichera trijuga* (Sapindaceae)	Siddharthnagar, India
20.	*C. siwalika* (Subram.) Ellis (1961) =*Helminthosporium siwalikum* Subramanian (1956)	*Helicteres* sp. (Sterculiaceae) *Lantana* sp. (Verbenaceae)	India
21.	*C. solanii* Sharma, Chaudhary & Kamal (2002)	*Solanum indicum* (Solanaceae)	Nichlaul forest, India
22.	*C. trematicola* Sharma, Chaudhary & Kamal (2002)	*Trema orientalis* (Ulmaceae)	Gorakhpur, India
23.	*C. trictioides* Meenu, Kharwar & Bhartiya (1998)	*Triumfetta rhomboidea* (Tiliaceae)	Kathmandu, Nepal
24.	*C. ulmacearum* Singh, A & S.K. and Kamal (2000)	*Trema orientalis* (Ulmaceae)	Gorakhpur, India
25.	*C. viticola* Sharma, Chaudhary & Kamal (2002)	*Cayratia carnosa* (Vitaceae)	Gorakhpur, India
26.	*C. woodfordiana* Meenu, Singh, A. and S.K. (1997)	*Woodfordia fruiticosa* (Lythraceae)	Nepal

Conidiophora macronematosa, mononematosa, oriunda singulata vel 2-5 fasciculata, recta vel curvata, simplicia, cylindricata, laevia, crassitunicata, usque 11 septata, 0-7 proliferationes successivas cylindricas, cellulae basali inflati, pallide vel atro olivaceo brunnea, 192-555 x 7-12.5 µm. *Cellulae conidiogenae* integratae, terminales, monotreticae, cicatrix incrassatae. *Conidia* solitaria primo in apice conidiophori et dein proliferationis cujusque successivas oriunda, sicca, acrogena, simplicia, laevia, tenuitunicata, recta vel curvata, obclavata vel obclavatocylindricata, ad apicem subacuta vel subobtusa, ad basim truncata, 5-9 pseudoseptata, hila non incrassata raro incrassata, pallide vel atro olivaceo lutea, 60-172 x 5.5-15.5 µm.

In foliis vivis *Buchanania lanjan* Spreng. (Anacardiaceae), Nichlaul, India; Mar. 1999; Nidhi; GPU Herb No. NS-1230 isotypus; HCIO No. 13889 holotypus.

Infection spots amphigenous, irregular, dark brown on upper surface, brown on lower surface. *Colonies* hypophyllous, effuse, grey. *Mycelium* of hyphae internal, branched, smooth walled, subhyaline, light olivaceous. *Stromata* absent. *Conidiophores* macronematous, mononematous, arising singly from hyphae, straight to curved, simple, cylindrical, smooth, thick walled, upto 11 septate with 0-7 successive cylindrical proliferations, basal cell swollen, light to dark olivaceous brown, 192-555 x 7-12.5 µm. *Conidiogenous cells* integrated, terminal, monotretic, scar thickened. *Conidia* arising singly through a pore at the apex of conidiophore which after the first conidium has fallen often proliferates through the apical pore forming a conidium at the apex of proliferation, dry, acrogenous, simple, smooth, thin walled, straight to curved,

Table - 2 : Comparison of morphotaxonomic features of *Corynespora buchananiae* sp. nov. with *C. lanneicola*

Corynespora sp.	Stromata	Conidiophores	Conidia
C. lanneicola Deighton & Ellis	Present	Arising singly or in a fascicle of 6 from the cells of stromata, erect, simple, dark blackish brown, septate with upto 3 successive proliferation, 120-160 x 7-11 µm	Formed singly, straight to slightly curved, obclavate, straw coloured to brown, 4-5 pseudoseptate, 40-48 x 10-15 µm.
C. buchananiae sp. nov.	Absent	Arising singly, erect, simple, straight to curved, light to dark olivaceous brown, septate with 0-7 successive proliferation, 192-555 x 7-12.5 µm.	Formed singly, straight to curved, obclavate to obclavatocylindrical, light to dark olivaceous yellow, 5-9 pseudoseptate, hilum unthickened, rarely thickened, 60-172 x 5.5-15.5 µm.

obclavate to obclavatocylindrical, apex subacute to subobtuse, base truncate, 5-9 pseudoseptate, hila unthickened, rarely thickened, some conidia producing germ tube from base and apex, light to dark olivaceous yellow, 60-172 x 5.5-15.5 µm.

It is evident from the literature that no species of *Corynespora* has yet been described on the host genus but one species i.e. *C. lanneicola* Deighton and Ellis (1957) has been described on the host family. The present collection is, therefore, compared with *C. lanneicola* (Table-2) in their morphotaxonomic features. From the comparative data provided in the table, it can be gathered that the present species is at great variance from earlier described species in the absence of stromata, number of proliferations and size of conidiophores. Conidial size also differs significantly. Therefore, this proposal of a new taxon, to accommodate it, is found to be worthwhile.

Corynespora crotalariicola Sharma and Srivastava sp. nov. (Fig.2)

Maculae amphigenae, circulares vel subcirculares, extensae per totum folii, griseo brunneae ad infernae, atro brunneae ad supernae. *Coloniae* hypophyllae, effusae, griseae. *Mycelium* ex hyphis internum, septatis, ramosis, pallide olivaceae. *Stromata* absentia. *Conidiophora* macronematosa, mononematosa, oriunda singulata vel 2 fasciculata, recta vel leniter flexuosa, simplicia, cylindricata, laevia, crassitunicata, 4-7 septata, 0-2 proliferationes successivas cylindricas, cellulae basali inflati, pallide vel atro olivaceo brunnea, 136-265 x 5.5-10 µm. *Cellulae conidiogenae* integrate, terminales, monotreticae,

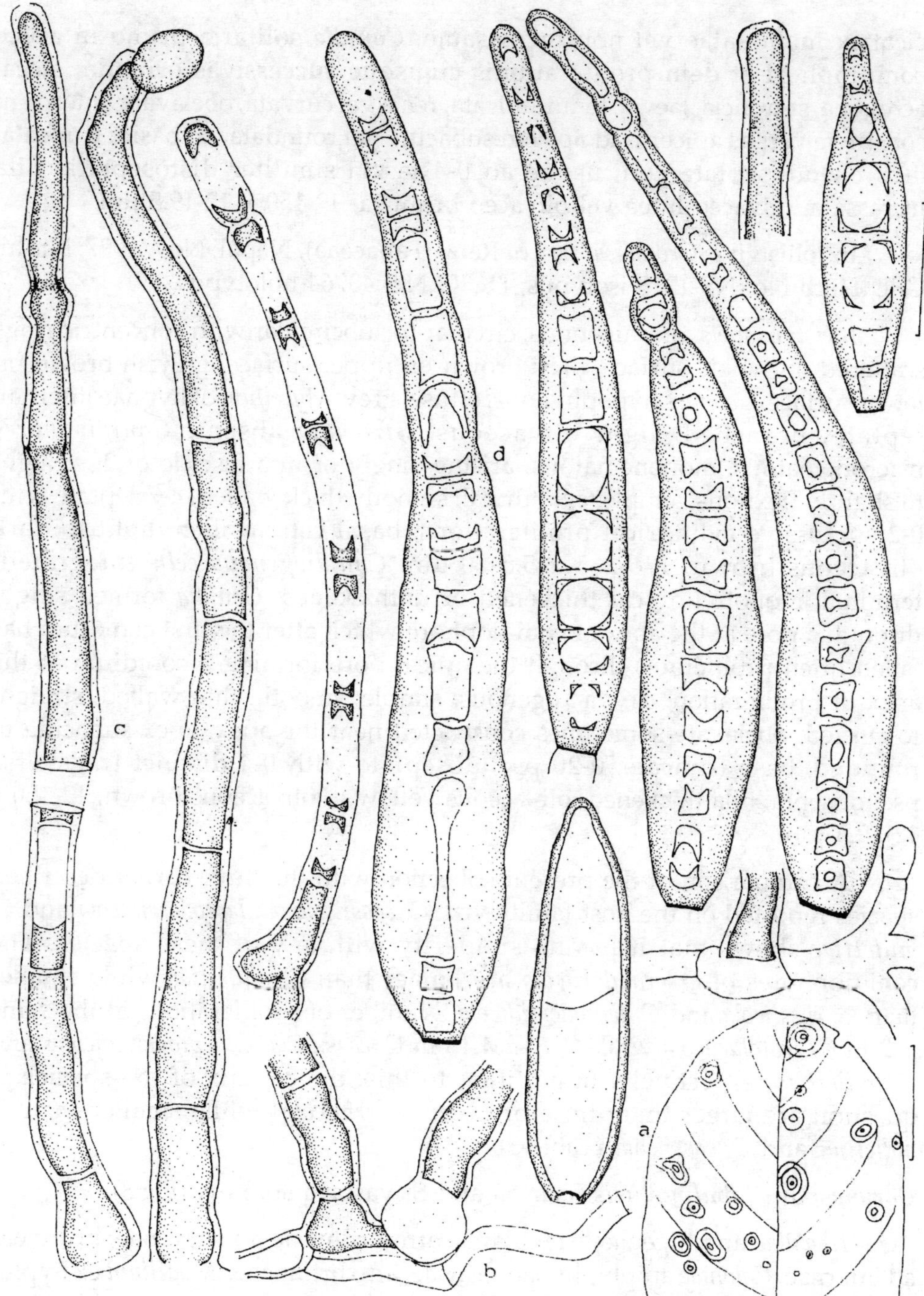

Fig.2: *Corynespora crotalariicola* sp. nov. (a) Leaf spots, (b) Mycelium, (c) Conidiophores, (d) Conidia; Bars: a=20mm, b-d 20 µm.

cicatrix incrassatae vel non incrassatae. *Conidia* solitaria primo in apice conidiophori et dein proliferationis cujusque successivas oriunda, sicca, acrogena, simplicia, laevia, tenuitunicata, recta vel curvata, obclavata, interdum constricta versus apicem, ad apicem subacuta vel roundata, ad basim truncata, 0-20 pseudoseptata cum usque ad 0-4 fascia similiter distoseptata, hila incrassata, olivaceo lutea vel olivaceo brunnea, 12-150 x 12-19.5 μm.

In foliis vivis *Crotalaria sericea* Retz. (Fabaceae), Nepal; Nov. 1997; Nidhi; GPU Herb No. NS-1218 isotypus; HCIO No. 43764 holotypus.

Infection spots amphigenous, circular to subcircular with concentric rings scattered on whole surface, dark brown on upper surface, greyish brown on lower surface. *Colonies* hypophyllous, effuse, grey. *Mycelium* of hyphae internal, septate, branched, light olivaceous. *Stromata* absent. *Conidiophores* macronematous, mononematous, arising singly or in a fascicle of 2, straight to slightly flexuous, simple, cylindrical, smooth, thick walled, 4-7 septate with 0-2 successive cylindrical proliferations, basal cell swollen, light to dark olivaceous brown, 136-265 x 5.5-10 μm. *Conidiogenous cells* integrated, terminal, monotretic, scar thickened or unthickened. *Conidia* formed singly through a pore at the apex of conidiophore which after the first conidium has fallen often proliferates through the apical pore forming a conidium at the apex of proliferation, dry, acrogenous, simple, smooth, thin walled, straight to curved, obclavate sometimes constricted near the apex, apex subacute to rounded, base truncate, 0-20 pseudoseptate with 0-4 distinct transverse pseudosepta, hila thickened, olivaceous yellow to olivaceous brown, 12-150 x 12-19.5 μm.

A comparison of the present collection with the three earlier described species reported on the host family viz., *C. cassiicola, C. leptoderridicola* and *C. millettiae* shows that it has no similarity with any of these species. The conidiophores of *C. crotalariicola* are longer than *C. millettiae* while smaller than *C. cassiicola* and *C. leptoderridicola*. Number of proliferations in the same is 2 in *C. crotalariicola* while 9, 6 and 18 in *C. cassiicola, C. leptoderridicola* and *C. millettiae* respectively. In addition to this, the conidia of the species in question are larger than those of *C. leptoderridicola* while smaller than *C. millettiae* and *C. cassiicola* (Table-3).

Corynespora jabalpurensis Sharma and Srivastava sp. nov. (Fig.3)

Maculae amphigenae, irregulares, atro brunneae ad supernae, brunneae ad infernae. *Coloniae* amphiphyllae, effusae, atro brunneae. *Mycelium* ex hyphis internum, ramosis, subhyalinae. *Stromata* absentia. *Conidiophora* macronematosa, mononematosa, oriunda singulata ex hyphis, recta, simplicia, cylindricata, laevia, crassitunicata, 1-7 septata, pallide vel atro olivaceo

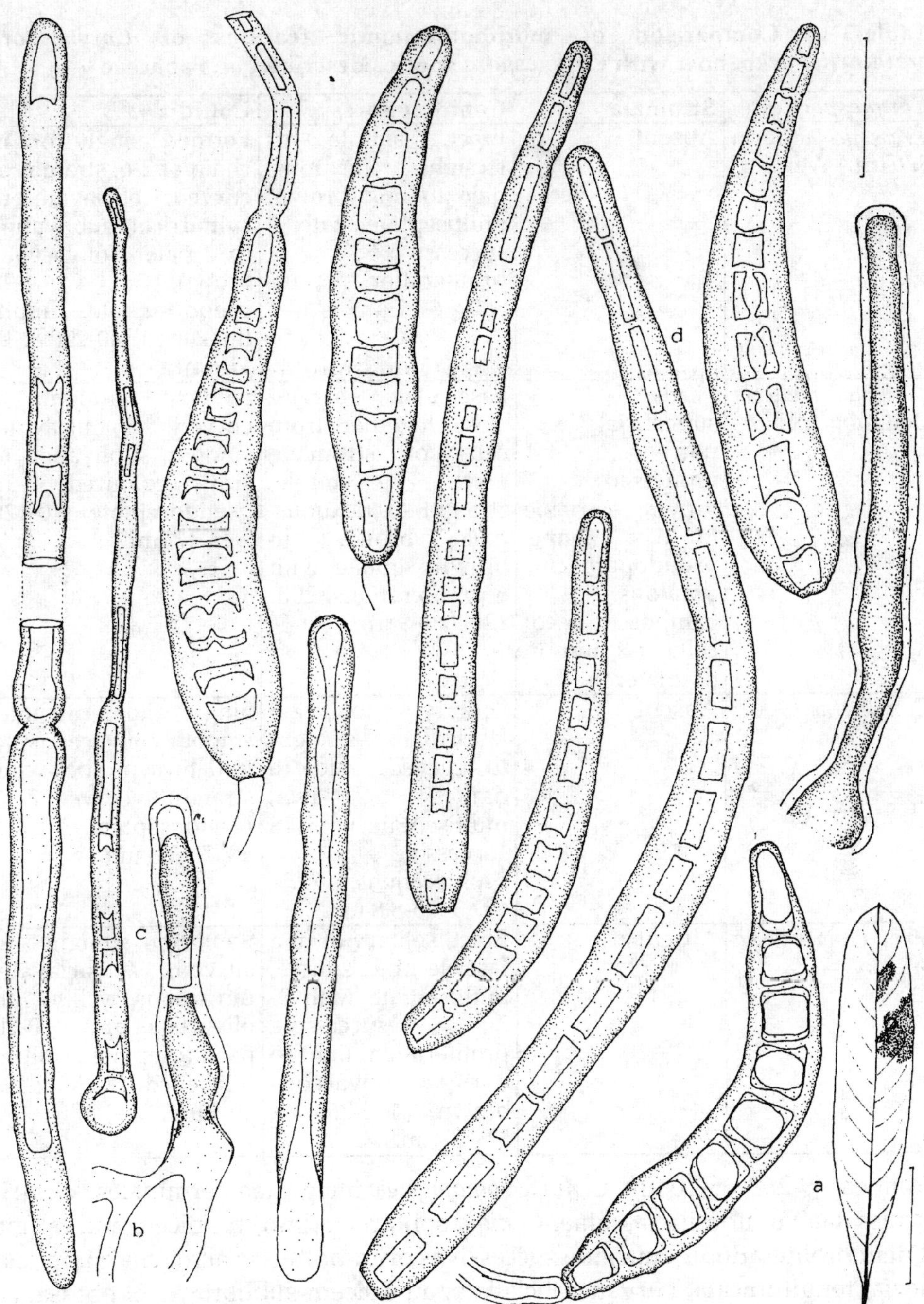

Fig.3: *Corynespora jabalpurensis* sp. nov. (a) Leaf spots, (b) Mycelium, (c) Conidiophores, (d) Conidia; Bars: a=20mm, b-d 20 μm.

Table-3 : Comparison of morphotaxonomic features of *Corynespora crotalariicola* sp. nov. with *Corynespora* species described on Fabaceae

Corynespora sp.	Stromata	Conidiophores	Conidia
C. cassiicola (Berk. & Curt.) Wei	Absent	Erect, simple or straight to flexuous, pale to mid brown, multiseptate with 9 successive proliferation, 110-850 x 4-11 μm.	Formed singly or in chain of 2-6, straight or curved, obclavate to cylindrical, subhyaline to pale olivaceous brown, 4-20 pseudoseptate, hilum thickened, 40-220 x 9-11 μm.
C. *leptoderridicola* Deighton & Ellis	Partly superficial, partly immersed, brown, variable in shape, pseudoparenchymatous, composed of cells 4-2 μm in diameter.	Fasciculate, arising close together from cells of stromata, erect, simple, straight to flexuous, pale brown to brown, septate with 6 proliferations, 200-1200 x 6-9 μm.	Solitary, straight to curved, obclavate, smooth, subhyaline to straw coloured, 6-16 pseudoseptate, 70-120 x 14-17 μm.
C. *millettiae* Guo	Absent	Solitary to 2-3 fasciculate, straight to curved, mid to dark brown, multiseptate with 18 successive proliferations, 77.5-492.5 x 5-8.8 μm.	Single to catenate, smooth, olivaceous to mid brown, obclavate, straight to curved, 2-15 pseudoseptate, 3-182.5 x 7.7-13.8 μm.
C. *crotalariicola* sp. nov.	Absent	Erect, solitary or in a fascicle of 2, simple, multiseptate with 0-2 successive proliferation, light to dark olivaceous brown, 136-265 x 5.5-10 μm.	Solitary, straight to curved, obclavate, olivaceous yellow to olivaceous brown, 0-20 pseudoseptate, hilum thickened, 12-150 x 12-19.5 μm.

brunnea, 49-260 x 5-10 μm. *Cellulae conidiogenae* integratae, terminales, cicatrix incrassatae vel non incrassatae. *Conidia* solitaria primo in apice conidiophori et dein proliferationis cujusque successivas oriunda, sicca, acrogena, simplicia, laevia, tenuitunicata, curvata, obclavata, ad apicem subobtusa vel obtusa, ad basim truncata, 7-15 pseudoseptata, hila non incrassata, pallide olivaceo lutea, 85-181 x 9-15 μm.

In foliis vivis *Dodonaea viscosa* Jacq. (Sapindaceae), Jabalpur, India; Dec.

1997; Nidhi; GPU Herb No. NS-1205 isotypus; HCIO No. 43752 holotypus.

Infection spots amphigenous, irregular, dark brown on upper surface, brown on lower surface. *Colonies* amphiphyllous, effuse, dark brown. *Mycelium* of hyphae internal, branched, subhyaline. *Stromata* absent. *Conidiophores* macronematous, mononematous, arising singly from hyphae, straight, simple, cylindrical, smooth, thick walled, 1-7 septate, basal cell may or may not be swollen, light to dark olivaceous brown, 49-260 x 5-10 µm. *Conidiogenous cells* integrated, terminal, scar thickened or unthickened. *Conidia* arising singly through a pore at the apex of conidiophore which after the first conidium has fallen often proliferates through the apical pore forming a conidium at the apex of proliferation, dry, acrogenous, simple, smooth, thin walled, curved, obclavate, apex subobtuse to obtuse, base truncate, 7-15 pseudoseptate, hila unthickened, light olivaceous yellow, 85-181 x 9-15 µm.

Two species of *Corynespora* have earlier been described on the host family viz., *C. pruni* (Berk. & Curt.) Ellis (1976) and *C. schleichericola* Singh *et al.* (2000). Present collection is, therefore, compared with these two so as to assess the differences warranting its description as a new species (Table-4).

The comparative account gives a clear view of *C. jabalpurensis* being distinctly and significantly different from the earlier described species in not having any proliferations in the conidiophores. Moreover, the conidiophores

Table-4 : Comparison of morphotaxonomic features of *Corynespora jabalpurensis* sp. nov. with *Corynespora* species reported on Sapindaceae

Corynespora sp.	Stromata	Conidiophores	Conidia
C. pruni (Berk. & Curt.) Ellis	Absent	Simple, multiseptate with 0-3 successive proliferation, brown to blackish brown, 60-280 x 7-11 µm.	Solitary to catenate, obclavate, olivaceous, brown, paler near apex, very dark at the scar, 4-9 pseudoseptate, hilum thickened, 50-130 x 10-16 µm.
C. schleichericola Singh *et al.*	Absent	Branched, 2-6 septate with 2-5 proliferation, light brown to brown, 77.5-166 x 3.5-4 µm.	Solitary to catenate, obclavate, light olivaceous, 1-12 pseudoseptate, hilum unthickened, 25.5-66 x 3.8-8.5 µm.
C. jabalpurensis sp. nov.	Absent	Simple, septate, proliferation absent, light to dark olivaceous brown, 49-260 x 5-10 µm	Solitary, obclavate, light olivaceous yellow, 7-15 pseudoseptate, hilum unthickened, 85-181 x 9-15 µm

are unbranched and much larger in size than that of *C. schleichericola* in which they are branched and smaller. Conidia are catenate in the earlier described species as against non catenate condition found in the present one. Hila thickened in *C. pruni* while unthickened in *C. jabalpurensis*. Size of conidia in the present collection is also different from the earlier described species. The differences are so marked that the present collection is recognized as a new taxon of species rank.

Corynespora luffae-cylindricae Sharma & Srivastava sp. nov. (Fig.4)

Maculae amphigenae, circulares vel irregulares, griseo brunneae ad supernae, margin atro griseae, pallide brunneae ad infernae. *Coloniae* amphiphyllae, effusae, griseae. *Mycelium* ex hyphis internum, septatis, laevibus, tenuitunicatis, ramosis. *Stromata* absentia. *Conidiophora* macronematosa, mononematosa, oriunda singulata ex hyphis, simplicia, raro ramosa, recta vel leniter curvata, cylindricata, laevia, crassitunicata, multiseptata, 0-6 proliferationes successivas cylindricas, pallide vel atro olivaceo brunnea, 62-472 x 5-9 µm. *Cellulae conidiogenae* integratae, terminales, cicatrix incrassatae vel non incrassatae. *Conidia* solitaria, primo in apice conidiophori et dein proliferationis cujusque successivas oriunda, sicca, acrogena, simplicia, laevia, tenuitunicata, recta vel curvata, obclavata vel obclavatocylindrica, ad apicem subobtusa vel roundata, ad basim truncata, 5-17 pseudoseptata, hila incrassata vel non incrassata, pallide olivaceo lutea, 57-279 x 8-19 µm.

In foliis vivis *Luffa cylindrica* Mill. (Cucurbitaceae), Gorakhpur, India; Sept. 1998; Nidhi; GPU Herb No. NS-1223 isotypus; HCIO No. 43769 holotypus.

Infection spots amphigenous, circular to irregular, grayish brown on upper surface with dark grey margin, light brown on lower surface. *Colonies* amphiphyllous, effuse, grey. *Mycelium* of hyphae internal, septate, thin walled, branched. *Stromata* absent. *Conidiophores* macronematous, mononematous, arising singly from hyphae, simple to rarely branched, straight to slightly curved, cylindrical, smooth, thick walled, multiseptate with 0-6 successive cylindrical proliferations, light to dark olivaceous brown, 62-472 x 5-9 µm. *Conidiogenous cells* integrated, terminal, monotretic, scar thickened or unthickened. *Conidia* arising singly through a pore at the apex of conidiophore which after the first conidium has fallen often proliferates through the apical pore forming a conidium at the apex of proliferation, dry, acrogenous, simple, smooth, thin walled, straight to curved, obclavate to obclavatocylindrical, apex subobtuse to rounded, base truncate, flat, rounded, 5-17 pseudoseptate, hila thickened or unthickened, light olivaceous yellow, 57-279 x 8-19 µm.

So far no species of *Corynespora* has been described on the host genus.

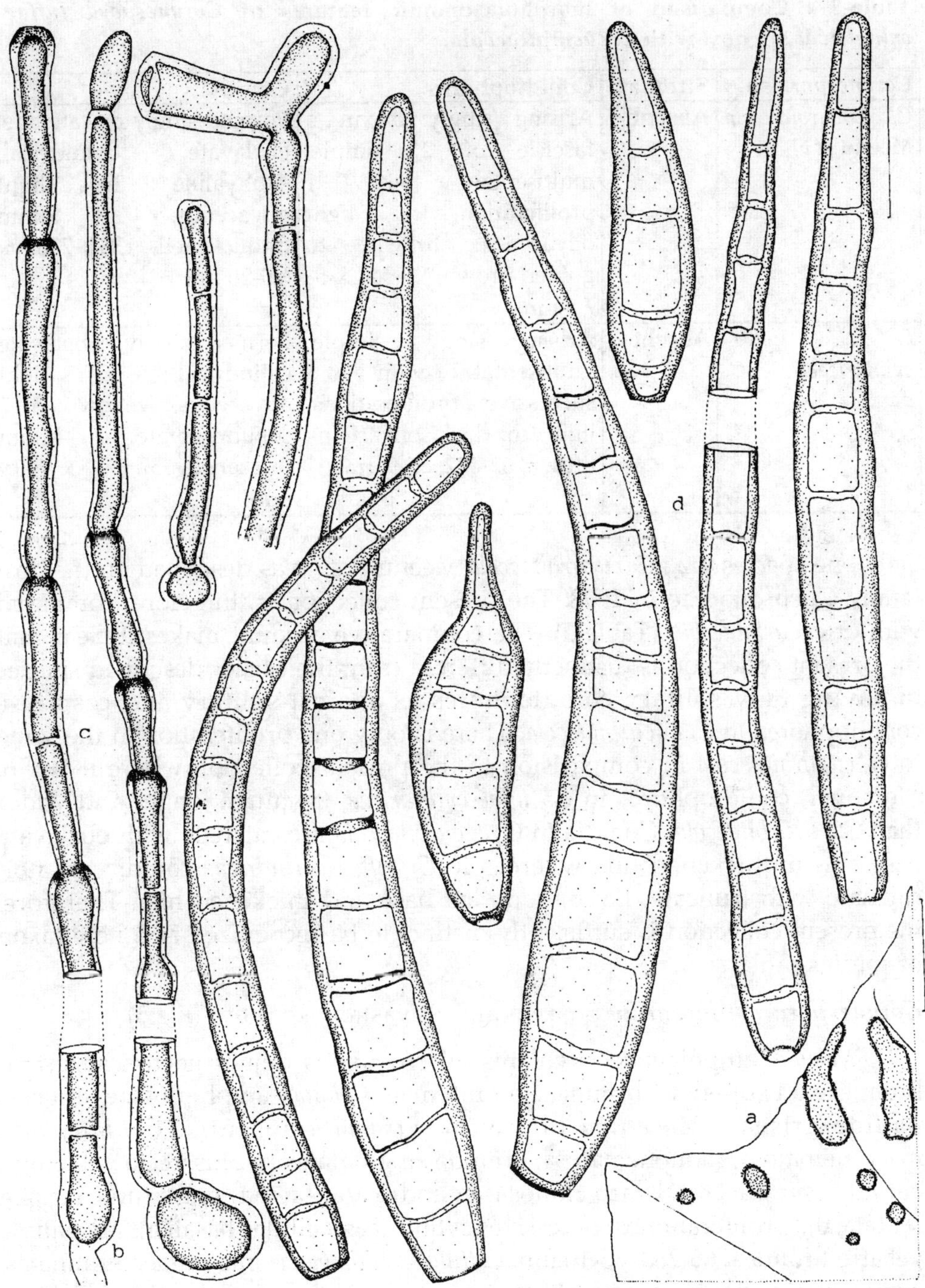

Fig.4: *Corynespora luffae-cylindricae* sp. nov. (a) Leaf spots, (b) Mycelium, (c) Conidiophores, (d) Conidia; Bars: a=20mm, b-d 20 μm.

Table-5 : Comparison of morphotaxonomic features of *Corynespora luffae-cylindricae* sp. nov. with *C. cucurbitaecola*.

Corynespora sp.	Stromata	Conidiophores	Conidia
C. cucurbitaecola Meenu *et al.*	Absent	Arising singly or in a fascicle of 2, simple, multiseptate with 1 proliferation, light olivaceous brown to golden brown, 90-347 x 5-7.5 µm	Formed singly or catenate, obclavate, cylindrical, subhyaline to light olivaceous, hilum unthickened, 38.5-230 x 6.5-20 µm
C. luffae-cylindricae sp. nov.	Absent	Arising singly, simple, multiseptate with 0-6 successive proliferations, light to dark olivaceous brown, 62-472 x 5-9 µm.	Formed singly, obclavate cylindrical, light olivaceous yellow, 5-17 pseudoseptate, hilum thickened, 57-279 x 8-19 µm.

Only one species viz., *C. cucurbitaecola* Meenu *et al.* was described on the host family Cucurbitaceae in 1998. The present collection is, therefore, compared with *C. cucurbitaecola* (Table-5). The comparative account makes it clear that the present collection is distinctly different from the earlier described species in having only solitary conidiophores as against solitary and cespitose conidiophores in *C. cucurbitaecola*. There is only one proliferation in the same in *C. cucurbitaecola* in comparision to the present collection where it is 0-6. Moreover, conidiophores in *C. luffae-cylindricae* are much longer and wider than *C. cucurbitaecola*. Conidia in *C. cucurbitaecola* are catenate with obclavate base and unthickened hila whereas in *C. luffae-cylindricae* conidia are non catenate with truncate, flat or rounded base and thickened hila. Therefore, the present collection is sufficiently distinct to be recognized as a new taxon of species rank.

Corynespora meliacearum Sharma and Srivastava sp. nov. (Fig.5)

Maculae amphigenae, circulares vel irregulares, deinde necroticae, griseo brunneae ad supernae, brunneae ad infernae. *Coloniae* amphiphyllae, effusae, pallide griseae. *Mycelium* internum. *Stromata* absentia. *Conidiophora* macronematosa, mononematosa, oriunda singulata ex hyphis, recta vel leniter curvata, simplicia vel raro ramosa, cylindricata, laevia, crassitunicata, 2-8 septata, 0-3 proliferationes successivas cylindricas, cellulae basali inflati, pallide vel atro brunnea, 65-283 x 5-10 µm. *Cellulae conidiogenae* integratae, terminales, monotreticae, cicatrix incrassatae vel non incrassatae. *Conidia* solitaria, primo in apice conidiophori et dein proliferationis cujusque successivas oriunda, sicca, acrogena, simplicia, laevia, tenuitunicata, recta, curvata, leniter flexuosa,

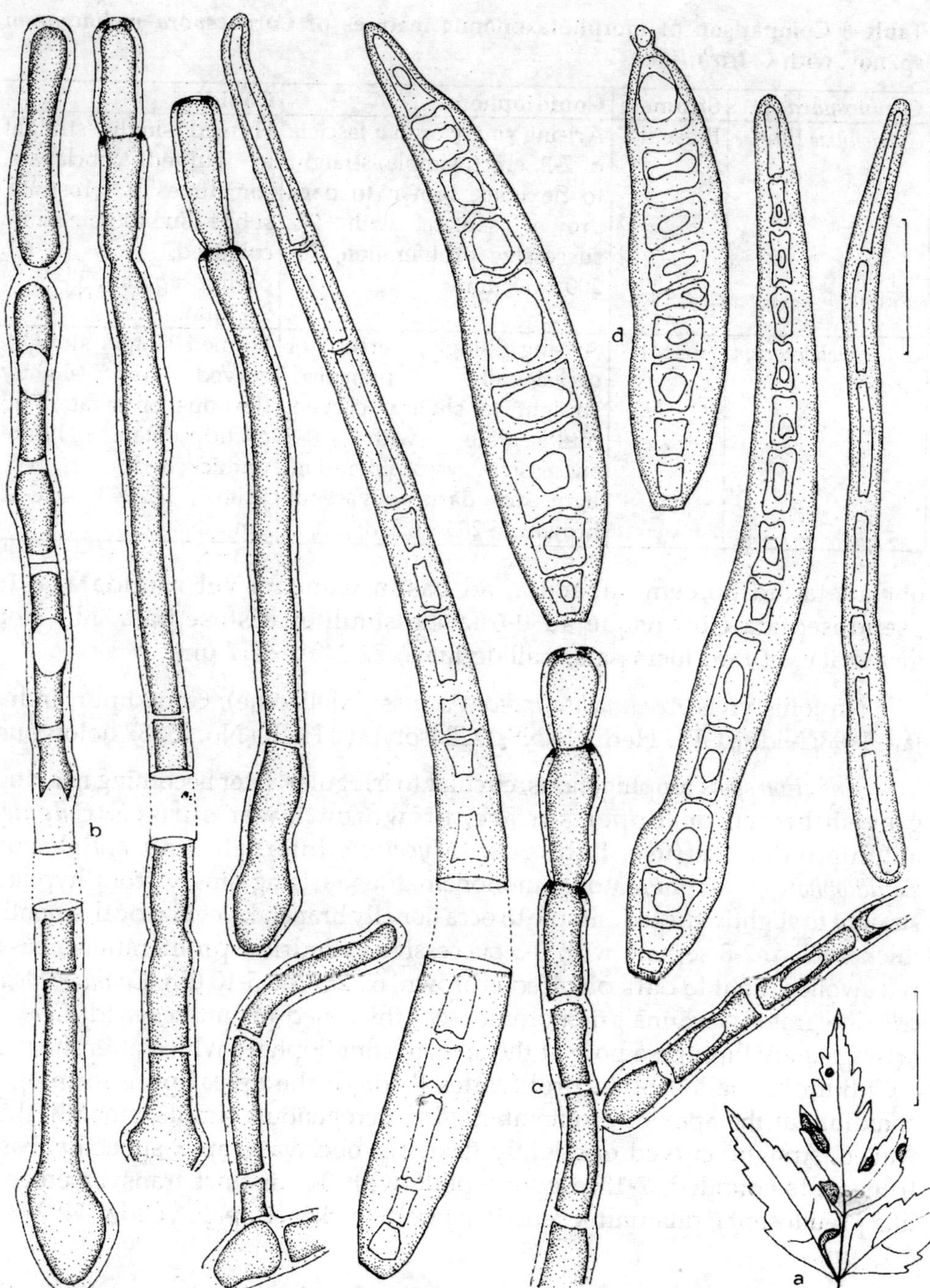

Fig.5: *Corynespora meliacearum* sp. nov. (a) Leaf spots, (b) Mycelium, (c) Branched conidiophore, (d) Conidia; Bars: a=20mm, b-d 20 µm.

Table-6 Comparison of morphotaxonomic features of *Corynespora meliacearum* sp. nov. with *C. trichiliae*

Corynespora sp.	Stromata	Conidiophores	Conidia
C. trichiliae Ellis	Present	Arising singly or in a fascicle of 2-3, erect, simple, straight to flexuous, brown to dark brown, septate with 0-2 successive proliferation, 75-200 x 6-8 µm.	Formed singly, straight to curved, obclavate, sometimes rostrate, subhyaline to pale straw coloured, 4-6 pseudoseptate, 53-74 x 9-11 µm.
C. meliacearum sp. nov.	Absent	Arising singly, erect or occasionally branched, straight to slightly curved, multiseptate with 0-3 successive proliferation, light to dark olivaceous brown, 65-283 x 6.5-10 µm.	Formed singly, straight, curved or slightly flexuous, obclavate, 7-15 pseudoseptate hilum unthickened, pale yellow, 72-243 x 6-17 µm.

obclavata, ad apicem subacuta, ad basim truncata vel roundata, 7-15 pseudoseptata cum usque ad 0-6 fascia similiter distoseptata, hila non incrassata vel raro incrassata, pallide lutea, 72-243 x 6-17 µm.

In foliis vivis *Azadirachta indica* A. Juss. (Meliaceae), Gorakhpur, India; Jan. 1998; Nidhi; GPU Herb No.NS-1228 isotypus; HCIO No. 43887 holotypus.

Infection spots amphigenous, circular to irregular, later becoming necrotic, greyish brown on upper surface, brown on lower surface. *Colonies* amphiphyllous, effuse, light grey. *Mycelium* internal. *Stromata* absent. *Conidiophores* macronematous, mononematous, arising singly from hyphae, straight to slightly curved, simple to occasionally branched, cylindrical, smooth, thick walled, 2-8 septate with 0-3 successive cylindrical proliferations, basal cell swollen, light to dark olivaceous brown, 65-283 x 6.5-10 µm. *Conidiogenous cells* integrated, terminal, monotretic, scar thickened or unthickened. *Conidia* arising singly through a pore at the apex of conidiophore which after the first conidium has fallen often proliferates through the apical pore forming a conidium at the apex of proliferation, dry, acrogenous, simple, smooth, thin walled, straight, curved or slightly flexuous, obclavate, apex subacute, base truncate to rounded, 7-15 pseudoseptate with 0-6 distinct transverse band like pseudosepta, hila unthickened or rarely thickened, pale yellow, 72-243 x 6-17 µm.

No species of *Corynespora* has yet been described on the host genus. Of course *C. trichiliae* Ellis (1960), described on the host family, is different from the present fungus as it is characterized by non fasciculate and branched conidiophores with a greater dimension than *C. trichiliae*. Number of pseudosepta and dimension of conidia of *C. meliacearum* are also very different from the earlier described species. Therefore, the present fungus is treated as a new taxon of species rank (Table-6).

Corynespora melongenae Sharma & Srivastava sp. nov. (Fig.6)

Maculae amphigenae, circulares vel irregulares, fusca brunneae ad supernae, pallide brunneae ad infernae. *Coloniae* amphiphyllae, effusae, griseae. *Mycelium* ex hyphis internum, septatis, ramosis. *Stromata* absentia. *Conidiophora* macronematosa, mononematosa, oriunda singulata ex hyphis, recta vel leniter curvata, simplicia, cylindricata, laevia, crassitunicata, 2-3 septata, 0-3 proliferationes successivas cylindricas, pallide vel atro olivaceo brunnea, 124-316 x 7-9 µm. *Cellulae conidiogenae* integratae, terminales, cicatrix incrassatae vel non incrassatae. *Conidia* solitaria, primo in apice conidiophori et dein proliferationis cujusque successivas oriunda, sicca, acrogena, simplicia, laevia, tenuitunicata, recta vel leniter curvata, ellipsoidea vel obclavata, 1-2 constricta, ad apicem subobtusa vel roundata, ad basim truncata, 0-8 pseudoseptata cum usque ad 0-1 fascia similiter distoseptata, hila non incrassata, olivaceo brunnea, 28.5-165 x 12-19 µm.

In foliis vivis *Solanum melongenae* Linn. (Solanaceae), Gorakhpur, India; Jan. 1999; Nidhi; GPU Herb No. NS-1203 isotypus; HCIO No. 43747 holotypus.

Infection spots amphigenous, circular to irregular, blackish brown on upper surface with dark brown centre, light brown on lower surface. *Colonies* amphiphyllous, effuse, grey. *Mycelium* of hyphae internal, septate, branched. *Stromata* absent. *Conidiophores* macronematous, mononematous, arising singly from hyphae, straight to slightly curved, simple, cylindrical, smooth, thick walled, 2-3 septate with 0-3 successive cylindrical proliferations, basal cell may or may not be swollen, light to dark olivaceous brown, 124-316 x 7-9 µm. *Conidiogenous cells* integrated, terminal, scar thickened or unthickened. *Conidia* arising singly through a pore at the apex of conidiophore which after the first conidium has fallen often proliferates through the apical pore forming a conidium at the apex of proliferation, dry, acrogenous, simple, smooth, thin walled, straight to slightly curved, ellipsoidal or obclavate, apex subobtuse to rounded, base truncate, 0-8 pseudoseptate with 0-1 distinct transverse band like pseudosepta, hila unthickened, olivaceous brown, 28.5-165 x 12-19 µm.

Perusal of the literature shows that two species of *Corynespora* i.e., *C. cassiicola* (Berk. & Curt.) Wei (1950), the type species and *C. solanii* Sharma *et al.* (2002) have been reported on the host genus. The present collection is, therefore, compared with the same (Table -7).

The table shows that *C. melongenae* is different from *C. cassiicola* and

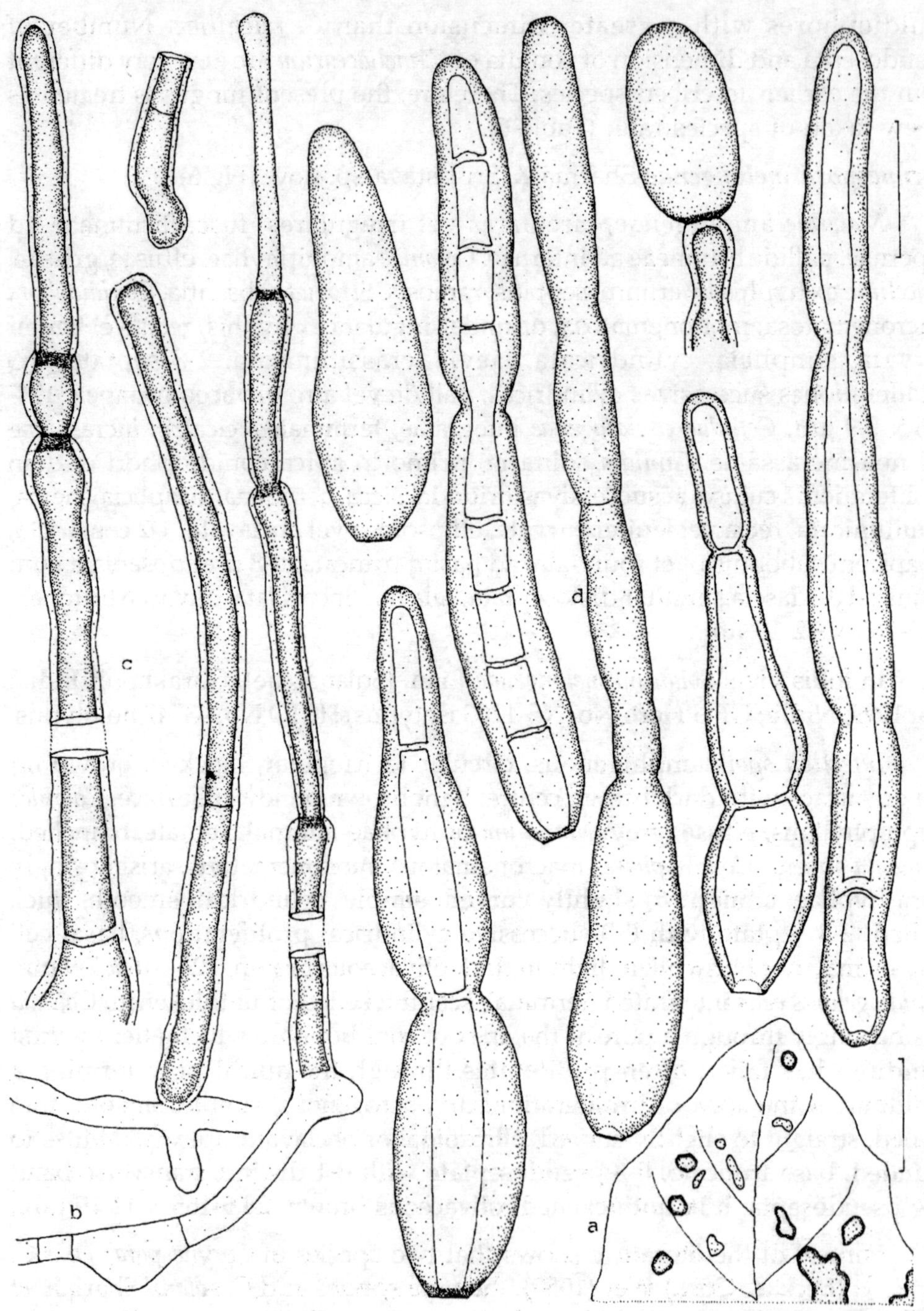

Fig.6: *Corynespora melongenae* sp. nov. (a) Leaf spots, (b) Mycelium, (c) Conidiophores, (d) Conidia; Bars: a=20 mm, b-d 20 µm.

Table-7 Comparison of morphotaxonomic features of *Corynespora melongenae* sp. nov. with *Colynespora* species on solanaccae

Corynespora sp.	Stromata	Conidiophores	Conidia
C. cassiicola (Berk. & Curt.) Wei	Absent	Erect, simple or straight to flexuous, pale to mid brown, multiseptate with 9 successive proliferation, 110-850 x 4-11 µm.	Formed singly or in chain of 2-6, straight or curved, obclavate to cylindrical, subhyaline to pale olivaceous brown, 4-20 pseudoseptate, hilum thickened, 40-220 x 9-11 µm.
C. solanii Sharma *et al.*	Absent	Erect, simple or occasionally branched, straight to flexuous, multiseptate with 0-1 proliferation, light olivaceous brown, 115-401 x 5-8.5 µm.	Formed singly or catenate, straight or rarely curved, cylindrical, olivaceous yellow, 1-17 pseudoseptate, hilum unthickened, 80.6-276 x 8-10 µm.
C. melongenae sp. nov.	Absent	Erect, simple, straight to sightly curved, multiseptate with 0-3 successive proliferations, light to dark olivaceous brown, 124-316 x 7-9 µm.	Formed singly, straight to slightly curved, obclavate, olivaceous brown, 0-8 pseudoseptate, hilum unthickened, 28.5-165 x 12-19 µm.

C. solanii in the dimension of conidiophores. The same in the present species are smaller than those of the earlier described species. The conidiophores have 0-3 proliferation in *C. melongenae* while it is 9 in *C. cassiicola* and only 0-1 in *C. solanii.* Moreover, conidiophores are strictly unbranched in the present collection whereas branched in the two earlier described species. In addition to this, shape of the conidia as well as its size and number of pseudosepta are also very different from earlier described species. Therefore, the proposal of a new species to accommodate the present collection is called for.

Corynespora nana var ***bifurcata*** Sharma & Srivastava var. nov. (Fig.7).

Infection spots amphigenous, irregular, dark brown on upper surface, light brown on lower surface, seminecrotic to necrotic. *Colonies* amphiphyllous, effuse, grayish brown. *Mycelium* of hyphae internal, septate, smooth, thin walled, branched subhyaline to light olivaceous. *Stromata* absent. *Conidiophores* macronematous, mononematous, arising singly from hyphae or in fascicles of 2, straight to curved, simple, cylindrical, smooth, thick walled, 0-5 septate with upto 2 successive cylindrical proliferation, basal cell swollen, light brown, 53-420 x 7-9 µm. *Conidiogenous cells* integrated, terminal, monotretic, scar

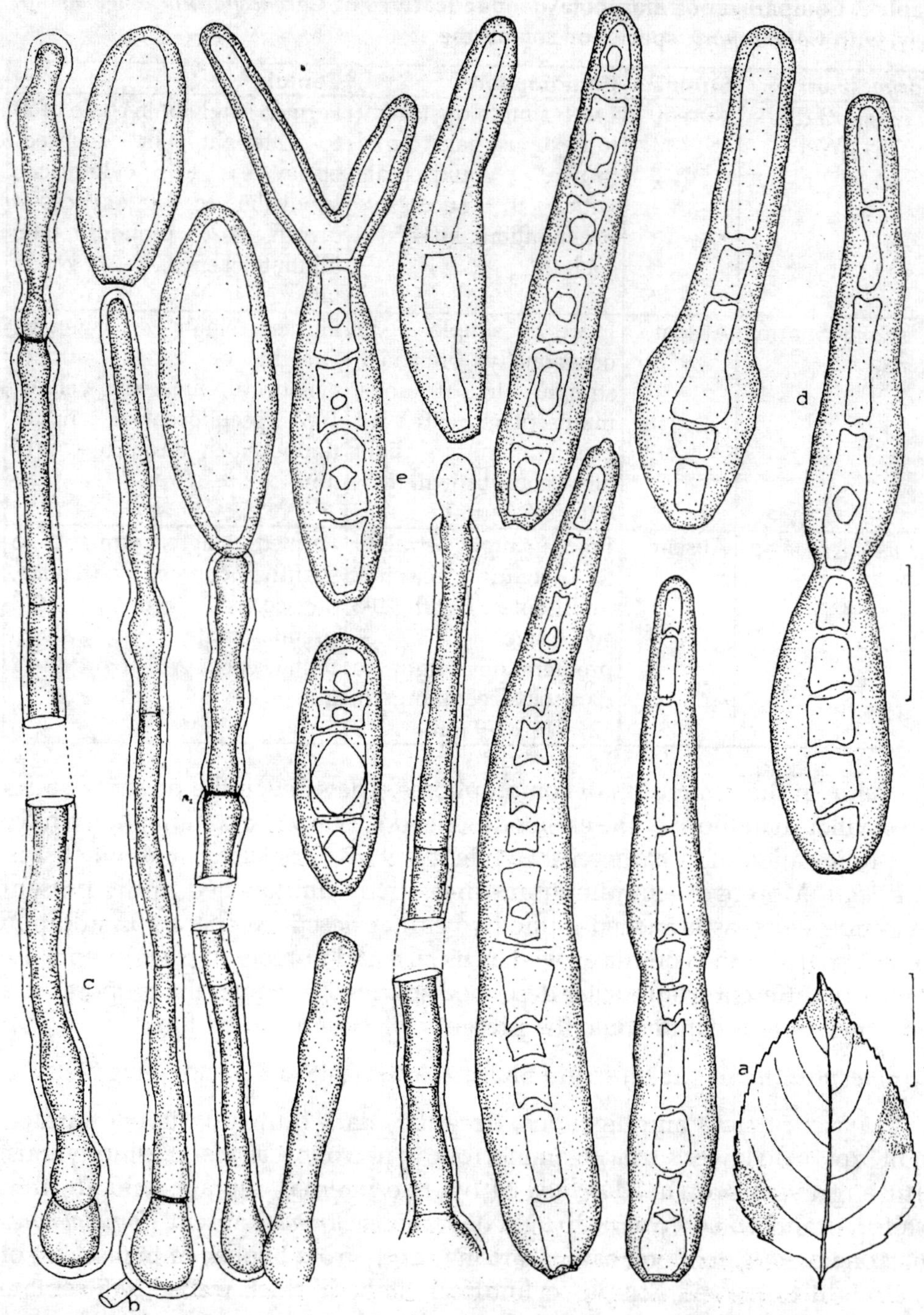

Fig.7: *Corynespora nana* var. *bifurcata* var. nov. (a) Leaf spots, (b) Mycelium, (c) Conidiophores, (d) Conidia; (e) Bifurcated conidium; Bars: a=20mm, b-e=20 µm.

unthickened. *Conidia* arising singly through a pore at the apex of conidiophores which after the first conidium has fallen often proliferates through the apical pore forming a conidium at the apex of proliferation, dry, acrogenous, simple, sometimes bifurcated near the apex, smooth, thin walled, straight to slightly curved, obclavate to ellipsoidal, apex subobtuse to rounded, base truncate, 1-11 pseudoseptate with 0-1 distinct transverse band like pseudosepta, hila unthickened, subhyaline to light olivaceous yellow, 37.5-124 x 12.5-18 µm.

On living leaves of *Lantana camara* Linn. (Verbenaceae); Gorakhpur, India; Nidhi; GPU Herb No. NS-1209 isotypes; HC10 No. 43756 holotype.

This fungus differs from *C. nana* Meenu and Kamal (1998) described earlier on *Lantana indica*. Conidiophores in the present species are larger than *C. nana*. The same are straight or curved in the present collection as against flexuous form in *C. nana*. Conidia in the present species are bifurcated and vary in shape from ellipsoidal to obclavate while these characteristics are not found in *C. nana*. However, due to otherwise similarities between the two fungi, the fungus in question has been described as a new variety, *C. nana* var. *bifurcata* var. nov. (Table-8).

Corynespora oleacearum Sharma and Srivastava sp. nov. (Fig.8)

Maculae amphigenae, circulares vel irregulares, pallide brunneae ad supernae, brunneae ad infernae. *Coloniae* hypophyllae, effusae, griseae. *Mycelium* internum. *Stromata* absentia. *Conidiophora* macronematosa, mononematosa, oriunda singulata ex hyphis, recta vel curvata, simplicia,

Table-8 : Comparison of morphotaxonomic features of *Corynespora nana* var. *bifurcata* var. nov. with *C. nana*

Corynespora sp.	Stromata	Conidiophores	Conidia
C. nana Meenu & Kamal	Absent	Erect, simple, straight to slightly flexuous, multiseptate with 1 proliferation, light olivaceous brown, 200-284 x 6-7 µm.	Formed singly or in chain of 1, smooth, straight to slightly curved, obclavate, 4-14 pseudoseptate, hilum unthickened, subhyaline to light olivaceous brown, 49.5-110 x 9-18.5 µm.
C. nana var. *bifurcata* var. nov.	Absent	Erect, solitary or in fascicle of 2, simple, multiseptate with 0-2 successive proliferations, light brown, 53-420 x 7-9 µm	Formed singly, straight to curved, ellipsoidal to obclavate, bifurcated, 1-11 pseudoseptate, hilum unthickened, subhyaline to light yellow, 37.5-124 x 12.5-18 µm.

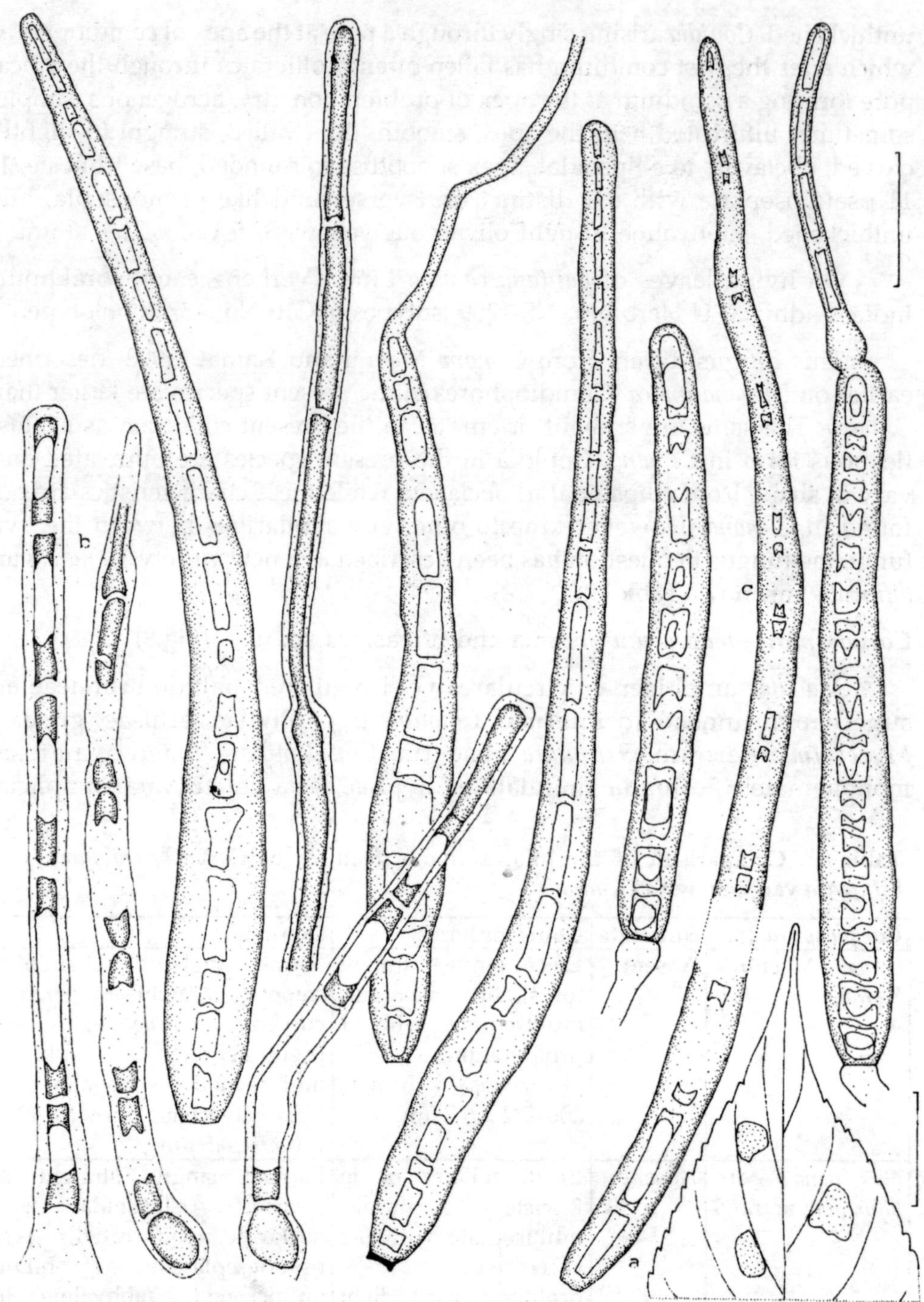

Fig.8: *Corynespora oleacearum* sp. nov. (a) Leaf spots, (b) Conidiophores, (c) Conidia; Baras: a=20mm, b-c 20 µm.

cylindricata, laevia, crassitunicata, 3-6 septata, 0-1 proliferationes cylindricas, cellulae basali inflati, pallide vel atro olivaceo lutea, 96-141 x 6-8 µm. *Cellulae conidiogenae* integrate, terminales, monotreticae, cicatrix non incrassatae. *Conidia* solitaria, primo in apice conidiophori et dein proliferationis cujusque successivas oriunda, sicca, acrogena, simplicia, laevia, tenuitunicata, recta vel curvata, obclavata vel cylindricata, ad apicem subobtusa vel roundata, ad basim truncata, 8-22 pseudoseptata cum usque ad 0-2 fascia similiter distoseptata, hila non incrassata, raro incrassata, pallide olivaceo lutea, 90-272.5 x 9-14.5 µm.

In foliis vivis *Nyctanthes arbor-tristis* Linn. (Oleaceae), Mahrajganj, India; Mar. 1997; Nidhi; GPU Herb No. NS-1220 isotypus; HCIO No. 43766 holotypus.

Infection spots amphigenous, circular to irregular, light brown on upper surface, brown on lower surface. *Colonies* hypophyllous, effuse, grey. *Mycelium* internal. *Stromata* absent. *Conidiophores* macronematous, mononematous, arising singly from hyphae, straight to curved, simple, cylindrical, smooth, thick walled, 3-6 septate with 0-1 cylindrical proliferation, basal cell swollen, light to dark olivaceous yellow, 96-141 x 6-8 µm. *Conidiogenous cells* integrated, terminal, monotretic, scar thickened. *Conidia* arising singly through a pore at the apex of conidiophore which after the first conidium has fallen often proliferates through the apical pore forming a conidium at the apex of proliferation, dry, acrogenous, simple, smooth, thin walled, straight to curved, obclavate to cylindrical, apex subobtuse to rounded, base truncate, sometimes germ tube appears from both the ends, 8-22 pseudoseptate with 0-2 distinct transverse pseudosepta, hila rarely thickened, light olivaceous yellow, 90-272.5 x 9-14.5 µm.

From the survey of literature it is evident that a species of *Corynespora* has not yet been reported on the host genus. But two species viz., *C. ligustri* Guo (1984) and *C. jasminiicola* Meenu *et al.* (1998) have been reported on the host family. Therefore, the present species is compared with these earlier described species (Table -9) so as to ascertain its distinct identity.

It is clear from the data presented in the table that this collection is different from the earlier described species. While the number of proliferation in conidiophores of *C. oleacearum* is only upto one, it is upto 16 in *C. ligustri* and 12 in *C. jasminiicola.* The same are much smaller in size than those in the earlier described species. The size of conidia in the present collection is also very large from that of the pre described species. Moreover, the hila of the conidia are rarely thickened in *C. oleacearum* whereas these are thickened in *C. ligustri* and unthickened in *C. jasminiicola.* Therefore, the present collection is treated as a new taxon of species rank.

Table-9 : Comparison of morphotaxonomic features of *Corynespora oleacearum* sp. nov. with *Corynespora* species on Oleaceae

Corynespora sp.	Stromata	Conidiophores	Conidia
C. ligustri Guo	Absent	Solitary to 2-3 fasciculate, simple, straight to curved, 1-18 septate with 16 successive proliferation, brown, 40-477.5 x 6.3-10 µm.	Formed singly or catenate, straight to curved, obclavate to cylindrical, 4-20 pseudoseptate, hilum thickened, brown, 25-225 x 7.5-30 µm.
C. jasminiicola Meenu & Kamal	Absent	Solitary, straight to curved, simple to branched, 5-12 septate with 1-2 cylindrical proliferations, golden brown to dark golden brown, 177.5-355 x 6-8.5 µm.	Formed singly, smooth, straight to slightly curved, obclavate, 2-18 pseudoseptate, hilum unthickened, light olivaceous yellow, 39.5-176 x 10-21 µm.
C. oleacearum sp. nov.	Absent	Solitary, simple, multiseptate with 0-1 proliferations, light to dark olivaceous yellow, 96-141 x 6-8 µm.	Formed singly, obclavate, 8-22 pseudoseptate, hilum thickened or unthickened, light olivaceous yellow, 90-272.5 x 9-14.5 µm.

Corynespora robusta Sharma & Srivastava sp. nov. (Fig.9)

Maculae amphigenae, irregulares, brunneae ad supernae, pallide brunneae ad infernae. *Coloniae* amphiphyllae, effusae, griseae. *Mycelium* ex hyphis internum, pallide olivaceae. *Stromata* absentia. *Conidiophora* macronematosa, mononematosa, oriunda singulata vel 2 fasciculata, recta vel curvata, simplicia, cylindricata, laevia, crassitunicata, 2-7 septata, 0-7 proliferationes successivas cylindricas, pallide vel atro olivaceo brunnea, 80-407 x 6-10 µm. *Cellulae conidiogenae* integratae, terminales, monotreticae, cicatrix incrassatae. *Conidia* solitaria, primo in apice conidiophori et dein proliferationis cujusque successivas oriunda, sicca, acrogena, simplicia, laevia, tenuitunicata, recta vel curvata, obclavata vel obovata, ad apicem subobtusa vel roundata, ad basim truncata vel roundata, 1-19 pseudoseptata, hila non incrassata, pallide vel atro olivaceo lutea, 16-171.5 x 12-20 µm.

In foliis vivis *Justicia betonica* Linn. (Acanthaceae), Gorakhpur, India; Nov. 1998; Nidhi; GPU Herb No. NS-1221 isotypus; HCIO No. 43767 holotypus.

Infection spots amphigenous, irregular, brown on upper surface, light brown on lower surface. *Colonies* amphiphyllous, effuse, grey. *Mycelium* of

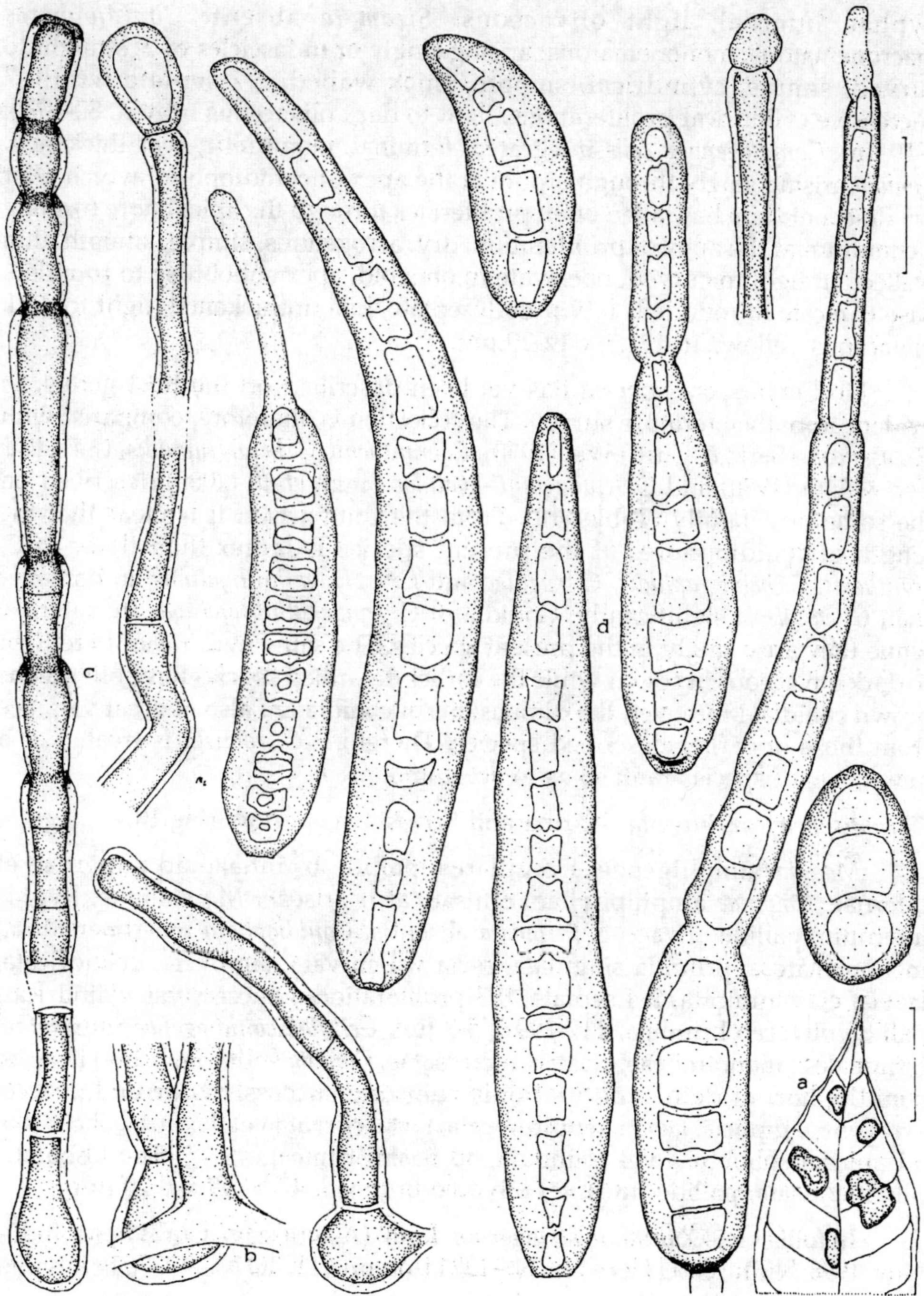

Fig.9: *Corynespora robusta* sp. nov. (a) Leaf spots, (b) Mycelium, (c) Conidiophores, (d) Conidia; Bars: a=20mm, b-d 20 µm.

hyphae internal, light olivaceous. *Stromata* absent. *Conidiophores* macronematous, mononematous, arising singly or in fascicles of 2, straight to curved, simple, cylindrical, smooth, thick walled, 2-7 septate with 0-7 successive cylindrical proliferations, light to dark olivaceous brown, 80-407 x 6-10 µm. *Conidiogenous cells* integrated, terminal, monotretic, scar thickened. *Conidia* arising singly through a pore at the apex of conidiophore which after the first conidium has fallen often proliferates through the apical pore forming a conidium at the apex of proliferation, dry, acrogenous, simple, smooth, thin walled, straight to curved, obclavate to obovoid, apex subobtuse to rounded, base truncate to rounded, 1-19 pseudoseptate, hila unthickened, light to dark olivaceous yellow, 16-171.5 x 12-20 µm.

No *Corynespora* species has yet been described on the host genus, as evident from the literature survey. The collection is, therefore, compared with *C. cassiicola* (Berk. & Curt.) Wei (1950), *C. eranthemi, C. hemigraphidis, C. ruelliae* Yen & Lim (1980) and *C. lepidagathii-hyalinae* Singh *et al.* (2000) described on the same host family (Table -10). From the comparison it is clear that the length of conidiophores of the present species is larger than those of *C. eranthemi, C. hemigraphidis, C. ruelliae* and *C. lepidagathii-hyalinae* and smaller than *C. cassiicola* significantly. Conidia of *C. lepidagathii-hyalinae* are catenate while they arise singly in the present species. The same in *C. robusta* are light to dark olivaceous in colour while the earlier described species have olivaceous brown conidia. Moreover, the dimensions of conidia are also at great variance from those of earlier described species. Therefore, *C. robusta* is treated as a new taxon of species rank so as to accommodate it.

Corynespora xanthiigena Sharma and Srivastava sp. nov. (Fig.10)

Maculae amphigenae, irregulares, pallide brunneae ad supernae et infernae. *Coloniae* amphiphyllae, effusae, atro griseae. *Mycelium* ex hyphis internum, pallide olivaceae. *Stromata* absentia. *Conidiophora* macronematosa, mononematosa, oriunda singulata, recta vel curvata, simplicia, cylindricata, laevia, crassitunicata, 2-4 septata, 0-3 proliferationes successivas cylindricas, pallide olivaceo brunnea, 117-299 x 5-7 µm. *Cellulae conidiogenae* integratae, terminales, monotreticae, cicatrix incrassatae. *Conidia* solitaria, primo in apice conidiophori et dein proliferationis cujusque successivas oriunda, sicca, acrogena, simplicia, laevia, tenuitunicata, recta, curvata vel flexuosa, obclavata, ad apicem subobtusa vel roundata, ad basim truncata, 5-17 pseudoseptata, hila incrassata, pallide lutea vel olivaceo brunnea, 47.5-199 x 8-20 µm.

In foliis vivis *Xanthium strumarium* Linn. (Asteraceae), Gorakhpur, India; Nov. 1998; Nidhi; GPU Herb No. NS-1224 isotypus; HCIO No. 43770 holotypus.

Infection spots amphigenous, irregular, light brown on both the surfaces. *Colonies* amphiphyllous, effuse, dark grey. *Mycelium* of hyphae internal, light olivaceous. *Stromata* absent. *Conidiophores* macronematous, mononematous,

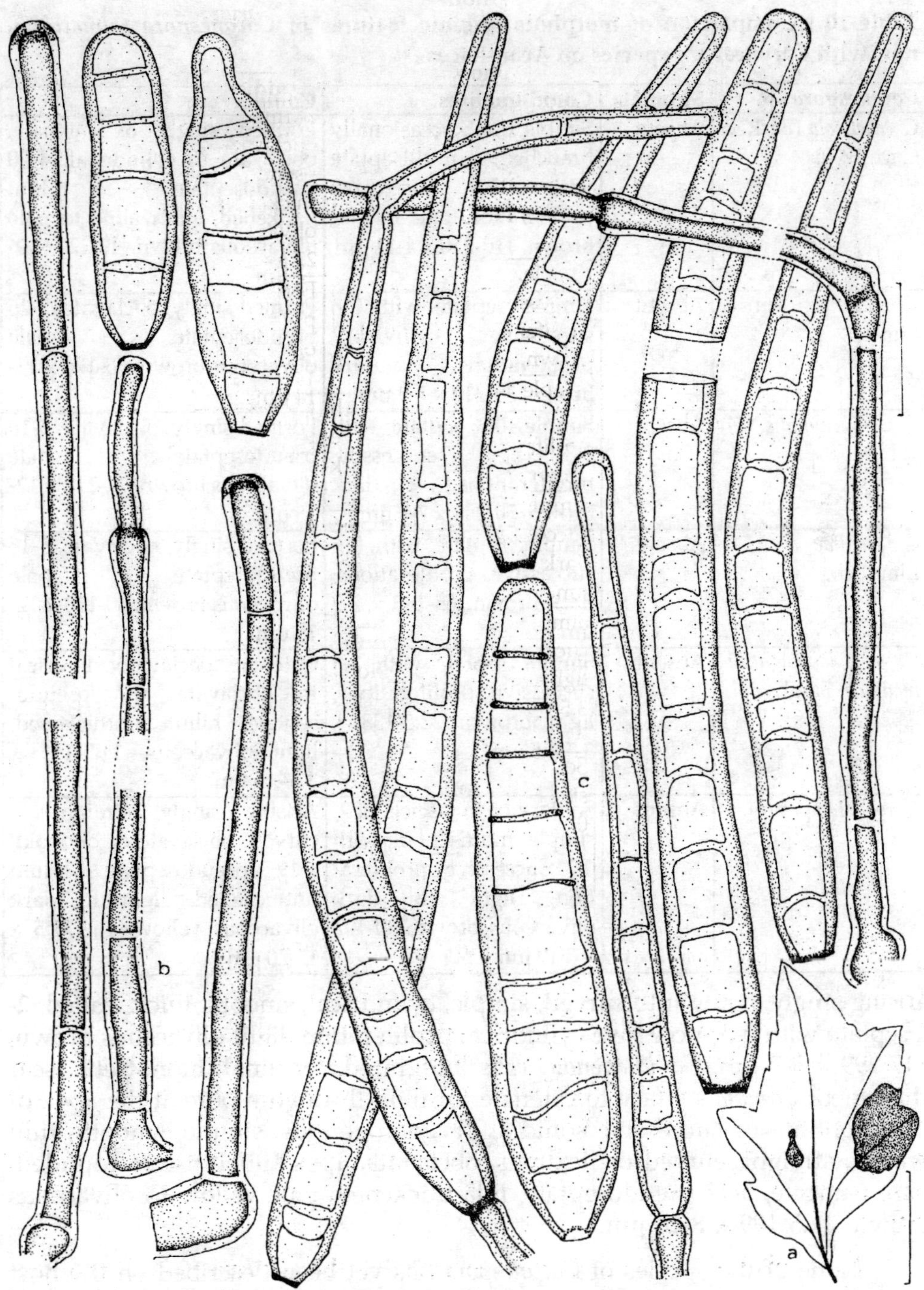

Fig.10: *Corynespora xanthiigena* sp. nov. (a) Leaf spots, (b) Conidiophores, (c) Conidia; Bars: a=20mm, b-c 20 μm.

Table-10 : Comparison of morphotaxonomic features of *Corynespora robusta* sp. nov. with *Corynespora* species on Acanthaceae

Corynespora sp.	Stromata	Conidiophores	Conidia
C. cassiicola (Berk. & Curt.) Wei	Absent	Simple or occasionally branched, multiseptate with 9 successive proliferation, pale to mid brown, 110-850 x 4-11 µm	Formed singly or catenate, obclavate to cylindrical, 4-20 pseudoseptate, hilum thickened, subhyaline to pale olivaceous brown, 40-220 x 9-22 µm.
C. eranthemi Yen & Lim	Absent	Simple, septate with 0-8 successive cylindrical proliferations, dark brown, 35-210 x 7-9 um.	Formed singly, obclavate, 5-25 pseudoseptate, pale olivaceous brown, 65-176 x 11-14 µm.
C. hemigraphidis Yen & Lim	Absent	Simple, 0-3 septate with 0-3 successive proliferations, dark brown, 50-244 x 7-9 µm.	Formed singly, obclavate, 5-16 pseudoseptate, pale olivaceous brown, 72-218 x 12-15 µm.
C. ruelliae Yen & Lim	Absent	Simple, septate with 0-6 successive proliferations, dark brown, 65-145 x 7-9 um.	Formed singly, obclavate, 5-16 pseudoseptate, pale olivaceous brown, 60-150 x 12-15 um.
C. lepidagathii-hyalinae Singh *et al.*	Absent	Simple, septate with 1-3 successive proliferations, light brown, 22.5-133 x 3.5-4.5 µm.	Catenate, obclavatocylindrical to obclavate, 0-7 pseudo-septate, hilum unthickened, light olivaceous, 11.5-151 x 3.5-4.5 um.
C. robusta sp. nov.	Absent	Solitary or in fascicle of 2, simple, multiseptate with 0-7 successive prolifera-tion, light to dark olivaceous brown, 80-407 x 6-10 µm.	Arising singly, straight to curved, obclavate or obovoid, 1-19 pseudoseptate, hilum unthickened, light to dark olivaceous yellow, 16-171.5 x 12-20 µm.

arising singly, straight to curved, simple, cylindrical, smooth, thick walled, 2-4 septate with 0-3 successive cylindrical proliferation, light olivaceous brown, 117-299 x 5-7 µm. *Conidiogenous cells* integrated, terminal, monotretic, scar thickened. *Conidia* solitary to catenate forming through a pore at the apex of conidiophore or successive conidia, dry, acrogenous, simple, smooth, thin walled, straight, curved or flexuous, obclavate, apex subobtuse to rounded, base truncate, 5-17 pseudoseptate, hila thickened, pale yellow to olivaceous brown, 47.5-199 x 8-20 µm.

None of the species of *Corynespora* has yet been described on the host genus but two species viz., *C. matuszakii* Morgan-Jones (1988) and *C. elephantopi* Singh *et al.* (2000) have been reported on the host family. So the present species is compared with these in table-11 to justify its novel identity.

Table-11 : Comparison of morphotaxonomic features of *Corynespora xanthiigena* sp. nov. with *C. matuszakii*

Corynespora sp.	Stromata	Conidiophores	Conidia
C. matuszakii Morgan-Jones.	None or very rudimentary	Arising singly, simple, cylindrical or narrowly clavate, smooth, thick walled with occasionally one or very rarely 2 terminal cylindrical proliferation, brown, upto 23 x 6-8 µm.	Formed singly or catenate, usually straight very rarely slightly curved, cylindrical or sometimes narrowly clavate and gradually attenuated towards the apex, cigar shaped, pale brown to straw coloured or mid brown, 2-10 distoseptate, 56-260 x 10-12.5 µm.
C. elephantopi Singh *et al.*	Absent	Unbranched, cylindrical, 1-6 successive proliferation, light olivaceous to olivaceous brown, 92-378 x 3.5-4.2 µm	Curved, rarely straight, obclavate, sub hyaline to light olivaceous, thin walled, apex sub-obtuse to obtuse, 0-13 pseudoseptate, hila unthickened, 14-96 x 3.5-4.2 µm
C. xanthiigena sp. nov.	Absent	Arising singly, simple, multiseptate with 0-3 successive proliferation, light olivaceous brown, 117-299 x 5-7 µm	Formed singly or catenate, straight to curved, obclavate, 5-17 pseudoseptate, hilum thickened, pale yellow to olivaceous brown, 47.5-199 x 8-20 µm.

The data in the table shows that in *C. matuszakii* stromata may or may not be present while in *C. xanthiigena* it is totally absent. The conidiophores of *C. xanthiigena* are larger than *C. matuszakii* and shorter than those of *C. elephantopi*. In the present collection conidia are smaller than those of *C. matuszakii* and larger than *C. elephantopi* to greater extent. Moreover, the same have an unthickened hila in *C. elephantopi* as against thickened hila in *C. xanthiigena*. Therefore, the proposal of a new speices to accommodate the present collection is called for.

Acknowledgements

The authors are thankful to Prof. Kamal for his guidance and critical suggestions and to the Head, Department of Botany, D. D. U. Gorakhpur University, for providing library and laboratory facilities. Our thanks are also due to the Curator, H. C. I. O., Indian Agricultural Research Institute, New Delhi for depositing the holotype material and providing accession number thereof.

References

Cooke, M.C. (1896). New melon disease. *Gdnrs. Chron.*, Ser. (3) **20**: pp 271-272.

Deighton, F. C. (1990). Observations on *Phaeoisariopsis*. Mycol. Res. **94**: 1096 - 1102.

Ellis, M.B. (1957). Some species of *Corynespora*. Mycological Papers, **65**: 1-15.

Ellis, M.B. (1960). Dematiaceous Hyphomycetes I. Mycological Papers, **76**: 1-36.

Ellis, M.B. (1961). Dematiaceous Hyphomycetes III. Mycological Papers, **82**: 1-55.

Ellis, M.B. (1971). Dematiaceous Hyphomycetes : 372-376.

Ellis, M.B. (1976). More Dematiaceous Hyphomycetes : 376-377.

Guo, Y.L. (1984). Four new species of genus *Corynespora*. Acta. Mycol. Sinica, **3**: 161-169

Güssow, H.T. (1906). Uber eine neue Krankheit an Gurken in England (*Corynespora mazei*, Güssow gen. et sp. nov.) Z. Pflkrankh, **16:** 10-13.

Kawamura, E. (1931). On the causal fungus of ringspot of cowpea (*Cercospora vignicola* sp. nov.) Fungi I, 14-20.

Lindau, G. (1910). Robenhorsts Kryptogamen - Flora, 2 Aufl., 1 (Pilze), I, pp. 805-807.

Liu, S.T. (1948). Seed borne diseases of Soybeans. Bot. Bull. Acad. Sinica. **2,** pp. 69-80.

Meenu and Kamal (1998a). New species of *Corynespora*. Mycol. Res. **102** : 344-346.

Meenu, Kharwar, R.N. and Bhartiya, H.D. (1998b). Some new forms of genus *Corynespora* from Kathmandu valley of Nepal. Indian Phytopath. **51** : 146-151.

Meenu, Singh, A. and Singh, S.K. (1997). Some new species of *Corynespora*. Indian Phytopath. **50:** 17-24.

Morgan-Jones, G. (1988). Notes on Hyphomycetes LX. *Corynespora matuszakii*, an undescribed species with narrow, cylindrical, catenate conidia and highly reduced conidil cell lumina. Mycotaxon. **33** : 483-487.

Munjal, R.L. and Gill, H. S. (1962). Some Dematiaceous fungi from India - I. Indian Phytopath. **15:** 269-279.

Rao, P.N. (1961). Two new species of Fungi Imperfecti on *Ficus hispida* L. Science and Culture, **27:** 414-415.

Saikia, V.N. and Sarbhoy, A.K. (1980). Hyphomycetes of North East India-II. *Indian Phytopath.* **33:** 466-470.

Sharma, N., Chaudhary, S. and Kamal (2002). Three new species of genus *Corynespora*. Indian Phytopath. **55**(2) (Accepted).

Singh, A. and S.K. and Kamal (2000a). Three new species of *Corynespora* from India. J. Mycol. Plant Pathol. **30**: 44-49.

Singh, A., Bhalla, K., Dubey, R. and Singh, S.K. (2000b). Additions to *Corynespora* from India.l J. Indian Bot. Soc. **79:** 185-190.

Solheim, W.G., Stevens, F.L. and Pierce, A.S. (1933). Fungi from Bombay. Indian Journal of Agricultural Sciences. **3:** 912-916.

Sutton, B.C. and Pascoe, I.G. (1988). Some Dematiaceous Hyphomycetes from Branches and Phyllodes of *Acacia* in Australia. Australian Systematic Botany. **1** : 127-138.

Swart, H.J. (1985). Australian leaf inhabiting fungi XVIII. *Corynespora acaciae* sp. nov. Trans. Br. Mycol. Soc. **84**: 175-177.

Tai, F.L. (1936). Notes on Chinese fungi VII. Bull. Chin. Bot. Soc., **2:** 45-66.

Teng, S.C. (1939). A contribution to our knowledge of the higher fungi of China. Academia Sinica, 614 pp.

Wei, C.T. (1950). Notes on *Corynespora*, Mycol. Pap. **34** : 1-9.

Yen, J.M. and Lim, G. (1980). Etude Sur Les Champignons Parasites Du Sud Est Asiatique XXXIX-Les *Corynespora* de Malaisie, Cryptog. Mycol. **1:** 83-90.

33

Variation in Sugarcane Red Rot Pathogen *Colletotrichum falcatum* Went

R. Viswanathan*, P. Malathi, and P. Padmanaban

Plant Pathology Section, Sugarcane Breeding Institute, ICAR, Coimbatore – 641007.

Red rot of sugarcane is one of the oldest recorded crop diseases in India. Severe epidemics occurred in different parts of the country during the last century led to elimination of several elite sugarcane varieties. Origin of new variants of the pathogen was responsible for the breakdown of the released resistant varieties. Loss of many sugarcane varieties with high yield and quality in the past focused the need to study the pathogenic variation in the pathogen. The fungal pathogen exhibited greater variation in cultural characters, conidial morphology and germination and appressorial characters. Based on cultural characters the isolates were categorized as light and dark isolates in the past. Differential host interaction studies showed a significant variation in pathogenic behavior and virulence. Detailed studies conducted in the institute showed greater cultural and serological variation among the strains of the fungus. *C. falcatum* pathotypes from tropical region exhibited higher virulence than those from subtropical region in the country. Recently molecular tools are being attempted to characterize the fungal variation.

Key-words : Sugarcane, *Colletotrichum falcatum*, cultural, pathogenic and serological variation

Sugarcane is an important commercial crop grown over an area of about 4.2 m ha in India. It is being cultivated in almost all the states except those in mountain ranges. Annual production of sugarcane in the country is about 300 MT. About 500 sugar mills utilize nearly 50% of sugarcane produced in the country, producing 18 MT of white sugar. About 40% of cane is being utilized by gur (jaggery) and khandsari making units. Sugarcane has contributed in developing Indian economy by way of sourcing Rs. 15,000 crores every year to the exchequer as cess and taxes. Further, location of sugar industries in rural areas resulted in rural development and employment generation to a significant extent. Stable pricing of sugarcane also helped the farmers to realize the farm income every year.

The seed piece transmissible fungal disease red rot, is one of the oldest known diseases of sugarcane and one which has had a far-reaching influence

*Corresponding Author E-mail: rasaviswanathan@yahoo.co.in

on the economics of sugar production in several countries. The disease is caused by the fungus *Colletotrichum falcatum* Went (Prefect state: *Glomerella tucumanesis* (Speg.) Arx and Muller). It was first reported and described from Java in 1893 and within a decade it was reported in several other countries; India (1901), West Indies (1904), Hawaii (1908), USA (1910), Australia (1925), etc. It was the cause of decline and eventual failure of the "Noble" canes in cultivation at that time in Louisiana in 1926, bringing the industry almost to the brink to ruin. The disease was first observed in India by Barber (1901) in the cultivar Red Mauritius in the Godavari delta of Andhra Pradesh. Butler (1906) the Imperial Mycologist critically studied this devastating disease in relation to the causal organism and its portals of entry into the cane stalk. He coined the name 'red rot' for the disease and mentioned that "in India, red rot in many places is the greatest obstacle to successful cane cultivation. In Madras, Bombay and Bihar the area under thick cane has, in certain districts, periodically shrunk as a result of an accumulation of this disease, to expand again only when the diseased cane has been replaced from outside" (Butler, 1918). The havoc of red rot epidemic in Gangetic plains during 1938-41 resulted in the widespread failure of Co 213, the chief commercial variety of the tract. Several sugar factories in the badly affected areas could crush only one-third of their normal 'crush' of cane owing to poor supplies. Sudden and complete failure of Co 213, the cane variety which had faithfully served the cane sugar industry in India for nearly one decade, over such a wide area in a tract where it had flourished for several years led to detailed studies on pathogenic variation (Chona, 1980). The sub-continent faced many such red rot epidemics in the past resulting in elimination of many popular sugarcane varieties from cultivation (Viswanathan and Samiyappan, 1999; 2000). These epidemics happened when the ruling varieties previously rated as resistant succumb to the disease owing to the development of new variants.

Background

Edgerton and Moreland (1920) first observed pathogenic variability in *C. falcatum.* The sudden failure of certain sugarcane varieties in Louisiana, notably POJ 213 due to red rot was attributed to the appearance of certain new virulent strains of the pathogen (Abbott, 1933). From his extensive stuides, Abbott (1938) concluded that the isolates obtained from red rot affected canes fell into two types, those which in culture has light coloured aerial mycelium and were relatively highly sporulating and those which had a dark-coloured mycelium and were poor sporulators. The large-scale epidemics in North Indian sugarcane belt were attributed to the appearance of new light strains with abundant sporulation and highly pathogenic to particular variety. Chona and Padwick (1942) found that the sudden failure of the predominant variety

Co 213 in India in 1936-39 was associated with the occurrence of light-coloured, highly sporulating strains on this variety. The successive failures of varieties like BO 11, Co 331, Co 453 and CoS 443 were attributed to the appearance of new strains (Rafay, 1950; Rafay and Singh, 1957). Sugarcane pathologists accepted the existence of pathogenic variation in *C. falcatum* in 1950s. In the absence of any race nomenclature system in *C. falcatum*, different workers have designated isolates/pathotypes in different manner. The strains were classified as A, B, C, D, E, F, G, H and I by Rafay *et al.* (1957). Kirtikar and his associates and other workers in Uttar Pradesh designated strains/isolates as R 103, R 115, R 117, R 121, R 123, R 135, R 141, R 146, R 158, R 183, R 185, R 191, R 200 etc. (Kirtikar *et al.*, 1972, Singh *et al.*, 1984; Pandey and Sakal, 1974; Gupta *et al.*, 1980). Recently sugarcane pathologists in India proposed nomenclature of CF01, CF02, CF03 etc to denote *C. falcatum* strains.

Cultural variation

Abbott (1935) described for the first time cultural variation in *C. falcatum.* He has grouped large number of isolates from USA, representing various sugarcane belts into two groups readily distinguished by the colour and texture of the mycelial growth. The colony turf of one group was dark grey in colour and had a velvety surface, while that of the other varied from almost white to light ashy grey in colour and was cottony or subfloccose in texture. For convenience in designation the two groups were referred to as 'dark' and 'light' types, respectively. The dark types fruit sparingly, acervuli generally being absent or limited to the periphery of the colony, although there are exceptions. The light types fruit abundantly, the surface of the colony often becoming pink and slimy from the abundant production of conidia. As the cultures age, whitish patches of mycelium usually appeared over the turf of the dark type colonies, a character seldoms shown by the light types. He described that physiologic specialization in *C. falcatum,* although proof is lacking that physiologic forms of the fungus exist in the same sense as have been described for the cereal rust fungi. He found that the differences in the appearance of the colony, do not appear to be correlated with other morphologic characters. In his observation, in making measures of conidia he established that significant differences in spore size occur, however, both dark and light types fall in both the large and the small spore groups. Rafay and Padmanaban (1941) identified three distinct types in culture of *C. falcatum* obtained in isolations from North Bihar in 1939-40 and characterized them as white to pale grey to cottony fluccose colonies without slimy pink masses of conidia.

Srinivasan (1962) commented after his extensive studies on cultural variation in the fungus as the range of variation in the conidial state of the fungus appears to be considerable. In the cultures at one end are the rare isolates, which are moniliaceous in morphology, producing their conidia indefinitely from vegetative hyphae and at the other extreme are those that produce strong tuberculoid stromata subtending the acervuli. In between are those, which are most common and develop the typical melanconiaceous acervuli, which may or may not bear setae. Another observation he made was all the wild cultures possessed light coloured mycelium and relatively good sporulation and virulence on first isolation and none of the isolates belonged to dark type. However, many of the isolates soon changed into non-sporulating avirulent dark types while others retained their original character as light isolates. This happened irrespective of whether the isolates were carried through successive cultural transfers or passed through host. He concluded existence of two distinct types of isolates viz. those that are relatively stable in their morphological and pathogenic characters and those that are unstable and are soon replaced by the dark type. However the stable isolates are relatively so since it occasionally gives ' dark type' variant sectors in the plates. Dark type variants were more frequent in the older parts of the lesions and relatively less so in the advancing margin of the lesion. Loss of sporulation in dark-type variant sectors in the plates was observed. Dark type variants were more frequent in the older parts of the lesions and relatively less so in the advancing margin of the lesion. Loss of sporulation in dark-type variants was irreversible even when they were passed through the host. He reported that the light-coloured, highly virulent ones generally prevalent in the red rot epiphytotic areas of North India such as eastern UP, Northern Bihar and pockets of Punjab, and the dark- coloured, poorly sporulating and relatively avirulent strains generally prevalent more particularly in areas in South India where red rot is not serious. He further classified four different morphological types based on the spore-bearing structures, and occurrence of setae in the acervuli. He described the most frequent type as aerial mycelium floccose, white in young cultures, sporulation in one two weeks; acervuli appearing black at first, surrounded by setae, later developing salmon orange, slimy, conidial masses, conidia falcate, 14.0-32.0 x 4.6-9.0 µm.

Chona and Hingorani (1950) studied the cultural conditions under which dark types arise from light types and have shown that this was an irreversible variation. Chattopadhyay and Sarkar (1960) studied *C. falcatum* isolates in West Bengal and divided them into 6 distinct morphological groups. Chona and Srivastava (1960) reported that isolates of *C. falcatum* were usually unstable in culture and there was no correlation between growth rate and morphology, physiological behaviour and virulence. Sharma (1970) reported cultural character of 3 isolates based on the colour of the mycelium as light and dark.

Pandey and Sakal (1974) observed great variability and apparent ability of red rot flora to produce new and possibly more virulent strain from time to time. Satyanaranaya and Rao (1995) reported that all the three major pathotypes viz. Cf419, Cf997 and Cf671 have completed by 11days after inoculation in 100 mm Petri plates. They found no appreciable difference either in cultural characteristics or in size of conidia of the three pathotypes. Jothi (1989) on the basis of cultural characters viz., colour, texture and sporulation, divided 30 isolates into 5 groups. The isolates showed tremendous variation in their radial growth on oatmeal agar, growth in liquid medium and sporulation. All the isolates were able to produce acervuli in culture and the diameter ranged from 0.639 to 1.54 mm, setae were found in all the isolates and in some cases it was rare. Number of setae ranged from 3 to 20 per acervulus and length ranged from 90 μm to 220 μm. Conidiophore length ranged from 120 μm to 330 μm. Subtle differences in their favourable pH requirement of the isolates were noticed. Similarly variation in sporulation by the isolates at differenct temperature regimes was found. There were also variations in the response of the isolates towards different fungicides.

Cultural characteristics of nine major *C. falcatum* pathotypes in the country were taken up recently. The results revealed a clear variation in mycelial growth among the strains. Similar results were obtained for mycelial dry matter production, where Cf767 and Cf7717 recording the highest mycelial dry weight and Cf64 recording the least. Regarding other characteristics Cf1148 and Cf 7717 from subtropics were producing thicker mycelium and they were darker as compared to other pathotypes (Fig. 1). The observations indicate that these pathotypes could be regarded as dark isolates as described by the previous workers. Among the pathotypes, Cf671 was fast in initiating acervuli production and this pathotype recorded higher sporulation along

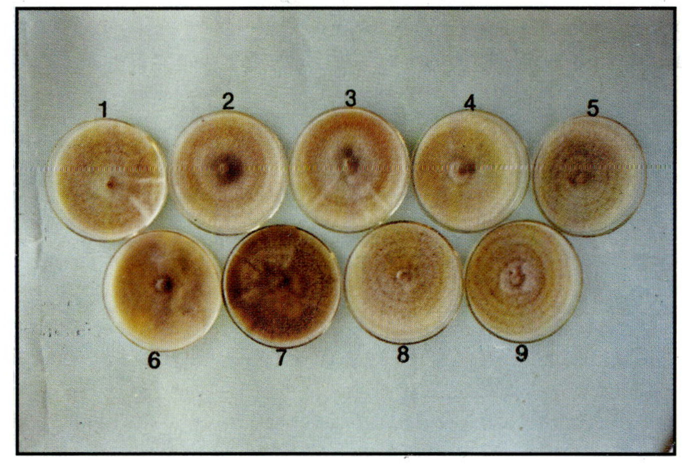

Fig. 1. Cultural variation in *C. falcatum* pathotypes (Seven days after planting on oatmeal agar)
1. Cf 671 , 2. Cf 90063, 3. Cf 9201 (Tamil Nadu)
4. Cf 419, 5. Cf 997 (Andhra Pradesh)
6. Cf 1148, 7. Cf 7717, 8. Cf 64, 9. Cf 767 (Subtropical India)

with Cf92061. Although similar conditions were provided conidial germination varied from 20 to 100 per cent among the pathotypes. The reasons for such variations in germination are not known. Most of the pathotypes recorded uni and biploar germination and few pathotypes showed sub-polar and horizontal germ tubes additionally. Most of the pathotypes produced globose shaped appressoria with variation in lobes (Table 1).

Table 1. Cultural variation in *Colletotrichum falcatum*

Cf Pathotypes	Mycelial characters				Acervuli initiation (day)	Sporulation (x 10^5/9mm disc)	Conidial germination		Appressorium	
	Growth		Texture	Colour			Per cent	Type	Presence (+/-)	Shape
	Diameter (mm)	Dry weight (mg)								
Cf671	8.25	90	More fluffy, thin	Whitish grey (L)	Fourth	+++	51	Uni, bi, sub-polar & horizontal	+	Globose, slightly lobed
Cf90063	6.50	85	More fluffy, thin	Greyish white (L)	Fifth	++	55	Uni, bi, sub-polar & horizontal	+	Globose, bilobed
Cf92061	6.25	100	More fluffy, thin	Greyish white (L)	Sixth	+++	46	Uni and bi-polar	+	Globose, bilobed
Cf419	7.25	90	Fluffy, thin	Whitish grey (L)	Sixth	+	27	Uni and bi-polar	+	Globose
Cf997	6.25	95	Fluffy, thin	Greyish white (L)	Seventh	++	100	Uni, bi- and sub-polar	+	Globose, lobed
Cf1148	4.75	100	Less fluffy, thick	Grey (D)	Sixth	+	82	Uni and bi-polar	+	Trilobed
Cf7717	8.50	110	Less fluffy, thick	Dark grey (D)	Sixth	++	97	Uni and bi-polar	+	Globose
Cf64	6.75	75	More fluffy, thin	Whitish grey (L)	Fifth	++	20	Uni and bi-polar	+/-	Globose, lobed
Cf767	7.25	110	More fluffy, thin	Greyish white (L)	Fifth	++	81	Uni, bi- and sub-polar	+	Labed

Mycelial growth in diameter, colour and texture were recorded on fifth day after inoculation on oat meal agar medium Mycelial dry weight was recorded on oatmeal broth 20 days after inoculation

Sporulation was recorded 8 days after inoculation

Spore germination and appressorial characters were recorded 15 days after inoculation

Sporulation

\+ - 2.08 to 16.67 x 10^5/ 9 mm disc

++ - 16.67 to 29.17 x 10^5/ 9 mm disc

+++ - 29.17 to 41.67 x 10^5/9 mm disc

D : Dark type L : Light type

Stroma formation

Chona *et al.* (1964) reported a stroma forming strain of *C. falcatum* from infected sugarcane cultivars from Karnal. Initially the cultures had dark grey cottony mycelium with scanty sporulation. The black stromatic bodies started appearing after 4-5 days of sub-culturing and became quite prominent in 10 days old cultures. The stromatic structures were hard and generally found on the edges of the slant and near the top portion where the depth of the mycelium was very shallow. They consist of fertile hyphae, some of which perform true function as conidiophores and bear typical falcate conidia of *C. falcatum* at their tips. A large majority, however, became dark brown to blackish in colour due to the accumulation of protoplasmic contents and cover the fertile portion in an irregular manner, transforming the whole mass into a big stroma measuring 4-7 mm in diameter. Sometimes a few black pimple-like extension of the stroma (eruptions) grows out and tends to release spores. Later a pink ooze of typical *C. falcatum* spores comes out of the stroma and gradually increase in size. These stromatic bodies withstood a dry heat upto 85^{0}C for one hour and moist heat upto 60^{0}C for 10 min without loss of viability, whereas, normal thermal death point of the spores and the mycelium of the most of the strains has been found to vary from 51 to 54^{0}C for 5 min and such structures may play an important role in the perpetuation of red rot inoculum.

Teleomorph

Chona and Srivastava (1952) reported the occurrence of *Physalospora tucumanensis* Speg. perfect stage of red rot pathogen in culture. They reported that the perithecial stage was responsible for sudden and rapid development of repeated epiphytotics. However, occurrence of teleomorph under field conditions was not recorded so far.

Conidial characteristics

Abbott (1938) reported that length of conidia ranged from 16 to 40 µm while Went (1893) reported 25 µm. Chona and Srivastava (1960) studied 32 isolates and reported that the length of the conidia being a minimum of 10 µm and with maximum of 36 µm. Various reports are available on the variation of conidial morphology and dimension. Two types of conidia, one with 28.9 x 3.1 µm and the other with 30.1 x 5.6 µm (Sharma, 1970), conidial size of 17.1 to 25.8 x 4.5 to 6.8 µm (Pandey and Sakal, 1974), a new biotype with size ranging from 33.0 to 37.4 µm x 4.4 to 4.95 µm (Gupta *et al.*, 1980), varying length of 23 to 30 µm (Khirbat *et al.*, 1980), sickle shaped or falcate conidia measuring 16-40 x 5-7 µm in size with oil globule in the middle (Agnihotri,

1983) and 15-42.6 x 3.7-8.6 µm (Jothi, 1989) were reported. Khirbat *et al.* (1986) reported that among the five isolates one isolate RH3 was distinctly different from the other four isolates by its growth rate and longer conidial length.

The *C. falcatum* isolates showed marked variation in conidial germination. On germination some isolates produced short, long and branched promycelia in which lengh ranged from 21-230 µm. Certain isolates did not produce promycelia and appressoria attach on the conidial wall itself. Light type isolates were more virulent than dark types. Conidial germination was a maximum of 90.3 per cent in a light isolate and minimum of 22.6 per cent in a dark isolate. Germination in light types ranged from 22.6 to 63.3 per cent whereas, it was from 35.0 to 90.3 in light isolates. However, Chona and Srivastava (1960) results were contrast to the finding. They identified unipolar, bipolar, horizontal or from the middle region germination. Jothi (1989) found four types of germination with the 30 isolates tested. Variation in promycelium length was also observed. She found appressoria were lobed, globose, triangular or irregular. Singh and Singh (1989) also reported similar findings.

Differential host interaction

Earlier workers gave emphasis on cultural and morphological variability rather than pathogenic variability based on differential reaction on common set of differential varieties. Later efforts were made to identify pathotypes based on differential varieties (Satyanarayana and Satyanarayana, 1976; Alexander *et al.*, 1979; Khirbat *et al.*, 1980; Beniwal *et al.*,1989; Padmanaban *et al.*, 1996). Alexander *et al.* (1979) classified *C. falcatum* isolates of India into five major pathotypes (Table 2)

Two isolates of red rot fungus collected from sugarcane varieties Co 419 and Co 997 from coastal Andhra Pradesh did not differ in respect of their

Table 2. Major pathotypes of *C. falcatum* in India

Sl. No.	Major Pathotypes	Endemic zone	Resistant varieties
1	MP 1	Tamil Nadu	Co 6806, Co 62198
2	MP 2	Coastal Andhra Pradesh	Co 975, Co 6907, Co 8014, CoA 7601
3	MP 3	Eastern UP, North Bihar	BO 90, BO 91, BO 99, BO 102
4	MP 4	Western UP, Haryana, Punjab	Co 6914, Co 7314, CoS 767
5	MP 5	Kerala	Co 6906, Co 7704

Source: Alexander *et al.* (1979)

morphological characters but differed in their pathogenicity on differential hosts viz., Co 419, Co 6907 and Co 62175 indicating the existence of two pathotypes in the light race of the pathogen (Satyanarayana and Sathyanarayana, 1976; Sathyanarayana and Rao 1984). Further studies of Satyanarayana and Rao (1995) with an additional strain from CoC 671 indicated the existence of three distinct pathotypes in Andhra Pradesh viz. Cf419, Cf671 and Cf997. They have further confirmed the existence of these three pathotypes by inoculating mixture of the three pathotypes.

Detailed pathogenic variability studies on red rot pathogen in Haryana were carried out by Khirbat *et al.* (1980) and Beniwal *et al.* (1989). Khirbat *et al.* (1980) reported existence of three pathotypes (RH-1, RH-2, RH-3) on the basis of differential reaction on three differential varieties Co 312, Co 62399 and CoL 9. Beniwal *et al.* (1989) also reported existence of three distinct pathotypes, namely pathotype 1, 2 and 3 based on differential reaction on varieties Co 7717 and Co 1148 by plug and nodal methods. Satyavir *et al.* (1995) identified two more new pathotypes from Haryana. Singh (2000) reported three new races Cf07 (CoJ 64 isolates from Narayangarh, Yamunangar, Shahabad), Cf08 (isolate from CoJ 64, Dhuri, Punjab) and Cf09 (isolates from CoS 767) in addition to the three established races Cf01, Cf02 and Cf03 from sub-tropical region. The major limitation of pathogenic variability studies is non-availability of a common set of differential varieties/species for the identification of pathotypes. At SBI, Padmanaban *et al.* (1996) conducted differential host interaction studies involving 50 sugarcane clones from *Saccharum* spp. and hybrids and isolates collected from different endemic locations in the country at three locations. Their studies indicated the presence of two pathotypes, which could be considered as distinct races of *C. falcatum* in Northwest zone comprising Western UP, Haryana and Punjab. Among the pathotypes of East Coast zone in Tamil nadu Cf671 and Cf8001 appeared to be similar, while Cf86062 seemed to be different. The isolates from Andhra Pradesh were grouped into three major pathotypes of Cf419, Cf997 and Cf671 as reported by Satyanarayana and Rao (1995). The study tentatively identified five possible races of *C. falcatum* namely Cf1148 and Cf7717 in the subtropics and Cf419, Cf997 and Cf671 in the tropics based on differential interaction. Thirteen host clones for subtropical India and 15 host clones for tropical region were identified as differential hosts to identify the pathotypes of respective regions. Recently Prakasam *et al.* (2000) reported identification of five races of the pathogen based on differential host studies viz. Cf1148, Cf7717, Cf671, Cf90063 and Cf92061. They have identified 11 host differentials viz. Co 7717, Co 1148, Co 6914, Co 8346, Co 8336, CoS 767, 80R41, CoC 671, CoC 90063, CoC 92061 and N2-5101 to differentiate these races.

In the all India Coordinated Research Programme on Sugarcane, differential host interaction is being studied with 13 sugarcane differentials viz., Baragua (*S. officinarum*), Khakai (*S. sinense*), SES 594 (*S. spontaneum*), Co 419, Co 975, Co 997, Co 1148, Co7717, Co 62399, CoS 767, CoJ 64, CoC 671 and BO 91 at eight loctions in subtropical region and four locations in tropical region. Ten isolates from Punjab were grouped into 6 pathotypes. The isolates from CoJ 64, Co 1148, CoJ 79 and CoJ 82 were separated upto 4 pathotypes, and the isolates from CoJ 84, CoJ 85, CoJ 86, CoJ 87 and CoPant 84211 constituted two pathotypes. Results of the study from Uchani, Haryana indicated that new isolates from CoJ 64 were distinctly different from the pathotype Cf03 and designated as Cf07. The new isolates from CoS 767 were tentatively grouped as pathotype Cf09. Results of studies at Karnal also revealed similar results. However, isolate from CoJ 64 (Dhuri) was found distinctly different in the pathogenic behaviour constituting a separate pathotype tentatively designated as Cf08. In the other regions in the sub-tropics, four pathotypes in Lucknow, three in Shahjahanpur, Seorahi and Pusa were identified. At Shahjahanpur the new isolates were categorized into three distinct pathotypes i) Isolate from CoS 767, ii) Isolates from CoS 90265, CoS 770 and CoPant 84211; iii) isolates from CoJ 64 and BO 22. At Pusa the isolates were grouped into three distinct pathotypes viz. i) Isolates from BO 70 and CoS 767, ii) Isolate from CoS 687 iii) Isolates from Co 1148 and CoLk 8001 (Anon, 2000). In the tropics, a new pathotype from 85A561 designated as Cf10 in addition to the existing Cf04, Cf05 and Cf06. Results from Cuddalore showed most unreliable results and they could not group the isolates into distinct pathotypes based on differential interaction. Pathogen isolates collected from Thiruvalla, Kerala did not show much pathogenic variability.

Virulence

Chona (1956) reported that a light coloured, highly sporulating type of red rot fungus predominated in the isolation made from red rot affected canes from endemic areas of UP and Bihar, whereas all the previous isolates prior to 1938 are of a dark type with sparse sporulation. The new, light type isolate is much more virulent than old dark type. By the virtue of its profuse production of spores it is naturally well suited for quick and widespread dissemination. In this change of red rot flora in the epidemic areas resulting in the predominance of a highly sporing, highly virulent strain, combined with secondary infection was found an explanation for the failure important commercial cane varieties. A great deal of difference in the virulence of various isolates was observed but there do not appear to exist any highly specialized physiologic forms like those of the wheat rust fungus, with definite specificity between an isolate and a particular cane variety or set of varieties. Most important feature of the red rot epidemic, as the development and predominance of a new more virulent strain may result in the failure of any commercial variety and may thus upset the varietal set up of a tract which takes quite a few years to build up.

Collection of isolates over several years by various workers indicated appearance of several new strains with marked physiologic differences as judged by their virulence on different cultivars (Kirtikar *et al.*, 1964; Pandey and Sakal, 1974; Anon, 1990). Kirtikar *et al.* (1964) studied 3 isolates for its comparative virulence on 25 varieties and found that in nature some intermediate types in between well-known dark and light coloured strains had arisen which showed greater virulence. Alexander and Rao (1972) conducted a study with the same set of genotypes using the same isolates by inoculating simultaneously at Coimbatore and Karnal and they attributed the diferences to environmental factors. Kar *et al.* (1965) concluded that the virulence is not necessarily associated with morphology. Fontenot *et al.* (1973) studied the relative pathogenicity of isolates of *C. falcatum* from sugarcane seed pieces from 15 locations and reported that they varied in pathogenicity to 3 sugarcane varieties in Brazil. Agnihotri and Budhraja (1975) explained the differences between leaf and midrib isolates and the isolates collected from sugarcane stalk tissues. Differences in virulence in different pathotypes of red rot fungus from varieties CoLK 8001, Co 8529 and Co 1148 was reported from Shahjahanpur, Pusa and Seorahi, respectively (Anon, 1990) .

Jothi (1989) studied virulence pattern of the 6 isolates on 16 sugarcane varieties whose reaction to *C. falcatum* is known. The results confirmed that the virulence of an isolate is not necesarily connected with the morphology of the isolate. The varieties behaved differently to the isolates (Table 3).

Table 3 Comparative virulence of *C. falcatum* pathotypes

Sugarcane Variety	Isolates					
	Cf997	Cf83013	Cf1148	Cf997	Cf658	Cf513
BO 91	-	*	*	-	-	-
Co 1148	-	*	**	-	**	*
Co 6914	*	**	***	**	-	***
Co 7314	-	-	**	-	*	*
Co 7704	*	-	**	*	*	*
Co 8314	*	-	**	-	*	***
CoS 767	-	***	***	-	*	***

-No white spot;* - Less white spots;**- More white spots;*** -Abundant white spots
Source: Jothi, (1989)

Comparison of tropical and sub-tropical pathotypes for their virulence on a set of sugarcane clones/cultivars indicated that pathogen isolated orginated from tropical India were more virulent than the sub-tropical ones. The pathotype isolated from CoC 671 was found to be the most virulent among all pathotypes tested and that from Co 1148 was the least virulent one (Padmanaban *et al.,* 1996). Further, pathotypes isolated from CoC 671, CoC 8001, CoC 85061, CoC 86062, CoC 91061 and CoC 92061 which have succumbled to red rot in the severe epidemics in Tamil Nadu state in the last decade were identical in their virulence. Studies of Viswanathan *et al.*(1997) also found a gradual reduction in red rot resistance year after year from 1985 to 1995 in the progenies developed in pre-zonal varietal trial at the Institute. Higher susceptibility of most of the clones tested in the later years was due to use of more aggressive pathotypes such as Cf671 and Cf92061. Though Cf1148 pathotype from subtropics was prevalent for over 20 years back, the variety was not withdrawn from cultivation since its virulence was comparatively less. On the other hand higher aggressiveness of pathotypes Cf671 and Cf92061 on the cultivars CoC 671 and CoC 92061 caused extensive loss to sugarcane cultivation in many parts of Tamil Nadu, Andhra Pradesh, parts of Karnataka and Gujarat. The results were in contrast to the earlier work at SBI which showed that isolates of Eastern UP and North Bihar are more virulent (Alexander *et al.*,1986). This study clearly established that acquisition of higher virulence in the pathotypes led to severe epidemics of the disease and loss of sugarcane varieties.

Recently a confirmatory study was carried out at the Institute to compare the virulence of the pathotypes on a set of sugarcane varieties with seven major pathotypes revealed that pathotype Cf90063 from CoC 90063 variety was the most virulent among the seven pathotypes from tropics and subtropics (Table 4). The studies clearly indicated that a new virulent strain which is more virulent to Cf671 and Cf92061 have emerged recently. The result also indicates that virulent strains emergy continuously in the field time to time depending on the variety under attack and environmental conditions, which suggests the need for a constant survey of emergence of new variants. Results of AICRP on sugarcane revealed emergence of new virulent pathotypes in different regions. Isolates from CoPant 84211 and CoJ 86 in Punjab, CoS 767 and CoJ 64 in Haryana, CoS 767 in Shahjahanpur and 85A261 in Andhra Pradesh exhibited higher virulence over the existing pathotyes (Anon, 2000), Prakasam *et al.* (1998) evaluated 44 sugarcane differentials of sugarcane for virulence/aggressivenss with two pathotypes from sub-tropics and three from tropics. They found the differences between differentials, pathotypes and the interaction were statistically significant and the differentials possessed both horizontal (HR) and vertical (VR) resistance. Statistical compartmentalization of the resistance revelaed a higher amount of VR in sugarcane genotypes than HR.

Physiological considerations

A successful pathogen requires potential warfare tools in the form of toxin and hydrolytic enzymes for penetration and infection in the host. *C. falcatum* produces pectic enzymes viz. pectin methyl esterase (PME) and polygalacturanase (PG) and cellulase Cx *in vitro* which might help in the entry of the pathogen into the host by dissolving the cell wall in association with the pectolytic enzymes. Evidence has accumulated to suggest that pectic enzymes play an important role in the development of rot and wilt diseases. The pectolytic enzymes in general solubilized the middle lamella and loosened cells get autolysed or the cells may be attacked by the invading pathogen. It is possible that weakly pathogenic strains have a reduced capacity to produce PG. This may results in decreased virulence as a result of slow invasion of the host by the weakly virulent strain because of weak PG activity. Singh and Hussain (1964) and Manocha and Vasudeva (1964) reported that the most virulent strain of *C. falcatum* produced large quantity of PG and small amount of PME in liquid culture. Jothi (1989) reported that the virulent isolate has shown higher activity of PME, PG and Cx as compared to five moderately

Table 4 Comparative virulence of *C. falcatum* pathotypes on sugarcane varieties

Variety	Red rot reaction to pathotyes (plug method)						
	Cf671	Cf90063	Cf92061	Cf1148	Cf7717	Cf419	Cf997
Co 419	MS	S	S	S	S	S	MS
Co 997	HS	HS	HS	HS	S	S	S
Co 1148	MS	MS	MS	MS	MR	MS	S
Co 7704	MR	MS	MS	MR	MS	R	MR
Co 7717	MR	HS	MS	HS	HS	S	S
Co 8021	MR	MR	R	R	MR	MR	R
Co 86032	MR	MS	MS	S	MS	MS	MS
Co 86249	MR	MR	R	R	MR	R	R
Co 93009	R	R	R	R	MR	R	MR
CoS 767	R	S	R	MR	MS	R	MR
CoC 671	S	S	S	S	MR	S	S
CoC 90063	MS	S	S	MS	MS	MS	MR
CoC 92061	S	HS	HS	HS	HS	HS	S
BO 91	MR	MR	R	R	MR	R	MR

virulent or less virulent isolates. Moderately virulent strains produced significantly less PG and weakly virulent the least. The virulent strain produced equal quantity of PME and cellulase. The isolates also varied in their capacity to produce endotoxins. Studies of Mohanraj *et al.* (1995) revealed that *C. falcatum* produces toxic metabolite of anthroquinone compound, which may facilitate pathogen infection and spread in the stalk. Further studies conducted at SBI clearly proved that a virulent strain produces the toxic metabolite in higher quantities as compared to less virulent strain (Malathi, Unpublished).

Emergence of virulent strain

Studies conducted at SBI and elsewhere indicate that newly emerging pathotypes are much more virulent than the existing ones. Some of the new isolates from CoC 90063 (Tamil Nadu), CoJ 64 (Haryana) and CoS 767 (UP) are highly virulent and in artificial inoculations such isolates knock down most of the genotypes. Earlier Srinivasan (1962) observed that varieties of sugarcane that encourage the dominance of light types or in other words, retard the appearance of dark variants, although they may possess a reasonable degree of resistance, should be dangerous in that they may tend to select out virulent types, while others that encourage the dark types may actually be safer even if they should prove to be more susceptible. Possible evidence for this apparently paradoxial situation may be that as long as the highly susceptible *S. officinarum* clones were dominant varieties in cultivation in the earlier decades in the last century, epiphytotics of red rot were relatively rare, while with the introduction of a multitude of hybrids, outbreaks of the disease have become distressingly more frequent. It would appear that in the absence of absolute immunity, atleast some of the relatively resistant varieties may tend to select out the most virulent variants while more rapidly eliminating the unstable variants. He felt that a given type is able to survive and multiply into an agent of an epiphytotic will depend, among other factors, largely on its stability, as also in its innate virulence. A highly virulent type may be short-lived if it is unstable, while one of even moderate virulence may be potentially dangerous if it has low variant propensity.

In accordance with the theory that heterokaryosis may be a factor in increasing virulence (Buxton, 1960), Carlvalho (1968) proposed a hypothesis that *C. falcatum* biotypes with different nutritional requirements, enzyme system and other properties, reflecting differences in pathogenicity could, in the presence of a variety resistant to each biotype individually, overcome this resistance by germination of their conidia followed by hyphal fusions to proudce heterokaryotic hyphae in which different nuclei complement each other. Consequently, we might have a strain more pathogenic than any individual biotype. This combination of nuclei would be present in the multi - nucleate appressoria; the several nuclei pass through the invasion tubes to the interior of the plant and the resulting hyphae colonize the host.

Serological variation

Important feature of serological reaction is their specificity and sensitivity to closely related substances, which can be distinguished by these methods whereas by other tests they appear to be identical. Pathotypes on isolation from the host clone can be identified for its distinctness based on diferential interaction on host clones. However few pathotypes isolated from the same area almost exhibited similar reaction on host clones suggesting that some other techniques are required to distinguish the pathogen isolates. Differentiation of fungal isolates or anastomosis groups was done by comparing serological relation in *Rhizoctonia solani* (Adam Jr. and Butler, 1979) and *Fomes* spp. (Madhosingh, 1964).

Polyclonal antisera were produced against *C. falcatum* pathotypes viz Cf687, Cf997, Cf1148 and Cf86062 as reported earlier (Viswanathan *et al.*,1999;2000). Ouchterlony's double diffusion technique (DDT) was performed to compare precipitin patterns on gel diffusion plates (1.25% Agarose, Sigma) (Ouchterlony, 1960).

Cf1148 antiserum when compared against antigen from Cf1148, Cf7717 and Cf64 (Haryana) common precipitin bands formed in homologous and heterologous reactions. Cf7717 shared an antigen with other two isolates and was clearly distinguishable from them based on reaction pattern (Fig. 2a). In field testing at Karnal (differential host studies) similar variation in these pathotypes was observed (Padmanaban *et al.*, 1996). The isolate Cf86062 shared another antigen with Cf658, which was, absent in all other 4 antigens. There was a double spur formation against Cf671, Cf92061, and Cf658 on either side. The isolates Cf8001 and Cf86062 were almost similar in their reaction and Cf671 and Cf92061 behaved similarly, although the former had a spur. The Cf658 isolate was clearly distinct from all other isolates though the isolate Cf86062 was found more closer to it, one specific antigen was lacking in Cf86062 which was the dominant antigen of Cf671 and Cf92061. These isolates fell into 3 serotypic groups (Fig. 2b). The Cf658 isolate was the first isolate collected from Tamil Nadu state and it was comparatively less virulent than all other isolates on differential interaction. In the same study Cf671 was found more virulent and Cf8001 behaved similar in its pathogenicity to Cf671 (Padmanaban *et al.*,1996) whereas, in serological tests both of them exhibit clear difference. In this context serological test demonstrated the distinction of isolates even though they behave similar in their pathogenicity.

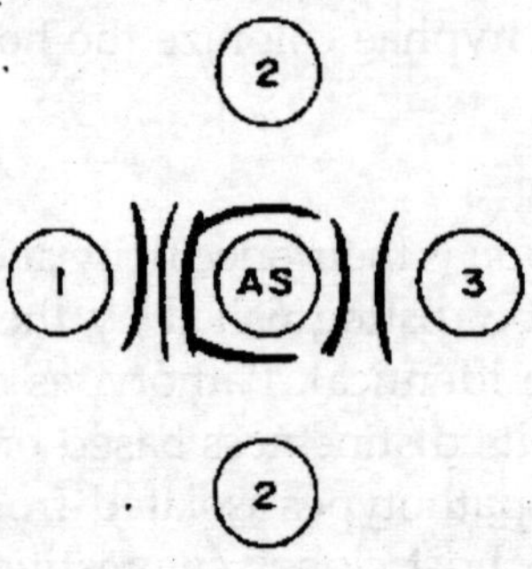

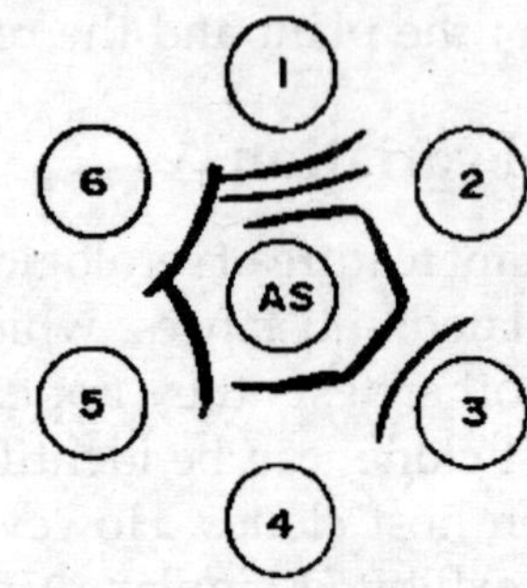

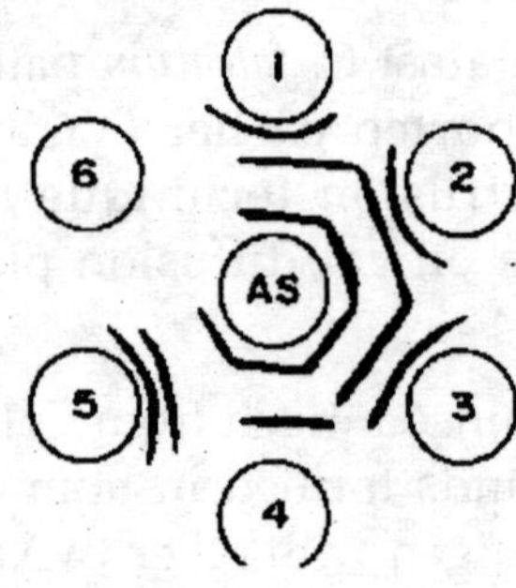

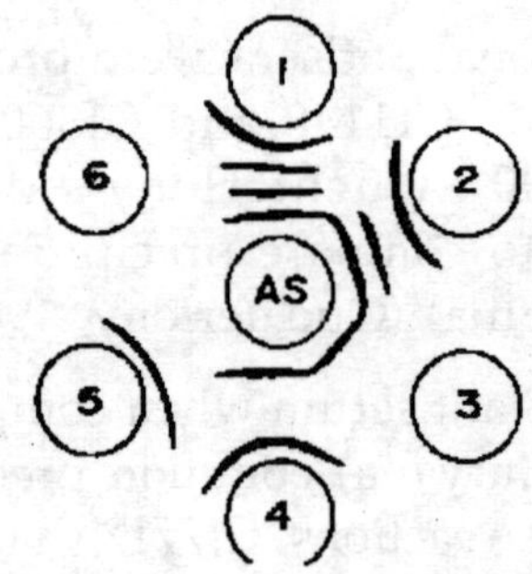

Fig. 2

a. Haryana pathotypes
Central well: As Cf1148
Outer well
1. An Cf1148
2. An Cf7717
3. An Cf64

b. Tamil Nadu pathotypes
Central well: AS Cf86062
Outer well
1. An Cf658
2. An Cf 8001
3. An Cf86062
4. An Cf 85061
5. An Cf92061
6. An Cf671

c. Andhra Pradesh pathotypes
Central well: As Cf997
Outer well
1. An Cf997
2. An Cf419
3. AnCf83070
4. An Cf671
5. An Cf80R41
6. Buffer

d. Bihar pathotypes
Central well: As Cf687
Outer well
1. An Cf687
2. An Cf1148
3. An Cf70
4. An Cf802
5. An CfLk8001
6. Buffer

Both Cf671 and Cf92061 showed similar serological reaction suggests further confirmation whether both are different or maintained separately since their hosts are different. Homologous reaction of Cf997 from Andhra Pradesh showed 3 clear precipitants, and Cf997 shared these antigens with all other isolates except Cf80R41. The isolates could be broadly grouped into two groups one comprising Cf419, Cf997, Cf671 and Cf8370 and other Cf80R41 (Fig. 2c). The pathotypes Cf419, Cf997 and Cf671 although behaved differently in field testing, serologically not much variation was noticed and this calls for more detailed studies on their variation.

Cf687 antiserum was compared with antigen from all 5 isolates (Bihar). The isolates could be classified into 3 groups viz. CfLk 8001 the first, Cf70 and Cf802 the second and Cf687 and Cf1148 the third serotype (Fig 2 d). Since Cf70 showed a discrete band pattern antigen from Cf70 and other antigens were cross-reacted with antiserum of Cf687 isolate. In all the 4 test Cf70 showed a single antigen which makes it as a distinct serotype among Bihar isolates. As in group testing (Fig 2d) CfLk8001 here also showed an arcing band. Triangular pattern of common band was observed (Fig. 3a). Cross reaction of Cf802 and Cf70 showed a hexagonal porecipitin band. As like group testing in homologous combination Cf802 produced arcing band however, here there were two sharp arcs noticed (Fig. 3b). Cf70 shared single antigen with Cf1148 and common band was circular (fig 3c). Precipitin band patterns between Cf687 and Cf70 were as in group testing (Fig. 3d). The results indicated that varying number of antigens in Bihar isolates one in Cf70, 2 in CfLk8001, 3 in Cf802 and Cf1148 and 4 in Cf687, which distinguished all the 5 serotypes. Results clearly established 3 serotypes among the Bihar isolates. Here also similarity of Cf687 and Cf1148 in serological test matches with the pathogenicity in sugarcane.

The purpose of the study was to differentiate the isolates of the red rot pathogen based on serological reaction. As such morphology of an isolate is not useful to distinguish with other. Sometimes disease reaction on the host was almost similar with the new isolate when compared to the reference isolate. In such a situation two or more isolates can give same reaction on host clones. Serological technique was preferred since this technique is based on the nature of antigens in each isolate. We thought once antiserum is raised against a reference isolate of a region, whenever, a new isolate is suspected it could be eaisly compared with reference serum and grouping can be done accordingly. Ultimate aim of this work was to minimize the duplication of isolates maintained to screen sugarcane genotypes in many research centres.

Clean precipitin bands were seen in homologous and heterologous combinations under the concentration of antiserum and antigen used.

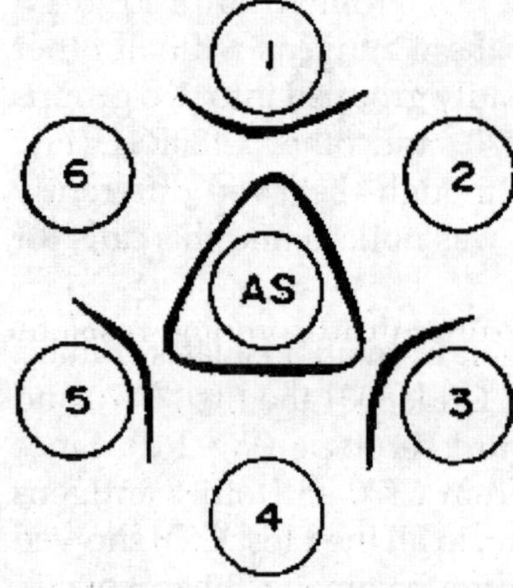

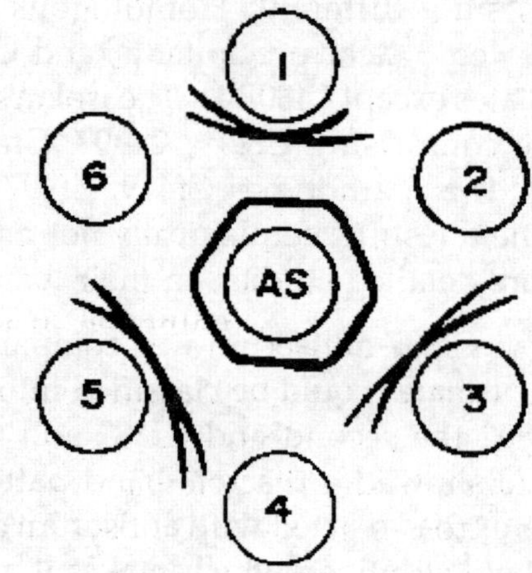

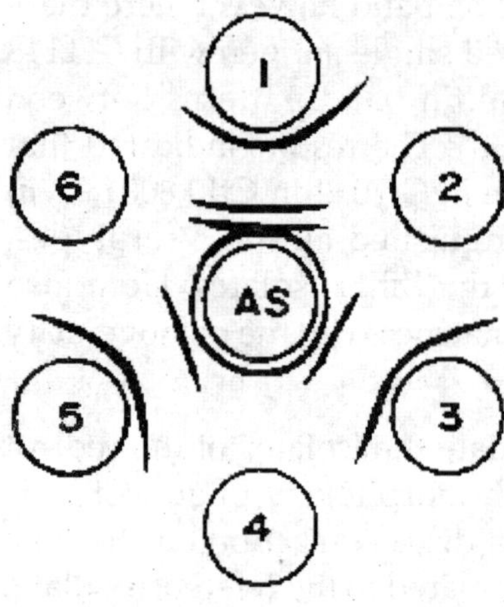

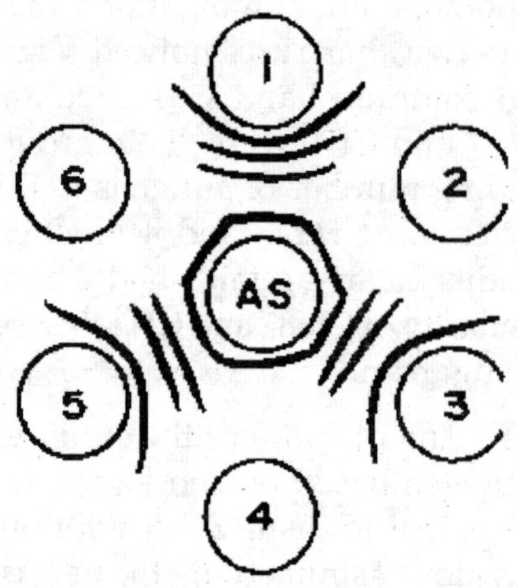

Fig. 3

CfLk8001 Vs Cf70 Outerwell 1,3,5: An CoLk8001

Cf802 Vs Cf70 Outerwell 1,3,5: An Cf1148

Cf687 Vs Cf70 Outerwell 1,3,5: An Cf687

All the 4 treatments outerwell 2,4,6: An Cf70

All the 4 treatments central well: As Cf687

AS: Antiserum

An. Antigen

Precipitin in all the tests suggests that all the isolates share certain common antigens among them. The results proved that distinctions brought out in serological tests was also observed in pathogenicity and virulence tests (Alexander and Jothi, 1995). In general, multiple precipitin formation was noticed in most of the cases which may be due to use of crude fungal extracts as antigen. However, we are able to group the isolates of each zone based on multiple precipitin bands. Similarly in other pathogens like *R. solani* (Adams Jr. and Butler, 1979), *Fomes* spp. (Modhosingh, 1964) and *Erwinia* spp. (Zeitoun and Wilson, 1966) isloates have been grouped as serotypes based on DDT precipitin patterns. In this antiserum to major isolate prevailing in a zone was prepared and tested with antigens from other isolates of the zone.

Using 3 antisera with their respective antigens an indirect-enzyme-linked immunosorbent assay (ELISA) revealed some variation in isolates of Cf1148, Cf7717 and Cf 671. Here the abosrbance values were taken for comparison. In homologous combination the values were higher, clear variation existed between Cf1148, and Cf7717 isolates. In the same study antigen from respective infected host also revealed similar variation (Viswanathan *et al.*,1999; 2000). It is also found that this serological technique may not aid in identifying virulence of the isolates and we have to depend more on field reaction for the same. Further all the isolates show definite variation in the serological studies. We isolate the cultures from a variety and maintain on them for further recovery. So, the variation exhibited by all the isolates may be due to sugarcane varietal influence. This suggests that sugarcane varieties may bring about adaptive changes in *C.falcatum* pathotypes. Srinivasan (1965) also expressed similar views.

Molecular variation

The DNA based isolate characterization methods such as restriction analysis, restriction fragment length polymorphism (RFLP) and random amplified polymorphism, using the polymerase chain reaction (RAPD-PCR) have been used for isolate characterization of many other species of *Colletotrichum.* Though simple restriction analysis studies alone have enabled the grouping of isolates of some fungi it has not been found to resolve the variability within isolates of *C. falcatum.* A review of published literature indicates that restriction analysis is sufficient to resolve variability only within isolates of prokaryotes and other organisms with smaller genome size. This suggests that the genome size of *C. falcatum* is relatively large rendering restriction analysis unsuitable to even broadly group the isolates of *C. falcatum.*

Studies of Mohanraj *et al.* (2002) brought out the wide genetic diversity between the pathotypes Cf1148 and Cf7717 and most of the observed genetic variability among the other pathotypes lies within these limits. This diversity is relatively in good correspondence to the differential reaction of the pathotypes (Beniwal *et al.*,1989; Padmanaban *et al.*,1996). Pathotypes from tropics viz. Cf671, Cf85061 and Cf92061 showed close genetic relationship as resolved by RAPDs and they also showed corresponding pathogenic homogenicity expressed as their cross infectivity. Surprisingly the pathotypes Cf90063 and Cf767 showed a genetic similarity. Another important finding was that the two pathotypes exhibiting closer reaction were the most virulent in tropic and subtropic regions, respectively. The sugarcane cultivar CoS 767 has been resistant to red rot for a long time in the sub-tropics and it became susceptible only in the late nineties. The fact that the tropical pathotype Cf90063 is able to infect the cv CoS 767 and that the RAPD profiles of isolates Cf90063 and Cf767 are relatively close suggests a considerable genetic as well as pathogenic similarity between these two isolates. In addition to PCR, a combination of different DNA based techniqeus is recently available to study variability among fungal pathogen.

In the above context the change of red rot flora in the country over a period of time and constant change in the host varieties becomes relevant. Under such situations the rapid recognition of groups of the prevailing pathogen variants, using molecular markers would enable identification of cane cultivars tolerant to them for effective depolyment and disease management.

Origin of new strains

Hybridization

The discovery of the perfect stage of the fungus suggested the possibility of hybridization; however further studies proved the fungus to be homothallic thus ruling out this hypothesis (Chona, 1980). Besides these established methods, the possibility of new physiologic races being produced by hybridization cannot be ruled out. Even though ascospores are homothallic, it does not rule out their heterozygocity.

Mutation

Patches of dark mycelium arise in slant or Petri dish cultures of the light type generally when the cultures gets old which in turn, on subculturing, gives rise to stable, dark type, sparsely sporulating mutants. The changes are, probably, due to spontaneous mutations. The pathogen has been observed to be highly mutable on artificial culture media (Chona and Hingorani, 1950). The frequency and spectrum of mutation are markedly enhanced by irradiation and by treatment with radiometric chemicals like nitrogen mustard. The mutants show pronounced morphological changes in shape and size of spores, colour of the colony and spore pigment (Agnihotri, 1983). Singh *et al.*, (1995) reported that gamma radiation on the pathogen causes changes in the morphology and pathogenicity. They found change of colony colour from light grey to white, grey to dark and dark grey to light grey with yellow pigmentation and mild pathogenic irrespective of the change in colony colour. Induced biochemical mutants showed variation in virulence ranging from mildly pathogenic to pathogenic. The work by different investigators showed that mutation is responsible for producing new races of *C. falcatum* but its frequency in nature is not known.

Heterokaryosis

The heterokaryotic mycelium has hyphae with 1-3 nuclei in intermediate and upto 8 in terminal hyphae. Thus the heterokaryotic mycelium, which develops from the appressoria, may result in genetically different hyphae and different nuclei need not necessarily be located in the same hypha. Through this heterokaryotic process nuclei located in different hyphae may supplement each other in their enzyme systems using transport of nutritive substances through the mycelium to other parts of the colony where hyphae are deficient. This complementation of enzyme systems and nutrients is an important factor in determining the pathogenicity since the research by earlier workers (Singh and Hussein, 1963; Manocha and Vasudeva, 1964) have established a relation between pectic/ cellulolytic enzymes and nitrogen sources and pathogenicity of the pathogen. Singh and Payak (1968) reported several types of hyphal fusions, namely, side-to-side, end-to-end and peg-to-peg, the latter occuring at several different points between the same pair of hyphae. They also reported some rare types

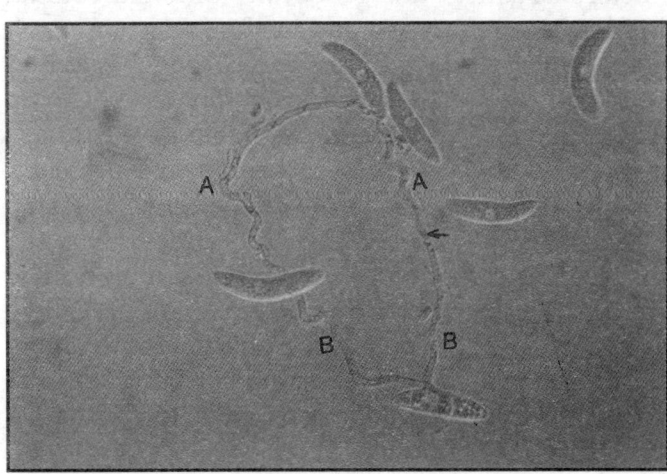

Fig. 4. Hyphal fusion between conidial germ tubes of *C. falcatum* A and B indicate germ tubes from two different conidia. : site of germ tube fusion.
Note : Conidium B is with dense matrix and germ tubes from the conidium contain many nuclei

of fusions, such as those between two conidia and between a conidium and an appressorium. They further reported that by intermixing two different isolates of *C. falcatum,* a new isolate that differed in pathogenicity and cultural characteristics is obtained. Recent studies at SBI also revealed fusion between two conidia or germ tubes of two conidia is very common (Fig. 4). The conidia of *C. falcatum* were generally uninucleate and the nucleus is located in the central portion. When the conidia become physiologically active, a nuclear division occurs that precedes the formation of a central septum, before the formation of the germination tube. In certain rare cases conidia with 2, 3, and 4 nuclei which differed morphologically from the normal with the characteristic falciform shape may also occur. Carvalho (1968) explained that these conidia may become multinucleate as a result of heterokaryosis. Soon after germination of the conidia, fusions of the germination tubes or of the hyphae may occur. At sometimes successive fusion between several germination tubes of different conidia in which nuclear migration occurs. Anastomosis seems to be influenced by the nuclei of different hyphae located side by side. At these points, after division, probably mitotic, of the nuclei, a swelling is formed on each hypha. These swellings grow, one touch the other, the walls fuse and one of the nuclei elongate and passes through connection tube. After nuclear migration, the hyphae becomes heterokaryotic (Parmeter *et al.*, 1963).

The conidia of *C. falcatum* are mostly uninucleate although base of the the conidiophore may be heterokaryotic. Thus genetically different conidia can occur in the same acervulus. Heterokaryosis and nuclear selection result in racial adaptation of *C. falcatum* to the varieties of sugarcane through selection of more pathogenic biotypes. The frequent cultural instability of *C. falcatum* with the formation of sectors is the result of nuclear selection through separation of the heterokaryotic nuclei because of environmental pressure. Heterokaryosis does not exclude other mechanisms of variation such as sexual process, mutations and parasexuality. On the contrary, it is undoubtedly the simultaneous occurence of these processes of variation that results in new biotypes making possible and giving significance to heterokaryosis. A large scale cultivation of a given variety would result in the selection of the more pathogenic biotypes within the population which would complement each other by heterokaryosis during the pre-penetration stage, thus establishing a race adopted to the variety.

Adaptation and influence of host

When a pathogen develops the capacity to carry out a biochemical process that it could not perform originally, it is said that the fungus has adopted and this phenomenon is called as adaptation. By this process the pathogen can metabolise protoplasm of hitherto unfavourable hosts, but the mechanism of adaptation is not fully understood. It may be due to adaptation of new type of cytoplasm, the pathogen acquiring tolerance to toxic materials and change in virulence to the host plant.

Morphologically the red rot fungus is practically indistinguishable from *C. graminicolum* Ces. Wilson, which is a parasite of sorghum, Johnoson grass and related grasses. A red rot disease of sorghum stalks caused by this fungus produces internal stalk symptoms very similar to those of red rot of sugarcane. Isolates from leaves of sorghum and Johnson grass that produced typical red rot symptoms when inoculated into sugarcane leaves and stalks. In cross inoculation experiments with isolates of *C. falcatum* from sugarcane and *C. graminicolum* from species of sorghum and related grasses showed that cultures from sugarcane were highly pathogenic to sugarcane and rarely so to the sorghum group, while those from the sorghum group were non- pathogenic to sugarcane and highly parasitic on sorghum. Thus, although some strains of the fungus from either sugarcane or the sorghum group of grasses may infect plants of the other grass group, they are considered as distinct species. Further evidence of this is the failure of *C. graminicolum* isolates to produce perithecia under conditions where those of *C. falcatum* were induced to do so (Abbott and Hughes, 1961).

Srinivasan (1962) studied in detail on host influence on emegence of new variants. Some varieties of sugarcane have been shown to induce rapid development and dominance in infested tissues of the dark, avirulent type of variant, while others appear to encourage the dominance of the virulent parental clone. Occasionally a more virulent variant than the parental clone has also been appeared. He found maintenance of cultural type even after 15 transfers in oatmeal in two types. Wheras certain other isolates within three transfers gave raise to poorly sporulating dark types. Passing the different stable and unstable cultures through host cultivar showed that (i) those of the cultures, which had proved unstable in culture also, gave predominantly the dark type variant on re-isolation. However, there were a few reisolates of the wild type of culture particularly from the advancing margin of the lesion (ii) the characters of the cultures re-isolated from the more stable isolates varied (iii) more dark-type variants appeared in the older parts of the lesion than the younger parts. Repeated passage through the host varieties showed that passing the dark variants through the four varieties of host did not result in further change and virulence and sporulation were not restored even after five such passages. Host juice affected the stability of the isolates grown on it. There were dark variants appearing in culture on the juice of certain varieties than on that of others. The living host also had a similar effect. However, he could not explain how the host influences the rate and direction of variation. He concluded that the dark type variants on account of loss of sporulation as also of pathogenic ability, would tend to get lost in nature soon after they arise.

Resistance evaluation

Earlier workers have observed that it is not easy to obtain and maintain varieties resistant to all multitudinous races and biotypes of single species of a pathogen. It is even more difficult to combine resistance to all races of all-important pathogens with all the desired plant characters and with quality and quantity of the product. It may be necessary to sacrifice some characters to get the desired resistance. The breeder and the pathologist will have to make all out efforts to maintain a constant flow of red rot resistant varieties to prevent the outbreaks of epidemics. The cultivator should also lend his cooperation by creating a disease free environment. This will enhance the scope of selection of even moderately suscepitble varieties and lengthen the potential of varieties released for cultivation.

Abbott (1938) has stated that varieties classed as resistant or susceptible to one had not changed their classification when tested with large number of isolates. However, Chona (1954) felt testing the clones with highly virulent one is likely to hold good. Kar (1964) indicated that varietal classification based on red rot resistance show considerable variations according to their behaviour with different isolates. He reported that changes in parasitic population are responsible for the deterioration of varieties indicating that physiologic specialization in the fungus is of a higher degree than recognized hitherto. The classification made for varieties with any virulent isolate may not necessarily is the same on that with others belonging to the same morphological groups. The isolates appearing in nature from time to time can accordingly be termed as virulent with reference to varieties. Currently a mixture of *C. falcatum* pathotypes is used for screening sugarcane genotypes in different locations.

Studies of Mohanraj *et al.* (2002) showed that the choice of isolates for evaluation of red rot reaction has to be based on their genetic relationship and differential host reaction rather than on a regional basis of their location of collection. Further molecular biological characterization of the newly emerging red rot pathotypes would be a constant requirement to continuously update the availability of cane varieties with matching resistance to the pathotypes for effective management of sugarcane red rot.

Conclusion

Despite concerted effort to establish a uniform classification system for *C. falcatum* pathotypes, several constraints still exist. For instance, the environment, the nature of inoculum, timing of inoculum etc may vary considerably from one centre to another. In addition, the subjectivity during disease evaluation may lead to misclassificaiton of the same strain by different pathologists.

Characterization of *Colletotrichum* species has relied on a number of criteria, including morphology, optimal growth temperature, vegetative compatibility, benomyl sensitivity and recently, molecular methods. Studies on vegetative compatibility offer another approach to determining genetic relatedness in anamorphic populations of *Colletotrichum* species. Studies with respect to vegetative compatibility grouping has not been attempted so far in *C. falcatum* and studies in this line of work in combination with serogrouping may give information on the relation among the different pathotypes collected from a same variety/location.

DNA based markers have increased the potential to detect and measure variability among individual strains. The markers are almost unlimited in number and are not affected by the environment. The distinct patterns obtained for each strain can be taken as specific fingerprints that describe them. Several researchers have proposed the use of molecular markers as a fast and safe alternative to differentiate and characterize fungal isolates. Preliminay work on molecular grouping of *C. falcatum* carried out at SBI gave a possible grouping of pathotypes. Further work with large number of sampling is to be done to finger print the pathogenic variation more clearly.

The points which needs careful further investigation are: how the new strains arise in nature; how these get spread over large areas so quickly; how can these be detected quickly and what in it that makes a strain more virulent; or a variety susceptible. In 1957, Abbott pointed out that the success achieved in detecting susceptible or resistant varieties will depend on the adequacy of sampling red rot flora and what is considered to be an adequate sample will depend on the pathologist's experience with the disease in a given area. Various workers have established the constantly changing nature of the pathogenic flora. Hence emphasis must be given on the constant monitoring of changes and exercising proper care in releasing sugarcane varieties which may be fairly resistant to red rot. This would be possible only if the inoculum used in resistance test is representative of the prevailing flora. Information on variation in differential host interaction, vegetative compatibility grouping, serotypic variation and molecular variation need to be combined to draw clear picture on the existing *C. falcatum* variation. Scientific advancements in molecular biology and bioinformatics would immensely help in the future to resolve variation in the fungal pathogen.

Acknowledgements

The authors are grateful to Dr. N. Balasundaram, Director of the institute for the encouragement and providing support to contirbute the chapter.

References

Abbott, E.V. (1933). Physiologic forms of *Colletotrichum falcatum* Went. Phytopathology, **23**:557-559.

Abbott, E.V. (1935). Physiologic specialization in *Colletotrichum falcatum. Proc. ISSCT*, **5**: 730-736.

Abbott, E.V. (1938). Red rot of sugarcane. Tech. Bull *USDA*, 641, p96.

Abbott, E.V. (1957). Some aspects of testing sugarcane varieties for red rot resistance. Proc. ISSCT, **9**: 1066-1070.

Abbott, E.V. and Hughes, C.G. (1961). Red rot. In: *Sugarcane* Diseases of the World Vol., (Eds) J.P. Martin, E.V. Abbott and C.G. Hughes, Elsevier, Amsterdam, pp 262-287.

Adams, C.G. Jr. and Butler, E.E. (1979). Serological relationship among anastomosis groups of *Rhizoctonia solani.* Phytopathology, 69: 629-633.

Angihotri, V.P. (1983). Diseases of sugarcane. Oxford & IBH Publishing Co, New Delhi, 363 pp.

Agnihotri, V.P. and Budhraja, T.R. (1975). Pathogenic potentialities of mid rib isolates of *Colletotrichum falcatum* in inciting stalk rot of sugarcane. Intern. Sugar J., **81**: 263-265.

Alexander, K.C. (1979). Progress in resistance breeding for red rot and smut. Maharashtra Sugar, **4**: 89-91.

Alexander, K.C. and Jothi, R. (1995). Current status of race flora of *Colletotrichum falcatum* Went. Proc. National Seminar on Strategies for Research and Management of Red Rot, Part II. (ed. V.P. Agnihotri, O.K. Sinha and R.P. Singh) IISR, Lucknow. pp 273-284.

Alexander, K.C., Kandasami, P.A., Rao, T.C.R., Mohanraj, D. and Rao, M.M. (1979). A new approach to genetic defense against red rot disease of sugarcane. Indian Sugar, **28**:695-696.

Alexander, K.C. and Rao, M.M. (1972). Comparative evaluation of genotypes in the centres in India for resistance to red rot and smut. Proc. Diamond Jubilee Symp. of Sugarcane Breeding Institute, Coimbatore.

Alexander, K.C., Rao, M.M., Mohanraj, D., Seshardri, K., Padmanaban, P., Prakasam, N. and Kshirsagar, S.S. (1986). A new approach to improvement of red rot resistance in sugarcane. Indian J. Mycol. Pl. Pathol., **16**: 108-112.

Anonymous, (1990). Report of Principal Investigator (Plant Pathology), All Indian Coordinated Research Project on Sugarcane, pp 6-18. IISR, Lucknow.

Anonymous, (2000). Report of Principal Investigator (Plant Pathology), All India Coordinated Research Project on Sugarcane, pp6-18. IISR, Lucknow pp1-9.

Barber, C.A. (1901). Sugarcane diseases in Godawari and Ganjam districts. Madras Dept. Land Records and Agri. Bull., **512**: (43) 181-194.

Beniwal, M.S., Satyavir and Virk, K.S. (1989). Pathogenic variability in *Colletotrichum falcatum* incitant of red rot of sugarcane. Indian Phytopathol. **42**: 95-98.

Butler, E.J. (1906). Fungal Diseases of Sugarcane in Bengal. Mem. Dept. Agri. India Bot. Ser. **1**: 2-24.

Butler, E.J. (1918). Fungi and Diseases in Plants. Thacker Spink & Co., Kolkata.

Buxton, E.W. (1960). Heterokaryosis, saltation and adaptation. In: Plant Pathology Vol II, eds. J.G. Horsfall and A.F. Dimond, Academic Press, New York, pp 359-405.

Carvalho, P.C.T. de. (1968). Heterokaryosis in *Colletotrichum falcatum*. Proc. ISSCT, **13**: 1251-1261.

Chatopadhyay, S.E. and Sarkar, A. (1957). Strains of *Colletotrichum falcatum* in West Bengal, India. J. Mycol. Res., **3** : 25-28.

Chona, B.L. (1954). Relative resistance of sugarcane varieties to red rot. Indian J. Agric. Sci. **24**: 301-315.

Chona, B.L. (1954). Chairman address. Proc. ISSCT, **9**: 975-986.

Chona, B.L. (1980). Red rot of sugarcane and sugar industry- A review. Indian Phytopathol. **33**: 191-206.

Chona, B.L. and Bajaj, B.S. and Sharma, R. (1964). A new stroma forming strain of *Colletotrichum falcatum* Went. Proc. 4th All Indian Conf. Sugarcane Res. & Dev. Worker, 620-622.

Chona, B.L. and Hingorani, M.K. (1950). Mutation in *Colletotrichum falcatum* Went the causal organism of sugarcane red rot. Phytopathology, **40**: 221-227.

Chona, B.L. and Padwick, W.G. (1942). More light on the red rot epidemic. Indian Farming, **3**: 70-73.

Chona, B.L. and Srivasatava, D.N. (1952). The perithecial stage of *Colletotrichum falcatum* Went in India. Indian Phytopath, **5**: 158-160.

Chona, B.L. and Srivastava, D.N. (1960). Variation in *Colletotrichum falcatum* Went the causal organism of red rot of sugarcane. Indian Phytopathol, **13**: 158-160.

Edgerton, C.W. and Moreland, C.C. (1920). Effect of fungi on the germination of sugarcane. *Louisiana* Agri. Sta. Bull., 169.

Gupta, S.C., Singh, M.P. and Upadhyay, U.C. (1980). A new biotype of *Colletotrichum falcatum* Went. Curr. Sci., **49**: 600-601.

Jothi, R. (1989). Studies on variation in red rot pathogen *Colletotrichum falcatum* Went on sugarcane, Ph.D. Thesis, Bharathiyar University, Coimbatore, 245 pp.

Kar, K., Rana, O.S. and Gupta, S.C. (1965). An intermediate virulent isolate of *Colletotrichum falcatum* in Eastern Uttar Pradesh. Indian Sugar., **15**: 161-163.

Khirbhat, S.K., Satyavir and Beniwal, M.S. (1980). Physiological and pathological variability in sugarcane red rot pathogen *Colletotrichum falcatum* Went in Haryana. Indian Phytopathol., **33**: 296-299.

Kirtikar, Rana, O.S. and Gupta, S.C. (1964). A new strain of *Glomerella tucumanensis* Arx and Muller. Indian Sugar, **9**: 27-31.

Kirtikar, Singh, G.P. and Singh, R.S. (1972). Reaction of cane varieties to red rot in Western tracts of U.P. Indian Sugar, **22**: 429-430.

Madhosingh, C., (1964). A serological comparison of isolates of *Fomes roseus and Fomes subroseus*. Can. J. Bot., **42**: 1677-1683.

Manocha, M.S. and Vasudeva, R.S. (1964). Factors affecting the pathogenicity of *Colletotrichum falcatum* Went causing red rot of sugarcane. Indian J. Agric. *Sci.* 34.

Mohanraj, D., Kumaresan S. and Sreenivasan, T.V. (2002). Molecular characterization of isolates of the sugarcane red rot pathogen *Colletotrichum falcatum* Went. Indian Phytopathol. (In Press).

Ouchterlony, O. (1960). Interpretation of comparative immune patterns obtained by diffusion- in gel technique. In: Immuno chemical approaches to problems in microbilogy (ed. M. Heidelberger, O.J. Plescia and R.A., Day). Rutgers Univ. Press, New Delhi, p; 5-19.

Padmanaban P, Mohanraj, D., Viswanathan, R., Rao, M. M., Prakasam, N., Jothi, R. and Alexander, K.C. (1996). Differential interaction of sugarcane clones to pathotypes of *Colletotrichum falcatum*

Went. Sugar Cane, 4: 16-20.

Pandey, L.N. and Sakul, R. (1974). New pathogenic strain of *Glomerella tucumanensis* (Speg.) Arx and Muller in Uttar Pradesh. Indian Sugar., 24: 707-709.

Parmeter, J.R., Snyder, W.C. and Reichle, R.E. (1963). Heterokaryosis and variability in plant pathogenic fungi. Rev. Phytopathol., **1**: 51-76.

Prakasam, N., Padmanaban, P. and Mohanraj, D. (2000). Identification of red rot pathotypes using differential host reaction. Annual Report for 1999-2000, Sugarcane Breeding Istitute, Coimbatore, p 44.

Prakasam, N., Padmanaban, P., Mohanraj, D. and Nallathambi, P. (1998). Studies on differential hosts under CCT. Annual Report for 1997-98, Sugarcane Breeding Institute, Coimbatore, pp 35-36.

Rafay, S.A. (1957). Another strain of *Physalospora tucumanensis.* Curr. Sci. 19: 385-386.

Rafay, S.A. and Singh, V.B. (1957). A new strain of *Glomerella tucumanensis.* Curr. Sci., **26**: 19-20.

Rafay, S.A. and Padmanaban, S.Y. (1941). Strains of *Colletotrichum falcatum* Went. Curr. Sci., **10**: 25-26.

Satyanarayana, V. and Rao, A.M. (1995). Investigations on the race flora of *Colletotrichum falcatum,* incitant of red rot disease of sugarcane. In: Proc. National Semianr on Sugarcane Production constraints and strategies for Research and Management of red rot. Part II (Eds) V.P. Agnihotri, O.K. Sinha and R.P. Singh, IISR, Lucknow. pp 285-293.

Satyanarayana, Y. and Rao, A.M. (1984). Evidence on the existence of two physiological different strains in the light race isolates of *Physalospora tucumanensis* Speg. in Andhra Pradesh. Sugar Cane, **3**: 6.

Satyanarayana, Y. and Satyanarayana, B. (1976). Variation in the light race isolates of red rot fungus in Andhra Pradesh. Cooperative Sugar, **7**: 389-392.

Satyavir, Virk, K.S. and Anil Kumar (1995). Resistance of sugarcane genotypes against red rot pathogen. In: Proc. National Seminar on Sugarcane Production Constraints and Strategies for Research and Management of Red Rot. Part II (Eds) V.P. Agnihotri, O.K. Sinha and R.P. Singh, IISR, Lucknow, pp 244-258.

Sharma, M.N. (1970). Occurrence of a new strain of *Physalospora tucumanensis.* Sci & Cult. **36**(1): 52-54.

Singh, G.P. and Hussain, A. (1963). Role of enzymes in red rot disease of sugarcane. Curr. Sci. **32**: 81-83.

Singh, G.P. and Rana, O.S. (1968). A new strain of *Glomerella tucumanensis* red rot pathogen in UP. Indian Sugar, **18**: 537-540.

Singh, K. and Singh R.P. (1989). Red rot. In: Diseases of Sugarcane: Major Diseases (Eds.) C. Ricaud, E.T. Egan, A.G. Gillaspie Jr. and C.G. Hughes, Elsevier, Amsterdam, pp 169-188.

Singh, M., Rao, G.P., Singh, S.B., Verma, K.P., Singh, H.N. and Singh, R.B. (1995). Effect of gamma radiation on cultural and pathogenic variability of *Colletotrichum falcatum.* In: Proc. National Seminar on Sugarcane Production Constrains and Strategies for Research and Management of Red *Rot.* Part II (Eds) V.P. Agnihotri, O.K. Sinha and R.P. Singh, IISR, Lucknow, pp 364-368.

Singh, M.P., Upadhyay, U.C., Verma K.P. and Singh. H.N. (1984). Pathogenic variabillity in red rot isolates of sugarcane in UP. Indian Sugar, **33**: 669-671.

Singh, N. (2000). Identification of pathotypes/races of red rot pathogen. Annual Report for 1999-2000, Sugarcane Breeding Institute, Coimbatore, p 64.

Singh, P. and Payak, M.M. (1968). Observations on karyology, hyphal fusions and production of a new isolate in *Colletotrichum falcatum.* Acta Phytopath. Acad. *Sci.* (Haryana), **3**: 143-155.

Srinivasan, K.V. (1962). Some observations on variation in the red rot pathogen, *Glomerella tucumanensis* (Speg.) Arx and Muller. Proc. ISSCT, **11**: 795-802.

Srinivasan, K.V. (1965). Towards the ideals of red rot resistance- Ends and means. Proc. ISSCT, **12**: 1108-1117.

Vasudeva, R.S. (1957). Report of Mycology and Plant Pathology, Sci. Rep. Indian Agric. Res. Inst., New Delhi for 1954-55, pp 87-101.

Viswanathan, R., Padmanaban, P. and Mohanraj, D. (1997). Growing virulence of red rot pathogen of sugarcane in Tamil Nadu. Indian Sugar, 47: 23-30.

Viswanathan, R., Padmanaban, P., Mohanraj, D. and Jothi, R. (1999). Rapid detection of red rot (*Colletotrichum falcatum*) in sugarcane and evaluation of host reaction by ELISA. In: Sugarcane Pathology vol. I Fungal Diseases (Eds.) G.P. Rao, A. Bergamin Filho, J. C. Autrey and R.L. Megraey, Science Publishers, Enfiled (NH), USA, pp 36-45.

Viswanathan, R., Padmanaban, P., Mohanraj, D. and Jothi, R. (2000). Indirect-ELISA technique for the detection of the red rot pathogen in sugarcane (*Saccharum* spp hybrid) and resistance screening. Indian J. Agric. Sci., **70**: 308-311.

Viswanathan, R. and Samiyappan, R. (1999). Red rot disease in sugarcane: a major constraint for the Indian sugar industry, Sugar Cane, **5**: 9-12.

Viswanathan, R. and Samiyappan, R. (2000). Red rot disease in sugarcane: Challenges and prospects. Madras Agric. J., **87**: 549-559.

Zeituoun, F.M. and Wilson, E.E. (1966) Serological comparison of *Erwinia nigrifluens* with certain other *Erwinia* species. Phytopathology, **56**: 381- 1385.

34

Morphology and Pathogenicity of Isolates of *Ceratocystis paradoxa* Causing Pineapple Disease of Sugarcane in Maharashtra

A.S.Patil[1] and G.P.Rao[2]

[1] *Department of Plant Pathology and Agril. Microbiology, Vasantdada Sugar Institute, Manjari, Pune 412307, Maharashtra*

[2] *Division of Plant Pathology, Sugarcane Research Station, Kunraghat, Gorakhpur-273 008, Uttar Pradesh*

Pineapple disease caused by *Ceratocystis paradoxa* in sugarcane is becoming alarming in many parts of Maharashtra. It causes significant losses to sugarcane by internal sett rotting. Twenty-five isolates of *C. paradoxa* were collected from different regions of Maharashtra and comparative morphology and taxonomy of these isolates were compared. On the basis of microscopic features, pathogenic behaviour and cultural characteristics, three isolate groups of *C. paradoxa* (CP1,CP2,CP3) were selected for detailed study. In CP1, the colony at the initial stages showed greyish white colour, while CP2 and CP3 were of little olivaceous black colour. The isolates exihibited diversity in cultural characteristics, morphology of micro- and macroconidia , conidial germination and pathogenecity of isolates . Detailed physiological studies showed greater cultural variation among the isolates of the fungus. Pathogenecity on different host interaction studies further confirmed variation in pathogenic behaviour and virulence of these isolates.

Keywords: *Ceratocystis paradoxa*, sugarcane, fungus diversity, Maharashtra

Pineapple disease caused by the fungus *Certocystis paradoxa* (Dade) Moreau is important disease of sugarcane and it causes serious losses in almost all the sugarcane growing countries of the world. Sugarcane, a cash crop, is cultivated in almost all states of India. It is playing a major role in the economy of Maharashtra. Among the different reports on diseases of sugarcane in Maharashtra, the pineapple disease is very important. Initially the incidence of this disease was considered minor but later it has been found increasing in different parts of Maharashtra state and now-a-days, it is cause of major concern. The pineapple disease of sugarcane was reported first in Maharashtra by Patil and Hapase in 1984. In the present chapter, we have discussed the

e-mail: gprao_gor@hotmail.com

economic importance, stages of infection and symptomatology of this important disease. The comparison of morphological characters, cultural behaviour and physiological and pathogenicity of different prominent isolates collected from various places in Maharashtra have been studied.

Herrera *et al.* (1981) reported that the pathogen developed better in red soil than black. In India, the disease has been reported from Maharashtra, Uttar Pradesh, Andhra Pradesh, Karnataka, Tamil Nadu and Kerala (Agnihotri, 1990).

Economic importance

Pineapple disease is capable of causing serious losses through the failure of infected cuttings to germinate. Ahmed (1960) noted the reduction of CCS% in all the infected cane varieties but deterioration in juice quality of Co 658 and Co 527 was not significant. Hughes (1949) reported that pineapple disease of sugarcane considered as a minor disease may cause serious losses to susceptible canes under certain conditions. Padmanabhan and Ahmed (1970) observed that the disease spreads very rapidly throughout the length of the cane resulting in the ultimate death of the plant. They also reported considerable losses arise due to lowering of sucrose content. Manzo (1970) reported that *Ceratocystis paradoxa* usually attacks only seed pieces.

Symptoms

Typical symptoms developed after 2 to 3 weeks of planting. The pathogen enters mainly through cut ends and proliferated rapidly in the parenchymatous tissues of the internodes. The affected tissues first develop a reddish colour, which turns to brownish black in later stages. Fibrovascular bundles in canes are attacked by the pathogen. Cavities are formed inside the severely affected internodes and the fungus sporulates luxuriantely on them. The presence of fungus in sugarcane prevents their rooting. The infected tissues breakdown in the center and later become hollow. Freshly affected tissues of sugarcane emit rotten pineapple like odour. In most of the cases, seed setts decayed before the bud sprouting of the shoot attained a height of 6-12 cms. It was observed that if the shoot roots develop before the attack, the plant survives but its growth is drastically impaired. Symptoms were observed in the standing crop also. The fibro vascular system remains intact and cane tops dried in artificially inoculated conditions. It was observed that in almost all the cases the infection was more towards the top than the bottom of the canes. The pathogen could not infect a particular variety equally showing their pathogenic capabilities (Plate- 1a).

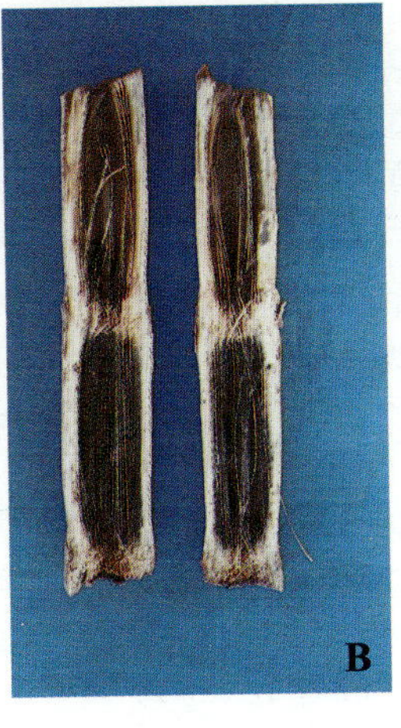

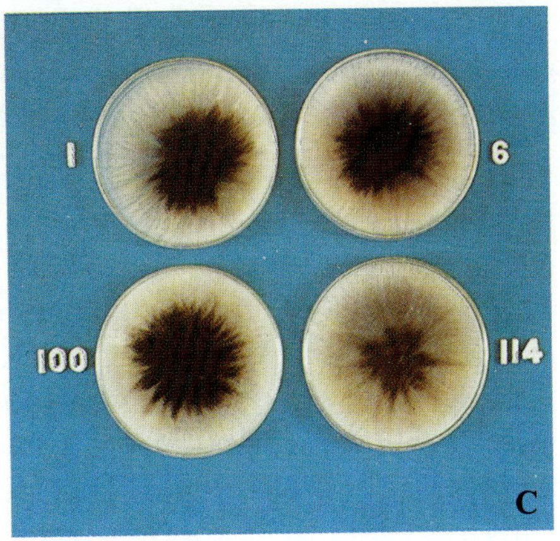

Plate 1a : Natural symptoms of Pineapple disease *Ceratocystis paradoxa;* A : The infected cane setts show brownish black colour due to pathogen. Parenchymatous tissues become hallow leaving fibro vascular bundles intact - Co 7219.; B : Rotting of setts due to entry of Pathogen. Internal tissues become deep black to olivaceous black in colour profused sporulation of pathogen observed naturally - CoC 671.; C : Culture of isolates : Pure culture of isolates of *Ceratocystis paradoxa* was obtained on Potato dextrose agar medium. 1 = CP1 isolate from Co 7219; 6 = CP6 isolate from Co 740; 100 = CP2 isolate from 7527; 114 = CP3 isolate from CoC 671.

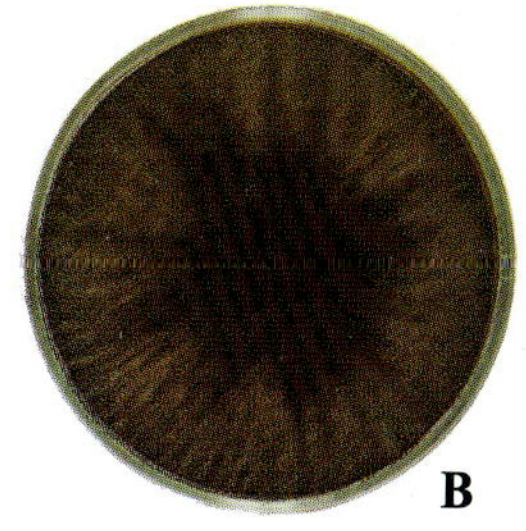

Plate 1b : Pineapple disease on Pineapple fruits; A : Natural infection of pineapple disease *Ceratocystis paradoxa* (Dade Moreau on Pineapple fruits; B : Culture of *Ceratocystis paradoxa* isolated from pineapple fruits. (i) Isolate CP1 (Co 7219); (ii) Isolate CP2 (Co 7527)

In natural conditions also, same types of symptoms were observed in standing canes due to entry of pathogen in canes having rats or bores damage, or some similar injury. Excessive dry conditions also accelerated the damage.

The first symptom was a light red discoloration of the internal tissue of the sugarcane. Later the cut end changed to black. The black colour being due to the production of spores of the fungus (Thomson, 1957). The black colour frequently extended into the inner nodes in the form of a sooty core and a characteristic black central cylinder was seen (Plate 1a, A & B). The freshly infected setts emitted a rotten pineapple like colour. The odour was due to ethyl acetate formed by the metabolic activity of the fungus *Ceratocystis paradoxa* (Agnihotri, 1990). The presence of *Ceratocystis paradoxa* in sugarcane setts prevents their rooting (Byther and Moore, 1974). In about 80 percent cases seed pieces decay before the bud sprouts or the shoot attains a height of 6-12 cm (Went, 1896; Bell, 1936; McMartin, 1937). The pathogen causes considerable damage in standing crops also (Padmanabhan and Ahmed, 1970; Manzo, 1975; Natrajan and Subba Raja, 1976; Singh *et al.*, 1991).

Isolate Diversity of *Ceratocystis paradoxa*

25 isolates (CP) of pineapple disease were collected from Co 740 (7 Nos.), Co 7219 (9 Nos.), CoC 671 (6 Nos.), Co 7527 (2 Nos.). and Co 8021 (1 Nos.), collected from different localities of Maharashtra (Table 1). Comparative morphology and taxonomy of CP- isolates are listed in Table 2.

Table 1 : Morphological grouping of 25 CP-Isolates and their classification

Sr. No.	Group – I Co 740 CP –1	Group –II Co 7219 CP-2	Group – III COC 671 CP-3	Group – IV CO 7527 CP-4	Group-V CO 8021 CP-5
1	CP-1	CP-8	CP-4	CP-2	CP-3
2	CP-5	CP-11	CP-6	CP-14	-
3	CP-7	CP-12	CP-10	-	-
4	CP-13	CP-15	CP-16	-	-
5	CP-18	CP-17	CP-24	-	-
6	CP-19	CP-20	CP-9	-	-
7	CP-122	CP-21	-	-	-
8	-	CP-23	-	-	-
9	-	CP-25	-	-	-
Total	7	9	6	2	1

Table 2 : Comparative study of morphology and taxonomy of *Caratocystis paradoxa* isolates

Isolate	Width of Mycelium	Microconidiophore	Microconidiophore	Microconidia	Macroconidia	Colour of macroconidia	Host source	Locality
C. paradoxa	-	100 µm inlenght	-	10-15x3.5-5.0	20-80 x 4 µm	Hyline to olive green or black	Sugarcane	Java
C. paradoxa	3.5-7 µm	-	-	-	16-19 x 10-12 µm	dark brown	Sugarcane	Java
C. paradoxa	-	-	-	7 x µm	25 x 8 µm	-	-	-
C. paradoxa	-	upto 20 µm long	short but, hyphal	6-24 x 2-5.5. µm	10-25 µm x 7.25 – 20 µm	brown	Sugarcane	Java
C. paradoxa	-	125 µm long x 7.5 µm wide at base	-	3.25 x 2.5-5 µm	9.5-25.0 x 7.20 µm	brown to reddish brown	Sugarcane	-
C. paradoxa	3.5-7 µm	-	-	6-12 x 4.5-6 µm	15-21 x 9-12 µm	brown black	Sugarcane	Coimbatore
C. paradoxa	3.2-6.5 µm	80-110 µm terminal cell	-	7.5-15.5 x 3.5 – 5.5 µm	15.2 x 8.5 12.5 µm	olivaecous green to greenish black	Sugarcane (Co 7219)	Pune
CP-Isolate of *Ceratocystis* sp. Group								
CP-1	4.37 – 7 µm (4.87 µm)	52.5-101.5 µm (75.68 µm)	49-87.5 µm *70 µm)	8.75-12.25 x 4.5-7 µm (10.5 x 6.47 µm)	10.5-21x7 – 12.25 µm 15.84 x 9.81 µm	Brown or black	Sugarcane Co 7219	Pune VSI
CP-2	6.5-8.75 µm (5.35 µm)	49-94.5 ? m (68.6 µm)	87.5-133 µm (116.37 µm)	8.75-14x4.37-7 µm (10.93x5.68 µm)	14-21x10.5-14 (17.7 x 11.48 µm)	Brown or black	Sugarcane Co 7527	Sangli
CP-3	3.5-8.75 µm (6.34 µm)	87.5-105 µm (94.5 µm)	59.5-87.5 µm (73.93 µm)	14-18.37x 3.5-7.87 µm (17.17 x 5.35 µm)	15.75-28x9.62-14 µm 15.84 x 9.81 µm	Brown	Sugarcane CoC 671	Kolhapur
CP-4	3.5-8.75 µm (5.79 µm)	105-217 µm (143.5 µm)	59.5-87.5 µm (95 µm)	11.37-17.5x5.25-7 µm (13.45 x 6.01 µm) (15.53 x 8.53 µm)	12.25-19.25 x 9-10.5 µm 15.84 x 9.81 µm	Brownish black	Sugarcane Co 7219	Jalgaon
CP-5	3.5-6.12 µm (3.44 µm)	147-178.5 µm (157 µm)	108.5-129.5 µm (122.5 µm)	10.5-19.25x 5.27 µm (13.58x5.79 µm)	12.25-17.5 x 7-10.5 µm	Brown	Sugarcane Co 740	Jalgaon

Isolation from Pineapple Fruits

During survey, the affected pineapple fruits *Ananas comosus* were also collected for comparison with sugarcane CP isolates. The causal organism *Ceratocystis paradoxa* a condial state of *Thielaviopsis paradoxa* was isolated on P D A from the pineapple fruits. Pathogenicity was proved on healthy pineapple fruits under laboratory condition. Cross inoculation studies of both the isolates, i. e. from pineapple fruits and from sugarcane setts were also made for their virulence studies. It is revealed from the results that there were clear difference in development of symptoms and virulence on pineapple fruits (Plate-1b), as shown by the three isolates of sugarcane.

IDENTITY OF THE PATHOGEN

25 isolates of *Ceratocystis paradoxa* causing pineapple disease of sugarcane (Table 1) were isolated on potato dextrose agar medium from the samples of different varieties collected during survey in the state. On the basis of microscopic features, pathogenic behaviour and cultural characteristics, only three isolate groups of *Ceratocystis paradoxa* were selected for the detailed study. The morphological and taxonomical characteristics of these three isolates are presented in Tables -2 and –3. The different CP – isolates included were as under :

CP 1 isolate from Sugarcane var. Co 7219

CP 2 isolate from Sugarcane var.Co 7527

CP 3 isolate from Sugarcane var. CoC671

External morphological characters of CP isolates

The growth of the fungus on the P.D.A. was very rapid. In three days, all the isolates showed maximum growth (up to 84 mm of colony diameter) on petriplates (solid medium). The colour of the colonies of the three isolates showed variations. In CP1, the colony at the initial stages showed grayish white color, while CP2 and CP3 were of little olivacious black colour (Plate – 1a). However, with age the colonies of the three isolates showed black colour. The colour of CP2 and CP3 colonies was darker than CP1 isolate. Colony of CP1 isolate showed aerial to sub-aerial grayish white mycelium. The growth was circular and growth pattern of radiating rays. Smooth periphery of the colony was also observed. In CP2 and CP3 isolates, the mycelium was sub-aerial to submerged, grayish to dark black in colour. Circular outline with radiating rays pattern of growth was observed.

Microscopic observations (Plate 2 & 3)

The fungus *Ceratocystis paradoxa* produced two types of spores, microspore (microconidia) and macrospores (macroconida). Microscopic observation for the morphology of pathogen revealed that the change in color of colonies was due to the production of masses of macroconidia on hyphae. The proportion of macrospores to microspores was much higher in all the sugarcane isolates of *Ceratocystis paradoxa*. Excellent mycelial growth was observed in all the three isolates on PDA medium. The comparative morphological observations of the 3 isolates are presented in Table -3 and 4.

Table 3 : Measurement of Mycelium, Conidiophore and conidia in micron (CP1, CP2, CP3)

	Observations of 8 microscopic fields									
	(1)	(2)	(3)	(4)	(5)	(6)	(7)	(8)	Mean	Range
CP1-Co7219										
Mycelium (width)	5.25	3.50	4.90	3.50	7.00	3.50	4.38	7.00	4.88	4.37-7.00
Microconidiophore (L)	52.50	94.50	101.50	77.00	77.00	70.00	73.50	59.50	75.69	52.5-101.50
Macroconidiophore (L)	49.00	87.50	77.00	66.50	63.00	70.00	77.00	87.50	72.19	49.5-87.50
Microconidia (Length)	10.50	8.75	8.75	10.50	12.25	12.25	10.50	10.50	10.50	8.75-12.25
(breadth)	7.00	5.25	5.25	4.55	7.00	8.75	7.00	7.00	6.47	4.55-8.75
Macroconidia (length)	17.50	16.62	17.50	12.25	10.50	15.75	17.50	14.00	15.20	10.50-17.50
(breadth)	12.25	10.50	10.50	7.00	7.88	10.50	10.50	8.75	9.73	7.00-12.25
CP2-Co 7527										
Mycelium (Width)	4.38	6.13	8.75	5.25	5.25	5.25	4.38	3.50	5.36	3.50-8.75
Microconidiophore (L)	133.00	112.00	119.00	129.50	115.50	122.50	112.00	87.50	116.38	87.5-133.5
Macroconidophore (L)	94.50	49.00	52.50	77.0	77.00	52.50	77.00	70.00	68.69	49.0-94.00
Microconidia (Length)	14.00	10.50	10.50	12.25	10.50	10.50	10.50	8075	10.94	8.75-14.00
(breadth)	5.25	7.0	5.25	6.13	5.25	7.00	5.25	4.38	5.69	4.38-7.00
Macroconidia (length)	14.00	14.00	19.25	19.25	19.25	19.25	15.75	21.00	17.72	14.0-21.00
(breadth)	11.38	10.50	14.00	10.50	10.50	12.25	10.50	12.25	11.48	10.5-14.00
CP3 CoC 671										
Mycelium (Width)	8.75	6.13	7.00	3.50	5.25	6.13	7.00	7.00	6.34	3.50-8.75
Microconidiophore (L)	91.00	94.50	91.00	94.50	94.50	98.00	105.0	87.50	94.50	87.5-105.50
Macrocnidiophore (L)	87.50	63.00	59.50	84.00	77.00	84.00	73.50	63.00	73.94	59.50-87.50
Microconidia (length)	17.50	17.50	18.38	14.00	17.50	18.38	17.50	16.63	17.17	14.00-18.37
(breadth)	7.00	7.88	4.38	4.38	4.38	7.00	4.38	3.50	5.36	3.50-7.87
Macroconidia (length)	17.50	17.50	15.75	19.25	15.75	21.00	28.00	17.50	19.03	15.75-28.00
(breadth)	14.00	13.13	12.25	10.50	9.63	14.88	14.00	12.25	12.58	9.62-14.00

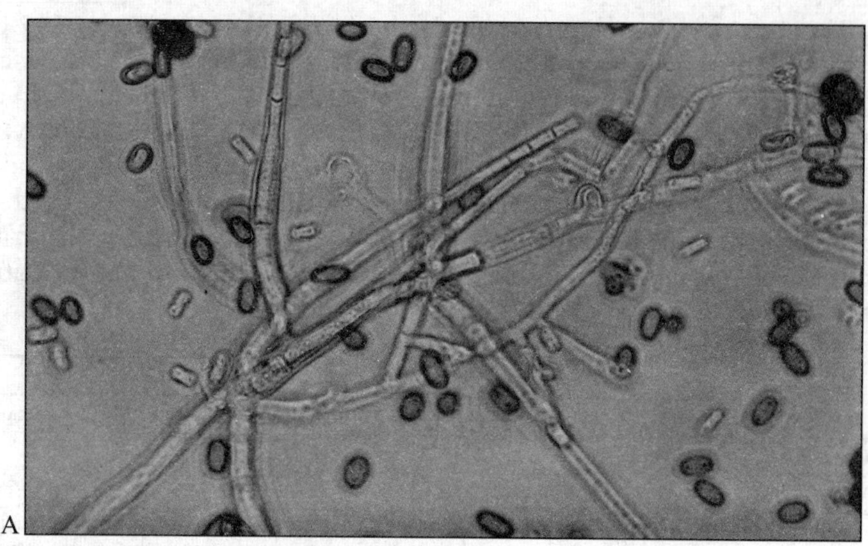

A

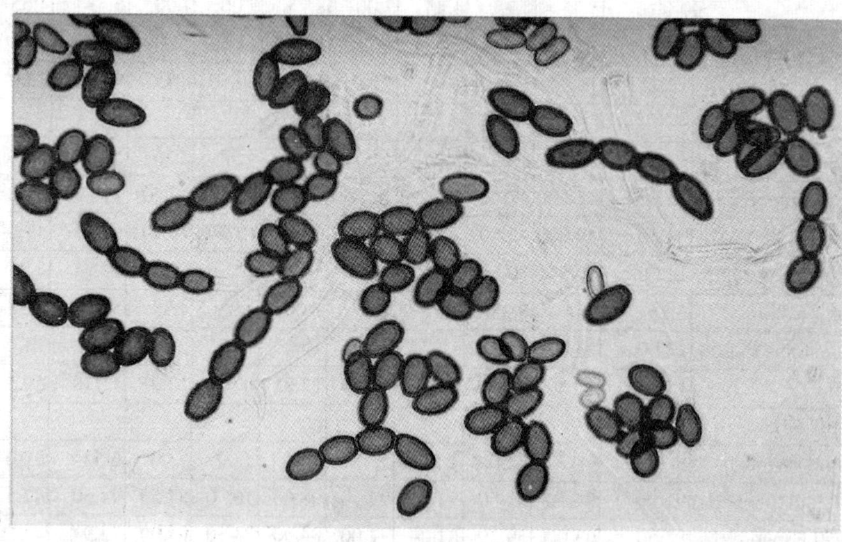

B

Plate 2 : Microscopic observations of *Ceratocystis paradoxa* (45 X); A: Microphotographs shows hyaline micro conidiophore and endogenous origin of conidia with rectangular hyaline microconidia. Macroconidiophore shows elongated basal cells and short apical cells; B : Macroconidia are thick walled and dark brown to deep brown in colour.

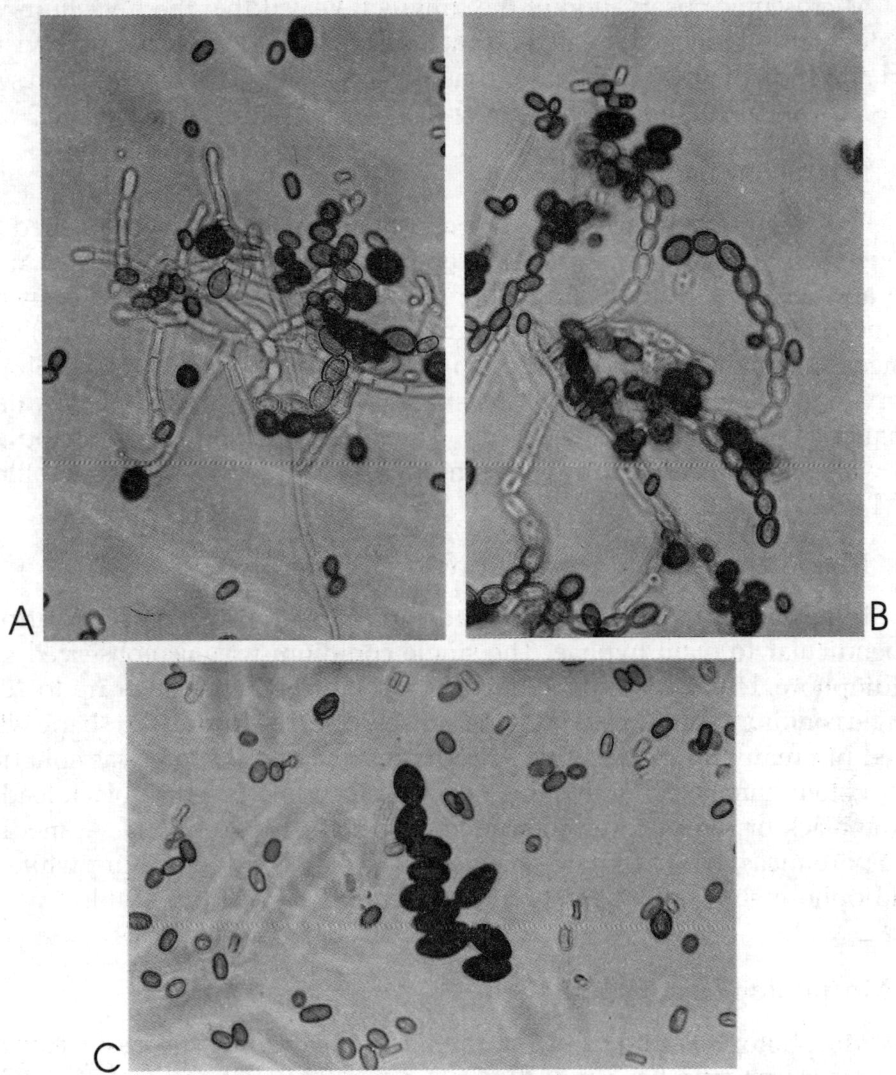

Plate 3 : Microscopic observations of CP1 isolate *Ceratocystis paradoxa* var. *sacchari* sp. nov. on Co 7219. A : Hyaline septate basal cells with elongated terminal cell, septate mycelium, Hyaline microconidiophore with endogenous origin of conidia; B : Macroconidia seen at the tip. Spores hyaline when attached with conidiophore. Hyaline septate microconidia endogenously produced in long chains; C : Microconidia light brown double walled oval to elliptical in shape. Initially hyaline turn brown on maturity. Macroconidia are thick walled dark brown to black or olivaceous black in colour. Egg shaped and bigger in size than microconidia.

i. Mycelium

Microscopic observation of the fungus revealed that the mycelium was septate, highly branched, hyaline, sometimes light brown colour with visible septa and smooth walls. Sometimes constrictions in the walls of mycelium were observed. The width of the mycelium was between 3.4 to 6. 34 µm (Plate 2 & 3).

ii. Microconidia

The microconidia were produced endogenously and were pushed out in long chains on conidiophore. The conidiophores showed short cells at the base and a long terminal cell at the end, hyaline septate and unbranched produced laterally. On P. D. A. medium, the microspore (microconidia) were formed first and the colony showed gray color to dark brown in colour. Similar observations were reported by Wismer (1951), Robertson (1962), Ogawa and Gilpatrick (1978) and Floribai (1985). They were rectangular in shape and thin walled and measured in between 8. 75 –17.17 ×4.55 – 7.0 µm (Table -3 and 4, Plate-2 & 3).

iii. Macroconidia (chlamydospores)

Macroconidia were produced on short lateral conidiophores perpendicular to main hyphae. The single conidium was also observed on a conidiophore. Usually they were found on chains of any number up to 20 on a single conidiophore. They were generally elliptical or truncate in shape when united but many times they are pyriform and the apical spore was spherical. Their colour changed from hyaline olive to green to a greenish black leading for the black or sooty within the rotted cane stalks, and on P. D. A. medium too. Spore measurement ranged between 10.5 – 19.5 × 7 –12.73 µm while the conidiophore showed the range in between 68.5 –73.93 µm (Table-3 and 4, Plate 2 & 3).

iv. Morphology of the pathogen

Morphological studies of the three CP isolates of *Ceratocystis paradoxa* from sugarcane varieties Co 7219 (CP1), Co 7527 (CP2) and CoC 671 (CP3) revealed considerable variation among them. On the basis of the morphological and taxonomical chlaracters, pathogenesis, virulence behaviour, their nutritional requirements etc., these three isolates of pineapple disease were categorized into two distinct groups as CP isolate group 1 and CP2 and CP3 as isolate group 2. After detailed studies, these two groups were identified and confirmed as *Ceratocystis paradoxa* (Dade) C. Moreau of CP2 and CP3 isolate group (Plate-2) and second group of CP1 isolate as *Ceratocystis paradoxa* var. *sacchari* sp. nov. (Plate 3). The same has been confirmed by CMI, England,

(Herb, IMI No 322884). The pineapple disease herein reported constitutes a new record on Co 7219, Co 740, CoC 671 and Co 7527 in Maharashtra State of India.

Pathogencity of Isolates

Two methods were employed to prove the pathogencity of the three CP isolates

i) Sett inoculation

ii) Soil inoculation

i) Sett inoculation

After 45-60 days, microscopic examination revealed the presence of pathogen in the parenchymatous tissues, reisolation from such artificially infected tissues yielded the same organisms as one originally isolated from natural infected setts of pineapple disease. Thus it proved the pathogenic nature of the three isolates (Plate-4).

ii) Soil inoculation technique

Microscopic observation of the setts from the soil inoculated pots showed the presence of pathogen in all cases. Reisolation of infected tissue (artificially) yielded the same pathogen as one originally isolated from naturally infected tissue of sugarcane. It was interesting to note that the rate of infection was rather higher or faster in sett inoculation method than soil inoculation.

Cross inoculation studies of three isolates

Two experiments were conducted to observe comparative virulence of the three isolates of *Ceratocystis paradoxa* (CP1, CP2, and CP3). These were i) Comparative virulence of three isolates on different sugarcane varieties, and ii) Comparative virulence of isolates among themselves only.

Experiment 1

Table - 5 shows the results of observations and % infection in terms of linear spread on the setts. CP1 isolate showed medium 37% and maximum 100% infection. CP2 isolate showed the range 27-100% infection and CP3 isolate showed 37-100% on different varieties of sugarcane. The maximum infection of 100% was observed on the varieties from where the isolate was sub cultured, i.e. CP1 showed 100% infection on Co 7219, CP2 and CP3 showed same reaction on varieties Co 7527 and CoC 671, respectively (Plate 5).

Table 4 : Comparative morphological observations of three isolates of *Ceratocystis paradoxa*

	Characters	Isolate Number 1 CP 1	Isolate No. 2 CP2	Isolate No. 3 CP3
1.	Morphology of mycelium	Hyaline, Sometimes light brown colour, highly branched septate, septa visible	Hyaline highly branched septate (close septa) walls smooth	Hyaline septate branched smooth walls
2.	Width of mycelium (μm)	3.5-8.75 Mean – 5.79	3.5-8.75 Mean 5.35	3.5 – 8.75 Mean 6.34
3.	Growth on (P.D.A.) mm	55	60	55
4.	Morphology of Microconidiophore	Hyaline septate basal cell small terminal cell elongated one conidiophore at the tip or a chain of 4-5 macroconidia seen at the tip. Spore hyaline when attached with conidiophore	Hyaline septate narrow at the tip, septate elongated terminal cell, one or more spores at the end of conidiophore	Hyaline septate arise laterally from hyphae narrow at the base and terminal cell long but broad at the tip
5.	Size of macrocornidiophore (length in μm)	52.5-119 Mean – 95	49-94.5 Mean 68.5	59.5-87.5 Mean – 73.93
6.	Morphology of microconidiophore	Hyaline septate microconidia endogenously produced in long chains elongated terminal cell.	Hyaline septate endogenously produced microconidia in the form of chains, unbranched, produced laterally	Colourless septate elongated terminal cell. No. of microconidia in the form of chains endogenous in origin, lat. conidiophore in position
7.	Size of microconidiophore (length in μm)	105-217 Mean – 143.5	187.5-133 Mean 116.37	87.5-105 Mean – 94.5
8.	Size of macroconidia (in μm) length x Breadth)	12.25-19.25 x 7-10.5 15.53 x 8.53	14.00-21 x 10.50 – 14.0 17.70 x 11.48	15.75-28.00 x 9.62 – 14.00 19.25 x 12.73
9.	Size of micrononidia (in μm) L x B	11.37-1.5 x 5.25-7. 13.36 x 6.01	8.75-14.00 x 4.37 – 7.00 10.93 x 5.68	14.0-18.37 x 3.50 – 7.87 17.17 x 5.35
10.	Morphology of microconidia	Light brown to dark brown double walled oval or elliptical in outline, initially hyaline turns brown on maturity, some spores are spherical in outline	Dark brown to black in colour, oval with tapering ends, double walled produced in chains 3-5 spores	Dark brown in colour produced in chain, double walled, oval in outline tapering at the ends, some spores to spherical dark brown in colour.
11.	Colour and shape of microconidia	Hyaline at maturity, light brown rectangular in outline, more in numbers.	Hyaline rectangular light brown at maturity	Hyaline rectangular, light brown at maturity, produced abundantly

Experiment 2

Setts of varieties Co 7219, Co 7527 and CoC 671 were inoculated with all the three isolates viz. CP1, CP2 and CP3.Observations were recorded after 90 days. The length of the lesion recorded down and % linear spread was calculated (Table 6).

It is observed that the isolates on their respective host variety showed maximum % infection in comparison with others. Other two isolates also showed infection on the same variety. The maximum infection was caused by CP3 isolate, i. e. 84. 02 % on all the three varieties.

CP1 isolates also infect upto 73.34 % and the least infection was seen by CP2, i. e. 71.75%. In the present study two methods were applied for proving the pathogenicity of isolates. In sett inoculation, setts were directly smeared

Table 5 : Cross inoculation studies of three Isolates (*Ceratocystis paradoxa*)

Serial	Variety	Mean percentage infection		
		CP 1	**CP 2**	**CP3**
1	Co 87044	43.1	91.1	57.6
2	Co 8014	100.0	100.00	100.
3.	Co 84036	88.2	100.0	100.0
4	Co 8011	97.8	100.0	100.0
5	Co 91061	44.9	27.0	37.4
6	Co 90063	37.3	39.1	63.5
7	Co 8504	88.9	100.0	90.0
9	CoC 671	90.0	98.8	100.0
10	Co 8021	80.1	31.4	1000
12	Co 7714	71.9	72.6	80.1
13	Co 7219	100.0	100.0	66.7
14	Co 7527	41.0	10.0	57.9
	Mean	72.7	73.8	76.8
Mean percentage infection				
Isolate	**CO 7219**	**CO 7527**	**COC 671**	**Mean**
CP 1	88.51	68.84	62.67	73.43
CP 2	56.48	85.24	73.54	71.78
CP 3	74.28	98.33	79.45	84.13
Control	0.00	0.00	0.00	0.00
CD at 5%	Isolate A	84.13		
–	Variety B	22.20		
	Isolate x Variety (A x B)	22.30		

with the cultures of pathogen and in soil inoculation the pathogen was mixed with healthy sterilized soil. Healthy setts were planted in the sick soil. It was observed that the rate of infection was faster in sett inoculation due to direct contact of pathogen with the sugarcane tissues, slow rate of infection was observed in sick soil due to continuous interaction of micro-organisms and soil factors. Omamor (1985) reported that *Ceratocystic paradoxa* being a soil borne fungus is transmitted mainly through rain-splash, wind and soil erosions.

The cross inoculation studies of three isolates of *Ceratocystis paradoxa* with different varieties revealed that all the three isolates infect to all the varieties with variation in their virulence. The artificial symptoms produced by CP1, CP2 and CP3 isolates on different sugarcane varieties are shown in Plate 7 as well as on pineapple fruits .Seven varieties were tested with three CP isolates (Co 740, Co 419, Co 86032, Co 7201, CoM 88121, CoM 7125, Co 8011).

Artificial infection of *Ceratocystis paradoxa* on standing cane

The data presented in Table 6 indicated that the linear spread of infection in 30 cultivars varied from 5 to 32.5 cms. The % disease infection of different varieties were compared with standard variety Co 7219. The maximum

Table : 6: Linear Spread of Infection Lesion in Standing Canes by Artificial Inoculation of *Ceratocystis paradoxa* (1991-92)

Name of Cultivar	Linear spread of infection lesion (cms)	% disease incidence against standard (Co 7219)	Name of Cultivar	Linear spread of infection lesion (cms)	% disease incidence against standard (Co 7219)
Co 7219	32.50	100.00	Co 8128	13.33	41.01
CoC 671	32.50	100.00	Co 85014	12.50	38.46
Co 7708	32.50	100.00	Co 8011	12.50	38.46
Co 85019	32,12	99.48	CoM 7620	12.50	38.46
Co 740	28.50	87.69	Co 7527	32.00	98.50
Co 85016	27.00	83.07	Co 85001	11.00	33.84
KHS 3347	23.00	70.76	Co 419	9.50	2.23
Co 7712	23.00	70.76	MS 8102	8.50	26.15
Co 8147	21.50	66.15	KHS 3296	8.50	26.15
Co 85010	20.66	63.56	CoR 8001	8.50	26.15
80 R 41	19.33	56.47	Co 7804	8.00	24.61
Co 85009	18.33	56.40	CoM 7125	7.00	21.53
Co 8021	16.00	49.23	CoM 7616	5.00	15.38
Co 85003	15.00	46.15	Co79218	13.75	423.30

infection was noticed in CoC 671 and Co 7708 while minimum infection was observed in CoM 7616 (15%). The fibro-vascular bundles remained intact and cane tops dried in artificially inoculated varieties. It was observed that in almost all the cultivars, infection was more towards the top than bottom of the canes. The isolate could not infect a particular variety equally showing their pathogenic capabilities. On the basis of these observations, the isolate was found to be most virulent with maximum linear spread 32.50 cms.

It is concluded from varietal screening that the single eye budded setts rotted completely, while two and three eye budded setts were comparatively less infected and the linear spread of the infection was also slow. It was noted that three eye budded setts were more appropriate for varietal screening against pineapple disease. Out of the thirty-two varieties only 7 were found resistant, 4 moderately resistant, 5 susceptible while remaining 16 were highly susceptible in 1991-92. The reactions of varieties were different in the following year, i. e. 1992-93, where none of them were found resistant. Only three were moderately resistant and rests all were rated as susceptible to highly susceptible.

Natarajan and Subbaraja (1976) reported 7 resistant, 15 moderately resistant and 23 susceptible varieties for the 45 sugarcane clones they screened. Ahmed *et al.* (1977) found 17 resistant out of 20 clones they tested, while only one was susceptible.

Waraitch and Kumar (1981) found only two of the thirty clones they tested to be resistant. Same authors in 1982 reported the linear spread in comparatively resistant clones was narrow while in susceptible variety it was quite broad and was accompanied with maceration of parenchymatous tissues. In our study artificial inoculation done with the help cork borer revealed that infection was more towards the top than the bottom of the cane and this is similar to the observation of Natarajan and Subbaraja (1976) on standing cane.

VIABILITY

In viability studies, it was observed that the fungus remained viable for a period of twelve months in natural conditions indicating that it was capable of survival for sufficiently long time to initiate the disease in the subsequent season. Inoculated stants of P D A were kept at 5 °C in refrigerator and subculturing was done at every two months and the fungus was found viable for 12 months.

Viability in sick soil

Since *Ceratocystis paradoxa* is a soil borne pathogen, its survival was seen in the soil. Suspension of spores was mixed in the sterilized soil. Isolation of the soil was done at every two months by soil dilution method. It was observed that fungus could survive well in the soil for more than 15 months and infect the sugarcane setts in next growing season.

It was observed that the fungus remained viable in culture for a period of twelve months. In soils it survived well for more than fifteen months, while in plant debris it survived for about six months but the growth was hampered due to contamination (bacteria and saprophytes). Badgar (1966) showed that non incubated conidia of *Thielaviopsis paradoxa* showed a high percentage of survival at 32 and 100% relative humididty, despite a rapid loss of viability at other humidities tested. Schreiber (1969) found that *Ceratocystic ulmi* could be recovered from American elms seedling inoculated with a spore suspension of the fungus in the initial stages and the recovery decreased after 16 to 22 weeks when it could not be recovered.

Germination

To observe germination, conidium from 7 days old culture was taken. 0. 1 % solutions of glucose, fructose, sucrose, lactose, galactose and mannose were used as sources of germination. Temperature was maintained at 28°C. Germination observed in distilled water served as control. The macroconidia took very long time to germinate, i. e. 10 to 13 hours. Microconidia also showed germination after 10 hours of incubation at 28°C. The type of germination in macroconidia was unipolar and both unipolar and bipolar types were observed in microconidia. It was observed that the germination was by the polarized growth of the newly deposited inner layer of the conidium wall.

Better germination of microconidia as compared to macroconidia has been observed in the different sources. Good germination of macroconidia was observed in 0.1 % fructose solution and poor germination in 1 % mannose, galactose and lactose solution while microconidial germination was good in 0.1% fructose (Table 7). It was noted that the germination was by the polarized growth of the newly deposited inner layer of the conidial was as reported by Hawes (1980).

Cultural Behaviour

Since there is no universal medium suitable for all the fungi, a wide choice for media for the purpose at hands is of fundamental importance to the pathogen in the perpetuation of the species, epiphytotics and speed of the disease etc. affords many interesting problems in the fungus physiology.

Table-7: Germination of Conidia *Ceratocystis paradoxa*

Source	Concentration %	Percentage Germination							
		CP1		CP2		CP3		Mean value	
		Micro conidia	Macro conidia	Micro conidia	Macro conidia	Micro conidia	Macro conidia	Micro conidia	Micro conidia
1. Sucrose	0.1	11.66	23.74	13.50	12.38	6.78	10.28	10.65	15.47
2. Glucose	0.1	14.16	15.88	7.50	18.74	8.77	11.42	10.14	15.35
3. Fructose	0.1	40.15	51.78	18.60	27.50	18.84	20.42	25.86	33.23
4. Lactose	0.1	11.66	18.83	6.66	10.00	10.00	11.47	9.44	13.43
5. Galactose	0.1	10.65	10.83	2.69	10.00	9.58	14.20	7.64	11.68
6. Mannose	0.1	8.33	33.33	11.25	33.33	0.00	8.33	6.53	25.00
7. Distilled water		11.07	16.11	13.33	13.44	11.05	13.77	11.82	14.44

Non-synthetic media

The fungus made excellent growth and sporulation in almost all the vegetable media tried viz., potato dextrose, host decoction, oat meal, beans, carrot, corn meal, cane juice except onion agar where least growth was observed. Maximum growth was obtained in host decoction and cane juice agar. Oat meal and host decoction supported good growth where extremely big macrospores, dark black in colour were observed (Plate 6a). Onion agar supported least growth but spores were found in long chains. The shape and size of the conidia were observed normal in all vegetable media (Plate 6a)

Synthetic media

The fungus does not grow profusely on all the media tried except glucose aspargine agar where growth and sporulation both were very profuse (Table 8-10). Sabouraud's agar also showed moderate growth. Asthana and Hawker and Richard's solution could not support fungal growth. Changed concentration of dextrose in Asthana and Hawker medium ranging from 10 gms / 1000 ml to 20 gms / 1000 ml were analyzed and left the plates for incubation but altered medium also could not support growth of *Ceratocystis paradoxa*. In Czapek-Dox, nitrogen concentration was altered and the range was kept in between 2 gms / 1000 ml to 4.5 gms / 1000 ml. At 28°C after incubation period, the medium could not support good growth of fungus. (Table 8-10).

Natural media (vegetable and plant part extract host decoction), stimulated the growth of fungus. Maximum growth (diameter) was observed on oat meal and host decoction media. Synthetic media failed to stimulate growth of fungus. Very thin but well spread growth was observed on Richard's and Czapek Dox media. Certain sugars like glucose and amino acid (aspargin)

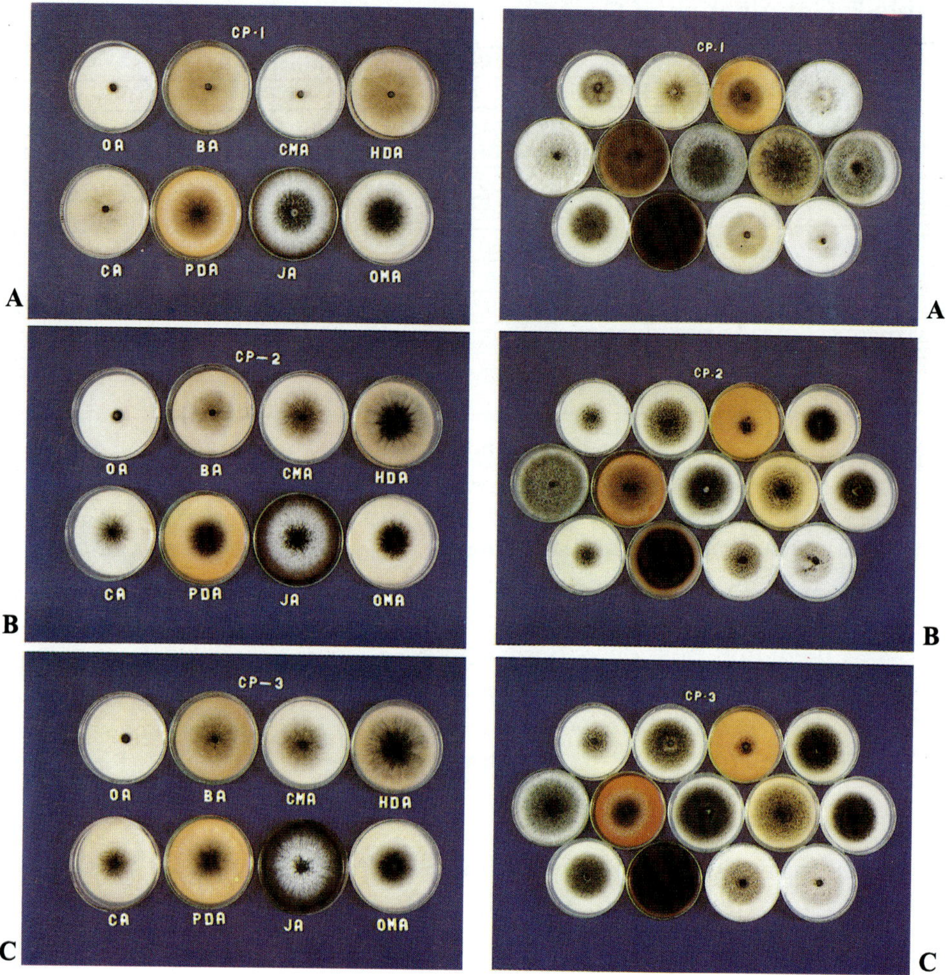

Plate 6a : Cultural behaviour of three isolates of Pineapple disease; A : CP1 isolate from Co 7219; B : CP2 isolate from 7527; C : CP3 isolate from CoC 671; All the isolates showed good growth on all non-synthetic media except OA and CA where growth was scanty; MEdia : OA - Onion agar; BA - Beans agar; CMA - Corn meal agar; HDA - Host decoction agar; CA - Carrot agar; PDA - Potato dextrose agar; JA - Cane juice agar; OMA - Oat meal agar

Plate 6b : Nitrogen nutrition of three isolates of pineapple disease on Glucose aspargine agar medium; A : CP1 isolate (Co 7219); B : CP2 isolate (Co 7527); C : CP3 isolate (CoC 671); First row: L-Phenylanine; Cystine; Tryptophan; Alanine; Second row : Glutamine; Histidine; Proline; Glycine; L-Glutomic acid; Third row: Methionine; Cystine hydroch., Tryosine, Control

Table-8: Cultural Behaviour on Various Synthetic Media *Ceratocystis paradoxa* CP1 (Co 7219)

Media	Colonly Dia mm	Sporulation	Growth characters	Microconidia (length & breadth in μm)	Macroconidia (length & breadth in μm)
1. Asthana & Hawker (A & H) agar	12	+	Very thin growth submerged white mycelium or hyaline, smooth grwoth	8.75-10.5 x 3.5 - 4.2 (9.91 x 3.73)	14 - 17.50 x 7.00 - 9.62 (15.75 x 8.45)
2. Czapek-Dox (C.Dox) agar	15	++	Very scanty, thing growth, submerged grayish mycelium, circular smooth margin	8.75 - 10.50 x 2.62 - 3.50 (9.91 x 3.20)	14.00 - 17.50 x 5.25 7.00 (15.75 x 6.12)
3. Glucose aspargine agar (G.A.)	87	++	Profuse, sub aerial to aerial, growth, mycelium in glucose whitish gray in colour, circular radiating growth, smooth margin	8.75 - 10.50 x 2.62 - 4.20 (9.91 x 3.44)	14.00 - 15.75 x 7.00 (14.58 x 7.00)
4. Richard' agar (R.A.)	13.5	+	Very scanty thin growth, submerged creeping, white mycelium, smooth margin	10.50 x 1.75 - 3.50 (10.50 x 2.62)	14.00 - 17.50 x 6.12 - 7.00 (15.75 x 6.70)
5. Brown's agar (B.A.)	15	+	Very thing growth, sub aerial creeping mycelium, white in colour, smooth margin	8.75 - 10.50 x 3.50 - 4.20 (9.91 x 3.73)	14.00-15.75 x 6.15 - 7.00 (15.16 x 6.70)
6. Sabouraud's agar (S.A.)	40	++	Mycelium blackish green in colour, aerial growth observed, smooth margin	8.75 - 10.50 x 3.50 (9.91 x 350)	14.00 x 7.00 - 8.75 (14.00 x7.58)
7. Nutrient agar (N.A.)	60	++	Aerial growth, mycelium is white in color, circular growth, smooth margin	8.75 - 10.50 x 3.5 - 4.55 (9.91 - 4.08)	14.00 - 17.50 x 6.12 - 7.00 (16.75 x 6.70)

Table-9: Cultural behaviour on various synthetic media *Ceratocystis paradoxa* CP2 (Co7527)

Media	Colony Dia mm	Sporulation	Growth characters	Microconidia (length & breadth in µm)	Macroconidia (length & breadth in µm)
1. Asthana & Hawker (A & H) agar	10	+	Very scanty, thin growth, submerged grayish mycelium, circular smooth margin	10.50 - 12.25 x 2.62 - 3.50 (11.08 x 3.20)	14.00 - 17.50 x 7.00 - 8.75 (16.33 - 8.16)
2. Czapek-Dox (C.Dox) agar	12	++	Very scanty thin growth, white mycelium smooth margin	7.00 - 10.50 x 2.62 0 3.5 (8.58 - 3.20)	15.75 - 17.50 x 8.75 - 10.50 (16.91 x 9.91)
3. Glucose aspargine agar (G.A.A.)	15	++++	Excellent profuse growth, mycelium grayish black in colour, circular radiating growth. Smooth margin, sub aerial to aerial grwoth	8.75 - 10.50 x 1.75 - 3.50 (9.91 x 2.21)	17.50 - 19.25 x 10.5 - 14.00 (18.08 x 11.06)
4. Richard' agar (R.A.)	12	+	Very poor scanty growth, submerged mycelium, smooth margin	8.75 - 10.5 x 3.50 - 4.20 (9.91 x 3.73)	14.00 - 21.00 x 7.00 - 8.75 (17.50 x 7.58)
5. Brown's agar (B.A.)	12	+	Very loss growth, thin creeping, submerged, circular with smooth margin	8.75 - 10.50 x 2.62 - 350 (9.91 x 3.20)	14.00 - 21.00 x 7.00 - 8.75 (17.50 x 7.58)
6. Sabouraud's agar (S.A.)	35	++	Thin, creeping submerged growth, white mycelium with circular smooth margin	17.50 x 8.75 - 10.50 (17.50 x 9.91)	8.85 - 10.50 x 2.62 - 3.5 (9.33 - 3.20)
7. Nutrient agar (N.A.)	642	++	Very thin creeping growth, white sub aerial mycelium, circular smooth margin	10.50 - 12.25 x 3.50 (11.08 x 3.50)	14.00 - 15.75 x 7.00 - 8.75 (14.58 x 7.58)

Table-10: Cultural behaviour on various synthetic media *Ceratocystis paradoxa* CP3 (CoC 671)

Media	Colony Dia mm	Sporulation	Growth characters	Microconidia (length & breadth in µm)	Macroconidia (length & breadth in µm)
1. Asthana & Hawker (A & H) agar	12	+	Very thin scanty growth, submerged, creeping circular, smooth margin	8.75 - 10.00 x 4.20 - 5.25 (9.25 - 4.70	14.70 - 1750 x 7.00 (16.00 x 7.00)
2. Czapek-Dox (C.Dox) agar	15	++	Very thin scanty growth, white mycelium, circular growth, smooth margin	7.00 - 8.75 x 3.50 - 4.20 (7.93 x 3.73)	10.50 - 12.25 x 4.20 - 575 (11.31 x 5.13)
3. Glucose aspargine agar (G.A.0	85	++++	Excellent profuse growth, mycelium grayish black circular aerial growth with smooth margin	10.50 x 3.50 - 4.20 (10.5 x 3.70)	11.20 - 14.00 x 6.30 - 7.00 (12.48 x 6.80)
4. Richard' agar (R.A.)	12	+	Very thin scanty growth, white mycelium creeping submerged circular growth	7.00 - 10.50 x 3.50 - 5.25 (8.75 - 4.20)	14.00 x 7.00 - 8.75 (14.00 x 8.10)
5. Brown agar (B.A.)	20	+	Very less growth, thin creeing submerged, circular with smooth margin	8.75 - 10.50 x 2.62 - 350 (9.91 x 3.20)	14.00 - 21.00 x 7.00 - 8.75 (17.50 x 7.58)
6. Sabouraud agar (S.A.)	28	++	Scanty growth, very less greenish black mycelium creeping growth	12.25 - 14.00 x 3.85 - 5.25 (13.41 x 4.40)	12.25 - 15.75 x 7.80 - 8.75 (14.00 x 7.58)
7. Nutrient agar (N.A.)	38	++	Good growth, aerial mycelium white in colour, circular with smooth margin	10.50 - 14.00 x 3.50 - 4.20 (12.25 x 3.80)	12.25 - 14.00 x 7.00 - 78.75 (12.0 x 7.70)

Table- 11: Effect of different aspargine concentration of growth and sporulation of *Ceratocystis paradoxa* (CP1 - Co 7219)

Quantity of medium	Concent-ration	gm's/1000 ml	Growth in mm (solid medium)	Growth mg (liquid medium)	Growth character
1. 250 ml	0.5%	1.25 gms	70	390	Growth was seen in almost all the concentration in liquid medium as well as solid medium. Initially mycelium was whitish gray in colour, aerial and was submerged. After 3 days showing black masses of conidia. Sporulation was profuse in all the concentration.
2. 250 ml	1.0%	2.5 gms	85	450	
3. 250 ml	1.5%	3.75 gms	85	435	
4. 250 ml	2.0%	5.0 gms	80	425	

induced excellent growth. Glucose aspargine agar medium was selected as basal medium (pH5) for nutritional physiology (Singh and Smalley, 1969). Profuse mycelial growth was observed both in liquid as well as solid medium. Proportion of microconidia was more in comparison to macroconidia. However the finding of the present study were similar to those of Anil Kumar and Nambiar (1990).

The following conclusions can be drawn from the cultural behaviour of the pathogen.

i). Thin walled microconidia are less fastidious in their nutritional requirements for the germination and vegetative growth. They utilize simple sugar like glucose or amino acids at low concentration. Thick walled macroconidia do not germinate under similar conditions of nutrition.

ii). Thick walled spores are more important in the survival of the pathogen as they are more tolerant to aging and drying.

The findings of present study are different from Sastry *et al.* (1989) and Floribai (1985) who observed good growth on all the synthetic media. Singh and Smalley (1969) also reported Glucose aspragine medium to culture the fungus *Ceratocystis paradoxa* that supports the present finding too.

Glucose Aspargine Medium

In glucose aspargine medium (Singh and Smalley, 1969), fungus showed good growth and sporulation. To standardize the growth on this medium, different concentrations of aspargine were used ranging from 5 gms / 1000 ml, 10 gms / 1000 ml, 15 gms / 1000 ml and 20 gms / 1000 ml in liquid medium. The maximum growth was found in 10 gms / 1000 ml in the form of dry mycelium weight (Table 11).

The growth was profuse and sporulation was very good so the Glucose Aspargine medium was selected as basal medium for all physiological experiments.

Physiological Studies

It is essential to study the physiological aspects of fungi in respect to nutritional and environmental factors which influence their growth and reproduction. The *in vitro* studies of the pathogen were conducted to find out the growth and sporulation of the pineapple disease fungus, *Ceratocystis paradoxa*. In the present study physiological aspects of the fungus in respect to environment and nutritional factors were studied.

Among the nutritional factors such as carbon, amino acids and stimulatory factors like vitamins and trace elements were analyzed to have better understanding on the physiology and disease development.

The influence of temperature and pH factor on the growth and sporulation was studied independently keeping all other factors constant.

ENVIRONMENTAL FACTORS

i. Temperature

Temperature is most important single factor that influences the germination, growth, reproduction and indeed all the activities of an organism. The range of temperature, in which fungi will grow varies to some extent with various species. Most fungi, however, either do not grow or grow very slowly at freezing point and at the other extreme, are unable to grow at temperature above 30 – 35 °C. The optimum temperature for growth are usually between 20 – 30 °C. The cardinal temperature and the temperature range very from organism to organism.

Considering the importance of temperature growth relation, present experiment was set up with three isolates. The results obtained in the present study clearly substantiate 'Kelps-hypothesis' (1900) that the temperature range for growth is wider (5-35°C) than that for sporulation, which was in this case between 20 °C and 30 °C. The minimum and maximum temperature for growth were 5°C and 37 °C, respectively with an optimum between 25 °C and 30 °C. There was no marked difference in spore shape, size and colour which remained normal (Table 12).

Table -12: Effect of temperature on growth and sporulation of 3 isolates of *Ceratocystis paradoxa* (CP1, CP2, CP3)

	ISOLATE CP1			ISOLATE CP2			ISOLATE CP3		
Sr. Temp	**Diam**	**Spor-ula-tion**	**Growth**	**Col. Diam**	**Spor-ula-tion**	**Growth**	**Col. Diam**	**Spor-ula-tion**	**Growth**
1. 5°C	10	+	Poor growth with scanty creepy mycelium	8	+	Very poor scanty growth, creeping mycelium	12	+	Poor scanty growth indistinct
2. 10°C	28	++	Faint, very much indistinct,	30	++	Poor growth, circular submerged	35	++	Very poor submerged growth
3. 15°C	40	++	Hyaline, thin matted, translucent olivacious, black grwoth	48	++	Hyaline, think ` rayish black growth	45	++	Hyaline to ` rayish black, thin growth
4. 20°C	70	++	Aerial, dense at center, uniform olivacious black in colour	70	+++	Aerial dense growth at the center uniform	72	+++	Submerged to aerial growth more dense at the center
5 28°C	90	++++	Very good growth, sub aerial, olivacious black	87	++++	Very good growth, sub aerial to submerged, dense at the center,	90	++++	Luxuriant growth, submerged, dense at center, distinct black zone due to sporulation
6 35°C	8	+	Thin scanty submerged growth, ` rayish white sub aerial mycelium	10	+	Very thin scanty growth submerged	12	+	Scanty, hyaline growth, little dense at the center
7. 40°C	-	-	-	-	-	-	-	-	-

SPORULATION + Less (10 spores / microscopic field)
++ Moderate (10 - 10 spores / microscopic filed)
+++ Good (20 - 30 spores / microscopic field)
++++ Excellent above 30 spores/microscopic field

Ogawa (1970) reported that *Ceratocystis paradoxa* grew well at tropical temperature of 20-28°C. In the present study the fungus was found growing in the range of 5-35°C (optimum temperature). The similar results were found by Westejein and Okaer (1972) who stated that *Ceratocystis paradoxa* grew well at 20-30°C.

Milanes *et al.* (1986) observed optimum temperature, for mycelial growth of *Thielaviopsis* (*Ceratocystis paradoxa*) *in vitro* measured by diameter of colonies was 28°C, maximum 34°C and minimum 24°C and optimum temperature for infection were 22-26°C. Sastry *et al.* (1988) stated that optimum temperature found for growth of *Ceratocystis paradoxa* was 28°C. Nambiar *et al.* (1989) stated max. decay due to *Thielaviopsis paradoxa* during the months of July to Nov. due to higher humidity and comparatively moderate temperature (26 - 28°C).

The results of the present study also are in the line where fungus was found growing in between 20-30°C with an optimum temperature 28°C for maximum growth (Table 15).

ii. Influence of pH

Growth and sporulation of fungi are profoundly influenced by pH of the medium and this may be due to the effect of pH on enzymatic activity. Growth and sporulation of fungi may be inhibited or prevented by the media which are too acidic or too alkaline and this is due to the differential activity of the hydrogen ions at different pH levels. Most fungi tolerate a wide range of hydrogen ion concentration of the medium and they grow best at neutral reactions (i. e. pH 7) or even slightly on the acid side of the neutral. Growth is usually stopped on the acid side at pH 3 and on the alkaline side at pH 8-9 and the range for maximum growth as narrower and it varies with organism and strains. Extensive investigation have carried out by several workers in the past on the limiting pH values for the optimum and minimum growth on various fungi. The maximum, optimum and minimum pH for a fungus vary but generally an acid medium is favorable for the growth of many fungi and alkaline for bacteria.

The fungus generally grows maximum over a certain range of initial pH values of the medium and will fail to grow at high and low extremes under given conditions.

The present experiment was carried out on liquid Glucose aspargine medium. The results of the experiment showed that *Ceratocystis paradoxa* can tolerate a wide range of pH. Growth was observed from pH 3 to 9. The fungus showed good growth between 3 to 7 but pH 5 was found to be the optimum.

Sporulation was found at all pH levels i. e. pH 3-9. The fungus appeared to tolerate an acid reaction in the experiment. In the pH levels, the spores were agglutinated and were not liberated in culture filterate. Similar results were reported by Olutiola and Cole (1977). The fungus also appeared to tolerate acid reaction.

Cochrane (1958) reported that an initial pH 5-7 was satisfactory for the majority of fungi. Adisa (1983) found optimum pH for growth and sporulation of *Ceratocystis paradoxa* was 6-8. Natarajan and Subbaraja (1976) stated that growth *in vitro* was maximum in pH range 6.0 to 7.5. Chi (1949) reported the range of pH for growth in between pH 3 and 9 with the most suitable range between pH 5and 7. The present study was in agreement with the general observation on *Ceratocystis paradoxa.*

Nutritional Factors

Some of the nutritional factors that influence organism are different sources of carbon, nitrogen, amino acids and stimulatory factors like vitamins and trace elements, etc.

i. Carbon Source

Carbon is the chief constituent of all living organisms and almost half of the dry weight of the fungus consists of carbon only. As the organic compounds differ widely from each other in their structure, composition and properties, fungi are extremely selective in their ability to utilize such sources.

Bilgrami and Verma (1978) stated that majority of fungi utilize starch, a polysaccharide with high efficiency through amylase activity. Since starch is a polymer of maltose units the efficient utilization of starch indicated sufficient activity of enzyme and maltose. Lilly and Barnett (1951) reported that the fungi producing amylase are able to utilize starch efficiently. Similar observations were made by Hiramath and Hedge (1983) in case of *Dreshlera oryzae*. In absence of carbon source fungus failed to grow and is highly dependent upon these sources for growth.

For the carbon nutrition, a detailed study was undertaken with three isolates of *C. paradoxa*. Detailed observations were made on dry mycelial weight, sporulation and change in pH after growth of fungus. The similar observations were taken in solid medium also after 3 days of incubation, the observations were recorded on radial growth and sporulation with colony characters. The results are presented in Tables 13 and 14.

Table-13 : Comparitive carbon requirements of three isolates of *Ceratocystis paradoxa* (CP1, CP2, CP3) "on liquid medium"

Sr. No. Test Source	ISOLATE CP1			ISOLATE CP2			ISOLATE CP3			Average Mycelia Dry Weight (mg)
	Myceli al Dry Weight (mg)	Sporu- lation	pH	Myceli al Dry Weight (mg)	Spor- ulation	pH	Myce- lial Dry Weight (mg)	Sporul- ation	pH	
A.MONOSACCHARIDES										
1. Xylose	530	++++	6.34	580	++++	6.07	520	++++	5.95	543
2. Fructose	520	++++	6.36	520	++++	6.19	7.30	++++	6.23	590
3. Glucose	450	++++	6.21	470	++++	5.96	560	++++	5.85	510
4. Galectose	430	++++	6.06	540	++++	5.94	560	++++	5.93	453
5. Mannose	380	++++	6.21	410	++++	6.17	570	++++	576	443
6. Arabinose	290	++++	6.28	660	++++	5.89	380	++++	5.76	443
7. Dulcitol	260	++++	6.11	270	++++	5.89	380	++++	5.18	237
8. Mono inosital	260	++++	5.85	350	++++	5.82	120	++++	5.58	243
9. Rhamnose	200	++++	5.33	170	++++	5.32	130	++++	5.61	167
10. Ribose	140	++++	4.24	140	++++	4.05	50	++++	4.17	110
B. DISACCHARIDES										
11. Cellobiose	530	++++	6.06	490	++++	5.96	440	++++	5.872	487
12. Maltose	500	++++	6.24	510	++++	6.08	540	++++	6.16	517
13. Sucrose	420	++++	6.05	460	++++	6.23	540	++++	5.92	473
14. Lactose	350	++++	6.24	430	++++	6.31	400	++++	5.81	393
15. Trehalose	330	++++	6.24	600	++++	5.85	340	++++	6.21	423
16. Melibiose	260	++++	6.27	290	++++	6.34	200	++++	6.01	250
C. POLYSACCHARIDES										
17. Dextrin	460	++++	6.32	510	++++	6.36	510	++++	6.22	493
18. Starch	450	++++	6.42	500	++++	5.75	430	++++	5.44	460
19. Salicin	130	++++	3.46	140	++++	3.09	180	++++	3.40	150
D. OLIGOSACCHARIDES										
20 Raffinose	300	++++	6.30	260	++++	6.23	300	++++	6.22	287
E. SUGAR ALCOHOLS										
21. Mannitol	230	++++	6.00	230	++++	5.87	150	++++	5.59	203
22. Sorbitol	210	++++	6.31	420	++++	6.28	160	++++	6.03	263
F. CONTROL	140	++++	6.11	130	++++	6.36	60	++++	5.58	110
SE	10.35			27.49		5.00				
SED	14.62			38.84		7.06				
CD 5%	29.56			78.49		14.28				

Table - 14: Comparitive carbon requirements of three isolates of *Ceratocystis paradoxa* (CI1, CP2, CP3) "in solid medium"

Sr. No.	Test Source	ISOLATE CP1		ISOLATE CP2		ISOLATE CP3	
		Dia mm	Growth Characters	Dia mm	Growth Characters	Dia mm	Growth Characters
		A. MONO SACCHARIDES					
1.	Xylose	85	Very rich growth, black mass in center, good sporulation	76	Poor growth, creeping scanty white mycelium	83	Thin hyaline creeping growth, white mycelium
2.	Fructose	85	Very good growth, circular whitish mycelium, central part black due to sporulation	85	Rich growth, circular white mycelium, central black mass due to sporulation	86	Very good growth, whitish gray mycelium, good sporulation
3.	Glucose	90	Profuse aerial growth, white mycelium	80	Very good growth, grayish black mycelium	80	Very good growth, black masses, due to sporulation
4.	Galactose	73	Thin growth, circular white mycelium	80	Sub aerial white mycelium	80	Very good growth, black masses, due to sporulation
5.	Mannose	85	Profuse aerial mycelium	70	Very good growth, greyish black colour	83	Very good growth, greyish black colour
6.	Arabinose	80	Very thin scanty growth, white mycelium	85	Very thin scanty growth white mycelium	65	Very thin scanty growth, white mycelium
7.	Dulcitol	80	Very poor growth, thin mycelium	80	Very poor growth, white mycelium	82	Very poor grwoth, white mycelium
8.	Meso inosital	78	Very thin poor growth white mycelium	58	Very thin poor growth, white mycelium	78	Very thin poor growth, white mycelium
9.	Rhamnose	80	Very thin growth, scanty white mycelium less sporulation	40	Very thin scanty growth, no black colour due to sporulation	73	Thin creeping growth, less sporulation, circular growth
10.	Ribose	53	Very thin growth, white mycelium, less sporulation	60	Scanty growth white mycelium, less sporulation	75	Moderate growth, thin creeping white mycelium less sporulation

B. DISACCHARIDES							
11.	Cellobiose	90	Very good growth, sub aerial mycelium, good sporulation	90	Very good sub-aerial mycelium, good sporulation	70	Very good growth, sub aerial mycelium, good sporulation
12.	Maltose	90	Profuse luxuriant growth, good sporulation	90	Good growth thin, moderate sporulation	90	Very good growth sub aerial mycelium, black masses due to sporulation
13.	Sucrose	85	Very good growth, profuse sporulation, sub aerial to aerial mycelium	88	Luxuriant growth, profuse sporulation, sub-aerial mycelium, good sporulation	70	Thin circular growth, white mycelium, aerial sporulation
14.	Lactose	85	Thin scanty white mycelium, less sporulation	73	Thin growth circular, white mycelium, aerial mycelium	80	Thin creeping growth, white mycelium, sub-aerial less sporulation
15.	Trehalose	83	Scanty thin growth, white mycelium, less sporulation	90	Scanty thin growth, white mycelium, less sporulation	65	Thin creeping growth less sporulation
16.	Melibiose	78	Very thin growth, scanty white mycelium, circular less sporulation	60	Very thin scanty growth, circular, less sporulation	65	Thin creeping growth, less sporulation
C. POLY SACCHARIDES							
17.	Dextrin	90	Luxuriant circular growth very thick sporulation	90	Very good growth, thick excellent sporulation	90	Very good growth circular excellent and thick sporulation
18.	Starch	90	Very rich luxuriant growth, sporulation in center	85	Profuse sporulation, more in the center	85	Very good growth, sporulation in the center

19.	Salicin	60	Very thin scanty growth creeping white mycelium	65	Very thin creeping growth, white mycelium	70	Very thin growth, white mycelium, white mycelium, creeping growth
D.OLIGO SACCHARIDE							
20	Raffinose	88	Scanty thin growth	75	Good growth but thin good sporulatin	73	Good growth but thin sub acrial to sub merged
E. SUGAR ALCOHOLS							
21	Mannitol	90	Very thin scanty growth, white mycelium less sporulation	80	Very thin growth, white mycelium less sporulation	80	Very thin white mycelium, less sporulation
22	Sorbitol	78	Very thin growth, white mycelium, poor sporulation	60	Very poor growth, white mycelium	65	Very scanty white mycelium, less sporulation
F.	**Control**	74	Scanty white mycelium very thin, no sporulation	60	Scanty white mycelium, very thin growth, no sporulation	75	Scanty growth, white mycelium, less sporulation
	SE SED CD5%		1.36 1.84 3.73		4.07 5.75 11.60		2.92 4.13 8.34

CP1 Isolate

Among all the monosaccharides, xylose was the best carbon source for the isolate followed by fructose and glucose. In solid medium also, maximum growth was seen in xylose, fructose and glucose. Galactose and rhamnose were moderately utilized whereas arabinose, dulcitol meso-inositol and sorbitol were utilized to some extent but ribose was found to be less utilized as carbon source among the monosaccharides. Sporulation was good in all the sources and variation in pH after growth was also not very much except in the case of ribose where it was drastically less (pH 4.24) in comparison with control (pH 6.11) (Table 13 and 14).

Among the disaccharides, cellobiose was best source for this isolate followed by maltose, sucrose, lactose, trehalose, melibiose. In solid medium, the maximum growth was seen in cellobiose followed by maltose and sucrose. Among polysaccharides, dextrin and starch were good carbon sources but salicin was found very poor source for this isolate in terms of mycelial weight. In solid medium, the maximum growth was seen in dextrin and starch. Raffinose was also moderately utilized and mannitol, sorbitol (Sugar alcohol) was found moderate source in comparison to control (Table -16). In solid medium, the colony showed big diameter but very thin scanty growth was observed with absolutely no sporulation (Table 13).

CP 2 Isolate

Among the monosaccharides, arabinose was best-utilized source of carbon followed by xylose, galactose, sorbitol, mannose, mono-inositol. Ribose was found to be best-utilized source of carbon for this isolate. In solid medium, the maximum growth was seen in arabinose followed by xylose, galactose, fructose and glucose. Ribose showed less growth.

Among the disaccharide, trehalose was the best source for CP2 isolate where maximum mycelial weight was found followed by maltose, cellobiose, sucrose, lactose, melibiose. Rhamnose was found very poor source of carbon for the mycelial growth of this isolate. The same results were obtained by solid medium observations also.

Among polysaccharide, dextrin and starch proved to be very good sources of carbon while salicin was very poor source of carbon. Raffinose and mannitol were utilized moderately by isolate. Dextrin and starch showed good growth on solid medium also.

In all the test sources, the sporulation was found very good. Agglutinated spores were found in masses. Macroconidia were more in comparison to microconidia. pH also showed variation in between pH 6 to pH 6.36 except in salicin and ribose where it dropped to pH 5 (Table 13 and 14).

CP 3 Isolate

Fructose among all the monosaccharides was found to be the best source of carbon for the growth and sporulation for this isolate followed by galactose, mannose, glucose and xylose. Arabinose meso-inositol and sorbitol were moderately utilized. Ribose was found as very poor carbon source for this isolate than cellobiose, 1actose, melibiose. Rhamnose was the poorest monosaccharide for the growth of this isolate.

Dextrin and starch were very good in comparison to salicin among polysaccharides. Raffinose and mannitol were moderate in utilization in comparison to control.

In all the cases the sporulation did occur and pH variation was found in the range of 5.61 to 6.23 except in ribose and salicin where it was observed 4.17 and 3.40 respectively.

Fructose was the best source of carbon followed by xylose, maltose, galactose, dextrin, glucose along with sporulation except ribose and salicin where growth was very poor. The control (without carbon) also showed poor sporulation but growth was very scanty and hyaline (Table 13 and 14).

Bhaskaran (1971) observed maximum growth in glucose followed by sucrose and fructose. Natrajan and Subbaraja (1976) observed maximum growth in glucose followed by sucrose and fructose. In the present study all the carbon sources were utilized by fungus to grow well. All the three isolates grew well in glucose and fructose followed by arabinose, dextrin, maltose, and starch. Chi (1949) reported that sucrose, maltose, and soluble starch, are the most favorable sources of carbohydrates for the growth of fungus *Ceratocystis paradoxa*.

Sastry *et al.* (1988) stated starch as the best source of carbon for fungal growth. Olutiola (1983) stated Galactose as a best carbon source for the growth of *Ceratocystis paradoxa*. Gautam (1983) found fructose and glucose media as best source. Adisa (1988) reported that all 6-carbon sources were very well utilized by *Ceratocystis paradoxa*. The present study reveals that all carbohydrates were utilized by fungus (Adisa, 1988), but in monosaccharide glucose and fructose were judged the best and dextrin and starch were very good in disaccharide's. Sugar alcohols were found poor source for the growth of *Ceratocystis paradoxa*.

The better growth of the fungus *Ceratocystis paradoxa* on glucose and fructose could be due to quicker hydrolysis of these compounds than sucrose (Natarajan and Subbaraja, 1976). Sucrose is disaccharide and it is utilized well because pathogen can convert sucrose easily into glucose and fructose before utilization. In the present study starch was also found good sources of carbon nutrition.

ii. Nitrogen Sources

Nitrogen is the one of essential elements used by fungi for functional as well as structural purposes. It is an essential element playing a major role in the growth and sporulation of fungi like other organisms. A particular source of nitrogen is, however, not equally suitable for every fungus because various organisms are very specific in the choice of nitrogen for their growth and reproduction. Generally, the pathogenic fungi grow more luxuriantly when organic nitrogen is available in the substrate then when inorganic nitrogen is present.

The results of our study indicated that ammonium nitrate and ammonium chloride were utilized by all three isolates but the pH was drastically altered after the growth of fungus (Table 15 and 16). Sodium nitrate, ammonium sulphate, ammonium dihydrogen phosphate, was also utilized. Urea and

Table-15: Comparative nitrogen requirements of three isolates of *Ceratocystis paradoxa* (CP1, CP2, CP3) "On Liquid Medium"

		ISOLATE CP1			ISOLATE CP2			ISOLATE CP3			
Sr. No.	Test Source	Mycelial Dry Weight (mg)	Sporulation	pH	Mycelial Dry Weight (mg)	Sporulation	pH	Mycelial Dry Weight (mg)	Sporulation	pH	Average Mycelial Dry Weight (mg)
1.	Ammonium nitrate	290	++++	1.92	300	++++	1.91	300	++++	1.89	297
2.	Ammonium chloride	290	++++	1.85	230	++++	1.95	250	++++	1.87	257
3.	Sodium nitrate	260	++++	4.32	140	++++	4.06	120	++++	5.50	173
4.	Ammonium sulphate	250	++++	2.43	190	++++	2.56	190	++++	2.80	210
5.	Ammonium dihydrogen phosphate	220	++++	2.68	190	++++	290	180	++++	2.49	197
6.	Urea	190	++++	1.52	160	++++	7.59	230	++++	6.90	193
7.	Potassium nitrate	150	++++	4.21	80	++++	4.00	160	++++	4.35	130
8.	Control	70	++++	3.37	90	++++	3.31	40	++++	3.05	67
	SE	10.74			7.93			7.20			
	SED	15.17			11.21			10.17			
	CD 5%	32.54			24.04			21.81			

Sporulation : ++++ Excellent; +++ Good; ++ Moderate; + Poor

potassium nitrate were found very poor sources of nitrogen where very less growth was observed in the three isolates except CP3 where urea showed moderate growth.

From the results it may be concluded that the fungus preferred nitrate nitrogen in the present study except potassium nitrate which showed very poor growth. On solid medium all the sources showed good growth except urea where less growth was observed (Plate 16).

In view of the variations in the different isolates, it was considered desirable to carry out some investigations to determine the suitable sources of nitrogen for the growth and reproduction of all the three isolates of *Ceratocystis paradoxa* isolated from the sugarcane plant.

Nitrates have been reported to be excellent sources for imperfect fungi and ascomycetes (Thind and Randhawa, 1957; Misra and Mahmood, 1960). Inorganic and organic forms are taken up rapidly during growth phase of fungi. In the present study ammonium nitrate and ammonium chloride were

Table-16 : Compraitive nitrogen requirements of three isolates of *Ceratocystis paradoxa* CP1, CP2, CP3) "in solid medium"

		ISOLATE CP1		ISOLATE CP2		ISOLATE CP3	
Sr. No.	**Test Source**	**Dia mm**	**Growth Characters**	**Dia mm**	**Growth Characters**	**Dia mm**	**Growth Characters**
1.	Ammonium nitrate	83	Profuse aerial mycelium, white good growth	90	Submerged greyish mycelium, full growth	85	Submerged greyish mycelium, full growth
2.	Ammonium chloride	90	Profuse growth, white mycelium, less sporulation	90	Profuse vegetative growth, white mycelium	90	Profuse vegetative growth, white mycelium
3.	Sodium nitrate	83	Scanty well spread thin growth, less sporulation	43	Scanty well spread thin growth, less sporulation	73	Scanty thin submerged growth, less sporulation
4.	Ammonium sulphate	90	Good growth, whitish gray aerial mycelinm	90	Submerged greyish mycelium, full growth	90	Good growth, greyish black mycelium, good sporulation
5.	Ammonium dihydrogen phosphate	90	Aerial white mycelium,well spread growth, good sporulation	90	Sub merged greyish mycelium, full growth	44	Submerged greyish mycelium, good growth
6.	Urea	38	Very scanty growth, less sporulation	47	Very thin scanty growth, less sporulation	44	Very thin scanty growth, less sporulation
7.	Potassium nitrate	51	Scanty growth submerged less sporulation	58	Scanty but well spread sub merged growth, less sporulation	53	Scanty thin well spread growth
8.	Control	50	Very scanty poor growth with less sporulation	50	Scanty poor growth, less sporulation	60	Scanty very poor grwoth, less sporulation
	SE		0.89		2.85		0.80
	SED		1.25		4.02		1.13
	CD 5%		2.60		8.63		2.42

very well utilized by all three isolates followed by ammonium sulphate, ammonium dihydrogen phosphate and urea. Potassium nitrate could not support good growth.

Durairaj (1965) reported that in ammonium oxalate the carbon source is increased by the oxalate apart from ammonium salts as nitrogen sources. In the present study all inorganic sources of ammonium salt were utilized well by the fungus due to addition of nitrogen sources in the form of ammonium salts.

The results are contradictory to Tandon (1967) who stated potassium nitrate was superior to sodium nitrate, calcium nitrate and magnesium nitrate.

Robbins (1937) reported that *Ceratocystis fimbricata* and *Ceratocystis ulmi* require ammonium or organic nitrogen and was unable to assimilate nitrate nitrogen. The results of present study also support the same view. Fungus could grow well only in presence of organic source (aspargine) in synthetic medium (Glucose aspargine medium).

The capacity to use nitrate by the fungi actually depends upon the nitrate reduces enzyme activity. Sastry *et al.* (1988) reported that two inorganic nitrogen sources did not support the growth of *Ceratocystis paradoxa* out of 7, which were tested. The all the organic nitrogen sources supported good growth of *Ceratocystis paradoxa.* Gautam (1983) reported ammonium oxalate as good nitrogen source to support maximum secretion of macerating enzymes.

Jamaluddin *et al.* (1975) reported the role of pH on the utilization of potassium nitrate (nitrogen source) otherwise nitrite does not the support growth of many fungi which is attributed to known toxicity of this substance (Bilgrami and Verma, 1978).

iii. Amino Acid Sources

Following are the results of amino acid studies on CP1, CP2, CP3 isolates:

CP1 Isolate

The maximum growth was observed with aspargine followed by histidine, glycine, cystine, tyrosine, proline, glutamic acid, serine, valine and glutamine. All the amino acids were utilized well by the fungus except tryptophan, where the growth was observed least. In all the cases, sporulation did occur. Very good agglutinated spores were observed and in most of the cases, pH of the basal medium altered drastically. In control the growth was very poor, thin and hyaline, but sporulation was observed there also (Table 17, Plate 6b).

Table-17: Comprative amino acids requirements of three Isolates of *Ceratocystis paradoxa* (CP1, CP2, CP3) "on liquid medium"

Sr. No.	Test Source	ISOLATE CP1			ISOLATE CP2			ISOLATE CP3			
		Myce-lial Dry Weight (mg)	Sporu lation	pH	Myce-lial Dry Weight (mg)	Sporu-lation	pH	Myce-lial Dry Weight (mg)	Sporu-lation	pH	Average Myce-lial Dry Weight (mg)
1.	Aspargine	400	++++	6.55	390	++++	6.87	400	++++	7.33	397
2.	Histidine	330	++++	4.85	280	++++	5.24	210	++++	5.04	273
3.	Glycine	280	++++	6.78	290	++++	7.05	300	++++	7.13	290
4.	Alanine	270	++++	6.80	260	++++	6.94	200	++++	7.79	243
5.	Cystine	250	++++	2.32	200	++++	2.11	420	++++	2.18	290
6.	Tyrosine	240	++++	2.98	280	++++	3.21	190	++++	2.99	237
7.	Proline	230	++++	6.51	260	++++	6.39	140	++++	6.55	210
8.	L(+) glutamic acid	210	++++	7.19	210	++++	7.09	140	++++	7.38	187
9.	L(-) Serine	210	++++	6.58	230	++++	6.32	230	++++	7.02	223
10.	Valine	210	++++	5.70	300	++++	5.67	120	++++	4.87	210
11.	Glutamine	200	++++	4.00	320	++++	5.98	260	++++	4.11	253
12.	Cysteine hydrochloride	180	++++	3.60	80	++++	3.56	170	++++	3.00	130
13.	Methionie	140	++++	3.14	110	++++	3.07	130	++++	3.06	12.3
14.	Leucine	130	++++	4.74	110	++++	4.81	140	++++	4.59	127
15.	L-aspartic acid	130	++++	6.77	240	++++	6.76	190	++++	5.67	187
16.	L-phenylanine	130	++++	3.14	130	++++	3.25	120	++++	3.33	123
17.	Tryptophane	120	++++	3.50	140	++++	3.23	210	++++	3.27	140
18.	Control	70	++++	3.37	90	++++	3.31	40	++++	3.05	67
	SE	11.94			9.62			5.83			
	SED	16.87			13.59			8.23			
	CD 5%	34.44			27.76			16.81			

Sporulation; ++++ Excellent; +++ Good; ++ Moderate; + Poor

CP2 Isolate

Aspargine showed maximum growth followed by valine, glycine, tyrosine, histidine, alanine, proline, leucine, glutamine, glutamic acid, cystine and L Serine. Cystine hydrochloride was least utilized by this isolate. Tryptophan and L-phenylanine also showed less growth in comparison to other amino acids (Table 17).

Sporulation took place in almost all the cases. In control also, sporulation did take place but the mycelia growth was very less due to lack of nitrogen sources in the basal medium.

CP3 Isolate

Maximum growth was observed in Cystine and Aspargine which were followed by Glycine, Glutamic acid, L-Serine, Histidine, Tryptophan, Alanine, and L-Aspartic acid. Other amino acids were utilized moderately. Good sporulatiion was seen in almost all the test sources. Valine and L-phenylanine were least utilized sources of amino acids (Table 17).

Control showed very less mycelial growth but sporulation did occur. In the test sources pH altered drastically except Aspargine, Glycine, Glutamic acid, Serine and Alanine.

Growth on solid medium

All the three isolates grew very well in the solid medium. It is revealed from the results that the CP1 isolate grew very well in all the test sources of amino acids except Tyrosine, Cystine, L-phenylanine, Tryptophan. For CP2 isolate, growth was submerged and sporulation was seen in the centre in the from of black masses except L-phenylanine and Tryptophan, the growth was good in all the sources.

CP3 isolate also showed submerged growth with black masses in the center due to sporulation Tryptophan showed very poor growth. L-phenylanine also showed less growth in solid medium (Plate -6b).

In the present study maximum growth and sporulation of *Ceratocystis paradoxa* was noticed in aspargine followed by glycine, histidine, cystine and glutamine out 17 amino acids sources. Mathiasen-Kaarik (1960), Bhaskaran (1971), Olutiola and Cole (1977) have observed the same findings and stated that maximum growth and sporulation of *Ceratocystis paradoxa* occurred with aspargine as the nitrogen source.

In present study most of the amino acids were utilized well. Bhaskaran

(1971) stated that histidine, leucine, Tryptophan and valine were commonly utilized by *C. paradoxa*. Crisan *et al* (1983) noted that cystine and cysteine were inhibitory to the growth and development of *C. paradoxa*. In present study cystine was found poor source for the growth *C. paradoxa*. Sastry *et al.* (1988) reported peptone as good source of nitrogen because it has a mixture of some amino acid as contaminants. Lilly (1963) stated that an amino acid or mixture of amino acids my be good source of nitrogen.

Olutiola (1983) observed aspargine as the best sources of nitrogen when galactose was used as sole source of carbon. Such better growth of fungi on organic nitrogen sources has been attributed to the extra carbon component and the possible presence of certain substances which may serve as growth factors (Olutiola, 1972).

iv. Vitamins Sources

Vitamins play an important role in promoting the growth of various fungi, and fungi widely differ in their ability to synthesize vitamins necessary for their growth and reproduction. Previous workers have shown that most fungi, which require biotin or thiamin, need more of the vitamin for adequate sporulation than they do for vegetative growth. Cochrane (1958), Robbins and Kavanagh (1942) have summarized the relation of fungi to vitamins with 113 filamentous fungi of different genera.

Vitamins in minute quanitites are effective in many of the biochemical reactions of the cell, which reflects their role as catalysts. Many of the vitamins perform their catalytic role as coenzyme or constituent parts of coenzyme (Bilgrami and Verma (1978).

Observations regarding dry weight, sporulation and growth characters of the fungus recorded at the end are assembled in Tables 18 & 19; Plate 7a).

CP1 Isolate

The effect of vitamins on the growth was enormous. In comparison to control (Without vitamin), all the test sources showed marked difference in the dry mycelial weight. Good sporulation was recorded in all the treatments. The maximum growth was observed with vitamin B complex followed by vitamin B1, biotin, folic acid, pyridoxine, riboflavin, thiamine, ca-penthanate, ascorbic acid, and niacin. The least effective Vitamin niacin also showed better growth than control (Table 18 & 19; Plate 7a).

In solid medium, the fungus grew very well in all the treatments. White colony with fluffy growth was observed more dense in the centre.

Table-18: Comparative vitamin nutritional requirement of three isolates of *Ceratocystis Paradoxa* CP1, CP2, CP3)" on liquid medium"

Sr. No.	Test Source	ISOLATE CP1 Myceli al Dry Wei-ght (mg)	ISOLATE CP1 Sporula tion	ISOLATE CP1 pH	ISOLATE CP2 Mycelial Dry Weight (mg)	ISOLATE CP2 Sporula tion	ISOLATE CP2 pH	ISOLATE CP3 Mycelia l Dry Weight (mg)	ISOLATE CP3 Sporula tion	ISOLATE CP3 pH	Average Mycelial Dry Weight (mg)
1.	B Complex	1170	++++	7.53	440	++++	7.07	580	++++	7.25	730
2.	Vitamin B12	1160	++++	7.38	8.10	++++	7.28	610	++++	7.24	860
3.	Biotin	900	++++	7.46	470	++++	6.35	560	++++	7.25	643
4.	Folic acid	880	++++	7.47	720	++++	7.43	400	++++	6.84	667
5.	Pyridoxine	880	++++	7.53	230	++++	5.12	350	++++	6.70	487
6.	Riboflavin	770	++++	7.49	430	++++	6.76	360	++++	6.96	520
7.	Thiamine	750	++++	765	480	++++	7.15	510	++++	7.07	580
8.	Ca.penthanate	730	++++	7.52	620	++++	7.15	430	++++	6.80	593
9.	Ascorbic acid	560	++++	7.56	420	++++	6.74	710	++++	7.43	563
10.	Niacin	550	++++	7.46	300	++++	6.94	380	++++	6.47	410
11.	Control	460	++++	7.48	430	++++	6.28	400	++++	6.26	430
	SE	7.86			4.55			56.34			
	SED	11.11			6.43			79.60			
	CD 5%	23.18			13.41			166.05			

Sporulation ++++ Excellent; +++ Good; ++ Moderate; + Poor

CP2 Isolate

Vitamins B12 was the best vitamin for the growth of mycelium followed by folic acid, Ca-penthanate, thiamine, biotin, B-complex, riboflavin, ascorbic acid, niacin and pyridoxine. Only in the case of pyridoxine and niacin, the growth was less in comparison with control (without vitamin). Rest all the treatments showed enhancement of growth due to the addition of vitamins. All the treatments showed good sporulation.

In solid medium, very good but submerged growth was observed in all the treatments. Black masses were seen in the center due to macrospores production. Control also showed good growth and sporulation (Plate 7a).

CP3 Isolate

Ascorbic acid was the best source of vitamin for this isolate followed by vitamin B12, B-compex, biotin, thiamine, ca-penthanate, folic acid, niacin, riboflavin, pyridoxine. In comparison to control, in all other treatments enhancement of growth was observed except pyridoxine, riboflavin and niacin.

In solid medium, very good growth was observed in all the treatments.

Table-19: Comparative vitamin requirements of three isolates of *Ceratocystis paradoxa* (CP1, CP2, CP3) "in solid medium"

Sr. No.	Test Source	ISOLATE CP1		ISOLATE CP2		ISOLATE CP3	
		Col. Dia mm	Growth Characters	Col. Dia mm	Growth Characters	Dia mm	Growth Characters
1.	B Complex	85	Cottony fluffy growth, white aerial mycelium, good growth	77	Submerged growth white mycelium at the periphery, central part black due to sporulation	75	Submerged good growth, white mycelium, good sporulation
2.	Vitamin B12	85	Cotton fluffy aerial white mycelium, dense growth in the center	75	Good growth, greyish white mycelium, black masses due to sporulation	70	Good growth, whitish gray mycelium, good sporulation
3.	Biotin	85	Cottony white fluffy mycelium in the center, circular full growth	75	Submerged white mycelium, wavy outline, central black masses due to sporulation	79	Submerged growth, white mycelium, black masses in the center due to sporulation
4.	Folic acid	85	Cottony white fluffy growth, circular but wavy margin	75	Submerged growth white mycelium, black masses due to sporulation	75	Submerged growth white mycelium, black masses in the center due to sporulation
5.	Pyridoxine	84	White fluffy cottony mycelium, sporulation only in the center	77	Submerged regular growth, sporulation in the center	79	Submerged regular white mycelium, black masses in the center due to sorulation
6.	Riboflavin	75	Aerial fluffy white mycelium, circular good growth	63	Submerged whitish gray mycelium, black masses due to sporulation	55	Sub aerial greyish white mycelium, central black masses due to sporulation
7.	Thiamine	79	Cottony white mycelium, good growth in the center	70	Submerged good growth, sporulation only in the center	75	Submerged good growth, sporulation only in the center
8.	Ca-penthanate	84	White cottony fluffy mycelium, dense in the center	75	Submerged growth, circular sporulation, only in the center	75	Submerged good growth, sporulation only in the center

9.	Ascorbic Acid	81	White fluffy aerial mycelium, circular smooth outline	77	Submerged white mycelium, central black masses due to spo9rulation	68	Submerged good growth, sporulation only in the center
10.	Niacin	82	White cottony fluffy mycelium, circular wavy outline	78	Submerged growth, central black masses due to sporulation	73	Submerged good growth, sporulation only in the center
11.	Control	85	Good growth, white cottony, mycelium, full growth	75	Submerged mycelium, central black masses due to sporulation	75	Submerged good growth, sporulation only in the center
	SE SED CD 5%	0.45 0.63 1.32		0.84 1.18 2.47		0.71 1.00 2.68	

Sporulation was also very good. The growth was submerged and due to microspore production, black masses were seen in the central part of petriplates.

In the present study there was marked stimulation of growth and sporulation of *Ceratocystis paradoxa* in different vitamin sources in liquid medium. Maximum growth was observed in vitamin B12 followed by B-complex where effective growth was observed in comparison to control (Table 18, 19 and Plate 7a).

Bilgrami and Verma (1978, 1981) stated that several fungi are known to respond with stimulated growth factors because in many such cases identical growth response could not be induced by addition of specific purified vitamin or other nutrients. Jamaluddin (1979) discussed the vitamin requiements of *Ceratocystis paradoxa* (pineapple fruits).

Trace Elements

Through the classical researches of Raulin (1869), Steinberg (1939, 1950), Blank (1941), the importance of trace elements came to be recognized in fungi, Blank and Talley (1941) took *Aspergillus-niger* as their experimental fungus for their exhaustive study on trace element nutrition.

With a view to find enhancement effect or retarding effect of trace elements on the growth and sporulation of fungus, an experiment was done. All the isolates showed very rich growth in all the trace elements as well as control (Plate 7b).

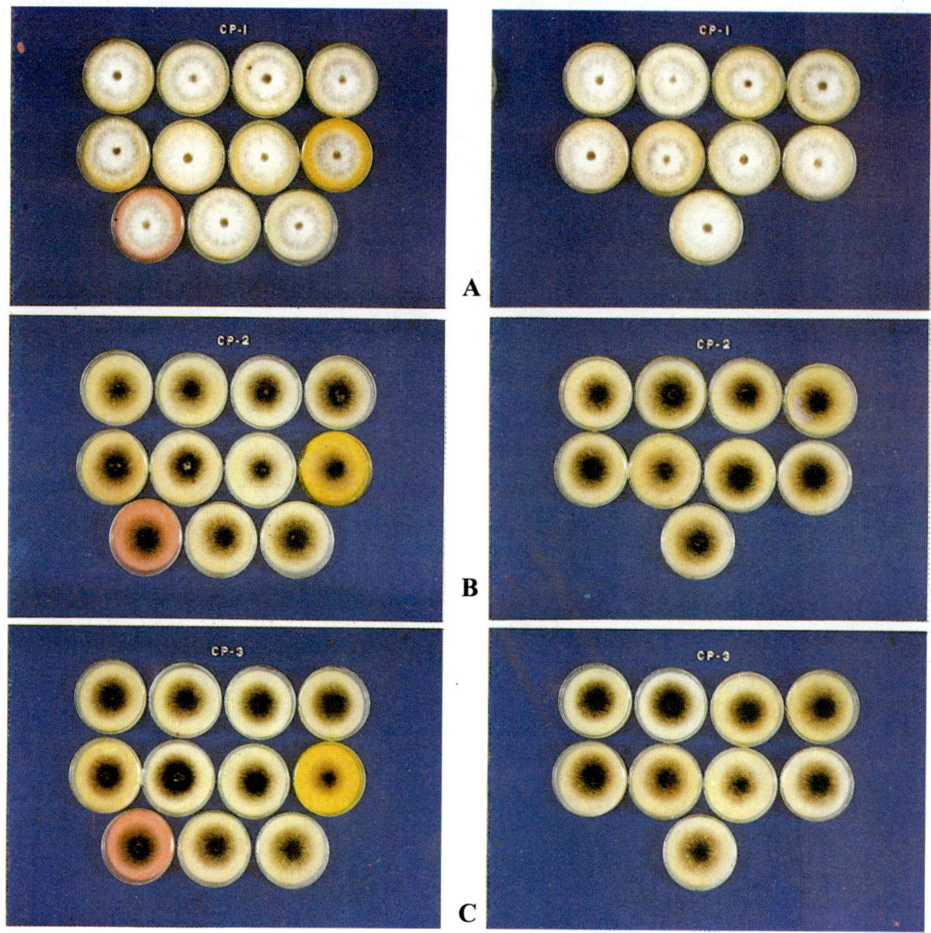

Plate 7a : Effect of vitamins on the growth of CP1, CP2 and CP3 isolates of *Ceratocystis paradoxa*; A : CP1 isolate (Co 7219); B : CP 2 isolate (Co 7527); C : CP3 isolate (CoC 671); All isolates showed profused growth on all test sources. The isolates grew equally well without vitamins also in solid medium (Control; Vitamin sources from left to right : First row : Ascorbic acid; B-Complex; Biotin; Ca-penthanate; Second row : Folic acid; Pyridoxin; Thiamine; Riboflavin; Third row : Vitamin B-12; Niacin; Control

Plate 7b : Effect of trace elements on the growth of three isolates of *Ceratocystis paradoxa*; A : CP1 isolate from Co 7219; B : CP2 isolate from 7527; C : CP3 isolate from CoC 671; All the isolates showed very rich growth in all the trace elements as well as control. Trace elements sources from left to right : First row : Boron; Calcium; Cobalt; Copper; Second row : Iron; Molybdenum; Lead; Zinc; Third row : Control

CP1 Isolate

The growth was enhanced by addition of trace elements for this isolate. The maximum growth was recorded when cobalt was added in the basal medium followed by copper, iron, lead, zinc, boron, calcium and molybdenum. The growth was much more in all the treatments in comparison to control. Sporulation was very good in all the treatments.

In solid medium also the growth was very profuse. Colonies showed cottony fluffy white mycelium towards periphery and slightly grayish in the center due to macrospores production. The growth in control was also equally good, but in liquid medium the growth was much more in comparison to solid medium.

CP2 Isolate

Boron showed maximum growth for this isolate followed by iron, calcium, molybdenum, zinc, lead, cobalt, copper. In comparison to control, the growth was less in cobalt and copper. Rest all showed good growth or equally good growth.

In solid medium also, boron showed maximum growth in colony diameter. Sporulation was very good. The growths of colonies were submerged. Black masses of macrospores were seen in the center of the plates.

CP3 Isolate

Calcium was found best among the various trace elements tried in the experiment followed by lead, zinc, boron, molybdenum, cobalt, iron, copper. Copper and cobalt showed less growth in the treatment in comparison to control. Iron and molybdenum also showed less growth in comparison to control.

In solid medium, good growth was observed. Black masses were seen in the center due to sporulation. Growth was less in the case of molybdenum.

In the present study marked influence of trace elements on the growth of *Ceratocystis paradoxa* is observed in liquid medium. There was not much difference in the solid medium in comparison to control. Cobalt showed maximum growth in terms of dry mycelium weight followed by lead, calcium, and boron which were also growth stimulators for *Ceratocystis paradoxa*.

References

Ahmed, H. U., Khan, M. A. and Mian, M. I. H. (1977). Screening sugarcane clones for resistance to pineapple disease. Sugarcane Pathol. News letter, 18: 8-8.

Ahmed, J. (1960). Preliminary study on varietal resistance of standing cane to pineapple disease and the influence of the disease on the juice quality. Agric. Jour., **47**: 8-15.

Anil Kumar and Nambiar K.K.N. (1990). Effect of certain physical and chemical factors on the germination of endoconidia and chlamydospores of *Thielaviopsis paradoxa* India Phytopathology, **43** (3): 460-462.

Adisa, V. A. (1988). Some nutritional requirements of spoilage molds of pineapple fruits. Cryptogamie, Mycologic, **9** (2): 111-116.

Agnihotri, V. P. (1990). Diseases of sugarcane. Oxford IBH Publishing Co., 363p.

Badgar, A. M. (1966). Influence of relative humidity on fungi causing brown rot of boxed bananas. Phytopathology, **55**: 688-692.

Bilgrami, K. S. and Verma, R. N. (1978). Physiology of fungi. Vikas Publ. House, New Delhi, 274-297.

Bilgrami, K. S. and Verma, R. N. (1981). Physiology of fungi. Vikas Pub. House, New Delhi pp. 274-297.

Bhaskaran, T. L. (1971). Studies on storage disease of banana. M. Sc (Agr.) Thesis, Tamil Nadu Agric. Univ. (Unpublished).

Bell, A.F. (1936). Pineapple disease. A cause of germination in cane. Cane Growers Quat. Bull., **4**: 46-48.

Byther, R.S. and Moore, P.H. (1974). Inhibition of sugarcane rooting by *Ceratocystis paradoxa*. Can. Jour. Bot., **52**: 761-766.

Blank, L. M. and Talley, P.J. (1941). The carbon utilization of carbohydrates activity of *Phymatotrichum omnivorum*. Am. J. Bot., **28**: 564-569.

Chi, Chien-Chen. (1949). A preliminary report on the study of pineapple disease of sugarcane in Taiwan. Jour. of Sugarcane Res., **3**: 71-102.

Cochrane, V. W. (1958). Physiology of fungi. John Wiley and Sons Inc. New York Chapman and Hall Ltd. London.

Crisan, A., Fabian, A., Soimu, T. and Fabian, N. (1983). Role of sulphur in metabolism of some plant pathogenic fungi. Clui Bapoca, Romanai Univ. "Sabes-Bolvai". 167-175.

Durairaj, V. (1965). Growth of *Colletotriehium Capsici* in pure culture. J. Indian Bot. Soc., **35**: 409-413.

Floribai, (1985). Pineapple disease in India M. SC. Thesis submitted to Tamil Nadu Agric Univ., Coimbatore (Unpublished).

Gautam. S. P. (1983). Effect of inhibitors on 'in vitro' activity of cellulotic enzymes by *Thielaviopsis paradoxa.* Nat. Acad. Sci. Letters, **6** (7): 215-217.

Hawes, C. R. (1980). Conidium germination in the chalara state of *Ceratocystis adiposa*: A light and electron microscope study. Trans. Bri. Mycol. Soc., **74** (2): 321-328.

Hiremath, P. C. and Hegde, R. K. (1983). Physiological studies of five isolates of *Drecaslera oryzae* (Breda da Hann). Rice Mysore J. Agri. Sci. **17**: 41-48.

Herrea, J., Mogling, R. and Veldes, M.A. (1981). Influence of some agro technological parameters in the planting of sugarcane on pineapple disease in Cuba. Centro Agucar, **8**: 3-9.

Hughes, C.O. (1949). The production and testing of sugarcane breeding. Tech. Commun. Bur. Sug. Expt. Stno., 17-69.

Jamaluddin, Tandon, N. P. and Tandon, R. N. (1975). *Ceratocystis* of pineapple and its chemical control. Proc. of the Nat. Acad. of Sci., **45** (4) : 217-220.

Jamaluddin, (1979). Vitamin requirements of *Ceratocystis paradoxa* (from pineapple). Proc. of the Nat. Acad. of Sci., **49** (1): 53-55.

Lilly, V. G. and Barnett, H. L. (1951). Physiology of fungi, McGrew Hill Book Co. Inc. USA.

Lilly, V. G. (1963). The relation of fungus physiology to physiology of diease V. V. Vir. Univ. Ag. Exp. Stru. Bull., **488**: 33-66.

Liu, J. and Monllor, A.C. (1972). Effect of temperuture and moisture on various apects of development, growth and pathogenicity of *Thielaviopsis paradoxa* from Sugarcane in Puerto Rico. J. Agri. Univ. Puerto Rico Rio Piedrus, **56** : 162-170

Milanes, V., Virellas, P. and Herrera Isla, L. (1986). Effect of temp. on the growth and pathogenic activity of *Thielaviopsis paradoxa* (de-seynes) Moreau. Centro agricola (1986) **13** (2): 24-28. (RPR 1988 Vol. 67 No. 372).

Misra, A. P. and Mahmood, M. (1960). Effect of carbon, nitrogen nutrition on growth world. Elsevier Publishing Co., Amsterdam. 542p. and sporulation of *Colletorichum capsisci* (Syd.) J. Indian Bot. Soc. **39**: 314-321.

Mathiesen-Kaarik, A. (1960). Growth and sporulation of Ophiostoma and some other blueing fungi on synthetic media. Symb. Bot. Upsaliena., **16**:168.

Martin, J. P., Abbott, E. V. and Hughes, C. G. (1961). Sugarcane diseases of the world. E.V. Abbott *et al.*, Elsvier Pub. Co., UK.

McMartin, A. (1937). Pathological condition affecting growth of sugarcane plant cutting from Natal, S. Africa sugar Jour., **21**: 353-359.

Manzo, S.K. (1975). Pineapple disease in standing sugarcane in Nigeria. Sugarcane Pathologists Newsletter, **13/14**:3-4.

Natarajan, S. and Subba Raja, K. T. (1976). Infection by *Ceratocystis paradoxa* Moreauon on standing canes of some sugarcane clones. Sugarcane Path. Newsletter, **17**: 25-28.

Nambiar, K.K.N., Anil Kumar., Shetty, K. and Joshi, Y. (1989). Seasonal effect on infection by coconut stem bleeding pathogen *Thielaviopsis paradoxa*. Current Sci., **58** (1): 34-35.

Ogawa, J. M. and Gilpatrick, J. D. (1978). Variation in fungal growth on various preparation of potato dextrose agar media. Plant Dis. Reptr., **62**: 437-441.

Ogawa, J. M. (1970). Post harvest diseases of bananas in China. F. A. O. Pl. Prot. Bull., **18** : 31-42."

Omamor, I. B. (1985). Black rot of sprouted seeds of the oil palm Elais guineensis. Trans. Brit. Mycol. Soc., **84** (1): 159-161.

Olutiola, P.O. and Cole, O.O. (1977). Some environmental and nutritional factors affecting growth and sporulation of *C. paradoxa*. Mycologia, **69**: 524-532.

Olutiola, O and Patrick (1983). Galactose as a carbon scource for the growth of an oil palm [*Elacis Guineensis jacq.*] isolate of *Ceratocysits paradoxa*. Indian bot. Soc., **62**: 220-225.

Olutiola, P.O. (1972). Aspects of carbohydrate physiology of *Rhynchosporium secalis.* Ph.D. Thesis, Univ. of Lancaster, Lancaster, Great Britain-107 pl 1-11, with legends figures 1-* L11 with legends.

Patil, A. S. and Hapase, D. G. (1984). A new record of pineapple disease of sugarcane in Maharashtra State. Proc. D.S.T.A. 34th Ann. Conv; A 189-A 194.

Padmanabhan, D. and Ahmed, J. N. (1970). Studies on the infection of *Ceratocystis paradoxa* (de Seynes) Moreau on the standing canes in sugarcane varities. Indian sugars, **20**: 605-612.

Robertson, J. S. (1962). Dry basal rot, a new disease of oil palms caused by *Ceratocystis paradoxa*. Trans. Brit. Mycol. Soc., **45**: 474-478.

Robbins, W. J. (1937). The assimilation by plants of various form of nitrogen. Amer. J. Bot., **24**: 243-250.

Robbins, W. J. and Kavanagh, K. (1942). Vitamins deficiencies of the filamentous fungi. Bot. Club., **8**: 411-471.

Schreiber R. Lowrence (1969). Viability of *Ceratocystis ulmi* in young seedlings of American Elm and the effects of extracts from their tissues on conidial germination Phylopathology, **60**: 31-35.

Singh, D. and Smalley, B. E. (1969). Changes in amino acid and sugar constituents of the xylem sap of American Elm following Inoculation with *Ceractocystis ulmi.* Phytopathology, **59**: 891-896.

Sastry, M. N. L., Kulkarni, B. G., Hegde R. K. and Hiremath, P. C. (1989). "Studies on the disease caused by *Thielaviopsis* state of *Ceratocystis paradoxa* (Dade) Morean on arecanut, coconut and sugarcane I. Morphological, cultural and cross inoculation studies of three isolated of *Ceratocystis paradoxa*. Mysore J. of Agri. Sci., **23** (1): 44-48.

Sastry, M. N. L., Kulkarni, B. G. P. Hegde, R. K. and Hiremath, P. C. (1988). Nutritional and physiological studies of three isolates of *Ceratocystis paradoxa*. Mysore Jour. Agric. Sci., **22** (4): 475-478.

Singh, H., Patil, A. S. and Hapase D.G. (1991). Pineapple disease of Sugarcane –a Review, Bharatiya Sugar, **16** (1): 187-192.

Thind, R.N. and Bilgrami, K. S. (1957). Nitrogen nutrition of *Phyllosticata artocarpina proc. natn.* Acad. Sci. India, **27B**: 269-273.

Tandon, R. N. (1967). Presidential address of Indian Sci. Congress, **54**: 1-14.

Thompson, G.M. (1957). The disease of sugarcane in Natal. Expt. Sta. S. African Sug. Assoc. Bull., **9**:1.

Wismer,. C.A. (1951). Controlling pineapple disease of sugarcane. Hawaiian Planters Res., **1**: 23-53.

Waraitch, K. S. and Kumar, B. (1981). Relative behaviour of various sugarcane clones to Ceratocystis paradoxa Dade) C. Moreau, causal organism of organism of pineapple disease. Sugarcane Pathol. Newsletter, **26**: 38-40.

Went, F.A.F.C. (1896). Notes on sugarcane diseases. Ann. Bot., **10** : 583-600.

Wenteijin, G. and Okafer (1972). Comparison of *Cassava yam* and potato dextrose agars as fungal culture media. Neth. J. Pl. Path., **77**: 134-139.

35

Fungal Diversity of Indoor Environment of Occupational Importance in India: An Overview

J. K. Misra[1], Iqbal Hashim[1], S. N. Bahoray[1] and V.K. Mishra[2]

[1]*Mycological Research Unit, Department of Botany, Sri Jai Narain P. G. College, Lucknow-226 001*

[2]*Department of Botany, K. M. Agrawal College, Kalyan - 421 301*

In our country, where a large number of occupational environs exist, only thirty-two have been investigated so far for their indoor mycoflora. Not more than a dozen occupational places such as bakery, cobbler shops and tanneries, flour mill, food storage places/grain godowns, ginnery, hospital wards, mechanical and textile industries, library, pig farm, poultry farm, saw mill and so on have been seriously surveyed. Many other places have only cursorily been looket into. We know almost two hundred fungal forms from such a small number of places searched for culturable fungi. Furthermore, only a few fungi have been implicated to deteriorate the materials and affect human lives indoor suggesting their significance. Briefly, methodology used by indoor mycologists has been mentioned. Thus, the chapter embodies the consolidated data available for indoor mycology, particularly of occupational environs in India. The need for furhter research has been stressed.

Key words : Indoor, fungi, biodiversity, occupational environs, allergy

Aeromycology-the study of fungi in the air, their impact on crop plants and the interaction of environmental factors with the microfungal population in the air has predominantly been the province of plant pathologists/aerobiologists for many a years (Mehta, 1940, 1952; Padmanabhan *et al.*, 1953; Sreeramulu and Seshavataram, 1962; Sreeramulu and Ramalingam, 1963, 1964, 1965, 1966; Sreeramulu, 1967, 1969, 1970, 1974; Tilak and Srinivasulu, 1967, 1971; Talde, 1968, 1969; Mishra and Srivastava, 1969, 1971; Sreeramulu and Ramakrishna, 1970; Kulkarni, 1971; Sreeramulu and Vittal, 1971; Mishra, 1972; Sharma and Sinha, 1973; Gaikwad, 1974; Tilak and Kulkarni, 1975, 1978, 1980; Pande, 1976; Dhaware, 1976; Sinha, 1976; Mane, 1978; Reddi, 1978; Tilak and Chakre, 1978; Chakre, 1979; Vishwe, 1979; Bhalke, 1980; Lakhe, 1980; Tilak, 1980; Mane and Tilak, 1980; Tilak and Shastri, 1980; Shastri, 1981; Tilak and Babu, 1981; Sumbali and Badyal,

1991; Benagi and Kulkarni, 1995; Patel *et al.*, 1995; Bagwan and Meshram, 2000) until people further realized the importance of suspended fungal spores in causing damages to human lives and other materials in storage (Kathapalia, 1960). On realizing these facts investigations started to assess the role of fungi in causing allergenic and other disorders among the human beings. In this connection the first important publication is that of Cunningham (1873) who studied the bio-content of air over Calcutta jail. The reports that are in print mainly pertain to the outdoor air, which include list of fungi that could be recovered from the air, phenology of the fungal spores in the air and the effect of environmental factors on fungi. Only some have emphasized the allergenic significance of suspended fungal spores, but mostly without undertaking investigations in that direction (Rajan *et al.*, 1952; Gupta *et al.*, 1960; Nair, 1963; Chaubal and Deodikar, 1964; Sreeramulu and Ramalingam, 1964; Sandhu *et al.*, 1964, 1979; Baruah and Chetia, 1966; Ramalingam, 1966, 1971; Agarwal *et al.*, 1969; Shivpuri and Agarwal, 1969; Mukerji *et al.*, 1969; Reddi, 1970; Mishra and Kamal, 1971; Tilak *et al.*, 1972, 1980; Chakraverty and Nandi, 1972; Chitaley and Bajaj, 1973, 1974, 1975; Vishnu -Mittre and Khandelwal, 1973; Agarwal and Shivpuri, 1974; Tilak and Bhalke, 1976, 1978; Pathak *et al.*, 1978; Sharma and Govil, 1979; Bhati and Gaur, 1979; Rati and Ramalingam, 1979; Sandhu and Randhawa, 1979; Wadhwani, 1979; Paradkar and Munshi, 1980; Tilak, 1980; Sulia and Khan, 1980; Sinha *et al.*, 1981, 1984; D'Silva and Freitas, 1982; Vittal and Ponnusamy, 1982; Rai and Singh, 1988; Bhat and Rajasab, 1988; Tilak and Rao, 1988; Jain and Mishra, 1988; Khandelwal *et al.*, 1988; Verma and Khare, 1988; Ghani, 1993; Agashe and Sudha, 1995; Shah, 1997; Prasad and Pramila, 1997; Raha and Bhattachaya, 1997; Majumdar and Bhattacharya, 2000; Verma *et al.*, 2000; Sahay *et al.*, 2001; Mishra, 2002; Saravanan and Vittal, 2002). Furthermore, the information for fungi in the indoor air is very limited in the sense that very a few places have been looked into for the fungal forms and very little has been done to assess their significance in either damaging the health of the inhabitants or deteriorating the material indoors. The indoor occupational environs, that are many in our country, have not also seriously been studied for the fungal components present therein. Therefore, this chapter aims at consolidating the available information and literature on indoor mycology of occupational environs besides emphasizing the need for further studies.

Methodology

To study the indoor mycoflora almost all described aerobiological techniques can be used, with suitable modifications, to achieve the goal of the investigation. Further, the individual researcher has to choose the techniques according to his own convenience and the availability of the equipment and facilities. The common equipment used for aerobiological/aeromycological studies are:

a. Gravity Slide Sampler-Durham Sampler

b. Rotating Arm Sampler-Rotorod Sampler

c. Filtration Sampler-Molecular Filter Membrane

d. Volumetric Sampler-Tilak, Hirst and Burkard Samplers

For indoor aeromycological investigations, we used Rotorod and Tilak air samplers to find out, both quantitatively and qualitatively, the suspended fungal spores in the working environments. Gravity culture plate method was also used to get culture-able fungal forms. These methods in combination have, to a greater extent, given satisfactory result. However, other investigators have used different samplers in combination with gravity culture plate method.

1. ***Rotorod air sampler*** : We used battery operated Rotorod air sampler (Manufactured by T. Mohan Rao, Pune) for our work on saw and flour mill. This sampler has a small constant speed motor that whirls thin-sticky-coated brass rods about its axis at a constant high speed. The collecting arms of the sampler are made up of brass, having 0.159 cm cross sectional area, are square in shape and slightly bent inwards. The vertical arms are 6 cm long and 4 cm apart from the rotating axis. This portable, efficient and sensitive sampler has been used by our group of researchers both for indoor and outdoor surveys of airspora of Cobbler shops and Tanneries, Flour mill, Saw mills and Ginneries, besides that of Mechanical and Textile Industries (Mishra, V. K., personal communication). The sampler was run for 30-60 minutes at various places of survey. The arms were covered with cello tape. Jelly was applied on the tape as an adhesive for the spores. After definite exposure the tapes were removed from the arms and mounted on clean glass slides over which molten jelly was poured and covered with glass cover slips. After drying, the slides were examined under a microscope for identification of the spores.

2. ***Tilak air sampler*** : Tilak air sampler is a volumetric air sampler which samples known quantity of air per unit time and provides qualitative and quantitative data of the bio-particles present in the air during the sampling period/unit time. Since it is run by electricity, it cannot be used everywhere, but wherever we used this equipment, it gave us satisfactory and reliable result.

3. ***Culture plate method*** : Ten cm sterile glass or plastic disposable Petri plates having any suitable semi-solid medium such as Czapeck- Dox agar, Potato dextrose agar, Sabouraud agar and so on, supplemented with

suitable antibiotics (Penicillin G and Streptomycin sulfate) can be exposed for varying period of time at different corners and at a center of the working place to get the culture-able spores settled onto the plates. The plates after exposure are incubated at suitable temperature, preferably 27± 1°C, for a few days to observe the fungal colonies and isolate them on suitable agar slants for further morphological/clinical studies. At those working environs where fungal load was very high and it was difficult to observe and isolate the fungal colonies from the Petri plates, Rose Bengal dye was used to restrict the fungal growth and facilitate isolation. For further details of methods, etc. one should refer to Tilak (1982, 1987).

Mycology of indoor environment and environs of occupational importance

The mycology of indoor environment has not been much studied (Tilak and Kulkarni, 1972; Shankara and Ramalingam, 1988) and the studies related to indoor environment of occupational importance are very a few (Baruah, 1961; Sreeramulu, 1961; Tilak and Kulkarni, 1972, 1975; Tilak *et al.*, 1972; Tilak and Vishwe, 1975; Sandhu *et al.*, 1977; Jayaprakash *et al.*, 1978; Tilak and Chakre, 1979; Rati *et al.*, 1980; Singh, 1981; Chaturvedi *et al.*, 1981; Jayaprakash and Ramalingam, 1981, 1983; Mehta and Sandhu, 1983; Gupta and Singh 1983; Shukla and Misra, 1984, 1986; Vittal and Glory, 1985; Harindranath *et al.*, 1985; Chakraverty and Sinha, 1985; Tilak and Saibaba, 1986; Shukla, 1987; Tripathi, 1987; Verma and Khare, 1987; Tilak and Pillai, 1988; Mishra, 1988; Misra, 1988; Misra *et al.*, 1988, 1989; Misra and Shukla, 1989; Rasool, 1989; Santra and Chanda, 1989; Singh *et al.*, 1990a, 1990b, 1994, 1995; Misra and Jamil, 1991; Pandit and Singh, 1992; Verma and Chile, 1992a, 1992b; Ghani *et al.*, 1993; Ghani, 1994; Singh and Singh, 1994, 1999; Nadimuthu and Vittal, 1995; Surekha and Reddy, 1996; Rafiyuddin *et al.*, 1997; Agashe and Anuradha, 1998; Pugalmaran and Vittal, 1999; Singh *et al.*, 2000; Sharma and Dutta, 2001; Sahney and Purwar, 2001; Kulshrestha and Chauhan, 2001; Begum *et al.*, 2001; Udayprakash and Vittal, 2002). More than thirty working environments have been looked into for the fungal flora (Table 1). A perusal to the tabulated information indicate that mainly less than a dozen real working places such as bakery, cobbler shops and tanneries, flour mill, food storage places/grain godowns, ginnery, hospital wards, mechanical and textile industries, library, pig farm, poultry farm, saw mill and so on have been seriously surveyed; others have not received desired attention. However, surveyed environs have showed that there exists enough diversity in the fungal flora of indoor environments. The number of publications bears the testimony of what we feel that little has been done. As can be seen from the table (Table 1) that the number of publications on indoor flora ranges from 1-14. It is only the library environment that has been studied by various

Table 1. Occupational environs in India studied for their indoor fungi

Environ	Reference
1. Agarbathi factory	Jayaprakash *et al.*, 1978; Jayaprakash and Ramalingam, 1981
2. Animal care facilties	Udayprakash andVittal, 2002
3. Bakery	Jayaprakash *et al.*, 1978; Jayaprakash and Ramalingam, 1981; Singh *et al.*,1990a; Singh and Singh, 1994; Rafiyuddin *et al.*, 1997 Sharma and Dutta, 2001
4. Cane sugar mills (Sugar factory)	Mehta and Sandhu, 1983, Pandit and Singh, 1992
5. Cattle shed	Sreeramulu, 1961
6. Cinema hall	Singh, 1981
7. Cobbler shop and Tanneries	Shukla and Misra, 1984; Shukla, 1987; Misra and Shukla, 1989
8. Coffee curing works	Jayaprakash *et al.*, 1978; Jayaprakash and Ramalingam, 1981
9. Cow shed	Baruah, 1961; Jayaprakash *et al.*, 1978; Jayaprakash and Ramalingam, 1981
10. Fire wood depot	Jayaprakash *et al.*, 1978; Jayaprakash and Ramalingam, 1981
11. Flour mill	Gupta and Singh, 1983; Misra, 1988; Misra and Zafar, 1991; Jayaprakash *et al.*, 1978; Jayaprakash and Ramalingam, 1981
12. Food storage places	Santra and Chanda, 1989
13. Fruit shop/market	Chakraverty and Sinha, 1985; Jayaprakash *et al.*, 1978; Jayaprakash and Ramalingam, 1981
14. Ginnery	Shukla and Misra, 1986
15. Grain shops/storage godown	Tilak and Chakrre, 1979; Jayaprakash and Ramalingam, 1981; Chakraverty and Sinha, 1985; Pugalmaran and Vittal, 1999; Sharma and Dutta, 2001
16. Hospital ward	Tilak *et al.*, 1980; Jayaprakash and Ramalingam, 1981; Chakraverty and Sinha, 1985; Santra andChanda, 1989; Verma and Chile, 1992a, b; Singh *et al.*, 1994; Kulshrestha and Chauhan, 2001; Sharma and Dutta, 2001
17. Hostel kitchen	Jayaprakash *et al.*, 1978; Jayaprakash and Ramalingam, 1981
18. Industry-Mechanical & Textile	Mishra, 1998; Mishra, 2001

19. Jute mill	Santra and Chanda, 1989
20. Library	Tilak and Vishwe, 1975; Paradkar andMunshi, 1980; Chaturvedi *et al.*, 1981; Vittal and Glory, 1981; Tripathi, 1987; Verma and Khare, 1987; Tilak and Pillai, 1988; Santra and Chanda, 1989; Ghani *et al.*, 1993; Nadimuthu and Vittal, 1995; Singh *et al.*, 1990; Singh and Singh, 1995; Agashe and Anuradha, 1998; Sahney and Purwar, 2001
21. Match factory	Jayaprakash and Ramalingam, 1981
22. Oil mill	Jayaprakash *et al.*, 1978; Jayaprakash and Ramalingam, 1981
23. Paper mill	Jayaprakash and Ramalingam, 1981; Sharma and Dutta, 2001
24. Pig farm	Begum *et al.*, 2001
25. Poultry shed/farm	Jayaprakash *et al.*, 1978; Rati *et al.*, 1980; Jayaprakash and Ramalingam, 1981; Tilak and Saibaba, 1986; Surekha and Reddy, 1996; Singh *et al.*, 2000; Sharma and Dutta, 2001
26. Rice mill	Jayaprakash *et al.*, 1978; Jayaprakash and Ramalingam, 1981
27. Saw mill	Misra, 1988, Misra *et al.*, 1989
28. Lecture hall/ School room	Chakraverty and Sinha, 1985
29. Scientific laboratory	Santra and Chanda, 1989; Ghani *et al.*, 1994
30. Silk filature	Harindranath *et al.*, 1985
31. Snuff depot	Jayaprakash *et al.*, 1978; Jayaprakash andRamalingam, 1981
32. Working environment	Rasool, 1989 (Not seen in original)

workers and there are 14 research papers that have appeared. This further indicates that we follow the trend of research, rather than investigating newer and significant areas that deserve attention. Vittal and Glory (1985) have reported 43 species of fungi representing 23 genera from a library in Madras. Occurrence of *Histoplasma capsulatum* from a library is their significant finding as this fungus is a causal organism of Histoplasmosis-a very serious lung disease. Why other investigators have not found this fungus posses a serious question and points towards the need for further search in libraries of different part of our country. Similarly, many places such as the agarbathi factory, cinema halls, food storage places, ginnery, mechanical and textile industries, jute mill, pig farm, saw mill and so on have a single publication on their indoor fungal flora (Table 1). This strongly suggests that these environments need more surveys with a view to find out both fungi suspended in the air and their possible correlation with the ailments of the workers of such places.

Since our researches so far on indoor mycology is limited only to the report of fungal forms in the air, it is now imperative to find out the allergenic potentialities of fungi that are the dominants ones and also of those that occur rarely. To bridge this gap we, the mycologists, have to develop collaboration with the physicians.

Two recent papers-one of Kulshretha and Chauhan (2001) for fungi of some hospitals in Agra city and the that of Begum *et al* (2001) from indoor air of pig farm at Khanapara (Assam) are pointers of the seriousness of mycologists to look into the fungal flora of various significant places, but it would have been more meaningful if there would have been attempt to find out some fungi specific to the places/substrate of that environment. For example, in pig farm we expect some coprophilous fungi other than the Mucorales. But to achieve this objective specific isolating techniques have to be adopted.

Fungal diversity, as evident from the tables (Tables 2 and 3), is quite encouraging and it would not be an exaggeration that one can find more variety of fungi if little more serious attempts are made using specific media to recover as many fungal forms as possible from the indoor air of occupational environs.

Conclusions

It is evident from the foregoing observations that there is enormous diversity among the fungal forms recovered from very a few places surveyed. Only assumptions have been made by most of the workers that Aspergilli, Penicilli and various species of *Cladosporium* that dominate the flora can cause allergy and damage the materials indoor. But now serious attempts have to be made to verify and justify the facts that fungal components of indoor air are damaging both to the inhabitants and materials of the indoor environment. Hence collaborative research is warranted.

Where We Need to Search Further

The data available and presented here are indicative of the fact that we need to know more about the aeromycoflora of a large number of places where people work and stay for a longer period during their working hours. For example there is no data about the fungi, as far as we know, of the places where air conditioning is done in India, particularly the hotels. There are also a number of places where a lot of chemicals are drained/poured during the working time and those chemicals may permit the growth of fungi that may be the causal agent of many allergenic reactions to those who work there, for example, the dark rooms of photographers. There would be a large number of

Table 2. Indoor mycoflora of places of occupational importance

	Occupational Environs																														
SN	Fungus	1	2	3	4	5	6	7	8	9	10	11	12	13	14	15	16	17	18	19	20	21	22	23	24	25	26	27	28	29	30
1	*Absidia corymbifera (Cohn.) Sacc.& Trottu*	-	+	-	-	-	-	-	-	-	-	-	-	+	-	-	-	-	-	-	-	-	-	-	-	-	-	-	-	-	-
2	*A. glauca Hagem*	-	-	-	-	-	-	-	-	-	-	-	-	-	-	-	-	-	+	+	-	-	-	-	-	-	-	-	-	-	-
3	*A. ramosa (Lindt.) Lendner*	-	-	-	-	-	-	-	-	-	-	-	-	-	-	-	+	+	-	-	-	-	-	-	-	-	-	-	-	-	-
4	*Acremonium sp.*	-	-	-	-	-	-	-	-	-	-	-	-	-	-	-	+	+	-	-	-	-	-	-	-	-	-	-	-	-	-
5	*Alternaria alternata (Fr.) Keissler*	-	-	+	-	+	-	-	-	+	+	+	-	+	+	-	+	+	+	+	-	-	-	+	+	-	+	-	+	-	-
6	*A. citri Ell. & Pierce*	-	-	-	-	-	-	-	-	-	-	-	-	-	-	-	-	-	-	-	-	-	-	-	+	-	-	-	-	-	-
7	*A. longissima Deighton & MacGarvie*	-	-	-	-	-	-	-	-	-	-	-	-	-	-	-	+	+	-	-	-	-	-	-	-	-	-	-	-	-	-
8	*A. padwickii (Ganguly) M. B. Elilis*	-	-	-	-	-	-	-	-	-	-	-	-	-	-	-	-	-	-	+	-	-	-	-	-	-	-	-	-	-	-
9	*A. solani Sorauer*	-	-	-	-	-	-	-	-	-	+	+	-	-	+	-	-	-	+	-	-	-	-	+	+	-	-	-	-	-	-
10	*A. tenuis Nees*	-	-	-	-	-	-	-	-	-	-	-	+	-	-	-	-	-	-	-	-	-	-	-	-	-	-	-	-	-	-
11	*A. tenuissima (Kunze ex Pers.) Wiltshire*	-	-	+	-	-	-	-	-	-	-	-	-	-	-	-	+	+	-	+	-	-	-	-	-	-	-	-	-	-	-
12	*Alternaria sp.*	-	-	-	-	-	-	-	-	-	-	-	-	+	+	-	-	-	-	+	-	-	-	-	+	-	-	-	-	-	-
13	*Aspergillus awamori Nakazava*	-	-	-	-	-	-	-	-	-	-	-	-	-	-	-	-	-	-	+	-	-	-	-	-	-	-	-	-	-	-
14	*A. candidus Link*	-	-	+	-	+	-	-	-	+	-	-	+	+	-	-	-	-	-	+	-	-	-	+	+	-	+	-	-	-	-
15	*A. carbonarius (Bainier) Thom*	-	-	-	-	-	-	-	-	+	-	-	-		-	-	-	-	-	-	-	-	-	-	-	-	+	-	-	-	-
16	*A. cervinus (Massee) Neill.*	-	-	-	-	-	-	-	-	-	-	-	-	-	-	-	-	-	-	+	-	-	-	-	-	-	-	-	-	-	-
17	*A. carneus (Van Tieghem) Bloch.*	-	-	-	-	+	-	-	-	+	-	-	-	-	-	-	-	-	-	-	-	-	-	-	-	-	+	-	-	-	-
18	*A. clavatus Desmazieres*	-	-	-	-	-	-	-	-	-	-	-	-	+	-	-	-	-	-	+	-	-	-	-	+	-	-	-	-	-	-
19	*A. flavipes (Bainier & Sartory) Thom & Church*	-	-	-	-	+	-	-	-	-	-	-	-	+	-	-	-	-	-	+	-	-	-	-	-	-	-	-	-	-	-
20	*A. flavus Link ex Fr.*	+	+	+	-	+	+	+	+	+	+	+	+	+	+	+	+	+	+	+	-	+	+	+	+	+	+	-	+	-	+

21	A. fumigatus Fres.	+	-	–	+	+	+	+	+	+	-	+	-	+	+	+	+	+	+	+	+	+	+	+	+	+	+	-	-	+	+
22	*A. glaucus* Link	-	-	-	-	-	-	-	-	-	+	+	-	+	+	-	-	-	-	+	-	-	-	+	+	-	-	-	+	-	-
23	*A. heteromorphus* Batista & Maia	-	-	-	-	-	-	-	-	-	-	-	-	-	-	-	-	-	-	+	-	-	-	-	-	-	-	-	-	-	-
24	*A. japonicus* Saito	-	+	–	-	+	-	-	-	+	-	-	+	-	-	-	-	-	-	+	-	-	-	-	-	-	+	-	-	-	-
25	*A. melleus** Yukawa	-	-	-	-	+	-	-	-	+	-	-	+	-	-	-	-	-	-	+	-	-	-	-	-	-	+	-	-	-	-
26	*A. nidulans** (Eidam) Wint.	-	-	–	-	-	-	-	-	+	-	-	+	+	-	-	+	+	-	+	-	-	+	-	-	-	+	-	-	-	-
27	*A. niger** van Tiegh.	-	+	–	-	+	-	-	-	+	+	+	+	+	+	-	+	+	+	+	-	-	+	+	+	-	+	-	+	-	-
28	*A. niveus* Bloch.	-	-	-	-	+	-	-	-	+	-	-	-	-	-	-	-	-	-	-	-	-	-	-	-	-	+	-	-	-	-
29	*A. ochraceous* Wilhelm	-	-	–	-	-	-	-	-	-	-	-	-	+	+	-	-	-	-	+	-	-	-	-	-	-	-	-	-	-	-
30	*A. oryzae* (Ahlb.) Cohn	-	-	-	-	-	-	-	-	-	+	-	-	+	+	-	-	-	-	+	-	-	-	-	-	-	-	-	-	-	-
31	*A. parasiticus* Speare	-	-	-	-	-	-	-	-	-	+	+	-	+	+	-	-	-	-	-	-	-	-	-	-	-	-	+	+	-	-
32	*A. restrictus* G. Smith	-	-	-	-	-	-	-	-	-	-	-	-	+	+	-	-	-	-	-	-	-	+	-	+	-	-	-	-	-	-
33	*A. sulphureus* (Fres.) Thom & Church	-	-	-	-	-	-	-	-	-	-	-	-	-	-	-	-	-	-	+	-	-	-	-	-	-	-	-	-	-	-
34	*A. sydowii* (Bainier & Sartory) Thom & Church	-	-	–	-	+	-	-	-	+	-	-	-	+	-	-	+	+	-	-	-	-	-	-	-	-	+	-	-	-	-
35	*A. tamarii** Kita	-	-	-	-	+	-	-	-	+	-	-	+	+	-	-	-	-	-	-	-	-	-	-	-	-	+	-	-	-	-
36	*A. terreus** Thom	-	-	-	-	+	-	-	-	+	+	-	+	+	+	-	+	+	+	+	-	-	+	-	+	-	+	-	+	-	-
SN	**Fungus**	1	2	3	4	5	6	7	8	9	10	11	12	13	14	15	16	17	18	19	20	21	22	23	24	25	26	27	28	29	30
37	*A. ustus* (Bain) Thom & Church	-	-	-	-	+	-	-	-	+	-	-	-	-	-	-	-	-	-	+	-	-	-	-	-	-	+	-	-	-	-
38	*A. variecolor* (Berk. & Br.) Thom & Raper	-	-	-	-	-	-	-	-	-	-	-	-	-	-	-	+	+	-	-	-	-	-	-	-	-	-	-	-	-	-
39	*A. versicolor* (Vuil) Tiraboschi	-	-	+	-	+	-	-	-	+	-	-	+	+	-	-	+	+	-	+	-	-	+	-	-	-	+	-	-	-	-
40	*A. wentii* Wehmer.	-	-	+	-	-	-	-	-	-	-	-	-	-	-	-	+	+	-	-	-	-	-	-	-	-	-	-	-	-	-
41	*Aspergillus* sp.	-	-	-	-	-	-	-	-	-	-	-	-	-	-	-	-	-	-	+	-	-	-	-	+	-	-	-	-	-	-
42	*Aureobasidium pullulans* (de Bary) Arnaud	-	-	-	-	-	-	-	-	-	-	-	-	+	-	-	+	+	-	+	-	-	-	-	-	-	-	-	-	-	-
43	*Botryotrichum* sp.	-	-	-	-	-	-	-	-	+	-	-	-	-	-	-	-	-	-	-	-	-	-	-	-	-	+	-	-	-	-
44	*Botrytis* sp.	-	-	-	-	-	-	-	-	-	-	-	-	-	+	-	-	-	-	-	-	-	-	-	-	-	-	-	-	-	-

45	*Calcarisporium sp.*	-	-	-	-	-	-	-	-	-	-	-	-	-	+	-	-	-	-	-	-	-	-	-	-	-	-	-	-	-	-
46	*Candida albicans (Robin) Berkhout*	-	-	-	-	-	-	-	-	-	-	-	-	+	+	-	+	-	-	+	-	-	-	-	-	-	-	-	-	-	-
47	*Cephalosporium sp.*	-	-	-	-	-	-	-	-	-	-	-	-	-	-	-	+	+	-	-	-	-	-	-	-	-	-	-	-	-	-
48	*Chaetomium globosum Kunze : Fr.*	-	-	-	-	+	-	-	-	-	-	-	-	+	-	-	-	-	-	+	-	-	-	-	-	-	-	-	-	-	-
49	*Chaetomium sp.*	-	-	-	-	-	-	-	-	-	-	-	-	-	+	-	-	-	-	-	-	-	-	-	-	-	-	-	-	-	-
50	*Choanephora cucurbitarum (Berk. & Rav.) Thaxt.*	-	-	-	-	-	-	-	-	-	-	-	-	+	-	-	-	-	-	-	-	-	-	-	-	-	-	-	-	-	-
51	*Chrysonilia sitophila (Mont.) von Arx*	-	-	-	-	+	-	-	-	-	-	-	-	-	-	-	-	-	-	-	-	-	-	-	-	-	-	-	-	-	-
52	*Chrysosporium pruinosum (Gil.& Abb.) Carmichael*	-	-	-	-	-	-	-	-	-	-	-	-	-	-	-	-	-	+	+	-	-	-	-	-	-	-	-	-	-	-
53	*Cladosporium buteicolla Cooke*	-	-	-	-	-	-	-	-	-	-	-	-	-	-	-	-	-	-	-	-	-	-	+	+	-	-	-	-	-	-
54	*C. cladosporioides (Fresen.) de Vries*	-	+	+	-	-	-	-	-	-	-	-	+	+	-	-	+	+	-	+	-	-	-	-	-	-	+	-	-	-	-
55	*C. epiphyllum Persoon*	-	-	-	-	-	-	-	-	+	-	-	-	-	-	-	-	-	-	-	-	-	-	-	-	-	+	-	-	-	-
56	*C. herbarum* (Pers.) Link ex. S. F. Gray*	-	-	-	-	-	-	-	-	+	-	+	-	-	+	-	+	+	+	-	-	-	-	+	+	-	+	-	+	-	-
57	*C. lignicolum Corda*	-	-	-	-	-	-	-	-	+	-	-	-	-	-	-	-	-	-	-	-	-	-	-	-	-	+	-	-	-	-
58	*C. macrocladium*	-	-	+	-	-	-	-	-	-	-	-	-	-	-	-	-	-	-	-	-	-	-	-	-	-	-	-	-	-	-
59	*C. psoraleae M. B. Ellis*	-	-	-	-	-	-	-	-	-	-	-	-	-	-	-	-	-	-	-	-	-	-	-	+	-	-	-	-	-	-
60	*C. sphaerospermum Penzig*	-	-	-	-	+	-	-	-	-	-	-	-	-	-	-	-	-	-	-	-	-	-	-	-	-	-	-	-	-	-
61	*C. tenuissimum Cooke*	-	-	-	-	-	-	-	-	-	-	-	-	-	-	-	-	-	-	-	-	-	-	+	+	-	-	-	-	-	-
62	*Cladosporium sp.*	-	-	+	-	-	-	-	-	-	-	-	-	+	+	-	-	-	-	+	-	-	+	-	+	-	-	-	-	-	-
63	*Circinella simplex van Tiegh.*	-	-	-	-	-	-	-	-	-	-	-	-	+	-	-	-	-	-	+	-	-	-	-	-	-	-	-	-	-	-
64	*C. umbellata van Tiegh. & Le Monn.*	-	-	-	-	-	-	-	-	-	-	-	-	+	-	-	-	-	-	-	-	-	-	-	-	-	-	-	-	-	-
65	*Colletotrichum sp.*	-	-	-	-	-	-	-	-	-	-	-	-	-	-	-	+	-	-	-	-	-	-	-	-	-	-	-	-	-	-
66	*Corynascus sepedonium*	-	-	-	-	-	-	-	-	-	-	-	-	-	-	-	-	-	-	+	-	-	-	-	-	-	-	-	-	-	-

67	*Corynespora cassiicola (Berk. & Curt.) Wei*	-	-	-	-	-	-	-	-	-	-	-	-	-	-	-	-	-	-	+	-	-	-	-	-	-	-	-	-	-	-
68	*Cunninghamella echinulata Thaxt.*	-	-	-	-	-	-	-	-	-	-	-	-	+	+	-	-	-	-	+	-	-	-	-	-	-	-	-	-	-	-
69	*Curvularia clavata Jain*	-	-	-	-	-	-	-	-	-	-	-	-	-	-	-	+	+	-	-	-	-	-	-	-	-	-	-	-	-	-
70	*C. geniculata (Tracy & Earle) Boedijn*	-	-	-	-	-	-	-	-	-	-	-	-	-	-	-	-	-	-	-	-	-	-	-	+	-	-	-	-	-	-
71	*C. lunata (Wakker) Boedijn*	-	-	-	-	+	-	-	-	-	-	+	+	-	+	-	+	+	+	+	-	-	-	-	+	-	-	-	+	-	-
72	*C. lunata var. aeria* (Batista, Lima & Vas.) Ellis*	-	-	-	-	-	-	-	-	+	-	-	-	-	-	-	-	-	-	-	-	-	-	-	-	-	+	-	-	-	-
73	*C. pallescens Boedijn*	-	-	+	-	-	-	-	-	-	-	-	-	+	-	-	-	-	-	-	-	-	-	-	+	-	-	-	-	-	-
74	*C. tetramera (Mc Kinney) Boedjn*	-	-	-	-	-	-	-	-	+	-	-	-	-	-	-	-	-	-	-	-	-	-	-	-	-	+	-	-	-	-
75	*C. tuberculata Jain*	-	-	-	-	-	-	-	-	-	-	-	-	-	+	-	-	-	+	+	-	-	-	-	-	-	-	-	-	-	-
76	*C. brachyspora Boedijn*	-	-	-	-	-	-	-	-	-	-	-	-	+	-	-	-	-	-	+	-	-	-	-	-	-	-	-	-	-	-
SN	**Fungus**	1	2	3	4	5	6	7	8	9	10	11	12	13	14	15	16	17	18	19	20	21	22	23	24	25	26	27	28	29	30
77	*Curvularia sp*	-	-	+	-	-	-	-	-	-	-	-	-	+	+	-	-	-	-	+	-	-	+	-	+	-	-	-	-	-	-
78	*Doratomyces microspora (Sacc.) Morton & Smith*	-	-	-	-	-	-	-	-	-	-	-	-	-	-	-	-	-	-	-	-	-	-	-	-	-	+	-	-	-	-
79	*Drechslera australiensis Subram. & Jain ex Ellis*	-	-	-	-	-	-	-	-	-	-	-	-	+	-	-	-	-	-	-	-	-	-	-	-	-	-	-	-	-	-
80	*D. halodes (Drechsler) Subram. & Jain*	-	-	-	-	-	-	-	-	-	-	-	-	-	-	-	-	-	-	+	-	-	-	-	-	-	-	-	-	-	-
81	*D. hawaiiensis (Bugnicourt) Subram. & Jain ex Ellis*	-	-	-	-	+	-	-	-	+	-	-	-	+	-	-	-	-	-	+	-	-	-	-	-	-	+	-	-	-	-
82	*Drechslera sp.*	-	-	-	-	-	-	-	-	-	-	-	-	-	+	-	+	+	-	+	-	-	-	-	-	-	-	-	-	-	-
83	*Emericella nidulans var. lata (Thom & Raper) Subram.*	-	-	-	-	+	-	-	-	+	-	-	-	-	-	-	-	-	-	-	-	-	-	-	-	-	+	-	-	-	-
84	*Emericella nidulans var. nidulans (Eidam) Vuillemin*	-	-	-	-	+	-	-	-	-	-	-	-	-	-	-	-	-	-	-	-	-	-	-	-	-	-	-	-	-	-

85	*Epicoccum sp.*	-	-	-	-	-	-	-	-	-	-	-	-	-	-	-	-	-	-	+	-	-	-	-	-	-	-	-	-	-	-
86	*Fusariella sp.*	-	-	-	-	-	-	-	-	-	-	-	-	-	-	-	-	-	-	-	-	-	-	-	+	-	-	-	-	-	-
87	*Fusarium annifosporem*	-	-	-	-	-	-	-	-	-	-	-	-	-	-	-	-	-	-	-	-	-	-	-	+	-	-	-	-	-	-
88	*F. equiseti (Corda) Sacc.*	-	-	-	-	-	-	-	-	-	-	-	-	-	-	-	-	-	-	-	-	-	-	+	+	-	-	-	-	-	-
89	*F. lateritium Nees*	-	-	-	-	-	-	-	-	-	-	-	-	-	-	-	-	-	-	+	-	-	-	-	-	-	-	-	-	-	-
90	*F. moniliforme Sheldon*	-	-	+	-	-	-	-	-	-	-	-	-	-	+	-	+	+	-	-	-	-	-	+	+	-	-	-	-	-	-
91	*F. oxysporum Schlech. Ex Fr.*	-	-	-	-	+	-	-	-	+	-	-	-	+	+	-	+	+	-	+	-	-	-	+	+	-	+	-	-	-	-
92	*F. roseum Link*	-	-	-	-	+	-	-	-	+	-	-	-	-	-	-	-	-	-	-	-	-	-	-	-	-	+	-	-	-	-
93	*F. semitectum Berk. & Rav.*	-	-	-	-	-	-	-	-	-	-	-	-	-	-	-	-	-	-	+	-	-	-	-	-	-	-	-	-	-	-
94	*F. solani (Mart.) App. & Wr.*	-	-	-	-	-	-	-	-	-	-	-	-	-	-	-	-	-	-	-	-	-	-	-	-	-	-	-	-	-	-
95	*F. stilboides Wr.*	-	-	-	-	-	-	-	-	-	-	-	-	-	-	-	-	-	-	+	-	-	-	-	-	-	-	-	-	-	-
96	*F. trichothecoides Wollenweber*	-	-	-	-	-	-	-	-	+	-	-	-	-	-	-	-	-	-	-	-	-	-	-	-	-	+	-	-	-	-
97	*Fusarium sp.*	-	-	-	-	-	-	-	-	-	-	-	-	+	-	-	-	-	-	+	-	-	+	-	+	-	-	-	-	-	-
98	*Geotrichum sp.*	-	-	+	-	-	-	-	-	-	-	-	-	+	+	-	-	-	-	-	-	-	+	-	+	-	-	-	-	-	-
99	*Gliocladium sp.*	-	-	-	-	-	-	-	-	-	-	-	-	-	-	-	-	-	-	-	-	-	-	-	-	-	+	-	-	-	-
100	*Helminthosporium australiense Bugnicourt*	-	-	-	-	-	-	-	-	-	-	-	-	-	-	-	+	-	-	-	-	-	-	-	-	-	-	-	-	-	-
101	*H. hawaiiense Bugnicourt*	-	-	-	-	-	-	-	-	+	-	-	-	-	-	-	-	-	-	-	-	-	-	-	-	-	+	-	-	-	-
102	*H. nodulosum Berkeley & Curtis ex Sacc.*	-	-	-	-	+	-	-	-	-	-	-	-	-	-	-	-	-	-	-	-	-	-	-	-	-	-	-	-	-	-
103	*H. oryzae Breda de Haan*	-	-	-	-	-	-	-	-	-	+	-	-	-	+	-	-	-	+	-	-	-	-	-	-	-	-	-	-	-	-
104	*H. sativum Pammel, King & Bakke*	-	-	-	-	-	-	-	-	-	+	+	-	-	+	-	-	-	-	-	-	-	-	-	-	-	-	-	+	-	-
105	*Helminthosporium sp.*	-	-	+	-	-	-	-	-	-	-	-	+	+	-	-	-	-	-	+	-	-	-	+	+	-	-	-	-	-	-
106	*Histoplasma capsulatum Darling*	-	-	-	-	-	-	-	-	-	-	-	-	-	-	-	-	-	-	+	-	-	-	-	-	-	-	-	-	-	-
107	*Humicola grisea Traaen*	-	-	-	-	-	-	-	-	-	-	-	-	+	-	-	-	-	-	-	-	-	-	-	-	-	-	-	-	-	-
108	*Humicola sp.*	-	-	+	-	-	-	-	-	-	-	-	-	+	+	-	-	+	-	+	-	-	+	-	-	-	-	-	-	-	-
109	*Isaria sp.*	-	-	-	-	-	-	-	-	-	-	-	-	-	+	-	-	-	-	-	-	-	-	-	-	-	-	-	-	-	-
110	*Meria sp.*	-	-	-	-	-	-	-	-	-	-	-	-	-	+	-	-	-	-	-	-	-	+	-	+	-	-	-	-	-	-

111	Microsporum gypseum (Bodin) Guiart & Grigorakis)	-	-	-	-	-	-	-	-	-	-	-	-	-	-	-	-	-	-	+	-	-	-	-	-	-	-	-	-	-	-
112	*Monilia sitophila* (Montagne) Sacc.	-	-	-	-	-	-	-	-	+	-	-	+	+	-	-	-	-	-	+	-	-	-	-	-	-	-	-	-	-	-
113	*Monodictys levis* (Wiltshire) Hughes	-	-	-	-	-	-	-	-	-	-	-	-	+	-	-	-	-	-	+	-	-	-	-	-	-	-	-	-	-	-
SN	**Fungus**	1	2	3	4	5	6	7	8	9	10	11	12	13	14	15	16	17	18	19	20	21	22	23	24	25	26	27	28	29	30
114	*Mucor hiemalis** Wehmer	-	-	-	-	+	-	-	-	+	-	-	-	-	-	-	-	-	+	+	-	-	-	-	-	-	+	-	-	-	-
115	*Mucor racemosus* Fresenius	-	-	-	-	-	-	-	-	-	-	-	-	+	-	-	+	+	-	+	-	-	-	-	-	-	-	-	-	-	-
116	*M. varians* Povah	-	-	+	-	-	-	-	-	-	-	-	-	-	-	-	-	-	-	-	-	-	-	-	-	-	-	-	-	-	-
117	*Mucor* sp.	-	-	+	-	-	-	-	-	-	-	-	+	+	+	-	-	-	-	+	-	-	+	+	+	-	-	-	-	-	-
118	*Myrothecium verrucaria* (Alb. & Schw.) Ditm. Ex Fr.	-	-	-	-	-	-	-	-	-	-	-	-	-	-	-	-	-	-	+	-	-	-	-	-	-	-	-	-	-	-
119	*Neurospora sitophila*	-	-	+	-	-	-	-	-	-	-	-	-	-	-	-	-	-	-	-	-	-	-	-	-	-	-	-	-	-	-
120	*Nigrospora sphaerica** (Sacc.) Mason	-	-	-	-	-	-	-	-	+	-	-	-	+	-	-	-	-	-	+	-	-	-	-	-	-	+	-	-	-	-
121	*Nigrospora* sp.	-	-	+	-	-	-	-	-	-	-	-	+	+	+	-	+	+	-	+	-	-	+	+	+	-	-	-	-	-	-
122	*Oidiodendron* sp.	-	-	+	-	-	-	-	-	-	-	-	-	-	-	-	-	-	-	-	-	-	+	-	-	-	-	-	-	-	-
123	*Oidium* sp.	-	-	+	-	-	-	-	-	-	-	-	-	-	-	-	-	-	-	-	-	-	-	-	-	-	-	-	-	-	-
124	*Paecilomyces fusisporus** Saksena	-	-	-	-	-	-	-	-	-	-	-	-	-	-	-	+	+	-	-	-	-	-	-	-	-	+	-	-	-	-
125	*P. varioti* Bain	-	-	+	-	+	-	-	-	+	-	-	-	+	-	-	-	-	-	+	-	-	-	-	-	-	+	-	-	-	-
126	*Penicillium camemberti* Thom	-	-	-	-	-	-	-	-	-	-	-	-	-	-	-	-	-	-	-	-	-	-	-	+	-	-	-	-	-	-
127	*P. chrysogenum* Thom	-	-	+	-	-	-	-	-	-	-	-	-	+	-	-	+	+	-	+	-	-	-	-	-	-	-	-	-	-	-
128	*P. citreo-viride* Biourge	-	-	-	-	-	-	-	-	-	-	-	-	-	-	-	-	-	-	+	-	-	-	-	-	-	-	-	-	-	-
129	*P. citrinum* Thom	-	-	+	-	+	-	-	-	+	-	-	-	+	-	-	-	-	-	+	-	-	+	-	-	-	+	-	-	-	-
130	*P. cyclopium* Westling	-	-	+	-	-	-	-	-	-	-	-	-	-	-	-	-	-	-	-	-	-	-	-	-	-	-	-	-	-	-
131	*P. expansum* (Link) Thom	-	-	+	-	-	-	-	-	-	-	-	-	+	-	-	-	-	-	-	-	-	-	-	-	-	-	-	-	-	-
132	*P. fellutanum* Biourge	-	-	-	-	-	-	-	-	-	-	-	-	+	-	-	-	-	-	-	-	-	-	-	-	-	-	-	-	-	-
133	*P. frequentans* Westling	-	-	-	-	-	-	-	-	-	-	-	-	+	-	-	-	-	-	-	-	-	-	-	-	-	-	-	-	-	-
134	*P. funiculosum* Thom	-	-	-	-	-	-	-	-	-	-	-	-	+	-	-	-	-	-	-	-	-	-	-	-	-	+	-	-	-	-

135	P. herquei Bainier & Sartory	-	-	-	-	-	-	-	-	-	-	-	-	-	-	-	-	-	-	+	-	-	-	-	-	-	-	-	-	-	-
136	P. implicatum Biourge	-	-	-	-	-	-	-	-	-	-	-	-	-	-	-	-	-	-	+	-	-	-	-	-	-	-	-	-	-	-
137	P. lanosum Westling	-	-	-	-	-	-	-	-	-	-	-	-	-	-	-	-	-	-	-	-	-	-	-	+	-	-	-	-	-	-
138	P. lapidosum Raper & Fennell	-	-	-	-	-	-	-	-	-	-	-	-	-	-	-	-	-	-	-	-	-	-	-	+	-	-	-	-	-	-
139	P. miczynskii Zaleski	-	-	-	-	-	-	-	-	-	-	+	-	-	+	-	-	-	+	-	-	-	-	-	-	-	-	-	+	-	-
140	P. nigricans (Bain.) Thom	-	-	-	-	-	-	-	-	-	-	-	-	-	+	-	+	+	-	-	-	-	-	-	-	-	-	-	-	-	-
141	P. oxalicum Currie & Thom	-	-	+	-	-	-	-	-	-	-	-	-	+	-	-	-	-	-	+	-	-	-	-	-	-	-	-	-	-	-
142	P. porulum	-	-	-	-	-	-	-	-	-	-	-	-	-	-	-	-	-	-	-	-	-	-	-	+	-	-	-	-	-	-
143	P. puberulum Bainier	-	-	-	-	-	-	-	-	-	-	-	-	-	-	-	-	-	-	-	-	-	-	+	+	-	-	-	-	-	-
144	P. purpurogenum Stoll.	-	-	-	-	-	-	-	-	-	-	-	-	+	-	-	-	-	-	+	-	-	-	-	-	-	-	-	-	-	-
145	P. restrictum Gilman & Abbott	-	-	-	-	-	-	-	-	-	-	-	-	-	-	-	-	-	-	+	-	-	-	-	-	-	-	-	-	-	-
146	P. rugulosum Thom	-	-	-	-	-	-	-	-	-	-	-	-	-	-	-	-	-	-	+	-	-	-	-	-	-	-	-	-	-	-
147	P. thomii Maire	-	-	-	-	-	-	-	-	-	-	-	-	+	-	-	-	-	-	-	-	-	-	-	-	-	-	-	-	-	-
148	P. vinaceum Gilman & Abbott	-	-	-	-	-	-	-	-	-	-	-	-	+	-	-	-	-	-	-	-	-	-	-	-	-	-	-	-	-	-
149	Penicillium sp.	-	-	+	-	-	-	-	-	-	-	-	+	+	+	-	-	-	-	+	-	-	+	+	+	-	-	-	-	-	-
150	Periconia sp.	-	-	-	-	-	-	-	-	+	-	-	-	-	-	-	-	-	-	-	-	-	-	-	-	-	+	-	-	-	-
151	Periconia byssoides Persoon ex Schweinitz	-	-	-	-	+	-	-	-	-	-	-	-	-	-	-	-	-	-	-	-	-	-	-	-	-	-	-	-	-	-
152	Pestalotia pezizoides de Not	-	-	-	-	+	-	-	-	-	-	-	-	-	-	-	-	-	-	-	-	-	-	-	-	-	+	-	-	-	-
153	Pestalotia sp.	-	-	-	-	-	-	-	-	-	-	-	-	-	+	-	-	-	-	+	-	-	-	-	-	-	-	-	-	-	-
154	Phialocephala sp.	-	-	-	-	-	-	-	-	-	-	-	-	-	-	-	-	-	-	-	-	-	+	-	-	-	-	-	-	-	-
155	Phoma betae (Oudem.) Frank	-	-	-	-	-	-	-	-	-	-	-	-	-	-	-	+	-	-	-	-	-	-	-	-	-	-	-	-	-	-
SN	**Fungus**	1	2	3	4	5	6	7	8	9	10	11	12	13	14	15	16	17	18	19	20	21	22	23	24	25	26	27	28	29	30
156	P. herbarum Westd.	-	-	-	-	-	-	-	-	-	-	-	-	-	-	-	-	-	-	-	-	-	-	-	-	-	+	-	-	-	-
157	P. hibernica Grimes, O' Connor & Cummins	-	-	-	-	-	-	-	-	-	-	-	-	-	-	-	+	-	-	-	-	-	-	-	-	-	-	-	-	-	-
158	P. jolyana Pirozynski & Morgan-Jones	-	-	-	-	-	-	-	-	-	-	-	-	-	-	-	-	-	-	-	-	-	-	-	-	-	+	-	-	-	-

159	*Pithomyces sp.*	-	-	-	-	-	-	-	+	-	-	-	-	-	-	+	+	-	-	-	-	-	-	-	-	+	+	-	-	-	-
160	*Phytophthora infestans (Mont.) de Bary*	-	-	-	-	-	-	-	-	-	-	-	-	-	-	-	-	-	-	-	-	-	-	-	+	-	-	-	-	-	-
161	*Rhizoctonia sp.*	-	-	+	-	+	-	-	-	+	-	-	-	-	-	-	+	+	-	-	-	-	-	-	-	-	+	-	-	-	-
162	*Rhizopus nigricans Ehrenb.*	-	-	-	-	-	-	-	-	-	+	+	-	-	+	-	+	+	-	+	-	-	-	-	-	-	-	-	-	-	-
163	*R. oryzae Went & Gerlings*	-	-	-	-	-	-	-	-	-	+	-	-	-	-	-	-	-	-	-	-	-	-	-	-	-	-	-	-	-	-
164	*R. stolonifer (Ehrenberg:Fr.) Vuillemin*	-	+	+	-	+	-	-	-	-	-	-	-	+	-	-	-	-	-	+	-	-	-	-	-	-	-	-	-	-	-
165	*Rhizopus sp.*	-	-	+	-	-	-	-	-	-	-	-	-	+	+	-	-	-	-	+	-	-	+	+	+	-	-	-	-	-	-
166	*Scolecabasidium constrictum Abbott*	-	-	-	-	-	-	-	-	-	-	-	-	+	-	-	-	-	-	-	-	-	-	-	-	-	-	-	-	-	-
167	*Scopulariopsis brevicaulis (Sacc.) Bainier*	-	-	-	-	+	-	-	-	-	-	-	-	+	-	-	-	-	-	-	-	-	-	-	-	-	-	-	-	-	-
168	*Scytalidium state of Hendersonula toruloides*	-	-	+	-	-	-	-	-	-	-	-	-	-	-	-	-	-	-	-	-	-	-	-	-	-	-	-	-	-	-
169	*Sepedonium sp.*	-	-	-	-	-	-	-	-	-	-	-	-	-	-	-	-	-	-	-	-	-	+	-	-	-	-	-	-	-	-
170	*Sphaeropsis malorum Berk.*	-	-	-	-	-	-	-	-	-	-	-	-	-	-	-	-	-	-	+	-	-	-	-	-	-	-	-	-	-	-
171	*Sporidesmium sp.*	-	-	-	-	-	-	-	-	-	-	-	-	-	+	-	-	-	-	-	-	-	-	-	-	-	-	-	-	-	-
172	*Sporotrichum sp.*	-	-	-	-	-	-	-	-	-	-	-	-	-	-	-	+	+	-	-	-	-	-	-	-	-	-	-	-	-	-
173	*Staphylotrichum sp.*	-	-	-	-	-	-	-	-	-	-	-	-	-	+	-	-	-	-	-	-	-	-	-	-	-	-	-	-	-	-
174	*Stachybotrys sp.*	-	-	-	-	-	-	-	-	-	-	-	-	-	-	-	-	-	-	-	-	-	+	-	-	-	-	-	-	-	-
175	*Stemphylium sp.*	-	-	-	-	-	-	-	-	+	-	-	-	-	-	-	-	-	-	+	-	-	-	-	-	-	+	-	-	-	-
176	Sterile mycelium	-	-	-	-	+	-	-	-	+	-	-	-	-	-	-	-	-	-	-	-	-	-	-	-	-	+	-	-	-	-
177	*Stigmina sp.*	-	-	-	-	-	-	-	-	+	-	-	-	-	+	-	-	-	-	-	-	-	-	-	-	-	-	-	-	-	-
178	*Syncephalastrum racemosum (Cohn) Schroeter*	-	-	-	-	+	-	-	-	+	-	-	-	+	-	-	-	-	-	+	-	-	-	-	-	-	+	-	-	-	-
179	*Syncephalastrum sp.*	-	-	-	-	-	-	-	-	-	-	-	-	-	-	-	-	+	-	-	-	-	-	-	-	-	-	-	-	-	-
180	*Talaromyces stipitatus C.R.Benj. apud Stolk & Samson*	-	-	-	-	+	-	-	-	-	-	-	-	-	-	-	-	-	-	-	-	-	-	-	-	-	-	-	-	-	-
181	*Torula ellisii Yadav & Lal*	-	-	-	-	+	-	-	-	-	-	-	-	-	-	-	-	-	-	-	-	-	-	-	-	-	+	-	-	-	-
182	*Torula sp.*	-	-	-	-	-	-	-	-	-	-	-	-	+	+	-	+	+	-	-	-	-	+	-	+	-	-	-	-	-	-

183	*Trichoderma harzianum* Rifai.	-	-	-	-	-	-	-	-	-	-	-	-	+	-	-	-	-	-	-	-	-	-	-	-	-	-	-	-	-	-
184	*T. koningii* Oudemans	-	-	-	-	+	-	-	-	-	-	-	-	-	-	-	-	-	-	-	-	-	-	-	-	-	-	-	-	-	-
185	*T. lignorum* (Tode) Herz.	-	-	-	-	+	-	-	-	+	-	-	-	-	-	-	-	-	-	-	-	-	-	-	-	-	+	-	-	-	-
186	*T. longibrachiatum* Rifai.	-	-	-	-	-	-	-	-	-	-	-	-	+	-	-	-	-	-	-	-	-	-	-	-	-	-	-	-	-	-
187	*T. viride* Pers.	-	-	+	-	-	-	-	-	-	-	-	-	+	-	-	+	+	-	+	-	-	-	-	-	-	-	-	-	-	-
188	*Trichoderma* sp.	-	-	-	-	-	-	-	-	-	-	-	-	+	+	-	-	-	-	-	-	-	+	+	+	-	-	-	-	-	-
189	*Trichothecium roseum* (Persoon) Link ex S. F. Gray	-	-	+	-	+	-	-	-	+	-	-	-	-	-	-	-	-	-	-	-	-	-	-	-	-	+	-	-	-	-
190	*Trichothecium* sp.	-	-	-	-	-	-	-	-	-	-	-	-	-	+	-	-	-	-	+	-	-	-	-	-	-	-	-	-	-	-
191	*Trichurus* sp.	-	-	-	-	-	-	-	-	-	-	-	-	-	-	-	+	+	-	-	-	-	-	-	-	-	-	-	-	-	-
192	*Ulocladium* sp.	-	-	-	-	-	-	-	-	-	-	-	-	-	+	-	-	-	-	-	-	-	-	-	-	-	-	-	-	-	-
193	*Verticillium* sp.	-	-	-	-	-	-	-	-	+	-	-	-	-	-	-	-	-	-	-	-	-	+	-	+	-	+	-	-	-	-

+ = Present
- = Absent

1= Agarbathi factory
2= Animal care facilities
3= Bakery
4= Cane sugar mills (Sugar factory)
5= Cobbler shop & Tanneries
6= Coffee curing works
7= Cow shed
8= Fire wood depot
9=Flour mill
10. Food storage places
11. Fruit shop/market
12= Ginnery
13= Grain shops/storage godown
14= Hospital ward
15= Hostel kitchen
16= Industry-Mechanical
17= Industry-Textile
18= Jute mill
19= Library
20= Match factory
21=Oil mill
22= Paper mill
23= Pig farm
24= Poultry farm
25= Rice mill
26= Saw mill
27= Scientific laboratory
28= Silk filature
29= Snuff depot
30= Working environment

Table 3. Common fungal spores recovered from indoor air of occupational environs by various workers

Fungal Spores			
1.	*Alternaria*	22.	*Hendersonula*
2.	*Aspergillus*	23.	*Heterosporium*
3.	*Basidiospores*	24.	Hyphal fragment
4.	*Beltrania*	25.	*Hysterium*
5.	*Bispora*	26.	*Nigrospora*
6.	*Botryodiplodia*	27.	*Penicillium*
7.	*Cephaliophora*	28.	*Periconia*
8.	*Cercospora*	29.	*Pithomyces*
9.	*Chaetomium*	30.	*Pleospora*
10.	*Cladosporium*	31.	Rust spores (mainly Urediniospores)
11.	*Corynespora*		
12.	*Curvularia*	32.	Smut spores
13.	*Dictyoarthrinium*	33.	*Sordaria*
14.	*Didymosphaeria*	34.	*Spegazzinia*
15.	*Diplodia*	35.	*Sporormiella*
16.	*Drechslera*	36.	*Stigmina*
17.	*Epicoccum*	37.	*Tetraploa*
18.	*Fusarium*	38.	*Torula*
19.	*Ganoderma*	39.	*Trichoconis*
20.	*Harknessia*	40.	*Trichothecium*
21.	*Helminthosporium*	41.	Unidentified spores

other places also, which although are not working places, but we should know the fungi that might be growing or can grow and cause health problems to the dwellers. Some such places may be slum-dwellings, houses made of mud/ soil (non-cemented houses) or rooms, ill-ventilated old houses and so on. Investigations undertaken for some such places will be rewarding.

References

Agarwal, M. K., Mukerji, K. G. and Shivpuri, D. N. (1969). Studies on the allergenic fungal spores of Delhi, Indian metropolitan area, Botanical aspects. J. Allergy, **44**: 193-203.

Agarwal, M. K. and Shivpuri, D. N. (1974). Fungal spores-Their role in respiration allergy. Advances in Pollen Spore Research, **1**: 78-128.

Agashe, S. N. and Anuradha, H. G. (1998). Aeromycological studies of a Library in Bangalore. Indian J. Aerobiol.,**11**: 24-26.

Agashe, S. N. and Sudha, P. (1995). Studies on circadian periodicity of fungal spores in Bangalore city. Indian J. Aerobiol., **8**: 17-23.

Bagwan, N. B. and Meshram, B. M. (2000). Relationship between fungal air-spora of apple storehouses and incidence of post-harvest fruit rots. Indian J. Aerobiol., **13**: 29-32.

Baruah, H. K. and Chetia, M. (1966). Air spora and allergenic human diseases. A study of certain fungal spores and pollen grains of Gauhati. J. Expt. Biol., **4**: 236-238.

Baruah, H. K. (1961). The air spora of cowshed. J. Gen. Microbiol., **25**: 483-491.

Begum, J., Ahmad, K. and Bora, K. N. (2001). A study of mycoflora in the indoor air of a Pig farm at Khanapara (Assam). Indian J. Aerobiol., **14**: 36-38.

Benagi, V. I. and Kulkarni, Srikant. (1995). Epidemiology and aerobiology of rust of groundnut caused by *Puccinia arachidis* Speg. Indian J. Aerobiol., **8**: 1-6.

Bhalke, S. P. (1980). *Studies in Air Spora Over Some Fields*. Ph. D. Thesis, Marathwada University, Aurangabad.

Bhat, M. M. and Rajasab, A. H. (1988). Incidence of viable fungal airspora in Gulbarga during 1984-86: Two year survey. Indian J. Aerobiol., **1**: 10-16.

Bhati, H. S. and Gaur, R. D. (1979). Studies on aerobiology—atmospheric fungal spores. New Phytol., **82**: 519-527.

Chakraverty, R. and Nandi, P. (1972). The seasonal periodicity of *Cladosporium*, a common allergen in the air of Calcutta and suburbs. Trans. Bose Res. Inst., **35**: 40-45.

Chakraverty, R. and Sinha, S. (1985). The incidence of *Aspergillus parasiticus* in the indoor and outdoor environments of Calcutta, India. Grana, **24**: 133-135.

Chakre, O. J. (1979). *Studies in the Microbial Pollution of Air at Aurangabad*. Ph. D. Thesis, Marathwada University, Aurangabad.

Chaturvedi, M., Srivastava, V., Ikram, S., Pathak, N. C. and Goyal, R. (1981). Mycoflora in the Library and Herbarium of the National Botanical Research Institute- a preliminary study. New Botanist, **8**: 63-71.

Chaubal, P. D. and Deodikar, G. B. (1964). Air borne spores around Poona. J. Poona University, **26**: 123-136.

Chitaley, S. D. and Bajaj, A. (1973). Air spora of Nagpur at high altitudes-I. Botanique, **4**: 27-34.

Chitaley, S. D. and Bajaj, A. (1974). Air spora of Nagpur at high altitudes-II. Botanique, **5**: 43-52.

Chitaley, S. D. and Bajaj, A. (1975). Air spora of Nagpur at high altitudes-III. Botanique, **6**: 59-68.

Cunningham, D. D. (1873). *Microscopic Examination of Air*. Govt. Printer, Calcutta, 58 pp.

D'Silva, A. M. and Freitas, Y. M. (1982). A survey of the aerial mycoflora of Bombay. Indian J. Microbiol., **22**: 27-33.

Dhaware, A. S. (1976). *Taxonomy of Some Ascomycetes and Air-spora Studies Over Paddy Field*. Ph. D. Thesis, Marathwada University, Aurangabad.

Gaikwad, Y. B. (1974). *Studies in Airspora and Taxonomy of Pyrenomycetes*. Ph. D. Thesis, Marathwada University, Aurangabad.

Ghani, T. (1993). *Study of Aeromycospora of Bhagalpur*. Ph. D. Thesis, Bhagalpur University, Bhagalpur.

Ghani, T. (1994). Incidence of air-borne fungi inside different scientific laboratories at Bhagalpur. Indian J. Aerobiol. **7**: 8-12.

Ghani, T., Kumar, S. and Kale, S. R. (1993). Aeromycoflora inside and outside a College Library. Indian Intl. J. Environ. Hlth. Res., **3**: 49-58.

Gupta, B. N. and Singh, K. P. (1983). Mycological studies in small flour mills. Biol. Mem., **8**: 95-102.

Gupta, K. D., Singh, I. C., Agarwal, M. K. and Shivpuri, D. N. (1960). Survey of allergenic aerial mold spores at Jaipur. Indian J. Chest Dis., **2**: 237.

Harindranath, N., Omprakash and Subba Rao, P. V. (1985). Prevalence of occupational asthma in silk filatures. Ann. Allergy, **55**: 511-515.

Jain, A. K. and Mishra, Rajeev. (1988). Air borne pollen grains, fungal spores and other biocomponents at Gwalior. Indian J. Aerobiol., **1**: 30-34.

Jayaprakash, K. B and Ramalingam, A. (1981). *Aspergillus fumigatus* in the air of environments at Mysore. Indian J. Bot., **4**: 17-23.

Jayaprakash, K. B and Ramalingam, A. (1983). Hazardous species of *Aspergillus ochraceus* group in the air of working environments at Mysore. *Proc. Indian Acad. Sci.* (Plant Science), **92**: 387-392.

Jayaprakash, K. B., Rathi, E. and Ramalingam, A. (1978). *Aspergillus flavus* in the air of working environments. Curr. Sci., **47**: 920-921.

Kathapalia, Y. P. (1960). Deterioration and conservation of Paper-I Biological deterioration. Indian Pulp and Paper, **18**: 117-125.

Khandelwal, A., Wadhwani, K. and Srivastava, A. K. (1988). A preliminary report of airspora at two different altitudes at Lucknow. Indian J. Aerobiol., **1**: 53-55.

Kulkarni, R. L. (1971). *Studies in the Air Spora of Some Fields at Aurangabad.* Ph. D. Thesis, Marathwada University, Aurangabad.

Kulshrestha, A. and Chauhan, S. V. S. (2001). Aeromycoflora of some hospitals of Agra city. Indian J. Aerobiol., **14**: 33-35.

Lakhe, D. G. (1980). *Studies in the Air Spora of Some Fields at Udgir.* Ph. D. Thesis, Marathwada University, Aurangabad.

Majumdar, M. R. and Bhattacharya, K. (2000). Aeromycoflora at the foot hills of Eastern Himalaya. Indian J. Aerobiol. **13**: 14-19.

Mane, D. A. & Tilak, S. T. (1980). Aerobiological approach to green ear disease of Bajra-II. All India II Symposium, Life Sciences, Abs. p. 41.

Mane, D. A. (1978). *Studies in the Air Spora over Some Fields at Aurangabad.* Ph. D. Thesis, Marathwada University, Aurangabad.

Mehta, K. C. (1940). Further studies on cereal rusts of India Part-I. Scientific Monograph 14. Imperial Council of Agri. Res., New Delhi.

Mehta, K. C. (1952). Further studies on cereal rusts of India Part-II. Scientific Monograph 18. Imperial Council of Agri. Res., New Delhi, 368 pp.

Mehta, S. K. and Sandhu, R. S. (1983). Immunological significance of *Aspergillus fumigatus* in canesugar mills. Ar. Env. Health, **38**: 41-46.

Mishra, R. R. and Kamal (1971). Aeromycology of Gorakhpur III- Seasonal variation in air fungal flora. Mycopath. et Mycol. Appl., **45**: 301-310.

Mishra, R. R. and Srivastava, V. B. (1971). Aeromycology of Gorakhpur. Spore content over a Paddy field. Mycopath. et Mycol. Appl., **44**: 283-288.

Mishra, R. R. and Srivastava, V. S. (1969). Aeromycology of Gorakhpur. V. Air spora over Wheat and Barley fields. *Proc. 50th. Indian Sci. Cong. Bot Sect.* Abstract No. 99.

Mishra, R. R. (1972). Aeromycology of Gorakhpur-IV. Periodical fluctuations of air-spora. *Mycopath. et* Mycol. Appl., **48**: 213-222.

Mishra, V. K. (1988). *The Study of Industrial Aeromycoflora of Bombay and Their Role in Human Allergy.* Ph. D. Thesis, University of Bombay, Bombay, 106 pp.

Mishra, V. K. (2002). Immunopharmacological studies of fungal extracts in guinea pigs. *National Conference on Fungal Diversity and Biotechnology,* KV Pendharkar College, Dombivli (East), 2-4 Feb. (Abstract) 34.

Misra, J. K. (1988). *Fungi in the Environs of Occupational Importance and Their Possible Role in Causing Allergenic Diseases Among Workers*. A final report submitted to the Department of Environment, Govt. of India, New Delhi, 132 pp.

Misra, J. K., and Jamil, Zafar. (1991). Fungi in the indoor environment of flour mill in Lucknow: Allergenic potentialities of some Aspergilli on humans. Grana, **30**: 398-403.

Misra, J. K., and Shukla, Poonam. (1989). An ecological study of air-borne mycoflora in cobbler's shops as affected by some meteorological factors. Trans. Mycol. Soc. Japan, **30**: 9-23.

Misra, J. K., Tewari, Rashmi and Jamil, Z. (1989). Occurrence, distribution and phenology of indoor fungal flora of Saw mills in Lucknow and their allergenic significance. Biol. Mem., **15**: 73-78.

Misra. J. K., Shukla, P., Tewari, R. and Jamil, Z. (1988). *Aspergillus terreus* Thom: A potent allergenic fungus. Biol. Mem., **14**: 150-152.

Mukerji, K. G., Agarwal, M. K. and Saxena, A. S. (1969). Where from the allergenic fungal spores are coming. Aspects Allergy App. Immunol., **2**: 181-189.

Nadimuthu, N. & Vittal, B. P. R. (1995). A volumetric survey of culturable molds in library. Indian J. Aerobiol. **8**: 12-16.

Nair, P. K. K. (1963). An analysis of atmospheric pollen, fungal spores and other vegetable matter at Vellore, Madras State (India). Indian J Med. Res., **51**: 447-452.

Padmanabhan, S. Y., Gangulay, D. and Balkrishnan, M. S. (1953). *Helminthosporium* disease of Rice II. Sources and development of seedling infection. Indian Phytopath., **6**: 96-105.

Pande, B. N. (1976). *Studies in Air-spora Over Some Fields*. Ph. D. Thesis, Marathwada University, Aurangabad.

Pandit, T. and Singh, A. B. (1992). Prevalance of air-borne fungi in sugar factory environment. Indian J. Aerobiol., **5**: 145-152.

Paradkar, S. A. and Munshi, S. K. (1980). Book deteriorating fungi in the air of Nagpur. J. Palynol. **16**: 83-84.

Patel, H. R., Shekh, A. M., Valand, G. B. & Dave, A. J. (1995). Aerobiology of late leaf spot fungi of groundnut. Indian J. Aerobiol., **8**: 7-11.

Pathak, S. K., Shukla, O. P., Saxena, K. C. and Jamil, Z. (1978). Allerginicity of air borne fungal spores of Lucknow urban area. Indian J Experim. Biol., **16**: 723-724.

Prasad, K. V. N. and Pramila, R. (1997). Patient's assessment of inhalant allergens by using 'Personal' sampler. Indian J. Aerobiol., **10**: 31-37.

Pugalmaran, M. and Vittal, B. P. R. (1999). Fungal diversity in the indoor and outdoor environments and dust of grain storage godowns. Indian J. Aerobiol. **12**: 24-29.

Rafiyuddin, M., Girisham, S. and Reddy, S. M. (1997). Aeromycoflora of bakery elaborating mycotoxins—A preliminary study. Indian J. Aerobiol. **10**: 9-12.

Raha, Subrata and Bhattacharya, K. (1997). Aeromycoflora of residential area at two distinct biozones of West Bengal, India. Indian J. Aerobiol. **10**: 38-42.

Rai, Bharat and Singh, A. K. (1988). A comparative aeromycological study of unpolluted and polluted air localities. Indian J. Aerobiol., **1**: 1-9.

Rajan, B. S. V., Nigam, S. S. & Shukla, K. K. (1952). A study of the atmospheric fungal flora of Kanpur. Proc. Indian Acad. Sci., **35**: 33-37.

Ramalingam, A. (1966). Biopollutants of air in India . Second International Symposium of Plant Pathology, New Delhi, Abstract, 87.

Ramalingam, A. (1971). Air spora of Mysore. Proc. Indian Acad. Sci., **74**: 227-240.

Rasool, S. K. (1989). *Studies on Airborne Fungal Spores in Indoor Working Environments*. M. Phil Dissertation, University of Madras, Chennai.

Rati, E. and Ramalingam, A. (1979). Toxic strains among airborne isolates of *Aspergillus flavus* Link. Indian J. Experim. Biol., **17**: 97-98.

Rati, E., Jayaprakash, K. B. and Ramalingam, A. (1980). Air spora of a poultry-shed at Mysore. Indian J. Microbiol., **20**: 6-12.

Reddi, C. S. (1970). A comparative survey of atmospheric pollen and fungus spores at two places, twenty five miles apart. Acta Allergol. Kbh., **25**: 189-215.

Reddi, C. S. (1978). Vertical profiles of spore concentration within and above a *Sorghum* crop. Phytopath. Z., **93**: 35-40.

Sahay, R. R., Mishra, K. B. and Singh, A. B. (2001). Assessment of Bioaerosols around agricultural processing area at Bodh Gaya (Bihar). Indian J. Aerobiol., **14**: 1-7.

Sahney, Manju and Purwar, Aparna. (2001). Aeromycoflora inside a Library of Allahabad University. Indian J. Aerobiol., **14**: 20-23.

Sandhu, D. K. and Randhawa, I. S. (1979). Studies on the air borne fungal spores in Amritsar, India: Their role in keratomycosis. Mycopathologia, **68**: 47-52.

Sandhu, D. K., Shivpuri, D. N. and Sandhu, R. S. (1964). Studies on the air-borne spores in Delhi—their role in respiratory allergy. Ann. Allergy, **22**: 374.

Sandhu, R. S., Khan , Z. U. and Randhawa, H. S. (1977). Natural occurrence of *A. fumigatus* in cane sugar mills. Sabouraudia, **15**: 263-272.

Sandhu, R. S., Mehta, S. K., Khan, Z. U. and Singh, M. M. (1979). Role of *Aspergillus* and *Candida* sp.in allergic bronchopulmonary mycoses. A comparative study. Scan. J. Resp. Dis., **60**: 235-242.

Santra, S. C. and Chanda, S. (1989). Airborne fungal flora in indoor environments of the Calcutta metropolis, India. Grana, **28**: 141-145.

Saravanan, T. and Vittal, B. P. R. (2002). A preliminary study on the vertical distribution of airborne moulds. *National Conference on Fungal Diversity and Biotechnology*, KV Pendharkar College, Dombivli (East), 2-4 Feb., (Abstract) p.14.

Shah, Ashok. (1997). Allergenic bronchopulmonary Aspergillosis: A clinical profile. Indian J. Aerobiol., **10**: 7-8.

Shankara, P. and Ramalingam, A. (1988). Influence of threshing on air pollution of intramural environments. Indian J. Aerobiol., **1**: 22-25.

Sharma, D. and Dutta, B. K. (2001). Observations on the aeromycoflora of indoor working environments of Greater Silchar. Indian J. Aerobiol., **14**: 8-15.

Sharma, J. K. and Sinha, S. (1973). Spore content of air over a Sorghum field near Agra, India. Sorghum-News Letter, **16**: 96.

Sharma, N. J. and Govil, C. M. (1979). Survey of air spora of Jaipur city. J. Indian Bot. Soc., **58**: 215-225.

Shastri, S. D. (1981). *Air Spora Over Some Fields.* Ph. D. Thesis, Marathwada University, Aurangabad.

Shivpuri, D. N. and Agarwal, M. K. (1969). Studies on allergenic fungal spores of Delhi, India Metropolitan area. Clinical aspects. J. Allergy, **44**: 204.

Shukla, P. (1987). *Aeromycoflora of Cobbler Shops and Tanneries in Uttar Pradesh and Their Allergenic Potentialities.* Ph. D. Thesis, University of Garhwal, Srinagar, U. P., 231 pp.

Shukla. P and Misra. J. K. (1984). Fungi in the environs of occupational importance: I. Cobbler Shop. Bio. Mem., **9**: 95-97.

Shukla. P and Misra. J. K. (1986). Fungi as potent biopollutant in the indoor air of Ginneries in Lucknow. Biol. Mem., **12**: 101-106.

Singh, A. and Singh, A. B. (1994). Airborne fungi in a bakery and the prevalence of respiratory dysfunction among workers. Grana, **33**: 349-358.

Singh, A. & Singh, A. B. (1999). *Aspergillus* spp. as an important occupational risk factor among susceptible individuals. Aerobiologia, **15**: 233-240.

Singh, A., Gangal, S. V. and Singh, A. B. (1994). Airborne fungi in the hospitals of metropolitan Delhi. Aerobiologia, **10**: 11-21.

Singh, A. B., Bhatnagar, A. K. and Gangal, S. V. (1990a). Prevalence of Aspergilli in a bakery environment. Indian J. Aerobiol. **3**: 15-21.

Singh, A. B., Chatterji, M., Singh, B. P. and Gangal, S. V. (1990b). Airborne viable fungi in library: before and after agitation of books. Indian J. Aerobiol. **3**: 32-38.

Singh, A., Ganguli, M. and Singh, A. B. (1995). Fungal spores are important components of library air. Aerobiologia, **11**: 231-237.

Singh, N. I. (1981). Microbiology of the air inside the cinema hall. Proc. Nat. Conf. Env. Bio., 199-206.

Singh, W. M., Singh, Kh. R., Sharma, M. H. and Singh, N. I. (2000). Airborne fungi from indoor air of a poultry farm at Lamangdong, Bishnupur district, Manipur. Indian J. Aerobiol. **13**: 20-24.

Sinha, S. (1976). Expanding horizons of microbial aerobiology. Indian Phytopath., **29**: 119-128.

Sinha, S., Chakraverty, R. and Mishra, A. K. (1981). Aeromycology of Calcutta. Aspects Allergy App. Immunol., **14**: 67-71.

Sinha, S., Chakraverty, R. and Mishra, A. K. (1984). Aspergilli in air over Calcutta. Indian J. Mircobiol. **24**: 35-38.

Sreeramulu, T. (1961). Concentration of fungus spores in the air inside cattle shed. Acta Allergol. Kbh.,**16**: 337-346.

Sreeramulu, T. (1967). Aerobiology in India. A review. J. Scient. Ind. Res., **26**: 474-481.

Sreeramulu, T. (1969). The diurnal and seasonal periodicity of certain plant pathogens in the air. Trans. Brit. Mycol. Soc., **42**: 177-184.

Sreeramulu, T. (1970). Conidial dispersal in two species of *Cercospora* causing Tikka leaf spots on Ground nut (*Arachis hypogea*). J. Agri. Sci., **40**: 173-178.

Sreeramulu, T. (1974). Air-spora of crop fields and its applications. *J. Palynology*, **6**: 31-38.

Sreeramulu, T. and Ramakrishna, V. (1970). Spore concentration in air within and above fields of Sugarcane, Rice, Sorghum and Banana. *Proc. 58th. Indian Sci. Cong. Bot.*, Abstract 52.

Sreeramulu, T. and Ramalingam, A. (1963). Spore content of air over Paddy fields-II. Changes in the field near Vishakhapatnam from November 1959 to January 1960. Proc. Nat. Acad. Sci., India B., **33**: 423-428.

Sreeramulu, T. and Ramalingam, A. (1964). Some short period changes in the atmospheric spore contents associated with changes in the weather and other conditions. Proc. Indian Acad. Sci., **59**: 154-172.

Sreeramulu, T. and Ramalingam, A. (1965). Vertical gradients in airborne spores three meters above ground level. *Proc. 51 & 52 Indian Sci. Cong. Delhi, Sec. Bot.*, Abstract, 324.

Sreeramulu, T. and Ramalingam, A. (1966). A two-year study of the air-spora of Paddy field near Vishakapatnam. Indian J. Agril. Sci., **36**: 112-132.

Sreeramulu, T. and Seshavataram. (1962). Spore content of air over Paddy fields I. Changes in a field near Pentapadu from 21 Sept. to 31st Dec. 1957. Indian Phytopath., **15**: 61-74.

Sreeramulu, T. and Vittal, B. P. R. (1971). Periodicity in the uredospore content of air within and above Sugarcane field. J. Indian Bot. Soc., **50**: 39-44.

Sulia, S. B. and Khan, K. R. (1980). Air spora of Bangalore city market and its relation to the occurrence of market disease. Advances in Pollen Spore Research *V-VII*. 157-169.

Sumbali, Geeta and Badyal, Kusum. (1991). Relationship between fungal airspora of fruit shops and incidence of fruit rots. Indian Phytopath., **44**: 214-218.

Surekha, M. and Reddy, S. M. (1996). Aeromycoflora of poultry shed in relation to incidence of mycotoxigenic fungi. J. Palynol., **32**: 135-141.

Talde, U. K. (1968). *Studies in Taxonomy of Some Pyrenomycetes and Discharged Spores*. Ph.D.Thesis, Marathwada University, Aurangabad.

Talde, U. K. (1969). Air spora of Parbhani. Indian J. Microbiol., **9**: 55-58.

Tilak, S. T. (1982). *Aerobiology*. Vaijayanti Prakashan, Aurangabad, 211pp.

Tilak, S. T. (1987). *Air Monitoring (Practical Manual)*. Vaijayanti Prakashan, Aurangabad, 110pp.

Tilak, S. T. and Kulkarni, R. L. (1978). Rust and Smut spores content of air above Sugarcane field. Biovigyanam, **4**: 103-108.

Tilak, S. T. and Babu, M. (1981). Aerobiological approach to leaf spot disease of Bajra. Indian J. Bot., **4**: 87-90.

Tilak, S. T. and Bhalke, S. P. (1976). Aeromycology at Aurangabad II. Ascospores. Biovigyanam, **5**: 175-177.

Tilak, S. T. and Bhalke, S. P. (1978). Aeromycology at Aurangabad II. Deuteromycetes. Indian J. Bot. **1**: 113-115.

Tilak, S. T. and Chakre, O. J. (1978). Atmospheric concentration of *Claviceps fusiformis* over Bajra fields in relation to their environmental factors. *Fourth International Palynological Conference 1976-77, Lucknow*, Abstract, p. 181-182.

Tilak, S. T. and Chakre, O. J. (1979). Indoor microbial pollution of air and its relevance to storage diseases of food grains. *Symposium on Environmental Pollution and Toxicology, Hissar*: 45-50.

Tilak, S. T. and Kulkarni, R. L. (1972). Microbial content of air inside and outside the caves at Aurangabad. Curr. Sci., **23**: 850-851.

Tilak, S. T. and Kulkarni, R. L. (1975). Atmospheric concentration of *Rhizopus* spores in air. Biovigyanam, **1**: 91-94.

Tilak, S. T. and Kulkarni, R. L. (1980). Some additions to the fungal flora of the air. Indian Phytopath., **34**: 69-71.

Tilak, S. T. and Pillai, S. G. (1988). Fungi in Library: an aerobiological survey. Indian J. Aerobiol. **1**: 92-94.

Tilak, S. T. and Rao, K. S. R. (1988). Atmospheric concentration of the basidiospores of *Ganoderma* Karst. Indian J. Aerobiol., **1**:17-21.

Tilak, S. T. and Saibaba, M. (1986). Atmospheric fungal spore concentration in poultry shed and its relevance to disease incidence. Indian Vet. J., **63**: 356-361.

Tilak, S. T. and Shastri, S. D. (1980). Aerobiology of the vegetable market-II. *All India Symposium Life Sciences, Nagpur*, Abstract, 47.

Tilak, S. T. and Srinivasulu, B. V. (1967). Air spora of Aurangabad. Indian J. Microbiol., **7**: 167-170.

Tilak, S. T. and Srinivasulu, B. V. (1971). Air spora of Aurangabad- II Ascospores. Indian Phytopath., **24**: 740-742.

Tilak, S. T. and Vishwe, D. B. (1975). Microbial content of air inside Library. Biovigyanam, **1**: 187-190.

Tilak, S. T. (1980). Aerobiological Research and Training in India. A key note address. *Second All India Symposium Life Sciences, Nagpur*, 1-7.

Tilak, S. T., Sharma, B. R. N., Sengupta, S. R. and Kulkarni, R. L. (1972). Studies in the microbiological deterioration of paintings at Ajanta and Ellora. Studies Museumology, **8**: 20-25.

Tilak, S. T., Talib, S. H. and Babu, M. (1980). Allergenicity of certain air borne fungal spores of hospital ward. Environment India, **3**: 27-30.

Tripathi, R. N. (1987). Fungal airspora inside the Central Library of Gorakhpur University. Water, Air Soil Pollution. **34**: 125-134.

Uday Prakash, N. K. and Vittal, B. P. R. (2002). Mold diversity in the atmosphere of animal care facilities. *National Conference on Fungal Diversity and Biotechnology*, KV Pendharkar College, Dombivli (East), 2-4 Feb. (Abstract) 15-16.

Verma, K. S. and Chile, S. (1992a). Fungi in the Medical College of the Jabalpur city and the allergenic behaviour of some species. J. Indian Bot. Soc., **71**: 247-249.

Verma, K. S. and Chile, S. (1992b). The airspora of hospital ward in relation to bronchial allergy. J. Palynol., **28**: 99-106.

Verma, K. S. and Khare, K. (1987). Fungal agents which damage paper materials in Library. In: *Atmospheric Biopollution*. Ed. N. Chandra. p. 193-197.

Verma, K. S. and Khare, K. (1988). Aeromycology at Jabalpur—A preliminary study. Indian J. Aerobiol., **1**: 56-58.

Verma, S., Sen, S. and Kaur, M. (2000). Survey of microbial aeroallergens at Solan-Himachal Pradesh. Indian J. Aerobiol. **13**: 10-13.

Vishnu-Mittre and Khandelwal, A. (1973). Air borne pollen grains and fungal spores at Lucknow during 1969-1970. Palaeobotanist, **22**:177-185.

Vishwe, D. B. (1979). *Studies in the Airspora of Aurangabad*. Ph. D. Thesis, Marathwada University, Aurangabad.

Vittal, B. P. R. and Glory, A. L. (1985). Airborne fungus spores of a library in India. Grana, **24**: 129-132.

Vittal, B. P. R. and Ponnusamy, P. (1982). Species composition of the atmospheric fungal flora of Madras. Indian J. Microbiol., **22**: 79-81.

Wadhwani, K. (1979). Air borne fungal flora of Lucknow (University area). J. Palynol., **15**: 105-110.

36

Aerial Fungal Biodiversity and Allergic Diseases

A.B.Singh* and Terlika Pandit

Aerobiology and Allergy Laboratory, Centre for Biochemical Technology Delhi University, Delhi-110007

India is a vast country inhabiting one fifth of the worlds' population. Today more than 20% of the more than 100 million Indian population suffer from allergic disorders such as allergic rhinitis, bronchial asthma and atopic dermatitis etc. Although fungi are ubiquitous in nature and are present in both outdoor and indoor environments, but greater attention is currently paid to indoor fungi as children and adults, both spend most of their time indoors. In this paper we have focussed on aerial fungal biodiversity and their role in allergic disorders. The sources of fungal spores in the outdoor air are mostly vegetation, cereal crops and garbage storage and in the indoor environment are mattresses, humidifiers, damp walls, carpets, dustbins, and indoor plants, decaying vegetation and organic wastes that provide conducive environment for fungal growth. The harmful effects of inhaling dust from fungal spores, grain, flour and hemp was reported as early as 1713 by Ramazzini. For effective diagnosis and therapy of allergic patients it is essential to know the quantitative and qualitative prevalence of fungi in different environments. Continuous efforts by workers have lead to the development of a vast array of sampling instruments. Every principle for separating particles from air, from sedimentation, filtration, inertial impaction, impingement in liquids, to thermal and electrical precipitation, has been applied to the collection of microorganisms.

In Indian environment, both outdoors and indoors a rich aerial fungal diversity is reported. The most predominant types in outdoor air are *Cladosporium*, Aspergilli-Penicilli, *Curvularia, Rhizopus* and *Helminthosporium* while some of the allergenic fungi reported in indoor air are Aspergilli-Penicilli, *Cladosporium, Penicillium nigricans, Aspergillus niger, Aspergillus flavus* and smuts spores. However, in different parts of the world there is an urgent need to organize all the information available in the form of seasonal calendars and all allergic fungal types enlisted and made readily available for the use of common man and physicians as a diagnostic tool. Important airborne fungi of different work environments have been studied and enlisted for their seasonal and annual variations.

Key-words : Fungal biodiversity, aerial microspora, allergic diseases

India is a vast country inhabiting one fifth of the worlds' population. The country is blessed with multireligious, multicultural, multilinguistic and multi-ethnic populations. Most of the population (< 70%) lives in villages

*Corresponding Author Email : singha49@hotmail.com

and their chief occupation is farming and agriculture. Since independence (early 1950s') there has been a lot of migration of people from rural areas to urban settlements for the sake of job opportunities, particularly in Metros of India. Today more than 20% of the 100 million Indian population suffers from allergic disorders such as allergic rhinitis, bronchial asthma and atopic dermatitis etc.

Bioparticulate matters such as fungal spores, pollens, insect debris, house dust mites, animal danders, chemicals, foods etc. have long been known to cause allergic disorders (Kino and Shing, 1978; Gravesen, 1979; Kang *et al.*, 1979; Shivpuri, 1980; Lacey and Crook, 1981; Salvaggio and Aukrust, 1981; Peterson *et al.*, 1983; Aalberse *et al.*, 1989; Burge, 1990). The seasonal and annual variations of these bioaerosols have been extensively studied in different parts of the world including India (Collins-Williams *et al.*, 1973; Al -Doory *et al.*, 1982; Long *et al.*, 1987) as their knowledge is of paramount importance for an effective diagnosis and therapeutic management of the allergic diseases. Many reports from India have provided information on prevalence of airborne fungi in ambient air (Ramalingam, 1971; Agarwal and Shivpuri, 1974; Tilak and Kulkarni, 1980; Gaur and Kasana, 1981; Maribhat and Rajasab, 1988; Pandit and Singh, 1991; Gupta *et al.*, 1993). Subsequent studies from all over the world established pollen grains and fungal spores as the major causative agents for respiratory allergic disorders (Feinberg, 1946; Nilsson *et al.*, 1977; Naranjo, 1958; Shivpuri *et al.*, 1960; Leuschner, 1974; Lewis, 1984; Gupta and Chanda, 1989; Singh and Singh, 1994, Rawat *et al.*, 2000).

Although fungi are ubiquitous in nature and are present in both outdoor and indoor environments, but greater attention is paid to indoor fungi as children and adults, both spend most of their time indoors. In this review we will focus on aerial fungal biodiversity and their role in allergic disorders. The sources of fungal spores in the outdoor air are cereal crops, the indoor environment are mattresses, humidifiers, damp walls, carpets, dustbins, and indoor plants, decaying vegetation and organic wastes, produce conducive environment for fungal growth.

The air inside buildings often is contaminated with particles and chemicals that adversely affect the health of the occupants. These pollutants are brought indoors from the outside or arise from sources indoors. Over the past several years the problem of indoor air pollution has been reviewed as is evident by a number of articles and books, with much attention to chemical and physical pollutants, e.g., nitrogen oxides, formaldehyde, asbestos fibers, and radon gas. But scant attention, has been paid to pollutants of biological origin and to their effects on subjects exposed to biopollutants.

Biological pollutants, often cause symptoms such as respiratory allergy, mucus membrane irritation, headache and fatigue. These symptoms are also often associated with what is termed as **"sick building syndrome"**. Biological aerosols have been the predominant cause of complaints in 1-5% of problem office buildings as investigated by the U.S. National Institute for Occupational Safety and Health (NIOSH). Airborne biological contamination may be a larger problem in homes where there is a greater variety of source materials and very different types of activities that contribute to the presence of microorganisms, and plant and animal matter.

History of Fungal Allergy

The harmful effects of inhaling dust from grain, flour and hemp was reported as early as 1713 by Ramazzini. But Floyer first implicated fungi as a cause of asthma in 1726. One and a half-century later, Blackley (1873) confirmed the relationship of fungal spores and allergy when he suffered **"bronchial catarrah"** after inhaling *Chetomium* and *Penicillium* spores. Half a century later van Leewan (1924) suggested that asthma was caused by 'climate allergens'. Feinberg (1935) reported respiratory allergic reactions to fungi in his patients and attributed outdoor environment as a source of fungi. Since then researches carried out all over the world have established the role of fungi as an important cause of allergic disorders as is evident from the spurt in literature on the subject during the last two decades (Agarwal and Shivpuri, 1974; Gravesen, 1979; Ibanez *et al.*, 1988; Aukrust, 1992; Singh and Singh, 1993, Pandit and Singh, 1992, 1994). More than 10-20% of Indian population is known to suffer from major allergic disorders (Vishwanathan *et al.*, 1965).

Many of the diseases that are now attributed to fungal infection were described as far back as Greek and Roman times. Hippocrates in his writings described the white patches of candidiasis (thrush) in debilitated patients. The Roman attributed dermatophytosis (ringworm) to insect bite and called the infection "tinea", meaning small insect larvae. The Greek, on the other hand, called dermatophytosis "herpes" (from the word herpo – to creep).

But it was only in the early 20th century that reports started pouring in about the relationship of fungi with allergies (Crook and Lacey, 1988; Feinberg, 1935, 1944). Since then several workers have established the role of fungi as causative agents in allergic diseases (Agarwal and Shivpuri, 1974; Gravesen, 1979; Jimenez-Diaz *et al.*, 1960; Salvaggio and Aukrust, 1981).

Liberation of Fungal Spores

More than 80,000 species of fungi are known (Haksworth *et al.*, 1983) to occur and some of them are present in much higher concentration in air as compared to pollen grains. Fungi are cosmopolitan in nature and possess highly evolved spore liberation mechanism. The spores are generally small (3 to > 100 μ) and remain suspended in the atmosphere for a long time – from a few hours to several days. While suspended, they are inhaled by humans and induce different symptoms of respiratory disorders in hypersensitive individuals.

Fungal spores are present in both outdoor and indoor environments and contribute the major part of suspended bioparticulate matter of the air. They are liberated in very high concentrations and can remain suspended in air for extended period of time. Their presence, profile of species, concentration etc. depends on various factors such as temperature, humidity, wind direction, sunshine, substrate precipitation and other seasonal and climatological factors. Dispersal of basidiospores depends on the moisture content of air, as their dispersal requires higher humidity in air (Zoberi, 1964) while spores of fungi like *Cladosporium* spp., *Alternaria* spp., *Drechslera* spp. etc. are wind dispersed so they are present in lower concentration when humidity is high.

Enumeration of Fungi by Air - Sampling Techniques

Continuous efforts by workers have led to the development of a vast array of sampling instruments developed since the time, Pasteur opened vials of sterile broth and learned that airborne microorganisms were more abundant in cities than that on a mountain. Many of the sampling devices developed and prevalent today are enumerated in Table 1. Some of them are described as follows.

Gravimetric Samplers

This is based on the principle that bioparticulates settle down on a surface due to gravitational force. The Durhams gravity-sampling device (Durham, 1946) consists of two horizontal discs with a diameter of 22.1 cm and 8.1 cm. The upper disc protects the slide from rain and sun. The slides are exposed daily, at a fixed hour, coated with adhesive glycerine jelly. After exposure, the slides are mounted in a drop of molten glycerine jelly, for various allergen types trapped. The results are expressed as number / unit area. However, this sampler is no more used in developed world, but developing countries still report the use of gravity settlement method for fungal spores/colonies. Opening plates of solid nutrient media, the settle plate method is inefficient. Even in

Table 1: Common samplers employed for sampling spores in different laboratories:

Samplers	Mode of Collection	Culture/ particulate	Battery operated/ power operated	Sampl-ing rate L/Min	Suitable for indoor/ outdoor sampling
Burkard personal slide sampler	Suction	Particulate	Battery/ power operated	10	Indoor
Burkard personal petriplate sampler	Suction	Culture	Battery/ power operated	10	Indoor
Rotorod aeroallergen model sampler	Intermittent rotating impactor	Particulate	Power operated	47	Indoor/ outdoor
Andersen six stage sampler	Suction	Culture	Power operated	28.3	Ideal for Indoor

still air the number of large particles collected on the surface will be overestimated and smaller particles that have a slower settling velocity will be underestimated.

Impaction Sampler

Rotorod sampler – developed by Perkins (1957) has leucite rods of 1-3 mm, coated with adhesive silicon grease are used to collect airborne particles. Collection quality depends on speed of sampler and width of sampling rod. It is a lightweight portable sampler operated by DC power. The exposure time can be adjusted according to requirement. There are three models available in the Rotorod Aeroallergen Models (40s, 85s, 95s) and the spore catch obtained by these three models is almost similar (Frenz and Lince, 1997).

These are used to study both culturable and non-culturable spore types on continuous basis for diurnal, seasonal and annual variations (Fig. 1).

Fig. 1. Rotorod Sampler (Aeroallergen Model)

Hirst Trap

The method requires suction of certain volume of air according to a known velocity and for a chosen duration on a trapping. There are several instruments currently in use. The Hirst spore trap (Hirst, 1952) is the most commonly used in UK. In this, trapping of bioparticles is done on slides coated with glycerine jelly and slides are replaced each day with fresh slides.

Burkard Seven Day Volumetric Sampler

The Hirst trap was later modified to Burkard trap in which slides were replaced with a drum, which can rotate and run continuously for seven days with a definite speed. Burkard continuous 7-day sampler uses an adhesive-coated tape mounted on a rotating drum. The drum is connected to a timer and rotate at constant speed. The tape is changed every 7 days. Exposed tape is cut in 7 strips corresponding to 7 days and mounted on a slide. This is a very good sampler to study diurnal or seasonal trends for fungal spores as well as pollen grains.

Burkard Slide Sampler

Burkard slide sampler is a compact battery operated sampler, it is 10cm in height and 8cm in diameter (Fig. 2 b) It has a rectangular orifice at the top and a slit on the side to insert microslide. The microslide is coated with glycerine jelly. The sampler sucks in 10L of air per minute. The particles get impacted on the slide in the form of a band. The slide is then mounted in glycerine jelly and scanned for fungal spore count under the microscope.

Burkard Petriplate Sampler

The Burkard personal petriplate sampler is similar except that it has a stage to hole the petriplate and a sieve to cover the petriplate and on top a lid to cover the sieve. On the cover is an opening from where the air is sucked in. These samplers are most convenient for outdoor and spot sampling and places where power connection is not available (Fig. 2 a).

Burkard petriplate sampler is similar to above sampler except that it has a stage to hold a petriplate containing nutrient agar media instead of slide. These two samplers are appropriate for spot sampling.

Andersen Sampler

The best device to obtain culturable fungal spore count is the Andersen six-stage volumetric sampler. It uses six petriplates in which media is kept

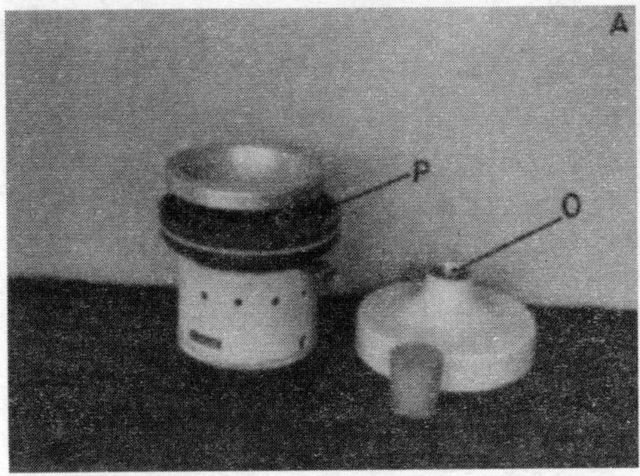

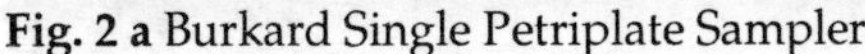

Fig. 2 a Burkard Single Petriplate Sampler

Fig. 2 b Burkard Slide Sampler

under different sieve size present in decreasing order of pore size. Each sieve has 400 pores. The sampler sucks in 28.3L of air per minute. The air passes from the orifice to all the six petriplates before passing out. The particles of similar aerodynamic dimensions are impinged on the same plate (Fig. 3). Now even two stage Andersen sampler is available which is quite efficient and less time consuming in number of plates to be examined.

The total number of colonies were corrected to the nearest count with the help of the correction factor given by Andersen (1958) and the counts were expressed as colony forming units per cubic meter of air (CFU/m^3).

Every principle for separating particles from air, from sedimentation, filtration, inertial impaction, impingement in liquids, to thermal and electrical precipitation, has been applied to the collection of microorganisms. Changes in aerosol concentration with time can be followed with the rotating slit or slit-to-agar sampler in which a large petri dish of medium is placed on a turn table beneath a stationary slit inlet (Fig. 3).

Fig. 3 Andersen Six Stage Volumetric Sampler

Identification and Analysis of Fungi

Analysis of the catch can be carried out by the following three methods:

Culture Analysis:

This includes the tally of colonies produced in culture or semisolid media. This technique of analysis is employed to identify the culturable fungal spores on the basis of their colony characteristics which otherwise, can not be identified under the microscope.

The culture plates, after exposure, are incubated at appropriate temperature and impacted spores are allowed to grow for a couple of days till colonies start forming. Each colony represents one spore and considered colony-forming unit.

Direct microscopy:

The microscopic identification of distinctive particles is an approach validated by years of practical applications of both gravimeteric and volumetric samplers. A variety of particles, including certain basidiospores, ascospores and spores of rust, smuts, downy mildew and primitive green plants are recognizable, which, fail to grow on most laboratory media. Therefore, it is most dependable way to identify most of the pollen and fungal spores.

After suspended particles have been collected on or in a suitable medium the viable microorganisms, those that will multiply when provided the appropriate conditions, contained in or on these particles can then be counted and identified. Environmental scientists looking for nonviable particles also use the techniques used to extract viable cells and particles carrying them from the air. The most efficient methods of removing suspended particles from the air, e.g., filtration through fine pore matrices, might be adequate for resistant forms of microorganisms, such as spores, but can damage less environmentally resistant, vegetative cells. The absence of these sensitive cells from a sample could cause one to mistakenly conclude that they were not present in the environment sampled. The total number of cells present can be estimated by microscopic examination, sometimes with the help of stains or fluorescent tags.

The correction factor was calculated as per the given formula:

$$\frac{\text{Total no. of fungal colonies}}{\text{Total volume of air sampled}} \times 1000$$

The total number of spores are expressed as spores per cubic meter (spores/m^3) of air. The formula used was as follows:

$$\frac{\text{Total no. of spores}}{\text{Total volume of air sampled}} \times 1000$$

Other limitations are that a considerable amount of experience is needed to identify particles by their morphology and to distinguish them from debris, and labelled antibodies for only a few clinically important microorganisms are readily available.

Immunoassay:

Immunochemical analysis following descending elution offers an analytic approach to dust without potential or defined form (eg. Dander, seed pomace, arthropod effluvia etc.). If micronic aerosols do carry pollen allergens, these fractions are also accessible to immunoassay in bulk samples obtained by filtration.

Merits and Demerits

One advantage in sampling for airborne microorganisms is that the amount of material needed to measure the concentration of viable air contaminants is much lower than that needed to gravimetrically quantitate nonviable particles. Where only one or a cluster of cell lands on an appropriate solid nutrient medium, or lawn of host cells, a macroscopic fungal or bacterial colony, or viral plaque, will develop. These isolates can then be identified very exactly using tests of biochemical and immunological reactions conducted on subcultures of the original material. The researcher, therefore, is not limited to the original amount of the sample but can culture as much material as is needed for the various tests used for identification.

Global Scenario

Aerial Distribution of fungi

Lacey (1981) found *Cladosporium* , *Aspergillus, Penicillium, Epicoccum, Fusarium, Drechslera*, as the most common fungal forms. *Cladosporium* being the most dominant type contributing 20-30% to the total airspora (Gregory and Sreeramulu, 1958; Lacey, 1962). From USA also *Cladosporium* has been reported as the most dominant fungal genus on the West Coast along with *Alternaria* (Al-Doory, 1967; Al-Doory *et al.*, 1987; Prince and Meyer, 1976). *Alternaria* also exceeded *Cladosporium* in warm dry regions of the world (Sandhu *et al.*, 1964; Agarwal and Shivpuri, 1974; Hariri *et al.*, 1978). In many countries basidiospores have been reported as predominant (Salvaggio and Seabury, 1971; Tarlo *et al.*, 1979; Mac Donald and O'Driscoll, 1980; Hasnain *et al.*, 1958; Jelks, 1989). Ogunala (1975) reported the prevalence of *Cladosporium, Curvularia, Fusarium, Aspergillus, Penicillium, Pithomyces, Aureobasidium, Rhizopus* and *Epicoccum* from Nigeria. From China yeast,

Aspergillus, Penicillium, Hormodendron, Mucor, Curvularia species were reported as dominant types (Chen *et al.*, 1978; Long *et al.*, 1987; Lu *et al.*, 1969). *Cladosporium* species has also been found to be the dominant type in the air of Australia followed by *Epicoccum, Nigrospora, Neurospora, Penicillium* (Tilak, 1981). In Spain, Denmark and Netherlands the dominant types that contributed more than 70% of the total catch were *Cladosporium, Alternaria, Aspergillus* and *Penicillium* (Beaumont *et al.*, 1985; Calvo *et al.*, 1980; Larsen, 1981).

Epidemiology of allergic diseases

Various surveys carried out all over the world have established the presence of fungi in air. About 10 to 20% of world population is known to suffer from major allergic disorders such as allergic rhinitis, bronchial asthma, atopic dermatitis and urticaria. The incidence of asthma alone was reported to be 1.5% to 6.0% in the populations in Britain, USA, Canada, France, Finland, New Zealand and Australia (Smith , 1983). However, allergic rhinitis varies from 6 - 12% in these countries (Smith and Salvin, 1988). The recent International study on asthma and allergies in childhood (ISAAC), based on questionnaire survey across the world, has reported as high as 20% prevalence of asthma alone in UK, Australia, New Zealand, Costa Rica, Peru and Singapore (Anonymous, 1998).

Meggs and his coworkers conducted a telephonic survey to determine the self reported prevalence of allergy in North Carolina, USA, and observed 35% of the individuals were having allergies (Meggs *et al.*, 1996). Observations compiled by Sly (1999) based on the local survey of 1962 to 1965 for allergic rhinitis in subjects with an average age of 18 years in United States disclosed a prevalence of 15 - 28%, while the national survey of 1976 - 1980 showed prevalence of allergic rhinitis to be 26%. Annual health surveys indicated increase in prevalence of asthma from 3.1% in 1980 to 5.4% in 1994 and prevalence among impoverished inner city children has been much higher (Sly, 1999).

In Australia, a study of a large random population of children in seven regions in New South Wales during 1991 - 93 observed high prevalence of airway hyper-responsiveness (22 - 24%) and diagnosed asthma in 24 - 38% of population (Peat *et al.*, 1995). They have also observed that prevalence of these symptoms was higher in coastal region and far west regions of South Wales as compared to the inland regions (Peat *et al.*, 1995).

An increase in seasonal allergic rhinitis has also been reported from different countries of Europe (Blenkinsopp and Blenkinsopp, 1989). In UK, general practitioners have observed a four-fold increase in allergic rhinits over the past 20 years. Burr *et al.* (1989) conducted two epidemiological surveys, 15 years apart, for respiratory allergy in 12 year old school children in South Wales. Their questionnaire survey was conducted along with the peak flow measurement and exercise provocation test. They observed 6% to 12% increase in prevalence of asthma, 5% to 16% for eczema and 9% to 15% for hay fever (allergic rhinitis). Ninan and Russel (1992) conducted a study to investigate the prevalence of respiratory symptoms and atopy in Aberdeen (UK) school children based on analysis of questionnaire completed in 1964 and 1989 and revealed, that diagnosis of asthma rose from 4.1% to 10.2% and prevalence of hay fever increased from 3.2% to 11.9%. Eczema also became more common (5.3% to 12 %) over the years.

The increase has also been observed in other countries. Aberg *et al.* (1995) assessed the prevalence of asthma, allergic rhinitis and eczema using two questionnaire studies, 12 years apart, with school children in Sweden. They observed an increase in prevalence of one or more symptoms that was 23.8% to 32.5% in Goteborg at southwestern coast of Sweden and 29.9% to 44.8% in Kirung, a mining town. In Istanbul, incidence of allergic rhinitis was 20.6% as reported by Dotterud *et al.* (1994). Whereas, only after five years Keles *et al.*, (1999) reported 62% and 51% prevalence of rhinitis in 1994 and 1996, respectively in Istanbul, which is very high as compared to previously available figures. A recent review on epidemiological studies in France, Charpin and his coworkers have observed that hay fever affects 6% of the children in France and 18% of the teenagers (Charpin *et al.*, 2000).

In Saudi Arabia, AL Dawood (2000) conducted an epidemiological survey in school children and reported that 9.5 % of the children had questionnaire-diagnosed asthma. Behbehani and his colleagues (2000) conducted one of the first epidemiological survey in Kuwait among 3,110 school children of 13 - 14 years of age. It revealed that 15.5% to 18.1% children had asthma within last 12 months and prevalence rate of allergic rhinitis was 17.1% and they have also observed that prevalence of asthma is higher in boys as compared to girls (Behbehani *et al.*, 2000). Thus an increase in the allergic diseases all over the world is evident with the spurt of publications on the epidemiology of respiratory disorders.

Indian Scenario

Fungal aerobiology

Airborne surveys of fungal allergens have been reported from different parts of India as well. The dominant forms include *Cladosporium* sp., *Aspergillus* sp., *Nigrospora*, *Alternaria*, *Curvularia*, basidiospores, ascospores,

Helminthosporium (Atluri and Subba Reddi, 1971; Maribhat and Rajasab, 1988; Ramalingam, 1971).

In India many reports provide information of prevalence of fungi in ambient air (Ramalingam, 1971; Agarwal and Shivpiuri, 1974; Tilak and Kulkarni, 1980). *Alternaria* is reported as the dominant fungal type from Delhi and Dehradun (Agarwal 1970; Sandhu, 1964). *Aspergillus, Penicillium, Cladosporium, Epicoccum* and *Drechslera* were reported as dominant types (Anonymous, 1998) from Lucknow from North India.

Studies carried out from western India by Tilak and coworkers (1980) have reported deuteromycetes as the most dominant types contributing 70% of the total spore catch. Recent survey from Pune and Aurangabad (Anonymous, 1998) have revealed *Cladosporium, Aspergillus, Penicillium, Curvularia, Rhizopus* as the dominant fungal types.

Studies carried out at Gaya, Gauhati, Calcutta, Gorakhpur and Imphal (Anonymous, 1998) revealed *Aspergillus, Penicillium, Cladosporium,* ascospores, rust and smut spores were the dominant types from eastern India.

Epidemiology

The first ever epidemiological study, carried out in India about four decades ago, revealed that nearly 10% of Indian population suffer from major allergic disorders (Vishwanathan, 1964). After this, epidemiological surveys were conducted only in late nineties. Shaikh reported from a survey of over two thousand patients attending clinics in Mumbai over a 6 year period (Shaikh, 1997, 1998). The study concluded that the incidence of allergy was 25.3% and more than 80% of these patients were below 40 years of age. Nasobronchial allergies i.e. asthma and/or rhinitis are reported to be commonest allergic manifestations with 73.4% of the patients belonging to this group. The incidence of atopic dermatitis was 3.6%. In a recent study carried out in school children in Delhi, prevalence of asthma was observed in 11.9% and asthmatics constituted 3.4 % of the total subjects (Chhabra *et al.*, 1998). The cumulative prevalence of asthma was 15.3%, which is very high as compared to 1% incidence reported earlier from Delhi (Anonymous, 1961).

In a recent study, based on the survey conducted on more than 50,000 populations in five cities in India under the All India Coordinated Project on Aeroallergens and Human Health, sponsored by the Ministry of Environment and Forests, 20% to 30% people have been found to suffer from various respiratory allergic disorders (Anonymous, 2000). The survey is based on medical questionnaire, Pulmonary Function Test (PFT) and skin tests on selected group of patients. Survey included urban and rural populations, adults, school children and industry workers as well. In Delhi, asthma was observed in 10.8% and allergic rhinitis was reported in 11.1% of 11,000 surveyed populations. Whereas, at Chandigarh 13.3% children were having allergic rhinitis and asthma was diagnosed in 5.6% of the children surveyed. However, 7.5% adults were patients of nasobronchial allergy. At Thiruvananthapuram, upper respiratory disorders are reported in 12 - 25% population and they are maximum in 5 – 25 years age group, whereas, current asthma was observed in 3% to 12% between different age groups. However, symptoms of asthma are reported in 5 – 10.9% urban and 4 - 13% rural population at Thiruvananthapuram. At Kolkata, 7.2% of the atopic population was suffering from respiratory and nasal allergy (Anonymous, 2000).

Fungal Aerobiology

In Outdoor Environment

Airborne survey of fungal allergens has been reported from different parts of India.

From Delhi and Dehradun, *Alternaria* is reported to be very high followed by *Cladosporium* in the studies concluded using gravitaional method (Agarwal, 1970; Sandhu, 1964). At Solan and Lucknow, 17 and 40 types, of fungal spores, respectively were recorded from ambient air. Aspergilli-Penicilli followed by *Cladosporium, Helminthsporium, Epicoccum and Drechslera* are the dominant types (Anonymous, 1998) Studies carried out at Gaya, Gauhati, Calcutta , Gorakhpur and Imphal (Anonymous, 1998) revealed Aspergil-Penicilli, *Cladosporium*, ascospores, rust and smuts spores, *Nigrospora, Periconia, Arthrium, Ganoderma* and *Rhizopus* as the major types from east India.

From western India, Tilak and coworkers (1980) have reported that Duteromycetes contributed 70% of the total fungal air spora. Recent survey has reported 18 spore types at Aurangabad and 22 at Pune. *Cladosporium*, Aspergilli-Penicilli, *Curvularia, Rhizopus and Helminthsporium* were the abundant types (Anonymous, 1998)

Dominant forms at Visakhapatnam, among 34 types were Aspergilli-Penicilli, *Cladosporium, Curvularia*, basidiospores, uredospores. At Chennai 50 types were identified, apart from the common genus, *Periconia, Curvularia and Ganoderma* are the dominant types (Anonymous, 1998). At Trivendrum, *Tetraploa* was another dominant fungal spore prevalent in the air among other common types.

At central India, surveys were conducted at Gwalior, Aurangabad and Nagpur and 15, 18 and 25 fungal types were recorded, respectively. Prominent types were Aspergili-Penicilli, *Cladosporium, Helminthosporium, Curvularia, Drechslera, Pyricularia, Syncephalastrum and* uredospores etc. (Anonymous, 1998). Recent studies conducted by Anonymous (2000) in occupational and extramural sites revealed the following results :

A total of 51 fungal types were identified from Banglore out of which 22 species belonged to *Aspergillus* and 2 to *Penicillium* group. Cladosporoides was the other dominant type. While in Gwalior 19 fungal types were reported out of which *Cladosporium herbarum* contributed the most (56.4%) followed by *Aspergillus flavus* and *A. niger* both contributing 8.6% each and *Fusarium roseum* 6.6%. The other dominant types were *Penicillium chrysogenum* and *Botrytis ceneria.* 9500 and 8750 colonies/m^3 were recorded in Vishakapatnam from outdoor and indoor environments, respectively. In all 48 different fungal types were recorded out of which the dominant types were *Cladosporium, Aspergillus niger, A. versicolor, Penicillium nigricans* and *A. fumigatus.* A total of 18,567 and 25,830 col/m^3 were recorded in Vishakapatnam from inside and outside, respectively. The dominant fungal types were *Aspergillus niger, Cladosporium, Aspergillus versicolor, Aspergillus fumigatus, Mucor* etc.

In Indoor Environments

Besides the outdoor environment, indoor and work environments are also greatly influenced by fungi especially occupational sites employing organic raw materials eg. granary, poultry, flour mills, bakery, sugar factory etc (Horejski *et al.*, 1960 ; Roby and Sneller, 1970; Lacey and Crooks, 1988). Survey conducted at working environments of bakery, poultry, sugar factory and libraries in Delhi revealed, Aspergilli-Penicilli and smuts spores as significant contributors in indoor air (Singh *et al.*, 1990 ; Singh , 1993; Pandit, 1994)

Poultry:

A total of 17 fungal forms were reported from Guwhati out of which *Cladosporium herbarum* was the most dominant followed by *Aspergillus schari* and *Penicillium sp.* In Gwalior 21 fungal types were reported from the Poultry. The most dominant being *Aspergillus niger* (24.6%) followed by *Cladosporium* sp. (22.3%), *Penicillium sp.*(10.7%), *Aspergillus flavus* (9.1%). *Botrytis, Fusarium, Aspergillus glaucus, Aspergillus fumigatus*, and *Curvularia* were the other dominant species. While in Bangalore *Aspergillus sp.* (531 CFU) were the most dominant contributors to the environment followed by *Penicillium* sp. (301 CFU) and *Cladosporium* sp. (169 CFU). *Mucor* and *Rhizopus* were among the other dominant types. A total of 14,164 and 12,837 colonies were recorded inside and outside, respectively from a Poultry farm in Vishakapatnam. A total of 54 fungal types were reported out of which majority belonged to

Aspergillus and *Penicillium* group. The other dominant fungi recorded were *Cladosporium, Penicillium nigricans, Trichoderma* etc.

Cattle shed:

A total of 51 fungal types were identified in Vishakapatnam out of which 22 species belonged to *Aspergillus* and 2 to *Penicillium* group. Cladosporoides was the other dominant type.

Residential Houses:

A total of 8909 and 9327 CFU/m^3 were recorded in Vishakapatnam from inside and outside, respectively for the 47 types identified. The dominant fungal types were *Cladosporium, Penicillium nigricans, Aspergillus versicolor,* and *Aspergillus oryzae.* While in Solan in Shimla sampling conducted in a wet house revealed *Penicillium* as the most dominant types contributing 30.7% followed by *Aspergillus* sp. (15.4%), *Alternaria* sp. (10.5%). In the same place sampling conducted in a mud house revealed *Aspergillus* sp. at 35.1% as the most dominant contributor followed by *Penicillium* (26.9%). The other dominant types were *Alternaria, Cladosporium, Curvularia* and *Fusarium.*

Garbage dump:

A total of 11,040 and 10,630 col/m^3 were recorded in Vishakapatnam from inside and outside, respectively. The dominant fungal types were *Aspergillus niger, Cladosporium, Aspergillus versicolor, Aspergillus fumigatus, Mucor* etc.

Jaggery godown:

A total of 12,855 and 10,570 col/m^3 were recorded in Vishakapatnam from inside and outside, respectively. The dominant fungal types were *Aspergillus niger, Cladosporium, Aspergillus versicolor, Penicillium brefeldianum, Curvularia,* etc.

Library:

A survey conducted in a library in Gwalior, *Cladosporium species* (30.3%) inside and (38.2%) outside was reported as the most dominant fungal type followed by *Aspergillus niger* (20.5%) inside and (15.2%) outside. Other dominant types included *A. teneus* and *Fusarium* sp. Another survey conducted in Banglore City library also showed *Cladosporium species* (51.6%)

Dairy farm:

A total of 9000cfu/m^3, 9800cfu/m^3, 8500cfu/m^3, 940cfu/m^3, 10,110cfu/m^3, were recorded from reception, processing area, packaging area, cold storage and outside air, respectively in Vishakapatnam. The dominant fungal types were *Aspergillus niger, Cladosporium, Curvularia, Aspergillus nigricans* in reception; *Aspergillus niger, Cladosporium, P.nigricans* in processing area and packaging area; *Aspergillus niger, Cladosporium* in outside air. Very few colonies were recorded in the cold storage.

Immunodetection - Common Diagnostic Tests:

Bio Assay – Assessment of allergenicity to fungi

In some cases bioassays are conducted on elutes of air samples or bulk samples of dust from a suspended environment. These tests use a living substrate for analysis, e.g., the *Limulus* amoebocyte assay for bacterial endotoxins. The common scratch test used by allergists is another example in which concentrated extracts of dust or pollen are applied to abraded skin or injected under the skin to test a person for sensitivity to specific antigens.

Animal dander, vegetable dusts, and the excretions or secretions from arthropods are less routinely determined in air samples of suspended particles. Although some antigen-specific immunosorbent assays are available, identification of these dusts is usually based on observation of their sources in the environment.

Among the newest analytical methods are the enzyme-linked immunosorbent assay (ELISA) (Kauffman *et al.*, 1983; Sepulveda *et al.*, 1979) and the radioallergosorbent test (RAST) used to measure the presence in human serum of antibodies that indicate exposure to specific microorganisms. The RAST technique measures the amount of circulating allergen specific immunoglobulin E (IgE) in blood serum. This *in vitro* analogue of a skin test is not as accurate and is more expensive to perform, but is safer for the patient because only the blood sample and not the person is exposed to the antigen. The ELISA test uses an enzyme label rather than the radioactive tag of the RAST technique to measure IgE, and is therefore safer and is also less expensive.

In some cases sensitivity to an environmental allergen is diagnosed using a provocation test in which the potential allergen is given orally or by inhalation under carefully controlled conditions.

Table 2 : Common diagnostic tests for allergies *in vivo* and *in vitro*

In vivo	Skin test - Prick - Intradermal Inhalation challenge test
In vitro	ELISA RAST Histamine release Gel diffusion CRIE

Skin Test with Antigenic Extracts:

Patients of nasobronchial allergy are selected for skin test based on their medical case history. Skin tests are performed with antigens of fungi of different species.

Skin Prick Test

Lyophilised extracts from antigen are reconstituted in 1: 10 (w/v) concentration in 50% glycerinated buffer. Skin prick test is carried out for patients with different extracts. A drop of each antigen is placed on precleaned volar surface of the arm with the applicator and pricked with 23G needle at an angle of 45°. The skin is lifted and needle withdrawn after the prick. Excess antigen is wiped off with tissue paper. Alongwith antigen, diluent and histamine (1mg/ml) are also used as negative and positive controls. Based on the wheal diameter skin reactions are graded as shown in Table 3.

Intradermal Test:

0.01ml of antigenic extract is injected intradermally on precleaned volar surface of the fore-arm and anterio-lateral part of the upper arm to make a bleb of 2-3 mm. Phosphate buffer saline and histamine (100 µg/ml) are also injected as negative and positive controls, respectively. To ensure that each antigen tested is devoid of non-specific irritants, 20 healthy volunteers should be included in the test. The skin reactions are graded as per the criteria given by Shivpuri and Agarwal (1969).

Skin test is an important tool in allergy diagnosis. The wheal and flare reactions to crude fungal extracts have been documented in patients with respiratory allergy. The commonly employed methods for skin test are skin prick test and intradermal skin test.

Agarwal and Shivpuri (1974) carried out skin tests for various fungal

Table 3: Criteria for grading intradermal skin tests:

Grade	Size of Weal	Size of Erythema	Pseudopodia
-ve	Same as buffer saline	Same as buffer saline	Nil
1+	2.5 times or more than –ve control	10-20mm	Nil
2+	3 times or more than –ve control	20-30mm	Nil
3+	4 times or more than –ve control	>30mm	1-2, small
4+	4-6 times or more than –ve control	>35mm	several

Criteria for grading skin prick tests:

Grade	Size of Weal	Remarks
-ve	Same as buffer	Patients either reacted to saline or did not react to saline
1+	2-3 mm	Allowance was made to the size of the weal in recording the reaction
2+	4-5 mm	
3+	5-8 mm	Erythema was present but grading was done based on weal size
4+	>8 mm with pseudopodia	

extracts and reported 2-34% fungal sensitivity in the Indian population.

Establishing Clinical Allergic Disease

Serum samples: Blood samples were collected from the patients showing markedly positive (2+ and above) skin reactions to antigens. Blood was allowed to clot and expressed sera were centrifuged at 5000 rpm at 4° C for 30 minutes. The serum was stored at –70°C in small aliquots for future use.

Blood was also collected from healthy volunteers with due consent who were not having allergic symptoms and family history of allergy. Sera obtained from these volunteers were used as control.

The major target organs affected by IgE-mediated reactions define the clinical syndrome in a particular patient. Diagnosis of allergic disease is a combination of history, physical findings and judicious use of laboratory testing, including allergy testing. The history is fundamental to any diagnosis of allergic disease. What are the symptoms? Do they occur in a certain place

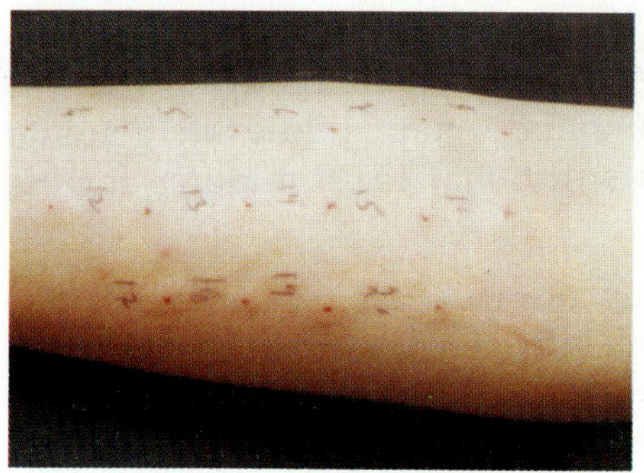
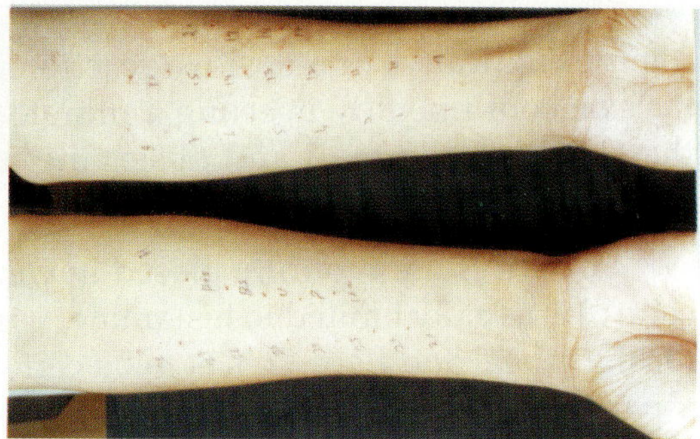

Fig. 4 Skin Prick Test

(i.e. home, workplace), time of year, time of day, after ingesting certain things (i.e. food, drugs, etc), after exercise, when stressed, etc. are all vitally important questions to be explored with every patient. Past medical history of previous allergic (and/or other immunologic) conditions can be associated with current illness. Family history of similar reactions as well as atopy in general is significant. The incidence of allergic diseases is estimated at about 10-15 % of the population (1 in 8 to 10). If one parent is truly atopic, the incidence is up to 1 in 4 and if both parents are atopic, the incidence rises to 1 in 2.

Physical examination is important because there are characteristic findings of allergic vs non-allergic conditions in a variety of organ systems including nose, eyes, lung and skin. Ruling out other causes for symptoms (i.e. mechanical obstruction for nasal congestion, pneumonia for shortness of breath, infection for irritated eyes and nonatopic inflammation for a skin rash) typically occur at this point.

Laboratory evaluation must be directed and cost effective. There are relatively few indications for obtaining a total serum IgE in the evaluation of routine allergic disease. However, directed assessment of allergen-specific IgE is very helpful in differentiating atopic from nonatopic individuals who have characteristic symptoms. Indeed, it is estimated that up to 50% of individuals who have nasal symptoms compatible with allergic rhinitis are nonatopic. Management strategies are dependent upon this distinction.

Allergy testing is commonly done by two methods - skin testing and in vitro determination of allergen specific IgE in the blood (RAST). Skin testing takes advantage of the allergen-specific IgE bound to mast cells in the skin. For safety, testing usually begins with percutaneous ("prick") testing. The allergen is applied and a needle passed through the drop to prick the skin. In highly sensitive individuals, this will result in the formation of a white, raised area (wheal) surrounded by a red, erythematous zone (flair) within 15-20 minutes indicative of a positive reaction. When indicated, negative prick tests may be confirmed by injecting a minute amount of more dilute allergen into the skin (intradermal) and observing an additional 15-20 minutes for the appearance of a wheal and flair.

Controls are typically placed in conjunction with the skin tests. Positive control is usually limited to histamine, which is the major mediator responsible for the wheal and flair. This is to be certain that the skin reaction can occur in a particular patient when histamine is released from a degranulated mast cell. Such reactions are most commonly suppressed by the recent use of antihistamines. The negative control (most commonly saline) is used to confirm that the skin mast cells do not nonspecifically degranulate from the mechanical manipulation of the skin (i.e. pricking or injecting). This is called dermatographism and is a contraindication to allergy skin testing, at least at that particular time.

In vitro testing is indicated when the histamine controls are negative (i.e. recent use of long acting antihistamines) or if the individual has a chronic skin condition that would make proper reading of skin tests impossible (i.e. dermatitis, dermatographism, etc). The principle is the use of a method that can detect allergen-specific IgE molecules in patient blood samples.

This method, while useful, has certain drawbacks. First, IgE is a very minor component of the circulating plasma. Allergen-specific IgE, even in highly atopic individuals, is only a small portion of the total IgE. Thus, very sensitive methods must be employed to detect allergen specific IgE in plasma. In contrast, it is estimated that as little as 50 crosslinked IgE pairs may be sufficient to activate a mast cell. This results in the sensitivity of RAST being only about 70% of skin testing. In addition, the half life of IgE in the blood is only 2-3 days, making determination of pollen-specific IgE by RAST more difficult out of season since the plasma level goes down during that time. In contrast, mast cell bound IgE is apparently stable for a more prolonged time (up to months). Finally, RAST testing is, on a per allergen basis, considerably more expensive than skin testing.

Interpretation of positive allergy testing results must be incorporated with historical and physical findings. It is only in this context that the diagnosis of allergic disease can be made and specific therapies initiated.

Treatment of Allergic Diseases

There are three common methods for treatment of allergic patients, Pharmacotherapy, Immunotherapy, Allergen Avoidance.

Pharmacotherapy

When mast cell-bound IgE is crosslinked, the mast cell membrane becomes activated. This "destabilizes" the membrane, resulting in the release of mediators from the large basophilic granules in the mast cell (a process called degranulation). The major constituent of the large granules is histamine. Histamine has many effects including itching, sneezing, bronchospasm and increased vascular permeability that can result in loss of intravascular volume (anaphylaxis) and/or extravasation of plasma into skin (wheal), fat layers (angioedema), nose or lung. Most histamine-induced effects occur when histamine binds to target tissue via histamine receptors. There are three distinct types of histamine receptors (H1,H2 and H3). Most allergic reactions are via H1 receptor binding.

Allergen Immunotherapy

The major therapy for most allergic diseases is the use of antihistamines, which are H1 receptor blockers. They do not stop histamine release from mast cells, rather they compete with histamine for the H1 receptor on target cells. It then follows that antihistamines are less effective after histamine has been released than if they block the receptor prior to mast cell degranulation.

Allergen Immunotherapy finally, a word should be said about allergen immunotherapy or allergy shots. They are still the only attempt at definitive therapy for IgE-mediated diseases. Through the administration of these allergens, clinical sensitivity to allergen-mediated reactions gradually diminishes.

Multiple theories have been advanced to explain the mechanism of action of allergen immunotherapy. Most researchers agree that three major events commonly occur in patients who receive a course of allergen immunotherapy - first, the production and release of many of the pro-inflammatory mediators (particularly cytokines) are diminished. This may be via a direct effect on mast cells and eosinophils or an immunoregulatory effect mediated by specific populations of lymphocytes.

Second, it is common to find increasing amounts of allergen-specific IgG circulating in the plasma of patients receiving allergen immunotherapy. Such IgG could also bind to the specific allergen and prevents its interaction with mast cell-bound IgE.

It is demonstrated that, after an initial rise, allergen-specific IgE levels in the plasma fall with allergen immunotherapy. This is thought to be due to active immunoregulatory mechanisms that alter how a specific individual responds to a particular allergen.

Not all mechanisms are likely to be active in every treated patient. Many researchers are trying to determine exactly which mechanism(s) is (are) active in a specific patient so allergen immunotherapy can be better tailored to the individual. Also work is ongoing to better chemically define the treating allergens, make allergen immunotherapy safer and even safely increase the interval between injections.

According to a recent WHO report in 1998 on allergen immunotherapy control studies have shown that allergen immunotherapy is an effective treatment for patients with allergic rhinitis/ conjunctivitis, allergic asthma and allergic reactions from stinging insects.

The treatment of allergic diseases is based on allergen avoidance, pharmacotherapy, allergen immunotherapy and educating the patients. Wherever found appropriate, immunotherapy should be used in combination with all forms of therapy with the goal that the allergic patient will become as symptom free as medically possible.

Allergen immunotherapy is indicated for patients who have demonstrated evidence of specific IgE to clinically relevant allergens.

Allergen avoidance

Avoidance of allergens can be useful in the management of patients with allergy. It is not only effective but relatively inexpensive and without any side effects. Although a little inconvenience is there due to required control procedures, but it is certainly helpful.

Since fungi thrive in hot and humid conditions therefore indoor humidity must be maintained below 50%. Bathrooms should be cleaned regularly (at least twice a week with bleach). As far as possible, avoid keeping things in corners of bathrooms that might retain moisture and water seepage and broken and damaged pipes should be repaired. Patients should stay away from basements where humidity and stuffiness is more. The walls and wooden door and furniture should be properly painted and varnished so that there is no mold growth on them. Exhaust fans should be used to take out the humid and stale air from the rooms.

Dustbins should be cleaned from time to time, as vegetable and food thrown in them are a very rich source of mold growth. If using a humidifier, it should be cleaned regularly. So the less the allergy causing material the patient inhales, the more comfortable he will be and the less medication will be required.

Conclusions

It is thus concluded that there is a rich fungal biodiversity in Indian environment both outdoors and indoors. However, in different parts of the world there is an urgent need to organize all the information available in the form of seasonal calendars and all allergic fungal types enlisted and made readily available for the use of common man and physicians as a diagnostic tool. Important airborne fungi of different work environments have been studied and enlisted for their seasonal and annual variations.

References

Aalberse, R.C., Bessot, J.C. and Bjorksten, B. (1989). Dust mite allergens and asthma - A world wide problem. J. Allergy Clin Immunol, **83**: 416-427.

Aberg, N, Hesselmar, B, Aberg, B and Eriksson, B. (1995). Increase of asthma, allergic rhinitis and eczema in Swedish school children between 1979 and 1991. Clin Exp Allergy, **25 (9)**: 815 - 819.

Agarwal,M.K. (1970). Studies on the allergenic fungal spores of Delhi Atmosphere. Ph.D. Thesis, University of Delhi.

Agarwal, M.K. and Shivpuri, D.N. (1974). Fungal spores: Their role in respiratory Allergy. Adv. Pollen – Spore Res, 1:78-128.

Al Dawood, K. (2000). Epidemiology of bronchial asthma among schoolboys in Al-Khobar city, Saudi Arabia: Cross sectional study. Croat Med J, **41(4)**: 437-441.

Al-Doory,Y. Domson,J.F. and Best,J. (1982). Further studies on the airborne fungi and pollens of Washington D.C. metropolitan area. Ann Allergy. **49**:265-269.

Andersen, A.A. (1958). New sampler for the collection, sizing and enumeration of viable airborne particles.

Anonymous, (1961). Morbidity survey on contributory health scheme beneficiaries, New Delhi. Indian Council of Medical Research, New Delhi.

Anonymous, (1998). World wide variations in the prevalence of asthma symptoms: the International Study of Asthma and Allergies in Childhood (ISSAC). Eur Respir J, **12 (2)**: 315 - 335.

Anonymous, (2000). All India Coordinated Project on Aeroallergens and Human Health. Report. Ministry of Environment and Forests, New Delhi.

Aukrust,L. (1992). Introduction to Mold Allergy. Clin Rev Allergy, **10** :147-152.

Behbehani N A, Abal A and Syabbalo N C, Abd Azeem A, Shareef E and Al-Momen J. (2000). Prevalence of asthma, allergic rhinitis and eczema in 13- to 14- years- old children in Kuwait: an ISAAC study. International Study of Asthma and Allergies in Childhood. Ann Allergy Asthma Immunol, **85 (1)**: 58-63.

Blackley, C.H. (1873). Hay Fever: Experimental research on the causes, treatment of catarahus aestivus, Baillere Tindall & Cox, London.

Blenkinsopp, A and Blenkinsopp, J. (1989). Continuing education. Pharm J, **20**:5.

Burge, H. Bioaerosols (1990). Prevalence and health effects in the indoor environments. J Allergy Clin Immunol, **86** : 687-701.

Burr, M.L., Butland, B.K., King, S. and Vaughan-Williams, E. (1989). Changes in asthma prevalence: two surveys 15 years apart. Arch Dis Child, **64 (10)** : 1452-1456.

Chhabra, S.K., Gupta, C.K., Rajpal, S. and Chhabra, P. (1998). Prevalence of asthma in schoolchildren in Delhi. J Asthma, **3**: 291-296.

Charpin, D, Raherison, D, Dutau, H and Taytard, A. (2000). Epidemiology of respiratory allergies: current data. Rev Mal Respir; **17 (1 pt 2)**: 139 - 158.

Collins-Williams, C. Kuo, H.K. Garey, D.N. Davidson, S. Collin-Williams, D. Fitch, M. and Fischer, J.B. (1973). Atmospheric mold counts in Tronto, Canada. Ann.Allergy, **31**:69-71.

Crook, B. and Lacey, J. (1988). Environment of airborne microorganism in work environment. Environ Tech Letters, **9**: 515-520,

Dotterud J., Kuammem, B. and Bolle, R., (1994). A survey of atopic diseases among school children in Sor-Varanger community. Acta Dermato G Venereio, **74**: 124 - 28.

Durham,O.C. (1946). J.Allergy, **17**:79.

Fienberg, S.M. (1935). Mold allergy. Its importance in Asthma and Hayfever. Wisconsin Med J, **34**: 254-262.

Fienberg, S.M. (1944). Probability of allergic reactions in fungus sensitive individuals: Preliminary experiments. J. Allergy, **15**: 271.

Feinberg, S.M. (1946). Allergy in Practice. 2nd Ed. The year book Publish. Inc., Chicago.

Floyer, J. (1726). Violent asthma after visiting a wine cellar. A treatise on asthma. 3rd Ed. London.

Frenz, D.A. and Naomi,L.L. (1997). A comparison of pollen recovery by three models of the Rotorod sampler. Ann Allergy Asth & Immunol. **79**:256-258.

Gravesen,S. (1979). Fungi as a cause of allergic diseases. Allergy, **34**:135-154.

Gaur, R.D. and Kasana, M.S. (1981). Studies on Aerobiology of Modinagar. J Ind Bot Soc., **60**:266-277.

Gupta, S. and Chanda S. (1989). Aeropalynological survey in subtropical Eastern Himalayas, Kurseong. Grana, **28**:219-221.

Gupta,S.K. , Perira,B.M.J. and Singh,A.B. (1993). Survey of culturable and non-culturable fungi at different sites in Delhi metropolis. Asian Paci j Allergy Immunol., **11**:19-28.

Hawksworth,D.L., Sutton,B.C. and Ainsworth,G.C. (1983). Ainsworth and Bisby's Dictionary of the Fungi, 7th Edition. Pub. Commonwealth Agriculture Bureau, KEW, Surrey.

Hirst, J.M. (1952). An automatic volumetric spore trap. Ann Appl Biol. **39**: 252-263.

Horejski, M., Sach, J., Tomsikova, A. and Med, A. (1960) A syndrome resembling farmers' lung in workers inhaling spores of Aspergilli and Penicilli moulds. Thorax, **15**: 212-217.

Ibanez, M.D., Horner, W.E., Liengswangwong, V., Sastre, J. and Lehrer, S.B. (1988). Identification and analysis of basidiospores allergens from puffballs. J Allergy Clinc Immunol, **82**: 787-795.

Jimenez –Diaz, C, Aber, J.M, Ortiz, F, Lahoz, F, Garcia Puente,L.M and Canto,G. (1960). Acta Allergol Kbh., **15**:139-149.

Kang, B., Velody, D., Homburger, H. and Yuninger, J.W. (1979) Analysis of indoor environment and atopic allergy in urban population with bronchial asthma. J Allergy Clin Immunol, **63**: 80-89.

Kino,T. and Singh, S.O.J (1978). Allergy Clin Immunol, **61**:10-16.

Kauffman, H.F., Beaumont, F., Meurs, H., van Der Heide,S. and de Vries,K. (1983). Comparison of antibody measurements against *Aspergillus fumigatus* by means of double diffusion and enzyme linked immunosorbent assay. J Allergy Clin Immunol, **72**:255-261.

Lacey, J. and Crook, B. (1988). Fungal and actimomycetes spores as pollutants of workplace and occupational allergens. Ann Occup Hyg, **32**: 515-533.

Leuschner, R.M. (1974). Investigation on airborne pollen in Basel and Davos (Switzerland) in connection with pollinosis. Grana (Abst) **14**:40-40.

Lewis, W.H. (1984). Pollen Allergy. In: Allergy: Theory and Practice. P.E. Korenblat and H.J. Wedner, eds., Grune and Startton, Orlando, pp.353-369.

Long,R. Yin,R. He,H. Li,Z. and Liu,D. (1987). Molds in air in southern part of Chengdu (China). J

West China Univ Med Sci., **18** : 60-63.

Maribhat,M. and Rajasaab, A.H. (1988). Airspora of commercial location at Gulbarga. In J Aerobiol, **1**:59-65.

Meggs, W.J., Dunn, K.A., Bloch, R.M., Goodman, P.E. and Davidoff, A.L. (1996). Prevalence and nature of allergy and chemical sensitivity in a general population. Arch Environ Health, **51 (4):** 275 - 282.

Naranjo, P. (1958). Etiological agents of respiratory allergy in tropical countries of central and south America. J Allergy, **29**:363-374.

Nilsson, S., Praglowski, J. and Nilsson, N. (1977). Atlas of airborne pollen grains and spores in northern Europe. Almquist and Wiksell International, Stockholm, Sweden.

Ninan, T.K. and Russel, G. (1992). Respiratory symptoms and atopy in Aberdeen schoolchildren: evidence from two surveys 25 years apart. Br Med J, **304**: 873 - 875.

Pandit, T. (1994). Aerobiological and Clinico-immunological studies of fungi of certain occupational sites. Ph.D. Thesis submitted to University of Delhi.

Pandit, T, and Singh, A.B. (1992). Prevalence of Airborne Fungi in a Sugar Factory Environment. Ind J Aerobiol. Special Vol., pp. 145-155

Pandit, T. and Singh, A.B. (1994). *Saccharomyces cerevisiae* (Yeast) : A Potential Aeroallergen for the Workers of Sugar Industry. Ind J Aerobiol., **7(1&2)** : 13-19.

Peat, J.K, Toelle, B.G., Gray, E.J., Haby, M.M., Belousova, E., Lellis, C.M. and Woodlcock, A.J. (1995). Prevalence and severity of childhood asthma and allergic sensitization in seven climatic regions of New South Wales. Med J Aust, **163 (1)** : 22 - 26.

Perkins, W.A. (1957). The rotorod sampler. The second semiannual report. Aerosl Lab. CML Standford University, **186**: 66.

Peterson, P.K. McGlave, P. Ramsay, N.K.C. Rhame, F.S. Cohen, E. PerryII G.S. Goldman, A.I. and Kersey, J. (1983). Prospective study of infectious diseases following bone marrow transplantation: Emergence of *Aspergillus* and Cytomegalovirus as the major causes of mortality. Infection Control, **4**:81-89.

Ramalingam, A. (1971). Proc Indian Acad Sci, **B74**: 227-240.

Rawat, A., Singh, A., Singh A. B., Gaur, S.N., Kumar, L., Roy, I. and Ravindran, P. (2000). Clinical immunological evaluation of Cedrus deodara pollen : A new allergen from India. Allergy, **55**: 1-7.

Roby,R.R., and Sneller, M.K. (1979). Ann Allergy. **43** : 286-288.

Salvaggio, J. and Aukrust, L. (1981). Mold induced asthma. J. Allergy Clin Immunol, **68**: 327-346.

Sandhu,S.K., Shivpuri,D.N. and Sandhu,R.S. (1964). Studies on the airborne fungal spores in Delhi. Ann Allergy, **22**:374-384.

Sepulveda,R. Longbottom,J and Pepys, J. (1979). Enzyme-linked immunosorbent assay (ELISA) for IgG and IgE antibodies to protein and polysaccharide antigens of *Aspergillus fumigatus*. Clin. Allergy, **9**:359-372.

Shaikh, W.A. (1997). Allergies in India: An analysis of 1619 patients attending an allergy clinic in Bombay. Int Rev Allergol Clin Immunol; **3 (2)**: 101.

Shaikh, W.A. (1998). Allergies and asthma in India: An analysis of 2467 patients seen over a six-year period. Ind J Clin Pract, **8 (12)**: 23.

Shivpuri, D.N., Vishwanathan R. and Dua L.K. (1960). Studies on pollen allergy in Delhi area I - pollination calendar. Ind J Med Res **48**:15-20.

Shivpuri, D.N. (1971). Proc 26th National Conff. T.B. and Chest Dis: 382.

Shivpuri, D.N. (1980). Clinically important pollen, fungal and insect allergens for naso-bronchial allergy patients in India. Asp Allergy Appl Immunol XIII : 19-23.

Shivpuri, D.N. and Parkash D. (1967). A study in allergy to *Prosopis juliflora* (Kabuli keekar). Ann Allergy **25**:643-648.

Shivpuri, D.N., Singh A.B. and Babu C.R. (1979). New allergenic pollens of Delhi state, India and their clinical significance. Ann Allergy, **42**:49-52.

Singh, A. (1993). Aerobiological and immunological studies on fungal spores of different occupational sites. Ph.D. Thesis submitted to Delhi University.

Singh, A., Prakash, D and Singh, A.B. (1998). Sensitization to different species of *Aspergillus* in bakery workers and general population. Asian Pacific J Allergy Immunol, **16**: 5-15.

Singh, A., Singh A.B., Bhatnagar A.K. and Gangal S.V. (1990). Prevalence of Aspergilli in bakery environment. Ind J Aerobiology **3**: 15-21.

Singh, A.B. (1994). FTR, Ministry of Environment and forests, New Delhi.

Singh, A.B. (1998). Airborne fungi of Allergenic significance in work environment. In Recent trends in Mycoses, 9-17.

Singh, A.B. and Babu, C.R. (1980). Pollen types in the atmosphere of Delhi. Phytomorphology, **30**:180-189.

Singh, A.B., Chatterjee, M., Singh B.P. and Gangal, S.V. (1990). Airborne viable fungi in library: before and after agitation of books. Ind J Aerobiol, **3** : 32-38.

Singh, A.B. and Malik, P. (1992). Pollen aerobiology and Allergy: An integrated Approach. Indian J. Aerobiology, **5**: 1-19.

Singh, A.B., Malik P., Parkash D. and Gupta C.K. (1992). Biological standardization of pollen allergens from India. Asia Pacific J Allergy Immunol **10**:103-109.

Singh, A.B., Malik P., Parkash D. and Gangal S.V. (1993). Identification of specific IgE binding Castor Bean (*Ricinus communis*) pollen obtained from different source materials. Grana, **32**:376-380.

Singh, A.B. and Singh, A., (1993). Fungi as allergenic agents. In fungal Ecology and biotechnology. Eds B. Rai, D.K. Arora, K. Dubey and P.D. Sharma, 195-208.

Singh, A.B. and Singh, A. (1994). Pollen Allergy - A Global Scenario. In Recent Trends in Aerobiology, Allergy and Immunology; ed. S.N.Agashe, Oxford IBH, New Delhi.

Singh, B.P., Singh A.B., Nair P.K.K. and Gangal, S.V. (1987a.) Survey of airborne pollen and fungal spores in DehraDun, India. Ann Allergy **59** : 229-234.

Singh, B.P., Singh A.B. and Parkash D. (1987b). Skin reactivity to airborne pollen and fungal antigens in patients of Naso Bronchial Allergy of Hill Regions (India). In: Atmospheric Bio Pollution, N. Chandra, ed., pp.125-134.

Singh, B.P., Singh A.B. and Gangal S.V. (1992). Calendars of different states, India. CSIR Centre for Biochemicals, Pub., Delhi, India.

Singh, K. and Shivpuri, D.N. (1971). studies on yet unknown allergenic pollen of Delhi state metropolitan: botanical aspects. Ind J Med Res **59**: 1392-1410.

Singh, N. and Devi, K.K. (1992). Aerobiology and allergic human diseases in Manipur.II. Airborne pollen grains of Imphal, Imphal District. Ind J Aerobiol Special volume:49-60.

Sly, R.M. (1999). Changing prevalence of allergic rhinitis and asthma. Ann Allergy Asthma Immunol, **82 (3):** 233 - 248.

Smith, J.M. (1983). Epidemiology and natural history of asthma, allergic rhinits and atopic dermatitis. In Allergy: Principles and Practice. Eds E. Middleton, C E Reed and E F Ellis, 2nd edition, St. Louis, C.V. Mosbey, USA, pp.633 - 658.

Smith, L.J. and Salvin R. G. (1988). Epidemiologic considerations in Atopic Diseases. In: Allergic Diseases from Infancy to Adulthood. Eds CW Bierman and D S Pearlman,. 2nd Ed, Saunders Co., pp.121 - 129.

Solomon, W.R. (1984). Sampling airborne allergens. Ann Allergy, **52** (3): 140-149.

Tilak, S.T. and Kulkarni,R.L. (1980). Addition in the fungal flora of the air. Ind Phytopathol., **34**:69-71.

Vishwanathan, R. (1964). Definition, incidence, etiology and natural history of asthma. Ind J Chest Dis, **6**:108 -124.

Vishwanathan, R. Modi,R.K. Prasad,S.S. and Sinha,S.P. (1965). Bronchial asthma and chronic bronchitis. J Ind Med. Assoc., **45**:480-480.

von Priquet, C. (1906). Allergy Munch Med. Worchen Schr, **53**: 1457.

WHO Position Paper. (1998). Allergen immunotherapy: therapeutic vaccines for allergic diseases. Eur J All and Clin Immunol., **5(2)**:1-42. Eds. (J. Bousquet, R.F.Lockey, H.-J. Malling.

Zoberi,M.H. (1964). Trans Br Mycol Soc. **47**:109-114.

37

Aeromycology in Relation to Foodgrain Storage

Indira Kalyanasundaram

Retired Professor, Department of Botany, University of Madras
30, Kasturi Avenue, MRC Nagar, Chennai- 600 028

A review of the literature on Aeromycology, and especially the work done in India between the late 1960's and early 1980's, makes it clear that the fungi that cause losses of foodgrain in storage are common members of the air mycoflora. They consist of a few xerophilic species of *Aspergilus* and *Penicillium*, and are often abundant in the air over crop fields, where they generally show an increase towards the harvest season. The studies made in the author's laboratory on rice crop in the vicinity of Madras leave no doubt that the air spora serve as a major source of contamination of grains with storage fungi. The air spora in turn can be traced to the surface soil, which becomes hot and dry during the periods of harvest and threshing. However, there are some regional differences, and in some regions where the soil and air are rather low in storage fungi, the storage structures would provide the source of contamination. In any case, a close monitoring of the air mycoflora especially during ripening of the crop, would indicate the need for preventive measures to protect the grain from fungal damage during storage.

Keywords : Aeromycology, Aspergilli, crop-field air, crop-field soil, foodgrain spoilage, storage fungi.

It is estimated that in developing countries like India, over 30% of the annual harvest of foodgrains is lost during storage. This is evident from the recent exhortation of the Supreme Court to the Government, in response to a public interest litigation during the starvation deaths in Orissa, that the food in its godowns should reach the poor and not be allowed to rot.

The causes of storage losses are listed as rodents, insects and fungi. Of these, the first two agents seem to claim more attention in our country, perhaps because they are readily visible, while the last-named — the fungi—are generally ignored. The Food Corporation of India does not consider it necessary to appoint Mycologists to periodically check the mycological status of its stored grains.

Yet, it is the unseen fungi that can cause the most extensive damage to stored grain whenever there is an increase in grain moisture above a critical level: an event that occurs all too often under our climatic conditions and storage practices. So, it is up to us Mycologists to raise our voices, to create an awareness of the importance of fungi in the deterioration of foodgrains during storage.

Field Fungi *vs* Storage Fungi

The fungi that commonly cause storage losses have ben termed as **storage fungi,** in contrast to **field fungi** which are said to infect and colonise grains in the standing crop in the field (Christensen, 1957). Ecologically these two groups are separated by their moisture requirements. The storage fungi are characterised by their ability to tolerate — or even to prefer — substrates with a very low water activity and are hence regarded as xerophilic. In culture they are osmophilic and can be selectively isolated on a high osmotic medium. They include a few species of *Aspergillus* and *Penicillium,* the most common Aspergilli being *A. flavus, A. niger, A. nidulans, A. candidus* and a few species of the glaucus group, while the Penicillia are variable. The field fungi can only grow in high-moisture substrates and hence disappear from the grains when these are dried to a low moisture content for storage. The common genera of field fungi are *Alternaria, Helminthosporium, Cladosporium, Fusarium* and *Pullularia.* According to Chirstensen and Kaufmann (1969), the storage fungi invade the grain only during storage, presumably from the equipment associated with storage.

'Storage Fungi' in the field

Lately, however, the term storage fungi has fallen into disuse, perhaps because of the realisation, even in the 1970's , that these fungi do occur even in freshly harvested grain, albeit at a low level, in India (Dhanraj *et al.,* 1972; Ayachi, 1974; Siradhana *et al.,* 1978; Bhadriah and Rama Rao, 1979; Agarwal *et al.,* 1980; Mehrotra and Dwivedi, 1980) and elsewhere (Hill and Lacey, 1979). More recently, they have been finding that the fungi can not only infect grains in the field but can even produce their deadly toxins *in situ.* Initially reported in maize from the U.S.A (Zuber *et al.,* 1976), aflatoxin formation in the standing crop has been reported even from India, besides maize, in diverse cereals like sorghum (Bhadraiah and Rama Rao, 1983), ragi (Shankar and Rai, 1985) and rice (Usha *et al.,* 1993).

These observations make it clear that the source of storage fungi on grain in many cases rests with the airborne spores over the field, rather than the storage equipment. The term storage fungi however, need not be given up, as it is a convenient term to designate the handful of osmoshilic Aspergilli and Penicillia that are usually found in stored grain all over the world. It is here that aeromycology comes into the picture. And I have chosen this topic for my article in this commemorative volume, in honour of some of the pioneering studies made by Prof. Kamal and his associates, which will be cited in due course. The general emphasis here is on Indian work.

Aeromycology of Crop Fields in India

Aerobiology is a fairly recent discipline, and following the classical work by Gregory (1961) , several aeromycological studies have been made, even in India. The reports up to 1966 have been ably reviewed by Sreeramulu (1967), himself a pioneer. In the present context, our interest is mainly in the mycoflora over food crop fields and grain processing areas; and in some cases, of the areas over which the air spora of the field would ultimately land, namely the plant surface. The early relevant studies include those of Mishra and Srivastava (1970, 71, 74), Mishra and Tewari (1975), Kamal and Singh (1974,75), Rati and Ramalingam (1976), Shenoi and Ramalingam (1976) and Verma and Kamal (1979, 80). Most of these authors have studied diurnal and seasonal fluctuations in fungal numbers, both total and differential. Some of them have compared the qualitative pattern over fields of different crops, others have tried to estimate the total number of species. Most of the authors have encountered Aspergilli and Penicillia, though not necessarily as dominant members of the air spora. Kamal and Singh (1974) and Verma and Kamal (1969), found some species of *Aspergillus* to show a marked seasonal periodicity, while a few species occurred consistently all through the year.

Aspergilli and Penicillia in the Air spora

Taken as a whole, it has been found that of the 200-odd fungal species recorded in the air over crop-fields, *Aspergillus* and *Penicillium* are represented by about 50 species each. A list of the *Aspergillus* species recorded in the air up to 1980, mostly over crop-fields, is presented in Table 1. It may be noted that species of the glaucus group, represented by *Aspergillus amstelodami, A. chevalieri* and *A. repens* (often clubbed together for convenience as *A. glaucus*), are rarely recorded by Indian authors, either in grain or in the air. Yet these are the most common of the grain mycoflora but they are true osmophils and will not show up unless the medium is high in sugar or salt, hence they are missed by most Indian workers.

It should be mentioned, however that the numbers cited above have emerged despite the fact that very few of the authors have used selective media for Aspergilli or Penicillia. In these specific isolations,some of the common storage fungi have been recorded, such as *Aspergillus niger, A. nidulans, A. flavus, A.fumigatus, A.sydowi* and *A. tereus* (Kamal and Singh, 1974; Rati and Ramalingam, 1973, 1976; Verma and Kamal, 1979, 1980). On the other hand, some authors have specifically pointed out the paucity of Aspergilli in the air

Table 1 : Qualitative distribution of Aspergilli in the atmosphere

Species/ group	**Regions where recorded (References in brackets)***
Aspergillus aculeatus	(1,2,5,6,10)
A. alliaceus	(7)
A. amstelodami	(7,8,9,10)
A. aureolatus	(10)
A. awamori	(10)
A. caespitosus	(7)
A. caespitosus	(2,6,7,8,10)
A. carbonarius	(5)
A. carneus	(5,7)
A. chevalieri	(7,8,9,)
A. clavatus	(6,7,9)
A. echinulatus	(8)
A. foetidus	(10)
A. flavipes	(4,5,6,10)
A. flavus	(1,2,3,4,5,6,7,9,10)
A. fumigatus	(1,5,6,7,8,9,10)
A. glaucus group	(5)
A. gracilis	(8)
A. japonicus	(4,6)
A. mangini	(7)
A. melleus	(7,10)
A. multicolor	(10)
A. nidulans	(1,2,3,4,5,7,8,9,10)
A. nidulans group	(10)
A. niger	(1,2,3,4,5,6,7,8,9,10)
A. niveus	(5,6,9)
A. ochraceus	(3,6,8,9,10)
Aspergillus ocharaceus group	(5,10)
A. oryzae	(4,8,10)
A. puniceus	(10)
A. penicillipormis	(10)
A. quadrilineatus	(7)
A. repens	(7,8)
A. restrictus	(7,8,10)
A. ruber	(7,8,10)
A. regulosus	(5,6,7,10)
A. schlerotiorum	(7,10)
A. stromatoides	(10)
A. sulphureus	(10)
A. sydowi	(1,2,3,5,6,7,10)
A. tamarii	(5,6)
A. terricola	(10)

A. terreus	(1,2,3,4,5,6,7,9,10)
A. ustus	(5,6,7,9, 10)
A. variecolor	(5,6,10)
A. versicolor	(4,6,7,8,9,10)
A. versicolor group	(4,5)
A. wentii	(10)

*1 Mishra and Srivastava, 1971, paddy field, Gorakhpur.
2& 3. Mishra and Srivastava, 1972, wheat field and barley field, Gorakhpur.
4. Palanichamy, 1979, Sorghum field, Madras.
5. Kamal and Singh, 1974, Sugarcane field, Gorakhpur.
6. Verma and Kamal, 1979, pigeon pea field, Gorakhpur.
7. Kramer *et al.*, 1960, Kansas.
8. Hudson, 1969, Cambridge.
9. Calvo *et al.*, 1980, Barcelona.
10. Rati and Ramalingam, 1976 , Mysore.

over crop-fields, and suggested that the storage fungi are probably acquired from the environment of the storage structures (Mehrotra and Dwivedi, 1980). Thus there can be regional differences in crop-field air mycoflora even within the same state, as seen from the studies made in Gorakhpur and in Allahabad.

Storage Fungi over Fields of Rice

It is pertinent to present here the results of some studies made some years ago in our laboratory, in order to pinpoint the sources of contamination of paddy with storage fungi. In this investigation, the fluctuations of the air, soil and plant surface mycoflora were assessed in relation to the growth cycle of the rice crop in a single experimental rice field for five successive cropping seasons, using a high osmotic medium for selective isolation of storage fungi (Banumathy,1982). A quantitative increase in numbers of these fungi was

Table 2. Storage fungi in rice-field air at different stages of development of the crop, in five successive cropping seasons.

	Stages of plants growth	*No. of propagules in hundreds/m² in (Cropping seasons)					Stage Total
		I	II	III	IV	V	
1	Vegetative	7.5	5.0	3.0	9.0	6.2	30.7
2	Panicle initiation	10.6	1.6	4.5	3.7	5.9	26.3
3	Ripening	10.1	7.5	3.1	5.9	6.7	33.3
4	Harvest	32.0	19.8	13.9	5.3	21.6	92.6
	Season total	60.2	33.9	24.5	23.9	40.4	182.9

F fatio in significant at 5% level for stages.

CD = 6.4; $\overline{1\ 2\ 3}$

Table 3. Mycoflora of threshing ground air during THRESHING in the five successive cropping seasons.

Kinds of fungi	* No. of propagules/m2 in (Cropping Seasons)					Average (of I, IV and V)	Winnowing I
	I	II	III	IV	V		
Total	10,919	α	α	9,858	7,332	9,360	15,038
'Other fungi'	7,379	α	1,404	1,1420	1,747	3,515	4,711
Storage fungi	3,540	α	α	8,408	5,585	5,845	10,327
Aspergilli	467	2,543	α	7,800	2,652	3,640	1,934
Penicillia	3,073	α	α	608	2,933	2,205	8,393
Break-up							
Aspergillus glauces	31	140		624	499	385	109
A. flavus	31	733	187	327	312	223	749
A. nidulans	78	1,127		5,741	1,544	2,454	31
A. niger	296	281		406	172	291	811
A. terreus	31	125	203	327	109	156	234
Aspergillus spp.	0	47	47	375	16	130	0
Penicillium sp. (1)	3,073	α	α	608	2,933	2,205	8,393

*The Petri dish exposure method was used, and for the sake of uniformity and to facilitate graphic representation, the results were quantified as the theoretical number of propagules settling over a square meter area in a period of 1 to 3 min.

α -Too numerous to count

seen concomitant with the growth cycle of the crop with a significant increase at harvest, with the result that during harvest and threshing, the air of the rice field and threshing-ground were literally teeming with the propagules of storage fungi (Tables 2 and 3).

In a situation like this, one can naturally expect a considerable proportion of these spores to have settled on the grain surface, and more would be likely to settle during harvest and threshing operations. even when a mechanical thresher is used, as was the case in seasons II to V in this investigation, the air within the thresher is teeming with storage fungi (Table 4). Thus the 'grain dust' that flies around during these operations is heavily loaded with fungal spores and is in fact an occupational hazard for the workers, likely to cause respiratory ailments. In fact, a great and immediate increase in fungal spore concentration during harvesting operations have been reported by several authors (Sreeramulu and Ramalingam, 1966; Mishra and Srivastava, 1972; Kamal and Singh, 1974, 75; Verma and Kamal, 1979).

Before making generalised conclusions on the basis of results obtained in coastal Madras in the above experiment, we wanted to see whether the

Table 4 : Air mycoflora inside mechanical thresher before/ after operation

Kinds of Fungi	No. of propagules/m² in (Cropping seasons)					
	II	III	IV		V	
	After threshing	Before threshing	Before threshing	After threshing	Before threshing	After threshing
Total	14,367	α	12,168	1,950	5,226	9,594
'Other fungi'	7,956	0	780	780	1,794	1,326
Storage fungi	6,411	α	11,388	1,170	3,432	8,2687
Aspergilli	2,667	α	11,388	936	2,184	7,722
Penicillia	3,744	α	0	234	1,248	546
Break-up						
Aspergillus glaucus	234	α	3,744	0	546	3,276
A. flavus	78	624	780	0	0	156
A. nidulans	1,825	α	4,680	702	1,404	3,900
A. niger	234	α	1,560	78	234	234
A. terreus	187	156	624	78	0	156
Aspergillus spp.	109	0	0	78	0	0
Penicillium sp. (1)	3,744	α	0	234	1,248	546

pattern of mycoflora over rice fields at harvest would be the same in other regions. So a comparative assessment was made over a paddy field in an interior district situated in the Western Ghats (Kodagu district of Karnataka). The results showed a paucity of storage fungi in soil as well as the air, of rice field and threshing ground. Yet the stored grain mycoflora was qualitatively comparable with that of the coastal area, suggesting that the storage structures could have been the sources of contamination (Banumathy, 1982; Kalyanasundaram and Jayaraman, 1996).

There is thus a strong regional influence with regard to the importance of air mycoflora as a source of contamination of foodgrains. If the quantitative fungal load on paddy is any indication, as we have seen from some samples from coastal Andhra Pradesh, our observations for Madras might hold good for other coastal regions also. (Kalyanasundaram and Rao, 1988). In some of the interior areas the situation seems to be different. Mehrotra and Dwivedi (1980) from Allahabad found a domiantion of *Cladosporium* and *Alternaria*, rather than *Aspergillus*, in the air over wheat fields, and suggested the environment of storage structures to be the source of storage fungal contamination of grains.

Soil as the Source of Storage Fungi in the Air

There is a tendency to point towards decaying organic matter as a source of fungal spores in the air. As far as the storage fungi in rice are concerned, our studies point strongly to the soil of the rice-field itself. The abundance of *Aspergillus* and *Penicllium* in soils including crop-field soils is well recorded (Das, 1963; Dutta and Ghosh, 1965; Sundaram, 1969; Kamal and Bhargava, 1972; Kamal and Gupta, 1973; Singh and Ram, 1979; Upadhyay and Rai, 1979; Behera and Dash, 1980). In the rice-field, the soil which is puddled at transplantation, dries up gradually with the growth of the crop and has to be absolutely dry at harvest, in order to facilitate the harvesting operations. We have observed an increase in populations of storage fungi in the rice-field soil during the growth cycle of the crop, even though the increase was not as consistent as that of the air mycoflora. In fact the difference in number between cropping seasons was significant rather than between growth stages, suggesting a strong climatic influence (Banumathy, 1982: Table 5). In Tamil Nadu, a portion of the field itself is levelled and beaten to a hard stiffness for the threshing operations. In the hot tropics, temperatures of dry surface soils at midday can go as high as 80^{0}C, and under these conditions most soil fungi die out, leaving only those xerophillic species that are commonly encountered in stored grain (Griffin, 1972). There is thus a striking parallel in qualitative pattern between the mycoflora of hot dry surface soils and of stored grain. In

Table 5 : Storage fungi in rice-field soil at different stages of development of the crop,in five successive cropping seasons

Stages of plant growth	No. of colonies in thousands/g in (Cropping seasons)					Stages Total
	I	II	III	IV	V	
1. Vegetative	12.0	15.5	13.0	21.0	21.7	83.2
2. Panicle inititation	13.3	8.2	11.9	18.5	30.3	82.2
3. Ripening	14.5	21.7	12.8	30.3	18.9	98.2
4. Harvest	16.9	26.9	12.7	30.3	30.4	117.2
Season total	56.7	72.3	50.4	100.1	101.3	380.8
Average	14.2	18.1	12.6	25.0	25.3	

F ratio is significant at 5% level for seasons.

CD = 7.25 I_2 II_2 III_2 IV_2 V_2

I_3 II_3 III_3 IV_3 V_3

I_4 II_4 III_4 IV_4 V_4

those areas where the air at harvest does not abound in storage fungi, the temperature of surface soils perhaps does not rise to the levels that cause selective survival of these fungi.

Spoilage of Grain by Storage Fungi

It thus turns out that in those areas where the air over the fields contains an abundance of storage fungi, the grain is already laden heavily with storage fungi even as it goes into storage. Under ideal conditions of storage these fungi would gradually die out over a long period of time. However, any inadvertent increase in grain moisture, as often happens in bulk storage especially in the open, to levels of 13 to 18% in the case of cereal grains, would lead to a rapid growth of these fungi resulting in either total spoilage of grain, or an insidious fungal growth and the formation of mycotoxins, depending on the level of moisture. The species of fungi that cause the spoilage would depend on the moisture level, as the individual species of storage fungi differ somewhat in their optimum moisture requirement (Jayaraman and Kalyanasundaram, 1989).

Aeromycology in Foodgrain Protection

From the foregoing account it becomes clear that the mycoflora of the air over fields of foodgrain would have to be monitored, especially during ripening of the grain, in order of estimate the amount of storage fungi likely to be present on the grain at harvest. An excess of these fungi in the air would call for protective measures to be initiated before the grain goes into storage, so as to minimise the surface contamination. This also holds true for other grain-processing areas such as mills, warehouses and smaller storage structures, where an excess of storage fungi would not only lead to ultimate spoilage of grain but also endanger the health of worker (Ghose *et al.*, 1977). Milled fractions of grains are some times seen to be more heavily contamined than the grains themselves (Jalaluddin and Sinha, 1971; Augustine *et al.*, 1984; Jayaraman and Kalyanasundaram, 1976; Vincent and Kalyanasundaram, 1999), and these in turn affect other processed foods such as bread and pasta (Paramasivan and Kalyanasundaram, 1986) and pasta (Augustine *et al.*, 1984).

The protective measures for grain and the grain environments should include the fitting of air filters, a de-dusting process for grain prior to storage, fumigation with chemical fumigants such as methyl bromide, phosphine or ethylene dibromide, or even with indigenous fumigants like incense (Kalyanasundarm and Jayaraman, 1996) or preservatives of chemical or plant origin, such as propionic acid, citral etc. (Mallick and Nandi, 1982; Asthana *et al.*, 1984; Kumar *et al.*, 1993). Some preventive measures are effective even long before harvest. Recently a foliar spray of azoxystrobin during plant growth has been found to effectively protect groundnuts from *Aspergillus flavus* infection and alflatoxin formation (Brenneman and Wilson, 2001). Clearly, more intensive research in this direction is called for, so that the surplus food in our public godowns reaches the needy instead of being allowed to rot.

References

Agarwal, G.P., Thakur, M.K. and Awasthy, S. (1980). Studies on the wheat grain storage in Madhya Pradesh . "Fungi associated with different varieties of freshly harvested wheat grains". Nat. Acad. Sci. Letter, **3**: 195-197.

Asthana, A., Dubey, N.K. and Dixit, S.N. (1984). Fungitoxicity of citral and eugenol against some storage fungi. Indian J. Microbiol., **24**: 199-201.

Augustine, B., Paravathi, M. and Kalyanasundaram, I. (1984). Potential mycotoxigens in wheat and wheat products. J. Fd Sci.Technol., **21**: 1-4.

Ayachi, S.S., (1974). Mycoflora of wheat grains: II. A Study on mycoflora associated with fresh wheat grains. J.N.K.V.V. Res. J., **8**: 161-162.

Banumathy, N. (1982). Storage fungi in rice: sources of inoculum. Ph.D. thesis, University of Madras, India.

Behera, N. and Dash, M.C. (1980). Seasonal dynamics of microfungal population in some crop fields of Sambalpur, India, Trop. Ecol., **20** :171-177.

Bharaiah, B. and Rama Rao, P. (1979). Seed mycoflora of four sorghum varieties. Natn Acad. Sci. Letters, **2**: 2150259.

Bhadraiah, B. and Rama Rao, P. (1983). Aflatoxin production in preharvest, freshly harvested and stored sorghums. *In* Bilgrami, K.S. (ed.) *Mycotoxins in food and feed.* Proceedings of all India Symposium, pp. 49-54. Allied Press, Bhagaplur.

Brenneman, T.B. and Wilson, D.M. (2001). Effects of azoxystrobin, tebuconazole and flutolanil on *Aspergillus* populations and aflatoxin formation in peanut. Phytopathology, **91** (6: Suppl.): S 11.

Calvo, A., Guarro, J., Suarez, G. and Ramirez, C. (1980). Air-borne fungi in the air of Barcelona (Spain) III. The genus *Aspergillus* Link. Mycopathologia, **71** :41-43.

Christensen, C.M. (1957). Deterioration of stored grains by fungi. Bot. Rev, **23**: 108-131.

Christensen, C.M. and Kanfmann, H.H. (1969). *Grain storage. The role of fungi in quality loss.* University of Minnesota Press, Minneapolis.

Das, A.C. (1963). Ecology of soil fungi of rice fields. 1. Succession of fungi on rice roots. 2. Association of soil fungi with organic matter. Trans. Br. Mycol. Soc., **46**: 431-443.

Dhanraj, K.S., Girish, G.K. and Krishnamurthy, K. (1972), Grain storage fungi associated with wheat in Punjab. Bull . Grain Technol., **10**: 297-298.

Dutta, B.G. and Ghosh, G.R. (1955). Soil fungi from Orissa (India.)IV. Soil fungi of paddy fields. Mycopathol. Mycol. Appl., **25**:316-322.

Ghosh, S.K., Metha, P.K., Patel, J.G., Kashyap, S.K. and Chatterjee, S.K. (1977). Isolation of aflatoxin positive *Aspergillus flavus* strain from rice mill atmosphere. Indian J. Microbiol., **17**:138-140.

Gregory, P.H. (1961). *The microbiology of the atmosphere.* Leonard Hill (Books) Ltd., London.

Griffin, D.M. (1972). *Ecology of soil fungi.* Chapman and Hall, London.

Hill, R. and Lacey, J. (1979). The ecology of *Penicillium* species on barley grain. Phytopathology, **69**: 528.

Hudson, H.J. (1969). Aspergilli in the air-spora at Cambridge. Trans. Br. Mycol. Soc., **52**:153-159.

Jalaluddin, M.and Sinha, S. (1971). Microorganisms associated with stored foodgrains and milled products of Dacca city godowns. Mycopathol. Mycol. Appl., **43**:61-64.

Jayaraman, P. and Kalyanasundaram, I. (1989). Moisture relations of storage fungi in rice. Indian Phytopathology, **42**:67-72.

Jayaraman, P. and Kalyanasundaram, I. (1996). Aflatoxin in rice bran and its products. *In* Toteja, G.S., S. Diwakar and B.N. Saxena (eds). *Surveillance, prevention and control of food contaminants*, pp. 131-138. Indian Council of Medical Research, Delhi.

Kalyanasundaram, I. and Jayaraman, P. (1996). Aflatoxigenic fungi in rice and rice bran: prevention and control . In Toteja, G.S., S. Diwakar and B.N. Saxena (eds). *Surveillance, prevention and control of food contaminants*, pp. 144-153. ICMR, Delhi.

Kalyanasundaram, I. and Rao, G.J. (1988). Storage fungi in rice in India. Kavaka, **14**:67-76.

Kamal and Bhargava, K.S. (1972). Studies on soil fungi from teak forests of Gorakhpur. II. A contribution to Indian Aspergilli. Mycopathol. Mycol. Appl., **47**:59-72.

Kamal and Gupta, M.L. (1973). Studies on Aspergilli from soils of Gorakhpur. IV. The genus *Aspergillus* in relation to other microfungi in Usar soils. Proc. Natn. Acad. Sci. India, **43**:229-236.

Kamal and Singh, N.P. (1974). An investigation on mycoorganic content of air over sugarcane field at Gorakhpur (U.P.) I –Aspergilli. Proc.Natn. Acad. Sci. India, **44**:134-138.

Kamal and Singh, N.P. (1975). Fungal airspora over cowpea and maize fields at Gorakhpur. Proc. Natn. Acad. Sci. India, **45**: 53-60.

Kumar, S., Sinha, R.K. and Prasad, T. (1993). Propionic acid as a preservative of maize grains in traditional storage in India. J. Stored Prod. Res., **29**: 89-93.

Mallick, A.K. and Nandi, B. (1982). Deterioration of stored rough rice *Oryza sativa* cultivar Ratna: 4. Preservation and palatability of citral and propionic acid treated grains. Acta Agric Scand., **32**: 177-188.

Mehrotra, B.S. and Dwivedi, P.K. (1980). Fungi associated with wheat in the field. Int. Biodeterior. Bull., **16**: 37-42.

Mishra, R.R. and Srivastava, V.B. (1970). Leaf surface mycoflora of *Triticum aestivum*. Bull. Torrey Bot. Club, **97**: 384-387.

Mishra, R.R. and Srivastava, V.B. (1971). Leaf surface fungi of *Oryza sativa* Linn. Mycopathol. Mycol. Appl., **44**:289-294.

Mishra, R.R. and Srivastava, V.B. (1972). Aeromycology of Gorakhpur. V. Airspora over wheat and barley fields. Mycopathol. Mycol. Appl., **47**:349-355.

Mishra, R.R. and Srivastava, V.B. (1974). Leaf surface microflora of *Hordeum vulgare L.* Acta Soc. Bot. Pol., **43**: 203-212.

Mishra, R.R. and Tewari, R.P. (1975). Impact of air fungal population on the microbiology of leaf surface environment. Trop Ecol., **16**: 49-54.

Palanichamy, T. (1979). Studies on the phylloplane mycoflora of sorghum. M.Phil Dissertation, University of Madras, India.

Paramasivan, P. and Kalyanasundaram, I. (1986). Sources of fungal contamination of bread. J. Fd. Sci. Technol., **23**: 168-171.

Rati, E. and Ramalingam, A. (1976). Air-borne Aspergilli at Mysore. Aspects Allergy Appl Immun., **9**:139-149.

Shankar, S.R. and Rai, P.V. (1985). Isolation and characterization of aflatoxins from a standing crop and stored ragi. Mysore J. Agric. Sci., **19**:180-182.

Shenoi, M.M. and Ramalingam, A. (1976). Airspora of a sorghum field at Mysore.J. Palynol., **12**:43-54.

Singh, B.N. and Ram, H. (1979). Seasonal distribution of *Penicillium* and *Aspergillus* in karail and pond soils. Indian J. Microbiol. **19**:14-16.

Siradhana, B.S., Dange, S.R.S. Rathore, R.S. and Singh, S.D. (1978). Seed mycoflora of maize with reference to mycotoxins. Curr. Sci., **47**:59-60.

Sreeramulu, T. (1967). Aerobiology in India. J. Sci. Ind. Res., **26**: 474-480.

Sreeramulu, T. and Ramalingam, A. (1966). A two year study of airspora of a paddy field near Visakhapatnam. Indian J. Agric. Sci., **30**: 111-132.

Sundaram, B.M. (1969). Fungal flora of rice field soils. Ph. D. thesis, University of Madras, India.

Upadhyay, R.S. and Rai, B. (1979). Ecological survey of Indian soil fungi with special reference to Aspergilli, Penicillia and *Trichoderma*. Rev. Ecol. Biol. Sol., **16:** 39-40.

Usha, C.M., Patkar, K.L. Shetty, H.S. Kennedy, R. and Lacey, J. (1993). Fungal colonization and mycotoxin contamination of developing rice grain. Mycol. Res., **97**:795-798.

Verma, A.K. and Kamal. (1979). An investigation on mycoflora of air over Arhar (*Cajanus cajan* (Linn.) Millsp.) fields. I- Aspergilli. Natn. Acad. Sci. Letters, **2**: 370-374.

Verma, A.K. and Kamal. (1980). Air-spora studies over Arhar (*Cajanus cajan* (L) Millsp). field—Periodicity in the spores of *Drechslera, Epicoccum* and *Penicillium.* Natn. Acad. Sci. Letters, **3**:3-17.

Vincent, T. and Kalyanasundaram, I. (1999). Prevalence of aflatoxin in rice bran, and some associated factors. Indian Phytopathol., **152**:129-133.

Zuber, M.S., Calvert, O.H., Lillehoj, E.B. and Kwolek, W.F. (1976). Preharvest development of aflatoxin B_1 in corn in the United States. Phytopathology, **66**: 1120-1121.

38

Aeromycospora of Palode Rural Market in Thiruvananthapuram District, Kerala

Thripthi Krishna Mohan, P.S. Jothish and T.S. Nayar*

Tropical Botanic Garden and Research Institute, Palode,
Thiruvananthapuram - 695 562, Kerala

Aeromycospora of rural market at Palode in Thiruvananthapuram district, Kerala were surveyed using Burkard Personal Slide Sampler for one year from July 1998 to June 1999. Sixteen fungal genera were identified. *Cladosporium, Aspergillus/Penicillium, Nigrospora,* Basidiospores and Ascospores were found throughout the year. They together constituted 90.6% of the total airborne fungal spores. The maximum spore incidence was recorded in wet season (June to November) and minimum in hot season (March to May). Total spore concentration at indoor and outdoor environments did not show significant difference.

Key-words : Aeromycospora, Fungi, rural market

Fungal spores are ubiquitous in air. Airborne fungal spores are potential to create health hazards and, therefore, survey of aeromycospora is an essential prerequisite in understanding the quality of air. Aeromycospora originate from plants, animals, soil, manure and decaying plants and animals.

Although concentration of airborne fungi has been extensively studied in the extramural and work environments in India and abroad (Janaki Bai and Subba Reddi, 1981, Nair *et al.*, 1986, Lacey and Crook, 1988, Vittal and Krishnamoorthi, 1988, Dames and Cadman, 1994, Halwagy, 1994, Levetin *et al.*,1995, Li and Kendrick, 1995, Simeray *et al.*, 1995, Majumdar and Bhattacharya, 2000) information on their occurrence in market environment is rather scanty. Places where fruits, vegetables and grocery are marketed contain abundant fungal spores. There is relationship between the occurrence of airborne fungal spores and that of market diseases in market environment (Sullia and Khan, 1980). Spoilage of stored vegetables is an important source of hazardous fungi in such environment.

*Corresponding Author E-mail : thripthikm@rediffmail.com

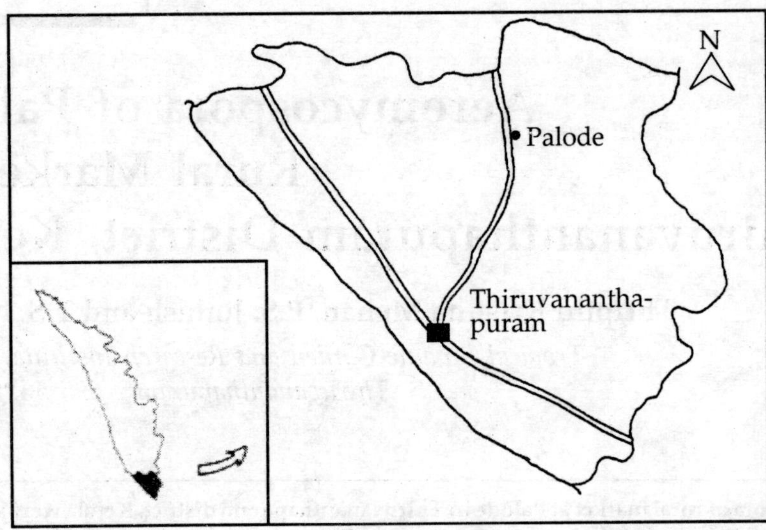

Fig. 1. Map of Thiruvananthapuram district and location of sampling site

The main objective of the present study was to record the prevalence of fungal spores at Palode market, a rural site in Thiruvananthapuram district of Kerala.

Study area

Palode is a forlorn village at the foothills of the Western Ghats situated about 35 km north-east of Thiruvananthapuram city (Fig. 1). There is a vegetable market at Palode in an open area spreading about 1.25 acres, which functions two days a week. About thousand people on average flock there on market days since this is the only market in an area of about 18 sq km around.

Vegetation in the area includes plantations of *Cocos nucifera, Tectona grandis, Hevea braziliensis* and *Artocarpus integrifolius* besides herbaceous weeds.

Survey was conducted twice a month at an interval of two weeks on market days for one year from July 1998 to June 1999. Burkard Personal Slide Sampler was used for the survey. Crowded market area was treated as the indoor and area outside the market at a distance of 100 meter was considered as the outdoor environments. Exposures were made for ten minutes between 8 AM and 10 AM, the duration when human activity was at its peak.

The exposed slides were scanned and spores were photographed using a Nikon Labophot 2 with HFX-DX and FX-35 DX dark box. Results are expressed as spores m^{-3} of air. Majority of the spores were identified upto the generic level. Spores of *Aspergillus/Penicillium* were categorised under one group. Basidiospores other than *Ganoderma* and Smuts/Rusts were grouped under 'other basidiospores'.

In Kerala, three seasons are distinguished: wet season (June to November; 21°C to 30°C), dry season (December to February; 19°C to 33°C) and hot season (March to May; 23°C to 37°C). Temperature, relative humidity, precipitation and wind velocity were recorded at the sites.

Observations

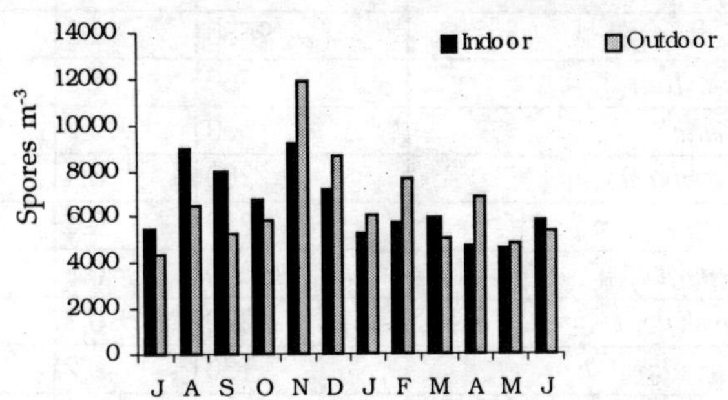

Fig. 2. Total Annual distribution at Palode rural market

Total annual spore distribution in the market environment shows prevalence of a rich fungal spora. Indoor environment (area inside the market) registered highest incidence of fungal spores in November (9190 spores m^{-3}) and August (9000 spores m^{-3}) whereas the outdoor environment showed highest incidence in November (11860 spores m^{-3}) (Fig. 2). Maximum spore incidence was observed in wet season and minimum in hot season.

The present study identified 16 fungal genera. *Cladosporium, Aspergillus/ Penicillium*, 'other basidiospores', *Nigrospora*, Ascospores, *Ganoderma* and Smuts/Rusts were found throughout the year. They together constituted 90.6% of the total airborne fungal spores. Other taxa included *Fusarium, Torula, Alternaria, Corynespora, Bispora, Drechslera, Curvularia, Torula* and *Spegazzinia* which together made up 3%. The remaining 6.4% could not be identified. Total annual concentration, percentage contribution and statistical significance of aeromycospora at indoor and outdoor environments are presented in Table 1.

Cladosporium was the main constituent of the aeromycospora with a total contribution of 17460 spores m^{-3} (22.5%). Season based comparison shows that though recorded round the year, it showed abundance in dry-hot seasons with maximum concentration in February (3610 spores m^{-3}). The wet season recorded its minimum occurrence with least concentration in October (140 spores m^{-3}) (Fig. 3a).

Aspergillus/Penicillium constituted the second dominant aeromycospora with a total contribution of 16870 spores m^{-3} (21.7%). This group dominated the wet season with peaks in October (2680 spores m^{-3}) and September (2410 spores m^{-3}). However, minimum concentration was observed in dry-hot

Table 1. Total annual concentration, percentage contribution and statistical significance of aeromycospora recorded from Palode rural market

Spores	Indoor		Outdoor		
	Spores m^{-3}	%	Spores m^{-3}	%	Significance
Deuteromycetes					
Cladosporium	17460	22.5	20830	26.6	NS
Aspergillus/Penicillium	16870	21.7	13490	17.2	NS
Nigrospora	9660	12.5	10280	13.1	NS
Fusarium	760	1.0	270	0.3	NS
Torula	560	0.7	290	0.4	NS
Alternaria	280	0.4	350	0.5	NS
Corynespora	270	0.4	240	0.3	NS
Bispora	240	0.3	290	0.4	NS
Drechslera	220	0.3	220	0.3	NS
Curvularia	130	0.2	260	0.3	NS
Tetraploa	10	0.01	-	-	
Spegazzinia	10	0.01	-	-	
Colletotrichum	-	-	10	0.01	
Basidiomycetes					
Other basidiospores	10600	13.7	13360	17.1	NS
Ganoderma	6040	7.8	6440	8.2	NS
Smuts/Rusts	2360	3.1	1480	1.9	NS
Ascomycetes					
Ascospores	7180	9.3	5280	6.7	NS
Unidentified	4950	6.4	5220	6.7	

NS – Not Significant at $p>0.05$

seasons with lowest concentrations in January, February and March (Fig. 3b). *Nigrospora* made a significant contribution of 9660 spores m^{-3} (12.5%) though it was present in low concentration year round except in November (Fig. 3c).

'Other basidiospores' constituted 10600 spores m^{-3} (13.7%) of total spore concentration. They were numerous in wet season with maximum concentration in November (2060 spores m^{-3}) and August (2050 spores m^{-3}) and minimum from January to May in dry-hot seasons (Fig. 3d). *Ganoderma* contributed 6040 spores m^{-3} (7.8%) and exhibited highest concentration in November (1680 spores m^{-3}) and August (1580 spores m^{-3}) whereas in January (30 spores m^{-3}) and March (60 spores m^{-3}) it showed minimum concentration. Smuts/Rusts were recorded year round in low concentration though they showed maximum concentration in June (600 spores m^{-3}) in the wet period.

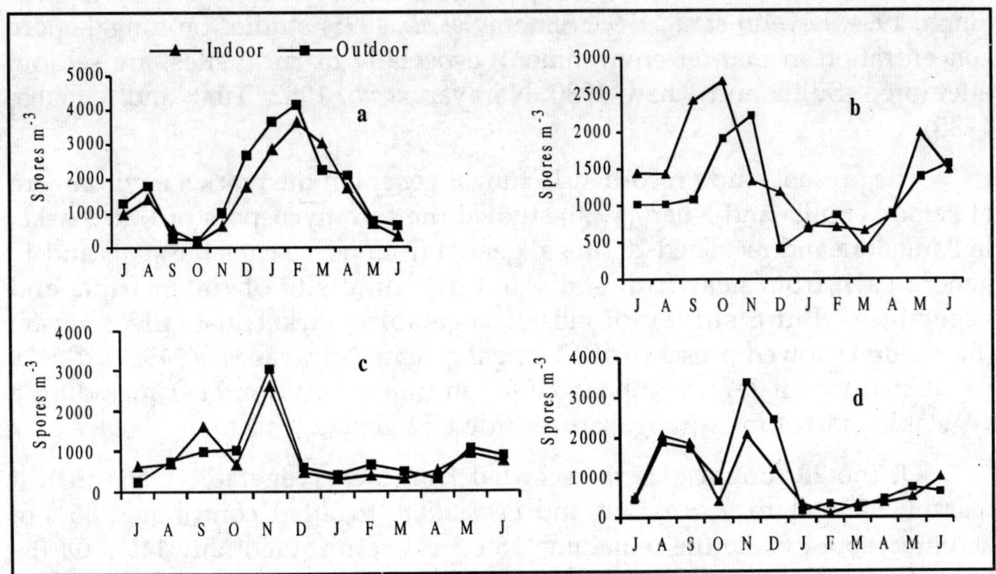

Fig. 3. Seasonal occurrence of dominant fungi: (a) *Cladosporium*; (b) *Aspergillus/Penicillium*; (c) *Nigrospora*; (d) Other basidiospores

The contribution of Ascospores to total spore concentration was 7180 spores m^{-3} (9.3%). The concentration remained more or less same year round though highest concentration was recorded in December (1650 spores m^{-3}).

The following taxa were caught rather intermittently: *Fusarium* with maximum concentration in August and June (180 spores m^{-3} each), *Torula* with continuous occurrence from December to February, *Alternaria* with continuous occurrence from December to April, *Corynespora* with maximum concentration in January (70 spores m^{-3}), *Drechslera* with continuous occurrence from October to January and *Curvularia* with its presence in September (10 spores m^{-3}), October (20 spores m^{-3}), December (50 spores m^{-3}), January (30 spores m^{-3}) and April (20 spores m^{-3}).

Concentration of total aeromycospora and that of each spore at indoor and outdoor environments were tested for significant differences using analysis of variance. They were found not significant ($p>0.05$) (Table 1). No difference in seasonal occurrence of aeromycospora was noticed at indoor and outdoor environments. (Figs. 3a.b,c,d).

Discussion

Though there is huge published literature on aeromycospora of different working environments (Lacey and Crook, 1988, Miller *et al.*, 1988, Singh and

Singh, 1994, Levetin *et al.*, 1995, Simeray *et al.*, 1995) studies on fungal spore concentration in market environment, especially of rural sites, are seldom attempted (Sullia and Khan, 1980, Narayan *et al.*, 1982, Tilak and Saibaba, 1985).

The present study recorded 16 fungal genera in the market environment at Palode. Sullia and Khan (1980) studied the aeromycospora of K.R. market in Bangalore and recorded 20 fungal genera from near vegetable stalls and 19 genera each from near fruit stalls and dumping site of rotten fruits and vegetables. But a survey of closed vegetable market in Malleeswaram (Bangalore) showed presence of 47 fungal genera (Narayan *et al.*, 1982). Study conducted by Tilak and Saibaba (1985) in indoor environment including a vegetable market in Aurangabad recorded 52 genera.

Of the 20 fungal genera recorded from near vegetable stalls in K.R market , Bangalore, *Aspergillus* and *Penicillium* together contributed 86% of aeromycospora while the remaining 18 genera contributed only 14%. Of the 52 fungal genera recorded from the vegetable market in Aurangabad, *Cladosporium, Torula, Aspergillus* and *Alternaria* together made up 52% of the aeromycospora. Remaining 48 genera constituted only 37%. Narayan *et al.* (1982) have not discussed the percentage contribution of the fungal genera recorded in their study in Malleeswaram market, Bangalore. In the present study *Cladosporium, Aspergillus/Penicillium*, 'other basidiospores', *Nigrospora*, Ascospores and *Ganoderma* together made up 87.5% of the aeromycospora. The remaining 10 genera contributed 6.1%. However, comparison of data from Palode on the concentration of total aeromycospora and that of each spore with data on those of Sullia and Khan (1980), Narayan *et al.* (1982) and Tilak and Saibaba (1985) could not be made since they did not provide details on concentration of aeromycospora.

The higher counts of most of the spores in the vegetable market in Aurangabad were observed in winter season (dry season of the present study) (Tilak and Saibaba, 1985) whereas the higher counts were observed at Malleeswaram in Bangalore in dry summer months (dry-hot seasons) (Narayan *et al.*, 1982). However, our study showed maximum spore concentration in wet season. Tilak and Saibaba (1985) attributed abundance of diseased samples for the higher counts in winter whereas Narayan *et al.* (1982) concluded that it was due to greater dispersal of dry powdery spores. The abundance of spores in wet season in our study could possibly be due to the conidia of *Aspergillus/Penicillium* and basidiospores.

The present study reports high concentration of *Cladosporium* in dry-hot seasons. No report exists on the seasonal occurrence of *Cladosporium* spores in the market environment. However, the aeromycological study conducted in the extramural environment in Durban (Dames and Cadman, 1994), Kuwait (Halwagy, 1994) and India (Janaki Bai and Subba Reddi, 1981, Majumdar and Bhattacharya, 2000) showed high concentration of *Cladosporium* in dry-hot seasons. Dominance of *Cladosporium* in air over all other spores as seen at Palode, has been reported in many working environments (Levetin *et al.*, 1995, Simeray *et al.*, 1995, Bagwan and Meshram, 2000). As observed in the present study, Tilak and Saibaba (1985) also recorded high prevalence of *Aspergillus/ Penicillium* in wet season in the vegetable market in Aurangabad. Sullia and Khan (1980) reported that 70% of aeromycospora in K.R. market in Bangalore was constituted by *Aspergillus* and *Penicillium.*

Basidiospores were recorded in high concentration (see Table 1) in the present study. They were found in low concentration in studies conducted by Narayan *et al.* (1982) and Tilak and Saibaba (1985). While there was total lack of *Nigrospora* in K.R. market (Sullia and Khan, 1980) and Malleeswaram (Narayan *et al.*, 1982) in Bangalore, it was observed round the year at Palode.

Dry season with high temperature and low relative humidity with minimum or no rainfall might have favoured maximum incidence of *Cladosporium* and other hyphomycetes such as *Torula, Alternaria, Corynespora, Drechslera* and *Curvularia* at Palode. Pasanen *et al.* (1991) observed that it was the water from condensation present on surfaces, which favoured the growth of Aspergillaceae. Our observation that spores of *Aspergillus/Penicillium* were in maximum numbers in the wet season corroborates this statement. Wet season favoured the development of sporophores and release of basidiospores.

Main sources of fungal spores in the market and surrounding areas at Palode were the decaying and rotten vegetable matters and other garbage heaped inside the market for longer periods. They provided substrate for the growth of fungal propagules. Temperory sheds with thatched coconut leaves could act as substratum in wet season for the growth of spores, especially those of *Aspergillus/ Penicillium.* The plantation around and the leaf litters accumulated beneath could function as hosts for plant pathogenic and saprobic fungal spores captured from this market environment.

Both the indoor and outdoor environments being open areas with a distance of only 100 metres apart, there is free flow of air from one environment to the other. This might be the main reason why no significant difference was observed in total spore concentration and in concentration of each spore type at indoor and outdoor areas.

Of the 16 fungal genera identified in the present study the allergenic nature of *Cladosporium, Aspergillus/Penicillium, Alternaria, Curvularia, Nigrospora* and Basidiospores has already been established (Agarwal and Shivpuri, 1974; Salvaggio and Aukrust; 1981, Tilak, 1990-1991, Vijay *et al.*, 1991).

Effective disposal of solid waste on one hand and continuous aeromonitoring on the other may improve the air quality of this rural market area.

Acknowledgement

Financial support from the Ministry of Environment and Forests, New Delhi in conducting this study is gratefully acknowledged. First author thanks CSIR, New Delhi for the award of a Senior Research Fellowship.

References

Agarwal, M. K. and Shivpuri, D. N. (1974). Fungal spores – their role in respiratory allergy. Advances in Pollen-Spore Research, **1**: 78-128.

Bagwan, N. B. and Meshram, B M. (2000). Relationship between fungal air-spora of apple store houses and incidence of post-harvest fruit rots. Indian Journal of Aerobiology, **13 (1&2)**: 29-32.

Dames, J. F. and Cadman, A. (1994). Airspora of Durban : a subtropical, coastal South African city. II Fungal spore component. Grana, **33** : 346-348.

Halwagy, M. H. (1994). Fungal airspora of Kuwait city, Kuwait, 1975 - 1987. Grana, **33**: 340-345.

Janaki Bai, A. and Subba Reddi, C. (1981). The fungus airspora of Visakhapatnam. Proceedings of Indian National Science Academy, **B47 (5)**: 731–747.

Lacey, J. and Crook, B. (1988). Fungal and actinomycete spores as pollutants of the work place and occupational allergens. Annals of Occupational Hygiene, **32** : 515–533.

Levetin, E., Shaughnessy, R., Fisher, E., Ligman, B., Harrison, J. and Brennan, T. (1995). Indoor air quality in schools : exposure to fungal allergens. Aerobiologia, **11**: 27-34.

Li, D.W. and Kendrick, B. (1995). A year - round outdoor aeromycological study in Waterloo, Ontario, Canada. Grana, **34** : 199-207.

Majumdar, M. R. and Bhattacharya, K. (2000). Aeromycoflora at the foothills of Eastern Himalayas. Indian Journal of Aerobiology, **13** : 14-19.

Miller, J. D., Laflamme, A. M., Sobul, Y., Lafontaine, P. and Greenhalgh, R. (1988) Fungi and fungal products in some Canadian houses. International Biodeterioration, **24** : 103-120.

Nair, P. K. K., Joshi, A. P. and Gangal, S. V. (Eds.) (1986). Airborne Pollen, Spores and other Plant materials of India - A Survey. Report of the All India Co-ordinated project on Aerobiology. CSIR Centre for biochemicals, Delhi and National Botanic Research Institute, Lucknow.

Narayan, M. C. J., Ravichandran, V. and Sullia, S. B. (1982). Aeromycology of the atmosphere of Malleeswaram market, Bangalore. Acta Botanica Indica, **10**: 196-200.

Pasanen, A. L., Kalliokoski, P., Pasanen, P., Jantunen, M, J. and Nevalainen, A. (1991). Laboratory studies on the relationship between fungal growth and atmospheric temperature and humidity. Environment International, **17** : 225-228.

Salvaggio, J. and Aukrust, L. (1981). Postgraduate course presentations : mould-induced asthma. Journal of Allergy and Clinical Immunology, **68** : 327-346.

Simeray, J., Mandin, D. and Chaumont, J.P. (1995). Variations in the distribution of fungal spores in the atmosphere of bake houses. Impact on the study of allergies. Grana, **34**: 269-274.

Singh, A. and Singh, A.B. (1994). Airborne fungi in a bakery and the prevalence of respiratory dysfunction among workers. Grana, **33**: 349-358.

Sullia, S. B. and Khan, K. R. (1980). Airspora of Bangalore city market and its relation to the occurrence of market diseases. Advances in Pollen Spore Research, **V-VII** : 157-169.

Tilak, S. T. and Saibaba, M. (1985). Indoor aero-mycology. Acta Botanica Indica, **13** : 203-211.

Tilak, S. T. (1990-1991). Fungal spores and Allergy. – Journal of Palynology Silver Jubilee Commemoration Volume: 369-386.

Vijay, H. M., Comtois, P., Sharma, R. and Lemieux, R. (1991). Allergenic components of *Ganoderma applanatum*. Grana, **30**: 167-170.

Vittal, B. P. R. & Krishnamoorthi, K. (1988). A census of the airborne mold spores in the atmosphere of the city of Madras, India. Annals of Allergy, **60** : 99-101.

39

Fungal Diversity and Diseases of Forest in Central India

R.K. Verma* and Yogendra Singh
Tropical Forest Research Institute, PO-RFRC, Jabalpur-482 021

Diversity of fungi in important forest trees and diseases caused by them of central India is reviewed. Forest fungal flora of Central India is listed which include 48 fungi in teak forest, 15 in sal forests, 46 in tree legumes, 18 in bamboos and 90 in miscellaneous forests. The most damaging diseases of above forests are vascular wilts, damping-off, root rots, cankers, rusts and foliar diseases. Wood decay in standing trees and woods by fungi is also included. Control measures of different diseases of teak, sal, tree-legumes, neem, bamboos, Khamer, *Eucalyptus* etc., are given in detail which include chemical, biological, cultural and integrated management approaches.

Key words : Diversity, fungi, forest trees, diseases, management

The forest cover of the country, as per the present assessment is 63.73 million ha. Constituting 19.39% of the geographic area, out of which 37.74 million ha (11.48%) is dense forest, 25.50 million ha (7.76%) open forest and 0.49 million ha (0.15%) mangroves. Madhya Pradesh accounts for the largest forest cover of the country i.e. 20.68% followed by Arunachal Pradesh (10.80), Orissa (7.38%), Maharashtra (7.32%) and Andhra Pradesh (6.94%). The seven North-Eastern states together comprise 25.70% of the total forest cover.

The total recorded forest area of the country as reported by the State/ Union Territory forest departments is 76.52 million ha. The forest cover as per the last assessment of the FSI, 1999, however, was 63.34 million ha. In most of the states, the forest cover is less than the recorded forest area. However, in the states of Arunachal Pradesh, Jammu and Kashmir, Manipur, Meghalaya, Mizoram, Nagaland, Sikkim and the union territory of Andaman and Nicobar Islands, the forest cover is more than the recorded forest area.

The Indian region is one of the most diverse biogeographic regions of the world, embracing a wide range of topography from perpetually snow

*Corresponding Author E-mail : rkverma28@rediffmail.com

covered high Himalayan ranges to plains at sea level, low lying swamps and mangroves, inland systems, tropical evergreen rain forests, fertile alluvial plains, hot deserts and high altitude cold deserts. There are almost rainless areas, to the world's highest rainfall areas. The climate ranges from tropical and sub-tropical in Indo-Gengetic plains and peninsular regions to temperate and arctic in the Himalayan region. The habitat types vary from the humid tropical Western Ghats to the hot deserts of Rajasthan, from cold deserts of Ladakh to the long, warm coastline stretches, peninsular India.

A great variety of climatic and altitudinal variations coupled with varied ecological habitats have contributed immensely to the rich vegetational wealth, and varied flora and fauna generating thus a very unique biodiversity.

A significant feature of the Indian flora is the presence of species from surrounding countries like-Malaya, Tibet, China, Japan, Europe, Africa, etc. and even from far widely separated countries like - U.S.A. and Australia.

National Forest Policy 1988 emphasizes on increasing forest productivity for restoring ecological balance and conservation of country's natural heritage and biodiversity on one hand and to meet the demands for timber, fuel, fiber, paper and pulp and fodder on the other. In achieving this aim, losses due to various abiotic and biotic factors for example, diseases must not cross the acceptable limits. Therefore, to develop and protect forests for maximum productive and protective functions, are the prime objectives of forestry. Protection of forest from diseases constitutes one such objective.

Diseases are known to be the most destructive agents affecting the planting stock in forest nurseries and consequently the plantation yield thus directly affecting the forest productivity. An account of Impact of Diseases in Tropical Forest is given in the Proceedings of the IUFRO Symposium (Nair *et. al.*, 1996).

Tropical forests of central India constitute a rich fungal diversity. The diversity of fungi in different forests is not properly documented till date. Proper documentation of these fungi will help in development of suitable biotechnological application packages to increase forest productivity. Keeping in view the wide diversity and diseases of forestry species, the work done on fungal diversity and diseases of important forestry trees is reviewed in this chapter.

Diversity of Fungi in Forests of Central India

India is one of the 12-mega diversity countries. With only 2.4% of the global land area, India possesses more than 45,500 plant species, representing about 11% of the world's biota. This is only a rough estimation, since many of the organisms, especially in lower groups like-bacteria, fungi, algae, lichens, bryophytes, etc. have yet to be described and many remote geographical areas have to be adequately surveyed. In terms of plant diversity, India ranks 10th in the world and 4th in Asia. Nearly fungi constitute 32% of India's biowealth (18.23%) and angiosperms (13.50%). Plant species in India represent more than 10% of the known plant species of the world.

Some scattered reports on rhizosphere and forest litter mycoflora are available from central India (Dadwal *et al.*, 1986; Soni and Jamaluddin, 1990) Fungi occurring in forests (three main types namely, teak, sal, bamboo, and miscellaneous) of central India on trees, associated plants, dead plant parts, leaf litter, soils, etc. are listed in Table 1.

Fungal Diseases of Important Tree Species

1. Diseases of Tree Legumes

Tree legumes include an extensive range of multipurpose, widely adapted species which are useful components of agricultural and silvicultural systems (NFT 1989). They are important throughout the tropics as sources of forage, firewood, charcoal, green manure and timber (NFT 1989, Hughes and Styles 1989). Over the last 15 years there has been a rapid explosion of reports of diseases affecting tree legumes.

A. Diseases of Acacias:

(i) *Acacia nilotica*

A. nilotica (L.) Willd is an extremely valuable source of fuel, gum, small timber, fodder, tanin (12-20%), and honey in the Indian subcontinent. It is very drought tolerant, withstands extreme temperature, and grows fast under favourable soil irrigation. Being resistant to termites and impervious to water, the wood is also used for railroad ties, tool handles, carts, and oars. It is also good for carving and boat building. Important diseases of A. *nilotica* are listed in Table 2.

Wilt and root rot due to *Fusarium solani* has also been reported on *A. nilotica*. *Botryodiplodia theobromae* has been found associated with this root disease (Bagchee 1945, 1958). *Haplosporella bakeriana* causes disease in young plantation (1-5 years) and 25% plants show die back symptoms, (Verma *et. al.*, 1997).

Table 1. Diversity of fungi in forests of Central India

S.No.	Name of fungi	Name of fungi
1.	**Teak forests**	
	1. *Acaulospora scrobiculata* Trappe 2. *Aspergillus niger* van Tiegh. 3. *A. nidulans* (Eidam) Wint. 4. *A.flavus* Link ex Fries 5. *A.terreus* Thom 6. *Arthrinum phaeospermum* (Cda.) Ell. 7. *Auricularia measenterica* (Dus) Farl 8. *Aureobasidium* sp. 9. *Calospora tectoniae* Tilak & Ramachandra Rao 10. *Chrysosporium* sp. 11. *Cephalosporium curtipes* Sukapur & Thirum 12. *Cercospora tectonae* Stevens 13. *Curvularia lunata* (Walker) Boedjin 14. *Dothiorella* sp. 15. *Fusarium moniliform* Shelden 16. *Fusarium pallidoroseum* (Cke.) Sacc. 17. *Ganoderma applanatum* (Pers.) Pat. 18. *Gigaspora* sp. 19. *Glomus aggregatum* Schenck & Smith emend Koske 20. *G.caledonium* (Nicol. & Gerd.) Trappe & Gerd. 21. *G.etunicatum Becker* & Gerd. 22. *G.fasciculatum* (Thaxter) Gerd. & Trappe emend Walker & Koske 23. *G. intraradices* Schenck & Smith 24. *G. macrocarpum* Tulasne & Tulasne 25. *G.microcarpum* Tulasne & Tulasne 26. *G. mosseae* (Nicol. & Gerd.) Gerd. & Trappe	27. *Hymenochaete rubiginosa* (Schrad) Levl., 28. *H. nigircans* (Lev.) Pat. 29. *Hypoxylon mediterraneum* (de Not.) JH Miller 30. *Helicobasidium compactum* Boed 31. *Macrophomina phaseolina* (Tassi) Goid. 32. *Moorella speciosa* Rao & Rao 33. *Leptosphaeria compressa (Rem.) Hohn.* 34. *Myriangium tectonae* Tendulkar 35. *Nitsehkia broomeiana* (Berk,) Nann. 36. *Olivia tectonae* (Recid) Thirum 37. *Penicillium* sp. 38. *Phoma exigua* Desm 39. *P.macrostoma* Mont. 40. *Phyllosticta tectonae* Syd. 41. *Rhizopus* sp. 42. *Scutellospora heterogama* (Nicol & Gerd) Walker & Sander 43. *S.pellucida* (Nicol & Schenck) Walker & Sander 44. *Sclerocystis coremiodes* Berk & Broome 45. *Sphaceloma tectonae* Wani & Thirum 46. *Trichoderma viride* Pers. ex Fr. 47. *T.pseudokoningii* Rifai 48. *Uncinula tectonae* Salm.

2.	**Sal forests**	
	1. *Amanita fulva* Schaeff. ex Pers. 2. *Comparophyllus niveus* (Scop. ex Fr.) Karst 3. *Cladosporium* sp. 4. *Cylindrocladium parvum* Anderson 5. *Fomes caryophylli* (Rac.) Bres. (=*F.fastulosus* (Lev.) Cooke 6. *Gloniella shoreae* Sivanesan 7. *Hymenochaete rubiginosa* (Schrad) Lev.	8. *H. nigricans* (Lev.) Pat. 9. *Hypoxylon mediterraneum* (de Not. JH Miller 10. *Lembosea shoreae* RK Verma, Soni, Tiwari & Jamaluddin 11. *Lenzitis melanophyllus* Dev 12. *Macrophoma shoreae* Bagchee 13. *Polyporus shoreae* Wahef. 14. *P. agariceus* Berk. 15. *Xylaria hypoxylon* (L) Grev.
3.	**Legumes**	
	1. *Botryodiplodia theobromae* Pat. 2. *Camptomeris albiziae* (Petch) Mason 3. *Camptomeris albiziicola* Thirum, & Narasimhan) Bessey 4. *Cercospora albiziae* Kar & Mandal, 5. *Cercospora pongamiae* Kar & Mandal, 6. *Cercospora sissoo* Syd., 7. *Colletotriochm dematium* (Pers. ex Fr.) Grove 8. *Colletotrichum gloeosporioides* Penzig, 9. *Colletotrichum* sp. 10. *Cyllindrosporium acaciae* Anahosur 11. *Diatrype albiziae* Rehm. 12. *Diplodia prospopidina* 13. *Diplodia dalbergiae* Died. 14. *Dothiorella indica* Soni, Dadwal & Jamaluddin 15. *Erysiphe acaciae* Blumer. 16. *Eutypella pongamiae* Agarwal & Gupta 17. *F. solani* f *dalbergiae* Gardon 18. *Fomes badius* Ber 19. *Fomes fustuosus* (Levl.) Cke. 20. *Fusarium pallidoroseum* (CKe.) Sacc. 21. *Fusarium oxysporum* Fr., 22. *Fusicladium pongamiae* Syd (= Asperisporium pongamiae (Syd.) Deighton 23. *Ganoderma applanatum* (Pers.) Pat.	24. *Ganoderma lucidum* (Leyss) Karst, 25. *Haplosporella bakeriana* (Petrak) Sacc. 26. *Haplosporella beaumontiana* Ahmad 27. *Helminthosporium dalbergiae* Ellis 28. *Maravalia achroa* (Syd.) Arth. & Cammus, 29. *Microstroma acaciae* Syd. 30. *Nectria flavistroma* 31. *Oidium* sp. 32. *Phoma nivea* (Syd) Srivastava & Bhargava 33. *Phomopsis dalbergiae* Sahni. 34. *Phyllactenia dalbergiae* Pirozynski 35. *Phyllosticta pongamiae* Syd. 36. *Polyporus gilvus* Fr. 37. *Ravenelia acaciae* (Mundk & Thirum. 38. *Ravenelia clemensiae* Syd. 39. *Ravenelia sessilis* Berk, 40. *Ravenelia tandonii* Syd.

	41. *Ravenelia hobsonii* Cooke 42. *Rhytidrysterium rufulum* (spreng.) Speg.	43. *Septoria prosopodis* Stev. & Pierce 44. *Sphaceloma pongamiae* Wani & Thirum. 45. *Uredo sissoo* Syd. & Butler
4.	**Bamboos**	
	1. *Alternaria alternata* (Fr.) Keissler 2. *Arthinum phaeospermum* (Cda) Ell 3. *Botryodiplodia theobromae* Pat. 4. *Curvularia lunata* (Walker) Boedjin 5. *Colletotrichum gloeosporioides* Penz. 6. *Dimerosporium strigosum* Pers. ex Fr. 7. *Dusturella divina* (Syd.) Mundkar & Kheswala 8. *Exerohilum rostratum* (Drech.) Leonard & Suggs. 9. *Glomerella cingulata* (Stonem) Spould & Schrenk 10. *Fusarium equiseti* (Corda) Sacc.	11. *Fusarium pallidoroseum* (Cke.) Sacc. 12. *Hypocrea dendrocalamae* Patil & Patil 13. *Ganoderma lucidum* (Leyess) Karst 14. *Metasphaeria dendrocalmae* Subedar & Rao 15. *Puccinia melanocephala* Syd 16. *Periconia saraswatipurensis* Bilgrami 17. *Rhizoctonia solani* Kuhn 18. *Sarocladium oryzae*
5.	**Miscellaneous forests**	
	1. *Aecidium rhytismoideum* Berk & Br 2. *Acremonium alabamense* Shukla, Agarwal & Jain 3. *A.sagarensis* Shukla, Agarwal & Jain 4. *Alternaria alternata* (Fr.) Keissler 5. *Botryosphaeria quercum* (Schwein) Sacc. 6. *Aspergillus niger* van Tiegh. 7. *A. fulvus* Link ex Fries 8. *A. fumigatus* (Fres.) Rai, Tewari & Agarwal 9. *A. nidulans* (Eidam) Wint. 10. *A. terreus* Thom 11. *Bartilinia robillardes* Tassi 12. *Botryodiplodia theobromae* Pat 13. *Bovilla pongamiae* Ramachandra Rao 14. *Chaetomium globosum* Kunzze & Schm. 15. *Calosphaeria lantatae* Tilak & Nefere 16. *Cercocladospora adinae* Agarwal & Singh, 17. *Cercospora adiniicola* Kar & Mandal	48. *Endophragmiella angustispora* Hughes 49. *Excipularia naraspurensis* Subram 50. *Ganoderma lucidum* (Leyss) Karst 51. *Haplosporella diospyri* Sheshadri, 53. *Hendersonula toruloidea* Nattras 54. *Leptosphaeria gmelinae* Jamaluddin, Dadawal & Soni, 55. *Macrophomina phaseolina* (Tassi) Goid). 56. *Maublancia indica* Pande and Patwardhan, 57. *Massarina terminaliae* Hasija,m Rajak, Prasad & Pandey 58. *Monochaetia jabalpurensis* Agarwal & Hasija

5.	**Miscellaneous forests**	
	15. *Calosphaeria lantatae* Tilak & Nefere 16. *Cercocladospora adinae* Agarwal & Singh, 17. *Cercospora adiniicola* Kar & Mandal 18. *Cercospora cassiae-siameae* Chiddarwar 19. *Cercospora catappe* P. Henn. 20. *Cercospora eucalypti* Cke & mass. 21. *Cercospora holopteleae* Chiddarwar 22. *Cercospora leucosticta* Ell. & Ev. 23. *Cercospora lythracearum* Hald & Wolf 24. *C.prasadi* Mathu & Mathur 25. *Circinotrichum poonensis* Pirozinski & Patel 26. *Cladosporium herbarum* (Pers.) Link 27. *Coleophoma empetri* (Rostr.) Ptr. 28. *Colletotrichum capsici* (Syd) Butler & Bisby 29. *Colletotrichum gloeosporioides* Penzig. 30. *Colletotrichum arjunae* Sahni 31. *Colletotrichum holopteleae* Hasija 32. *Coniella granati* (Sacc.) syd. & Petrak 33. *Corticium rolfsii* Curzi 34. *C. salmonicolor* B & Br. 35. *Curvularia lunata* (Walker) Boed. 36. *Cylindrocladium scoparium* Morg 37. *Cytospora agarwalii* Soni, Dadwal & Jamaluddin 38. *Daldinia concentrica* (Bolt ex Fr.) Ces de Not. 39. *Diatrypella veruucaeformis* (Her.) Nits. 40. *Dictyosporium interanedium* Subram. 41. *Dothiorella indica* Soni, Dadwal & Jamaluddin 42. *Excipularia naraspurensis* Subram 43. *Elsinoe kamatii* Tewari 44. *Fusariella atrovirens* Sacc. Hasija, Rajak, Prasad & Pandey	45. *Fusariella atrovirens* Sacc. 46. *F.solani* (Mart) Appel & Wr. 47. *Fusarium oxysporum* Schlecst ex Fr. 48. *Monochaetia jabalpurensis* Agarwal & Hasija 49. *Muragenella eucalypti* Sutton & ND Sharma 50. *Mucor circinelloides* van Tiegh. 51. *Mycosphaerella holopteleae* Naphade, 52. *Nicrodiplodia zizyphae* Subedar & Rao 53. *Oidium azadirachtae* Narayanaswamy & Ramakr. 54. *Paecilomyces variotii* Bain 55. *Pestalotiopsis versicolor* (Speg.) Stey 56. *Pestalotiopsis terminaliae* Agarwal & Hasija 57. *Phaeoseptoria eucalypti* (Hansf.) Walker 58. *Phoma tropica* Schcavider & Boerem, 59. *Phoma nebulosa* 60. *Phomopsis* sp. 61. *Phyllactenia suffulta* (Rsbenth) Sacc., var. *gmelinae* Patil 62. *Phyllactenia terminaliae* Ramkr. TS, 63. *Phyllosticta eucalypti* Thrum,., 64. *Pseudocercospora subsessilis* (H&P Syd.) Deigthon,

65. *Pseudocercospora microphora* (Seshadri) Deighton 66. *Pseudopeltrakia kambakkamensis* subram. & Ellis 67. *Pyrenochaeta terrestris* 68. *Pythium* sp. 69. *Rhizoctonia solani* Kuhn 70. *Rhizopus stolonifer* (Ehrenb. Ex Fr.) Linder 71. *Schizophyllum commune* Fr. 72. *Sclerotium rolfsii* Sacc. 73. *Selenophoma kamatii* Kalani	74. *Sirosporium suttoni* Manoharacharii and Rao 75. *Sphaceloma lagerostromiae* Wani & Thirum 76. *Sphaceloma terminaliae* Bitanc. 77. *Thrysterina multispora* Sivaneasan, Rajak & Gupta 78. *Thielavia appendiculata* Gilman 79. *Trichoderma harzianum* Rifai 80. *Uredo terminaliae* P. Henn 81. *Verticillium* sp.

Ganoderma root rot in *Acacia arboretum*

Severe disease incidence due to attack of *Ganoderma lucidum* was noticed in *Acacia arboretum* (including *A. nilotica*). The arboretum was established in 1982 with 18 different *Acacia* spp. after clear felling of teak forest in Seoni District of Madhya Pradesh. Out of 18 only 2 species, namely *A. greggii* and *A. verek* showed disease resistance (Harsh *et. al*, 1993).

(ii) *Acacia catechu* Willd (*Khair*)

Khair is moderate sized deciduous tree. The wood is largely used for extraction of *cutch* and *katha*. The former is largely used for dying and tanning, while the latter is mainly used as an ingredient for chewing betel leaves. It is also used for fuel and charcoal. Important diseases of *Acacia catechu* are listed

Table 2. Diseases of *A. nilotica* with their causal organisms and control measures

S. No.	Name of the disease	Causal organism	Control
1.	Wilts	*Fusarium oxysporum, F.pallidoroseum*	Soil drenching with Bavistin @ 0.5%
2.	Leaf spot	*Cylindrosporium acaciae*	Spraying with 0.2% Dithane M45
3.	Leaf rust	*Ravenelia acaciae*	Spraying with Dithane Z78 @ 0.2%
4.	Root rot	*Ganoderma lucidum*	As in *D. sissoo*
5.	Die back	*Haplosporella bakeriana*	--
6.	Heart rot	*Fomes badius*	As in Sal

Table 3 Diseases of *A. catechu* from central India

S.No.	Name of disease	Causal organism
1.	Powdery leaf spot	*Microstroma acaciae*
2.	Leaf rust	*Ravenelia tandonii*
3.	Powdery mildews	*Erysiphe acaciae, Phyllactenia acaciae*
4.	Root rots	*Ganoderma lucidum, Polyporus gilvus*
5.	Heart rot	*Fomes badius, Fomes fustuosus, Ganoderma applanatum.*

in Table 3.

Of the diseases listed above *Ganoderma lucidum* causes serious root rot mortality in khair (Bakshi, 1976). Symptomatology, infection and management strategy is same as described for *Ganoderma* root rot of sissoo.

B. Diseases of Albizias

(i) *Albizia lebbek* (East Indian walnut, kokko, woman's tongue):

Leguminous as well, this is a very adaptable tree that produces small timber and fuel, and is good for reforesting dry, alkaline soils. It is one of the best known trees in India, and for which, has spread through much of the southern hemisphere. The wood is dense, coppices well, and carries a high calorific value at 5, 200 kcal. Furniture, beautification, fodder (20% protein), and erosion control make up its other main uses.

Seed and seedling diseases

Rhizoctonia solani caused destructive damping-off of seedlings in nurseries of *A. lebbek* in India (Mehrotra, 1989). Leaf rusts caused by *Ravenelia* spp. are important seedling diseases of *Albizia* spp. in India (Sharma and Bhardwaj, 1988).

Foliage diseases

Camptomeris albiziae caused leaf blotch and foliar necrosis of *A. lebbek.* A similar leaf blotch, *Camptomeris albiziicola,* has been recorded on *A. lebbek* in India.

Cercospora glauca and *C. albiziae* caused leaf spots of *A. lebbek. Phylliosticta albizinae* may seriously defoliate young plantations of *A. lebbek* in India (Bakshi, 1976). *Rhizoctonia solani* causes leaf web blight of *A. lebbek* (Mehrotra, 1990).

Many rusts belonging to the genera *Ravenelia* and *Uredo* are recorded on this tree (Bakshi, 1976); however, in most cases, their importance has not been determined.

Stem diseases

Fusarium solani caused greyish-black cankers on 15-20 year old trees of *A. lebbek* in India. The pathogen invades through wounds and later develops a stratified canker through repeated killing of the cambium (Bakshi, 1976). In severe infections, the canker may extend 2-4 m almost girdling the stem and causing drying of the crown. Trees may snap at the region of the canker. *Haplosporella beaumontiana* causes die back disease in young plantation (1-5 years) and 21% plants show die back symptoms (Verma *et al.*, 1997).

(ii) *Albizia procera*

Albizia procera Benth is one of the most important tree legumes raised in nurseries and plantations of India. The wood is extensively used for building, furniture and agricultural implements. Important diseases of *A. procera* occurring in central India are listed in Table 4.

Table 4 Diseases of *A. procera* with their causal organisms in nurseries and plantations of central India.

S.No.	Name of disease	Causal organism
1.	Leaf spots	*Cercospora albiziae, Camptomeris albiziae*
2.	Anthracnose	*Colletotrichum dematium*
3.	Leaf and pod rust	*Ravenelia sessilis, R. clemensiae*
4.	Root rots	*Ganoderma lucidum, Ganoderma applanatum*

Leaf spot

Cercospora albiziae attacks leaflets of young plants of safed siris. The infection spots which occur on both sides of leaf are brownish - whitish grey in the center surrounded by narrow, brown to blackish ring. In some cases, the center of spot gives a deep greyish powdery appearance (Kar and Mandal, 1969).

Anthracnose

Colletotrichum dematium produces small brown - dark brown spots on leaves which later on increase in size and form lesions.

Control: Spraying with Captafol or Blitox 50 @ 0.2% has been found effective in controlling the disease.

Leaf and pod rust

Symptom: Yellowish brown to black pustules are produced on leaves and pods. Profuse development of rust pustules, which almost plasters the leaf surfaces, may affect the metabolic activity of plants. The disease is severe on young plants. Seedlings when heavily attacked are deformed, crumpled and may die subsequently. *Mycovellosiella ravenaliae* M.D. Mehrotra and R.K. Verma a hyper parasite of *Ravenelia clemensiae* (leaf rust pathogen of *A. sprocera*) is also reported (Mehrotra and Verma, 1993).

Control: Spraying with Dithane Z78 or Sulfex @ 0.2%

Root rots

Ganoderma lucidum and *Ganoderma applanatum* cause root rots in *A. procera*. Symptomatology, infection and management strategy for the disease is same as described for *Ganoderma* root rot of sissoo.

Dalbergia sissoo

Sissoo (*Dalbergia sissoo* Roxb. ex DC. Fabaceae) is large deciduous tree with loose spreading crown. Livestock eats the young branches and foliage of this timber tree. After teak, it is the most important cultivated timber tree in India, planted on roadsides, and as shade trees for tea plantations. Sissoo makes first class cabinetry and furniture. It is used for plywood, agricultural, and musical instruments, skis, carvings, boats, floorings, etc. The leaves are used for fodder. This is a moderately fast growing tree that adapts well to semi arid conditions and produces first-class firewood (roughly 5,000 kcal per kg). Unfortunately sissoo is known to suffer from a number of diseases. Major diseases of sissoo are listed in Table 5.

Among diseases listed above some are responsible for huge losses regularly and are described in detail.

Table 5 Diseases of *D. sissoo* with their causal organism in Central India

S.No.	Name of disease	Causal organism
1.	Leaf spots	*Botryodiplodia theobromae, Cercospora sissoo, Colletotrichum dematium, gloeosporioides, Phoma nivea, Phomopsis dalbergiae*
2.	Leaf rusts	*Maravalia achroa, Uredo sissoo*
3.	Wilts	*Fusarium oxysporum, F. solani* f *dalbergiae*
4.	Powdery mildew	*Phyllactinia dalbergiae*
5.	Root rot	*Ganoderma lucidum*
6.	Stem & twig disease	*Helminthosporium dalbergiae*

Leaf rusts

Maravalia twig and leaf rust is widespread in the country and has also been reported in serious form in forest nurseries (Mehrotra and Pandey, 1993). The disease appear in early March in nurseries and young plantations with formation of yellow uredospores on young and juvenile leaves. Later in the season teleospores are produced. The incidence of the disease in nurseries is almost hundred per cent and extent of damage to seedlings is alarming, particularly under shade and humid conditions. The infected seedlings show leaf malformation and become weak and stunted.

Control: The disease may be controlled by application of Bayleton (0.08%) or Zineb 0.2%.

Uredo sissoo attacks sissoo of all ages in nurseries, plantations and natural forests. Uredia appear in November on the lower surface of leaves as minute, powdery, densely scattered pustules surrounded by numerous paraphyses. Rust sori mature by January when infected leaves fall off along with uninfected leaves during normal leaf fall.

Control : The disease may be controlled by 3-4 applications of Zineb (0.2%).

Powdery mildew

Symptoms appear as white powdery coating on leaves. Lower surface of sissoo leaves is attacked resulting in stunting distortion, chlorosis and defoliation.

Control : Dusting sulphur powder keeps the disease in check.

Wilts

Wilt disease is prevalent throughout the country in nurseries and plantations raised in unsuitable sites. The disease is absent in natural forests and in plantation on suitable sites.

Symptoms: Wilt disease is systemic in nature and manifests itself during July to September. Symptom of the disease includes yellowing and death of leaves in acropetal succession up the tree. Eventually the entire tree presents a yellow appearance. In later stages, leaves drop off rendering the branches increasingly bare. Affected trees die with in a few months. In nurseries the disease results in pre emergence rot and post emergence damping-off of seedlings, In diseased roots vascular browning is another pronounced symptom. Xylem vessels are plugged with hyphae and infiltrated with jelly-like substances, which hinder sap flow to the crown resulting in wilt symptom.

Control:

1. Wilt of sissoo is common in plantations raised on stiff and clayey soil with inadequate drainage. Therefore, proper selection of site with loose texture and good drainage can keep the plantations healthy, free from wilt disease (Bakshi *et al.*, 1972).
2. Soil drenching with Bavistin 0.2% has been reported to control the disease.
3. Application of biocontrol agents (*Trichoderma* spp.) with AM fungi and *Rhizobium* has been found to control the disease effectively in nurseries (Singh *et al.*, 2002).

Root rot

Root rot in sissoo due to *Ganoderma lucidum* is common both in plantations and natural forests. Trees of advanced age are generally attacked by the pathogen. The disease is prevalent in both light and heavy textured soils, however, the pathogen spreads rapidly on roots in light soils and trees are killed in a short period.

Disease spread and damage: Infection usually comes from the residual stumps and roots of the previous forest cover. *G.lucidum* infects roots through intact as well as injured surfaces. The pathogen attacks and kills the bark and cause white spongy rot in the sapwood. Young plants are killed soon after infection. Mature trees are also killed when most of the roots become infected. The affected trees exhibit a stag headed appearance in which condition they may continue for a few years before being killed. Infected plants can be located easily because of the presence of fruit bodies of the pathogen, which commonly develop on affected plants, usually at the base.

Control:

1. **Mechanized plantations:** The disease risk is minimized to a large extent by raising mechanized plantations where stumps and roots are mostly removed. However, if any mortality occur because of infection from left over infected roots serving as source of infection, the dead trees should be removed along with decayed roots.
2. **Removal of stumps:** If it is not possible to raise mechanized plantation, the incidence of the disease can be minimized, if the stumps and roots present in lines (line planting) are removed completely.
3. **Isolation trenches:** In sissoo plantations, the lateral spread of the disease can be checked by isolating the diseased and dead plants from the adjacent healthy plants by digging trenches about 0.3 m wide, 0.7 m deep and 1.5m long across the line. It may be advisable to dig an additional trench,

which should include one apparently healthy plant (as a buffer) immediately after the first trench. Such plant may harbour initial infection but the inoculum potential may not be up to the extents of producing disease symptom. This will provide an additional safe guard in containing the disease.

4. **Planting resistant species:** In a reforested stand where the *Ganoderma lucidum* root rot mortality is apprehended, it may be advisable not to plant susceptible species like sissoo and khair, in the first rotation but to raise resistant species, like *Ailanthus* and Semul having rotation age of about 10 years and 40 years.
5. **Mixed stands:** The age-old practice to control root rot mortality is to raise mixed stands in which the resistant species are mixed with susceptible ones. It is therefore, recommended to plant sissoo in intimate mixtures with resistant species, like semul and ailanthus, which will break the continuity of a pure line of sissoo and thereby prevent rapid spread of the disease in the plantation.

D. *Leucaena*

Leucaena leucocephala is a leguminous species. Fuelwood, poles, and pulpwood are among the species' most useful byproducts. In addition, the tree's rapid growth makes it an excellent choice for live fences and erosion control. It is also the most intensively researched, widely used and best known forage tree. Many disease caused by fungi have been recorded on *Leucaena* species throughout the tropics.

Foliage diseases

Camptomeris leaf spot caused by *Camptomeris leucaenae,* reduces forage production and quality of *L. leucocephala*. The pathogen is specific to *Leucaena,* characterised by chlorotic patches on upper surfaces of leaflets and blotches of profuse sporulation in crowded black pustules on lower surfaces. Coalescence of lesions results in chlorosis and defoliation. The protein content of severely affected leaves was reduced by 18%.

Stem diseases

Gummosis, the exudation of gum from the main trunk, stems and branches of *Leucaena* spp., with associated leaf senescence, dieback and stunted growth, is described as the most serious disease of *L. leucocephala* (NAS, 1984). Yet, little is known of its pathology and economic importance (Lenne, 1991). Some researchers consider gummosis to be a physiological phenomenon (Hegde, 1984).

Root rots

Root rots caused by *Ganoderma* and *Fusarium* spp. have been reported on *L. leucocephala* (Raina 1980, 1983, Pathak, 1986). *Ganoderma lucidum* causes wilting, drying of apical meristems, stem blackening, defoliation, root rot and tree death. Seven year-old trees had 10-15% mortality at moist sites (Pathak, 1986).

E. *Prosopis juliflora* (mesquite):

P. juliflora, is best suited for very warm, dry climates. This fast-growing species makes excellent fuelwood and charcoal. Additional useful byproducts include timber, forage, and honey.

A large number of fungi have been recorded on *P. juliflora* but no information is available on their importance. These include leaf spots caused by *Pleospora scirrhioides* and *Septoria prosopodis;* dieback caused by *Diplodia dalbergiae, D. prosopidina, Nectria flavistroma* (Browne, 1968).

F. *Pongamia pinnata* (karanj)

Pongamia pinnata (L) Pierre is distributed through out India. Wood is used as fuel. The seeds yield a thick oil used for burning and for skin diseases. Important diseases of *P. pinnata* are given in Table 6.

Leaf spots

Leaf spots caused by *Cercospora pongamiae, Colletotrichum gleoeosporioides* and *Phyllosticta pongamiae* are commonly found in nurseries of central India. These disease may be controlled by application of Dithane M-45 or Blitox @ 0.2%.

Table 6 Diseases of *Pongamia pinnata* with their causal organisms in nurseries and plantation of central India.

S.No.	Name of disease	Causal organism
1.	Leaf spots	*Cercospora pongamiae, Colletotrichum gleoeosporioids, Phyllosticta pongamiae*
2.	Leaf blight	*Fusicladium pongamiae (=Asperisporium pongamiae)*
3.	Leaf rust	*Ravenelia hobsonii*
4.	Powdery mildew	*Oidium* sp.
5.	Anthracnose	*Sphaceloma pongamiae*
6.	Dieback	*Haplosporella beaumontiana*
7.	Pod rots	*Colletotrichum* sp., *Dothiorella indica*
8.	Root rot	*Ganoderma lucidum*

Fusicladium leaf blight

The disease has been reported to occur in severe form from Jabalpur Madhya Pradesh (Jamaluddin *et al.*, 1981). It appears in the nursery beds during rainy season, particularly in the months of August when there is sufficient moisture in the atmosphere. The severity increases with subsequent rainfall.

Symptoms: Mostly younger leaves of plants are heavily infected. Infection usually starts from the margin or tip of the leaves, initially scattered brown - dark brown lesions form on lower side of leaf lamina which coalesce to form large pustules as the disease advances. These pustules produce heavy sporulation. The upper surface of the leaves become chlorotic. In severe cases leaf curling may also occur.

Dieback

Haplosporella beaumontiana causes dieback disease in young plantation (1-5 years) and 11% plants showed die back symptoms (Verma *et al.*, 1997).

Control: The diseases can be control by the application of Bavistin 0.2%.

Pod rot

The pod rot of *Pongamia pinnata* caused by a complex of three fungi, namely *Dothiorella* sp., *Colletotrichum* state of *Glomerella cingulata* and *Phoma* sp. has been reported from Madhya Pradesh (Jamaluddin *et al.*, 1983).

Symptoms: The symptoms caused by *Dothiorella* sp. appeared in the form of black dots on both the surfaces of pods, which coalesced to form big patches of irregular size. Few black minute dots were also produced by *Phoma* sp. on pod surface. Infection by *Colletotrichum* sp. started as small, pin head sized yellowish dots, usually one mm diam. These dots later on coalesced to form big sized light brown oval-irregular zones on the surface.

Control: Dithane M45 and Difolatan @0.3% may control the disease under field condition.

2. Diseases of Teak (*Tectona grandis*)

Teak is indigenous to India, Burma, Thailand, Indonesia and other islands in the far east. It is a large deciduous tree with a tall and cylindrical bole and rounded crown. It is highly valued for its timber quality. The major diseases of teak occurring in central India are given in Table 7.

Table 7 Diseases of teak with their causal organisms in nurseries and plantation in central India.

S.No.	Name of disease	Causal organism
1.	Leaf spots	*Cercospora tectonae, Dothiorella* sp., *Phyllosticta tectonae, Phoma exigua*
2.	Wilt	*Fusarium moniliforme*
3.	Rust	*Olivea tectonae*
4.	Powdery mildrew	*Uncinula tectonae*
5.	Root rot	*Helicobasidium compactum*
6.	Basal canker	*Fusarium pallidoroseum*

Leaf spots

Leaf spot caused by *Cercospora tectonae* is common in Central India. The pathogen produces dull brown to greyish-brown or white angular spots on teak leaves. Later on spots coalesce to form large blotches. Spray of Dithane M-45 (0.2% has been recommended to control the disease.

Leaf rust

The leaf rust disease (*Olivea tectonae*), is widely distributed throughout the country. The disease is especially common in the nurseries. It is also found in plantations. Leaves are commonly attacked usually from October to February. At first Uredia develop followed later by telial sori which may be intermixed with Uredia or produced in separate sori. Both sori are orange yellow and occur on the lower surface. The upper surface of the leaf may present gray appearance corresponding to the position of sori on the lower surface. Infection causes premature defoliation, which results in retardation of plant growth in early age. The rust was earlier known only in the uredial stage and described as *Uredo tectonae* by Raciborski in 1900. Later on based on telial stage, Thirumalachar placed the rust under *Olivea.*

Control: The teak rust can be controlled by application of sulfur based fungicides such as sulfex @ 0.2%.

Powdery mildew

Powdery midlew is common on teak in central India. It is wide spread in teak nurseries and forests. The disease begins to appear soon after the rains. The upper surface of the infected leaves are covered with white mycelium containing black fruiting bodies of the fungus and present bluish-gray mealy appearance. In the teak forest, nearly all trees may become infected. Due to heavy coating of leaves by fungus growth, photosynthesis activity is retarded which may affect growth. The mycelium sends in haustoria which are restricted in the upper epidermal cells only. Conidia which develop abundantly on the leaf surface are wind dispersed to cause fresh infection.

Control: The disease can be controlled by spraying with Calixin or Sulfex @ 0.2%.

Root rot

The disease is known to occur both in nursery and also in newly established plantation up to age of two years particularly on sites with poor drainage. The disease declines with the maturity of plants.

The fungus forms an extensive mycelium in the soil, hyphae from which run up to the roots of plants, penetrate the tissue and form a cushion - like covering, brownish-violet when dry and dark purple-brown when wet, on the base of the stems, often girdling them. The symptom includes decline of the foliage, production of sprouts followed by death of the plant.

Control

1. Well drained suitable sites should be selected for nurseries. In plantations where the disease may be severe mostly on low lying and poorly drained soil, trenching should be done to isolate diseased patches.
2. Dead and dying trees should be removed to check the spread of the disease.

Basal canker

The disease was noticed in one-year plantation at Hosangabad district of Madhya Pradesh. The disease mostly occurs in waterlogged areas and may cause 20-30% damage. The plants started drying from top to the base. (Dadwal and Jamaluddin, 2001).

Control: The root shoot should be treated with 0.1% Bavistin before planting in the field. After shoot development one spray with 0.2% Dithane M-45 during Oct./Nov. is advisable.

3. Diseases of Sal (*Shorea robusta* Gaertnf)

Sal, a fairly large deciduous tree with a majestic shining foliage is one of the most important trees of South Asia, found extensively in India and also in parts of Bangladesh and Myanmar. It is the most important timber tree of Central and northern India. Sal grows gregariously and tends to form a dense crop (Tewari, 1995). Important diseases of sal as reported from central India are given in Table 8.

Table 8 Diseases of Sal (*Shorea robusta*) with their causal organisms in forests of Central India.

S.No.	Name of Disease	Causal organisms
1.	Root rot	*Polyporus shoreae*
2.	Heart rot	*Hymenochaete rubiginosa*, *Fomes caryophylli*, *F. fastuosus*
3.	Stem canker and witches' Broom	*Macrophoma shoreae*
4.	Leaf spot	*Cladosporium* sp., *Lembosea shoreae*

Polyporus shoreae Root rot

Polyporus shoreae is a serious root parasite on sal and occurs throughout its range of distribution (Jamaluddin, 1991). It is not commonly recorded on other hosts except rarely on *Shorea assamica* Dyer (Bagchee, 1959) and *Pometia pinnata* Forst. It is sporadic in drier types of sal, becoming common in locations with higher rainfall.

Infection and symptoms: The fungus infects the roots first at the distal end. The infection progresses up the root to the collar region and spreads from the disease tree to adjoining healthy trees by root contact. Affected trees don't exhibit symptoms till some roots are attacked. The remaining healthy roots are unable to meet the water requirement of the crown, particularly in big trees, which show symptoms of top dying. Dying of branches gradually extends downwards till the death of the tree in a few year. Commonly such trees become wind blown due to decay of roots. Small light yellow mycelial felts of the fungus develop on bark of roots. In some cases the trees may put forth epicormic branches. The sporophores of the fungus are usually formed on affected trees at the base or on exposed roots of wind blown trees during the rainy season. Sporophores are sessile, rarely sub stipitate, dimediate, sometimes funnel shapped, soft when fresh becoming brittle when dry, single or imbricate, upper surface is brown to black, glabrous, hymenial surface is brown to dark brown (Bakshi, 1971).

Control:

1. **Controlled burning:** In wet sat forest, controlled burning is an effective method of reducing soil moisture and thereby minimizing infection by the fungus.
2. **Dry marking:** The dead trees and also trees in advanced stages of the disease should be removed and the timber utilized as it remains sound.

Heart rot

Sal is one of the most naturally durable timbers in service and is not easily subjected to decay. However, some fungi such as *Hymenochaete rubiginosa, Fomes caryophylli* and *Fomes fastuosus* cause heart rot of economic significance in sal. An appraisal of loss of timber due to decay carried out by Bakshi and Singh 1970 revealed that out of 704 trees examined 70% showed heart rot and total loss of timber was 10%.

Infection and Symptoms: In high sal forests, heart rot establish in trees from trunk as well as butt infections, while in coppice forests, they establish mainly on the stool from where they progress up the tree when the heart wood develops in coppice shoot. Wounds caused due to fire injury, frost injury, breaking of branches, lightning, browsing by animals constitute the main avenues for infections. Seventy to eighty per cent infections in sal may occur through branch stubs. In advanced stages, developments of sporophores on the host indicates decay in all cases. Punk knots (soft and punk cylindrical cores of decayed wood) are a characteristic symptom of attack by *Fomes caryophylli* in sal. Dead branches, knots and branch stubs, which constitute infection courts, may also indicate decay.

Two other heart rot causing fungi namely *Trametes cubensis* (Mont.) Sacc. and *T. incerta* (Curry) Cooke, have also been recorded but they are not as common as the fungi listed above.

Control: Losses due to heart rot in sal may be prevented or minimized through management practices and silvicultural measures. In areas where frost is common, the injury to the young crop may be prevented by adjustment of the canopy. Fire protection measures including control burning should be adopted. Thinning and improvement felling should always be carried out with a view to stand improvement when deformed, crooked, injured, diseased and dead trees should also be removed from the stand. During cultural operations, all trees showing injury due to fire, frost and also those exhibiting sporophores and punk knots should be removed.

Stem canker and Witches' Broom

Symptoms: The symptom of the disease includes knoted twigs, prominent cankers with terraced callus and formation of witches' broom.

Pycnidia of *Macrophoma shoreae* are embedded in the cracked bark, dark brown to black, spherical-oblong and produce one-celled spores.

Control: Removal of diseased twigs and spray of Dithane M45 @ 0.2% can minimize the disease.

4. Diseases of *Eucalyptus*

Eucalyputs L. Herit is indigenous to Australia. Because of the wide adaptations as regards site conditions and fast growth in exotic locations, *Eucalyptus* have been introduced in many countries including India. There are about 500 species mostly Australian. Many are being cultivated in India. *Eucalyptus teriticornis* Sm. is being raised in large-scale plantations in most states. *E.grandis* (Hill) Maiden is suited to high altitude and high rainfall locations. *E.globulus* Labill introduced in the Nilgiris in the 19th century has naturalized in the area. *E. citriodora* Hook is also planted on a large scale. Apart from meeting demand for timber and fuelwood, *Eucalyptus* is planted to meet the increasing demand of industrial wood, particularly for the paper industry. However, some diseases have become limiting factor and it is reported to be victim of several fungal diseases both at seedling stage and standing tree (Bakshi 1976). Important diseases of *Eucalyptus* as reported from different parts of central India are given in Table 9.

Table 9 Diseases of *Eucalyptus* with their causal organisms in nurseries and plantations.

S.No.	Name of disease	Causal organisms
1.	Leaf spot and seedling blight	*Cylindrocladium scoparium*
2.	Leaf spots	*Alternaria alternata, Curvularia lunata, C. prasadi, Cercospora eucalypti, Phyllosticta eucalypti, Phaeoseptoria eucalypti.*
3.	Damping off	*Pythium* sp., *Verticillium* sp.
4.	Twig and stem cankers	*Cytospora agarwalii, Dothiorella indica, Hendersonula toruloidea*
5.	Pink disease	*Corticium salmonicolor*
6.	Wilt	*Fusarium oxysporum*
7.	Root rots	*Macrophomina phaseolina, Sclerotium rolfsii, Ganoderma lucidum.*

Cylindrocladium leaf spot and seedling blight

Symptoms: The pathogen infects leaves, young twigs and also the main stems of one-year-old plants. Infection on leaves results in premature defoliation where as infection of twigs and main stem results in wilting of plants. Symptoms appear as dots like spots with slight discoloration, which deepens to become grayish black. Necrotic tissues present a water soaked appearance. The disease is severe in warm and humid weather when the white mycelial growth with profuse sporulation develops on affected parts. The incidence of disease declines in dry weathers.

Control: The disease may be controlled through spraying with chemicals like Dithane M45 or Fytolan @0.2%. Other leaf spot diseases caused by *Alternaria alternata, Curvularia lunata, Cercospora eucalyptii, Phyllosticta eucalypti* and *Phaeoseptoria eucalyptii* may be controlled by spraying with Dithane M45, Fytolan or Blitox 50 @ 0.2%.

Pink disease

Corticium salmonicolor, the cause of pink disease is widely distributed throughout the world and has a wide host range.

Damage: As exotics in India, species of *Eucalyptus* have not developed any resistance against the pathogen, which occurs naturally on a wide range of hosts in the natural forests. All the species of *Eucalyptus,* which are widely planted in India, are highly susceptible to the disease which may appear in epidemic form under condition of high rainfall and high temperature during the summer. Eucalyptus is susceptible to the disease at a very early age. Young plants may show repeated dieback and mortality, while those which have escaped infection during the early age, may at a later period become infected resulting in development of cankers.

Symptoms: Branches and the main stems of young plants are girdled due to infection during the later part of rainy season. The affected plants are killed Plants escaping infection at an early age put up a good height and growth, however, such trees may subsequently become infected on the main bole resulting in splitting and cracking of bark and exposure of wood. Gum exudes from the resultant cankers. Trees may continue to grow as the attack is localized and girdling is usually not complete.

Control:

1. In locations where disease incidence due to *Corticicum salmonicolor* is very high particularly in young plants up to four years of age, 2-3 prophylactic spray with suitable fungicides during the most vulnerable period of infection may afford protection to plants against infection.
2. Infected plants should be removed.

Ganoderma root rot

Different species of *Eucalyptus* are susceptible to attack by *Ganoderma lucidum* root rot. Attacked seedings are killed. In trees, the fungus causes decay in roots due to which plants loose footholds in soil and become wind blown. Roots and stumps of the previous crop decayed by the pathogen serve as infection centers for Eucalyptus planted under reforestation programmes.

Control: *Ganoderma* root rot in *Eucalyptus* may be minimized or kept under control by mechanized plantations, extraction of stumps, digging isolation trenches, planting resistant species and mixed stands.

Charcoal root rot

Symptoms: *Macrophomina phaseolina* causes a serious root and stem rot in juvenile seedlings of *Eucalyptus* in nurseries (Jamaluddin *et al.*, 1987). Initially the leaves become yellowish turn brown and dry up. In advanced stage death of the plants takes place. Infected roots and the stem exhibit charred appearance due to development of black dots like sclerotia and pycnidia on the surface.

Control: The disease may be minimized by soil drenching with Bavistin 0.1%

Twigs and stem diseases

These diseases have been reported by Jamaluddin *et al.* (1987) from Madhya Pradesh.

Symptoms: Twigs and stems of *Eucalyptus* suffer from canker, blight and dieback symptoms, which result in the death of the affected branches, and in some cases of the whole plant. Symptoms include the development of large number of epicormic branches on the affected plants.

Control: The disease can be successfully controlled or minimized by removing the affected twigs, branches or whole plant according to the nature of infection or damage caused therein.

5. Diseases of Neem

Neem (*Azadirachta indica* A Juss.) is widely cultivated all over India. Recently it is being raised by forest department and other agencies on large scale for its diverse multipurpose uses, particularly for its medicinal and pesticidal properties. Important diseases of neem occurring in central India are listed in Table 10.

Leaf spot pathogen *Pseudocercospora subsessilis* occurs widely and cause heavy losses through out the natural distribution of this tree species (Tewari, 1992). Both leaf spot and tip blight diseases may be controlled by application of Fytolan 0.2% and Blitox 0.2% (Jamaluddin *et al.*, 1994).

Wilt caused by *Fusarium oxysporum* and *F. solani* affect mainly the seedlings of 1-2 months age. Symptoms include yellowing of leaves, defoliation

Table 10. Diseases of neem in central India

S.No.	Name of disease	Causal organism
1.	Leaf spot	*Pseudocercospora subsessilis, Cercospora leucosticta, Leptosphaeria azadirachtae.*
2.	Tip blight	*Colletotrichum gloeosporioides*
3.	Wilt	*Fusarium oxysporum, F.solani.*
4.	Powdery mildew	*Oidium azadirahtae*
5.	Web blight	*Rhizoctonia solani*
6.	Root rot	*Ganoderma lucidum*

and wilting of plants completely. The disease may be controlled by soil drenching with Bavistin 0.05%. Leaves of plants infected with powdery mildew pathogen appear to be coated with white powdery fungal growth. Disease appears usually after the rainy season. In severe infection plants become shrivelled and defoliated. Disease can be controlled by application of Sulfex 0.2%

Leptosphaeria azadirachtae causing twig and stem canker in young plantations of neem has been reported from Orissa (Verma *et al.*, 2001).

6. Diseases of *Gmelina arborea* (Khamer)

Gmelina arborea Roxb. occurs scattered throughout greater part of India in deciduous and moist deciduous forests. The wood is excellent for furniture. It is also used for ship building. Diseases occurring on this species are given in Table 11.

Diseases caused by *Phoma tropica, Leptosphaeria gmelinae, Alternaia, alteranata* and foot rot by *F oxysporum* causing serious damage in nurseries have been reported (Jamaluddin *et al.*, 1988). Top dying and mortality in

Table 11. Diseases of *Gmelina arborea* from central India

S.No.	Name of disease	Causal organism
1.	Leaf spots	*Phoma tropica, Leptosphaeria gmelinae, Alternaria alternata, Helicomina microphora, Microdiplodia zizyphae.*
2.	Anthracnose	*Colletotrichum capsici*
3.	Powdery mildew	*Phyllactenia suffulta va. gmelinae*
4.	Foot rot	*Fusarium oxysporum*
5.	Top dying	*Phomopsis* sp.

provenence trial plantation caused by *Phomopsis* sp.was studied in detail (Harsh, *et al.* 1992). Diseases can be controlled effectively with the application of Bavistin and Dithane M45 @ 0.2%.

7. Diseases of Bamboos

Bamboos are common man's timber having diversified uses. It is used in house building, for making thatches and scaffoldings. In paper industry it is still the largest source of pulp wood in India. It suffers from diseases, which attack the rhizome, roots, stem, culm foliage, flowers and seeds. Important diseases of bamboos occurring in central India are listed in Table 12.

The diseases listed in Table 12 have been recorded mainly from four widely distributed species belonging to two genera, namely *Bambusa arundinacea, B. nutans, B. vulgaris* and *Dendrocalamus strictus.*

Culm rot and sheath blight

This disease is of particular significance as it is being reported to occur in moderate to severe form from different part of the country. Heavy mortality in *Bambusa nutans* plantations in coastal areas of Orissa due to this disease by *Sarocladium oryzae* has been reported (Jamaluddin *et al.*, 1992). The disease causes heavy losses to the farmers in coastal belts of Orissa. Same pathogen has been reported to cause a serious dieback of *B.balcoa, B. tulda* and *B. vulgaris* from Bangladesh (Boa and Rahman, 1987, Rahman, 1988) and is suspected that it has spread to West Bengal in areas adjoining Bangladesh.

Table 12 Diseases of bamboos with their causal organisms

S.No.	Name of disease	Causal organism
1.	Leaf spots	*Alternaria alternata, Curvularia lunata, Colletotrichum gloeosporioides, Exserohilum rostratum Botryodiplodia theobromae*
2.	Wilt	*Fusarium oxysporum*
3.	Collar rot	*Rhizoctonia solani*
4.	Leaf rust	*Dusturella divina, Puccinia melanocephala*
5.	Culm rot	*Fusarium equiseti*
6.	Root and clum rot	*Fusarium pallidoroseum*
7.	Culm rot and sheath blight	*Sarocladium oryzae*
8.	Culm sheath rot	*Glomerella cingulata*
9.	Root rot	*Ganoderma lucidum*

Control

Leaf spot diseases of bamboos listed above may be controlled or minimized with the application of Dithane M45 of Fytolan @ 0.2%. wilt may be controlled by soil drenching with Bavistin @ 0.05%, leaf rust by spraying Dithane Z-78 0.2%; culm rot and sheath blight by spraying Bavistin 0.1% and Dithane M45 0.3%. Root rot caused by *G. lucidum* may be minimized or kept under control by removing infected rhizome and fruit bodies of the pathogen.

8. Diseases of Other trees

Adina cardifolia

Cercocladospora adinae Agarwal and Singh, *Cercospora adiniicola* Kar and Mandal are reported from Jabalpur *Marssonia poonensis* Patwardhan and Pande

Anogeissus latifolia (Roxb) Bedd

Pestalotiopsis versicolor (Speg.) Steyaert and *Monochaetia jabalpurensis* on leaves were recorded from Jabalpur, M.P. *Marssonina poonensis* Patwardhan and Pande was recorded from MS.

Cassia siamea Lam.

Cercospora cassiae-siameae Chiddarwar is recorded on leaf from Pune, MS

Diospyros melanoxylon Roxb.

Aecidium rhytismoideum Berk & Br is reported from Chanda, MS.

Hardwickia binnata

Wood damage caused by *Schizophyllum commune.*

Holoptelea integrifolea Planch

Leaf diseases caused by *Cercospora holopteleae* Chiddarwar from Khandala, *Mycosphaerella holopteleae* Naphade from Pune, MS and *Colletotrichum holopteleae* Hasija from Jabalpur, MP are reported.

Lagerstroemia parviflora Roxb

Cercospora lythracearum Heald & Wolf from Khandala, MS and *Sphaceloma lagerstroemiae* Wani & Thirum. from Pune, MS are reported.

Sterculia urens

Excipularia naraspurensis Subram. causes bark disease on this tree.

Terminalia arjuna W & A:

Colletotrichum arjunae Sahni from Jabalpur and *Sphaceloma terminaliae* Bitanc from Khandla, MS are reported.

T.bellarica Roxb.:

Phyllactenia terminaliae Ramkr. TS from Khandla, MS *Pestalotiopsis terminaliae* Agarwal & Hasija from Jabalpur.

T. chebula Retz.:

Uredo terminaliae P. Henn., *Phyllactenia terminaliae* and *Cercospora catappe* P. Henn. from Mahabaleswar, MS.

T. chenulata Heyne ex Roth.:

Phyllactenia terminaliae from Sinhgad, MS and *Cercospora catappe* from Mahabaleshwar, MS.

Syzyzium cumini (L) Skeels:

Elsinoe kamatii Tewari from Mahabaleshwar, *Maublancia indica* Pande and Patwardhan from MS, *Selenophoma kamatti* Kalani from Pune, MS are reported on leaves.

No disease is reocrded on *Ailanthus excelsa* Roxb and *Bridelia retusa* Spreng from central India.

Future Research Needs

Since India is one of the 12 megadiversity countries of the world and nearly 18% of the countries bio-wealth is constituted by fungi alone, the structure and dynamics of microbial community by fungi in soil are believed to influence many facets of forest productivity including the production of quality seedlings and plantations yield and development of diseases. Considering the importance of microorganisms in forestry, only limited information is available on what types of microorganisms exists in forest and forest soils and how they influence productivity. Therefore, there is need to intensify researches on identification of microorganism using recombinant DNA techniques and molecular phyllogeny. This will help in better understanding the microbial community, which may oepn up the possibilities of manipulation of specific organisms to improve quality of seedllings and forest yield and better manage the diseases.

The extent of knowledge of diseases of trees in the tropics is generally poor. Knowledge of diseases is mainly restricted to records in host lists with limited information on importance and pathogenic variability, and even doubts regarding the true status of the 'pathogen'. Conflicting reports of the importance of diseases have been made from the same country. There are many important species for which little or no pathological information is presently available. Limited work has been done to assess losses caused by known diseases and the economics of control.

References

Bagchee, K. (1959). The fungal diseases of sal (*Shorea robusta* Gaertnh) V. The heart rot of sal caused by *Trametes incerta* (Currey) Cke. Indian Forest Record (NS) Mycol, **2** : 61-69.

Bakshi, B.K. (1971). Indian Polyporaceae on trees and timber 246pp. Indian Council Agricultural Research, New Delhi.

Bakshi, B.K. (1976). Forest Pathology: Principles and Practice in Forestry, 400 pp Delhi, India.

Boa, E.R. and Rahman, M.A. (1987). Bamboo blight and the bamboos of Bangladesh. Bulletin *1 Forest Pathology Series, Forest Research Institute, p43.*

Browne, F.G. (1968). Pests and diseases of forest plantation trees: an annotated list of the principal species occurring in the British Commonwealth. Clarendon Press, Oxford University Press, Oxford.

Dadwal, V.S., Soni, K.K. and Jamaluddin. (1986). Rhizosphere microflora of teak (*Tectona grandis*) *Indian Journal of Forestry*, **9**: 59-62.

Dadwal, V.S. and Jamaluddin. (2001). A note on basal canker of teak (*Tectona grandis*) in plantations. Indian forester **127**: 365-366.

Harsh, N.S.K., Jamaluddin and Tiwari, C.K. (1992). Top dying and mortality in provenence trial platation of *Gmelina arborea.* J.Trp. Forestry, **8**: 55-61.

Harsh, N.S.K., Soni, K.K. and Tiwari, C.K. (1993). Gandoerma root rot in *Acacia arboretum* in Madhya Pradesh. European Journal of Forest Pathology, **23** *: 252-255.*

Hegde, N.G. (1984). Gummosis in *Leucaena leucocephala.* Leucaena Research Reports **5** : 31 - 32.

Hughes, C.E. and Styles, B.T. (1989). The benefits and risks of woody legume introductions. In: Stirton, C.H. and Zarucchi, J.L. (eds), Advances in Legume Biology. Monograph Systematic Botany, Missouri Botanical Garden, pp. 505-531.

Jamaluddin. (1991). Status of *Polyporus shoreae* causing root rot in dry and wet sal forest. Jounal of Tropical Forestry, **7** : 342-344.

Jamaluddin, Dadwal, V.S. and Soni, K.K. (1981). Studies on Fusicladium leaf spot in *Pongamia pinnata*. Indian Forester, **107** : 500-504.

Jamaluddin, Dadwal, V.S. and Soni, K.K. (1983). Studies on pod rot of *Pongamia pinnata* and its control. Indian Journal of Forestry, **6** : 287-288.

Jamaluddin, Soni, K.K. and Dadwal, V.S. (1987). Some noteworthy diseases of *Eucalyptus* in Madhya Pradesh. Indian Journal of Forestry, **10** : 55-57.

Jamaluddin, Dadwal, V.S. and Soni, K.K. (1988). Some new and noteworthy diseases of *Gmelina arborea* from MP. Journal of Tropical Forestry, **4** : 297-299.

Jamaluddin, Verma, R.K. and Nath, V. (1994) Diseases occurring on different provenances of neem in central India and their control. Indian Journal of Tropical biodiversity, **2** : 466-468.

Jamaluddin, Gupta, B.N., Bohidar, S.C. and Dadwal, V.S. (1992). Mortality of bamboo (*Bambusa nutans* Walt) in coastal areas of Orissa. Journal of Tropical Forestry, **8** : 252-261.

Kar, A.K. and Mandal, M. (1969). New *Cercospora* species from West Bengal I. Transactions of British Mycological Society, **53** : 337-360.

Lenne, J.M. (1991) Diseases of Leucaena species. Tropical Pest Management, **37**: 281-289.

Mehrotra, M.D. (1990) *Rhizoctonia solani,* potentially dangerous pathogen of khasi pine and hardwoods in forest nurseries. European Journal of Forest Pathology, **20**: 329-338.

Mehrota, M.D. and Pandey, P.C. (1993). *Maravalia* twig and leaf rust of sissoo and its management. Indian Forester, **119** : 955.

Mehrotra, M.D. and Verma, R.K. (1993). *Mycovellosiella ravenaliae* a hyper parasite of *Ravenelia clemensiae* from India. Mycological Research, **97** : 1020.

NAS, (1984) Leucaena: Promising Forage and Tree Crop for the Tropics. Second Edition. National Academy Press, Washington, DC 115 pp.

Nair, K.S.S., Sharma, J.K. and Varma, R.V. (1996). The Proceedings of the IUFRO Symposium 23-26 November 1993 Pecehi, India, 521 p.

NFT, (1989) Why nitrogen fixing trees? NFT Highlights 89-03, NFTA, Hawaii.

Pathak, P.S. (1986). Mortality in *Leucaena* due to *Ganoderma lucidum.* Leucaena Research Reports **7**, 65.

Raina, A.K. (1983). Performance of *Leucaena* in the indian arid zones. 3. *Fusarium* gummosis and *Ganoderma* root rot. *Leucacna* Research Reports **4**: 35-36.

Raina, A.K. (1980) Epidemiology of *Ganoderma* root rot- a case study. Woodpower 80. In: Proceedings of the International Symposium of Forestry and Wood Utilization, Part 3 ((Abstract), June 1980.

Rahman, M.A. (1988). Perspective bamboo blight in Bangladesh. Indian Forester, **114**: 726-736.

Sharma, R.C. and Bhardwaj, L.N. (1988). Forest nursery diseases and their management. In: Ram Parkash (ed.) Advances in Forestry in India, International Book Distributors, Dehra Dun, India, Vol. II, pp. 91-118.

Soni, K.K. and Jamaluddin. (1990). *Eucalyptus* litter decomposition in tropical dry deciduous forests of Madhya Pradesh. Indian Forester **116**:286-291.

Singh, Y., Verma, R.K. and Jamaluddin. (2002). An integrated approach to control *Fusarium* wilt of sissioo. Indian Forester, **128** : 432-438.

Tewari, D.N. (1995). A Monograph on Sal (*Shorea robusta* Gaertn.f), 277 p. International Book distributros, Dehra Dun, India.

Tewari, DN (1992). A Monograph on neem (*Azadirachta indica* A Juss.), 279 p. International Book distributors Dehra Dun.

Verma R.K., Jamaluddin and Soni, K.K. (1997). Die back of leguminous trees in young plantations caused by *Haplosporella* in central India. Indian Forester, **123(7)**: 652-655.

Verma, R.K., Soni, K.K., Tiwari, C.K. and Jamaluddin, (2001). Two new ascomycetes from India. International Symp on Diversity of fungi in Asia. Gorakhpur Univ. 9-11 Feb.

40

Diversity of Important Fungal Biocontrol Agents and Their Commercial Application in Agriculture

Nasim Ahmad* and O. M. Bambawale

National Centre for Integrated Pest Management
Pusa Campus, New Delhi 110 012

Synthetic chemical pesticides were first developed during the mid-twentieth century and in due course became the only means of plant protection available to the farmers. Since early eighties, the deleterious effects of their over use and abuse were noticed in the form of pest resistance, resurgence of minor pests, disturbance of ecosystem due to destruction of natural enemies, environmental pollution and finally direct health problems in mammals. Now, most of the Universities and Research Institutions advocate the principles of Integrated Pest Management (IPM). An important component of IPM is biological control and now several microbial biopesticides are available which have a high potential to at least partly replaced the use of synthetic molecules. Amongst the microbials, the fungal control agents form a sizable part and we have an array of fungi, which can be utilized in the management of plant diseases, insect pests, plant nematodes and weeds. This review provides a broad perspective on the range of diversity of fungal biocontrol agents available for commercial exploitation, commercial products now in use, niches in which these work and limitation in their use with suitable examples. The fungal bioagents are expected to hold high promise in addressing some of our key pest problems in near future.

Key words : Diversity, Systematics, Fungal Biocontrol Agents, Commercial Application, Agriculture

Biological control of pathogens, that is, the total or partial destruction of pathogen populations by other organisms, occurs routinely in nature. There are for example, several diseases in which the pathogen cannot develop in certain areas either because the soil, called suppressive soil, contains microorganisms antagonistic to these pathogens or because the plant that is attacked by a pathogen has also been naturally inoculated with antagonistic microorganisms before or after the pathogen attacked. Sometimes, the antagonistic microorganisms may consist of avirulent strains of the same pathogen that destroy or inhibit the development of the pathogen, as happens in *hypovirulence* and cross protection. In some cases, even higher plants reduce the amount of inoculum either by trapping available pathogens (trap plants) or by releasing into the soil, substance toxic to the pathogens. Agriculturists

*Corresponding Author E-mail : nasimc@rediffmail.com

have increased their efforts to take advantage of such natural biological control against several plant diseases. Biological antagonisms, although subject to numerous ecological limitations, are expected to become an important part of the control measures employed against many more diseases.

Numerous kinds of antagonistic microorganisms have been found to increase in suppressive soils; most commonly, however, pathogen and disease suppression has been shown to be caused by fungi, such as *Trichoderma, Penicillium,* and *Sporidesmium,* or by bacteria of the genera *Pseudomonas, Bacillus,* and *Streptomyces.* Suppressive soil added to conducive soil can reduce the amount of disease by introducing microorganism antagonistic to the pathogen. For example, soil amended with soil containing a strain of a *Streptomyces* species antagonistic to *Streptomyces scabies*, the cause of potato scab, resulted in potato tubers significantly free from potato scab. In this chapter we will discuss potential fungal biocontrol agents of diseases and insect pests on economically important crops and their commercial exploitation in Indian agriculture.

Biological control practices for direct protection of plants from pathogens involve the deployment of antagonistic microorganisms at the infection court before or after infection takes place. Although hundreds, possibly thousands, of microorganisms have been shown to interfere with the growth of plant pathogens in the laboratory, greenhouse, or field, and to provide some protection from the diseases caused by them, so far only six microorganisms have been registered and are commercially available for use. The six microorganisms include three fungi: *Gliocladium virens,* for control of seedling diseases of ornamental and bedding plants; *Trichoderma harzianum,* for control of several soilborne plant pathogenic fungi and *Trichoderma harzianum / T. polysporum,* for the control of wood decays. The other three commercially available microorganisms are bacteria: *Agrobacterium radiobacter* K-84, for use against crown gall; *Pseudomonas fluorescsens,* for use against *Rhizoctonia* and *Pythium* damping-off of cotton; and *Bacillus subtilis,* used as a seed treatment. Although the actual use of these biological control products is rather limited, it is expected that these and other such products will find wide acceptance and will fill a real need in the not too distant future (Adams, 1990).

1. Fungal Antagonists of Plant Pathogens

Reducing Amount of Pathogen Inoculum through Antagonistic Microorganisms

Soil-borne Pathogens

The mycelium and resting spores (oospores) or sclerotia of several phytopathogenic soil fungi such as *Pythium, Phytophthora, Rhizoctonia, Sclerotinia,* and *Sclerotium* are invaded and parasitized (mycoparasitism) or are lysed (mycolysis) by several fungi, which as a rule are not pathogenic to plants. Several non-plant pathogenic fungi, including some oomycetes, chytridiomycetes, and hyphomycestes, and some *Pseudomonas* and actinomycetous bacteria infect the resting spores of several plant pathogenic fungi. Among the most common mycoparasitic fungi are *Trichoderma* sp., mainly *T. harzianum* and *T. viride.* The latter fungus has been shown to parasitize mycelia of *Rhizoctonia* and *Sclerotium,* to inhibit the growth of many other fungi, for example, *Pythium, Phytophthora, Fusarium,* and *Heterobasidion* (Fomes), and to reduce the diseases caused by most of these pathogens. Other common mycoparasitic fungi are *Laetisaria arvalis* (*Corticium sp.*), a mycoparasite and antagonist of *Rhizoctonia* and *Pythium.* Also, *Sporidesmium sclerotivorum, Gliocladium virens,* and *Coniothyrium minitants,* all destructive parasites and antagonists of *Sclerotinia sclerotiorum* effectively control several of the *Sclerotinia* diseases. *Talaromyces flavus* parasitizes *Verticillium* and controls *Verticillium* wilt of eggplant. Also, some *Pythium* species parasitize species of *Phytophthora* and other species of *Pythium.* Several yeasts, for example, *Pichia gulliermondii,* also parasitize and inhibit the growth of plant pathogenic fungi like *Botrytis and Penicillium* (Andrew, 1992).

Aerial Pathogens

Many other fungi have been shown to antagonise and inhibit numerous fungal pathogens of aerial plant parts. For example, *Chaetomium sp.* and *Athelia bombacina* suppress *Venturia inaequalis* ascospores and conidia production in fallen and growing leaves, respectively. *Tuberculina maxima* parasitizes the white pine blister rust fungus *Cronartium ribicola; Darluca filum* and *Verticillium lecanii* parasitize several rusts; *Ampelomyces quisqualis* parasitizes several powdery mildews; *Tilletiopsis* sp. parasitizes the cucumber powdery mildew fungus *Spaerotheca fuliginea;* and *Nectria inventa* and *Gonatobotrys simplex* parasitize two pathogenic species of *Alternaria.*

Mechanisms of Action

The Mechanisms by which antagonistic microorganisms affect pathogen populations are not always clear, but they are generally attributed to one of four effects:

(1) Direct parasitism or lysis and death of the pathogen.

(2) Competition with the pathogen for food.

(3) Direct toxic effects on the pathogen by antibiotic substances released by

the antagonist.

(4) Indirect toxic effects on the pathogen by volatile substances, such as ethylene, released by the metabolic activities of the antagonist.

Many of the antagonistic microorganisms mentioned above are naturally present in crop soils land exert a certain degree of biological control over one or many plant pathogens regardless of human activities. Humans, however, have been attempting; increase the effectiveness of antagonists either by introducing new and larger populations of antagonists, for example, *Trichoderma harzanium* and *Pasteuria penetrans*, in fields where they are lacking, and /or by adding soil amendments that serve as nutrients for, or otherwise stimulate growth of, the antagonistic microorganisms and increase their inhibitory activity against the pathogen. Unfortunately, although both approaches are effective in the laboratory and in the greenhouse, neither has been particularly successful in the field. New microorganisms added to the soil of the field cannot compete with the existing microflora and cannot maintain themselves for very long. Also, soil amendments, so far, have not been selective enough to support and build up only the populations of the introduced or existing antagonists. Thus, their potential for eventual disease control is quite limited. There are several cases of successful biological control of plant pathogens when the antagonistic microorganism is used for direct protection of the plant from infection by the pathogen.

Biocontrol of *Heterobasidion annosum* by *Peniophora gigantea*

Heterobasidion annosum, the cause of root and butt rot of conifers, infects freshly cut pine stumps and then spreads into the roots. Through the root contacts, it then spreads into the roots of the standing trees, which it kills. If the stump surface is inoculated with oidia of the fungus *Peniophora gigantea* immediately after the tree is felled, *Peniophora* occupies the cut surface and spreads through the stump into the lateral roots. There, it successfully competes with and excludes or replaces the pathogenic *Heterobasidion* in the stump, thereby protecting nearby trees. The oidia are applied to the cut surface either as a water suspension or as a powder or they are added to the lubricating oil placed on the chain saw and are thus deposited on the surface as it is cut.

Biocontrol of Chestnut Blight with Hypovirulent Strains of the Pathogens

Chestnut blight, caused by the fungus *Cryphonectria* (*Endothia*) *parasitica*, is controlled naturally in Italy, and artificially in France, through inoculation of cankers, caused by the normal pathogenic strains of the fungus, with hypovirulent strains of the same fungus. The hypovirulent strains carry virus like double-stranded RNAs (dsRNAs) that apparently limit the pathogenicity of the virulent strains. These dsRNAs apparently pass through mycelial anastomoses from the hypovirulent to the virulent strains, the latter are rendered hypovirulent, and the development of the canker slows down or stops. Hypovirulent strains are being tested in the United States also, but so far control of chestnut blight has been limited to experimental trees. It appears that the U.S. strains are more variable than the European ones and require multiple hypovirulence factors, therefore, the interaction is more strain-specific (Anagnostakis, 1982).

Biological Control of Soilborne Diseases

The principal fungi used as biological control agents against soilborne diseases include the two mentioned above as being used commercially, namely, *Gliocladium virens* and *Trichoderma harzianum*. They are used in potting mixes, are mixed with soil, or are used and solid matrix in seed priming treatments. They are effective against damping-off diseases of ornamentals and vegetables caused by *Pythium*, *Phytophthora*, *Sclerotium*, and some other fungi. In addition, *Sporidesmium sclerotiorum*, *Conithyrium minitants*, *Talaromyces flavus*, and others give experimental control of some diseases caused by *Sclerotinia*, *Rhizoctonia*, and *Verticillium*. Some species of *Pythium*, such as *Pythium nunn* and *P. oligandrum*, protect potted ornamentals and vegetable from plant pathogenic species of *Pythium, and Talaromyces flavus* and binucleate Rhizoctonias protect plants from the pathogenic *Rhizoctonia solani*. Experimental control of several other soilborne diseases has been obtained with many other fungi. Finally, *Fusarium* wilts of several crops, such as celery, cucumber, and sweet potato, caused by the respective formae specialis of *Fusarium oxysporum* have been reported to be successfully controlled by inoculating transplants or cutting with nonpathogenic strains of the same fungus. Some of these strains have been isolated from the vascular tissue of host plants that remained healthy while nearby plants had been killed by the wilt-inducing strains of the fungus. It is believed that the nonpathogenic strains not only compete with the pathogenic ones in the rhizosphere and for infection sites, but they also enhance the resistance of the host toward the pathogenic strains (Bruchi, 1975).

Biological Control of Diseases of Aerial Plant Parts with Fungi

Many filamentous fungi and yeasts have been shown to be effective antagonists of fungi infecting the aerial parts of plants, for example, inoculation of postbloom, dead tomato flowers with conidia of *Cladosporium herbarum* or *Pencillium* sp. almost completely suppressed subsequent infection of developing fruits by *Botrytis cinerea*. Similarly, spraying spores of common bark saprophytes, such as *Cladosporium* sp. and *Epicoccum* sp., and of the soil fungus *Trichoderma* on pruning cuts of fruits trees has prevented infection by canker causing pathogens such as *Nectria galligena* and *Leucostoma* (*cytospora*) sp. Sprays with *Trichoderma* in the fields also reduced *Botrytis* rot of strawberries and of grapes at the time of harvest and in storage. Several foliar diseases have also been reduced significantly (by more than 50 percent) when the leaves were sprayed with spores of common phylloplane fungi, for example, *Alternaria, Cochliobolus, Septoria, Colletotrichum*, and *Phoma*; or with spores of hyperparasites, for example, the cucumber powdery mildew fungus *Sphaerotheca fuligiena* with spores of *Ampelomyces quisqualis*, or *Tilletiopsis*, the wheat leaf rust fungus *Puccinia recondita* with spores of *Darluca filum*, and the carnation rust fungus with *Verticillium lecanii*. None of the above, nor any of the numerous other known cases of fungal antagonism by fungi, is used as yet for practical control of any disease of aerial plant parts.

Biological control of Post-harvest Diseases

Post-harvest rots of several fruits could be reduced considerably by spraying with spores of antagonistic fungi and saprophytic yeasts at different stages of fruit development, or by dipping the harvested fruit in the inoculum. For example, yeast treatments reduced post-harvest rotting of peach and apple. Also, significant reduction of citrus green mold (caused by *Penicillium digitatum*) was obtained by treating the fruit with the antagonist *Trichoderma viride*, whereas pre-harvest and post-harvest *Botrytis* rot of strawberries was reduced by several sprays of *Trichoderma* spores on strawberry blossoms and young fruit. *Penicillium* rot of pineapple was reduced considerably by spraying the fruit with non-pathogenic strains of the pathogen. Similarly, several antagonistic yeasts protected grapes and tomatoes from *Botrytis, Penicillium*, and *Rhizoctonia* rots. In addition, the yeast *Candida oleophila* was approved for post-harvest decay control in citrus and apples under the trade name Aspire (Harvey, 1978).

Products for Biological Control of Soilborne Pathogens

Presently, there are over 80 products for biocontrol of pathogens worldwide, a significant improvement over the past ten years. Most of these products are formulations either of the fungi *Trichoderma-Gliocladium* or the bacteria *Pseudomonas* and *Bacillus*. However, not all of these products are registered as biocontrol agents, but are marketed as plant growth promoters, plant strengtheners, or soil conditioners. These designations have enabled the products to get to the marketplace with less stringent toxicology or efficacy testing than would be required for plant protectants. However, some countries still have a regulatory framework in place for "plant strengthener." For example, in Germany in 1998, 208 plant strengtheners were registered with der Biologischen Bundesanstalt, up from 25 in 1991. These products include inorganic compounds such as SiO_2, $NaHCO_3$, organic constituents such as compost, homeopathic compounds, and some containing microorganisms such as *Trichoderma harzianum, Bacillus subtilis, Pseudomonas,* and *Pythium oligandrum.* In the following section, we highlight some of the most important products in use.

Coniothyrium minitans

This mycoparasite destroys sclerotia of *Sclerotinia sclerotiorum* and *S. minor.* The only biological control agent (BCA) registered as a biopesticide in Germany, *Coniothyrium minitans* is used for the control of Sclerotinia wilt of lettuce in greenhouse and rape in the field. It is marketed as a wettable granule, called Contans® WG, in Germany. This fungus has been extensively tested on glasshouse lettuce in England. The fungus was produced on a number of solid substrates such as barley, bran-vermiculite, millet, oats, peat-bran, and wheat and tested in a sequence of glasshouse lettuce crops. *C. minitans* reduced the sclerotial populations at the soil surface, survived at least 39 weeks at a density of 10^4-10^5 CFU/g, and spread to infect sclerotia in control plots (McQuilken and Whipps, 1995). In another series of glasshouse trials on lettuce, *C. minitans* outperformed *Gliocladium virens,* infecting over 80% of the sclerotia (Budge *et al.*, 1995). *C. minitans* is also registered in Switzerland and marketed in Hungary under the name of Koni®. Sales reach 30 metric tons per year, but registration is still pending in the United States and most European countries.

Gliocladium virens (= *Trichoderma virens*)

This product was developed by the Biocontrol of Plant Disease Laboratory of the USDA-ARS in Beltsville, MD. Isolated in the late 1980s from a soil in Maryland, the fungus is widely distributed in soil worldwide. It was developed for control of *Pythium ultimum* and *Rhizoctonia solani* in soilless mixes (Lumsden and Locke, 1989). *G. virens* isolate GL-21 was first formulated as an alginate prill (GlioGard®) by W. R. Grace Co. A granular fluid (SoilGard®) was later developed for greenhouse applications, and is presently marketed by Thermo Triology Corp., Columbia, MD. The fungus produces two fungitoxic compounds, Glioviren and Gliotoxin compounds (Howell and Stipanovic, 1995; Lumsden *et al.*, 1992). The lessons learned about commercialization and regulation of this product have been recounted in numerous articles (Lumsden *et al.*, 1996; Mintz and Walter, 1993; Whipps and Lumsden, 2001).

Trichoderma harzianum Strain T-22

This strain was produced in the late 1980s by protoplast fusion between T-95, a rhizosphere-competent strain of *T. harzianum* originally isolated from a Columbian soil, and *T. harzianum* T-12 from New York soil (Hadar *et al.*, 1984; Statz *et al.*, 1988). This strain has been extensively tested in greenhouse and field trials (Harman, 2000). This rhizosphere-competent strain can colonize all parts of the root system and persists for a long period when applied as a seed treatment, greenhouse soil drench or granules, or in-furrow drench or granules. T-22 is marketed by Bioworks, Geneva, NY, as a granular formulation (RootShield®) or a water-suspendable drench containing conidia (PlantShield®). The product has been shown to reduce Fusarium crown and root rot of tomatoes grown in potting mix-containing T-22 and transplanted into the field (Dantoff *et al.*, 1995). In greenhouse trials, T-22 controlled *R. solani* in poinsettia, geraniums, and *Catharanthus*, and *Pythium* on geraniums, impatiens, and petunias. The control it provides is equal to that by fungicides (Harman *et al.*, 2000), with which it is mostly compatible, but it must be applied as a preventive before disease occurs. *T. harzianum* has multiple mehanisms of action, including mycoparasitism via production of chitinases, β 1-3 glucanases and β 1-4 glucanases (Lorito *et al.*, 1996), antibiotics (Sivasithamparam and Ghisalberti 1998), competition (Elad *et al.*, 1999), solubilization of inorganic plant nutrients (Altomare *et al.*, 1999), induced resistance (Baileye and Lumsden, 1998), and inactivation of the pathogen's enzymes involved in the infection process (Elad *et al.*, 1999; Elad and Kapat, 1999). Retail sales in 1999 exceeded $3 million, making it one of the first biocontrol agents in North America to achieve this level of success. Other strains of *T. harzianum* are also marketed in a number of products, including T -35 or Trichodex® from Israel.

Gliocladium catenulatum Strain JI446

Originally isolated from a Finnish field soil, this isolate is the active ingredient in Primastop®, a product marketed by Kemira Agro Oy, Finland. Primastop® received EPA registration in July 1998 for 55 different crops, but only for greenhouse and indoor use. Target pests include damping-off, seed rot, root rot, and wilt pathogens. It is sold as a wettable powder that can be applied to the soil, roots, or foliage. In glasshouse trials with ornamental bedding plants, application by incorporation into the growing mix or drench reduced damping-off caused by *Pythium* and *Rhizoctonia*. In some cases, *G. catenulatum* was as effective as the fungicides propamocarb or tolclofos (McQuilken and Mohammad, 1997).

Nonpathogenic *Fusarium oxysporum* Strain Fo47

Originally isolated from a *Fusarium* suppressive soil in France, this isolate has been investigated for over 25 years by the laboratory of Claude Alabouvette, lNRA, Dijon. It is effective against Fusarium wilt diseases on carnation (Postma and Rattink, 1992), tomato (Fuchs *et al.*, 1999), and Fusarium crown and root rot on tomato (Lemanceau and Alabouvette, 1991). Mechanisms of action include competition for carbon (Lemnceau *et al.*, 1993), direct competition with pathogenic strains (Epavier and Alabouvette 1994), and induction of host defenses (Duiff *et al.*, 1998; Fuchs *et al.*, 1997). It is marketed by Natural Plant Products, Nogueres, France, as a liquid formulation for soilless culture such as tomato in rockwool, and as a clay formulation for mixing in cyclamen potting mixes. It is also marketed in Germany for use on cyclamens.

Products for Biological Control of Foliar Pathogens

Ampelomyces quisqualis

The fungus *Ampelomyces quisqualis* is the first organism reported to be a hyperparasite of powdery mildews (Yarwood, 1932) and antagonizes fungi in the orders Erysiphales, Mucorales, and Perisporiales (Falk *et al.*, 1995; Kiss, 1997; Sundheim , 1982; Sundheim and Tronsmo, 1988). In studies of its mode of action, *A. quisqualis* was shown to colonize hyphae and conidiophores of several species of powdery mildew fungi and formed pycnidia within the conidiophores of its hosts (Hashioka and Nakai, 1980). Falk *et al.* (1995) and Kiss (1997) further showed that *A. quisqualis* parasitized cleistothecia of *Uncinula necator* and *Blumeria graminis*, respectively, albeit at low rate of colonization.

Attempts have been made to exploit *A. quisqualis* as a biocontrol agent. Under greenhouse or field conditions, this antagonist was reported to be effective only under very high humidity. Jarvis and Slingby (1977) therefore, proposed the use of water sprays in combination with *A. quisqualis* to alleviate its need for high humidity. However, water sprays alone reduced the severity of *S. fuliginea* under greenhouse conditions, providing moderately good control. In the early 1980s, *A. quisqualis* was shown to be tolerant to some fungicides and could therefore be used in an integrated approach against *S. fuliginea* on greenhouse cucumber, when relative humidity remained high (Sundheim, 1982). A mixture of *A. quisqualis* with 2% paraffin oil was proposed to control cucumber powdery mildew in the field (Philipp *et al.*, 1990).

The use of *A. quisqualis* to control powdery mildew has been studied (Philipp and Hellstern, 1986; Sztejnberg *et al.*, 1989). The spores of the fungus were shown to germinate into the hyphae of powdery mildew leading to the collapse of the pathogen (Philipp and Hellstern, 1986; Sztejnberg *et al.*, 1989). A formulated product was registered in the late 1980s in Australia, but the high humidity requirements of *A. quisqualis* have hampered its efficacy.

Recently, Ecogen, Inc. developed a formulation (AQ-10®), sold as water-dispersible granules, based on a new strain that reportedly tolerates lower humidities. Good control of powdery mildew of cucumber with AQ-10® was claimed when disease pressure was moderate (Feldman *et al.*, 1993). Early attempts to introduce AQ-10® into the grape industry failed because of poor formulation and efficacy (Hofstein *et al.*, 1996), but now its use is recommended as part of an integrated program and as a preventative measure only. AQ-l0® is registered for a number of crops including grapes, vegetables, and other fruits, recommended in conjunction with a wetting/dispersing agent, AddQ, to overcome humidity requirements of the fungus.

Trichoderma harzianum Strain T -39

Trichoderma harzianum strain T-39, developed at the Volcani Center in Israel and marketed as TRICHODEX, 20P by Makhteshim Ltd. (Be'er Sheva, Israel), is targeted at *Botrytis cinerea*. The reported mode of action of *T. harzianum* T-39 is competition for nutrients and interference with the production of lytic enzymes by the pathogen (Elad and Kopat, 1999) thus in addition to slowing the germination of the pathogen's conidia, T39 also prevents the penetration of the host tissue and the maceration process.

Although Trichodex was developed primarily for the grape market (Elad, 1994), its efficacy as a biocontrol agent on greenhouse crops has been tested intensively under commercial conditions (Elad *et al.*, 1996, Elad and Shtienberg, 1995). Trichodex has effectively controlled *Botrytis* diseases in greenhouse crops in Israel (Elad *et al.*, 1993) and other countries (Elad *et al.*, 1995) where it is registered for agricultural use. It is the first such product to be introduced commercially to greenhouses. As with most biocontrol agents, the efficacy of *T. harzianum* T-39 can be influenced by ambient environmental conditions. For this reason, the use of Trichodex in combination or alternating with chemicals has been investigated (Elad *et al.*,1999). Also examined was the integration of biological and chemical controls aided by the use of a forecaster to predict *Botrytis* outbreaks. This integrated system for disease management saved 60% of the chemical sprays, which were replaced by Trichodex.

Ulocladium atrum

No biological product based on this organism is currently available, but some European companies have expressed interest in light of promising results of biocontrol of necrotrophs obtained in different laboratories. For instance, applications of spore suspensions of *U. atrum* on greenhouse tomatoes consistently reduced infections by *B. cinerea* (Fruit and Nicot, 1999). Similar results have been obtained on other greenhouse crops such as cyclamen, geranium, and roses (Gerlagh *et al.*, 2001; Kohl *et al.*, 2000). The antagonism of *Ulocladium atrum* against *Botrytis* spp. appears to involve competition in necrotic tissues since no toxins or cell wall degrading enzymes were found (Kohl *et al.*, 1997). Interestingly, among the four antagonists tested, *U. atrum* colonized dead onion leaves the best and also had the strongest antagonism on leaves exposed to field conditions (Kohl *et al.*, 1995). This antagonist also showed a high ecological competence for the habitat of above-ground necrotic plant tissues. It germinated and colonized the substrate under various environmental conditions in the presence of *Botrytis* spp. and other naturally occurring saprophytic fungi (Kohl *et al.*, 1999). Microbial suppression of sporulation on necrotic tissues lower spore load in the crop, which will, in turn, slow progression of disease epidemics, as shown for *Botrytis* spp. in field grown onions (Kohl and Fokkema, 1998).

The preceding example is very valuable in demonstrating how ecological processes regulate fungal populations. Nutrient competition can be an excellent strategy for microbial suppression of unspecialized necrotrophs such as *Botrytis* spp. On the other hand, the research has somewhat undermined the value of antibiosis and overshadowed the importance of parasitism interfering with the saprophytic phase of a fungus. More recently, Kessel (1999) used *U. atrum* against *B. cinerea* in cyclamen and found that the antagonist was as effective as commercial fungicides in suppressing the disease. This success was attributed to the ability of *U. atrum* to exclude *B. cinerea* from necrotic tissue that was a mutual substrate.

Registration and Commercialization

Although biocontrol agents are arguably more effective in protected crops, their commercialization can be jeopardized if the market seems too small to justify the expenses inherent in their registration (Fravel *et al.*, 1999). This paradox is largely responsible for the limited number of biofungicides currently available. Meanwhile, there is a flourishing industry of insect biological control agents that dominates the green- house market in Europe. Why are microbial agents for disease control lagging behind. Undoubtly, the registration process is one of the main reasons. Registration of insect predators is not subject to the long and costly steps of human and environmental toxicology, formulations, quality assurance of the formulated product, genetic stability, etc. required for microbial pest control agents. The initial investment required to bring beneficial insects to the market is more moderate and therefore, not so discouraging to small, specialized companies. There are currently 26 natural enemy producers in Europe, with approximately 65 worldwide (Van Lenteren, 2000), and the number of insect biological control agents has grown from 2 in 1970 to almost 100 today. These figures attest to the viability of the greenhouse market with regard to biofungicides. Although multinational agrochemical companies have shied away from both insect and microbial pest control agents, smaller companies that have already profited from establishing their niche in the greenhouse market can recognize the potential of the largely unexploited field of disease biocontrol. In fact, promotion of disease biocontrol in the greenhouse environment is attractive from several standpoints. First, as discussed in several reviews (Belanger *et al.*, 1998; Elad *et al.*, 1999; Whipps and Lumsden, 2001), is increased and reproducible efficacy of bicontrol agents under controlled environments. The success enjoyed by insect biocontrol in greenhouses demonstrates convincingly that biological control of pests is easier to achieve when growers can manipulate environmental conditions to best promote the development of the biocontrol agents. Therefore, since the biofungicides industry is still in its infancy, it is important to disseminate positive information to gain credibility among growers.

The second advantage of biologicals over chemicals lies in the registration process. Registration of a biofungicide is notoriously frustrating and legally cumbersome as attested by numerous recent reviews (Cross and Polonenko, 1996; Harman, 2000; Hofstein *et al.*, 1996). However, regulators in the United States, Canada, and Europe tend to be favorably disposed toward biological pesticides and encourage companies to register such products. For example, the cost of registration of biofungicides is lower than for chemical pesticides, or can be completely waived. Biological products automatically enter a fast-track review process that speeds up registration. Canada and the United States have recently implemented a joint review process of biological products whereby a registration dossier receives speedier analysis, and approval, when granted, allows commercialization in both countries simultaneously.

2. Entomopathogenic Fungi

There are approximately 750 species of fungi found throughout the five major taxa comprising the division Eumycota. Eumycota have been identified as being entomogenous or having confirmed or suspected entomogenous spp. (Anon, 1979). The major taxa according to Ainsworth *et al.*, (1973) are the Mastigomycotina (zygosporic fungi), the Ascomycotina (sac fungi), the Basidiomycotina (club fungi) and the Deuteromycotina, better known as Fungi Imperfecti. The potential entomogenous fungi and their commercial products are given in the Table 1 and 2 respectively.

Entomophthora

Entomophthora was utilized in integrated control programs in the U.S.S.R. to control the pea aphid (Belova, 1978). Extensive ecological studies, conducted

Table : 1. Current entomogenous fungi being considered as potential microbial insecticides

Fungus	Infective Stage	Insect host	Habitat
Entomophthora	Conidia or resting spores	Aphids	Foliage
Achersonia	Conidia	Whitefly	Foliage
Beauveria	Conidia	Beetles Caterpillars	Foliage
Hirsutella	Conidia	Mites	Foliage
Metarhizium	Conidia	Froghoppers, Beetles	Foliage
Nomurae	Conidia	Caterpillars	Foliage
Verticillium	Conidia	Aphids Whitefly Scales	Glasshouse Foliage

from 1962 to 1969, showed that rainfall was a key factor in the development of an Entomophthora mycosis in the pea aphid. By dividing the country into regions (based on hydrothermal coefficients), epizootics could be accurately predicted thereby resulting in reduced or no treatment for control of aphids.

The use of sprinklers to increase the realtive humidity in glasshouses, resulted in the initiation of *E. fresenii* epizootics in *Aphis fabae* (Dedryver, 1979). The same fungus introduced into field populations of *A. fabae* in England, however, failed to spread (Wilding, 1981). Aqueous suspensions of conidia, hyphal bodies, and resting spores of *E. thaxteriana* and *E. virulenta*, sprayed in glasshouses for aphids and mites (Chudare, 1978), provided rapid control possibly due to the presence of a toxin. A suspension of *E. aphidis* (6 x10^6 conidia/ml), obtained from bran and water, gave 76 to 100% mortality of several aphid species (Wilding, 1981).

The use of the resting spores for the field application of *Entomophthora* genrally has been unsuccessful. These programs involved the use of *E. obscura*

Table 2. Trade names of commercial or experimental preparations of fungi formulated as microbial insecticides

Fungal species	Trade name	Producer
Aschersonia aleyrodis	Aseronija	All Union Inst. (U.S.S.R.)
Beauveria bassiana	Biotrol FBB	Nutrilite Products (U.S.)
	Boverin	Glavmikrobioprom (U.S.S.R.)
	BAG-6178	Abbott Labs. (U.S.)
Hirsutella thompsonii	Mycar	Abbott Labs. (U.S.)
Metarhizium anisopliae	Biotrol FMA	Nutrilite Products (U.S.)
	Metaquino	CODECAP (Brazil)
Verticillium lecanii	Vertalec	Tate and Lyle (England)
	Mycotol	Tate and Lyle (England)

against cereal aphids in France (Latge and Perry, 1980) and *E. virulenta* against green peach aphid in the U.S. (Soper, 1978). A major limitation to the performance of these species in the field appears to be their inability to spread.

Beauveria

During the last 15 years, special attention has been given in the U.S.S.R. and eastern Europe to the use of *Beauveria bassiana* as a component of an integrated pest management strategy against the Colorado potato beetle, *Leptinotarsa decemlineata* and to a lesser extent, the codling moth *Cydia pomonella.* An excellent summary of this research has been published by Ferron (1981).

Large-scale studies against *Leptinotarsa decemlineata* in the U.S.S.R. were begun by Pospelov (1944) and followed by mass production and field application (along or combined with insecticides) by numerous workers (Ferron, 1978; Yevlakhova, 1978; Telenga *et al.*, 1964; Pristavko, 1996). Generally, two treatments made at 15-day intervals with 1.5 kg/ha concentrated boverin (30 x 10^9 conidia /g) combined with a reduced dosage of chlorophos were used against larval stage. A similar treatment scheme was developed in Poland, combining *Beaurveria bassiana, Paecilomyces farinosus,* and a parasitic nematode. Results, however, were no better than with *B. bassiana* alone (Fedorko *et al.*, 1977). *Beauveria bassiana* plus *Bacillus thuringiensis* and *Beauveria bassiana* plus Despirol (Drapatyi, 1979) also have been used in the U.S.S.R. with varied results. In France, two to four application of *B. bassiana* at 10^4 blastospores/hectare significantly reduced larval populations of *L. decemlineata* (Fargues *et al.*, 1980). Conidia at 2.0 to 4.5 x 10^9/g compared to 2.0 to 2.5 x 10^9 blastospores/g provided superior control of *L. decemlineata* larvae in Romania after 12 days (Beratlief, 1979). In the U.S., *B. bassiana* applied to the soil, reduced pupal populations of colorado potato beetle (Watt and Lebru, 1984). Generally, high doses of boverin will work as a prophylactic measure against younger larvae of *L. decemlineata* but an insecticide must be included to control older larvae.

Strategies for the use of boverin in the Ukraine, usually with reduced dosage of chemical insecticides (malathion), were successful by reducing the following spore population after treatment of the summer generation of codling moth, *Cydia pomonella* (Sikura and Smietnik, 1967). Introduced starins of *Beauveria bassiana* have appeared most effective (Lappa *et al.*, 1977). Where fungicide applications do not interfere, three applictions of boverin at 1kg (24 x 10^9 spores/g) per hectare were effective in controlling *C. pomonella* in the U.S.S.R. In France conidia of *B. bassiana* applied at 6 x 10^9 conidia per tree, against late instar larvae of *C. pomonella*, gave a 50% reduction in population and also reduced the following spring population (Ferron and Vincent, 1978).

In the Peoples Republic of China, *Beauveria bassiana* has been used for many years against the European corn borer (Chiang and Huffaker, 1976). Likewise, a specific mutant has been effective in experimental trails against *Ostrinia nubilalis* in France (Riba, 1984). In the U.S., field trials with conidial dust have been conducted against lygus bugs (Dunn and Mechalas, 1963) and caterpillars of cob crops (Ignoffo *et al.*, 1979). Soil-inhabiting weevils have also been treated with *B. bassiana*. (Walstad *et al.*, 1970; Muller-Koglar and Stein, 1970; McCoy *et al.*, 1985).

The other closely related *Beauveria* species, *B. brongiartii* has been tested extensively against soil-inhabiting scarbeid beetle, primarily *Melolontha melolontha* in Western Europe (Ferron, 1971; Keller, 1982). Soil treatment with 20 x 10^9 conidia/ m^2 caused an epizootic and the muscardine appeared to persist in the soil for long periods of time (Ferron and Lutte, 1997). Blastospores preparations were inferior to conidia and sublethal dosages of insecticides combined with conidia did not enhance control (Keller *et al.*, 1979).

Metarhizium

The green muscardine, *Metarhizium anisopliae* var. *anisopliae* and var. *major* like the white muscardine fungi have been studied for over 100 years. However, promising field results with this fungus have occurred only within the early seventies (Ferron, 1978).

The rhinoceros beetle, *Oryctes rhinoceros*, a serious pest in coconut growing regions of the South Pacific, has been successfully controlled in Fiji, Tonga and Wester Samoa by *Metarhizium anisopliae* var. *major* (Latch, 1965; Lach and Falloon, 1976; Peterson, 1977). This species was propagated on grain and introduced to breeding sites of the larvae. About 50 g of conidial preparation /m^2 of soil gave 80% mortality after15 to 24 months and improved coconut yields by 25% within one year.

The use of Metaquino in Brazil (the conidial formulation of *M. anisopliae*) was successfully used to control spittel bug on sugarcane and in pastures for nearly eight years. As much as 50,000 ha in a single province were aerially treated at 50 l/ha with 6 x 10^{11} to 1.2 x 10^{12} conidia/l. Although *M. anisopliae* occurs in natural epizootics, introducing the fungus early in the season appears to keep spittel bugs below damaging levels. *M. anisopliae* has also been effective in field trials against mosquitoes. In Australia, it has reduced pasture cockchafer populations when applied as a spore bait (Coles and Pinnock, 1981). In China, *M. anisopliae* was pathogenic to *Colasposoma* when applied to the soil at a rate of 75 kg/ ha (Xiong and Wu 1981). From 50 to 60% mortality of pecan weevil also has been obtained in the U.S. with applications of *M. anisopliae* (Gottwald and Tedders, 1983).

Nomuraea

Epizootics of *Nomuraea rileyi* noctuid caterpillars occur regularly in agro-ecosystems in the U.S. and other countries (Ignoffo *et al.*, 1976; Allen *et al.*, 1971). Extensive research has been conducted on the survival of the fungus in soybean fields and a model has been proposed for predicting the level of infection in velvetbean caterpillar population in Florida (Kish and Allen, 1978).

Early close planting of soybeans and cotton, that influences microclimatic field conditions, also enhanced the development of *N. rileyi*. The prophylactic use of *N. rileyi* has been tested using high conidial doses with encouraging results. A dust formulation of *N. rileyi* with 12.4 x 10^{10} conidia /g applied early in the season at 1.1 x 10^{13}/ acre to soybeans caused a premature epizootic (Ignoffo *et al.*, 1976). These studies suggest that a prophylactic treatment 2 to 4 weeks before economic levels of injury are anticipated, may provide control of defoliating caterpillars. Data corroborating these suggestions have been published using infected larvae of *Helioithis virescens* in soybeans in North Carolina (Sprenkle *et al.*, 1975).

Verticillium

Verticillium (=*Cephalosporium*) *lecanii,* reported from scale insects for some time, has become important as a microbial insecticide for use in glasshouses (Hall 1981). In the U.S.S.R., *V. lecanii* has been successfully cultivated on a synthetic medium containing grain or bread cubes. A spore preparation containing 41 x 10^7 active spores /g either undiluted or as a 10% concentration (diluted with talc or water) was applied to citrus leaves infested with *Coccus hesperidum.* In the greenhouse, 85 to 100% mortality was achieved within 3 weeks. Although *V. lecanii* killed all stages of the brown soft scale, it was unable to complete its development because of lack of nutrients. Hence the fungus, unable to survive and spread within the scale population, must be repeatedly introduced.

Hall and Burges (1979) achieved excellent control of *Myzus persicae,* a major pest; of Chrysnathemums in glasshouses with *Verticillium lecanii.* Conidia (10^5 to 10^7 /ml) and blastospores (10^7 to 10^8/ml), suspended in phosphate buffer containing 0.02% Triton X-100 as a wetting agent, sprayed to run-off gave consistent results when sprayed in the evening and when plants were covered with polyethylene sheets for humidity. In similar trials on Chrysanthemums, *Macrosiphoniella sanborni* and *Brachycaudus helychrysi* were also controlled from slow to high population densities during all seasons (Hall 1976). Although differences in species susceptibility and effect of spore concentration were detected in Laboratory assay, no differences appeared in glasshouse trials (Hall, 1980).

In trials on cucumbers, *Verticillium lecanii* was also effective against *Aphis gossypii* (within 14 days of treatment), and the whitefly, *Trialeurodes vaporariourum* (Yendoe and Roberts, 1970; Hall, 1982).

Aschersonia

Aschersonia aleyrodis has been recognized for nearly a century as an effective biological control agent of citrus whitefly particularly in Florida. From 1958 to 1973, 11 strains of several species of *Aschersonia* were introduced from India, China, Japan, Vietnam, the U.S., and Cuba for use in biological control of the citrus whitefly *(Dialeurodes citri)* in the U.S.S.R. The citrus whitefly was accidentally imported in the U.S.S.R. in 1957 and had no natural enemies. Citrus plantings in Azerbaijan were sprayed with *Aschersonia* to control whitefly during the period 1961 to 1964. Approximately 80% larval mortality was obtained (under favorable environmental conditions) and the introduced fungus adapted well and spread to new citrus plantations. A strain of *A. placenta* of Vietnamese origin proved to be particularly virulent, producing 90% larval mortality. This isolate was applied, as an aqueous suspension containing 10^6 to 10^8 spores/ml at the rate of 900 to 1000 l/ha.

Tests with *Aschersonia* to control the greenhouse whitefly *(Trialeurodes vaporariorum)* began in 1973 in the U.S.S.R. Moldavian, Cuban, and Japanese isolates of *Aschersonia* were the most effective against larvae, but effectiveness was highly dependent upon humidity. Best results (83% mortality by the Cuban isolate) were obtained in the greenhouse on cucumbers when humidity reached 85 to 95% (Solovey and Koltsov, 1976; Primak and Chizhik, 1975). Aqueous suspensions containing 5×10^6 to 5×10^8spores/ml were the most effective rates, and applications had to be repeated after 10 to 14 days. In Bulgaria, good control *of T. vaporariorum* on cucumbers was obtained using *A. aleyrodis* by spraying the larvae on the underside of leaf. Conidial preparations gave 71.3% mortality compared to 14.3% mortality when dried and ground mycelium was used (Spasova, 1974). Field tests for the control of *Dialeurodes citri* on citrus with *Aschersonia* sp. in Japan were inhibited by fungicides (Uchida, 1970). In Holland, *A. aleyrodis* gave acceptable control of *T. vaporariorum* larvae on greenhouse cucumbers in combination with the parasite *Encarsia* (Ramakar, 1983; Ramakar and Samson, 1984).

Hirsutella

Hirsutella thompsonii causes seasonal epizootics in citrus rust mite populations lasting 2 to 3 weeks during the summer when mite densities reach injurious levels in Fiorida (McCoy, 1981; McCoy *et al.*, 1976). Non-selective pesticides can reduce the incidence of *H. thompsonii* through direct and indirect action resulting in a greater dependence on chemical control for the mites. By using a selective fungicide for phytopathogenic fungi and raising the spray threshold for the citrus rust mite, the incidence of *H. thompsonii* in mite populations was significantly increased thereby decreasing the number of spray applications from 4 to 2/year. During the last 12 years, *H. thompsonii* was mass-produced, formulated as a mycoacaricide, and applied by government agencies and industry for control of the citrus rust mite on citrus and other mite pests of turf, coconut, and greenhouse crops.

The fungus, in early tests, was applied as fragmented-mycelial slurry to citrus trees (30 l/tree; 1.0 to 10.0% w/v) with a high pressure sprayer with excellent results in Fiorida (McCoy *et al.*, 1971; McCoy and Selhime, 1977). In China, laboratory-produced mycelium was applied at dosages of 0.5 to 1.0 g/l to citrus. In greenhouses Gerson *et al.* (1979) sprayed conidial-bran wettable powder (2.5 x 10^6 conidia/ml) on groundnut plants infested with the carmine spider mite; however, extreme temperatures (35°C) and low humidity prevented infection. The inoculum, however, survived and subsequent mite exposure resulted in infection. Gardner *et al.* (1982) obtained up to 99% mortality of *Tetranychus urticae* in laboratory bioassays with *H. thompsonii,* however, rates of 1.2 to 9.6 g/l applied to plants in the greenhouse were ineffective.

In 1976, Abbott Laboratories produced the first commercial formulation of *H. thompsonii* and labeled it Mycar. Initial small-scale field trials with a wettable powder at 2.5, 5.0, and 10.0 g/500 ml (1.9 x 10^5 spores/g) and about 30 litre provided good control of mites in Florida. Additional field studies with both a wettable powder and dust formulation at 2 to 4 lb and 25.0 lb/acre, respectively, provided good control of mites (McCoy and Couch, 1982). Mycar was also effective in field trials against the citrus rust mite in the U.S.S.R (Litvinova *et al.,* 1979), Brazil and Colombia. In the U.S. *Hirsutella thompsonii* was registered for control of eriophid mites on citrus and turf. After the sale of several hundred kilograms of technical material, commercial production was terminated in 1985. Numerous factors such as the sensitivity of the primary conidia (infective unit) to available water and fungal survival in the field influenced its reliability as a commercially acceptable control practice. Subsequently, limitations in storage and transportation to maintain fungal stability resulted in a discontinuation of commercial sale in the U.S.

3. Nematophagous Fungi

Plant pathogenic nematodes are also parasitized by other microorganisms. For example, the dagger nematode *Xiphenema* and the cyst nematodes *Heterodera* and *Globodera* are parasitized by the nematophagous fungi *Catenaria auxiliaris, Nematophthora gynophila, Verticillium chlamydosporium and Hirsutella sp.*, whereas the root-knot nematode *Meloidogyne* sp. is parasitized by the fungi *Dactylella, Arthrobotrys, Paecilomycses,* and *Hirustella sp.*

The genus *Arthobotrys* is known as facultative nematophagus fungus. *Verticillium* species are known to have biocontrol potential for plant pathogenic nematodes. *Verticillium* species either attack free living nematodes or colonizes cysts and eggs or young sessile females of *Heterodera* species. The genus *Paecilomyces* (Type: *Paecilomyces variotii)* was introduced by Bainier (1907) who described it as being closely related to *Penicillium.* Amongst the other species, *Paecilomyces lilacinus* is known to be well established egg parasite of nematodes and has tremendous potential as biocontrol agent and their products are commercially patented (Bicon) against several nematode species.

4. Biological Control of Weeds

Microorganism used as biocontrol agents of weeds are generally fungi pathogenic to specific weeds, to which they cause significant damage or death. Weed pathogens are isolated from infected weeds locally or elsewhere in the world. They are then grown and multiplied in culture, and are tested for their efficiency in infecting and damaging or killing the specific weed in the laboratory, the greenhouse, and in the field. They are also tested for their specificity, that is, inability to infect cultivated hosts of related plant genera and families.

Of the several hundred, perhaps thousands, of pathogen-weed host combinations tested, several dozen of them were shown to be quite effective. Some of the weed pathogens are now sold commercially. For example, the fungus *Colletotrichum gloeosporioides*, sold as Collego, is effective against the northern jointvetch weed (*Aeschynomene virginica*) growing in rice and soybean fields in several states; *Phytophthora palmivora*, sold as DeVine, is effective against the milkweed vine weed (*Morrenia adorata*) growing in citrus orchards of florida. Two other weed pathogens are expected to be marketed in the near future: *Alternaria cassiae*, to be sold as Casst, is effective against the weed sickledpod (*Cassia obtusifolia*) growing in peanut and soybean crops in the southeastern United States and another strain of *Colletotrichum gloeosporioides*, to be sold as BioMal, is effective against the weed roundleaf mallow (*Malva pusilla*) growing in small-grain fields in North America. A few others, among the many promising pathogens of weeds that are likely to be used as biocontrol agents against these weeds in the near future inlclude the following: *Phomopsis amaranticola* against pigweed (*Amaranthus* spp.), *Colletotrichum dematium* f. sp. *crotalariae* against showy crotalaria (*Crotalaria spectabilis*), *Alternaria helianthi* against cocklebur (*Xanthium pennsylvanicum*), *Alternaria macrospora* against spurred anoda (*Anoda cristata*), *Colletotrichum coccodes* against velvetleaf (*Abutilon theophrasti*), and *Cercospora rodmanni* against water hyacinth (*Eichhornia crassipes*). Many other fungi (*Ascochyta, Bipolaris, Fusarium, Phoma, Puccinia, Sclerotinia,* ect.) have been studied extensively as mycoherbicides of various weeds.

Acknowledgement

Authors are grateful to Director, NCIPM for providing facilities to prepare this manuscript. Authors are also thankful to Mr. Abdul Ali Jafri and Mr. Rajeev Kumar Gautam, NCIPM for help in preparing the manuscript. Thanks are also due to Mr. Kuldeep Sharma for typing the manuscript.

References

Adams, P. (1990). The potential of mycoparasites for biological control of plant diseases. Annu. Rev. Phytopathol., **28** : 59-72.

Ainsworth, G.C., Sparrow, F.K., and Sussman, A.S. (1973). Eds., The fungi: an Advanced Treatise, Vol. 4a, Academic Press, New York, 592.

Ainsworth, G.C., Sparrow, F.K., and Sussman, A.S. (1973). Eds., The Fungi: An Avcanced Treatise, Vol., 4b, Academic Press, New York, 504.

Allen , G. E. Greene, G. L., and Whitcomb, W.H. (1971). An epizootic of *Spicaria rileyi* on the velvet bean caterpillar, *Anticarsia gemmatalis* in Florida, Fla. Entomo., **54** : 189.

Altomare C., Norvell, W. A., Bjorkman, T. and Harman, G. E. (1999). Solubilization of phosphates and micronutients by the plant growth promoting and biocontrol fungus *Trichoderma harzianum* Rifai 1295-22. Appl. Environ. Microbiol., **65** : 292-33

Anagnostakis, S. L. (1982). Biological control of chestnut blight. Science, **215** : 466-471.

Andrew, J.H. (1992). Biological control in the phyllosphere. Ann. Rev. Phytopathol., **30** : 630-635.

Anon. (1979). Microbial insect control agents, in Microbial Processes: Promising Technologies for Developing countries, national Academy of Science, Washington, D.C.

Bailey B. A. and Lumsden, R. D. (1998). Direct effects of *Trichoderma* and *Gliocladium* on plant growth and resistance to pathogens. In: *Trichoderma* and *Gliocladium* , ed. Kubicek, C.P., Harman, G.E. **2**:185-204, Taylor and Francis.

Bainier, G. (1907). Mycotheque de l'ecole de Pharmacie XI *Paecilomyces* Bull. Trimest. Soc. Mycol. Fr; **23**:26-27.

Barron, G. L. (1969). Isolation and maintenance of endoparasitic nematophagus Hyphomycetes. Can. J. Bot. **47** : 1899-1902.

Belanger, R.R., Dik A.J., and Menzies, J.G. (1998). Powdery mildews-recent advances toward integrated control. In: Plant-microbe interaction and biological control ed. Boland, G.J. and Kuykendall, L.D., pp. 89-109, New York, Marcel Dekker.

Belova, R.N. (1978). Development of the technology of boverin production by the submersion method, in proc. First Joint US/USSR Conf. On Production, Selection, and Standardization of Entomopthoghenic fungi of the US/USSR Joint working Group on the Production of substances by Microbial Means, Ignoffo, C.M., Ed., 102.

Beratlief, Z. (1979). Investigation on the entomopathogenic fungus *Beauveria bassiana* and its action on the Colorado beetle and the bet weevil, An. Inst. Cercet. Prof. Plant. Inst. Ont. Cercet. Agric. Bucharest. **15** : 233.

Bruchi, G.W. ed. (1975). Biology and control of soilborne plant pathogens. APS Press, St. Paul, Minnesota.

Budge, S. P., McQuilken M.P., Fenlon JS and Whipps, J. M. (1995). Use of *Coniothyrium minitans* and *Gliocladium virens* for biological control of *Sclerotinia sclerotiorum* in glasshouse lettuce. Biol. Control, **5** : 513-22

Chiang , H. C. and Huffaker, C. B. (1976). Insect pathology and microbial control of insects in the People's Republic of China in Proc. Ist Inst. Coloq. On Invertebr. Pathol., Queens's University Press Kingston, Canada.,42.

Chudare, Z.P. (1978). Applicton of *Entomophthora,* in Proc. First Joint US/USSR, Conf. On Production, Selection, and Standardization of Entomopathogenic fungi of the US/USSR Joint Working Group on the Production of substances by Microbial Means, Ignoffo, C.M., Ed., 254.

Coles, R.B. and Pinnock, D. E. (1981). Control of the pasture cockchafer with the fungal pathogen *Metarhizium anisopliae,* Proc. 3rd August. Conf. Grasslands, 191.

Cross, J.V. and Polonenko, D.R. (1996). An industry perspective on registration and commercialization of biocontrol agents in Canada. Can. J. Plant Pathol., **18** : 455-62

Dantoff, L.E., Nemec, S. and Pernezny, K. (1995). Biological control of Fusarium crown and root rot of tomato in Florida using *Trichoderma harzianum* and *Glomus intraradices.* Biol. Control, **5**:427-31

Dedryver, C.A. (1979). Intiating an epizootic in a glasshouse with *Entomophthora fresenii* on *Aphis fabae* through inoculum introduction and control of the relative humidity, Entomophaga, **24** : 443.

Drapatyi, N. A. and Beilakh, G. (1979). A mixture of beauverin and despiro tested against the Colorado betle, Kartofel Ovoshchi, 5 : 37.

Duiff, B.J., Pouhair, D., Olivian, C., Alabouvette, C. and Lemanceau P. (1998). Implication of systemic induced resistance in the suppression of Fusarium wilt of tomato by *Pseudomonas fluorescens* WCS417r and by non-pathogenic *Fusarium oxysporum* Fo47. Eur. J. Plant Pathol., **104**:903-10

Dunn, P.H. and Mechalas, B.J. (1963). The potential of *Beauveria bassiana* (Balsmo) Vuillemin as a microbial insecticide, J. Insect Pahtol., **5** : 451.

Elad, Y., David, D.R., Levi t, Kapat, A. and Kirshner, B. (1999). *Trichoderma harzianum* T-39-mechanims of biocontrol of foliar pathogens. In Modern fugicides and antifugal compounds II, ed. H Lyr, pp. 459-67. Andover, Hants, UK: Intercept.

Elad, Y and Shtienberg, D. (1995). *Botrytis cinerea* in greenhouse vegatable; chemical, cultural, physiological and biological controls and their integration. Integration Pest Management Rev. **1**:15-29

Elad, Y. (1994). Biological control of grape grey mould by *Trichoderma harzianum*. Crop Prot. **13**:35-38

Elad, Y., Belanger, R.R. and Kohl, J. (1999). Biological control of diseases in the phyllosphere. In: Integrated pest and diseases management in green house crops, ed. Albajes *et al*. Dordrecht: Kluwer pp. 338-52

Elad, Y., Gullino, M. L., Shtienberg, D. and Aloi C. (1995). Coping with tomato grey mould under Mediternean conditon. Crop Prot. **14**:105-9

Elad, Y. and Kapat, A. (1999). The role of *Trichoderma harzianum* protease in the biocontrol of *Botrytis cinerea*. Eur. J. Plant Pathol., **105**:177-89

Elad, Y., Malathrakis Ne and Dik A.J. (1996). Biological control of *Botrytis*-incited disease and powdery mildew in greenhouse crops. Crop Prot., **15** : 229-40.

Elad, Y., Zimand, G., Zaqs, Y., Zuraiel, S. and Chet, I. (1993). Use of *Trichoderma harzianum* in combination or alternation with fungicides to control cucumber grey mould (*Botrytis cinerea*) under commercial greenhouse conditions. Plant Pathol., **42** : 324-32

Epavier, A. and Alabouvette, C. (1994). Use of ELISA and GUS-transformed strains to study competition between pathogenic and nonpathogenic *Fusarium oxysporum* for root colonization. Biocontrol Sci. Technol., **4** : 35-47

Falk, S.P., Gaadoury, D.M., Cortes, I.P., Pearson R.C. and Seem R.C. (1995). Parasitism of *Uncinula necator* cleistothecia by the mycoparasitse *Ampelomyces quisqualis*. Phytopathology, **85**:749-800

Fargues , J., Cugier, J.P. and van de Weghe, P. (1980). Experimentation parcelles du champignon *Beauveria bassiana* contre *Leptinotarsa decemlineata*, Acta Ecol., 1 : 49.

Fedorkoi, A., Bajan, Ç., Kmitowa,K. and Wojciechowska, M. (1977). Effect of a joint introduction into the soil of several Entomopthogenic microorganisms on the level of reduction of the Colorado beetle, Pol. Ecol. Stud., 3 : 135.

Feldman, K., Keren-Zur M., Hofstein, R. and Fridlender, B. (1993). *Ampelonmyces quisqualis*, an important component of an IPM programme for the control of powdery mildew. Int. Congr. Plant. Pthol. 6th, Abstr. 3.2.11,p. 58

Ferron, P. and Vincent, J.J. (1978). Preliminary experiments on the use of *Beauveria bassiana* against *Carpocapsa pomonella*, Mitt. Biol. Bunderrsat. Land Fortwirsch. Berlin-Dahlem, 180, 120.

Ferron, P. (1971). Modification of the development of *Beauveria tenella* mycosis in *Melolontha melolontha* larvae by means of reduced doses of organophosphorus insecticides, Entomol. Exp. Appl. **14** : 457.

Ferron, P. (1978). Biological control of insect pests by entomogenous fungi, Annu. Rev. Entomol., 23 : 409.

Ferron, P. (1981). Pest control by the fungi *Beauveria* and *Metarhizium*, in Microbial control of Pests and Plant Disease 1970-1980, Burges, H. D., Ed., Academic Press, New York, 465.

Ferron, P. and Lutte, (1977). Microbiologique contre le hanneton commun *Melolontha melolontha*, Meded. Fac, Landouwwet. Rijksuniv. Gent., **42** : 1323.

Fravel, D.R., Rhodes, D.J. and Larkin, R.P. (1999). Production and commercialization of biocontrol products. In: Inegrated pest and diseases management in green house crops, ed. Albajes et al. Dordrecht: Kluwer, pp. 365-76

Fruit, L. and Nicot., P. (1999). Biological control of *Botrytis cinerea* on tomato stem wounds with *Ulocladium atrum*, integrated control in Glasshouses, IOBC Bulletin 22, pp. 81-84

Fuchs, J.G., Moenne- Loccoz J.G. and Defago G. (1999). Ability of nonpathogenic *Fusarium oxysporum* Fo47 to protect tomato against Fusarium wilt Biol. Control, **14** : 105-10

Fuchs, J.G. and Moenne-Loccoz Defago G. (1997). Nonpathogenic *Fusarium oxysporum* strain Fo47 induces resistance to Fusarium wilt in tomato. Plant Dis., **81** : 492-96

Gardner, W. A., Oetting, R.D. and Storey, G.K. (1982). Susceptibility of the two-spotted spider mite, Tetranychus urticae to the fungal pathogen *Hirsutella thompsonii*, Fla. Entomol., **65** : 458.

Gerlagh, M., Amsing, J.J., Molhoek, W.M.L., Bosker, van Zessen, A.I., Lombaers-van der Plas C.H. and Kohl, J. (2001). The effect of treatment with *Ulocladium atrum* on *Botrytis cinerea* attack of geranium (*Pelagronium zonale*) stock plants and cuttings. Eur. J. Plant Pathol. In press.

Gerson, U., Kenneth, R., and Muttath, T.I. (1979). *Hirsutella thompsonii*, a fungal pathogen of mites. II. Host-pathogen interactions, Annl . Appl. Biol., **91** : 21.

Gottwalsd, T.R. and Teders, W.L. (1983). Suppression of pecan weevil populations with entomopathogenic fugi, Enviorn. Entomol, **12** : 471.

Gotwald, T.R. and Tedders, W.L. (1984). Colonizaiton transmission and longevity of *Beauveria bassiana* and *Metarhizium anisopliae* on pecan weevil larvae, Curculio caryae in the soil, environ. Entomol., **13** : 557.

Hadar, Y., Harman, G.E. and Taylor, A.G. (1984). Evaluation of *Trichoderma koningii* and *T. harzianum* from New York soils for biological control of seed rot caused by *Pythium* spp. Phytopathology, **74**:106-10

Hall, R.A. and Burges, H.D. (1979). Control of aphids in glasshouse with the fungus *Verticillium lecanii*, Ann. Appl. Niol., **93** : 235.

Hall, R.A. (1976). *Verticillium lecanii* on the aphid, *Macrosiphoniella sanborni*, J. Inertebr. Patho, **28** : 389.

Hall, R.A. (1980). Control of aphid by the fungus, *Verticillium lecanii*, I. Effect of spore concentration, Enthol. Exp. Appl. **27** : 1.

Hall, R.A. (1981). The fungus *Verticillium lecanii* as a microbial insecticide of aphids and scales, in Microbial control of Pests and Plant Diseases 1970-1980, Burges, H. D., ED., Academic Press, New York, 483.

Hall, R.A. (1982). Control of whitelfy, *Trialeurodes vaporarioum* and cotton aphid, *Aphis gossypii* in glasshouses by two isolates of the fungus *Verticillium lecanii*, Ann. Apl. Biol., **101** : 1.

Harman, G.E. (2000). Myths and dogmas of Biocontrol: changes interceptions derived from research on *Trichoderma harzianum* T-22. Plant Dis., **84** : 377-93

Harvey, J.M. (1978). Reduction of losses in fresh market fruits and vegetables. Annu. Rev. Phytopathol. **16** : 321-324.

Hashoika, Y. and Nakai, Y. (1980). Ultrastracture of pycnidal development and mycoparasitism of *Ampelomyces quisqualis* parasite on Erysiphalse. Trans. Mycol. Soc. Jap., **21** : 329-38

Hofstein, R., Daoust, R.A. and Aeschlimann J.P. (1996). Constraints to the development of biofugicides: the example of "AQ-10", a new product for controlling powdery mildews. Entomophagia, **41** : 455-60

Ignoffo, C.M., Garcia, R., Alyoshina, O.A. and Lappa, N.V. (1979). Laboratory and field studies with boverin: a mycoinsecticidal preparation of *Beuveria bassiana* produced in the Soviet Union, J. Econ. Entomol. **72** : 562.

Ignoffo, C.M., Marston, N.L., Putler, B., Hosteter, D.L., Thomas, G.D. Biever, K.D. and Dickerson, W.A. (1976). Natural biotic agenst controlling insect pests of Missouri soyaneans, world Soyabean Rep., 561.

Ignoffo,C.M., Marston, N.L., Hostetter, D.L., Puttler, B. and Bell, J.V. (1976). Natural and induced epizootics of *Nomuraea rileyi* in soybean caterpillars, J. invertebr. pathol, **27** : 191.

Jarówaja, N. (1970). The genus *Arthrobotrys* Corda Acta mycologia, **6** : 337-406.

Jarvis, W.R. and Slingsby, K. (1977). The control of powdery mildew of greenhouse cucumber by water sprays and *Ampelomyces quisqualis*. Plant Dis. Reptr., **61** : 728-30

Keller, S. (1982). The importance of fungal diseases in the regulation of cockchafer populations Mitt. Schweiz. Entomol. Ges., **55** : 392.

Keller, S., Keller, E. and Ramser, E. (1979). Result of a field trial for the microbial control of the cockchafer *Melolontha melolontha* with the fungus *Beauveria tenella*, Mitt. Schweix. Entomol. Ges., 2, 35.

Kessel, G. (1999). Biological control of Botrytis spp. by *Ulocladium atrum*, an ecological analysis. Ph.D. thesis. Wageningen Univ., the Netherlands. 155 pp.

Kish, L.P. and Allen, G. E. (1978). The biology and ecology of *Nomuraea rileyi* and a program for predicting its incidence on Anticarsia gemmatalis in soybean, Fla. Agric. Exp. Stn. Bull., 795.

Kiss, L. (1997). Graminicolous powdery mildew fungi as new natural hosts of *Ampelomyces parasites*. Can. J. Bot., **75** : 680-83

Kohl, J., Belanger R.R. and Fokkema N.J. (1997). Interaction of four antagonistic fungi with Botrytis aclada in dead onion leaves: a comparative microscopic and ultrstrctural study. Phytopathology, **87** : 634-42

Kohl, J. and Fokkema N.J. (1998). Biological control of nectrotrophic foliar fungal pathogens. In: Plant-microbe interaction and biological control ed. Boland, G.J. and Kuykendall, L.D., pp. 49-88., New York, Marcel Dekker.

Kohl, J., Gerlagh M. and Grit, G. (2000). Biocontrol of *Botrytis cinerea* by *Ulocladium atrum* in different production systems of cyclamen. Plant Dis, **84** : 569-73

Kohl, J., Lombaers-van der Plas, C.H., Molhoek, W.M.L., Kessel, G.J. and Goossen-van de Geijn, H.M. (1999). Competitive ability of the antagonists *Ulocladium atrum* and *Gliocladium roseum* at temperature favourable for Botrytis spp. development. Biocontrol, **44** : 329-46

Kohl, J., Molhoek W.M.L., van der Plas C.H. and Fokkema N. J. (1995). Suppression of sporulation of Botrytis spp. as a valid biocontrol strategy. Eur. J. Plant Pathol., **101**:251-59

Lappa, N.V., Goral, V.M., and Drozda, V.F. (1977). Effectiveness of boverin and pecilomin in the control of the codling moth Zasch . Rast. (Kiev), **6** : 24.

Latch, G.C..M. and Falloon, R.E. (1976). Studies on the use of *Metarhizium anisoplae* to control Orycetes rhinoceros. Entomophaga, **21** : 39.

Latch, G.C.M. (1965). *Metarhizium anisopliae* strains in New Zealand and their possible use for controlling pasture inhabiting insects, J. Agric. Res., **8** : 384.

Latge, J.P. and Perry, D.F. (1980). The Utilization of an *Entomophthora obscura* resting spore preparation in biological control experiments against cereal aphids, IOBC Bull., **111** : 19.

Lemaceau P. and Alabouvette C. (1991). Biological control of *Fusarium* diseases by fluorescent *Pseudomonas* and non-pathogsenic *Fusarium.* Crop Prot. **10**(4) : 279-286

Litvinova, M. N., Aleshina, O. A. and Deysadze, T. A. (1979). Report on testing of an entomopathogenic preparation based on *Hirsutella thompsonii* obtained from the U.S.A, Sci. Res. Rep. All Union Sci. Res. Inst. Bacterial Prep., 6.

Lorito, M., Woo S.L., D. Ambrosio M., Harman, G.E and Hayes, C.K. (1996). Synergistic interaction between cell wall degrading enzymes and membrane affecting compounds. Mol. Plant-Microbe Interact., **9** : 206-13

Lumsden, R.D. and Locke, J.C. (1989). Biological control of damping-off caused by *Pythium ultimum* and *Rhizoctonia solani* with *Gliocladium virens* in soilless mix. Phytopathology **79** : 361-66

Lumsden, R.D., Ridout, CJ, Vendemia, M.E., Harrioson, DJ, Waters, R.M. and Walter, J.F. (1992). Characterization of Major secondary metabolites produced in soilless mix by a formulated strain of the biocontrol fugnus *Gliocladium virens.* Can. J. Plant Pathol. **38** : 1274-80

Lumsden, R.D., Walter, J.F. and Baker, C.P. (1996). Development of *Gliocladium virens* for damping -off disease control. Can. J. Plant Pahtol, **18** : 463-68

McCoy, C.W. and Couch, T. L. (1982). Microbial control of the citrus rust mite with the mycoacaricide Mycar, Fla. Entomol., **65** : 116,.

McCoy, C.W. and Selhime, A.G. (1977). The fungus pathogen *Hirsutella thompsonii* and its potential use for control of the citrus rust mite in Florida, Proc. Int. Citrus Congr., **2**, 521.

McCoy, C.W. (1981). Pest control by the fungus *Hirsutella thompsonii* , in Microbial Control of Pests and Plant Diseases 1970-1980. Burges. H. D., Ed., Academic Press, New York, 499.

McCoy, C.W., Beavers, G.M. and Taran , C.A. (1985). Susceptibility of *Artipus floridanus* of different isolates of *Beauveria bassiana*, Fla. Entomol., **68** : 402,.

McCoy, C.W., Brooks, R.F., Allen, J.C. and Selhime, A.G. (1976). Management of arthropod pests and plant diseases in citrus agroecosystem, Proc. Tall Timbers con. Ecol. Anim. Control Habitat managem., **6** : 10.

McQuiken, M.P. and Mohammad, O. (1997). Evaluation of commercial formulation of *Gliocladium catenulatum* (JI 1446) for biocontrol of damping-off in bedding plants. Meded Rijksfac. Landbouwwet Univ. Gent., **62** : 987-92

McQuilken, M.P and Whipps, J.M. (1995). Production, survival and evaluation of solid substrate inocula of *Coniothyrium minitans* against *Sclerotinia sclerotoirum*. Eur. J. Plant Pathol. **101**:101-10

Mints, A.S. and Walter, J.F. (1993). A private industry approach: development of GlioGard™ for disease control in horticulture. In pest Management: Biologically Based Technologies, ed. RD Lumsden, JL Vaughn, pp. 398-403. Washington, DC: Am. Chem. Soc.

Muller-Kogler, E. and Stein, W. (1970). Gewachshausveruache mit *Beaveria bassiabna* (Bals) Vuill. Zur infection von Sitona lineatus (L.) (Colept., Curcul.) im Boden, Z, Angew, Entomol, **65:** 59.

Peterson , G.D. (1977). Research on the control of the coconut palm rhinoceros beetle in Fiji, Tonga, Western samoa, Review of project results, U. N. Dev. Programme FAO, **21** : 109.

Philipp, W.D., Beuther, E, Hermann, D, Keinkert, E and Oberwalder, C, (1990). Formulation of the powdery mildew hyperparasite *Ampelomyces quisqualis* Ces. Z. Pflanzenkr. Pflanzenschts **97**:120-32

Philipp, W.D. and Hellstern, A. (1986). Biologiische Mehltauberkampfung mit *Ampelomyces quisqualis* ber reduzierter Luftfeuchtigkeit. Z. Pflanzenker. Pflanzensschutz, **93** : 384-91

Pospelo, V.P. (1944). Microbiological method of controlling agricultural pests, Proc. Leningrad Acad. Agric. Sci.; abstracted in Rev. Appl. Entomol. Ser. A, **29** : 52.

Postma, J and Rattink, H. (1992). Biological control of Fusarium wilt of carnation with a non-pathogenic isolate of *Fusarium oxysporum* Can. J. Bot., **70** : 1199-205

Primak, T. A. and Chizhik, R. I. (1975). The basis for possible use of Aschersonia aleyrodis in the control of glasshouse whitefly, Zakhyst Rosl. Resp. Mizhvid. Temat. Nauk. Zb. **22** : 53.

Pristavko, V.P. (1966). Processes pathologiques consecutif a Iaction de *Beauveria bassiana* (Bals.) Vuill. Associe a de fibles doses de DDT chez Leptinotars decemlineata Say, Entomophaga, **11**: 311.

Ramaker, P.M. C. (1983). *Aschersonia aleyrodis*, a selective biological insecticides , IOBC/WPRS, Bull., VI 167.

Ramaker, P.M.J. and Samson, R.A. (1984). *Aschersonia aleyrodis*, a fungal pathogen of whitefly. II. Application as a biological insecticide in glasshouses, Z. Angew. Entomol., **97** : 1. 1971.

Rattink, H. (1992). Biological control of Fusarium wilt diseases of carnation by a non-pathogenic isolate of *Fusarium oxysporum* . Acta Hortic., **307** : 37-42

Riba, G. (1984). Application en essias parcellaries de plein champ d'un mutant artificiel du champignan entomopathogene *Beauveria bassiana* contre la pyrale du mais, Ostrinia nubilalis, Entomophaga.

Samson, R.A. (1974). Paecelomyces and some allied hyphomycetes. Stud. Mycology **6** : 3.

Sikura, A.I. and Smietnik, A.I. (1967). The effectiveness of beauverin in conjunction with some insecticides to control *Carpocapsa pomonella*. Zashch. Râst. (Kiev). **5** : 20.

Sivasithamparam, K and Ghisalberti, E.L. (1998). Secondary metabolism in *Trichoderma* and *Gliocldium*. In: *Trichoderma* and *Gliocladium*, ed. Kubicek, C.P., Harman, G.E. 101, 1:139-91, Taylor and Fransis.

Solovey, Y. F. and Koltsov, P.D. (1976). Effect of the entomopathogenic fungus Aschersonia on the orange whitefly, Milkol, Fitopatol., **10** : 425.

Soper, R.S. (1978). Development of *Entomophthora* species as possible microbial insecticides, in Proc. First Joint US/USSR Conf. On Production, Selection and standardization of Entomopathogenic Fungi of the US/USSR Joint working Group on the production of substances by Microbiological Means, Ignoffo, C.M., Ed., 270.

Spasova, P. (1974). Use of the entomopahtogenic fugus *Aschersonia* against *Trialeurodes vaporariorum*, Gradinarstvo, **16** : 13.

Sprenkel, R.K. and Brooks, W.M. (1975). Artificial dissemination and lepizootic intiation of *Nomuraea rileyi* an entomogenous fungus of lepidopterous pests of soybeans. J. Eco. Entomol., **68** : 847.

Statz, T.E., Harmann, G.E. and Weeden N.F. (1998). Protoplasts preparation and fusion in two biocontrol strains of *Trichoderma harzianum*.; Mycologia, **80**:141-50

Sundheim, L, (1982). Control of cucumber powdery mildew by the hyperparasite *Ampelomyces quasqualis* and fungicides. Plant Pathol. **31** : 209-114

Sundheim, L. and Tronsmo, A. (1988). Hyper parasites in biological control. In: biocontrol of plant diseases, ed. K.G. Mukerji, K.L. Garg, **1**:53-69. Boca Ration : CRC Press.

Sundheim, L. (1982). Effects of four fungi on condial germination of the hyper parasite *Ampelomyces quisqualis*. Plant Pathol. **31**:340-47

Sztejnberg, A, Galper, S, Mazar, S and Liker, N. (1989). *Ampelomyces quisqualis* for biological and integrated control of powdery mildews in Israel. J. Phytopathol., **124** : 285-89

Telenga, N.A., Sikura., A.I. and Smetnik, A.I. (1964). The use of the biological material boverin in conjunction with insecticides to control *Leptinotarsa decemlineata* Say, Zashch,. Rast. (Kiev), **4** : 3.

Uchida, M. (1970). Studies on the use of the parasitic fungus *Aschersonia* sp. for controlling citrus whitefly, *Dialeurodes citri*, Bull., Kanagawa Hortic. Exp. Stn., **18** : 66.

van Lenteren, J.C. and Tommasini, M.G. (1999). Mass production, storage, shipment and quality control of natural enemies. In: Integrated pest and diseases management in green house crops, ed. Albajes et al. Dordrecht: Kluwer, pp. 276-94

van Lenteren, JC, (2000). A greenhouse without pesticides fact or fantasy ? Crop Prot. **19** : 375-84

Walstad, J.E. (1970). Anderson, R.F., and Stambaugh, W. J., Effects of environmental conditions on two species of muscardine fungi (*Beauveria bassiana* and *Metarhizium anisolipae*). J. Invertebr. Pathol., **16** : 221.

Watt, B.A. and Lebru, R.A. (1984). Soil effects of *Beauveria bassiana* on pupal populations of the Colordo potato beetle, Environ, Entomol., **13** :15.

Whipps, J.M. and Lumsden R.D. (2001). Commercial use of fungi as plant disease biological control agents: status and prospects. In: Fungal biocontrol agents - progress, problems and potential eds. Butt et al. Wallingford CAB.

Wilding, N. (1981). Pest control by Entomophthorales, in Microbial control of pest and Plant Disease 1970-1980, Burges, H.D., Ed., Academic Press, New York, 539.

Wilding, N. (1981). The effect of introducing aphid-pathogenic entomophthoraceae into field populations of *Aphis fabae*, An. Appl. Biol., **99** : 11.

Xiong, D.Z. and Wu, G. (1981). The application of *Metarhizium anisopliae* to control *Colasposoma metallicum*, Zhiwu Baohu, **3** : 36.

Yarwood, C.E. 1932. *Ampelomyces quisqualis* on clover mildew. Pbytopathology 22:31 (Abstr.)

Yenddoi, W.G. and Roberts, D.W. (1970). Is microbial control with entomogenous fungi possible? Proc. 4th Int. colloq on Insect Pathol., College Park, Md., 28,.

Yevlakhova, A. (1978). Basic trends in the use of entomopathogenic fungi in the USSR, in Proc. First Joint US/USSR Conf. On production, selection and Stadnardization of Entomopathogenic Fungi, Ignoffo, C.M., Ed., 20.

41

Commercialization of Biocontrol Agents: Problems and Prospects

H. B. Singh, A. Singh and C.S. Nautiyal

National Botanical Research Institute, Lucknow-226 001

The present review deals with the status of biological control as an alternative to the current practices of plant disease management. The detailed list of biocontrol products available worldwide has been also tabulated. The importance of commercialization of biological control agents, e.g. PGPR and fungal antagonists, the problems and prospects of its industralization are discussed.

Key words : biological control, plant disease, management, commercialization

Though biological control of plant pathogens, as it stands now, is an established sub-discipline in plant pathology, its beginning can be traced back to 1920s, when Stanford (1926) and Millard and Taylor (1927) demonstrated that control of potato scab, caused by *Streptomyces scabies* was due to the biocontrol activity of *Bacillus* sp (cf. Pal and Jalali, 1998) Weindling (1934, 1937) carried out his pioneer work on *Trichoderma* and *Gliocladium*, the two antagonistic fungi which were later to hold the top position in the totem pole of biological control. These works were followed by blurred efforts in the area of biological control. The last few years, however, have witnessed a burst in terms of activities in research efforts in this direction, and several review articles (Papvizas and Lumsden, 1980, Schroth and Hancock, 1981, Nautiyal, 2000 and Whipps, 2001) and books (Cook and Baker, 1983; Papavizas, 1981) discuss the use of specific microorganisms for the biocontrol of plant diseases. In the last decade, more than 60 biocontrol products have appeared on commercial market. Still they are just a small fraction of total number and sales of chemical fungicides in the field. In a 1993 report, the sale of biofungicides represented less than $2.5 million in sales whereas the sales of chemical fungicides were in excess of $ 29 billion. Optimistic estimates projected that the sales of biocontrol products could reach $ 15 million by the year 2000 (Powell and Justum, 1993). Therefore, it is preposterous to believe that the commercialized biocontrol agents whether in the form of biopesticides would be able to replace the chemical pesticides overnight as the foremost method for controlling plant pests viz. insects, fungi and bacteria. According

*Corresponding Author E-mail : hbs1@rediffmail.com

to Schroth and Hancock (1982) the biocontrol of phytopathogens is a fascinating, challenging but elusive and frustrating area of study for plant scientists and microbiologists alike.

The emergence of biological control as a viable and eco-friendly alternative to chemical pesticides has much to do with the well-recognized observation that a substantial amount of biological control does operate in nature. In most of the cases, the biocontrol agents are microorganisms commonly found in nature and are well known antagonists. It has also been observed that the induction of suppressiveness is related to factors that enhance their activity in a particular agricultural habitat. Most of the reports using biological control agents for disease management have been restricted to the study of different antagonists in controlling one or more diseases of a particular crop in micro-plot. However, very few reports are there for large scale and successful application of biocontrol agents in disease management. There are still fewer examples of commercialization of biological control formulations despite the origin of concept of use of biocontrol agents for increasing plant yield over 100 years ago.

Need for commercialization of biopesticides in India

Our research in the field of agriculture immediately after independence has been mainly basic. It was only after the green revolution that the research in agricultural disciplines started to move towards applied areas and the scientists started to categorize the effective researches on the basis of their reach to the farmers, how easily it could be applied in the field and the cost-benefit ratio etc. In our country and elsewhere N, P, K has been in use since ancient times through the use of cow dung, leftovers of the previous crops etc and the disease management practices were in vogue with the use of simple techniques of rouging and burning the diseased plant, use of $CuSO_4$ etc. but it was only after the commercial production of that chemical fungicides. Same is the case with biological control agents. The scientists have screened a large number of antagonists, selected the effective ones and tried them in field for controlling the diseases in micro-plots. In this process crude formulations of the effective antagonists were developed.

The next logical step of the above sequence should have been commercialization of the biocontrol products. The commercialization of biocontrol products is also necessary to overcome the negative and harmful effects of the chemicals, which are caused by their prevalence in our current schemes of disease management. This is more so in case of developing countries where the chemicals banned in most of the developed countries are still in use e.g. Pentachloronitrobenzene (PCNB) (a highly toxic fungicide is still used for disease management in our country whereas it is banned in USA). It is anticipated that the successful development and commercialization of the biocontrol fungi will effectively reduce the use of hazardous chemicals and ultimately it may eliminate them. This will help us to provide our future generations with safer and cleaner soil, water and air than it is now.

As mentioned earlier there are approximately 61 biocontrol products being produced on a commercial basis all over the world. The products are restricted in their use in the sense that they can control only one or two pathogens and have been tried on relatively few crops before being released in to the market. The efforts made in the direction of commercialization of biocontrol agents may be categorized in to three groups based on both the increasing level of difficulty and decreasing potential of repeat sales. The first group proposed to apply the antagonist directly and precisely to the infection court when and where needed, much like the use of soothing balm in medicine. This is done to preempt or swamp the pathogen by the application of high population of antagonist organism (Cook *et al.*, 1996) e.g. seed-applied antagonists for protection of germinating seeds against damping-off and antagonists applied to fruits for protection against fruit decay in storage and bacteria growing at below 0^0 C for protection of plants against frost injury (Lindow, 1983). Nearly all microbes commercialized to date fall in to this group with the exception of *Gliocladium virens* GL-21 (Lumsden and Walter, 1995) which is applied in soil in close proximity to seed or plant or with sufficiently high inoculum density to assure instantly high population in the infection court. This group of bioagents forms the first logical step in their commercialization for use against plant diseases and is important for the group's significant contribution in increasing the experience and familiarity with microbes within the industry, growers and society.

Plant Growth Promoting Rhizobacteria (PGPR) form the backbone of second group which includes antagonists being applied at one place e.g. on seeds with the intent that they will spread with the growth of the plants and protect their roots and shoots (Kloepper *et al.*, 1980; Nautiyal, 2000)). This group of antagonists may include cases exemplifying the strategy of inundative and augmentative application where the antagonist is expected to multiply, persist for part or all of the life of the plant and increase an inadequate population of a related or similar antagonist is already present in the rhizosphere (Cook *et al.*, 1996; Nautiyal, 2000). For example, a toxigenic *Aspergillus flavus* was used to spread from colonized wheat grains scattered on the soil surface on to the floral parts of cotton plants where they forestall the establishment of toxigenic strains of *A. flavus* (Cotty and Bayman, 1993). This category of examples presents a more challenging approach to biocontrol as the antagonists involved are subjected to greater vulnerability to the effects of competition and environmental conditions. No antagonist of this group has been commercialized to date.

The third category belongs to the antagonists which are released in to the infection court once or several times (Cook *et al.*, 1996). Most of the successful examples of biological control of perennial weeds and insects with microbes belong to this category. The reason for the success of this group of antagonists is that they are pathogens dependent on the target pest for their ability to multiply and persist on the host for a longer period of time. Despite all the success stories of effective biological control of plant diseases and pests the rate of commercialization of the biocontrol agents has been slow and therefore their application in the field is restricted. Of the 33 commercial biocontrol products listed by Fravel *et al.* (1999) over half of them have applications in nursery or greenhouses and many were developed specifically for soil-borne plant pathogens e.g. *Pythium* and *Rhizoctonia*. The world's total greenhouse area is 307,000 hectare including both plastic and glass (Gullino *et al.*, 1999) whereas the total land area under cultivation in 1998 was 1.51 billion hectare (FAOSTAT, 2000). The use of biocontrol is more prevalent in greenhouses and protected structures than in open field. This is another area of concern, which must be looked into before moving towards commercialization.

There are various problems associated with the slow rate of commercialization of the biocontrol products. Weller (1988) has reviewed the more important reasons for failure or inconsistent performance of the biological control in the rhizosphere with introduced microbes. They are:

(i) low disease pressure for an effective test

(ii) the treatment favours the increased damage from a non-target diseases

(iii) colonization of the roots affected by the introduced strain or the loss of ecological competence by the strain is variable and

(iv) the expression of production of antibiotics, wherever it is necessary for the effective action of antagonist, is either too late or too low in quantity to be effective for controlling the disease.

However, these limitations can be sorted out by selection of effective strains of antagonist and improvement in their activity through modern biotechnological methods. Some other challenges related to commercialization of biocontrol products are a mixture of institutional problems, technical limitations and unrealistic expectations from these products. However, we will have to overcome these obstacles if we are serious about the development of commercial products for biological control.

The rhizosphere competence of the formulation developed as a result of years of research varies with changing soil conditions (Ahmad and Baker, 1987). After isolating a rhizosphere competent strain of an effective biocontrol agent/antagonist all the research must be directed towards the development and production of the strain. The field trials are necessary supplements for the development of a product, which need to be carried out under different soil conditions in different fields, on different crops and various pathogenic fungi in order to develop a broad-spectrum product, which can effectively control a large number of pathogenic fungi under different conditions.

Another constraint is the selection of suitable substrate for the development of effective formulation. A number of substrates have been tried for the development of biocontrol formulations (Table 1). It would be

Table 1. Different substrates used for mass production of biocontrol fungi in laboratory conditions

S. No.	Substrates
1	Powdered rye grass seed
2	Diatomaceous earth granules + molasses
3	Wheat bran formulation
4	Wheat bran saw dust formulation
5	Molasses - yeast medium
6	Wheat bran peat formulation
7	Sorghum seeds
8	Sugarcane and Maize straw
9	Tapioca rind or thippi substrate
10	Talc formulation
11	Coffee fruit skin and cherry husk
12	Alginate-Wheat bran
13	Coconut coir pith
14	Alginate pellets
15	Vermiculite wheat bran and cellulose granule formulation
16	Distilled waste of aromatic plants
17	Soy flour and molasses
18	Sugarcane pressmud
19	Used tea leaves
20	Pine needles
21	Banana pseudostem
22	Pine apple peelings
23	Compost
24	Rye grains

economically viable to use the agricultural wastes viz., tea, coffee waste, rice hay, distilled wastes from oil yielding plants, wheat bran etc. We have recently developed an economical method for mass multiplication of biocontrol fungi for field application (Singh and Johri, 2000; Singh, 2002).

Number of commercialized products have been developed and are available in the world market for disease management but they are either location specific or disease specific or both. Not many products are available which are broad-spectrum one and can be used on large scale. As of today, a large number of diseases are still not covered by the commercialized biocontrol formulations. Thus, there is need for moving ahead with selecting the effective strains of biocontrol fungi and developing them as broad-spectrum formulations for use against a large number of soil-borne diseases in varying soil conditions. A large chunk of biocontrol products commercialized all over the world for managing soil borne fungal diseases of plants are *Trichoderma*-based e.g. in our country, twenty-nine manufacturers are producing *Trichoderma*-based formulations.

Economics is the key as well as an important barrier to commercialization of microbial agents for use in agriculture. This is manifested by the fact that the cost of development and registration (the policy varying in different countries) has to be paid by a very limited market e.g. the market represented by a single disease on single crop. In USA, providing the toxicological data to EPA, greatly reduces the prospect for commercialization (Cook, 1996). Another major obstacle is absence of infrastructure for scale up and commercialization of biocontrol products in most of the countries (Cook, 1993). A system similar to that developed for release of new cultivars by breeders must be developed for commercialization and dissemination of biocontrol agents to the farmers.

The ideal conditions required for development of a biocontrol agents and its subsequent commercialization are:

Selection of suitable strain

The relative ease of finding the right antagonist strain under laboratory and field conditions is an added advantage but it does ensure that the agent would work consistently and effectively in field. The search for an effective strain involves years of screening, testing and/or involvement of luck. For example Kim *et al.* (1997) screened 2000 strains of spore forming aerobic bacteria, from which strain L324-92 was selected. The discovery of phloroglucinol producing rhizosphere competent bacteria came after several years of research on phenazine-producing strains of fluorescent pseudomonads. It now forms the basis for the expectation that the right strain for biological control of take all disease of wheat is under development. The development of T-22 strain of *Trichoderma harzianum* by Harman and his associates took more than a decade and it was only in last decades of last century that the sale of its product, Root Shield, started to pick up from scratch (Harman, 2000).

Genetic engineering of effective strains or those strains, which do not have the desired traits, must also be tried. The demonstration by Huang *et al.* (1997) of the activity against *Rhizoctonia* root rot in addition to the ability of phloroglucinol producing strain to produce phenazine-1-carboxylate shows that it may be possible to produce strains with ability to control more than one disease and possibly more than one crop. This may broaden the market potential needed to justify costs of development and registration of commercialized biological control products. There are some of the biological control agents like Plant Growth Promoting Rhizobacteria (PGPR) and some strains of *Trichoderma* and *Gliocladium,* which besides controlling various soil borne diseases also enhance the yield of host crop (Windham *et al.*, 1986, MacKenzie and Starman, 1995, Kalra *et al.*, 2000; Nautiyal, 2000). It would be a boon to the industries associated with development of biocontrol products, if a strain that posses both these properties and can be produced successfully at large scale.

Extensive performance trials should be conducted before accepting one strain for advancement towards commercialization. The release of new wheat cultivars is done after conducting experiments for 100 site years (number of sites x number of years) so as to produce enough data. The industries interested in developing a strain to a commercialized product conduct their own trials extensively but only after being sure about the potential of the strains.

Shelf life and storage

For the establishment of a product containing microbial antagonists in the market it is essential that it has long shelf life and can be stored at room temperature for at least one crop season and preferably well into the next season. Mathre *et al.* (1999) suggests that a shelf life of 18-months is preferable for a commercial microbial product with a requirement of storage at room temperature. The last condition is essential, as the end users can not afford to buy costly equipment to keep the products at any temperature other than room temperature. Besides, it will render the biocontrol product economically unviable due to high cost of maintenance.

Application Technology

Generally most biocontrol agents will be applied with the same equipment currently used to apply other seed treatment products, foliar sprays or granular formulations. Development of a new application technique will not be acceptable easily unless it is more economically viable than the existing techniques and/or fit into current production scheme. For example, the so-called 'hopper box' treatment of seeds whereby the seed is treated as a part of the operations of filling the drill or planting the seeds may well replace the current practice of treatment of the seed warehouse, by treating only the seeds that will be actually planted. Thus no seeds will be treated which are not going to be used.

Scale-up and quality control

The research on biological control all over the world is initially carried out on small scale e.g. *in vitro* screening by dual culture, production on substrates in Erelnmeyer flasks, autolavable bags etc. For example, the conditions required for growth of a bacterium in 250-ml Erlenmeyer flasks is much different from that which involves the production of 10,000 liters of cells depending mainly on the fermenter used. Likewise the growth of fungal biocontrol agents in flasks produces a small quantity of formulation which can be used in microplots for disease control. When the production of biocontrol agents is scaled up for commercialization (several tons of products instead of several grams/kilograms in laboratory) the basic problem is that of maintaining the quality of the material produced, which is outside the capability of most researchers. This can be sorted out by team work where public sector researchers collaborate with the scientist of industry. When growing in large quantities the microbial agents may develop contamination or mutations at any of the various steps involved in the production.

There is another aspect regarding the use of fermentation technology for mass production of biocontrol agents. Technologically liquid state fermentation of submerged cultivation of biocontrol agents is more superior in the sense that it is easier to control the process parameters and to guarantee homogenous conditions for development of the agent. However, solid state fermentation has its own advantages. The advantages associated with solid state fermentation are:

(i) simplicity of the technique

(ii) smaller investment cost

(iii) larger quantity of product per unit substrate

(iv) reduction of energy consumption in comparison to liquid fermentation

(v) the conditions are similar to those found in nature

(vi) production of air conidia and spores that are more resistant and more durable

(vii) better yields of pure and homogenous spores

Diego *et al.* (2001) compared static and dynamic solid state fermentation of *Trichoderma resei* with the application of electromagnetic fields. They found more spores per unit substrate were produced in dynamic state fermentation than those in static state fermentation. The minds from industrial and scientific fields will have to standardize and choose the better method for commercialization.

Cook *et al.* (1996) have mentioned four adverse non-target effects of microbes used as introduced agents for biological control of pests and diseases. These are:

1. Competitive displacement of a beneficial microorganism
2. allergenicity
3. toxigenicity
4. pathogenicity

Of the four non-target effects all but allergenicity are also desired target effects depending on the mechanism involved in biological control. Since these non-target effects have been identified for all categories of biocontrol agents including insect and weed pathogens, there is a need to examine the potential of these effects on various aspects of biological control especially for antagonists of plant pathogens. Competitive displacement is one non-target effect of biological control which must be done away with. This kind of non- target effect would show up in the field as a negative rather that positive response of the treated crop. Allergenicity is another issue which was identified as worker safety issue but not as a public health issue (Cook *et al.*, 1996) and can be managed by exercising caution in workplace e.g. wearing of dust masks by the workers.

Toxigenicity becomes a major problem in biological control where the control mechanism involves the use of a secondary metabolite for inhibition of the target organism. However this should not raise concern to workers unless the toxicant produced by the antagonist is a component of the product handled and applied to the soil or plant. Probably the greatest hazard with secondary metabolites is their potential for toxigenicity to plants. This would be an agronomic problem but not an environmental or human health issue as is the case with the use of chemical pesticides.

Pathogenicity to non-target organisms is a major problem when the biocontrol agent used is an insect or weed pathogen. For antagonists the concern would be for microbes with potential as opportunistic pathogens to humans e.g. *Aspergillus ochraceus* and clinical strains of *Burkhoderia cepacia*. Analysis of the elimination of these organisms should be done during the screening process.

Every effort should be made to guarantee the safety of environment and human health while developing a microorganism on a commercial basis for controlling the plant diseases.

Success stories

The number of commercial biocontrol formulations is about sixty (Table 2). The preliminary trials for diseases management using biocontrol

Table 2. List of commercialized products

Products	Biocontrol agents	Name of the manufacturer/distributor
(1)	(2)	(3)
Actigard	Acibenzolar-S-methyl	Syngenta Crop Protection, Greensboro
Actinovate*	*Streptomyces lydicus*	Natural Industreis, Houston, USA
AQ10 biofungicide	*Ampelomyces quisqualis* isolate M-10	Ecogen, Inc. Israel
Anti-Fungus	*Trichoderma* spp.	Grondortsmettingen De Cuester, Belgium
Aspire	*Candida oleophila* I-182	Ecogen, Inc. Israel
Bactophyte	*Bacillus subtilis*	Russia
Biofungus	*Trichoderma* spp.	Grondortsmettingen De Cuester n. v., Belgium
Bas-derma	*T. viride*	Basarass Biocontrol Res. Lab., India
Binab T	*Trichoderma harzianum* (ATCC 20476) and *Trichoderma polysporum* (ATCC 20475)	Bio-Innovation AB, UK
Bioderma	*Trichoderma viride/T. harzianum*	Biotech International Ltd., India
Biofox C	*Fusarium oxysporum* (Nonpathogenic)	S. I. A. P. A., Italy
Bioject Spot-Less	*Pseudomonas aureofaciens*	Eco Soil Systems, USA
Bio-Save10LP, 110	*Pseudomonas syringae*	CTT Corp., USA

(1)	(2)	(3)
Bio-Save11	*Pseudomonas syringae*	EcoScience Corp., USA
Blight Ban A506	*Pseudomonas fluorescence*	Plant Health Technologies, USA
Blue Circle	*Burkholderia cepacia*	EcoScience Corp., USA
Cedomon	*Pseudomoans chloraphis*	BioAgri AB, Sweden
Coniothyrin	*Coniotyrium minitans*	Russia
Companion	*Bacillus subtilis* GB03, *B. lichniformis, B. megaterium*	Growth Products, USA
Contans, Intercept WG	*Coniothyrium minitans*	Prophyta Biologischer Pflanzenschutz GmbH', Germany
Conquer	*Pseudomonas fluorescens*	Mauri Foods, Australia and Sylvan Laboratories, USA
Deny (Formerly PERCEP)	*Burkholderia cepacia* and *Pseundomonas cepacia* type Wisconsin	Stine Microbial Products, Shwanee, KS
DiTera	Fermentation product of killed *Myrothecium verrucaria*	Valent Biosciences, USA
Ecofit	*T. viride*	Hoechst Schering Afgro Evo Ltd., India
Epic	*Bacillus subtilis*	Gustafson Inc., USA
Fusaclean	*Fusarium oxysporum* (Nonpathogenic)	Natural Plant Protection, France
Kali sena	*Aspergillus niger*	Cadilla Phrma., India
Galtrol-A	*Agrobacterium radiobacter* strain 84	AgBioChem Inc., USA
Intercept	*Pseudomonas cepacia, Burkhoderia cepacia*	Soil Technologies Corp., USA
HiStick N/T	*Bacillus subtilis*	MicroBio Group, UK
Kodiak, Kodiak HB, Kodiak AT	*Bacillus subtilis*	Gustaffson Inc., USA
Koni	*Coniothyrium minitans*	BIOVED Ltd., Hungary
Messenger	*Erwinia amylovora* HrpN (Hairpin protein)	BIOVED Ltd., Hungary
Mycostop	*Streptomyces griseoviridis* strain K61	Kemira Agro. Oy, Finland
Nogall, Diegall	*Agrobacterium tumefaciens*	Bio-Care Technology Pvt. Ltd., Australia
Norbac 84 C	*Agrobaceterium radiobacter*	New BioProducts Inc., USA
Paecil	*Paecilomyces lilacinus*	Technological Innovation Corporation Pvt. Ltd., Australia
Phagus	Bacteriophage	Natural Plant Protection, France
Polyversum	*Pythium oligandrum*	Biopreparatory Ltd., Czech Republic
PSSOL	*Pseudomonas colanacearum* (nonpathogenic)	Natural Plant Protection, France

(1)	(2)	(3)
Prestop, Primastop	*Gliocladium catenulatum*	Kemira Agro. Oy, Finland
Rhizo-Plus	*Bacillus subtilis* FZB24	KFZB Biotech GmbH, Germany
Root shield, Plant Shield, T-22 Planter Box	*Trichoderma harzianum* Rifai strain KRL-AG (T-22)	Bioworks Inc., USA
Root Pro, RootProtato	*Trichoderma harzianum*	Efal Agr, Israel
Rotstop	*Phlebia gigantean*	Kemira Agro. Oy, Finland
Serenade	*Bacillus subtilis*	AgrQuest Inc., USA
Soilgard	*Gliocladium virens* strain GL-21	Thermo Trilogy,USA
Supresivit	*Trichoderma harzianum*	Borregaard and Reitzel, Czech Republic
System 3	*Bacillus subtilis* and chemical pesticides	Helena Chemicals Co., USA
T-22 G, T-22 HB	*Trichoderma harzianum* strain KRL-AG2	THT Inc., USA
Trichodex, Trichopel	*Trichoderma harzianum*	Makhteshim Chemical Works Ltd., USA
Trichopel, Trichoject, Trichodowels, Trichoseal	*Trichoderma harzianum* and *Trichoderma viride*	Agrimm Technologies Ltd., New Zealand
Trichopel	*Trichoderma harzianum* and *Trichoderma viride*	Agrimm Technologies Ltd., NewZealand
Trichoderma 2000	*Trichoderma* sp.	Myocontrol Ltd., Israel
Tri-control	*Trichoderma* spp.	Jeypee Biotechs, India
Trieco	*Trichoderma viride*	Ecosense Labs Pvt. Ltd., Mumbai, India
TY	*Trichoderma* sp.	Mycocontrol, Israel
Yield Shield*	*Bacillus pumillus* GB34	Gustafson Inc., USA
Vaminoc	Mycorrhizal fungi	Biological Crop Protection Ltd., USA
Victus	*Pseudomonas fluorescens*	Sylvan Spawn Laboratory, USA
* EPA Registration Pending		

formulations involve the mass production of biocontrol fungi on different substrates. A number of diseases have been controlled by the use of biological control agents by different workers. Our group has developed a process for successful mass production of biocontrol fungi on agri-waste and filed a US patent for the process (Singh *et al.*, 1999). Our group has also filed a US patent on a novel strain of *Trichoderma harzianum* having growth promotory effect, potential to control a large number of soil-borne fungal pathogens and also nematode inhibitory effect (Kalra *et al.*, 2000). Different scientists have tried number of substrates for producing bioagents *en* masse (Table 1). There has been a number of plant diseases controlled by using biocontrol formulations developed on these wastes e.g. wilt of chickpea, collar rot of *Mentha* spp., corm rot of gladiolus etc.

Future prospects

What is the future of biological control against the plant pathogens causing various diseases all over the world? To convert the prognosis for the future of biological control the major question is how the technology can be moved from laboratory to the commercial growers. More scientific efficacy trials with proper replication and statistical analysis are needed under commercial or near-commercial conditions. This data should be in the public domain; its ready accessibility to growers and extension personnel will be more persuasive than reliance on company advertisements. The reasons for the lack of data on biological control are not far to seek. Some are:

(a) Many of the data are proprietary and found only in company reports or in registration submissions,

(b) Universities and research institutions have moved away from routine testing of products and results are published in research station bulletins which are not accessible through citation services

(c) Many trials may show lack of effectiveness and are not published.

(d) Trials showing the effectiveness of biocontrol products are not published in scientific journals because of perceived lack of the innovation.

More studies are required on the epidemiology and ecology of the pathogen, which is specially lacking for soil borne pathogens. In addition several other questions must be answered before harping on the successes of biological control in comparison to chemical control methods. For example,

(a) What is the method of introduction of pathogen and its spread?

(b) What is the relationship between the population density and damage?

(c) What are the effects of various environmental conditions on the efficacy of introduced microorganism?

Every option, which we have, doing nothing being one of them, has environmental consequences. But we have to find a way out of this quagmire to assume that our commercial products for biocontrol of plant diseases have the effectiveness and safety as their chief traits. Only then, the commercialization of biocontrol agents can be assumed to be successful.

Acknowledgement

The authors are grateful to Dr. P. Pushpangadan, Director, National Botanical Research Institute, Lucknow for providing necessary facilities and encouragement.

References

Ahmad, J.S. and Baker, R. (1987). Rhizosphere competence of *Trichoderma harzianum*. Phytopathology, **77**: 182-189.

Cook, R.J. (1993). Making greater use of introduced microorganism for biological control of plant pathogens Ann. Rev. Phytopathol, **81:** 53-80.

Cook, R.J. and Baker, K.F. (1983). The Nature and Practice of Biological Control of Plant Pathogens. St. Paul, Minn: Am. Phytopathol. Soc, 539 pp.

Cook, R.J., Bruckart, W.L. and Coulson, J.R. (1996). Safety of microorganisms intended for pest and plant disease control Framework for Scientific Evaluation. Biol. Control. **7**:333-351.

Cotty, P.J. and Bayman, P. (1993). Competitive exclusion of toxigenic strains of *Aspergillus flavus* by an atoxigenic strains. Phytopathology, **83**: 1283-1287.

Diego, S.M.M., Morla, A.G., Sofia, M.C. and Pulles, D.C. (2001). Growing of *Trichoderma resei* on static and dynamic solid substrate fermentation. (Abstract). In: International Conference on New Horizons in Biotechnology. Trivandrum, India. April 18-21, 2001. p. 74.

FAOSTAT. (2000). Food and Agriculture Organization of United Nations. http://apps.fao.org/

Fravel, D.R., Rhodes, D.J. and Larkin, R.P. (1999). Production and commercialization of biocontrol products. In: Integrated Pest Management in Greenhouse Crops. [Eds. R. Albajes, M. Lodovica Gullino, J. C. Van Lenteren, and Y. Elad]. Kluwer Academic Publishers, Dorcrecht, 545 pp. pp. 365-376.

Gullinos, M.L., Albajes, R. and van Lenteren, J.C. (1999). Setting the stage: characteristics of protected cultivation and tools for sustainable crop production. In: Integrated Pest Management in Greenhouse Crops. [Eds. R. Albajes, M. Lodovica Gullino, J. C. Van Lenteren, and Y. Elad]. Kluwer Academic Publishers, Dorcrecht, 545 pp. pp. 1-15.

Harman G.E. (2000). Myths and Dogmas of Biological Control: Change in Perceptions desired from Research on *Trichoderma harzianum* T-22. Plant Disease, **84 (4):** 377-393.

Huang,Z. Thomashow, L.S., Mavrodi, D.V., Raijmakers, Weller, D.M. and Cook, R.J. (1997). Transgenic strains for biological control of plant root diseases. Patent Application Filed 12 December 1997.

Kalra, A., Singh, H.B., Pandey, R., Patra, N.K., Kumar, S., Gupta, M.L., Dhawan, OP and Kumar, S. (2000). A process for the isolation of a novel strain of *Trichoderma harzianum* useful as nematode inhibitor, fungicide and growth promoter. Indian Patent Filed NF 101/2000.

Kim, D.S., Cook, R.J. and Weller, D.M. (1997). *Bacillus* sp. L324-92 for biological control of three root diseases of wheat grown with reduced tillage. Phytopathology, **87**: 551-558.

Kloepper, J.W., Schroth, M.N. and Miller, T.D. (1980). Effects of rhizosphere colonization by plant growth promoting Rhizobacteria on potato plant development and yield. Phytopathology, **70**: 1078-1082.

Lindow, S.E. (1983). Role of bacterial ice nucleation in frost injury to plants. Annuual Review of Phytopathology, **21**: 363-384.

Lumsden, R.D. and Walter, J.F. (1995). Development of biocontrol fungus *Gliocladium virens*: Risk Assessment and approval for horticultural use. In: Benefits and Risks [eds. MT Hokkanen and JM Lynch]. Cambridge University, Cambridge, UK.

MacKenzie, A.J. and Starman, T.W. (1995). Enhanced root and shoot growth of *Chrysanthemum* cuttings propagated with the fungus *Trichoderma harzianum*. Hort. Science, **30** (3): 496-498.

Mathre, D.E., Cook, R.J. and Callan, N.W. (1999). From discovery to use: Traversing the world for commercializing biocontrol agents for plant disease control. Plant Disease, **83** (11): 972-983.

Nautiyal, C.S. (2000). Plant beneficial rhizosphere competent bacteria. Proc. Natl. Acad. Sci., India, **70**: 107-123.

Pal, V. and Jalali, I. (1998). Rhizosphere bacteria for biocontrol of plant disease. Indian Journal of Microbiology, **38** : 187-204.

Papavizas, G.C. (1981). Biological control in crop production. Totowa, NJ: Allanhel, Osmun. 461 pp.

Papavizas, G.C. and Lumsden, R.D. (1980). Biological control of soil-borne fungal propagules. Annual Review of Phytopathology, **18**: 389-413.

Powell, K.A. and Justum, A.R. (1993). Technical and commercial aspects of biocontrol products. Pestic. Sci., **37**: 315-321.

Schroth, M.N. and Hancock, J.H. (1981). Selected topics in biological control. Annual Review of Phytopathology, **35**: 453-476.

Schroth, M.N. and Hancock, J.G. (1982). Disease suppressive soil and root colonizing bacteria. Science, **216**: 1376-1381.

Singh, H.B. (2002). Production of Biofungicide on Agriculture wastes. In : Directory of Rural Technologies Part-III. (Eds. M.C. Kunnumkal and B.R. Sant). National Institute of Rural Development, Hyderabad. pp. 434-437.

Singh, H.B. and Johri, J.K. (2000). Gramin Kshetra mein paudh rog niyantrana hetu jaivik niyantrana samagriva takniki nirman (Hindi). National Symposium on Gramin Prodyodiki, Aug. 23-24, 2000. RRL Trinuvanantapuram, P. 16.

Singh, H.B., Kalra, A., Patra, N.K., Kumar, S., Pandey, R., Khanuja, S.P.S. and Shashney, A.K. (1999). A process for the preparation of novel growth media from distillation and other plant wastes for mass multiplication of biocontrol fungi. Indian Patent No. 235/D/99 US No. 09/281784 Dt. 31.3.99.

Weindling, R. (1934). Studies on a lethal principle effective in parasitic action of *Trichoderma lignorum* on *Rhizoctonia solani* and other soil fungi. Phytopathology, **24**: 1153-1179.

Weindling, R. (1937). Isolation of toxic substances from the culture filtrates of *Trichoderma* and *Gliocladium*. Phytopathology, **27**: 1175-1177.

Weller, D.M. (1988). Biological control of soil borne plant pathogens in the rhizosphere with bacteria. Ann. Rev. Phytopathol, **26**: 379-407.

Windham, M.T., Elad, Y., Baker, R. (1986). A mechanism for increased plant growth by *Trichoderma* spp. Phytopathology, **76**: 518-521.

Whipps (2001). Microbial interactions and biocontrol in the rhizosphere. Journal of Experimental Botany, **52**: 487-511.

42

Distribution of Ectomycorrhizal Fungi in Association with Economically Important tree species in Southern India

V. Mohan

Division of Forest Protection, Institute of Forest Genetics and Tree Breeding, Coimbatore-641 002

Mycorrhizal fungi are usually beneficial symbionts for plant growth. They enhance the uptake of nutrients, especially phosphorus, increase the absorptive surface area of the roots and act as bio-control agents against soil-borne pathogens. This symbiotic mycorrhizal association is very essential to forest trees. Among the different mycorrhizal fungi, the ectomycorrhizal fungi associate with most of the gymnosperms especially the conifers and some of the angiosperms. Species of *Eucalyptus* are native to Australia and have successfully been introduced in different parts of the country by the State Forest Departments, Forest Development Corporations, Research Institutes, Industries etc. Similarly, species of Acacias and Casuarinas are also introduced to India and they grow on a variety of soils and different agro-climatic conditions. These tree species are much useful for pulp wood, pole, cheap timber etc. An attempt has been made to investigate the status of different ecotomycorrhizal fungi in association with *Acacia* spp. (*Acacia auriculiformis, A. holosericea, A. mangium*), *Casuarina* spp. (*Casuarina equisetifolia, C. junghuhniana*) and *Eucalyptus* spp. (*Eucalyptus camaldulensis, E. globulus, E. teretricornis*) in different plantations of Kerala, Tamilnadu and Pondicherry and southern India. The study revealed the occurrence of different ectomycorrhizal fungi viz., *Lycoperdon perilatum, Pisolithus tinctorius, Scleroderma* sp., and *Thelephora ramarioides* in association with different tree species. The significance of the results of the findings is presented and discussed.

Key words : Ectomycorrhizal fungi, *Acacia, Casuarina, Eucalyptus, Lycoperdon perlatum, Pisolithus tinctorius, Scleroderma* sp., *Thelephora ramarioides*

Forestry has the main objective of developing and protecting forests for their maximum productivity. With increasing demand for wood, fodder, fuel and timber, plantation forestry has gained importance in recent years. Among different plantation species, *Eucalyptus* is an economically important one extensively cultivated in different parts of the country for its pulp wood. Casuarinas are another important fast growing species of multiple end uses, is now widely planted in Southern India by the state forest departments, industries, NGOs and farmers. Similarly, species of acacias viz., *A. auriculiformis, A. holosericea* and *A. mangium* are being cultivated in different parts of Southern India due to its high value of pulpwood, cheap timber, pole and fuel. Also, the acacias are well suited for afforestation of wasteland areas. These tree species grow on a variety of soils such as red soils, laterite soils, gravely brown soils, sand dunes, shallow to deep, infertile to fertile soils and pH from acid to highly alkaline soils and can tolerate periodic water logging.

Mycorrhizal fungi are beneficial symbionts for plant growth. They enhance the uptake of nutrients, especially phosphorus (Hayman, 1982), the resistance against pathogens (Marx, 1969) and increase the absorptive surface area of the roots. Among the different mycorrhizas, the ectomycorrhizal fungi occurring in about 10% of the world's flora include most of the gymnosperms especially the conifers and some of the angiosperms. Report on distribution of ectomycorrhizal fungi in tropical angiosperm tree species is very limited (Sampangiramaiah and Ganesh Bhatta, 1996; Vijayakumar *et al.*, 1999). Hence, an attempt was made to investigate the distribution of different ectomycorrhizal fungi in association with species of difference trees viz., *Acacia, Casuarian* and *Eucalyptus* in various experimental plantations such as Progeny trial cum Seeding Seed Orchards, Provenance trials, Clonal Seed Orchard etc developed by the Institute of Forest Genetics and Tree Breeding and the State Forest Departments in Tamilnadu, Kerala and Pondicherry. The present paper gives a detailed account on the status of the distribution of different ectomycorrhizal fungi associated with economically important tree species in Sourthern India.

Periodical surveys were undertaken to study the natural occurrence of ectomycorrhizal fungi in association with various economically important forest tree speices such as *Acacia auriculiformis, A. holosericea, A. mangium, Casurina equisetifolia, C. junghuhniana, Eucalyptus camaldulensis, E. globulus* and *E. tereticornis* in experimental plantations of Tamilnadu, Kerala and Pondicherry. A total of 10 different sites were selected for *Eucalyptus camaldulensis* and *E. tereticornis*, 2 sites for *E. globulus* and 5 different sites for acacias and casuarinas. Data on the number of basidiomata in different plantations was collected at periodical intervals over a period of one year.

Root and rhizosphere soil samples underneath the fungal basidiomata were also collected and preserved for further studies. The root samples were processed for studying the mycorrhizal association by adopting the methods provided by (Agerer, 1987-1997; Mohan *et al.*, 1993). Soil samples were analysed for its nutrient status such as pH, avaialble nitrogen, phosphorus and potassium. Pure culture of these ectomycorrhizal fungal types was attempted from young, fresh, unopened basidiomata.

Distribution of Ectomycorrhizal Fungi in Different Plantations

Data on the distribution of basidiomata of different ectomycorrhizal fungi in association with trees of Acacias (3 species) was collected and the results are presented in Table 1. Total of four different ectomycorhizal fungi such as *Lycoperdon perlatum, Pisolithus tinctorius, Scleroderma* sp. and *Thelophora ramarioides* were recorded. Among them, the ectomycorrhizal fungus, *T. ramarioides* was found dominant and distributed in association with most of the tree species except *E. globulus.* The basidiomata of *P. tinctorius* were recorded in association with tree species of *E. camaldulensis* and *E. tereticornis* in different sites. Similarly, the basidiomata of *Scleroderma* sp. and *L. perlatum* were recorded in association with *E. globulus* and *E. tereticornis* and the same were not observed in plantations of other tree species.

Frequency distribution of the ectomycorrhizal fungus, *P. tinctorius* was studied in 10 different plantations of *E. camaldulensis* and *E. tereticornis.* The data was presented in Table 2. Maximum distribution of this fungus was found in *E. tereticornis* (80%), followed by *E. camaldulensis* (60%). There was no distribution of this fungus in *E. globulus* plantations at Ootacamund and Kodaikanal.

The frequency distribution of the ectomycorrhizal fungus, *T. ramarioides* was also studied in plantations of *Eucalyptus* species and the results are

Table 1. Distribution of different ectomycorrhizal fungi in association with acacias, casuarinas and eucalypts in different plantation sites in Southern India.

Tree species	Ectomycorrhizal fungi			
	Lp	Pt	Sc	Tr
Acacia auriculiformis	(-)	(-)	(-)	(+)
Acacia holosericea	(-)	(-)	(-)	(+)
Acacia mangium	(-)	(-)	(-)	(+)
Casuarina equisetifolia	(-)	(-)	(-)	(+)
Casuarina junghuhniana	(-)	(-)	(-)	(+)
Eucalyptus camaldulensis	(-)	(+)	(-)	(+)
Eucalyptus globulus	(+)	(-)	(+)	(-)
Eucalyptus tereticornis	(+)	(+)	(+)	(+)

Lp = *Lycoperdon* sp. (-) = ABSENT
Pt = *Pisolithus tinctorius*
Sc = *Scleroderma* sp. (+) = PRESENT
Tr = *Thelephora ramarioides*

Table 2: Frequency distribution of *Pisolithus tinctorius* in association with *Eucalyptus* spp. in different plantation sites in Sourthern India.

Sites	Tree Speices		
	Eucalyptus camaldulensis	*E. globulus*	*E. tereticornis*
Chennai	(+)	(-)	(+)
Kodaikanal	(-)	(-)	(-)
Neyveli	(+)	(-)	(+)
Ootacamund	(-)	(-)	(-)
Sadivayal	(+)	(-)	(+)
Sethumadai	(-)	(-)	(+)
Pondicherry	(+)	(-)	(+)
Panampally	(+)	(-)	(+)
Trivandrum	(+)	(-)	(+)
Walayer	(-)	(-)	(+)
Site frequency (%)	60	0	80

presented in Table 3. It was found that both *E. camaldulensis* and *E. tereticornis* had equal distribution (50%) of this fungus. There was no distribution of *T. ramarioides* in *E. globulus* plantations at Ootacamund and Kodaikanal over different periods of observations.

Table 3. Frequency distribution of *Thelephora ramarioides* in association with *Eucalyptus* spp. in different plantation sites in Southern India.

Sites	Tree Speices		
	Eucalyptus camaldulensis	*E. globulus*	*E. tereticornis*
Chennai	(-)	(-)	(-)
Kodaikanal	(-)	(-)	(-)
Neyveli	(+)	(-)	(+)
Ootacamund	(-)	(-)	(-)
Sadivayal	(+)	(-)	(+)
Sethumadai	(-)	(-)	(+)
Pondicherry	(+)	(-)	(+)
Panampally	(+)	(-)	(+)
Trivandrum	(+)	(-)	(+)
Walayer	(-)	(-)	(+)
Site frequency (%)	50	0	50

Table 4: Frequency distribution of *Thelephora ramarioides* in association with acacias and casuarinas in different plantation sites in Sourthern India.

Trees speices	Sites					Frequency (%)
	1	2	3	4	5	
Acacia auriculiformis	(-)	(+)	(+)	(+)	(+)	80
A. holosericea	(+)	(+)	(+)	(+)	(+)	100
A. mangium	(-)	(-)	(+)	(+)	(+)	60
Casuarina equisetifolia	(+)	(+)	(+)	(+)	(+)	100
C. junghuhniana	(+)	(+)	(-)	(+)	(+)	80

(1) = Marakanam (TN)
(2) = Neyveli (TN)
(3) = Sadivayal (TN)
(4) = Panampally (Kerala)
(5) = Pondicherry (Pondicherry)

Data on the frequency distribution of the ectomycorrhizal fungus, *T. ramarioides* was collected in different plantations of Acacias and Casuarinas (Table 4). The results indicated that the fungus had association with all the species of Acacias and Casuarinas but variation in freuqency distribution was noticed in *A. auriculiformis* and *C. junghuhniana* plantations. A *mangium* had only 60% frequency distributiion of this fungus.

It was observed that most of the ecomycorrhizal fungi emerging within a fortnight of the first showers during the months of June and July and continued appearing up to October and November. The basisdiomata of *L. perlatum*, *P. tinctorius* and *Scleroderma* sp. were distributed in irregular patterns at varying distances from the tree base in plantations with or without leaf litter. But the ectomycorrhizal fungus *T. ramarioides* was formed as a circle around the *Casuarina* trees in most of the plantation sites (Plate 1 C).

Data on mean number of basidomata of *T. ramarioides* occurring in different plantations is presented in Table 5. The results showed that maximum number of basidiomata was recorded during July to September, followed by October to December months. Less number of basidiomata was observed during April to June months. Among the different tree plantations, *A. holosericea* harboured large number of basidiomata (3965) followed by *C. equisetifolia* (3060) during July to September months.

The data on mean number of basidiomata of *P. tinctorius* occurring in different plantations of *E. camaldulensis* and *E. tereticornis* is presented in Table 6. It was found that large number of basidiomata was recorded during July to September followed by October to December months. Less number of basidiomata was recorded during April to June months. Among different tree plantations, *E. tereticornis* had maximum number of basidiomata of this fungus during different periods of observation.

Table 5: Mean number of basidiomata of *Thelephora ramarioides* occurring in different plantations in Southern India.

Trees species	Period of observation			
	Jan.-Mar. 99	Apl-Jun.99	Jul-Sep. 99	Oct-Dec. 99
	Mean Number of Basidiomata			
Acacia auriculiformis	325	0	2800	1200
Acacia holosericea	510	90	3965	1975
Acacia mangium	260	0	2110	875
Casuarina equisetifolia	490	75	3060	1400
Casuarina junghuhniana	310	65	2430	1105
Eucalyptus camaldulensis	225	50	1905	865
E. tereticornis	295	60	1855	800

The absence of basidiomata of *P. tinctorius* and *T. ramarioides* in *E. globulus* plantations at Octacamund and Kodaikanal may be due to edaphic and climatic factors. The fungus, *P. tinctorius* has been earlier reported in association with *Pinus* spp. (Marx, 1977) in the United States, *Hopea odorata* and *Shorea* spp. (Suhardi and Tasimin, 1988) in Indonesia, *Pinus kesiya, P. merkusii, P. caribaeae* (Nopamornbodi, 1988) in Thailand, *Eucalyptus tereticornis* (Natarajan *et al.*, 1988) in India and *Acacia auricultiformis* (Sampangiramaiah, Ganesh Bhatta, 1996) in India. This fungus is being first time reported from India in association with *Acacia mangium, A. holosericea* and *Eucalyptus camaldulensis* tree species in low fertile soils of Andhra Pradesh (Vijayakumar *et al.*, 1999).

Similarly, the ectomycorrhizal fungus, *Thelephora ramarioides* was well distributed in all the plantation sites surveyed (Tables 3 and 4). Among them,

Table 6: Mean number of basidiomata of *Pisolithus tinctorius* occurring in different plantations in Sourthern India.

Tree species	Period of observation			
	Jan.-Mar. 99	Apl-Jun.99	Jul-Sep. 99	Oct-Dec. 99
	Mean Number of basidiomata			
Eucalyptus camaldulensis	350	150	750	550
E. tereticornis	600	230	1350	850

PLATE

(A) Basidiomata of *Pisolithus tinctorius* in *Eucalyptus camaldulensis* plantation.
(B) Basidiomata of *Pisolithus tinctorius.*
(C) Basidiomata of *Thelephora ramarioides* in *Casuarina equisetifolia* plantation.

maximum distribution was observed in *A. holosericea* and *C. equisetifolia* (100%) followed by *A. auriculiformis* and *C. junghuhniana* (80%) and less distribution in *A. mangium* (60%), *E. camaldulensis* and *E. tereticornis* (50%). This ectomycorrhizal fungus has been earlier reported in association with *Casuarina equisetifolia* in coastal belts in South East Asian countries by Corner (1968). This fungus has been first time reported in association with *Casuarina equisetifolia* plants in the Coromandel Coast near Madras (Tamilnadu), India (Natarajan *et al.*, 1988). In the present study, this fungus is being first time reported from India in association with experimental trials of *Acacia holosericea, A. mangium, Eucalyptus camaldulensis* and *E. tereticornis* trees in South India.

Another fungus *Scleroderma* sp. was recorded in assocation with *E. tereticornis* and *E. globulus* trees in limited plantation sites during the period of observation (Table 1). In India, Bakshi (1966) recorded *Scleroderma verrucosum* as the ectomycorrhizal symbiont in *Eucalyptus*. Thapar and Singh (1967) have reported the natural occurrence of *Scleroderma verrucosum* and *S. cepa* in several *Eucalyptus* species from North India. Bakshi (1974) also reported the association of *Scleroderma bovista, S. cepa, S. geaster* and *S. verrucosum* in *Pinus* spp. Sharma and Mishra (1988) recorded *Scleroderma auranticum* as the symbiont in *Pinus kesiya* in Shillong region of India. Similarly, Mohan (1991) reported the association of *Scleroderma citrinum in Pinus patula* plantations in the Nilgiri Hills, Tamil Nadu, Southern India. The association of *Scleroderma* sp. with *Shorea* spp. was reported by Suhardi and Tasimin (1988). Nopamornbodi (1988) reported the association of *Scleroderma flavidum* with *Pinus kesiya* and *S. auranticum* with *Pinus merkusii.* Vijayakumar *et al.*, (1999) also reported this fungus in assocation with *Acacia* species in Andhra Pradesh.

Root samples collected underneath the basidiomata of different fungi in association with different tree species were processed for identification of the fungal association. All the samples had ectomycorrhizal association in the roots showing fungal mantle and Hartig-net structures.

Conclusion

The acreage under species of Acacias, Casuarinas and Eucalypts is increasing both in forest plantations and agricultural lands. Mycorrhization of forest crops has attracted considerable attention over the last few years because of their role as biofertilizer improving host growth as well as contributing to disease suppression (Marx, 1969, 1973; Sampangiramaiah and Bagyaraj, 1989). Hence, the natural association of *P. tinctorium, Scleroderma* sp. and *T. ramarioides* with acacias, casuarinas and eucalypts is of considerable significance and needs further study and exploitation.

References

Agerer, R. (1987-1997). Colour Atlas of Ectomycorrhizae. 1st-8th edn. Schwabisch Gmund, Germany:Einhorn-Verlag.

Bakshi, B.K. (1966). Mycorrhiza in *Eucalyptus*. Indian Forester, **92:** 19-21.

Bakshi, B.K. (1974). Mycorrhiza and its role in forestry. P.L. 480 Project Report, Forest Research Institute and Colleges, Dehra dun, India. 89 pp.

Corner, E.J.H. (1968). A monograph of *Thelephora* (Basidiomycetes). Nova Hedwigia, Beih., **27**: 1-110.

Hayman, D.S. (1982). Practical aspects of Vesicular-Arbuscular mycorrhiza. In : Advances in Agricultural Microbiology (Ed.) N.S. Subba Rao. Oxford and IBH Publishing Co., New Delhi, India pp. 325-373.

Marx, D.H. (1969). The influence of ectotrophic mycorrhizal fungi on the resistance of pine roots to pathogenic infections. I. Antagonism of mycorrhizal fungi to root pathogenic fungi and soil bacteria. Phytopathology, **59**: 158-163.

Marx, D.H. (1973). Mycorrhizae and feeder root diseases. In: *Ectomycorrhizae - their ecology and physiology*. (Eds. G.C. Marka and T.T. Kozlowski). Academic Press, London. pp. 351-382.

Marx, D.H. (1977). Tree host range and world distribution of the ectomycorrhizal fungus *Pisolithus tinctorius*. Can. J. Microbiol., **23** : 217-223.

Mohan, V. (1991). Studies on ectomycorrhizal association in *Pinus patula* plantations in the Nilgiri Hills, Tamilnadu. Ph.D. thesis, University of Madra, Madras, India. 260 pp.

Mohan, V., Natarajan, K. and Ingleby, K. (1993). Anatomical studies on ectomycorrhizas. I. The ectomycorrhizas produced by *Thelephora terrestris* on *Pinus patula*. Mycorrhiza, **3:** 39-42.

Natarajan, K., Mohan, V. and Kaviyarasan, V. (1988). On some ectomycorrhizal fungi occurring in Southern India. Kavaka., **16:** 1-7.

Nopamornbodi, O. (1988). Status of mycorrhizal research in Thailand. In: *Mycorrhizae for Green Asia. Proc. Ist Asian Conf. on Mycorrhizae.* (Eds.) A. Mahadevan, N. Raman and K. Natarajan. University of Madras, Madras, India. pp-27.

Sampangiramaiah, K. and Bagyaraj, D.J. (1989). Root diseases and mycorrhizae-a review. Journal of Phytopathological Research, **2:** 1-6.

Sampangiramaiah, K. and Ganesh Bhatta, S.M. (1996). Natural association of *Pisolithustinctorius* with *Acacia auriculiormis* in India. In: *Mycorrhizae: Biofertilizers for the furture.* (Eds.) Alok Adholeya and Sujan Singh. *Proc. 3rd National Conf. on Mycorrhizae.* TERI, New Delhi, India. pp. 65-66.

Sharma, G.D. and Mishra, R.R. (1988). Production of mass inoculum and inoculation techniques of ectomycorrhizal fungi in sub-tropical pine (*Pinus kesiya*). In: *Mycorrhizae for Green Asia. Proc. Ist Asian Conf. on Mycorrhizae.* (Eds.) A. Mahadevan, N. Raman and K. Natarajan, University of Madras, Madrasm, India. pp. 319-321.

Suhardi, I. and Tasimin, Y.F. Status of mycorrhizal research in Indonesia. In: *Mycorrhizae for Green Asia. Proc. Ist Asian Conf. on Mycorrhizae.* (Eds.) A. Mahadevan, N. Raman and K. Natarajan. University of Madras, Madras, India. pp. 14-15.

Thapar, H.S. and Singh, B. (1967). Mycorrhiza in *Eucalyptus*. Indian Forester, **93**: 756-759.

Vijaya Kumar, R., Prasada Reddy, B.V. and Mohan, V. (1999). Distribution of ectomycorrhizal fungi in forest tree species of Andhra Pradesh, Southern India - A New record. Indian Forester, **125** (5) : 496-502.

43

Arbuscular Diversity of VAM Fungi

D. Vyas*, O.P. Dwivedi, R.K. Yadav and K.M. Vyas

Lab. of Microbial Technology and Plant Pathology,
Department of Botany,Dr. H.S. Gour University
Sagar (M.P.) 470 003

The structural diversity of Vasicular-arbuscular mycorrhizas which are now also known as AM fungi has been assessed. Two major structural classes of VAM the *Arum* and the *Paris*-types are discussed in detail. *Paris*-type occurred more frequently in the plant kingdom. In Bryophytes presence of aseptate hyphal coils has been reported. Both Arum-type and Paris-type of VA mycorrhizal structure have been recovered from the roots of Filicales (true fern). However, extensive hyphal coiling of *Arum*-type was reported in Pteridophytes. Paris-type of hyphal coiling dominates in the Gymnosperms. Arum-type and Paris-type of VAM are widely scattered throughout, both in mono and dicotyledones families of Angiosperms. Our recent work suggests that wild or ornamental plants belonging to genus *Lantana* of Verbenaceae family also shows both types (*Arum* and *Paris*) of VAM structures. Similarly, in wheat of Gramineae family both Arum and Paris types of VAM structures are observed. The presence or absence of intercellular space and difference in wall structure of cortical cells of host may probably be responsible for the occurrence of different types of endophyte structure of VA mycorrhizas.

Key words : VA mycorrhiza, *Arum* and *Paris*-type, diversity.

Mycorrhizas (fungus roots) are symbiotic association between specialized soil fungi and plants. Seven types of mycorrhizas are recognized (Brundrett, 2002). Of these, Vesicular-arbuscular mycorrhizas (VAM also called arbuscular mycorrhizae), are the most wide spread in plants. The VAM fungi are placed in the Zygomycetes, order Glomuales in the genera *Aculospora, Glomus, Scutellospora, Gigaspora, Paraglomus* and *Arachaespora* (Morton and Redcket, 2001). These fungi are considered to be primitive due to their relatively simple spore, their lack of sexual reproduction and because there are relatively few species of these fungi and they associate with a wide diversity of plants (Morton,1990). These fungi are incapable of growth without plants. Ribosomal genome diversity within these fungi is consistent with the absence of sexual reproduction and makes it difficult to define the species and indivisuals (Hosny *et al.* 1999;Pringle *et al.*,2000). The functional diversity of these fungi is likely to be much greater than is suggested by the common observations on a number of currently recognized species (Brundrett, 1991; Abbott *et al.*,1992).

*For Correspondence : e-mail: dvyas1@rediffmail.com; dvyas@sancharnet.in

The study of VAM fungi has been impeded by their obligate biographic nature and by difficulties in the identification of spores, especially in the field material. In this situation, molecular techniques have enormous potential and there have been a number of important recent advances. Both mitochondrial DNA and the multistory nuclear ribosomal RNA (r-RNA) gene have been used for species identification and phylogenetic analysis of fungi (Bruns *et al.*, 1991). In mycorrhizal symbiosis formed between plant roots and the arbuscular mycorrhizal (AM) fungi, the Glomales is of great interest to botanists because of its potential influence on ecosystem processes, its role in determining plant diversity in natural communities and the ability of these fungi to induce a wide variety of growth responses in coexisting plant species. Little attention, has however, been paid to the ecological role of diversity of AM fungi, athough the, AM fungi have been shown to play a significant role in the floristic diversity and structure of annual and perennial plant communities. (Grime *et al.* , 1987; Gange *et al.* , 1990; Sanders and Koide, 1994).

The lack of basic information about the AM fungi stems from a few fundamental difficulties encountered in their identification, culture and taxonomy, coupled with the great difficulty in manipulating AM fungi in natural ecosystems without greatly modifying the environment in other ways. In this article, we outline briefly as to why the study of AM fungal diversity warrants greater attention.

Understanding the real, rather than the potential significance of AM fungal diversity in natural communities has posed a greater problem. For a given plant community we need to know how diverse the AM fungal community is, which plant roots are colonized by which AM fungi, whether there is seasonality in the patterns of colonization, whether any specificity between plant and AM fungi occurs and what will be the effect of those AM fungi on plants and the ecosystem. The difficulty of identifying AM fungi in the roots of plants has always been an obstacle to their study in natural communities, though the identification of AM fungi in roots based on morphological observations has been successful in pot experiments (Abbot, 1982). The usefulness of this technique for field investigations appears to be limited because the hyphal morphology of one AM fungal isolate is likely to vary with host species (Gallaud, 1905; Gerdemann, 1965; Lackie *et al.*, 1987).

Structural Diversity in AM Symbiosis

In the whole plant kingdom, two major structural classes of VAM fungi are found (Gallaud, 1905). These are (1) the *Paris*–type, defined by the absence of the intercellular phase and presence of extensive intracellular hyphal coils

(Fig.1). (2) the *Arum*–type, defined on the basis of an extensive intercellular phase of hyphal growth in the root cortex and development of terminal arbuscules on intracellular hyphal branches (Fig. 2). Arbuscules are intercalary structures on the coils. These two structural classes of vesicular arbuscular mycorrhizas, were named *Arum* and *Paris*-type, after the plants in which they were first described. In both classes, the initial epidermal penetration results in intracellular hyphae, often forming coils (pelotons) in the hypodermis where present and in outer cortex. The *Arum* type mycorrhiza was characterized by the existence of extensive intercellular hyphae underlying

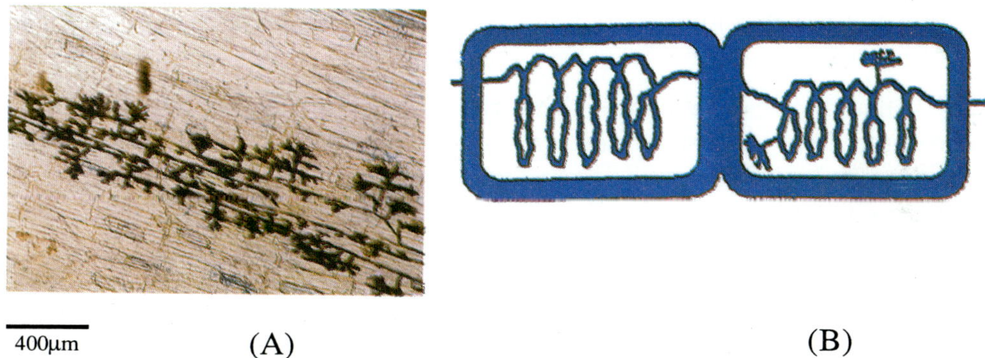

Fig.1 : (A) *Paris*-type structure in the cortex of *Triticum aestivum* L. (Bar x 400)
(B) Schematic Diagram of *Paris*- type structure.

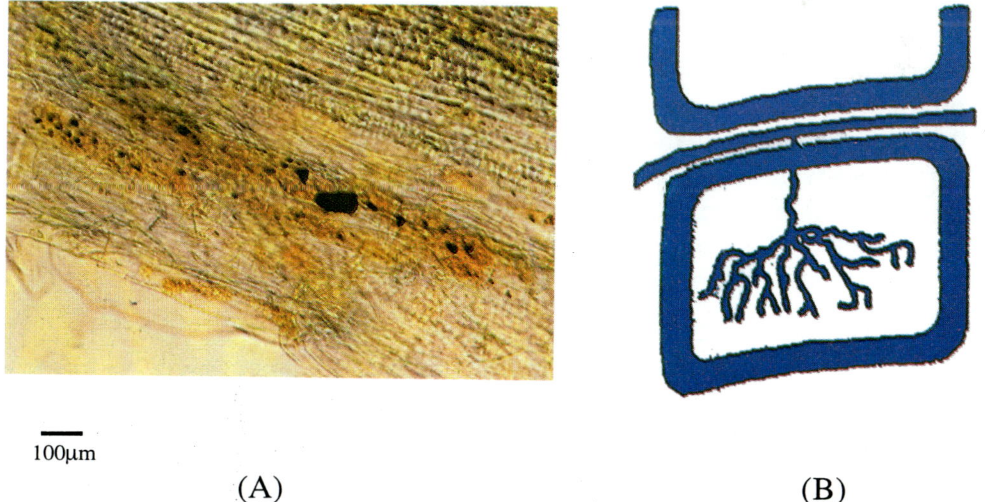

Fig.2 : (A) *Arum*-type structure in the cortex of *Allium cepa* L. (Bar x 100).
(B) Schematic Diagram of *Arum* -type structure.

this initial phase of colonization. Arbuscules are formed normally as terminal structures on so called trunk hyphae. Vesicles can be intercellular or intracellular and are not formed by all VA mycorrhizal fungi. Accordingly, the term "arbuscular mycorrhiza"(AM) is coming into use in place of VAM.

There have been many reports showing that in *Paris*-type, arbuscules are relatively few in numbers, small or absent altogether. *Paris* type occurs more frequently in the plant kingdom than *Arum* type and predominate in ferns, gymnosperms and many wild angiosperms. The cultivated herbs that are the subject of much experimental work are mostly *Arum* type. In 41 angiosperm families there are records of only *Paris* type VA mycorrhizas and while *Arum*-type is found in 30 families. Another 21 families have examples of both classes or intermediates between them.

Gallaud (1905), defined the *Paris* type VA mycorrhiza by the absence of intercellular hyphae. Instead, there are extensive coils of intracellular hyphae within cortical cells, form which arbuscules are normally formed as intercalary structures. One of the cells illustrated contained no arbuscules, but small and indistinct arbuscules were present in adjacent cells. Vesicles, where present in *Paris* types, are intracellular. A good amount of experimental work has been done with temperate crop plants that have *Arum* type mycorrhizas (e.g. *Allium*, *Trifolium* and *Vicia* etc.).

On the basis of an earlier classic study also (Janse, 1897); the two classes are clearly distinguishable, although the fine details of arbuscules are not recognized. However, the lack of taxonomic knowledge of these fungi led to some confusion, the emphasis was mainly on similarities with the intracellular hyphal structures–especially the hyphal coils, found in other endomycorrhizas. The situation was further confused by Peyronel (1923, 1924), who studied both types of structures and concluded the fungi that produced hyphal coils in *Paris* types were secondary colonists in roots which were already colonized by the true VA mycorrhizal fungi, Peyronel believed that these secondary fungi did not produce arbuscules or vesicles and that the hyphae were septate to a degree that does not occur in the fungi from VA mycorrhizas.

The confusion over whether *Paris* types are "genuine" VA mycorrhizas should have been laid to rest by experimental investigations which showed conclusively that VA mycorrhizal fungi isolated from *Arum* type hosts produce *Paris* type structures in other hosts and *vice versa*. Work by others suggests that the Solanaceae includes representatives of both classes. A similar investigation made was by Gerdmann (1965), who showed that the same Glomalean isolates produce a *Paris* type VA mycorrhiza in *Liriodendron* (tulip tree) and an *Arum* type in *Zea*. Jacquelinet – Jeanmougin and Giainnazzi –

Pearson (1983) like wise showed that the development of *Paris* type structures in *Gentiana* was produced by Glomalean fungi that formed *Arum* type in *Allium*. The Gentianaceae, which includes achlorophyllous (Mycoheterotrophic *sensu* Leake, 1994) genera, is a classic *Paris* type family, as first shown by Janse (1897); Gay *et al.* (1982) and McGee(1985).

Control of many major mycorrhizal structures by the genome of the host was also emphasized by Bonfante–Fasolo and Fontana (1985), Daniels–Hetrick *et al.* (1985) and Smith (1995). Nevertheless, despite the impressive experimental evidence, it is still possible that the fungal genome might also exert some control over the structures.

There have been numerous studies of the development of *Arum* type structures, including those by Mosse and Hepper (1975) with *Trifolium*, Bonfante –Fasolo and Scannerini (1977) with *Ornithogalum*, Holley and Peterson (1979) with *Phaseolus*, Morton (1985) with *Zea*, Alexander *et al.* (1989) with *Allium*, *Phaseolus* and *Zea* and Weber, *et al.*(1995) with plants in the Apocynaceae. Detailed work on development of *Paris* type (or near *Paris* type) includes that by Kinden and Brown (1975, 1976) with *Liriodendron*, Bonfante–Fasolo and Fontana (1985) with *Ginkgo*, Yawney and Schultz (1990) and Cooke *et al.* (1992,1993) with *Acer*, Weber *et al.* (1995) with plant in the Apocynaceae. Our interest in structural diversity in VA mycorrhizas arose from considerations of differences in mycorrhizal efficiency i.e. the variation in nutritional benefits and costs to the plant that are associated with differences in amounts of inorganic nutrients such as P or Zn transferred to the plant in return for organic carbon (e.g. Sugar) transferred to the fungus. Previously, it has been widely assumed that bi-directional transfer of inorganic nutrients and organic carbon occurs solely across arbuscules. Indeed in their taxonomic G-revision of the Glomales, Morton and Benny (1990) took this assumedly unique role of the arbuscules as a fundamental feature of the order.

DIVERSIFICATION OF AM FUNGI:

The AM symbiosis is ancient, the fossil record suggests that it occurred in the first land plants (Pyrozynski and Dlape, 1989; Remy *et al.*,(1994). However, morphological diversification is low, with only 152 recognized species in six genera (Walker and Trappe, 1993). Law (1985) pointed out that considering the age of the symbiosis this diversification is low, and he suggested that this is typical of mutulistic symbioses, as there are few external pressures which would favour the selection of new traits. However, the low divergence of AM fungi seems at odds with their functional diversity.

Until recently, methods for studying AM fungal diversity and the ecology of the symbiosis have been wholly reliant on the morphology of the spore phase. Spore of AM fungi are relatively large, easy to extract from soil and have several morphological characters that allow species identification to be determined by experienced personnel. However, the use of spore data alone for the assessment of ecosystem diversity and ecology has long been recognized as unsatisfactory (Walker *et al.*,1982; Klironomos *et al.*, 1993; Rolden–Fajardo, 1994). The relationship between the morphological diversity of AM fungi and their genetic and functional diversity has not been established. According to Walker (1992) many described taxa are "Workable" in that they are recognizable and are found repeatedly in different parts of the world albeite in diverse environments. For example, several species of AM fungi which were reported in a semi – arid site in Australia (Mc Gee, 1989) also occurred in a species rich meadow in the north of England (Sanders, 1993), both ecosystems comprising completely different soils, vegetation and climate. This indicates that AM fungi are very plastic in their environmental adaptability.

Distribution of the Two Classes in the Plant Kingdom

1. Bryophytes and Pteridophytes

The absorbing organs of a wide range of lower land plants have long been known to be mycotrophic and these groups were studied by Janse (1897) and Gallaud (1905). Among many other early workers, Rayner (1927) and Harley (1969) have given good account of early work, and Bonfante-Fasolo (1984) has reviewed developmental aspects. Bryophytes are colonized by a range of fungal endophytes and previous uncertainties about the fungal taxa are now being addressed (Duckett & Read, 1995). VA mycorrhizal Glomalean fungi occur as endophytes in many Brophytes and they are very widespread in Pteridophytes. Where the cells of the absorbing organs have ready access to the soil environment, as some thalloid structures lacking epidermal-type layers, there can be direct penetration of external hyphae between cells (i.e. formation of an inter-cellular phase) but in the intracellular phase extensive aseptate hyphal coils are common, and arbuscules (where present) are intercalary rather than terminal on hyphae. A good example is the gametophyte of *Phaeoceros* (Bryophyta: Anthocerophyta), which has been studied in detail by Ligrone (1988). The prothalli (gametophytes) of different *Lycopodium* spp. (Pteridophytes) show variation in .mycorrhizal structures (the absorbing organs are not roots !), as first discussed by Rayner, (1927). Whereas *L. cernuum* had intracellular hyphae in the peripheral region, there were intercellular hyphae in the central tissues. In other species, the hyphae are entirely intracellular. Interestingly, the prothalli of *L. cernuum* contain photosynthetic cells while those of the second group are achlorophyllous. These examples aside, and allowing for the simple anatomy of the absorbing organs, bryophytes and' lower' Pteridophytes (non- Fillicales) mostly have V A structures that are *Paris*- types. However, absence of arbuscules seems quite common, as found for the thalloid hepatic *Cryptothallus* by Pocock and Duckett (1984), for *Lycopodium* by Duckett & Ligrone (1991) and for *Psilotum* (Pteridophyte) by Peterson *et al.* (1981); Janse (1897).

Both *Arum*-type and *Paris*-type VA mycorrhizal structures have been recorded in roots of the Filicales (true ferns.). Extensive *Arum*-type intercellular hyphae and arbuscules were first described in *Angiopteris* (Marattiaceae: Marattiales) by Janse (1897) and Gallaud (1905). This structure was later confirmed by Boullard (1958), who showed that in contrast, *Danaea* (same family) forms the *Paris*-type. The position in the Ophioglossales is also not clearcut. Janse (1897) described intercellular hyphae in '*Ophioderma*' (*Ophioglossum*) but Gallaud (1905) included mycorrhiza in *Ophioglossum* as a *Paris*-type, stating that coils are absent but the hyphae are always intracellular. Descriptions and illustrations by Boullard (1958) show *Paris*-type features including coils, vesicles and the remnants of arbuscules in *Ophioglossum* and the related genus *Botrychium*. As far as records indicate the representatives of all families in the leptosporangiate Filicales that have been examined appear to have *Paris*-type structures (Boullard, 1958; Cooper, 1976; Berch and Kendrick, 1982; Harley, 1969).

Interestingly, Remy *et al.* (1994) have now unequivocally demonstrated the presence of arbuscules in the early Devonian fossil *Aglaophyton*, which shows features of both bryophytes and vascular plants. It would be very interesting to know if this plant was a *Paris*-or *Arum*-type: it ought to be possible to detect the occurrence of intercellular hyphae or intracellular coils in the fossils. The structures described by Kidston and Lang (1921) from the same fossil flora (Rhynie chert) included, coiled intracellular hyphae, suggesting that they were *Paris*-types.

2. Gymnosperms

There have been many studies of VA mycorrhizas in gymnosperms, and in all cases excepting possibly one, they can be classified as *Paris*-types, irrespective of whether or not the investigators have used this term. Examples include Podocarpaceae: (Janse, 1897; Shibata, 1902; Gallaud, 1905: Baylis *et al.*, 1963); Taxaceae: (Prat, 1926; Strullu *et al.*, 1981); Taxodiaceae (Gallaud, 1905; Konoe, 1957). Bonfante-Fasolo (1984) has given a general survey. The possible exception is *Ginkgo* (Ginkgoaceae) in which Bonfante-Fasolo and Fontana (1985) found rare intercellular hyphae and vesicles and abundant intracellular hyphal coils and intercalary arbuscules in the inner cortex. Accordingly, *Ginkgo* should perhaps strictly be included in a 'near-*Paris*' or 'Intermediate' class; it is certainly not an *Arum*-type. We note here that the presence of rare intercellular hyphae in other examples classified as *Paris*-types might compromise the validity of the classification. More such examples will be mentioned. The methodology of the investigation might be a problem in that the presence of rare intercellular hyphae may be obscured by the dense intracellular coils and missed altogether when root squashes are examined, rather than sections. Also, standard staining methods are not uniformly successful. Early workers, such as Janse (1897) and Gallaud (1905), and others including Bonfante-Fasolo and Fontana (1985) did not fall into these traps.

3. Angiosperms

Arum-type and *Paris*-type VA mycorrhizas are widely scattered throughout both in monocotyledonous and dicotyledonous families in the angiosperms.

Table 1 lists families in each class, giving numbers of genera that have been studied in each family, and using the classification by Cronquist (1981), In some cases the names in the early literature had to be updated using Mabberley (1989) for this purpose. While it is not established that all species in one genus are all within the same class, we have found only one case (*Ranunculus*) where different species within a single genus appear to form VA mycorrhizas of both classes, and another (*Zea*) where there may be major differences between cultivars. These are discussed below Table 1 relies heavily on the surveys made by Janse (1897), Gallaud (1905), Johnston (1949) and Stelz (1968), with examples from other papers as listed. Not all authors have used Gallauds classification, and we have placed families in the *Arum*-type and *Paris*-type classes on the basis of records of presence or absence of intercellular hyphae in the cortex; and in the latter class, presence of hyphal coils.

Generally, all examples included have been claimed to be VA (or arbuscular) mycorrhizas by the authors cited, except of course where the work was done before these terms were introduced. Fortunately, early workers usually included detailed illustrations which have helped confirm the basis of the classification. We have not been deterred by reports of total absence of arbuscules, since all such cases in Table 1 (though not identified there) included statements of presence of vesicles (sometimes rare) and hyphae that are mainly aseptate. Some potentially useful lists have been excluded because, although they include plants with extensive intracellular hyphae, they do not specify whether intercellular hyphae are absent, Examples are based on surveys by

Table 1. Family groupings of angiosperms having distinct *Arum*-type and *Paris*-type VA mycorrhizas and those with both types and/or intermediate types

c	Arum-types	Paris-types	Both types (B) and/or intermediate types (I)
Monocots	Agavaceae	Burmanniaceae	Araceae
	Araceae	Cannaceae	Gramineae
	'Liliaceae'	Dioscoraceae	
	Alliaceae	Heliconiaceae	Arecaceae
	Asphodelaceae	'Liliaceae'	Pandanaceae
	Anthericaceae	Colchicaceae	
	Convallariaceae	Liliaceae .	
	Hyacinthaceae	Trilliaceae	
	Hypoxidaceae	Uvulariaceae	
	Ruscaceae	Mararitaceae	
	Zingiberaceae	Thismiaceae,	
		Triuridaceae	
Dicots	Alangiaceae	Aceraceae	Apocynaceae
	Anacardiaceae	Annonaceae	Asclepiadaceae
	Asclepiadaceae	Araliaceae	Burseraceae
		Aristolochiaceae	Caprifoliaceae
	Balsaminaceae	Bombacaceae	Euphorbiaceae
	Begoniaceae	Caricaceae	Flacourtiaceae
	Boraginaceae	Casuarinaceae	Labiatae
	Campanulaceae	Cecropiaceae	Leguminosae
	Combretaceae	Cornaceae	Caesalpinoideae
	Compositae	Cunoniaceae	Mimosoideae
	Cucurbitaceae	Gentianaceae	Papilionoideae
	Elaeocarpaceae		
	Guttiferae	Grossulariaceae	Meliaceae
	Malvaceae	Hamamelidaceae	Menyanthaceae
	Oleaceae	Hippocastanaceae	Ranunculaceae
	Proteaceae	Lecythidaceae	Rubiaceace
	Rosaceae	Linaceae	Rutaceae
	Staphyleaceae	Loganiaceae	Scrophulariaceac
	Symplocaceae	Magnoliaceae	Solanaceae
	Thymeleaceae	Malpighiaceae	Sterculiaceae
	Turneraceae	Melastomaceae	Verbenaceae
	Urticaceae	Moraceae	
	Vitaceae	Myrsinaceae	
		Myrtaceae	
		Polygalaceae	
		Rubiaceae	
		Sapindaceae	
		Saxifragaccac	
		Theaceae	
		Ulmaceae	
		Umbelliferae	
		Violaceae	

Thapar and Khan (1973) of forest trees (both, gymnosperms and angiosperms) in India, and by Louis (1990) of coastal plants in Singapore. As already noted, Louis was conscious of the possible functional implication of absence of arbuscules.

In many cases, Table 1 gives records for only one species or genus per family, which is a hazardous basis for generalizing but where different workers have made examinations of the same species or genera they have almost always come to the same conclusion about structures. Examples to the contrary are discussed below. Thus, classic *Arum*-type families include Compositae and Malvaceae, whereas *Paris*-type families include Aceraceae and Gentianaceae. This list is not exhaustive, and it does not include a number of early surveys published in languages other than English that we have not seen. Nevertheless, it shows that *Paris*-types are indeed very common, as suggested by Brundrett and Kendrick (1990 b).

The Liliaceae as defined by Cronquist seems at first sight a particularly difficult family, with representatives of both classes (Table 1). It includes variations of *Paris*-types that have distincitive intracellular hyphal swellings (bobbits). These were described in detail by Widden (1996) in *Clintonia* and *Medeola.* Rather similar features in *Colchicum* were noted by Gallaud (1905). The swellings are a consistent feature of these plants. The classification of Liliaceous plants and the different structural classes can be distributed quite neatly (bearing in mind the limited numbers of genera surveyed) between the families listed by Dahlgren *et al.* (1985) (Table 1). Table 2 gives more details, and shows the distinct distribution within the Super order Liliiflorae *sensu* Dahlgren *et al.* (1985), which includes taxa from outside the Liliaceae *sensu* Cronquist (1981). *Ornithogalum* (Hyacinthaceae) was described as an *Arum*-type by Gallaud. A detailed study of *O. umbellatum* by Bonfante-Fasolo and Scannerini (1977) showed extensive intracellular hyphal coils in the outer cortex with intercellular hyphae in the inner cortex. This difference appears in other plants. Extensive ramifications of intercellular hyphae between epidermis and cortex in *Thysanotus* (Anthericaceae) are another example of a variant in the *Arum*-type class (McGee, 1988). Structural variation within *Paris*-type dicots is illustrated by the presence of lobed intracellular hyphae in *Linum catharticum* (Gay *et al.*, 1982).

Families with genera of both classes are also listed in Table 1 (last column: 'B'). Examples include the Apocynaceae, Euphorbiaceae and Solanaceae (this last being a favorite for experimental studies). Some plants contain *Paris*-type features (especially extensive hyphal coils) but also many intercellular hyphae: see Table 1: last column ' I '. The Apocynaceae again includes genera in this category (Weber *et al.*, 1995). Obviously, as with the gymnosperm *Ginkgo*, discussed above, there is a complication where rare intercellular hyphae have been recorded in roots that are otherwise distinctly *Paris*-type. *Theobroma* (cacao) is one such example (Laycock, 1945; Nadarajah, 1980). Gallaud (1905) wrestled with these issues when he found that some *Ranunculus* spp. fell into the *Paris* class, whereas others had *Arum*-type features. This appears to be the sole example where species of one genus were put into both classes by a single investigator. The most complicated family is the Leguminosae, where the situation is not clarified when it is divided into the different sub-families.

The Gramineae are also complicated (Table 1) and we have not attempted to sort out distribution of the VA mycorrhizal classes into the graminaceous sub-families. Nicolson's (1959) survey of grasses in Britain suggested that *Paris*-types are very common. Greny (1973) made a detailed anatomical morphological study of several important members of the Gramineae and concluded that *Zea, Avena, Triticum, Hordeum, Lolium* and *Holcus* all have *Arum*-type mycorrhizas whereas *Phleum, Alopecurus, Agrostis* and *Agropyrum* have *Paris* type. By contrast, Johnston (1949) described *Zea* as having *Paris* type mycorrhizas. Although he did not refer to Gallaud's classification, his survey of plants in the West Indies split them on the basis of (a) the presence of an extensive intercellular phase in the outer and inner cortex, or (b) the presence of extensive intracellular hyphal coils in the cortex. Apart from the distinction between outer and inner cortex, these are essentially the same as Gallaud's definitions. Since the many other workers who have used *Zea* have shown it to contain extensive intercellular hyphae, this poses a problem, and it must be emphasized that a number of families are in the' B' category in Table 1, as a result of Johnston's survey of plants from the West Indies. Currently we have studied the mycorrhizal colonization in different cultivars of wheat (*Triticum aestivum*) *viz.*HI-1502, HI-1500, GW-1172, HD-2781, HW-2004, MACS-3208, MPO-1126, A-9-30-1, HI-1501, HD-4694, HD-4672, C-306 and Sujata. Interestingly, we have observed only *Paris*-type of endomycorrhizal association. The cortical cells having dense coiled hyphae with intercalary arbuscules.

Geographical factors aside, there may be no need to have doubts about Johnston's powers of observation, because a number of investigators have noted that *Zea* has hyphal coils in the outer cortex (Winter, 1951; Meloh, 1963; Gerdemann, 1965; Greny, 1973; Kariya and Toth, 1981; Morton, 1985). Hence in this case there might be plasticity, which might be exhibited in different maize cultivars. *Triticum* and *Hordeum* can also have extensive hyphal coils (Greny, 1973). Hyphal coils in the outer cortex occur in other genera, including tobacco (Hildebrand and Koch, 1936) and some legumes, including *Medicage* and *Trifolium* spp. (Abbott, 1982). They may reflect differences in cortical

Table 2: Distribution of *Arum*-types and *Paris*-types within the Super-order Liliiflorae *sensu* Dahlgren *et al.* (1985) (for simplicity, only name of genera are listed).

	***Arum*-types**	*!Unexpected End of Formula***Paris*-types**
O. Asparagales		
Convallariaceae*	*Convallaria*	
	Maianthemum	
	Polygonatum	
	Smilacina	
Ruscaceae*	*Ruscus*	
Agavaceae	*Agave*	
Hypoxidaceae*	*Curculigo*	
Asphodelaceae*	*Aloe*	
Anthericaceae*	*Anthericum.*	
Hyacinthaceae *	*Hyacinthoides*	
	Muscari	
	Ornithogalum	
	Scilla	
Alliaceae*	*Allium*	
0. Burmanniales		Thismia
Thismiaceae		
O. Dioscorales		Tamus
Dioscoraceae		Dioscorea
Trilliaceae*		Paris
		Trillium
		Medeolat[ø]
O. Liliales		Colchicumt [ø]
		Wurmbea
Liliaceae *sensu*		Erythronium
*stricto**		Disporumt [ø]
Uvulariaceae*		
		Clintoniat[ø]
		Uvulariat[ø]

*Included in the Liliaceae *sensu* Cronquist (1981).

[ø] Hyphal swellings: see text.

structure and function. In addition to formation in the outer cortex, hyphal coils are frequently formed during penetration fungal hyphae through cells of hypodermes of both *Arum*-types and *Paris*-types, as shown by several early workers, including Gallaud (1905); Smith *et al.* (1989). We were then in ignorance of the earlier work. Hypodermis, though frequently overlooked, is present in many plants (Peterson, 1988; Perumalla *et al.*,1990).

Weber and his collaborators have published a valuable series of papers that survey development of VA mycorrhizas in the Apocynaceae,

Asclepiadaceae, Gentianaceae and Loganiaceae (Table 1). These families are all members of the Order Gentsanales. As already noted, the achlorophyllous members of the Gentianaceae are classic *Paris*-types. However, the other family in the *Arum*-types, both types or intermediates the Apocynaceae contains all of these (Table 1; Weber *et al.*, 1995). Weber and his colleagues have proposed that *Paris*-types are phylogenetically more advanced than *Arum*-types, reflecting weakened fungal vigour in some hosts and hence an evolutionary tendency towards successful mycobeterotrophy (Imhof and Weber, 1997). Given the frequency of *Paris*-types, and their dominance, in lower plants. We are not convinced by this argument. Much of information in Tables 1 and 2 has been obtained with field grown plants and hence plants of uncertain age. Preliminary results by Soderstrom *et al.* (1996) have suggested that up to *Petroselinum crispum* (Parsley) and *Daucus carota* (carrot) both in the Umbelliferae display some *Arum*-type features, with *Paris*-type features predominating afterwards. These developmental effects deserve further study in relation to root anatomy and the ageing and turn-over of individual fungal colonization units. Recently we observed (data unpublished) that wild species of *Lantana camara* and cultivated ornamental species *L. indice* and *L. aculeato* have both type of endophyte hyphal arrangement i.e. hyphal branches (*Arum*-type) and hyphal coiling (*Paris*-type). Similar observation are also obtained with the members of Euphorbiaceae family (e.g. *Euphorbia pulcherima*, *E. geniculata* and *E. hirta*).

References

Abbott,L.K.(1982).Comparative anatomy of vesicular arbuscular mycorrhizas formed on subterranean clover. Aust.J. Botany, **30**: 485-499.

Abbott,L.K., Robson, A.D. Gazey, C.(1992). Selection of inoculent vesicular mycorrhizal fungi. In: Norris J.R.,Read,D.J., and Verma,A.K.,eds.Methods in Microbiology, Techniques for the study of mycorrhiza. London,UK:Academic Press. **24** :1-21.

Alexander, T., Toth, R., Meier,R. and Weber,H.C.(1989).Dynamics of arbuscule deveopment and degeneration in onion, bean and tomato with references to vesicular-arbuscular mycorrhizae in grasses. Can. J. Bot., **67**:2505-2513.

Baylis, G.T.S.,McNabb, R.F.R. and Morrison, T.M. (1963). The mycorrhizal nodues of Prodocarps. Trans.Brits. mycol. Soc., **46** :378-384.

Berch, S.M., and Kendrick, B. (1982). Vesicular-arbuscular mycorrhizae of Southern Ontario. Mycologia., **74**:769-776.

Bonfante-Fasolo, P. (1984). Anatomy and morphology of VA mycorrhiza. Boca Raton FL, USA:CRC Press, 5-33.

Bonfante-Fasolo, P. and Fontana, A. (1985).VAM fungi in *Gingko biloba* roots: their interaction at cellular level. Symbiosis, **1**:53-63.

Bonfante-Fasolo, P. and Scannerini, S. (1977).A cytological study of the vesicular-arbuscular mycrrhiza in *Ornithogalum umbellatum, L. Allionia*, **22** : 6 - 20.

Boullard, B. (1958).*La mycotrophie chez pteridophytes: sa frequence, ses caracteres, sa significtion.* Bordeaux: E. Drouillard.

Brundrett, M.C.(1991).Mycorrhizas in natural ecosystems.In:Macfayden A, Begon M, Fitter AH,eds. Advances in ecological research, London,UK:Academic Press., **21** : 171-313.

Brundrett, M.C. (2002).Coevolution of roots and mycorrhizas of land plants.New Phytol., **154** : 275-304.

Brundrett, M.C. and Kendrick, B. (1990b).The roots and mycorrhizas of herbaceous woodland plants.II. Structural aspects of morphology. New Phytol., **114** : 469-479.

Cooke, M. A., Widden, P. and O'Halloran, I. (1992). Morophology, incidence an fertilization effect on the vesicular–arbuscular mycorrhiza of *Acer saccharum* in a Quebec hardwood forest. Mycologia., **84** : 422-430.

Cooke, M. A., Widden, P. and O'Halloran, I. (1993). Development of vesicular –arbuscular mycorrhiza in sugar maple (*Acer saccharum*) and effects of base-cation amendment on vesicle and arbuscule formation. Can. J. Bot., **71**:1421-1426.

Cooper, K.M. (1976). A field survey of mycorrhizas in New Zealand ferns. New Zealand J. Bot.,**14** :169-181.

Cronquist, A.(1981).An integrated system of flowering plants. New York:Columbia University Press.

Dahlgren, R.M.T., Clifford,H.T. and Yeo, P.F.(1985).The families of the monocotyledon: structure,evolution and taxonomy. Berlin : Springer-Verlag.

Daniels-Hetrick,B.A.,Bloom, J.,Feyerherm, S.M. (1985). Root colonization of *Glomus epigaeum* in nine host species. Mycologia, **77**:825-828.

Duckett, J.G. and Ligrone, R. (1991). A light and electron microscopy study of fungal endophytes in the sporophyte and gametophytes of *Lycopodium cernuum* with observations on the gametophytes-sporophytes junction. Can. J. Bot., **70**:58-72.

Duckett, J.G., Read, D.J. (1995). Ericoid mycorrhizas and rhizoids ascomycetes association in liverworts share the same mycobiont: isolation of partner and resysnthesis of the associations *in vitro*.New Phytol., **129**:439-447.

Gallaud, I. (1905). Etudes sur les mycorrhizes endotrophes. Revue Generale de Botanique 17:5-48,66-83, 123-136,223-239,313-325,425-435,479-500.

Gange, A.C., Brown, V.K. and Farmer, L.M. (1990). A test of mycorrhizal benefit in an early successional plant community. New Phytologist, **115**:85-91.

Gay, P.E., Grubb,P.J. and Hudson, H.J.(1982).Seqasonal changes in the concentrations of nitrogen, phosphorus and potassium, and in the density of mycorrhiza, in biennial and matrix-forming perennial species of closed chalk-land turf. J. Ecology, **70** : 571-593.

Gerdmann, J.W.(1965). Vesicular-arbuscular mycorrhizas formed on maize and tulip tree by *Endogone fasiculata*. Mycologia., **57** :562-575.

Greny, A. (1973). Etude anatomo-morphologique des endomycorrhizes constitue pal le Mais, l' Avoine, le Ble,'l' Orge et diverses Graminees prairials et adentices. Momoire rpresente au Conservatoire des Arts et Metiers poiur obtenir le Diplome d' Ingenieu r C.N.A.M.

Grime, J.P, Mackey, J.M, Hillier, S.M. and Read, D.J. (1987). Floristic diversity in a model system using experimental microcosms Nature, **328** :420-422.

Harley, J.L.(1969). The biology of mycorrhiza. 2nd edn. London:Leonard H ill.

Hilberbrand, A.A. and Koch, L.W. (1936). A microscopical study of infection of the roots of strawberry and tobaco seedling by micro-organisms of the soil. Can. J. Bot., **57**:11-25.

Holley, J. E. and Peterson, R.L. (1979). Development of a vesicular-arbuscular mycorrhiza in bean roots. Cand. J. Bot., **57** :1960-1978.

Hosney, M., Hijri, M.,Passerieux, E. and Dulieu, H.(1999). rDNA units are highly polymorphic in *Scutellospora castanea* (Glomlaes Zygomycetes) *Gene* **266**: 61-71.

Imhof, S. and Weber, H. C. (1997). Root anatomy and mycotrophy (AM) of the achlorophyllous *Voyria truncata* (Standley) Standley & Steyermrk (Gentianaceae). Botanica Acta., **110** :127-134.

Jacquelient-Jeanmougin, S. and Gianinazzi-Pearson, V. (1983). Endomycorrhizas in the Gentianaceae. I. The fungus associated with *Gentiana lutea* L. New Phytol., **95**: 663-666.

Janse, J.M. (1897).Les endophytes radicaux de quelques plants Javanaises, Annales du jardin Botanique de Buitenzorg **14** : 53-201.

Johnston, A.(1949). Vesicular-arbuscular mycorrhiza in sea-island cotton and other tropical plants.Tropical Agri.(Trinidad), **26** :118-121.

Kariya,N. and Toth, R.(1981).Ultrastructure of the mycorrhizal association formed between *Zea diploperennis* and *Glomus fasciculatum*. Mycologia.,**73** : 1027-1039.

Kidston, R. and Lang, W.H. (1921).On the old red sandstone plants showing structure from the Rhynie chert bed ,Aberdeenshire. V. The thallophyta occurring in the peat bed, the succession of the plant through a vertical section of the bed, and the condition of accumulation and preservation of the peat. Trans. Royal Soc. Edinburgh., **52** :855-902.

Kiden, D.A. and Brown, M.F. (1975). Electron microscopy of vesicular-arbuscular mycorrhizae of yellow poplar. I. Characterzation of endophytic structure by scanning electron sterescopy.Cand. J. Microbiol., **21**:989-993.

Kiden,D.A. and Brown, M.F.(1976). Electron microscopy of vesicular-arbuscular mycorrhizae of yellow poplar.IV. Host endophytes interactions during arbuscular deterioration. Cand. J. Microbiol., **22**:64-75.

Klironomos, J.N., Moutoglis, P., Kendrick, B. and Wilden, P.(1993). A comparison of spatial heterogeneity of vesicular- arbuscular fungi in two maple forest soils. Canad. J. Botany, **11** : 1472 1480.

Konoe, R. (1957). Uber das vorkommen der Wurzelpilze bei Metasequoia und den nachst verwantden in Pflanzen. J. Inst. Polytechnics, Osaka City Univ., **8** : 179-184.

Lackie, S.M.,Garriock, M.L., Peterson, R.L. and Bowley, S.R. (1987). Influence of host plant on the morphology of the vesicular arbuscular mycorrhizal fungus *Glomus versiforme* (Daniels and Trappe)Berch.., Symbiosis : 147-158.

Laycock,D. H.(1945). Preliminary investigation into the function of the endothrophic mycorrhiza of *Theobromas cacao* L. Trop. Agric.(Trinidad). **22** : 77-80.

Law, R. (1985). Evolution in a mutulistic environment. In ;Boucher DH ed The biology of mutualism: Ecology and evolution , London: Croon Helm, 145-170.

Leake, J.R.(1994). The biology of myco-heterotrophic (saprophytic) plants. New Phytol., **127**:171-216.

Ligrone, R. (1988). Ultrastructure of a fungal endophytes in *Phaeoceros laevis* L. Prosh (Anthocerophyta). Botanical Gazette (Chicago)., **149** : 92-100.

Louis, I. (1990). A mycorrhizal survey of plant species colonizing costal reclaimed land in Singapore. Mycologia., **82** :772-778.

Mabberley, D.J.(1989). The plant book. Cambrigde : Cambridge University Press. McGee PA. 1985. Lack of spread of endomycorrhizas of *Centaurium* (Gentianaceae). New Phytol., **101**: 451-458

McGee, P.A. (1988). Growth response and morphology of mycorrhizal of *Thysanotus* (Anthericaceae:monocotyledonae). New Phytol., **109L**:459-463.

McGee, P.A. (1989). Variation in propagule number of vesicular arbuscular mycorrhizal fungi in a semi-arid soil. Mycological Res., **92**:28-33.

Melok, K.A. (1963). Untersuchungen Zur Biologie der endotrophen Mykorrhiza bei *Zea mays* L. und *Avena sativa* L. *Archiv fiir* Mikrobiologie., **46** : 369-381.

Morton, J. B.(1985). Variation in mycorrizal and spore morphology of *Glomus occultum* and *Glomus diaphanum* as influenced by plant host and soil environment. Mycologia, **77**: 192-204.

Morton, J.B.(1990).Evolutionary relationships among arbuscular mycorrhizal fungi in the Endogone. Mycologia.**82**:192-207.

Morton, J.B. and Benny,G.L.(1990). Revised classification of arbuscular mycorrhizal fungi (Zygomycetes):a new order, Glomales, tow new suborder,Glomineae and Gigasporineae, and two new families, Acaulosporaceae and Gigasporaceae,with an emendation of Glomaceae. Mycotaxon, **37**:477-491

Morton, J.B. and Redecker,D.(2001).Tow new families of Glomales, Archaeosporaceae and paraglomaceae, with tow new genera Archaeospora and Paraglomus, based on concordant molecular and morphological characters.Mycologia., **93**:181-195.

Mosse, B. and Hepper,C.M.(1975). Vesicular-arbuscular infection in root organ cultures. Physiological Plant Pathol., **5**:215-223.

Nadarajah,P. (1980). Species of Endogonaceace and mycorrhizal association of *Elaeis guineesis* and *Theobroma cacao*. In: Mikola< P.ed Tropical mycorrhis. Oxford: Cleredon Press.233-237.

Nicolson, T.H.(1959). Mycorrhiza in the Gramineae. I. Vesicular-arbuscular endophytes, with special reference to the external phase. Tran. of the British Mycol. Soc., **42**:421-438.

Perumalla, C. J., Peterson, C. A. and Enstone, D.E. (1990). A survey of angiosperm species to detect hypodermal Casparian bands. I. Roots with a uniseriate hypodermis and epidermis. Botan. J. Linnean Soc. Lond.,**193** : 93-112.

Peterson, C.A.(1988).exodermal Casparian bands: their significance for ion uptake by roots. Physiologia Plantarum, **72**:711-720

Peterson,R.L.,Howarth,M.J.,Whittier,D.P.(1981).Interractions between a fungal endophytes and gametophyte cells in *Psilotum nudum*. Cand. J. Bot., **59** : 711-720.

Peyronel, B. (1923). Prime ricerche sulle micorize endotrofiche sulla micoflora radicoa normale delle fanerogame.Rivista di Biologia **5** : 463-485.

Peyronel, B. (1924). Prime ricerche sulle micorize endotrofiche sulla micoflora radicoa normale delle fanerogame.Rivista di Biologia, **6** : 17-53.

Pocock, K. and Duckett, J. G.(1984). A comparative ultrastructural analysis of the fungal endophytes in the *Cryptopthallus mirabilis* Malm. And other British thalloid hepatics J. Bryol., **13**:227-233.

Prat, H. (1926). Etude des Mycorrhizes du Taxus baccata amnales des Science Naturelles, Botanique et Biologie Vegetale, Paris, **8**:141-162.

Pringle,A., Moncalve, J.M. and Vilgalys, R. (2000). High levels of varition in ribosomal DNA sequences within and among spores of a natural population of the arbuscular mycorrhizal fungui *Acaulospora colossica*. Mycologia., **92**:259-268.

Pyrozynski, K. A., Dalpe, Y.(1989).Geological history of the Glomaceae with particular references to mycorrhizal symbiosis. Symbiosis. 7:1-36.

Rayner, M. C. (1927). Mycorrhiza. An account of non-pathogenic infection by fungi in vasicular plants and bryophytes. New Phytol. Reprint No. 15 London :Weldon and wesley.

Remy, W., Taylor, T.N., Hass, H. and Kerp, H. (1994). Four-hundred-milliom-year-old vesicular-arbuscular mycorrhizae. Proceedings of the National Academy of Science, USA, **91**:11841-118443.

Rolden-Fajardo, B.E. (1994). Effect of indigenous arbuscualr mycorrhizal endophytes on the development of six wild plants colonising a semi-arid area in south–east Spain. New Phytol., **127**:115-122.

Sanders, I.R.(1993). Temporal infectivity and specificity of vesicular arbuscular mycorrhizas in co-existing grassland species. Oecologia., **93**:349-355.

Sanders, I.R. and Koide, R.T.(1994). Nutrient acquisition and community structure in co-occurring mycotrophic and two mycotrophic old-field annuals. Functional Ecology., **8**:77-84.

Shibata, K. (1902). Cytologische Studien uber die endotrophen mykorrhizen. Jahrbucker fur die wissenschaftliche Botanik, **37** : 643-684.

Smith, S.E. (1995). Discoveries, discussion and directions in mycorrhizal research. In : Verma A, Hock B, eds.Mycorrhiza: Structure Function Molecular Biology and Biotechnology, Berlin Springer-Verlag, Berlin :3-24.

Smith, S.E.,Long,C.M. and Smith,F.A.(1989). Infection of roots with dimorphics hypodermis : possible effects on solute uptake. Agriculture, Ecosystem and Environment., **29**:403-407.

Soderstrom,B.,Dickson, S., Smith, F.A. and Smith, S.E.(1996). *Arum* and *Peris*-type of endomycorrhiza in some plants –AM fungus species combinations.In Szaro, TM, Bruns TD, eds.Program and Abstracts, First International Congress, on Mycorrhizae Berkely, California: University of California. 111.

Stelz, T. (1968). Mycorrhizes et vegetation des pelouses calcaires These pour l' obtention du Doctorat, Faculte' des Sciences de l' Universite de Rouen, France.

Strullu, D.G.,Gourret,J.P.,Garrec,J.P. and Fourcy, A. (1981). Ultrastructure and electron-probe microanalysis of the mentacrometic vacuolar granules occurring in *Taxus* mycorrhizas.New Phytol. **87**:537-545.

Thaper,H.S. and Khan, S.N. (1973). Studies on endomycorrhiza in some forest species. Proceedings, Indian National Science Academy. **B39**:687-694.

Tiemann,C., Demuth, K. and Weber, H.C. (1994a). Zur VA-Mycorrhiza von *Gelsemium rankinii* and *G. Sempervirens* (Loganiaceae). Beitrage zur Biologie der Pflanzen **68** : 311-321.

Tiemann,C., Demuth, K. and Weber, H.C. (1994b). Zur symbiose von *Cynanchlum vincetoxicum* L. PERS.,*Asclepias curassavica* L. und *Ceropegia woodii* SCHL.(Asclepidaceae) mit mycorrhiza-pilsen (VAM). Flora **189** : 1-6.

Untch, Wber, H.C.(1995). Strukturen der Mycorrhiza (AM) bei *Ceropegia dicho toma* Haw., *Ceropegia fusca* Bolle und *Periploca laevigata* Ait. (Asclepiadeceae).Beitrage zur Biologie dar Pflanzen., **69** :129-140.

Walker, C. (1992). Systemics and taxonomy of the arbuscular endomycorrhizal fungi (Glomales) – a possible way forward. Agronomie., **12**:887-897.

Walker, C., Mize, C.W. and McNabb, H.S. (1982). Population of endogonaceous fungi at two locations in central Iowa. Cand. J. Bot., **60** : 2518-2529.

Walker,C. and Trappe, J.M. (1993). Names and epithests in the Glomales and Endogonales. Mycological Res. 97:339-344.

Weber, H.C., Klahr, A. and Morron-Heimbuch, M. (1995). Anatomical structures of the VA mycorrhiza in the Apocynaceae (Genetianales). Botanica Acta., **108**:525-534.

Widden, P. (1996). The morphology of vesicular-arbuscular mycorrhizae in *Clintonia borealis* and *Medeola virginiana.* Cand. J. Bot. **74:** 679-685.

Winter, A.G.(1951). Untersuchungen uber die Verbreitung und Bedeutung der Mykorrhizen bei kultivierten Gramineen und einigen anderen landwirtschaftlichen Nutzpflanzen. Phytopathologische Zeitschrift. **17**:421-432.

Yawney, W.J. and Schultz, R.C. (1990). Anatomy of vesicular-arbuscular endomycorrhizal symbiosis between sugar maple (*Acer saccharum* Marsh) and *Glomus etunicatum* Becker and Gerdmann. New Phytol., **114** : 47-57.

44

Deuteromycotina - A Hidden Group of Fungi

C. Manoharachary

Mycology and Plant Pathology Lab., Department of Botany, Osmania University, Hyderabad - 500 007, A.P.

Fungi belonging to Deuteromycotina represent the asexual phases of Ascomycotina and Basidiomycotina which are world wide in distribution and contribute significantly through their biotechnological products to the human welfare. Global market is now in a position to market commercialized fungal products. Diversified habitats are explored for the study of fungi and their association. Conidial ontogeny was used to identify the Fungi-Imperfecti. Large number of fungi need to be brought and studied in culture.

Keywords : Deuteromycotina, Diversity, Taxonomy, Biotechnology.

Moulds are known to colonize diversified habitats and multiply in micro-ecological niches. The survival time of fungi vary from few months to many years. Fungi are ubiquitous and cosmopolitan in distribution. The habitats which get colonized by fungi include water, soil, air, litter, herbivore dung plant and animal parts and extremophilic habitats. The colonization of fungi in diversified habitats is of much significance as their secondary metabolities are of much relevance to the human welfare. Fungi are known to produce antibiotics, organic acids, toxins, dyes, alkaloids, immuno-modules, fermented products, food and feed. Mycoflora differs in its composition from an ecological niche to the other. Nutrient status, pH, moisture, temperature, composition of the substrate, altitude, rainfall, relative humidity, type of habitat and other related parameters have greater role in the occurrence, distribution and seasonal variation among different groups of fungi.

Diversity and Classification

The world of fungi, which includes moulds and yeasts, is vast. Some 76000 species are known; but it is estimated to comprise 1,500,000 species. One of the important characters that distinguishes fungi from plants is the

Email : cmchary@rediffmail.com

composition of their cell wall that is made, for the most part, of chitin, a type similar to that of Arthropoda. Being heterotrophic, they live as parasites, saporphytes and symbionts.

India has a rich diversity of fungi and forms an important geographical region for fungal distribution (Subramonian, 1962). The knowledge of fungal diversity is based on observation, taxonomy, nomenclature, classification and identification. Taxonomy is based on morphotaxonomic characterization and also through application of molecular tools. Vegetative characters and reproductive features help in the differentiation and segregation of genera and species. However, strain differentiation is difficult. Therefore RFLP, RAPD, ELISA and PCR have become important tools in the segregation of strains and biotypes.

Species is the basic unit of classification. Naming is a reference system and proper name has to be given to the type specimen representing the taxon. This constitutes an essential part of the name of the species. Classification deals with the placement of the species in a hierarchial system. This is mainly based on morphological and biological resemblance. The species name along with the genus forms binomial. Naming of fungi regulates the proposing names, their validity, legitimacy and always maintaining a single correct name for each fungus. Saccardo introduced term Deuteromycetae which otherwise called the Fungi Imperfecti. Deuteromycotina accomodates fungi which exists in conidial state or asexual state and whose perfect states for many remain unidentified or yet to be discovered. These fungi possess their perfect states either in Ascomycotina or Basidiomycotina. The asexual state is called anamorph and perfect state is called Teleomorph. Hawksworth *et al.,* (1995) have proposed a classification under mycota having eleven phyla and excluded Fungi Imperfecti. Saccardo (1880, 1884) classified the fungi Imperfecti based on conidial shape, septation, colouration and fruit bodies as amerospore, didymospore, phragmospore, dictyospore, scoleospore, helicospore and staurospore. Hughes (1953) has stated that conidiogenesis paves the way for classifying Fungi Imperfecti into eight modes viz., Blastosporae, Botryoblastosporae, Sympodulosporae, Aleuriosporae or Annellosporae, Phialosporae, Arthrosporae, Meristem arthrosporae and porosporae. Tubaki (1958) added a ninth mode. Since then some intermediaries have been discovered and till today the conidiogenesis remain as an important criterion. Thus Deuteromycetes has been recognized as a unified taxonomic system comprising Hyphomycetes and Coelomycetes. Based on scientific scrutiny it has been observed that the two basic modes of conidiogenesis namely Blastic and Arthric are important in the characterization of fungi. Taking into account of the complex phenonena of conidiogenesis, it is certainly an important step towards a more natural classification of anamorphic fungi. The mitosporic fungi forms a dominant group of fungal wealth. It is an artifical group of fungi that can live and multiply without a sexual phase. From the point of number, the group is represented by 2400 genera and 20000 species. Most of these fungi are observed *in situ* and can be isolated on nutritive medium. Approximately 8000 fungal species representing this group are reported from India.

Biotechnology

Deuteromycetes contribute to the biodegradation and recycling of organic matter. They are cellulolytic (*Chrysosporium*) and lignolytic (*Sporotrichum*) and play vital role in waste water treatment and in the biodegration of pesticides. Certain Fungi are important in cheese porduction (e.g., *Penicillium* spp), organic acid production (species of *Aspergillus* and *Penicillium*) and *Penicillium chrysogenum* is used in the production of Antibiotic (*P. chrysogenum*) and biofertilizer or cattle feed. Penicillins, griseofulvin, cephalosporin and cyclosporin are the abiomedicines produced by some members of the Fungi Imperfecti. Cyclosporin A is the powerful immunosuppresant. Forty percent of industrial enzymes are of fungal origin. Marketable success has been achieved in the filed of biocontrol using *Beauvaria bassiana* as bio-insecticide, *Athrobotrys* as predator of nematodes, *Trichoderma* in the control os soilborne and root-borne pathogens, *Colletotrichum gloeosporioides* as bioherbicide and several others.

In the field of medicine and veterinary science, some members of Fungi Imperfecti are knwon to cause mycoses (*Aspergillus, Histoplasmosis* (Histoplasma), superficial mycoses (*Scytalidium, Phialophora*), ring worm, pulmonary or epidermal allergies. Several fungal toxins induce various sicknesses in humans and animals. The best known example are aflatoxins/ mycotoxins produced by *Aspergillus* and *Penicillium.* Deuteromycetes cause a large number of plant diseases manifesting themselves in various symptoms (wilts, anthracnose, leaf spots, blue mould etc.). Practically, all foods and organic matter can be altered to the point of toxicity by Deuteromycetes. Even non-organic matter like glass is damaged by *Penicillium citrinum,* metals such as aluminum by *Aspergillus,* paints by *Phoma* and others.

During the survey of microfungi from 1970 - 2001 (Manoharachary and Rao, 1993; Manoharachary, 1995) colonizing diversified habitats of Andhra Pradesh State, the author has collected around 1200 speices. Table 1 gives the list of some important fungi Imperfecti that the author has encountered and Table 2 enlists conidial fungi isolated from fresh water bodies. Table 3 lists out some biotechnological product from fungi.

Table – 1. New additions to the Fungi of India

Species	Species
Acladium curtisii (Berk:) Ellis	*Helicorhoidon botryoidecum* (Cooke) Hughes
Acrimoniuloa fagi Ellis:	*Helicosporium griseum* (Bon) Sacc. c.f. Moore
Acrodictys bambusicola Ellis:	*Helicosporium nematosporum* Linder
Acrodictys fulginosa Sutton	*Helicosporium viridae* (Corda) Hughes: c.f. Pirozynski
Bahusakala cokei Ellis:.	*Hermatomyces tucumanensis* Spegazzini
Beltraniopsis esenbeckiae Batista and Benzerra	*Idriella angustispora,* Morgan-Jones
Berkleasmium tropicale Morris	*Isariopsis alborosella* (Desm) Saccardo
Brachydesmiella biseptata Hughes	*Lacellinopsis spirallis,* Ellis
Brachysporiella setosa Berk & Curt	*Megalodochium inflatum,* Holubova-Jechova
Corynespora obclavata Dyko and Sutton	*Podosporium rigidium* Schweinitz
Craspidodihymum condensatum (Wallr) Mason and Hughes, c.f. Remington	*Spadicoides constricta* Wang and Sutton
Dendriphiopsis state of *Amphisphaeria incrustans* Ellis and Everh	*Spiropes dorycarpus* (Mont) Ellis
Dictyosporium elegans Corda	*Sporidesmum bambusae* Ellis
Didymobotryum rigidum (Berk and Br) Saccardo	*Sporidesmium nodipus* (Penz and Sacc) Hughes
Diplococcum spicatm Grove	*Sporidesmium pedunculatum* (Peck) Ellis
Exosporium monanthotaxis Pirozynski	*Tripospermum myrti* (Lind) Hughes
Helicoma mulleri Corda	*Virgariella atra* Hughes
Helicomyces tenuis Spegazzini	*Xemosporium berkeleyi* (Curtis) Pirozynski
	Xenosporium thaxterii (Linder) Pirozynski

Conclusions and Future outlook

(1) Fungi belonging to Deuteromycotina are an important group from the point of biotechnology, industry, plant diseases and environment management.

(2) There is a necessity of inventorizing the fungi of this group with special reference to their ecological niches, relevance, strain differentiation and others.

(3) Biodiversity status of Fungi Imperfecti has to be recast.

Table-2 . Conidial fungi in fresh water from India

Actinospora megalospora Ingold	*Flagellospora penicilloides* Ingold
Alatospora acuminata Ingold	*Helicomyces* sp.
Anguillospora logissima (de Wild) Ingold	*Isthamotricladia laeensis* Matsushima
Anguillospora crassa Ingold	*Lemonniera aquatica* de Wild
Articulospora tetracladia Ingold	*Lunulospora curvula* Ingold
Articulospora angulata Tubaki	*Lunulospora cymbiformis* Miura
Brachiosphaera tropicalis Nawawi	*Phalangispora constricta* Nawawi & Webster
Camposporium pallucidum (Grove) Hughes	*Speiropsis irregularis* Petersen
Condylospora spumigena Nawawi	*Speiropsis pedatospora* Tubaki
Centrospora acerina (Hartig) Newhall	*Tetracladium marchalianum* deWild
Clavariopsis aquatica de Wild	*Tetracladium setigerum* (Grove) Ingold
Clavariana aquatica de Wild	*Tetrachaetum elegans* Ingold
Clavariopsis brachycladia Tubaki	*Tetraploa aristata* Berk & Br.
Clavatospora filiformis Nawawi	*Tricladium attenuatum* Iqbal
Clavatospora stellata (Ingold & Cox) Nilsson ex. Marvanova & Nilsson	*Tricladium curvisporium* Descals
Dactylella appendiculata Anastasious	*Tricladium fuscum* Nawawi
Dactylella aquatica (Ingold) Ranzoni	*Tricladium splendens* Ingold
Dendrospora moniliformis Descals	*Triscelophorus monosporus* Ingold
Diplocladiella scalaroides Arnuad ex. Ellis	*Trisulcosporium* sp.
Flabellospora tetracladia Nawawi	*Tripospermum myrti* (Lind) Hughes
Flabellospora verticillata Alasoadura	*Varicosporium elodea* Kegel
Flagellospora curvula Ingold	*Weisneriomyces javanicus* Koerders

Table-3. Biotechnology products from conidial fungi

.		
Cephalosporium acremonium	-	Cephalosporine
Penicillium griseofulvum	-	Griseofulvin
Aspergillus niger	-	Citric Acid, Lactase
Aspergillus oryzae	-	Kojic acid, α-amylase
Fusarium moniliforme	-	Gibberellins
Trichoderma reese	-	Cellulase
Beauvaria bassiana	-	Insect killer
Penicillium chrysogenum	-	Bisol
Fusarium graminearum	-	Quorn
Trichoderma harzianum	-	Biocontrol agent
Colletotrichum gloesporioides	-	Collego

(4) Aspects related to hot spots, biogeographical mapping, distribution and culturing of Deuteromycotina needs to be given emphasis.

(5) Species 2000 programme has to be implemented with reference to the Fungi Imperfecti.

(6) Many taxa of Fungi-Imperfecti need to be studied with reference to their Teleomorphs.

References

Hawksworth, D.L., Kirk., P.M. , Sutton, B.C. and Pejer, D.N. (1995). Ainsworth & Bisbys Dictionary of Fungi (Revised), 8th edition, CAB, International, U.K. p. 616.

Hughes, S.J. (1953). Conidiophores conidia and classification. Can. J. Bot., **31** : 577 - 659.

Manoharachary, C and Rao, N.K. (1993). New additions to the Fungi of India. Asian J. Plant Sci., **5** (1) : 1-25.

Manoharachary, C. (1995). Aspects and prospects of water-brone conidial fungi from India. Adv. Biol. Res., **4** (1&2) 160-183.

Saccardo, P.A. (1880). Conspectus generum fungorum italiae inferiorum nempe ad Sphaeropsideas - Melanconieas et Hyphomycetas Pertinentium Systemate Sporologic dispositorn Michella Commentorium Mycologicum 2, 1-38.

Saccardo, P.A. (1884). Sylloge Fungorum omnium hucusan cognitorum vol. 3. Pavia, 860 p.

Subramanian, C.V. (1962). A classification of the Hyphomycetes. Curr. Sci., **31** : 409 - 411.

Tubaki, K. (1963). Taxonomic study of Hyphomycetes. Ann. Rep. Inst. for Ferm. Osaka., **1**, 24-54.

45

Some Limitations in the Systematics of Fungal Classification

D.K. Agarwal

Division of Plant Pathology, IARI, New Delhi - 110012

Taxonomic studies serve as the basis for accurate classification and identification of fungi. Fungal identification is essential and it needs to be systematically arranged. Taxonomy of fungi has become a need based study in view of its role in industry, agriculture, medicine applied biology, pest management and biotechnology. The limitations in the fungal taxonomy are discussed with scientific accuracy.

Key words : Fungal taxonomy, importance, limitations

Taxonomy is a science of methods, observations identification, naming and interpretation of thoughts. The main principle of fungal taxonomy is the diagnosis and the act of identifying an organism or a disease from signs and symptoms. Although stated simply, the process is often not so simple- at least in terms of modern agriculture. According to Ham (1951) a physician is responsible for the care of the patient, similarly mycologist or pathologist is responsible for identification and control of the disease. This implies the need for accurate diagnosis which depends on a professional evaluation of the person and his environment. In mycology and plant pathology the patients are plants or perhaps several thousands of them in many acres of land if we are dealing with commercial agriculture. The plant diagnostician often plays a different role from that of the physician. The latter bases his diagnosis on his own personal observations and history of the patient, plus whatever additional information he can obtain from a pathologist in a medical laboratory. The mycologist and plant pathologist are usually expected to make the diagnosis based on diseased plant specimens submitted to him from fields, forests, orchards and more often that too in poor condition. This diagnosis can be more accurate when it is made at the site where the plant is grown by some one who is knowledgeable about the host species. Although the principles of diagnosis in plant pathology, medicine and veterinary sciences are much the same, the plant pathologists have some unique handicaps. The aim of the diagnosis is to provide sufficient understanding to preserve or protect host plants in some herbarium or to study them in glasshouse conditions.

Limitations

Newer techniques in diagnosing fungi will undoubtedly be developed in the years ahead. The prospect for rapid and accurate identification of pathogens or fungi by procedures perhaps yet undreamed of may be available to our successors. The act of identification requires, much knowledge, common sense and skill. We must cultivate and demonstrate an attitude of identifying a disease, since it is a great service for the nation whether it is a crop or human being.

Pound (1968) noted with alarm the growing disparity of funding between basic and production research and emphasized the need for balance in both areas. Support and appreciation in all aspects of plant pathology was neither achieved nor maintained. It is indeed difficult to "manage fungi" without being able to recognize them or the damage they cause. Many departments are much concerned and are seeking to improve these opportunities but there is, perhaps, more talk than progress. The excitement of tracking down a causal agent through scientific detective work was not emphasized. This was a treasure that was almost lost as many mycological researches are fading away and given no importance. We should see that the students are trained in all these important practical aspects which some of us have no intention of doing ourselves. The modern taxonomic concept has consistently emphasized the importance of developmental criteria and morphology of fungal organisms both for obligate parasites and saprophytes growing on various substrates in nature. A mycologist bears the responsibility of the degree of taxonomic sophistication in categorising his fungus in an available system of classification. The classical work of Brefeld (1881) who devised pure culture technique and which has been successfully used in the conidiogenesis of various fungi and their developmental stages is now considered to be of immense importance. According to Simmons (1967), "It is not essential to base our major systematic divisions on the most obvious characters of early growth but the mature sporulating conditions instead. In a majority of cases we can compare any number of molds under precisely the same conditions and in all stages of their growth and development". Several important factors viz. host specificity, variation in morphological characters due to environmental factors, spore morphology, spore germination, development of conidia and conidiophores, correlation of imperfect and perfect stages and the phenomenon of heteroecism are some of the problems in taxonomic studies of fungi. These problems get further aggravated if there are no proper laboratory and instrumentation facilities. The use of modern biochemical, physiological character and molecular techniques as tools in taxonomic studies of fungi in their different facets of life cycles has added another milestone in the history of taxonomy.

Speciation in several groups of fungi viz. rusts, smuts, powdery mildews etc. which are based specifically on hosts on which they thrive, unfortunately conflicts the minds of taxonomists. If host specificity is alone taken as a criteria for separating taxa it may result in considerable confusion. The situation is worse where it is not possible to identify the host due to lack of floral parts or when a fungus is growing in artificial culture medium.

Taxonomic Approaches

Myxomycota

In Myxomycota, plasmodial forms have been used as a criteria for establishing or verifying relationships among the various genera and species for taxonomic treatment. Alexopoulos (1961) showed that plasmodial forms viz. phaneroplasmodia, aphanoplasmodia and protoplasmodia are peculiar to the Physarales, a few species of *Stemonites, Comatrichia* and *Lamproderma* and some species of *Echinostelium, Licea* and *Clastoderma,* respectively. On the basis of these developmental studies in culture Alexopoulos (1969) transferred *Clastoderma* from the Stemonitales to Echinosteliales and suggested that *Barbeyella* does not belong to the Stemonitales. This shows the importance of experimental work before new taxa are made or described and major revisions are undertaken. Still the plasmodia of large number of Myxomycota have not been grown in culture. Hence, it is essential to base taxonomic, treatment of the Myxomycota almost entirely upon the character of the fruiting stage, which is relatively constant, most commonly observed, and may be preserved indefinitely. The realization by many modern mycologists that parasitic organisms vary immensely in their pathogenic traits appears to be basic to the principle that these organisms should be classified, at the species level, mainly on morphological criteria and that physiological characteristics, such as pathogenicity, should be represented by specialized form.

Mastigomycotina Zygomycotina

Oomycetes on the basis of physiology in general, show a similar level of organisation to both Chytridiomycetes and Zygomycetes but otherwise appear unrelated to either of the classes. Numerous examples can be cited where on the basis of nutritional requirements, wide diversity has been reported not only in the families of an order but among the various genera and their species. Hence, it seems doubtful, that any fixed criteria can be applicable to all fungi, in general.

In Zygomycotina, the structure and biochemical composition of the cell wall have for long been important taxonomic criteria. Occurrence of chitin or cellulose seems to be a distinguishing character of the main group. Sarbhoy

(1975) explored the possibilities of grouping the generic and specific characters of the family Zygomycetes. Spores which are produced in slime as in *Spinellus, Mucor ramannianus* and *Piptocephalis freseniana,* lack the prominent projections found on dry spores of *Cunninghamella, Mucor plumbeus, Piptocephalis virginiana* and species of *Chaetocladium* (Mehrotra and Sarbhoy, 1960). Presence of capitate spines seen in *Gilbertella* and *Choanephora* may be used as an additional character for identification of the family Choanephoraceae. The results of Young (1969) are interesting because the genus *Chaetocladium* which has only three recognized species (i.e. C. *jonesii,* C. *brefeldi* and C. *hesseltineii)* can easily be separated into two groups on the basis of the shape of their spines. The advantage has been taken of the surface ornamentation of the zygospore and appendages of the suspensors. The zygospore surface is smooth in Radiomycetaceae and Kickxellaceae, punctate, or pitted in Dimargaritaceae and echinulate in some other families of Zygomycetes. They form mycelial clusters in *Mortierella,* finger-like in *Absidia,* dichotomously branched in *Phycomyces,* acuminate in *Radiomyces,* coarsely roughened in *Thamnidium, Chaetocladium* and *Helicostylum* and smooth in *Mucor, Rhizopus* and other members of Mucorales. More work on these lines would be rewarding in characterizing various taxonomical1y difficult members of this or other orders. Many workers have revised and relegated several species of Zygomycotina, but stil1 confusion prevails due to incomplete information on morphology, physiology and reproduction of reported members.

In recent years biochemical tests are being used in differentiating various pathogenic fungi. These tests and to divide the pathogenic fungi into two different categories on the basis of their utilization of various carbohydrates viz. *(i)* tests to assess what particular compound(s) that the fungi assimilate or degrade during growth process *(ii)* tests to assess the constituents of fungi. Sarbhoy (1965, b, 1969) reported that sorbose, mannose and galactose were not at all suitable for the growth of *Zygorrhynchus heterogamus, Z. exponens* and Z. *californiensis:* Lilly and Barnett (1953) and Mehrotra and Singh (1970) also found sorbose to be toxic to several Mucorales.

Yeasts and Filamentous Fungi

Assimilation tests are much variable in the taxonomy of yeasts but of lesser importance in the study of filamentous fungi. George and Camp (1957) showed that the differentiation of species of fungi which are very much alike can be obtained on the basis of such tests e.g. utilization of histidine in distinguishing *Trichophyton megninii* from *T. rubrum* and *T. gallinae* or of nicotinic acid to separate *T. equinum* from *T. mentagrophytes.* Scholer and Mul1er (1966) studied forty six strains of *Absidia corymbifera,* twenty four strains of *Mucor pussillus,* four of *M. mieheie* on the basis of biochemical tests (assimilation of forty one carbon sources, three nitrogen and vitamin dependance) found several strains of *A. racemosa* and *A. italina,* led to the concept of one single species. In *M. pussillus* and *M. mieheie,* which are closely related morphologically, biochemical patterns proved to be similar to each other but differed in different environmental and nutritional conditions, the principle morphological characters between the two species of *Pierella* remain unchanged (Mehrotra and Singh, 1970). Nutritional tests are of utmost importance for the identification when one deals with virtually morphologically identical organisms. In the identification of species of *Nocardia,* the assessment of capacity to attack variety of compounds such as casein, tyrosine, xanthine, gelatine and various sugars is of paramount importance (Gordon and Mihen, 1959). Sarbhoy (1965) reported significant differences among the different species of *Zygorrhynchus* while studying the utilization of potassium nitrate. Out of the six species, three viz. Z. *exponens,* Z. *moelleri* and Z. *heterogamus* could utilize nitrate while Z. *macrocarpus* and Z. *vuilleimini* failed to utilize it. Similar results were obtained with carbon sources as well. It is clear that some of these differences in the nutritional requirements noticed in the various species of *Zygorrhynchus* can be regarded as an additional taxonomic criteria for differentiating these species.

Obligate Parasites

In obligate parasites such as rusts, smuts, powdery and downy mildews, pathogenic specialization appears to have reached its greatest heights. In these organisms, a morphological pointer has been recommended for similar species which may generally be subdivided into pathogenic entities "adapted in parasitism to particular host plants, but the range of the adaptation may differ in various organisms whether it is to family, tribe or genus. This type of situation is somewhat remarkably special in Erysiphales which usually occur in the conidial state but their taxonomy is mainly based on characters of cleistothecial stage. The species are successful leaf parasites despite the absence of spermatia. This observation requires necessary taxonomic elaboration. The mycelium in powdery mildews is superficial on the surface of the leaf rather than internal as in most of the leaf parasites. Conidia are produced profusely throughout the growing season so that there is ample opportunity for multiple spore colonization of single leaves. Plasmogamy is readily accomplished between the colonies of different genetic types by means of fusions of ascogonia and hyphae having appropriate types of nuclei. In such cases the term antheridia has not been used for cells, fusing with ascogonia in any of the Ascomycota. Preliminary studies have shown that there is ample variation in the conidia and conidiophores, which can serve as useful criteria for the speciation in the genus *Oidium.* Therefore, there is a definite need to correlate the conidial and cleistothecial stages wherever feasible and also to work out a satisfactory scheme of classification of these fungi where perfect stages are not known (Munjal, 1970).

Pleomorphism in fungi complicates the taxonomic problems specially in rusts and Ascomycetes. Narasimhan (1965) drew our attention to the lack of information of life cycles of heteroecious rusts. The confusion of identity of rusts of even major crop plants like sugarcane, bajra etc. clearly established that there are two rusts of sugarcane, caused by *Puccinia erianthi* Padwick and Khan and *P. kuehnii* (Krueg.) Butler. The former has urediniospores with uniformly thick wall (epispore), while the latter has an apically thickened wall. The rust of bajra commonly recorded as *Puccinia penniseti* Zimm. has been wrongly named and actually consists of two varieties viz., *P. substriata* var. *penicillare* (Speg.) Ramachar and Cummins which occurs in North India and in which case brinjal *(Solanum melongena* L.) does not act as the alternate host. The other variety namely *P. substriata* var. *indica* Ramachar and Cummins occur in South India and has brinjal as the alternate host. The two varieties are morphologically distinct. Smuts are also beset with taxonomic problems both at generic and specific level. Despite the excellent monograph of Duran and Fischer (1961), it is still difficult for us to agree to their disposition of *Neovossia* as a synonym of *Tilletia.* Khanna *et al.* (1966) using electron microscope showed that teliospore of *Neovossia indica* has a peridium which is absent in *Tilletia carries* and *T. foetida.* Besides this, several workers who worked with *Neovossia* species have shown that in this group, the teliospore on germination produces a very large number of sporidia, usually bitunicate and only rarely fuse to form 'H' shaped structures. These characters make *Neovossia* quite distinct from *Tilletia.* At the species level the fungi causing loose smut of wheat and barley *(Ustilago tritici* and *U. nuda);* covered smuts of oats and barley *(U. kolleri* and *U. hordei)* and flag smut of wheat and *Agropyron* (*Urocystis tritici* and *U. agropyri)* are morphologically indistinguishable from each other. As suggested by Fischer (1953), the correct names of these smut fungi are *Ustilago nuda, U. hordei* and *Urocystis agropyri,* respectively. Pathologically these are distinct biologic forms. The plant pathologists, however, continue to use wrong names on the basis of long and continuous usage, a plea which is not tenable. The problem can be easily solved by using the trinomial classification such as *Ustilago nuda tritici* or more appropriately *Ustilago nuda* var. *tritici* which is in consonance with the recommendations of the International Code of Botanical Nomenclature.

Higher Fungi

Recently in the literature on the taxonomy of higher fungi in relation to modern generic concepts more emphasis is placed on features such as presence or absence of clamp connections, thick walled hyphae in contrast to thin walled, the arrangement of hyphae in a tissue such as the hymenophore of a bolete, and colour changes on contact with chemicals of the cell wall, cell content and to some extent materials excreted by the cells. (Smith, 1968). It is now well known that by their use we are arriving at a more meaningful species concepts. However, when it comes to postulating relationships we find such proposals as the one grouping *Mycobonia* with *Polyporus* (Singer, 1963) and *Stereum taxodii* with some other genus such as *Echinodontium tinctorum* (Gross, 1964).

Fungi Imperfecti

In fungi imperfecti, a taxonomist faces more problems than in any other group. The study of *Helminthosporium.* species show that several of them belong to *Corynespora, Exosporium, Sporidesmium* etc. In genera like *Alternaria, Cladosporium, Cercospora* and *Curvularia* the speciation is made difficult due to highly variable nature of conidia and conidiophores. Notwithstanding the fact that in *Alternaria tenuis* group, one or more species inhabit the same spot. Even single spore cultures show a lot of variation in shape and size of conidia. A number of new species of *Alternaria, Cladosporium, Cercospora, Curvularia* and other related genera have been described on morphological characters based on host specificity alone and not studied in culture. Variations in the morphological characters due to environmental factors is well known. Munjal (1970) stated that creation of new species based on morphological characters observed on host alone has added a lot of confusion in speciation. This is somewhat unwarranted. In *Cercospora,* because of the fact that they have restricted host range and only a few of them have been cultured on artificial media, too much dependence on host has clouded our concept of the true taxonomic relationships. In case of mistaken host identity shifting of the fungal species also takes place. A re-examination of *Cercospora* species in some cases has lead to their transfer to related genera like *Pseudocerospora, Eriocercospora, Mycovellosiella, Stenella, Phaeoisariopsis, Phaeoramularia, Passalora, Stigmina, Cercosporidium,* etc. There are nearly 2,000 species of *Cercospora* and about half of them are quite unlike the type of the genus *C. apii.* Dieghton (1967, 1973). It is by now generally recognized that in *Cercospora* spp. similar to *C. apii,* morphological characters such as length and width of conidia, width of scars, length and degree of geniculations and branching of the conidiospores are all liable to be modified under environmental influences. In conclusion, it can be suggested that much valuable information would be available if more of the "species" of true *Cercospora* already described are compared by careful experimental work. Many more cross inoculation studies, field observations and cultural characters are needed, with adequate controls by repeating the studies of Cercosporas occurring on other host plants. Such type of work will contribute much more to our knowledge than the automatic description, as a "new species", of a morphologically indistinguishable *Cercospora* where sole characteristic is that it has been collected on a different host plant. Any one working on a disease caused by *Cercospora* should be interested in whether the inoculum is likely to come from a particular host that he is studying or equally well have come from other related or even quite unrelated host species in the vicinity. The present use of the host index for the identification of such genera is apt to blind one to such possibilities.

In Coelomycetes, the genera like *Colletotrichum, Pestalotiopsis, Phoma, Phyllosticta* and *Septoria* etc. constitute important pathogenic groups. In the genus *Colletotrichum,* von. Arx (1957) has contributed significantly in understating speciation problem. Cultural characters together with sporemorphology serve as better criteria for taxonomic speciation than spore morphology alone. An examination of *Pestalotia* species known to occur in India have revealed that these mostly belong to genus *Pestalotiopsis.* Species differentiation in the group presents difficulty because the colour scheme advocated by Guba is not suitable and the size of appendages is also a variable character. The position regarding *Phoma* and *Phyllosticta* is also arbitrary .The characters used for their separation so far do not solve the problem very satisfactorily.

The occurrence of several or a group of fungi simultaneously on the same substrate often lead to erroneous conclusions. Many species of *Meliola* and their related genera are frequently parasitized by other fungi. Much confusion in morphology and taxonomy has resulted from failure to discriminate between the members of Meliolinae and its numerous parasites. The pycnidia and conidiophores and even the perithecia of the parasites on Meliolinae have been repeatedly described as the organs of *Meliola.* Due to such errors, *Helminthosporium, Arthrobotryum* etc. have been described as the conidial stage of *Meliola.*

Modern Approaches

In modern taxonomy the use of simple fluorescent antibody technique in the identification of plant pathogenic fungi using fluorescein iso-thio-cyanate by ultraviolet light and dark ground microscopy is likely to be of much value in near future. Preece (1963) studied the following fungi viz. *Mycosphaerella citrulina, M. pinoides, Phytophthora parasitica, P. cactorum, Fasarium culmorum, F. avenaceum, Aspergillus flavus* and *A. niger* with the help of fluorescent antibody technique and found the results quite interesting and helpful in the identification of these fungi upto their generic level. The technique has some implications in plant diseases caused by fungi but the impermeable nature of the undamaged leaf to stains and reagents, promotes its use in the study of taxonomy of fungi. Serological investigations of *Fusarium oxysporum, F. moniliforme* and *F. solani* have already shown that they can be clearly differentiated from each other. It was also confirmed that *F. oxysporum* is more closely related to *F. moniliforme* than to *F. solani,* a relationship already believed to exist from morphological and cultural evidence (Booth, 1966). This evidence naturally provides a taxonomist more detailed information about the molecular structure of the pathogen involved. Electrophoretic separation of the cell components in starch or polycrylamide gels enables us to characterize large molecules because of the molecular sieving properties of the gels, in terms of their size and shape. The application of these techniques has given information of considerable significance. It is hoped that in near future the plant pathologists will be able to establish different serotypes in this and its allied genera.

Development of new approaches employing genetic, cytological and biochemical evidences and the opportunity of handling data by computers has greatly stimulated the taxonomist's field. The computerized data should be comprehensive in coverage and the degree of yield of information available should be high. There should be a place in the system for every fungus so that one can easily locate that place whenever one wishes to store new information or to retrieve what already is there.

In conclusion, it should be reiterated that taxonomy with its numerous principles and problems is a dynamic discipline, and it plays a dominant role in the development of biology. No doubt, we have some problems which can be solved by broadening our knowledge in this subject by utilizing newer techniques and sincere devotion towards the development of this science. This is a traditional field richly endowed with accomplishments, yet juvenile, large and versatile enough to offer challenging opportunities to contribute towards its progress.

References

Alexopoulos, C.J. (1961). Gross morphology of the *Plasmodium* and its possible significance in the relationship among Myxomycetes. Mycologia, **52:** 1-20.

Alexopoulos, C.J. (1969). The experimental approach to the taxonomy of the Myxomycetes, Mycologia, **61:** 219-239.

Arx, J.A. von (1957). Die Arten den gattung *Colletotrichum* Cda. Phytopath. Z., **29:** 28-68

Brefeld, O. (1881). Botanische Untersuchungen Uber Schimmel pilze, Heft. IV: 76-78.

Booth, C. (1966). Fruit bodies in Ascomycetes : In *"The Fungi"* (G.C. Ainsworth and A.S. Sussman Eds.), Vol. 2, Chapter 5, 133, A.P. New York, USA.

Deighton, F.C. (1967). Studies on *Cercospora* and allied genera –II. *Passalora, Cercosporidium* and some species of *Fusicladium* on *Euphorbia*. Mycol. Pap. **112 :** 80 pp.

Deighton, F.C. (1973). Studies on *Cercospora* and allied genera –IV. *Cercosporella* Sacc., *Pseudocercosporella* Gen. Nov. & *Pseudocercosporidium* Gen. Nov. Mycol. Pap. **133:** 62 pp.

Duran, R. and Fischer, G.W. (1961). The Genus *Tilletia*. Washington State Univ., Pullman U.S.A.

Fischer, G.W. (1953). *Manual of North American Smut Fungi*. Ronald Press, New York.

George, L.K. and Camp, L.B. (1957). Studies on nutrition and physiology of *Trichophyton* species. J. Bact.,**74:** 121-124.

Gordon, R.E. and Mihen, J.M., (1959). A comparison of *Nocardia asteroides* and *Nocardia brasiliensis* J. Gen. Microbiol., **20:** 129-135.

Gross, Henry L. (1964). The Echinodontiaceae, Mycopathol. Mycol. Appl., **24:** 1-26.

Ham, T.H. (1951). A Syllabus of Laboratory examination and Clinical diagnosis. Cambridge, Mass., Harvard Univ. Press, p. 496.

Khanna, Asha, Payak, M.M. and Mehta, S.C. (1966). Teliospore morphology of some smut fungi- I. Electron Microscopy. Mycologia, **58:** 462-469.

Lilly, V.G. and Barnett, H.L. (1953). The utilization of sugars by fungi. West Virginia Univ. Bull. 3621, U.S-A.

Mehrotra, B.S. and Sarbhoy, A.K. (1960). A new species *Chaetocladium* from India. Mycologia, **52:**464-470.

Mehrotra, B.S. and Singh, S.N. (1970). Nutritional studies on Mucraceous genus *Syncephalis*. Zentbl. Bakt. Parasitkunde Abt., **2.124:** 434-440.

Munjal, R.L. (1970). Some comments on the taxonomy of phytopathogenic fungi. *Pl. Dis. Probl.* International Symposium on Plant Pathology. Published by Indian Phytopath. Soc. IARI, New Delhi, pp. 191-193.

Narasimhan, M.J. (1965). Study of heteroeceous life cycles of rusts in India –I. Indian Phytopath, **18:** 107-115.

Pound, F.T. (1968). The importance of taxonomic criteria for classification of fungi. Mycologia, **60** : 402-406.

Preece, T.F. (1963). Fluorescent antibody a new technique for identification of fungi. *Trans. Brit.* Mycol. Soc., **46:** 523-525.

Sarbhoy, A.K. (1965). Nutritional studies on six members of Mucorales-VII, Mycopath. Mycol. Appl. **24:** 161-164.

Sarbhoy, A.K. (1965b). Nutritional studies on the genus *Zygorrhynchus*, Can. J. Microbiol.**11:** 297-308.

Sarbhoy, A.K. *(1969)*. Recent trends in Typification of some phytopathogenic fungi, Indian Phytopath. **5:** 11-14.

Sarbhoy, A.K. (1975). *Advances in Mycology and Plant Pathology*. Published by Prof. R.N. Tandon's Birthday Celebration Committee, Editors S.P. Raychaudhuri *et. al.*, pp. 15- 30.

Scholer, Von, H.J. and Muller, E. (1966). Biochemical and physiological approach for identification of Mucorales. Path. Microbiol., **29:** 730-741.

Simmons, E.G. (1967). Typification of *Alternaria, Stemphyllium* and *Ulocladium*, Mycologia, **59 :** 67-92.

Singer, ROLF (1963). The Agaricales in Modern Taxonomy. J. Cramer, Weinheim. p. 915.

Smith, Alexander H. (1968). The Agaricales, Mycologia, **60:** 742-755.

Young, T.W.K. (1969). Asexual spore morphology in species of *Piptocephalis*, New Phytol, **68:** 359-362.